CHAPTER 8a
Special Topic: Drugs and the Mind

http://books.nap.edu/html/marimed/ - *Report on Medical Marijuana, Institute of Medicine*

http://www.ca.org/ - *Cocaine Anonymous*

http://www.nida.nih.gov/infofax/ecstasy.html - *Ecstasy, National Institute on Drug Abuse, NIH*

http://www.nida.nih.gov/infofax/heroin.html - *Heroin, National Institute on Drug Abuse, NIH*

http://www.niaaa.nih.gov/ - *National Institute on Alcohol Abuse and Alcoholism, NIH*

http://www.collegedrinkingprevention.gov/ - *College Drinking, National Institute on Alcohol Abuse and Alcoholism, NIH*

http://www.factsontap.org/ - *Facts on Tap: Alcohol and Your College Experience*

http://www.samhsa.gov/ - *Substance Abuse and Mental Health Services Administration*

CHAPTER 9
Sensory Systems

http://www.preventblindness.org/ - *Prevent Blindness America*

http://www.aoa.org/ - *American Optometric Association*

http://www.nei.nih.gov/ - *National Eye Institute, NIH*

http://www.nonoise.org/ - *Noise Pollution Clearinghouse*

http://www.cdc.gov/niosh/topics/noise/ - *Noise and Hearing Loss Prevention, CDC*

http://www.nidcd.nih.gov/ - *National Institute of Deafness and other Communication Disorders, NIH*

CHAPTER 10
The Endocrine System

http://www.hormone.org/ - *The Hormone Foundation*

http://www.diabetes.org/ - *American Diabetes Association*

http://www.thyroidmanager.org/ - *Thyroid Disease Manager*

http://www.endocrineweb.com/ - *Endocrine Disorders and Endocrine Surgery*

CHAPTER 11
Blood

http://www.redcross.org/ - *American Red Cross*

http://www.bloodbook.com/ - *Blood Book, Blood Information*

http://www.labtestsonline.org/ - *Lab Tests Online*

CHAPTER 12
The Circulatory System

http://www.heartsavers.org/ - *National Heart Savers Association*

http://www.nemahealth.org/ - *National Emergency Medicine Association*

http://www.heart1.com/ - *Complete Source for Cardiac Care*

CHAPTER 13
Body Defense Mechanisms

http://www3.niaid.nih.gov/ - *National Institute of Allergy and Infectious Diseases, NIH*

http://www.a-o-t-a.org/ - *American Organ Transplant Association*

http://www.transplantliving.org/ - *Transplant Living, Your Prescription for Transplant Information*

http://www.aarda.org/ - *American Autoimmune Related Diseases Association*

http://www.autoimmune-disease.com/ - *Autoimmune Diseases Online*

CHAPTER 13a
Special Topic: Infectious Disease

http://www.cdc.gov/ncidod - *National Center for Infectious Diseases, CDC*

http://www.nfid.org/ - *National Foundation for Infectious Diseases*

http://www.fda.gov/oc/opacom/hottopics/anti_resist.html - *Antibiotic Resistance, FDA*

http://www.cdc.gov/drugresistance/community - *Antibiotic Resistance, CDC*

http://www.bt.cdc.gov/ - *Emergency Preparedness and Bioterrorism, CDC*

http://www.fda.gov/oc/opacom/hottopics/bioterrorism.html - *Bioterrorism, FDA*

http://www.who.int/en/ - *World Health Organization*

CHAPTER 14
The Respiratory System

http://www.cdc.gov/ncidod/dbmd/respirat.htm - *Bacterial Respiratory Diseases, CDC*

http://www.nlm.nih.gov/medlineplus/lungdiseases.html - *Medline Respiratory Diseases, NIH*

http://www.who.int/tb/en/ - *Tuberculosis Campaign, World Health Organization*

http://www.lungusa.org/ - *American Lung Association*

http://www.nhlbi.nih.gov/index.htm - *National Heart, Lung, and Blood Institute, NIH*

http://www.cdc.gov/nchstp/tb/default.htm - *Division of Tuberculosis Elimination, CDC*

CHAPTER 14a
Special Topic: Smoking and Disease

http://www.quitsmokingsupport.com/ - *Quit Smoking Support*

http://www.tobacco.org/ - *Tobacco News and Information*

http://www.quitsmoking.com/ - *The Quit Smoking Company*

http://www.cdc.gov/tobacco/ - *Office on Smoking and Health, CDC*

CHAPTER 15
The Digestive System

http://www.digestive.niddk.nih.gov/ - *National Digestive Disease Information Clearinghouse, NIH*

http://www.healthyteeth.org/ - *Healthy Teeth Information*

http://www.ada.org/ - *American Dental Association*

http://www.dentalhealthfoundation.org/ - *Dental Health Foundation*

CHAPTER 15a
Special Topic: Nutrition and Weight Control

http://www.eatingdisorderinfo.org/ - *The Alliance for Eating Disorders Awareness*

http://www.anred.com/ - *Anorexia Nervosa and Related Eating Disorders*

http://www.nationaleatingdisorders.org/ - *National Eating Disorders Association*

http://www.overeatersanonymous.org/ - *Overeaters Anonymous*

http://www.thebodypositive.org/ - *The Body Positive*

http://www.healthyeating.net/ - *Healthy Eating Nutritional Website*

http://www.nal.usda.gov/fnic/foodcomp/index.html - *Nutrient Data Laboratory, U.S. Department of Agriculture* http://www.mypyramid.gov/ - *U.S. Department of Agriculture*

http://www.nutrition.gov/home/index.php3 - *Access to all online government information on nutrition*

http://www.cdc.gov/nccdphp/dnpa/obesity/ - *Overweight and Obesity, CDC*

CHAPTER 16
The Urinary System

http://www.kidney.niddk.nih.gov/index.htm - *National Kidney and Urologic Diseases Information Clearinghouse, NIH*

http://www.urologychannel.com/ - *The Urology Channel, Your Urology Community*

CHAPTER 17
Reproductive Systems

http://www.cdc.gov/reproductivehealth/WomensRH/index.htm - *Women's Reproductive Health, CDC*

http://www.plannedparenthood.org/ - *Planned Parenthood Federation of America*

http://www.nationalbreastcancer.org/ - *National Breast Cancer Foundation*

http://www.nabco.org/ - *National Alliance of Breast Cancer Organizations*

http://www.breastcancer.org/ - *Breast Cancer Education*

http://tcrc.acor.org/ - *Testicular Cancer Resource Center*

http://www.tc-cancer.com/ - *Testicular Cancer Information and Support*

http://www.prostate-cancer.org/ - *Prostate Cancer Research Institute*

http://www.4npcc.org/ - *National Prostate Cancer Coalition*

http://www.prostatecancerfoundation.org/ - *Prostate Cancer Foundation*

CHAPTER 16a
Special Topic: Sexually Transmitted Diseases and AIDS

http://www.ashastd.org/ - *American Social Health Association*

http://www.projectinform.org/ - *HIV Education, Project Inform*

http://www.unaids.org/ - *Joint United Nations Program on HIV/AIDS*

http://www.hivandhepatitis.com/ - *HIV and Hepatitis Information*

http://www.aidsinfo.nih.gov/ - *Aids Information, NIH*

CHAPTER 18
Development throughout Life

http://www.childbirth.org/ - *Pregnancy and Childbirth Information*

http://www.nacd.org/ - *National Association for Child Development*

http://www.ffcd.org/ - *Foundation for Child Development*

http://www.marchofdimes.com/ - *March of Dimes, Saving Babies Together*

http://www.reproductive-health.org/ - *American Foundation for Urologic Disease*

http://www.aoa.gov/ - *Administration on Aging*

http://www.nia.nih.gov/ - *National Institute on Aging, NIH*

http://www.ncoa.org/ - *The National Council on the Aging*

CHAPTER 19
Chromosomes and Cell Division

http://www.ndss.org/ - *National Down Syndrome Society*

http://www.ndsccenter.org/ - *National Down Syndrome Congress*

http://www.turner-syndrome-us.org/ - *Turner Syndrome Society of the United States*

http://www.fivepminus.org/ - *Family Support Group for Children with Cri du Chat Syndrome*

http://www.fragilex.org/ - *The National Fragile X Foundation*

CHAPTER 19a
Stem Cells—The Body's Repair Kit

http://stemcells.nih.gov/ - *Resource for Stem Cell Research, NIH*

CHAPTER 20
Genetics and Human Inheritance

http://www.genetests.org/ - *Genetic Testing Information*

http://www.nlm.nih.gov/medlineplus/prenataltesting.html - *Medline Prenatal Testing, NIH*

CHAPTER 21
DNA and Biotechnology

http://www.humancloning.org/ - *Human Cloning Foundation*

http://www.cloninginformation.org/ - *Americans to Ban Cloning*

http://www.isscr.org/ - *International Society for Stem Cell Research*

http://www.thecampaign.org/ - *The Campaign to Label Genetically Engineered Foods*

CHAPTER 21a
Special Topic: Cancer

http://www.cancer.org/ - *American Cancer Society*

http://www.cancer.gov/ - *National Cancer Institute*

http://www.cancercare.org/ - *Cancer Care Non Profit*

http://www.nccf.org/ - *National Childhood Cancer Foundation*

CHAPTER 22
Evolution and Our Heritage

http://www.becominghuman.org/ - *The Institute of Human Origins*

http://www.amnh.org/ - *American Museum of Natural History*

http://www.aaalac.org/ - *Association for Assessment and Accreditation of Laboratory Animal Care*

CHAPTER 23
Ecology, the Environment, and Us

http://www.environmentalhealthnews.org/ - *Environmental Health News*

http://yosemite.epa.gov/oar/globalwarming.nsf/content/index.html - *Global Warming Information, Environmental Protection Agency*

http://www.globalwarming.net/ - *Global Warming International Center*

http://www.thehungersite.com/ - *World Hunger*

http://www.worldhunger.org/ - *World Hunger Education Service*

CHAPTER 24
Human Populations, Limited Resources, and Pollution

http://www.census.gov/ - *U.S. Census Bureau*

http://www.un.org/popin - *United Nations Population Information Network*

http://www.prb.org/ - *Population Reference Bureau*

http://www.census.gov/main/www/popclock.html - *U.S. Census Bureau*

Brief Contents

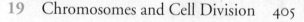
i

BIOLOGY OF HUMANS

BIOLOGY OF HUMANS
CONCEPTS, APPLICATIONS, AND ISSUES

SECOND EDITION

JUDITH GOODENOUGH
University of Massachusetts, Amherst

BETTY MCGUIRE
Cornell University

ROBERT A. WALLACE

PEARSON
Prentice Hall

Upper Saddle River, NJ 07458

Library of Congress Cataloging-in-Publication Data

Goodenough, Judith.

Biology of humans : concepts, applications, and issues / Judith Goodenough, Betty McGuire, Robert A. Wallace. — 2nd ed.

  p. cm.

ISBN 0-13-178999-6

1. Human biology.   I. McGuire, Betty. II. Wallace, Robert A. III. Title.

QP34.5.G658 2007

612—dc22   2006030153

Executive Editor: Gary Carlson
Editor in Chief, Science: Dan Kaveney
Project Manager: Crissy Dudonis
Senior Media Editor: Patrick Shriner
Editor in Chief, Development: Carol Trueheart
Development Editor: Moira Nelson
Production Editor: Debra A. Wechsler
Executive Managing Editor: Kathleen Schiaparelli
Assistant Managing Editor: Beth Sweeten
Managing Editor, Media: Rich Barnes
Media Production Editor: Dana Dunn
Marketing Manager: Mandy Jellerichs
Manufacturing Buyer: Alan Fischer
Manufacturing Manager: Alexis Heydt-Long
Director of Creative Services: Paul Belfanti
Creative Director: Juan R. López
Art Director: Kenny Beck
Interior Design: Anne DeMarinis
Cover Design: Kenny Beck

Senior Managing Editor, Art Production and Management:
  Patricia Burns
Manager, Production Technologies: Matthew Haas
Managing Editor, Art Management: Abigail Bass
Art Development Editor: Jay McElroy
AV Production Editor: Rhonda Aversa
Art Studio: Imagineering Media Services, Inc.
Manager, Formatting: Allyson Graesser
Formatter: Joanne Del Ben
Director, Image Resource Center: Melinda Reo
Manager, Rights and Permissions: Zina Arabia
Interior Image Specialist: Beth Brenzel
Cover Image Specialist: Karen Sanatar
Image Permission Coordinator: Joanne Dippel
Photo Researcher: Yvonne Gerin
Editorial Assistant: Lisa Tarabokjia
Cover Image: Rafting on McKenzie River, Oregon,
  USA/John Giustina/Photodisc/Getty Images, Inc.

© 2007, 2005 Pearson Education, Inc.
Pearson Prentice Hall
Pearson Education, Inc.
Upper Saddle River, NJ 07458

Pearson Prentice Hall™ is a trademark of Pearson Education, Inc.

Printed in the United States of America

10 9 8 7 6 5 4 3 2 1

ISBN 0-13-178999-6

Pearson Education LTD., *London*
Pearson Education Australia PTY, Limited, *Sydney*
Pearson Education Singapore, Pte. Ltd.
Pearson Education North Asia Ltd., *Hong Kong*
Pearson Education Canada, Ltd., *Toronto*
Pearson Educatión de Mexico, S.A. de C.V.
Pearson Education—Japan, *Tokyo*
Pearson Education Malaysia, Pte. Ltd.

About the Authors

Judith Goodenough

Judith received her B.S. from Wagner College (Staten Island, NY), and her doctorate from New York University. She has 30 years of teaching experience at the University of Massachusetts, Amherst, specializing in introductory level courses. The insights into student concerns and problems gained from more than 25 years of teaching Human Biology and 20 years of team-teaching The Biology of Social Issues have helped shape this book. In 1986, Judith was honored with a "Distinguished Teaching Award." In addition to teaching, she coordinates the introductory biology laboratories at UMass. Judith has written articles in peer-reviewed journals, contributed chapters to several introductory biology texts, and written numerous laboratory manuals. With the author team of McGuire and Wallace, she wrote *Perspectives on Animal Behavior.*

Judith Goodenough

Betty McGuire

Betty received her B.S. in Biology from Pennsylvania State University, where she also played varsity basketball. She went on to receive an M.S. and Ph.D. in Zoology from the University of Massachusetts at Amherst, and then spent two happy years as a postdoctoral researcher at the University of Illinois, Champaign-Urbana. Her field and laboratory research emphasize the social behavior and reproduction of small mammals. She has published numerous research papers, co-authored the text *Perspectives on Animal Behavior* and several introductory biology study guides, and served as an Associate Editor for *Mammalian Species,* a publication of the American Society of Mammalogists. Betty taught Human Biology, Introductory Biology, Vertebrate Biology, and Animal Behavior at Smith College. She now teaches Mammalogy and The Vertebrates: Structure, Function, and Evolution at Cornell University.

Betty McGuire

Robert A. Wallace

The late Robert Wallace received a B.A. in Fine Art and Biology from Harding University, an M.A. in Muscle Histochemistry from Vanderbilt University, and a Ph.D. in Behavioral Ecology from the University of Texas at Austin. He subsequently taught at a number of colleges and universities in the United States and Europe, including the Richard Bland College of William and Mary, the University of Maryland-Overseas Division, Duke University, and the University of Florida. He is the author of seven previously published biology textbooks, including *Biology: The World of Life,* and two mass-market science books, *The Genesis Factor* and *How They Do It,* as well as numerous scientific articles on a variety of subjects. Robert was also a Fellow of the Explorers Club of New York and the Royal Geographical Society of London. He received the Orellana Medal from the government of Ecuador in recognition of his work with the medicinal plants of vanishing tribes in that country.

Humans are curious by nature. This book, intended for nonscience majors, was written to stimulate that natural curiosity, inspiring an appreciation for the intricacy of human biology and the place of humans in the ecosphere. Once awakened, however, curiosity demands satisfaction in the form of solid information. Toward this end, we provide students with a conceptual framework for understanding how their bodies work and for dealing with issues relevant to human health in the modern world. All along, we sustain the student's interest and curiosity by continuously illustrating the connections between biological concepts and issues of current social, ethical, and environmental concern. Our central belief is that the application of biological concepts to familiar experiences is the key to helping students see the excitement of science and its importance in their lives.

Strengths We Built On This edition builds on the strengths of clarity, liveliness, consistency, currency, and relevance that characterized the first edition. The writing is engaging, the explanations straightforward, and the pedagogical framework meticulously constructed. Great care has been taken to keep the level of coverage consistent throughout. All features—titles, outlines, headings, illustrations and legends, tables, vocabulary lists, summaries, questions, and so on—are designed to help students identify important facts and ideas, understand them, and appreciate why they matter.

As in the first edition, the text continually uses applications of scientific knowledge to engage students' interest, bring concepts to life, and illustrate the social relevance of human biology. This strategy is especially apparent in the seven Special Topics chapters (described shortly) and the dozens of Issues Boxes distributed throughout. Many discussions of applications have been added or enhanced. For example, the essay on asbestos in Chapter 3 now mentions the effects of asbestos contamination at Ground Zero on many of the surviving first responders from 9/11. As another example, Chapter 8a now has a discussion of the medical uses of marijuana. Newspaper headlines prompted the addition of five new essays: SOCIAL ISSUE *Scientific Misconduct* (Chapter 1); ENVIRONMENTAL ISSUE *Lead Poisoning* (Chapter 11); HEALTH ISSUE *The Fear That Vaccines Cause Autism* (Chapter 13); SOCIAL ISSUE *Bird Flu, Will It Become a Pandemic?* (Chapter 14); ENVIRONMENTAL ISSUE *Environmental Estrogens* (Chapter 17); and SOCIAL ISSUE *Forensic Science, DNA, and Personal Privacy* (Chapter 21).

Because students deserve information that is not only relevant but *current*, the text has been updated throughout. The chapter on Nutrition and Weight Control presents the new USDA food guide pyramid and recommendations for healthy living. The statistics on smoking, sexually transmitted diseases, and cancer are the latest available. The essays on health, social, and environmental issues have also been updated. Examples include the *Global Warming* essay in Chapter 23 and the discussion of shock wave lithotripsy in the essay on kidney stones in Chapter 16. (At the time of the first edition, the lithotripsy procedure was deemed wonderful because surgery wasn't needed. However, in spring 2006, reports revealed that it appears to increase risk of diabetes and hypertension.)

Practical Goals and Special Features

The principal goal of this textbook is to give *a clear presentation* of the fundamental concepts of human anatomy, physiology, development, genetics, evolution, and ecology. The second goal is to *apply these concepts* in ways that will both interest and benefit the student. For example, the explanation of neurotransmitters is followed immediately by a discussion of the roles of neurotransmitters in Alzheimer's disease, depression, and Parkinson's disease. When students understand why a topic is relevant, they have a reason to want to learn about it. Thus, the chapters on organ systems explain how a healthy system functions, how that system might malfunction, steps to take to avoid a malfunction, and the ways in which medical science can help when systems are compromised or fail. Discussions of topics that students are likely to encounter in the media on an almost daily basis—smoking, contraception, STDs, cancer, antibiotic-resistant bacteria—help them see connections between classroom activities and the concerns of daily life. Connections between classroom topics and environmental issues help students develop a global perspective concerning questions they once thought of as being strictly personal.

Much of the information offered in the text is practical: What type of exercise is of greatest benefit to the heart? How can a person cope safely with insomnia? What can be done to protect against unwanted pregnancy and prevent the spread of sexually transmitted diseases? The body each of us is born with is a most intricate machine, but it does not come equipped with an owner's manual. In a sense, this book can be the student's owner's manual. Studying and applying the lessons to their individual lifestyles and health issues can help your students live longer, happier, and more productive lives.

The third goal is to help students *develop reasoning skills*, so they can make use of their newly acquired knowledge in situations they face in daily life. The "stop and think" and "what would you do" questions distributed within the chapters, as well as the "applying the concepts" questions at the end of each chapter are all written with this goal in mind. The fourth goal is to help students understand how the choices they make can *affect society and the planet, as well as their own quality of life*. Much of the material learned in human biology has a bearing on social and environmental issues that are important to us all. This text will help instructors heighten students' awareness of their impact on the biosphere and prepare students to be responsible citizens of their country and the world. Society is currently immersed in many pressing biological debates—concerning, for example, the cloning of human cells, stem cell research, genetically modified foods, gene therapy, organ transplants, the definition of death, and the prevention and treatment of HIV infections—and students need the tools to understand these issues and make informed decisions about them.

Every aspect of the book is designed with these four goals in mind, but the following special features are particularly effective.

SPECIAL TOPIC CHAPTERS

The text contains seven Special Topic chapters: Chapter 8a, Drugs and the Mind; 13a, Infectious Disease; 14a, Smoking and Disease; 15a, Nutrition and Weight Control; 17a, Sexually Transmitted Diseases and AIDS; 19a, Stem Cells—the Body's Repair Kit; and 21a, Cancer. Each of these short chapters builds on the "pure biology" presented in the immediately preceding chapter to cover issues likely to be of personal interest and therefore likely to motivate learning in students. The discussions they contain are more thoroughly developed than would be possible in a boxed essay. Even if you do not include these special topics in your reading assignments, we believe the issues are so pertinent to students that they will read the special chapters of their own volition, or at least refer to them occasionally as guides to a healthier lifestyle.

"STOP AND THINK" QUESTIONS

The "stop and think" questions scattered throughout each chapter are intended to promote active learning. They invite the student to pause in order to think about the information that was just presented and apply it to a new and interesting situation. These periodic checks allow the student to determine whether he or she has followed and understood the basic chapter content.

"WHAT WOULD YOU DO?" QUESTIONS

The "what would you do?" questions, which are also positioned throughout each chapter, challenge the student to form an opinion or to take a stand on a particular issue that society faces today, as well as to identify the criteria used in reaching that opinion or decision. These questions help students see the relevance of biology to real-life problems and foster the practice of thinking through such complicated issues as the fluoridation of water, routine screening for prostate cancer, the use of animal organs to save human lives, the export of pesticides to developing countries, and strategies for slowing the growth of human populations. When the subject of one of these questions is controversial, the text presents examples of arguments from both sides, as well as evidence in support of competing arguments. There frequently is no right answer to these questions. They simply demonstrate the broad implications of the topics discussed. Instructors may choose to use these questions to begin a classroom presentation, to stimulate in-class discussion, or as a way of sending students out of the classroom still thinking about the topic of the day's lecture. Additional critical-thinking questions are provided on the Companion Website.

ISSUE ESSAYS

Three categories of boxed essays use the basic scientific content of the chapters to explore issues having broader impact on individual health, society, and the environment. Instructors might use these essays to arouse students' interest in lecture topics or to provide them with information needed for informed in-class discussions. **Health Issue** essays deal primarily with personal health topics. They provide current information on certain health problems that students, their families, or their friends might encounter. Examples of topics discussed in Health Issue essays are acne, osteoporosis, treatments for the common cold, and heartburn. These essays give students a better understanding of topics that commonly arise during visits to the doctor. The **Social Issue** essays explore ethical and social issues related to the topics in a chapter. For example, they explore questions concerning such subjects as anabolic steroids, gene testing, possible pandemic flu, and the use of primates in research. Finally, the **Environmental Issue** essays deal with ways in which human activities alter the environment or, conversely (sometimes simultaneously), ways in which the environment influences human health. Among the topics discussed in Environmental Issues essays are biodiversity, acid rain, noise pollution, and global warming.

ENTICING ILLUSTRATION PROGRAM

Users of the first edition–instructors and students alike–were unreservedly enthusiastic in praising the illustrations for their appeal and helpfulness. The visual program consists of simple but elegantly rendered illustrations that have been carefully designed for effective pedagogy. Their very beauty stimulates learning. This is particularly true of the many vibrant, three-dimensional anatomical figures, whose realistic style and appropriate depth and detail make them easy for students to interpret and use for review.

Within each category of illustrations—from molecular models to depictions of human tissues and organs—the figures are consistent in plan and style throughout the text. Numerous key figures pull concepts together to present the "big picture." For example, the chapters on organ systems have figures that show both the anatomical structure and the function of all major components of the organ system. These figures, such as Figure 14.2, of the respiratory system, help keep students aware of the important relationship between structure and function. Pairing micrographs with drawings lends authority to the drawings, helps students interpret the micrographs, and in the end enhances the pedagogical value of both. Reference figures help students locate particular structures within the body. Figure 4.2, for instance, which illustrates types of connective tissue, not only pairs micrographs with drawings but also shows examples of where each type of connective tissue can be found. Many illustrations provide surrounding context for the anatomical structure being examined. For example, Figure 8.3, a sagittal section through the brain, shows the surrounding head, including facial features, to provide orientation and perspective. Flowcharts walk students through a process one step at a time, so they can follow the progress of a discussion visually after reading an explanation in the body of the text. Similarly, difficult concepts are reviewed using step-by-step figures that break the concepts down into simple components. Such figures help the student understand how events, processes, and ideas fit together. For example, Figure 18.2 guides the student through the process of early development, from ovulation to implantation. Liberal use of "voice balloons" makes all categories of illustrations easier to interpret and review.

Finally, color is used in the visual program as an effective means of organizing information. Where appropriate, color delineates the steps in a process. For example, in Figure 13.12, subtle differences in background shading distinguish different steps in the antibody-mediated immune response. In Figure 12.12, a depiction of the electrical activity of the heart, color indicates progression from one step in a heartbeat to the next.

Organization and Pedagogy

The text begins with a discussion of the chemistry of life, proceeds through cells, tissues, organs, and organ systems, and ends with discussions of genetics, populations, and ecosystems. As teachers ourselves, we understand the difficulty of covering all the topics in a human biology text in one semester. Instructors are inevitably forced to make difficult decisions concerning what to include and what to leave out. We also know that there are many equally valid ways of organizing the material. For this reason, the chapters in this text are written so as not to depend heavily on material covered in earlier chapters. The independence of each chapter allows the instructor to tailor the use of this text to his or her particular course. At the same time, cross-references are given where they may be helpful to direct students to relevant discussions in other chapters.

The pedagogical features that provide a consistent framework for every chapter have been designed not only to help students understand the information presented in their human biology course, but to help them study more effectively. Some of the most important of these elements are described below.

CHAPTER OUTLINES

Each chapter begins with an outline constructed from the chapter's two major levels of headings. Because it identifies the chapter's important concepts and the relationships between them, this feature provides a conceptual framework on which students can mentally organize new information as they read. It also serves as a meaningful review of chapter contents that students can use when studying for exams.

HEADINGS AND SUMMARIES

The two major levels of headings are worded as sentences stating the main point of the discussion that follows. Thus, students immediately see the big picture and therefore have an easier time focusing on the accompanying explanation. The "Highlighting the Concepts" sections that summarize each chapter are organized using the chapter's highest level of heading. Relevant page numbers are included in this summary to guide students back to the full text discussion of any topic they may wish to review.

KEY TERMS AND GLOSSARY

We have held the use of technical language to a minimum, because this text is intended for students who are not science majors. Important terms are set in bold type where they are formally introduced and are listed as key terms at the end of the chapter. New terms of lesser importance are set in italics. The Key Terms list also provides chapter page numbers indicating where each term is defined. A Glossary at the end of the book contains definitions for all the key terms.

END-OF-CHAPTER QUESTIONS

The questions provided at the end of each chapter are designed in several formats to meet several purposes. Some, specifically the "Reviewing the Concepts," are intended simply as content review; others—particularly those under the heading "Applying the Concepts"—require critical thinking and challenge the students to apply what they have learned to new situations. Review questions that require a written answer are followed by the number of the page containing the relevant discussion. These questions encourage students to review and understand the relevant material instead of simply memorizing a few salient facts. Answers to the multiple-choice and fill-in-the-blank questions are provided in an appendix, as are hints for answering the "Applying the Concepts" questions. The hints, which help students identify the information needed to answer each question, are intended to guide students in their thinking process instead of simply providing a quick answer.

ICONS

 WEB Tutorial icons displayed throughout the book direct the student to the text's Companion Website at www.prenhall.com/goodenough/. The Web tutorials employ animations and interactive exercises to teach processes and concepts that may be challenging for the student or that are difficult to understand from words and illustrations alone. Quizzes are offered to help students assess their understanding of a topic after viewing the animation or completing the exercise.

A second type of icon accompanies many of the tables in the text, identifying those tables as being presented for reference only. These icons, along with specific table titles, help students decide what material should be studied for exams and what material is provided primarily for reference or as a source of illustrative examples.

RELEVANT WEBSITES

The front inside text cover lists 160 relevant and useful Websites for students who want to explore certain topics in greater detail. Because the text discusses many health issues that students or their families and friends may be dealing with, the list of Websites includes resources that can provide support for people with various health problems, including cancer, drug and alcohol abuse, and smoking.

Supplements

Carefully designed packages of supplemental material for the student and instructor are available to further facilitate learning and teaching.

FOR INSTRUCTORS

Instructor Resource Center on CD/DVD This multiple-disc resource is designed to fully support individual teaching approaches and invigorate lectures has been upgraded with the highest-quality media assets available. Every figure, photo, and table in the text has been enhanced for optimal projection results and is provided in a variety of JPEG and PowerPoint® options. A full set of Lectures that can be adapted to individual classroom needs is offered in PowerPoint as well. The array of animations on our IRC on CD/DVD is truly unparalleled. The more traditional set of 2-D animations created to support the text has been thoroughly revised and improved. In addition and for the first time, a set of 3-D animations and simulations from Prentice Hall's visually-stunning and scientifically precise "BLAST" series is also offered with this text. Both types of animations are pre-loaded into PowerPoint in PC and Macintosh versions.

Transparencies   The set of transparencies consists of approximately 400 illustrations from the text on full-color acetates.

Instructors' Guide   This manual contains material that will assist instructors in the preparation of lectures. Its contents include lecture outlines and hints, suggestions for classroom activities, answers to the end-of-chapter questions, and suggestions for assignments.

Test Item File   The printed Test Item File contains over 1000 questions instructors can use to prepare exams. Types of questions include multiple choice, true/false, short answer, fill-in-the-blank, matching, and labeling.

TestGen EQ Computerized Testing Software   The questions from the Test Item File are also available as part of the TestGen EQ Testing Software, a text-specific testing program that can be networked for administering tests. The program allows instructors to view and edit these test questions and also to add their own questions to create an unlimited variety of quizzes and tests.

FOR THE STUDENT:

Companion Website   At **www.prenhall.com/goodenough/**, the Companion Website offers a wealth of student learning tools. Each chapter offers an *Online Study Guide* designed to help students gain focus with their limited study time. Study features include concise summaries with pop-up windows displaying key figures and tables, learning objectives, key term flashcards, interactive exercises, and interactive practice quizzes loaded with specific useful feedback. *Web Tutorials* use animation, interactivity, and quizzes to provide assistance with the most challenging and important concepts in each chapter. Icons alerting the student to the availability of specific tutorials have been included throughout the book. There is an access card for the Companion Website in the front of all new copies of the text. Students with used books can go to the web address above to find out about getting access to the Companion Website.

Student Study Guide (printed study aid) by Claudia Douglass, Central Michigan University   The Student Study Guide is designed to help students master the topics and concepts covered in the textbook. The study guide contains several types of review questions, including multiple choice, fill-in-the-blank, critical thinking, matching, and labeling. It also includes study objectives, summaries, key terms, and student study tips.

Understanding Human Biology: Laboratory Exercises by Mimi Bres and Arnold Weisshaar, Prince George's Community College In both content and approach to learning, this lab manual is an appropriate ancillary to *Biology of Humans*. It emphasizes developing an understanding of the basic principles of human anatomy and physiology, genetics and evolutionary change, ecology, and the impact of human actions on the environment.

Acknowledgments

We are truly fortunate to be part of a team of dedicated and talented people. Gary Carlson, our executive editor, has enthusiastically supported us at every turn. He has been involved at almost every level, and the result is clear evidence of his effectiveness. Gary and Carol Trueheart, editor in chief of development, helped us develop a revision plan that built on the strengths of the first edition while selectively introducing new content and paring back needless detail. Our developmental editor, Moira Nelson, helped us implement that plan. She is one of the best in the business. She was a source of great ideas for new directions and always willing to help. During the academic year, most of our writing took place on weekends. Moira always checked her email and quickly addressed any of our concerns—often including amusing quips that eased the stress.

Jane Loftus did an excellent job of copyediting the book: keeping our writing style, while improving accuracy and consistency.

We are especially pleased to work with Debra Wechsler as our production editor again on this edition, because of both her professional skills and her friendly personality. By helping us prioritize, she kept us to the schedule and still allowed us to meet the needs of our families. Throughout the project she was calm, patient, and helpful.

The art program was designed and developed by two medical illustrators: Jay McElroy from Prentice Hall with the help of Jack Haley of Imagineering, Inc. Their art development has created an accurate and pedagogically sound illustration program. We especially enjoyed working with Jay McElroy on art development. He was always responsive to questions and suggestions and somehow managed to keep a wonderful demeanor throughout this complex and demanding project. Imagineering Media Services skillfully rendered the art. Photo Researcher Yvonne Gerin persistently worked to find the best and most original photographs. Kenny Beck and Anne DeMarinis created an elegant design that highlights and effectively integrates the many graphic and pedagogical elements. Formatter, Joanne Del Ben, skillfully implemented the design.

Project manager, Crissy Dudonis, did a skillful job of coordinating development and production of the ancillary materials that accompany our text. She brought together lecture and study aids to create a supplemental package that is highly effective for instructors and students. Crissy also worked with Patrick Shriner, our excellent media editor, on the selection and development of media supplements.

We owe a great deal to our late friend and colleague, Bob Wallace. We are thankful for all he taught us. His wife, Jayne Wallace, provided information about medicines in the rain forest and has continued to support our efforts.

FROM JUDITH GOODENOUGH

I thank my family and friends who supported and encouraged me at every stage of this project. My husband, Steve, was a cheerleader and convinced me that I would complete this project. Without his witty quips, I would have lost my sanity. He reassures me that I'll always be his first wife. My daughters, Aimee and Heather, inspire me and continually remind me that the people you love should always come first. The willingness of my mother, Betty Levrat, to help in any way, allowed me to focus on writing.

Margaret Ludlam cheered me up and helped me cope when things got tense. Margaret's willingness to pick up the slack at UMass allowed me to take vacation days to work at home during the semester. Lee Estrin, one of "the group," dear friends who have always provided moral support and advice, was always willing to visit for an "I need a break weekend," even when I couldn't actually stop working completely.

FROM BETTY MCGUIRE

I thank my husband, Willy Bemis, for support and encouragement throughout this project, and my children, Kate and Owen, for (usually) waiting patiently for me to complete just one more sentence or paragraph. Dora, Jim, and Cathy McGuire were endlessly encouraging and understanding, as they have been throughout my life. Lowell Getz, my friend and research colleague, waited with good humor as one after another of our papers took a back seat to a book chapter.

We must thank the dozens of reviewers who so carefully read chapters. They were immensely helpful in shaping the book and in catching errors. Of course, any errors that remain are our responsibility, and we hope you will inform us if you find any.

Judith Goodenough
Betty McGuire

REVIEWERS OF THE FIRST TWO EDITIONS

Richard Adler, *University of Michigan, Dearborn*
John Aliff, *Georgia Perimeter College*
Merrilee Anderson, *Mt. Aloysius College*
Bert Atsma, *Union County College*
Tamatha Barbeau, *Francis Marion University*
Erwin Bautista, *University of California, Davis*
Susan Bear, *Saint Joseph's University*
Virgina Bliss, *Framingham State University*
Robert Boyd, *Auburn University*
Christopher Brey, *Rutgers University*
Betheny Bush, *Fresno City College*
Alessandro Catenazzi, *Florida International University*
Tony Contento, *Iowa State University*
Christine Curran, *University of Cincinnati*
Michael A. Davis, *Central Connecticut State University*
Jeff W. Dawson, *University of Cambridge*
Claudia Douglass, *Central Michigan University*
Don Drewiske, *University of Wisconsin, Green Bay*

Meredith Durmowizc, *Villa Julie College*
Lucinda H. Elliott, *Shippensburg University*
Fledia Ellis, *Gulf Coast Community College*
Beth Erviti, *Greenfield Community College*
Jane Finan, *Lewis-Clark State College*
Deb Firak, *McHenry County College*
Michael P. Franklin, *California State University, Bakersfield*
Paul Gannon, *El Paso Community College*
Donald Glassman, *Des Moines Area Community College*
Sheldon R. Gordon, *Oakland University*
Margarit Gray, *Anderson College*
Debra Grieneisen, *Harrisburg Area Community College, Wildwood*
Gretel Guest, *Alamance Community College*
Esther Hager, *Passaic County Community College*
Craig Hanke, *University of Wisconsin, Green Bay*
Clare Hoffman Hays, *Metropolitan State College of Denver*
P.C. Hirsch, *East Los Angeles College*
Donald B. Hoagland, *Westfield State College*
Nina Holland, *University of California, Berkeley*
John Hoover, *Millersville University*
Robert J. Huskey, *University of Virginia*
Janice Ito, *Leeward Community College*
R. Kent Johnson, *San Jose City College*
Carol Kasper, *MacMurray College*
Jon Kelty, *Central Michigan University*
Mary Jo Kenyon, *North Dakota State University*
Sherrene D. Kevan, *Wilfrid Laurier University*
Peter King, *Francis Marion University*
William Kleinelp, *Middlesex Community College*
Robert Krebs, *Cleveland State University*
Dominic Lannutti, *El Paso Community College*
Lee Lee, *Montclair State University*
Maureen Leupold, *Gennessee Community College*
James Loats, *Black Hawk College*
Patricia Matthews, *Grand Valley State University*
Ann Maxwell, *Angelo State University*
Marjorie McConnell, *Ramapo College*
Linda McNally, *Davidson College*
Daniel Meinhardt, *University of Wisconsin, Green Bay*
Thomas Mester, *State Center Community College District, Clovis Center*
David P. S. Mork, *St. Cloud State University*
James Murphy, *Monroe Community College*
Angelo Nishimoto, *Leeward Community College*
Nelson Ngoh, *University of Bridgeport*
Robert K. Okazaki, *Weber State University*
Nancy L. Pencoe, *State University of West Georgia*
Polly Phillips, *Florida International University*
Shawn Phippen, *Valdosta State University*
Kim H. Ponto, *Central Ohio Technial College*
Richard Rausch, *Clark College*
Mary Reese, *Mississippi State University*
Bernice Richards, *Northern Essex Community College*
John Rinehart, *Eastern Oregon University*
Beverly A. Roe, *Erie Community College*

Susan Rohde, *Triton College*
Elaine Rubenstein, *Skidmore College*
Matthew Schmidt, *Empire State College and Stony Brook University*
Mark A. Schneegurt, *Wichita State University*
Harold F. Sears, *University of South Carolina, Union*
Richard Shippee, *Vincennes University*
Jeff Simpson, *Metropolitan State College of Denver*
Paulette Snyder, *Metropolitan State College of Denver*
Karen Stine, *Ashland University*
Steven Stocking, *San Joaquin Delta College*
Ken Thomas, *Northern Essex Community College*
Chad Thompson, *SUNY, Westchester Community College*
Janis Thompson, *Lorain County Community College*
Michael Troyan, *Pennsylvania State University*
Anthony Uzwiak, *Rutgers University*
Ann Vernon, *St. Charles Community College*
Mary A. Vetter, *Luther College, University of Regina*
Susan Wadkowski, *Lakeland Community College*
Doris Ward, *Bethune–Cookman College*
Jennifer Warner, *University of North Carolina, Charlotte*
Rosemary Waters, *Fresno City College*
James J. Watrous, *Saint Joseph's University*
Lura Williamson, *University of New Orleans*
Brian D. Wisenden, *Minnesota State University, Moorhead*
Sabrina D. Woodford, *University of Cambridge*
Amie Yenser, *The Pennsylvania State University, Hazleton*
Scott Zimmerman, *University of Wisconsin, Stout*

ACCURACY REVIEWERS

William Bassham
L. Leann Kanda

MEDIA REVIEWERS

James J. Watrous, Lead Advisor, *Saint Joseph's University*
Christine Curran, *University of Cincinnati*
Michael Davis, *Central Connecticut State University*
Claudia Douglass, *Central Michigan University*
Lucinda Elliott, *Shippensburg University*
Deb Firac, *McHenry County College*
Eric Gillock, *Fort Hays State University*
Margarit Gray, *Anderson College*
Clare Hoffman Hays, *Metropolitan State College of Denver*
Donna Huffman, *Calhoun Community College*
George Jacob, *Xavier University*
William Kleinelp, *Middlesex County College*
Mary Katherine Lockwood, *University of New Hampshire*
James Mulrooney, *Central Connecticut State University*
Paulette Snyder, *Metropolitan State College of Denver*
Mary Vetter, *University of Regina*
Susan Wadkowski, *Lakeland Community College*
Jennifer Warner, *University of North Carolina, Charlotte*
Cheryl Watson, *Central Connecticut State University*
Arnold Weisshaar, *Prince Georges Community College*
Brian Wisenden, *Minnesota State University, Moorhead*

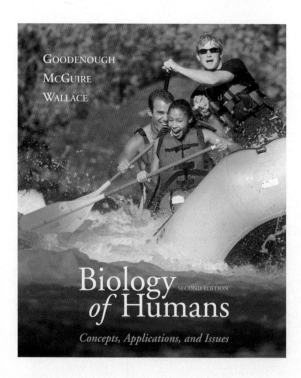

BIOLOGY OF HUMANS: CONCEPTS, APPLICATIONS, AND ISSUES is designed to motivate students to learn basic biological concepts. It does this by providing more examples and applications to students' personal health, their environment, and society than any other book. It also equips students with tools for developing reasoning skills so the information they learn can be applied to life choices.

Biology is a visual science. For that reason graphics have been meticulously designed to help students learn biological concepts. Biology is not a native language for students. For that reason the chapters have been written, rewritten, read by hundreds of reviewers, and carefully developed so that concepts are clearly stated. The combination of exceptional writing and graphics gives students the opportunity to enter into and begin to understand the world of human biology.

Writing For Comprehension

- Accessible writing smoothly integrates just the right amount of detail with intriguing applications to keep students engaged
- Excellent metaphors and analogies abound to anchor concepts to what students already know
- Topic heads are written as declarative sentences to give students an overview and framework for review and understanding
- Occasionally a chapter ends with an *exploring further* box that makes connections between two or more chapters

Applications

- Seven SPECIAL TOPICS Chapters highlight important areas that warrant detailed explanation and exploration, but remain optional to instructors
- Interesting applications are woven into the text at every opportunity
- Contemporary, important issues are addressed in special boxes entitled *Health Issue, Social Issue,* and *Environmental Issue*
- Each chapter ends with questions focused on *Applying the Concepts*

Reasoning Skills

- Periodic interludes entitled *stop and think* and *what would you do?* ask the student to reflect and reason as well as to consider bioethical issues
- Most boxes end with provocative questions that invite students to ponder broader aspects of an issue
- *Applying the Concepts* questions invite critical thinking

Helping Students Understand Human Biology

"The writing style is easy to read and understand. I do choose textbooks that are readable for the students, so this text is similar to others I have used in that respect. I think the students will be engaged and motivated. I like the neutral, factual presentation of the material without being judgmental or emotional about health issues."

—JANICE ITO,
LEEWARD COMMUNITY COLLEGE

There is some truth to the saying "You are what you eat." Rest assured, however, that no matter how many hamburgers you eat, you will never become one. Instead, the hamburger becomes *you*. The transformation is possible largely because of the activities of the digestive system. Like an assembly line in reverse, the digestive system takes the food we eat and breaks the complex organic molecules into their chemical subunits. The subunits are molecules small enough to be absorbed into the bloodstream and delivered to body cells. Food molecules ultimately meet one of two fates: they may be used to provide energy for daily activities or they may provide materials for growth and repair of the body. Imagine that you just ate a hamburger on a bun. The starch in the bun may fuel a jump for joy; the protein in the beef may be used to build muscle; and the fat may become myelin sheaths that insulate nerve fibers. Without the digestive system, this food would be useless to us, because it could never reach the cells of our body.

A simple experiment can demonstrate the effectiveness of venous valves. Allow your hand to hang by your side until the veins on the back of your hand become distended. Place two fingertips from the other hand at the end of one of the distended veins nearest to the knuckles. Then, leaving one fingertip pressed on the end of the vein, move the other toward the wrist, pressing firmly and squeezing the blood from the vein. Lift the fingertip near the knuckle and notice that blood immediately fills the vein. Repeat the procedure, but this time lift the fingertip near the wrist. You will see the vein remain flattened, because the valves prevent the backward flow of blood.

"The text is easy to read, and clearly defines new terms and concepts. It has distilled the chapter to the most relevant material without removing important details. The examples will be familiar to readers, and suggestions are given to relate concepts to known situations. This is very well done."

—MERRILEE ANDERSON,
MOUNT ALOYSIUS COLLEGE

"The writing style perfectly addresses the non-science student population nicely."

—KENNETH THOMAS,
NORTHERN ESSEX COMMUNITY COLLEGE

"I really enjoy the writing style. It is informative and accurate but casual. It will engage the students. This style, in addition to the Special Topics sections and the wonderful companion website, is the main reason I switched to Goodenough et al."

—TAMATHA R. BARBEAU,
FRANCIS MARION UNIVERSITY

Contraction of skeletal muscle squeezes veins. Virtually every time a skeletal muscle contracts, it squeezes nearby veins. This pressure pushes blood past the valves toward the heart. The mechanism propelling the blood is not unlike the one that causes toothpaste to squirt out of the uncapped end of the tube regardless of where the tube is squeezed. When skeletal muscles relax, any blood that moves backward fills the valves. As the valves fill with blood, they extend further into the lumen of the vein, closing the vein and preventing the flow of blood from reversing direction (Figure 12.6b). Thus, the skeletal muscles are always squeezing the veins and driving blood toward the heart.

"The writing style of the Goodenough text has a very student-friendly tone...it is relaxed, easy to read, much like a conversation."

—LINDA McNALLY,
DAVIDSON COLLEGE

"Once again, I like the writing style of Goodenough et al.'s book. I feel that Drs. Goodenough, McGuire, and Wallace's extensive background and experience in teaching biology are reflected in their direct, concise style of presenting the material. Again, as mentioned previously, I have had numerous positive comments about how students like the textbook. These comments are really the first from non-majors students about their textbook and I never had such feedback from the previous and most recent textbooks used for our human biology course."

—ROBERT OKAZAKI,
WEBER STATE UNIVERSITY

Clear and Engaging Writing Motivates Students

8a

SPECIAL TOPIC

Drugs and the Mind

Psychoactive Drugs Alter Communication between Neurons

Drug Dependence Causes Continued Drug Use

Alcohol Depresses the Central Nervous System
- The rate of alcohol absorption depends on its concentration
- Alcohol is distributed to all body tissues
- The elimination of alcohol from the body cannot be increased
- Alcohol has many health-related effects

Marijuana's Psychoactive Ingredient Is THC
- Marijuana binds to THC receptors in the brain
- Long-term marijuana use has many effects on the body
- Legalization of medical marijuana is controversial

Stimulants Excite the Central Nervous System
- Cocaine augments the neurotransmitters dopamine and norepinephrine
- Amphetamines augment the neurotransmitters dopamine and norepinephrine

Hallucinogenic Drugs Alter Sensory Perception

Sedatives Depress the Central Nervous System

Opiates Reduce Pain

M addie had decided to keep her wedding small and was excited to be planning it herself. Even so, she was beginning
to feel overwhelmed by all the details. After weeks of ly-
awake nights running through a long mental checklist of errands to
run and decisions still to be made, she realized that what she nee
more than anything was a decent night's sleep. Her doctor consente
prescribe a mild sedative that helped her get more rest and feel calm
the days leading up to the celebration.

The wedding went as smoothly and happily as she had drea
But a few weeks later, when she was beginning to feel settled in he
life, Maddie stopped taking the sleeping pills and was distressed t
that her sleeplessness and anxiety flared up worse than ever. Ala
she consulted her doctor. Apparently, her nervous system had dev
a dependency on the drug, even in the limited period over wh
had taken it. She and her doctor made a plan that included subs
the original sedative for a similar one and reducing her use of it
ual incr   idn't work she might have to seek more

We may think of alcohol as adding to the festivity of an occasion, but
in fact its effect on the brain is that of a depressant.

7 SPECIAL TOPIC CHAPTERS

focus on topics
of high student interest.

Circulation influences skin color in non-albinos, as well.
When well-oxygenated blood flows through vessels in the dermis,
the skin has a pinkish or reddish tint that is most easily seen in
light-skinned people. Intense embarrassment can increase the
blood flow, causing the rosy color to heighten, particularly in the
face and neck. This response, known as blushing, is impossible to
stop. Other intense emotions may cause color to disappear temporarily from the skin. A sudden fright, for example, can cause a
rapid drop in blood supply, making a person pale. Skin color may
also change in response to changing levels of oxygen in the blood.
Compared to well-oxygenated blood, which is bright red, poorly
oxygenated blood is a much deeper red that gives the skin a bluish
appearance. Poor oxygenation is why the lips appear blue in extremely cold conditions. When it is cold, your body shunts blood
away from the skin to the body's core, which conserves heat and
keeps vital organs warm. This shunting reduces the oxygen supply
to the blood in the small vessels near the surface of the skin. The
oxygen-poor blood seen through the thin skin of the lips makes
them look blue. When you do not get enough sleep, the amount of
oxygen in your blood may be slightly lower than usual, causing the
color to darken. In some people, the darker color is visible through
the thin skin under their eyes as dark circles.

exploring further . . .

In Chapter 19, we learned about the cell cycle. In Chapters 20 and 21, we
learned about genes, their inheritance, and their regulation, and also considered how mutations affect gene functions. In Chapter 21a, we will use this
information to understand cancer, a family of diseases in which mutations in
genes that regulate the cell cycle cause a loss of control over cell division.

APPLICATIONS are woven
into the text more than
any other textbook.

THESE BOXES connect
a chapter to preceding
and following chapters.

SOCIAL ISSUE

The Ethics of Radiation Research on Humans

In the United States, between about 1945 and 1971, a number of people were exposed to radiation as part of research into its effects and uses (Figure 2.A). These people included terminally ill hospital patients, mentally delaye[...] ers, and scientists. Did [...] their families understa[...] periments? Did anyon[...] words, did the particip[...] ing for them, give the[...]

In one set of radia[...] itary's security concern[...] fully informing study p[...] 1945 and 1947, 18 te[...] ceived injections of a s[...] was not disclosed to t[...]

of such questions, terminally ill people are generally no longer used in research, except when the purpose is to test an experimental treatment for their illness.

The principle of informed consent was

able, although in others, informed consent was properly obtained. In 1974, the U.S. Department of Health, Education, and Welfare established regulations for all research with human subjects. However, it was not until 1991 that these requ-

ENVIRONMENTAL ISSUE

Asbestos: The Deadly Miracle Material

During the first three-quarters of the twentieth century, over 30 million tons of asbestos were used for various purposes in the United States. About half a million people are estimated to have died as a result. What is asbestos? Why do we use it? And why do we sometimes die from it?

Asbestos is a fibrous silicate mineral, found in many forms in nature, that is strong, flexible, and resistant to heat and corrosion (Figure 3.A). Be-

Asbestosis is the most common disease caused by exposure to asbestos. It results from the dangerous interaction between asbestos and lysosomes. Apparently, cells responsible for cleaning the respiratory passages engulf small particles of asbestos inhaled into the lungs; lysosomes inside the cleaning cells then fuse with the vesicles containing the asbestos particles. Unfortunately, the lysosomal enzymes cannot break down the asbestos particles. Instead, the particles destabi-

present in many buildings constructed before the ban went into effect. This point was made painfully clear when asbestos was detected in the debris and dust at the site where the World Trade Centers collapsed on September 11, 2001. Asbestos had been used as insulation in parts of the towers.

What can be done about the asbestos already present in our schools and workplaces? It is generally recommended and often required that [...]losed by other [...] a sealant. Ex[...]hod is best in [...]sealing or re-

HEALTH ISSUE

Correcting Vision Problems

Although most people wear glasses to correct vision problems, almost 20 million Americans use contact lenses instead. The contact lens sits on the cornea over a layer of tears. The outer surface of the lens is the corrective surface, and the inner surface fits snugly on the cornea. There are two types of contact lenses. Rigid gas-permeable lenses are made of slightly flexible plastic. These lenses are tough and durable, but some people find that they irritate the eyes. Soft lenses are made of a more malleable plastic. Although soft lenses are gentler on the eyes, they become scratched and damaged more easily.

Contact lenses must be cleaned and disin-

frequent insertion and removal of lenses would reduce trauma to the cornea. However, a serious safety concern with contact lenses that are left in overnight or for extended periods of time is ulcerative keratitis, a condition in which the cells of the cornea may be rubbed away by the contact lens, sometimes leading to infection and scarring. If not promptly treated, ulcerative keratitis can lead to blindness. Another problem with leaving contact lenses in overnight seems to be a reduced oxygen supply to the cornea, which causes it to swell.

Many people who are tired of depending on glasses or contact lenses have opted to undergo laser eye surgery. This procedure is popu-

instead of in the entire cornea, eliminating some of the complications caused by the deeper corneal flaps. LASEK is used mostly for people who are poor candidates for LASIK because their corneas are thin or flat. In yet another procedure, photorefractive keratectomy (PRK), a surgeon uses short bursts of a laser beam to shave a microscopic layer of cells off the corneal surface and flatten it. A computer calculates and controls the laser exposure.

The shape of the cornea can also be altered using a surgical procedure that does not involve lasers. A surgeon can place corneal ring segments, two tiny crescent-shaped pieces of plastic, in the cornea to flatten it. This 15-minute

Rich in Applications and Connections

Art development editors and authors evaluated each illustration to determine how it could best be rendered to teach the concept or process at hand. A color palette was selected for its visual impact as well as its ability to provide contrast, and ultimately to project well as PowerPoint slides or acetates. Colors consistently code for recurring elements in the art program. Photomicrographs complement figures where appropriate.

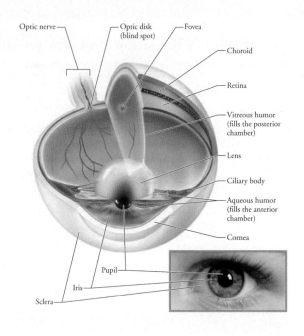

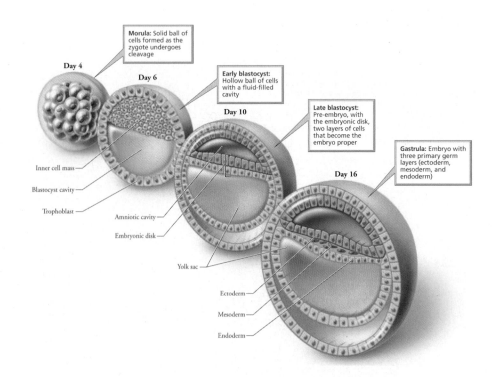

"The illustrations are one of the things that made me choose this text. They show much of the detail that is described in the text, and in a way that complements the text. They are bright and colorful, making them appealing to the eye. Different versions of the pictures also help illustrate sequences of events."

—DON DREWISKE,
UNIVERSITY OF WISCONSIN, GREEN BAY

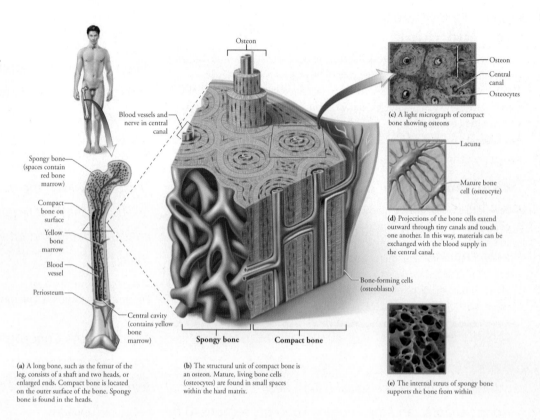

Osteon

Blood vessels and nerve in central canal

Spongy bone (spaces contain red bone marrow)

Compact bone on surface

Yellow bone marrow

Blood vessel

Periosteum

Central cavity (contains yellow bone marrow)

Spongy bone **Compact bone**

Osteon
Central canal
Osteocytes

(c) A light micrograph of compact bone showing osteons

Lacuna

Mature bone cell (osteocyte)

(d) Projections of the bone cells extend outward through tiny canals and touch one another. In this way, materials can be exchanged with the blood supply in the central canal.

Bone-forming cells (osteoblasts)

(e) The internal struts of spongy bone supports the bone from within

(a) A long bone, such as the femur of the leg, consists of a shaft and two heads, or enlarged ends. Compact bone is located on the outer surface of the bone. Spongy bone is found in the heads.

(b) The structural unit of compact bone is an osteon. Mature, living bone cells (osteocytes) are found in small spaces within the hard matrix.

Art That Teaches

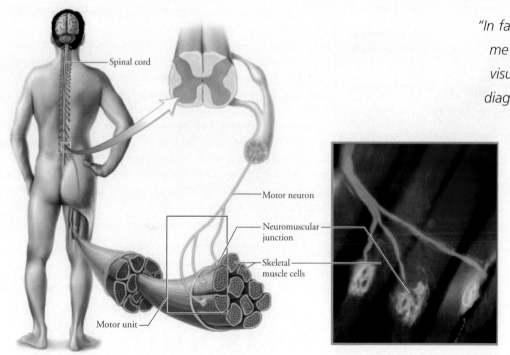

Spinal cord

Motor neuron

Neuromuscular junction

Skeletal muscle cells

Motor unit

stop and think

It takes an ATP molecule to break the cross-bridges so that new ones may be formed. Without ATP, cross-bridges cannot be broken, and the muscle becomes stiff. When a person dies, ATP is no longer produced. How does this explain the stiffening of muscles, known as rigor mortis, that begins 3 to 4 hours after death?

what would you do?

Few people would question the justifiability of providing drugs to elevate neurotransmitter function to someone who is depressed or suicidal in order to help the person live a normal life. However, researchers believe that the levels of key neurotransmitters also affect personality traits, such as shyness or impulsiveness. If so, we may someday be able to design our own personalities. Should minor personality problems be treated with drugs? Should a personality "flaw" be treated? What do you think?

"Stop and Think Questions: 'Again, these questions engage students and I would definitely bring them up in lecture.'"

—MARY REESE,
MISSISSIPPI STATE UNIVERSITY

"I especially liked the questions in the 'what would you do' boxes that will require students to go through critical thinking and integrating knowledge learnt on this topic."

—ALESSANDRO CATENAZZI,
FLORIDA INTERNATIONAL UNIVERSITY

"Applying the Concepts — this section is great. Students can use this information in small groups to study and to reinforce integration of the material."

—CLAUDIA DOUGLASS,
CENTRAL MICHIGAN UNIVERSITY

"I would consider the social issues, health issues, special topics chapter, and the applying concepts sections to be strengths. I especially like the applying concepts section because this section allows the students to apply critical thinking and demonstrate their ability and understanding of the concepts for that chapter."

—SHAWN PIPPEN,
VALDOSTA STATE UNIVERSITY

"The 'Reviewing the Concepts' gives students a variety of question styles to practice their knowledge and 'Applying the Concepts' provides excellent opportunities to practice critical thinking skills."

—LINDA MCNALLY,
DAVIDSON COLLEGE

what would you do?

There are various reasons that many foods today are intentionally irradiated, including (1) removal of harmful microorganisms (pathogens), insect pests, and parasites; (2) delaying spoilage and increasing shelf life; and (3) sterilizing the material in which the food is packaged. The food does not become radioactive as a result. Supporters of the practice note that test animals fed on irradiated food show no adverse effects. Opponents, however, point to the environmental risks of building and operating food irradiation plants. Opponents also note that carefully controlled experiments are still needed to verify that irradiated food is safe for people of all ages and nutritional states. Several foods, including wheat flour, poultry, and white potatoes, have been approved for irradiation in the United States. These irradiated foods have a distinctive logo (Figure 2.5) on their packaging. Would you eat irradiated food? Explain your answer.

Critical Thinking
and Reasoning Skills

Instructor Support

Instructor Resource Center on CD/DVD

(0-13-230266-7)

Multiple-disc package of media support that includes:

- JPEGs of all textbook figures (labeled and unlabeled), photos, and tables with labels and colors fully optimized for projection
- Image Gallery PowerPoint® presentation of all JPEGs for each chapter
- Suggested Lecture PPT presentations for each chapter
- 200+ Instructor Animations preloaded into PPT files in PC and Macintosh versions
- Prentice Hall BLAST (Biology Lecture Animation and Simulation Tool) 3-D modules vividly depicting biological concepts, preloaded into PPT files in PC and Macintosh versions
- Classroom Response System (CRS) "Clicker" questions for each chapter in PPT files
- Instructor's Resource Guide and Instructor's Test Item File in an editable Word format

—See pages xxvi & xxvii for more details

Transparency Pack (0-13-179015-3)

Includes 400 illustrations from the text on full-color acetates.

Instructor's Guide (0-13-179013-7)

By Jennifer Warner, University of North Carolina at Charlotte. This tool provides tips for teaching non-majors by making the material relevant, interesting, and interactive. Each chapter includes: Lecture Activity suggestions that include objectives, procedures, and assessment suggestions; Objectives that identify goals for students and instructors; and Answer Key to all end-of-chapter questions.

Test Item File (0-13-179014-5)

More than 2000 questions were carefully prepared by a team of educators and coordinated by Jennifer Warner, University of North Carolina, Charlotte.

Test Gen-EQ (0-13-179016-1)

This text-specific testing program is networkable for administering tests. It allows you to edit existing test items or add your own questions to create a nearly unlimited number of quizzes and tests.

Understanding Human Biology:

Laboratory Exercises

By Mimi Bres and Arnold Weisshaar, Prince George's Community College

(0-13-179009-9)

- Written in a style which is easily accessible to students with no previous background in biology
- Focuses on collaborative, small group activities that encourage student interactions and maximize laboratory resources
- Focuses on understanding basic principles of human anatomy and physiology, genetics and evolutionary change, ecology, and the impact of human actions on the environment
- Exercises are designed to run smoothly, even in large laboratory sections with 25–40 students.
- Equipment and supplies needed are cost effective and easily accessible to large and small schools
- Each lab exercise is suitable for completion in two or three hour-long periods and can be divided to accommodate 90-minute sessions

Student Support

Study Guide (0-13-179001-3)

By Claudia Douglass of Central Michigan University
This printed guide offers learning objectives; key concepts; chapter summaries; labeling exercises; as well as short-answer, fill-in-the-blank, and critical-thinking questions. Each chapter concludes with a multiple-choice Practice Test to help students gauge their understanding.

Study Lecture Notebook (0-13-17900-5 CD)

A lecture notebook for students that contains a selection of figures in the main text.

Companion Website
www.prenhall.com/goodenough

A full array of student learning features including a comprehensive Online Study Guide, interactive Chapter Quizzes, engaging Web Tutorials, and student versions of Prentice Hall's unique BLAST 3-D animated modules vividly depicting important biological concepts. There is an access code for the Companion Website in the front of new copies of this textbook. Others can purchase access online.

—See pages xxiv & xxv for more details

Full Support for
Instructors and Students

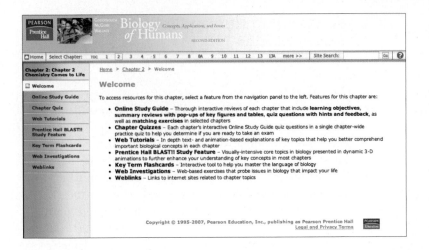

Companion Website—
www.prenhall.com/goodenough

This 24/7 study tool is designed to help students make the most of their limited study time. There is an access code in front of new copies of the textbook. Others can purchase access online. The Companion Website's most beneficial study features are the Chapter Quizzes and Online Study Guide (see below), as well as the Web Tutorials (see facing page).

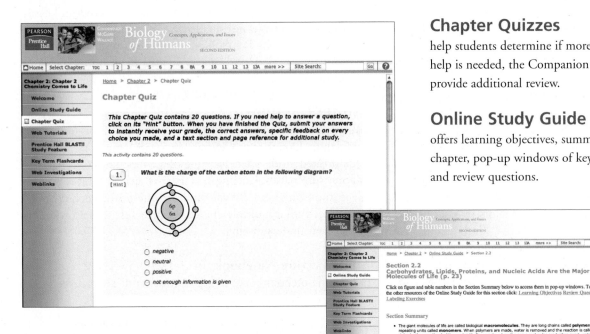

Chapter Quizzes

help students determine if more study is required. If more help is needed, the Companion Website's other features provide additional review.

Online Study Guide

offers learning objectives, summaries of each section of each chapter, pop-up windows of key figures, interactive exercises, and review questions.

Website
Maximizes Student Study

Web Tutorials review concepts from the text in an engaging tutorial format. 71 Tutorials guide you through biological concepts using animation, instructional narration, and review quizzes that help students master the most challenging topics. Icons have been placed throughout the book to indicate that relevant Web Tutorials are available.

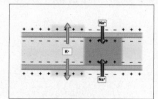
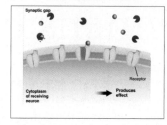
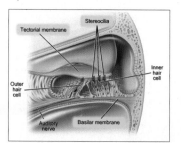

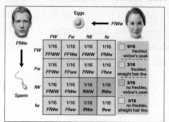

For Instructors
Instructor Resource Center on CD/DVD
Our multiple-disc package includes all the tools you need to build engaging lectures—including JPEGs of every figure, photo, and table in the text with labels and colors optimized for projection, a full complement of PowerPoint® presentations, and Prentice Hall BLAST 3-D animations. Download resources quickly with our new browser-based Instructor Resource Center on CD/DVD.

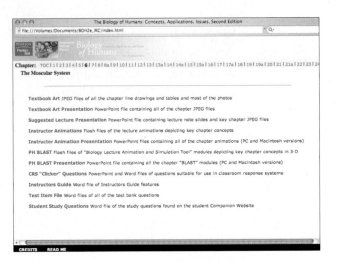

Fully Optimized JPEGs of Every Textbook Figure, Photo, and Table
Images that project poorly in lecture halls can diminish the impact of your lectures. We have taken added steps to ensure that the stunning figures in this textbook look every bit as good projected in your lecture hall as they do in the book.

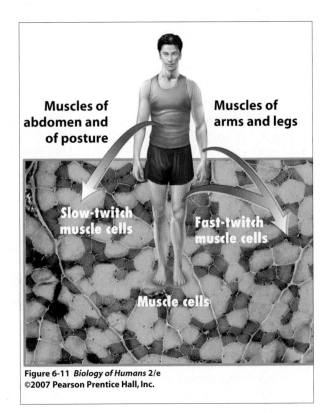

Figure 6-11 *Biology of Humans 2/e*
©2007 Pearson Prentice Hall, Inc.

A New Dimension
Added to Your Lectures

BLAST

BIOLOGY LECTURE ANIMATION & SIMULATION TOOL

Prentice Hall's BLAST

Available only with our books—BLAST offers unparalleled vivid 3-D animation and scientific precision to enliven lectures. BLAST is designed to support individual teaching approaches and class needs.

Chapter 2
Chemical Bonds
Protein Structure
Nucleic Acid Structure and Function
Enzymes

Chapter 3
Cell Size and Function
Osmosis
Diffusion and Transport
Diffusion

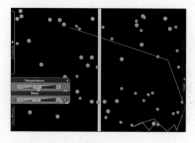

Animal Cell Overview
Structure of ATP
Formation and Breakdown of ATP in
  the Cell
Harvesting Energy Overview
Glycolysis
Krebs (Citric Acid) Cycle
Electron Transport Chain

Chapter 4
Feedback Loops
Negative Feedback: Body Temperature
Positive Feedback: Labor

Chapter 6
How Muscle Works

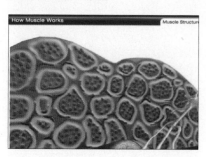

Chapter 7
Action Potential
Signal Transmission at Synapses

Chapter 10
Cell Signaling: Types
Cell Signaling: Amplification

Chapter 12
Cardiac Cycle Overview
Electrical Coordination of Cardiac Cycle
Details of Cardiac Cycle

Chapter 13
Innate (Nonspecific) Immunity
Adaptive Immunity

Chapter 13a
Antibiotic Resistance in Bacteria

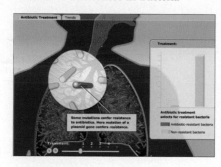

Chapter 16
Kidney Anatomy and Function

Chapter 17a
HIV Structure
HIV Life Cycle
AIDS Onset
AIDS Treatment Strategy

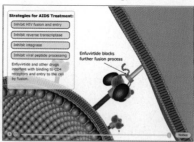

Chapter 19
DNA Packaging
Mitosis
Meiosis
Mitosis and Meiosis Compared
Genetic Variation from Meiosis

Chapter 19a
Stem Cells

Chapter 20
Single-Trait Crosses
Two-Trait Crosses

Chapter 21
DNA Replication
Transcription and Translation
Electrophoresis and DNA Fingerprinting

Chapter 21a
Cell Cycle Control and Cancer

Chapter 22
Homologous Limb Development

Chapter 23
Energy and Nutrient Cycles

BLAST

Contents

4 Body Organization and Homeostasis 62

PART II | CONTROL AND COORDINATION OF THE BODY

5 The Skeletal System 82

6 The Muscular System 97

Neurons: The Matter of the Mind 110

The Nervous System 124

SPECIAL TOPIC

8a Drugs and the Mind 142

9 Sensory Systems 152

The Endocrine System 174

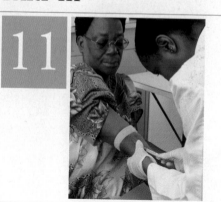

MAINTENANCE OF THE BODY

PART III

Blood 197

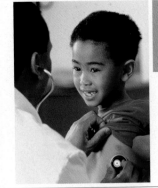

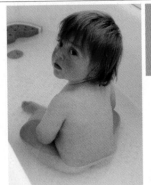

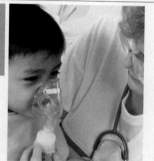

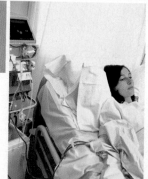

The Urinary System   329

REPRODUCTION

PART IV

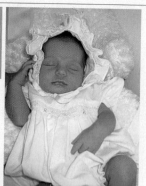

Reproductive Systems   349

17a

SPECIAL TOPIC

Sexually Transmitted Diseases and AIDS 370

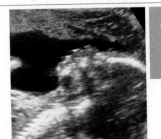

18

Development throughout Life 382

GENETICS AND DEVELOPMENT

PART V

Chromosomes and Cell Division 405

SPECIAL TOPIC

Stem Cells—The Body's Repair Kit 424

EVOLUTION AND ECOLOGY

PART VI

Special-Interest Essays

ENVIRONMENTAL ISSUE ESSAYS

HEALTH ISSUE ESSAYS

SOCIAL ISSUE ESSAYS

1

Humans in the World of Biology

All Living Things Share Basic Characteristics

Living Organisms Are Classified by Evolutionary Relationships

Life Has Many Levels of Organization

The Scientific Method Gathers Information for Drawing Logical Conclusions

- Inductive and deductive reasoning help solve problems
- Clinical trials follow strict guidelines
- Epidemiological studies look for patterns in populations

Critical Thinking Helps Us Make Informed Decisions

 **ENVIRONMENTAL ISSUE**   Medicinal Plants and the Shrinking Rain Forest

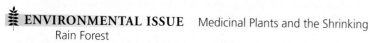 **SOCIAL ISSUE**   Scientific Misconduct

The dugout canoe, with four on board, turns from the muddy river and approaches the slippery bank. A tribesman cheerfully steps off into mud up to his knees and, laughing at his friend, who is pushing with a pole from the back, shoves the canoe as far as possible onto the bank and ties it to a tree.

The tribesman helps you to the grassy area, and you peer through the tangled, viney vegetation into the dark interior of the forest. Already you are soaked with sweat. Still, you are eager to do what you came for—explore this part of the Amazon rain forest, searching for medicinal plants. You know that two chemicals from the rosy periwinkle of Madagascar are being used to treat leukemia and Hodgkin's disease (Figure 1.1). You also know that the Amazon rain forest, home to millions of species of plants, is being destroyed at an alarming rate. You want to learn its secrets before they disappear forever.

To learn those secrets, you need to talk to the native people about the plants they use for healing. You have spent months preparing—learning about the culture of the native people, cataloging the information already available on local plants and their uses, learning how to collect and preserve the plants, and gathering tools and supplies. A physician helped prepare simple descriptions and photographs of diseases, which will help you communicate with local healers. If these people recognize the symptoms you describe, they will show you the plants they use to treat them. In laboratories at home, chemists and biologists will study the plants you have collected to determine their medicinal value. ∎

Humans are a small, but important, part of the diverse life on the earth.

FIGURE **1.1** The rosy periwinkle (*Catharanthus roseus*) is a source of two anticancer drugs.

The rain forest is an excellent place to begin our consideration of human biology. Its incredible diversity of life—the bewildering variety of plants, the brilliant colors of the flowers, the strange songs of unseen birds—reminds us that humans are just a small, although important, part of life on the earth. We share certain characteristics with all other living things and have characteristics that set us apart from them.

As we look around, we can see that life has many levels of organization: individual, population, community, ecosystem, and biosphere. In most of this book, we will focus on the human individual—how the individual body functions, and the biological principles that govern those functions. However, we will also examine many of the larger health, social, and environmental issues that we as individuals must be aware of, for the very reason that they can affect us all. Our actions affect the health of the rain forest. At the same time, the rain forest is a source of biological substances that may be used to improve human health.

All Living Things Share Basic Characteristics

As you continue exploring the rain forest, how do you know if something unfamiliar to you is alive? In some cases, the question is easy to answer. Although the leaves around you have different shapes and sizes, a brief examination assures you that they are leaves. And the tree whose trunk you are exploring is clearly a tree, but what about the gray material adhering to the trunk? How can you tell if it is alive?

Defining life might seem to be easy, but it is not. In fact, no single definition would satisfy all life scientists. For example, if we say you can tell something is alive if it reproduces, someone is likely to note that a page with a wet ink spot can fall on top of another page and reproduce itself almost exactly. If we say you can tell something is alive if it grows, what should we conclude about crystals? They grow, but they are not alive. And so it goes. It seems there is no single defining feature of life.

No single definition applies to all forms of life, so we find that instead of defining life, we can only characterize it. That is, we can only list the traits associated with life. Most biologists agree that, in general, the following statements characterize life.

1. **Living things contain nucleic acids, proteins, carbohydrates, and lipids.** The same set of slightly more than 100 elements is present in various combinations in everything on the earth—living or nonliving. However, living things can combine certain elements to create molecules that are found in all living organisms: nucleic acids, proteins, carbohydrates, and lipids. The nucleic acid DNA (deoxyribonucleic acid) is especially important because DNA molecules can make copies of themselves, an ability that enables organisms to reproduce (Figure 1.2a). The molecules of life are discussed in Chapter 2.

2. **Living things are composed of cells. Cells** are the smallest units of life. Some organisms have only a single cell; others, such as humans, are composed of trillions of cells (Figure 1.2b). All cells come from preexisting cells. Because cells can divide to form new cells, reproduction, growth, and repair are possible. Cells are discussed in Chapter 3 and cell division in Chapter 19.

3. **Living things reproduce.** Living things have ways of generating new individuals that carry some of the genetic material of the parents. Some organisms, such as bacteria, reproduce simply by making new and virtually exact copies of themselves. Other organisms, including humans, reproduce by combining genetic material with another individual (Figure 1.2c). Reproduction is discussed in Chapter 17.

4. **Living things use energy and raw materials. Metabolism** refers to the sum total of all chemical reactions that occur within the cells of living things. Through metabolic activities, organisms extract energy from various nutrients and transform it to do many kinds of work. Metabolism maintains life and allows organisms to grow (Figure 1.2d). Chemical reactions involved in the transformation of energy are discussed in Chapter 2.

5. **Living things respond.** A boxer weaves and ducks to avoid the blow of an opponent. A chameleon takes aim at and captures its prey (Figure 1.2e). For a living thing to respond, it must first detect a stimulus and then have a way to react. As you will learn, your sensory organs detect stimuli. Your nervous system processes sensory input, and your skeletal and muscular systems enable you to respond. The skeletal and muscular systems are discussed in Chapters 5 and 6. The nervous system is discussed in Chapter 8, and sensory organs in Chapter 9.

6. **Living things maintain homeostasis. Homeostasis** is the relatively constant and self-correcting internal environment of a living organism. We generally find that life can exist only within certain limits and that living things tend to behave in ways that will keep their body systems functioning within those limits. For example, if you become too cold, you shiver (a metabolic response). Shivering produces heat that warms your body. Alternatively, if you become too hot, your sweat glands will be activated to cool you down. In addition to these and other physiological responses, the sensation of being hot or

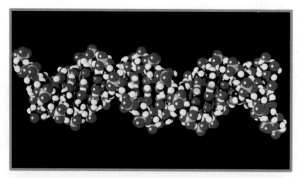

(a) Living things contain nucleic acids, proteins, carbohydrates, and lipids.
This is a computer-generated model of the nucleic acid DNA, which carries genetic information.

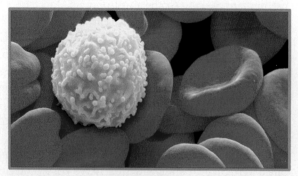

(b) Living things are composed of cells.
These are red blood cells (disks) and a white blood cell (sphere).

(c) Living things reproduce.
All organisms reproduce their own kind.

(d) Living things use energy and raw materials.
This mother orangutan is sharing bananas with her offspring. This food will provide energy and raw materials for the maintenance, repair, and growth of their bodies.

(e) Living things respond.
This chameleon sees and catches its prey.

(f) Living things maintain homeostasis.
This child cools himself, a behavioral response to maintain a relatively constant body temperature.

(g) Populations of living things evolve and have adaptive traits.
The orchid is adapted to live perched on branches of trees. It uses other plants for support so that it can receive enough sunlight to produce its own food by photosynthesis.

FIGURE **1.2** Characteristics of life

cold may motivate you to behave in ways that cool you down or warm you up (Figure 1.2f). We will discuss homeostasis in Chapter 4, where we make an initial survey of body systems.

7. **Populations of living things evolve and have adaptive traits.** Members of a population possessing beneficial genetic traits will survive and reproduce better than members of the population that lack these traits. As a result of this process, called *natural selection*, each of the amazing organisms you see around you has **adaptive traits**—that is, traits that help it survive and reproduce in its natural environment. Most plants in the rain forest have shallow root systems, because the topsoil is thin and nutrients are near the surface. As a result, tall trees have acquired, through evolution, supports like cathedral buttresses to hold them up, while vines climb over both roots and trees to reach the light. Many plants do not grow in the ground at all but live high above it in the canopy. These plants, called epiphytes, the group to which orchids belong, are rooted on the surfaces of other plants (Figure 1.2g). Rain forest animals also have adaptations—the ability to fly or climb, for example—that enable them to reach the plants for food. Evolution is discussed in Chapter 22.

stop and think

Scientists have discovered pools of water on the surface of a small moon of Saturn called Enceladus. These pools of water may be the most likely place in the solar system to find extraterrestrial life. If samples of water from Enceladus were brought back to scientists on earth, what characteristics of life could they look for to determine whether the samples contain anything that is or was once alive?

Living Organisms Are Classified by Evolutionary Relationships

At least 10 million species of organisms live on the earth. Scientists organize, or classify, living organisms in a way that shows evolutionary relationships among them. This means, for the most part, that organisms with the greatest similarity are grouped together.

Several classification systems have been proposed. One system favored by many biologists recognizes three domains. Two of the domains, Bacteria and Archaea, consist of the various kinds of prokaryotes, all of which are single-celled organisms that lack a nucleus or other internal compartments. All other organisms, including humans, belong to the third domain, Eukarya. Organisms in domain Eukarya have eukaryotic cells, which contain a nucleus and complex internal compartments called organelles. Domain Eukarya is subdivided into four kingdoms, as shown in Figure 1.3. Within each kingdom, organisms are further categorized into groups whose members share characteristics that distinguish them from members of other groups in the kingdom. The kingdoms are first divided into large groups, and those are subdivided into smaller groups to show successively closer relationships.

As humans, we belong to a subdivision of the animal kingdom called vertebrates (animals with a nerve cord protected by a true backbone), and more specifically to the group called mammals. Two characteristics that make us mammals are that we have hair and that we feed our young milk produced by mammary glands. However, we are further defined as belonging to the primates, along with lemurs, monkeys, and apes, because we share a suite of features that includes forward-looking eyes and a particularly well-developed brain. Humans, monkeys, and apes also have opposable thumbs (a thumb that can touch the tips of the other four fingers).

Domain Bacteria	Domain Archaea	Domain Eukarya			
Unicellular prokaryotic organisms	Unicellular prokaryotic organisms; most live in extreme environments	Eukaryotic cells that contain a membrane-bound nucleus and internal compartments			
		Kingdoms			
		Protists	**Fungi**	**Plants**	**Animals**
		Protozoans, algae, slime molds	Molds, mushrooms	Mosses, ferns, seed plants	Invertebrates and vertebrates

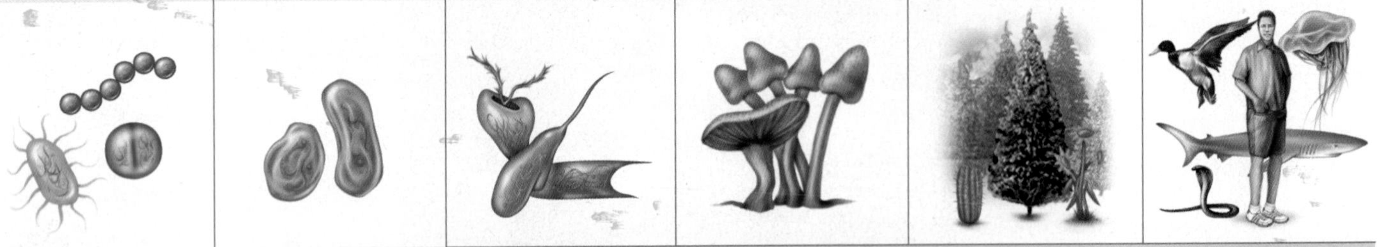

FIGURE **I.3** The three domains and four kingdoms of life

FIGURE **1.4** Social interactions are an important thread in the fabric of human life.

Smaller details, such as tooth structure and skeletal characteristics, serve to divide the primates into smaller groupings.

What most sets humans apart from all other living species? Human characteristics include a large brain size relative to body size and a two-legged gait. But nothing distinguishes humans more than culture. *Culture* may be regarded as a set of social influences that produce an integrated pattern of knowledge, belief, and behavior (Figure 1.4). Other animals, of course, have social interactions, from various forms of cooperation and mating behavior to territoriality, competition, and social hierarchies. But in these cases, social influences are much less pronounced than in the corresponding human interactions. Consider, for example, our rituals—weddings, graduation, burial of the dead—and the way our lives are enriched with art, music, and dance.

Keep in mind, however, that there is not one human culture but many. If you were a scientist following native guides through the Amazon rain forest, you would quickly learn that their culture is different from ours. In fact, there are many different cultures in the rain forest. Separate groups of people in the same environment do not adapt to it in the same way. If you were to describe your culture to someone from a rain forest tribe your description may elicit anything from astonishment to howls of laughter. Even so, you would notice that some things, especially love of children and the need to belong to a group, seem the same almost everywhere. Ultimately, whatever the similarities or differences between human populations, culture powerfully defines the human condition.

Life Has Many Levels of Organization

As we study human biology, we will see that life can be organized on many levels (Figure 1.5). Cells, the smallest unit of life, are composed of molecules. A multicellular organism may consist of different tissues, groups of similar cells that perform specific functions. Organs also may consist of different types of tissue that work together for a specific function. Two or more organs working together to perform specific functions form an organ system. Humans are described as having 11 organ systems, as we will see in Chapter 4.

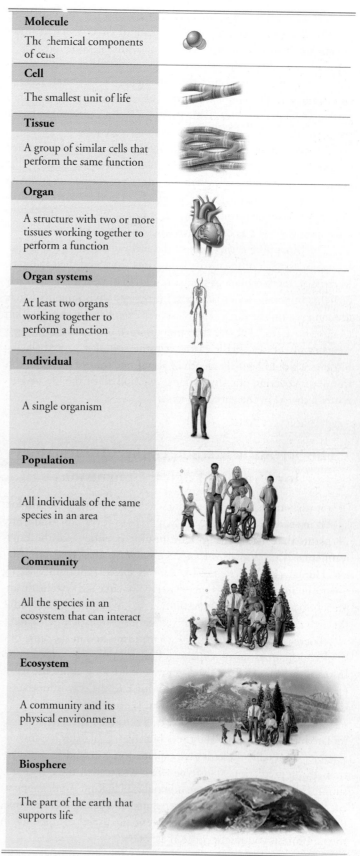

Molecule	
The chemical components of cells	
Cell	
The smallest unit of life	
Tissue	
A group of similar cells that perform the same function	
Organ	
A structure with two or more tissues working together to perform a function	
Organ systems	
At least two organs working together to perform a function	
Individual	
A single organism	
Population	
All individuals of the same species in an area	
Community	
All the species in an ecosystem that can interact	
Ecosystem	
A community and its physical environment	
Biosphere	
The part of the earth that supports life	

FIGURE **1.5** Levels of organization of life

Life can also be organized at levels beyond the individual organism. A **population** is all the individuals of the *same species* (individuals that can interbreed) living in a distinct geographic area. A population might be the yellow-bellied marmots living in a meadow or the four-eyed butterfly fish living in a coral reef. A **community** is all living species that can potentially interact in a particular geographic area. For example, a community might be *all the species* that live and interact in an alpine meadow.

An **ecosystem** includes all the living organisms in a defined area and their physical environment. The size of the locality that defines the ecosystem varies with the interest of the person studying it. An ecosystem can be defined as the whole earth, a particular forest, or even a single rotting log within a forest. Whatever its size, an ecosystem is viewed as being relatively self-contained.

The **biosphere** is that part of the earth in which life is found. It encompasses all of the earth's living organisms and their habitats. In essence, the biosphere is the narrow zone in which the interplay of light, minerals, water, and gases produces environments where life can exist on the earth. The biosphere extends only about 11 km (7 mi) above sea level and the same distance below, to the deepest trenches of the sea. If the earth were the size of a basketball, the biosphere would have the depth of about one coat of paint. In this thin layer covering one small planet, we find all of the life *we* currently know of in the entire universe.

The Scientific Method Gathers Information for Drawing Logical Conclusions

I t is not surprising that there are scientists exploring the rain forest in search of any secrets it may have to reveal, including any plants that may prove to have healing qualities (see the Environmental Issue essay *Medicinal Plants and the Shrinking Rain Forest*). Humans are an irrepressibly curious species, constantly asking questions about the things they observe. **Science** is a systematic approach to answering those questions, a way of acquiring knowledge through carefully documented investigation and experimentation.

There is no such thing as *the* scientific method in the sense of a single, formalized set of steps to follow for doing an experiment. Instead, the **scientific method** is a way of learning about the natural world by applying certain rules of logic to the way information is gathered and conclusions are drawn. It often begins with an observation that raises a question. Next, a possible explanation is formulated to answer the question, but that explanation must be testable. Generally, the tentative explanation will lead to a prediction. If the prediction holds true when it is tested, the test results support the explanation. If the original explanation is not supported, an alternative explanation is generated and tested. For example, if you were a scientist looking for potentially useful compounds in the rain forest, you might proceed as follows:

1. **Make careful observations and ask a question about the observations.** If, while walking through the rain forest, you came across a plant—which you later identify as a lemon ant plant—growing alone, with no other plants growing around them, you might record and describe that observation (Figure 1.6). Considering the density of different plants that you see on most of the forest floor, the relative isolation of this group of plants strikes you as odd, and you make up your mind to discover the reason. You begin your study of the situation by collecting information about the plant and its surroundings. This includes, among other things, measuring the area of the patch, the approximate number of plants in it, and the size of the cleared area around the plant. You also look for another patch, make the same measurements, and compare them to see whether there is some correlation between the size of the patch and the area of the clearing. Your measurements suggest that the larger the patch, the larger the cleared area.

2. **Develop a testable hypothesis (possible explanation) as a possible answer to your question.** Now you start to formulate ideas about why other plants do not grow near patches of lemon ant plants. Your first guess is that the lemon ant plants are producing a chemical that inhibits the growth of other plants. This statement is your **hypothesis**–a testable explanation for your observation. A hypothesis that cannot be tested is worthless because it cannot be supported or refuted. However, a hypothesis that is tested and shown to be incorrect can still be useful in developing alternative hypotheses.

3. **Make a prediction based on your hypothesis and test it with a controlled experiment.** Now you must plan an experiment to test your hypothesis. You might predict, for example, that if your hypothesis is correct, the chemical alone, in the absence of the intact plant, will inhibit the growth of other plants. You therefore select one lemon ant plant and liquefy its tissues in alcohol, using a blender, to produce an extract of all the chemicals in the plant that are soluble in alcohol. This extract can be used to treat the seeds of other plants to determine its effects on their growth.

FIGURE **1.6** A lemon ant plant harbors small ants considered delicious by local inhabitants. Few or no other plants grow around a lemon ant plant.

ENVIRONMENTAL ISSUE

Medicinal Plants and the Shrinking Rain Forest

The healing powers of many plants have been known for centuries. Historically, such knowledge was gained by trial and error and passed along by word of mouth. For example, many cultures have long known that tea made from willow bark relieves pain and reduces fever. Much later, scientists learned that willow bark contains salicylic acid. They isolated the compound and developed it into the drug we know today as aspirin. Similarly, digitalis, a heart medication, was discovered after a patient with an untreatable heart condition was seen to benefit from an herbal drink provided by a local gypsy. The potion contained purple foxglove, which, like willow bark, is frequently mentioned in ancient texts as a healing herb. Broccoli, a more familiar plant, contains the anticancer chemical sulforaphane.

Medicinal plants are still important today. More than 25% of the prescription medicines sold in the United States contain chemicals that came from plants. Many more healing chemicals first discovered in plants used medicinally by native people are now routinely synthesized in laboratories (Table 1.A).

Although medicinal plants can be found almost everywhere, a particularly rich source is the tropical rain forest. The American Cancer Society has identified approximately 3000 plants that have anticancer properties, and 70% of them are found in the rain forest. As a result, there are now more than 100 drug companies funding projects to learn about medicinal plants from native people living in tropical rain forests.

Most plants that have proved to be medically useful are found in the tropics, regions where the human population is growing rapidly. Unfortunately, the forests in these regions are being cut to create living space and foster economic development. In Madagascar, home of the rosy periwinkle (Figure 1.1), which is the source of two cancer drugs, humans have destroyed 90% of the vegetation. Experts estimate that roughly nine-tenths of the original tropical rain forest will have been destroyed by 2030. Considering that 155,000 of the known 250,000 plant species are from tropical rain forests, and that less than 2% of the known plant species have been tested for medicinal value, we have no way of knowing what potential new medicines are being destroyed.

TABLE 1.A   SELECTED MEDICINAL PLANTS AND THEIR USES

SOURCE PLANT	COMPOUND	MEDICINAL USE
Willow bark and meadowsweet	Salicylic acid (active ingredient in aspirin)	Pain relief, fever reduction
Foxglove	Digitalin	Treatment of heart conditions (increase intensity of heart contraction, slows heart rate)
Quinine tree	Quinine	Malaria preventive
Opium poppy	Morphine	Pain relief
Eucalyptus tree	Menthol	Decongestant
Rauvolfia root	Reserpine	Treatment of high blood pressure
Belladonna	Atropine	Dilates pupils for eye exams
Pacific yew	Taxol	Treatment of ovarian and breast cancer
Rosy periwinkle	Vincristine	Treatment of Hodgkin's disease (a type of cancer)
	Vinblastine	Treatment of leukemia

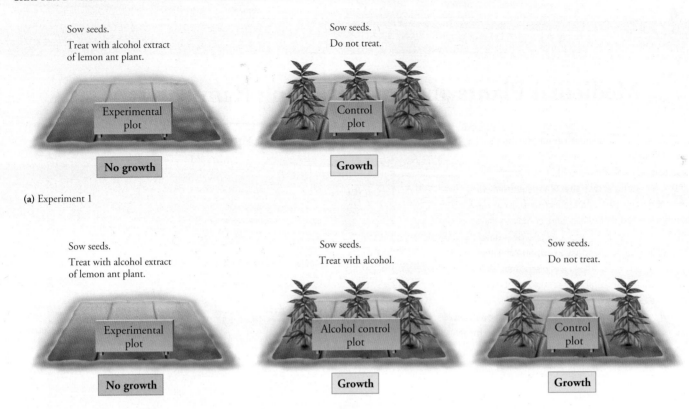

(a) Experiment 1

(b) Experiment 2

FIGURE 1.7 In a controlled experiment, subjects are divided into groups that are exactly alike except for the variable. In both experiments shown here, the plant seeds are the subjects. (a) In experiment 1, the variable is the extract of the lemon ant plant. (b) In experiment 2, a second control is added to the design to determine whether the alcohol used to prepare the extract was responsible for the inhibition of growth.

If possible, your experiment must be designed in such a way that there can be only one explanation for the results. In a **controlled experiment**, for example, the subjects (in this case, plant seeds) are randomly divided into two groups, one called "the control group" and one called "the experimental group" (Figure 1.7a). Both groups are treated in the same way except for the *one* factor, called the **variable**, whose effect the experiment is designed to reveal. In this case, the variable is the extract of plant chemicals. To discover whether those chemicals inhibit the growth of other plants, you prepare two identical plots of ground and seed them both with the same variety of plants. You then spread the plant extract over the experimental plot, but you leave the control plot untreated. Sure enough, no new plants grow in the area that you treated with the alcohol extract, but plants grow well in the untreated areas.

Do these results support your hypothesis? They seem to ... but are you certain that the alcohol in the extract is not what is preventing the seeds in the experimental plot from growing? In the first experiment, without intending to, you actually had two variables: the alcohol and the extract itself. In a scientific study, additional variables that have not been controlled for and may have affected the outcome are called *confounding variables*. When there are confounding variables, we cannot say for sure which variable or variables caused the effect. So you experiment again, this time with two control groups and an experimental group. In other words, you use three plots, as identical to one another as possible. You till the soil, plant the seeds, and water each plot. Next, you treat one plot with the alcohol extract; you treat another plot with alcohol only; and you leave the third plot untreated (Figure 1.7b). Then you wait. Sure enough, plant growth is retarded only in the area treated with the extract. Plants grow in the plot treated only with alcohol and in the untreated plot. Your experimental results indicate that the chemicals in the plant tissue are inhibiting the growth of other plants.

4. **Draw a conclusion based on the results of the experiment.** Next, you arrive at a conclusion. Your conclusion, in this case, may be that the lemon ant plant produces a chemical that inhibits the growth of other plants.

Note, however, that although your results support your hypothesis and by extension your conclusion, they do not *prove* the hypothesis. A hypothesis can be proven wrong when the results do not support it, but results that are consistent with a hypothesis do not prove that it is correct. There may be more than one possible explanation for the results you obtained. For instance, perhaps the lemon ants that live on the plant deposit a chemical on its surface that washes off with rain and inhibits plant growth in the surrounding area. The ant-produced

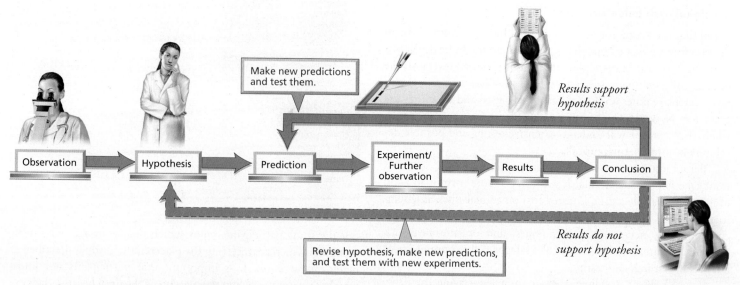

FIGURE **1.8** The scientific process consists of observation, creating testable hypotheses, experimentation, drawing conclusions, revising hypotheses, and designing new experiments.

chemical might also have gotten into your extract, along with the chemicals produced by the plant itself. Thus, even successful experiments often lead to revised hypotheses and further experimentation (Figure 1.8).

What if the plant extract had not inhibited the growth of other plants? This would mean that your hypothesis needs to be rethought and that you may have to find another explanation. Each new or revised hypothesis calls for new experiments (with new controls) and new conclusions. Maybe the inhibitory chemical is not soluble in alcohol. Maybe the chemical is a gas. Maybe it does not cause the inhibition of growth but removes nearby plants in other ways. Essentially, you must now look for new variables. You might want to go back to the forest and note whether there are other insects living on the plant and affecting the surroundings in some way. Or maybe the lemon ant plant was not *actively* inhibiting the growth of other plants but was better at obtaining something necessary for plant growth, such as water or a specific nutrient. How could you test these ideas?

Another requirement of scientific inquiry is that experiments must be repeated and yield similar results. Other scientists following the same procedure should obtain an outcome that is similar. Note, however, that it can be very difficult to identify all the factors that might affect the outcome of an experiment. In our rain forest experiment, for example, rainfall, the amount of light, and soil characteristics might all influence the results. Each time an experiment is repeated, additional factors may be identified as important.

The testing and refinement of a hypothesis represents one level of the scientific process. As time passes, related hypotheses that have been confirmed repeatedly can be fit together to form a **theory**—a wide-ranging explanation for some aspect of the physical universe. Because of its breadth, a theory cannot be tested by a single experi-

ment but instead emerges from many observations, hypotheses, and experiments. Nevertheless, a theory, like a hypothesis, leads to additional predictions and continued experimentation. Among the few explanations that have been tested thoroughly enough to be considered theories are the cell theory of life (which says all cells come from preexisting cells) and the theory of evolution by natural selection (which you will learn about in Chapter 22).

■ Inductive and deductive reasoning help solve problems

As you examined the lemon ant plant, asked questions about it, and devised experiments to answer those questions, you made use of two types of logical reasoning: inductive and deductive.

In **inductive reasoning,** facts are accumulated through observation until the sheer weight of the evidence allows some logical general statement to be made. An example is the way you collected as much information as you could about the lemon ant plant, other plants in the area, the soil, the light, and anything else that might have a bearing on your question. You then used those specific facts to draw the general conclusion that the lemon ant plant could be interfering with the growth of other plants in the area. You used inductive reasoning to develop a testable hypothesis.

Deductive reasoning begins with a general statement that leads logically to one or more deductions, or conclusions. The process can usually be described as an "if-then" series of associations. *If* this plant is making something that interferes with the growth of other plants, *then* there should be a cleared area around this plant; *then* this plant should have a way of producing some chemical; *then* that chemical should be detectable; *then* the plant should have a delivery system to transport the chemical to the soil; and so on. You used deductive reasoning to make predictions that would support your hypothesis and to decide whether the results of your experiment supported or refuted the hypothesis.

■ Clinical trials follow strict guidelines

Recall that the reason for your interest in chemicals from plants in the rain forest was the hope that some of them might be developed into drugs to treat human ailments. You interviewed several native healers, collected many species of plants they use for healing, and made extensive notes on how the plants are used and what they are good for. Now, you will preserve the plants for the trip home, where their extracts will undergo extensive testing and possible development into pharmaceuticals.

Before testing a new drug or treatment on humans, scientists must take steps to ensure that it will not do more harm than good (Table 1.1). Usually a drug is tested first on animals such as laboratory rodents. Most medical advances, including vaccinations, chemotherapy, new surgical techniques, and organ transplants, began with animal studies. Strict rules safeguard the care and use of animals in research and testing.

If no ill effects are discovered in animals receiving the drug, then studies on humans, called *clinical trials*, may begin. In all phases of clinical testing, the studies are done on people who volunteer. In phase I, the drug is screened for safety on fewer than 100 healthy people. At this stage, researchers hope to learn whether they can safely give the drug to humans, the effective dosage range, and side effects.

If the drug is found to be safe, it is tested further. In phase II, a few hundred people with the target disease are given the drug to see if it works for its intended purpose. If it does, the new drug will be compared with alternative treatments in phase III trials. Thousands of participants are involved in phase III of testing. The Food and Drug Administration (FDA) approves only those drugs or treatments that have passed all phases of testing.

what would you do?

The job of the FDA is to ensure the safety and effectiveness of new drugs and treatments. It must balance the patients' desires for access to new treatments against the government's desire to protect patients from treatments that may be unsafe and ineffective. The drug approval process is painstakingly slow. Do you think that the FDA should bypass certain steps of the approval process to make new drugs available to critically ill patients who may not be able to wait? If you do, what criteria should be used to decide the degree of illness that would warrant treatment with a drug that was not yet approved? Who should be held responsible if early access to a drug of unknown safety causes a patient to suffer serious side effects or premature death? If you were ill and there was a drug for your illness in clinical trials, would you participate in those trials? What factors would influence your decision?

Clinical studies must be properly designed. Recall that a well-designed experiment has both an experimental group and a control group. In a drug trial, the experimental group receives the drug under consideration. The control group receives a **placebo**, an innocuous, nondrug substance made to look like the drug being tested. Study participants are randomly assigned to either the control group or the experimental group and do not know whether they are receiving the treatment or a placebo. When neither the researchers

TABLE I.I   TESTS PERFORMED ON A NEW DRUG BEFORE IT IS APPROVED BY THE FOOD AND DRUG ADMINISTRATION (FDA)

Tests on laboratory animals

Is the drug safe for use on animals?

Clinical trials

Phase I   Is the drug safe for humans?

Phase II   Does the drug work for its intended purpose?

Phase III   How does the new drug compare with other available treatments?

nor the study participants know which people are receiving treatment and which are receiving the placebo, the study is described as being *double blind*. It is important that participants not know whether they are receiving the placebo or the drug because their expectations about the drug could affect the way they respond. Researchers should not know which people are in the experimental or control groups because their expectations or desire for a particular result could affect the interpretation of the data (see the Social Issue essay *Scientific Misconduct*).

Finally, it is extremely important that study participants give their **informed consent** to participate. An informed consent document lists all the possible harmful effects of the drug or treatment and must be signed before a person can take part in the study. Examples of experiments on humans in which informed consent was not obtained or was obtained under questionable circumstances are described in the Social Issue essay in Chapter 2, *The Ethics of Radiation Research on Humans*.

■ Epidemiological studies look for patterns in populations

Human health can also be studied without clinical trials. In an *epidemiological study*, for example, researchers look at patterns that occur within large populations. Thus, an epidemiological study to investigate the effects of air pollution on asthma (a condition in which airway constriction causes breathing problems) would look for a correlation of some kind between the variable of interest (air pollution) and its suspected effects (worsening of asthma). If the researchers' hypothesis is that air pollution aggravates asthma, they might predict and then look for evidence that the number of people admitted to hospitals for asthma-related problems increases with increased levels of air pollution.

Several studies have in fact shown that the number of hospital admissions for asthma and other respiratory problems do increase with the level of air pollution. In one study, the residents of Brisbane City, Australia, were followed for 8 years. During this time, the levels of ozone, nitrogen dioxide, sulfur dioxide, and particulate matter—all components of air pollution—were measured every 30 minutes at nearby government sites. The number of admissions to hospital emergency rooms for respiratory symptoms was also recorded. At the end of the study, the records showed that when the levels of ozone and particulate matter increased, so did the number of hospital admissions for asthma.

Scientific Misconduct

In 2005, there was a global gasp of outrage at the news that the breakthrough stem cell research of South Korean Hwang Woo-Suk was fraudulent. According to the University of Seoul, Hwang's research team did not create the lines of human stem cells they had claimed to. The results published in the journal *Science* in 2004 and 2005 were false. As explained in Chapter 19a, *Stem Cells—The Body's Repair Kit,* stems cells are relatively unspecialized cells that scientists hope to transform into specialized cells of various types to cure certain diseases or repair certain types of damage in the body.

The news that Hwang's research was fraudulent made headlines because if he had actually been successful, his work would have represented a major advance in a controversial area of study. Even so, it is unlikely that any of us were directly affected by the deceit. However, around the same time that Hwang's scandal was revealed, several prestigious medical journals announced that they had published inaccurate or misleading studies on the painkiller Vioxx and on cancer treatments. Scientific fraud in these kinds of medical research is potentially of direct concern to each of us.

Scientific fraud in clinical trials can take a number of forms. In some cases, the results are fabricated. For example, in early 2006 a paper by Dr. Jon Sudbo attributing the use of anti-inflammatory drugs to a reduced risk of mouth cancer, published in the British medical journal *The Lancet*, was retracted upon the discovery that none of the 908 volunteers in the clinical trial existed.

Clinical trials sometimes reveal undesirable and harmful side effects of a drug. Failure to reveal those side effects is another form of scientific fraud. It is alleged that the company that manufactures Vioxx failed to report the results of trials that showed the drug increased the risk of heart attack. It is also alleged that the company that manufactures the antidepressant Paxil failed to divulge that adolescents in clinical trials of Paxil were more likely than other adolescents to develop suicidal tendencies as a result of taking the drug.

Clearly, scientific fraud in clinical trials can harm your health and your wallet. It could mean that the prescription drug you pay top dollar for is not more effective than a placebo. Worse, it can lead to Food and Drug Administration approval of drugs that are too dangerous to be marketed.

The scientific method provides an important safeguard against scientific fraud in basic research by stipulating that scientists who repeat each others' experiments should be able to obtain the same results. This was Hwang's downfall. His deceit was discovered when his results could not be replicated. Other instances of scientific fraud have also been detected this way. Unfortunately, controlled clinical trials are expensive—they can easily cost $150 million dollars—and for this reason, repeated trials are not conducted as often as they should be.

Respectable science journals use peer review to evaluate the quality of research submitted for publication. In this process, experts in the same field of research read the paper, evaluate it, and send a confidential report to the editor. Peer review can and does uncover problems with experimental design or with the way data were handled, but it cannot detect fabricated data. However, some editors of medical journals have recently banded together to announce that they are on the lookout for fraud and will scrutinize with greater care the material submitted to them for publication.

Conversely, we would also expect the rate of asthma attacks to go down as air pollution levels decrease. The 1996 Olympic Games in Atlanta, Georgia, provided an opportunity to test this idea. To reduce traffic congestion in the city during the games, Georgia officials provided 24-hour public transportation, increased park-and-ride services, closed the city to private automobiles, and promoted alternative work schedules and telecommuting. As a result, the level of air pollution was lower during the games than during the 4 weeks preceding and the 4 weeks following the games. Consistent with our prediction, the frequency of asthma attacks in children during the interval of lower air pollution dropped by 11% to 44%, depending on the hospital being monitored. Note, however, that although these results are certainly in line with our predictions, they do not prove our hypothesis, because alternative explanations are possible.

One alternative explanation that might be suggested for the decrease in asthma attacks during the Olympic Games is that the games took place during a time of year when mold counts are usually low, and molds can also trigger asthma attacks. Thus, mold was another variable that could have changed over the course of the study. The next question is whether the number of asthma attacks among children dropped because the mold counts were lower. Probably not. The levels of mold did not change between the time of the Olympic Games and the weeks preceding and following the games. Furthermore, the daily mold count and the number of asthma attacks did not correlate at any time during the study in the Atlanta area.

Critical Thinking Helps Us Make Informed Decisions

Few of us perform controlled experiments in our everyday lives, but all of us must evaluate the likely strength of scientific claims. We encounter them in many forms—as advertisements, news stories, and anecdotes told by friends. We often must make decisions on the basis of these claims, but how can we decide whether they are true? Critical-thinking skills can help us analyze the information and make prudent decisions.

The key to becoming a critical thinker is to ask questions. The following list is not exhaustive, but it may help guide your thinking process.

1. **Is the information consistent with information from other sources?** The best way to answer this question is to gather as much information as possible, from a variety of sources. Do not passively accept a report as true. Do some research.

2. **How reliable is the source of the information?** Investigate the source of the information to determine whether that person or group has the necessary scientific expertise. Is there any reason to think the claim may be biased? Who stands to gain if you accept it as true? For example, the Food and Drug Administration is probably a more reliable source of information on the effectiveness of a drug than is the drug company marketing the drug. If a claim is controversial, listen to both sides of the debate and be aware of who is arguing on each side.

3. **Was the information obtained through proper scientific procedures?** Information gathered through controlled experiments is more reliable than anecdotal evidence, which cannot be verified. For example, your friend might tell you that his muscles have gotten larger since he started using some special exercise equipment. But you cannot be sure unless measurements were taken before and after he began to use it. Even if your friend can prove his muscles are bigger, there is no guarantee that exercising with this equipment will build *your* muscles.

4. **Were experimental results interpreted correctly?** Consider, for example, a headline advertising capsules containing fish oils: "Fish Oils Increase Longevity." It may be tempting to conclude from this headline that you will live longer if you take fish oil supplements, but in fact the headline is referring to an experiment in which *dietary* fish oil was altered in *rats*. Rats fed a diet high in fish oils lived longer than did rats on a diet low in fish oil. The claim that taking fish oil supplements will increase longevity is not a valid conclusion based on the experiment. First, the study was done on rats, not on humans. Second, the amount of dietary fish oil, not the amount of fish oil from capsules that supplement dietary fish oil, was the variable. Supplements of fish oil may not have the same effect as dietary fish oils. It could be that taking fish oil supplements would boost the amount of fish oil in your body to unhealthy levels.

5. **Are there other possible explanations for the results?** Suppose you learn that the fish oil headline is based on a study showing that people who eat fish at least three times a week live longer than those who eat fish less frequently. In this case, the data indicate that there is a correlation between fish in the diet and length of life. However, a correlation between two factors does not prove that one *caused* the other. The difference in longevity may be due to other differences in the lifestyles of the two groups. For instance, people who eat fish may exercise more frequently or have less stressful jobs or live in areas with less pollution.

You will be asked to make many decisions in your life. Some will affect your community and even beyond. Should we eliminate genetically modified food? Should stem cell research be permitted? Should we vaccinate the general public against smallpox? Should companies polluting the atmosphere be taxed? We will raise these and similar questions throughout this text. Others will appear daily in the news media.

Although you may never be one of the lawmakers deciding these issues, you are a voter who can help choose the lawmakers and voice your opinions to the lawmakers, who will decide. Scientists can provide facts that may be useful as we all struggle to answer the complex questions facing society, but they cannot provide simple answers. As scientific knowledge grows and our choices become increasingly complex, each of us must stay informed and review the issues critically.

HIGHLIGHTING THE CONCEPTS

All Living Things Share Basic Characteristics (pp. 2–4)

1. Life cannot be defined, only characterized.
2. Living things contain nucleic acids, proteins, carbohydrates, and lipids; are made of cells; reproduce, metabolize, respond, and maintain homeostasis. Populations of living things evolve.
   WEB TUTORIAL 1.1 Signs of Life

Living Organisms Are Classified by an Evolutionary Relationships (pp. 4–5)

3. Classifications of living organisms reflect the evolutionary relationships among them. One classification system recognizes three domains: Bacteria, Archaea, and Eukarya. Domain Eukarya consists of four kingdoms: protists, fungi, plants, and animals.
4. Humans are classified as animals, vertebrates, mammals, and primates.
5. Humans share traits with other species, but they also have many features not shared by most other animals, among the most important of which is culture.

Life Has Many Levels of Organization (pp. 5–6)

6. Human biology can be studied at different levels. Within an individual, the levels of increasing complexity are molecules, cells, tissues, organs, and organ systems. Beyond the level of the individual are populations, communities, ecosystems, and the biosphere.

The Scientific Method Gathers Information for Drawing Logical Conclusions (pp. 6–11)

7. The scientific method consists of observation, formulation of a question and hypothesis (a testable explanation for the observation), experimentation (performed with controls), and drawing a conclusion, which may lead to further experimentation.
8. As evidence mounts in support of related hypotheses, the hypotheses may be organized into a theory, which is a well-supported explanation of nature.
9. Inductive reasoning uses a large number of specific observations to arrive at a general conclusion. Deductive reasoning, in contrast, uses "if-then" logic to progress from the general to the specific.
10. There are strict rules concerning experiments on humans and other animals. The drug approval process usually begins with studies on animals. If no ill effects in animals are observed, the drug is then tested on humans. Phase I trials determine whether the drug is safe for

humans, phase II whether it works for its intended purpose, and phase III whether it is more effective than existing treatments.

11. The design of a human experiment often includes an experimental group that receives the treatment and a control group that receives a placebo, which is a non-drug substance presumed to have no effect on the condition being treated. In a double-blind experiment, neither the study participants nor the researchers know who is receiving the real treatment. Study participants must sign an informed consent document indicating that they were made aware of the possible harmful consequences of the treatment.

12. Epidemiological studies examine patterns within populations to find a correlation between a variable and its suspected effects.
WEB TUTORIAL 1.2   The Scientific Method
Critical Thinking Helps Us Make Informed Decisions (pp. 11–12)

13. Critical thinking consists of asking questions, gathering information, and evaluating evidence carefully before drawing conclusions.

KEY TERMS

cell   *p. 2*
metabolism   *p. 2*
homeostasis   *p. 2*
adaptive trait   *p. 4*
population   *p. 6*

community   *p. 6*
ecosystem   *p. 6*
biosphere   *p. 6*
science   *p. 6*
scientific method   *p. 6*

hypothesis   *p. 6*
controlled experiment   *p. 8*
variable   *p. 8*
theory   *p. 9*
inductive reasoning   *p. 9*

deductive reasoning   *p. 9*
placebo   *p. 10*
informed consent   *p. 10*

REVIEWING THE CONCEPTS

1. List seven features that characterize life.   *pp. 2, 4*
2. What characteristics make humans classified as mammals? As primates?   *p. 4*
3. Distinguish between a population, a community, an ecosystem, and the biosphere.   *p. 6*
4. What is a hypothesis? How does it differ from a theory?   *pp. 6, 9*
5. Define a controlled experiment.   *p. 8*
6. Differentiate between inductive and deductive reasoning.   *p. 9*
7. What is a placebo?   *p. 10*
8. What is meant by a double-blind experiment?   *p. 10*
9. Describe the procedure in an epidemiological study.   *pp. 10–11*

10. A theory is
    a. a testable explanation for an observation.
    b. a conclusion based on the results of an experiment.
    c. a wide-ranging explanation for natural events that has been extensively tested over time.
    d. the factor that is altered in a controlled experiment.
11. The maintenance of physical and chemical conditions inside the body within tolerable ranges is called _____.
12. A(n) _____ is a testable explanation for an observation.
13. A trait that increases the chance that an organism will survive and reproduce in its natural environment is described as being _____.

APPLYING THE CONCEPTS

1. A native you met in the rain forest told you that one of the plants you collected brings relief to people who are having difficulty breathing. You suspect that it might be a good treatment for asthma, a condition in which constriction of airways causes breathing problems. You are able to isolate a component of this plant as a drug. Tests on animals show that it is effective. Tests on humans show that the drug can be given safely to humans. Design an experiment to test the hypothesis that this drug eases breathing during an asthma attack.

Additional questions can be found on the companion website.

2. There have been several, as yet unproven, claims that the MMR (measles, mumps, rubella) vaccine and vaccines containing mercury cause autism in children. Is there a link between vaccines and autism? Design an epidemiological study to answer this question. What populations of children would you compare?

3. Find an article or advertisement that makes a scientific claim. Use your critical-thinking skills to evaluate the claim.

2

Chemistry Comes to Life

In later years, familiarity with a few basic chemical principles will help these children understand the biology of their own bodies.

Basic Chemistry Helps Us Understand Human Biology

- Atoms contain protons, neutrons, and electrons
- Elements combine to form compounds
- The atoms or ions of a compound are held together by chemical bonds
- Water is essential to life

Carbohydrates, Lipids, Proteins, and Nucleic Acids Are the Major Molecules of Life

- Carbohydrates supply energy to cells
- Lipids store energy and form cell membranes
- Proteins provide structure and speed up chemical reactions
- The nucleic acids are DNA and RNA
- ATP is a nucleotide that releases energy

 **SOCIAL ISSUE**   The Ethics of Radiation Research on Humans

ENVIRONMENTAL ISSUE   What Is Happening to the Rain?

Chris arranged the pH meter and a variety of pH indicator strips in the middle of an otherwise empty table, hoping to pique the interest of his fourth graders when they came in from morning break. The class was about to learn a game he had made up called "Acid, Base, or Neutral?" This was Chris's first teaching job since receiving his degree in education, and he wasn't really sure how the game would go over; but he thought it was the kind of activity that 10-year-olds could get excited about. If so, he hoped the same excited feeling would come back to them when they encountered acids and bases again in more advanced science classes.

Chris's students spent the rest of the morning testing the pH of every liquid they could find, including many that Chris had bought on his way to school. Soon they began to make more confident (and accurate) guesses about what liquids would have lower pH values than others—and thus be stronger acids—and they were very impressed when Chris told them that their stomachs secrete a liquid whose value is even lower than the pH of lemon juice. They were also surprised to learn that saliva's value was higher than the value for water. Chris told them that the different chemical reactions in their bodies work best at different pH's and that the body has ways of providing the proper pH for each one.

Chemistry is the branch of science concerned with the composition and properties of material substances, including the substances from which our bodies are built. Chemical concepts, such as pH, contribute important insights into how our bodies work. Chapter 2 reviews several of these fundamental concepts, laying a foundation for discussions throughout the rest of the book. ■

Basic Chemistry Helps Us Understand Human Biology

The world around you contains an amazing variety of physical substances: the grass or concrete on which you walk, the water you drink, the air you breathe, and even this book that you are reading. All of these substances that make up our world are called matter. Fundamentally, **matter** is anything that takes up space and has mass. All forms of matter are made up of atoms.

■ Atoms contain protons, neutrons, and electrons

Atoms are units of matter that cannot be broken down into simpler substances by ordinary chemical means. Each atom is composed of even smaller, subatomic particles, such as protons, neutrons, and electrons. These subatomic particles are characterized by their location within the atom, their electrical charge, and their mass.

Each atom has a nucleus at its center and a surrounding "cloud" of electrons. As you can see in Figure 2.1, the nucleus contains protons and neutrons. Electrons occupy the space around the nucleus in more or less concentric regions called shells. The shell closest to the nucleus can hold up to 2 electrons. The next shell out can hold up to 8 electrons. Atoms with more than 10 electrons have additional shells. Neutrons, as their name implies, have no electrical charge; they are neutral. Protons, in contrast, have a positive charge. As a result, the nucleus of an atom has a positive charge. Electrons have a negative charge. Most atoms have the same number of positively charged protons and negatively charged electrons. As a result, they are "neutral," having no net charge.

Protons have essentially the same mass as neutrons. Their mass is measured in special units called atomic mass units, or amu. Electrons are much smaller than either neutrons or protons and have almost no mass. Table 2.1 summarizes the basic characteristics of protons, neutrons, and electrons.

TABLE 2.1 REVIEW OF SUBATOMIC PARTICLES			
PARTICLE	LOCATION	CHARGE	MASS
Proton	Nucleus	1 positive unit	1 atomic mass unit
Neutron	Nucleus	None	1 atomic mass unit
Electron	Outside the nucleus	1 negative unit	Negligible

ELEMENTS

An **element** is a "pure" form of matter containing only one kind of atom. You are probably familiar with many elements, such as gold, silver, iron, and oxygen. The earth is made up of a little over 100 elements. Only about 20 elements are found in the human body, which consists mostly of carbon, oxygen, hydrogen, and nitrogen. Each element consists of atoms containing a certain number of protons in the nucleus. For example, all carbon atoms have 6 protons. The number of protons in the atom's nucleus is called the *atomic number.*

The periodic table lists the elements and describes many of their characteristics. Figure 2.2 depicts a simplified periodic table. Note that each element has a name and a one- or two-letter symbol. The symbol for the element carbon, for example, is C, and that for chlorine is Cl. Besides an atomic number, each atom also has an atomic weight. Recall that each proton and neutron has an approximate mass of 1 atomic mass unit, or amu. The mass of an electron is so small that it is usually considered zero. Because electrons have negligible mass, and protons and neutrons each have an atomic mass of 1, the *atomic weight* for any atom equals the number of protons plus the number of neutrons. Oxygen has an atomic weight of 16, indicating that it has 8 protons (we know this from its atomic number) and 8 neutrons in its nucleus.

ISOTOPES AND RADIOISOTOPES

All the atoms of a particular element contain the same number of protons; they can, however, have different numbers of neutrons. Such differences result in atoms of the same element having slightly different atomic weights. Atoms that have the same number of protons but differ in the number of neutrons are called **isotopes**. Over 300 isotopes occur naturally on the earth. The element carbon, for example, has three isotopes. All carbon atoms have 6 protons in the nucleus. Most carbon atoms also have 6 neutrons, but some have 7 or 8. The isotopes of carbon thus have atomic weights of 12, 13, and 14, respectively, depending on the number of neutrons in the nucleus. These isotopes are written ^{12}C (the most common form in nature), ^{13}C, and ^{14}C.

Radiation is energy moving through space. Examples include radio waves, light, heat, and the excess energy or particles given off by unstable atoms. Some elements have both stable and unstable isotopes. Unstable, radiation-emitting isotopes are called **radioisotopes**. About 60 occur naturally, and many more have been made in laboratories.

Radiation can be dangerous or useful. When dangerous, it can damage the body directly or indirectly. Direct damage to a person

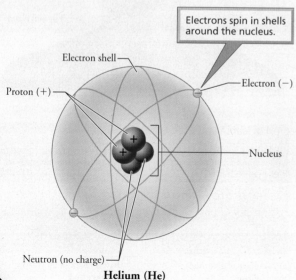

Electrons spin in shells around the nucleus.

Electron shell

Proton (+)

Electron (−)

Nucleus

Neutron (no charge)

Helium (He)

FIGURE **2.1** An atom of helium, showing protons and neutrons in the nucleus and electrons occupying a region around the nucleus.

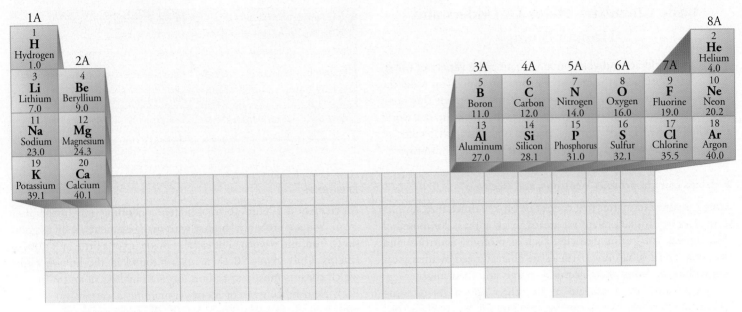

FIGURE **2.2** A simplified version of the periodic table, showing elements with atomic numbers 1 through 20

absorbing harmful radiation may include a low white blood cell count, development of some cancers, and damage to organs and glands (Figure 2.3). In other cases, radiation may not produce any noticeable injury to the person who was exposed, but it may alter the hereditary material in the cells of his or her reproductive system, possibly causing defects in the individual's offspring.

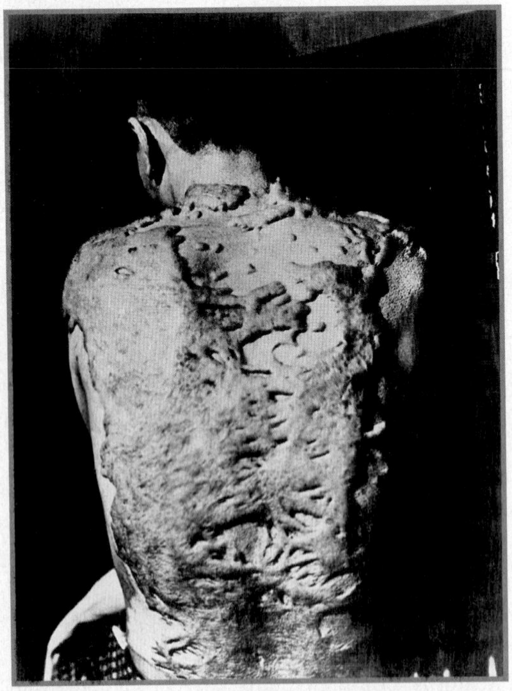

In stark contrast to the harmful effects of radiation are its medical uses. Medical professionals use small doses of radiation to generate visual images of internal body parts in order to diagnose irregularities in the body's structure or function. Radioactive iodine, for example, is often used to identify disorders of the thyroid gland. This gland, located in the neck, normally accumulates the element iodine, which it uses to regulate growth and metabolism. Small doses of iodine-131 (^{131}I), a radioactive isotope, can be given to a patient suspected of having metabolic problems. The radioactive iodine is taken up by the patient's thyroid gland and detected by medical instruments, as shown in Figure 2.4. The small

FIGURE **2.3** An example of direct damage caused by radiation. This individual survived the atomic explosion at Nagasaki, Japan, on August 9, 1945. Damage to the reproductive cells of people exposed to radiation has proved more difficult to recognize. Children of bomb survivors are still being monitored.

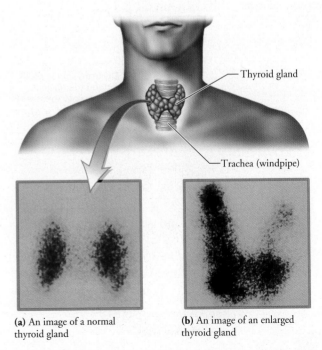

(a) An image of a normal thyroid gland

(b) An image of an enlarged thyroid gland

FIGURE **2.4** Radioactive iodine can be used to generate images of the thyroid gland. Such images may be used to diagnose metabolic disorders.

amount of radioactive iodine used in imaging does not damage the thyroid gland or surrounding structures. However, larger doses can be used to kill thyroid cells when the gland is enlarged and overactive. A more familiar medical use of small doses of radiation is the x-ray.

Radiation can also be used to kill cancer cells. Cancer cells divide more rapidly and have higher rates of metabolic activity than do most normal cells. For these reasons, cancer cells are more susceptible to the destructive effects of radiation. Furthermore, when medical professionals aim a beam of radiation at a tumor, to kill the cancer cells inside it, they also take steps to shield the surrounding healthy tissue. Because radiation can be used either to kill or to heal, research into its effects and uses often raises ethical concerns, particularly when human subjects are used for experiments (see the Social Issue essay, *The Ethics of Radiation Research on Humans*).

what would you do?

There are various reasons that many foods today are intentionally irradiated, including (1) removal of harmful microorganisms (pathogens), insect pests, and parasites; (2) delaying spoilage and increasing shelf life; and (3) sterilizing the material in which the food is packaged. The food does not become radioactive as a result. Supporters of the practice note that test animals fed on irradiated food show no adverse effects. Opponents, however, point to the environmental risks of building and operating food irradiation plants. Opponents also note that carefully controlled experiments are still needed to verify that irradiated food is safe for people of all ages and nutritional states. Several foods, including wheat flour, poultry, and white potatoes, have been approved for irradiation in the United States. These irradiated foods have a distinctive logo (Figure 2.5) on their packaging. Would you eat irradiated food? Explain your answer.

■ Elements combine to form compounds

Two or more elements may combine to form a new chemical substance called a **compound**. A compound's characteristics are usually different from those of its elements. Consider what happens when the element sodium (Na) combines with the element chlorine (Cl). Sodium is a silvery metal that explodes when it comes into contact with water. Chlorine is a deadly yellow gas. In combination, however, they form a crystalline solid called sodium chloride (NaCl)—plain table salt (Figure 2.6).

■ The atoms or ions of a compound are held together by chemical bonds

The atoms (or, as we will see below, ions) in a compound are held together by chemical bonds. There are two types of chemical bond: covalent and ionic. In both, atoms behave as if they are trying to attain a full outer electron shell. Recall that these shells are the regions surrounding the nucleus where the electrons are most likely to be found. Figure 2.7 depicts the first two shells as concentric

(a) The element sodium is a solid metal.

(b) Elemental sodium reacts explosively with water.

(c) The element chlorine is a yellow gas.

(d) When the elements sodium and chlorine join, they form table salt, a compound quite different from its elements.

FIGURE **2.6** The characteristics of compounds are usually different from those of their elements.

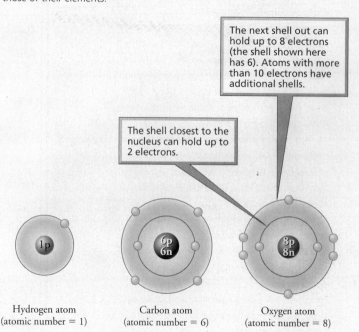

The next shell out can hold up to 8 electrons (the shell shown here has 6). Atoms with more than 10 electrons have additional shells.

The shell closest to the nucleus can hold up to 2 electrons.

Hydrogen atom (atomic number = 1)

Carbon atom (atomic number = 6)

Oxygen atom (atomic number = 8)

FIGURE **2.7** Atoms of hydrogen, carbon, and oxygen. Each of the concentric circles around the nucleus represents a shell occupied by electrons.

The Ethics of Radiation Research on Humans

In the United States, between about 1945 and 1971, a number of people were exposed to radiation as part of research into its effects and uses (Figure 2.A). These people included terminally ill hospital patients, mentally delayed school children, prisoners, and scientists. Did the research subjects or their families understand the risks of these experiments? Did anyone warn them? In other words, did the participants, or someone speaking for them, give their informed consent?

In one set of radiation experiments, the military's security concerns took precedence over fully informing study participants. Between 1945 and 1947, 18 terminally ill volunteers received injections of a substance whose identity was not disclosed to them. The substance was plutonium, an element used to make one type of atomic bomb. The rationale for the study was that many workers might have been exposed to plutonium during construction of the bomb and many more might be exposed at nuclear facilities in the years ahead, yet little was known about the effects of plutonium on humans. Plutonium is metabolized differently by different species, so experiments on the effects of plutonium on other animals could not be applied to humans. Also, although several thousand workers at atomic facilities had probably been exposed to plutonium, the precise levels of exposure were unknown because the exposures had been accidental. The researchers therefore believed it necessary to inject known quantities of the element into human subjects and monitor its movement through the body in order to discover whether plutonium is rapidly excreted by humans or held in their tissues for years.

The scientists believed the research was necessary to set safety standards for plutonium exposure. Nevertheless, their failure to fully inform the volunteers in these experiments raised ethical questions. Is it ethical to use terminally ill volunteers in potentially harmful experiments? If terminally ill patients were considered convenient subjects because any plutonium remaining in the body could be measured at a not-too-distant autopsy, could the need for an autopsy interfere with the doctor-patient relationship by making a quick death desirable? And what if some of the patients had been misdiagnosed and were not close to death after all? Because

of such questions, terminally ill people are generally no longer used in research, except when the purpose is to test an experimental treatment for their illness.

The principle of informed consent was legally established in the United States during the 1930s. However, for years afterward, some U.S. research groups violated the principle and did not obtain informed consent from their human subjects. In a number of radiation studies, as we have seen, the consent was questionable, although in others, informed consent was properly obtained. In 1974, the U.S. Department of Health, Education, and Welfare established regulations for all research with human subjects. However, it was not until 1991 that these regulations were fully operational in all government agencies. Today, researchers working with human subjects must obtain the informed consent of all participants. Nevertheless, new ethical questions continue to arise. ✹

ETHICS IN RADIOACTIVITY EXPERIMENTS ON HUMANS

 In some of the radiation studies cited in newspapers in late 1993 and in an earlier congressional report, known as the Markey Report, subjects did not freely consent to the experiments. In other studies it is doubtful whether informed consent was obtained. But in some of the studies informed consent was truly given. Here are examples from each category.

Date	Experiment
Possible Infliction of Harm or No Informed Consent	
1945–47	18 supposedly terminal patients were injected with high doses of plutonium to learn whether the body absorbed it.
1946–47	6 hospital patients were injected with uranium salts to determine the dose that produced injury to the kidneys.
1963–71	67 prison inmates had their testicles exposed to x-rays to measure radiation damage to sperm production.
Questionable Consent	
1946	17 mentally delayed teenagers at the Fernald School in Waltham, Massachusetts, ate meals with trace amounts of radioactive iron to learn about iron absorption in the body.
1953–57	11 comatose brain cancer patients were injected with uranium to learn whether it is absorbed by brain tumors.
1954–56	32 mentally delayed teenagers at the Fernald School drank milk with trace amounts of radioactive calcium to learn whether oatmeal impeded its absorption by the body.
Informed Consent	
1945	10 researchers and workers at Clinton Laboratory, in Oak Ridge, Tennessee, voluntarily exposed patches of their skin to radioactive phosphorus.
1951	14 researchers at Hanford Nuclear Reservation voluntarily exposed patches of their skin to gaseous tritium.
1963	54 hospital patients volunteered to take trace amounts of radioactive lanthanum to measure effects on the large intestine.

FIGURE **2.A** A sample of radiation experiments performed on people

circles around the nucleus. A full innermost shell contains 2 electrons. However, as you can see in the figure, a hydrogen atom has only 1 electron. A full second shell contains 8 electrons. Atoms with more than 10 electrons have additional shells. When atoms form bonds, they lose, gain, or share the electrons in their outermost shell.

COVALENT BONDS

A **covalent bond** forms when two or more atoms *share* electrons in their outer shells. Consider the compound methane (CH_4). Methane is formed by the sharing of electrons between one atom of carbon and four atoms of hydrogen. Notice in Figure 2.8a that the

outer shell of an isolated carbon atom contains only four electrons, even though it can hold eight. Also note that hydrogen atoms only have one electron, although the first shell can hold two electrons. A carbon atom can fill its outer shell by joining with four atoms of hydrogen. By forming a covalent bond with the carbon atom, each hydrogen atom fills its own outer shell. We see, then, that the covalent bonds between the carbon atom and hydrogen atoms of methane result in filled outer shells for all five atoms involved.

A **molecule** is a chemical structure held together by covalent bonds. Recall that compounds are formed by two or more elements, so molecules that contain only one kind of atom are not considered compounds. Oxygen gas, formed by the joining of two

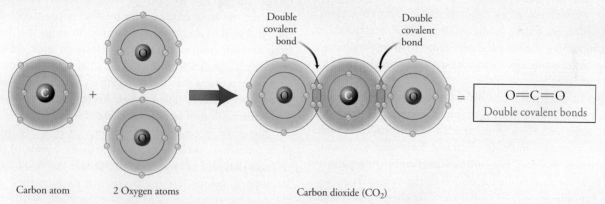

Carbon atom   **4 Hydrogen atoms**   **Methane (CH_4)**

(a) The molecule methane (CH_4) is formed by the sharing of electrons between one carbon atom and four hydrogen atoms. Because, in each case one pair of electrons is shared, the bonds formed are single covalent bonds.

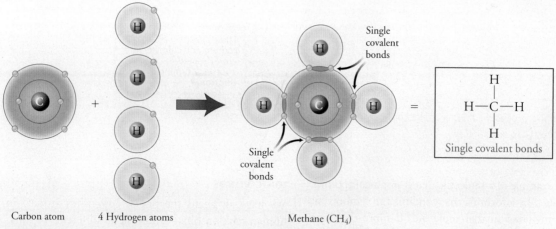

Carbon atom   **2 Oxygen atoms**   **Carbon dioxide (CO_2)**

(b) The oxygen atoms in a molecule of carbon dioxide (CO_2) form double covalent bonds with the carbon atom. In double bonds, two pairs of electrons are shared.

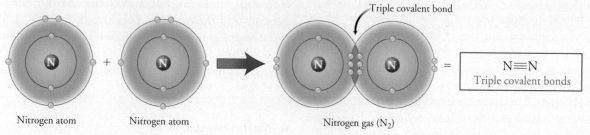

Nitrogen atom   **Nitrogen atom**   **Nitrogen gas (N_2)**

(c) The nitrogen atoms in nitrogen gas (N_2) form a triple covalent bond in which three pairs of electrons are shared.

FIGURE **2.8** Covalent bonds form when electrons are shared between atoms. Shown here are examples of single, double, and triple covalent bonds. For each example, the structural formula is shown on the far right.

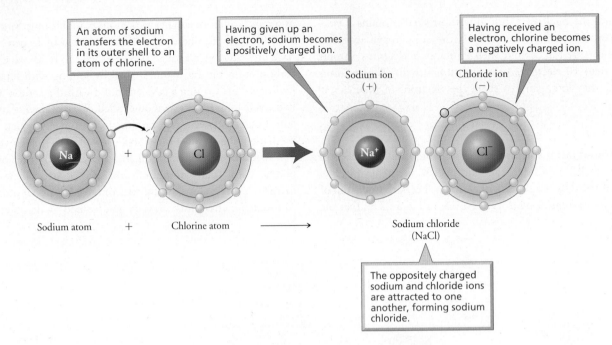

FIGURE **2.9**   An ionic bond involves the transfer of electrons between atoms. Such a transfer creates oppositely charged ions that are attracted to one another.

oxygen atoms, is an example of a molecule that is not a compound. Molecules are described by a formula that contains the symbols for all of the elements included in that molecule. If more than one atom of a given element is present in the molecule, a subscript is used to show the precise number of that kind of atom. For example, the molecular formula of sucrose (table sugar) is $C_{12}H_{22}O_{11}$, showing that one molecule of sucrose contains 12 atoms of carbon, 22 atoms of hydrogen, and 11 atoms of oxygen. Numbers placed in front of the molecular formula indicate more than one molecule. For example, three molecules of sucrose are described by the formula $3C_{12}H_{22}O_{11}$.

As shown in the methane molecule in Figure 2.8a, the bond between each of the four hydrogen atoms and the carbon atom consists of a single pair of electrons. Such a bond is called a single bond, so the methane molecule contains four single covalent bonds. Sometimes, however, two atoms share two or three pairs of electrons. These bonds are called double and triple covalent bonds, respectively. Carbon dioxide, CO_2, produced by chemical reactions inside our cells, has double covalent bonds between the carbon atom and each oxygen atom (Figure 2.8b). The nitrogen atoms in nitrogen gas, N_2, are joined by triple covalent bonds (Figure 2.8c).

Covalent bonds in molecules are sometimes depicted by a structural formula. Notice in the box at the right of Figure 2.8a that one straight line is drawn between the carbon atom and each hydrogen atom in the structural formula for the methane molecule. The single line indicates a single covalent bond resulting from a pair of shared electrons. The double lines between the carbon and oxygen atoms in the carbon dioxide molecule indicate a double covalent bond, or two pairs of shared electrons (see Figure 2.8b). Three lines drawn between the two nitrogen atoms in gaseous nitrogen depict a triple covalent bond, or three pairs of shared electrons (see Figure 2.8c).

IONIC BONDS

We have all heard the phrase "opposites attract" in reference to human relationships—and it is no different for ions. An **ion** is an atom or group of atoms that carries either a positive (+) or a negative (−) electrical charge. Electrical charges result from the *transfer* (as opposed to sharing) of electrons between atoms. Recall that a neutral atom has the same number of positively charged protons and negatively charged electrons. An atom that loses an electron has one more proton than electrons and therefore has a positive charge. An atom that gains an electron has one more electron than protons and has a negative charge. Oppositely charged ions are attracted to one another. An **ionic bond** results from the mutual attraction of oppositely charged ions.

Ions form because of the tendency of atoms to attain a complete outermost shell. Consider, again, the atoms of sodium and chlorine that join to form sodium chloride. As shown in Figure 2.9, an atom of sodium has one electron in its outer shell. An atom of chlorine has seven electrons in its outer shell. Sodium chloride is formed when the sodium atom transfers the single electron in its outer shell to the chlorine atom. The sodium atom now has a full outer shell, but having lost an electron, it has one more proton than electrons and therefore has a positive charge (Na^+). The chlorine atom, having gained an electron to fill its outer shell, has one more electron than protons and has a negative charge (Cl^-). These oppositely charged ions are attracted to one another, and an ionic bond forms.

■ Water is essential to life

Water is such a familiar part of our everyday lives that we often overlook its unusual qualities. Unique properties of water include its virtuosity as a dissolving agent, its high heat capacity, and its

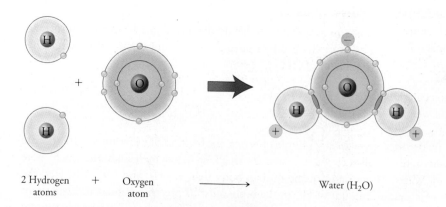

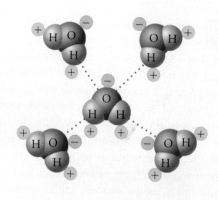

2 Hydrogen + Oxygen ⟶ Water (H_2O)
atoms   atom

(a) Water is formed when an oxygen atom covalently bonds (shares electrons) with two hydrogen atoms. As a result of unequal sharing of electrons, oxygen carries a slight negative charge and the hydrogen atoms a slight positive charge.

(b) The hydrogen atoms from one water molecule are attracted to the oxygen atoms of other water molecules. This relatively weak attraction (shown by dotted lines) is called a hydrogen bond.

FIGURE **2.10** The hydrogen bonds of water

high heat of vaporization. As it turns out, water's unusual qualities can be traced to its polarity (tendency of its molecules to have positive and negative regions) and the hydrogen bonds between its molecules.

POLARITY AND HYDROGEN BONDS

In covalently bonded molecules, electrons may be shared equally or unequally between atoms. When the sharing of electrons is unequal, different ends of the same molecule can have slight opposite charges. Unequal covalent bonds are called *polar*, and molecules with unevenly distributed charges are called polar molecules. In water (H_2O), for example, the electrons shared by oxygen and hydrogen spend more time near the oxygen atom than near the hydrogen atom. As a result, the oxygen atom has a slight negative charge; each hydrogen atom has a slight positive charge; and water molecules are polar. The hydrogen atoms of one water molecule are attracted to the oxygen atoms of other water molecules. The attraction between a slightly positively charged hydrogen atom and a slightly negatively charged atom nearby is called a **hydrogen bond** (Figure 2.10a). In the case of water, the hydrogen bonding occurs between hydrogen and oxygen. However, sometimes hydrogen bonds form between hydrogen and atoms of other elements.

Hydrogen bonds are weaker than either ionic or covalent bonds. For this reason, they are illustrated by dotted lines rather than solid lines in Figure 2.10b. Even though individual hydrogen bonds are very weak, collectively they can be significant. Hydrogen bonds maintain the shape of proteins and our hereditary material, DNA, and they account for some of the unique physical properties of water.

Covalent bonds, ionic bonds, and hydrogen bonds are summarized in Table 2.2.

PROPERTIES OF WATER

Life depends on the properties of water. Let's consider some of the ways that the polarity and hydrogen bonding of water give water the properties that make it so vital to life.

Because of the polarity of its molecules, water interacts with many substances. This interactivity makes it an excellent solvent, easily dissolving both polar and charged substances. Ionic compounds, such as NaCl, dissolve into independent ions in water. The sodium ions and chloride ions separate from one another in water because the sodium ions are attracted to the negative regions of water molecules and the chloride ions are attracted to the positive regions. Because of its excellence as a solvent, water serves as the body's main transport medium. As the liquid component of blood, it carries dissolved nutrients, gases, and wastes through the circulatory system. Metabolic wastes are excreted from the body in urine, another watery medium.

TABLE 2.2	REVIEW OF CHEMICAL BONDS		
TYPE	**BASIS FOR ATTRACTION**	**STRENGTH**	**EXAMPLE**
Covalent	Sharing of electrons between atoms; the sharing between atoms may be equal or unequal	Strongest	CH_4 (methane)
Ionic	Transfer of electrons between atoms creates oppositely charged ions that are attracted to one another	Strong	NaCl (table salt)
Hydrogen	Attraction between a hydrogen atom with a slight positive charge and another atom (often oxygen) with a slight negative charge	Weak	Between a hydrogen atom on one water molecule and an oxygen atom on another water molecule

Water also helps prevent dramatic changes in body temperature. This ability comes from its *high heat capacity*, which simply means that a great deal of heat is required to raise its temperature. Hydrogen bonds hold multiple water molecules together, so a large amount of heat is required to break these bonds (a higher temperature corresponds to an increase in the movement of the molecules). About 67% of the human body is water. Thus, if a person weighs 68 kg (150 lb), water makes up about 45 kg (100 lb) of the body weight. Because humans, as well as other organisms, are made up largely of water, they are well suited to resist changes in body temperature and to keep a relatively stable internal environment. Water in blood also helps redistribute heat within our bodies. Our fingers don't usually freeze on a frigid day because heat is carried to them by blood from muscles where the heat is generated.

Another property that helps prevent the body from overheating is water's *high heat of vaporization*, which means that a great deal of heat is required to make water evaporate (that is, change from a liquid to a gas). Water's high heat of vaporization is also due to its hydrogen bonds, which must be broken before water molecules can leave the liquid and enter the air. Water molecules that evaporate from a surface carry away a lot of heat, cooling the surface. We rely on the evaporation of water in sweat to cool the body surface and prevent overheating (Figure 2.11). Unfortunately, water vapor in the air can inhibit the evaporation of sweat.

stop and think

Sharp increases in body temperature can cause heat stroke, a condition that may damage the brain. Explain why heat stroke is more likely to occur on a hot humid day than on an equally hot dry day.

FIGURE **2.11** The evaporation of water in sweat cools the surface of this runner's body. Water has a high heat of vaporization, so when water molecules in sweat evaporate, they carry away a lot of heat.

ACIDS AND BASES

Sometimes a water molecule dissociates, or breaks up, forming a positively charged hydrogen ion (H^+) and a negatively charged hydroxide ion (OH^-):

$$H-O-H \rightleftharpoons H^+ + OH^-$$
$$\text{Water} \qquad \text{Hydrogen} \qquad \text{Hydroxide} \\ \text{ion} \qquad \qquad \text{ion}$$

In any sample of water, the fraction of water molecules that are dissociated is extremely small, so water molecules are much more common in the human body than are H^+ and OH^-. In fact, the amount of H^+ in the body must be precisely regulated. Even slight changes in the concentration of H^+ can be disastrous, disrupting chemical reactions within cells. Substances called acids and bases influence the concentration of H^+ in solutions.

Acids and bases are defined by what happens when each is added to water. An **acid** is anything that releases hydrogen ions (H^+) when placed in water. A **base** is anything that produces hydroxide ions (OH^-) when placed in water. Hydrochloric acid (HCl), for example, dissociates in water to produce hydrogen ions (H^+) and chloride ions (Cl^-). Because HCl increases the concentration of H^+ in solution, it is classified as an acid. Sodium hydroxide (NaOH), on the other hand, dissociates in water to produce sodium ions (Na^+) and hydroxide ions (OH^-). Because NaOH increases the concentration of OH^- in solution, it is classified as a base. The OH^- produced when NaOH dissociates reacts with H^+ to form water molecules and thus reduces the concentration of H^+ in solution. Therefore, acids increase the concentration of H^+ in solution, and bases decrease the concentration of H^+ in solution.

THE pH SCALE

We often want to know more than simply whether a substance is an acid or a base. For example, how strong an acid is battery acid? How strong a base is household ammonia? Questions such as these can be answered by knowing the pH of these solutions and understanding the pH scale (Figure 2.12). The **pH** of a solution is a measure of hydrogen ion concentration. The **pH scale** ranges from 0 to 14, with a pH of 7 being neutral (the substance does not increase H^+ or OH^-), a pH of less than 7 being acidic, and a pH of greater than 7, basic. Usually, the amount of H^+ in a solution is very small. For example, the concentration of H^+ in a solution with a pH of 6 is 1×10^{-6} (or 0.000001) moles per liter (a mole, here, is not a small animal, but a unit of measurement that indicates a specific number of atoms, molecules, or ions). Similarly, the concentration of H^+ in a solution with a pH of 5 is 1×10^{-5} (or 0.00001) moles per liter. Technically, pH is the negative logarithm of the concentration of H^+ in a solution. According to the pH scale, the lower the pH, the greater the acidity, or concentration of H^+, in a solution. Each reduction of pH by one unit represents a tenfold increase in the amount of H^+. So a solution with a pH of 5 is 10 times more acidic than a solution with a pH of 6. And a solution with a pH of 4 is 100 times more acidic than one with a pH of 6. Some of the characteristics of acids and bases, including their values on the pH scale, are summarized in Table 2.3.

BUFFERS

Most biological systems must keep their fluids within a narrow range of pH values. Substances called **buffers** keep their pH values from changing dramatically. Buffers remove excess H^+ from solution when concentrations of H^+ increase. Buffers add H^+ when concentrations of H^+ decrease. For example, an important buffering system that keeps the pH of blood at about 7.4 is the carbonic acid–bicarbonate system. When carbon dioxide is added to water it forms carbonic acid (H_2CO_3), which dissociates into hydrogen ions and bicarbonate ions (HCO_3^-):

$$CO_2 + H_2O \rightleftharpoons H_2CO_3 \rightleftharpoons H^+ + HCO_3^-$$

Carbon dioxide   Water   Carbonic acid   Hydrogen ion   Bicarbonate ion

The buffering action of carbonic acid and bicarbonate results from the fact that when levels of H^+ decrease in the blood, carbonic acid dissociates, adding H^+ to solution. When levels of H^+ increase in the blood, the H^+ combines with bicarbonate and is removed from solution. Such action is essential because even slight changes in the pH of blood—say, a drop from 7.4 to 7.0 or an increase to 7.8—can cause death in a few minutes. (The critical link between pH and life is illustrated by the impact of acid rain on our environment and health. See the Environmental Issue essay, *What Is Happening to the Rain?*)

In the human body, almost all biochemical reactions occur around pH 7 and are maintained at that level by powerful buffering systems. An important exception occurs in the stomach, where hydrochloric acid (HCl) produces pH values from about 1 to 3. In the stomach, HCl kills bacteria swallowed with food or drink and promotes the initial breakdown of proteins. These activities require an acid stomach, and the stomach has several ways of protecting itself from the acid. However, sometimes stomach acid backs up into the esophagus, and "heartburn" is the uncomfortable result. Taking an antacid can ease the discomfort of heartburn. Antacids consist of weak bases that temporarily relieve the pain of stomach acid in the esophagus by neutralizing some of the hydrochloric acid.

Carbohydrates, Lipids, Proteins, and Nucleic Acids Are the Major Molecules of Life

Most of the molecules we have discussed so far have been small and simple. Many of the molecules of life, however, are enormous by comparison and have complex architecture. Some proteins, for example, are made up of thousands of atoms linked together in a chain that repeatedly coils and folds upon itself. Exceptionally large molecules, including many important biological molecules, are known as **macromolecules**.

Macromolecules that consist of many small, repeating molecular subunits linked in a chain are called **polymers**. The small molecular subunits that form the building blocks of the polymer are called **monomers**. We might think of a polymer as a pearl necklace, with each monomer representing a pearl. As we shall see, a protein is a polymer, or chain, of amino acid monomers linked together. And glycogen, the storage form of carbohydrates in animals, is a polymer of glucose monomers.

Polymers form through **dehydration synthesis**. In this process, the reaction that bonds one monomer covalently to another releases a water molecule: One of the monomers donates OH, and the other donates H. The reverse process, called **hydrolysis,** which the body uses to break many polymers apart, requires the

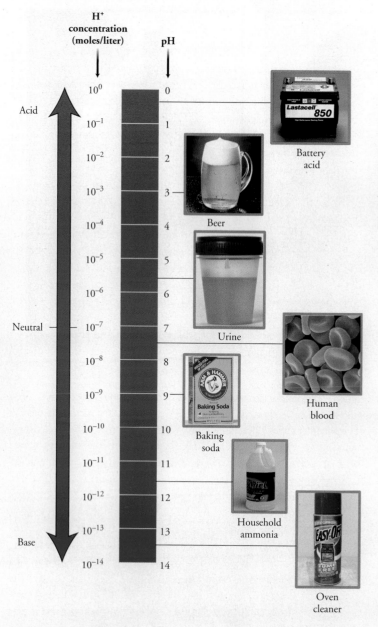

FIGURE **2.12** The pH scale and the pH of some body fluids and other familiar substances

TABLE 2.3 REVIEW OF THE CHARACTERISTICS OF ACIDS AND BASES

CHARACTERISTIC	ACID	BASE
Behavior in water	Releases H^+	Releases OH^-
pH	Less than 7	Greater than 7
Example	HCl (hydrochloric acid)	NaOH (sodium hydroxide)

ENVIRONMENTAL ISSUE

What Is Happening to the Rain?

Statues that have withstood not only the scrutiny of art critics but also the ravages of time are part of our cultural heritage. Some have stood for centuries over city streets and sidewalks, gazing at generations of passersby. But now the statues have had to be moved, and many that were not moved have been virtually destroyed (Figure 2.B). Rain has dissolved away their features, because in recent years it has become a solution of acid.

Acid rain is usually defined as rain with a pH lower than 5.6, the pH of natural precipitation. It is caused largely by the burning of fossil fuels in cars, factories, and power plants. Only a small percentage of the pollutants that cause acid rain are released into the atmosphere by volcanic eruptions and other natural processes. The sulfur dioxide and nitrogen oxides produced by these activities react with water in the atmosphere to form sulfuric acid (H_2SO_4) and nitric acid (HNO_3). These acids fall to the earth as rain, snow, or fog, with pH values sometimes as low as 1.5.

The effects of acid rain on the environment have been devastating (Figure 2.C). On land, acid rain has been linked to the decline of forests. Trees, for example, become stressed and more susceptible to disease when their nutrient uptake is disrupted by increased acidity in the soil. In aquatic environments, acid rain has been linked to the decrease and sometimes total elimination of populations of fish and amphibians. Embryos of spotted salamanders, for example, develop abnormally under somewhat acidic conditions and die when pH values are less than about 5. Acid rain also causes mercury and aluminum to leach from soils and rocks. These harmful elements eventually make their way into lakes and streams, where they contaminate aquatic life.

Acid rain is harmful to human health, as well. The pollutants that cause acid rain form fine particles of sulfate and nitrate that are easily inhaled. Once inside us, they cause irritation and respiratory illnesses such as asthma and bronchitis.

Because most acid rain is caused by human activity, it is within our power to reduce, if not eliminate, the problem. Many of the devastating effects could be reversed by tighter controls on industrial emissions. The installation of antipollution devices in our cars also would be helpful. And, of course, we could opt for walking or biking instead of driving and could carpool with others when driving is necessary. ⚘

(a) A figure, known as "The Vampire," on a cathedral in Paris around 1920

(b) The same figure, about 1990

FIGURE **2.B** Acid rain is caused by the release into the atmosphere of nitrogen and sulfur oxides during the burning of fossil fuels. Such rain is slowly eating away our works of art. The damage to this French statue took place over a period of about 70 years.

(b) In some areas, acid rain has reduced or eliminated populations of aquatic organisms. Acidic conditions have been shown, for example, to disrupt the development of spotted salamanders or to kill the embryos outright.

(a) Acid rain has destroyed parts of our forests.

FIGURE **2.C** Effects of acid rain

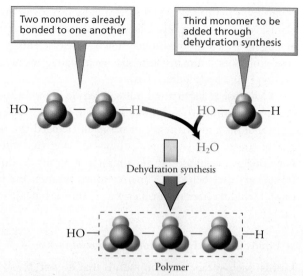

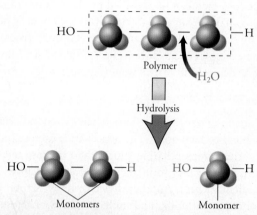

(a) Polymers are formed by dehydration synthesis, in which a water molecule is removed and two monomers are joined.

(b) Polymers are broken down by hydrolysis, in which the addition of a water molecule disrupts the bonds between two monomers.

FIGURE **2.13** Formation and breaking apart of polymers

addition of water across the covalent bonds. The H from the water molecule attaches to one monomer and the OH to the adjoining monomer, thus breaking the covalent bond between the two. Hydrolysis plays a critical role in digestion. Most foods consist of polymers too large to pass from our digestive tract into the bloodstream and on to our cells. Thus, the polymers are hydrolyzed into their component monomers. The monomers can then be absorbed into the bloodstream for transport throughout the body. Dehydration synthesis and hydrolysis are summarized in Figure 2.13.

■ Carbohydrates supply energy to cells

The **carbohydrates** known commonly as sugars and starches provide fuel (energy) for the human body. Carbohydrates are compounds made entirely of carbon, hydrogen, and oxygen, with each molecule having twice as many hydrogen atoms as oxygen atoms. Sugars and starches can be classified by size into monosaccharides, disaccharides, and polysaccharides. Some common carbohydrates are described in Table 2.4.

TABLE 2.4	REVIEW OF SOME COMMON CARBOHYDRATES		
CARBOHYDRATE	**MOLECULAR FORMULA**	**SOURCE**	**MONOMERS**
Monosaccharides			
Glucose	$C_6H_{12}O_6$	Blood, fruit, honey	
Fructose	$C_6H_{12}O_6$	Fruit, honey	
Galactose	$C_6H_{12}O_6$	From hydrolysis of lactose (milk sugar)	
Disaccharides			
Sucrose	$C_{12}H_{22}O_{11}$	Sugar cane, maple syrup	Glucose, fructose
Maltose	$C_{12}H_{22}O_{11}$	From hydrolysis of starch; ingredient in beer	Glucose
Lactose	$C_{12}H_{22}O_{11}$	Component of milk	Glucose, galactose
Polysaccharides			
Starch	*	Potatoes, corn, some grains	Glucose
Glycogen	*	Stored in muscle and liver cells	Glucose
Cellulose	*	Cell walls of plants	Glucose

* These complex carbohydrates consist of chains containing hundreds of glucose molecules joined to each other in long strings.

MONOSACCHARIDES

Monosaccharides, also called simple sugars, are the smallest molecular units of carbohydrates. They contain from three to seven carbon atoms and, in fact, can be classified by the number of carbon atoms they contain. A sugar that contains five carbons is pentose; one with six carbons is hexose; and so on. Glucose, fructose, and galactose are examples of six-carbon sugars (Table 2.4). Monosaccharides can be depicted in several ways (Figure 2.14).

DISACCHARIDES

Disaccharides are double sugars that form when two monosaccharides covalently bond to each other. The disaccharide sucrose (table sugar) consists of the monosaccharides glucose and fructose (Figure 2.15). Two glucose molecules form the disaccharide maltose, an important ingredient of beer. Another disaccharide is lactose, the principal carbohydrate of milk and milk products. Lactose is formed by the joining of glucose and galactose.

POLYSACCHARIDES

Polysaccharides are complex carbohydrates that form when monosaccharides (most commonly glucose) join together in long chains. Most polysaccharides store energy or provide structure. In plants, the storage polysaccharide is **starch**; in animals it is **glycogen**, which can be broken down to release energy-laden glucose molecules. Humans store glycogen mainly in the cells of the liver and muscles (Figure 2.16a).

Cellulose is a structural polysaccharide found in the cell walls of plants (Figure 2.16b). Humans lack the enzymes necessary to digest cellulose, and as a result, it passes unchanged through our digestive tract. (Enzymes are discussed later in this chapter.) Nonetheless, cellulose is an important form of dietary fiber (roughage) that helps fecal matter move through the large intestines. Including fiber in our diet may reduce the incidence of colon cancer.

■ Lipids store energy and form cell membranes

Lipids, such as fats, are compounds that do not dissolve in water. They are insoluble because they are nonpolar (having no electrical charges) and water is polar. Because of this difference, water shows no attraction for lipids and vice versa, so water and lipids do not mix. Three types of lipids that are important to human health are triglycerides, phospholipids, and steroids.

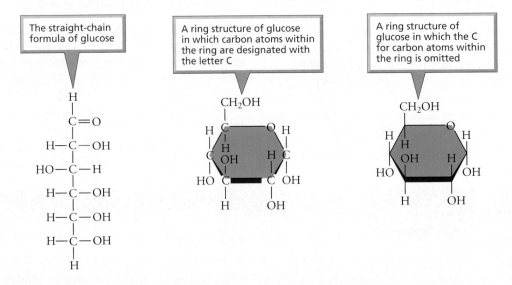

FIGURE **2.14** Monosaccharides are simple sugars, generally having a backbone of three to six carbon atoms. Many of these carbon atoms are also bonded to hydrogen (H) and a hydroxyl group (OH). In the fluid within our cells, the carbon backbone usually forms a ring like structure. Here, three representations of the monosaccharide glucose ($C_6H_{12}O_6$) are shown.

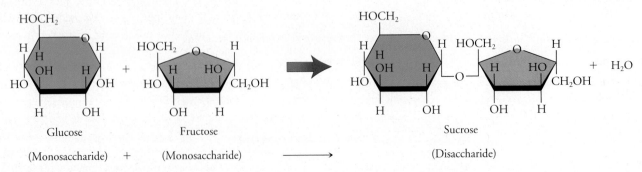

FIGURE **2.15** Disaccharides are built from two monosaccharides. Here, a molecule of glucose and one of fructose combine to form sucrose.

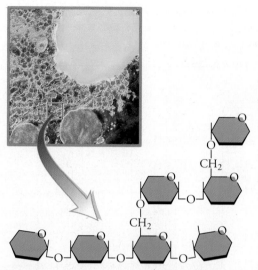

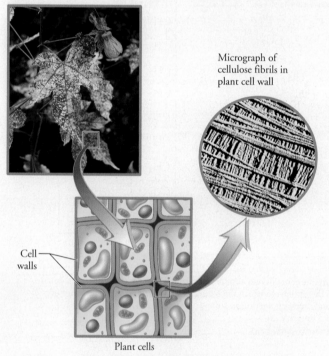

(a) Glycogen is the storage polysaccharide in animals. Granules of glycogen are stored in cells of the liver.

Micrograph of cellulose fibrils in plant cell wall

Cell walls

Plant cells

(b) Cellulose is a structural polysaccharide found in the cell walls of plants.

FIGURE **2.16** Polysaccharides may function in storage (as in glycogen) or provide structure (as in cellulose).

TRIGLYCERIDES

Fats and oils are **triglycerides**, compounds made of one molecule of glycerol and three fatty acids. Fatty acids are chains of carbon atoms also bonded to hydrogens and having the acidic group COOH at one end. The fatty acids bond to glycerol through dehydration synthesis (Figure 2.17a). Triglycerides are classified as saturated or unsaturated, depending on the presence or absence of double bonds between the carbon atoms in their fatty acids (Figure 2.17b). Saturated fatty acids have only single covalent bonds link-

ing the carbon atoms. They are described as "saturated with hydrogen," because their carbon atoms are bonded to as many hydrogen atoms as possible. Saturated fats are made from saturated fatty acids. Butter, a saturated fat, is solid at room temperature because its fatty acids can pack closely together. Fatty acids with one or more double bond between carbon atoms are described as unsaturated—that is, not saturated with hydrogen—because they could bond to more hydrogen atoms if the double bonds between their carbon atoms were broken. The double bonds cause "kinks" in the fatty acids and prevent molecules of unsaturated fat from packing tightly into a solid. Thus, unsaturated fats, such as olive oil, are liquid at room temperature.

Fats and oils provide about twice the energy per gram as carbohydrates or proteins do. This high energy density makes fat an ideal way for the body to store energy for the long term. Our bulk would be much greater if most of our energy storage consisted of carbohydrates or proteins, given their relatively low energy yield compared with fat. (The carbohydrate glycogen, described earlier, functions in short-term rather than long-term energy storage. Glycogen reserves take us from one meal to the next rather than one day to the next.)

In preparation for long-term energy storage, excess triglycerides, carbohydrates, and proteins from the foods we consume are converted into small globules of fat that are deposited in the cells of adipose tissue. There, the fat remains until we skip a meal or two; then our cells break it down to release the energy needed to keep vital processes going. In addition to long-term energy storage, fat serves a protective function in the body. Thin layers of fat surround major organs such as the kidneys, cushioning the organs against physical shock from falls or blows. Fat also serves as insulation and as a means of absorbing lipid-soluble vitamins from the intestines and transporting them to the cells that use them. Despite the importance of fats and oils to human health, in excess they can be very dangerous, particularly to our circulatory system.

PHOSPHOLIPIDS

Phospholipids are made of a molecule of glycerol bonded to two fatty acids and a negatively charged phosphate group. Another small molecule of some kind, though usually polar, called the variable group, is linked to the phosphate group. This general structure provides phospholipids with two regions having very different characteristics. As you will notice in Figure 2.18a, the region made up of fatty acids is nonpolar; it is described as a hydrophobic, or "water-fearing," tail. The other region consists of the hydrophilic, or "water-loving," head. The tails, being hydrophobic, do not mix with water. The heads, being hydrophilic, interact readily with water. The hydrophilic heads and hydrophobic tails of phospholipids are responsible for the structure of plasma (cell) membranes. In the membrane surrounding a cell, phospholipids are arranged in a double layer, called a bilayer (Figure 2.18b), with the hydrophilic heads of each layer facing away from each other. That way, each surface of the membrane consists of hydrophilic heads in contact with the watery solutions inside and outside the cell. The hydrophobic tails of the two layers point toward each other and help hold the membrane together.

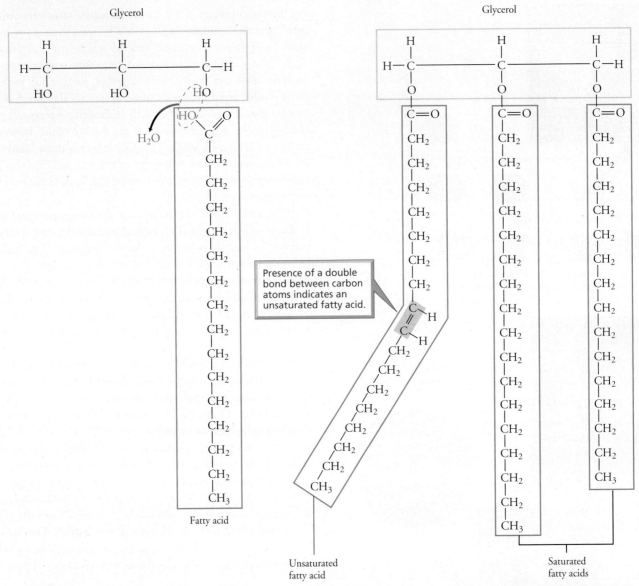

(a) A fatty acid bonds to glycerol through dehydration synthesis.

(b) This triglyceride contains one unsaturated fatty acid (note the presence of a double bond between the carbon atoms) and two saturated fatty acids (note the absence of any double bonds between the carbon atoms).

FIGURE **2.17** Triglycerides are composed of a molecule of glycerol joined to three fatty acids.

STEROIDS

A **steroid** is a type of lipid made up of four carbon rings attached to molecules that vary from one steroid to the next. Cholesterol, one of the most familiar steroids, is a component of the plasma membrane and is also the foundation from which steroid hormones, such as estrogen and testosterone, are made. Cholesterol, estrogen, and testosterone are shown in Figure 2.19. A high level of cholesterol in the blood is considered a risk factor for heart disease.

■ Proteins provide structure and speed up chemical reactions

A **protein** is a polymer made of one or more chains of amino acids. In many proteins, the chains are twisted, turned, and folded to pro-

duce complicated structures. Thousands upon thousands of different proteins are found in the human body, contributing to structural support (as does collagen, a major component of bone), transport (as does hemoglobin, which carries oxygen in the blood), movement (as do the actin and myosin filaments in muscles), and regulation of chemical reactions (as do enzymes, which speed up reactions). Despite their great diversity in structure and function, all proteins are made from a set of only about 20 kinds of amino acids.

AMINO ACIDS

Amino acids are the building blocks of proteins. They consist of a central carbon atom bound to a hydrogen atom (H), an amino

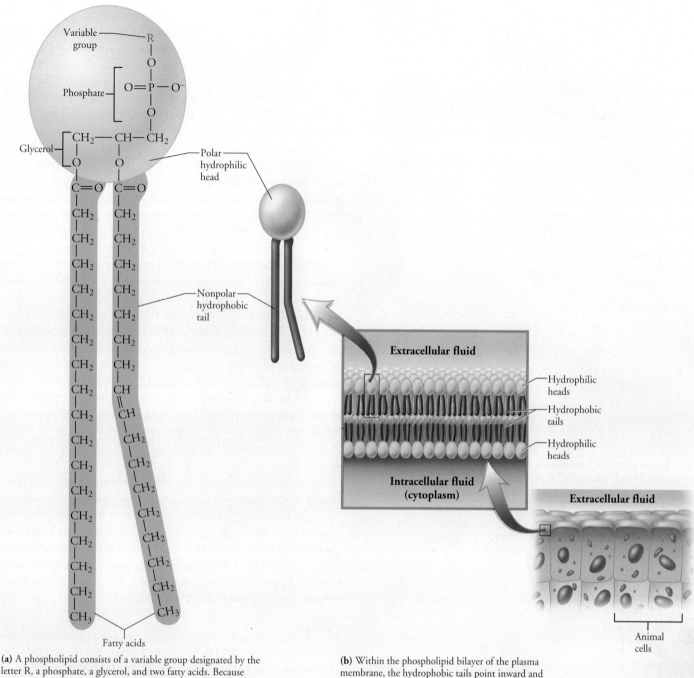

(a) A phospholipid consists of a variable group designated by the letter R, a phosphate, a glycerol, and two fatty acids. Because the variable group is often polar and the fatty acids nonpolar, phospholipids have a polar hydrophilic (water-loving) head and a nonpolar hydrophobic (water-fearing) tail.

(b) Within the phospholipid bilayer of the plasma membrane, the hydrophobic tails point inward and help hold the membrane together. The outward-pointing hydrophilic heads mix with the watery environments inside and outside the cell.

FIGURE **2.18** Structure of a phospholipid. Phospholipids are the main components of the plasma membrane encasing a cell and separating its internal and external watery environments.

group (NH_2), an acidic carboxyl group (COOH), and a side chain, often designated by the letter R (Figure 2.20). Amino acids differ from each other only in their side chains. Some amino acids, called nonessential amino acids, can be synthesized by our bodies. Other amino acids, called *essential amino acids*, cannot be synthesized by our bodies and must be obtained from the foods we eat.

The amino acids that form proteins are linked by bonds called peptide bonds, which are formed through dehydration synthesis. A peptide bond links the carboxyl group (COOH) of one amino acid to the amino group (NH_2) of the adjacent amino acid, as shown in Figure 2.21. Chains containing only a few amino acids are called **peptides**. Dipeptides contain two amino acids, tripeptides contain three amino acids, and so on. Chains containing 10 or more amino acids are called **polypeptides**. The term *protein* is used for polypeptides with at least 50 amino acids.

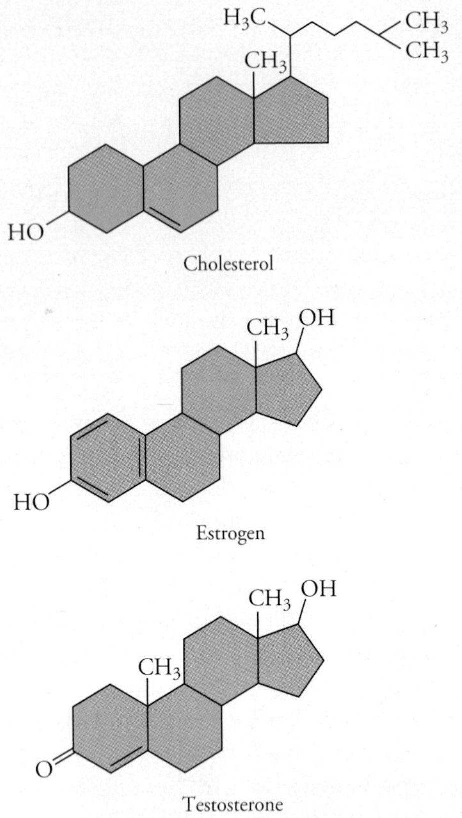

FIGURE 2.19 The steroid cholesterol is a component of cell membranes and is the substance from which steroid hormones such as estrogen and testosterone are made. All steroids have a structure consisting of four carbon rings. Steroids differ in the groups attached to these rings.

Cholesterol

Estrogen

Testosterone

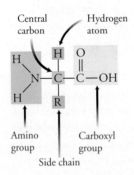

FIGURE 2.20 Structure of an amino acid. Amino acids differ from one another in the type of R group (side chain) they contain.

PROTEIN STRUCTURE

Proteins have four levels of structure: primary, secondary, tertiary, and quaternary. The **primary structure** of a protein is the particular sequence of amino acids. This sequence, determined by the genes, dictates a protein's structure and function. Even slight changes in primary structure can alter a protein's shape and ability to function. The inherited blood disorder sickle cell anemia provides an example. This disease primarily affects populations in central Africa, but also occurs in 1 in 500 African Americans. It results from the substitution of one amino acid for another during synthesis of the protein hemoglobin, which carries oxygen in our red blood cells. This single substitution in a molecule that contains hundreds of amino acids creates a misshapen protein that then alters the shape of red blood cells. Death can result when the oddly shaped cells clog the tiny vessels of the brain and heart.

The **secondary structure** of proteins consists of patterns known as pleated sheets and helices, which are formed by certain kinds of bends and coils in the chain, as a result of hydrogen bonding. Alterations in the secondary structure of a protein normally found on the surface of nerve cells can transform the protein into an infectious agent known as a prion. Prions have been implicated in a number of diseases, including Creutzfeldt-Jakob disease in humans and "mad cow disease" in cattle (see Chapter 13a).

The **tertiary structure** is the overall three-dimensional shape of the protein. Hydrogen, ionic, and covalent bonds between different side chains may all contribute to tertiary structure. Changes in the environment of a protein, such as increased heat or changes in pH, can cause the molecule to unravel and lose its three-dimensional shape. This process is called **denaturation**. Even minor change in the shape of a protein can result in loss of function.

Finally, some proteins consist of two or more polypeptide chains. Each chain, in this case, is called a subunit. **Quaternary structure** is the structure that results from the assembled subunits. The forces that hold the subunits in place are largely the attractions between oppositely charged side chains. The four levels of protein structure are summarized in Figure 2.22.

ENZYMES

Life is possible because of enzymes. Without them, most chemical reactions within our cells would occur far too slowly to sustain life. **Enzymes** are substances—almost always proteins, but occasionally RNA molecules—that speed up chemical reactions without being consumed in the process. Typically, reactions with enzymes proceed 10,000 to 1,000,000 times faster than the same reactions without enzymes.

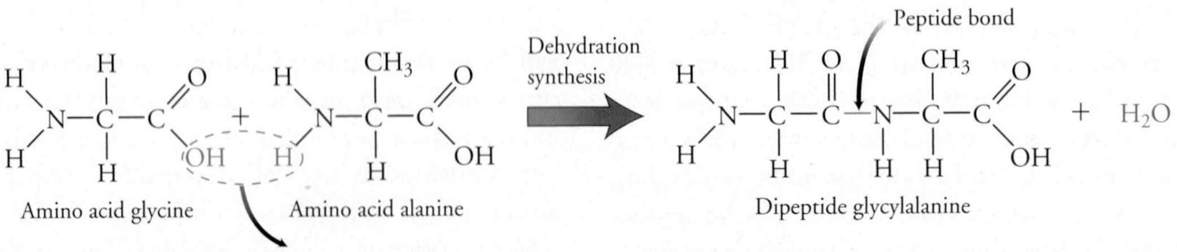

Amino acid glycine   Amino acid alanine

Dipeptide glycylalanine

FIGURE 2.21 Formation of a peptide bond between two amino acids through dehydration synthesis. The carboxyl group (COOH) of one amino acid bonds to the amino group (NH_2) of the adjacent amino acid, releasing water.

(a) Primary structure is the specific sequence of amino acids. Each amino acid is depicted here as a bead within the polypeptide chain.

(b) Secondary structure, such as the helix shown here, results from the bending and coiling of the chain of amino acids.

(c) Tertiary structure is the three-dimensional shape of proteins.

(d) Some proteins have two or more polypeptide chains, each chain forming a subunit. Quaternary structure results from the attractive forces between two or more subunits.

FIGURE **2.22** Levels of protein structure

The basic process by which an enzyme speeds up a chemical reaction can be summarized by the following equation:

$$E + S \rightarrow ES \rightarrow E + P$$

Enzyme   Substrate   Enzyme-substrate complex   Enzyme   Product

The particular substance that an enzyme works on is called its substrate. For example, the enzyme sucrase speeds up the reaction in which sucrose is broken down into glucose and fructose. In this reaction, sucrose is the substrate, and glucose and fructose are products. Similarly, the enzyme maltase speeds up the breakdown of maltose (the substrate) into molecules of glucose (the product). From these examples you can see that an enzyme's name may resemble the name of its substrate. These particular examples are decomposition reactions, in which a substance is broken down into its component parts. Enzymes also increase the speed of many synthesis reactions.

During reactions promoted by enzymes, the substrate binds to a specific location, called the **active site**, on the enzyme, to form an **enzyme–substrate complex**. This binding orients the substrate molecules so they can react. The substrate is converted to one or more products that then leave the active site, allowing the enzyme to bind to another substrate molecule. The entire process occurs very rapidly. One estimate suggests that a typical enzyme can con-

vert about 1000 molecules of substrate into product every second. Figure 2.23 summarizes the steps of an enzymatic reaction.

Enzymes are very specific in their activity; each is capable of binding to and acting on only one or at most a few particular substrates. Maltase, for example, acts only on maltose and not on sucrose, a structurally similar compound. This specificity is due to the unique shape of each enzyme's active site. The enzyme's active site and the substrate fit together like pieces of a jigsaw puzzle.

Sometimes enzymes need *cofactors*, nonprotein substances that help them convert substrate to product. Some cofactors permanently reside at the enzyme's active site. Other cofactors bind to the active site at the same time as the substrate. Some cofactors are the organic (carbon-containing) substances we know as vitamins. Organic cofactors are called *coenzymes*. Other cofactors are inorganic (non-carbon-containing) substances such as zinc or iron.

Enzyme deficiencies sometimes occur and affect our health. Lactase deficiency is a relatively common example. Lactase is the enzyme needed to digest the lactose in milk products, breaking it down to its two component monosaccharides, glucose and galactose. Infants and children usually produce adequate amounts of lactase, but many adults do not. For them, consumption of milk and milk products can lead to diarrhea, cramps, and bloating, caused by undigested lactose passing into the large intestine where it feeds resident bacteria. The bacteria, in turn, produce gas and lactic acid that irritate the bowels. The milk industry has responded to this problem (often called "lactose intolerance") by marketing lactose-reduced milk, but tablets and caplets that contain the enzyme lactase are also available.

■ The nucleic acids are DNA and RNA

In our discussion of protein structure we mentioned that genes determine the protein's primary structure, which is the sequence of amino acids. Genes, our units of inheritance, are segments of long polymers called deoxyribonucleic acid, or DNA. DNA is one of the two types of nucleic acids.

NUCLEOTIDES

The two nucleic acids in our cells are **deoxyribonucleic acid (DNA)** and **ribonucleic acid (RNA)**. Both are polymers of smaller units called **nucleotides**, joined into chains through dehydration synthesis. Every nucleotide monomer consists of a five-carbon (pentose) sugar bonded to one of five nitrogen-containing bases and at least one phosphate group (Figure 2.24). The five nitrogen-containing bases are adenine, guanine, cytosine, thymine, and uracil. Some of the bases (cytosine, thymine, and uracil) have a single ring made of carbon and nitrogen atoms. The others (adenine and guanine) have two such rings. The sequence of bases in DNA and RNA determines the sequence of amino acids in a protein. DNA, as we said above, is the nucleic acid found in genes. RNA, in various forms, converts the genetic information found in DNA into proteins.

DNA AND RNA

Key differences in the structures of RNA and DNA are summarized in Table 2.5. RNA is a single strand of nucleotides. The

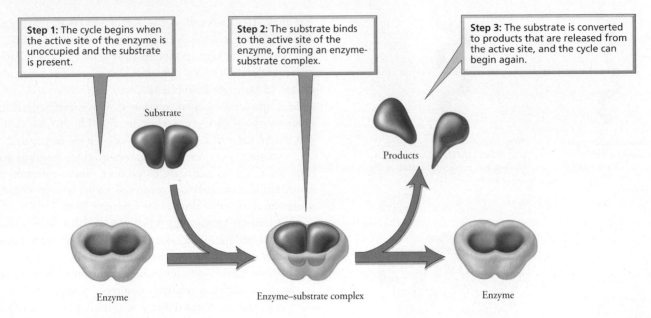

Step 1: The cycle begins when the active site of the enzyme is unoccupied and the substrate is present.

Step 2: The substrate binds to the active site of the enzyme, forming an enzyme-substrate complex.

Step 3: The substrate is converted to products that are released from the active site, and the cycle can begin again.

Substrate

Products

Enzyme

Enzyme–substrate complex

Enzyme

(a) A decomposition reaction involving an enzyme

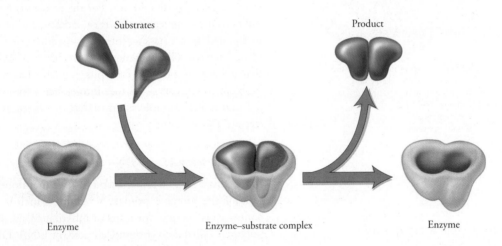

Substrates

Product

Enzyme

Enzyme–substrate complex

Enzyme

(b) A synthesis reaction involving an enzyme

FIGURE **2.23** The working cycle of an enzyme

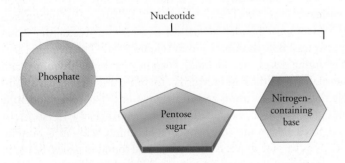

Nucleotide

Phosphate

Pentose sugar

Nitrogen-containing base

FIGURE **2.24** Structure of a nucleotide. Nucleotides consist of a five-carbon (pentose) sugar bonded to a phosphate molecule and one of five nitrogen-containing bases. Nucleotides are the building blocks of nucleic acids.

five-carbon sugar in RNA is ribose. The nitrogen-containing bases in RNA are cytosine, adenine, guanine, and uracil (Figure 2.25). In contrast, DNA is a double-stranded chain. Its two parallel strands, held together by hydrogen bonds between the nitrogen-containing bases, twist around one another to form a double helix. The five-carbon sugar in DNA is deoxyribose. The nitrogen-containing bases in DNA are adenine, thymine, cytosine, and guanine (Figure 2.26).

■ ATP is a nucleotide that releases energy

At this moment within your cells, many molecules of the nucleotide **adenosine triphosphate (ATP)** are each losing a phosphate group. As the phosphate group is lost, energy that the

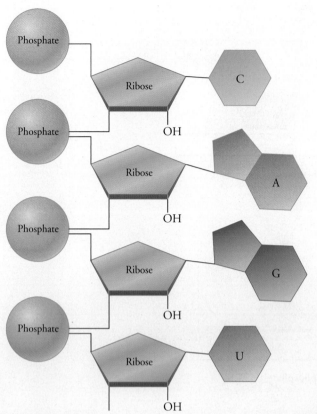

FIGURE **2.25** RNA is a single-stranded nucleic acid. It is formed by the linking together of nucleotides composed of the sugar ribose, a phosphate group, and the nitrogen-containing bases cytosine (C), adenine (A), guanine (G), and uracil (U).

FIGURE **2.26** DNA is a nucleic acid in which two chains of nucleotides twist around one another to form a double helix. The two chains are held together by hydrogen bonds between the nitrogen-containing bases. Each nucleotide of DNA contains the pentose sugar deoxyribose, a phosphate group, and one of the following four nitrogen-containing bases: adenine (A), thymine (T), cytosine (C), and guanine (G).

TABLE 2.5 REVIEW OF THE STRUCTURAL DIFFERENCES BETWEEN RNA AND DNA		
CHARACTERISTIC	**RNA**	**DNA**
Sugar	Ribose	Deoxyribose
Bases	Adenine, guanine, cytosine, uracil	Adenine, guanine, cytosine, thymine
Number of strands	One	Two; twisted to form double helix

molecule stored by holding on to the phosphate group is released. Your cells trap that energy and use it to perform work. It is because of this activity that you are able to sit up and read this book. ATP consists of the sugar ribose, the base adenine, and three phosphate groups, attached to the molecule by phosphate bonds. It is formed from adenosine diphosphate (ADP) by covalent bonding of a phosphate group to the ADP in an energy-requiring reaction. The energy absorbed during the reaction is stored in the new phosphate bond. It is an unstable bond and easily broken when the cell requires energy (Figure 2.27).

ATP is often described as the energy currency of cells. All energy from the breakdown of molecules such as glucose must be channeled through ATP before it can be used by the body.

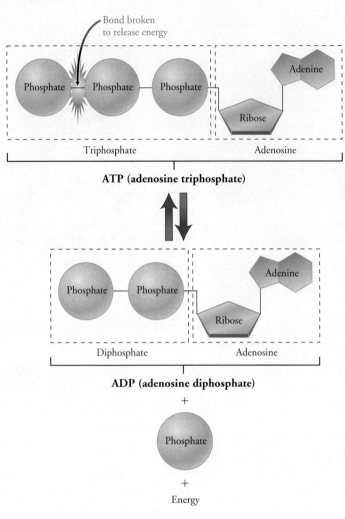

FIGURE **2.27** Structure and function of adenosine triphosphate (ATP). This nucleotide consists of the sugar ribose, the base adenine, and three phosphate groups. The phosphate bonds of ATP are unstable. When cells need energy, the last phosphate bond is broken, yielding adenosine diphosphate (ADP), a phosphate molecule, and energy.

HIGHLIGHTING THE CONCEPTS

Basic Chemistry Helps Us Understand Human Biology (pp. 15–23)

1. Atoms consist of subatomic particles called protons, neutrons, and electrons. Protons have a positive charge. Neutrons have no charge. Both are found in the nucleus and have atomic masses equal to 1. Electrons have a negative charge, weigh almost nothing, and are found around the nucleus in shells.

2. Each element is made of atoms containing the same number of protons. The atomic number of an element is the number of protons in one of its atoms. The atomic weight for any atom equals the number of protons plus the number of neutrons.

3. Isotopes are atoms that have the same number of protons but different numbers of neutrons. Some isotopes, called radioisotopes, emit radiation.

4. Elements combine to form compounds. The characteristics of compounds are usually different from those of the elements that form them.

5. When atoms join to form compounds, chemical bonds form between them. Covalent bonds form when atoms share electrons. Atoms that have lost or gained electrons have an electric charge and are called ions. The attraction between oppositely charged ions is an ionic bond.

6. Sometimes the sharing of electrons in a covalent bond is unequal, resulting in a polar covalent bond and sometimes a polar molecule. Hydrogen bonds are weak attractive forces between the charged regions of polar molecules.

7. Water is a polar molecule, and the polarity results in hydrogen bonding. Both make water an important component of the human body. Water is an excellent solvent because of its polarity. Its hydrogen bonds give water a high heat capacity, which helps the body's internal temperature remain constant, and a high heat of vaporization, so that evaporation of perspiration from the body surface prevents overheating.

8. Acids increase the concentration of H^+ in solution. Bases decrease the concentration of H^+ in solution. The strengths of acids and bases are measured on the pH scale. Biological fluids must remain within a narrow pH range. Buffers prevent dramatic changes in pH.

WEB TUTORIAL 2.1 Atoms, Ions, and Bonding
WEB TUTORIAL 2.2 Water and Chemistry

Carbohydrates, Lipids, Proteins, and Nucleic Acids Are the Major Molecules of Life (pp. 23–34)

9. A polymer is a large molecule made of many smaller molecules, called monomers. Polymers form through dehydration synthesis (removal of a water molecule) and are broken apart by hydrolysis (addition of a water molecule).

10. Carbohydrates are sugars and starches that provide fuel for the human body. Monosaccharides (simple sugars) are the smallest monomers of carbohydrates. Disaccharides ("double sugars") consist of two monosaccharides joined through dehydration synthesis. Polysaccharides ("many sugars") are complex carbohydrates formed when large numbers of monosaccharides join through dehydration synthesis.

11. Lipids, such as triglycerides, phospholipids, and steroids, are nonpolar molecules that do not dissolve in water. Triglycerides (fats and oils) are made of glycerol and three fatty acids and function in long-term energy storage. Phospholipids are important components of plasma membranes. Steroids include cholesterol, a component of the plasma membrane that also serves as the foundation for steroid hormones.

12. Proteins are polymers made from a set of about 20 amino acids linked together through dehydration synthesis to form chains. There are four levels of protein structure. Primary structure is the specific sequence of amino acids in a protein. Secondary structure results from the bending and coiling of the amino acid chain into pleated sheets or helices. Tertiary structure is the three-dimensional shape of the proteins. Some proteins consist of two or more polypeptide chains, each comprising a subunit. Attractive forces between the subunits of these proteins produce quaternary structure.

13. Enzymes are proteins, or sometimes RNA molecules, that speed up chemical reactions without being consumed in the process. The substrate binds to the enzyme at the active site, forming an enzyme–substrate complex, and is then converted to products that leave the active site.

14. Deoxyribonucleic acid (DNA) and ribonucleic acid (RNA) are polymers of nucleotides. A nucleotide consists of a five-carbon sugar bonded to one of five nitrogen-containing bases and a phosphate group. Genes are stretches of DNA that determine the sequence of amino acids during protein synthesis. DNA is double-stranded, its sugar component is deoxyribose, and its nitrogen-containing bases are adenine, guanine, cytosine, and thymine. RNA is a single-stranded molecule that also plays a major role in protein synthesis. The sugar in RNA is ribose, and the nitrogen-containing bases are adenine, guanine, cytosine, and uracil.

15. Adenosine triphosphate (ATP), the energy currency of cells, is a nucleotide made of the sugar ribose, the base adenine, and three phosphate groups. When cells require energy, one of the unstable phosphate bonds is broken and energy is released.

WEB TUTORIAL 2.3 Monomers and Polymers
WEB TUTORIAL 2.4 Lipid Structure and Function
WEB TUTORIAL 2.5 Protein Structure
WEB TUTORIAL 2.6 Nucleic Acids

KEY TERMS

chemistry *p. 14*
matter *p. 15*
atom *p. 15*
element *p. 15*
isotope *p. 15*
radioisotope *p. 15*
compound *p. 17*
covalent bond *p. 19*
molecule *p. 19*
ion *p. 20*
ionic bond *p. 20*
hydrogen bond *p. 21*
acid *p. 22*

base *p. 22*
pH *p. 22*
pH scale *p. 22*
buffer *p. 23*
macromolecule *p. 23*
polymer *p. 23*
monomer *p. 23*
dehydration synthesis *p. 23*
hydrolysis *p. 23*
carbohydrate *p. 25*
monosaccharide *p. 26*
disaccharide *p. 26*
polysaccharide *p. 26*

starch *p. 26*
glycogen *p. 26*
cellulose *p. 26*
lipid *p. 26*
triglyceride *p. 27*
phospholipid *p. 27*
steroid *p. 28*
protein *p. 28*
amino acid *p. 28*
peptide *p. 29*
polypeptide *p. 29*
primary structure *p. 30*
secondary structure *p. 30*

tertiary structure *p. 30*
denaturation *p. 30*
quaternary structure *p. 30*
enzyme *p. 30*
active site *p. 31*
enzyme–substrate complex *p. 31*
deoxyribonucleic acid (DNA) *p. 31*
ribonucleic acid (RNA) *p. 31*
nucleotide *p. 31*
adenosine triphosphate (ATP) *p. 32*

REVIEWING THE CONCEPTS

1. What is an atom? *p. 15*
2. Compare protons, neutrons, and electrons with respect to their charge, mass, and location within an atom. *p. 15*
3. What is an isotope? *p. 15*
4. Describe the damage to the body caused by radiation. *pp. 15–16*
5. How is radiation used to diagnose or cure illness? *pp. 16–17*
6. Describe ionic, covalent, and hydrogen bonds. Give an example of each. *pp. 17, 19–21*
7. What characteristics of water make it a critical component of the body? *pp. 21–22*
8. Define acids and bases according to what happens when they are added to water. *pp. 22–23*
9. What is the pH scale? *pp. 22–23*

10. How are polymers formed and broken? Give three examples of polymers and their component monomers. *pp. 23, 25–26*
11. Name two important energy-storage polysaccharides and one important structural polysaccharide. *p. 26*
12. Describe the structure of a phospholipid. What important roles do they play in the human body? *pp. 27, 29*
13. What are saturated and unsaturated triglycerides? What are the functions of triglycerides in the body? *pp. 27, 28*
14. Describe the four levels of protein structure. *pp. 30, 31*
15. Compare the structures of RNA and DNA. *pp. 31–32, 33*
16. Describe the structure and function of ATP. *pp. 32–34*
17. Compare and contrast the ways in which proteins and nucleic acids are used in the body. *pp. 28, 31*

18. Hydrogen bonds
    a. are stronger than either ionic or covalent bonds.
    b. form between a slightly positively charged hydrogen atom and a slightly negatively charged atom nearby.
    c. maintain the shape of proteins and DNA.
    d. b and c
19. Water
    a. is a nonpolar molecule and therefore an excellent solvent.
    b. has a high heat capacity and therefore helps maintain a constant body temperature.
    c. has a low heat of vaporization and therefore helps prevent over-heating of the body.
    d. makes up about 25% of the human body.
20. Acids
    a. release hydrogen ions when placed in water.
    b. produce hydroxide ions when placed in water.
    c. have a pH greater than 7.
    d. prevent normal digestive activities in the stomach.
21. Carbohydrates
    a. consist of chains of amino acids.
    b. supply our cells with energy.
    c. contain glycerol.
    d. function as enzymes.
22. Triglycerides
    a. have one molecule of glycerol and three fatty acids.
    b. are poor sources of energy.

    c. are saturated when there are two or more double bonds linking carbon atoms.
    d. are major components of plasma membranes.
23. Enzymes
    a. speed up chemical reactions and are consumed in the process.
    b. are always proteins.
    c. are usually nonspecific and therefore capable of binding to many different substrates.
    d. have locations, known as active sites, to which the substrate binds.
24. Changes in temperature or pH can cause a protein to lose its three-dimensional shape and become nonfunctional. This process is called _____.
25. DNA and RNA are polymers of smaller units called _____.
26. The _____ structure of a protein is the precise sequence of amino acids.
27. _____ are arranged in a double layer (bilayer) that forms the plasma membrane of cells.
28. In plants, the storage polysaccharide is _____, and in animals it is _____.
29. The nucleotide _____ is the energy currency of cells.
30. Buffers prevent dramatic changes in _____.
31. _____ bonds form when outer-shell electrons are shared between atoms.
32. _____ is the process through which monomers are linked together through the removal of a water molecule to form polymers. _____ is the process that breaks polymers apart by the addition of water.

APPLYING THE CONCEPTS

1. A friend eyes your lunch and begins to lecture you on the perils of high-fat foods. You decide to acknowledge the health risks of eating a high-fat diet but also to point out to her the important roles that lipids play in your body. What will you say?
2. Bill claims that eating fruits and vegetables is overrated because humans lack the enzyme needed to digest cellulose and thus cellulose passes unchanged through our digestive tract. Is Bill correct that we gain nothing from eating food that contains cellulose?
3. For years your mother has been telling you to drink several large glasses of water each day. Having just read this chapter, you now un-

derstand the critical roles that water plays in your body. You vow to explain during your next phone call home what you have learned. What will you say to your mother? Also, why do you think dehydration might be dangerous?
4. Dental x-rays are taken using small doses of radiation and are used to detect tooth decay, injuries to the roots of teeth, and problems with the bones supporting the teeth. A dental professional covers the patient with a lead apron that runs from the neck to the abdomen and then leaves the room to take the x-ray. Why are such precautions necessary?

Additional questions can be found on the companion website.

Our Cells Are Highly Structured

- Eukaryotic cells are structurally more complex than prokaryotic cells
- Cells are very small
- The plasma membrane has diverse functions
- Organelles are specialized compartments within cells
- The cytoskeleton provides support and movement

Our Cells Use Cellular Respiration and Fermentation to Generate ATP

- Cellular respiration requires oxygen
- Fermentation does not require oxygen

ENVIRONMENTAL ISSUE   Asbestos: The Deadly Miracle Material

3
The Cell

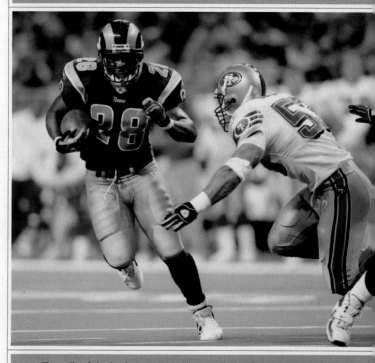

The cells of the body cooperate like a well-coordinated team in which each member is suited to a particular role. At the same time, all cells have certain basic features in common.

Morinville quarterback Joe Norton found receiver Charlie Mendoza in the right corner of the end zone with 15 seconds remaining in the fourth quarter. Norton, playing with a strained ankle in the second half, passed for 25 yards on the winning drive, taking advantage of Mendoza's legendary speed and agility. Mendoza rushed for 90 yards and two touchdowns, sealing his place as Most Valuable Player in the championship game. Rob Douglas launched a stunning 47-yard field goal with one minute remaining in the third quarter and kept Beaumont at bay with a series of punts and kickoffs into the end zone and inside the 5-yard line.

Coach Anderson's satisfaction was palpable. "The offensive line deserves tremendous credit," he said. "With their blocking and Norton and Mendoza's performances, no one could have beat us today, even though the score was close. And we couldn't have done it without superb defense. We held Beaumont at key times in the second half, and the two sacks on Tony Jones sealed it. You fellas deserve immense credit. It's a historic win."

As any coach knows, a team cannot function without the coordinated efforts of players in a wide range of specialized positions. By the same token, each player's body cannot contribute to the team's performance without the coordinated actions of a vast number of cell types. Our bodies contain over 10 trillion cells that specialize in more than 100 different functions. To throw a pass, muscle cells contract in response to electrical signals from the brain. To dodge a tackle, cells of the adrenal glands release hormones such as norepinephrine that trigger near-instantaneous reactions. Meanwhile, kidney cells continuously remove wastes from the burning of cellular fuels. Neurons coordinate the perceptions and intellect of each player.

In this chapter we discover how cells work by examining the structures that all cells have in common. We begin with the plasma membrane, the outer boundary of a cell, and work inward. After examining the various organelles—small compartments within cells that have specialized functions—we explore the ways in which cells obtain the energy they need to do their work and run the body. ■

Our Cells Are Highly Structured

The **cell theory** is a fundamental organizing principle of biology, guiding the way biologists think about and study living things. It states that (1) a cell is the smallest unit of life; (2) cells make up all living things, from unicellular to multicellular organisms; and (3) new cells arise from preexisting cells.

There are two basic types of cells—eukaryotic cells and prokaryotic cells—depending on the presence or absence of internal compartments enclosed by membranes. These membrane-bound compartments are called **organelles**, or "little organs." **Eukaryotic cells** have membrane-bound organelles. **Prokaryotic cells** do not.

■ Eukaryotic cells are structurally more complex than prokaryotic cells

The cells of plants, animals, and all other organisms except bacteria and archaea are eukaryotic. All the cells that make up your body, therefore, are eukaryotic, having extensive internal membranes (Figure 3.1). One of the membrane-bound organelles in eukaryotic cells is a well-defined nucleus containing DNA. (Our red blood cells are an interesting exception to the "rule" that eukaryotic cells have a well-defined nucleus and other membrane-bound organelles. As described in Chapter 11, when red blood cells mature, they extrude their nucleus and most organelles, leaving more space for hemoglobin, the protein that transports oxygen.)

Prokaryotic cells are limited to bacteria and another group of microscopic organisms called archaea. You are probably already aware of bacteria, some of which inhabit your body. Archaea may be less familiar to you. They include species that inhabit extreme environments such as the Great Salt Lake or the hot sulfur springs of Yellowstone National Park. Prokaryotic cells are much simpler and typically much smaller than eukaryotic cells. Most are sur-

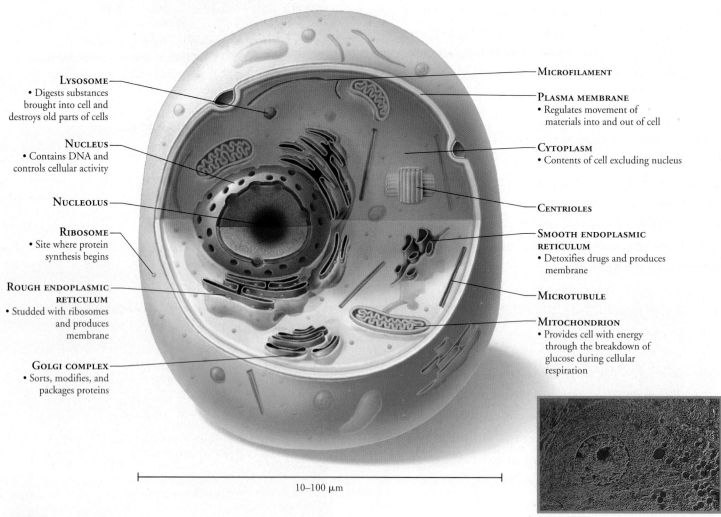

LYSOSOME
• Digests substances brought into cell and destroys old parts of cells

NUCLEUS
• Contains DNA and controls cellular activity

NUCLEOLUS

RIBOSOME
• Site where protein synthesis begins

ROUGH ENDOPLASMIC RETICULUM
• Studded with ribosomes and produces membrane

GOLGI COMPLEX
• Sorts, modifies, and packages proteins

MICROFILAMENT

PLASMA MEMBRANE
• Regulates movement of materials into and out of cell

CYTOPLASM
• Contents of cell excluding nucleus

CENTRIOLES

SMOOTH ENDOPLASMIC RETICULUM
• Detoxifies drugs and produces membrane

MICROTUBULE

MITOCHONDRION
• Provides cell with energy through the breakdown of glucose during cellular respiration

10–100 μm

FIGURE **3.1** Eukaryotic cells, such as the generalized animal cell shown here, have internal membrane-bound organelles.

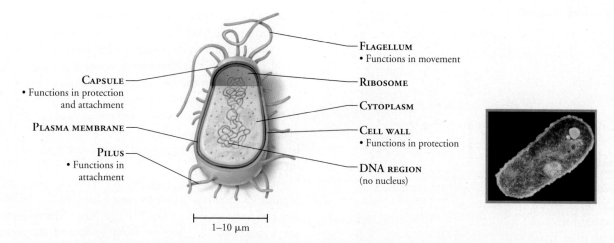

FLAGELLUM
• Functions in movement

CAPSULE
• Functions in protection
and attachment

RIBOSOME

CYTOPLASM

PLASMA MEMBRANE

CELL WALL
• Functions in protection

PILUS
• Functions in
attachment

DNA REGION
(no nucleus)

1–10 μm

FIGURE **3.2** Prokaryotic cells, such as a bacterium, lack internal membrane-bound organelles.

rounded by a rigid cell wall (among eukaryotes, plant cells have cell walls, but animal cells do not). As shown in Figure 3.2, the DNA of prokaryotic cells is not enclosed in a nucleus. Table 3.1 reviews the major differences between eukaryotic and prokaryotic cells.

TABLE 3.1 REVIEW OF FEATURES OF EUKARYOTIC AND PROKARYOTIC CELLS

FEATURE	EUKARYOTIC CELLS	PROKARYOTIC CELLS
Organisms	Plants, animals, fungi, protists	Bacteria, archaea
Size	10–100 μm across	1–10 μm across
Membrane-bound organelles	Present	Absent
DNA form	Coiled, linear strands	Circular
DNA location	Nucleus	Cytoplasm
Internal membranes	Many	Rare
Cytoskeleton	Present	Absent

■ Cells are very small

Most eukaryotic and prokaryotic cells are so small you need a microscope to see them (Figure 3.3). Their small size is dictated by a physical relationship known as the *surface-area-to-volume ratio*. This relationship says that as a cell gets larger, its surface area increases much more slowly than its volume (Figure 3.4). Nutrients enter a cell and wastes leave a cell at its surface. A large cell would have difficulty moving all the nutrients it needs and all the wastes it produces across its inadequate surface and would die. A small cell, on the other hand, has sufficient surface for the uptake and removal of substances and would survive.

■ The plasma membrane has diverse functions

We begin our examination of the cell at its outer surface—the **plasma membrane**. This remarkably thin outer covering controls the movement of substances both into and out of the cell. Because the concentrations of substances in a cell's interior are critically balanced, molecules and ions are not permitted to move randomly in and out.

Both prokaryotic and eukaryotic cells have a plasma membrane, but only eukaryotic cells also contain internal membranes

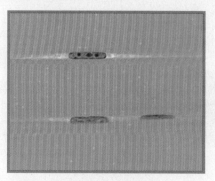

(a) Striated muscle cells viewed with a light microscope.

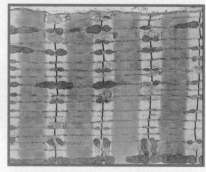

(b) Striated muscle cells viewed with a transmission electron microscope.

(c) Striated muscle cells viewed with a scanning electron microscope.

FIGURE **3.3** Micrographs are photographs taken through a microscope. Here, a striated muscle cell has been photographed using three different types of microscope. Electron microscopes use beams of electrons to produce images with finer details than is possible with light microscopes.

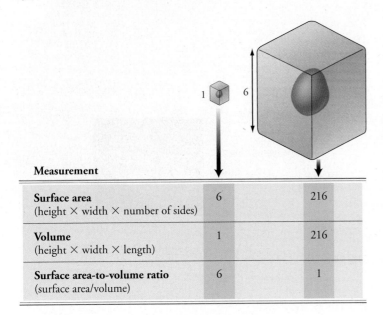

Measurement	1	6
Surface area (height × width × number of sides)	6	216
Volume (height × width × length)	1	216
Surface area-to-volume ratio (surface area/volume)	6	1

FIGURE **3.4** Cells must remain small in size because the ratio of surface area to volume decreases rapidly as cell size increases.

that divide the cell into many compartments. Each compartment contains its own assortment of enzymes and is specialized for particular functions. In general, the principles described for the plasma membrane also apply to the membranes inside the cell.

STRUCTURE OF THE PLASMA MEMBRANE

The plasma membrane is made of lipids, proteins, and carbohydrates. Recall from Chapter 2 that phospholipids are the major components. These molecules, with their hydrophilic (water-loving) heads and hydrophobic (water-fearing) tails, form a double

layer, called the lipid bilayer, at the surface of the cell (Figure 3.5). The hydrophilic heads on the side of the membrane facing outside the cell interact with the **extracellular fluid** (also known as interstitial fluid), which is the watery solution outside cells. The hydrophilic heads on the side of the membrane facing inside the cell interact with the **cytoplasm**, which is the jellylike solution inside the cell. The cytoplasm contains all contents of the cell between the plasma membrane and the nucleus. Within the lipid bilayer, the hydrophobic tails point toward each other and hold the plasma membrane together.

Interspersed in the phospholipid bilayer are proteins, as seen in Figure 3.5. Some proteins are embedded in the membrane, and some of these span the bilayer completely. Other proteins are simply attached to the inner or outer surface of the membrane. In many cases, the surface proteins are attached to an exposed portion of an embedded protein. Molecules of cholesterol are also scattered throughout the bilayer.

As you can see in Figure 3.5, carbohydrates attach only to the outer surface of the plasma membrane. Some of these carbohydrates are attached to lipids, forming substances called glycolipids. Most of the carbohydrates, however, are attached to proteins, forming glycoproteins. We will see that glycoproteins often function in cell recognition.

The structure of the plasma membrane is often described as a **fluid mosaic**. The proteins are interspersed among the lipid molecules like tiles of different colors within a mosaic. Many of the proteins are able to move sideways through the bilayer to some degree, giving the membrane its fluid quality.

FUNCTIONS OF THE PLASMA MEMBRANE

The plasma membrane performs several vital functions for the cell. First, by imposing a boundary between the cell's internal and

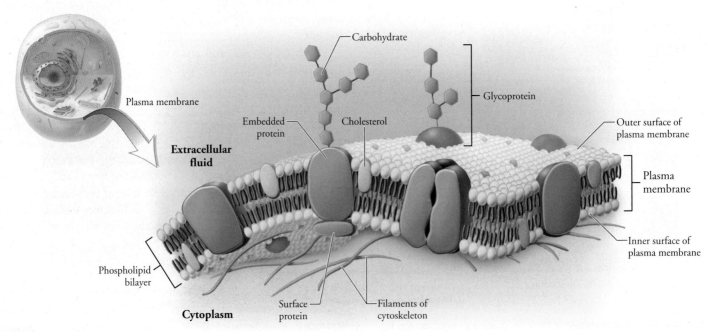

FIGURE **3.5** The structure of the plasma membrane of a cell according to the fluid mosaic model

external environment, the plasma membrane maintains the cell's structural integrity. Second, the structure of the plasma membrane regulates the movement of substances into and out of the cell, permitting entry to some substances but not others. For this reason, the membrane is often described as being **selectively permeable**. You will read more about the transport of materials across the plasma membrane later in this chapter.

The plasma membrane also functions in cell-cell recognition. Cells distinguish one type of cell from another by recognizing molecules—often glycoproteins—on their surface. Membrane glycoproteins differ from one species of organism to another and among individuals of the same species. Even different cell types within an individual have different membrane glycoproteins. This variation allows the body to recognize foreign invaders such as bacteria. Your own body, for example, would recognize such invaders because the bacteria lack the surface molecules found on your cells. The bacteria, in turn, "read" the different surface molecules of your cells in order to settle preferentially on some kinds of cells but not others.

Another important function of the plasma membrane is communication between cells. Such communication relies on *receptors*, specialized proteins in the plasma membrane (or inside the cell) that bind particular substances that affect the activities of the cell. For example, hormones secreted by one group of cells may bind to receptors in the plasma membranes of other cells. The receptors then relay a signal to proteins inside the cell, which transmit the message to other nearby molecules. Through a series of chemical reactions, the hormone's "message" ultimately initiates a response by the recipient cell, perhaps causing it to release a certain chemical.

Lastly, the plasma membrane plays an important role in binding pairs or groups of cells together. *Cell adhesion molecules* (CAMs, for short) extend from the plasma membranes of most cells (Figure 3.6) and help attach the cells to one another, especially during the formation of tissues and organs in an embryo.

TABLE 3.2 REVIEW OF PLASMA MEMBRANE FUNCTIONS
Maintenance of structural integrity of the cell
Regulation of movement of substances into and out of the cell
Recognition between cells
Communication between cells
Sticking cells together to form tissues and organs

The functions of the plasma membrane are reviewed in Table 3.2.

stop and think

Of the five functions of the plasma membrane summarized in Table 3.2, which might explain the difficulty of transplanting tissues and organs successfully from one body to another? Why would rejection of such transplants occur? Under what circumstances might one body accept a tissue graft or organ from another?

MOVEMENT ACROSS THE PLASMA MEMBRANE

The movement of substances into and out of the cell is controlled in several ways. Even so, some materials cross the plasma membrane freely, by simple diffusion.

SIMPLE DIFFUSION *Concentration* is the number of molecules of a substance in a particular volume. A *concentration gradient* is a difference in the relative number of molecules or ions of a given substance in two adjacent areas. The random movement of a substance from a region of higher concentration to a region of lower concentration is **simple diffusion**. The end result of simple diffusion is an equal distribution of the substance in the two areas. Consider what happens when someone is cooking bacon in the kitchen. At first, the smell of bacon is localized in the kitchen. Soon, however, the smell permeates adjoining rooms, too, as odor molecules move from where they are more concentrated (the kitchen) to where they are less concentrated (other parts of the house). Eventually the odor molecules are equally distributed, but they still move randomly in all directions. Likewise, when a substance diffuses across a membrane from a region of higher concentration to a region of lower concentration, the movement of its molecules does not stop once the concentration has equalized. Instead, the molecules continue to move randomly back and forth across the membrane. The rate of movement in each direction, however, is now the same. Substances such as carbon dioxide and oxygen diffuse through the lipid bilayer of our cells (Figure 3.7).

FACILITATED DIFFUSION Water-soluble substances are repelled by lipids and so cannot move through the lipid bilayer by simple diffusion. If they are to cross a cell membrane, their transport must be assisted, or "facilitated," by certain proteins within the membrane. Some of these proteins bind to a particular water-soluble substance and carry it across. Other proteins form

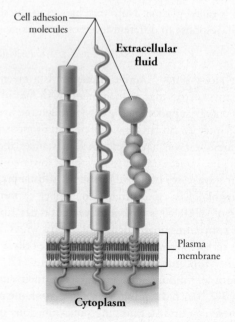

Cell adhesion molecules

Extracellular fluid

Plasma membrane

Cytoplasm

FIGURE **3.6** Cell adhesion molecules (CAMs). These uniquely shaped molecules hold cells together to form tissues and organs and are involved in cellular interaction and movement.

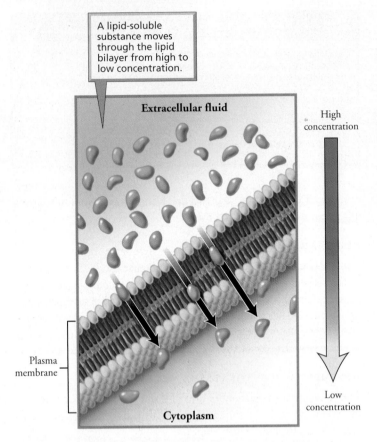

A lipid-soluble substance moves through the lipid bilayer from high to low concentration.

Extracellular fluid

High concentration

Plasma membrane

Low concentration

Cytoplasm

FIGURE **3.7** Simple diffusion is the random movement of a substance from a region of higher concentration to a region of lower concentration.

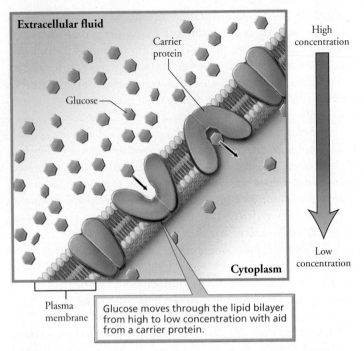

Extracellular fluid

Carrier protein

High concentration

Glucose

Low concentration

Cytoplasm

Plasma membrane

Glucose moves through the lipid bilayer from high to low concentration with aid from a carrier protein.

WEB TUTORIAL **3.3** FIGURE **3.8** Facilitated diffusion is the movement of a substance from a region of higher concentration to a region of lower concentration with the aid of a membrane protein that acts as a channel or a carrier protein.

channels through which certain water-soluble substances can move. **Facilitated diffusion** is the movement of a substance from a region of higher concentration to a region of lower concentration with the aid of a membrane protein. Molecules of glucose, for example, enter fat cells by facilitated diffusion. In this example, a carrier protein helps move glucose from outside to inside the fat cell (Figure 3.8).

OSMOSIS **Osmosis** is a special type of diffusion in which water moves across a plasma membrane or any other selectively permeable membrane from a region of higher water concentration to a region of lower water concentration. The movement of water occurs in response to a concentration gradient of a dissolved substance (solute). Consider what happens when a substance such as table sugar (the solute) is dissolved in water (the solvent) in a membranous bag through which water, but not sugar, can move. Keep in mind that when solute concentration is low, water concentration is high, and when solute concentration is high, water concentration is low. If the membranous bag is placed into a **hypertonic solution**, meaning a solution whose solute (sugar) concentration is higher than that inside the bag, more water moves out of the bag than in, causing the bag to shrivel (Figure 3.9a). If, on the other hand, the bag is placed into an **isotonic solution**, one with the same solute (sugar) concentration as inside the bag, there is no net movement of water in either direction, and the bag maintains its normal shape (Figure 3.9b). When the bag is placed

into a **hypotonic solution,** in which the concentration of solute (sugar) is lower than that inside the bag, more water moves into the bag than out, causing the bag to swell and possibly burst (Figure 3.9c).

The bag in our example can be replaced by a body cell, such as a red blood cell, as shown in Figure 3.9. The solution that the bag was placed into can be replaced by the fluid surrounding red blood cells, which is called plasma. Note the changes in the shape of red blood cells in response to different levels of solute concentration in the plasma.

ACTIVE TRANSPORT **Active transport** is a mechanism that moves substances across plasma membranes with the aid of a carrier protein and energy supplied by the cell (through the breakdown of ATP; see Chapter 2). So far in our discussion of movement across plasma membranes, we have described substances moving from regions of higher concentration to regions of lower concentration. However, in most cases of active transport, substances are moved from regions of lower concentration to higher concentration, as shown in Figure 3.10. This type of movement is described as going "against the concentration gradient" and occurs when cells need to concentrate certain substances. For example, the cells in our bodies contain higher concentrations of potassium ions (K^+) and lower concentrations of sodium ions (Na^+) than their surroundings. Through active transport, proteins in the plasma membrane help maintain these conditions, pumping potassium ions into the cell and sodium ions out of the cell. In this example, both potassium and sodium are moving from regions of lower concentration to regions of higher concentration.

The bag gains and loses the same amount of water and maintains its shape.

98% water
2% sugar

The bag loses more water than it gains and shrivels.

The bag gains more water than it loses and swells.

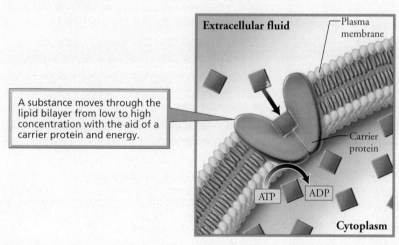

(a) *Hypertonic* solution
(90% water 10% sugar)

(b) *Isotonic* solution
(98% water 2% sugar)

(c) *Hypotonic* solution
(100% water, distilled)

WEB TUTORIAL 3.4 FIGURE **3.9** Osmosis is diffusion of water across a selectively permeable membrane. The drawings show what happens when a membranous bag through which water but not sugar can move is placed in solutions that are (a) hypertonic, (b) isotonic, or (c) hypotonic to the solution inside the bag. The width of the arrows corresponds to the amount of water moving into and out of the bag. The photographs show what happens to red blood cells when placed in the three kinds of solutions. Red blood cells are normally shaped like flattened disks, as in part b.

ENDOCYTOSIS Most small molecules cross the plasma membrane by simple diffusion, facilitated diffusion, or active transport. Large molecules, single-celled organisms such as bacteria, and droplets of fluid containing dissolved substances enter cells through endocytosis. In **endocytosis**, a region of the plasma membrane engulfs the substance to be ingested and then pinches off from the rest of the membrane, thus enclosing the substance in a saclike structure called a **vesicle**. The vesicle then travels into the cell. Two types of endocytosis are phagocytosis ("cell eating") and pinocytosis ("cell drinking"). In **phagocytosis**, cells engulf large particles or bacteria (Figure 3.11a). In **pinocytosis**, they engulf droplets of fluid (Figure 3.11b), thus bringing all of the substances dissolved in the droplet into the cell.

EXOCYTOSIS The process by which large molecules leave cells is **exocytosis** (Figure 3.12). In a cell that produces hormones, for example, the hormones are enclosed in membrane-bound vesicles that travel through the cell toward the plasma membrane. When the vesicle reaches the plasma membrane, the vesicle membrane fuses with the plasma membrane and the vesicle releases the hormone outside the cell. Nerve cells also release chemicals by exocytosis.

Table 3.3 reviews the ways in which substances move across the plasma membrane.

stop and think

People with kidney disorders may need dialysis to remove waste products from their blood. The blood is passed through a long coiled tube submerged in a tank filled with a fluid called dialyzing fluid. The tubing is porous, allowing small molecules to diffuse out of the blood into the dialyzing fluid. Urea is a waste product that must be removed from the blood. Glucose molecules, on the other hand, should remain in the blood. What should the concentrations of urea and glucose be like in the dialyzing fluid to achieve these goals?

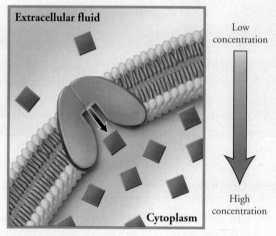

Extracellular fluid

Plasma membrane

A substance moves through the lipid bilayer from low to high concentration with the aid of a carrier protein and energy.

Carrier protein

ATP   ADP

Cytoplasm

Extracellular fluid

Low concentration

High concentration

Cytoplasm

FIGURE **3.10** Active transport is the movement of molecules across the plasma membrane, often from an area of lower concentration to one of higher concentration with help from a carrier protein and energy, usually in the form of ATP.

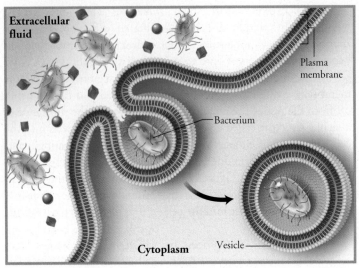

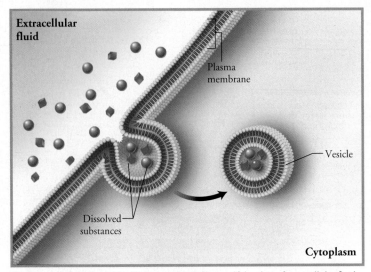

(a) Phagocytosis ("cell eating") occurs when cells engulf bacteria or other large particles.

(b) Pinocytosis ("cell drinking") occurs when cells engulf droplets of extracellular fluid and the dissolved substances therein.

FIGURE **3.11** Endocytosis—phagocytosis or pinocytosis—occurs when a localized region of the plasma membrane surrounds a bacterium, large molecule, or fluid containing dissolved substances and then pinches off to form a vesicle that moves into the cell.

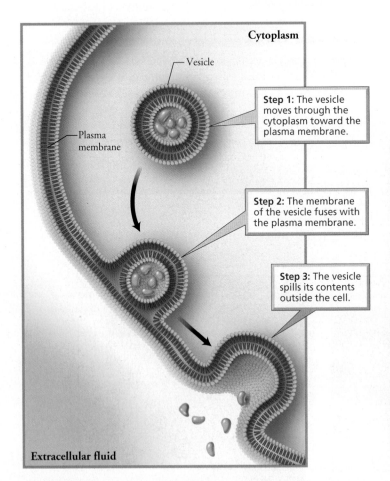

Step 1: The vesicle moves through the cytoplasm toward the plasma membrane.

Step 2: The membrane of the vesicle fuses with the plasma membrane.

Step 3: The vesicle spills its contents outside the cell.

FIGURE **3.12** Cells package large molecules in membrane-bound vesicles, which then leave by exocytosis.

TABLE 3.3 REVIEW OF MECHANISMS OF TRANSPORT ACROSS THE PLASMA MEMBRANE

MECHANISM	DESCRIPTION
Simple diffusion	Random movement from region of higher to lower concentration
Facilitated diffusion	Movement from region of higher to lower concentration with the aid of a carrier or channel protein
Osmosis	Movement of water from region of higher water concentration (lower solute concentration) to lower water concentration (higher solute concentration)
Active transport	Movement often from region of lower to higher concentration with the aid of a carrier protein and energy usually from ATP
Endocytosis	Materials are engulfed by plasma membrane and drawn into cell in a vesicle
Exocytosis	Membrane-bound vesicle from inside the cell fuses with the plasma membrane and spills contents outside of cell

■ Organelles are specialized compartments within cells

Inside the eukaryotic cell, the primary role of membranes is to create separate compartments where specific chemical processes critical to the life of the cell are carried out. The membrane-bound compartments, or organelles, distributed in the jellylike cytoplasm of the cell, have different functions—just like the different offices within a large company, some of which are responsible for production, some for purchasing, and others for shipping. The compartmentalization allows segregated combinations of molecules to carry out specific tasks (refer again to Figure 3.1).

Some organelles give directions for the manufacture of cell products. Others make or modify the products or transport them. Still other organelles process energy or break down substances for use or disposal.

NUCLEUS

The **nucleus** of the cell contains almost all the cell's genetic information (Figure 3.13). The DNA within the nucleus (the chromatin in Figure 3.13 consists of DNA and its associated proteins) controls cellular structure and function because it contains a code for the production of proteins. All our cells contain the same genetic information. The characteristics of a particular cell—that make it a muscle cell or a liver cell—are determined in large part by the specific directions it receives from its nucleus.

A double membrane called the **nuclear envelope** surrounds the nucleus and separates it from the cytoplasm, as shown in Figure 3.13. Communication between the nucleus and cytoplasm occurs through openings in the envelope called **nuclear pores**. The traffic of selected materials across the nuclear envelope allows the contents of the cytoplasm to influence the nucleus and vice versa.

The genetic information within the nucleus is organized into **chromosomes**, threadlike structures made of DNA and associated proteins. The number of chromosomes varies from one species to another. For example, humans have 46 chromosomes (23 pairs), house mice have 40 chromosomes, and domestic dogs have 78. Individual chromosomes are visible with a light microscope during cell division, when they shorten and condense (Figure 3.14a). At all other times, however, the chromosomes are extended and not readily visible. In this dispersed state, the genetic material is called *chromatin* (Figure 3.14b). The chromatin and other contents of the nucleus constitute the *nucleoplasm*. We will discuss chromosomes and cell division in Chapter 19.

The **nucleolus**, a specialized region within the nucleus (see Figure 3.13), forms and disassembles during the course of the cell cycle (see Chapter 19). It is not surrounded by a membrane but is

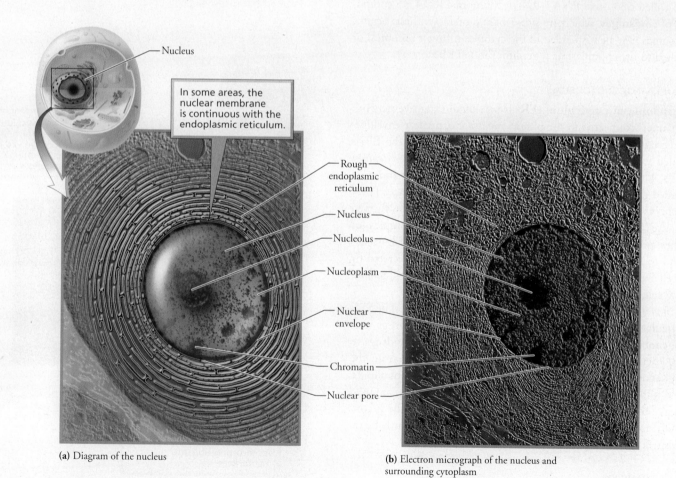

Nucleus

In some areas, the nuclear membrane is continuous with the endoplasmic reticulum.

Rough endoplasmic reticulum

Nucleus

Nucleolus

Nucleoplasm

Nuclear envelope

Chromatin

Nuclear pore

(a) Diagram of the nucleus

(b) Electron micrograph of the nucleus and surrounding cytoplasm

FIGURE **3.13** The nucleus contains almost all the genetic information of a cell.

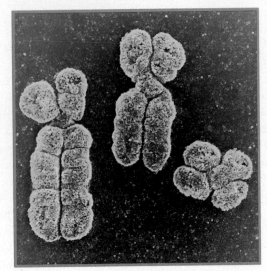

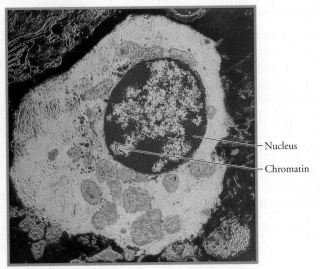

—Nucleus

—Chromatin

(a) Individual chromosomes are visible during cell division when they shorten and condense.

(b) At all other times, the chromosomes are extended and not readily visible. In this dispersed state, the genetic material is called chromatin.

FIGURE **3.14**   Chromosomes are composed of DNA and associated proteins.

simply a region where DNA has gathered to produce a type of RNA called ribosomal RNA (rRNA). Ribosomal RNA is a component of **ribosomes**, which are sites where protein synthesis begins. Ribosomes may be suspended in the cytoplasm (free ribosomes) or attached to the endoplasmic reticulum (bound ribosomes).

ENDOPLASMIC RETICULUM

The **endoplasmic reticulum (ER)** is part of an extensive network of channels connected to the plasma membrane, nuclear envelope, and certain organelles (Figure 3.15). In some regions, the ER is studded with ribosomes and is called *rough endoplasmic reticulum* (RER). The amino acid chains made by the attached ribosomes are threaded through the RER's membrane to its internal space. There the chains are processed and modified, enclosed in vesicles formed from the RER membrane, and transferred to the Golgi complex for further processing and packaging. Proteins made by ribosomes bound to ER will be incorporated into membranes or secreted by the cell eventually. Proteins produced by free ribosomes will remain in the cell.

Smooth endoplasmic reticulum (SER) lacks ribosomes. The SER (particularly in liver cells) detoxifies alcohol and other drugs such as phenobarbital. Typically, enzymes of SER modify the drugs to make them more water soluble and easier to eliminate from the body. Another function of SER is the production of phospholipids. These phospholipids, along with proteins from the RER, are used to make the RER membrane. Because the RER membrane is continually used to form vesicles for shipping, it must constantly be replenished.

GOLGI COMPLEX

Resembling a stack of dinner plates, the **Golgi complex** consists of flattened sacs made of membrane and it is the protein processing and packaging center of the cell (Figure 3.16). Protein-filled vesi-

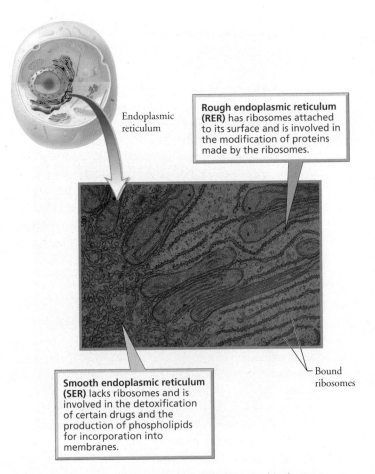

Endoplasmic reticulum

Rough endoplasmic reticulum (RER) has ribosomes attached to its surface and is involved in the modification of proteins made by the ribosomes.

Smooth endoplasmic reticulum (SER) lacks ribosomes and is involved in the detoxification of certain drugs and the production of phospholipids for incorporation into membranes.

—Bound ribosomes

FIGURE **3.15**   The endoplasmic reticulum (ER), shown in this electron micrograph, is continuous with the plasma and nuclear membranes and consists of two regions: rough ER and smooth ER.

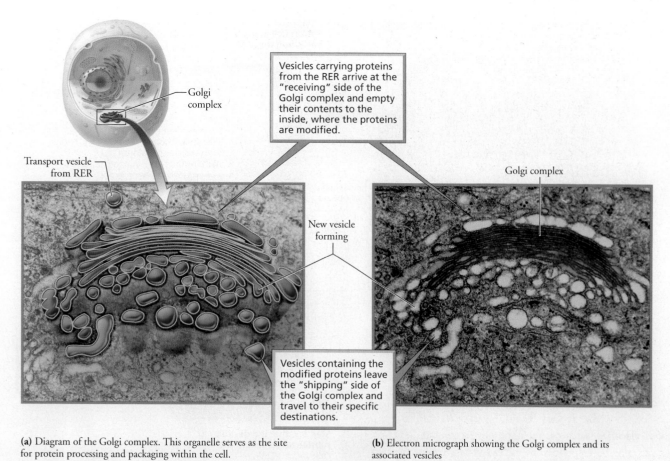

(a) Diagram of the Golgi complex. This organelle serves as the site for protein processing and packaging within the cell.

(b) Electron micrograph showing the Golgi complex and its associated vesicles

FIGURE **3.16** The Golgi complex

cles from the RER arrive at the "receiving side" of the Golgi complex, fuse with its membrane, and empty their contents inside. The Golgi complex then chemically modifies many of the proteins as they move, by way of vesicles, from one membranous disk in the stack to the next. When the processing is finished, the Golgi complex sorts the proteins, much as a postal worker sorts letters, and sends them to their various destinations. Some of the proteins emerging from the "shipping side" are packaged in vesicles and sent to the plasma membrane for export from the cell or to become membrane proteins. Other proteins are packaged in lysosomes.

LYSOSOMES

How does the cell dismantle worn-out parts or digest large objects that it takes in by phagocytosis? If it simply released digestive enzymes into its cytoplasm, it would soon destroy itself. Instead, intracellular digestion occurs mainly within lysosomes. (Intracellular digestion is digestion within a cell. The digestion of food in our gastrointestinal tract is a different phenomenon; it occurs outside of cells and is described in Chapter 15.) **Lysosomes** are roughly spherical organelles consisting of a single membrane packed with about 40 different digestive enzymes. The enzymes and membranes of lysosomes are made by the RER and then sent to the Golgi complex for further processing. Eventually, enzyme-filled lysosomes bud from the Golgi complex and begin their diverse roles in digestion within the cell. These roles include the destruc-

tion of foreign invaders, such as bacteria, and the dismantling of worn-out organelles.

Consider, for example, what happens when a cell engulfs a bacterium. You can follow this process in Figure 3.17. During the process of phagocytosis (Step 1), a vesicle encircles the bacterium. A lysosome released from the Golgi complex then fuses with the vesicle (Step 2), and the lysosome's digestive enzymes break the bacterium down into smaller molecules. These diffuse out of the vesicle into the cytoplasm, where they can be used by the cell (Step 3). Indigestible residues may be expelled from the cell by exocytosis (Step 4), or they may be stored in vesicles inside the cell (Step 5).

Lysosomes also dismantle obsolete parts of the cell itself. Old organelles and macromolecules are broken down into smaller components that can be used again (see Figure 3.17). An organelle called a mitochondrion, for example, lasts only about 10 days in a typical liver cell before being destroyed by lysosomes. Such "housecleaning" keeps the cell functioning properly and promotes the recycling of essential materials.

The absence of a single kind of lysosomal enzyme can have devastating consequences. Molecules that would normally be broken down by the missing enzyme start to collect in the lysosomes and cause them to swell. Ultimately, the accumulating molecules interfere with cell function. These lysosomal storage diseases are inherited and progress with age.

Tay-Sachs disease is a lysosomal storage disease caused by the absence of the lysosomal enzyme hexosaminidase (Hex A), which

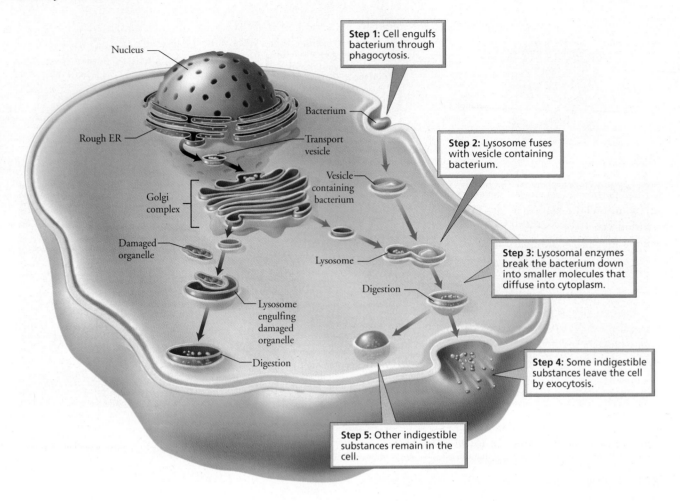

Nucleus

Rough ER

Transport vesicle

Golgi complex

Damaged organelle

Bacterium

Vesicle containing bacterium

Lysosome

Digestion

Lysosome engulfing damaged organelle

Digestion

Step 1: Cell engulfs bacterium through phagocytosis.

Step 2: Lysosome fuses with vesicle containing bacterium.

Step 3: Lysosomal enzymes break the bacterium down into smaller molecules that diffuse into cytoplasm.

Step 4: Some indigestible substances leave the cell by exocytosis.

Step 5: Other indigestible substances remain in the cell.

FIGURE **3.17** Lysosome formation and function in intracellular digestion. Lysosomes, released from the Golgi complex, digest a bacterium engulfed by the cell (see pathway on right). Lysosomes also digest obsolete parts of the cell itself (see pathway on left).

breaks down lipids in nerve cells. When Hex A is missing, the lysosomes swell with undigested lipids (Figure 3.18). Infants with Tay-Sachs disease appear normal at birth but begin to deteriorate by about 6 months of age as abnormal amounts of lipid accumulate in the nervous system. Typically, motor skills that had developed, such as crawling and grasping, progressively decline. Senses, such as sight and hearing, also deteriorate. Recurrent seizures are common in 2-year-olds. By the age of 4 or 5, Tay-Sachs causes paralysis and death. At present there is no cure for this disease. However, there is a blood test to detect individuals that carry the gene for Tay-Sachs. Called carriers, these individuals do not have the disease but could pass the gene to their offspring.

what would you do?

Imagine that you and your spouse want to start a family but that both of you are carriers of Tay-Sachs disease and could pass the gene to your children. The possible outcomes for any child you might conceive are as follows: the child may not have the gene for Tay-Sachs and may be healthy, may have the disease and die in early childhood, or may be a carrier as you are. Your parents urge adoption. Your spouse prefers not to adopt but to use prenatal screening to check if your fetus has the disease. What would you do?

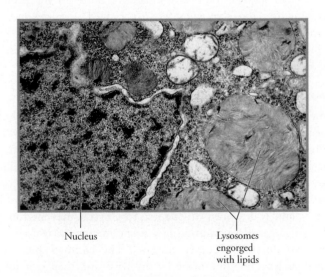

Nucleus

Lysosomes engorged with lipids

FIGURE **3.18** An electron micrograph of a nerve cell from the brain of a person with Tay-Sachs disease. Note the lysosomes engorged with undigested lipids.

Certain environmental factors cause disease by interfering with lysosomes. In the Environmental Issue essay, *Asbestos: The Deadly Miracle Material*, we describe the impact of asbestos on lysosomes and health.

MITOCHONDRIA

Most cellular activities require energy. Energy is needed to transport certain substances across the plasma membrane and to fuel many of the chemical reactions that occur in the cytoplasm. Specialized cells such as muscle cells and nerve cells require energy to perform their particular activities. The energy needed by cells is provided by **mitochondria** (singular, mitochondrion), the organelles within which most of cellular respiration occurs. *Cellular respiration* is a four-phase process in which oxygen and an organic fuel such as glucose are consumed and energy in the form of ATP is released. The first phase takes place in the cytoplasm. The remaining three phases occur in the mitochondria. The process of cellular respiration will be described later in this chapter.

The number of mitochondria varies considerably from cell to cell and is roughly correlated with a cell's demand for energy. Most cells contain several hundred to thousands of mitochondria. Like the nucleus, but unlike other organelles, mitochondria are bounded by a double membrane (Figure 3.19). The inner and outer membranes create two separate compartments that serve as sites for some of the reactions in cellular respiration. The infoldings of the inner membrane of a mitochondrion are called *cristae*, and these are

the sites of the last phase of cellular respiration. Finally, mitochondria contain ribosomes and a small percentage of a cell's total DNA (the rest being found in the nucleus). Mitochondria have ribosomes and DNA because they are likely descendants of once free-living bacteria that invaded or were engulfed by ancient cells (see Chapter 22).

Table 3.4 reviews the functions of organelles.

TABLE 3.4	REVIEW OF MAJOR ORGANELLES AND THEIR FUNCTIONS
ORGANELLE	**FUNCTION**
Nucleus	Contains almost all the genetic information and influences cellular structure and function
Rough endoplasmic reticulum (RER)	Studded with ribosomes (sites where the synthesis of proteins begins); produces membrane
Smooth endoplasmic reticulum (SER)	Detoxifies drugs; produces membrane
Golgi complex	Sorts, modifies, and packages products of RER
Lysosomes	Digest substances imported from outside the cell; destroy old or defective cell parts
Mitochondria	Provide cell with energy through the breakdown of glucose during cellular respiration

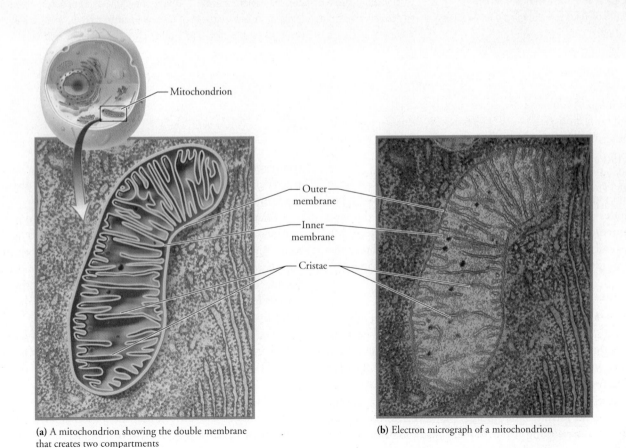

(a) A mitochondrion showing the double membrane that creates two compartments

(b) Electron micrograph of a mitochondrion

FIGURE **3.19** Mitochondria are sites of energy conversion in the cell.

ENVIRONMENTAL ISSUE

Asbestos: The Deadly Miracle Material

During the first three-quarters of the twentieth century, over 30 million tons of asbestos were used for various purposes in the United States. About half a million people are estimated to have died as a result. What is asbestos? Why do we use it? And why do we sometimes die from it?

Asbestos is a fibrous silicate mineral, found in many forms in nature, that is strong, flexible, and resistant to heat and corrosion (Figure 3.A). Because of these properties, asbestos has been used widely in construction—as an insulator on ceilings and pipes, for example, or to soundproof and fireproof the walls of schools.

The very properties that make asbestos an ideal building material—its fibrous nature and durability—also can make it deadly. For example, when fibers of asbestos insulation are dislodged, small, light particles become suspended in the air and can be inhaled into the lungs. There is some evidence that the different forms of asbestos differ in the time they persist in lung tissue. Nevertheless, particles of at least one form appear resistant to degradation and remain in the lungs for life (Figure 3.B).

Asbestosis is the most common disease caused by exposure to asbestos. It results from the dangerous interaction between asbestos and lysosomes. Apparently, cells responsible for cleaning the respiratory passages engulf small particles of asbestos inhaled into the lungs; lysosomes inside the cleaning cells then fuse with the vesicles containing the asbestos particles. Unfortunately, the lysosomal enzymes cannot break down the asbestos particles. Instead, the particles destabilize the membranes of the lysosomes, causing massive release of enzymes, and the enzymes destroy the cells of the respiratory tract. Irreversible scarring of lung tissue is the result, eventually interfering with the exchange of gases in the lungs. People with asbestos-damaged lungs experience chronic coughing and shortness of breath. These symptoms become more severe over time and may cause death from impaired respiratory function.

At present, there is no effective treatment for asbestosis. The focus, therefore, has been on prevention. In the United States, the use of asbestos for insulation and fireproofing, or for any new purposes, is banned. However, asbestos is still present in many buildings constructed before the ban went into effect. This point was made painfully clear when asbestos was detected in the debris and dust at the site where the World Trade Centers collapsed on September 11, 2001. Asbestos had been used as insulation in parts of the towers.

What can be done about the asbestos already present in our schools and workplaces? It is generally recommended and often required that exposed asbestos be removed, enclosed by other building materials, or covered with a sealant. Experts should determine which method is best in any given situation. Moreover, the sealing or removal of asbestos should be done by experts, because the greatest risk of asbestos exposure occurs when asbestos is handled improperly. Workers at risk of exposure to asbestos—plumbers, electricians, insulation workers, and carpenters, to name a few—should insist upon frequent testing of the air in their workplace, as was done at the World Trade Center site and in surrounding neighborhoods. Workers at Ground Zero were instructed to wear respirators and other protective equipment (Figure 3.C). So far,

■ The cytoskeleton provides support and movement

Traversing the cytoplasm of the cell is a complex network of fibers called the **cytoskeleton**. The fibers are divided into three types: **microtubules** are the thickest; **microfilaments** are the thinnest; and **intermediate filaments** are the diverse group in between. Microtubules and microfilaments are seen to disassemble and reassemble, whereas intermediate filaments tend to be more permanent.

MICROTUBULES

Microtubules are straight, hollow rods made of the protein tubulin. Some microtubules provide support to the cell. Others, near the plasma membrane, maintain cell shape. Microtubules also form tracks along which organelles or vesicles travel (Figure 3.20). For example, secretory vesicles that have budded from the Golgi complex make their way to the plasma membrane by moving along a microtubule track. Finally, microtubules play a role in the separation of chromosomes during cell division. A microtubule-organizing center, located near the nucleus, contains a pair of **centrioles**, each

composed of nine sets of triplet (three) microtubules arranged in a ring (Figure 3.21). Centrioles may function in cell division and in the formation of cilia and flagella.

Microtubules serve as the working parts of two types of cell extensions called cilia (singular, cilium) and flagella (singular, flagellum). **Cilia** are numerous, short extensions on a cell, resembling hairs and moving with the back-and-forth motion of oars. They are found, for example, on the surfaces of cells lining the respiratory tract (Figure 3.22a), where they sweep debris trapped in mucus away from the lungs. Smoking destroys these cilia and hampers cleaning of respiratory surfaces. A **flagellum** resembles a whip and moves in an undulating manner. Flagella are much longer than cilia. The only cell with a flagellum in humans is the sperm cell (Figure 3.22b).

Cilia and flagella differ in length, number per cell, and pattern of movement. Nevertheless, they have a similar arrangement of microtubules, called a 9 + 2 pattern (Figure 3.22c), at their core. This arrangement consists of nine pairs of microtubules arranged

monitoring of workers involved in the rescue and recovery effort has revealed that many are experiencing persistent respiratory symptoms, and some show reduced ability to move air in and out of their lungs. There also has been an increased incidence of sarcoidosis, another potentially fatal lung scarring disease. It is too early to know if any of these health effects can be specifically attributed to asbestos. However, doctors believe that the effects are generally related to the dust, ash, and smoke inhaled by the workers in the aftermath of the attacks on September 11, 2001. 🌲

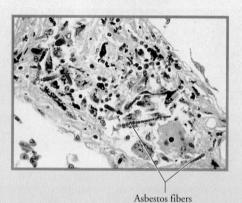

FIGURE **3.A** The deadly miracle material asbestos. The qualities of asbestos that make it so useful for construction—strength, durability, and flexibility—also make it hazardous to our health. Particles of asbestos released into the air may be inhaled into the lungs, where they cause asbestosis (scarring of lung tissue).

Asbestos fibers

FIGURE **3.B** Asbestos fibers in lung tissue. The destruction of lung tissue characteristic of asbestosis results from the release of lysosomal enzymes inside the cells responsible for cleaning the respiratory tract.

FIGURE **3.C** A worker at Ground Zero, New York City. Proper protective equipment, such as a respirator, can safeguard those exposed to asbestos and other contaminants in the workplace.

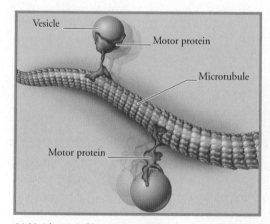

(a) Vesicles carried by proteins travel along a microtubule.

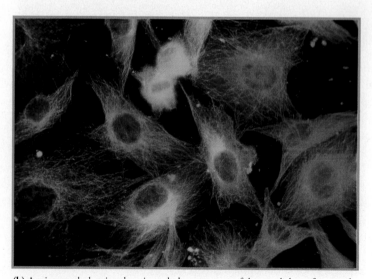

(b) A micrograph showing the microtubule component of the cytoskeleton for several cells. The round area in the center of each cell is the nucleus.

FIGURE **3.20** Microtubules serve as tracks along which organelles or vesicles move.

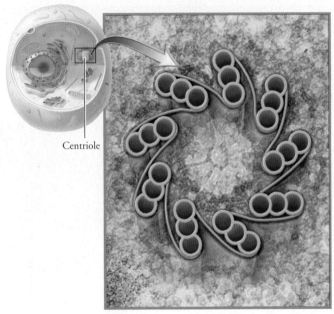

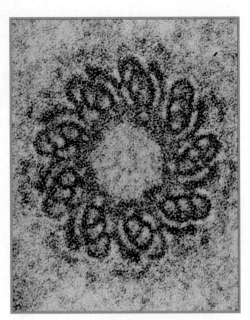

Centriole

(a) Diagram of a centriole. Each centriole is composed of nine sets of triplet microtubules arranged in a ring.

(b) Electron micrograph showing the microtubules of a centriole

FIGURE **3.21** Centrioles may help organize microtubules.

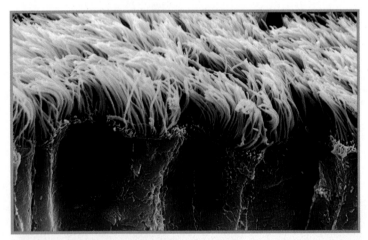

(a) Cilia on cells lining the respiratory tract

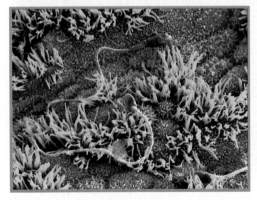

(b) Sperm cells in a fallopian tube

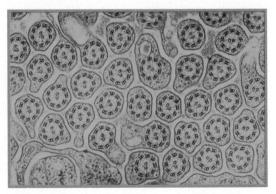

(c) Several cilia in cross section showing the 9+2 arrangement of microtubules. Flagella (not shown) have a similar arrangement of microtubules

FIGURE **3.22** Microtubules are responsible for the movement of cilia and flagella.

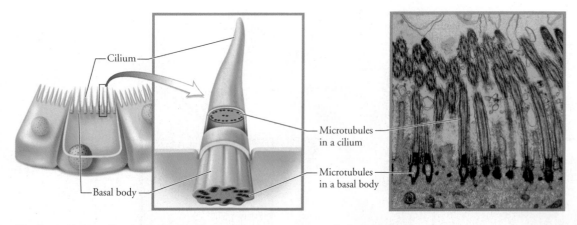

Cilium

Microtubules
in a cilium

Microtubules
in a basal body

Basal body

FIGURE **3.23** Basal bodies, anchoring cilia to cells. Basal bodies contain an arrangement of microtubules identical to that found in centrioles (nine sets of triplet microtubules).

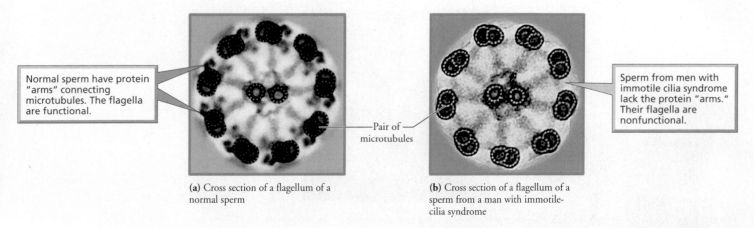

Normal sperm have protein "arms" connecting microtubules. The flagella are functional.

Sperm from men with immotile cilia syndrome lack the protein "arms." Their flagella are nonfunctional.

Pair of microtubules

(a) Cross section of a flagellum of a normal sperm

(b) Cross section of a flagellum of a sperm from a man with immotile-cilia syndrome

FIGURE **3.24** Loss of function of cilia and flagella

in a ring with two microtubules at the center. The microtubules of a cilium or flagellum are anchored to the cell by a *basal body*. Like the centrioles described above, basal bodies contain nine triplets of microtubules arranged in a ring (Figure 3.23).

Immotile cilia syndrome is a genetic disorder characterized by severe loss of function in cilia and flagella throughout the body. People with immotile cilia syndrome experience chronic respiratory problems, and males may be sterile because of immotile sperm. Loss of ciliary and flagellar function appears to be caused by the absence of armlike structures that normally link the nine pairs of microtubules in the 9 + 2 pattern (Figure 3.24).

MICROFILAMENTS

Microfilaments are solid rods made of the protein actin. These fibers are best known for their role in muscle contraction, where they slide past thicker filaments made of the protein myosin. They are also the structures that produce localized movements of plasma membrane, as when a cell puts out extensions called *pseudopodia* (*pseudo*, false; *pod*, foot) that enable the cell to creep upon a surface (Figure 3.25). Finally, microfilaments play a role in cell division, forming a band that contracts and pinches the cell in two.

INTERMEDIATE FILAMENTS

Intermediate filaments are a diverse group of ropelike fibers helping to maintain cell shape and anchoring certain organelles in place. Their protein composition varies from one type of cell to another.

Our Cells Use Cellular Respiration and Fermentation to Generate ATP

Living requires work, and work requires energy. Logic tells us, therefore, that living requires energy.

We get our energy from the food we eat. Our digestive system (discussed in Chapter 15) breaks down complex macromolecules, such as carbohydrates, proteins, and fats, into their simpler components, such as glucose, amino acids, and fatty acids. These simpler molecules are then absorbed into the bloodstream and carried to our cells, where some of the energy stored in the molecules' chemical bonds is used to make ATP, the energy-rich molecule that our cells use to do their work. (Some energy is also given off as heat.)

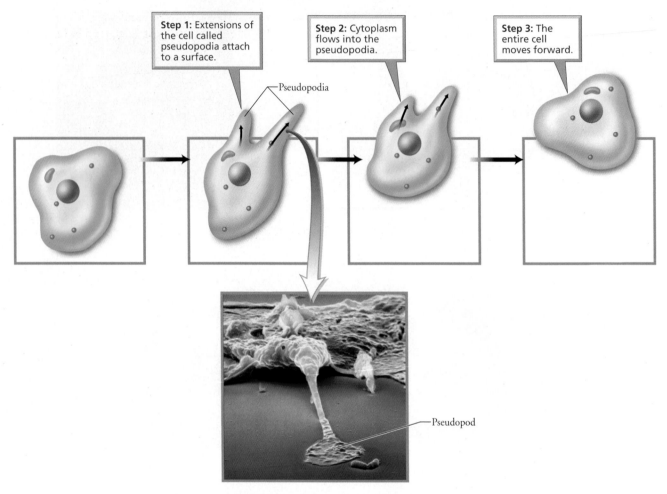

Step 1: Extensions of the cell called pseudopodia attach to a surface.

Step 2: Cytoplasm flows into the pseudopodia.

Step 3: The entire cell moves forward.

Pseudopodia

Pseudopod

FIGURE **3.25** Formation of pseudopodia. Microfilaments form pseudopodia, extensions of the cell that produce movement across a surface.

Cells have two ways of breaking glucose molecules apart for energy: cellular respiration and fermentation. Cellular respiration requires oxygen; fermentation does not.

All the chemical reactions that take place in a cell constitute its *metabolism*. These chemical reactions are organized into metabolic pathways. Each pathway consists of a series of steps in which a starting molecule is modified, eventually resulting in a particular product. Specific enzymes speed up each step of the pathway. Cellular respiration and fermentation are examples of *catabolic pathways*, pathways in which complex molecules are broken down into simpler compounds, releasing energy. *Anabolic pathways*, on the other hand, build complex molecules from simpler ones, and consume energy in the process.

■ **Cellular respiration requires oxygen**

Cellular respiration is the oxygen-requiring pathway in which glucose is broken down by cells. It is an elaborate series of chemical reactions whose final products are carbon dioxide, water, and energy. In a laboratory beaker, glucose and oxygen can be combined to produce those products in a single step. However, under those circumstances, the glucose burns, and all the energy is lost as heat. The process employed by cells in which glucose is broken down in

a series of steps enables them to obtain much of the energy in a usable form—specifically, as a high-energy chemical bond in ATP. Recall from Chapter 2 that ATP is formed from ADP (adenosine diphosphate) and inorganic phosphate (Pi) in a process that requires energy.

Cellular respiration has four phases: (1) glycolysis, (2) the transition reaction, (3) the citric acid cycle, and (4) the electron transport chain. All four phases are occurring continuously within cells. Glycolysis takes place in the cytoplasm of the cell. The transition reaction, the citric acid cycle, and the electron transport chain take place in mitochondria. You will see that some of these phases consist of a series of reactions in which the products from one reaction become the substrates (raw materials) for the next reaction. You will also see that the transfer of electrons from one atom or molecule to another is a key feature of the process our cells use to capture energy from fuel. As the electrons are passed along a chain of intermediate compounds, their energy is used to make ATP.

GLYCOLYSIS

The first phase of cellular respiration, called **glycolysis** (*glyco*, sugar; *lysis*, splitting), begins with glucose, a six-carbon sugar, being split into two three-carbon sugars. These three-carbon sugars are then converted into two molecules of pyruvate (Figure 3.26), another three-carbon compound. Glycolysis occurs in several steps, and each step requires a different, specific enzyme. During the first steps, two molecules of ATP are consumed, because energy is needed to prepare glucose for splitting. During the remaining steps, four molecules of ATP are produced, for a net gain of two ATP.

Glycolysis also produces two molecules of nicotine adenine dinucleotide (NADH), which are generated when electrons are donated to the coenzyme NAD^+. Glycolysis does not require oxygen and releases only a small amount of the chemical energy stored in glucose. Most of the energy remains in the two molecules of pyruvate. These molecules move from the cytoplasm into the inner compartment of the mitochondrion.

TRANSITION REACTION

Once inside the inner compartment of the mitochondrion, pyruvate reacts with a substance called coenzyme A (CoA) in a reaction called the transition reaction. The *transition reaction* results in the removal of one carbon (in the form of carbon dioxide, CO_2) from each pyruvate (Figure 3.27). The resulting two-carbon molecule,

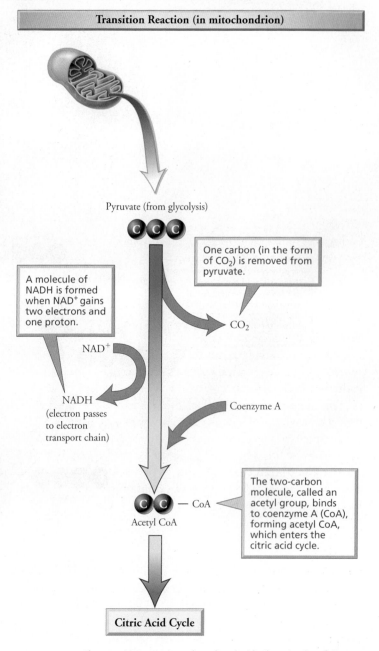

FIGURE **3.26** Glycolysis is a several-step sequence of reactions in the cytoplasm. Glucose, a six-carbon sugar, is split into two three-carbon molecules of pyruvate.

FIGURE **3.27** The transition reaction takes place inside the mitochondrion and is the link between glycolysis and the citric acid cycle.

called an acetyl group, then binds to CoA to form acetyl CoA. A molecule of NADH is also produced from each pyruvate.

CITRIC ACID CYCLE

Still in the inner compartment of the mitochondrion, acetyl CoA reacts with a four-carbon compound in the first of a cyclic series of eight chemical reactions known as the **citric acid cycle**, named after the first product (citric acid, or citrate) formed along its route (Figure 3.28). The cycle is also called the Krebs cycle, after the scientist Hans Krebs, who described many of the reactions.

Rather than considering each of the chemical reactions in the citric acid cycle, we will simply say that it completes the loss of electrons from glucose and yields two molecules of ATP (one from each acetyl CoA that enters the cycle) and several molecules of NADH and $FADH_2$ (flavin adenine dinucleotide). NADH and $FADH_2$ are carriers of high-energy electrons. The NADH and $FADH_2$ produced in glycolysis, in the transition reaction, and in the citric acid cycle enter the electron transport chain, the final phase of cellular respiration. The citric acid cycle also produces CO_2 as waste.

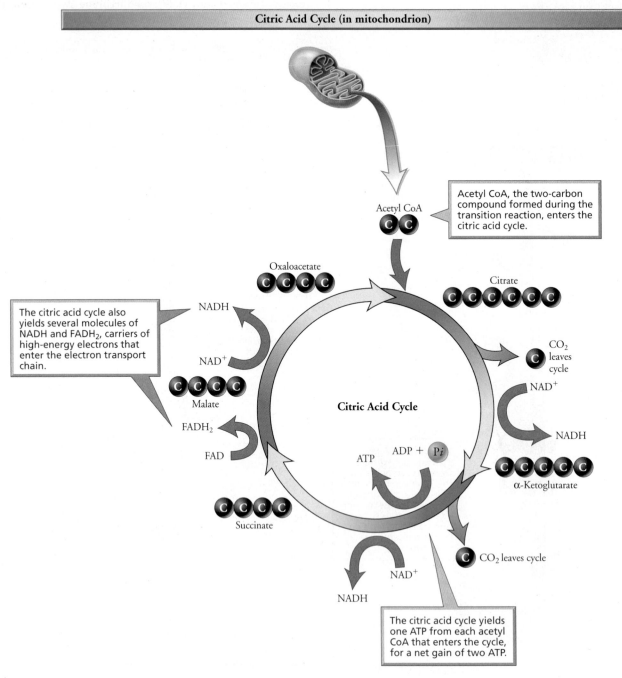

FIGURE **3.28** The citric acid cycle is a cyclic series of eight chemical reactions that occurs inside the mitochondrion and yields two molecules of ATP and several molecules of NADH and $FADH_2$ per molecule of glucose.

ELECTRON TRANSPORT CHAIN

During the final phase of cellular respiration, the molecules of NADH and $FADH_2$ produced by earlier phases pass their electrons to a series of carrier proteins embedded in the inner membrane of the mitochondrion. These proteins are known as the **electron transport chain** (Figure 3.29). (Recall that the inner membrane of the mitochondrion is highly folded, providing space for thousands of sets of carrier proteins.) During the transfer of electrons from one protein to the next, energy is released and used to make ATP. Eventually, the electrons are passed to oxygen, the final electron acceptor, which then combines with two hydrogen ions to form water. Oxygen has a critical role in cellular respiration. When oxygen is absent, electrons accumulate in the carrier molecules, halting the citric acid cycle and cellular respiration. When oxygen is present, however, and accepts the electrons, respiration continues. The electron transport chain produces 32 molecules of ATP per molecule of glucose.

Certain deadly poisons, such as cyanide, inhibit cellular respiration by blocking the flow of electrons in the electron transport chain. Symptoms of cyanide poisoning in humans include gasping, convulsions, coma, and death, in cases of significant exposure. Cyanide is present in several plants—including, it might surprise you to learn, the roots of the bitter cassava, a tropical plant from which tapioca is made. The roots are heated to destroy the cyanide, so you can eat your pudding with near total peace of mind!

The results of cellular respiration are summarized in Figure 3.30. Basic descriptions of each phase can be found in Table 3.5. Altogether, cellular respiration produces 36 molecules of ATP per molecule of glucose: 2 ATP from glycolysis, 2 ATP from the citric acid cycle, and 32 ATP from the electron transport chain.

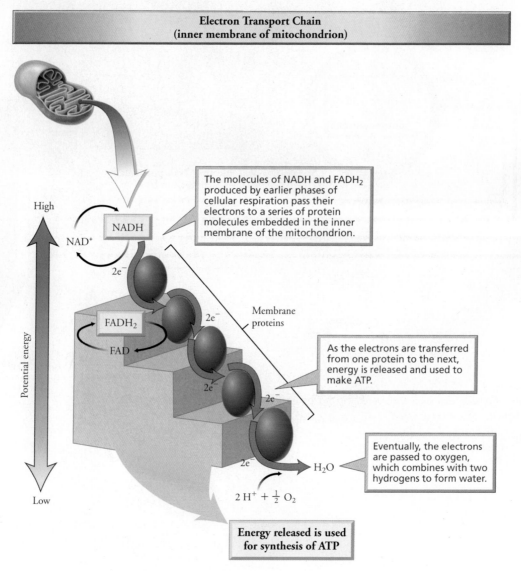

**Electron Transport Chain
(inner membrane of mitochondrion)**

The molecules of NADH and $FADH_2$ produced by earlier phases of cellular respiration pass their electrons to a series of protein molecules embedded in the inner membrane of the mitochondrion.

High

NADH

NAD^+

$2e^-$

Membrane proteins

FADH$_2$

$2e^-$

FAD

$2e^-$

$2e^-$

As the electrons are transferred from one protein to the next, energy is released and used to make ATP.

$2e^-$

H_2O

Eventually, the electrons are passed to oxygen, which combines with two hydrogens to form water.

$2 H^+ + \frac{1}{2} O_2$

Potential energy

Low

Energy released is used for synthesis of ATP

FIGURE **3.29** The electron transport chain is the final phase of cellular respiration. This phase yields 32 ATP per molecule of glucose.

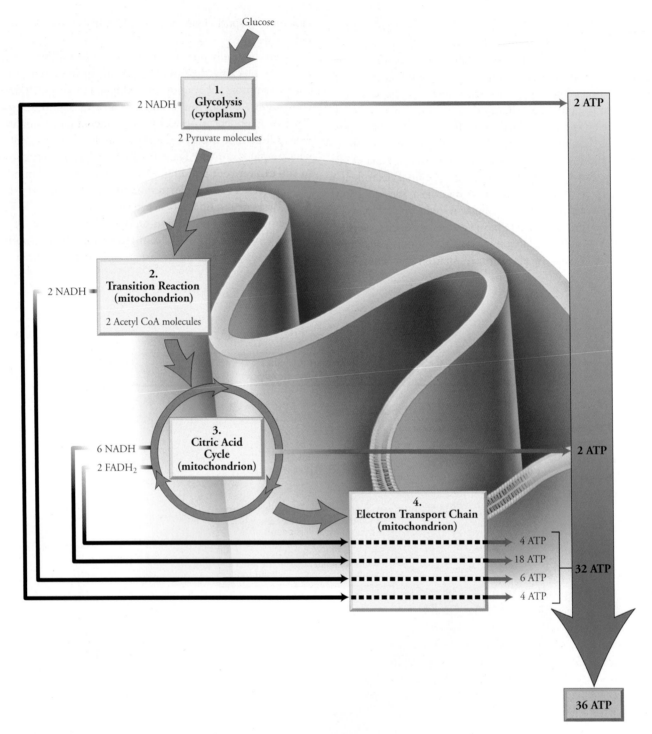

FIGURE 3.30 Summary of cellular respiration. Cellular respiration produces 36 ATP per molecule of glucose (2 ATP from glycolysis, 2 ATP from the citric acid cycle, and 32 ATP from the electron transport chain).

TABLE 3.5	REVIEW OF CELLULAR RESPIRATION		
PHASE	**LOCATION**	**DESCRIPTION**	**MAIN PRODUCTS**
Glycolysis	Cytoplasm	Several-step process by which glucose is split into 2 pyruvate	2 pyruvate 2 ATP 2 NADH
Transition reaction	Mitochondria	One CO_2 is removed from each pyruvate; the resulting molecules bind to CoA, forming 2 acetyl CoA	2 acetyl CoA 2 NADH
Citric acid cycle	Mitochondria	Cyclic series of eight chemical reactions by which acetyl CoA is broken down	2 ATP 2 $FADH_2$ 6 NADH
Electron transport chain	Mitochondria	Electrons from NADH and $FADH_2$ are passed from one protein to the next, releasing energy for ATP synthesis	32 ATP H_2O

■ Fermentation does not require oxygen

You have seen that cellular respiration depends on oxygen as the final electron acceptor in the electron transport chain. Without oxygen, the transport chain comes to a halt, blocking the citric acid cycle and stopping cellular respiration. Is there a way for cells to harvest energy when molecules of oxygen are scarce? The answer is yes, and the pathway is fermentation.

Fermentation is the breakdown of glucose without oxygen. It begins with glycolysis, which, as you recall, occurs in the cytoplasm and does not require oxygen. From one molecule of glucose, glycolysis produces two molecules each of pyruvate, the electron carrier NADH, and ATP. The remaining reactions of fermentation also take place in the cytoplasm, transferring electrons from NADH to pyruvate or a derivative of pyruvate. This transfer of electrons is critical because it regenerates NAD^+, which is essential for the production of ATP through glycolysis. Recall that in cellular respiration, oxygen is the final electron acceptor in the electron transport chain, whereas in fermentation it is pyruvate or a pyruvate derivative. Fermentation therefore nets only 2 molecules of ATP compared with the 36 molecules of ATP produced by cellular respiration. In short, fermentation is a very inefficient way for cells to harvest energy.

LACTIC ACID FERMENTATION

Different types of fermentation found in different organisms differ in the waste products formed from pyruvate. The fermentation that takes place in the human body is *lactic acid fermentation*, in which NADH passes electrons directly to pyruvate to produce lactate or lactic acid as a waste product (Figure 3.31). During strenuous exercise, the oxygen supply in our muscle cells runs low. Under these conditions, the cells increase lactic acid fermentation to ensure the continued production of ATP. The muscle pain we often experience after intense exercise is caused in part by the accumulation of lactic acid due to fermentation. In time, the soreness disappears as the lactic acid moves into the bloodstream, which carries it to the liver.

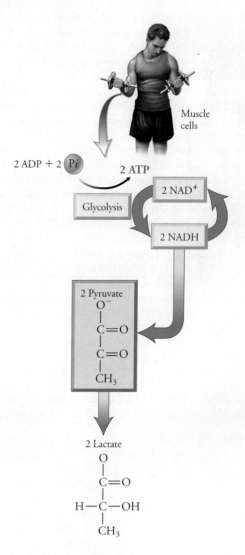

FIGURE **3.31** Lactic acid fermentation. Fermentation occurs in the cytoplasm, does not require oxygen, and yields only two molecules of ATP per molecule of glucose. The symbol P*i* stands for inorganic phosphate (PO_4^{-2}).

HIGHLIGHTING THE CONCEPTS

Our Cells Are Highly Structured (pp. 38–53)

1. There are two main types of cell. Prokaryotic cells, unique to bacteria and archaea, lack membrane-bound organelles. Eukaryotic cells, found in all other organisms, have membrane-bound organelles.

2. As a cell grows, its volume increases more than its surface area; therefore, a cell that is too large will encounter problems due to inadequate surface area. For that reason, most cells are very small.

3. The plasma membrane is made of phospholipids arranged in a bilayer with proteins and molecules of cholesterol interspersed throughout and carbohydrates attached to the outer surface. The structure of the plasma membrane is often described as a fluid mosaic.

4. The plasma membrane maintains the cell's integrity, regulates movement of substances into and out of the cell, functions in cell-cell recognition, promotes communication between cells, and binds cells together to form tissues and organs.

5. Substances cross the plasma membrane in several ways. Some cross by random movement from higher to lower concentration (simple diffusion); others need help from a carrier or channel protein (facilitated diffusion). Water also moves across the plasma membrane from higher to lower concentration (osmosis). Substances being concentrated by cells cross the plasma membrane from lower to higher concentration with the help of ATP and a carrier protein (active transport). Finally, cells may engulf outside materials by surrounding them with plasma membrane (endocytosis) or may release substances to the external environment by fusing internal vesicles with the plasma membrane and spilling their contents to the outside (exocytosis).

6. Within a eukaryotic cell, membranes delineate specialized compartments, or organelles, within which specific processes occur. The nucleus contains almost all the genetic information and thus holds the code for the cell's structure and many of its functions. The nucleolus is a region within the nucleus that makes ribosomal RNA. Ribosomes are the sites where protein synthesis begins. Endoplasmic reticulum (ER) functions in membrane production and may be studded with ribosomes (rough endoplasmic reticulum, RER) or free of ribosomes (smooth endoplasmic reticulum, SER). The Golgi complex sorts, modifies, and packages products of the RER. Lysosomes digest bacteria engulfed by cells and break down old or defective components within cells. Mitochondria process energy for cells.

7. The cytoskeleton, a complex network of fibers throughout the cytoplasm of the cell, consists of three categories of fibers, all made of proteins: microtubules, microfilaments, and intermediate filaments. Microtubules are hollow rods of tubulin that function in cell movement (cilia and flagella), support, and the movement within cells of chromosomes, organelles, and vesicles. Microfilaments are rods of actin that function in muscle contraction, movement (pseudopodia), and cell division. Intermediate filaments, which are different kinds of ropelike fibers, maintain cell shape and anchor organelles.

WEB TUTORIAL 3.1 Eukaryotic Cells
WEB TUTORIAL 3.2 Membrane Structure
WEB TUTORIAL 3.3 Passive and Active Transport
WEB TUTORIAL 3.4 Diffusion and Osmosis
WEB TUTORIAL 3.5 Endocytosis and Exocytosis

Our Cells Use Cellular Respiration and Fermentation to Generate ATP (pp. 53–59)

8. Cells require energy to work. We get this energy from the food we eat. Carbohydrates, fats, and proteins that we consume are each broken down by our digestive system into smaller units such as simple sugars, fatty acids, and amino acids. These simpler substances are absorbed into the bloodstream and carried to our cells, where the energy stored in the substances' chemical bonds is transferred for the cell's use to the chemical bonds of ATP. (Some energy is also given off as heat.)

9. Cells use two catabolic pathways—cellular respiration and fermentation—to break down the carbohydrate glucose and store its energy as ATP. Cellular respiration requires oxygen and yields 36 molecules of ATP per molecule of glucose. Fermentation does not require oxygen and yields only 2 molecules of ATP per molecule of glucose.

10. Cellular respiration has four phases—glycolysis, the transition reaction, the citric acid cycle, and the electron transport chain—all occurring continuously within cells. Glycolysis occurs in the cytoplasm and splits glucose into pyruvate while producing NADH (a carrier of high-energy electrons) and a net gain of 2 ATP. The molecules of pyruvate move from the cytoplasm into the inner compartment of the mitochondrion, where they pass through a few preparatory steps, known as the transition reaction, before entering the citric acid cycle. The citric acid cycle is an eight-step cycle that completes the breakdown of glucose into carbon dioxide. The citric acid cycle also yields 2 ATP and carriers of high-energy electrons (NADH and FADH$_2$). In the electron transport chain, the carriers of high-energy electrons produced during glycolysis, the transition reaction, and the citric acid cycle pass their electrons to a series of proteins embedded in the inner membrane of the mitochondrion; oxygen is the final electron acceptor. Energy released during the transfer of electrons yields 32 ATP.

11. Fermentation is the breakdown of glucose without oxygen. It occurs in the cytoplasm and begins with glycolysis, the splitting of glucose into pyruvate. The remaining chemical reactions involve the transfer of electrons from NADH to pyruvate or a derivative of pyruvate. Compared with cellular respiration, fermentation is an inefficient way for cells to harvest energy because it yields only 2 ATP as compared with the 36 ATP of cellular respiration.

WEB TUTORIAL 3.6 Breaking Down Glucose for Energy

KEY TERMS

cell theory   *p. 38*
organelle   *p. 38*
eukaryotic cell   *p. 38*
prokaryotic cell   *p. 38*
plasma membrane   *p. 39*
extracellular fluid   *p. 40*
cytoplasm   *p. 40*
fluid mosaic   *p. 40*
selectively permeable   *p. 41*
simple diffusion   *p. 41*
facilitated diffusion   *p. 42*

osmosis   *p. 42*
hypertonic solution   *p. 42*
isotonic solution   *p. 42*
hypotonic solution   *p. 42*
active transport   *p. 42*
endocytosis   *p. 43*
vesicle   *p. 43*
phagocytosis   *p. 43*
pinocytosis   *p. 43*
exocytosis   *p. 43*
nucleus   *p. 45*

nuclear envelope   *p. 45*
nuclear pore   *p. 45*
chromosome   *p. 45*
nucleolus   *p. 45*
ribosome   *p. 46*
endoplasmic reticulum (ER)   *p. 46*
Golgi complex   *p. 46*
lysosome   *p. 47*
mitochondrion   *p. 49*
cytoskeleton   *p. 50*
microtubule   *p. 50*

microfilament   *p. 50*
intermediate filament   *p. 50*
centriole   *p. 50*
cilia   *p. 50*
flagellum   *p. 50*
cellular respiration   *p. 54*
glycolysis   *p. 54*
citric acid cycle   *p. 56*
electron transport chain   *p. 57*
fermentation   *p. 59*

REVIEWING THE CONCEPTS

1. How do prokaryotic and eukaryotic cells differ? *pp. 38–39*
2. Describe the structure of the plasma membrane. *p. 40*
3. List five functions of the plasma membrane. *pp. 40–41*
4. What is the difference between simple and facilitated diffusion? Give an example of each. *pp. 41–42*
5. Describe endocytosis and exocytosis. *p. 43*
6. What is contained within the nucleus of a cell? *pp. 45–46*
7. List two functions of lysosomes. *pp. 47–48*
8. What are lysosomal storage diseases? *pp. 47–48*
9. Name the three types of fibers that make up the cytoskeleton. Describe their structures and functions. *pp. 50–53*
10. What is the basic function of cellular respiration? *pp. 53–54*
11. How do cellular respiration and fermentation differ with respect to the number of ATP molecules produced and the requirement for oxygen? *p. 59*
12. Prokaryotic cells
    a. have internal membrane-bound organelles.
    b. are usually larger than eukaryotic cells.
    c. lack internal membrane-enclosed organelles.
    d. have linear strands of DNA within a nucleus.
13. The plasma membrane
    a. is selectively permeable.
    b. contains lipids that function in cell-cell recognition.
    c. has cell adhesion molecules that prevent cells from sticking together.
    d. is made of nucleic acids.
14. Facilitated diffusion is
    a. the random movement of a substance from a region of higher concentration to a region of lower concentration.
    b. the movement of water across the plasma membrane.
    c. the movement of molecules across the plasma membrane against a concentration gradient with the aid of a carrier protein and energy supplied by the cell.
    d. the movement of a substance from a region of higher concentration to a region of lower concentration with the aid of a membrane protein.

15. Almost all the genetic information of a cell is found in the
    a. endoplasmic reticulum.
    b. Golgi complex.
    c. nucleus.
    d. mitochondria.
16. Ribosomes
    a. are found on smooth endoplasmic reticulum.
    b. are sites where protein synthesis begins.
    c. process and modify proteins.
    d. break down foreign invaders and old organelles.
17. Mitochondria
    a. process energy for cells.
    b. lack ribosomes and DNA.
    c. are bounded by a single membrane.
    d. function in cell digestion.
18. Microtubules
    a. are found in eukaryotic cilia and flagella.
    b. are made of the protein actin.
    c. form pseudopodia.
    d. pinch a cell in two during cell division.
19. Glycolysis
    a. occurs in the mitochondria.
    b. requires oxygen.
    c. splits glucose into pyruvate.
    d. nets 32 molecules of ATP per molecule of glucose.
20. _____ is the jellylike solution within a cell that contains everything between the nucleus and the plasma membrane.
21. _____ is called "cell drinking."
22. The _____ is a specialized region within the nucleus that is involved in the production of rRNA.
23. _____ is the final electron acceptor in the electron transport chain.
24. Our muscle cells may switch from cellular respiration to _____ when oxygen is low.

APPLYING THE CONCEPTS

1. Joe's kidneys are failing, and he needs a kidney transplant. Luckily, he has an identical twin who wishes to donate one of his kidneys. Why would a kidney from his identical twin be better than one from another family member or a complete stranger? Explain your answer with reference to the plasma membrane of Joe's cells.
2. Would you expect to find more mitochondria in muscle cells or bone cells? Explain your answer.

3. During cell division, the nucleus and cytoplasm of a cell split into two daughter cells. Cancer is characterized by uncontrolled cell division and is often treated with chemotherapy. How might some drugs used in chemotherapy halt cell division by affecting the cytoskeleton? Which of the cytoskeletal elements (microtubules, microfilaments, intermediate filaments) might be affected?

Additional questions can be found on the companion website.

4

Body Organization and Homeostasis

Cells are arranged in tissues, and tissues, in turn, form organs. The skin, our largest organ, helps protect underlying tissues and helps to regulate body temperature.

The Organization of the Human Body Increases in Complexity from Cells to Organ Systems

- Groups of similar cells form tissues
- Many tissues have specialized junctions between the cells
- Tissues combine to form organs that in turn form organ systems
- Most organs are housed in body cavities that are lined with membranes

The Skin Is an Organ System

- Skin protects, regulates body temperature, and excretes
- The skin has two layers
- Skin color is determined by pigment and blood flow
- Hair, nails, and glands are skin derivatives

Homeostasis Is a State of Relative Internal Constancy

- Homeostasis is maintained by negative feedback mechanisms
- The hypothalamus regulates body temperature

HEALTH ISSUE   Fun in the Sun? Sunlight and Skin Cancer

HEALTH ISSUE   Acne: The Misery, the Myths, and the Medications

The children greeted Jenna happily when they got home from the beach, where they had been with her younger brother Mark (their favorite uncle). But Jenna could see that Mark wasn't feeling well. She was sure he had followed her child-care instructions to the letter: restoring the sunblock on Allison and Andrew's shoulders every time they came out of the water, keeping hats on their heads when the sun was strong, giving them frequent sips of water, and reminding them to lie down occasionally under the umbrella. She was equally sure that he hadn't bothered to take the same precautions himself. He never let himself get dehydrated or overheated when training with the track team at his college; but while chasing over the sand after his niece and nephew, he probably never noticed that he was hot and perspiring, and the thought of taking a drink of water, wearing a hat, or resting in the shade would never have crossed his mind. No wonder he was complaining of a headache now and looking rather green around the gills. The human body does a fairly good job of adjusting so as to function optimally under a wide range of conditions, but it does need a little assistance from the owner.

This chapter describes the body's organization at four levels: cells, tissues, organs, and organ systems. It looks at the functions of the skin as an organ system and at the interactions between skin and other organ systems. It then discusses how all the body's systems interact to maintain relatively constant internal conditions, when they can, at every organizational level. ■

The Organization of the Human Body Increases in Complexity from Cells to Organ Systems

Think for a moment about the multitude of functions taking place in your body at this very instant. Your heart is beating. Your lungs are obtaining oxygen and eliminating carbon dioxide. Your eyes are forming an image of these words, and your brain is thinking about them. Your body can carry out these functions and more because its cells are specialized to perform specific tasks. But cell specialization is not enough. Specialized cells are organized into tissues, organs, and organ systems.

■ Groups of similar cells form tissues

A **tissue** is a group of cells of similar type that work together to serve a common function. Human tissues come in four primary types: epithelial tissue, connective tissue, muscle tissue, and nervous tissue. **Epithelial tissue** covers body surfaces, lines body cavities and organs, and forms glands. **Connective tissue** serves as a storage site for fat, plays an important role in immunity, and provides the body and its organs with protection and support. **Muscle tissue** is responsible for movement of the organism and of substances through the organism. **Nervous tissue** conducts nerve impulses from one part of the body to another. As you read this chapter, you will learn more about each of these types of tissue.

EPITHELIAL TISSUE

The three basic shapes of epithelial cells are suited to their functions. **Squamous** (skwa'-mus) **epithelium** is made up of flattened, or scale-like, cells. These cells form linings—in the blood vessels or lungs, for instance—where their flattened shape allows oxygen and carbon dioxide to diffuse across the lining easily. In blood vessels, the smooth surface of the lining reduces friction. **Cuboidal epithelium**, consisting of cube-shaped cells, and **columnar epithelium**, consisting of tall, rectangular, column-shaped cells, are specialized for secretion and absorption. Cuboidal cells are found in many glands and in the lining of kidney tubules. The small intestine is lined with columnar cells. These, like many examples of columnar cells, have numerous, small, fingerlike folds on their exposed surfaces, greatly increasing the surface area for absorption.

All three epithelial cell shapes—squamous, cuboidal, and columnar—can be either simple or stratified. Because stratified epithelium has many layers of cells, it often serves a protective role. Table 4.1 and Figure 4.1 summarize the types of epithelial tissue.

A **gland** is epithelial tissue that secretes a product. **Exocrine glands** secrete their products into ducts leading to body surfaces, cavities, or organs. Exocrine glands include the salivary glands of the mouth and the oil and sweat glands of the skin. **Endocrine glands** lack ducts and secrete their products, hormones, into intercellular spaces, from which the hormones diffuse into the bloodstream and are carried throughout the body.

Epithelial tissue is supported by a noncellular layer called the **basement membrane**. The basement membrane binds the epithelial cells to underlying connective tissue and helps the epithelial tissue resist stretching.

CONNECTIVE TISSUE

Connective tissue has many forms and functions. Sometimes described as the body's glue, its most common role is to bind and support the other tissues. However, certain connective tissues specialize in transport (blood) and energy storage (adipose tissue). Connective tissue is the most abundant and widely distributed tissue in the body.

All connective tissues contain cells embedded in an extracellular *matrix*. This matrix consists of protein fibers and a noncellular material called *ground substance*. The ground substance may be solid (as in bone), fluid (as in blood), or gelatinous (as in cartilage). It is secreted by the cells themselves or other cells nearby. Whereas all other types of tissue consist primarily of cells, connective tissue is made up mostly of its matrix. The cells are distributed in the matrix like pieces of fruit in a gelatin dessert.

The connective-tissue matrix contains three types of protein fibers in proportions that depend on the type of connective tissue. **Collagen fibers** are strong and ropelike and can withstand pulling

TABLE 4.1 REVIEW OF EPITHELIAL TISSUE			
SHAPE	**NUMBER OF LAYERS**	**EXAMPLE LOCATIONS**	**FUNCTIONS**
Squamous (flat, scale-like cells)	Simple (single layer)	Lining of heart and blood vessels, air sacs of lungs	Allows passage of materials by diffusion
	Stratified (more than one layer)	Linings of mouth, esophagus, and vagina; outer layer of skin	Protects underlying areas
Cuboidal (cube-shaped cells)	Simple	Kidney tubules, secretory portion of glands and their ducts	Secretion; absorption
	Stratified	Ducts of sweat glands, mammary glands, and salivary glands	Protects underlying areas
Columnar	Simple	Most of digestive tract (stomach to anus), air tubes of lungs (bronchi), excretory ducts of some glands, uterus	Absorbs; secretes mucus, enzymes, and other substances
	Stratified	Rare; urethra, junction of esophagus and stomach	Protects underlying areas, secretes

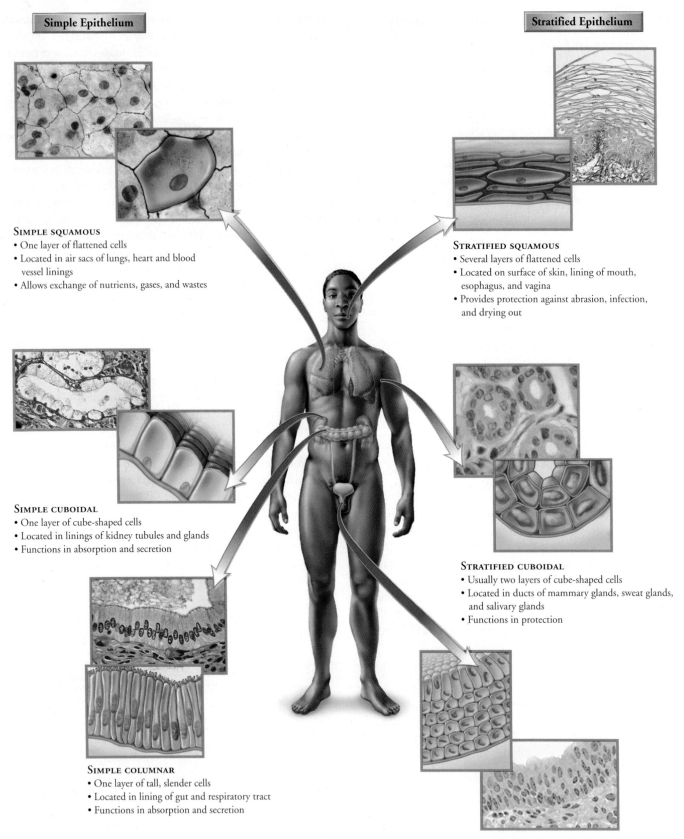

Simple Epithelium

Stratified Epithelium

SIMPLE SQUAMOUS
- One layer of flattened cells
- Located in air sacs of lungs, heart and blood vessel linings
- Allows exchange of nutrients, gases, and wastes

STRATIFIED SQUAMOUS
- Several layers of flattened cells
- Located on surface of skin, lining of mouth, esophagus, and vagina
- Provides protection against abrasion, infection, and drying out

SIMPLE CUBOIDAL
- One layer of cube-shaped cells
- Located in linings of kidney tubules and glands
- Functions in absorption and secretion

STRATIFIED CUBOIDAL
- Usually two layers of cube-shaped cells
- Located in ducts of mammary glands, sweat glands, and salivary glands
- Functions in protection

SIMPLE COLUMNAR
- One layer of tall, slender cells
- Located in lining of gut and respiratory tract
- Functions in absorption and secretion

STRATIFIED COLUMNAR
- Several layers of tall, slender cells
- Rare, located in urethra (tube through which urine leaves the body)
- Functions in protection and secretion

FIGURE **4.1** Types of epithelial tissue. These are named for the shape of the cell and the number of cell layers.

because of their great tensile strength. **Elastic fibers** contain random coils and can stretch and recoil like a spring. They are common in structures where great elasticity is needed, including the skin, lungs, and blood vessels. **Reticular fibers** are thin strands of collagen[1] that branch extensively, forming interconnecting networks suitable for supporting soft tissues (for example, they support the liver and spleen).

The protein fibers are produced by cells called **fibroblasts** in the connective tissue. Fibroblasts also repair tears in body tissues. For example, when skin is cut, fibroblasts move to the area of the wound and produce collagen fibers that help close the wound, cover the damage, and provide a surface upon which the outer layer of skin can grow back.

Table 4.2 and Figure 4.2 group the many types of connective tissue into two broad categories—connective tissue proper (loose and dense connective tissue) and specialized connective tissue (cartilage, bone, and blood)—and summarize the characteristics of each type. The characteristics of any specific connective tissue are determined more by its matrix than by its cells.

CONNECTIVE TISSUE PROPER   Loose and dense connective tissues differ in the ratio of cells to extracellular fibers. **Loose connective tissue** contains many cells but has fewer and more loosely woven fibers than are seen in the matrix of dense connective tissue (Figure 4.2). One type of loose connective tissue, *areolar* (ah-re'-o-lar) *connective tissue*, functions as a universal packing material between other tissues. Its many cells are embedded in a gelatinous matrix that is soft and easily shaped. Areolar connective tissue is found, for example, between muscles, permitting one muscle to move freely over the other. It also anchors the skin to underlying tissues and organs.

The second type of loose connective tissue is **adipose tissue**, which contains cells that are specialized for fat storage. Most of the

[1]Reticular fibers have the same subunits as collagen fibers, but they are assembled into a slightly different kind of structure.

body's long-term energy stores are fat. Fat also serves as insulation and, around certain organs, as a shock absorber.

Dense connective tissue forms strong bands because of its large amounts of tightly woven fibers. It is found in ligaments (structures that join bone to bone), tendons (structures that join muscle to bone), and the dermis (layer of skin below the epidermis).

SPECIALIZED CONNECTIVE TISSUE   Specialized connective tissue, as shown in Figure 4.2, comes in three types: cartilage, bone, and blood. **Cartilage** is tough but flexible. It serves as cushioning between certain bones and helps maintain the structure of certain body parts, including the ears. The cells in cartilage (chondrocytes) sit within spaces in the matrix called lacunae. The protein fibers and somewhat gelatinous ground substance of cartilage are responsible for the tissue's resilience and strength. Cartilage lacks blood vessels and nerves, so nutrients reach cartilage cells by diffusion from nearby capillaries. Because this process is fairly slow, cartilage heals more slowly than bone, which is a tissue with a rich blood supply.

The human body has three types of cartilage:

- **Hyaline cartilage**, the most abundant, provides support and flexibility. It contains numerous cartilage cells in a matrix of collagen fibers and a bluish white, gel-like ground substance. Known commonly as gristle, hyaline cartilage is found at the ends of long bones (look carefully at your next drumstick), where it allows one bone to slide easily over another. It also forms part of the nose, ribs, larynx, and trachea.

- **Elastic cartilage** is more flexible than hyaline cartilage because of the large amounts of wavy elastic fibers in its matrix. Elastic cartilage is found in the external ear, where it provides strength and elasticity.

- **Fibrocartilage** contains fewer cells than either hyaline or elastic cartilage. Like hyaline cartilage, its matrix contains collagen fibers. Fibrocartilage forms the outer part of the shock-absorbing disks between the vertebrae of the spine. It is made to withstand pressure.

TABLE 4.2   REVIEW OF CONNECTIVE TISSUE

TYPE	EXAMPLE LOCATIONS	FUNCTIONS
Connective tissue proper		
Loose, areolar	Between muscles, surrounding glands, wrapping small blood vessels and nerves	Wraps and cushions organs
Loose, adipose (fat)	Under skin, around kidneys and heart	Stores energy, insulates, cushions organs
Dense	Tendons; ligaments	Attaches bone to bone (ligaments) or bone to muscle (tendons)
Specialized connective tissue		
Cartilage (semisolid)	Nose (tip); rings in respiratory air tubules; external ear	Provides flexible support, cushions
Bone (solid)	Skeleton	Provides support and protection (by enclosing), and levers for muscles to act on
Blood (fluid)	Within blood vessels	Transports oxygen and carbon dioxide, nutrients, hormones, and wastes; helps fight infections

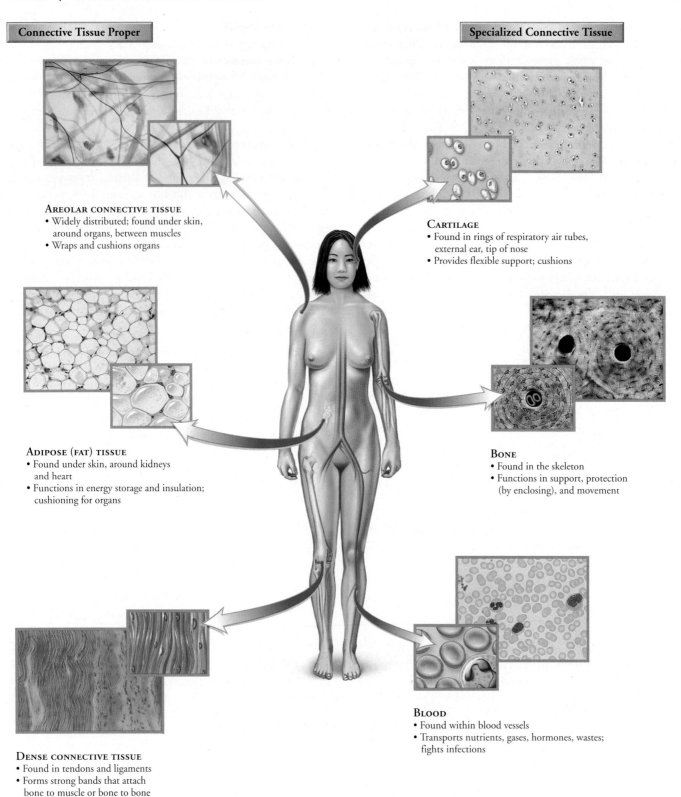

Connective Tissue Proper

AREOLAR CONNECTIVE TISSUE
• Widely distributed; found under skin, around organs, between muscles
• Wraps and cushions organs

ADIPOSE (FAT) TISSUE
• Found under skin, around kidneys and heart
• Functions in energy storage and insulation; cushioning for organs

DENSE CONNECTIVE TISSUE
• Found in tendons and ligaments
• Forms strong bands that attach bone to muscle or bone to bone

Specialized Connective Tissue

CARTILAGE
• Found in rings of respiratory air tubes, external ear, tip of nose
• Provides flexible support; cushions

BONE
• Found in the skeleton
• Functions in support, protection (by enclosing), and movement

BLOOD
• Found within blood vessels
• Transports nutrients, gases, hormones, wastes; fights infections

FIGURE **4.2** Types of connective tissue

Bone, in combination with cartilage and other components of joints, makes up the skeletal system. To many people's surprise, bone is a living, actively metabolizing tissue with a good blood supply that promotes prompt healing. Bone has many functions: protection and support for internal structures; movement, in conjunction with mus-cles; storage of lipids (in yellow marrow), calcium, and phosphorus; and production of blood cells (in red marrow). The matrix secreted by bone cells is hardened by calcium, enabling bones to provide rigid support. Collagen fibers in bone also lend it strength. You will read more about the structure and function of bones in Chapter 5.

Blood is a specialized connective tissue consisting of a liquid matrix, called *plasma*, in which so-called formed elements (cells and cell fragments called platelets) are suspended (Figure 4.2). The "fibers" in blood are soluble proteins, visible only when the blood clots. An important function of blood is to transport various substances, many of which are dissolved in the plasma. One kind of formed element, the *red blood cell*, transports oxygen to cells and also carries some of the carbon dioxide away from cells. The other two kinds of formed element are *white blood cells*, which help fight infection, and *platelets*, which help with clotting. Both white blood cells and platelets help protect the body. You will read more about blood in Chapter 11.

MUSCLE TISSUE

Muscle tissue is composed of muscle cells (called muscle fibers) that contract when stimulated. As shown in Figure 4.3, there are three types of muscle tissue: skeletal, cardiac, and smooth. Their characteristics are summarized in Table 4.3. You will read more about muscle tissue in Chapters 6 and 12.

- **Skeletal muscle tissue** is so named because it is usually attached to bones. When skeletal muscle tissue contracts, therefore, it usually moves a part of the body. Because skeletal muscle is under conscious control, it is described as voluntary muscle. Skeletal muscle cells are long cylinder-shaped cells, each containing several nuclei. In addition, skeletal muscle cells have striations, which are alternating light and dark bands visible under a light microscope. The striations are caused by the orderly arrangement of the contractile proteins actin and myosin.

- **Cardiac muscle tissue** is found only in the walls of the heart, where its contractions are responsible for pumping blood to the rest of the body. Cardiac muscle contractions are considered involuntary; we cannot *make* them occur by thinking about them. Cardiac muscle cells resemble branching cylinders and have striations and typically only one nucleus. Special junctions at the plasma membranes of these cells strengthen cardiac tissue and promote rapid conduction of impulses throughout the heart.

- **Smooth muscle tissue** is involuntary and is found in the walls of blood vessels and airways, where its contraction reduces the flow of blood or air. Smooth muscle is also found in the walls of organs such as the stomach, intestines, and bladder, where it aids in mixing and propelling food through the digestive tract and in eliminating wastes. The cells of smooth muscle tissue taper at each end, contain a single nucleus, and lack striations.

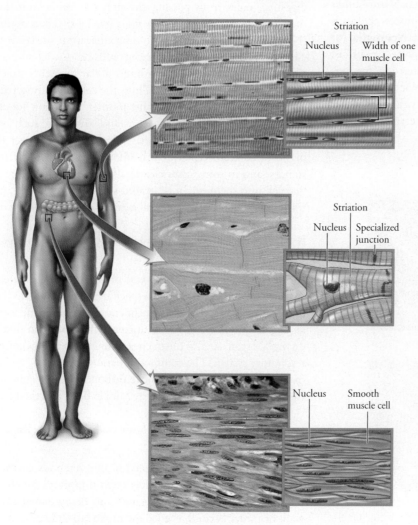

Striation
Nucleus
Width of one muscle cell

SKELETAL MUSCLE
- Long cylindrical striated cells with many nuclei
- Voluntary contraction
- Most are found attached to the skeleton
- Responsible for voluntary movement

Striation
Nucleus
Specialized junction

CARDIAC MUSCLE
- Branching striated cells, one nucleus
- Involuntary contraction
- Found in wall of heart
- Pumps blood through the body

Nucleus
Smooth muscle cell

SMOOTH MUSCLE
- Cells tapered at each end, one nucleus
- Involuntary contraction
- Found in walls of hollow internal organs, such as the intestines, and tubes, such as blood vessels
- Contractions in digestive system move food along
- When arranged in circle, controls diameter of tube

FIGURE **4.3** Types of muscle tissue

TABLE 4.3 **REVIEW OF MUSCLE TISSUE**

TYPE	DESCRIPTION	EXAMPLE LOCATIONS	FUNCTIONS
Skeletal	Long, cylindrical cells; multiple nuclei per cell; obvious striations	Muscles attached to bones	Provides voluntary movement
Cardiac	Branching striated cells; one nucleus; specialized junctions between cells	Wall of heart	Contracts and propels blood through the circulatory system
Smooth	Cells taper at each end; single nucleus; arranged in sheets; no striations	Walls of digestive system, blood vessels, and tubules of urinary system	Propels substances or objects through internal passageways

NERVOUS TISSUE

The final major type of tissue, nervous tissue, makes up the nervous system: brain, spinal cord, and nerves. Nervous tissue consists of two general cell types, neurons and accessory cells called *neuroglia* (Figure 4.4). **Neurons** generate and conduct nerve impulses, which they conduct to other neurons, muscle cells, or glands. Although neurons come in many shapes and sizes, most have three parts: the cell body, dendrites, and an axon. The cell body houses the nucleus and most organelles. Dendrites are highly branched processes that provide a large surface area for the reception of signals from other neurons. A neuron generally has one axon, a long process that usually conducts impulses away from the cell body. **Neuroglia** (or more simply, glial cells) support, insulate, and protect neurons. They increase the rate at which impulses are conducted by neurons and provide neurons with nutrients from nearby blood vessels. Recent studies indicate that glial cells communicate with one another and with neurons. You will read more about nervous tissue in Chapters 7 and 8.

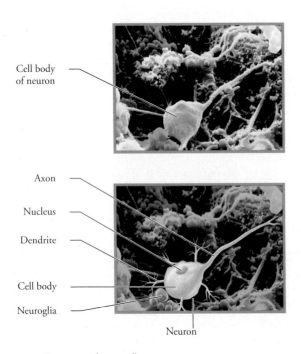

Cell body of neuron

Axon

Nucleus

Dendrite

Cell body

Neuroglia

Neuron

FIGURE **4.4** Neurons and neuroglia

■ Many tissues have specialized junctions between the cells

In many tissues, especially epithelial tissue, the cell membranes have structures for forming attachments between adjoining cells. There are three kinds of junctions between cells: tight junctions, adhesion junctions, and gap junctions. Each type of junction suits the function of the tissue. In *tight junctions* (Figure 4.5), the membranes of neighboring cells are attached so securely that they form a leakproof seal. Tight junctions are found in the lining of the urinary tract and intestine, where secure seals between cells prevent urine or digestive juices from passing through the epithelium. Less rigid than tight junctions, *adhesion junctions* (also called desmosomes) (Figure 4.5b) resemble rivets holding adjacent tissue layers together. The plasma membranes of adjacent cells do not actually touch but are instead bound together by intercellular filaments attached to a thickening in the membrane. Thus, the cells are connected but can still slide slightly relative to one another. Adhesion junctions are common in tissues that must withstand stretching, such as the skin and heart muscle. *Gap junctions* (Figure 4.5c) connect the cytoplasm of adjacent cells through small holes, allowing certain substances, and in some cases electrical charges, to flow directly from one cell into the next. In heart and smooth muscle cells, gap junctions help synchronize electrical activity and thus contraction.

■ Tissues combine to form organs that in turn form organ systems

An **organ** is a structure composed of two or more different tissues that work together to perform a specific function. Organs themselves do not usually function as independent units but instead form part of an **organ system**—a group of organs with a common function. For example, organs such as the trachea, bronchi, and lungs constitute the respiratory system. The common function of these organs is to bring oxygen into the body and remove carbon dioxide. Figure 4.6 shows the 11 major organ systems of the human body and their functions.

■ Most organs are housed in body cavities that are lined with membranes

Most of our organs are suspended in internal body cavities (Figure 4.7, page 71). These cavities have two important functions. First, they help protect the vital organs from being damaged when we walk or jump. Second, they allow organs to slide past one another and change shape. Sliding and changing shape are important when

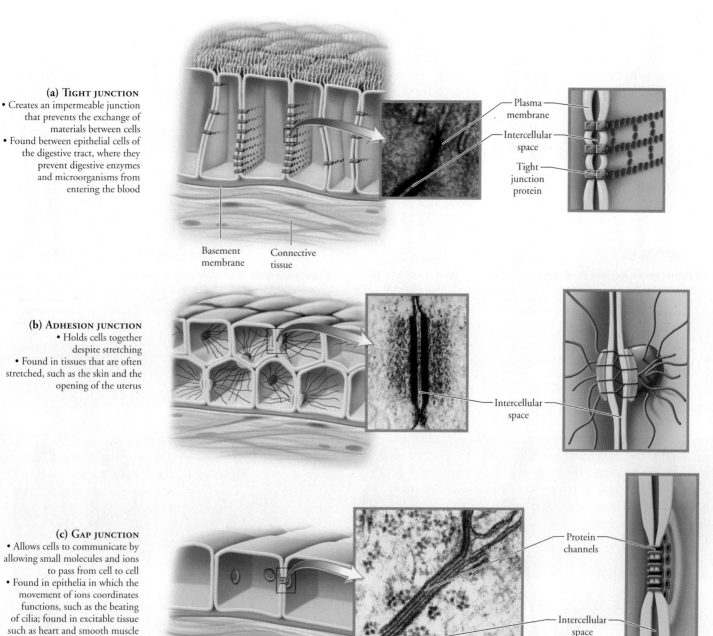

(a) Tight junction
- Creates an impermeable junction that prevents the exchange of materials between cells
- Found between epithelial cells of the digestive tract, where they prevent digestive enzymes and microorganisms from entering the blood

Plasma membrane
Intercellular space
Tight junction protein

Basement membrane
Connective tissue

(b) Adhesion junction
- Holds cells together despite stretching
- Found in tissues that are often stretched, such as the skin and the opening of the uterus

Intercellular space

(c) Gap junction
- Allows cells to communicate by allowing small molecules and ions to pass from cell to cell
- Found in epithelia in which the movement of ions coordinates functions, such as the beating of cilia; found in excitable tissue such as heart and smooth muscle

Protein channels

Intercellular space

FIGURE **4.5** Specialized cell junctions

the lungs fill with air, the stomach fills with food, the urinary bladder fills with urine, or our bodies bend or stretch.

There are two main body cavities—the ventral and dorsal cavities—each of which is further subdivided. The ventral (toward the abdomen) cavity is divided into the *thoracic* (chest) *cavity* and the *abdominal cavity.* The thoracic cavity is subdivided again into the pleural cavities, which house the lungs, and the pericardial cavity, which holds the heart. The abdominal cavity contains the digestive system, the urinary system, and the reproductive system. A muscle sheet called the diaphragm separates the thoracic and abdominal cavities. The dorsal (toward the back) cavity is subdivided into the *cranial cavity,* which encloses the brain, and the *spinal cavity,* which houses the spinal cord.

Body cavities and organ surfaces are covered with membranes—sheets of epithelium supported by connective tissue. Membranes form physical barriers that protect underlying tissues. The body has four types of membrane.

1. **Mucous membranes** line passageways that open to the exterior of the body, such as those of the respiratory, digestive, reproductive, and urinary systems. Some mucous membranes, including the mucous membrane of the small intestine, are specialized for absorption. Others, those of the respiratory system, for instance, secrete mucus that traps bacteria and viruses that could cause illness.

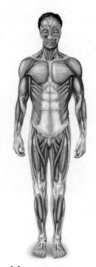
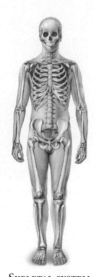
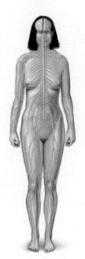
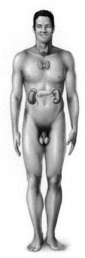

INTEGUMENTARY SYSTEM
- Protects underlying tissues from abrasion and dehydration
- Provides cutaneous sensation
- Regulates body temperature
- Immune function
- Synthesizes vitamin D
- Excretion

SKELETAL SYSTEM
- Attachment for muscles
- Encloses and protects organs
- Stores calcium and phosphorus
- Produces blood cells
- Stores fat

MUSCULAR SYSTEM
- Moves body and maintains posture
- Internal transport of fluids
- Generation of heat

NERVOUS SYSTEM
- Regulates and integrates body functions via neurons

ENDOCRINE SYSTEM
- Regulates and integrates body functions via hormones

CIRCULATORY SYSTEM
- Transports nutrients, respiratory gases, wastes, and heat
- Transports cells and antibodies for immune response
- Transports hormones
- Regulates pH through buffers

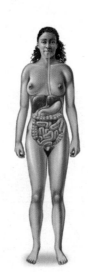
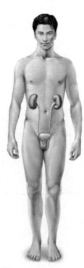
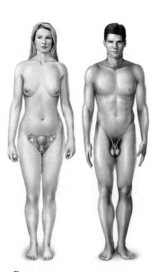

LYMPHATIC SYSTEM
- Returns tissue fluids to bloodstream
- Protects against infection and disease

RESPIRATORY SYSTEM
- Exchanges respiratory gases with the environment

DIGESTIVE SYSTEM
- Physical and chemical breakdown of food
- Absorbs, processes, stores, and controls release of digestive products

URINARY SYSTEM
- Maintains constant internal environment through the excretion of nitrogenous waste

REPRODUCTIVE SYSTEM
- Produces and secretes hormones
- Produces and releases egg and sperm cells and accessory secretions
- Forms placental attachment with fetus (females only)

FIGURE **4.6** The major organ systems of the human body

2. **Serous membranes** line the thoracic and abdominal cavities and the organs within them. They secrete a fluid that lubricates the organs within.

3. **Synovial membranes** line the cavities of freely movable joints, such as the knee. These membranes secrete a fluid that lubricates the joint, easing movement.

4. **Cutaneous membrane,** or skin, covers the outside of the body. Unlike other membranes, it is thick, relatively waterproof, and relatively dry.

We will consider the structure and function of the skin in the next section. Details about the structure and functions of the other organ systems are presented in subsequent chapters.

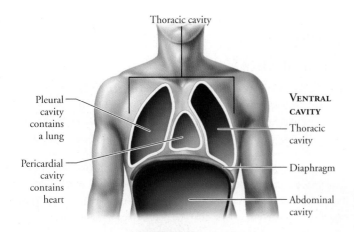
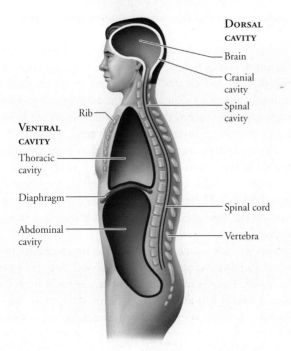

FIGURE **4.7** Body cavities. The internal organs are suspended in body cavities that protect the organs and allow organs to slide past one another as the body moves. Ventral means toward the abdomen, and dorsal means toward the back.

The Skin Is an Organ System

We have all been told that "beauty is only skin deep," but our skin does much more than just make us attractive.

■ Skin protects, regulates body temperature, and excretes

An integument is an outer covering, so the skin is sometimes called the **integumentary system**. It is the part of our body that makes direct contact with the environment. Few of us realize, however, the many other services performed by the skin and its derivatives—hair, nails, sweat glands, and oil glands.

One of the main functions of our skin is protection. It serves as a physical barrier that shields the contents of the body from invasion by foreign bacteria and other harmful particles, from ultraviolet radiation, and from physical and chemical insult. In addition to

offering this somewhat passive form of protection, skin contains cells called macrophages that have a more active way of fighting infection, as we will see in Chapter 13.

The skin has many other functions, as well. For example, because its outermost layer of cells contains the water-resistant protein keratin, the skin also plays a vital role in preventing excessive water loss from underlying tissues. It plays a role in temperature regulation, too. Although we perspire almost constantly, our sweat glands increase their output of perspiration dramatically during times of strenuous exercise or high environmental temperatures. The evaporation of this perspiration from the skin's surface helps rid the body of excess heat. Later we will see how changes in the flow of blood to the skin help regulate body temperature. The skin even functions in the production of vitamin D. Modified cholesterol molecules in the skin's outer layer are converted to vitamin D by ultraviolet radiation. The vitamin D then travels in the bloodstream to the liver and kidneys, where it is chemically modified to assume its role in stimulating the absorption of calcium and phosphorus from the food we eat.

The skin also contains structures for detecting temperature, touch, pressure, and pain stimuli. These receptors—components of the nervous system—help keep us informed about conditions in our external environment. Keep these many functions of the integumentary system in mind as you read on about the structure of the skin and its derivatives.

■ The skin has two layers

On most parts of your body, the skin is less than 5 mm (less than a quarter of an inch) thick, yet it is one of your largest organs. It represents about one-twelfth of your body weight and has a surface area of 1.5 to 2 m^2 (1.8 to 2.4 yd^2).

The skin has two principal layers, as shown in Figure 4.8. The thin, outer layer, the **epidermis**, forms a protective barrier against environmental hazards. The inner layer, the **dermis,** contains blood

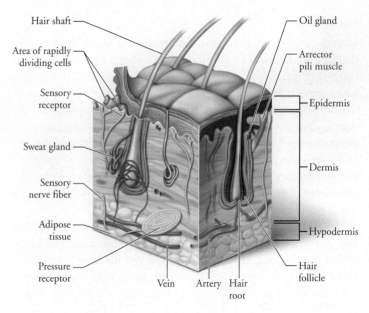

FIGURE **4.8** Structures of the skin and underlying hypodermis

vessels, nerves, sweat and oil glands, and hair follicles. Beneath the skin is a layer of loose connective tissue called the *hypodermis* or *subcutaneous layer*, which anchors the skin to the tissues of other organ systems that lie beneath.

THE EPIDERMIS

The outermost layer of skin, the epidermis, is itself composed of several layers of epithelial cells. The outer surface of epidermis, the part you can touch, is made up of dead skin cells. Thus, when we look at another person, most of what we see on the person's surface is dead. These dead cells are constantly being shed, at a rate of about 30,000 to 40,000 each minute. In fact, much of the dust in any room consists of dead skin cells. When you go swimming or soak in the bathtub for a long time, the dead cells on your skin's surface absorb water and swell, causing the skin to wrinkle. This is particularly noticeable where the layer of dead cells is thickest, such as on the palms of the hands and soles of the feet.

The skin does not get thinner as the dead cells are shed because they are continuously replaced from below. The deepest layer of epidermis contains rapidly dividing cells. As new cells are produced in this layer, older cells are pushed toward the surface. On their way, they flatten and die because they no longer receive nourishment from the dermis. Along this death route, keratin, a tough, fibrous protein, gradually replaces the cytoplasmic contents of the cells. It is keratin that gives the epidermis its protective properties. About 2 weeks to a month pass from the time a new cell is formed to the time it is shed.

Viruses called *papilloma viruses* sometimes cause the newly forming skin cells to multiply even faster than that which is normal for them, resulting in small, raised masses known as warts. Warts may be removed in various ways, such as cryosurgery with liquid nitrogen (essentially freezing the tissue to death), surgical removal, or chemical destruction with nonprescription drops. The best news, however, is that most warts eventually go away without any treatment.

THE DERMIS

Over most parts of the body, the dermis is a much thicker layer than the epidermis. The dermis lies just beneath the epidermis and consists primarily of connective tissue. In addition, it contains blood vessels, hair follicles, oil glands, the ducts of sweat glands, sensory structures, and nerve endings. Unlike the epidermis, the dermis does not wear away. This durability explains why tattoos—designs created when tiny droplets of ink are injected into the dermal layer—are permanent (Figure 4.9). The only way to remove a tattoo is by surgical means, including the use of lasers or "shaving" (abrading) of the skin. Because the dermis is laced with nerves and sensory receptors, getting a tattoo hurts.

Blood vessels are present in the dermis but not in the epidermis. Nutrients reach the epidermis by passing out of dermal blood vessels and diffusing through tissue fluid into the layer above. Such tissue fluid is probably quite familiar to you. Where skin is traumatized by, for example, a burn or an ill-fitting shoe rubbing against

your heel, this fluid accumulates between the epidermis and dermis, separating the layers and forming blisters.

stop and think

Burns—tissue damage caused by heat, radiation, electric shock, or chemicals—can be classified according to the depth to which the damage penetrates. First-degree burns are confined to the upper layers of epidermis, where they cause reddening and slight swelling. In second-degree burns, damage extends through the epidermis into the upper regions of the dermis. Blistering, pain, and swelling occur. Third-degree burns extend through the epidermis and dermis into underlying tissues. Severe burns, particularly if they cover large portions of the body, are life threatening. Given your knowledge of skin functions, what would you predict the immediate medical concerns to be when third-degree burns are present?

The lower layer of the dermis consists of dense connective tissue containing collagen and elastic fibers, a combination that allows the skin to stretch and then return to its original shape. Unfortunately, the skin's elasticity has limits. Pregnancy or substantial weight gain may exceed the ability of skin to return to its earlier size after stretching; instead, the dermis tears, and stretch marks appear. The resilience of our skin also decreases as we age. The most pronounced effects begin in the late forties, when collagen fibers begin to stiffen and decrease in number and elastic fibers thicken and lose elasticity. These changes, combined with reductions in moisture and the amount of fat in the hypodermis, lead to wrinkles and sagging skin.

Certain wrinkles, such as frown lines, are caused by the contraction of facial muscles. A controversial and popular treatment for these wrinkles is the injection of Botox, the toxin from the bacterium that causes botulism. When Botox is injected into facial muscles, they become temporarily paralyzed, so the muscle contractions that form the wrinkles cannot occur, and the skin

FIGURE **4.9** Tattoos—designs created when droplets of ink are injected into the dermis—are essentially permanent because, unlike the epidermis, the dermis is not shed.

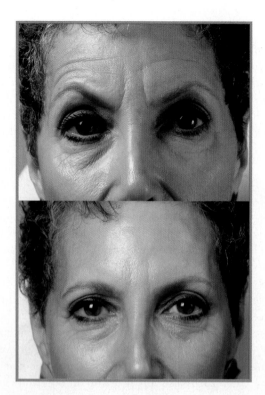

FIGURE **4.10** Botox injections before and after

smoothes out. The muscles regain the ability to contract over the next several months, however, and the injection has to be repeated (Figure 4.10).

THE HYPODERMIS

The **hypodermis**, a layer of loose connective tissue just below the epidermis and dermis, is not usually considered part of the skin. It does, however, share some of the skin's functions, including cushioning blows and helping to prevent extreme changes in body temperature, because it contains about half of the body's fat stores. In infants and toddlers, this layer of subcutaneous fat—often called baby fat—covers the entire body, but as we mature, some of the fat stores are redistributed. In women, subcutaneous fat tends to accumulate in the breasts, hips, and thighs. In both sexes, it has a tendency to accumulate in the abdominal hypodermis, contributing to the all-too-familiar pot belly, and in the sides of the lower back, forming "love handles."

Liposuction, a procedure for vacuuming fat from the hypodermis, is the most commonly performed cosmetic surgery in North America. The physician makes a small incision in the skin above the area of unwanted fat, inserts a fine tube, and moves the tube back and forth to loosen the fat cells, which are then sucked into a container. Liposuction is not a way of losing a lot of weight, because only a small amount of fat—not more than a few pounds—can be removed. However, it is a way of sculpting the body and removing bulges. Furthermore, because liposuction removes the cells that store the fat, fat does not usually return to those areas. In general, the procedure is safe, but it is not risk free. In some patients, it has produced blood clots that traveled to the lungs, causing death. People who are considering liposuction should choose their doctor carefully.

what would you do?

There are many social issues connected to liposuction. One is the procedure's increasing popularity among teenagers. What does this trend say about the value we place on physical beauty? Should there be age restrictions on liposuction? And what of allegations that some physicians who perform liposuction take unnecessary risks to increase profit? If you decided to have liposuction, what criteria would you use to choose a physician?

■ Skin color is determined by pigment and blood flow

Two interacting factors produce skin color: (1) the quantity and distribution of pigment and (2) blood flow. The pigment, called **melanin**, is produced by cells called **melanocytes** at the base of the epidermis. These cells, with their spiderlike extensions, produce two kinds of melanin: a yellow-to-red form and the more common black-to-brown form. The melanin is then taken up by surrounding epidermal cells, thus coloring the entire epidermis.

All people have about the same number of melanocytes. Differences in skin color are due to differences in the form of melanin produced and the size and number of pigment granules. A person's genetic makeup determines whether the yellowish red or the brown form of melanin is produced.

Albinism is an inherited condition in which an individual's melanocytes are incapable of producing melanin. It occurs in about 1 in 10,000 people and is not specific to a particular race. Individuals with this condition, known as albinos, lack pigment in their eyes, hair, and skin (Figure 4.11a). The eyes and skin of albinos have a pinkish color because the underlying blood supply shows through. Vitiligo, in contrast, is a condition in which melanocytes disappear either partially or completely from certain areas of the skin, leaving white patches in their wake (Figure 4.11b).

Circulation influences skin color in non-albinos, as well. When well-oxygenated blood flows through vessels in the dermis, the skin has a pinkish or reddish tint that is most easily seen in light-skinned people. Intense embarrassment can increase the blood flow, causing the rosy color to heighten, particularly in the face and neck. This response, known as blushing, is impossible to stop. Other intense emotions may cause color to disappear temporarily from the skin. A sudden fright, for example, can cause a rapid drop in blood supply, making a person pale. Skin color may also change in response to changing levels of oxygen in the blood. Compared to well-oxygenated blood, which is bright red, poorly oxygenated blood is a much deeper red that gives the skin a bluish appearance. Poor oxygenation is why the lips appear blue in extremely cold conditions. When it is cold, your body shunts blood away from the skin to the body's core, which conserves heat and keeps vital organs warm. This shunting reduces the oxygen supply to the blood in the small vessels near the surface of the skin. The oxygen-poor blood seen through the thin skin of the lips makes

(a)

(b)

FIGURE **4.11** Conditions caused by abnormalities in melanocytes. (a) Albinism, a genetic condition in which an individual's melanocytes cannot produce melanin, occurs throughout the world. Here, an Angolan mother holds her albino child. (b) Vitiligo is a condition in which melanocytes disappear from areas of the skin, leaving white patches.

them look blue. When you do not get enough sleep, the amount of oxygen in your blood may be slightly lower than usual, causing the color to darken. In some people, the darker color is visible through the thin skin under their eyes as dark circles.

Tanning is a change in skin color from exposure to the sun. Melanocytes respond to the ultraviolet (UV) radiation in sunlight by increasing the production of melanin. This is a protective response, because melanin absorbs some UV radiation, preventing it from reaching the lower epidermis and the dermis. The skin requires some ultraviolet radiation for the production of vitamin D, but too much can be harmful. See the Health Issue essay, *Fun in the Sun? Sunlight and Skin Cancer.*

■ Hair, nails, and glands are skin derivatives

Many seemingly diverse structures are derived from the epidermis: hair, nails, oil glands, sweat glands, and teeth. We will now consider the first four of these in terms of their structure and roles in everyday life. (Teeth are discussed in Chapter 15.)

HAIR

Hair usually grows all over the body, except on a few areas such as the lips, palms of the hands, and soles of the feet. Although hair is not essential for our survival, its presence or absence in certain places can have a strong psychological impact. Indeed, the strong emotional response of some individuals to the growth or loss of hair seems surprising when we learn that hair is essentially nothing more than dead cells filled with keratin and packed into a column.

What functions do these dead cells serve? An important one is protection. Hair on the scalp protects the head from UV radiation. Hair in the nostrils and external ear canals keeps particles and bugs from entering. Likewise, eyebrows and eyelashes help keep unwanted particles and glare (and perspiration and rain) out of the eyes. Hair also has a sensory role: receptors associated with hair follicles are sensitive to touch.

A hair consists of a shaft and a root (see Figure 4.8). The shaft projects above the surface of the skin, and the root extends below the surface into the dermis or hypodermis, where it is embedded in a structure called the hair follicle. Nerve endings surround the follicle and are so sensitive to touch that we are aware of even slight movements of the hair shaft. (Try to move just one hair without feeling it.) Each hair is also supplied with an oil gland that opens onto the follicle and supplies the hair with an oily secretion that makes it soft and pliant. In the dermis, a tiny smooth muscle called the arrector pili is attached to the hair follicle. Contraction of this muscle—which pulls on the follicle, causing the hair to stand up—is associated with fear and with cold. The tiny mound of flesh that forms at the base of the erect hair is sometimes called a goose bump.

A strand of hair, as we said above, is made up of dead, keratinized cells. The cuticle, or outermost layer, of the strand, consists of a single layer of thin, flat, scale-like cells that are packed with keratin. Exposure to the elements and abrasion can cause the cuticle to wear off at the hair's tip, making the underlying cortex frizz and thus producing dreaded split ends.

NAILS

Nails protect the sensitive tips of fingers and toes. Although the nail itself is dead and lacks sensory receptors, it is embedded in tissue so sensitive that we detect even the slightest pressure of an object on the nail. In this way, nails serve as sensory "antennas." They also help us manipulate objects, as when we undo a tight knot in a shoelace.

Like hair, nails are modified skin tissue hardened by the protein keratin. Nails differ from hair, however, in that they grow continuously (a hair stops growing after 2 to 6 years) but much more slowly.

GLANDS

Three types of glands—oil, sweat, and wax—are found in the skin. Although all three types develop from epidermal cells, they differ in their locations, structures, and functions.

Oil (sebaceous) **glands** are found virtually all over the body except on the palms of the hands and soles of the feet. They secrete sebum, an oily substance made of fats, cholesterol, proteins, and salts. The secretory part of these glands is located in the dermis, as shown in Figure 4.8. Sebum lubricates hair and skin and contains substances that inhibit growth of certain bacteria. Sometimes,

HEALTH ISSUE

Fun in the Sun? Sunlight and Skin Cancer

It is a sunny day, and the beach is packed with people. Some jump in the surf; some build sand castles, and others lie on blankets or towels. It is hardly a scene we would equate with disfigurement and death. Nonetheless, that is a connection we should make, because skin cancers are increasing at an alarming rate, largely due to our enjoyment of the sun.

The ultraviolet (UV) radiation of sunlight causes the melanocytes of the skin to increase their production of the pigment melanin. The melanin, taken up by surrounding epidermis, absorbs UV radiation before the radiation can travel any deeper into the body and damage the genetic information of the cells.

Unfortunately, this protective buildup of melanin is not instantaneous. Sunburn—damage to skin accompanied by reddening and peeling—is a typical sequel to the first day at the beach. Excessive exposure to sun destroys the cells at the surface of the skin, causing blood vessels in the dermis to dilate, which turns the skin red. The large-scale destruction of epidermis provokes an increased production of new cells. These push the burned cells off the skin surface and the skin peels, sometimes in small strands, but in severe cases, in large sheets.

In skin cancer, the genetic material in skin cells is altered by UV radiation so that the cells grow and divide uncontrollably, forming a tumor. Three types of skin cancer are caused by overexposure to the sun (Figure 4.A):

- Basal cell carcinoma arises in the rapidly dividing cells of the deepest layer of epidermis. It is the most common type of skin cancer.

- Squamous cell carcinoma arises in the newly formed skin cells as they flatten and move toward the skin surface. It is the second most common form of skin cancer.

- Melanoma is the least common and most dangerous type of skin cancer. It arises in melanocytes, the pigment-producing cells of the skin. Unlike basal or squamous cell carcinomas, melanomas, when left untreated, often metastasize (spread rapidly) throughout the body, first infiltrating the lymph nodes and later the vital organs. The survival rate in persons whose melanoma is found before it has metastasized is about 90% but drops to about 14% if the cancerous cells have spread throughout the body.

You can catch melanomas at an early stage if you carefully examine your skin while applying the ABCD mnemonic of the American Cancer Society:

A stands for asymmetry. Most melanomas are irregular in shape.

B stands for border. Melanomas often have a diffuse, unclear border.

C stands for color. Melanomas usually have a mottled appearance and contain colors such as brown, black, red, white, and blue.

D stands for diameter. Growths with a diameter of more than 5 mm (about 0.2 in.) are threatening.

Most skin cancers are treated either by excision, along with nearby tissue, or by radiation. Sometimes laser surgery, cryosurgery (freezing), or electrodesiccation (drying by electric current) are used to destroy cancerous cells. Many skin cancers occur on the face, and disfigurement caused by removal of large chunks of the nose or ears may necessitate reconstructive surgery.

The best way to avoid getting skin cancer is to avoid prolonged exposure to the sun. If you must be out in the sun, use a sunscreen and avoid exposure between 10:00 A.M. and 3:00 P.M., when the UV rays are the strongest. Sunscreens should have a sun protection factor (SPF) of at least 15. Apply your sunscreen about 45 minutes before going out into the sun, and reapply it after swimming or perspiring.

Always remember that sunscreens are not foolproof. Most block the higher-energy portion of the sun's UV radiation, known as UV-B, while providing only limited protection against the lower-energy portion, called UV-A. Whereas UV-B causes skin to burn, recent research suggests that exposure to UV-A weakens the body's immune system, possibly impairing its ability to fight melanoma. Ironically, by providing protection from sunburn, sunscreens have had the potentially devastating effect of enabling people to spend more time in the sun, possibly increasing their risk of developing melanoma.

Sunblocks such as zinc oxide, the white ointment used by lifeguards and others who spend long hours in the sun, differ from sunscreens in that they totally deflect ultraviolet rays. For this reason, sunblocks are recommended for particularly sensitive or already burned areas of the face, such as the nose and lips.

Avoid tanning salons. For many years, tanning salons claimed to use "safe" wavelengths of UV radiation because they did not use skin-reddening UV-B. But these "safe" wavelengths are actually UV-A. Given the apparent link between UV-A and increased risk of melanoma, the potential danger of these "safe" wavelengths is now obvious. 🏃

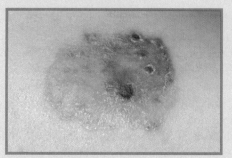

(a) Basal cell carcinoma

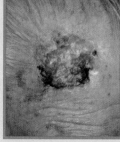

(b) Squamous cell carcinoma

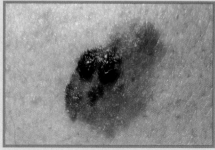

(c) Melanoma

FIGURE **4.A** Three skin cancers

however, the duct of an oil gland becomes blocked, causing sebum to accumulate and disrupt the gland's proper function. Next, bacteria invade the gland and hair follicle, resulting in a condition called acne. See the Health Issue essay, *Acne: The Misery, the Myths, and the Medications.*

As their name implies, **sweat glands** produce sweat, which is largely water plus some salts, lactic acid, vitamin C, and metabolic wastes such as urea. Although some wastes are eliminated through sweating, the principal function of sweat is to help regulate body temperature by evaporating from the skin surface. Wax glands are modified sweat glands found in the external ear canal.

Homeostasis Is a State of Relative Internal Constancy

To remain healthy, the organ systems of the body must constantly adjust their functioning in response to changes in the internal and external environment. We have already learned that the body's organ systems are interdependent, working together to provide the basic needs of all cells—water, nutrients, oxygen, and a normal body temperature. Just as city dwellers breathe the same air and drink the same city water, the cells of the body are surrounded by the same extracellular fluid. Changes in the makeup of that fluid will affect every cell.

One advantage of our body's multicellular, multi–organ system organization is its ability to provide a controlled environment for the cells. Although conditions outside the body sometimes vary dramatically, our organ systems interact to maintain relatively stable conditions within. This ability to maintain a relatively stable internal environment despite changes in the surroundings is called **homeostasis** (meaning "to stay the same"). But, conditions within the body never stay the same. As internal conditions at any level vary, the body's processes must shift to counteract the variation. Homeostatic mechanisms do not maintain absolute internal constancy, but they do dampen fluctuations around a set point to keep internal conditions within a certain range. Thus, homeostasis is not a static state but a dynamic one.

Illness can result if homeostasis fails. We see this in diabetes, a condition in which either the pancreas does not produce enough of the hormone insulin or the body cells are unable to use insulin. Normally, as a meal is digested and nutrients are absorbed into the bloodstream, the rising level of glucose in the blood stimulates the pancreas to release insulin. The general effect of insulin is to lower the blood level of glucose, returning it to a more desirable value. Without insulin, blood glucose can rise to a point that causes damage to the eyes, kidney, nerves, and blood vessels. A healthy diet, exercise, medication, and sometimes insulin injections can help people with diabetes regulate their blood glucose level.

Homeostasis depends on communication within the body. The nervous and endocrine systems are the two primary means of communication. The nervous system can bring about quick responses to changes in internal and external conditions. The endocrine system produces hormones, which bring about slower and longer-lasting responses to change.

stop and think

When you exercise, your breathing rate, blood pressure, and heart rate increase. Is this a violation of homeostasis?

■ Homeostasis is maintained by negative feedback mechanisms

Homeostasis is maintained primarily through **negative feedback mechanisms**—corrective measures that slow or reverse a variation from the normal value of a factor, such as blood glucose level or body temperature, and return the factor to its normal value. When the normal value is reached, the corrective measures cease; the normal value is the feedback that turns off the response. (In contrast, a positive feedback mechanism causes a change that promotes continued change in the same direction.)

Homeostatic mechanisms have three components (Figure 4.12):

1. A *receptor* detects change in the internal or external environment. A receptor, in this context, is a sensor that monitors the environment. When the receptor detects a change in some factor or event—some variable—it sends that information to the control center, the second of the three components.

2. A *control center* determines the factor's set point—the level or range that is normal for the factor in question. The control center integrates information coming from all the pertinent receptors and selects an appropriate response. In most of the body's homeostatic systems, the control center is located in the brain.

3. An *effector,* often a muscle or gland, carries out the selected response.

Consider how a negative feedback mechanism controls the temperature in your home during the frigid winter months. A thermostat in the heating system serves as both the temperature-sensing receptor and the control center. If it senses that the temperature inside your home has fallen below a certain programmed set point, the thermostat turns on the heating system (the effector). When the internal temperature has returned to the set point, the thermostat turns the heating system off. Thus, the temperature fluctuates around the set point but remains within a certain range. Now, let's apply these principles to see how homeostatic mechanisms regulate body temperature.

■ The hypothalamus regulates body temperature

The body's temperature control center is located in a region of the brain called the hypothalamus. Its set point is approximately 37°C (98.6°F), although it differs slightly from one person to the next. Body temperature must not vary too far from this mark, because even small temperature changes have dramatic effects on metabolism. Temperature is sensed at the body's outer surface by skin receptors and deep inside the body by receptors that sense the temperature of the blood. The hypothalamus receives input from both types of receptor. If the input indicates that body temperature is below the set point, the brain initiates mechanisms to increase

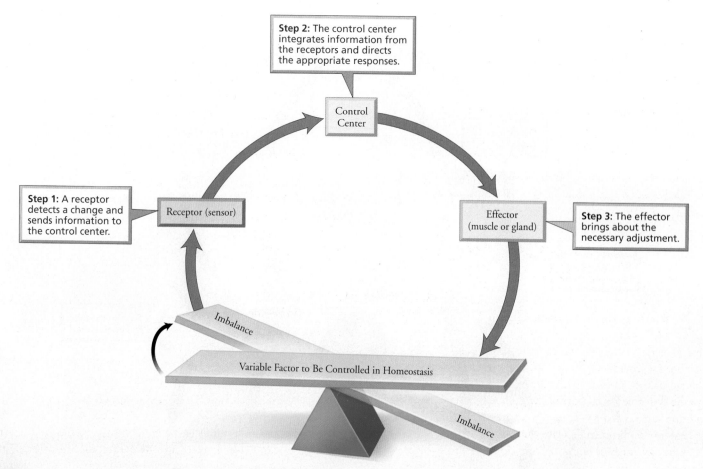

Step 2: The control center integrates information from the receptors and directs the appropriate responses.

Control Center

Step 1: A receptor detects a change and sends information to the control center.

Receptor (sensor)

Effector (muscle or gland)

Step 3: The effector brings about the necessary adjustment.

Imbalance

Variable Factor to Be Controlled in Homeostasis

Imbalance

WEB TUTORIAL 4.1

FIGURE **4.12** The components of a homeostatic control system by negative feedback mechanisms

heat production and conserve heat. When the input indicates that body temperature is above the set point, the brain initiates mechanisms that promote heat loss (Figure 4.13).

Let's consider what happens when we find ourselves in an environment where the temperature is above our set point, say 38°C (100.4°F). Thermoreceptors in the skin detect heat and activate nerve cells that send a message to the hypothalamus, which then sends nerve impulses to the sweat glands to increase their secretions. As the secretions (perspiration) evaporate and body temperature drops below 37°C (98.6°F), the signals from the brain to the sweat glands are discontinued. In this homeostatic system, the thermoreceptors in the skin are the receptors, the hypothalamus is the control center, and the sweat glands are the effectors. The system is a negative feedback mechanism because it produces an effect (cooling of the skin) that is relayed back (feeds back) to the control center, shutting off the corrective mechanism when the desired change has been produced.

Other mechanisms the brain may activate to lower body temperature include dilation of blood vessels in the dermis and relaxation of the arrector pili muscles attached to the hair follicles. The former response releases more heat to the surrounding air and explains the flushed appearance we get during strenuous exercise. The latter keeps damp, cooling hair lying close to the skin. The brain

may also initiate behavioral responses to a high body temperature, such as seeking shade or removing a sweatshirt.

Now let's consider what happens when body temperature drops below the set point. Subtle drops in body temperature are detected largely by thermoreceptors in the skin, which send a message to the hypothalamus in the brain. The brain then sends nerve impulses to sweat glands, ordering a decrease in their activity, and also sends messages to vessels in the dermis, telling them to constrict. This constriction reduces blood flow to the extremities, conserving heat for the internal organs and giving credence to the saying "cold hands, warm heart." Another response to decreasing body temperature is contraction of arrector pili muscles, which causes hairs to stand on end, thereby trapping an insulating layer of air near the body. This response, known as piloerection, is less effective in humans than in more heavily furred animals. The body also responds to cooling by increasing metabolic activity to generate heat and by the repeated contraction of skeletal muscles, known as shivering. Finally, behavioral responses, such as folding one's arms across one's chest, may be called upon to help combat a drop in body temperature.

Sometimes the mechanisms for lowering higher-than-normal temperatures fail, resulting in potentially deadly *hyperthermia*—abnormally elevated body temperature. Some marathon runners

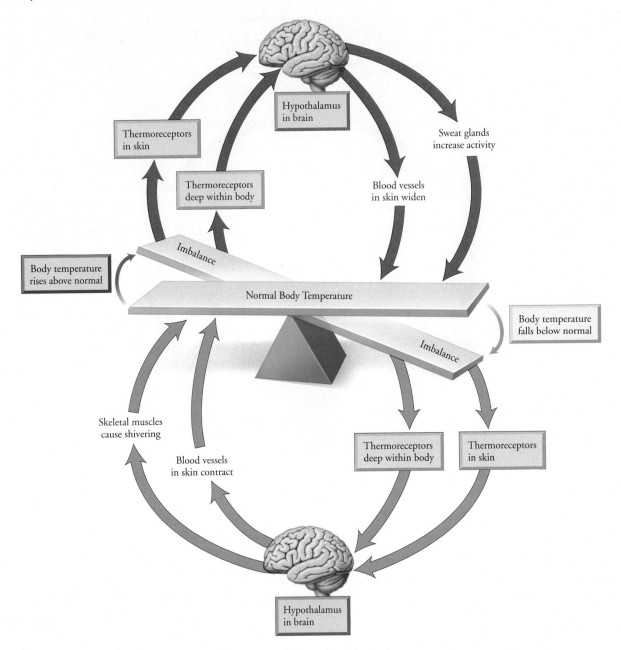

FIGURE **4.13** The homeostatic regulation of body temperature by negative feedback mechanisms

have died as a result of elevated core temperatures, as have people sitting in a hot tub heated to too high a temperature (say, up around 114°F). In both situations, perspiration could bring no relief; its evaporation would be prevented by high humidity in the case of the runners or by the surrounding water in the case of the hot-tubbers. Commonly called heat stroke, hyperthermia is marked by confusion and dizziness. If the core temperature reaches about 42°C (107°F), the heartbeat becomes irregular, oxygen levels in the blood drop, the liver ceases to function, and unconsciousness and death soon follow. Few people can survive core temperatures of 43°C (110°F).

If the body's temperature drops too far—to 35°C (95°F) or below—a condition called *hypothermia* results, disrupting nervous-system function and temperature-regulating mechanisms. People suffering from hypothermia usually become giddy and confused at first. When their temperature drops to 33°C (91.4°F), they lose consciousness. Finally, at a body temperature of 30°C (86°F), blood vessels are completely constricted and temperature-regulating mechanisms are fully shut down. Death soon follows. Hypothermia can be treated if detected early enough. In severe cases, dialysis machines may be used to artificially warm the blood and pump it back into the body.

stop and think

Frostbite is damage to tissues exposed to cold temperatures. Given what you know about the body's response to cold temperature, why are fingers and toes particularly susceptible to frostbite?

Acne: The Misery, the Myths, and the Medications

Acne and adolescence go hand in hand. In fact, about four out of five teenagers have acne, a skin condition that will probably annoy, if not distress, them well into their twenties and possibly beyond. What is acne, and what causes it? Is it really brought on by the chocolate, potato chips, and sodas so prevalent in the teenage diet? Why do males usually experience more severe acne than females do? Can acne be treated?

Simple acne is a condition that affects hair follicles associated with oil glands. When adolescence begins, oil glands increase in size and produce larger amounts of sebum. These changes are prompted by increasing levels of "male" hormones called androgens in the blood of both males and females; the androgens are secreted by the testes, ovaries, and adrenal glands. The changes thus induced in the activity and structure of oil glands set the stage for acne. It should come as no surprise, then, that acne occurs most often on areas of the body where oil glands are largest and most numerous: the face, chest, upper back, and shoulders.

Acne is the inflammation that results when sebum and dead cells clog the duct where the oil gland opens onto the hair follicle (Figure 4.B).

A follicle obstructed by sebum and cells is called a whitehead. Sometimes the sebum in plugged follicles mixes with the skin pigment melanin, forming a blackhead. Thus, melanin, not dirt or bacteria, lends the dark color to these blemishes. The next stage of acne is pimple formation, beginning with the formation of a red raised bump, often with a white dot of pus at the center. The bump occurs when obstructed follicles rupture and spew their contents into the surrounding epidermis. Such ruptures may occur naturally by the general buildup of sebum and cells or may be induced by squeezing the area. The sebum, dead cells, and bacteria that thrive on them then cause a small infection—a pimple or pustule—that will usually heal within a week or two without leaving a scar.

Eating nuts, chocolate, pizza, potato chips, or any of the other "staples" of the teenage diet does not cause acne. Also, acne is not caused by poor hygiene. Follicles plug from below, so dirt or oil on the skin surface is not responsible. (Most doctors do, however, recommend washing the face two or three times a day with hot water to help open plugged follicles.)

Treatments for acne fall into two main categories: topical (those applied to the skin) and oral (those taken by mouth). Most topical medicines must be used for several weeks before improvements are noticed. One topical preparation is a 5% solution of benzoyl peroxide sometimes sold over the counter. Applied once or twice a day as a cream after the affected area has been washed, benzoyl peroxide kills bacteria living in the follicles. Another topical option, tretinoin (retinoic acid, Retin-A), available only by prescription, is a derivative of vitamin A that helps thin the epidermis, reduce the stickiness of dead cells, and increase the rate at which the cells are sloughed off. These actions help keep follicles clear and speed the removal of existing whiteheads and blackheads.

Severe cases of acne may call for oral, or systemic, medicines, such as tetracycline or minocycline, that work by inhibiting the bacteria in the follicle. Isotretinoin (Accutane), another prescription-only derivative of vitamin A, is probably the most effective treatment for severe acne. This drug works by poisoning the oil glands, causing them to shrink and reduce their output of sebum. Some of the shrinkage is permanent, making Accutane different from other treatments in that it suppresses acne even after the treatment has stopped. 👥

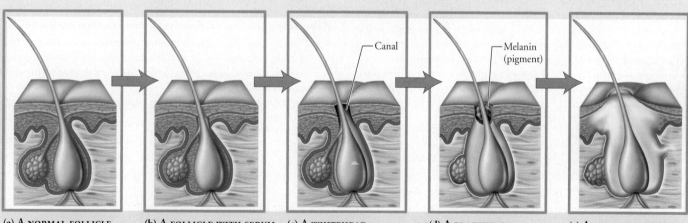

(a) A NORMAL FOLLICLE
Note the tiny hair and cells of the oil gland.

(b) A FOLLICLE WITH SEBUM
The canal becomes clogged with sebum, dead cells, and bacteria.

(c) A WHITEHEAD
Sebum, cells, and bacteria accumulate in the follicle.

(d) A BLACKHEAD
Sebum in the clogged follicle oxidizes and mixes with melanin.

(e) AN INFLAMED PIMPLE
The follicle wall ruptures, releasing the contents of a whitehead or blackhead into the surrounding epidermis.

Canal

Melanin (pigment)

FIGURE **4.B** The stages of acne

HIGHLIGHTING THE CONCEPTS

The Organization of the Human Body Increases in Complexity from Cells to Organ Systems (pp. 63–71)

1. Tissues are groups of cells that work together to perform a common function. There are four main types of tissue in the human body: epithelial (covers body surfaces, forms glands, and lines internal cavities and organs), connective (acts as storage site for fat, plays role in immunity and in transport, and provides protection and support), muscle (generates movement), and nervous tissue (coordinates body activities through initiation and transmission of nerve impulses).

2. All connective tissues contain cells in an extracellular matrix composed of protein fibers and ground substance. The types of connective tissue are connective tissue proper (loose and dense connective tissue) and specialized connective tissue (cartilage, bone, and blood).

3. Blood consists of formed elements (red blood cells, white blood cells, and platelets) suspended in a liquid matrix (plasma). The protein fibers are normally dissolved in the plasma and play a role in blood clotting. Red blood cells transport oxygen and carbon dioxide; white blood cells aid in fighting infections; and platelets function in blood clotting.

4. Muscle tissue is composed of muscle fibers that contract when stimulated, generating a mechanical force. There are three types of muscle tissue: skeletal, cardiac, and smooth. Skeletal muscle tissue is usually attached to bone, is voluntary, has cross-striations visible under a microscope, and has several nuclei in each cell. Cardiac muscle tissue is found in the walls of the heart, is involuntary, has cross-striations, and usually contains only one nucleus in each cell. Smooth muscle tissue is in the walls of blood vessels, airways, and organs. It is involuntary and lacks striations. A smooth muscle cell tapers at each end and has a single nucleus.

5. Nervous tissue consists of cells called neurons and accessory cells called neuroglia. Neurons convert stimuli into nerve impulses that they conduct to glands, muscles, or other neurons. Neuroglia increase the rate at which impulses are conducted by neurons and provide neurons with nutrients from nearby blood vessels. Neuroglial cells can communicate with one another and with neurons.

6. Three types of specialized junctions hold tissues together. Tight junctions prevent the passage of materials through the boundaries where cells meet. Adhesion junctions link cells by intercellular filaments attached to thickenings in the plasma membrane. Gap junctions have small pores that allow physical and chemical communication between cells.

7. An organ is a structure that is composed of two or more different tissues and has a specialized function.

8. Two or more organs that participate in a common function are collectively called an organ system. The human body has 11 major organ systems.

9. Internal organs are located in body cavities. There are two main body cavities: (1) the dorsal cavity, which is subdivided into the cranial cavity, where the brain is located, and the spinal cavity, where the spinal cord is located; (2) the ventral cavity, which is subdivided into the thoracic (chest) cavity and the abdominal cavity. The thoracic cavity is further divided into the pleural cavities, which contain the lungs, and the pericardial cavity, which contains the heart. Membranes line body cavities and spaces within organs.

The Skin Is an Organ System (pp. 71–76)

10. The integumentary system includes the skin and its derivatives such as hair, nails, and sweat and oil glands. It protects underlying tissues from abrasion and dehydration; regulates body temperature; synthesizes vitamin D; detects stimuli associated with touch, temperature, and pain; and body defense mechanisms.

11. The skin has two layers. The outermost layer, or epidermis, is composed of epithelial cells that die and wear away. The dermis, just below, is a much thicker, nondividing layer composed of connective tissue and containing nerves, blood vessels, and glands. Below the dermis is the hypodermis, a layer of loose connective tissue that anchors the skin to underlying tissues.

12. The epidermis is a renewing barrier. Cells produced in the deepest layer are pushed toward the skin surface, flattening and dying as they move away from the blood supply of the dermis and replace their cytoplasmic contents with keratin.

13. Skin color is determined, in part, by melanin, a pigment released by melanocytes at the base of the epidermis and taken up by neighboring cells on their way to the surface. Blood flow and blood oxygen content also influence skin color.

14. Hair is a derivative of skin. The primary function of hair is protection.

15. Nails are modified skin tissue hardened by keratin. They protect the tips of our fingers and toes and help us grasp and manipulate small objects.

16. Oil and sweat glands are derivatives of skin. Sebum, the oily substance secreted by oil glands, lubricates the skin and hair, prevents desiccation, and inhibits the growth of certain bacteria.

17. The integumentary system does not function alone but instead works closely with other organ systems.

Homeostasis Is a State of Relative Internal Constancy (pp. 76–79)

18. Homeostasis is the relative internal constancy maintained at all levels of body organization. It is a dynamic state, with small fluctuations occurring around a set point, and is sustained primarily through negative feedback mechanisms.

19. Homeostatic mechanisms consist of receptors, a control center, and effectors.

WEB TUTORIAL 4.1 Homeostasis

KEY TERMS

tissue   *p. 63*
epithelial tissue   *p. 63*
connective tissue   *p. 63*
muscle tissue   *p. 63*
nervous tissue   *p. 63*
squamous epithelium   *p. 63*
cuboidal epithelium   *p. 63*
columnar epithelium   *p. 63*
gland   *p. 63*
exocrine gland   *p. 63*
endocrine gland   *p. 63*
basement membrane   *p. 63*

collagen fibers   *p. 63*
elastic fibers   *p. 65*
reticular fibers   *p. 65*
fibroblast   *p. 65*
loose connective tissue   *p. 65*
adipose tissue   *p. 65*
dense connective tissue   *p. 65*
cartilage   *p. 65*
hyaline cartilage   *p. 65*
elastic cartilage   *p. 65*
fibrocartilage   *p. 65*
bone   *p. 66*

blood   *p. 67*
skeletal muscle tissue   *p. 67*
cardiac muscle tissue   *p. 67*
smooth muscle tissue   *p. 67*
neurons   *p. 68*
neuroglia   *p. 68*
organ   *p. 68*
organ system   *p. 68*
mucous membrane   *p. 69*
serous membrane   *p. 70*
synovial membrane   *p. 70*
cutaneous membrane   *p. 70*

integumentary system   *p. 71*
epidermis   *p. 71*
dermis   *p. 71*
hypodermis   *p. 73*
melanin   *p. 73*
melanocyte   *p. 73*
oil gland   *p. 74*
sweat gland   *p. 76*
homeostasis   *p. 76*
negative feedback mechanism
   p. 76

REVIEWING THE CONCEPTS

1. What are the four types of tissue found in the human body?   *p. 63*
2. Contrast the organization of epithelial and connective tissues. How are differences in the matrix of different types of connective tissue related to their functions?   *pp. 63, 65*
3. Why does bone heal more rapidly than cartilage   *p. 65*
4. What type of tissue is blood?   *p. 67*
5. Contrast skeletal, cardiac, and smooth muscle with respect to structure and function.   *p. 67*
6. What types of cells are found in nervous tissue? What are their functions?   *p. 68*
7. List the functions of the integumentary system.   *p. 71*
8. Describe the roles of pigments and blood flow in determining skin color.   *pp. 73–74*
9. What causes goose bumps?   *p. 74*
10. What are oil glands?   *pp. 74, 76*
11. Describe the functions of the sweat glands.   *p. 76*
12. What is homeostasis?   *p. 76*
13. Describe the body's homeostatic mechanisms for raising and lowering core temperature.   *pp. 76–77*

14. The four basic tissue types in the body are
    a. simple, cuboidal, squamous, columnar.
    b. neural, epithelial, muscle, connective.
    c. blood, nerves, bone, cartilage.
    d. fat, cartilage, muscle, neural.
15. The lining of the intestine is composed primarily of _____ cells.
    a. epithelial
    b. muscle
    c. connective tissue
    d. nerve
16. Cells that form pads that cushion the vertebrae are
    a. bone.
    b. muscle.
    c. epithelium.
    d. cartilage.
17. The mineral that makes bone hard is _____.
18. The maintenance of body processes within a relatively constant range is called _____.

APPLYING THE CONCEPTS

1. Thor, a ski champion at his college, tore the cartilage in his knee in a ski accident. He asked the doctor if he would be ready to compete in a month. Why would you expect the doctor's answer to be "No"?
2. The parathyroid glands secrete parathyroid hormone in response to decreased blood calcium. This hormone is in a negative feedback relationship with calcium. What effect will it have on blood calcium level?
3. Hannah has the flu. As her fever rises, she gets the chills. She shivers and covers herself in extra blankets. She takes some acetominophen and her fever breaks. As her body temperature returns to normal, she throws off the blankets, looks flushed, and perspires. Use the mechanisms of body temperature control to explain what is happening as Hannah's fever rises and falls.
4. Ehlers Danlos syndrome is a group of disorders caused by defects in genes that disrupt the production of collagen, which is one of the chief components of connective tissue. Explain why symptoms of Ehlers Danlos syndrome include joints that extend beyond their normal range and stretchy, saggy skin.

Additional questions can be found on the companion website.

5

The Skeletal System

The 206 bones of the human body provide support against gravity and protect internal organs. The joints, places where bones meet, determine the type of movement that is possible

Bones Function in Support, Movement, Protection, Storage, and Blood Cell Production

Bones Have a Hard Outer Layer of Compact Bone Surrounding Spongy Bone

Bone Is Living Tissue
- Most of the skeleton begins as a cartilage model
- Hormones regulate bone growth

Bone Fractures Are Healed by Fibroblasts and Osteoblasts

Bones Are Continuously Remodeled

We Divide the Human Skeleton into Two Parts
- The axial skeleton protects our internal organs
- The appendicular skeleton makes locomotion possible

Joints Are Junctures between Bones
- Synovial joints permit flexibility

HEALTH ISSUE   Osteoporosis: Fragility and Aging

At age 4 Jamal tried to fly. After swooping around the backyard with a small blanket for a cape, flapping his arms, he decided that he would have a better chance if he started with a little elevation. Starting from the second step of the ladder to his backyard slide did seem to work a bit better, so he decided to go up two more steps and try again. This too seemed to work, except that his outstretched hands stung a little when he hit the ground. He decided to go to the sixth step of the slide and dive with his cape extended. This time when he hit the ground, the pain made him scream. In a matter of seconds his mom was there to pick him up. A trip to the emergency room and an x-ray indicated that the wrist was sprained. The doctor gave him a lollipop, told him he was lucky he hadn't broken the bones in his wrist, and never, never to jump off of ladders again.

The fact that Jamal cannot fly and the fact that he could have broken his wrist in his fall are both related to the composition and design of human bones and muscles. As you read about the structure and properties of bone in this chapter, you may be surprised to learn that bone is a dynamic, living tissue and to discover the number of functions our skeletal system performs for us (even though it doesn't allow us to fly). In considering how bones grow, we will even learn a few of the reasons why some of us are taller than others. ■

Bones Function in Support, Movement, Protection, Storage, and Blood Cell Production

The **skeleton** is a framework of bones and cartilage that performs the following functions for the body:

1. **Support.** It provides a rigid framework that supports soft tissues. The leg bones and backbone hold our bodies upright, and the pelvic girdle supports the abdominal organs.
2. **Movement.** It provides places of attachment for muscles. Contraction of muscles allows bones to move at joints.
3. **Protection.** It shields our internal organs, such as the heart and lungs, which are enclosed within the chest cavity, and the brain, which lies within the skull.
4. **Storage of minerals.** It stores minerals, particularly calcium and phosphorus, that can be released to the rest of the body when needed.

5. **Storage of fat.** It stores energy-rich fat in yellow bone marrow (the soft tissue within some bones). The fat can be metabolized when the energy is needed.
6. **Blood cell production.** It produces blood cells in the red marrow of certain bones.

Bones Have a Hard Outer Layer of Compact Bone Surrounding Spongy Bone

The 206 bones of the human body come in a range of sizes and a variety of shapes. Most bones contain both compact tissue and spongy tissue in proportions that depend on the bone's size and shape.

Compact bone is very dense, with few internal spaces (Figure 5.1). It forms most of the shaft of long bones, such as those of the arms and legs, and in fact is what you see when you look at the outside of any bone. In the body, compact bone is covered by a glove-like membrane, the **periosteum**, that nourishes the bone. The periosteum contains blood vessels and nerves as well as cells that

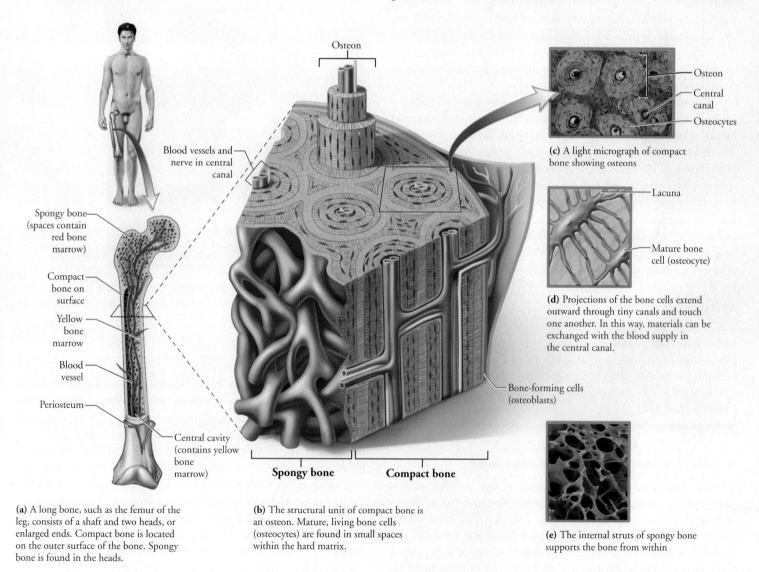

(c) A light micrograph of compact bone showing osteons

(d) Projections of the bone cells extend outward through tiny canals and touch one another. In this way, materials can be exchanged with the blood supply in the central canal.

(a) A long bone, such as the femur of the leg, consists of a shaft and two heads, or enlarged ends. Compact bone is located on the outer surface of the bone. Spongy bone is found in the heads.

(b) The structural unit of compact bone is an osteon. Mature, living bone cells (osteocytes) are found in small spaces within the hard matrix.

(e) The internal struts of spongy bone supports the bone from within

FIGURE **5.1** The structure of bone

function in bone growth and repair. When a bone is bruised or fractured, most of the pain results from injury to the periosteum.

Spongy bone is a latticework of thin struts of bone with open areas between. This internal network braces the bone from within. Spongy bone is largely found in small, flat bones, such as most of the bones of the skull, and in the heads (enlarged ends) and near the ends of the shafts of long bones. Some of the spongy bone in adults (for example, in the ribs, pelvis, backbone, skull, sternum, and long-bone ends) is filled with **red marrow**, where blood cells form. The cavity in the shaft of adult long bones is filled with **yellow marrow**, a fatty tissue used for energy storage.

Bone Is Living Tissue

Compact bone is highly organized living tissue containing a microscopic repeating structural unit called an **osteon** (shown in Figure 5.1). Each osteon consists of mature bone cells, called **osteocytes** (*osteo*, bone; *cyte*, cell), arranged in concentric rings around a central canal. Each osteocyte lies within a tiny cavity called a lacuna in the hardened matrix. From each lacuna, tiny canals connect with nearby lacunae and eventually with the central canal. The cells have projections that extend through the tiny canals to touch neighboring cells. Nutrients, oxygen, and wastes move from cell to cell at the points of contact, traveling to or from the blood vessels in the central canal.

Bone is, indeed, a living tissue, but its most noticeable characteristics result from its nonliving component, the matrix. Secreted by the bone cells, the matrix makes bone both hard and resilient. The hardness comes from mineral salts, primarily calcium and phosphorus, in the matrix. The resilience comes from strands of the strong elastic protein collagen woven through the matrix. Without the calcium and phosphorus salts, bone would be rubbery and flexible like a garden hose. Without the collagen, bone would be brittle and crumbly like chalk. In some disorders, bones do bend, causing bowlegs. An example is rickets, in which the amount of calcium salts in the bones is greatly reduced (Figure 5.2).

stop and think

Strontium-90 is a radioactive substance that enters the atmosphere after atomic explosions. Humans may ingest it in milk from cows that grazed on contaminated grass. The strontium-90 can then replace calcium in bone and kills nearby cells or alter their genetic information. Explain why exposure to strontium-90 can lead not just to bone cancer but also to disruption of blood cell formation.

■ Most of the skeleton begins as a cartilage model

During embryonic development, most of the skeleton is first formed of cartilage, a strong yet flexible connective tissue (Figure 5.3). Unlike mature bone cells, which cannot divide because they are enclosed in a solid matrix, cartilage cells are able to divide and multiply quickly. Thus, the cartilage model can grow as rapidly as the fetus does. In the third month of development, the cartilage begins to be replaced by bone.

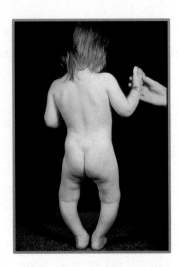

FIGURE **5.2** The legs of this child are bent because of rickets, a condition caused by insufficient vitamin D, which is needed for proper absorption of calcium from the digestive system. Calcium salts make the matrix of bone very hard. Therefore, in rickets, the bones become soft and somewhat pliant.

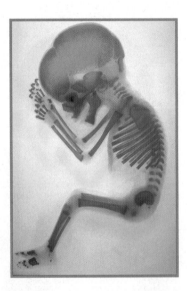

FIGURE **5.3** The fetal skeleton is first made of cartilage and gradually replaced by bone. This shows the cartilaginous model of a 16-week-old human fetus.

The transformation of a long bone, such as an arm or leg bone, from cartilage to bone begins with the formation of a collar of bone around the shaft of the cartilaginous model (Figure 5.4). The collar is produced by bone-forming cells called **osteoblasts** (*osteo*, bone; *blast*, beginning or bud). Osteoblasts produce the matrix of bone by secreting collagen (as well as other organic materials) and then depositing calcium salts on it. The bony collar supports the shaft as the cartilage within it breaks down, leaving the marrow cavity. Blood vessels then bring osteoblasts into the cavity, which fill the cavity with spongy bone. Unlike cartilage, osteoblasts cannot undergo cell division. Once they form the matrix around themselves, they are called osteocytes, which, as we have seen, are mature bone cells and the principal cells in bone tissue.

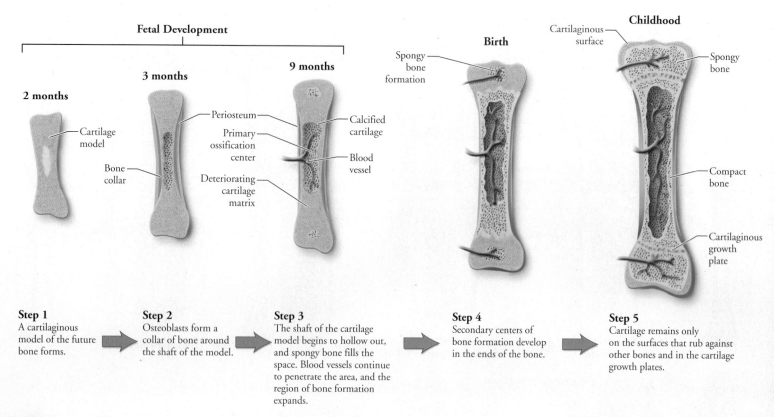

Step 1
A cartilaginous model of the future bone forms.

Step 2
Osteoblasts form a collar of bone around the shaft of the model.

Step 3
The shaft of the cartilage model begins to hollow out, and spongy bone fills the space. Blood vessels continue to penetrate the area, and the region of bone formation expands.

Step 4
Secondary centers of bone formation develop in the ends of the bone.

Step 5
Cartilage remains only on the surfaces that rub against other bones and in the cartilage growth plates.

FIGURE **5.4** Steps of bone formation in long and short bones of an embryo

At about the time of birth, bone growth centers form in the ends of long bones, and spongy bone begins to fill them. Two kinds of structures made of cartilage will remain. One is a cap of cartilage over each end of the bone, where one bone rubs against another in a joint. The second is a plate of cartilage, called an epiphyseal plate, that separates each end of the bone from its shaft. The epiphyseal plate is commonly called the **growth plate**. Cartilage cells within the growth plate divide, forcing the end of the bone farther away from the shaft. As bone replaces the newly formed cartilage in the region closer to the shaft, the bone becomes longer.

■ Hormones regulate bone growth

Parents proudly measure the growth of a child, inch by inch, on a growth chart. During childhood, bone growth is powerfully stimulated by growth hormone, released by the anterior pituitary gland. Growth hormone prompts the liver to release growth factors that produce a surge of growth in the growth plate. Thyroid hormones modify the activity of growth hormone to ensure that the skeleton grows with the proper proportions.

At puberty, many children experience a growth spurt during which the length of pant legs can seem to shrink almost weekly. These dramatic changes are orchestrated by the increasing levels of male or female sex hormone (testosterone and estrogen, respectively) produced at that stage. Initially the sex hormones stimulate the cartilage cells of the growth plates into a frenzy of cell division. But this growth generally stops toward the end of the teenage years (age 18 in females and 21 in males) because of later changes initiated by

the sex hormones. The cartilage cells in the growth plates start dividing less frequently. The plates become thinner as cartilage is replaced by bone. Finally, the bone in the ends fuses with the bone in the shaft.

stop and think

We have considered the mechanism by which long bones grow and the role of growth hormone in that process. Why would it be ineffective for a short, middle-aged person to be treated with growth hormone to stimulate growth?

Bone Fractures Are Healed by Fibroblasts and Osteoblasts

In spite of their great strength, bones occasionally break. Fortunately, bone tissue can heal. The first thing that happens when a bone breaks is bleeding (even with a simple fracture) followed by formation of a clot (hematoma) at the break. Within a few days, connective tissue cells called fibroblasts grow inward from the periosteum and invade the clot. The fibroblasts secrete collagen fibers that form a mass called a callus, which links the broken ends of the bone. Some of the fibroblasts then transform into cartilage-producing cells and secrete cartilage into the callus.

Next, osteoblasts from the periosteum invade the callus and begin to transform the cartilage into new bone material. As this transformation happens, the bony callus becomes thicker than the

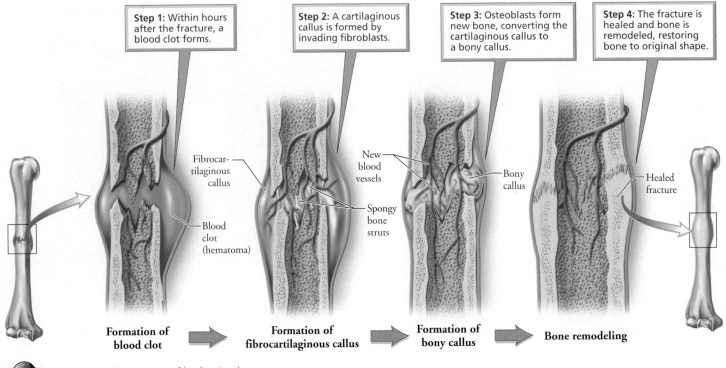

Step 1: Within hours after the fracture, a blood clot forms.

Step 2: A cartilaginous callus is formed by invading fibroblasts.

Step 3: Osteoblasts form new bone, converting the cartilaginous callus to a bony callus.

Step 4: The fracture is healed and bone is remodeled, restoring bone to original shape.

Fibrocartilaginous callus

Blood clot (hematoma)

New blood vessels

Spongy bone struts

Bony callus

Healed fracture

Formation of blood clot → **Formation of fibrocartilaginous callus** → **Formation of bony callus** → **Bone remodeling**

WEB TUTORIAL 5.2

FIGURE 5.5 The progress of healing in a bone

undamaged part of the bone and protrudes from it. In time, however, the extra material will be broken down, and the healed part of the bone will return to normal size (Figure 5.5).

Bones Are Continuously Remodeled

Even after reaching our full height, we continue to undergo a life-long process of bone deposition and absorption called remodeling. We have seen that bone is deposited by osteoblasts. Another kind of bone cell, called an **osteoclast**, breaks down bone, releasing minerals that are reabsorbed by the body. The process of remodeling is continuous but occurs at different rates in different parts of bones. For example, the end of the femur (thighbone) closest to the knee is completely replaced every 5 to 6 months, but the shaft of the same bone is replaced more slowly.

An important factor determining the rate and extent of bone remodeling is the degree of stress to which a bone region is subjected. Bone forms in response to stress and gets absorbed when it is not stressed. Weight-bearing exercise, such as walking or jogging, thickens the layer of compact bone tissue, leading to a stronger bone. Bones that are used frequently may actually change shape. For example, continual practice may enlarge the knuckles of pianists and the big toes of ballet dancers. On the other hand, a few weeks without stress can cause a bone to lose nearly a third of its mass. This is a concern for astronauts living in the weightless environment of space as well as for the person right here on earth who is using crutches with a leg in a cast.

If the breakdown process occurs faster than the deposition of new tissue, a bone becomes weak and easy to break. This imbalance occurs in **osteoporosis**, a condition in which there is a progressive loss in bone density. (See the Health Issue essay, *Osteoporosis: Fragility and Aging.*)

We Divide the Human Skeleton into Two Parts

Figure 5.6 uses color to distinguish what we call the axial skeleton from the appendicular skeleton. The **axial skeleton** (shown in orange) includes the skull, the vertebral column (backbone), and the bones of the chest region (sternum and rib cage). The **appendicular skeleton** (shown in light brown) includes the pectoral girdle (shoulders), the pelvic girdle (pelvis), and the limbs (arms and legs).

■ The axial skeleton protects our internal organs

In general, the axial skeleton protects and supports our internal organs. We will focus on only the major bones of the 80 that make up this portion of the skeleton, beginning with the bones of the skull.

THE SKULL

The skull is the most complex bony structure in the body. Its principal divisions are the cranium and the face (Figure 5.7).

THE CRANIAL BONES The **cranium** protects the brain, houses the structures of hearing, and provides attachment sites for

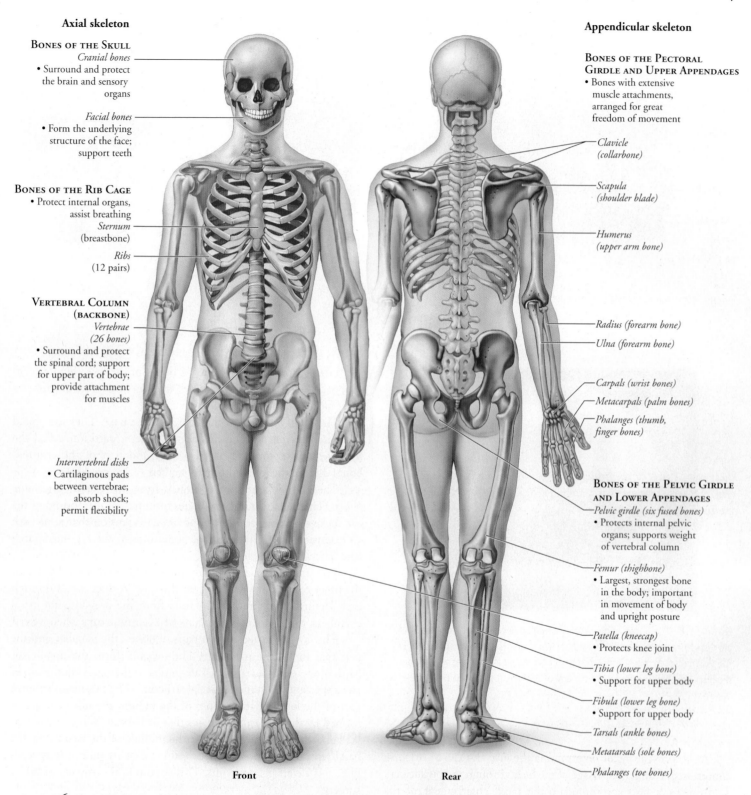

Axial skeleton

BONES OF THE SKULL
Cranial bones
- Surround and protect the brain and sensory organs

Facial bones
- Form the underlying structure of the face; support teeth

BONES OF THE RIB CAGE
- Protect internal organs, assist breathing

Sternum (breastbone)

Ribs (12 pairs)

VERTEBRAL COLUMN (BACKBONE)
Vertebrae (26 bones)
- Surround and protect the spinal cord; support for upper part of body; provide attachment for muscles

Intervertebral disks
- Cartilaginous pads between vertebrae; absorb shock; permit flexibility

Appendicular skeleton

BONES OF THE PECTORAL GIRDLE AND UPPER APPENDAGES
- Bones with extensive muscle attachments, arranged for great freedom of movement

Clavicle (collarbone)

Scapula (shoulder blade)

Humerus (upper arm bone)

Radius (forearm bone)

Ulna (forearm bone)

Carpals (wrist bones)

Metacarpals (palm bones)

Phalanges (thumb, finger bones)

BONES OF THE PELVIC GIRDLE AND LOWER APPENDAGES
Pelvic girdle (six fused bones)
- Protects internal pelvic organs; supports weight of vertebral column

Femur (thighbone)
- Largest, strongest bone in the body; important in movement of body and upright posture

Patella (kneecap)
- Protects knee joint

Tibia (lower leg bone)
- Support for upper body

Fibula (lower leg bone)
- Support for upper body

Tarsals (ankle bones)

Metatarsals (sole bones)

Phalanges (toe bones)

Front

Rear

FIGURE **5.6** The major bones of the human body. The bones of the axial skeleton are shown in orange; the bones of the appendicular skeleton are shown in light brown. Cartilage is shown in gray.

the muscles of the head and neck. It is formed from eight (or sometimes more) flattened bones. The single *frontal bone* forms the forehead and the front of the brain case. Behind it, extending from either side of the midline, the two *parietal bones* form the top and sides of the skull. The *occipital bone* lies at the back of the head and

surrounds the *foramen magnum*, the opening through which the spinal cord passes.

Before and shortly after birth, the bones of the cranium are connected by membranous areas called the *fontanels*, often referred to as soft spots. During birth, the fontanels allow the skull to be

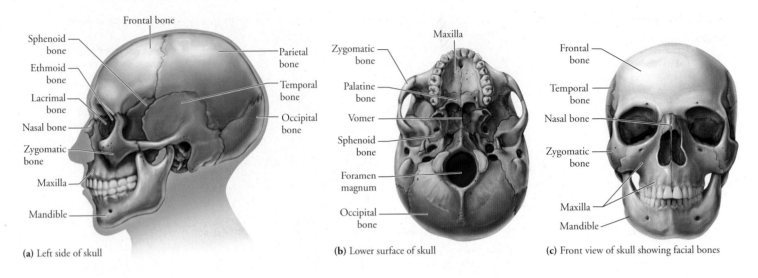

(a) Left side of skull

(b) Lower surface of skull

(c) Front view of skull showing facial bones

FIGURE **5.7** The major bones of the skull and face

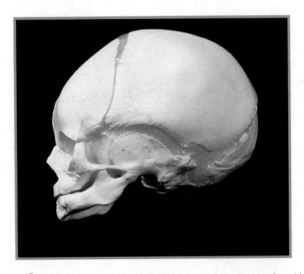

FIGURE **5.8** The bones of the skull of a human newborn are not fused but are instead connected by fibrous connective tissue. These "soft spots" allow the skull bones to move during the birth process, easing the passage of the skull through the birth canal. By the age of 2, the soft spots will be replaced by bone.

compressed, easing the passage of the head through the birth canal. The fontanels also accommodate the rapid enlargement of the brain during fetal growth and infancy (Figure 5.8). They are replaced by bone by the age of 2.

The cranium also contains the *temporal bones*, part of which form what we think of as our temples. The *sphenoid bone*, with its bow-tie shape, forms the cranium's floor. The *ethmoid bone*, the smallest bone in the cranium, separates the cranial cavity from the nasal cavity. The olfactory nerves from the nasal cavity, which are responsible for our sense of smell, communicate with the brain through tiny holes in the ethmoid.

THE FACIAL BONES The 14 bones of the face (Figure 5.7c) support several sensory structures and serve as attachments for most facial muscles.

The *nasal bones* form the bridge of the nose. They are paired and fused at the midline. Inside the nose, a partition called the *nasal septum* (composed of the vomer and part of the ethmoid bone) divides the left and right chambers of the nasal cavity.

Cheekbones are formed largely from the paired *zygomatic bones*. Flattened areas of these bones form part of the bottom of the eye sockets. Each zygomatic bone has an extension that joins with an extension from the temporal bone to form the zygomatic arch (the "cheekbone" itself).

The smallest facial bones are the two *lacrimal bones*, located at the inner corners of the eyes, near the nose. A duct passes through each lacrimal bone and drains tears from the eyes into the nasal chambers; this connection explains why our nose runs when we cry.

The jaw is formed by two pairs of bones, the maxillae and the pair that forms the mandible. The *maxillae* form the upper jaw. Most of the other facial bones are joined to them, giving the upper jaw a special importance in facial structure. The maxillae also form part of the hard palate, or roof of the mouth, the rest of which is formed by the two *palatine bones* that lie behind. When the maxillae fail to fuse to one another at the midline of the face below the nasal septum, the mouth cavity and nasal cavity do not fully separate, and a cleft palate results. This condition is easily corrected by surgery.

The lower jaw, called the *mandible*, is also formed from two bones connected at the midline. The mandible is connected to the skull at the temporal bone, forming a hinge called the temporomandibular joint. (Joints are discussed later in this chapter.) This joint allows the mouth to open and close. Emotional stress causes some people to clench or grind their teeth, sometimes unconsciously. This clenching often leads to headaches, toothaches, or even earaches. The condition is known as temporomandibular joint (or TMJ) syndrome.

Osteoporosis: Fragility and Aging

About 20 million Americans, most of them elderly white women, suffer from osteoporosis, a decrease in bone density that occurs when the destruction of bone outpaces the formation of new bone during bone remodeling. The net destruction causes bones to become thin, brittle, and susceptible to fracture. A fall, blow, or lifting action that would not bruise a person with healthy bones could easily cause a bone to fracture in a person with severe osteoporosis.

Osteoporosis is sometimes very apparent. The afflicted person becomes hunched and shorter as the vertebrae lose mass and compress (Figure 5.A). It can be startling to encounter a middle-aged or elderly person you haven't seen in a while and find that you suddenly tower above them!

As we have seen, bone remodeling occurs throughout life. Until we reach about age 35, bone is formed faster than it is broken down. Our bones are strongest and densest during our midthirties, after which they begin to lose density. Peak bone density is influenced by a number of factors, including sex, race, nutrition (dietary levels of calcium and vitamin D), exercise (we have seen the importance of weight-bearing exercise), and overall health. In men, bone mass is generally 30% higher than in women. The bones of African Americans are generally 10% denser than those of Caucasians and Asians. The degree to which our bones will become weakened as we age depends largely on how dense they were at their peak. Women are at greater risk for developing osteoporosis than are men. This difference is not just because women have less bone mass at peak but also because their rate of bone loss is accelerated for several years after menopause (the time in a woman's life when she stops producing mature eggs or menstruating). The acceleration occurs because menopause is followed by a sharp decline in the female hormone estrogen, and estrogen stimulates bone formation and is also important in the absorption of calcium from the intestines.

It is worth noting that intense physical exercise sometimes alters hormone levels in young women, and causes them to stop menstruating and placing their bones at risk. In one study, 20% of young female athletes who failed to ovulate during even one cycle out of the whole year suffered as much as a 4% loss in bone density.

A number of other factors have also been implicated in the onset of osteoporosis. For example, height is important. Short people are at greater risk, perhaps because they generally start with less bone mass. People with a good supply of body fat are less at risk because fat can be converted to estrogen. Heavy drinkers are at higher risk, because alcohol interferes with estrogen function. Smoking is also bad for bones, because it can reduce estrogen levels. We have already noted that people who do not take in enough calcium, or who cannot absorb it because of insufficient vitamin D, have thinner bones because calcium is necessary for bone growth. (Calcium-rich foods include milk products as well as broccoli, spinach, shrimp, and soybean products.) Certain drugs, such as caffeine (a diuretic), tetracycline, and cortisone can promote osteoporosis. Finally, as mentioned earlier, sedentary people have thinner bones than do active people.

We can expect osteoporosis and other afflictions associated with aging to become more common as life expectancy increases. Thus, it would seem prudent for each of us to start an early program of prevention by eating a diet rich in calcium and vitamin D and by engaging in regular weight-bearing exercise, such as walking or jogging.

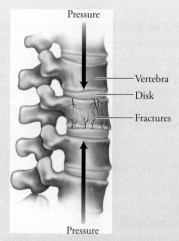

(a) Normal bending can place pressure on vertebrae that can cause small fractures.

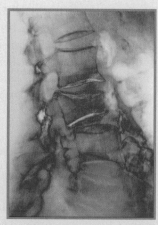

(b) Osteoporosis can be seen in this colored x-ray image of vertebrae as the pink regions in the bones.

(c) A loss of height and a stooped posture result as the weakened bones of the vertebrae become compressed.

FIGURE 5.A Osteoporosis is a loss of bone density that occurs when bone destruction outpaces bone deposition during the continuous process of bone remodeling. The bones become brittle and are easily fractured.

THE VERTEBRAL COLUMN

The **vertebral column**, known more familiarly as the backbone or spine, is a series of bones through which the spinal cord passes as it descends down the back (Figure 5.9). Each of these bones is called a **vertebra**. The vertebrae (plural of vertebra) are classified according to where they lie along the vertebral column. There are

- 7 *cervical* (neck) *vertebrae,*
- 12 *thoracic* (chest) *vertebrae,*
- 5 *lumbar* (lower back) *vertebrae,*
- 1 *sacrum* (formed by the fusion of five vertebrae), and
- 1 *coccyx* (or tailbone, formed by fusion of four vertebrae).

Scoliosis, which means "twisted disease," is an abnormal curvature of the spine to the left or right. The most common form has no known cause and affects over 1.5 million adolescents, primarily females. It usually begins during, and progresses through, the adolescent growth spurt. Treatment, if needed, may consist of a brace or surgery to straighten the spine.

The fusion of the sacral vertebrae gives additional strength to the backbone in the region where the sacrum joins the pelvic girdle. This reinforcement is necessary because of the great stress placed on the sacrum by the weight of the vertebral column and the powerful

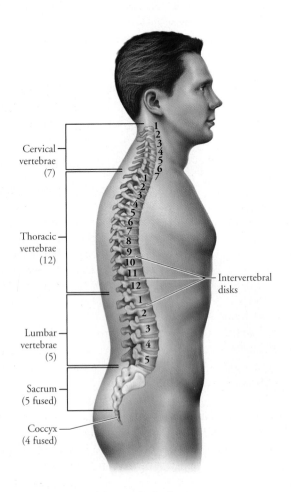

Cervical vertebrae (7)

Thoracic vertebrae (12)

Lumbar vertebrae (5)

Sacrum (5 fused)

Coccyx (4 fused)

Intervertebral disks

FIGURE **5.9** A side view of the vertebral column

movements of the leg. The coccyx, on the other hand, may have become fused for precisely the opposite reason. It serves no function and is regarded as a vestigial tail (*vestigial* means an evolutionary relic) whose bones may have joined as a side effect of growth without movement. (The upper part of the coccyx, however, is supplied with nerves and can be extremely painful when injured.)

Above the sacrum, the vertebrae are separated from one another by **intervertebral disks**, pads of fibrocartilage that help cushion the bones of the vertebral column. The smooth, lubricated surfaces of these disks help give the column flexibility. As the intervertebral disks become compressed over the years, the person may become shorter. This compression, alone or coupled with osteoporosis, can have a pronounced effect on an aging person's height.

Excessive pressure on the disks can cause various problems. For example, if it causes them to bulge inward, pressing against the spinal cord itself, they can interfere with muscle control and perception of incoming stimuli. On the other hand, a disk that presses outward against a spinal nerve branching from the spinal cord can be a source of great pain. The sciatic nerve, a large nerve that extends down the back of the leg, is one of the nerves most frequently affected in this way. Sciatica, the resulting inflammation, can be cripplingly painful.

Lower back pain is a particularly mysterious human ailment that accounts for more missed workdays than any medical problem besides colds. The source of the pain is notoriously difficult to pinpoint. A person with a slipped disk in the region may or may not have pain, whereas a person whose back looks perfectly normal may complain of terrible pain. Part of the diagnostic problem seems to be that the pain may originate from muscle rather than bone, and muscles do not show up on x-rays. Some cases may occur because of weak abdominal muscles that cannot counteract the pull of the powerful back muscles and therefore cause the vertebrae to misalign.

THE RIB CAGE

Twelve pairs of ribs attach at the back of the rib cage to the thoracic vertebrae (Figure 5.10). At the front, the upper 10 pairs of ribs are attached by cartilage either directly or indirectly to the sternum (breastbone). Their flexibility permits the ribs to take some blows without breaking and to move during breathing. The last two pairs of ribs do not attach to the sternum and are called floating ribs.

■ The appendicular skeleton makes locomotion possible

The appendicular skeleton consists of the pelvic and pectoral girdles and the attached limbs. A girdle is a skeletal structure that supports the arms, in the case of the **pectoral girdle**, or the legs, in the case of the **pelvic girdle**. The pectoral girdle connects the arms to the rib cage, and the pelvic girdle connects the legs to the vertebral column, enabling the body to move from one place to another.

THE PECTORAL GIRDLE

The pectoral girdle is composed of the *scapulae* (shoulder blades) and the *clavicles* (collarbones) (Figure 5.11). The clavicles are more

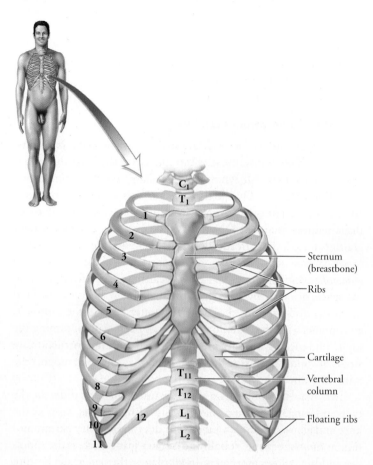

FIGURE **5.10** The bones of the rib cage

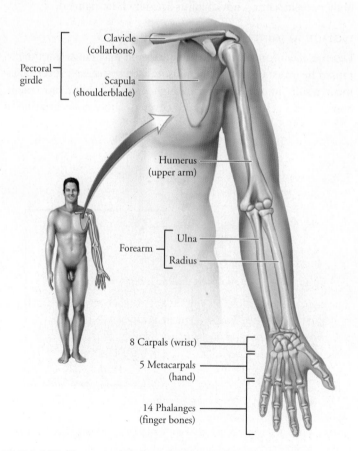

FIGURE **5.11** The pectoral girdle and arm

curved in males than in females (one way to tell the sex of a skeleton), and each forms a relatively rigid bridge between a scapula and the sternum. One corner of the roughly triangular scapula has a socket into which fits one end of the *humerus*, the upper arm bone. The other end of the humerus joins at the elbow the *radius* and *ulna*—the bones of the lower arm, or forearm. The elbow is formed by the extension of the ulna past the junction with the humerus. The radius and ulna meet the eight *carpals* at the wrist, and these join with the five *metacarpals*, which are the bones of the hand. The hand gives rise to three *phalanges* in each finger and two in each thumb.

The carpal tunnel is a narrow opening through the carpal bones that form the wrist, and through it passes a nerve that controls sensations in the fingers and in some of the muscles in the hand. *Tendons*, bands of connective tissue that attach muscles to bones, also pass through the carpal tunnel. Repetitive motion in the hand or wrist can cause these tendons to become inflamed and press against the nerve, resulting in numbness or tingling in the hand and pain that may affect the wrist, hand, and fingers. This condition, known as carpal tunnel syndrome, has been experienced by barbers, cab drivers, and pianists, but it is becoming increasingly common as people spend more time operating computer keyboards or playing video games. The computer industry has responded by redesigning their keyboards and games. Employers have responded, too, by providing adjustable work stations and rest and exercise periods.

what would you do?

There has been a rash of lawsuits against computer manufacturers and video game designers by people who have developed carpal tunnel syndrome as a result of endless hours spent using keyboards and video game controllers. If you developed carpal tunnel syndrome from using a keyboard or video game, would you sue the manufacturer? What criteria would you use to decide?

THE PELVIC GIRDLE

The pelvic girdle is much more rigid than the pectoral girdle, as you know if you have ever tried to shrug your hips. The two *pelvic bones* that make up the pelvic girdle (Figure 5.12) attach in back to the sacrum and then curve down and around to the front, where they join to a cartilage disk at the *pubic symphysis*. The male and female hips are easily distinguishable because the opening in the female is wider to facilitate childbirth.

The *femur*, or thighbone, rotates within a socket in the pelvis. At the knee, the femur joins the *tibia* (the shinbone). The *fibula* is a smaller bone running down the side of the tibia. The junction where the tibia joins the femur is covered by a *patella* (kneecap). At the ankle, the lower leg bones meet the *tarsals*, or ankle bones. These are connected to the *metatarsals*, or foot bones, which in turn are connected to the phalanges, the toe bones.

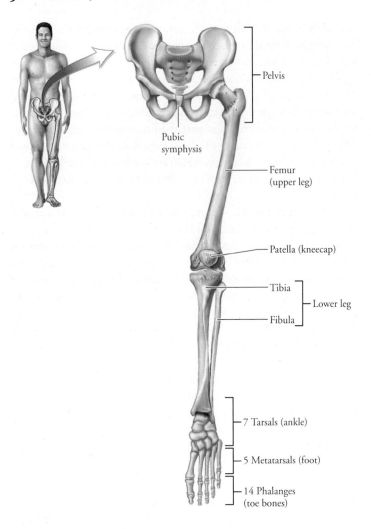

FIGURE **5.12** The pelvic girdle and leg

Joints Are Junctures between Bones

Joints, the places where bones meet, can be classified as fibrous, cartilaginous, or synovial, depending on their components and structure. Some joints allow no movement; others permit slight movement; and still others are freely movable.

Fibrous joints are held together by fibrous connective tissue. They have no joint cavity, and most do not permit movement. For example, the joints between the skull bones of an adult, called **sutures**, are virtually immovable because the bones are interlocked and held together tightly by fibrous connective tissue (Figure 5.13). In fact, the sutures can actually move ever so slightly. This ability is fortunate, because the movements serve as a shock absorber when you hit your head.

Cartilage, which is rather rigid, holds bones together in *cartilaginous joints*. Some cartilaginous joints are immovable, and others allow slight movement. We find such joints between vertebrae; in the attachment of ribs to the sternum; and in the pubic symphysis, the joint between the two pelvic bones. In a pregnant

woman, hormones loosen the cartilage of the pubic symphysis, allowing the pelvis to widen during childbirth.

■ Synovial joints permit flexibility

Most of the joints of the body are freely movable **synovial joints**. Because of these joints, muscles can maneuver the body into thousands of positions. All synovial joints share certain common features (Figure 5.14). One is that the surfaces that move past one another in the joints have a thin layer of hyaline cartilage. The cartilage reduces friction, so the bones slide over one another without grating and grinding. In addition, synovial joints are surrounded by a two-layered joint capsule. The inner layer of the capsule, the *synovial membrane*, secretes viscous, clear fluid (synovial fluid) into the space, or *joint cavity*, between the two bones. The synovial fluid lubricates and cushions the joint. The outer layer of the capsule is continuous with the covering membranes of the bones forming the joint. The entire synovial joint is reinforced with **ligaments**, strong straps of connective tissue that hold the bones together, support the joint, and direct the movement of the bones.

All synovial joints share these features but may differ in the type and range of motion they permit. **Hinge joints**, such as the knee and elbow, resemble a hinge on a door in that they permit motion in only one plane. A **ball-and-socket joint**, such as the shoulder and hip, allows movement in all planes: the ball at the head of one bone fits into a socket on another bone. Notice that you can swing your arm around in a complete circle. Some of the ways that body parts move at synovial joints are shown in Figure 5.15.

DAMAGE TO JOINTS

Damage to a ligament is called a **sprain** and may range from slight, caused by overstretching, to serious, caused by tearing. A torn ligament results in swelling and enough pain to inhibit movement. Sometimes, because ligaments are covered with a concentration of

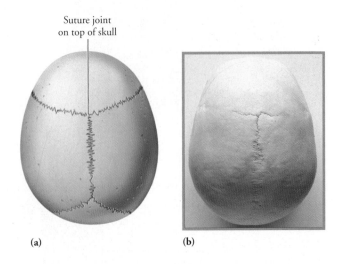

FIGURE **5.13** The bones of the skull are joined by suture joints. (a) Top view of the skull. (b) Photograph of suture joints in an adult skull.

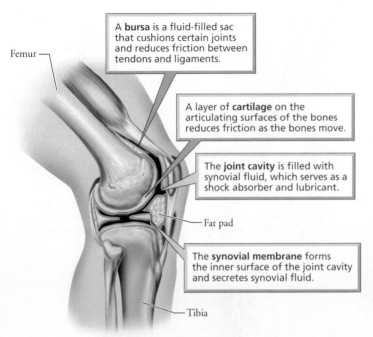

A **bursa** is a fluid-filled sac that cushions certain joints and reduces friction between tendons and ligaments.

A layer of **cartilage** on the articulating surfaces of the bones reduces friction as the bones move.

The **joint cavity** is filled with synovial fluid, which serves as a shock absorber and lubricant.

Femur

Fat pad

The **synovial membrane** forms the inner surface of the joint cavity and secretes synovial fluid.

Tibia

(a) Synovial joints, such as the knee shown here, permit a great range of movement.

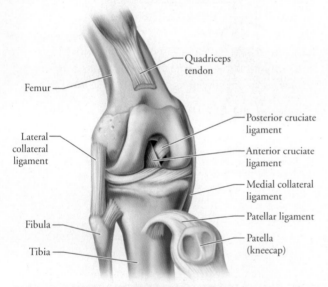

Quadriceps tendon

Femur

Lateral collateral ligament

Posterior cruciate ligament

Anterior cruciate ligament

Medial collateral ligament

Patellar ligament

Fibula

Tibia

Patella (kneecap)

(b) Ligaments hold bones together, support the joint, and direct the movement of the bones.

FIGURE **5.14** The knee is a synovial joint.

pain receptors that are very sensitive to stretching and swelling, the injury is not as severe as the pain would suggest. As with most musculoskeletal swelling, the initial treatment is likely to be an ice pack to reduce the swelling. Like tendons, ligaments have few blood vessels and heal slowly.

A common knee injury among athletes, especially gymnasts and football or soccer players, is a tear in the anterior cruciate ligament (ACL) (Figure 5.16). Why is this ligament so vulnerable?

When the knee is bent, the ACL acts as a restraining wire that restricts front-to-back twisting movement between the thighbone (femur) and the shinbone (tibia). An external blow to a bent knee—as may occur during a tackle or a hard landing—can stretch the ACL. If the force applied to the two bones is greater than the strength of the ligament, the ACL can tear.

In synovial joints and certain other locations where movement might cause friction between moving parts, the body has its own "ball bearings" in the form of fluid-filled sacs called **bursae** (singular, *bursa*; meaning pouch or purse). Bursae are lined with synovial membranes. They are found in places where skin rubs over bone as the joint moves and between tendons and bones, muscles and bones, and ligaments and bones.

Joint injury or repeated pressure can cause bursae to become inflamed and swell with excess fluid, a condition called bursitis (*-itis* means an inflammation). Bursitis is characterized by intense pain that becomes worse when the joint is moved and that cannot be relieved by resting in any position. Nonetheless, bursitis is not serious and usually subsides on its own within a week or two. In severe cases, a physician may drain some of the excess fluid to remove the pressure.

ARTHRITIS

Arthritis is a general term referring to joint inflammation. Some of the more than 100 kinds of arthritis are far more serious than others. Osteoarthritis is a degeneration of the surfaces of a joint, caused by wear and tear. Eventually the slippery cartilage at the ends of the affected bones begins to disintegrate until the bones themselves come into contact and grind against each other, causing intense pain and stiffness. Any joint surface that undergoes friction is bound to wear down, but osteoarthritis is most likely to occur in weight-bearing joints, such as the hip, knee, and spine. It is also occasionally seen in the finger joints or wrist.

Prevention is always better than treatment. One tip for preventing osteoarthritis is to control your weight to avoid overburdening your knees and hips. Another tip is to exercise. Lifting weights will strengthen the muscles that help support your joints. Stretching will give you a greater range of movement.

A far more threatening form of arthritis (and, unfortunately, one of the most common) is rheumatoid arthritis (discussed further in Chapter 13). Rheumatoid arthritis is marked by inflammation of the synovial membrane. The resulting accumulation of synovial fluid in the joint causes swelling, pain, and stiffness. Eventually, the constant irritation can destroy the cartilage, which may then be replaced by fibrous connective tissue that further impedes joint movement.

Rheumatoid arthritis differs from other types of arthritis in that it is apparently an autoimmune disease. That is, the body mistakenly rejects its own synovial membranes just as it would some invasive foreign matter. Rheumatoid arthritis can vary in severity over time, but it is a permanent condition. It normally affects the joints of the fingers, wrist, knees, neck, ankles, and hips. Sometimes the only effective treatment is to replace the damaged joint with an artificial one (Figure 5.17).

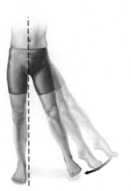

Flexion
Motion that *decreases* the angle between the bones of the joint, bringing the bones closer together

Extension
Motion that *increases* the angle between the bones of the joint

Adduction
Movement of a body part *toward* the body midline

Abduction
Movement of a body part *away from* the body midline

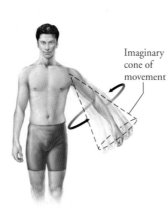
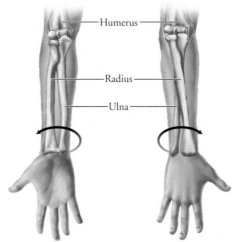

Humerus

Radius

Ulna

Imaginary cone of movement

Rotation
Movement of a body part around it own axis

Circumduction
Movement of a body part in a wide circle so that the motion describes a cone

Supination
Rotation of the forearm so that the palm faces up

Pronation
Rotation of the forearm so that the palm faces down

FIGURE **5.15** Types of movement at synovial joints

FIGURE **5.16** A common sports injury is a tear in the anterior cruciate ligament of the knee.

FIGURE **5.17** Replacement prosthetic of the hip

HIGHLIGHTING THE CONCEPTS

Bones Function in Support, Movement, Protection, Storage, and Blood Cell Production (p. 83)

1. The skeleton is a framework of bone and cartilage that supports and protects the internal organs and has a role in movement. It also serves as a storage site for minerals (calcium and phosphorus) and fat. Blood cells are produced in the red marrow of certain bones.

Bones Have a Hard Outer Layer of Compact Bone Surrounding Spongy Bone (pp. 83–84)

2. Compact bone is dense bone tissue found on the outside of all bones. It is covered by a membrane, the periosteum, that nourishes the bone cells. Spongy bone is a latticework of bony struts found in flat bones and near the knobby ends, or heads, of long bones.

3. The shaft of a long bone has a cavity filled with fatty yellow marrow. The spongy bone of certain bones in adults is filled with red marrow.

Bone Is Living Tissue (pp. 84–85)

4. The structural unit of bone, called an osteon, consists of a central canal surrounded by concentric circles made of bone cells (osteocytes) in a hard matrix. Cellular extensions of the osteocytes touch one another through tiny canals in the matrix, allowing exchange of materials between cells and the central canal. Bone matrix is hardened by calcium salts and strengthened by strands of collagen.

5. In an embryo, most of the skeleton forms first as cartilage that is gradually replaced by bone. In the replacement process, osteoblasts form a bony collar around the shaft of the bone. Next, cartilage within the shaft begins to break down and is replaced by bone, after which the cartilage in the heads (epiphyses) is replaced. Epiphyseal growth plates of cartilage remain and allow bones to grow in length until the person is fully grown.

6. Bone growth is stimulated by growth hormone from the anterior pituitary gland. Sex hormones (estrogen and testosterone) initially stimulate growth, but later they cause the growth plates to disappear, and growth ceases.
WEB TUTORIAL 5.1 Bone Growth

Bone Fractures Are Healed by Fibroblasts and Osteoblasts (pp. 85–86)

7. In bone-fracture repair, blood clot formation is followed by a cartilaginous callus that links the broken ends and is gradually replaced by bone.
WEB TUTORIAL 5.2 Bone Repair

Bones Are Continuously Remodeled (p. 86)

8. Bone is constantly being remodeled. Osteoblasts deposit new bone in response to stress, and osteoclasts break bone down when it is not stressed.

We Divide the Human Skeleton into Two Parts (pp. 86–91)

9. The skeleton is described as consisting of the axial skeleton, which protects and supports internal organs, and the appendicular skeleton, which allows you to move from place to place.

Joints Are Junctures between Bones (pp. 92–94)

10. Joints, the places where bones meet, are classified by composition and degree of movement. Some joints, such as the sutures between the skull bones, allow no movement.

11. Synovial joints are freely movable. They have cartilage on the adjoining bone surfaces, are surrounded by a synovial cavity filled with synovial fluid, and are held together by ligaments.

12. Arthritis is inflammation of a joint. Osteoarthritis occurs when the surface of a joint degenerates because of use. Rheumatoid arthritis is an autoimmune disease.

KEY TERMS

skeleton *p. 83*
compact bone *p. 83*
periosteum *p. 83*
spongy bone *p. 84*
red marrow *p. 84*
yellow marrow *p. 84*
osteon *p. 84*
osteocyte *p. 84*

osteoblast *p. 84*
growth plate *p. 85*
osteoclast *p. 86*
osteoporosis *p. 86*
axial skeleton *p. 86*
appendicular skeleton *p. 86*
cranium *p. 86*
vertebral column *p. 90*

vertebra *p. 90*
intervertebral disk *p. 90*
pectoral girdle *p. 90*
pelvic girdle *p. 90*
joint *p. 92*
sutures *p. 92*
synovial joint *p. 92*
ligament *p. 92*

hinge joint *p. 92*
ball-and-socket joint *p. 92*
sprain *p. 92*
bursae *p. 93*
arthritis *p. 93*

REVIEWING THE CONCEPTS

1. List six functions of the skeleton. *p. 83*
2. Compare compact and spongy bone. *pp. 83–84*
3. Describe the structure of a long bone. Where are the yellow and red marrow found? *pp. 83–84*
4. Diagram an osteon, including the osteocytes and central canal. *p. 84*
5. Describe the formation of bone in a fetus. Explain how bone growth continues after birth. *pp. 84–85*
6. Explain how a bone heals after it has been fractured. *pp. 85–86*
7. What is bone remodeling? Explain the role of osteoblasts and osteoclasts. How does stress affect remodeling? *p. 86*
8. Describe the axial and appendicular parts of the skeleton. *p. 86*
9. Describe a synovial joint. *p. 92*
10. The functional units of compact bone are called
    a. myofilaments.
    b. striations.
    c. osteons.
    d. osteocytes.
11. Which of the following persons would be expected to have the *densest* bones?
    a. an elderly woman whose favorite pastime is knitting
    b. a 16-year-old female who swims regularly
    c. a 33-year-old male African American who loves to drink milk and plays basketball regularly
    d. a 50-year-old-male workaholic with a desk job
12. The skull, vertebral column, sternum, and rib cage constitute the _____ skeleton.
13. Bone is hardened by the mineral _____.
14. The cells that form bone are called _____.
15. A _____ is a tear in a ligament that may cause pain after the injury.

APPLYING THE CONCEPTS

1. Edith is a 64-year-old woman who has been confined to a wheelchair since she was 35 because of a car accident that left her legs paralyzed. Her leg bones are very thin and weak. Why?

2. Makiri is a 19-year-old woman who is pregnant with her first child. Her diet consists primarily of junk food. During the fifth month of her pregnancy, she trips and breaks her arm in the fall. Tests show that her bones have become weak owing to a loss of calcium. What has caused Makiri's bone weakness?

3. A forensic scientist is examining skeletal remains that includes a femur (thighbone) and a tibia (lower leg bone). By measuring the length of the bones, she determines that these leg bones could have belonged to a tall child or a short young adult. What should the scientist look for to determine whether the bones were from a child or young adult?

Additional questions can be found on the companion website.

The Muscular System Moves Our Body Parts and Maintains Our Posture

Most Skeletal Muscles Work in Pairs

Sarcomeres Are the Contractile Units of Muscle

- Skeletal muscle contracts when actin filaments slide across myosin filaments
- Calcium ions and regulatory proteins control contraction
- Nerves stimulate muscle contraction
- Muscular dystrophy causes progressive muscle deterioration

The Strength of Muscle Contraction Depends on the Number of Motor Units Stimulated

The Strength of Contraction Increases If a Muscle Is Stimulated before It Has Relaxed

ATP for Muscle Contraction Comes from Many Sources

Slow-Twitch and Fast-Twitch Muscle Cells Differ in Contraction Speed and Duration

Aerobic Exercise Increases Endurance; Resistance Exercise Builds Muscle

SOCIAL ISSUE Building Muscle Fair and Square? Anabolic Steroid Abuse

6

The Muscular System

97

Weightlifting and other resistance exercises build muscle, but scientists know very little about how muscle building occurs. They do know a great deal about the structure of muscle tissues and how they contract to move our body parts, so that we can make our impact on the world.

Hiking on the mountain trail, Michael felt exhilarated by the fresh breeze, the scent of pine needles, and the warm sun. He had been walking for about an hour, his mind finally freed of worries. Between the demands of his job, his courses at the community college, and the stress of a recent breakup with Heather, he hadn't had a decent night's sleep for weeks. This hike, he thought to himself, was the first real exercise he had had since the Heather debacle. The only problem at the moment was that he was getting uncomfortably thirsty because he had forgotten his water bottle. Deciding to head home, he changed direction and suddenly felt a sharp pain in his left calf. It felt as if the whole muscle had locked up in a severe cramp that prevented him from taking another step. He knew that he now had to stretch his leg gradually to work the cramp out. What he didn't realize was that the cramp was caused by the very problems that had just been passing through his mind: his fatigue (from insomnia), lack of regular exercise, and dehydration. Fortunately, gentle stretching relieved the pain in a matter of minutes, and after a brief rest Michael was able to walk home.

In this chapter, we will explore muscle structure and the mechanism of contraction and learn why muscle contraction rarely results in a cramp. Then, we will see how nerves control contractions and study the energy sources that fuel muscle activity. ■

The Muscular System Moves Our Body Parts and Maintains Our Posture

There are three kinds of muscles—skeletal, cardiac, and smooth. Each has distinct qualities and functions, but they all have four traits in common.

1. **Muscles are excitable.** They respond to stimuli.
2. **Muscles are contractile.** They have the ability to shorten.
3. **Muscles are extensible.** They have the ability to stretch.
4. **Muscles are elastic.** They can return to their original length after being shortened or stretched.

This chapter will focus on skeletal muscle, the kind of muscle we usually think of when muscles are mentioned. Smooth and cardiac muscles are discussed in Chapters 4 and 12, respectively. Skeletal muscles allow you to smile with pleasure and scowl in anger. They allow you to move, some of us more gracefully than others. And they allow you to remain erect, maintaining your posture despite the pull of gravity. Unlike cardiac and smooth muscle, skeletal muscle is under voluntary control. We can contract it when we want to.

Most Skeletal Muscles Work in Pairs

Most of the major muscles we use for locomotion, manipulation, and other voluntary movements are attached to bones. Each end of the muscle is attached to a bone by a **tendon**, which is a band of connective tissue. A muscle is often attached to two bones on opposite sides of a joint. The muscle's **origin** is the end attached to the bone that remains relatively stationary during a movement. The muscle's **insertion** is the end attached to the bone that moves. Thus, the bones act as levers in working with skeletal muscles to produce movement.

Most muscles work in pairs or groups. Muscles that must contract at the same time to cause a certain movement are called *synergistic muscles*. Most muscles, however, are arranged in **antagonistic pairs**, which produce a movement when one of the pair contracts and the other relaxes (Figure 6.1). When a muscle in an antagonistic pair contracts, it shortens and pulls on one of the bones that meet at a given joint, causing movement of the bone in one direction. Any time such a muscle contracts and pulls on a bone, however, its partner, which has an opposing action, must relax. For the bone to then move back to its former position, the first muscle must relax and the other muscle in the pair must contract, thus pulling the bone in the opposite direction. The biceps muscle, located on the top of the upper arm (the muscle people like to show off), cooperates in this way with the triceps muscle, on the back of the arm (the one we feel when we do push-ups). Whereas the contraction of the biceps causes the arm to flex and bend at the elbow, the contraction of the triceps causes the arm to extend and straighten at the elbow. The body has more than 600 skeletal muscles. The major ones—those most prominent in giving shape to the body—are shown in Figure 6.2.

In the preceding description, we said that a contracting muscle pulls a bone. In actuality, the contraction of a muscle pulls on the tendon that connects that muscle to a bone. Excessive stress on a tendon can cause it to become inflamed, a condition called *tendinitis*. Most tendinitis is probably due to overuse, misuse (as when lifting improperly), or age. Unfortunately, tendons heal slowly, because they are poorly supplied with blood vessels. The most effective treatment is rest: if it hurts, do not use it.

The terms *muscle pull, muscle strain,* and *muscle tear* are all used almost interchangeably to describe damage to a muscle or its tendons caused by overstretching. This common sports injury may also include damage to small blood vessels in the muscle, causing bleeding and pain. Treatment includes applying ice to the injured area to reduce swelling and to keep the muscle stretched.

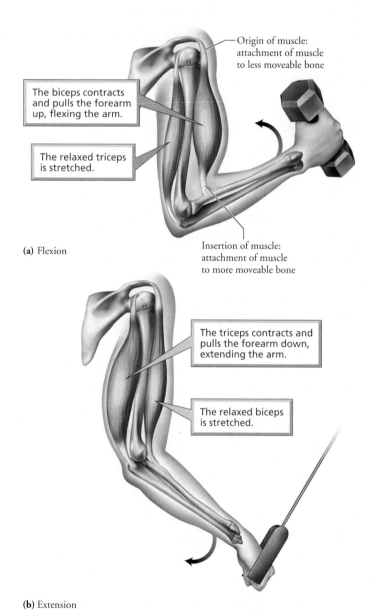

Origin of muscle: attachment of muscle to less moveable bone

The biceps contracts and pulls the forearm up, flexing the arm.

The relaxed triceps is stretched.

(a) Flexion

Insertion of muscle: attachment of muscle to more moveable bone

The triceps contracts and pulls the forearm down, extending the arm.

The relaxed biceps is stretched.

(b) Extension

FIGURE **6.1** The antagonistic action of the triceps and biceps muscles during flexion and extension. The origins and insertions of the muscles are shown.

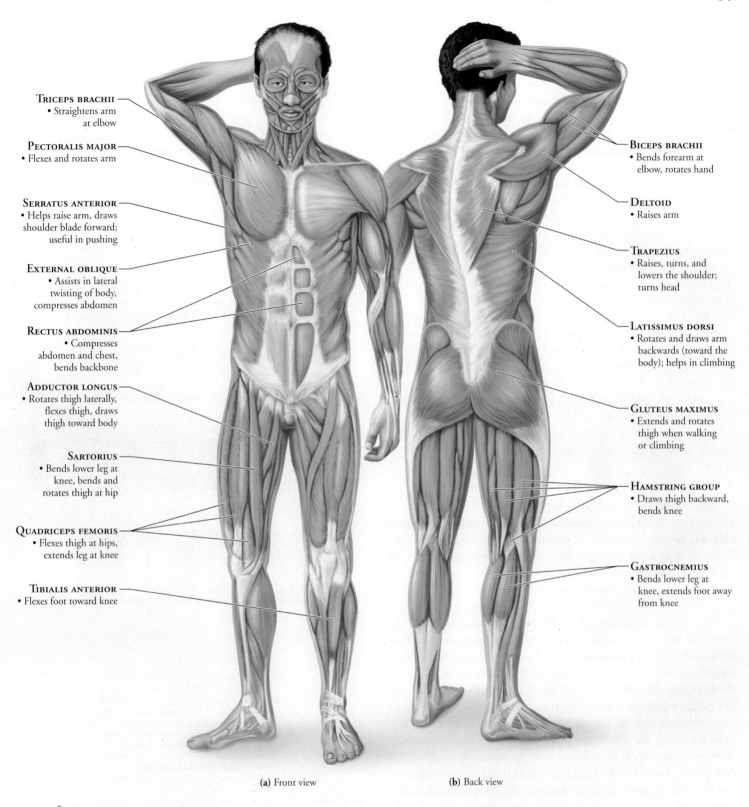

TRICEPS BRACHII
• Straightens arm at elbow

PECTORALIS MAJOR
• Flexes and rotates arm

SERRATUS ANTERIOR
• Helps raise arm, draws shoulder blade forward; useful in pushing

EXTERNAL OBLIQUE
• Assists in lateral twisting of body, compresses abdomen

RECTUS ABDOMINIS
• Compresses abdomen and chest, bends backbone

ADDUCTOR LONGUS
• Rotates thigh laterally, flexes thigh, draws thigh toward body

SARTORIUS
• Bends lower leg at knee, bends and rotates thigh at hip

QUADRICEPS FEMORIS
• Flexes thigh at hips, extends leg at knee

TIBIALIS ANTERIOR
• Flexes foot toward knee

BICEPS BRACHII
• Bends forearm at elbow, rotates hand

DELTOID
• Raises arm

TRAPEZIUS
• Raises, turns, and lowers the shoulder; turns head

LATISSIMUS DORSI
• Rotates and draws arm backwards (toward the body); helps in climbing

GLUTEUS MAXIMUS
• Extends and rotates thigh when walking or climbing

HAMSTRING GROUP
• Draws thigh backward, bends knee

GASTROCNEMIUS
• Bends lower leg at knee, extends foot away from knee

(a) Front view   (b) Back view

FIGURE **6.2** Some major muscles of the body

Sarcomeres Are the Contractile Units of Muscle

Skeletal muscle cells can be several centimeters long, which is enormously long compared to most of the body's other cells. Muscle cells in the thigh can be 30 cm (1 ft) in length. The fine structure of these cells provides numerous clues to the mechanism responsible for muscle contraction.

Skeletal muscle is also called **striated** (striped) **muscle**, because under the microscope, the cells are seen to have pronounced bands that look like stripes (Figure 6.3). The striations are caused by the orderly arrangement of many elongated **myofibrils**, which are specialized bundles of proteins, within the muscle cell. Each myofibril contains two types of long protein filaments called **myofilaments**: the thicker **myosin filaments** and a greater number of the thinner **actin filaments**. The bundles of myofilaments make up about 80% of the cell volume.

Along its length, each myofibril can be described as having contractile units called **sarcomeres**. The ends of each sarcomere are marked by dark bands of protein, called Z lines. Thousands of sarcomeres are arranged end to end along a muscle cell. Within each sarcomere, the actin and myosin filaments are arranged in a specific manner. One end of each actin filament is attached to a Z line. Myosin filaments lie in the middle of the sarcomere, their ends partially overlapping with surrounding actin filaments. The degree of overlap increases when the muscle contracts.

■ **Skeletal muscle contracts when actin filaments slide across myosin filaments**

The leading explanation of muscle contraction is called the **sliding filament model.** It says that a muscle contracts when the actin filaments slide along the myosin filaments, increasing the degree of overlap between actin and myosin and thus shortening the sarcomere. When many sarcomeres shorten, the muscle as a whole contracts (Figure 6.4).

To explain the sliding of the actin and myosin filaments, we must look at their molecular structures. A thin, actin myofilament is made up of two chains of spherical actin molecules resembling two strings of beads twisted around each other to form a helix. A thick, myosin filament is composed of myosin molecules shaped like golf clubs but with two heads side by side on the same shaft. In a myosin filament, the "shafts" of several hundred myosin molecules lie along the filament's length, and the "heads" protrude along each end of the bundle in a spiral pattern. The club-shaped ends of the myosin molecules, the so-called *myosin heads*, are the key to the movement of actin filaments and, therefore, to muscle contraction.

Muscle contraction results from the following cycle of interactions between myosin and actin (Figure 6.5):

- **Resting sarcomere.** At the start of each cycle, the myosin heads have already split a molecule of ATP to ADP and inorganic phosphate (P_i). The energy released from splitting the ATP causes the myosin heads to swivel in a way that extends them toward the Z bands at the ends of the sarcomere. This step is analogous to cocking a pistol.

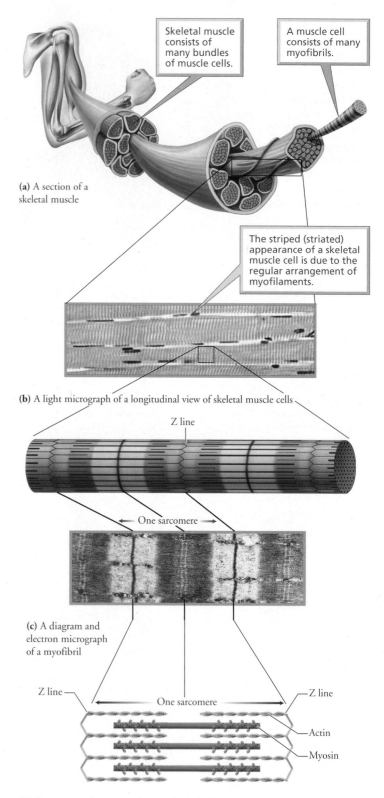

Skeletal muscle consists of many bundles of muscle cells.

A muscle cell consists of many myofibrils.

(a) A section of a skeletal muscle

The striped (striated) appearance of a skeletal muscle cell is due to the regular arrangement of myofilaments.

(b) A light micrograph of a longitudinal view of skeletal muscle cells

Z line

← One sarcomere →

(c) A diagram and electron micrograph of a myofibril

Z line

One sarcomere

Z line

Actin

Myosin

(d) A sarcomere, the contractile unit of a skeletal muscle, contains actin and myosin myofilaments.

WEB TUTORIAL 6.1

FIGURE **6.3** The structure of a skeletal muscle

- **STEP 1: Cross-bridge attachment.** The myosin heads attach themselves to the nearest actin filament. When the myosin head is bound to an actin molecule, it acts as a bridge between the thick and thin filaments. For this reason, myosin heads are also called **cross-bridges.**

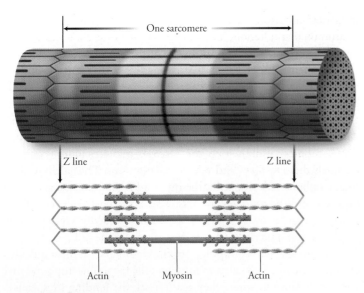

One sarcomere

Z line   Z line

Actin   Myosin   Actin

(a) Sarcomere relaxed

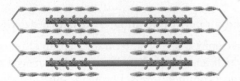

(b) Sarcomere contracted

FIGURE **6.4** Each myofibril is packed with actin filaments and myosin filaments. When a muscle contracts, actin filaments slide past myosin filaments. Movements of the heads of myosin filaments pull actin filaments toward the center of a sarcomere.

- **STEP 2: Bending of myosin head (the power stroke).** Binding to actin causes the myosin heads to swing forcefully back to their bent positions. The actin filaments, still bound to the myosin heads, are therefore pulled toward the midline of the sarcomere. This so-called power stroke is analogous to pulling the trigger on a pistol. The ADP and inorganic phosphate then pop off of the myosin head.
- **STEP 3: Cross-bridge detachment.** New ATP molecules now bind to the myosin heads, causing the myosin heads to disengage from the actin.
- **STEP 4: Myosin reactivation.** The myosin heads split the ATP and store the energy, causing the contraction cycle to begin again.

This cycle of events is repeated hundreds of times in a second.

stop and think

It takes an ATP molecule to break the cross-bridges so that new ones may be formed. Without ATP, cross-bridges cannot be broken, and the muscle becomes stiff. When a person dies, ATP is no longer produced. How does this explain the stiffening of muscles, known as rigor mortis, that begins 3 to 4 hours after death?

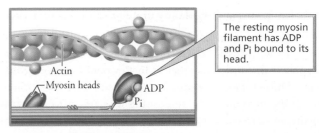

Actin

Myosin heads   ADP
                       P_i

The resting myosin filament has ADP and P_i bound to its head.

Resting sarcomere

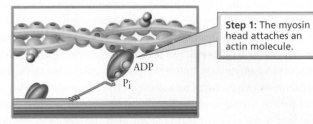

ADP
P_i

Step 1: The myosin head attaches an actin molecule.

Cross-bridge attachment

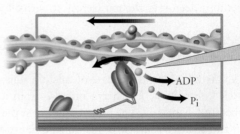

ADP

P_i

Step 2: The myosin head bends, causing the actin filament to slide across the myosin filament. This is the power stroke.

Bending of myosin head (the power stroke)

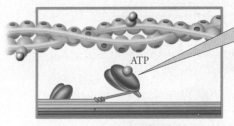

ATP

Step 3: Free ATP binds to myosin, causing it to release actin.

Cross-bridge detachment

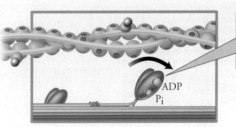

ADP
P_i

Step 4: ATP is split to ADP and P_i and myosin returns to its resting position.

Myosin reactivation

FIGURE **6.5** The sliding filament model of muscle contraction. (The regulatory proteins on the actin are described in the following subsection.)

■ Calcium ions and regulatory proteins control contraction

Muscle contractions are controlled by the availability of calcium ions. How? Muscle cells contain the proteins **troponin** and **tropomyosin,** which together form a troponin–tropomyosin complex. During muscle *relaxation*, the actin–myosin-binding sites where the myosin heads would otherwise attach to the actin filaments are covered over by the troponin–tropomyosin complex. Contraction occurs when calcium ions enter the sarcomere and bind to troponin, causing it to change shape. This change causes tropomyosin to shift position, which pulls the tropomyosin molecule away from the myosin-binding sites on the actin. The exposed binding sites enable the formation of cross-bridges (Figure 6.6).

Where do the calcium ions come from? Calcium ions are stored in the **sarcoplasmic reticulum**, an elaborate form of smooth endoplasmic reticulum found in muscle cells. The sarcoplasmic reticulum can be pictured as a sleeve of thick lace surrounding each myofibril in a muscle cell. (Recall that a myofibril is a bundle of actin and myosin.) Also scattered through the cell are a number of **transverse tubules** (T tubules), which are tiny, cylindrical inpocketings of the muscle cell's plasma membrane. The T tubules carry signals from motor neurons deep into the muscle cell to virtually every sarcomere.

■ Nerves stimulate muscle contraction

The stimulus that ultimately leads to the release of calcium ions, and therefore to muscle contraction, is a nerve impulse from a motor neuron (described in Chapter 7). The junction between the tip of a motor neuron and a skeletal muscle cell is called a **neuro-muscular junction** (Figure 6.7). When a nerve impulse reaches a neuromuscular junction, it causes the release of the chemical acetylcholine from small packets in the tip of the motor neuron. Acetylcholine then diffuses across a small gap onto the surface of the muscle cell, where it binds to special receptors on the muscle cell membrane, causing changes in the membrane's permeability and thus creating an electrochemical message similar to a nerve impulse. The message travels along the muscle cell's plasma membrane, into the T tubules, and then to the sarcoplasmic reticulum, causing channels there to open and release calcium ions. The calcium ions then combine with troponin, the myosin-binding sites on actin are exposed, and the muscle contracts. This is the chain of events that must happen in order for you to absent-mindedly scratch your head, and it happens a lot faster than it takes to describe.

When the nerve impulse stops, membrane pumps quickly clear the sarcomere of calcium ions, and the troponin–tropomyosin complexes move to where they again block the binding sites, causing the muscle to relax. The contraction of other muscles stretches the sarcomere back to its original length.

■ Muscular dystrophy causes progressive muscle deterioration

Muscular dystrophy is a general term for a group of inherited conditions in which muscles become increasingly damaged and weak. One of the most common forms, Duchenne muscular dystrophy, is caused by a defective gene for production of the protein dystrophin. The lack of dystrophin allows calcium ions to enter the muscle cell and rise to such high levels that other important proteins are destroyed. Muscle cells die as a result and are replaced by fat and connective tissue, so the skeletal muscles become progressively weaker. (The mode of inheritance and progression of Duchenne muscular dystrophy are described in Chapter 20.)

The Strength of Muscle Contraction Depends on the Number of Motor Units Stimulated

Skeletal muscles are stimulated to contract by motor neurons (nerve cells). Each motor neuron that brings an impulse from the brain to a muscle makes contact with a number of different cells in that muscle. The motor neuron and all the muscle cells it stimulates is called a **motor unit** (Figure 6.8). All the muscle cells in a given motor unit contract together. On average, there are 150 muscle cells in a motor unit, but this number is quite variable. Muscles responsible for precise, finely controlled movements, such as those of the fingers or eyes, have small numbers of muscle cells in each motor unit. In contrast, muscles for less precise movements, such as those of the hip or calf, have many muscle cells in a motor unit. A motor unit in tiny eye muscles may have only three muscle cells, whereas a motor unit in a calf muscle may have thousands.

The nervous system increases the strength of a muscle contraction by increasing the number of motor units being stimulated, a process called *recruitment*. The muscle cells of a given motor unit are generally spread throughout the muscle. Thus, if a single motor unit is stimulated, the entire muscle contracts, but only weakly. Although several of the same muscles are used to lift a table as to lift a

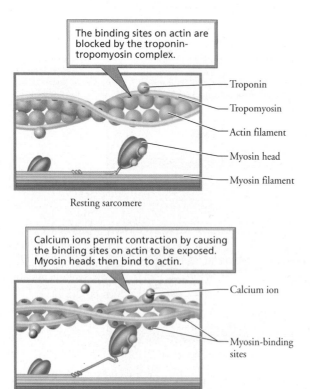

The binding sites on actin are blocked by the troponin-tropomyosin complex.

— Troponin
— Tropomyosin
— Actin filament
— Myosin head
— Myosin filament

Resting sarcomere

Calcium ions permit contraction by causing the binding sites on actin to be exposed. Myosin heads then bind to actin.

— Calcium ion
— Myosin-binding sites

Cross-bridge formation

FIGURE **6.6** Calcium ions initiate muscle contraction.

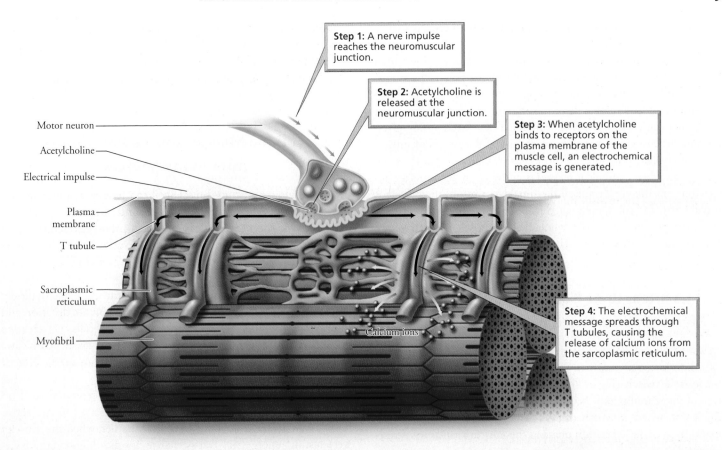

Step 1: A nerve impulse reaches the neuromuscular junction.

Step 2: Acetylcholine is released at the neuromuscular junction.

Step 3: When acetylcholine binds to receptors on the plasma membrane of the muscle cell, an electrochemical message is generated.

Motor neuron

Acetylcholine

Electrical impulse

Plasma membrane

T tubule

Sacroplasmic reticulum

Myofibril

Calcium ions

Step 4: The electrochemical message spreads through T tubules, causing the release of calcium ions from the sarcoplasmic reticulum.

FIGURE **6.7** The connection between a motor neuron and a muscle cell is called a neuromuscular junction.

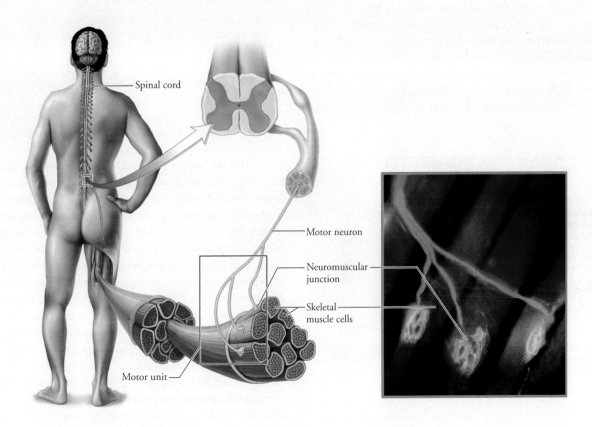

Spinal cord

Motor neuron

Neuromuscular junction

Skeletal muscle cells

Motor unit

FIGURE **6.8** A motor unit includes a motor neuron and the muscle cells it stimulates.

fork, the number of motor units summoned in those muscles is greater when lifting the table.

Our movements tend to be smooth and graceful rather than jerky because the nervous system carefully choreographs the stimulation of different motor units, timing it so that they are not all active simultaneously or for the same amount of time. Although an individual muscle cell is either contracted or relaxed at any given instant, the muscle as a whole can consist of thousands of muscle cells, and usually only some of those cells are contracted at the same time. Even when a muscle is relaxed, some of its motor units are active (but not the same units all the time). As a result, the muscle is usually firm and solid, even when it is not being used. This state of intermediate contraction of a whole skeletal muscle is called *muscle tone*. A muscle that lacks muscle tone is limp.

The Strength of Contraction Increases If a Muscle Is Stimulated before It Has Relaxed

If a whole skeletal muscle is artificially stimulated briefly in the laboratory, some of the muscle cells will contract, causing a **muscle twitch** (Figure 6.9a). The interval between the reception of the stimulus and the time when contraction begins is called the latent period. The contraction phase is quite short and is followed by a longer relaxation phase as the muscle returns to its resting state.

If a second stimulus is given before the muscle is fully relaxed, the second twitch will be stronger than the first. This phenomenon is described as **summation**, because the second contraction is added to the first (Figure 6.9b). Summation occurs because an increasing number of muscle cells are stimulated to contract.

When stimuli arrive even more frequently, the contraction becomes increasingly stronger as each new muscle twitch is added on. If the stimuli occur so frequently that there is no time for any relaxation before the next stimulus arrives, the muscle goes into a constant, powerful contraction called **tetanus** (Figure 6.9c). Tetanus

cannot continue indefinitely. Eventually, the muscle will become unable to produce enough ATP to fuel contraction, and lactic acid will accumulate (as discussed below). As a result, the muscle will stop contracting despite continued stimulation, a condition called *fatigue*.

ATP for Muscle Contraction Comes from Many Sources

A single contracting muscle cell can require as much as 600 trillion ATP molecules per second simply to form and break the cross-bridges producing the contraction. Even small muscles contain thousands of muscle cells. If ATP is the only source of energy for muscle contraction, where does all the ATP come from?

Our muscles have a number of sources of ATP and typically employ them in a certain sequence, depending on the duration and intensity of exercise: (1) ATP stores in muscle cells, (2) creatine phosphate, (3) anaerobic (without oxygen) metabolic pathways, and (4) aerobic (with oxygen) respiration (Figure 6.10). (Recall that cell respiration was discussed in Chapter 2.)

A resting muscle stores some ATP, but this reserve is used up quickly. During vigorous exercise, the ATP reserves in the active muscles are depleted in about 6 seconds. Earlier, when the muscles were resting, energy was transferred to another high-energy compound, called creatine phosphate, that is stored in muscle tissue. Creatine phosphate has a high-energy bond connecting the creatine and the phosphate parts of the molecule. A resting muscle contains about six times as much creatine phosphate as stored ATP and can release its stored energy when needed to convert ADP to ATP. This energy can add another 10 seconds to the exercise time.

$$\text{Creatine phosphate} \xrightarrow{\quad \text{ADP} \quad \text{ATP} \quad} \text{Creatine}$$

Activities such as diving, weight lifting, and sprinting, which require a short burst of intense activity, are powered entirely by ATP and creatine phosphate reserves.

Once the supply of creatine phosphate is diminished, ATP must be generated from either anaerobic metabolic pathways or aerobic respiratory pathways. The primary fuel for either pathway is glucose, and the glucose that fuels muscle contraction comes mainly from glycogen, which, as you learned in Chapter 2, is stored in muscle (and liver) cells and consists of a large chain of glucose molecules. When an active muscle cell runs short of ATP and creatine phosphate, enzymes begin converting glycogen to glucose. About 1.5% of a muscle cell's total weight is glycogen. Even so, long-term activity, as in an endurance sport, may deplete a muscle's glycogen reserves. The accompanying feeling of overwhelming fatigue is known to runners as "hitting the wall," to cyclists as "bonking," and to boxers as becoming "arm weary."

The circulatory system can supply enough oxygen for aerobic respiratory pathways to power *low* levels of activity, even if continued for a prolonged time. However, anaerobic pathways produce ATP two and one-half times faster than do aerobic pathways.

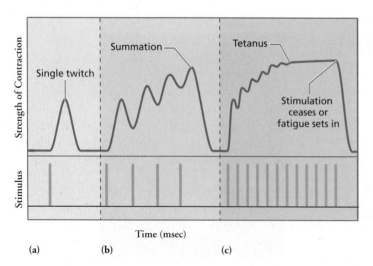

FIGURE **6.9** Muscle contraction shown graphically: (a) muscle twitch, (b) summation, (c) tetanus

6 seconds	10 seconds	30–40 seconds	End of exercise		After prolonged exercise
ATP stored in muscles	ATP formed from creatine phosphate and ADP	ATP generated from glycogen stored in muscles and broken down to form glucose			Oxygen debt paid back
		Oxygen limited • Glucose oxidized to lactic acid	Oxygen present • Heart beats faster to deliver oxygen more quickly • Myoglobin releases oxygen		Breathe heavily to deliver oxygen • Lactic acid used to produce ATP • Creatine phosphate restored • Oxygen restored to myoglobin • Glycogen reserves restored

FIGURE **6.10** Energy sources for muscle contraction

Therefore, during strenuous activity lasting 30 or 40 seconds, anaerobic pathways supply the ATP that fuels muscle contraction. Indeed, burstlike activities such as tennis or soccer rely nearly completely on anaerobic pathways of ATP production. Chapter 2 explained that in anaerobic respiratory pathways, the pyruvic acid produced in glycolysis is converted to lactic acid.

$$\text{Glycogen} \rightarrow \text{Glucose} \xrightarrow[\text{No oxygen required}]{\text{ADP} \backsim \text{ATP}} \text{Pyruvic acid} \rightarrow \text{Lactic acid}$$

During more prolonged muscular activity, the body gradually switches back to aerobic pathways for producing ATP. The necessary oxygen can come from either of two sources. One is the oxygen bound to hemoglobin in the blood supply. As activity continues, the heart rate increases, and blood is pumped more quickly; moreover, it is shunted to the neediest tissues. Another oxygen source is *myoglobin*, an oxygen-binding pigment within muscle cells. Aerobic pathways produce more than 90% of the ATP required for intense activity lasting more than 10 minutes. These pathways also produce nearly 100% of the ATP that powers a truly prolonged intense activity, such as running a marathon.

$$\text{Glycogen} \rightarrow \text{Glucose} \xrightarrow[\text{Oxygen required}]{\text{ADP} \backsim \text{ATP}} \text{Carbon dioxide} + \text{Water}$$

After prolonged exercise, a person continues to breathe heavily for several minutes. The extra oxygen relieves the **oxygen debt** that was created by the muscles' using more ATP than was provided by aerobic metabolism. Most of the oxygen is used to generate ATP to convert lactic acid back to glucose. In addition, the oxygen that was released by myoglobin is replaced, and glycogen and creatine phosphate reserves are restored.

stop and think

Some athletes take dietary creatine supplements to improve their performance. Creatine does seem to boost performance in sports that require short bursts of energy but not in those that require endurance. How might this difference be explained? Creatine does not increase muscle mass, yet it can enhance performance in sprint sports. How is this effect possible?

Slow-Twitch and Fast-Twitch Muscle Cells Differ in Contraction Speed and Duration

There are two general types of skeletal muscle cells: slow-twitch and fast-twitch (Figure 6.11). **Slow-twitch cells** contract slowly when stimulated, but with enormous endurance. These cells are dark and reddish because they are packed with the oxygen-binding pigment myoglobin and because they are richly supplied with blood vessels. Slow-twitch cells also contain abundant mitochondria, the organelles in which aerobic production of ATP occurs. Because they can produce ATP aerobically for a long time, slow-twitch cells are specialized to deliver prolonged, strong contractions.

In contrast, **fast-twitch cells** contract rapidly and powerfully, but with far less endurance. Fast-twitch cells have a form of an enzyme that can split ATP bound to myosin more quickly than the same enzyme in slow-twitch cells. Because they can make and break cross-bridges more quickly, they can contract more rapidly. In addition, compared with their slow-twitch cousins, fast-twitch cells have a wider diameter because they are packed with more actin and

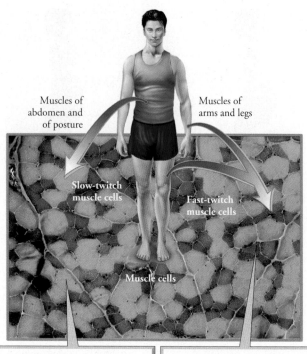

FIGURE **6.11** Slow- and fast-twitch muscle cells

Slow-twitch muscle cells:
• Designed for endurance
• Contract slowly
• Strong, sustained contractions
• Steady supply of energy
 – Many mitochondria (structures for aerobic production of ATP)
 – Many capillaries
 – Packed with the oxygen-binding pigment myoglobin

Fast-twitch muscle cells:
• Designed for rapid, powerful response
• Contract rapidly
• Short, powerful contraction because there is more actin and myosin than in slow-twitch cell
• Depend more heavily on anaerobic metabolic pathways to generate ATP, so fatigue rapidly

Aerobic Exercise Increases Endurance; Resistance Exercise Builds Muscle

Exercise can greatly influence the further development of muscle. Different kinds of exercise produce different results. In *aerobic exercise*, such as walking, jogging, or swimming, enough oxygen is delivered to the muscles to keep them going for long periods. This kind of exercise fosters the development of new blood vessels that service the muscles and development of more mitochondria to facilitate energy usage. Aerobic exercise also increases muscle coordination, improves digestive tract movement, and increases the strength of the skeleton by exerting force on the bones. It brings cardiovascular and respiratory system improvements that help muscles to function more efficiently. For example, it enlarges the heart so that each stroke pumps more blood.

Aerobic exercises, however, generally do not increase muscle size. Muscular development, the kind that helps you look good on the beach, comes mostly from *resistance exercise*, such as that which you develop from lifting heavy weights. To build muscle mass, you have to make your muscles exert that which develop more than 75% of their maximum force. These exercises can be very brief. Only three sets of six contractions each are needed to build bulk and increase strength. The bulk results from the existing muscle cells increasing in diameter, although some researchers suggest that heavy exercise splits or tears the muscle cells and that each of the parts then regrows to a larger size.

Nautilus and other muscle-building machines available at many gyms automatically adjust the resistance they offer to muscles during an exercise (Figure 6.12). This is a valuable feature because muscles are weaker at some parts of their range of motion than at others. The design of the exercise machine makes the lifting easier

myosin. This feature adds to their power. However, fast-twitch cells, rich in glycogen deposits, depend more heavily on anaerobic means of producing ATP. As a result, fast-twitch cells tire more quickly than slow-twitch cells.

The two kinds of cells are distributed unequally throughout the human body. The abdominal muscles do not need to contract rapidly, but they need to be able to contract steadily to hold our paunch in at the beach and to balance the powerful, slow-twitch back muscles so that we can stand upright. Fast-twitch cells are more common in the legs and arms, because our limbs may be called upon to move quickly.

People vary in the relative amounts of slow- and fast-twitch muscle cells they possess. Whereas the leg muscles of endurance athletes, such as marathoners, are made up of about 80% slow-twitch cells, those of sprinters are about 60% fast-twitch cells. To some extent these differences are genetic, so if you are a "fast-twitch person," you can build a certain level of endurance, but only to a degree. Endurance runners, on the other hand, dread those times when, at the end of a long race, some well-trained competitor, genetically endowed with lots of fast-twitch cells, sprints past them at the finish line.

FIGURE **6.12** Muscles get larger when they are repeatedly made to exert more than 75% of their maximum force.

Building Muscle Fair and Square? Anabolic Steroid Abuse

People have always appreciated strength, perhaps because they associate it with ability. Of course, such appreciation has led us not only to displays of strength itself but also to the display of a muscular body, that is, to bodybuilding. Unfortunately, some people are always looking for shortcuts to acquiring desirable traits and abilities.

One of the most dangerous behaviors of young people in sports today is the use of anabolic (building) steroids. Taken in large amounts, these synthetic hormones that mimic the male sex hormone testosterone stimulate the body to build muscle, often leading to a dramatic increase in strength (Figure 6.A). The steroids promote these improvements by stimulating protein formation in muscle cells and by reducing the amount of rest needed between workouts.

Anabolic steroids have more than 70 possible unwanted side effects that can range in severity from acne to liver cancer. Most common and most serious are the damage to the liver, cardiovascular system, and reproductive system. The cardiovascular problems, including heart attacks and strokes, may not show up for years, but the effects on the reproductive system are more immediate. The testicles of male steroid users often become smaller, and the user may become sterile and impotent (unable to achieve an erection). These effects occur because anabolic steroids inhibit natural testosterone production. Female steroid abusers develop irreversible masculine traits such as a deeper voice, growth of body hair, loss of scalp hair, smaller breasts, and an enlarged clitoris (during embryological development, the clitoris develops from the same structure that develops into the penis in a male). Among the other risks are injuries resulting from the intended effect of the drug—increased muscle strength—because that

increase is not accompanied by a corresponding increase in the strength of tendons and ligaments. Injuries to tendons and ligaments may take a long time to heal.

Steroid use can have psychological effects, too. For instance, it may promote aggression and a feeling of invincibility, often referred to as "roid rage." Users may also develop severe depression. In addition, steroids are addictive. Although anabolic steroids are available only by prescription, school age children report that it is not difficult to obtain them without a prescription. Commonly called 'roids, juice, or slop, steroids are swallowed as a pill or injected. Because most are obtained illegally through the black market or from foreign countries, their purity and quality are questionable at best.

We might ask, then, why steroid use is so attractive. One reason seems to be that steroids provide an easy route to the body image so prized by society. Nearly one-third of high school males who use steroids admit that they use the drugs to acquire a muscular, well-built look. Steroid use seems to be particularly widespread among high school senior boys. Perhaps they are unsure of themselves as they prepare to leave the nest and enter the "real world."

Among athletes, however, the reason for steroid use is clear: it gives the user a distinct competitive edge, and everyone wants to be a winner. Almost all international athletic committees prohibit steroid use. Indeed, more than a few medals have had to be returned when cheating was discovered. Allegations of steroid use have also caused scandal in professional sports. *Cheating* is the right word because the advantage obtained by taking steroids is clearly unfair.

Some athletes skirt the legal issues connected with steroid use by taking a legal drug called

androstenedione that is available without prescription. The liver converts this drug to testosterone. Although there are no laws prohibiting the use of androstenedione, the National Football League, the National Collegiate Athletic Association, and the International Olympic Committee have banned its use among their athletes. The long-term effects are largely unknown, but in the short term, androstenedione increases blood cholesterol levels, thereby increasing the risk of heart disease. It is also thought to increase the risk of prostate cancer.

The person who abuses steroids is the one immediately responsible for the behavior. Nevertheless, one cannot help but wonder how great an influence the ideals of a society that idolizes winners may have been in shaping this dangerous trend. ✸

FIGURE **6.A** Anabolic steroids are sometimes abused as an easy way to build muscle and strength.

where the muscle is weaker and more difficult in the range the muscle can handle. "Free weights," such as barbells and dumbbells, require more training to use safely than does an exercise machine.

what would you do?

The International Olympics Committee and the World Anti-Doping Agency are concerned about the future possibility of creating genetically modified athletes. Gene therapy techniques (see Chapter 21) may soon be developed for inserting genes into people to improve their athletic performance. For example, researchers exploring muscle deterioration have developed gene therapy that repairs and rebuilds muscles in old mice. Clinical trials of the technique will soon begin on humans who have muscle-wasting disorders. If you were a competitive athlete, would you use gene therapy to repair damaged muscle tissue? Would you use gene therapy to build muscle tissue and enhance your performance? Do you think any of these uses of gene therapy should be illegal?

One problem with building muscle is that you have to keep at it. If you train in the spring only, to look good at the beach, you have an uphill battle, because all the mass of muscle you built last year began to disappear 2 weeks after the training stopped.

A frequently asked question is, are men naturally stronger than women? The general answer is yes. Women have about 35% muscle mass, and men have 42%. One reason is that the level of the male hormone testosterone, which builds muscles, is generally higher in men than in women (see the Social Issue essay, *Building Muscle Fair and Square?*). Also, men are generally larger than women and so have more muscle mass to begin with. However, the individual muscle cells of men and women have the same strength.

HIGHLIGHTING THE CONCEPTS

The Muscular System Moves Our Body Parts and Maintains Our Posture (p. 98)

1. There are three types of muscle: skeletal, smooth, and cardiac. All types of muscle cells are excitable, contractile, extensible, and elastic.

Most Skeletal Muscles Work in Pairs (pp. 98–99)

2. Many of the skeletal muscles of the body are arranged in antagonistic pairs so that the action of one is opposite to the action of the other. One member of such a pair usually causes flexion (bending at the joint) and the other, extension (straightening at the joint).

3. Muscles are attached to bones by tendons. The muscle origin is the end attached to the bone that is more stationary during the muscle's movement, and the insertion is the end attached to the bone that moves.

Sarcomeres Are the Contractile Units of Muscle (pp. 100–102)

4. The entire muscle is wrapped in a connective tissue sheath. A muscle cell is packed with myofibrils, which are composed of myofilaments (the contractile proteins actin and myosin).

5. A muscle contracts when myosin heads in some of its cells bind to actin, swivel, and pull the actin toward the midlines of sarcomeres, causing the actin and myosin filaments to slide past one another and increasing their degree of overlap.

6. Contraction is controlled by the availability of calcium ions. Calcium ions interact with two proteins on the actin filament—troponin and tropomyosin—that determine whether myosin can bind to actin. Calcium ions are stored in the sarcoplasmic reticulum and released when a motor nerve sends an impulse.

7. Motor nerves contact muscle cells at neuromuscular junctions. When the nerve impulse reaches a neuromuscular junction, acetylcholine is released and causes a change in the permeability of the muscle cell's membrane that, in turn, causes calcium ions to be released from the sarcoplasmic reticulum.
WEB TUTORIAL 6.1 Muscle Structure and Function

The Strength of Muscle Contraction Depends on the Number of Motor Units Stimulated (pp. 102–104)

8. A motor neuron and all the muscle cells it stimulates are collectively called a motor unit.

The Strength of Contraction Increases If a Muscle Is Stimulated before It Has Relaxed (p. 104)

9. The response of a muscle cell to a single brief stimulus in the laboratory is called a twitch. If a second stimulus arrives before the muscle has relaxed, the second contraction builds upon the first. This phenomenon is known as summation. Frequent stimuli cause a sustained contraction, called tetanus.

ATP for Muscle Contraction Comes from Many Sources (pp. 104–105)

10. Each time a cross-bridge between myosin and actin forms and is broken, two ATP molecules are used. The sources of ATP are (1) stored ATP, (2) creatine phosphate, (3) anaerobic metabolic pathways, and (4) aerobic respiration.

Slow-Twitch and Fast-Twitch Muscle Cells Differ in Contraction Speed and Duration (pp. 105–106)

11. Slow-twitch muscle cells contract slowly but with enormous endurance. Fast-twitch muscle cells contract rapidly and powerfully but with less endurance.

Aerobic Exercise Increases Endurance; Resistance Exercise Builds Muscle (pp. 106–108)

12. Resistance exercise will increase muscle size. Forcing muscles to exert more than 75% of their maximal strength adds myofilaments to existing muscle cells and increases the cell diameter.

KEY TERMS

tendon *p. 98*

origin *p. 98*

insertion *p. 98*

antagonistic pair *p. 98*

striated muscle *p. 100*

myofibril *p. 100*

myofilament *p. 100*

myosin filament *p. 100*

actin filament *p. 100*

sarcomere *p. 100*

sliding filament model *p. 100*

cross-bridge *p. 100*

troponin *p. 102*

tropomysin *p. 102*

sarcoplasmic reticulum *p. 102*

transverse tubules *p. 102*

neuromuscular junction *p. 102*

motor unit *p. 102*

muscle twitch *p. 104*

summation *p. 104*

tetanus *p. 104*

oxygen debt *p. 105*

slow-twitch cells *p. 105*

fast-twitch cells *p. 105*

REVIEWING THE CONCEPTS

1. Why are most skeletal muscles arranged in antagonistic pairs? Give an example that illustrates the roles of each member of an antagonistic pair of muscles. *p. 98*

2. Describe a skeletal muscle, including descriptions of muscle cells, myofibrils, and myofilaments. *p. 100*

3. What is the sliding filament model of muscle contraction? *pp. 100–101*

4. What causes actin to move during muscle contraction? *p. 101*

5. Explain the roles of troponin, tropomyosin, and calcium ions in regulating muscle contraction. *p. 102*

6. Explain how the events that occur when an impulse from a motor nerve cell reaches a neuromuscular junction and calcium ions are released from the sarcoplasmic reticulum lead to muscle contraction. *p. 102*

7. Define a motor unit. What are the consequences of the differences in motor units' size? *pp. 102, 104*

8. Define muscle twitch, summation, and tetanus. *p. 104*

9. List the sources of ATP for muscle contraction, and explain when each source is typically called upon. *pp. 104–105*

10. Differentiate between slow- and fast-twitch muscle cells. *pp. 105–106*

11. What type of exercise can build muscles? *p. 106*

12. A single motor neuron and all the muscle cells it stimulates is called a
    a. sarcoplasmic reticulum.
    b. neuromuscular junction.
    c. motor unit.
    d. summation.

13. In a muscle, energy is stored in the form of
    a. creatine phosphate.
    b. ADP.
    c. myosin.
    d. glucose.

14. A muscle cell contracts when _____ filaments and _____ filaments slide past one another.

15. The contractile unit of a muscle is called a/an _____.

APPLYING THE CONCEPTS

1. Hakeem is a 22-year-old who ate some of his Aunt Sophie's canned tomatoes for dinner last night. This morning, his speech is slurred and he is having trouble standing or walking, so his brother rushes him to the emergency room. The doctor tells Hakeem that he probably has botulism, a type of food poisoning produced by a bacterium. The botulinum toxin prevents the release of acetylcholine at neuromuscular junctions. Explain how Hakeem's symptoms are related to this effect.

2. A new injury called "BlackBerry thumb" is becoming increasingly common because people send e-mails on their BlackBerries and text messages on their cell phones. BlackBerry thumb is a form of tendini-

   tis. What would cause BlackBerry thumb? What would the symptoms be? How would you treat it?

3. You are the CEO of a drug company. A research scientist approaches you with a plan to develop a new muscle-relaxing drug. The scientist explains that the drug works by flooding the muscle cell with calcium ions. Would you finance the development of this drug? Why or why not?

4. Chickens are ground-dwelling birds. They run to escape from predators, and they can fly only short distances. How do these activities explain the distribution of "white" and "dark" meat on a chicken? (Meat is skeletal muscle.)

Additional questions can be found on the companion website.

7

Neurons: The Matter of the Mind

The brain cells we use to plot and plan our way to victory in a chess match have the same design and mode of functioning as the cells that tell our muscles to contract or that carry information from our sensory organs to our brain. They are neurons, the basic functional units of the nervous system.

Neurons and Neuroglial Cells Are the Cells of the Nervous System
- Neuroglial cells support, protect, insulate, and nurture neurons
- Neurons can be sensory, motor, or associative

Neurons Have Dendrites, a Cell Body, and an Axon
- Axons and dendrites are bundled together to form nerves
- The myelin sheath increases the rate of conduction and helps in repair

The Nerve Impulse Is an Electrochemical Signal
- Ions move passively through ion channels
- The sodium-potassium pump uses ATP to transport sodium ions out and potassium ions in
- The inside of a resting neuron has a negative charge relative to the outside
- An action potential is a reversal and restoration of the charge difference across the membrane
- The sodium-potassium pump restores the original distribution of ions
- Action potentials are all-or-nothing events
- A neuron cannot fire during the refractory period

Synaptic Transmission Is Communication between Neurons
- Synaptic transmission involves the release of neurotransmitter and the opening of ion channels
- Neurons "sum up" input from excitatory and inhibitory synapses
- The neurotransmitter is quickly removed from the synapse
- Different neurotransmitters play different roles

ENVIRONMENTAL ISSUE Environmental Toxins and the Nervous System

A melia, age 12, has won every junior chess tournament in her state. She remembers every chess game she has ever played, categorizing them by opening, general strategy, and closing moves. Her IQ is extremely high, but competitive chess demands all her ability. Today she is taking part in a chess club exhibition, playing 15 games simultaneously with a 5-minute time limit for each game. Walking from board to board, she looks briefly at the chess pieces, makes her move, and presses the timer button on the clock. As she walks, she mentally retraces the moves in a difficult match she played 2 weeks ago. An hour passes, and after winning all 15 games and giving a brief interview to a television reporter, Amelia goes home to finish her calculus assignment.

Amelia is gifted almost beyond measure, yet her brain functions with the same biological materials we all possess. In this chapter, we will begin to explore the brain and other parts of the nervous system, starting with the cells from which they are built. We will discover the special features that allow these cells to communicate with one another. We will then proceed, in Chapter 8, "The Nervous System", to explore the larger divisions of the brain and spinal cord and identify their functions. ■

Neurons and Neuroglial Cells Are the Cells of the Nervous System

The *nervous system* integrates and coordinates all the body's varied activities. Its two primary divisions, which are the subject of Chapter 8, are (1) the central nervous system, which is made up of the brain and spinal cord, and (2) the peripheral system, which is made up of all the nervous tissue in the body outside the brain and spinal cord. Both these major divisions of the nervous system consist of two types of specialized cells. **Neurons** (nerve cells) are excitable cells that generate and transmit messages. **Neuroglial cells** (also called simply glial cells) are supporting cells that outnumber the neurons by about 10 to 1.

■ Neuroglial cells support, protect, insulate, and nurture neurons

The nervous system has several types of glial cells, each with different jobs to do. Some glial cells provide structural support for the neurons of the brain and spinal cord. Glial cells also provide a steady supply of chemicals called nerve growth factors that stimulate nerve growth. Without nerve growth factors, neurons die. Other glial cells form insulating sheaths around axons, which, as described shortly, are the long projections extending from certain neurons. This sheath, called the myelin sheath, has several important functions, also described below. Scientists now know that glial cells can communicate with one another and with neurons.

Unlike neurons, glial cells are able to reproduce. In fact, when neurons die, glial cells multiply to fill up the vacant space. A consequence of this trait is that the uncontrolled division of glial cells forms most brain tumors.

■ Neurons can be sensory, motor, or associative

The basic unit of the nervous system is the neuron, or nerve cell. Neurons, which are responsible for an amazing variety of functions, can be grouped into the three general categories depicted in Figure 7.1.

- **Sensory** (or afferent) **neurons** conduct information *toward* the brain and spinal cord. These neurons generally extend from sensory receptors, which are structures specialized to gather information about the conditions within and around our bodies.
- **Motor** (or efferent) **neurons** carry information *away from* the brain and spinal cord to an **effector**—either a muscle, which will contract, or a gland, which will secrete its product, as a response to information from a sensory or interneuron.

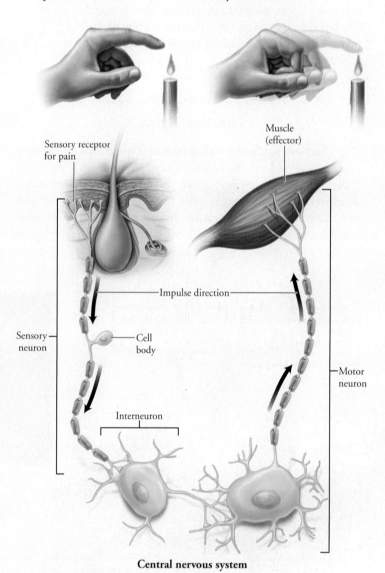

Central nervous system

FIGURE **7.1** Neurons may be sensory neurons, interneurons, or motor neurons. This diagram traces the pathway of an impulse from a sensory receptor to an interneuron and from there to a motor neuron and its effector. Sensory receptors detect changes in the external or internal environment. An interneuron usually receives input from many sensory neurons, integrates that information, and if the input is appropriate, stimulates a motor neuron. The motor neuron then causes a muscle or a gland (an effector) to respond.

- Association neurons, commonly called **interneurons**, are located between sensory and motor neurons. They are found only within the brain and spinal cord, where they integrate or interpret the sensory signals, thereby "deciding" on the appropriate response. Interneurons are by far the most numerous nerve cells in the body, accounting for over 99% of the body's neurons.

AMYOTROPHIC LATERAL SCLEROSIS

In a progressive disease called *amyotrophic lateral sclerosis* (ALS), also called Lou Gehrig's disease, motor neurons throughout the brain and spinal cord die and stop sending messages to skeletal muscles. Without stimulation from motor neurons, the muscles gradually weaken, and the person loses control over arms, legs, and body. The cause of death is respiratory failure, because the muscles that control breathing (the diaphragm and rib muscles) eventually die. Sensory neurons and interneurons are not affected by ALS, and therefore awareness and reasoning do not deteriorate.

Neurons Have Dendrites, a Cell Body, and an Axon

The very shape of a typical neuron is specialized for communicating with other cells (Figure 7.2) To begin with, a neuron has numerous short, branching projections called **dendrites** that create a huge surface for receiving signals from other cells. The signal then travels toward an enlarged central region called the cell body, which has all the normal organelles, including a nucleus, for maintaining the cell. When a neuron responds to an incoming signal, it transmits its message along the **axon**, a single long extension that carries messages away from the cell body. The message is sent along the axon to either another neuron or an effector, either a muscle or a gland. In some cases, the axon allows the neuron to communicate over long distances. For example, a motor neuron that allows you to wiggle your big toe has its cell body in the spinal cord, and an axon carries the message all the way to the muscles of your toe. The end of the axon has many branches specialized to release a chemical, called a neurotransmitter, that alters the activity of the effector. In summary, the dendrites and cell body are typically the *receiving* portions of the neuron, and the axon is the *sending* portion.

To appreciate the dimensions of a neuron, we will pretend for a moment that there is such a thing as a "typical neuron." We will choose as our example a motor neuron, one that carries a message from the spinal cord to a muscle. Imagine an enlarged cell body to be about the size of a tennis ball. The axon of this neuron would be about 1 mile (1.6 km) long but only about one-half inch (1.3 cm) in diameter. The dendrites, the shorter but more numerous projections of the neuron, would fill an average-sized living room.

■ Axons and dendrites are bundled together to form nerves

A **nerve** is a bundle of parallel axons, dendrites, or both from many neurons. Each nerve is covered with tough connective tissue and, depending on the type of neurons it contains, can be classified as sensory, motor, or mixed (a mixed nerve is made up of both sensory and motor neurons).

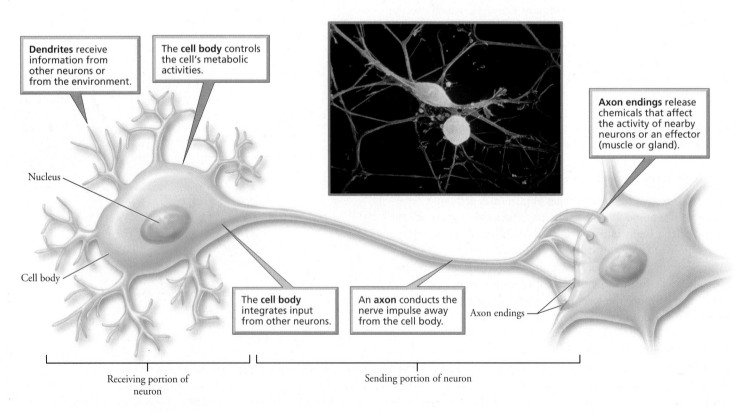

Dendrites receive information from other neurons or from the environment.

The **cell body** controls the cell's metabolic activities.

Axon endings release chemicals that affect the activity of nearby neurons or an effector (muscle or gland).

Nucleus

Cell body

The **cell body** integrates input from other neurons.

An **axon** conducts the nerve impulse away from the cell body.

Axon endings

Receiving portion of neuron

Sending portion of neuron

FIGURE **7.2** The structure of a neuron

■ The myelin sheath increases the rate of conduction and helps in repair

Most of the axons outside the brain and spinal cord, and some of those within, have an insulating outer layer called a **myelin sheath** (Figure 7.3), composed of the plasma membranes of glial cells. Outside of the brain and spinal cord, for example, glial cells known as **Schwann cells** form neurons' myelin sheaths. A Schwann cell plasma membrane wraps around the axon many times to provide a covering that looks somewhat like a jelly roll. The myelin sheath serves as a kind of living electrical tape, insulating individual axons and preventing messages from short-circuiting between neurons. It also allows the signal to move along the axon faster. The myelin sheath is kept alive by the nucleus and cytoplasm of the Schwann cell, which are squeezed to the periphery as the sheath forms.

A single Schwann cell encloses only a small portion, about 1 mm long, of an axon. The gaps between adjacent Schwann cells, where the axon is exposed to the extracellular environment, are called *nodes of Ranvier*. Their presence is very important to the speed at which a neuron transmits messages. With the myelin sheath in place, a nerve impulse "jumps" successively from one node of Ranvier to the next in a type of transmission called **saltatory conduction** (*saltare*, to jump), which is up to 100 times faster than signal conduction would be on an unmyelinated axon of the same diameter. Not surprisingly, the axons that are responsible for conducting signals over long distances are typically myelinated.

To get a sense of how this "jumping" mode of transmission increases the speed at which a message travels, think of the different ways a ball can be moved down the court during a basketball game. When only seconds are left in the game, dribbling the ball the length of the court would take too much time. Passing the ball through a series of players is faster. Likewise, an impulse passed from one node to the next, as occurs in myelinated nerves, moves faster than one traveling uniformly along the full length of the axon, as occurs in unmyelinated nerves.

The myelin sheath also plays a role in helping to repair a neuron when a nerve is a cut or crushed. When a neuron in the peripheral nervous system is cut, the part of the axon that has been separated from the cell body can no longer receive life-sustaining materials from the rest of the cell and begins to degenerate within a few minutes. The Schwann cells that wrapped the axon then begin to clear all the axon fragments from inside the myelin sheath. Finally, the axon stump begins "sprouting" from the cell body and regrows along the proper path, with the guidance of the empty myelin tube.

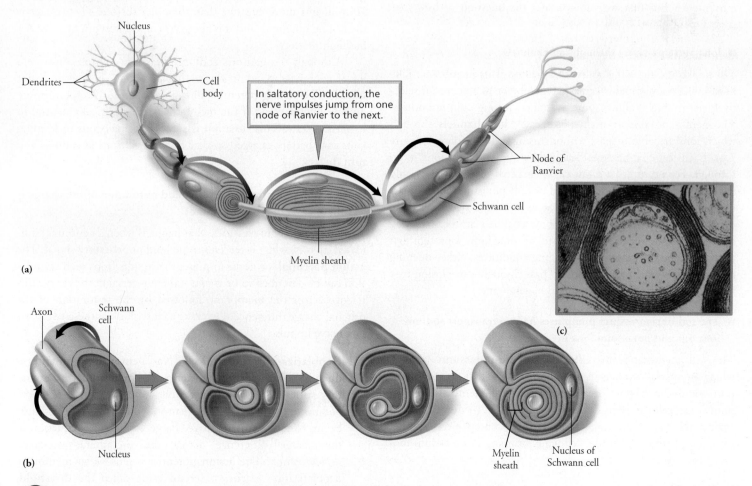

(a)

In saltatory conduction, the nerve impulses jump from one node of Ranvier to the next.

Nucleus

Dendrites

Cell body

Node of Ranvier

Schwann cell

Myelin sheath

(b)

Axon

Schwann cell

Nucleus

Myelin sheath

Nucleus of Schwann cell

(c)

WEB TUTORIAL 7.1 FIGURE **7.3** The myelin sheath. (a) An axon protected by a myelin sheath. The Schwann cells that form the myelin sheath are separated by nodes of Ranvier—areas of exposed axon that allow for saltatory conduction. (b) The myelin sheath forms from multiple wrappings of Schwann cell plasma membranes. (c) An electron micrograph of the cut end of a myelinated axon.

MULTIPLE SCLEROSIS

The importance of the myelin sheath becomes dramatically clear in people with multiple sclerosis, a disease in which the myelin sheaths in the brain and spinal cord are progressively destroyed. The damaged regions of myelin become hardened scars called scleroses (hence the name of the disease) that interfere with the transmission of nerve impulses. Short-circuiting between normally unconnected conduction paths delays or blocks the signals going from one brain region to another. Depending on the part of the nervous system affected, the result can be paralysis or the loss of sensation, including loss of vision.

The Nerve Impulse Is an Electrochemical Signal

The neuron membrane itself is specialized for communication. A nerve's message, which is called a nerve impulse, or action potential, is a electrochemical signal caused by sodium ions (Na^+) and potassium ions (K^+) crossing the neuron's membrane to enter and leave the cell. We will return later to the action potential to see the events that cause and result from these ion movements. But first we will consider the question of how ions cross a membrane.

■ **Ions move passively through ion channels**

The key to the signaling ability of a neuron is its membrane. Like most living membranes, the plasma membrane of a neuron is selectively permeable; it allows some substances through but not others. The membrane contains many pores, called **ion channels**, through which ions are able to pass without using cellular energy (Figure 7.4). Each ion channel specifically allows only certain ions to go through. For example, sodium channels allow the passage of only sodium ions. Potassium channels allow only potassium ions to pass through. Thus, ion channels function as molecular sieves, selecting which ions can enter and leave. Some channels are permanently open. Others are regulated by a "gate" of some kind, a protein that changes shape in response to changing conditions, either opening the channel—which allows ions to pass through—or closing it—which prevents ions from crossing the membrane.

■ **The sodium-potassium pump uses ATP to transport sodium ions out and potassium ions in**

The cell membrane also contains **sodium-potassium pumps**, which are special proteins in the cell membrane that actively transport sodium and potassium ions across the membrane. These pumps use cellular energy in the form of ATP to move the ions against their concentration gradients. Each pump ejects sodium ions (Na^+) from within the cell while bringing in potassium ions (K^+) (see Figure 7.4).

■ **The inside of a resting neuron has a negative charge relative to the outside**

It will be easier to understand the movement of ions during an action potential if we first consider a neuron that is not transmitting

an action potential—that is, a neuron in its resting state. As we will see, however, *resting* is hardly the word to describe what is going on at this stage. The membrane of a resting neuron maintains a difference in the electrical charges near the two membrane surfaces (the surface facing inside the cell and the surface facing outside the cell), keeping the inside surface more negative than the outside one. This charge difference across the membrane, called the **resting potential**, results from the unequal distribution of ions across the membrane.

In a resting neuron, sodium and potassium ions are unequally distributed across the plasma membrane. There are more sodium ions outside the membrane than inside. Furthermore, there are more potassium ions inside than outside. Potassium ions tend to leak out because they are more concentrated inside the axon. (Recall from Chapter 3 that substances tend to move from an area of higher concentration to one of lower concentration.) To a lesser extent, sodium ions leak in. However, sodium-potassium pumps maintain the resting potential by pumping out sodium ions while moving potassium ions back in. There are also some negatively charged ions within the cell that are too large to pass through the membrane.

The result of the unequal distributions described above is that the inner surface of a resting neuron's membrane is typically about 70 millivolts more negative than the outer surface. This voltage is the resting potential, and it is about 5% of the strength of a size AA flashlight battery.

Although the neuron's sodium-potassium pumps consume a lot of energy to maintain the resting potential, the energy is not wasted. The resting potential allows the neuron to respond more quickly than it could if the membrane were electrically neutral in its resting state. This situation is somewhat analogous to keeping your car's battery charged so that the car will start as soon as you turn the key.

■ **An action potential is a reversal and restoration of the charge difference across the membrane**

Now we are ready to consider what happens when a neuron is stimulated, that is, when it receives some kind of excitatory signal. The **action potential**, or nerve impulse, that results from such stimulation can be described in brief as a sudden reversal in the charge difference across the membrane followed by the restoration of the original charge difference. Let's take a closer look at these two parts of a nerve impulse.

1. **Depolarization: sodium ions (Na^+) enter the axon and reduce the charge difference across the membrane.** An excitatory stimulus causes the gates on sodium channels to open. Sodium ions then enter the neuron, and their positive charge begins to reduce the negative charge within. The reduction of the charge difference across the membrane is called *depolarization*. The action potential begins when membrane depolarization reaches a certain value called the **threshold**. When the threshold is reached, the gates on more sodium channels open. Enough sodium ions enter through the open gates to create a net positive charge in that region, as shown in Figure 7.5.

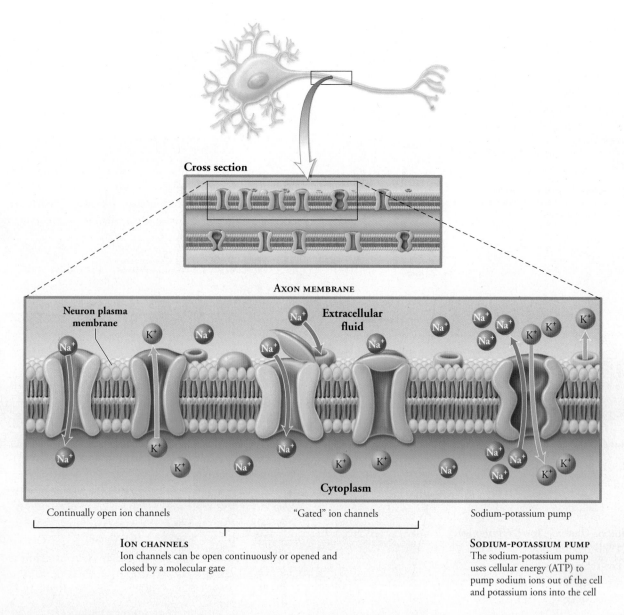

Cross section

Axon membrane

Neuron plasma membrane

Na⁺ K⁺ Na⁺ Na⁺ Extracellular fluid Na⁺ Na⁺ Na⁺ Na⁺ K⁺ K⁺ K⁺ Na⁺ Na⁺ Na⁺

Cytoplasm

Na⁺ K⁺ K⁺ Na⁺ Na⁺ K⁺ K⁺ Na⁺ K⁺

Continually open ion channels   "Gated" ion channels   Sodium-potassium pump

ION CHANNELS
Ion channels can be open continuously or opened and closed by a molecular gate

SODIUM-POTASSIUM PUMP
The sodium-potassium pump uses cellular energy (ATP) to pump sodium ions out of the cell and potassium ions into the cell

FIGURE **7.4** The plasma membrane of a neuron provides two general ways for ions to enter or leave the cell: (1) diffusion through a channel or (2) active transport by a pump.

2. **Repolarization: potassium (K^+) ions leave the axon and cause the membrane potential to return to close to its resting value.** About halfway through the action potential, the gates on potassium channels open. Potassium ions now leave the cell. The exodus of potassium ions with their positive charge causes the interior of the neuron to become negative once again relative to the outside. The outward flow of potassium ions returns the membrane potential to close to its resting value. The restoration of the charge difference across the membrane is called *repolarization*.

To repeat, the action potential is a reversal of the charge difference across the membrane caused by the inward flow of sodium ions followed immediately by the restoration of the original charge difference caused by the outward flow of potassium ions. These changes occur sequentially along the axon like a wave rippling away from the cell body, as will be discussed shortly.

■ **The sodium-potassium pump restores the original distribution of ions**

Notice that, at the end of an action potential, the charge distribution across the membrane has returned to the resting potential. However, there are slightly more sodium ions and slightly fewer potassium ions inside the cell than before. This alteration is corrected by the sodium-potassium pump, which restores the original distribution of sodium and potassium ions. The action of the sodium-potassium pump is slow. Therefore, it does not contribute directly to the events of the action potential.

■ **Action potentials are all-or-nothing events**

The reason the action potential is described as a *wave* of changes that travels down the neuron's plasma membrane is that the events do not occur simultaneously along the entire length of the axon. Instead, as sodium ions enter the cell at one location along the

RESTING NEURON

Plasma membrane is charged, with the inside negative relative to the outside.

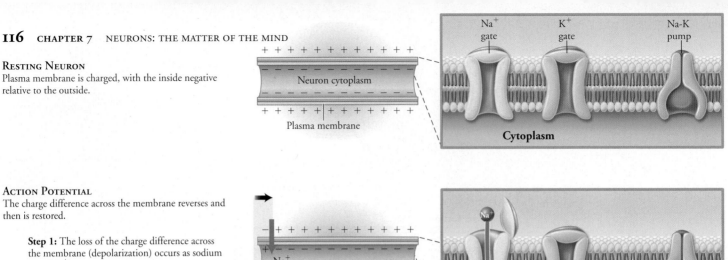

ACTION POTENTIAL

The charge difference across the membrane reverses and then is restored.

Step 1: The loss of the charge difference across the membrane (depolarization) occurs as sodium ions (Na^+) enter the axon.

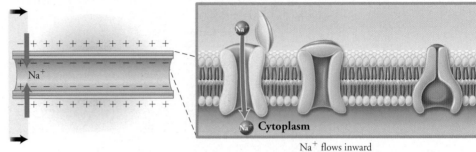

Na$^+$ flows inward

Step 2: The return of the membrane potential to near its resting value (repolarization) occurs as potassium (K^+) ions leave the axon.

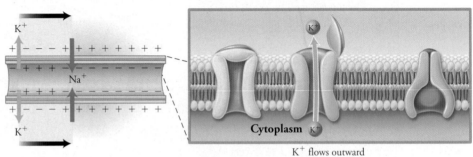

K$^+$ flows outward

RESTORATION OF ORIGINAL ION DISTRIBUTION

The sodium-potassium pump restores the original distribution of ions.

(a)

Sodium-potassium pump active here

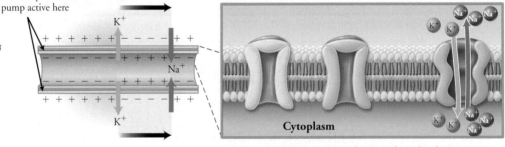

Na-K pump restores the original ion distribution

(b)

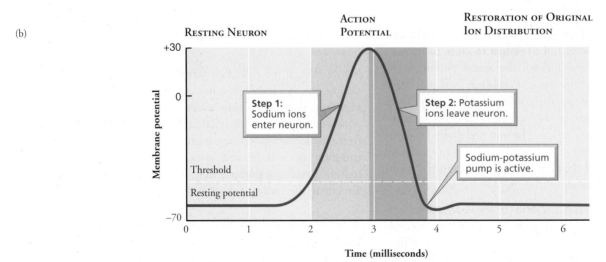

FIGURE **7.5** The resting state and the propagation of an action potential along an axon. (a) The sequential opening and closing of sodium-channel gates and potassium-channel gates produces the action potential. (b) The voltage across the membrane can be measured by electrodes placed inside and outside the axon. The graph shows the changes in voltage that accompany an action potential.

membrane, and the charge inside the membrane becomes less negative in that region of the cell, the change in charge causes the opening of the sodium channel gates in an adjacent part of the membrane. As a result of this sequential opening of gates, the change in charge travels down the length of the axon. Once started, action potentials do not diminish, just as the last domino in a falling row falls with the same energy as the first. Moreover, the intensity of the nerve impulse does not vary with the strength of the stimulus that triggered it. If an action potential occurs at all, it is always of the same intensity as any other action potential. This "all-or-nothing" aspect of nerve-cell conduction is similar to the firing of a gun in that the force of the bullet is not changed by how hard you pull the trigger.

■ A neuron cannot fire during the refractory period

Immediately after an action potential, the neuron cannot be stimulated again for a brief instant called the **refractory period**. During the refractory period, the sodium channels are closed and cannot be opened. Consequently, a new action potential cannot yet be generated. Because of the refractory period, a prolonged stimulus that is above threshold can cause only a series of discrete nerve impulses, not a single, larger, sustained impulse. For this reason, increasing the strength of a stimulus will increase the frequency of impulses, but only up to a point. The inability of the sodium gates to open during the refractory period is also the reason that the nerve impulses cannot reverse and go backward toward the cell body.

stop and think

"Red tides" in the ocean are caused by proliferation of dinoflagellates. These single-celled marine algae contain a chemical called saxitoxin (STX), extremely small concentrations of which prevent sodium channels in mammalian neurons from opening. The clams, scallops, and mussels that consume the dinoflagellates are insensitive to the toxin, but the STX accumulates in their tissues. What effect would you expect STX to have on nerve transmission in humans who accidentally consume tainted shellfish?

Synaptic Transmission Is Communication between Neurons

When a nerve impulse reaches the end of an axon, in almost all cases the message must be relayed to the adjacent cell across a small gap that the impulse itself cannot cross. To transmit the message to the adjacent cell requires a brief change of the medium of communication from an electrochemical signal to a chemical signal. Therefore, when an action potential reaches the end of the axon, a chemical is released from the axon's tip. That chemical, called a **neurotransmitter**, diffuses across the gap and conveys the message to the adjacent cell.

The junction between a neuron and another cell is called a **synapse**. The structure of a synapse between two neurons is shown in Figure 7.6. The gap between the cells is called the *synaptic cleft*. Recall that the axon branches near the end of its length. Each branch ends with a small bulblike swelling called a *synaptic knob*.

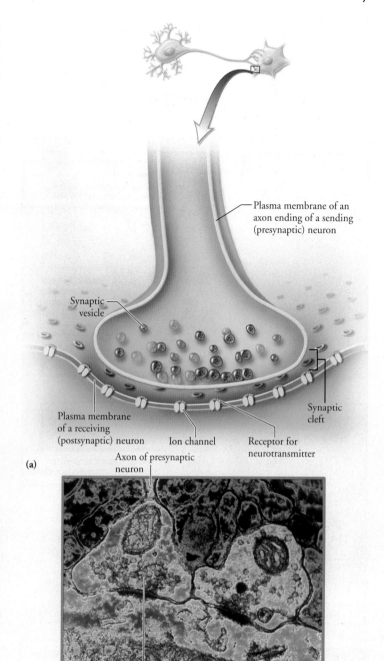

(a)

(b)

Labels: Plasma membrane of an axon ending of a sending (presynaptic) neuron; Synaptic vesicle; Plasma membrane of a receiving (postsynaptic) neuron; Ion channel; Receptor for neurotransmitter; Synaptic cleft; Axon of presynaptic neuron; Synaptic vesicles

FIGURE **7.6** Structure of a synapse. (a) The synaptic knob at the end of the axon on the presynaptic neuron is separated from the dendrite or cell body of the postsynaptic neuron by a small gap called a synaptic cleft. Within the synaptic knob are small sacs, called synaptic vesicles, filled with neurotransmitter molecules. (b) An electron micrograph of a synapse.

The neuron sending the message is the *presynaptic neuron* (meaning "before the synapse"). The neuron receiving the message is the *postsynaptic neuron* ("after the synapse").

■ Synaptic transmission involves the release of neurotransmitter and the opening of ion channels

Now let's consider the events that occur in the synapse as the message is sent from one neuron to the next (Figure 7.7).

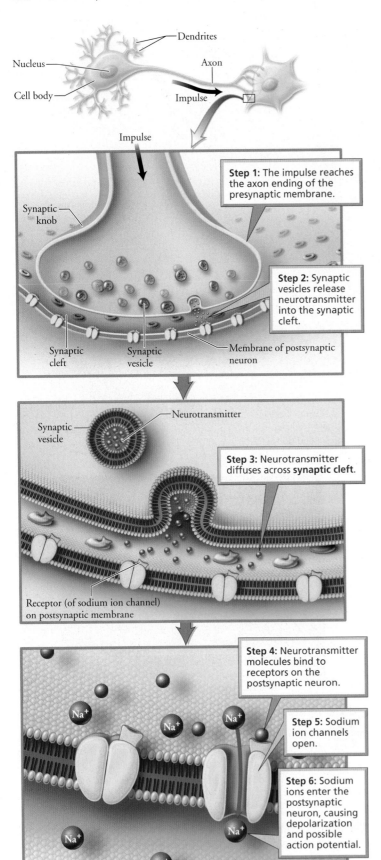

Nucleus

Cell body

Dendrites

Axon

Impulse

Impulse

Step 1: The impulse reaches the axon ending of the presynaptic membrane.

Synaptic knob

Step 2: Synaptic vesicles release neurotransmitter into the synaptic cleft.

Synaptic cleft

Synaptic vesicle

Membrane of postsynaptic neuron

Synaptic vesicle

Neurotransmitter

Step 3: Neurotransmitter diffuses across **synaptic cleft**.

Receptor (of sodium ion channel) on postsynaptic membrane

Step 4: Neurotransmitter molecules bind to receptors on the postsynaptic neuron.

Step 5: Sodium ion channels open.

Na+

Na+

Na+

Na+

Step 6: Sodium ions enter the postsynaptic neuron, causing depolarization and possible action potential.

Na+

Na+

FIGURE **7.7** Transmission across an excitatory synapse

1. **Synaptic knobs release packets of neurotransmitter.** Within the synaptic knobs of the presynaptic neuron, the neurotransmitter is contained in tiny sacs called *synaptic vesicles.* When a nerve impulse reaches the synaptic knob, the gates of calcium ion channels in the membrane there open. Calcium ions move into the knob, causing the membranes of the synaptic vesicles to fuse with the plasma membrane at the synaptic knob and to dump the enclosed neurotransmitter into the synaptic cleft.

2. **Neurotransmitter diffuses across the synaptic cleft and binds with receptors on the membrane of the postsynaptic neuron.** A receptor is a protein that recognizes a particular neurotransmitter, much as a lock "recognizes" a key. The only cells a neurotransmitter can stimulate are cells that have receptors specific for that particular neurotransmitter. Thus, only certain neurons can be affected by a given neurotransmitter.

3. **When a neurotransmitter binds to its receptor, an ion channel is opened, either exciting or inhibiting the postsynaptic neuron.** The binding of the neurotransmitter to a receptor causes the opening of an ion channel in the postsynaptic neuron. The response that is triggered as a result depends on the type of ion channel the receptor opens. It is really the receptor that determines which ion channels will open and what the effect of a given neurotransmitter will be.

An *excitatory synapse* is one where the binding of the neurotransmitter to the receptor opens sodium channels, allowing sodium ions to enter and increasing the likelihood that an action potential will begin in the postsynaptic cell. In contrast, an *inhibitory synapse* is one where the binding of the neurotransmitter decreases the likelihood that an action potential will be generated in the postsynaptic neuron. In this case, the cell's interior becomes more negatively charged than usual. As a result, the cell will require larger than usual amounts of an excitatory neurotransmitter in order to reach threshold.

■ Neurons "sum up" input from excitatory and inhibitory synapses

A neuron may have as many as 10,000 synapses with other neurons (Figure 7.8). Some of these synapses will have excitatory effects on the postsynaptic membrane. Others will have inhibitory effects. The summation of excitatory and inhibitory effects on a neuron at any given moment determines whether an action potential is generated. This integration of input from large numbers of different kinds of synapses gives the nervous system fine control over neuronal responses, just as having both an accelerator and a brake gives you finer control over the movement of a car.

■ The neurotransmitter is quickly removed from the synapse

After being released into a synapse, neurotransmitters are quickly removed, so their effects are temporary. If they were not removed, they would continue to excite or inhibit the postsynaptic membrane indefinitely. Depending on the neurotransmitter, disposal is accomplished in one of two ways. First, enzymes can deactivate a neurotransmitter. For example, the enzyme acetylcholinesterase removes the neurotransmitter **acetylcholine** from synapses where it

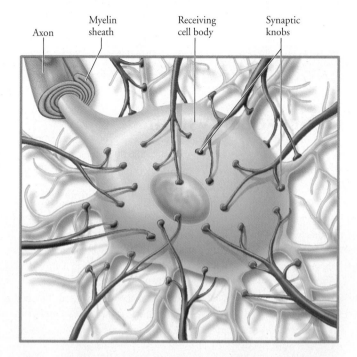

Axon | Myelin sheath | Receiving cell body | Synaptic knobs

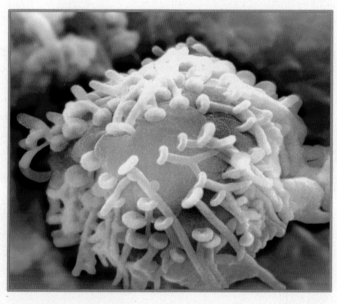

FIGURE **7.8** A neuron may have as many as 10,000 synapses at which it receives input from other neurons. Some synapses have an excitatory effect on the membrane of the postsynaptic neuron and increase the likelihood that the neuron will fire. Other synapses have an inhibitory effect and reduce the likelihood that the postsynaptic neuron will fire. The net effect of all the synapses determines whether an action potential is generated in the postsynaptic neuron. (The shape of the synaptic knobs shown in this electron micrograph is distorted as a result of the preparation process.)

has been released. Second, the neurotransmitter may be actively pumped back into the presynaptic knob. (See the Environmental Issue essay, *Environmental Toxins and the Nervous System.*)

■ Different neurotransmitters play different roles

As we have seen, neurotransmitters are the chemical means of communication within the nervous system. There are dozens of them, carrying messages between our different neurons and between neu-

rons and muscles or glands. The activities of neurotransmitters produce our thoughts and feelings and enable us to interact appropriately with the world around us. Some neurotransmitters produce different effects on different types of cells.

Acetylcholine and norepinephrine are neurotransmitters that act in both the peripheral and the central nervous systems. Both have either excitatory or inhibitory effects depending on where they are released. As we will see in Chapter 8, most internal organs receive input from neurons that release acetylcholine *and* from neurons that release norepinephrine. Norepinephrine stimulates most organs but inhibits certain others. Whatever its effect on any particular organ, acetylcholine will have the opposite effect.

Acetylcholine is also the neurotransmitter released at every neuromuscular junction (the junction of a motor neuron and a skeletal muscle cell), where it triggers contraction of voluntary (skeletal) muscles. *Myasthenia gravis* is an autoimmune disease in which the body's defense mechanisms attack the acetylcholine receptors at such junctions. As a result, a person with myasthenia gravis has little muscle strength. Repeated movements become feeble quite rapidly because the amount of acetylcholine released with each nerve impulse decreases after the neurons have fired a few times in succession. The low number of acetylcholine receptors makes people with myasthenia gravis extremely sensitive to even the slightest decline in acetylcholine availability. Drugs that inhibit acetylcholinesterase are prescribed to prevent the breakdown of acetylcholine, elevating the level of acetylcholine in the neuromuscular junction.

About 50 neurotransmitters are employed by the central nervous system for communication between the neurons in our brains. Why so many? One reason seems to be that different neurotransmitters are involved with different behavioral systems. Norepinephrine, for instance, is important in the regulation of mood, in the pleasure system of the brain, and in arousal. Norepinephrine is thought to produce an energizing "good" feeling. It is also thought to be essential in hunger, thirst, and the sex drive. Serotonin is thought to promote a generalized feeling of well-being. Dopamine helps regulate emotions. It is also used in pathways that control complex movements.

A change in the level of a neurotransmitter affects the behaviors controlled by neurons that communicate using that neurotransmitter. Neurotransmitter levels may change as a result of taking certain drugs (see Chapter 8a). Changes in neurotransmitter levels also cause certain diseases.

ALZHEIMER'S DISEASE

Alzheimer's disease is progressive and results in loss of memory, particularly for recent events, followed by sometimes severe personality changes (Figure 7.9). A formerly good-natured person may become moody and irritable. In Alzheimer's disease, the parts of the brain important in memory and intellectual functioning (hippocampus and cerebral cortex; see Chapter 8) lose large numbers of neurons. Some of the neurons in these regions use the neurotransmitter acetylcholine, which may decrease in level by as much as 90% in a person with Alzheimer's, possibly explaining the loss of memory and mental capacity. In addition, a brain affected by

ENVIRONMENTAL ISSUE

Environmental Toxins and the Nervous System

Many of us are exposed on a daily basis to substances that can harm our nervous system. The exposure may occur because of our job, our lifestyle, the foods we eat, or the medications we use. Because we are born with all the neurons we will ever have, the death of neurons can lead to irreversible loss of function. Some damage may take years to become apparent, but other effects are immediate.

Pesticides are chemicals for killing organisms that would otherwise damage crops and proper-

ty. For example, the widely used organophosphate insecticides Malathion, Parathion, and Diazinon are poisons that kill insects by excessively boosting the activity at certain synapses in the insect's body. They work by inhibiting acetylcholinesterase, the enzyme that breaks down the neurotransmitter acetylcholine. As a result, acetylcholine accumulates in the synapses and has a continuous effect.

Because they have the same effect in the human body, these insecticides accidentally poi-

son approximately 500,000 people around the world each year, primarily farm workers; 5,000 to 14,000 die as a result. One symptom of pesticide poisoning is muscle spasms (continuous involuntary contractions), because acetylcholine is the neurotransmitter that triggers contraction of voluntary muscles, such as those in the arms or legs.

Mercury is a naturally occurring heavy metal that damages the nervous system. In adults, symptoms of mercury poisoning include numbness in the arms and legs, sensory disturbance,

FIGURE 7.9 Former President Ronald Reagan died in June 2004 at the age of 93. He had Alzheimer's disease since at least 1994. His wife, Nancy Reagan, described Alzheimer's disease as "the long good-bye" because it robs people of their memories long before it takes their lives.

Alzheimer's disease is pocked with clusters of proteins, some between the neurons (amyloid plaques) and others within the neurons (neurofibrillary tangles). The amyloid plaques and neurofibrillary tangles are the prime suspects for the cause of the death of acetylcholine-producing neurons.

Hypothesizing that the loss of acetylcholine is responsible for some of the Alzheimer's symptoms, researchers and physicians have attempted to treat Alzheimer's disease with drugs meant to raise or at least maintain acetylcholine levels. Although such drugs (Aricept, Exelon, and Reminyl) do improve the memory and intellectual ability of some people with Alzheimer's, they do not help in all cases. Moreover, any improvement is rapidly lost when a person stops taking the drugs.

DEPRESSION

We all describe ourselves as feeling depressed at times. However, more than 19 million Americans experience depression that lasts

for weeks, months, or years and interferes with their ability to function in daily life. Clinical depression of this description can affect anyone regardless of age, sex, race, or income. It is thought to be related in some way to insufficient levels of the neurotransmitter serotonin as well as of dopamine and norepinephrine. Signs of depression are a loss of interest and pleasure in the activities and hobbies that were previously pleasurable; anxiety; sleep problems; decreased energy; and feelings of sadness, hopelessness, worthlessness, and guilt. Depression takes the joy out of life and complicates certain medical conditions such as heart disease, cancer, diabetes, epilepsy, and osteoporosis.

Many depressed people exhibit suicidal behavior or actually commit suicide (currently, the third-ranking cause of death for teenagers). Studies comparing the brains of people who committed suicide with those of people who died of other causes reveal structural and chemical differences. The brains of people who took their own lives indicate a problem in the production and activity of serotonin.

Depression can be treated successfully. Unfortunately, few of the millions of people suffering from depression recognize the symptoms and seek help. Antidepressant drugs affect the functioning of the neurotransmitters responsible for the problem: norepinephrine and serotonin. Older medications affect both neurotransmitters simultaneously. Newer ones, including Prozac, Zoloft, and Paxil, specifically affect serotonin functioning, increasing the level of serotonin in the synapse by reducing its rate of removal.

what would you do?

Few people would question the justifiability of providing drugs to elevate neurotransmitter function to someone who is depressed or suicidal in order to help the person live a normal life. However, researchers believe that the levels of key neurotransmitters also affect personality traits, such as shyness or impulsiveness. If so, we may someday be able to design our own personalities. Should minor personality problems be treated with drugs? Should a personality "flaw" be treated? What do you think?

lack of coordination, weakness and tremor, and slowed and slurred speech. Continued exposure can cause paralysis, convulsions, and death. Exposure of a pregnant woman to mercury can cause abnormalities in the nervous system of the fetus. Nursing mothers can pass mercury to their babies in breast milk.

During the nineteenth century, mercury was used in the hat-making industry, causing mental derangement in hat makers and giving rise to the saying "mad as a hatter." Today, mercury is still used in a variety of industrial processes, many of which release wastewater that carries mercury into natural bodies of water. Gaseous emissions, such as those from coal-fired power plants and the burning of municipal or medical waste, can release mercury into the air, later to be deposited on land or in water.

Bacteria in water convert the mercury they ingest to methyl mercury. Other organisms eat the bacteria and are themselves consumed by still others. As a result, mercury accumulates in the bodies of fish and other creatures higher in the food chain (see Chapter 23). Today, most exposure of humans to mercury results from eating mercury-contaminated food, especially fish. The Food and Drug Administration (FDA) advises pregnant women and women of child-bearing age not to eat any shark, swordfish, king mackerel, or tilefish because of their high mercury content, and to limit consumption of albacore (white) tuna to one serving a week and of all other fish to two servings a week. At the same time, the FDA encourages consumers to eat up to two servings a week of fish and shellfish that are lower in mercury, such as salmon, pollock, catfish, canned light tuna, and shrimp. 🌲

PARKINSON'S DISEASE

Parkinson's disease is a progressive disorder that results from the death of dopamine-producing neurons deep in the brain's movement control center (Figure 7.10). A person with Parkinson's disease moves slowly, usually with a shuffling gait and a hunched posture, and may suffer from involuntary muscle contractions that further interfere with intended movements. The contractions may cause tremors (involuntary rhythmic shaking) of the hands or head, because of muscles alternately contracting and relaxing, or they may cause muscle rigidity, because of muscles contracting continuously. This muscle rigidity may cause sudden "freezing" in the middle of a movement.

As the dopamine-producing neurons in the brain's movement control center die, dopamine levels begin to fall. Initially, the symptoms are subtle and are written off as part of the aging process. By the time the Parkinson's diagnosis is apparent, 80% of the neurons in this small area of the brain may have died.

Attempts to treat Parkinson's disease have focused on replacing dopamine or helping the brain get by with the remaining dopamine. Unfortunately, swallowing pills containing dopamine does not help, because dopamine is prevented from reaching the brain by the blood-brain barrier you will read about in Chapter 8. Instead, patients are given other substances that *can* reach the brain. The most common and effective treatment combines two drugs: L-dopa, an amino acid that the brain converts to dopamine, and carbidopa, which prevents dopamine from forming outside of the brain and causing undesirable side effects. However, this treatment does not stop the steady loss of dopamine-producing neurons, so it loses effectiveness as the disease progresses. Some patients are treated with drugs that enhance their levels of dopamine by inhibiting the enzyme that breaks dopamine down.

A controversial experimental treatment for Parkinson's disease, the transplantation of stem cells, shows some promise. These special cells are capable of differentiating into virtually any type of tissue. Although stem cells can be collected from certain adult tissues, the stem cells of an embryo are more versatile. However, the use of human embryonic stem cells has generated intense political debate (see Chapter 19a).

FIGURE **7.10** Actor Michael J. Fox and former heavyweight champion Muhammad Ali have Parkinson's disease, a progressive debilitating disease characterized by slowed movements, tremors, and rigidity. The symptoms result from insufficient amounts of the neurotransmitter dopamine, caused by the death of dopamine-making nerve cells in a movement control center of the brain. The Michael J. Fox Foundation for Parkinson's Research (www.michaeljfox.org) funds research on the early diagnosis and treatment of Parkinson's disease.

HIGHLIGHTING THE CONCEPTS

Neurons and Neuroglial Cells Are the Cells of the Nervous System (pp. 111–112)

1. The nervous system integrates and coordinates the body's activities. It is made up of the central nervous system (the brain and spinal cord) and the peripheral nervous system (nervous tissue outside the brain and spinal cord).

2. The nervous system has two types of specialized cells: neurons (nerve cells) and neuroglial cells.

3. Neuroglial cells outnumber neurons. They provide structural support for neurons, supply nerve growth factors, and form myelin sheaths around certain axons.

4. There are three general categories of neurons. Sensory (or afferent) neurons conduct information from the sensory receptors toward the central nervous system. Motor (or efferent) neurons conduct information away from the central nervous system to an effector. Association neurons (interneurons) are positioned between sensory and motor neurons and are located in the central nervous system.

Neurons Have Dendrites, a Cell Body, and an Axon (pp. 112–114)

5. Neurons are specialized for communicating with other cells. A "typical" neuron has a cell body containing the organelles that maintain the cell. Numerous branching fibers called dendrites conduct messages *toward* the cell body. A single long axon conducts impulses *away* from the cell body.

6. An axon may be enclosed in an insulating layer called a myelin sheath. Neuroglia called Schwann cells form the myelin sheath by wrapping their plasma membranes repeatedly around the axon. The myelin sheath greatly increases the rate at which impulses are conducted along an axon; it also plays a role in the regeneration of cut axons in the peripheral nervous system.

WEB TUTORIAL 7.1 Myelinated Neurons and Saltatory Conduction

The Nerve Impulse Is an Electrochemical Signal (pp. 114–117)

7. The message conducted by a neuron, called a nerve impulse or action potential, is caused by sodium ions (Na^+) and potassium ions (K^+) crossing the neuron's plasma membrane to enter and leave the cell.

8. Ion channels in the plasma membrane are small pores through which ions move without the use of cellular energy. Ion channels are usually specific to one or a few types of ions.

9. The sodium-potassium pump uses cellular energy in the form of ATP to pump sodium ions out of the cell and potassium ions into the cell against their concentration gradients.

10. In the resting state, a neuron has an electrical potential difference, called the resting potential, across its plasma membrane. The resting potential is generated by an unequal distribution of charges across the membrane that makes the neuron more negative inside than outside: there are many negatively charged proteins inside the cell. Sodium ions are in greater concentration outside the neuron, and potassium ions are in greater concentration inside.

11. The action potential begins when a region of the membrane suddenly becomes permeable to sodium ions. If enough sodium ions enter to reach threshold, an action potential is generated. The gates on sodium channels open, and many sodium ions enter the cell, making its interior in that region of the membrane temporarily positive (depolarization). Potassium ions then leave the cell, making the inside once again more negative than the outside (repolarization). The same events are repeated all along the axon in a wave of depolarization and repolarization called an action potential.

12. At the end of an action potential, the sodium-potassium pump moves sodium ions out of the neuron and potassium ions into the neuron, restoring the original ion distribution.

13. Once initiated, an action potential sweeps to the end of the axon without diminishing in strength.

14. During the brief refractory period immediately following an action potential, the neuron cannot be stimulated.

WEB TUTORIAL 7.2 The Nerve Impulse

Synaptic Transmission Is Communication between Neurons (pp. 117–121)

15. The point where one neuron meets another is called a synapse. The neuron sending the message (the presynaptic neuron) and the neuron receiving the message (the postsynaptic neuron) are separated by a small gap called the synaptic cleft.

16. The arrival of a nerve impulse at the axon's end causes calcium ions to enter the presynaptic cell there. These ions cause synaptic vesicles that store neurotransmitters to fuse with the plasma membrane of the presynaptic neuron and release their contents into the synaptic cleft. The neurotransmitter then diffuses across the gap and binds to receptors on the membrane of the postsynaptic neuron.

17. If the synapse is excitatory, sodium ions enter the postsynaptic cell and increase the likelihood that it will generate a nerve impulse. If the synapse is inhibitory, the charge difference across the membrane of the receiving neuron is increased, reducing the likelihood that the postsynaptic neuron will generate a nerve impulse.

18. Postsynaptic cells integrate excitatory and inhibitory input from many cells. If threshold is reached, an action potential is generated in the postsynaptic cell.

19. Neurotransmitter is quickly removed from the synapse either by enzymatic breakdown or by transport back into the presynaptic neuron.

20. Acetylcholine, epinephrine, and norepinephrine are neurotransmitters used in both the peripheral and central nervous systems. Acetylcholine causes voluntary muscle to contract. All three neurotransmitters can have excitatory or inhibitory effects depending on where they are released.

21. Many neurotransmitters are found in the brain. Different ones are active in different behavioral systems. Disturbances in brain chemistry affect mood and behavior. Alzheimer's disease, characterized by a progressive loss of memory, is associated with a loss of acetylcholine in certain parts of the brain. Reduction in the functioning of dopamine, norepinephrine, and serotonin can cause depression and suicidal behavior. Low levels of dopamine in another brain region cause Parkinson's disease, which is characterized by slow movements, tremors, and muscle rigidity.

WEB TUTORIAL 7.3 The Synapse

KEY TERMS

neuron *p. 111*
neuroglial cell *p. 111*
sensory neuron *p. 111*
motor neuron *p. 111*
effector *p. 111*
interneuron *p. 112*

dendrite *p. 112*
axon *p. 112*
nerve *p. 112*
myelin sheath *p. 113*
Schwann cell *p. 113*
saltatory conduction *p. 113*

ion channel *p. 114*
sodium-potassium pump *p. 114*
resting potential *p. 114*
action potential *p. 114*
threshold *p. 114*
refractory period *p. 117*

neurotransmitter *p. 117*
synapse *p. 117*
acetylcholine *p. 118*

REVIEWING THE CONCEPTS

1. List the three types of neurons and give their general functions. *pp. 111–112*
2. What are the functions of neuroglial cells? *p. 111*
3. Draw a typical neuron and label the following: cell body, nucleus, dendrites, and axon. *p. 112*
4. Explain how a myelin sheath is formed. What are the functions of the myelin sheath? *p. 113*
5. Describe the distribution of sodium ions and potassium ions during a neuron's resting state. Explain how the movements of these ions affect the charge difference across the membrane. *p. 114*
6. Why is the resting potential important? *p. 114*
7. What happens to sodium ions at the beginning of an action potential? *p. 114*
8. Describe the events that bring about the restoration of the charge difference across the membrane (repolarization). *p. 115*
9. How is the original ion distribution of sodium and potassium ions restored? *p. 115*
10. What is the refractory period? *p. 117*
11. Draw a synapse between two neurons. Label the following: presynaptic neuron, postsynaptic neuron, synaptic cleft, synaptic vesicles, neurotransmitter molecules, and receptors. *p. 117*
12. How do the events at an excitatory synapse differ from those at an inhibitory synapse? *p. 118*
13. How is the action of a neurotransmitter terminated? *pp. 118–119*
14. Choose the *incorrect* statement:
    a. Neurotransmitters diffuse across the myelin sheath.
    b. An inhibitory neurotransmitter makes it less likely that an action potential will be generated in the postsynaptic (after the synapse) neuron.
    c. Neurotransmitters are stored in synaptic vesicles.
    d. Most interneurons are found in the central nervous system.
15. In botulism, a type of food poisoning, the poison produced by bacteria in the spoiled food prevents the person's synaptic vesicles from fusing with the neuron's membrane. You would expect that this effect would
    a. cause excessive destruction of neurotransmitter in the synaptic cleft.
    b. destroy myelin.
    c. prevent the message of the presynaptic cell from reaching the postsynaptic cell.
    d. cause neurotransmitter to clog the synaptic cleft.
16. In a resting neuron
    a. potassium ions are more concentrated outside the membrane than inside.
    b. the inside is more negative than the outside.
    c. sodium ions are more concentrated inside than outside.
    d. action potentials are being generated.
17. You are a neurophysiologist trying to identify the function of a particular nerve cell. You notice that this nerve cell fires immediately before a person's pinky finger bends. You correctly conclude that this axon
    a. may be part of a sensory neuron carrying information from the finger toward the brain.
    b. may be part of a motor neuron carrying information from the brain toward the pinky.
    c. is without a doubt part of an interneuron.
    d. is dead.
18. The synaptic cleft
    a. is a chemical that allows two neurons to communicate with one another.
    b. is a gap between two neurons.
    c. is a gap between two Schwann cells forming the myelin sheath.
    d. allows saltatory conduction.
19. The _____ is an insulating layer formed by Schwann cells that increases the rate of an action potential.
20. The _____ uses cellular energy to move sodium ions out of the axon and potassium ions into the axon.
21. Parkinson's disease is caused by the loss of neurons that produce the neurotransmitter _____.

APPLYING THE CONCEPTS

1. Indira has taken the sedative Valium, a drug that has a molecular shape similar to that of the neurotransmitter GABA. When Valium binds to GABA receptors, it causes the same effects as the binding of GABA itself. GABA is an inhibitory neurotransmitter. On the basis of this information, explain why Valium has a calming effect on the nervous system.
2. A nerve gas (diisopropyl fluorophosphate) used during war blocks the action of acetylcholinesterase, the enzyme that breaks down acetylcholine in the synapse. What effects would you expect this gas to have at the synapses that use acetylcholine?
3. Mohammed has a kidney condition that raises the level of potassium ions in the extracellular fluid that bathes cells. What effect would you expect this condition to have on his ability to generate nerve impulses (action potentials)?
4. Kerry is experiencing a weakness in her legs. A physician tells her that she has Guillain-Barré syndrome, which is a progressive but reversible condition in which the myelin sheath is lost from parts of the nervous system. Explain why the loss of myelin would cause weakness in Kerry's legs.
5. Ouabain is a drug that causes a neuron to lose its resting potential. What effect would ouabain have on the ability of a neuron to generate action potentials?

Additional questions can be found on the companion website.

8

The Nervous System

Gasp! Gulp. Munch. The nervous system works in concert with other systems in the body to interpret and respond to events and objects in the world around us.

The Nervous System Consists of the Central and Peripheral Nervous Systems

Bone, Membranes, and Cerebrospinal Fluid Protect the Central Nervous System

The Brain Is the Central Command Center
- The cerebrum is the conscious part of the brain
- The thalamus allows messages to pass to the cerebral cortex
- The hypothalamus is essential to homeostasis
- The cerebellum is an area of sensory-motor coordination
- The brain stem controls many of life's basic processes and connects the brain and spinal cord
- The limbic system is involved in emotions and memory
- The reticular activating system filters sensory input

The Spinal Cord Transmits Messages to and from the Brain and Is a Reflex Center

The Peripheral Nervous System Consists of the Somatic and Autonomic Nervous Systems
- The somatic nervous system controls conscious functions
- The autonomic nervous system controls internal organs

Disorders of the Nervous System Vary in Health Significance
- Headaches have several possible causes
- Strokes occur when the brain is deprived of blood
- Coma is a lack of response to all sensory input
- Spinal cord injury results in impaired function below the site of injury

HEALTH ISSUE Meningitis: Bacterial-Type and West Nile Virus

HEALTH ISSUE To Sleep, Perchance to Dream

J ohn's fingers trembled as he rang Tegan's doorbell. How would their first date go? Would he know the right things to say? To do? Would she like the movie he had selected? His mind was racing. In the theater, he could hardly taste the popcorn they were sharing. He barely felt the ice from his cola when he spilled it on his lap.

Eventually his nervousness subsided, and he began to be caught up in the film. His pulse raced as the aliens attacked with missiles, guns, and chainsaws. He dodged in his seat as smashed cars and shattered glass appeared to fly off the screen. He laughed at the crisp one-liners.

On the way home, Tegan smiled and thanked him for a lovely evening. She said she had really enjoyed the witty script and amazing special effects. Could they see another film next week, he asked nervously? "Of course," she said, "perhaps the new Brad Pitt film at the Cineplex downtown." Awkwardly, they kissed goodnight.

Although they were probably not aware of it, John and Tegan's night at the movies was sponsored by their nervous systems. Emotion, foresight, language, reflexes, reasoning, and awareness of sensory stim- uli all depend on the actions of billions of neurons in the brain, spinal cord, and peripheral nerves. How can this array of cellular wires and switches control such a wide range of human responses?

In this chapter, we explore the organization of the nervous system and the structures responsible for its many functions. We also discuss some disorders of the brain and spinal cord and their effects on the human body and mind. ■

The Nervous System Consists of the Central and Peripheral Nervous Systems

I f you were to view the nervous system apart from the rest of the body, you would see a dense mass of neural tissue where the head should be, with a cord of neural tissue extending downward from it where the middle of the back should be (Figure 8.1). These structures are the brain and spinal cord, and they constitute the **central nervous system (CNS)**, which integrates and coordinates all voluntary and involuntary nervous functions. Connected to the brain and spinal cord are many communication "cables"—the nerves

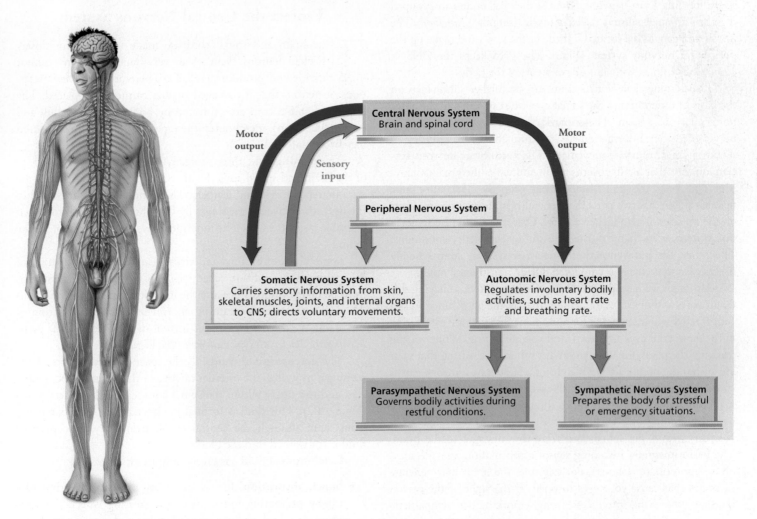

FIGURE **8.1** An overview of the nervous system. The various parts of the human nervous system have special functions but work together as an integrated whole.

HEALTH ISSUE

Meningitis: Bacterial-Type and West Nile Virus

Meningitis is an inflammation of the meninges, the protective coverings of the brain and spinal cord. Many types of bacteria and certain viruses can cause it. If examination of the cerebrospinal fluid shows bacteria to be the cause, the person is treated with antibiotics. If the cause is a virus, treatment includes medicines to alleviate pain and fever while the body's immune system fights the virus.

Bacterial meningitis is a serious, sometimes fatal disease. The initial symptoms, which generally (but not always) include a high fever,

headache, stiff neck, and vomiting, may develop within hours or over several days. The bacteria may spread to the blood, causing blood poisoning (septicemia), which may cause a skin rash that can start anywhere on the body. The rash initially looks like a cluster of tiny pinpricks. However, if they are not treated with antibiotics, the spots join and look like fresh bruises. The bacteria can also cause gangrene, which is the death of body tissues from a failure of the blood supply. The gangrene may require amputation of arms or legs. The bacteria can also damage the kidneys,

requiring dialysis or even a kidney transplant. About 1 case in 10 is fatal.

Freshmen college students housed in dormitories are at increased risk of getting bacterial meningitis because of their close living quarters. Part of the reason is the means by which the bacteria are spread. People can carry the bacteria in their throat without having any symptoms of illness and can spread the infection through coughing, sneezing, or intimate kissing. Upperclassmen are less susceptible, perhaps because they have built up immune defenses against the bacteria.

that carry messages to and from the CNS. The nerves branch extensively, forming a vast network. Some of their cell bodies are grouped together in small clusters called **ganglia** (singular, *ganglion*). The nerves and ganglia are located outside of the CNS and make up the **peripheral nervous system (PNS)**. The PNS keeps the CNS in continuous contact with almost every part of the body.

The peripheral nervous system can be further subdivided on the basis of function into the somatic nervous system and the autonomic nervous system. The **somatic nervous system** consists of nerves that carry information to and from the CNS, resulting in sensations and voluntary movement. The **autonomic nervous system**, on the other hand, governs the involuntary, unconscious activities that keep the body functioning properly. The autonomic nervous system has two parts that generally produce opposite effects on the muscles or glands they control. One, the **sympathetic nervous system,** is in charge during stressful or emergency conditions. The other, the **parasympathetic nervous system,** adjusts bodily function so that energy is conserved during nonstressful times.

Although we have described the nervous system as having different parts and divisions, remember that all the parts function as a coordinated whole. Imagine for a moment that you are meditating in the park; your eyes are closed, and you are resting. While you are relaxing, the parasympathetic nervous system is ensuring that your life-sustaining bodily activities continue. Suddenly, someone grasps your hand. Sensory receptors in the skin (which are part of the somatic nervous system) respond to the pressure and warmth of the hand by sending messages over sensory nerves to the spinal cord. Neurons within the spinal cord relay the messages to the brain. The brain integrates incoming sensory information and "decides" on an appropriate response. For example, the brain may generate messages that cause your eyes to open. If the sight of the person holding your hand produces strong emotion, the sympathetic nervous system may speed up your heartbeat and perhaps even your breathing.

Bone, Membranes, and Cerebrospinal Fluid Protect the Central Nervous System

The brain and spinal cord are made up of many closely packed neurons. Neurons are very fragile, and most cannot divide and produce new cells. Therefore, with few exceptions, a neuron that is damaged or dies cannot be replaced. The brain and spinal cord are protected by bony cases (the skull and vertebral column), membranes (the meninges), and a fluid cushion (cerebrospinal fluid).

The **meninges** are three protective connective tissue coverings of the brain and spinal cord (Figure 8.2). The outermost layer, the dura mater, is tough and leathery. The middle layer is anchored to the next lower layer of meninges by thin threadlike extensions that resemble a spider's web. The innermost layer is molded around the brain.

Some kinds of bacteria and viruses can cause inflammation of the meninges, a condition called *meningitis*. All cases of meningitis must be taken seriously because the infection can spread to the underlying nervous tissue, causing encephalitis, or inflammation of the brain. We discuss meningitis further in the Health Issue essay, *Meningitis: Bacterial-Type and West Nile Virus.*

The **cerebrospinal fluid** fills the space between layers of the meninges as well as the internal cavities of the brain, called ventricles, and the cavity within the spinal cord, called the central canal (Figure 8.2). This fluid is formed in the ventricles and circulates from them through the central canal. Eventually, cerebrospinal fluid is reabsorbed into the blood.

Cerebrospinal fluid has several important functions:

- **Shock absorption.** Just as an air bag protects the driver of a car by preventing impact with the steering wheel, the cerebrospinal fluid protects the brain by cushioning its impact with the skull during blows or other head trauma.

Vaccines are available against several forms of meningitis. The protection lasts at least 3 years, so a freshman who is vaccinated should not require a booster during the college years. Some colleges are now requiring that incoming freshmen be vaccinated against meningitis.

The West Nile virus, which is transmitted by infected mosquitoes, can cause both meningitis and encephalitis (brain inflammation). The first reported cases of West Nile virus in North America were in New York City in 1999. Since then, the disease has spread to nearly every state. The virus can infect certain vertebrates, including humans, horses, birds, and occasionally dogs and cats. Testing mosquitoes and dead birds, especially crows and starlings, for the virus is one way to track its spread. Because the symptoms are similar to those of the flu—fever, headache, and muscle and joint pain—many people who become infected are unaware of it; and most people under the age of 50 have few symptoms or none at all. However, older people have weaker immune systems. If they become infected, they are more likely to develop meningitis or encephalitis, either of which can cause brain damage, paralysis, or death.

You can protect yourself from West Nile virus by avoiding wet and humid places that harbor mosquitoes. If you must enter an area where mosquitoes are likely to be, wear light-colored clothing that covers your body and use insect repellent.

Human-to-human transfer of the West Nile virus is possible but less likely. Transfusion of blood contaminated with the virus has caused about two dozen cases, but this mode of transmission is now preventable, because there is a test that can screen blood donations for the virus. There have also been a few instances in which the virus has either spread from a mother to an infant through breast feeding or from a pregnant woman to her fetus. ♀♂

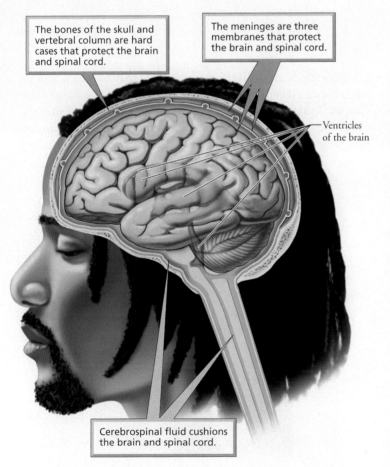

FIGURE **8.2** The central nervous system is protected by the meninges, the cerebrospinal fluid, and the bones of the skull and vertebral column.

- **Support.** Because the brain floats in the cerebrospinal fluid, it is not crushed under its own weight.
- **Nourishment.** The cerebrospinal fluid delivers nutrients and chemical messengers and removes waste products.

The CNS is also protected by the **blood-brain barrier**, a mechanism that selects the substances permitted to enter the cerebrospinal fluid from the blood. This barrier is formed by the tight junctions between the cells of the capillary walls that supply blood to the brain and spinal cord. Because the cells are held together much more tightly than are cells in capillaries in the rest of the body, substances in the blood are forced to pass through the cells of the capillaries instead of between the cells. Thus, the membranes of the capillary cells filter and adjust the consistency of the filtrate by selecting the substances that can leave the blood. The plasma membranes of the capillary walls are largely lipid. So, lipid-soluble substances, including oxygen and carbon dioxide, can pass through easily. Certain drugs, including caffeine and alcohol, are lipid soluble, explaining why they can have a rapid effect on the brain. However, the blood-brain barrier prevents many potentially life-saving, infection-fighting, or tumor-suppressing drugs from reaching brain tissue, which frustrates physicians.

The Brain Is the Central Command Center

In a sense, your brain is more "you" than is any other part of your body, because it holds your emotions and the keys to your personality. Yet, if you were to look at your brain, you probably would deny any relationship to it. The brain is the consistency of soft cheese and weighs less than 1600 g (3 lb), which is probably less than 3% of your body weight. Nevertheless, it is the origin of your secret thoughts and desires; it remembers your most embarrassing moment; and it keeps all your body systems functioning

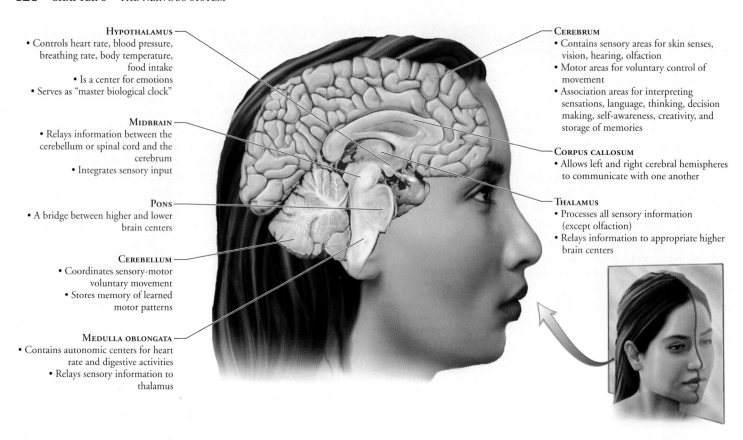

HYPOTHALAMUS
- Controls heart rate, blood pressure, breathing rate, body temperature, food intake
- Is a center for emotions
- Serves as "master biological clock"

MIDBRAIN
- Relays information between the cerebellum or spinal cord and the cerebrum
- Integrates sensory input

PONS
- A bridge between higher and lower brain centers

CEREBELLUM
- Coordinates sensory-motor voluntary movement
- Stores memory of learned motor patterns

MEDULLA OBLONGATA
- Contains autonomic centers for heart rate and digestive activities
- Relays sensory information to thalamus

CEREBRUM
- Contains sensory areas for skin senses, vision, hearing, olfaction
- Motor areas for voluntary control of movement
- Association areas for interpreting sensations, language, thinking, decision making, self-awareness, creativity, and storage of memories

CORPUS CALLOSUM
- Allows left and right cerebral hemispheres to communicate with one another

THALAMUS
- Processes all sensory information (except olfaction)
- Relays information to appropriate higher brain centers

FIGURE **8.3** A section through the brain from front to back, indicating the functions of selected structures

harmoniously while your conscious mind concentrates on other activities. Let's look at how its many circuits are organized to perform these amazing feats.

■ The cerebrum is the conscious part of the brain

The **cerebrum** is the largest and most prominent part of the brain (Figure 8.3). It is, quite literally, your "thinking cap." Accounting for 83% of the total brain weight, the cerebrum gives you most of your human characteristics.

The many ridges and grooves on the surface of the cerebrum make it appear wrinkled. Some furrows are deeper than others. The deepest indentation is in the center and runs from front to back. This groove, called the longitudinal fissure, separates the cerebrum into two hemispheres (Figure 8.4). Each hemisphere receives sensory information from and directs the movements of the opposite side of the body. In addition, the hemispheres process information in slightly different ways and are, therefore, specialized for slightly different mental functions.

The thin outer layer of each hemisphere is called the **cerebral cortex** (Figure 8.5). (*Cortex* means bark or rind.) The cerebral cortex consists of billions of neuroglial cells, nerve cell bodies, and unmyelinated axons and is described as **gray matter**. Although the cerebral cortex is only 1 to 2 mm (about 1/8 in.) thick, it is highly folded. These folds, or convolutions, triple the surface area of the cortex.

Beneath the cortex is the cerebral **white matter**, which appears white because it consists primarily of myelinated axons. Recall from Chapter 7 that myelin sheaths increase the rate of conduction along axons and are, therefore, found on axons that conduct information over long distances. The axons of the cerebral white matter allow

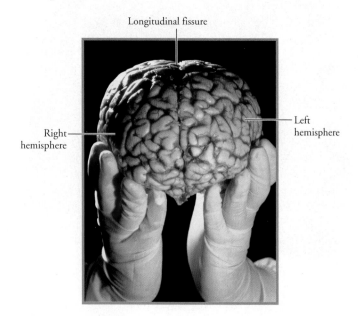

Longitudinal fissure

Right hemisphere

Left hemisphere

FIGURE **8.4** A photograph of the human brain from the front, showing the left and right cerebral hemispheres

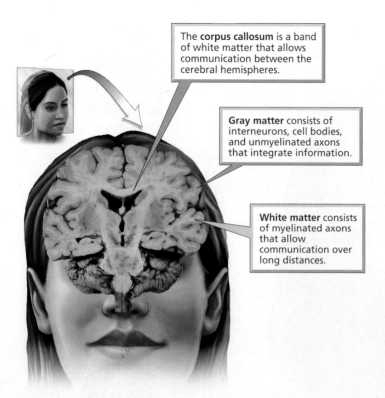

The **corpus callosum** is a band of white matter that allows communication between the cerebral hemispheres.

Gray matter consists of interneurons, cell bodies, and unmyelinated axons that integrate information.

White matter consists of myelinated axons that allow communication over long distances.

FIGURE **8.5** A cross section through the brain, showing the gray matter (the cerebral cortex), white matter, and corpus callosum

various regions of the brain to communicate with one another and with the spinal cord. A very important band of white matter, called the *corpus callosum*, connects the two cerebral hemispheres so they can communicate with one another.

Other grooves on the surface of the brain mark the boundaries of four lobes on each hemisphere: the frontal, parietal, temporal, and occipital lobes (Figure 8.6). Each of these lobes has its own specializations. Although the assignment of a specific function to a particular region of the cerebral cortex is imprecise, it is generally agreed that there are three types of functional areas: sensory, motor, and association.

SENSORY AREAS

Our awareness of sensations depends on the sensory areas of the cerebral cortex. The various sensory receptors of the body send information to the cortex, where each sense is processed in a different region. If you stood on a street corner watching a parade go by, you would hear the band play because information from your ears would be sent to the auditory area in the temporal lobe. You would see the flag wave because information from your eyes would be sent to the visual area in the occipital lobe. When you catch a whiff of popcorn, information is sent from the olfactory (smell) receptors in your nose to the olfactory area in the temporal lobe of the cortex. As you eat that popcorn, you know it is too salty because information from the taste receptors is sent to gustatory areas in the parietal lobe.

Still watching the parade, you know that you are standing in the hot sun and that your belt is too tight because information from touch, pain, and temperature receptors in the skin and from receptors in the joints and skeletal muscles is sent to the **primary**

somatosensory area. This region forms a band in the parietal lobes that stretches over the cortex from ear to ear (Figure 8.7). Sensations from different parts of the body are represented in different regions of the primary somatosensory area (of the hemisphere on the opposite side of the body). The greater the degree of sensitivity, the greater the area of cortex devoted to that body part. Thus, your most sensitive body parts, such as the tongue, hands, face, and genitals, have more of the cortex devoted to them than do less sensitive areas, such as the forearm.

MOTOR AREAS

If you decide to join the parade, the **primary motor area** (Figure 8.7) of the cerebral cortex will send messages to your skeletal muscles. This motor area controls voluntary movement. It forms a band in the frontal lobe, just anterior to the primary somatosensory area. The motor area is organized in a manner similar to the somatosensory area. Each point on its surface corresponds to the movement of a different part of the body. The parts of the body over which we have finer control, such as the tongue and fingers, have greater representation on the motor cortex than do regions with less dexterity, such as the trunk of the body.

Just in front of the motor cortex is the *premotor cortex*. It coordinates learned motor skills that are patterned or repetitive, such as typing or playing a musical instrument. The premotor cortex coordinates the movement of several muscle groups at the same time. When a pattern of movement is repeated many times, the proper pattern of stimulation is stored in the premotor cortex. For example, as a guitar player practices playing a particular song many times, the pattern of stimulation needed to play that song is stored in the premotor cortex. Then, each time the song is played, the premotor cortex will stimulate the primary cortex in the pattern needed to play that song, without requiring the musician to think about where on the strings the fingers should be placed.

ASSOCIATION AREAS

Next to each primary sensory area is an association area. These communicate with the sensory and motor areas, and with other parts of the brain, to analyze and act on sensory input. In particular, each sensory association area communicates with the general interpretation area, to recognize what the sensory receptors are sensing. The general interpretation area assigns meaning to sensory information by integrating the input from sensory association areas with stored sensory memories. For example, on a dark night, your eyes may detect a small moving object. If the object then rubs against your legs and purrs, your general interpretation area will assist you in recognizing it as the neighbor's friendly cat. However, if the object turns away from you and raises its tail, you will recognize it as a skunk.

Once the sensory input has been interpreted, the information is sent to the most complicated of all association areas, the **prefrontal cortex**. This most anterior part of the frontal lobe will predict the consequences of various possible responses to the information it receives and decide which response will be best for you in your current situation. The prefrontal cortex enables us to reason, plan for the long term, and think about abstract concepts. It also plays a key role in determining our personality.

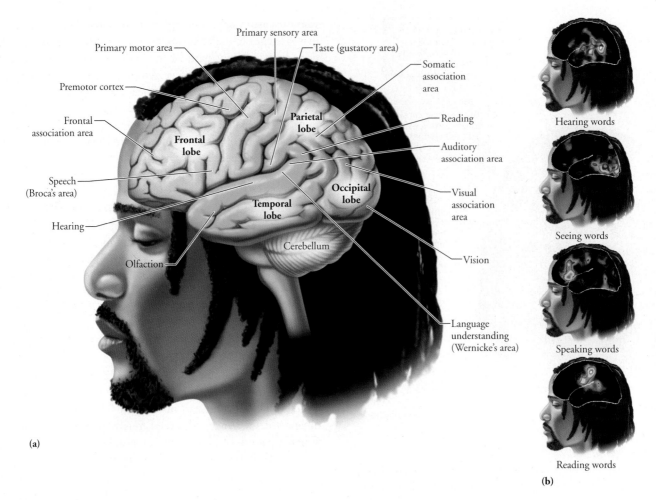

(a)

(b)

FIGURE **8.6** The cerebral cortex. (a) The cerebral cortex has four lobes. Some of the functions associated with each lobe are indicated. (b) These PET scans of the brain show regions of increased blood flow during different mental activities. The increased flow of blood shows which region becomes active when the cerebrum is engaged in hearing words, seeing words, speaking words, and reading words. Notice the relation between active regions of the cerebral cortex during these tasks and the cortical areas for various language skills shown in part (a).

■ The thalamus allows messages to pass to the cerebral cortex

The cerebral hemispheres sit comfortably over the **thalamus** (see Figure 8.3). This structure is often described as the gateway to the cerebral cortex because all messages to the cerebral cortex must pass through the thalamus first. The thalamus functions in sensory experience, motor activity, stimulation of the cerebral cortex, and memory. Sensory input from every sense except smell and from all parts of the body is delivered to the thalamus. The thalamus sorts the information by function and relays it to appropriate regions of the cortex for processing. Some regions of the thalamus also integrate information rather than just relaying it. At the thalamic level of processing, you have a general impression of whether the sensation is pleasant or unpleasant. If you step on a tack, for instance, you may experience pain by the time the messages reach the thalamus; however, you will not know where you hurt until after the message is directed to the cerebral cortex.

■ The hypothalamus is essential to homeostasis

Below the thalamus is the **hypothalamus** (*hypo,* "under"), a small region of the brain that is largely responsible for homeostasis—the

body's maintenance of a stable environment for its cells (discussed in Chapter 4). By coordinating the activities of the nervous and endocrine (hormonal) systems, the hypothalamus, shown in Figure 8.3, influences blood pressure, heart rate, digestive activity, breathing rate, and many other vital physiological processes. It keeps body temperature near the set point, and it regulates hunger and thirst and therefore the intake of food. Moreover, because the hypothalamus receives input from the cerebral cortex, it can make your heart beat faster when you so much as see or think of something exciting or dangerous—a rattlesnake about to strike, for instance.

The hypothalamus is part of the limbic system (discussed later in this chapter), so it is also part of the circuitry for emotions. Specific regions of the hypothalamus play a role in the sex drive and in the perception of pain, pleasure, fear, and anger.

The hypothalamus also contains an area called the suprachiasmatic nucleus, often described as the body's "master biological clock." Day after day, many of the physiological processes within the body, affecting many physical abilities and even a person's mental sharpness, fluctuate so predictably that the variations are called biological rhythms and are described as being controlled by biological clocks. The suprachiasmatic nucleus synchronizes all these bio-

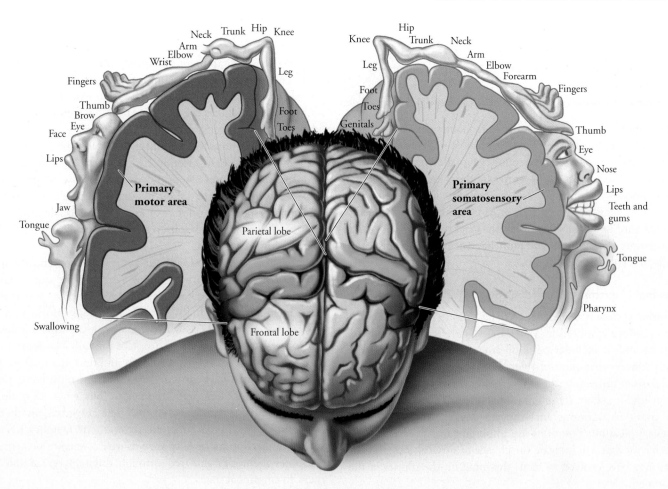

FIGURE **8.7** The primary motor and the primary somatosensory regions of the cerebral cortex are organized in such a way that each location on their surface corresponds to a particular part of the body. The general arrangement is similar in the two regions.

logical clocks and sets the timing for all the body's biological rhythms. Note that these fluctuations do not negate the concept of homeostasis; they simply remind us that the body's optimal internal state is not always the same but instead varies with the rhythm of our external environment and our daily activities.

■ **The cerebellum is an area of sensory-motor coordination**

The **cerebellum** (see Figure 8.3) is the part of the brain responsible for sensory-motor coordination. It acts as an automatic pilot that produces smooth, well-timed voluntary movements and controls equilibrium and posture. Sensory information concerning the position of joints and the degree of tension in muscles and tendons is sent to the cerebellum from all parts of the body. By integrating this information with input from the eyes and the equilibrium receptors in the ears, the cerebellum knows the body's position and direction of movement at any given instant.

The coordination of sensory input and motor output by the cerebellum involves two important processes: comparison and prediction. During every move you make, the cerebellum continuously compares the actual position of each part of the body with where it *ought* to be at that moment (with regard to the intended movement) and makes the necessary corrections. Try to touch the

tips of your two index fingers together above your head. You probably missed on the first attempt. However, the cerebellum makes the necessary corrections, and you will likely succeed on the next attempt. At the same time, the cerebellum calculates future positions of a body part during a movement. Then, just before that part reaches the intended position, the cerebellum sends messages to stop the movement at a specific point. Therefore, when you scratch an itch on your cheek, the hand stops before slapping your face!

■ **The brain stem controls many of life's basic processes and connects the brain and spinal cord**

The brain stem consists of the medulla oblongata, the midbrain, and the pons. The **medulla oblongata** is often called simply the medulla. This marvelous inch of nervous tissue contains reflex centers for some of life's most vital physiological functions—including the pace of the basic breathing rhythm, the force and rate of heart contraction, and blood pressure. The medulla (see Figure 8.3) connects the spinal cord to the rest of the brain. Therefore, all sensory information going to the upper regions of the brain and all motor messages leaving the brain are carried by nerve tracts running through the medulla.

stop and think

Why would a brain tumor that destroyed the functioning of nerve cells in the medulla lead to death more quickly than a tumor of the same size on the cerebral cortex?

The **midbrain** processes information about sights and sounds and controls simple reflex responses to these stimuli. For example, when you hear a unexpected loud sound, your reflexive response is to turn your head and direct your eyes toward the source of the sound.

The **pons**, which means "bridge," connects lower portions of the CNS with higher brain structures. More specifically, it connects the spinal cord and cerebellum with the cerebrum, thalamus, and hypothalamus. In addition, the pons has a region that assists the medulla in regulating respiration.

■ The limbic system is involved in emotions and memory

The **limbic system** is a collective term for a group of structures that help to produce emotions and memory (Figure 8.8). It is called a system because it includes parts of several brain regions and the neural pathways that connect them. The limbic system is defined on the basis of function rather than anatomy.

The limbic system is our emotional brain. It allows us to experience countless emotions, including rage, pain, fear, sorrow, joy, and sexual pleasure. Emotions are important because they motivate behavior that will increase the chance of survival. Fear, for example, may have evolved to focus the mind on the threats in the environment.

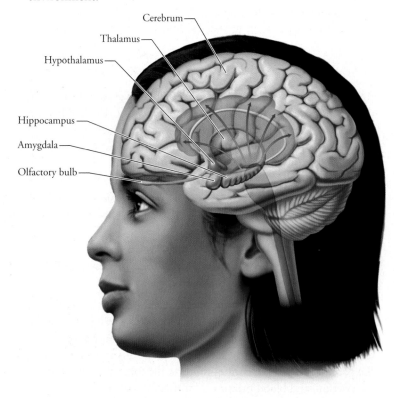

FIGURE **8.8** The limbic system and reticular activating system. The diagram shows the limbic system as a three-dimensional structure within the brain viewed from the left side. The reticular activating system is shown as upward arrows.

Connections between the cerebrum and the limbic system allow us to have *feelings* about *thoughts*. As a result, you may become excited at the thought of winning the lottery. They also allow us to have thoughts about feelings, thus keeping us from responding to emotions, such as rage, in ways that would be unwise. The limbic system includes the hypothalamus, as Figure 8.8 shows. In addition, it is connected to lower brain centers, such as the medulla, that control the activity of internal organs. Therefore, we also have "gut" responses to emotions.

Certain structures associated with our sense of smell (olfaction) are also part of the limbic system. As a result, we often have emotional responses to odors. For example, we usually consider the odor of baking cinnamon rolls to be pleasant and the odor of skunk to be repulsive.

You wouldn't be you without your memories, and the limbic system plays a role in forming them. Memory, the storage and retrieval of information, takes place in two stages. The first is **short-term memory**, which holds a small amount of information for a few seconds or minutes, as when you look up a phone number and remember it only long enough to place the call. The second stage, **long-term memory**, stores seemingly limitless amounts of information for hours, days, or years. Not all short-term memories get consolidated into long-term memories, but when they do, the **hippocampus** plays an essential role. The amygdala, another part of the limbic system that functions in long-term memory, has widespread connections to sensory areas as well as to emotion centers. It associates memories gathered through different senses and links them to emotional states.

■ The reticular activating system filters sensory input

The **reticular activating system (RAS)** is a very extensive network of neurons that runs through the medulla and projects to the cerebral cortex (Figure 8.8). The RAS functions as a net, or filter, for sensory input. Our brain is constantly flooded with tremendous amounts of sensory information, about 100 million impulses each second, most of them trivial. The RAS filters out repetitive, familiar stimuli—the sound of street traffic, paper rustling, the coughing of the person next to you, or the pressure of clothing. However, infrequent or important stimuli pass through the RAS to the cerebral cortex and, therefore, reach our consciousness. Because of the RAS, you can fall asleep with the television on but wake up when someone whispers your name.

In addition, the RAS is an activating center. Unless inhibited by other brain regions, the RAS activates the cerebral cortex, keeping it alert and "awake." Consciousness occurs only while the RAS stimulates the cerebral cortex. When sleep centers in other regions of the brain inhibit activity in the RAS, we sleep. In essence, then, the cerebrum "sleeps" whenever it is not stimulated by the RAS. Sensory input to the RAS results in stimulation of the cerebral cortex and an increase in consciousness, explaining why it is usually easier to sleep in a dark quiet room than in an airport terminal. Conscious activity in the cerebral cortex can also stimulate the RAS, which will, in turn, stimulate the cerebral cortex. Therefore, thinking about a problem may keep you awake all night.

The Spinal Cord Transmits Messages to and from the Brain and Is a Reflex Center

The other major component of the central nervous system besides the brain is the spinal cord. The **spinal cord** is a tube of neural tissue that is continuous with the medulla at the base of the brain and extends about 45 cm (17 in.) to just below the last rib. For most of its length, the spinal cord is about the diameter of your little finger. It becomes slightly thicker in two regions, just below the neck and at the end of the cord, because of the large group of nerves connecting these regions of the cord with the arms and legs. The central canal, filled with cerebrospinal fluid, runs the length of the spinal cord.

The spinal cord is encased in and protected by the stacked bones of the vertebral column (Figure 8.9). Pairs of spinal nerves (considered part of the peripheral nervous system) extend from the spinal cord through openings between the vertebrae to serve different parts of the body. The vertebrae are separated by disks of cartilage that act as cushions.

The spinal cord has two functions: (1) to transmit messages to and from the brain and (2) to serve as a reflex center. The transmission of messages is performed primarily by white matter, found toward the outer surface of the spinal cord. White matter consists of myelinated axons grouped into tracts[1]. Ascending tracts carry sensory information up to the brain. Descending tracts carry motor information from the brain to a nerve leaving the spinal cord.

The second function of the spinal cord is to serve as a reflex center. A reflex is an automatic response to a stimulus, prewired in a circuit of neurons called a **reflex arc**. The circuit consists of a receptor, a sensory neuron (which brings information from the receptors toward the CNS), usually at least one interneuron, a motor neuron (which brings information from the CNS toward an effector), and an effector (a muscle or a gland; Figure 8.10). Shaped somewhat like a butterfly in the central region of the spinal cord, the gray matter houses the interneurons and the cell bodies of motor neurons involved in reflexes.

[1]Within the CNS, bundles of axons are called tracts. Outside the CNS, bundles of axons are called nerves.

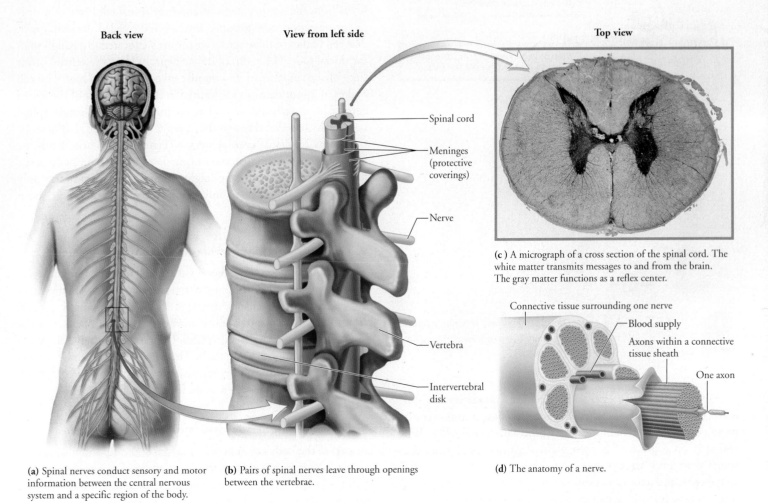

Back view **View from left side** **Top view**

Spinal cord

Meninges (protective coverings)

Nerve

Vertebra

Intervertebral disk

(**c**) A micrograph of a cross section of the spinal cord. The white matter transmits messages to and from the brain. The gray matter functions as a reflex center.

Connective tissue surrounding one nerve

Blood supply

Axons within a connective tissue sheath

One axon

(**a**) Spinal nerves conduct sensory and motor information between the central nervous system and a specific region of the body.

(**b**) Pairs of spinal nerves leave through openings between the vertebrae.

(**d**) The anatomy of a nerve.

FIGURE **8.9** The spinal cord is a column of neural tissue that runs from the base of the brain to just below the last rib. It is protected by the bones of the vertebral column.

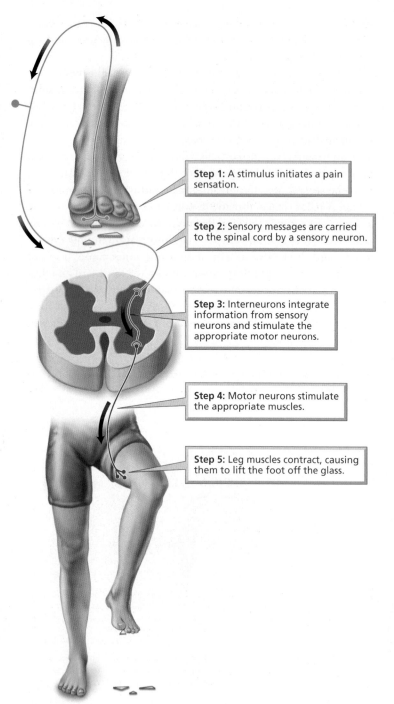

Step 1: A stimulus initiates a pain sensation.

Step 2: Sensory messages are carried to the spinal cord by a sensory neuron.

Step 3: Interneurons integrate information from sensory neurons and stimulate the appropriate motor neurons.

Step 4: Motor neurons stimulate the appropriate muscles.

Step 5: Leg muscles contract, causing them to lift the foot off the glass.

 FIGURE **8.10** A reflex arc consists of a sensory receptor, a sensory neuron, usually at least one interneuron, a motor neuron, and an effector.

Spinal reflexes are essentially "decisions" made by the spinal cord. They are beneficial when a speedy reaction is important to a person's safety. Consider, for example, the withdrawal reflex. When you step on a piece of broken glass, impulses speed toward the spinal cord over sensory nerves. Within the gray matter of the spinal cord, the sensory neuron synapses with an interneuron. The interneuron, in turn, synapses with a motor neuron that sends a message to the appropriate muscle to contract and lift your foot off the glass.

While the spinal reflexes were removing the foot from the glass, pain messages from the cut foot were sent to the brain through ascending tracts in the spinal cord. However, it takes longer to get a message to the brain than it does to get one to the spinal cord, because the distance and number of synapses to be crossed are greater. Therefore, by the time pain messages reach the brain, you have already withdrawn your foot. Nonetheless, once the sensory information reaches the conscious brain, decisions can be made about how to care for the wound.

The Peripheral Nervous System Consists of the Somatic and Autonomic Nervous Systems

The nerves and ganglia of the peripheral nervous system (PNS) carry information between the CNS and the rest of the body. The PNS consists of spinal nerves and cranial nerves.

The body has 31 pairs of **spinal nerves**, each of which originates in the spinal cord and services a specific region of the body (Figure 8.11a). One member of the pair serves a part of the right side of the body and the other serves the corresponding part of the left side. All spinal nerves carry both sensory and motor fibers. The fibers from the sensory neurons enter the spinal cord from the dorsal, or posterior, side, grouped into a bundle called the *dorsal root*. The cell bodies of these sensory neurons are located in a ganglion in the dorsal root. The axons of motor neurons leave the ventral (front side) of the spinal cord in a bundle called the *ventral root*. The cell bodies of motor neurons are located in the gray matter of the spinal cord. The dorsal and ventral roots join to form a single spinal nerve, which passes through the opening between the vertebrae.

The 12 pairs of **cranial nerves** (Figure 8.11b) arise from the brain and service the structures of the head and certain body parts, including the heart and diaphragm. Some cranial nerves carry only sensory fibers, others carry only motor fibers, and others carry both types of fiber.

■ The somatic nervous system controls conscious functions

The peripheral nervous system is subdivided into the somatic nervous system and the autonomic nervous system. The somatic nervous system carries sensory messages that tell us about the world around us and within us, and it controls movement. Sensory messages carried by somatic nerves result in conscious sensations, including light, sound, and touch. The somatic nervous system also controls our voluntary movements, allowing us to smile, stamp a foot, sing a lullaby, or frown as we sign a check.

■ The autonomic nervous system controls internal organs

As part of the body's system of homeostasis, the autonomic nervous system automatically adjusts the functioning of our body organs so that the proper internal conditions are maintained and the body is able to meet the demands of the world around it. The somatic nervous system sends information about conditions within the body to the autonomic nervous system. The autonomic nervous system then makes the appropriate adjustments. Its activities alter digestive activity, open or close blood vessels to shunt blood to areas that need it most, and alter heart rate and breathing rate.

Recall that the autonomic nervous system consists of two branches: the sympathetic and the parasympathetic nervous systems. The sympathetic nervous system gears the body to face an emergency or stressful situation, such as fear, rage, or vigorous exercise. Thus, the sympathetic nervous system prepares the body for fight or flight. In contrast, the parasympathetic nervous system adjusts body function so that energy is conserved during relaxation.

Both the parasympathetic and the sympathetic nervous systems send nerve fibers to most, but not all, internal organs (Figure 8.12). When both systems send nerves to a given organ, they have opposite, or antagonistic, effects on its function. If one system stimulates, the other system inhibits. The antagonistic effects are brought about

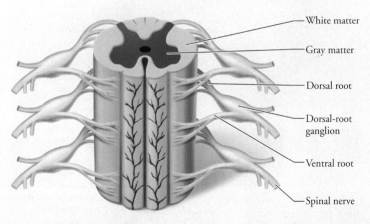

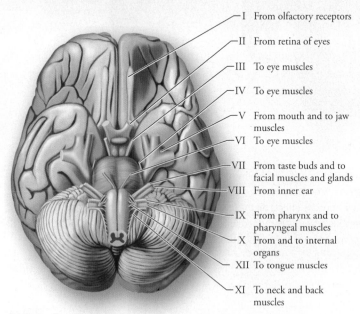

(a) View from front of body

- White matter
- Gray matter
- Dorsal root
- Dorsal-root ganglion
- Ventral root
- Spinal nerve

I	From olfactory receptors
II	From retina of eyes
III	To eye muscles
IV	To eye muscles
V	From mouth and to jaw muscles
VI	To eye muscles
VII	From taste buds and to facial muscles and glands
VIII	From inner ear
IX	From pharynx and to pharyngeal muscles
X	From and to internal organs
XII	To tongue muscles
XI	To neck and back muscles

(b) View of underside of brain

FIGURE **8.11** (a) Spinal and (b) cranial nerves. The 12 pairs of cranial nerves can be seen in this view of the underside of the brain. Most cranial nerves service structures within the head, but some service organs lower in the body. The descriptions indicate whether the neuron carries sensory information (toward the brain) or motor information (away from the brain).

by different neurotransmitters. Whereas sympathetic neurons release mostly norepinephrine at their target organs, parasympathetic neurons release acetylcholine at their target organs.

The sympathetic nervous system acts as a unified whole, bringing about all its effects at once. It is able to act in this way because its neurons are connected through a chain of ganglia. A unified response is exactly what is needed in an emergency. To meet a threat, the sympathetic nervous system increases breathing rate, heart rate, and blood pressure. It also increases the amount of glucose and oxygen delivered to body cells to fuel the response. In addition, it stimulates the adrenal glands to release two hormones, epinephrine and norepinephrine, into the bloodstream. These hormones back up and prolong the other effects of sympathetic stimulation. Lastly, it inhibits digestive activity, because digesting the previous meal is hardly a priority during a crisis.

The effects of the parasympathetic nervous system occur more independently of one another. After the emergency, organ systems return to a relaxed state at their own pace. Organs can respond to the parasympathetic nervous system independently because the ganglia containing the parasympathetic neurons that stimulate each organ are not connected in a chain near the spinal cord, as they are in the sympathetic nervous system. Instead, the ganglia of the parasympathetic nervous system are located near the individual organs.

Disorders of the Nervous System Vary in Health Significance

Disorders of the nervous system vary tremendously in severity and impact on the body. Some disorders, such as a mild headache, are often more of a nuisance than a health problem. Others, such as insufficient sleep, can cause more problems than a person might expect (see the Health Issue essay, *To Sleep, Perchance to Dream*). Still other disorders, such as stroke, coma, and spinal cord injury, can have devastating effects on a person's well-being.

■ Headaches have several possible causes

Excessive exercise may make your muscles hurt. However, thinking too much cannot cause a headache. The brain has no pain receptors, so a headache is not a brainache. Headaches can occur for almost any reason: they can be caused by stress or by relaxation, by hunger or by eating the wrong food, or by too much or too little sleep. The most common type of headache is a tension headache, affecting some 60% to 80% of people who suffer from frequent headaches. The pain of a tension headache is usually a dull, steady ache, often described as feeling like a tight band around the head. Migraine headaches are usually confined to one side of the head, often centered behind one eye. A migraine headache typically causes a throbbing pain that increases with each beat of the heart. It is sometimes called a sick headache because it may cause nausea and vomiting. Some migraine sufferers experience an aura, a group of sensory symptoms, different for different people, that occurs just before an attack. The aura may include visual disturbances (a blind

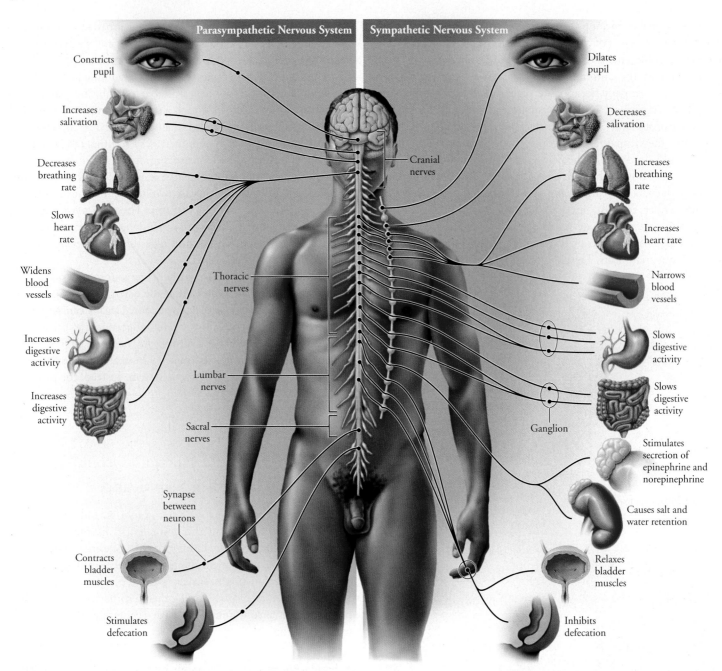

| Parasympathetic Nervous System | Sympathetic Nervous System |

Constricts pupil

Increases salivation

Decreases breathing rate

Slows heart rate

Widens blood vessels

Increases digestive activity

Increases digestive activity

Cranial nerves

Thoracic nerves

Lumbar nerves

Sacral nerves

Synapse between neurons

Contracts bladder muscles

Stimulates defecation

Dilates pupil

Decreases salivation

Increases breathing rate

Increases heart rate

Narrows blood vessels

Slows digestive activity

Slows digestive activity

Ganglion

Stimulates secretion of epinephrine and norepinephrine

Causes salt and water retention

Relaxes bladder muscles

Inhibits defecation

WEB TUTORIAL 8.3

FIGURE **8.12** The structure and function of the autonomic nervous system. Most organs are innervated by fibers from both the sympathetic and the parasympathetic nervous systems. When this dual innervation occurs, the two branches of the autonomic nervous system have opposite effects on the level of activity of that organ. A chain of ganglia links the pathways of the sympathetic nervous system, which therefore usually acts as a unit, with all its effects occurring together. In contrast, the ganglia of the parasympathetic nervous system are each near the organ they service, so parasympathetic effects are more localized.

To Sleep, Perchance to Dream

Not all sleep is the same. Indeed, each hour and a half during a typical night's sleep, you will cycle through a sequence of sleep stages. If the natural sleep pattern is disturbed, you may not feel well rested, even if you have "slept" for 8 hours. Stage 1 sleep occurs while you "drift off to sleep." During this period of transition from wakefulness to sleep, you become less aware of your surroundings, slowly tuning out lights and low-level noises (Figure 8.A). Within the next few minutes, stage 2 sleep usually occurs. During stage 2 sleep, you are unaware of your surroundings. Muscle tension is lower than during wakefulness, and breathing and heart rate also decrease. More than half the night's sleeping time is spent in periods of stage 2 sleep. Soon after entering stage 2, you sink even deeper into sleep—stages 3 and 4—and become increasingly difficult to awaken. After about 20 minutes of stage 4 sleep, you usually switch back to stage 3 followed by stage 2. Then, you begin an interval of paradoxical sleep, so named because the type of brain waves that occur are also seen during alert mental activity. It would therefore seem that the brain is quite active at this time, and yet it is more difficult to awaken a person in paradoxical sleep than at any other stage. Paradoxical sleep is also called rapid-eye-movement (REM) sleep, because your eyes move rapidly behind closed eyelids during this stage. In REM sleep, your heart and breathing rate can be quite variable, but almost every skeletal muscle in your body, except for those of eyes and ears, is virtually paralyzed. This paralysis may keep us from hurting ourselves or others.

The first bout of REM sleep usually occurs about 90 minutes after falling asleep and lasts about 10 minutes. The REM period of each successive cycle is a little longer. By morning, the interval of REM sleep may be about an hour long. As the bouts of REM lengthen, stages 3 and 4 are lost. So the second half of a night's sleep is not identical to the first half: there is much more REM in the second half. This is why taking several short catnaps does not provide the same quality of sleep as a single longer period of sleep, even if the total number of hours of sleep is identical.

REM sleep is the stage when dreams that have a storylike progression of events are apt to occur. About 80% of the people who are awakened during REM sleep and asked what was going on say they were dreaming. Only about 20% report that they were dreaming if awakened during other stages of sleep. Moreover, the dreams that occur outside of REM sleep are more likely to be a thought, an image, or an emotion than a story.

Unfortunately, millions of Americans cheat on sleep. Students pull "all-nighters" studying for exams. Many of them are holding down jobs as well. Parents juggle jobs and family. Workers cope with long shifts and long commutes. And when does *fun* fit into the schedule? People this busy may not have enough time for a good night's sleep.

After insufficient sleep, people usually can manage to get through the day as long as what they are doing is simple—walking, seeing, hearing. However, they cannot think clearly about complicated matters and reach a rational decision. Their attention span is short, and their ability to learn is affected. Sleepy students often sit through class in a daze and sometimes nod off.

Sleep deprivation can be hazardous to health. Drowsiness is a major cause of industrial accidents and traffic fatalities. In a national poll, 20% of American drivers reported having fallen asleep at the wheel within the past year. Driving on Friday night is a greater risk than on Monday night, because so many drivers have been sleep deprived all week.

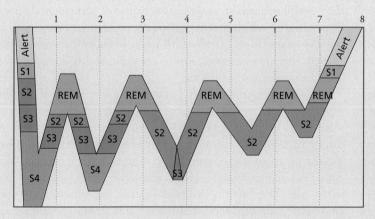

FIGURE **8.A** During a good night's sleep a person generally cycles through a sequence of sleep stages every 90 minutes.

continued →

Ironically, because of the pressures of life, sleep does not always come easily. Tens of millions of Americans will lie awake tonight suffering from insomnia. This sleep disorder occurs in different patterns—difficulty falling asleep, waking up during the night, or waking up earlier than desired.

Here are some suggestions for what to do if you occasionally have trouble sleeping. Establish a regular bedtime and a regular waking time,

but do not remain in bed if you cannot sleep. Relax before bedtime: read a book, watch television, take a warm shower—whatever helps you unwind. For at least 6 hours before bedtime, avoid caffeinated beverages, such as coffee, tea, and cola. Also, avoid drinking alcohol for 2 hours before bedtime. Establish a regular pattern of exercise, because mild exercise promotes sleep and reduces stress. However, do not exercise too close to bedtime. Lastly, avoid using

sleeping pills, even those sold over the counter in the drugstore. At best, they decrease the amount of time it takes to fall asleep by only 10 to 20 minutes and lengthen the night's sleep by only 20 to 40 minutes. Sleeping pills decrease the user's REM sleep, and the REM sleep that does occur is not normal. Furthermore, people quickly become less sensitive to the effects of sleeping pills and as a result tend to increase the dosage to a level that is dangerous. 👫

---

spot, zigzag lines, flashing lights), auditory hallucinations, or numbness. Migraines are set off by an imbalance in the brain's chemistry. Specifically, the level of one of the brain's chemical neurotransmitters, serotonin, is low. With too little serotonin, pain messages flood the brain. Cluster headaches are headaches that tend to occur in groups, occurring two or three times a day for days or weeks. More common in men than in women, cluster headaches cause severe pain that lasts for a few minutes to a few hours, often waking a person from a sound sleep.

what would you do?

In an experimental pain treatment for severe headaches, a tiny electrode is implanted in the skin and placed near the nerve responsible for the pain. The device, powered by a battery implanted near the collarbone, delivers continuous electric pulses intended to block the pain signals in the nerve and stop the pain. If you suffered from severely painful headaches, would you opt for this treatment? What criteria would you use to decide?

■ Strokes occur when the brain is deprived of blood

A stroke, also called a cerebrovascular accident, is the death of nerve cells in a region of the brain. The extent and location of the mental or physical impairment caused by a stroke depend on the region of the brain involved. If the left side of the brain is affected, the person may lose sensations in or the ability to move parts of the right side of his or her body because motor nerve pathways cross from one side of the brain to the other in the lower brain. Because the language centers are usually in the left hemisphere, the person may have difficulty speaking. When the stroke damages the right rear of the brain, some people show what is called the neglect syndrome and behave as if the left side of things, even their own bodies, does not exist. The person may comb only the hair on the right side of the head or eat only the food on the right side of the plate.

Neurons have a high demand for both oxygen and glucose. Therefore, when the blood supply to a portion of the brain is shut off, the affected neurons begin to die within minutes. Common causes of strokes include blood clots blocking a vessel, hemorrhage from the rupture of a blood vessel in one of the meninges, or the formation of fatty deposits that block a vessel. High blood pressure, heart disease, diabetes, smoking, obesity, and excessive alcohol intake increase the risk of stroke.

■ Coma is a lack of response to all sensory input

Although a comatose person seems to be asleep—with eyes closed and no recognizable speech—a coma is not deep sleep. A person in a coma is totally unresponsive to all sensory input and cannot be awakened. Although the cerebral cortex is most directly responsible for consciousness, damage to the cerebrum is rarely the cause of coma. Instead, coma is caused by trauma to neurons in regions of the brain responsible for stimulating the cerebrum, particularly those in the reticular activating system or thalamus. Coma can be caused by mechanical shock—as might be caused by a blow to the head— tumors, infections, drug overdose, or failure of the liver or kidney.

■ Spinal cord injury results in impaired function below the site of injury

The spinal cord is the pathway that allows the brain to communicate with the rest of the body. Therefore, damage to the spinal cord can impair sensation and motor control below the site of injury. The extent and location of the injury will determine how long these symptoms persist, as well as the degree of permanent damage. Depending on which nerve tracts are damaged, injury may result in the loss of sensation, paralysis, or both. If the cord is completely severed, there is a complete loss of sensation and voluntary movement below the level of the cut.

Restoring the ability to function to people with spinal cord injuries is an active area of research. Some researchers are trying to reestablish neural connections by stimulating nerve growth through treatments with nerve growth factors. Others are exploring the potential use of embryonic stem cells for treatment (discussed in Chapter 19a). Stem cells retain the ability to develop into nerve cells. Mice with spinal cord injuries that were treated with embryonic stem cells recovered some ability to move. Another approach to restoring the ability to move is to use computers to electronically stimulate specific muscles and muscle groups. The stimulation is delivered through wires that are either implanted under the skin or woven into the fabric of tight-fitting clothing. A small computer, which is usually worn at the wrist, directs the stimulation to the appropriate muscles. This technology has helped some people with a spinal cord injury to walk again. It has also helped some people by stimulating the diaphragm, a muscle important in breathing.

exploring further . . .

In Chapter 7, we learned that neurons communicate with one another using chemicals called neurotransmitters. Neurotransmitter molecules fit into receptors on the membrane of the receiving neuron and cause ion channels to open, either exciting or inhibiting the receiving neuron. Different neurotransmitters play roles in different behavioral systems.

In this chapter, we learned that different parts of the nervous system are specialized for different functions. The limbic system of the brain is a "pleasure center." The sympathetic nervous system prepares the body for emergency situations.

Next, in Chapter 8a, Special Topic: Drugs and the Mind, we will consider psychoactive drugs—those that affect a person's mental state. We will see that psychoactive drugs work by increasing or decreasing the effects of specific neurotransmitters and therefore affecting specific regions of the brain.

HIGHLIGHTING THE CONCEPTS

The Nervous System Consists of the Central and Peripheral Nervous Systems (pp. 125–126)

1. The nervous system is divided into the central nervous system (CNS), meaning the brain and spinal cord, and the peripheral nervous system (PNS), meaning all the neural tissue outside the CNS. Ganglia are clusters of nerve cell bodies located outside the CNS. The peripheral nervous system can be further subdivided into the somatic nervous system and the autonomic nervous system.

Bones, Membranes, and Cerebrospinal Fluid Protect the Central Nervous System (pp. 126–127)

2. The brain and the spinal cord are protected by the bony cases of the skull and vertebral column, by membranes (the meninges), and by a fluid cushion (cerebrospinal fluid).
3. The meninges are three protective layers of connective tissue that cover the brain and spinal cord. Bacteria and viruses can cause inflammation of the meninges, resulting in the condition called meningitis.
4. The cerebrospinal fluid, located between layers of the meninges, serves as a shock absorber for the brain, supports the brain, and provides nourishment to the brain.
5. The blood-brain barrier is a filter that allows only certain substances to enter the cerebrospinal fluid from the blood, thus protecting the brain and spinal cord from many potentially damaging substances.

The Brain Is the Central Command Center (pp. 127–132)

6. The brain serves as the body's central command center, coordinating and regulating the body's other systems.
7. The cerebrum is the thinking, conscious part of the brain. It consists of two hemispheres. Each hemisphere receives sensory impressions from and directs the movements of the opposite side of the body. The cerebrum has an outer layer of gray matter called the cerebral cortex and an underlying layer of white matter consisting of myelinated nerve tracts that allow communication between various regions of the brain.
8. The cerebral cortex has three types of functional areas: sensory, motor, and association. Our awareness of sensation depends on the sensory areas of the cerebral cortex. Motor areas of the brain control the movement of different parts of the body. Association areas communicate with the sensory and motor areas to analyze and act on sensory input.
9. The thalamus is an important relay station for all sensory experience except smell. It also plays a role in motor activity, stimulation of the cerebral cortex, and memory.
10. The hypothalamus is essential in maintaining a stable environment within the body. It regulates many vital physiological functions, such as blood pressure, heart rate, breathing rate, digestion, and body temperature. The hypothalamus also coordinates the activities of the nervous and endocrine systems through its connection to the pituitary gland. As part of the limbic system, the hypothalamus is a center for emotions. It also serves as a "master biological clock."
11. The primary function of the cerebellum is sensory-motor coordination. It integrates information from the motor cortex and sensory pathways to produce smooth movements. The cerebellum also stores memories of learned motor skills.
12. The medulla oblongata regulates breathing, heart rate, and blood pressure. It also serves as a pathway for all sensory messages to higher brain centers and motor messages leaving the brain.
13. The pons connects lower portions of the CNS with higher brain structures. It connects the spinal cord and cerebellum to the cerebrum, thalamus, and hypothalamus.
14. The limbic system, which includes several brain structures, is largely responsible for emotions. The hippocampus, which is part of the limbic system, is essential to converting short-term memory to long-term memory.
15. The reticular activating system is a complex network of neurons that filters sensory input and keeps the cerebral cortex in an alert state.

The Spinal Cord Transmits Messages to and from the Brain and Is a Reflex Center (pp. 133–134)

16. The spinal cord is a cable of nerve tissue extending from the medulla to approximately the bottom of the rib cage. The spinal cord has two functions: to conduct messages between the brain and the body and to serve as a reflex center.

WEB TUTORIAL 8.1 Reflex Arcs

The Peripheral Nervous System Consists of the Somatic and Autonomic Nervous Systems (pp. 134–135)

17. The PNS consists of spinal nerves, each originating in the spinal cord and serving a specific region of the body, and cranial nerves, each arising from the brain and serving the structures of the head and certain body parts such as the heart and diaphragm.
18. The PNS is divided into the somatic nervous system, which governs conscious sensations and voluntary movements, and the autonomic nervous system, which helps regulate our unconscious, involuntary internal activities.
19. The autonomic nervous system can be divided into the sympathetic and parasympathetic nervous systems, two branches with antagonistic actions. The sympathetic nervous system gears the body to face stressful or emergency situations. The parasympathetic nervous system adjusts body functioning so that energy is conserved during restful times.

WEB TUTORIAL 8.2 Cranial and Spinal Nerves
WEB TUTORIAL 8.3 The Autonomic Nervous System

Disorders of the Nervous System Vary in Health Significance (pp. 135–138)

20. Headaches can range from relatively mild (tension headaches) to severe (migraine and cluster headaches).

21. A stroke is caused by an interruption of blood flow leading to the death of nerve cells. The effects depend on the region of the brain affected.
22. Coma, a condition in which a person is totally unresponsive to sensory input, can be caused by a blow to the head, tumors, infections, drugs, or failure of the liver or kidney.
23. Because the spinal cord contains the pathways of communication between the brain and the rest of the body, damage to the spinal cord impairs functioning below the site of injury.

KEY TERMS

central nervous system   *p. 125*
ganglia   *p. 126*
peripheral nervous system   *p. 126*
somatic nervous system   *p. 126*
autonomic nervous system   *p. 126*
sympathetic nervous system
 p. 126
parasympathetic nervous system
 p. 126
meninges   *p. 126*

cerebrospinal fluid   *p. 126*
blood-brain barrier   *p. 127*
cerebrum   *p. 128*
cerebral cortex   *p. 128*
gray matter   *p. 128*
white matter   *p. 128*
primary somatosensory area
 p. 129
primary motor area   *p. 129*
prefrontal cortex   *p. 129*

thalamus   *p. 130*
hypothalamus   *p. 130*
cerebellum   *p. 131*
medulla oblongata   *p. 131*
midbrain   *p. 132*
pons   *p. 132*
limbic system   *p. 132*
short-term memory   *p. 132*
long-term memory   *p. 132*
hippocampus   *p. 132*

reticular activating system
 p. 132
spinal cord   *p. 133*
reflex arc   *p. 133*
spinal nerves   *p. 134*
cranial nerves   *p. 134*

REVIEWING THE CONCEPTS

1. Distinguish between the central nervous system and the peripheral nervous system. List the principal components of each.   *pp. 125–126*
2. Describe three ways the brain and spinal cord are protected.   *pp. 126–127*
3. What are the functions of cerebrospinal fluid?   *pp. 126–127*
4. What is gray matter? What is white matter?   *pp. 128–129*
5. Describe the three types of functional areas of the cerebral cortex.   *p. 129*
6. In what way are the organizations of the primary somatosensory area and that of the primary motor area of the cerebral cortex similar? In what way do these areas differ?   *p. 129*
7. List five functions of the hypothalamus.   *pp. 130–131*
8. What is the function of the cerebellum?   *p. 131*
9. Which functional system of the brain is responsible for emotions?   *p. 132*
10. Describe the two functions of the reticular activating system.   *p. 132*
11. List the two functions of the spinal cord and relate each function to the structure of the spinal cord.   *pp. 133–134*
12. You are cooking dinner and carelessly touch the hot burner on the stove. You remove your hand before you are even aware of the pain. Using the anatomy of a spinal reflex arc, explain how you could react before you were aware of the pain.   *p. 134*
13. What are the two divisions of the peripheral nervous system? What type of response does each control?   *pp. 134–135*
14. Compare and contrast the functions of the sympathetic and parasympathetic nervous systems.   *p. 135*
15. List some effects of sympathetic stimulation, and explain how these prepare the body for an emergency.   *p. 135*
16. You are watching a football game with your friends. A wide receiver makes an incredible catch and then runs 20 yards, skillfully dodging defensive players to make a touchdown. Your friend Joe says, "Amazing! How *does* he do that?" The receiver's outstanding sensory-motor coordination is largely due to the actions of his
    a. cerebellum.
    b. medulla.
    c. reticular activating system.
    d. hypothalamus.

17. As you sit here studying, you are unlikely to be aware of the pressure of your clothes against your body and the rustling of paper as other students turn pages. The part of the brain that "decides" that these are unimportant stimuli is the
    a. reticular activating system.
    b. cerebellum.
    c. hypothalamus.
    d. medulla.
18. Belinda was riding a bicycle without a helmet and was struck by a car. She hit the back of her head very hard in the fall. The physician is quite concerned because the medulla is located at the base of the skull. She explains to the parents that injury to the medulla could result in
    a. the loss of coordination so that the child may never regain the motor skills needed to ride a bicycle.
    b. the loss of speech.
    c. amnesia (the loss of all memory).
    d. death because many life-support systems are controlled here.
19. The neural center that regulates body temperature is the _____.
20. The region of the brain that regulates basic physiological processes such as breathing and heart rate is the _____.
21. The branch of the nervous system that prepares the body to respond to emergency situations is the _____.
22. The brain region responsible for intelligence and thinking is the _____.

APPLYING THE CONCEPTS

1. Your Aunt Rosa had a stroke; that is, some of the neurons in her brain died or were injured when a blood clot or hemorrhage reduced the blood supply to them. She can understand what you say to her, but she cannot speak to answer you. She has trouble moving her right arm. What region of the brain was affected by the stroke? How do you know?

2. Joe and Henry were both in car accidents, and both suffered spinal cord damage. Joe's injury was in the lower back, and Henry's was in the neck region. The degree of injury to the spinal cord is similar in both Joe and Henry. Would the resulting problems be equal in severity? Explain. Would either one or both require a respirator to breathe for him? Why? Describe some of the difficulties that you might expect Joe and Henry to have.

3. When you have a cold, you might take a decongestant to help you breathe. Some decongestants contain pseudoephedrine, which mimics the effects of the sympathetic nervous system. What side effects might you expect? Would you expect this medication to make you drowsy?

4. When you have dental work done, the dentist often administers a local anesthetic in the gums near the region that requires drilling. You are usually advised not to eat anything until the anesthetic wears off. This advice is given out of concern for your tongue, not your teeth. Why?

5. After Jorge's car accident he could remember events that took place before the accident, but he would quickly forget a conversation or a television show he just watched. What part of the brain was injured in the accident?

Additional questions can be found on the companion website.

8a

Drugs and the Mind

Psychoactive Drugs Alter Communication between Neurons

Drug Dependence Causes Continued Drug Use

Alcohol Depresses the Central Nervous System
- The rate of alcohol absorption depends on its concentration
- Alcohol is distributed to all body tissues
- The rate of elimination of alcohol from the body cannot be increased
- Alcohol has many health-related effects

Marijuana's Psychoactive Ingredient Is THC
- Marijuana binds to THC receptors in the brain
- Long-term marijuana use has many effects on the body
- Legalization of medical marijuana is controversial

Stimulants Excite the Central Nervous System
- Cocaine augments the neurotransmitters dopamine and norepinephrine
- Amphetamines augment the neurotransmitters dopamine and norepinephrine

Hallucinogenic Drugs Alter Sensory Perception

Sedatives Depress the Central Nervous System

Opiates Reduce Pain

We may think of alcohol as adding to the festivity of an occasion, but in fact its effect on the brain is that of a depressant.

Maddie had decided to keep her wedding small and was excited to be planning it herself. Even so, she was beginning to feel overwhelmed by all the details. After weeks of lying awake nights running through a long mental checklist of errands to be run and decisions still to be made, she realized that what she needed more than anything was a decent night's sleep. Her doctor consented to prescribe a mild sedative that helped her get more rest and feel calmer in the days leading up to the celebration.

The wedding went as smoothly and happily as she had dreamed. But a few weeks later, when she was beginning to feel settled in her new life, Maddie stopped taking the sleeping pills and was distressed to find that her sleeplessness and anxiety flared up worse than ever. Alarmed, she consulted her doctor. Apparently, her nervous system had developed a dependency on the drug, even in the limited period over which she had taken it. She and her doctor made a plan that included substituting the original sedative for a similar one and reducing her use of it in gradual increments. If that didn't work, she might have to seek more specialized treatment.

In this chapter, we will consider how drugs affect the nervous system and alter our state of mind. We will see why the use of some drugs leads to a need to continue using the drug. Then we will take a look at some of the more common mind-altering drugs. ■

Psychoactive Drugs Alter Communication between Neurons

A drug that alters one's mood or emotional state is often described as a psychoactive drug. The mind-altering effects of these drugs result from their ability to alter communication between nerve cells. As you learned in Chapter 7, neurons communicate with one another using chemicals called neurotransmitters. Neurotransmitters are released by one neuron, diffuse across a small gap, and bind to specific receptors on another neuron, triggering changes in the activity of the second neuron. Under normal conditions, the action of the neurotransmitter is stopped almost immediately because the neurotransmitter is either broken down by enzymes, is reabsorbed into the cell that released it, or simply diffuses away.

A psychoactive drug may alter this communication between neurons in any of several ways (Figure 8a.1). It may stimulate the release of a neurotransmitter, thereby enhancing the response of the receiving neuron. It may inhibit the release of a neurotransmitter and thus dampen the response of the receiving neuron. Or, it may increase and prolong the effect of a neurotransmitter by delaying its removal from the synapse. If the drug is chemically similar to the normal neurotransmitter, it may bind to the receptor and affect the activity of the receiving neuron in the same manner as the neurotransmitter. Alternatively, its binding to the receptor may prevent the neurotransmitter from acting at all.

One of the problems with using psychoactive drugs is that the user may develop some level of dependence on the drug. Before we consider the drugs themselves, let's consider the concepts of dependence and tolerance.

Drug Dependence Causes Continued Drug Use

Tolerance is a progressive decrease in the effectiveness of a drug in a given person. As tolerance to a particular drug develops, a person must take larger or more frequent doses to produce the same effect. Tolerance develops partly because the body steps up its production of enzymes that break down the drug and partly because of changes in the nerve cells that make them less responsive to the drug. Cross-tolerance occurs when tolerance to one drug results in a lessened response to another, usually similar drug. If a person abuses codeine, for instance, tolerance develops not only for codeine but also for other drugs that have similar effects on the nervous system, such as morphine and heroin.

A loose definition of dependence might be, "It is what causes a person to continue using a drug." More precisely, dependence is the state in which the drug is necessary for physical or psychological well-being. A person who is physically dependent on a drug experiences withdrawal symptoms when the drug use is stopped.

Certain drugs cause users to continue using them because the drugs stimulate the "pleasure" centers in the limbic system of the brain (see Chapter 8). An animal with electrodes implanted in the pleasure center will quickly learn to press a lever to stimulate this brain region. If permitted, it will self-stimulate repeatedly, sometimes hundreds of times an hour, until it is exhausted. When the

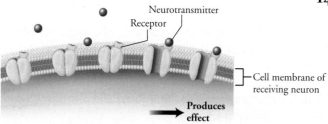

(a) The natural sequence of events: molecules of neurotransmitter released by one neuron diffuse across a gap and fit into receptors on the membrane of a receiving neuron, causing a response.

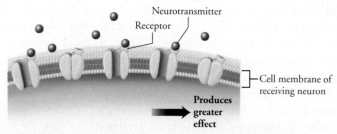

(b) A psychoactive drug may increase the number of neurotransmitter molecules released, increasing the response of the receiving neuron.

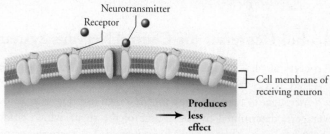

(c) A psychoactive drug may decrease the number of neurotransmitter molecules released, decreasing the response of the receiving neuron.

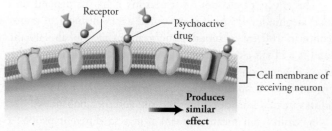

(d) A psychoactive drug may fit into the receptors for a neurotransmitter, causing a similar response by the receiving neuron.

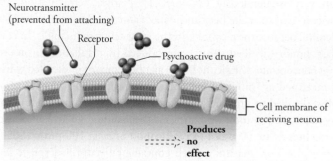

(e) A psychoactive drug may fit into the receptors for a neurotransmitter and prevent the neurotransmitter from entering the receptor, blocking the response by the receiving neuron.

 FIGURE **8a.1** Psychoactive drugs alter a person's mental state by affecting communication between neurons.

experiment is changed so that a small dose of a drug is released into the blood of an animal when it presses a lever, the animal will learn to press the lever to self-administer the drug—if the drug stimulates the pleasure center. Drugs that stimulate the pleasure center include cocaine, amphetamine, morphine, and nicotine. (Nicotine is discussed further in Chapter 14a.)

Several issues should be considered before a person takes a given drug. One is safety—both short term and long term. Are there health risks associated with the use of the drug? Is it a drug that leads to tolerance and dependence? Pregnant women have an additional issue to consider: the possible effects of the drug on the growing fetus. Most drugs can move from the mother's blood to the fetus's blood. In this case, the drug affects the fetus directly and causes effects similar to those experienced by the mother. Other drugs can harm the fetus by constricting blood vessels in the umbilical cord, shutting down the lifeline through which oxygen and nutrients reach the fetus.

In the following discussions of psychoactive drugs, we will consider their habit-forming potential as well as possible health risks associated with their use.

 Use this table to appreciate the relative strengths of various alcoholic beverages.

TABLE 8A.1 The ALCOHOLIC CONTENT OF SELECTED BEVERAGES

ALCOHOLIC BEVERAGE	PERCENT ALCOHOL (ETHANOL)
Light beer	4.5
Most beer	~5
Flavored malt beverages (coolers)	5
Ice beer	5.5–5.9
Dark beer (stout, porter, bock)	6–7
Malt liqour	8
French and German wines	8.5–10
American wine (most)	12–14
Sherry and port	18–21
Distilled liquor (vodka, gin, scotch, whiskey, rum, brandy, cognac)	Most 40% (80 proof); some 45% (90 proof) or 50% (100 proof)

Alcohol Depresses the Central Nervous System

In every alcoholic drink, the alcohol is ethanol. Ethanol is produced as a by-product of fermentation when yeast cells break down sugar to release energy for their own use. The taste of the beverage is determined by the source of the sugar, which comes from the fruit or vegetable that is fermented. For example, grapes are fermented to make wine, grain (barley, oats, rice, or wheat) to make beer, and malt to make scotch.

The effects of alcohol on a person's behavior depend on the blood alcohol level, which is measured as the number of grams of alcohol in 100 ml of blood. One gram of alcohol in 100 ml of blood is a blood alcohol level of 1%. The blood alcohol level in a person depends on several factors, including how much alcohol is consumed; how rapidly it is consumed (over what period of time); and its rate of absorption, distribution, and metabolism.

A "drink" can mean different things to different people. For some it is a can of beer, for others a glass of wine, and for others scotch on the rocks. The amount of pure ethanol in the beverage can vary tremendously (Table 8a.1). Natural fermentation, the process used to make beer and wine, cannot yield more than 15% alcohol, because the alcohol kills the yeast cells producing it. In contrast, liquor (distilled spirits) is produced by distillation, a process that concentrates the alcohol. The alcohol content of distilled spirits is measured as "proof." One degree of proof equals 0.5% alcohol. Most distilled spirits (vodka, gin, scotch, whiskey, rum, brandy, and cognacs) are 80 proof, which means they contain 40% alcohol. But some brands are 90 proof (45% alcohol) or even 100 proof (50%). Clearly, a 3-oz[1] martini, which contains only distilled spirits and therefore about 1.2 oz of alcohol, is more intoxicating than an 8-oz mug of beer, which contains about 0.4 oz of alcohol.

The differences in the alcohol content of beverages are reflected in the standard amounts in which they are served as drinks. Generally, the standard drink contains 0.5 oz (10 g) of pure ethanol. A standard drink is one bottle or can of beer, a glass of wine, or a jigger of 80 proof distilled spirits, such as gin, whisky, vodka, or scotch (Figure 8a.2)

■ The rate of alcohol absorption depends on its concentration

The intoxicating effects of alcohol begin when it is absorbed from the digestive system into the blood and is delivered to the brain. As a rule, the rate of absorption of alcohol depends on the concentration of alcohol in the drink: the higher the concentration, the faster the rate of absorption. So wine, beer, or distilled spirits diluted with a mixer will be absorbed more slowly than pure liquor. The choice of mixer also influences the rate of absorption. Carbonated beverages speed the rate of absorption because of the pressure of the gas bubbles.

Although very few substances are absorbed across the walls of the stomach, about 20% of the alcohol consumed is absorbed there. The remaining alcohol is absorbed through the intestines. Because alcohol can be absorbed from the stomach, a person begins to feel the effects of a drink quickly, usually within about 15 minutes. The presence of food in the stomach slows alcohol absorption because it dilutes the alcohol, covers some of the stomach lining through which alcohol would be absorbed, and slows the rate at which the alcohol passes into the intestines.

■ Alcohol is distributed to all body tissues

Ethanol is a small molecule that is soluble in both fat and water, so it is distributed to all body tissues. Therefore, overall body size affects one's blood alcohol level and the degree of intoxication. A large person would have a lower blood alcohol level and as a result

---

[1]In this chapter, the term *oz* (ounce) refers to a fluid ounce, which is equal to about 29 ml.

Type of Drink	Serving Size	Caloric Content
 6 oz Mixer Fruit juice	1 glass	35–105 calories
 8 oz Mixer Carbonated beverage	1 glass	~70–120 calories
 1.5 oz Distilled spirits (80 proof gin, whisky, vodka, scotch)	1 jigger	~100 calories
 5 oz Wine (12% alcohol)	1 glass	110–200 calories
 12 oz Most beer	1 bottle or can	140–150 calories

FIGURE **8a.2** A standard drink contains 0.5 oz of alcohol. Different types of alcoholic beverages vary in their alcohol content, so the size of a standard drink varies with its alcoholic content.

be less intoxicated than would a small, slender person after they both consumed the same amount of alcohol (Figure 8a.3).

■ The rate of elimination of alcohol from the body cannot be increased

Ninety-five percent of the alcohol that enters the body is metabolized (broken down) before it is eliminated. Most of that metabolism occurs in the liver, which converts alcohol to carbon

dioxide and water. The rate of metabolism is slow, about one-third of an ounce of pure ethanol per hour.

In practical terms, it takes slightly more than an hour for the liver to break down the alcohol contained in one standard drink—a can of beer or a glass of wine. Because alcohol cannot be stored anywhere in the body, it continues to circulate in the bloodstream until it is metabolized. Therefore, if more alcohol is consumed in an hour than is metabolized, both the blood alcohol level and the degree of intoxication increase (Figure 8a.4).

There is no way to increase the rate of alcohol metabolism by the liver and, therefore, no way to sober up quickly. A cup of coffee may slightly counter the drowsiness caused by alcohol, but it does not reduce the level of intoxication. Furthermore, because muscles do not metabolize alcohol, exercise does not help. So walking around the block will not make a person sober, nor will a cold shower.

If a man and a woman consume the same amount of alcohol, the woman will usually feel the effects before the man does. Part of the reason for this difference is anatomical. Women are usually smaller than men. In addition, on average, a woman's body contains a higher percentage of fat than a man's. Alcohol dissolves in fat more slowly than in water. As a result, the alcohol is diluted more slowly in a woman's body, and the effects are prolonged. But there's more to the story; the difference also has to do with differences in metabolism. Although the liver is the primary site for alcohol metabolism, some alcohol is broken down by an enzyme found in the stomach lining. Alcohol that is metabolized in the stomach never enters the bloodstream and cannot cause intoxication. Women have less of this enzyme in their stomach linings than do men. Consequently, a woman absorbs about 30% more of the alcohol in a drink than a man does. When weight differences between an average man and woman are also taken into account, 2 oz of liquor can have approximately the same effect on a woman as 4 oz would have on a man.

A small amount of alcohol, about 5%, is eliminated from the body unchanged through the lungs or in the urine. Alcohol eliminated from the lungs is the basis of the breathalyzer test that may be administered by law enforcement officers who want an on-the-spot sobriety check.

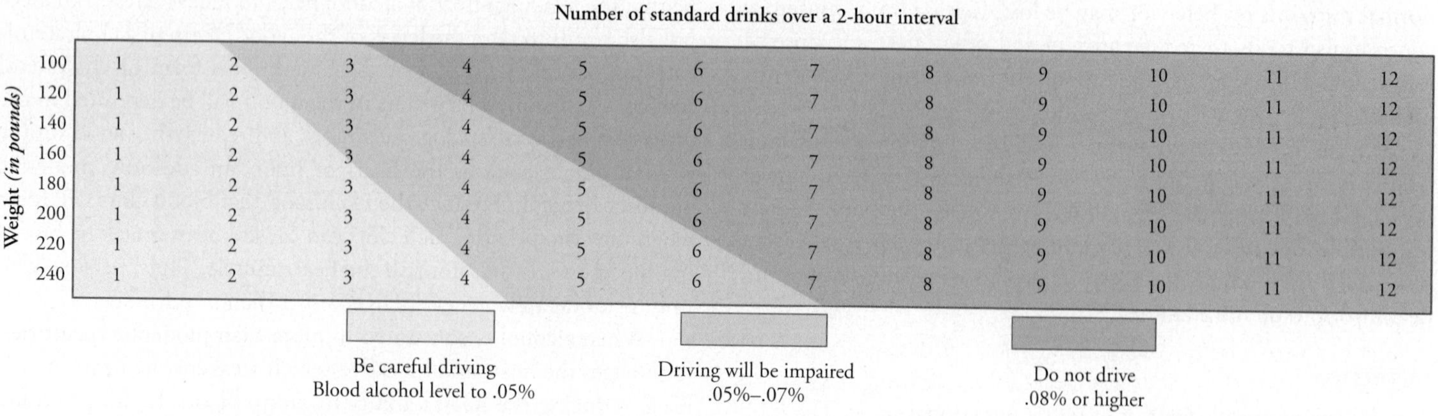

Number of standard drinks over a 2-hour interval

Weight (in pounds)												
100	1	2	3	4	5	6	7	8	9	10	11	12
120	1	2	3	4	5	6	7	8	9	10	11	12
140	1	2	3	4	5	6	7	8	9	10	11	12
160	1	2	3	4	5	6	7	8	9	10	11	12
180	1	2	3	4	5	6	7	8	9	10	11	12
200	1	2	3	4	5	6	7	8	9	10	11	12
220	1	2	3	4	5	6	7	8	9	10	11	12
240	1	2	3	4	5	6	7	8	9	10	11	12

Be careful driving
Blood alcohol level to .05%

Driving will be impaired
.05%–.07%

Do not drive
.08% or higher

FIGURE **8a.3** Alcohol consumption can impair driving. The blood level of alcohol depends on both the number of drinks consumed and body size. A small person has a higher blood alcohol level than a large person after consuming the same amount of alcohol. The blood level of alcohol determines the effect on the nervous system.

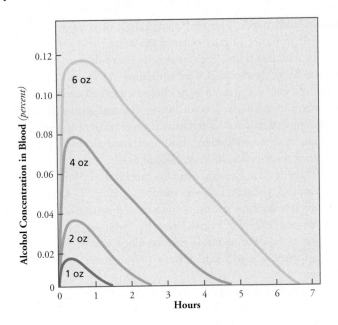

FIGURE **8a.4** Blood alcohol levels in a man at hourly intervals after drinking 1, 2, 4, or 6 oz of alcohol. Alcohol remains in the blood and has mind-altering effects until it is broken down in the liver. The liver breaks down alcohol at a slow constant rate of about one-third ounce of pure ethanol an hour. Therefore, it takes slightly more than an hour to metabolize the alcohol in one standard drink.

Source: Reproduced with permission from *The Encyclopedia Britannica*, © 2004 by Encyclopedia Britannica, Inc.

■ Alcohol has many health-related effects

Because alcohol has a negative effect on virtually every organ of the body, people do not have to be alcoholics for alcohol to impair their health.

THE NERVOUS SYSTEM

Many people believe that alcohol is a stimulant, but it is a depressant. In other words, it slows down the activity of all the neurons of the brain, beginning with the higher cortical, or "thinking," centers. Alcohol often is mistakenly thought to be a stimulant because it depresses the inhibitory neurons first, allowing the excitatory ones to take over. As alcohol removes the "brakes" from the brain, normal restraints on behavior may be lost. Release from inhibitory controls also tends to reduce anxiety and often creates a sense of well-being. However, discrimination, control of fine movements, memory, and concentration are gradually lost as well.

The brain centers for balance and coordination are affected next, causing a staggering gait. Numbed nerve cells send slower messages, resulting in slower reflexes. Eventually, the brain regions responsible for consciousness are inhibited, causing a person to pass out. Still higher concentrations of alcohol can cause coma and death from respiratory failure.

NUTRITION

Alcohol is high in calories but has little nutritional value. The high caloric value is part of the reason alcohol can make a person fat, but another part is that the body "prefers" to utilize alcohol over other nutrients. When alcohol is present in the body, it is metabolized for energy before fats are used. Unused fat is then stored in such places as the thighs, hips, and belly. (The term *beer belly* is often accurate!) Alcohol also robs the body in other ways. It decreases the absorption of certain vitamins, including folate, thiamine, B_{12}, and B_6. In addition, it causes the kidneys to pump important substances such as potassium ions, zinc, calcium, magnesium, and folate out of the body.

THE LIVER

Excessive alcohol consumption damages the liver, an organ that performs many vital functions in the body. Severe damage to the liver is a serious threat to life. Alcohol metabolism preempts fat metabolism in the liver, causing fats to accumulate in liver cells. Four or five drinks daily for several weeks are enough to cause fat accumulation to begin. At this early stage, however, the liver cells are not yet harmed, and with abstinence, they can be restored to normal. With continued drinking, the accumulating fat causes liver cells to enlarge, sometimes so much that the cells rupture or grow into cysts that replace normal cells. The fat also reduces blood flow through the liver, causing inflammation known as alcoholic hepatitis. Signs of alcoholic hepatitis include fever and tenderness in the upper abdominal region. Gradually, fibrous scar tissue may form, a condition known as cirrhosis, which further impedes blood flow and impairs liver functioning. Cirrhosis can lead to intestinal bleeding, kidney failure, fluid accumulation, and eventually death, if drinking continues. Indeed, cirrhosis of the liver, the ninth leading cause of death in the United States, is most often caused by alcohol abuse.

CANCER

A person who drinks heavily is at least twice as likely to develop cancer of the mouth, tongue, or esophagus than is a nondrinker. Evidence also exists that the risk of cancer from both drinking and smoking cigarettes is greater than the sum of the risks caused by either drinking or smoking alone.

HEART AND BLOOD VESSELS

Here's the good news. *Moderate* amounts of alcohol can be good for the heart. Teetotalers are more likely to suffer heart attacks than are persons who drink moderately, say a drink a day. One reason may be that the relaxing effect of alcohol helps to relieve stress. But alcohol also seems to raise the levels of the "good" form of a cholesterol-carrying particle—HDL—in the blood. This form of cholesterol reduces the likelihood that fats in the blood will be deposited in the walls of blood vessels, clog the vessels, and reduce the blood supply to vital organs such as the heart or brain. In addition, moderate amounts of alcohol reduce the likelihood that blood clots will form when they should not. Such clots can cause a heart attack by blocking blood vessels that nourish the heart muscle. Thus, persons who imbibe moderately generally live longer than nondrinkers.

When alcohol is consumed in more than moderate quantities, it damages the heart and blood vessels. It weakens the heart muscle itself, reducing the heart's ability to pump blood. It also promotes the deposit of fat in the blood vessels, making the heart work harder to pump blood through them. Consuming more than moderate quantities of alcohol may also elevate blood pressure substantially.

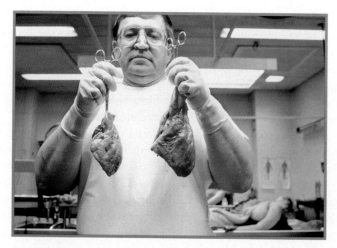

FIGURE **8a.5** The alcoholic's heart on the right is nearly twice the normal size.

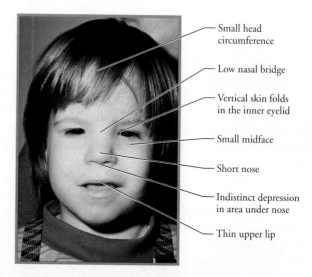

Small head circumference

Low nasal bridge

Vertical skin folds in the inner eyelid

Small midface

Short nose

Indistinct depression in area under nose

Thin upper lip

FIGURE **8a.6** Characteristic facial features of children with fetal alcohol syndrome

Together, these effects—damage to heart muscle, blood vessels clogged with fatty materials, and high blood pressure—can enlarge the heart to twice its normal size (Figure 8a.5).

ACCIDENTS

Accidents—at home or on the road—are more likely to happen when a person has been drinking. A blood alcohol level of only 0.04% to 0.05%, which can result from as few as two or three drinks, decreases peripheral vision, decreases the light sensitivity of the eye by 30% (equivalent to wearing sunglasses at night), slows recovery from headlight glare, and reduces reaction time by as much as 25%. At the same time, alcohol impairs the ability to concentrate and to judge distances and speed. Considering that a single mile of city driving requires a driver to make roughly 300 split-second decisions, such impairments can mean the difference between life and death.

EFFECTS ON FETAL DEVELOPMENT

Alcohol moves freely from a drinking mother's blood to the fetus. Soon the blood alcohol level of the fetus is the same as the mother's. If a pregnant woman consumes even one or two drinks a day, she increases the chance that her baby will be of lower than normal birth weight, and low birth weight is associated with complications after birth. Heavy drinking during pregnancy increases the risk of miscarriage and stillbirth.

Fetal alcohol syndrome (FAS) is a pattern of growth abnormalities and birth defects common among children of women who drink during pregnancy. These include mental retardation, growth deficiency, and characteristic facial features (Figure 8a.6), although all the characteristics of FAS are not always present in any one infant. After Down syndrome and spina bifida, alcohol is the third leading cause of birth defects associated with mental retardation. Of these three leading causes, alcohol use is the only one that is preventable.

FAS may be caused by chronic alcohol abuse throughout pregnancy or by binge drinking at critical times during fetal development. The fetus is particularly sensitive to chemicals, such as alcohol, during the first 3 months of pregnancy, when most organ systems are forming. The best advice for pregnant women is to avoid alcohol consumption entirely, because safe levels are not known.

ALCOHOLISM

Although alcohol is a legal drug, alcoholism is America's number one drug problem. Alcoholics can be young or old, rich or poor, and of any race, economic status, or profession. Because there is no such thing as a typical alcoholic, alcoholism can be difficult to identify.

Different alcoholics have different drinking patterns. Some binge and some chronically overindulge. But what they all share is a loss of control over their drinking. When an alcoholic takes the first sip of alcohol, he or she cannot predict how much or how long the drinking episode will continue.

Marijuana's Psychoactive Ingredient Is THC

Marijuana is the most widely used illegal drug in the United States today. It consists of the leaves, flowers, and stems of the Indian hemp plant, *Cannabis sativa*. The principal psychoactive ingredient (the component that produces mind-altering effects) is delta-9-tetrahydrocannabinol, or THC.

The effects of marijuana depend on the concentration of THC in the sample and on the amount consumed. In small to moderate doses, THC produces feelings of well-being and euphoria. In large doses, THC can cause hallucinations and paranoia. Anxiety may even reach panic proportions at very high doses.

Marijuana is not addicting in the sense of producing severe, unpleasant withdrawal symptoms. Nonetheless, a withdrawal syndrome has been identified. It includes symptoms such as restlessness, irritability, mild agitation, insomnia, nausea, and cramping. Withdrawal symptoms are not common, and if they do occur, are usually relatively mild and short-lived.

■ Marijuana binds to THC receptors in the brain

Researchers are just beginning to understand how marijuana brings about its effects. When THC binds to certain receptors on nerve cells, it triggers a cascade of events that ultimately lead to the "high" that the user experiences. (Figure 8a.7).[2] Once the receptors were discovered, researchers quickly set about to find the neurotransmitter that normally binds to them, reasoning that the receptors did not evolve millions of years ago just in case someone would someday decide to smoke marijuana. The researchers sorted through thousands of chemicals in pulverized pig brains to find one that would bind to THC receptors. The first one to be identified was a hitherto unknown chemical messenger they named *anandamide,* after the Sanskrit word meaning "internal bliss" (*ananda*). Anandamide functions as the brain's own THC. Its normal functions probably include the regulation of mood, memory, pain, appetite, and movement. In addition to mimicking anandamide, THC seems to stimulate the release of dopamine in the reward pathways of the brain.

■ Long-term marijuana use has many effects on the body

Marijuana is usually smoked. It should not be surprising, then, that the most clearly harmful effects of marijuana are on the respiratory system. However, the damage, it seems, is done by the residual materials in the smoke and not by the THC itself. People who smoke marijuana on a daily basis smoke fewer joints (marijuana cigarettes) than cigarette smokers smoke cigarettes. However, compared with a regular cigarette, a joint has 50% more tar, which contains cancer-causing chemicals. Also, marijuana smoke is usually inhaled deeply and held within the lungs. As a result, three times as much tar is deposited in the airways, and five times as much carbon monoxide is inhaled. Carbon monoxide prevents red blood cells from carrying needed oxygen to the cells of the body. The end result of prolonged smoking of marijuana is marked trauma to the respiratory system. Similar to cigarette smoke, marijuana smoke inflames air passages and reduces breathing capacity.

[2]There is a slightly different type of THC receptor found primarily on cells of the immune system.

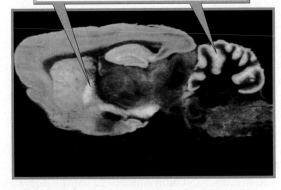

Marijuana receptors are most highly concentrated in regions involved in thinking and memory and in motor control areas.

FIGURE **8a.7** The areas with marijuana receptors appear yellow in this photograph showing a section of a rat's brain.

Another known effect of smoking marijuana is that it makes the heart beat faster, sometimes double its normal rate. In some people, marijuana also increases blood pressure. Either change increases the heart's workload and could pose a threat to people with preexisting cardiovascular problems, such as high blood pressure or atherosclerosis (fatty deposits in the arteries). Thus, although the risk is small, marijuana increases one's risk of immediate heart attack.

Some studies have shown that THC can interfere with reproductive functions in both males and females. At least some of these disturbances may result from the structural similarity between THC and the female hormone estrogen. Males who smoke marijuana often have lower levels of the male sex hormone testosterone and produce fewer sperm than do nonusers. If a man who frequently smokes marijuana is having difficulty becoming a father, quitting might boost his fertility. Testosterone and sperm levels return to normal when THC is cleared from the body.

The effects of marijuana on the female reproductive system are not clear. Although we know that THC interferes with ovulation in female monkeys, we know less about what it may do to human females. Some doctors, however, report menstrual problems and reproductive irregularities in women who smoke marijuana.

■ Legalization of medical marijuana is controversial

In response to public pressure to permit the use of marijuana for medicinal purposes, the Office of National Drug Control Policy recently funded a study by the Institute of Medicine (IOM), a branch of the National Academy of Sciences. The study team carefully evaluated what was known and what was not known about the medicinal use of marijuana—its helpful and harmful effects, its potential for abuse, and the likelihood that its use will lead to the use of other drugs.[3]

The study concluded that the THC in marijuana can relieve some symptoms of certain medical problems. It is an effective antinausea drug for cancer patients. It can also stimulate the appetite and combat the weight loss (wasting syndrome) that afflicts some people with AIDS. In addition, marijuana can relieve pain and so may help some patients with multiple sclerosis or phantom limb pain (pain that seems to originate in an amputated limb). Pressure within the eye is also reduced by marijuana. Thus, marijuana is often described as an effective treatment for glaucoma, a condition in which the pressure within the eyeball increases to the point where it can cause blindness. However, the study team found that this effect is short-lived and requires high doses. Thus, other existing therapies for glaucoma are better than marijuana.

One interesting conclusion of the IOM team is that marijuana's future as medicine does *not* involve smoking it. The team decided that the risks of *smoking* marijuana outweigh the benefits. Crude marijuana contains some 400 chemicals that when burned may release 2000 chemicals, some of which are known carcinogens. The researchers therefore recommend developing other means of delivering this drug to its targets in the body. Possibilities include inhalers, nasal

[3]Janet E. Joy, Stanley J. Watson, Jr., and John A. Benson, Jr., Editors. 1999. *Marijuana and Medicine, Assessing the Scientific Base.* National Academy Press, Washington DC; Alison Mack and Janet Joy for the Institute of Medicine. 2001. *Marijuana as Medicine? The Science Beyond the Controversy.* National Academy Press, Washington DC.

sprays or gels, rectal suppositories, or a pill that could be placed under the tongue and from there absorbed directly into the blood.

The team also concluded that there are no convincing data to support the notion that legalization of marijuana will lead to an increase in its nonmedical use or to general drug abuse. Will medical use of THC lead to addiction? *Probably* not. Few long-term marijuana users show withdrawal symptoms when they discontinue use, but some do.

Stimulants Excite the Central Nervous System

Stimulants are drugs that excite the central nervous system (CNS; see Chapter 8). Here we'll look at two types: cocaine and the amphetamines.

■ Cocaine augments the neurotransmitters dopamine and norepinephrine

Cocaine is extracted from the leaves of the coca plant (*Erythroxylon coca*), which grows naturally in the mountainous regions of South America. When cocaine powder is inhaled into the nasal cavity ("snorted"), it reaches the brain within a few seconds and produces an effect almost as intense as when a solution of cocaine is injected. Smoking the drug is an even more effective delivery route. Forms of cocaine that can be smoked—specifically, freebase and crack—are obtained by further extraction and purification. Crack is an extremely potent form of cocaine used by two-thirds of cocaine addicts (several hundred thousand) in the United States.

When ingested in one of these ways, cocaine brings about a rush of intense pleasure (euphoria), a sense of self-confidence and power, clarity of thought, and increased physical vigor. It does so

by increasing the levels of two "feel-good" neurotransmitters: dopamine and norepinephrine (Figure 8a.8). The euphoria is caused primarily by cocaine's effect on dopamine, a neurotransmitter used by nerve cells in the pleasure centers of the brain. Normally, dopamine is almost immediately reabsorbed into the nerve cell that released it, and its effect on the next nerve cell ceases. Cocaine, however, interferes with the reuptake of dopamine, thus increasing and prolonging dopamine's effect. Cocaine also increases the effects of another neurotransmitter, norepinephrine. Norepinephrine brings about the effects of the sympathetic nervous system, the part of the nervous system that prepares the body to face an emergency. Thus, cocaine also triggers the responses that ready the body for stress: increased heart rate and blood pressure, narrowing of certain blood vessels, dilation of pupils, a rise in body temperature, and a reduction of appetite.

The effects of cocaine are short-lived, lasting from 2 to 90 minutes, depending on how the cocaine enters the body. When the high wears off, it is generally followed by a "crash," a period of deep depression, anxiety, and extreme fatigue. These uncomfortable feelings often produce a craving for more cocaine. The higher the high, the lower the crash, and therefore, the more intense the craving. Crack, which is estimated to be about 75% pure compared with the 10% to 35% purity of street cocaine, causes higher highs and lower crashes and therefore is extremely addicting.

CARDIOVASCULAR RISKS

Cocaine may cause heart attack or stroke, emergencies in which an interruption of blood flow deprives heart cells or brain neurons of oxygen and nutrients. One way that cocaine blocks blood flow is by constricting arteries. Thus, one way it produces heart attack is

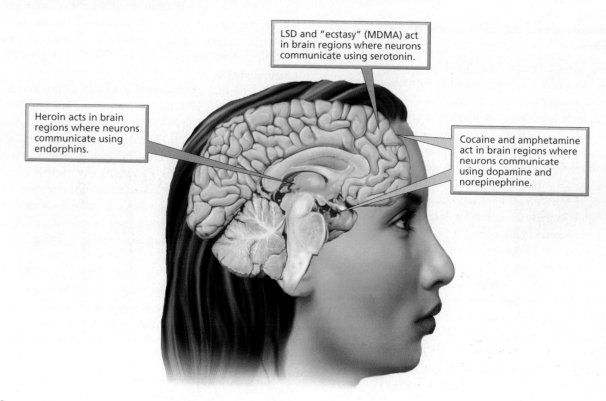

LSD and "ecstasy" (MDMA) act in brain regions where neurons communicate using serotonin.

Heroin acts in brain regions where neurons communicate using endorphins.

Cocaine and amphetamine act in brain regions where neurons communicate using dopamine and norepinephrine.

FIGURE **8a.8** The effects of certain drugs on the brain

by causing spasms in the arteries supplying the heart. Blood flow to the heart can also be disturbed by irregularities in heartbeat. By disabling the nerves that regulate heartbeat, cocaine can cause disturbed heartbeat rhythms that result in chest pain and heart palpitations (that uncomfortable feeling of being aware of your heart beating). Cocaine use may even cause the heart to stop beating. In addition, cocaine increases blood pressure, which may cause a blood vessel to burst.

RESPIRATORY RISKS

Stimulation of the CNS is always followed by depression. As the stimulatory effects of cocaine wear off, the respiratory centers in the brain that are responsible for breathing become depressed, or inhibited. Breathing may become shallow and slow and may stop completely. In short, cocaine use can cause respiratory failure that can lead to death.

Damage to the respiratory system as a result of cocaine use depends largely on how the drug is taken. If it is snorted, cocaine causes damage to the nerves, lining, and blood vessels of the nose. It can dry out the nose's delicate mucous membranes so that they crack and bleed almost continuously. Symptoms of a sinus infection—a perpetually runny nose and a dull headache spanning the bridge of the nose—are common. What is more alarming, the partition between the two nasal cavities may disintegrate. Because cocaine is a painkiller, considerable damage to the nose may occur before being discovered.

Smoking crack, on the other hand, damages the lungs and airways. The chronic irritation leads to bronchitis and may also cause lung damage from reduced oxygen flow into the lungs or blood flow through the lungs.

■ Amphetamines augment the neurotransmitters dopamine and norepinephrine

Amphetamines are synthetically produced stimulants that closely resemble dopamine and norepinephrine (the same neurotransmitters whose levels are increased by cocaine). There are many forms of amphetamine, including dextroamphetamine and methamphetamine, also known as crystal meth. Methamphetamine is illegally sold as a powder that can be injected, snorted, swallowed, or smoked. All amphetamines are stimulants of the CNS. Like cocaine, amphetamines make the user feel good—exhilarated, energetic, talkative, and confident. They suppress appetite and the need for sleep. Amphetamines are active for longer periods than cocaine—hours as opposed to minutes.

Amphetamines can be swallowed in a pill or injected intravenously. One crystalline form of methamphetamine called ice is smoked to produce effects similar to those of crack cocaine. Methamphetamine can produce hazardous effects, including blood vessel spasm, blood clot formation, insufficient blood flow to the heart, and accumulation of fluid in the lungs.

Amphetamines bring about their physical and psychological effects by causing the release of certain neurotransmitters, such as norepinephrine and especially dopamine. As a result, the drugs ele-

vate blood pressure, increase heart rate, and open up airways in the lungs (refer to Figure 8.12)

Tolerance to amphetamines develops (accompanied by simultaneous tolerance to cocaine) along with both physical and psychological dependence. Withdrawal symptoms include extreme fatigue, depression, and increased appetite. Also, the pleasurable feelings caused by amphetamine use can lead to a compulsion to overuse the drug.

Amphetamines cross the placenta and can affect the developing fetus. Women who use amphetamines during pregnancy are more likely to give birth prematurely and to have a low-birth-weight baby than are nonusers.

Hallucinogenic Drugs Alter Sensory Perception

The drugs that are classified as hallucinogenic are grouped together because of their similar effects, despite having very different structures. These effects include visual, auditory, or other distortions of sensation as well as vivid, unusual changes in thought and emotions.

The psychedelic drugs include some natural and some synthetic compounds. There are about six natural psychedelic drugs, the best known being mescaline, which comes from peyote cactus, and psilocybin, which is found in certain mushrooms. There are also synthetic ones, including LSD (lysergic acid diethylamide) and MDMA,[4] which is popularly known as ecstasy.

These drugs are thought to act by augmenting the action of the neurotransmitters serotonin, norepinephrine, or acetylcholine. For example, LSD, psilocybin, DMT (dimethyltryptamine), and bufotenin bind to the serotonin receptors in the brain, thereby mimicking the natural effects of serotonin. Ecstasy both binds to serotonin receptors and promotes the release of serotonin and dopamine. Mescaline, on the other hand, is structurally similar to norepinephrine.

The normal physiological reactions to psychedelic drugs are not especially harmful, but the distortions of reality the drugs produce may lead to behavior that is quite dangerous. *Bad trip* is the term most often used to describe an unpleasant reaction to a psychedelic drug. The symptoms may include paranoia, panic, depression, and confusion. The trip usually ends within 24 hours, when drug levels decrease. Brief recurrences of the sensory or emotional alterations produced by the drug are called flashbacks. Almost any aspect of the drug experience may be relived during a flashback. Most flashbacks are fleeting and merely annoying, but sometimes they can be frightening.

Tolerance for psychedelic drugs develops quickly. Furthermore, cross-tolerance among these drugs is the rule; that is, a person who has become tolerant of one psychedelic drug will be tolerant to others as well. Craving and withdrawal reactions are unknown, however, and laboratory animals offered LSD will not take it voluntarily.

---

[4]3,4-methylenedioxymethamphetamine

Ecstasy deserves additional consideration because its use is growing rapidly. It is sometimes called the "love drug" or "hug drug" because of users who say it puts them at peace with themselves and at ease with others. Also a stimulant, ecstasy causes increases in heart rate, blood pressure, and body temperature— sometimes to dangerous levels. People on ecstasy feel energetic enough to dance all night, sometimes suffering dehydration and heat stroke—which occasionally lead to death—as a result. (Deaths are rare, however.) The release of serotonin from neurons causes a euphoric high, but the temporary depletion of serotonin in the days following ecstasy use can bring depression and anxiety. Ecstasy can also cause nausea, vomiting, and dizziness. Another danger is that ecstasy pills may contain drugs other than MDMA. Among the drugs commonly found in these pills are the cough suppressant dextromethorphan, caffeine, ephedrine, and pseudoephedrine. Even the poison strychnine has been found.

Phencyclidine, or PCP, also known as "angel dust," and ketamine, also known as "Special K," are classified as psychedelic anesthetics. That is, they cause hallucinations and alter physical sensations. Unlike other psychedelic drugs, they do not work by raising serotonin levels.

Ketamine is gaining popularity as a "club drug." It can be snorted as a powder or swallowed as pills. Its effects range from a dreamy feeling to hallucinations to a deep, paralyzing out-of-body experience. High doses can cause seizures and coma.

Sedatives Depress the Central Nervous System

Sedatives are drugs that depress the CNS. They belong to the drug class known as benzodiazepines, two examples of which are Xanax and Ativan. Like alcohol, the depressant drugs affect inhibitory neurons first. As a result, low doses initially produce a relief from anxiety and a mild euphoria. Higher doses also inhibit excitatory neurons, and sleep follows.

The effects of different depressants are additive, so if you drink alcohol and also take a sedative, the depressant effects will be greatly intensified. Combining depressant drugs can easily lead to overdose. The respiratory system can, in fact, become so depressed that breathing stops. Administering a CNS stimulant—amphetamine, for example—cannot reverse an overdose of depressants. Although the drugs may have antagonistic (meaning opposite) effects, they work by binding to different receptors or by affecting different regions of the brain. Thus taking a stimulant and a depressant at the same time is like simultaneously stepping on the brake and the accelerator of a car. Hitting the accelerator does not remove the brake. To stop an overdose, it would be necessary to remove the drug from its receptors. A stimulant may temporarily arouse the person, but when it wears off, the resulting rebound depression added to the residual depression caused by the sedative could be fatal.

Continued use of sedatives leads to tolerance as well as both psychological and physical dependence. Cross-tolerance also develops, so tolerance to one sedative reduces the response to other sedatives. Tolerance develops because the liver produces more of the enzymes that break down the drug and, to some extent, because the nerve cells become less sensitive to its effects. As tolerance develops, overdose becomes a greater risk. Increased doses are required for the drug's sedating effect, but the neurons in the breathing center are still sensitive to a lethal dose.

A dangerous depressant that has been gaining popularity lately is GHB (gamma-hydroxybutrate). It is a clear liquid that is administered by being mixed with a beverage. This fast-acting sedative has the reputation of a "date rape" drug because men have used it to make women vulnerable to sexual assault. The most common negative effects of GHB are dizziness, nausea, vomiting, confusion, and sometimes blackouts. Overdoses can be fatal because GHB inhibits the respiratory system.

Opiates Reduce Pain

The opiates are natural or synthetic drugs that affect the body in ways similar to morphine, the major pain-relieving agent in opium. These and related drugs have two faces. On one hand, they have a high potential for abuse because tolerance and physical dependence occur. On the other hand, they are medically important because they alleviate severe pain.

Morphine and codeine, which come from the opium poppy, were among the first opiates used. Heroin is a synthetic derivative of morphine that is more than twice as effective. Heroin is usually injected intravenously, but there are now forms that can be smoked or inhaled into the nasal cavity. Regardless of how it enters the body, heroin reaches the brain very quickly, producing a feeling that is usually described in ecstatic or sexual terms. This rush of euphoria has made heroin the drug preferred by opiate abusers and spurs its continued use.

The opiates, including heroin, exert their effects by binding the receptors for the body's endogenous (natural, internally produced) opiates: compounds called endorphins, enkephalins, and dynomorphins. These are neurotransmitters whose functions include roles in the perception of pain and fear.

Heroin and other commonly abused opiates have effects similar to those of morphine: euphoria, pain suppression, and reduction of anxiety. They also slow the breathing rate. An overdose may cause the user to fall into a coma and stop breathing. Overdose is always a potential problem with opiates bought on the street for illegal use. The buyer has no idea of the strength of the drug. Street supplies are often diluted with sugar. If a heroin addict who is accustomed to a diluted drug injects heroin that is much more potent than he or she is used to, death due to overdose often occurs. Extremely constricted pupils in an unconscious person are a sign of heroin overdose.

Many of the problems associated with heroin use arise because heroin addicts often suffer from a general disregard for good health practices. Other problems are associated with the injections themselves. Frequent intravenous injection of any drug is associated with certain ailments. Because of the constant puncturing, veins can become inflamed. In addition, shared needles may spread disease-causing organisms, including the viruses that cause AIDS and hepatitis and the bacterium that causes syphilis.

9
Sensory Systems

Sensory Receptors Generate Electrochemical Messages in Response to Stimuli

Receptors Are Classified by the Type of Stimulus to Which They Respond

Receptors for the General Senses Are Distributed throughout the Body

- Mechanoreceptors detect touch and pressure
- Cold and heat receptors detect temperature change
- Muscle spindles and Golgi tendon organs detect body and limb position
- Pain is caused by any sufficiently strong stimulus

Vision Depends on the Eye

- The wall of the eyeball has three layers
- The eye has two fluid-filled chambers
- Sharp vision requires the image to be focused on the retina
- Light changes the shape of pigment molecules, which generate neural messages
- Vision in dim light depends on rods
- Color vision depends on cones

Hearing Depends on the Ear

- The ear collects and amplifies sound waves and converts them to neural messages
- Variations in the movements of the basilar membrane determine loudness and pitch
- Hearing loss can be conductive or sensorineural
- Ear infections can occur in the ear canal or in the middle ear

Balance Depends on the Vestibular Apparatus of the Inner Ear

Smell and Taste Are the Chemical Senses

 HEALTH ISSUE Correcting Vision Problems

 ENVIRONMENTAL ISSUE Noise Pollution

We generally rely on our own senses—sight, hearing, taste, touch, and so on—for identifying the resources and conditions our bodies require and for recognizing danger. When extra-sensitive olfaction is needed, however, we sometimes borrow the smell receptors of a dog.

Lani, a Labrador retriever, is a trained and certified professional employed by a mold-testing company. Her job is to sniff through buildings in search of hidden mold that may be growing behind deceptively clean-looking drywall or beneath solid floors. She has been trained to recognize the odors of 18 genera of undesirable molds. When she senses one, she sits and points her nose toward its source. Lani works faster and is easier to transport and deploy than any electronic mold-detecting device, and follow-up laboratory analyses show that she is accurate. Like dogs who sniff out bombs, drugs, stolen money, accident victims, or evidence of arson, she provides her human partners with information about the environment that their own senses are not able to perceive.

Although most humans cannot sniff out mold hidden under floorboards, our senses are nevertheless quite impressive and do a good job of telling us about our surroundings. In this chapter, we will explore the body's general senses (such as touch, pressure, vibration, temperature, and pain) and our special senses (vision, hearing, balance, smell, and taste). We will look especially closely at vision, hearing, and balance, examining how light, sound, and body position are reported by the intricate structures of the eye and ear. We also will explore the relationship between smell and taste. ■

Sensory Receptors Generate Electrochemical Messages in Response to Stimuli

Information about the external and internal world comes to us through our **sensory receptors**, structures that are specialized to respond to stimuli, or changes in the environment, by generating electrochemical messages. If a stimulus is strong enough, these messages eventually become nerve impulses (action potentials) that are then conducted to the brain.

Sensation is an awareness of a stimulus. Whether a sensation is experienced as sight, sound, or something else depends on which part of the brain receives the nerve impulses. Photoreceptors respond best to light, but they can respond to pressure. Regardless of the stimulus, nerve impulses go to the visual cortex of the brain, and we see light. This is why, if you press gently on your closed eyelids, you will have the sensation of seeing clouds of light. The pressure stimulates visual receptors, which send nerve impulses to the visual cortex. We hear sounds when nerve impulses go to the auditory cortex of the brain. Therefore, if you press on the flap of skin at the opening of your ear, sensory receptors in the inner ear will send nerve impulses to the auditory cortex, and you will have the sensation of hearing sounds.

Perception is the conscious awareness of sensations. It occurs when the cerebral cortex integrates sensory input (Figure 9.1). For example, light reflected from an apple strikes the eye, stimulating some of the photoreceptors. The brain interprets the pattern of input from the photoreceptors as seeing an apple.

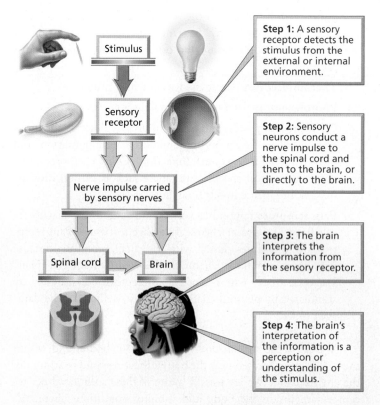

Step 1: A sensory receptor detects the stimulus from the external or internal environment.

Step 2: Sensory neurons conduct a nerve impulse to the spinal cord and then to the brain, or directly to the brain.

Step 3: The brain interprets the information from the sensory receptor.

Step 4: The brain's interpretation of the information is a perception or understanding of the stimulus.

FIGURE **9.1**   An overview of the steps involved in sensation and perception

Each type of sensory receptor responds best to one form of energy. Photoreceptors, for instance, respond best to light. The response of a sensory receptor is an electrochemical message (a change in the charge difference across plasma membrane) that varies in magnitude with the strength of the stimulus. For instance, the louder a sound, the larger the change in the charge difference across the membrane—up to a point. When the change reaches threshold level, an action potential (nerve impulse) is generated. Most types of sensory receptors gradually stop responding when they are continuously stimulated, a phenomenon called **sensory adaptation**. As receptors adapt in this way, we become less aware of the stimulus. For example, the musty smell of an antique store may be obvious to a person who just walked in, but the salesclerk working in the store no longer notices it. Some receptors, such as those for pressure and touch, adapt quickly. For this reason, we quickly become unaware of the feeling of our clothing against our skin. Other receptors adapt more slowly or not at all. For instance, the receptors in muscles and joints that report on the position of body parts, for instance, never adapt. Their continuous input is essential for coordinated movement and balance.

Receptors Are Classified by the Type of Stimulus to Which They Respond

Receptors are classified according to the stimulus to which they respond. Several classes of receptors are traditionally recognized:

1. **Mechanoreceptors** are responsible for the sensations we describe as touch, pressure, hearing, and equilibrium. They

respond to distortions in the receptor itself or in nearby cells. In addition, the body has mechanoreceptors that detect changes in blood pressure and others that indicate the body's position.

2. **Thermoreceptors** detect changes in temperature.

3. **Photoreceptors** detect changes in light intensity.

4. **Chemoreceptors** respond to chemicals. We describe the input from the chemoreceptors of the mouth as taste (gustation) and those from the nose as smell (olfaction). Other chemoreceptors monitor levels of specific substances such as carbon dioxide, oxygen, or glucose in our body fluids.

5. **Pain receptors** respond to very strong stimuli that usually result from physical or chemical damage to tissues. Pain receptors are sometimes classed with the chemoreceptors, because they often respond to chemicals liberated by damaged tissue, and occasionally with the mechanoreceptors, because they are stimulated by physical changes, such as swelling, in the damaged tissue.

We can sense stimuli both outside of our bodies and within them. Receptors located near the body surface respond to stimuli in the environment. We are usually aware of these stimuli. Other receptors are inside the body and monitor conditions there. Although we often are unaware of the activity of internal receptors, they play a vital role in maintaining homeostasis. In fact, they are key components of the feedback loops that regulate blood pressure, blood chemistry, and breathing rate. Internal receptors also cause us to feel pain, hunger, or thirst, thereby prompting us to attend to our body's needs.

The **general senses**—touch, pressure, vibration, temperature, body and limb position, and pain—arise from receptors in the skin, muscles, joints, bones, and internal organs. Although we are not usually aware of the general senses, they are important because they provide information about body position and help keep internal body conditions within the limits optimal for health. The **special senses** are vision, hearing, the sense of balance, or equilibrium, smell, and taste. These are what usually come to mind when we think of "the senses," largely because we are so dependent on them for perceiving and understanding the world. The receptors of the special senses are located in the head. Most of them reside within specific structures.

Receptors for the General Senses Are Distributed throughout the Body

The receptors for general senses are distributed throughout the body. Some monitor conditions within the body; others provide information about the world around us. Some of the receptors are free nerve endings; in other cases the nerve endings are encapsulated. Free nerve endings are branched tips of dendrites of sensory neurons. An encapsulated ending is one in which a connective tissue capsule encloses the tips of the dendrites.

■ **Mechanoreceptors detect touch and pressure**

Throughout life, we actively use touch as a way of learning about the world and of communicating with one another. The messages we get from pressure can be equally important. Some pressure receptors inform us of the need to loosen our belt after a big meal. Other pressure receptors monitor internal conditions, including blood pressure. As noted earlier, the receptors that respond to touch and pressure—to any stimulus that stretches, compresses, or twists the receptor membrane—are called mechanoreceptors.

Light touch, as when the cat's tail brushes your legs, is detected by several types of receptors (Figure 9.2). For example, free nerve endings wrapped around the base of the fine hairs on the skin detect any bending of those hairs. Free nerve endings and the special cells they end on (Merkel cells) form *Merkel disks*. They also sense light touch. When compressed, Merkel cells stimulate the free nerve endings in the associated Merkel disks to tell us that something has touched us. Merkel disks are found on both the hairy and hairless parts of the skin. *Meissner's corpuscles* are encapsulated nerve endings that tell us exactly where we have been touched. They are common on the hairless, very sensitive areas of skin, such as the lips, nipples, and fingertips.

The sensation of pressure generally lasts longer than does touch and is felt over a larger area. *Pacinian corpuscles*, which consist of onionlike layers of tissue surrounding a nerve ending, respond when pressure is first applied and therefore are important in sensing vibration. They are scattered in the deeper layers of skin and the underlying tissue. *Ruffini corpuscles* are encapsulated endings that respond to continuous pressure.

■ **Cold and heat receptors detect temperature change**

Thermoreceptors respond to changes in temperature. In humans, thermoreceptors are specialized free nerve endings found just below the surface of the skin. One kind responds to cold and another responds to warmth. They are widely distributed throughout the body but are especially numerous around the lips and mouth. You may have noticed that the sensation of hot or cold fades rapidly. This fading occurs because thermoreceptors are very active when temperature is changing but adapt rapidly when temperature is stable. As a result, the water in a hot tub may feel scalding at first, but very soon it feels comfortably warm.

■ **Muscle spindles and Golgi tendon organs detect body and limb position**

Whether you are at rest or in motion, the brain "knows" the location of all your body parts. It continuously scans the signals from muscles and joints to check body alignment and coordinate balance and movement. *Muscle spindles*—specialized muscle fibers wrapped in sensory nerve endings—monitor the length of a skeletal muscle. *Golgi tendon organs*—highly branched nerve fibers located in tendons (connective tissue bands that connect muscles to bones)—measure the degree of muscle tension. The brain combines information from muscle spindles and Golgi tendon organs with information from the inner ear (as we will see below) to coordinate our movements.

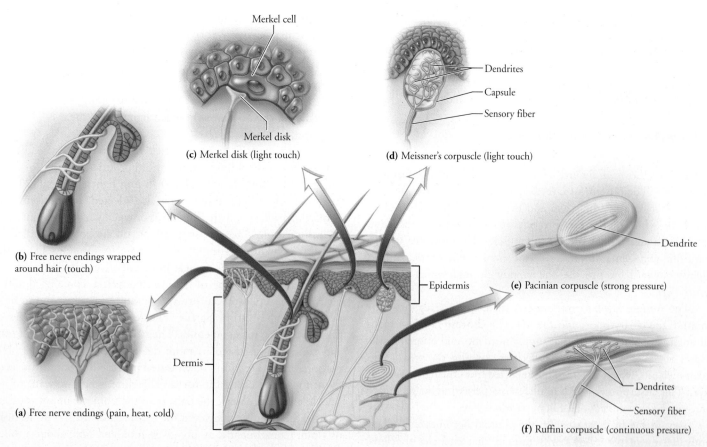

FIGURE **9.2** General sense receptors of the skin allow us to feel touch, pressure, temperature, and pain.

■ **Pain is caused by any sufficiently strong stimulus**

The receptors for pain are free nerve endings found in almost every tissue of the body. When tissue is damaged, cells release chemicals that alert the free nerve endings of the injury. The stimulated sensory neurons then carry the message to the brain, where it is interpreted as pain. Aspirin and ibuprofen reduce pain by interfering with the production of one of the released chemicals. Any stimulus strong enough to damage tissues, including heat, cold, touch, and pressure, will cause pain.

Many of our internal organs also have pain receptors. However, pain originating in an internal organ is sometimes perceived as pain in an uninjured region of the skin (Figure 9.3). This phenomenon is called **referred pain**. For example, the pain of a heart attack is often experienced as pain in the left arm. This pain probably occurs because sensory neurons from the internal organ and those from a particular region of the skin communicate with the same neurons in the spinal cord. Because the message is delivered to the brain by the same neurons, the brain interprets the input as coming from the skin.

Pain is an important mechanism that warns the body and protects it from injury. For example, pain usually prevents a person with a broken leg from causing additional damage by moving the limb. Nonetheless, few of us appreciate the value of pain while we are experiencing it. Furthermore, pain that persists long after the warning is needed can be debilitating.

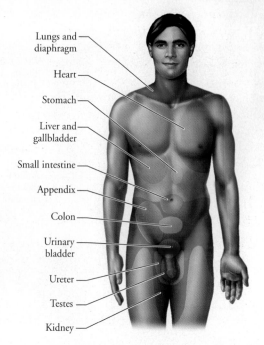

FIGURE **9.3** Referred pain. Pain from certain internal organs is sensed as originating in particular regions of the skin.

Vision Depends on the Eye

Humans are very visual creatures. We may not see detail as well as an eagle or movement as well as an insect, but we see far better than most other mammals and depend on vision in most of the activities that make up our daily lives.

■ The wall of the eyeball has three layers

The human eyeball is an irregular sphere about 25 mm (1 in.) in diameter. As shown in Figure 9.4, the wall of the eyeball consists of three layers. The outermost layer is a tough, fibrous covering with two distinct regions: the sclera and the cornea. The **sclera**, often called the white of the eye, protects and shapes the eyeball and serves as an attachment site for the muscles that move the eye. In the front and center of the eye, the transparent **cornea** bulges slightly outward and provides the window through which light enters the eye.

The middle layer of the eye has three distinct regions—the choroid, the ciliary body, and the iris. The **choroid** is a layer containing many blood vessels that supply nutrients and oxygen to the tissues of the eye. The choroid layer also contains the brown pigment melanin, which absorbs light reflected from the light-sensitive layer. This absorption of light results in sharper vision by helping to prevent excessive reflection of light within the eye.

Toward the front of the eye, the choroid becomes the **ciliary body,** a ring of tissue, primarily muscle, that encircles the **lens**, which focuses light on the retina. The ciliary body holds the lens in

place and controls its shape, which is important for focusing light on the light-sensitive layer of the eye.

In front of the ciliary body, the choroid becomes the iris. The **iris**, the colored portion of the eye that can be seen through the cornea, regulates the amount of light that enters the eye. It is shaped like a flat doughnut. The doughnut hole—the opening in the center of the iris through which light enters the eye—is called the **pupil**.

The iris contains smooth muscle fibers that automatically adjust the size of the pupil to admit the appropriate amount of light. The pupil becomes larger (dilates) in dim light and smaller (constricts) in bright light. Pupil size is also affected by emotions. The pupils dilate when you are frightened or when you are very interested in something. They constrict when you are bored. Candlelight creates a romantic setting partly because its dimness dilates the pupils, making the lovers appear more attentive and interested.

The innermost layer of the eye, the **retina**, contains almost a quarter-billion photoreceptors, the structures that respond to light by generating electrical signals. The retina contains two types of photoreceptors: rods and cones. The structure and function of rods and cones are discussed in more detail later in the chapter. The photoreceptors are most concentrated in a small region in the center of the retina called the **fovea**. Therefore, when we want to see fine details of an object, the light reflected from the object must be focused on the fovea. However, the fovea is only the size of the head of a pin. Consequently, at any given moment, only about a thousandth of our visual field is in sharp focus. Eye movements bring different parts of the visual field to the fovea. The **optic nerve** carries the message from the eye to the brain, where the message is interpreted. The region where the optic nerve leaves the retina has no photoreceptors. As a result, we cannot see an image that strikes this area. This area is, therefore, called the **blind spot**. To find your blind spot, cover your left eye and focus on the X with your right eye. You should see the circle in your peripheral vision. Place this book in front of you and slowly move your head toward it. As your head moves, the image of the circle becomes focused on different regions of the retina. When the circle disappears, its image is focused on the blind spot.

You are usually not aware of your blind spot because involuntary eye movements constantly move the position of the image on the retina, allowing the brain to "fill in" the missing parts as the visual information is processed by the brain. Table 9.1 summarizes the structures of the eye and their functions.

■ The eye has two fluid-filled chambers

In addition to having three outer layers, the eyeball can be divided into two fluid-filled cavities, or chambers (see Figure 9.4). The posterior chamber, located at the back of the eye between the lens and the retina, is filled with a jellylike fluid called **vitreous humor**. This fluid helps keep the eyeball from collapsing and holds the thin retina against the wall of the eye. The anterior chamber, located at the front of the eye between the cornea and the lens, is filled with a

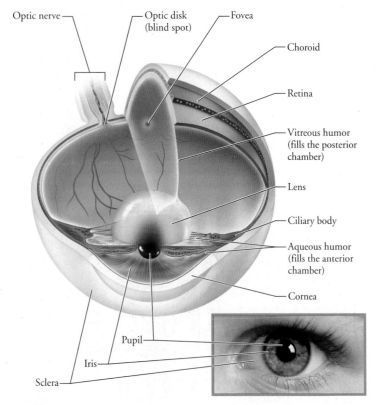

FIGURE **9.4** Structure of the human eye. Light enters the eye through the transparent cornea and then passes through the pupil. The lens focuses light on the light-sensitive retina, which contains the rods and cones.

TABLE 9.1 A REVIEW OF THE STRUCTURES OF THE EYE AND THEIR FUNCTIONS

STRUCTURE	DESCRIPTION	FUNCTION
Outer layer		
Sclera	Outer layer of the eye	Protects the eyeball
Cornea	Transparent dome of tissue forming the outer layer at the front of the eye	Refracts light, focusing it on the retina
Middle layer		
Choroid	Pigmented layer containing blood vessels	Absorbs stray light; delivers nutrients and oxygen to tissues of eye
Ciliary body	Encircles lens; contains the ciliary muscles	Controls shape of lens; secretes aqueous humor
Iris	Colored part of the eye	Regulates the amount of light entering the eye through the pupil
Pupil	Opening at the center of the iris	Opening for incoming light
Inner layer		
Retina	Layer of tissue that contains the photoreceptors (rods and cones); also contains bipolar and ganglion cells involved in retinal processing	Receives light and generates neural messages
Rods	Photoreceptor	Responsible for black and white vision and vision in dim light
Cones	Photoreceptor	Responsible for color vision and visual acuity
Fovea	Small pit in the retina that has a high concentration of cones	Provides detailed color vision
Other structures of the eye		
Lens	Transparent, semispherical body of tissue behind the iris and pupil	Fine focusing of light onto retina
Aqueous humor	Clear fluid found between the cornea and the lens	Refracts light and helps maintain shape of the eyeball
Vitreous humor	Gelatinous substance found within the chamber behind the lens	Refracts light and helps maintain shape of the eyeball
Optic nerve	Group of axons from the eye to the brain	Transmits impulses from the retina to the brain

fluid called **aqueous humor**. This clear fluid supplies nutrients and oxygen to the cornea and lens and carries away their metabolic wastes. In addition, the aqueous humor creates pressure within the eye, helping to maintain the shape of the eyeball. Unlike the vitreous humor, which is produced during embryonic development and is never replaced, aqueous humor is replaced about every 90 minutes. It is continuously produced from the capillaries of the ciliary body, circulates through the anterior chamber, and drains into the blood through a network of channels that surround the eye.

GLAUCOMA

If the drainage of aqueous humor is blocked, the pressure within the eye may increase to dangerous levels. This condition, called *glaucoma*, is the second most common cause of blindness (after cataracts, which are discussed shortly). The accumulating aqueous humor pushes the lens partially into the posterior cavity of the eye, increasing the pressure there and compressing the retina and optic nerve. This pressure, in turn, collapses the tiny blood vessels that nourish the photoreceptors and the fibers of the optic nerve. Deprived of nutrients and oxygen, the photoreceptors and nerve fibers begin to die, and vision fades. Unfortunately, glaucoma is progressive but painless, making detection difficult. Late signs include blurred vision, headaches, and seeing halos around objects. Because many people do not realize they have a problem until some vision has been lost, eye specialists recommend that everyone over 40 years of age be tested for glaucoma every year.

■ **Sharp vision requires the image to be focused on the retina**

Sharp, clear vision requires that the light rays entering the eye converge so that their focal point, the point where they are "in focus," is on the retina. The structures of the eye accomplish this focusing by bending the light rays to the necessary degree.

Most bending of light occurs as the light passes through the curved surface of the cornea. Because of the way that the curved cornea bends the light rays, the image created on the retina is upside down and backward. The cornea has a fixed shape, so it always bends light to the same degree and cannot make the adjustments needed to focus on objects at varying distances.

The lens, on the other hand, is elastic and can change shape to focus on both near and distant objects. Picture the lens as an underinflated round balloon. If you pull on the sides of such a balloon, it flattens. When you release its sides, the balloon assumes its usual, rounder shape. Similar pulling and releasing changes the shape of

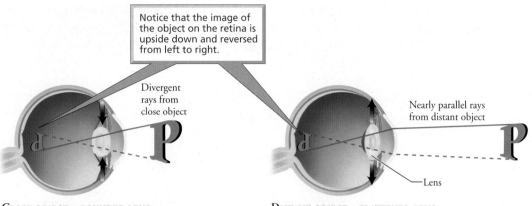

Notice that the image of the object on the retina is upside down and reversed from left to right.

Divergent rays from close object

Nearly parallel rays from distant object

Lens

CLOSE OBJECT – ROUNDED LENS:
• Ciliary muscles contract
• Ligaments to lens loosen
• Lens becomes rounded

DISTANT OBJECT – FLATTENED LENS:
• Ciliary muscles relax
• Ligaments to lens stretch
• Lens flattens

FIGURE **9.5** The lens changes shape so the eye can view objects at different distances.

the lens. Figure 9.5 shows that a rounder, thicker lens bends light to a greater degree, enabling the eye to focus on objects that are near. Changing the shape of the lens to change the bending of light is called **accommodation**.

These changes in lens shape are controlled by the ciliary muscle, which is attached to the lens by ligaments. Because the ciliary muscle is circular, its diameter becomes smaller when it contracts, much like pursed lips do. This contraction relaxes the tension on the ligaments, freeing the lens to assume the rounded shape needed to focus on nearby objects. Relaxation of the ciliary muscle increases the diameter of the ciliary muscle, increasing the tension on the ligaments and the lens. Consequently, the lens flattens and focuses light from more distant objects on the retina. As we age, the lens becomes less elastic and does not round up to focus on nearby objects as easily, explaining why we hold the newspaper farther away as we become older.

CATARACTS

A **cataract** is a cloudiness or opaqueness in the lens, usually because of aging. Cataracts are the most common eye problem affecting men and women older than 50. Typically, the lens takes on a yellowish hue that blocks light on its way to the retina. In the beginning, cataracts may cause a person to see the world through a haze that can limit activities and cause automobile accidents, early retirement, and falls that can result in hip fractures. As the lens becomes increasingly opaque, the fog thickens. Indeed, cataracts are the leading cause of blindness worldwide. When cataracts make it impossible for a person to perform everyday tasks, the clouded lens can be surgically removed and replaced with an artificial lens.

FOCUSING PROBLEMS

The three most common visual problems—farsightedness, nearsightedness, and astigmatism—are problems in focusing, and normal vision can be restored with corrective lenses (Table 9.2). In *farsightedness,* distant objects are seen more clearly than nearby ones because the eyeball is too short or the lens is too thin, causing images of nearer objects to be focused behind the retina (Figure 9.6). Although distant objects can be seen clearly, the lens cannot become round enough to bend the light sufficiently to focus on nearby objects. Corrective lenses that are thicker in the middle than at the edges (convex) cause the light rays to converge a bit before they enter the eye. The lens can then focus the image on the retina. Other solutions to vision problems are discussed in the Health Issue essay, *Correcting Vision Problems.*

About 25% of the American population are nearsighted; that is, they can see nearby objects more clearly than those far away. *Nearsightedness* (myopia) occurs when the eyeball is elongated or when the lens is too thick. This condition causes the image to focus in front of the retina. Nearsighted people see nearby objects clearly because the lens becomes round enough to focus the image on the retina. However, the lens simply cannot flatten enough to bring the focused image of distant objects to the retina, and so

TABLE 9.2 FOCUSING PROBLEMS			
PROBLEM	**DESCRIPTION**	**CAUSE**	**CORRECTION**
Farsightedness	See distant objects more clearly than nearby objects	Eyeball too short or lens too thin; lens cannot become round enough	Convex lens; increases corneal curvature
Nearsightedness	See nearby objects more clearly than distant objects	Eyeball too long or lens too thick; lens cannot flatten enough	Concave lens; decreases corneal curvature
Astigmatism	Visual image is distorted	Irregularities in curvature of cornea or lens	Lenses that correct for the asymmetrical bending of light

those objects appear blurred. Lenses that are thinner in the middle than at the edges (concave) can correct nearsightedness. These lenses cause the light rays to diverge only slightly before entering the eye.

Although genetics undoubtedly plays a role in the development of nearsightedness, frequent close work, such as reading or working at a computer, is also a cause. When you do a lot of close work, the frequent contractions of the ciliary muscles that change

the shape of the lens increase the pressure within the eye. This pressure can cause the eye to stretch and elongate, resulting in nearsightedness. Eye specialists recommend that you look up from the page or away from the computer screen at frequent intervals—particularly if nearsightedness runs in your family.

Irregularities in the curvature of the cornea or lens will cause distortion of the image because they cause the light rays to converge unevenly. This condition is called *astigmatism*. Vision can be restored to normal by corrective lenses that compensate for the asymmetrical bending of light rays.

■ Light changes the shape of pigment molecules, which generate neural messages

We said earlier that humans are very visual creatures, but it may surprise you to know that the eye contains 70% of all the sensory receptors in the body. The function of the eye's receptors, both rods and cones, is to respond to light by sending neural messages to the brain, where they are translated into images of our surroundings.

Our world is, in fact, so visual that few of us ever stop to wonder how vision works. So let's take a moment to consider the overall process (Figure 9.7). First, light waves in the visible spectrum (the spectrum of wavelengths our eyes can detect) strike an object.

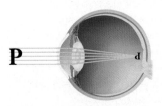

(a) Normal eye
- Close and distant object seen clearly
- Image focuses on retina

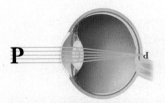

(b) Farsightedness
- Distant objects seen clearly
- Close objects out of focus
- Short eyeball causes image to focus behind retina

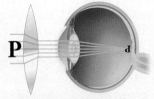

Convex lens
- Causes light rays to converge so that the image focuses on the retina

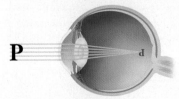

(c) Nearsightedness (myopia)
- Close objects seen clearly
- Distant objects out of focus
- Long eyeball causes image to focus in front of retina

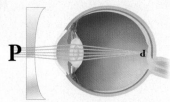

Concave lens
- Causes light rays to diverge so that the image focuses on the retina

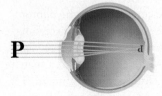

(d) Astigmatism
- Image blurred
- Irregular curvature of cornea or lens causes light rays to focus unevenly

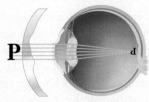

Uneven lens
- Focuses entire image on retina

FIGURE **9.6** Focusing problems such as farsightedness, nearsightedness, and astigmatism are caused when the image of an object is not focused on the retina. These vision problems can be corrected with the use of specific lenses.

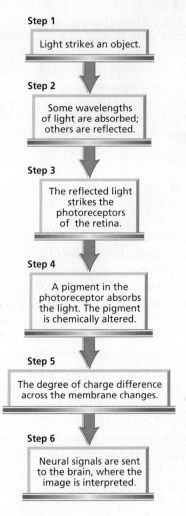

Step 1
Light strikes an object.

Step 2
Some wavelengths of light are absorbed; others are reflected.

Step 3
The reflected light strikes the photoreceptors of the retina.

Step 4
A pigment in the photoreceptor absorbs the light. The pigment is chemically altered.

Step 5
The degree of charge difference across the membrane changes.

Step 6
Neural signals are sent to the brain, where the image is interpreted.

FIGURE **9.7** An overview of the visual process

HEALTH ISSUE

Correcting Vision Problems

Although most people wear glasses to correct vision problems, almost 20 million Americans use contact lenses instead. The contact lens sits on the cornea over a layer of tears. The outer surface of the lens is the corrective surface, and the inner surface fits snugly on the cornea. There are two types of contact lenses. Rigid gas-permeable lenses are made of slightly flexible plastic. These lenses are tough and durable, but some people find that they irritate the eyes. Soft lenses are made of a more malleable plastic. Although soft lenses are gentler on the eyes, they become scratched and damaged more easily.

Contact lenses must be cleaned and disinfected regularly using special solutions. Infections, caused by bacteria, fungi, or *Acanthamoeba* (a type of amoeba), can develop behind the lens. Tears leave deposits of protein, fats, and calcium that help infectious organisms adhere to the contact lens. At best, these deposits are difficult to remove. The more infrequently the lenses are cleaned, the more difficult it becomes to remove the deposits and the more likely infection becomes. Extended-wear contact lenses of both types have been developed, some of which can be left in place for a month or more. Initially, eye specialists thought that less

frequent insertion and removal of lenses would reduce trauma to the cornea. However, a serious safety concern with contact lenses that are left in overnight or for extended periods of time is ulcerative keratitis, a condition in which the cells of the cornea may be rubbed away by the contact lens, sometimes leading to infection and scarring. If not promptly treated, ulcerative keratitis can lead to blindness. Another problem with leaving contact lenses in overnight seems to be a reduced oxygen supply to the cornea, which causes it to swell.

Many people who are tired of depending on glasses or contact lenses have opted to undergo laser eye surgery. This procedure is popularly known as LASIK, which stands for laser-assisted *in situ* keratomileusis. LASIK permanently changes the shape of the cornea. The procedure involves cutting a flap in the cornea, using pulses from a computer-controlled laser to reshape the middle layer of the cornea, and then replacing the flap. This treatment is used to correct for nearsightedness, farsightedness, and astigmatism.

Another procedure for reshaping the cornea is LASEK (laser-assisted subepithelial keratomileusis). It differs from LASIK in that the surgeon creates a flap in the epithelium only,

instead of in the entire cornea, eliminating some of the complications caused by the deeper corneal flaps. LASEK is used mostly for people who are poor candidates for LASIK because their corneas are thin or flat. In yet another procedure, photorefractive keratectomy (PRK), a surgeon uses short bursts of a laser beam to shave a microscopic layer of cells off the corneal surface and flatten it. A computer calculates and controls the laser exposure.

The shape of the cornea can also be altered using a surgical procedure that does not involve lasers. A surgeon can place corneal ring segments, two tiny crescent-shaped pieces of plastic, in the cornea to flatten it. This 15-minute procedure reduces nearsightedness. The ring segments are intended to be permanent but can be removed if necessary.

Lastly, a procedure called orthokeratology uses special hard contact lenses to produce long-lasting, but not permanent, changes in the cornea. At first, the lenses are worn for 8 hours a day. After the cornea has been reshaped for clearest vision, the lenses need only be worn a few hours every few days. If their use is discontinued, the cornea gradually returns to its natural shape.

At least some wavelengths are reflected off the object to the retina of the eye and are absorbed by pigment molecules in the photoreceptors. When light strikes a photopigment, it causes the components of the photopigment molecule to split apart. This change, in turn, causes a series of reactions that reduce the photoreceptor's release of inhibitory neurotransmitter. The neurotransmitter normally inhibits the cells involved in processing the information (discussed shortly), and so light increases the activity of these processing cells. The optic nerve carries these nerve signals from the retina to the back of the brain, where a perception of the world outside takes shape. In a sense, then, seeing is an illusion, because there are no pictures inside our heads.

The processing of electrical signals from the rods and cones begins before the signals leave the retina. The messages are first sent to bipolar cells in the retina, rare examples of neurons that have two processes, an axon and a single dendrite, extending from opposite sides of the cell body. Bipolar neurons direct the information to interneurons called ganglion cells (Figure 9.8). Together, bipolar cells

and ganglion cells convert the input from the retina into patterns, such as edges and spots.

■ Vision in dim light depends on rods

The **rods** are more numerous than the cones and are the photoreceptors responsible for black-and-white vision. Rods are exceedingly sensitive to light and are capable of responding to light measuring one 10-billionth of a watt—the equivalent of a match burning 50 mi (80.5 km) away on a clear, pitch-dark night! (Of course, the ideal conditions necessary to actually *see* a burning match at such a distance are impossible to attain.) The rods allow us to see in dimly lit rooms and in pale moonlight. By detecting changes in light intensity across the visual field, rods contribute to the perception of movements.

The pigment in rods, called **rhodopsin**, is packaged in membrane-bound disks, which are stacked like coins in the outer segment of the rod. Light strikes a rod, splits the rhodopsin, and

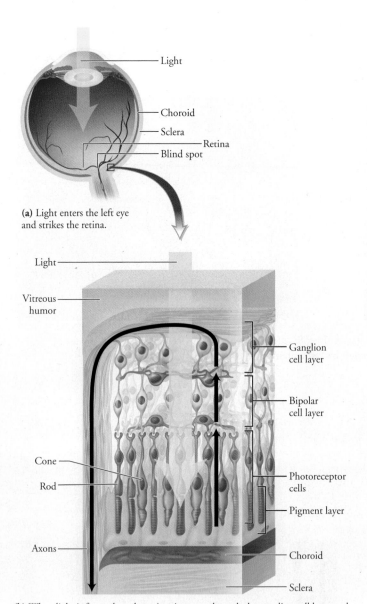

(a) Light enters the left eye and strikes the retina.

(b) When light is focused on the retina, it passes through the ganglion cell layer and bipolar cell layer before reaching the rods and cones. In response to light, the rods and cones generate electrical signals that are sent to bipolar cells and then to ganglion cells. These cells begin the processing of visual information.

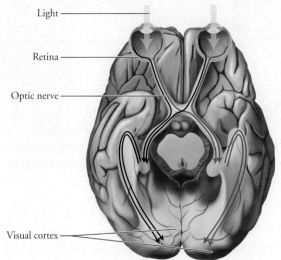

(c) The axons of the ganglion cells leave the eye at the blind spot, carrying nerve impulses to the brain (viewed from below) by means of the optic nerve.

FIGURE **9.8** Neural pathways of the retina convert light to nerve signals.

triggers events that reduce the permeability of the plasma membrane to sodium ions, which eventually leads to changes in the activity of bipolar cells and ganglion cells. In the dark, rhodopsin is resynthesized. The reason you have difficulty seeing when you first walk into a dark room from a brightly lit area is that the rhodopsin in your photoreceptors has been split. As rhodopsin is resynthesized, your vision becomes sharper (Figure 9.9).

■ Color vision depends on cones

The **cones** are the photoreceptors responsible for color vision. Unlike the rods, the cones produce sharp images, because each cone usually sends information to only one bipolar cell for processing.

What makes you see red? Actually, red, or any other color, is determined by wavelengths of light. White light, such as that emitted by an ordinary lightbulb or the sun, consists of all wavelengths. (Rainbows are created when the various wavelengths of white light are separated as they pass through tiny droplets of water in the air.) When light strikes an object, some of the wavelengths may be absorbed and others reflected; we see the wavelengths that are reflected. Thus, an apple looks red because it reflects mostly red light and absorbs most of the other wavelengths.

We see color because we have three types of cones, called blue, green, and red. The cones are named for the wavelengths they absorb best, not for their color. When light is absorbed, the cone is stimulated. Actually, each type of cone absorbs a range of wavelengths, and the ranges overlap quite a bit. As a result, colored light stimulates each type of cone to a different extent. For instance, when we look at a bowl of fruit, light reflected by a red apple stimulates red cones, light reflected by a ripe yellow banana stimulates both red and green cones, and light reflected from blueberries stimulates both blue and green cones. The brain then interprets color according to how strongly each type of cone is stimulated.

stop and think

Most bipolar cells receive input from several rods but from only one cone in the fovea. How might this difference explain why rods permit us to see at lower light intensities than do cones? How might it explain why the sharpest images are formed when the object is focused on the fovea?

COLOR BLINDNESS

Color blindness is a condition in which certain colors cannot be distinguished. Most people who are colorblind see some colors, but they tend to confuse certain colors with others. A lack, or a reduced number, of one of the types of cones causes the confusion. A person who lacks red cones sees deep reds as black. In contrast, a person who lacks green cones sees deep reds but cannot distinguish between reds, oranges, and yellows. Absence of blue cones is extremely rare.

People who are colorblind generally function normally in everyday life. They compensate for the inability to distinguish certain colors by using other cues, such as intensity, shape, or position. Red and green, the universal traffic colors for stop and go, are the most commonly confused colors. However, it is usually possible to pick out the brightest of the lights in a traffic signal, and all traffic

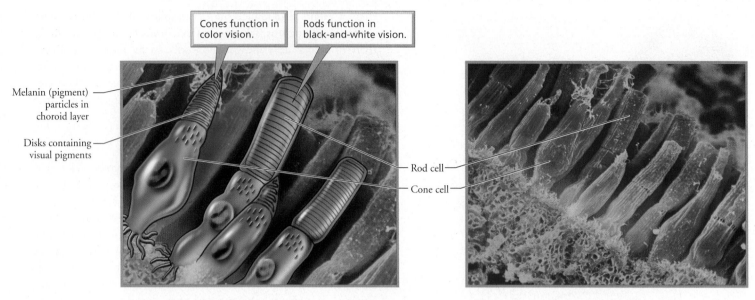

Cones function in color vision.

Rods function in black-and-white vision.

Melanin (pigment) particles in choroid layer

Disks containing visual pigments

Rod cell

Cone cell

FIGURE **9.9** Rods and cones, the photoreceptors in the retina, are named for their shapes. Their function is to generate neural messages in response to light. The outer portion of each photoreceptor is packed with membrane-bound disks that contain pigment molecules.

signals are arranged the same way—red on top and green on the bottom. School children are regularly tested for color blindness so that both students and teachers are aware of the condition. Color blindness may cause difficulty with classwork, such as reading maps or graphs that are presented in color. If a teacher is aware of the condition, steps can be taken to avoid frustration and confusion for the student.

Hearing Depends on the Ear

From a social perspective, hearing is perhaps the most important of our senses, because speech plays such an important role in the communication that binds society. Hearing can also make us aware of things in the external world that we cannot see—telling us whether the person approaching us on a pitch-black night is a friend or a foe, for instance. Hearing can also add to our understanding of events that we *can* see, as when a mother determines by the sound of the cry whether her baby is hungry or in pain. And hearing can enrich the quality of our lives, as when we listen to music or hear waves crashing on the shore.

In every instance, what we hear are sound waves produced by vibration. Vibrating objects, such as guitar strings, the surface of the stereo speaker, or vocal cords, move rapidly back and forth. The vibrations push repeatedly against the surrounding air, creating sound waves, as shown in Figure 9.10. The loudness of sound is determined by the amplitude of the sound wave, the distance between the top of the peaks and the bottom of the troughs. The pitch is determined by the frequency, the number of cycles (repetitions of the wave) per second. The more cycles, the higher the pitch. Sound waves travel through the air to reach our ears.

Greatest compression

Greatest expansion

One cycle

(a) Sound is caused by a vibrating object that causes pressure waves in the air (or water).

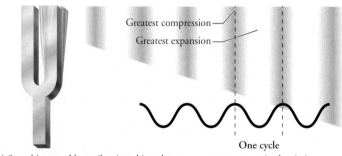

Low amplitude

Higher amplitude

(b) The amplitude (height) of the wave determines the loudness of the sound.

Low frequency

Higher frequency

(c) The frequency (cycles per second) of the waves determines the pitch of the sound.

FIGURE **9.10** A sound wave and the effects of amplitude (height) and frequency (cycles per second)

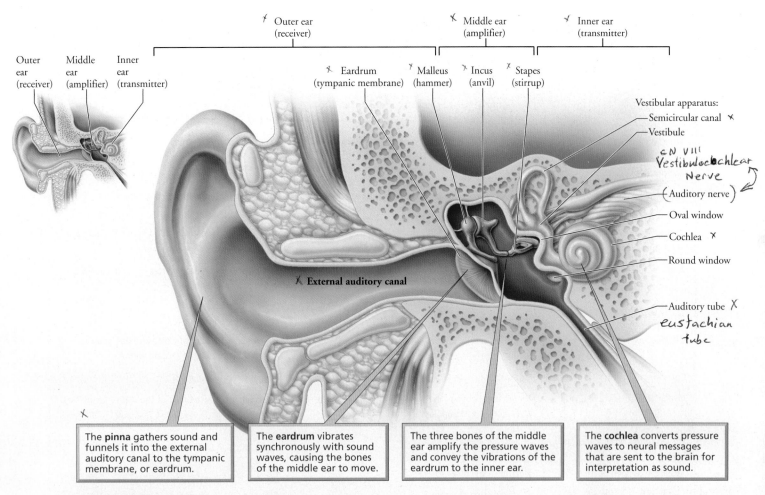

Outer ear (receiver)
Middle ear (amplifier)
Inner ear (transmitter)

Outer ear (receiver)
Middle ear (amplifier)
Inner ear (transmitter)

Eardrum (tympanic membrane)
Malleus (hammer)
Incus (anvil)
Stapes (stirrup)

Vestibular apparatus:
Semicircular canal
Vestibule
CN VIII
Vestibulocochlear Nerve
(Auditory nerve)
Oval window
Cochlea
Round window
Auditory tube
eustachian tube

External auditory canal

The **pinna** gathers sound and funnels it into the external auditory canal to the tympanic membrane, or eardrum.

The **eardrum** vibrates synchronously with sound waves, causing the bones of the middle ear to move.

The three bones of the middle ear amplify the pressure waves and convey the vibrations of the eardrum to the inner ear.

The **cochlea** converts pressure waves to neural messages that are sent to the brain for interpretation as sound.

FIGURE **9.11** Structure of the ear. The human ear has three parts: the outer ear (the receiver), the middle ear (the amplifier), and the inner ear (the transmitter).

■ **The ear collects and amplifies sound waves and converts them to neural messages**

The ear has three main parts—the outer ear, the middle ear, and the inner ear—as shown in Figure 9.11 and summarized in Table 9.3. The **outer ear** functions as a receiver. The part of the outer ear visible on the the head consists of a fleshy funnel called the **pinna**. The pinna gathers the sound and channels it into the **external auditory canal**, the canal that leads from the pinna to the eardrum. The pinna also accentuates the frequencies of the most important speech sounds, making speech easier to pick out from background noise. The pinnas' shape and placement at either side of the head help us determine the direction from which a sound comes.

The sheet of tissue that separates the outer ear from the middle ear is the eardrum, or **tympanic membrane**. It is as thin as a sheet of paper and as taut as the head of a tambourine. When sound waves strike the eardrum, it vibrates at the same frequency as the sound waves and transfers the vibrations to the middle ear.

stop and think

As we age, the tissues of the eardrum thicken and become less flexible. How do these changes partially explain why sensitivity to high-frequency sounds is usually lost as we grow older?

The **middle ear** serves as an amplifier. It consists of an air-filled cavity within a bone of the skull and is spanned by the three smallest bones of the body: the **malleus** (hammer), **incus** (anvil), and **stapes** (stirrup). Together, these bones function as a system of levers to convey the airborne sound waves from the eardrum to the **oval window**, a sheet of tissue that forms the threshold of the inner ear. The malleus is attached to the inner surface of the eardrum. Therefore, the vibrations of the eardrum in response to sound cause the malleus to rock back and forth. The rocking of the malleus, in turn, causes the incus and then the stapes to move. The base of the stapes fits into the oval window.

The force of the eardrum's vibrations is amplified 22 times in the middle ear. The magnification of force is necessary to transfer the vibrations to the fluid of the inner ear. The amplification occurs because the eardrum is larger than the oval window. This size difference concentrates the pressure against the oval window.

The air pressure must be nearly equal on both sides of the eardrum for the eardrum to vibrate properly. If the pressure is not equal, the eardrum will bulge inward or outward, and will be held in that position, causing pain and difficulty hearing. For instance, when a person ascends quickly to a high altitude, the atmospheric pressure is lower than the pressure in the middle ear, and the eardrum bulges outward. Unequal pressure usually is alleviated by

TABLE 9.3 REVIEW OF THE STRUCTURES OF THE EAR AND THEIR FUNCTIONS		
STRUCTURE	**DESCRIPTION**	**FUNCTION**
Outer ear		
Pinna	Fleshy, funnel-shaped part of the ear protruding from the side of the head	Collects and directs sound waves
External auditory canal	Canal between pinna and tympanic membrane	Directs sound to the middle ear
Middle ear		
Eardrum (tympanic membrane)	Membrane spanning the end of the external auditory canal	Vibrates in response to sound waves
Malleus (hammer), incus (anvil), and stapes (stirrup)	Three tiny bones of the middle ear	Amplify the vibrations of the tympanic membrane and transmit vibrations to inner ear
Auditory tube	A tube that connects the middle ear with the throat	Allows equalization of pressure in middle ear with external air pressure
Inner ear		
Cochlea	Fluid-filled, bony, snail-shaped chamber	Houses spiral organ (of Corti) and has openings called oval window and round window
Spiral organ (of Corti)	Contains hair cells	The organ of hearing
Oval window	Membrane between the middle and inner ear that the stapes presses against	Transmits the movements of the stapes to the fluid in the inner ear
Round window	Membrane at the end of the lower canal in cochlea	Relieves pressure created by the movements of the oval window
Vestibular apparatus	Fluid-filled chambers and canals	Monitors position and movement of the head
Vestibule (utricle and saccule)	Two fluid-filled chambers	Maintains static equilibrium (body and head stationary, information on position of head)
Semicircular canals	Three fluid-filled chambers oriented at right angles to one another	Maintain dynamic equilibrium (body or head moving)

the **auditory tube**, a canal that connects the middle ear cavity with the upper region of the throat. Most of the time, the auditory tube is closed and flattened. However, swallowing or yawning opens it briefly, allowing the pressure in the middle ear cavity to equalize with the air pressure outside the ear. The sensation of pressure suddenly equalizing is described as the ear "popping." It can occur whenever the pressure on the external eardrum changes quickly, as when you ride up or down in an elevator, take off or land in an airplane, or go scuba diving.

The **inner ear** is a transmitter. It generates neural messages in response to pressure waves caused by sound waves, and it sends the messages to the brain for interpretation. The inner ear contains two sensory organs, only one of which, the **cochlea** (*kok* le ah), is concerned with hearing. The other sensory organ, the vestibular apparatus, is concerned with sensations of body position and movement and will be discussed in the next section.

The cochlea is the about the size of a pea, and yet it can be considered the true seat of hearing. It is a bony tube about 35 mm (1.4 in.) long that is coiled about two and a half times, somewhat like the shell of a snail, as you can see in Figure 9.12. (*Cochlea* is

from the Latin for "snail.") The wider end of the tube, where the snail's head would be, has two membrane-covered openings. The upper opening is the oval window into which the stapes fits. The lower opening, called the **round window**, serves to relieve the pressure created by the movements of the oval window.

The internal structure of the cochlea is easier to understand if we imagine the cochlea uncoiled so that it forms a long straight tube. We would then see that two membranes divide the interior of the cochlea into three longitudinal compartments, each filled with fluid. The central compartment (the cochlear duct) ends blindly, like the finger on a glove, and does not extend completely to the end of the cochlea's internal tube. As a result, the upper and lower compartments (the vestibular canal and the tympanic canal) are connected at the end of the tube nearest the innermost end of the tube. The basilar membrane is the floor of the central compartment. The **spiral organ** (of Corti), the portion of the cochlea most directly responsible for the sense of hearing, is supported on the basilar membrane (Figure 9.12). The spiral organ consists of the **hair cells** and the overhanging **tectorial membrane**. Each hair cell has about 100 "hairs," slender projections from its upper surface

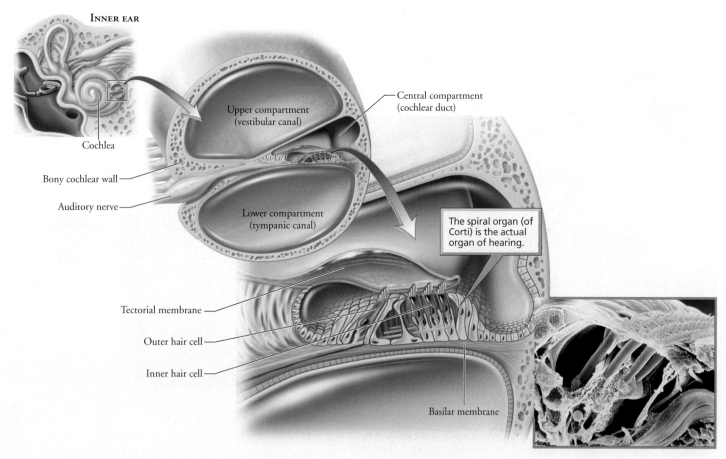

FIGURE **9.12** The cochlea houses the spiral organ (of Corti). The spiral organ consists of the hair cells and the overhanging gelatinous tectorial membrane. The spiral organs rests on the basilar membrane. The hairs are the receptors for hearing. In the micrograph, the hair cells are colored pink and their "hairs" are colored yellow.

that project into the tectorial membrane. The hair cells are arranged on the basilar membrane in rows that resemble miniature picket fences. The hair cells stimulate the nerve cells that carry nerve impulses to the brain.

When the stapes moves to and fro against the oval window, it sets up corresponding movements—pressure waves—in the fluid of the inner ear (Figure 9.13). Those pressure waves are then transmitted from the fluid of the cochlea's upper compartment to the fluid of the lower compartment, because the compartments are continuous. The movements of the fluid cause the basilar membrane to swing up and down; this swinging, in turn, causes the projections on the hair cells to be pressed against the tectorial membrane. The bending of the hairs ultimately results in nerve impulses in the auditory nerve.

■ **Variations in the movements of the basilar membrane determine loudness and pitch**

The louder the sound, the greater the pressure changes in the fluid of the inner ear and, therefore, the stronger the bending of the basilar membrane. Because hair cells have different thresholds of stimulation, more vigorous vibrations in the basilar membrane stimulate more hair cells. The brain interprets the increased number of impulses as louder sound.

How does the ear determine the pitch of a sound? Different regions of the basilar membrane vibrate in reponse to sounds of different pitch. In a sense, the cochlea is like a spiral piano keyboard. The basilar membrane varies in width and flexibility along its length. Near the oval window, the basilar membrane is narrow and stiff. Like the shorter strings on a piano, this region vibrates maximally in response to high-frequency sound, such as a whistle. At the tip of the tube in the cochlea, the basilar membrane is wider and floppier. Low-frequency sounds, such as a bass drum, cause maximum vibrations in this region of the basilar membrane. Thus, sounds of different pitches activate hair cells at different places along the basilar membrane. The brain then interprets input from hair cells in different regions as sounds of different pitch.

■ **Hearing loss can be conductive or sensorineural**

An estimated 28 million Americans have some degree of hearing loss, and 2 million of them are completely deaf. Hearing loss that is severe enough to interfere with social and job-related communication is among the most common chronic neural impairments in the United States.

There are two types of hearing loss: conductive loss and sensorineural loss. Conductive loss results when an obstruction anywhere along the route prevents sounds from being conducted

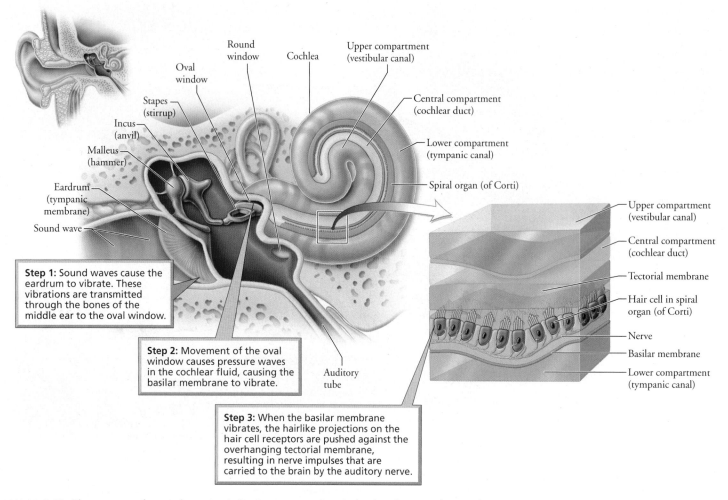

Step 1: Sound waves cause the eardrum to vibrate. These vibrations are transmitted through the bones of the middle ear to the oval window.

Step 2: Movement of the oval window causes pressure waves in the cochlear fluid, causing the basilar membrane to vibrate.

Step 3: When the basilar membrane vibrates, the hairlike projections on the hair cell receptors are pushed against the overhanging tectorial membrane, resulting in nerve impulses that are carried to the brain by the auditory nerve.

FIGURE **9.13** The sequence of events from sound vibration to a nerve impulse begins when sound enters the external auditory canal.

through the external auditory canal to the eardrum and over the bones of the middle ear to the inner ear. The external auditory canal can become clogged with wax or other foreign matter. In this case, the insertion of cotton-tip swabs usually aggravates the situation by further clogging the canal and should, therefore, be avoided. Other possible causes for conductive hearing loss are thickening of the eardrum, which might occur with chronic infection, or perforation of the eardrum, which might occur as a result of trauma. Excess fluid in the middle ear, which often occurs with middle ear infections, can also cause conductive hearing loss. A common age-related cause of conduction deafness is otosclerosis, a condition in which the bones of the middle ear become fused and can no longer transmit sound. In this case, the only way that sound can be transmitted to the inner ear is through the bones of the skull, which is much less efficient. Otosclerosis is routinely corrected by surgery.

Sensorineural deafness is caused by damage to the hair cells or to the nerve supply of the inner ear. A common cause is gradual loss of hair cells throughout life. Some of the loss is simply due to aging. After age 20, we lose about 1 Hz of our total perceptive range of 20,000 Hz every day. However, much of the damage to hair cells is caused by exposure to loud noise (see the Environmental Issue essay, *Noise Pollution*). Neural damage can be caused by infections,

including mumps, rubella (German measles), syphilis, and meningitis. Drugs such as those used for treating tuberculosis and certain cancers can also cause neural damage.

One of the first signs of sensorineural damage is the inability to hear high-frequency sounds. Because consonants, which are needed to decipher most words, have higher frequencies than do vowels, speech becomes difficult to understand. So beware if you find yourself asking, "Could you repeat that?"

The most effective way to deal with hearing loss is with a hearing aid. The basic job of a hearing aid is to amplify sound. One type of hearing aid presents amplified sound to the eardrum. Another type uses a vibrator or stimulator placed in the bone of the skull behind the ear. Sound is then conducted through the bones of the skull to the inner ear.

Hearing aids often are unable to help profoundly deaf people, but cochlear implants can pierce the silence for some of them. A cochlear implant does not amplify sounds as a hearing aid does; it compensates for parts of the inner ear that are not functioning properly. It is an array of electrodes that transforms sound into electrical signals that are delivered to the nerve cells near the cochlea. The implant is surgically inserted into the cochlea through the round window. A cochlear implant allows some people who were

ENVIRONMENTAL ISSUE

Noise Pollution

It is difficult to escape the din of modern life—noise from airports, city streets, loud appliances, stereos. Noise pollution threatens your hearing and your health. Exposure to excessive noise is to blame for the hearing loss of one-third of all hearing-impaired people. Loud noise damages the hairs on the hair cells of the inner ear. When the hairs are exposed to too much noise, they wear down, lose their flexibility, and can fuse together (Figure 9.A). Unfortunately, there is no way to undo the damage; you cannot grow spare parts for your ears.

The loudness of noise is measured in decibels (dB). The decibel scale is logarithmic. An increase of 10 dB generally makes a given sound twice as loud. The decibel ratings and effects of some familiar sounds are given in Table 9.A. Most people judge sounds over 60 dB to be in-

trusive, over 80 dB to be annoying, and over 100 dB to be extremely bothersome. The federal Occupational Safety and Health Administration (OSHA) has set 85 dB as the safety limit for 8 hours of exposure. The threshold for physical pain is 140 dB.

Hearing can be damaged by exposure to noise that is loud enough to make conversation difficult. The louder a sound, the shorter the exposure time necessary to damage the ear. Even a brief, explosively loud sound is capable of damaging hair cells. More commonly, however, hearing loss results from prolonged exposure to volumes over 85 dB. Your ears can endure sound at 90 dB for about 8 hours. For every 5 dB above that, it takes half as long for damage to begin. Thus, sound at 95 dB will damage your ears in only 4 hours. At 110 dB, the average rock concert or stereo headset at full blast

can damage your ears in as little as 30 minutes. If sounds seem muffled to you or if your ears are ringing after you leave a noisy area, you probably sustained some damage to your ears.

Hearing loss is expected in the elderly, but a surprising number of young people also have impaired hearing. The culprit is most likely noise—probably in the form of music. How can you protect yourself from bad vibes? Don't listen to loud music. Keep it tuned low enough that you can still hear other sounds. If you are listening with earphones, no one else should be able to hear the sound from them. When you cannot avoid loud noise, as when you are mowing the lawn, vacuuming, or attending a rock concert, wear earplugs to protect your hearing. You can buy them at most drugstores, sporting goods stores, and music stores. 🌲

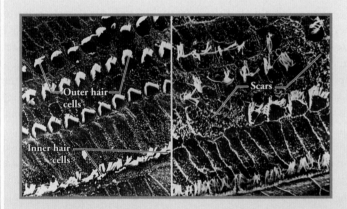

FIGURE **9.A** The hair cells of the inner ear can be permanently damaged by loud noise. The micrograph on the left shows a portion of the spiral organ (of Corti) from a normal guinea pig, showing the characteristic three rows of outer hair cells and a single inner row. The micrograph on the right shows the spiral organ (of Corti) from a guinea pig that had been exposed to loud (120 dB) sound. Note that scars have replaced some injured cells. On most of the hair cells that have survived, the hairs are no longer in an orderly pattern.

TABLE 9.A EFFECTS OF NOISE POLLUTION

SOUND SOURCE	LOUDNESS (dB)	EFFECT FROM PROLONGED EXPOSURE
Jet plane at takeoff	150	Eardrum rupture
Deck of aircraft carrier	140	Very painful; traumatic injury
Rock-and-roll band (at maximum volume)	130	Irreversible damage
Jet plane at 152 m (500 ft)	110	Loss of hearing
Subway, lawn mower	100	
Electric blender	90	Annoying
Washing machine, freight train at 15 m (50 ft)	80	
Traffic noise	70	Intrusive
Normal conversation	65	
Chirping bird	60	
Quiet neighborhood (daytime)	50	
Soft background music	40	Quiet
Library	30	
Whisper	20	Very quiet
Breathing, rustling leaves	10	
	0	Threshold of hearing

severely hearing impaired or totally deaf to converse with someone in person or on the phone. They also help deaf children to learn to speak well.

what would you do?

Many deaf people consider themselves to be a cultural group, not a group with a medical problem. Some people argue, for instance, that communication is the real challenge faced by the deaf and that the solution should therefore be a social, not a medical, one. If medical technology is available, should we use it to "fix" conditions such as deafness? If you had a child who was born deaf, would you want a cochlear implant for that child? Why or why not? (Assume that the child is not yet able to make that decision for himself or herself.)

■ Ear infections can occur in the ear canal or in the middle ear

An external ear infection is an infection in the ear canal. Swimmer's ear, the most common such infection, is precipitated by water becoming trapped in the canal and creating an environment favorable for the growth of bacteria. The first symptom of swimmer's ear is itching, usually followed by pain that can become intense and constant. Chewing food or touching the earlobe sharpens the pain. The usual treatment consists of antibiotic ear drops, heat, and pain medication. To prevent swimmer's ear, always be sure to drain any remaining water from the ear canal after swimming or bathing, by shaking the head if need be.

Middle ear infections usually result from infections of the nose and throat that work their way through the auditory tubes connecting the throat and middle ear. At least half of all children get an ear infection at some time; some children get four or five ear infections a year. Middle ear infections are more common in children than in adults because the auditory tubes curve and tilt downward in adults. In children, the tubes are nearly horizontal, allowing infectious organisms to travel through them more easily. The signs of a middle ear infection are a stabbing earache, impaired hearing, and a feeling of fullness in the ear, often accompanied by fever. A middle ear infection usually is treated with an antibiotic.

Balance Depends on the Vestibular Apparatus of the Inner Ear

The **vestibular apparatus**, a fluid-filled maze of chambers and canals within the inner ear, is responsible for monitoring the position and movement of the head. The receptors in the vestibular apparatus are hair cells similar to those in the cochlea. Head movements or changes in velocity cause the hairs on these cells to bend, sending messages to the brain. The brain uses this input to maintain balance.

The vestibular apparatus consists of the semicircular canals and the vestibule. The **semicircular canals** are three canals in each ear that contain sensory receptors and help us stay balanced as we move. They monitor any sudden movements of the head, including those caused by acceleration and deceleration, to help us maintain

our equilibrium when the body or head is moving (see "dynamic equilibrium" in Figure 9.14). At the base of each semicircular canal is an enlarged region called the *ampulla*. Within the ampulla is a tuft of hair cells. The hairlike projections from these cells are embedded in a pointed "cap" of stiff, pliable, gelatinous material called the *cupula*. When you move your head, fluid in the canal lags a little behind, causing the cupula to bend the hair cells and stimulate them.

The three semicircular canals in each ear are oriented at right angles to one another. Head movements cause the fluid in the canals to move in the direction opposite the direction of movement, in the same way that sudden acceleration slams passengers back against the car seat. Whereas nodding your head to say yes will cause the fluid in the canals parallel to the sides of the head to swirl, shaking it from side to side to say no will cause fluid in the canals parallel to the horizon to move. Tilting your head to look under the bed will cause fluid in the canals that are parallel to your face to move back and forth. Even the most complex movement can be analyzed in terms of motion in three planes.

stop and think

When you stop moving, there is a time lag before the fluid in the semicircular canals stops swirling. How does this time lag account for the sensation of dizziness after you have been twirling around for a while?

The **vestibule**, the other part of the vestibular apparatus, is important for static equilibrium, the maintenance of balance when we are not moving. The vestibule consists of the utricle and the saccule, two fluid-filled cavities that let us know literally which end is up (Figure 9.14). These cavities tell the brain the position of the head with respect to gravity when the body is not moving. They also respond to acceleration and deceleration but not to rotational changes as the semicircular canals do. Both the utricle and the saccule contain hair cells overlaid with a gelatinous material in which small chalklike granules of calcium carbonate are embedded. These granules, called *otoliths* make the gelatin heavier than the surrounding material and, therefore, make it slide over the hair cells whenever the head is moved. The movement of the gelatin stimulates the hair cells, which send messages to the brain, and the brain interprets the messages to determine the position of the head relative to gravity.

The utricle and saccule sense different types of movement. The utricle senses the forward tilting of the head, as well as forward motion, because its hair cells are on the floor of the chamber, oriented vertically when the head is upright. In contrast, the hair cells of the saccule are on the wall of the chamber, oriented horizontally when the head is upright. They respond when you move vertically, as when you jump up and down.

Motion sickness—that dreadful feeling of dizziness and nausea that sometimes causes vomiting—is thought to be caused by a mismatch between sensory input from the vestibular apparatus and sensory input from the eyes. For example, you may be one of the many people who get carsick from reading in the car. When looking down at a book, your eyes tell your brain that your body is stationary. However, as the car changes speed, turns, and hits bumps

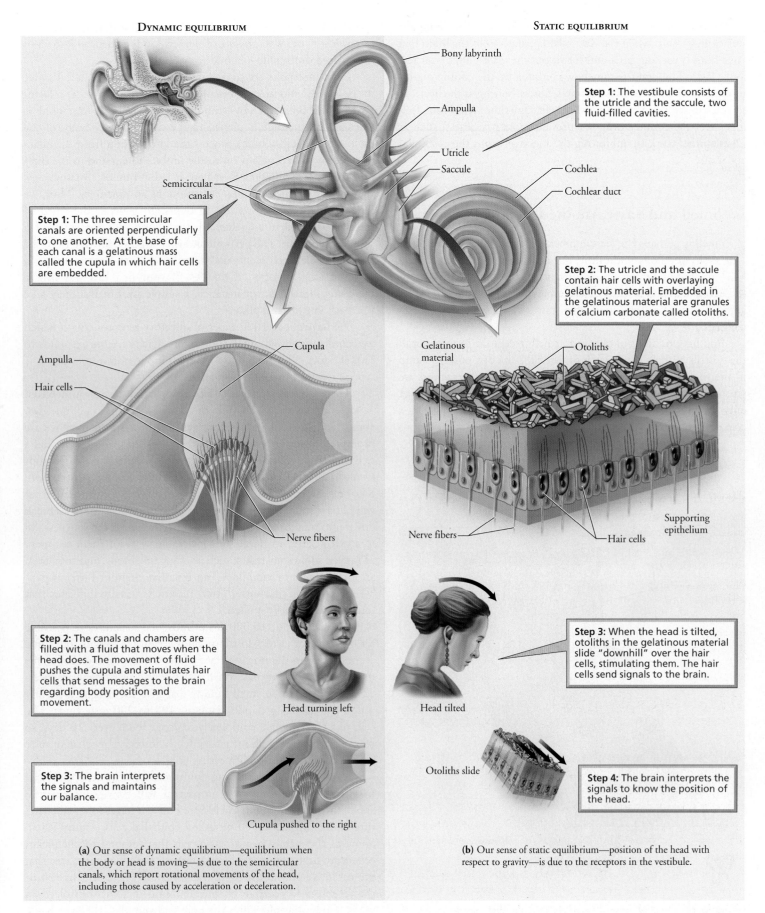

DYNAMIC EQUILIBRIUM

STATIC EQUILIBRIUM

Bony labyrinth

Ampulla

Utricle

Saccule

Cochlea

Cochlear duct

Step 1: The vestibule consists of the utricle and the saccule, two fluid-filled cavities.

Semicircular canals

Step 1: The three semicircular canals are oriented perpendicularly to one another. At the base of each canal is a gelatinous mass called the cupula in which hair cells are embedded.

Step 2: The utricle and the saccule contain hair cells with overlaying gelatinous material. Embedded in the gelatinous material are granules of calcium carbonate called otoliths.

Cupula

Ampulla

Hair cells

Gelatinous material

Otoliths

Supporting epithelium

Nerve fibers

Nerve fibers

Hair cells

Head turning left

Head tilted

Step 2: The canals and chambers are filled with a fluid that moves when the head does. The movement of fluid pushes the cupula and stimulates hair cells that send messages to the brain regarding body position and movement.

Step 3: When the head is tilted, otoliths in the gelatinous material slide "downhill" over the hair cells, stimulating them. The hair cells send signals to the brain.

Step 3: The brain interprets the signals and maintains our balance.

Cupula pushed to the right

Otoliths slide

Step 4: The brain interprets the signals to know the position of the head.

(a) Our sense of dynamic equilibrium—equilibrium when the body or head is moving—is due to the semicircular canals, which report rotational movements of the head, including those caused by acceleration or deceleration.

(b) Our sense of static equilibrium—position of the head with respect to gravity—is due to the receptors in the vestibule.

FIGURE **9.14** Dynamic and static equilibrium

in the road, your vestibular system is detecting motion. Similarly, seasickness results when the vestibular apparatus tells the brain that your head is rocking back and forth, but the deck under your feet looks level. The brain is somehow confused by the conflicting information, and the result is motion sickness. Staring at the horizon, so you can see that you are moving, can sometimes relieve the feeling. Over-the-counter drugs to prevent motion sickness, such as Dramamine, work by inhibiting the messages from the vestibular apparatus.

Smell and Taste Are the Chemical Senses

Smell is perhaps the least appreciated of our senses. You could easily get along without it, but life would not be as interesting. For one thing, about 80% of what we usually think of as the flavor of a food is really detected by our sense of smell. This relationship explains why food often tastes bland when the nose is congested.

You have millions of olfactory (smell) receptors, located not in the nostrils but in a small patch of tissue the size of a postage stamp in the roof of each nasal cavity (Figure 9.15). These receptors are sensory neurons that branch into long olfactory hairs that project outward from the lining of the nasal cavity and are covered by a coat of mucus, which keeps them moist and is a solvent for odorous

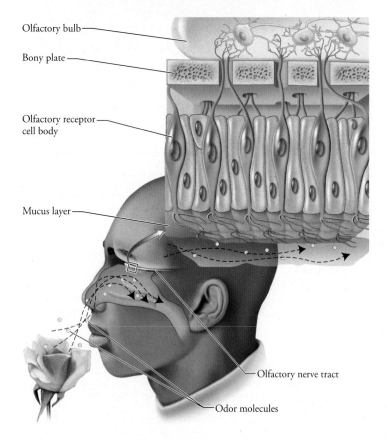

Olfactory bulb

Bony plate

Olfactory receptor cell body

Mucus layer

Olfactory nerve tract

Odor molecules

FIGURE **9.15** Our sense of smell resides in a small patch of tissue in the roof of each nasal cavity. Within each cavity are some 5 million olfactory receptors.

molecules. The hairs move, gently swirling the mucus. Olfactory receptors, by the way, are one of the few kinds of neuron known to be replaced during life—about every 60 days.

Odor molecules are carried into the nasal cavity in air, dissolve in the mucus, and bind to receptor sites on the hairs of the olfactory receptor cells, thereby stimulating the cells. If threshold is reached, the message is carried to the two olfactory bulbs of the brain. The olfactory bulbs process the information from the olfactory receptors and pass it on to the limbic system and to the cerebral cortex (see Chapter 8), where it is interpreted. Interestingly, the limbic system is a center for emotions and memory. Thus, we rarely have neutral responses to odors. In general, the smell of freshly baked bread is perceived as pleasing, but the smell of a skunk is repulsive. The perfume industry makes a fortune from the association between scents and sexuality. Odors can also trigger a flood of long-forgotten memories. If your first kiss was near a blooming lilac bush, for instance, a simple sniff of lilac may take you back in time and place.

We have about 1000 types of olfactory receptors, with which we can distinguish about 10,000 odors. Each receptor responds to several odors. Thus, the brain relies on input from more than one type of receptor to identify an odor.

There are four primary tastes: sweet, salty, sour, and bitter. These four tastes are enough to answer the important question about food or drink: Should we swallow it or spit it out? We are prompted to swallow if it tastes sweet, because sweetness implies a rich source of calories and thus energy. Salty tastes also prompt swallowing, because salts will replace those lost in perspiration. Sourness may pose a dilemma. Sour is the taste of unripe fruit that would have more food value later on. As fruit ripens, starches break down into sugars that create a sweet taste and mask the sourness. So it is often better to reject sour fruits and wait for them to ripen. However, some sour fruits, such as oranges, lemons, and tomatoes, are rich sources of vitamin C, an essential vitamin. Bitter is easy. We reject foods that taste bitter. Bitterness usually indicates that food is poisonous or spoiled.

We have about 10,000 **taste buds**, the structures responsible for our sense of taste. Most taste buds are on the tongue, but some are scattered on the inner surface of the cheeks, on the roof of the mouth, and in the throat. Most taste buds on the tongue are located in papillae, those small bumps that give the tongue a slightly rough feeling. Each papilla contains 100 to 200 taste buds. The cells of a taste bud are completely replaced about every 10 days, so you need not worry about permanently losing your sense of taste when you burn your mouth eating hot pizza.

A taste bud is the meeting point between chemicals dissolved in saliva and the sensory neurons that will convey information about them to the brain. Each taste bud is a lemon-shaped structure containing about 40 modified epithelial cells (Figure 9.16). Some of these cells are the taste cells that respond to chemicals; others are supporting cells. The taste cells have long *taste hairs* that project into a pore at the tip of the taste bud. These taste hairs bear the receptors for certain chemicals found in food. Dissolved in saliva, food molecules enter the pore and stimulate the taste hairs. Although taste cells are not neurons, they generate electrical signals, which are then sent to sensory nerve cells wrapped around the taste cell.

Each taste bud responds to all four basic tastes but is usually more sensitive to one or two of them than to the others. Because of the way the taste buds are distributed, different regions of the tongue have *slightly* different sensitivity to these tastes. In general, the tip of the tongue is most sensitive to sweet tastes, the back to bitter, and the sides to sour. Sensitivity to salty tastes is fairly evenly distributed on the tongue.

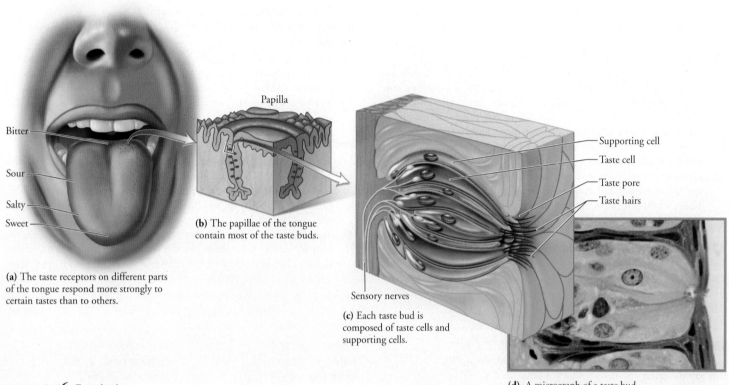

(a) The taste receptors on different parts of the tongue respond more strongly to certain tastes than to others.

(b) The papillae of the tongue contain most of the taste buds.

(c) Each taste bud is composed of taste cells and supporting cells.

(d) A micrograph of a taste bud

FIGURE **9.16** Taste buds

HIGHLIGHTING THE CONCEPTS

Sensory Receptors Generate Electrochemical Messages in Response to Stimuli (p. 153)

1. Changes in external and internal environments stimulate sensory receptors, which in turn generate electrochemical messages that eventually are converted to nerve impulses that are conducted to the brain.

Receptors Are Classified by the Type of Stimulus to Which They Respond (pp. 153–154)

2. There are five types of sensory receptors. Mechanoreceptors are responsible for touch, pressure, hearing, equilibrium, blood pressure, and body position. Thermoreceptors detect changes in temperature. Photoreceptors detect light. Chemoreceptors monitor chemical levels within the body and detect smells and tastes. Pain receptors respond to strong stimuli that cause physical or chemical damage to tissues.

3. Some sensory receptors are located near the body surface and respond to environmental changes. Other sensory receptors are located inside the body and monitor internal conditions.

Receptors for the General Senses Are Distributed throughout the Body (pp. 154–155)

4. The general senses are touch, pressure, temperature, sense of body and limb position, and pain.

5. Thermoreceptors, which are widely distributed throughout the body, respond to changes in temperature.

6. Muscle spindles, specialized muscle fibers with sensory nerve endings wrapped around them, and Golgi tendon organs, highly branched nerve fibers in tendons, help the brain monitor the location of all parts of the body.

7. Pain receptors, located in almost every tissue of the body, respond to any strong stimulus that results from physical or chemical damage to tissues.

Vision Depends on the Eye (pp. 156–162)

8. The wall of the eye has three layers. The sclera and the cornea make up the outer layer. The choroid, ciliary body, and iris make up the middle layer. The retina, which contains the photoreceptors (rods and cones), is the innermost layer.

9. The cornea and the lens work together to focus images on the retina. The lens can accommodate (change in shape) to focus on near or distant objects.

10. When pigment molecules in the rods or cones (photoreceptors) absorb light, a chemical change occurs in the pigment molecule. The resulting change in the permeability of the membrane of the photoreceptor generates a neural message that is carried by the optic nerve to the brain.

11. Color vision depends on the cones. The three types of cones are named for the color of light they absorb best—green, red, or blue—and allow us to see colors.

WEB TUTORIAL 9.1 The Human Eye

Hearing Depends on the Ear (pp. 162–168)

12. The ear is divided into three regions: the outer ear, the middle ear, and the inner ear. The outer ear, which functions as a receiver, consists of the pinna and the external auditory canal. The middle ear, which serves as an amplifier, consists of the tympanic membrane (eardrum), three small bones (malleus, incus, and stapes), and the auditory tube. The inner ear is a transmitter and consists of the cochlea and vestibular apparatus.

13. Hearing is the sensing of sound waves caused by vibration. Sound enters the outer ear and vibrates the eardrum (tympanic membrane). These vibrations move the malleus, which moves the incus and the stapes. The stapes conveys these vibrations to the inner ear by means of the oval window.

14. Pressure on either side of the eardrum is regulated by the auditory tube.

15. The cochlea is a coiled tube enclosed in bone. Its interior is divided into three longitudinal tubes. The middle tube contains the spiral organ (of Corti), which is lined with hair cells and is the portion of the cochlea most responsible for hearing. Sensory hair cells extend from the basilar membrane, which forms the floor of the spiral organ. Vibrations in the oval window cause fluid in the cochlea to move. Movement of the fluid causes the basilar membrane to vibrate and push hair cells into the overlying tectorial membrane, bending hair cells and initiating nerve impulses.

WEB TUTORIAL 9.2 The Human Ear

Balance Depends on the Vestibular Apparatus of the Inner Ear (pp. 168–170)

16. Balance is controlled by the vestibular apparatus, which consists of the semicircular canals and the vestibule. The semicircular canals monitor sudden movements of the head. The vestibule is made of two components, the saccule and the utricle, which tell the brain the position of the head with respect to gravity.

Smell and Taste Are the Chemical Senses (pp. 170–171)

17. Smell receptors are located in the nasal cavity. They are lined with cilia and coated with mucus. Odorous molecules dissolve in the mucus and bind to the receptors' hairs, stimulating the receptors. Information is passed to the olfactory bulbs and then to the limbic system and the cerebral cortex.

18. Taste buds are responsible for our sense of taste. They are on the tongue, inside the cheeks, on the roof of the mouth, and in the throat. They sense the four basic tastes of sweet, salty, sour, and bitter.

KEY TERMS

sensory receptors *p. 153*	choroid *p. 156*	cataract *p. 158*	oval window *p. 163*
sensory adaptation *p. 153*	ciliary body *p. 156*	rods *p. 160*	auditory tube *p. 164*
mechanoreceptor *p. 153*	lens *p. 156*	rhodopsin *p. 160*	inner ear *p. 164*
thermoreceptor *p. 154*	iris *p. 156*	cones *p. 161*	cochlea *p. 164*
photoreceptor *p. 154*	pupil *p. 156*	outer ear *p. 163*	round window *p. 164*
chemoreceptor *p. 154*	retina *p. 156*	pinna *p. 163*	spiral organ *p. 164*
pain receptor *p. 154*	fovea *p. 156*	external auditory canal *p. 163*	hair cells *p. 164*
general senses *p. 154*	optic nerve *p. 156*	tympanic membrane *p. 163*	tectorial membrane *p. 164*
special senses *p. 154*	blind spot *p. 156*	middle ear *p. 163*	vestibular apparatus *p. 168*
referred pain *p. 155*	vitreous humor *p. 156*	malleus *p. 163*	semicircular canals *p. 168*
sclera *p. 156*	aqueous humor *p. 157*	incus *p. 163*	vestibule *p. 168*
cornea *p. 156*	accommodation *p. 158*	stapes *p. 163*	taste buds *p. 170*

REVIEWING THE CONCEPTS

1. Define sensory adaptation and give an example. *pp. 153–154*

2. What are the five classes of receptors? Which would give rise to the general senses? Which would give rise to special senses? *pp. 153–154*

3. When stimulated, what do Merkel disks tell us? *p. 154*

4. What are the functions of the two distinct regions of the outer layer of the eye? *p. 156*

5. How is light focused on the retina? *pp. 157–158*

6. How is light converted to a neural message? *pp. 159–160*

7. How are sound waves produced? How are loudness and pitch determined? *p. 162*

8. What is the function of the eardrum (tympanic membrane)? *p. 163*

9. Why is it necessary for the force of the vibrations to be amplified in the middle ear? How is amplification accomplished? *p. 163*

10. What are the effects of unequal air pressure on either side of the eardrum? How does the auditory tube help regulate air pressure? *pp. 163–164*

11. How does the basilar membrane respond to pitch? *p. 165*

12. What are the two types of hearing loss? Explain how they differ. *pp. 165–166*

13. What are the two structures for equilibrium in the inner ear? Explain the structure of each and how it detects body position and motion. *pp. 165–170*

14. What causes motion sickness? How can it be counteracted? *pp. 169, 170*

15. Where are the olfactory receptors located? Explain their structure as it is related to their function. *p. 170*

16. What are the structures responsible for taste? Where are they found? *pp. 170–171*

17. What are the four primary tastes? *p. 171*
18. Describe the structure of a taste bud. *pp. 170–171*
19. Pacinian corpuscles sense
    a. pressure.
    b. light touch.
    c. warmth.
    d. pain.
20. The spiral organ (of Corti) is important in sensing
    a. body movement.
    b. sound.
    c. light.
    d. degree of muscle contraction.
21. The greatest concentration of cones is found in the
    a. sclera.
    b. lens.
    c. ciliary body.
    d. fovea.

22. Noah is a 4-year-old boy who has a middle ear infection. The cause of the infection could be
    a. bacteria that spread to the ear from a sore throat.
    b. excessive wax buildup in the ear canal.
    c. water that was trapped in the ear after bathing.
    d. noise pollution.
23. The blind spot of the eye is
    a. located in the cornea.
    b. the region where the optic nerve leaves the eye.
    c. the region of the eye where rods outnumber cones.
    d. the region where the ciliary muscle attaches.
24. The receptors responsible for color vision are the _____.
25. The _____ of the eye changes shape to focus.
26. The snail-shaped structure concerned with hearing is the _____.
27. The semicircular canals sense _____.

APPLYING THE CONCEPTS

1. Molly needs glasses for driving but not for reading. What is the name for her visual problem? What type of corrective lens would she need?
2. Sarah has a viral infection that is causing vertigo (dizziness). In which part of the ear is the infection?
3. Manuel is a guitar player in a rock band. He complains that he is having difficulty hearing high-pitched sounds. Explain the relationship between his hearing problem and his occupation.

Additional questions can be found on the companion website.

4. You are at a baseball game. After watching the ball be hit into the stands for a home run, you turn to your friend sitting next to you to ask for some peanuts. What changes in your eyes for your friend's face to come into focus?

10

The Endocrine System

The Endocrine System Communicates Using Chemical Messages
- Hormones are the messengers of the endocrine system
- Feedback mechanisms regulate the secretion of hormones

Hormones Influence Growth, Development, Metabolism, and Behavior
- Pituitary hormones often prompt other glands to release hormones
- Thyroid hormones regulate metabolism and decrease blood calcium
- Parathyroid hormone increases blood calcium
- The adrenal glands secrete stress hormones
- Hormones of the pancreas regulate blood glucose
- Hormones of the thymus gland promote maturation of white blood cells
- The pineal gland secretes melatonin

Other Chemical Messengers Act Locally

HEALTH ISSUE   Is It Hot in Here, or Is It Me? Hormone Replacement Therapy and Menopause

HEALTH ISSUE   Hormones and Our Response to Stress

HEALTH ISSUE   Melatonin: Miracle Supplement or Potent Drug Misused by Millions?

The endocrine system produces hormones, substances often described as the chemical messengers of the body. The messages they carry are varied but vital, with effects ranging from body height to alertness level, energy production, and resistance to disease.

Basketball was Kazuo's passion. He was only 5 feet 4 inches tall, but he made up in speed, skill, and desire what he lacked in size. Although most of the guards he played against were 6 feet or taller, he had done well so far, averaging eight points and eight assists per game. Tonight he would be matched against Conklin, the 6 foot 3 inch all-conference point guard for the Tigers.

It was a grueling game. With 6 seconds left to play and the score tied, and after being beaten all night by Conklin's use of his height advantage, Kazuo suddenly managed to reach around Conklin's back and knock the ball away. He felt a huge surge of energy as he dove for the ball, got back on his feet, and raced the rest of the way down the court to the basket. With Conklin in midair above him and almost rim high, Kazuo drove his pivot foot into the floor, sprang to the left, switched the ball to his left hand, and deftly spun the ball off the backboard and into the hoop. The buzzer rang just as he hit the floor, exhausted and exhilarated at the same time.

Kazuo's late-in-the-game surge of energy came in the form of epinephrine, a hormone released from his adrenal gland. In this chapter, you will learn about the adrenal and other endocrine glands and the hormones they secrete. You will see that organs communicate with each other and with other tissues by means of hormones and that hormones initiate both long-term changes, such as growth and development, and short-term changes, such as Kazuo's winning surge of energy. In the end, you will understand that proper functioning of the body depends on this system of internal communication. The endocrine system complements our body's other, more rapid system of internal communication, the nervous system (Chapters 7 and 8). Because the nervous and endocrine systems share the common function of regulating and coordinating the activities of all body systems, some consider these two systems to be one—the neuroendocrine system. ■

The Endocrine System Communicates Using Chemical Messages

Our bodies contain two types of glands, exocrine glands and endocrine glands (Figure 10.1). **Exocrine glands** secrete their products into ducts that empty either onto the body surface, into the spaces within organs, or into a body cavity. Oil (sebaceous) glands, for example, are exocrine glands that secrete oil into ducts that open onto the surface of the skin. Salivary glands, another example of exocrine glands, secrete saliva into ducts that open into the mouth. **Endocrine glands** are made of secretory cells that release their products, called hormones, into the fluid just outside the cells. Endocrine glands do not secrete their products into ducts. Instead, the hormones move from the cells that produced them to the fluid just outside the cells, where they diffuse directly into the bloodstream.

The endocrine system consists of endocrine glands and of organs that contain some endocrine tissue; these latter organs have other functions besides hormone secretion (Figure 10.2). The major endocrine glands are the pituitary gland, thyroid gland, parathyroid glands, adrenal glands, and pineal gland. Organs with some endocrine tissue include the hypothalamus, thymus, pancreas, ovaries, testes, heart, and placenta. Organs of the digestive and urinary systems, such as the stomach, small intestine, and kidneys, also have endocrine tissue. Our discussion here will focus on the major endocrine glands. We also will examine three organs with endocrine tissue: hypothalamus, thymus, and pancreas. The remaining organs containing endocrine tissue will be discussed in the chapters that cover the organs' other functions. For example, the kidneys are discussed in Chapter 16 on the urinary system, and the ovaries and testes are discussed in Chapter 17 on the reproductive system. Table 10.1 summarizes the glands and other organs of the endocrine system and the hormones they release.

■ Hormones are the messengers of the endocrine system

Hormones are the chemical messengers of the endocrine system. They are released in very small amounts by the cells of endocrine glands and tissues and enter the bloodstream to travel throughout the body. Although hormones contact virtually all cells, most affect only a particular type of cell, called a **target cell**. Target cells have *receptors,* protein molecules that recognize and bind to specific hor-

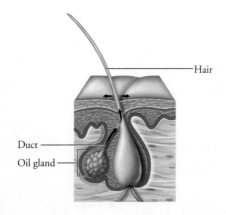

(a) Exocrine glands, such as this oil gland, secrete their products into ducts.

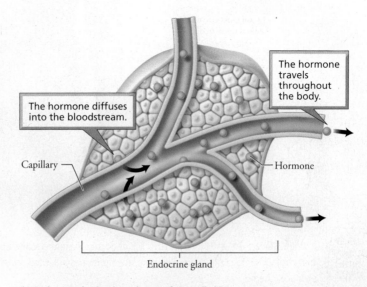

(b) Endocrine glands release their products, called hormones, into the fluid just outside cells. The hormones then diffuse into the bloodstream to be transported throughout the body.

FIGURE **10.1** Exocrine and endocrine glands

mones. Once a hormone binds to its specific receptor, it begins to exert its effects on the cell. Cells other than target cells lack the correct receptors and are unaffected by the hormone.

The mechanisms by which hormones influence target cells depend on the chemical makeup of the hormone. Hormones are classified as being either lipid soluble or water soluble. Lipid-soluble hormones include **steroid hormones**, a group of closely related hormones derived from cholesterol. The ovaries, testes, and adrenal glands are the main organs that secrete steroid hormones. Lipid-soluble hormones move easily through any cell's plasma membrane because it is a lipid bilayer. Once inside a target cell, a steroid hormone combines with receptor molecules in the cytoplasm (only target cells have the proper receptors). The hormone-receptor complex then moves into the nucleus of the cell, where it attaches to DNA and activates certain genes. Ultimately, such activation leads to the synthesis of specific proteins by the target cell. (The precise steps involved in protein synthesis are described in Chapter 21).

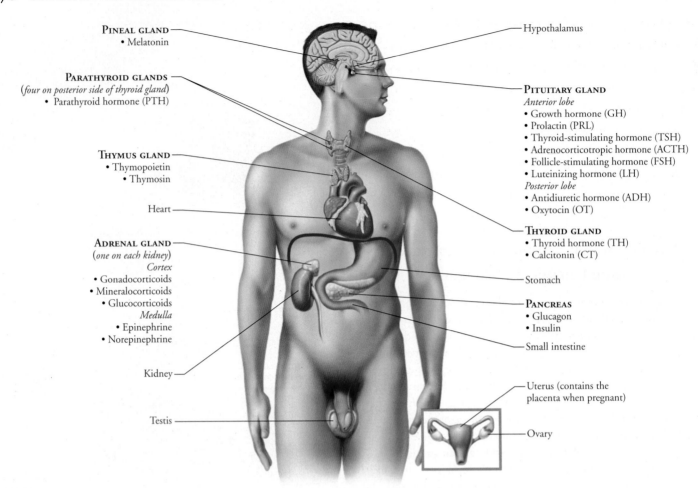

PINEAL GLAND
• Melatonin

PARATHYROID GLANDS
(*four on posterior side of thyroid gland*)
• Parathyroid hormone (PTH)

THYMUS GLAND
• Thymopoietin
• Thymosin

Heart

ADRENAL GLAND
(*one on each kidney*)
Cortex
• Gonadocorticoids
• Mineralocorticoids
• Glucocorticoids
Medulla
• Epinephrine
• Norepinephrine

Kidney

Testis

Hypothalamus

PITUITARY GLAND
Anterior lobe
• Growth hormone (GH)
• Prolactin (PRL)
• Thyroid-stimulating hormone (TSH)
• Adrenocorticotropic hormone (ACTH)
• Follicle-stimulating hormone (FSH)
• Luteinizing hormone (LH)
Posterior lobe
• Antidiuretic hormone (ADH)
• Oxytocin (OT)

THYROID GLAND
• Thyroid hormone (TH)
• Calcitonin (CT)

Stomach

PANCREAS
• Glucagon
• Insulin

Small intestine

Uterus (contains the placenta when pregnant)

Ovary

FIGURE **10.2** The endocrine system. The endocrine system is made up of endocrine glands and of organs that contain some endocrine tissue. Here, the hormones are listed under the endocrine gland or organ that produces them.

These proteins may include enzymes that stimulate or inhibit particular metabolic pathways. Figure 10.3 summarizes the mode of action of steroid hormones.

Water-soluble hormones, such as protein or peptide hormones, cannot pass through the lipid bilayer of the plasma membrane and therefore cannot enter target cells. Instead, the hormone—which in this situation is called the first messenger—binds to a receptor on the plasma membrane of the target cell. This binding activates a molecule—called the second messenger—in the cytoplasm. **Second messengers** are molecules within the cell that influence the activity of enzymes, and ultimately the activity of the cell, to produce the effect of the hormone. A system in which the second messenger is cyclic adenosine monophosphate (cAMP) is illustrated in Figure 10.4.

■ **Feedback mechanisms regulate the secretion of hormones**

Now that we have seen how hormones work at the cellular level, let's turn our attention to the factors that stimulate and regulate the release of hormones from endocrine glands. The stimuli that cause endocrine glands to manufacture and release hormones are other hormones, signals from the nervous system, and changes in the levels of certain ions or nutrients in the blood.

Recall from Chapter 4 that homeostasis keeps the body's internal environment relatively constant. Such constancy most often is achieved through **negative feedback mechanisms**, which, as we saw in Chapter 4, are homeostatic mechanisms in which the outcome of a process feeds back to the system, shutting the process down. Negative feedback mechanisms regulate the secretion of some hormones. Typically, a gland releases a hormone, and then rising blood levels of that hormone inhibit its further release. In an alternative form of negative feedback, some endocrine glands are sensitive to the particular condition they regulate rather than to the level of the hormone they produce. For example, the pancreas secretes the hormone insulin in response to high levels of glucose in the blood. Insulin prompts the liver to store glucose, which causes the blood level of glucose to decline. The pancreas senses the low glucose in the blood and stops secreting insulin. Secretion of hormones is sometimes regulated by **positive feedback mechanisms**, in which the outcome of a process feeds back to the system and stimulates the process to continue. For example, during childbirth, the pituitary gland releases the hormone oxytocin (OT), which stimulates the uterus to contract. Uterine contractions then stimulate further release of oxytocin, which stimulates even more contractions (we will revisit this example later). The feedback is described as positive be-

TABLE 10.1 REVIEW OF SOME ENDOCRINE GLANDS AND THEIR HORMONES

GLAND OR ORGAN	HORMONE	FUNCTION
Anterior lobe of pituitary	Growth hormone (GH)	Stimulates growth, particularly of muscle, bone, and cartilage Stimulates breakdown of fat
	Prolactin (PRL)	Stimulates breasts to produce milk
	Thyroid-stimulating hormone (TSH)	Stimulates synthesis and release of hormones from thyroid gland
	Adrenocorticotropic hormone (ACTH)	Stimulates synthesis and release of glucocorticoid hormones from adrenal glands
	Follicle-stimulating hormone (FSH)	Stimulates gamete development in males and females Stimulates secretion of estrogen by the ovaries
	Luteinizing hormone (LH)	Causes ovulation and stimulates ovaries to secrete estrogen and progesterone Stimulates cells of testes to develop and secrete testosterone
Posterior lobe of pituitary	Antidiuretic hormone (ADH)	Promotes water reabsorption by the kidneys
	Oxytocin (OT)	Stimulates milk ejection from the breasts Stimulates uterine contractions during childbirth
Thyroid	Thyroid hormone (TH)	Regulates metabolism and heat production Promotes normal development and functioning of nervous, muscular, skeletal, and reproductive systems
	Calcitonin (CT)	Decreases blood levels of calcium (stimulates absorption of calcium by bone)
Parathyroid	Parathyroid hormone (PTH)	Increases blood levels of calcium (stimulates breakdown of bone and rate at which calcium is removed from urine and absorbed from gastrointestinal tract)
Adrenal cortex	Gonadocorticoids (androgens, estrogens)	Amounts secreted by adults are so low that effects are probably insignificant
	Mineralocorticoids (aldosterone)	Increase sodium reabsorption by kidneys Increase potassium excretion by kidneys
	Glucocorticoids (cortisol, corticosterone, cortisone)	Stimulate glucose synthesis Inhibit the inflammatory response
Adrenal medulla	Epinephrine	Fight-or-flight response Short-term response to stress
	Norepinephrine	Fight-or-flight response Short-term response to stress
Pancreas	Glucagon	Increases blood glucose level (prompts liver to increase conversion of glycogen to glucose and formation of glucose from fatty and amino acids)
	Insulin	Decreases blood glucose level (stimulates transport of glucose into cells, inhibits breakdown of glycogen to glucose, prevents conversion of fatty and amino acids to glucose)
Thymus	Thymopoietin, thymosin	Promote maturation of white blood cells
Pineal	Melatonin	Reduces jet lag and promotes sleep

cause it acts to stimulate, rather than to inhibit, the release of oxytocin. Eventually, some change breaks the positive feedback cycle. In the case of childbirth, expulsion of the baby and placenta terminates the feedback cycle. When we discuss the various glands and their hormones in the sections that follow, we also will describe the feedback mechanisms by which they are regulated.

Sometimes the nervous system overrides the controls of the endocrine system. During times of severe stress, for example, the nervous system overrides the mechanism controlling the release of

the hormone insulin. Recall that the level of insulin in the blood is controlled by a negative feedback system. Because of this negative feedback system, levels of glucose in the blood are normally maintained within a relatively narrow range. However, under conditions of severe stress, such as those associated with strong emotional reactions, heavy bleeding, or starvation, the nervous system permits levels of blood glucose to rise much higher than normal. This response gives cells access to the large amounts of energy they need for coping with the stressful situation.

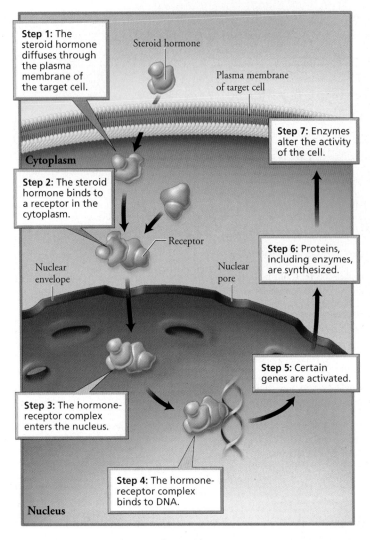

Step 1: The steroid hormone diffuses through the plasma membrane of the target cell.

Steroid hormone

Plasma membrane of target cell

Step 7: Enzymes alter the activity of the cell.

Cytoplasm

Step 2: The steroid hormone binds to a receptor in the cytoplasm.

Receptor

Step 6: Proteins, including enzymes, are synthesized.

Nuclear envelope

Nuclear pore

Step 5: Certain genes are activated.

Step 3: The hormone-receptor complex enters the nucleus.

Step 4: The hormone-receptor complex binds to DNA.

Nucleus

FIGURE **10.3** Mode of action of steroid hormones

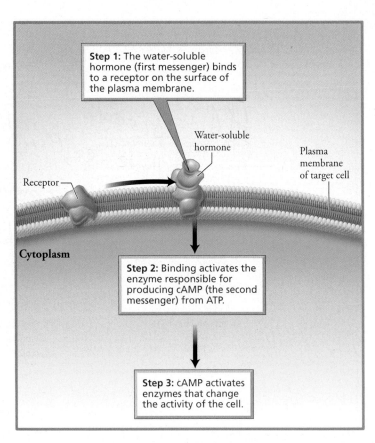

Step 1: The water-soluble hormone (first messenger) binds to a receptor on the surface of the plasma membrane.

Water-soluble hormone

Plasma membrane of target cell

Receptor

Cytoplasm

Step 2: Binding activates the enzyme responsible for producing cAMP (the second messenger) from ATP.

Step 3: cAMP activates enzymes that change the activity of the cell.

FIGURE **10.4** Mode of action of some water-soluble hormones: the second messenger system of cAMP. Because water-soluble hormones cannot cross the plasma membrane, they must affect the activities of target cells indirectly.

Hormones Influence Growth, Development, Metabolism, and Behavior

TUTORIAL 10.3

Now we'll take a look at the individual endocrine glands, describing the location and general structure, hormones, and hormonal effects of each. We also will consider disorders associated with each gland and its hormones. As we describe individual glands, keep in mind that the main function of the endocrine system—like that of the nervous system—is to regulate and coordinate other body systems and thereby maintain homeostasis. Recall also that whereas the nervous system is our rapid system of internal communication, the endocrine system is our more leisurely system, coordinating and regulating longer-term processes such as growth, development, and reproduction. We begin with the pituitary, a gland the size of a pea.

■ Pituitary hormones often prompt other glands to release hormones

The **pituitary gland** is suspended from the base of the brain, just above the roof of the mouth, by a short stalk (Figure 10.5). The gland consists of two lobes, the anterior lobe and the posterior lobe. These lobes differ in size and in their relationship with the **hypothalamus**—the area at the base of the brain to which the pituitary is connected by its stalk. The lobes of the pituitary secrete different hormones.

The anterior lobe is the larger one. Hormones of the hypothalamus control the secretion of hormones from the anterior lobe of the pituitary. Nerve cells in the hypothalamus synthesize and secrete hormones that travel by way of the bloodstream to the anterior lobe, where they stimulate or inhibit hormone secretion. Substances that stimulate hormone secretion are called *releasing hormones*. Those that inhibit hormone secretion are called *inhibiting hormones*. The anterior pituitary responds to releasing and inhibiting hormones from the hypothalamus by modifying its own synthesis and secretion of six hormones. These hormones are growth hormone (GH), prolactin (PRL), thyroid-stimulating hormone (TSH), adrenocorticotropic hormone (ACTH), follicle-stimulating hormone (FSH), and luteinizing hormone (LH).

The posterior lobe of the pituitary is very small, just larger than the head of a pin. It consists of neural tissue that releases hormones. In contrast to the *circulatory* connection between the hypothalamus and the anterior lobe, the connection between the hypothalamus and the posterior lobe is a *neural* one. As shown in Figure 10.6, nerve cells from the hypothalamus project directly into the posterior lobe. These nerve cells release oxytocin (OT) and antidiuretic hormone (ADH).

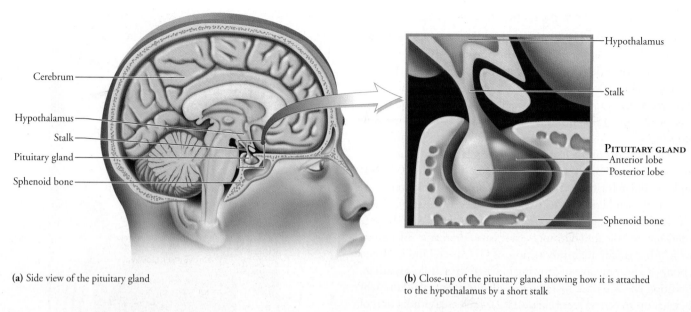

(a) Side view of the pituitary gland

(b) Close-up of the pituitary gland showing how it is attached to the hypothalamus by a short stalk

FIGURE **10.5**   Location and structure of the pituitary gland

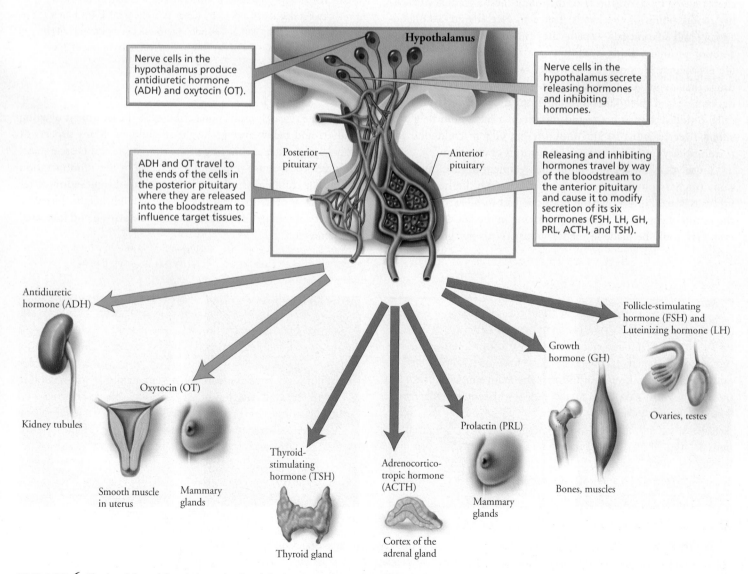

FIGURE **10.6**   The two lobes of the pituitary gland and the hormones they secrete

HORMONES OF THE ANTERIOR LOBE

The anterior lobe of the pituitary produces and secretes six major hormones. We will begin with **growth hormone (GH)**, the primary function of which is to stimulate growth through increases in cell size and rates of cell division. Cells of bone, muscle, and cartilage are most susceptible to GH, but cells of other tissues are affected as well. Growth hormone also plays a role in glucose conservation by making fats more available as a source of fuel.

Two hormones of the hypothalamus regulate the synthesis and release of GH. Growth hormone-releasing hormone (GHRH) stimulates the release of GH. Growth hormone-inhibiting hormone (GHIH) inhibits the release of GH. Levels of GH are normally maintained within an appropriate range. However, excesses or deficiencies of the hormone can have dramatic effects on growth. Abnormally high production of GH in childhood, when the bones are still capable of growing in length, results in **giantism**, a condition characterized by rapid growth and eventual attainment of heights up to 8 or 9 feet (Figure 10.7). Increased production of GH in adulthood, when the bones can thicken but not lengthen, causes acromegaly (literally, "enlarged extremities"). **Acromegaly** is characterized by enlargement of the tongue and a gradual thickening of the bones of the hands, feet, and face (Figure 10.8). Giantism and acromegaly usually are caused by a tumor of the anterior pituitary. Both conditions are associated with decreased life expectancy. Tumors can be treated with surgery, radiation, or drugs that reduce GH secretion and tumor size. Insufficient production of GH in childhood results in **pituitary dwarfism**. Typically, pituitary dwarfs are sterile and attain a maximum height of about 4 feet (Figure 10.9). Administering GH in childhood can treat pituitary dwarfism but not other forms of dwarfism.

In the past, the use of GH for the treatment of medical conditions (such as pituitary dwarfism) was extremely limited because GH was scarce, given that the hormone had to be extracted from the pituitary glands of cadavers. Beginning in the late 1970s, however, GH could be made in the laboratory. With this greater avail-

FIGURE **10.7** Robert Wadlow, a pituitary giant, was born in 1918 at a normal size but developed a pituitary tumor as a young child. The tumor caused increased production of GH. Robert never stopped growing until his death at 22 years of age, by which time he had reached a height of 8 feet 11 inches.

ability came research on its potential uses in the treatment of aging in adults and below-average height in children. Many aspects of normal aging, such as thinning of the skin, decreased muscle mass, and increased body fat, appear to be reversed by the administration of GH. For children who are of below-average height, administration of GH appears to add a few inches of adult height. Negative side effects of taking GH include fluid accumulation and increased blood pressure.

(a)

(b)

(c)

(d)

FIGURE **10.8** Acromegaly. Excess secretion of GH in adulthood, when the bones can thicken but not lengthen, causes acromegaly, a gradual thickening of the bones of the hands, feet, and face. The disorder was not apparent in this female at ages (a) 9 or (b) 16, but it became apparent by ages (c) 33 and (d) 52.

FIGURE **10.9** Pituitary dwarfism is caused by insufficient GH in childhood.

what would you do?

A recent review of studies in which children of below-average height were administered GH concluded that such therapy could yield an additional 2 inches of adult height. Opponents of this use of GH believe it is wrong to give a powerful and potentially harmful hormone to children who are basically healthy. Rather than administering GH to such children, opponents suggest working to increase societal acceptance of persons who are short. Opponents also point out that GH therapy is expensive (each additional inch of adult height gained costs about $35,000) and has negative side effects. Do you think that GH should be used to "treat" below-average height in children? What would you do if you were the parent of a healthy child who was destined to be very short? What if your child were a pituitary dwarf; would you approve the use of GH then?

Prolactin (PRL), another hormone secreted by the anterior lobe of the pituitary gland, stimulates the mammary glands to produce milk. Oxytocin, a hormone secreted from the posterior pituitary, causes the ducts of the mammary glands to eject milk. PRL interferes with female sex hormones, explaining why most mothers fail to have regular menstrual cycles and to conceive while nursing their newborn. (Lactation should not, however, be relied upon as a method for birth control, because the suppression of female hormones and ovulation lessens as mothers breast-feed their infants less frequently. Thus, a mother who breast-feeds infrequently may ovulate, and the released egg could be fertilized.)

Growth of a pituitary tumor may cause excess secretion of PRL, which may cause infertility in females, along with production of milk when birth has not occurred. In males, PRL appears to be involved in the production of mature sperm in the testis, but its precise role is not yet clear. Nevertheless, production of too much PRL, as might occur with a pituitary tumor, can cause sterility and impotence in men. Some hormones from the hypothalamus stimulate and others inhibit production and secretion of PRL.

The remaining hormones produced by the anterior lobe of the pituitary gland influence other endocrine glands. Hormones that influence another endocrine gland are called **tropic hormones**. Two such hormones secreted by the anterior lobe of the pituitary are thyroid-stimulating hormone and adrenocorticotropic hormone. **Thyroid-stimulating hormone (TSH)** acts on the thyroid gland in the neck to stimulate synthesis and release of thyroid hormones. **Adrenocorticotropic hormone (ACTH)** controls the synthesis and secretion of glucocorticoid hormones from the outer portion (cortex) of the adrenal glands (see Figure 10.2).

Two other tropic hormones secreted by the anterior lobe of the pituitary gland influence the gonads (ovaries in the female and testes in the male) and are therefore considered gonadotropins. **Follicle-stimulating hormone (FSH)** promotes development of egg cells and secretion of the hormone estrogen from the ovaries in females. In males, FSH promotes maturation of sperm. **Luteinizing hormone (LH)** causes ovulation, the release of a future egg cell by the ovary in females. LH also stimulates the ovaries to secrete estrogen and progesterone. These two hormones prepare the uterus for implantation of a fertilized ovum and the breasts for production of milk. In males, LH stimulates cells within the testes to produce the hormone testosterone.

HORMONES RELEASED FROM THE POSTERIOR LOBE

The posterior pituitary does not produce any hormones. However, neurons of the hypothalamus manufacture antidiuretic hormone and oxytocin. These hormones travel down the nerve cells into the posterior pituitary, where they are stored and released.

The main function of **antidiuretic hormone (ADH)** is to conserve body water by decreasing urine output. ADH accomplishes this task by prompting the kidneys to remove more water from the fluid destined to become urine. The water is then returned to the blood. Alcohol temporarily inhibits secretion of ADH, causing increased urination following alcohol consumption. The increased output of urine causes dehydration and the resultant headache and dry mouth typical of many hangovers. ADH is also called vasopressin. This name comes from its role in constricting blood vessels and raising blood pressure, particularly during times of severe blood loss. A deficiency of ADH may result from damage to either the posterior pituitary or the area of the hypothalamus responsible for the hormone's manufacture. Such a deficiency results in **diabetes insipidus**, a condition characterized by excessive urine production and resultant dehydration. Mild cases may not require treatment. Severe cases may cause extreme fluid loss, and death through dehydration can result. Treatment usually includes administration of synthetic ADH in a nasal spray. Diabetes insipidus (*diabetes*, overflow; *insipidus*, tasteless) should not be confused with **diabetes mellitus** (*mel*, honey). The latter is a condition in which large amounts of glucose are lost in the urine as a result of an insulin deficiency. Both conditions, however, are characterized by increased production of urine. (We will encounter diabetes mellitus again when we discuss the hormones of the pancreas.) Look again at what their names mean. Can you guess how physicians in the past distinguished between these two forms of diabetes?

Oxytocin (OT) is the second hormone produced in the hypothalamus and released by the posterior pituitary. The name oxytocin (*oxy*, quick; *tokos*, childbirth) reveals one of its two main

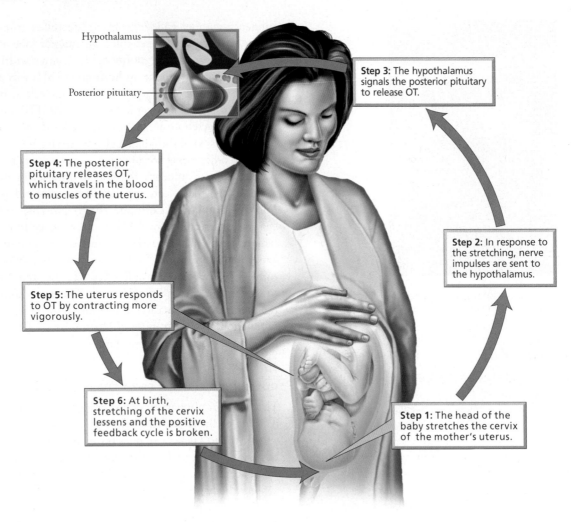

Hypothalamus

Posterior pituitary

Step 3: The hypothalamus signals the posterior pituitary to release OT.

Step 4: The posterior pituitary releases OT, which travels in the blood to muscles of the uterus.

Step 2: In response to the stretching, nerve impulses are sent to the hypothalamus.

Step 5: The uterus responds to OT by contracting more vigorously.

Step 6: At birth, stretching of the cervix lessens and the positive feedback cycle is broken.

Step 1: The head of the baby stretches the cervix of the mother's uterus.

FIGURE **10.10** The steps by which OT stimulates uterine contractions during childbirth

functions, stimulating the uterine contractions of childbirth (Figure 10.10). As described earlier, the control of OT during the induction of labor is an example of a positive feedback mechanism. Pitocin is a synthetic form of OT used to induce labor.

The second major function of oxytocin is to stimulate milk ejection from the mammary glands. Milk ejection occurs in response to the sucking stimulus of an infant (Figure 10.11). Recall that prolactin secreted by the anterior pituitary stimulates the mammary glands to produce, but not to eject, milk. Men also secrete OT, and there is some evidence that this hormone facilitates the transport of sperm in the male reproductive tract.

stop and think

Women who have just given birth are often encouraged to nurse their babies as soon as possible after delivery. How might an infant's suckling promote completion of, and recovery from, the birth process? Consider that the placenta (afterbirth) must still be expelled after the birth of the baby and that the uterus must return to an approximation of its prepregnancy form.

■ Thyroid hormones regulate metabolism and decrease blood calcium

The **thyroid gland** is a shield-shaped, deep red structure in the front of the neck, as shown in Figure 10.12a. (The color stems from its prodigious blood supply.) Within the thyroid are small spherical chambers called follicles (Figure 10.12b, c). Cells line the walls of the follicles and produce thyroglobulin, the substance from which **thyroid hormone (TH)** is made. Other endocrine cells in the thyroid, called parafollicular cells, secrete the hormone calcitonin (Figure 10.12c).

Nearly all body cells have receptors for TH. Not surprisingly, the hormone has broad effects. TH regulates the body's metabolic rate and production of heat. It also maintains blood pressure and promotes normal development and functioning of several organ systems. TH affects cellular metabolism by stimulating protein synthesis, the breakdown of lipids, and the use of glucose for production of ATP. The pituitary gland and hypothalamus control the release of TH. Falling levels of TH in the blood prompt the hypothalamus to secrete a releasing hormone. The releasing hormone

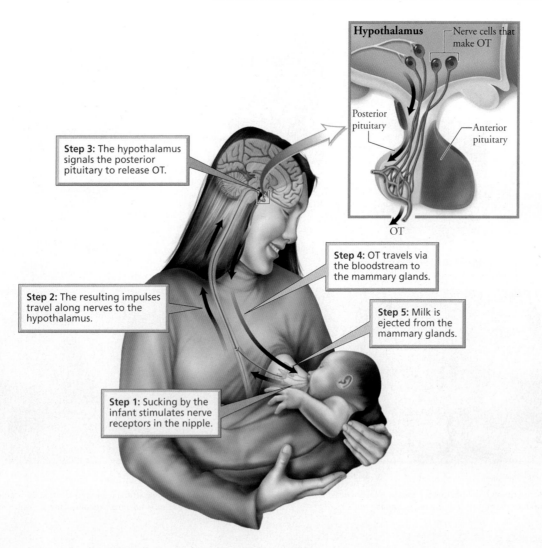

Hypothalamus — Nerve cells that make OT

Posterior pituitary

Anterior pituitary

OT

Step 3: The hypothalamus signals the posterior pituitary to release OT.

Step 2: The resulting impulses travel along nerves to the hypothalamus.

Step 4: OT travels via the bloodstream to the mammary glands.

Step 5: Milk is ejected from the mammary glands.

Step 1: Sucking by the infant stimulates nerve receptors in the nipple.

FIGURE **10.11** The steps by which OT stimulates milk ejection from the mammary glands

stimulates the anterior pituitary to secrete TSH, which, in turn, causes the thyroid to release more TH.

Iodine is needed for production of thyroid hormone. A diet deficient in iodine can produce a **simple goiter**, that is, an enlarged thyroid gland (Figure 10.13). When intake of iodine is inadequate, the level of TH is low, and the low level triggers secretion of TSH. TSH stimulates the thyroid gland to increase production of thyroglobulin. The lack of iodine prevents formation of TH from the accumulating thyroglobulin. In response to continued low levels of TH, the pituitary continues to release increasing amounts of TSH, which cause the thyroid to enlarge in a futile effort to filter more iodine from the blood. In the past, goiters were quite common, especially in parts of the midwestern United States (dubbed the Goiter Belt), where iodine-poor soil and little access to iodine-rich shellfish led to diets deficient in iodine. The incidence of goiter in the United States dramatically decreased once iodine was added to most table salt beginning in the 1920s. Simple goiter can be treated by iodine supplements or administration of TH.

Undersecretion of TH during fetal development or infancy causes **cretinism**, a condition characterized by dwarfism, and de-

layed mental and sexual development (Figure 10.14). If a pregnant woman produces sufficient TH, many of the symptoms of cretinism do not appear until after birth, when the infant begins to rely solely on its own malfunctioning thyroid gland to supply the needed hormones. Oral doses of TH can prevent cretinism, so most infants are tested for proper thyroid function shortly after birth. Undersecretion of TH in adulthood causes **myxedema**, a condition in which fluid accumulates in facial tissues. Other symptoms of undersecretion of TH include decreases in alertness, body temperature, and heart rate. Oral administration of TH can prevent these symptoms.

Oversecretion of TH causes **Graves' disease**, an autoimmune disorder in which a person's own immune system produces Y-shaped proteins called antibodies (discussed in Chapter 13) that in this case mimic the action of TSH. The antibodies stimulate the thyroid gland, causing it to enlarge and overproduce its hormones. The symptoms of Graves' disease include increased metabolic rate and heart rate, accompanied by sweating, nervousness, and weight loss. Many patients with Graves' disease also have *exophthalmos*, protruding eyes caused by the swelling of tissues in the eye orbits

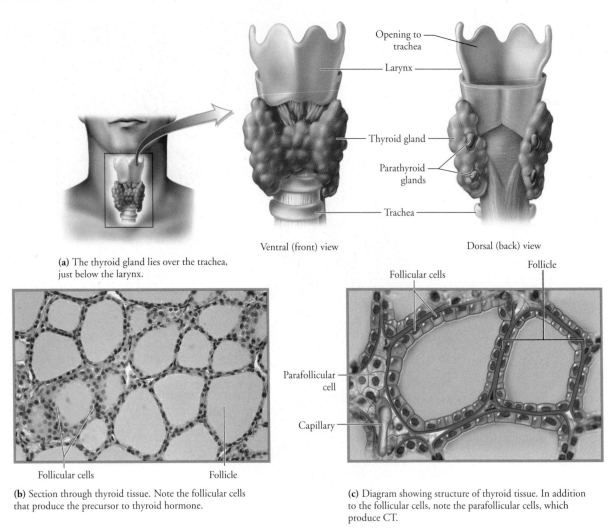

Ventral (front) view

Dorsal (back) view

(a) The thyroid gland lies over the trachea, just below the larynx.

Follicular cells

(b) Section through thyroid tissue. Note the follicular cells that produce the precursor to thyroid hormone.

(c) Diagram showing structure of thyroid tissue. In addition to the follicular cells, note the parafollicular cells, which produce CT.

FIGURE **10.12** Location and structure of the thyroid gland and parathyroid glands

FIGURE **10.13** Simple goiter. In response to an iodine-deficient diet, the thyroid gland enlarges, causing a goiter.

(Figure 10.15). Graves' disease may be treated with drugs that block synthesis of thyroid hormones. Alternatively, thyroid tissue may be reduced through surgery or the administration of radioactive iodine. The thyroid gland accumulates iodine; thus, ingestion of radioactive iodine (usually in capsules) selectively destroys thyroid tissue.

The **calcitonin (CT)** secreted by the parafollicular cells of the thyroid helps regulate the concentration of calcium in the blood to ensure the proper functioning of muscle cells (calcium ions bind to the protein troponin leading to changes in other muscle proteins, eventually causing muscle contraction; Chapter 6) and neurons (calcium causes the release of neurotransmitters into the synaptic cleft and therefore is critical in the transmission of messages from one neuron to the next; Chapter 7). When the level of calcium in the blood is high, CT stimulates the absorption of calcium by bone and inhibits the breakdown of bone, thereby lowering the level of calcium in the blood. CT also lowers blood calcium by stimulating an initial increase in the excretion of calcium in the urine. When the level of calcium in the blood is low, the parathyroid glands, which we discuss next, are prompted to release their hormone.

FIGURE **10.14** Cretinism, a condition characterized by dwarfism and delayed mental development, is caused by undersecretion of TH during fetal life or infancy.

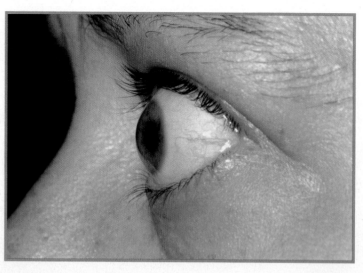

FIGURE **10.15** Exophthalmos, a disorder of the thyroid gland that is often associated with Graves' disease. Oversecretion of TH leads to an accumulation of fluid behind the eyes, causing the eyes to bulge out.

■ Parathyroid hormone increases blood calcium

The **parathyroid glands** are four small, round masses at the back of the thyroid gland (Figure 10.12a, dorsal view). These glands secrete **parathyroid hormone (PTH)**, also called *parathormone*. As mentioned above, CT from the thyroid gland lowers the level of calcium in the blood. In contrast, PTH increases levels of calcium in the blood. Low levels of calcium in the blood stimulate the parathyroid glands to secrete PTH, which causes calcium to move from bone and urine into the blood. PTH exerts its effects by stimulating (1) bone-destroying cells called osteoclasts that release calcium from bone into the blood, (2) the removal of calcium from the urine and its return to the blood, and (3) the rate at which calcium is absorbed into the blood from the gastrointestinal tract. PTH also inhibits bone-forming cells called osteoblasts and thereby reduces the rate at which calcium is deposited in bone. The feedback system by which CT and PTH together regulate levels of calcium in the blood is summarized in Figure 10.16.

Surgery on the neck or thyroid gland may damage the parathyroid glands. The resultant decrease in PTH causes decreased blood calcium that in turn produces nervousness and muscle spasms. In severe cases, death may result from spasms of the larynx and paralysis of the respiratory system. PTH is difficult to purify, so deficiencies are not usually treated by administering the hormone. Instead, calcium is given either in tablet form or through increased dietary intake. A tumor of the parathyroid gland can cause excess secretion of PTH. Oversecretion of PTH pulls calcium from bone tissue, causing increased blood calcium and weakened bones. High levels of calcium in the blood may lead to kidney stones, calcium deposits in other soft tissue, and decreased activity of the nervous system.

■ The adrenal glands secrete stress hormones

The body's two **adrenal glands**, each about the size of an almond, are located at the tops of the kidneys (*ad*, upon; *renal*, kidney). Each gland has two distinct regions (Figure 10.17). The outer region, the **adrenal cortex**, secretes more than 20 different steroid hormones, generally divided into three groups: the gonadocorticoids, mineralocorticoids, and glucocorticoids. The inner region, called the **adrenal medulla**, secretes epinephrine and norepinephrine.

The **gonadocorticoids** are male and female sex hormones known as **androgens** and **estrogens**, respectively. In both males and females, the adrenal cortex secretes both androgens and estrogens. However, in normal adult males, androgen secretion by the testes far surpasses that by the adrenal cortex. Thus, the effects of adrenal androgens in adult males are probably insignificant. In females, the ovaries and placenta also produce estrogen, although during menopause, the ovaries decrease secretion of estrogen and eventually stop secreting it. The gonadocorticoids from the adrenal cortex may somewhat alleviate the effects of decreased ovarian estrogen. One option for menopausal women suffering from estrogen deficiency is hormone replacement therapy. The advantages and disadvantages of this therapy are considered in the Health Issue essay, *Is It Hot in Here, or Is It Me? Hormone Replacement Therapy and Menopause.*

The **mineralocorticoids** secreted by the adrenal cortex affect mineral homeostasis and water balance. The primary mineralocorticoid is **aldosterone**, a hormone that acts on cells of the kidneys to increase reabsorption of sodium ions (Na^+) into the blood. This reabsorption prevents depletion of Na^+ and increases water retention. Aldosterone also acts on kidney cells to promote the excretion of potassium ions (K^+) in urine. **Addison's disease** is a disorder

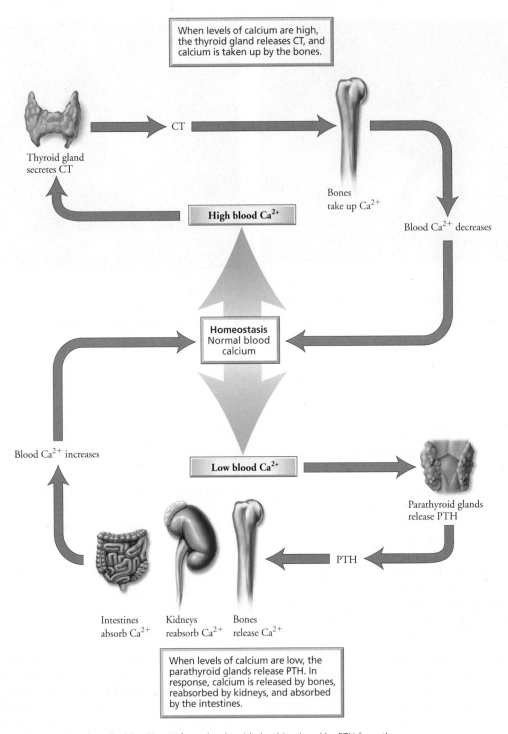

When levels of calcium are high, the thyroid gland releases CT, and calcium is taken up by the bones.

CT

Thyroid gland secretes CT

Bones take up Ca^{2+}

Blood Ca^{2+} decreases

High blood Ca^{2+}

Homeostasis
Normal blood calcium

Blood Ca^{2+} increases

Low blood Ca^{2+}

Parathyroid glands release PTH

PTH

Intestines absorb Ca^{2+}

Kidneys reabsorb Ca^{2+}

Bones release Ca^{2+}

When levels of calcium are low, the parathyroid glands release PTH. In response, calcium is released by bones, reabsorbed by kidneys, and absorbed by the intestines.

FIGURE **10.16** Regulation of calcium levels in the blood by CT from the thyroid gland (top) and by PTH from the parathyroid glands (bottom)

caused by the undersecretion of the glucocorticoid cortisol (see the next page) and aldosterone. This disease appears to be an autoimmune disorder in which the body's own immune system perceives cells of the adrenal cortex as foreign and destroys them. The resulting deficiency of adrenal hormones causes weight loss, fatigue, electrolyte imbalance, poor appetite, and poor resistance to stress. A peculiar bronzing of the skin also is associated with Addison's disease. Recall that the pituitary gland secretes ACTH, which stimulates the cortex of the adrenal glands to secrete its hormones. Thus,

Addison's disease can also be caused by inadequate secretion of ACTH by the pituitary. Addison's disease can be treated with tablets containing the missing hormones.

stop and think

High blood pressure can signal abnormal aldosterone secretion. Would high blood pressure be associated with the undersecretion or oversecretion of aldosterone?

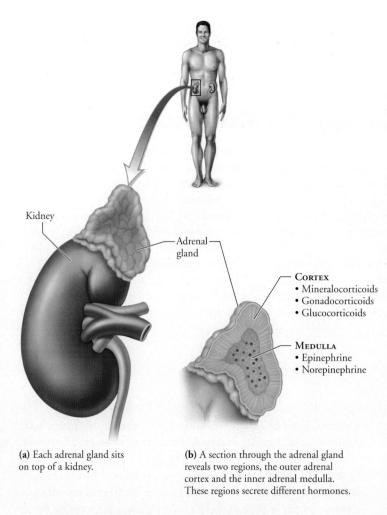

Kidney

Adrenal gland

CORTEX
• Mineralocorticoids
• Gonadocorticoids
• Glucocorticoids

MEDULLA
• Epinephrine
• Norepinephrine

(a) Each adrenal gland sits on top of a kidney.

(b) A section through the adrenal gland reveals two regions, the outer adrenal cortex and the inner adrenal medulla. These regions secrete different hormones.

FIGURE **10.17** Location and structure of an adrenal gland

Cushing's syndrome results from prolonged exposure to high levels of cortisol. Body fat is redistributed and fluid accumulates in the face (Figure 10.18). Additional symptoms include fatigue, high blood pressure, and elevated glucose levels. A tumor on either the adrenal cortex or the anterior pituitary may cause the oversecretion of cortisol that leads to Cushing's syndrome. (Recall that the anterior pituitary secretes ACTH, which stimulates the release of hormones from the adrenal cortex.) Tumors are treated with radiation, drugs, or surgery. Cushing's syndrome also may result from giving glucocorticoid hormones to treat asthma, lupus, or rheumatoid arthritis. Treatment in medically induced cases of Cushing's syndrome typically entails a gradual reduction of the glucocorticoid dose, ideally to the lowest level necessary to control the existing disorder.

The adrenal medulla produces **epinephrine** (adrenaline) and **norepinephrine** (noradrenaline). These hormones are critical in the **fight-or-flight response**, the body's reaction to emergencies, which is mounted by the sympathetic nervous system (Chapter 8). Imagine, for example, that you are walking home alone late at night and a stranger suddenly steps toward you from the bushes. Impulses received by your hypothalamus are sent by neurons to your adrenal medulla. These impulses cause cells in your adrenal medulla to increase output of epinephrine and norepinephrine. In response to these hormones, your heart rate, respiratory rate, and blood glucose levels rise. Blood vessels associated with the digestive tract constrict because digestion currently is not of prime importance. Vessels associated with skeletal and cardiac muscles dilate, allowing more blood, glucose, and oxygen to reach them. These substances also reach your brain in greater amounts, leading to the increased mental alertness needed for fleeing or fighting. We explore further how our bodies react to stress in the Health Issue essay, *Hormones and Our Response to Stress.*

The **glucocorticoids** are hormones secreted by the adrenal cortex that affect glucose levels. Glucocorticoids act on the liver to promote the conversion of fat and protein to intermediate substances that are ultimately converted into glucose. The glucocorticoids also act on adipose tissue to prompt the breakdown of fats to fatty acids that are released into the bloodstream, where they are available for use by the body's cells. Glucocorticoids further conserve glucose by inhibiting its uptake by muscle and fat tissue.

Glucocorticoids also inhibit the inflammatory response; such inhibition can be beneficial when the body is faced with the swelling and intense irritation associated with skin rashes such as that caused by poison ivy. One way glucocorticoids inhibit inflammation is by slowing the movement of white blood cells to the site of injury. Another way is by reducing the likelihood that other cells will release chemicals that promote inflammation. Unfortunately, these activities of glucocorticoids inhibit wound healing. Steroid creams containing glucocorticoids are therefore intended to be applied only to the surface of the skin and to be used for superficial rashes only. These creams should not be applied to open wounds. Some examples of glucocorticoids are cortisol, corticosterone, and cortisone.

(a) Patient diagnosed with Cushing's syndrome

(b) Same patient after treatment

FIGURE **10.18** Cushing's syndrome. Prolonged exposure to cortisol causes fluid to accumulate in the face. Most often, Cushing's syndrome is caused by the administration of cortisol for allergies or inflammation.

HEALTH ISSUE

Is It Hot in Here, or Is It Me?
Hormone Replacement Therapy and Menopause

Many endocrine disorders are characterized by little or no secretion of certain hormones. Cretinism, for example, is caused by the undersecretion of thyroid hormone during fetal life or infancy. In other cases, hormone secretion declines as part of the normal aging process. Estrogen, a hormone produced primarily by the ovaries, is secreted in decreasing amounts beginning when women are about age 30. This decrease eventually leads into menopause, the termination of ovulation and menstruation.

Menopause usually occurs when women are between 45 and 55 years of age. Undersecretion of hormones, whether caused by malfunctioning glands or normal aging, often is treated with hormone replacement therapy (HRT). In HRT, the deficient hormone is replaced using injections, pills, patches, or creams.

Hormone replacement has had many successes, but it is not without controversy, particularly when it is used to treat normal declines in hormone levels. As mentioned above, levels of

ovarian hormones decline as women age; menopause is the ultimate result. Some women face a host of uncomfortable symptoms in conjunction with the eventual end of ovulation and menstruation. Such symptoms include hot flashes, night sweats, vaginal dryness, stress incontinence (involuntary loss of small amounts of urine when laughing, coughing, or sneezing), heightened mood swings, and memory loss. Symptoms may last only a few months or a few years. Declining estrogen levels also have been linked to increased

---

■ Hormones of the pancreas regulate blood glucose

The **pancreas** is located in the abdomen just behind the stomach; it contains both endocrine and exocrine cells (Figure 10.19). The exocrine cells secrete digestive enzymes into ducts that empty into the small intestine. The role of the pancreas in digestion will be discussed in Chapter 15. The endocrine cells occur in small clusters called **pancreatic islets** (or islets of Langerhans). These clusters contain two major types of hormone-producing cells. One type produces the hormone glucagon; the other produces the hormone insulin.

Glucagon increases glucose in the blood. It does so by prompting cells of the liver to increase conversion of glycogen (the storage polysaccharide in animals) to glucose (a simple sugar, or monosaccharide). Glucagon also stimulates formation of glucose from lactic acid and amino acids. The liver releases the resultant glucose molecules into the bloodstream, causing a rise in blood sugar level.

In contrast to glucagon, **insulin** decreases glucose in the blood; insulin and glucagon thus have opposite or *antagonistic effects*. Insulin decreases blood glucose in several ways. First, insulin stimulates transport of glucose into muscle cells, white blood cells, and connective tissue cells. Second, insulin inhibits the breakdown of glycogen to glucose. Finally, insulin prevents conversion of amino and fatty acids to glucose. As a result of these actions, insulin promotes protein synthesis, fat storage, and the use of glucose for energy. Figure 10.20 summarizes the regulation of glucose in the blood by insulin and glucagon.

Insulin has dramatic effects on health. More than 120 million people worldwide, over 15 million of them in the United States, suffer from diabetes mellitus, a group of metabolic disorders characterized by an abnormally high level of glucose in the blood. There

are several types of diabetes mellitus. *Type 1 diabetes mellitus* used to be known as insulin-dependent diabetes. It also was called juvenile-onset diabetes, because it usually develops in people younger than 25 years of age. In this autoimmune disorder, which represents about 5% to 10% of all diagnosed cases of diabetes, a person's own immune system attacks the cells of the pancreas responsible for insulin production. Symptoms include nausea, vomiting, thirst, and excessive urine production. Treatment involves daily, sometimes multiple, injections of insulin. Exercise and careful monitoring of diet and blood glucose levels also are essential. Insulin cannot yet be taken orally, because it is a protein hormone that can be broken down in the digestive tract. However, work is under way to package insulin in "digestion-resistant" microcapsules. There is also the possibility that insulin could be administered as a nasal spray. Other research has focused on whether insulin-producing cells can be transferred into the pancreas of a diabetic.

Type 2 diabetes mellitus was also known as non-insulin-dependent diabetes. It was formerly called adult-onset diabetes, as well, because it usually develops after age 40, although it recently has begun showing up in younger people. Type 2 diabetes, which accounts for between 90% and 95% of diabetes cases, is characterized by decreased sensitivity to insulin. The lower sensitivity may result from a decrease in insulin receptors on target cells. Many people with type 2 diabetes are overweight. Treatment typically involves dietary restrictions, exercise, and weight loss. Oral medications that increase insulin sensitivity or production also may be given. About 40% of people with type 2 diabetes need insulin injections. This was one reason for dropping the name non-insulin-dependent diabetes.

risk of osteoporosis (increased bone loss leading to increased risk of fracture; see Chapter 5). Given all these effects, it is not surprising that for many years doctors administered ovarian hormones to millions of women at or past menopause. In most cases, doctors prescribed estrogen and progestin (a form of progesterone). Estrogen alone, called estrogen replacement therapy (ERT), was given if a woman's uterus had been surgically removed (this surgical procedure is known as a hysterectomy). Both HRT and ERT have some benefits. Both therapies relieve hot flashes and night sweats, and slow the loss of bone.

In recent years, research has revealed a dark side to HRT and ERT. In 2002, the U.S. govern-ment formally listed all forms of estrogen used in replacement therapies as "known human carcinogens." The discovery that estrogen replacement therapy increases a woman's risk of developing endometrial cancer (cancer of the lining of the uterus) led doctors to use lower doses of estrogen and to combine it with progestin, which protects the uterine lining against cancer. However, there also is the possibility that HRT and ERT may increase the risk in some women of breast cancer, heart attack, stroke, blood clots, and gall bladder disease. To make matters worse, early claims that HRT and ERT slowed the progression of Alzheimer's disease have not been substantiated by recent studies.

In view of the risks of HRT and ERT, many physicians are now urging women to consider the many other ways available to stave off osteoporosis. These physicians urge women to avoid smoking, to eat a diet rich in calcium, and to exercise regularly. Nonhormonal drugs may be better and safer for preventing bone fractures. Some alternative practitioners advise getting estrogen from dietary sources such as yams and soybeans. These foods are certainly weaker sources than prescription estrogen and may reduce the risks. Finally, if you choose to use HRT to relieve the effects of decreasing estrogen, then it is advised that you work closely with your physician to use the lowest effective dose for the shortest possible time. ♂♀

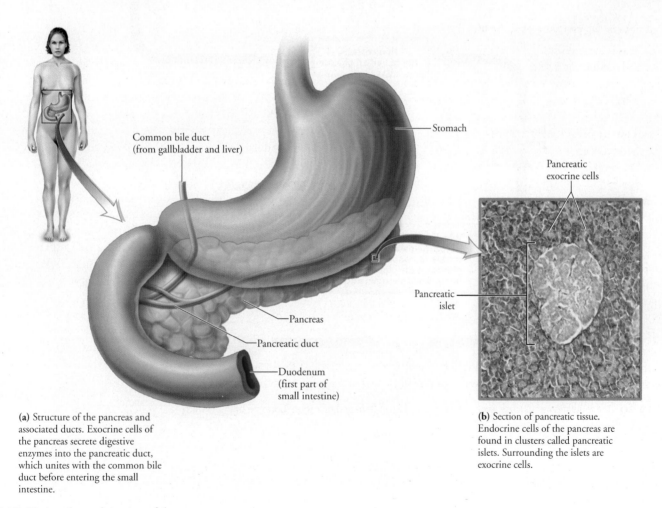

(a) Structure of the pancreas and associated ducts. Exocrine cells of the pancreas secrete digestive enzymes into the pancreatic duct, which unites with the common bile duct before entering the small intestine.

(b) Section of pancreatic tissue. Endocrine cells of the pancreas are found in clusters called pancreatic islets. Surrounding the islets are exocrine cells.

FIGURE **10.19** Location and structure of the pancreas

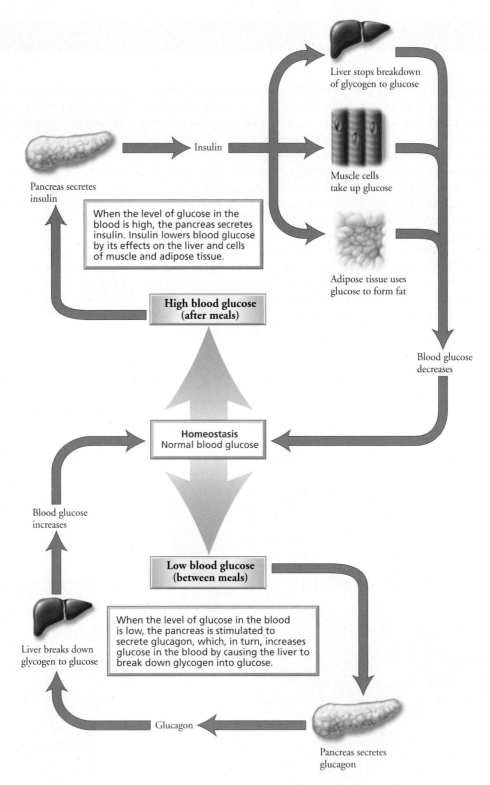

Pancreas secretes
insulin

Insulin

Liver stops breakdown
of glycogen to glucose

Muscle cells
take up glucose

Adipose tissue uses
glucose to form fat

When the level of glucose in the
blood is high, the pancreas secretes
insulin. Insulin lowers blood glucose
by its effects on the liver and cells
of muscle and adipose tissue.

**High blood glucose
(after meals)**

Blood glucose
decreases

Homeostasis
Normal blood glucose

Blood glucose
increases

**Low blood glucose
(between meals)**

When the level of glucose in the blood
is low, the pancreas is stimulated to
secrete glucagon, which, in turn, increases
glucose in the blood by causing the liver to
break down glycogen into glucose.

Liver breaks down
glycogen to glucose

Glucagon

Pancreas secretes
glucagon

FIGURE **IO.20** The regulation of glucose level in the blood by insulin (top) and glucagon (bottom), both of which are secreted by the pancreas

HEALTH ISSUE

Hormones and Our Response to Stress

Stress can be broadly defined as mental or physical tension. Rarely does a day go by that we are not subjected to stress. Awaiting the start of an exam, some personal performance, or an interview can be stressful. Our bodies usually can deal with everyday stresses and maintain the relative constancy of our internal environment. Sometimes, however, stress is extreme in its intensity and duration, and our coping mechanisms prove inadequate. At such times, stress triggers the hypothalamus to initiate the general adaptation syndrome (GAS), a series of physiological adjustments made by our bodies in response to extreme stress.

The GAS has three phases: alarm, resistance, and exhaustion. The alarm phase is also known as the fight-or-flight response. Recall that the fight-or-flight response is initiated by epinephrine from the adrenal medulla. The response immediately funnels huge amounts of glucose and oxygen to the organs most critical in responding to danger.

Sometimes the adjustments of the alarm phase are sufficient to end or escape whatever is causing the stress. At other times, stress is so intense and long lasting that the individual enters the resistance phase. Changes wrought by the resistance phase are more long term than are those of the alarm phase. Also, rather than being stimulated by nerve impulses from the hypothalamus, the resistance phase is initiated by the release of hormones from the hypothalamus. The released hormones stimulate the anterior pituitary to secrete several hormones. Some of these hormones stimulate other glands to secrete their hormones. Glucocorticoids from the adrenal cortex are the main hormones of the resistance phase. Two primary effects of glucocorticoids are to mobilize the body's protein and fat reserves and to conserve glucose for use by cells of the nervous system. The hormones of the resistance phase also result in conservation of body fluids.

The resistance phase is sustained by the body's fat reserves. This phase may last for weeks or months, but it cannot go on indefinitely. Sooner or later, lipid reserves are exhausted and structural proteins must be broken down to meet energy demands. Eventually, organs are unable to meet the heavy demands of the resistance phase, and they begin to fail. This is the exhaustion phase. Without immediate attention, death may result from collapse of one or more organ systems.

Stress can have dramatic effects on our health, especially when it is prolonged and uncontrollable. It depresses wound healing, increases our susceptibility to infections, and leads to disorders such as hypertension, irritable bowel syndrome, and asthma. Some studies have shown that stress puts people at greater risk for developing chronic diseases. Overall, prolonged stress appears to shorten the life span.

Given the connection between stress and health, it is important to reduce some of the stress in our lives. A first step is to be realistic in assessing the levels of stress associated with life events, such as starting college or getting married. Because not everyone is affected in the same way by the same life event, it is important to consider what you find stressful and then take positive steps to reduce your exposure to stress. It also is important to develop ways to cope with unavoidable stress. Commonly used means to alleviate the effects of stress include relaxation techniques and regular exercise. A more specialized option is biofeedback, a procedure that can be used to help a person recognize the symptoms of stress and learn how to control them. During a stress biofeedback session, a health care professional connects a patient to a machine that monitors one or more physiological indicators of stress, such as heart rate or muscle tension (Figure 10.A). The health care worker then discusses a stressful situation with the patient. The machine gives off signals when the conversation makes the patient begin to experience stress. For example, increased tension in muscles might prompt a clicking sound. The patient can then practice decreasing the muscle tension through deep breathing and relaxation. This decrease can be monitored by decreases in the frequency of clicks. Eventually, patients are able to recognize and cope with signs of stress without the help of the machine. ♂♀

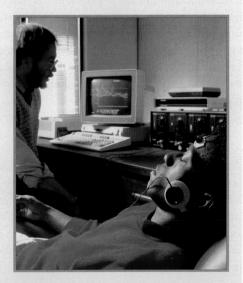

FIGURE **10.A** Biofeedback is one way in which people can learn to recognize the symptoms of stress and how to cope with them.

There are two other general categories of diabetes. When diabetes mellitus develops in women during gestation, it is called gestational diabetes. This condition occurs in 2% to 5% of all pregnancies. It usually begins in the second half of pregnancy. The placenta is the organ that supplies the growing fetus with nutrients and oxygen and carries away wastes and carbon dioxide (Chapter 18). The placenta also produces hormones, some of which may block the effects of insulin. Normally, such blocking effects are overcome by the pancreas's producing more insulin. Sometimes, however, the production of insulin is insufficient, and gestational diabetes results. This form of diabetes generally goes away after delivery because the placenta is expelled a few minutes after the baby. Treatment for gestational diabetes typically includes testing blood glucose levels, possibly taking insulin, eating a healthy diet, and engaging in regular physical activity. Lastly, "other specific types" of diabetes represent 1% to 2% of all diagnosed cases. These types include insulin deficiencies resulting from damage to the pancreas from disease, infection, or drugs.

Diabetes has serious complications. Diabetics are at increased risk for blindness, kidney disease, heart disease, high blood pressure, and atherosclerosis (buildup of fatty deposits in the arteries). Many diabetics suffer from gum disease and from damage to their nervous system that may include impotence and loss of sensation. Poor circulation and problems with nerves in the lower legs may make amputation of the lower limbs necessary. Indeed, more than half of lower limb amputations in the United States are done in people with diabetes. Given these serious complications, it is important to diagnose and treat diabetes early and, of course, to prevent it when possible. The risk factors for diabetes include obesity, high blood pressure, parent or sibling with diabetes, and a history of gestational diabetes.

Too much insulin sometimes results from a tumor of the pancreas. More typically, however, a diabetic person mistakenly injects too much insulin. The result is depressed levels of glucose in the blood. Initial symptoms of low blood glucose include anxiety, sweating, hunger, weakness, and disorientation. Because brain cells fail to function properly when starved of glucose, the initial symptoms may be followed by convulsions and unconsciousness. The consequences associated with severe depletion of blood glucose are known collectively as *insulin shock*. Insulin shock can prove fatal unless blood sugar levels are raised.

■ Hormones of the thymus gland promote maturation of white blood cells

The **thymus gland** lies just behind the breastbone, on top of the heart (see Figure 10.2). It is more prominent in infants and children than in adults because it decreases in size as we age. The hormones it secretes, such as **thymopoietin** and **thymosin**, promote the maturation of white blood cells called T lymphocytes. T lymphocytes, also known as T cells, are part of the body's defense mechanisms (Chapter 13).

■ The pineal gland secretes melatonin

The **pineal gland** is a tiny gland at the center of the brain (Figure 10.21). Its secretory cells produce the hormone **melatonin**. Levels

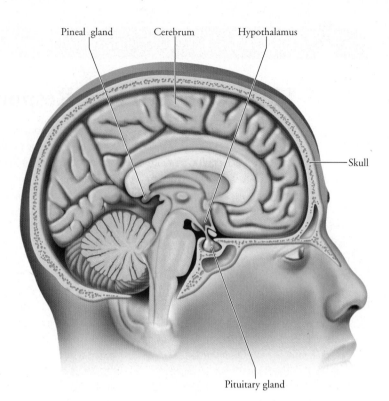

FIGURE **10.21** The pineal gland. Located at the center of the brain, the pineal gland secretes the hormone melatonin.

of circulating melatonin are greater at night than during daylight hours, because of input the pineal gland receives from visual pathways. Neurons of the retina, stimulated by light entering the eye, send impulses to the hypothalamus and ultimately the pineal gland, where they inhibit secretion of melatonin.

Research in the past few decades has suggested diverse roles for melatonin. Melatonin may, as it does in some nonhuman animals, inhibit production of the pigment melanin by melanocytes of the skin. Melatonin may also influence daily rhythms. Sleep and, for some people, seasonal changes in mood appear to be influenced by melatonin. Melatonin also may slow the aging process. The possible links between melatonin and sleep, daily rhythms, and aging have led to widespread over-the-counter purchasing of melatonin, a situation described as "melatonin mania." We examine the promise and possible risks of taking melatonin in the Health Issue essay, *Melatonin: Miracle Supplement or Potent Drug Misused by Millions?*

One disorder associated with too much melatonin is *seasonal affective disorder (SAD)*. This form of depression is associated with winter, when short day length results in overproduction of melatonin. (Recall that melatonin is secreted in the absence of light.) Too much melatonin causes symptoms such as lethargy, long periods of sleep, low spirits, and a craving for carbohydrates. The symptoms usually appear around October and end about April in the Northern Hemisphere. Three quarters of persons who suffer from SAD are female. Treatment of SAD often includes repeated exposure to very bright light for about an hour each day. The intense light inhibits melatonin production.

Melatonin: Miracle Supplement or Potent Drug Misused by Millions?

Melatonin, a hormone secreted by the pineal gland, has often been billed as a wonder drug. There are claims that melatonin can reduce jet lag, cure insomnia, boost immunity, and slow the aging process. In addition, melatonin is apparently nontoxic, even at extremely high doses. But which, if any, of these claims is well supported by scientific evidence? And even if melatonin can be helpful under certain circumstances, should this potent hormone be sold as an over-the-counter dietary supplement to self-prescribing consumers?

The claim that melatonin helps to alleviate jet lag is well supported by scientific evidence. Studies have shown that melatonin taken by mouth on the day of travel and continued for several days has the following effects in about half the people who take it to fight jet lag: (1) reduced fatigue during the day; (2) reduced time to fall asleep at night; and (3) more rapid development of a normal sleep pattern. Such benefits are usually most evident during eastward travel that crosses more than four time zones. For people who are not traveling, but who toss and turn and watch the clock on the nightstand at home, there is some evidence that melatonin can bring sleep. Research, beginning in the 1980s and continuing today, has shown that melatonin reduces the time it takes to fall asleep and may lengthen the time that people stay asleep. Although most agree that melatonin can be an effective sleep aid, scientists debate the precise mechanism by which melatonin produces its soporific effect.

What about the claims that melatonin can be used to boost immunity? Recall that the thymus gland, responsible for the production of infection-fighting white blood cells called T lymphocytes, is largest in infants and young children and shrinks as we age. Old mice also show shrinkage of the thymus. Such shrinkage can be reversed in geriatric mice by injections of melatonin. Indeed, the immune systems of injected animals showed signs of revitalization. Human trials on the potential immune-enhancing effects of melatonin in cancer patients are now underway. Although some promising results have been reported, there is insufficient evidence at present to support the claim that melatonin effectively boosts the human immune system.

Melatonin may slow aging through its effective scavenging of free radicals, molecular fragments that contain an unpaired electron. Free radicals have been implicated in many chronic diseases of old age, such as cancer and heart disease. Substances, such as melatonin, that destroy free radicals are called antioxidants. Free radicals are normally generated by some cells of the body for the purpose of destroying bacteria and old cells. Environmental agents such as drugs, toxins, and radiation also generate free radicals. What evidence is there to support melatonin as an effective antioxidant? Although numerous studies with other animals have supported the antioxidant properties of melatonin, well-designed studies with humans are lacking.

Should consumers hoping to cure jet lag or insomnia or to boost immunity and slow aging be gobbling down melatonin pills and lozenges purchased at health food stores? At present, melatonin is sold as a dietary supplement and thus is not subject to intense scrutiny by the Food and Drug Administration. No one knows the possible long-term effects of high doses. Another concern is that production of the hormone by companies not subject to regulation raises questions about the strength and purity of the final marketed product. Questions also remain about the benefits of melatonin. As a result, some people in the medical profession caution all consumers to wait until melatonin is subjected to further scrutiny and regulation. Others argue, however, that melatonin is probably safe for most people but that certain groups should avoid it until further information is available. For example, avoidance of melatonin is urged for pregnant or nursing women (the effects of high doses on fetuses or infants are unknown) and children (who normally produce melatonin in large amounts). Avoidance of melatonin is also indicated for people with autoimmune disorders or cancers of the immune system (melatonin may stimulate the immune system and worsen these conditions). Finally, people taking other medications should consult their physician before taking melatonin. Such consultation should help people avoid dangerous drug interactions. ♙

Other Chemical Messengers Act Locally

Now that we have surveyed the endocrine glands and their hormones, let's consider chemical messengers that act locally. Once secreted by a cell, these **local signaling molecules** exert their effects on adjacent target cells within seconds or milliseconds, much faster than the communication carried out by hormones traveling to distant sites within the body. Neurotransmitters, discussed in Chapter 7, are examples of chemicals that rapidly convey messages from one cell (a neuron) to a neighboring cell (often another neuron). Prostaglandins, growth factors, and nitric oxide (NO) are other examples of local signaling molecules.

Prostaglandins are lipid molecules continually released by the plasma membranes of most cells. Different types of cells secrete different prostaglandins. At least 16 different prostaglandin molecules function within the human body. These molecules have remarkably diverse effects, influencing blood clotting, regulation of body temperature, diameter of airways to the lungs, and the body's inflammatory response. Prostaglandins also affect the reproductive system. Menstrual cramps are thought to be caused by prostaglandins released by cells of the uterine lining. These prostaglandins act on the smooth muscle of the uterus, causing muscle contractions and cramping. Drugs such as aspirin inhibit the synthesis of prostaglandins and thus may lessen the discomfort of menstrual cramps. Prostaglandins also are found in semen, the fluid discharged from the penis at ejaculation. Once in the female reproductive tract, prostaglandins in semen cause the smooth muscles of the uterus to contract, perhaps helping the sperm continue their journey.

Growth factors are peptides or proteins that, when present in the fluid outside target cells, stimulate those cells to grow, develop, and multiply. For example, one growth factor causes precursor cells in the bone marrow to proliferate and differentiate into particular white blood cells. Another growth factor prompts endothelial cells

to proliferate and organize into tubes that eventually form blood vessels (fully formed blood vessels have an inner lining of endothelial cells, a middle layer of smooth muscle, and an outer layer of connective tissue; Chapter 12). The gas nitric oxide (NO) functions in the cellular communication that leads to the dilation of blood vessels. Basically, endothelial cells of the inner lining of blood vessels make and release NO, which signals the smooth muscles in the surrounding (middle) layer to relax, allowing the vessel to dilate. NO also functions as a neurotransmitter, carrying messages from one neuron to the next.

HIGHLIGHTING THE CONCEPTS

The Endocrine System Communicates Using Chemical Messages (pp. 175–178)

1. Endocrine glands lack ducts and release their products, hormones, into the spaces just outside cells. The hormones then diffuse into the bloodstream. Endocrine glands and organs that contain some endocrine tissue constitute the endocrine system, a system of internal communication that regulates and coordinates other organ systems and helps maintain homeostasis.

2. Hormones, the chemical messengers of the endocrine system, contact virtually all cells within the body. However, hormones affect only target cells, those cells with receptors that recognize and bind specific hormones.

3. Steroid hormones are lipid soluble. Steroids cross through the plasma membrane of target cells into the cytoplasm, where they combine with a receptor molecule, forming a hormone-receptor complex that moves into the nucleus of the cell. There the complex directs synthesis of specific proteins, including enzymes that stimulate or inhibit particular metabolic pathways.

4. Water-soluble hormones, many of which are peptides and proteins, cannot pass through the lipid bilayer of the plasma membrane. Thus, they exert their effects indirectly by activating second messenger systems. The hormone, considered the first messenger, binds to a receptor on the plasma membrane. This event activates a molecule in the cytoplasm, considered the second messenger, that carries the hormone's message inside the cell, changing the activity of enzymes and chemical reactions.

5. Endocrine glands are stimulated to manufacture and release hormones by chemical changes in the blood, hormones released by other endocrine glands, and messages from the nervous system. Hormone secretion is usually regulated by negative feedback mechanisms but sometimes by positive feedback mechanisms.
   WEB TUTORIAL 10.1 Modes of Action of Hormones
   WEB TUTORIAL 10.2 Hormonal Feedback Loops

Hormones Influence Growth, Development, Metabolism, and Behavior (pp. 178–193)

6. The pituitary gland has an anterior lobe and a posterior lobe. The anterior lobe is influenced by the hypothalamus through a circulatory connection. Nerve cells in the hypothalamus release hormones that travel by way of the bloodstream to the anterior lobe, where they stimulate or inhibit release of hormones. The anterior pituitary releases six hormones (growth hormone, GH; prolactin, PRL; thyroid-

stimulating hormone, TSH; adrenocorticotropic hormone, ACTH; follicle-stimulating hormone, FSH; and luteinizing hormone, LH). Four of the six hormones are tropic hormones that influence other endocrine glands (TSH, ACTH, FSH, LH). In contrast to the circulatory connection between the hypothalamus and the anterior lobe of the pituitary gland, the connection between the hypothalamus and the posterior lobe is neural. Nerve cells from the hypothalamus extend down into the posterior lobe, where they store and release oxytocin (OT) and antidiuretic hormone (ADH).

7. The thyroid gland, at the front of the neck, produces thyroid hormone (TH) and calcitonin (CT). Thyroid hormone has broad effects, including regulating metabolic rate, heat production, and blood pressure. CT maintains low levels of calcium in the bloodstream.

8. The parathyroid glands, four small masses of tissue at the back of the thyroid gland, secrete parathyroid hormone (PTH, or parathormone), an antagonist to CT. As such, PTH is responsible for raising blood levels of calcium by stimulating the movement of calcium from bone and urine to the blood.

9. Each of two adrenal glands sits on top of a kidney and has two regions. The adrenal cortex (outer region) secretes gonadocorticoids, mineralocorticoids, and glucocorticoids. The adrenal medulla (inner region) produces epinephrine (adrenaline) and norepinephrine (noradrenaline) that initiate the fight-or-flight response.

10. The pancreas secretes the hormones glucagon (increases glucose in the blood) and insulin (decreases glucose in the blood). Type 1 diabetes mellitus is an autoimmune disorder in which a person's own immune system attacks the insulin-producing cells of the pancreas, causing insulin deficiency. Type 2 diabetes mellitus is characterized by a decreased sensitivity to insulin that usually develops in middle age.

11. The thymus gland lies on top of the heart and plays an important role in immunity. Its hormones influence the maturation of white blood cells called T lymphocytes.

12. The pineal gland, at the center of the brain, secretes the hormone melatonin. Melatonin appears to be responsible for establishing biological rhythms and triggering sleep.
    WEB TUTORIAL 10.3 The Hypothalamus and Pituitary

Other Chemical Messengers Act Locally (pp. 193–194)

13. Some local chemical messengers convey information between adjacent cells, evoking rapid responses in target cells. Examples of local signaling molecules include neurotransmitters, prostaglandins, growth factors, and nitric oxide.

KEY TERMS

exocrine gland *p. 175*
endocrine gland *p. 175*
hormone *p. 175*
target cell *p. 175*
steroid hormone *p. 175*
second messenger *p. 176*
negative feedback mechanism *p. 176*
positive feedback mechanism *p. 176*
pituitary gland *p. 178*
hypothalamus *p. 178*
growth hormone (GH) *p. 180*
giantism *p. 180*
acromegaly *p. 180*
pituitary dwarfism *p. 180*
prolactin (PRL) *p. 181*

tropic hormone *p. 181*
thyroid-stimulating hormone (TSH) *p. 181*
adrenocorticotropic hormone (ACTH) *p. 181*
follicle-stimulating hormone (FSH) *p. 181*
luteinizing hormone (LH) *p. 181*
antidiuretic hormone (ADH) *p. 181*
diabetes insipidus *p. 181*
diabetes mellitus *p. 181*
oxytocin (OT) *p. 181*
thyroid gland *p. 182*
thyroid hormone (TH) *p. 182*
simple goiter *p. 183*

cretinism *p. 183*
myxedema *p. 183*
Graves' disease *p. 183*
calcitonin (CT) *p. 184*
parathyroid glands *p. 185*
parathyroid hormone (PTH) *p. 185*
adrenal glands *p. 185*
adrenal cortex *p. 185*
adrenal medulla *p. 185*
gonadocorticoids *p. 185*
androgen *p. 185*
estrogen *p. 185*
mineralocorticoids *p. 185*
aldosterone *p. 185*
Addison's disease *p. 185*
glucocorticoids *p. 187*

Cushing's syndrome *p. 187*
epinephrine *p. 187*
norepinephrine *p. 187*
fight-or-flight response *p. 187*
pancreas *p. 188*
pancreatic islets *p. 188*
glucagon *p. 188*
insulin *p. 188*
thymus gland *p. 192*
thymopoietin *p. 192*
thymosin *p. 192*
pineal gland *p. 192*
melatonin *p. 192*
local signaling molecules *p. 193*

REVIEWING THE CONCEPTS

1. How do endocrine glands differ from exocrine glands? Give examples of each. *p. 175*
2. Given that hormones contact virtually all cells in the body, why are only certain cells affected by a particular hormone? *p. 175*
3. How do lipid-soluble and water-soluble hormones differ in their mechanisms of action? *pp. 175–176, 178*
4. Compare negative and positive feedback mechanisms with regard to regulation of hormone secretion. Provide an example of each. *pp. 176–177*
5. How do the anterior and posterior lobes of the pituitary gland differ in size and relationship with the hypothalamus? *pp. 178–179*
6. List the hormones secreted by the anterior lobe of the pituitary and their functions. *pp. 179–181*
7. List the hormones released by the posterior lobe of the pituitary and their functions. *pp. 179, 181–182*
8. What are the effects of thyroid hormone? *pp. 182–184*
9. Describe the feedback system by which calcitonin and parathyroid hormone regulate levels of calcium in the blood. *pp. 184–186*
10. What are the major functions of the glucocorticoids, mineralocorticoids, and gonadocorticoids secreted by the adrenal cortex? *pp. 185–187*
11. What is the fight-or-flight response? Which hormones are critical in initiating this response? *p. 187*
12. What hormones are secreted by the pancreas? What are their functions? *pp. 188, 190*
13. Explain the differences between Type 1 and Type 2 diabetes mellitus. *p. 188*
14. What is the basic function of hormones secreted by the thymus gland? *p. 192*
15. What roles might melatonin play in the body? *pp. 192–193*
16. How do local signaling molecules differ from true hormones? *pp. 193–194*
17. Which of the following does *not* characterize the anterior lobe of the pituitary gland?
    a. releases oxytocin and antidiuretic hormone
    b. circulatory connection to the hypothalamus
    c. larger of the two lobes
    d. secretes growth hormone and prolactin

18. A diet deficient in iodine may produce
    a. cretinism.
    b. Graves' disease.
    c. Cushing's syndrome.
    d. goiter.
19. Which of the following does *not* characterize the adrenal medulla?
    a. inner region of the adrenal gland
    b. secretes epinephrine and norepinephrine
    c. secretes glucocorticoids
    d. secretes hormones involved in fight-or-flight response
20. Type 1 diabetes
    a. is more common than Type 2 diabetes.
    b. usually develops after age 40.
    c. is an autoimmune disorder.
    d. is characterized by decreased sensitivity to insulin due to decreased insulin receptors on target cells.
21. Overproduction of melatonin by the pineal gland may cause
    a. seasonal affective disorder.
    b. diabetes insipidus.
    c. acromegaly.
    d. Addison's disease.
22. _____ hormones combine with receptor molecules in the cytoplasm of target cells, whereas _____ hormones bind to receptors on the surface of target cells and activate second messengers.
23. Oversecretion of growth hormone in childhood causes _____. Oversecretion in adulthood causes _____.
24. The hormone _____ lowers blood levels of calcium, whereas the hormone _____ increases blood levels of calcium.
25. In males, androgens are produced by the testes and the _____.
26. The hormone _____ lowers glucose in the blood, whereas the hormone _____ increases glucose in the blood.

APPLYING THE CONCEPTS

1. Mary has an itchy rash on the surface of her skin, and Rick has cut his finger on glass. Would either person benefit from applying a steroid cream containing cortisone? Why? Why not?

2. Matt is a thin 20-year-old. He is weak, disoriented, and sweating profusely. His friend brings him to the emergency room of the local hospital and explains that Matt is diabetic. Which type of diabetes does Matt likely have? What might explain his current condition? What might be done to help him?

3. It is winter in Massachusetts and Theresa has felt "down" and lethargic since the fall. She has trouble getting out of bed in the morning, and once up, she craves carbohydrates. What might explain Theresa's symptoms? What might alleviate them, and why?

4. Velma tells her friend Carlos that he produces the female hormone estrogen. Is she correct? If yes, where is the estrogen produced in Carlos? Where is it produced in Velma?

Additional questions can be found on the companion website.

Blood Functions in Transportation, Protection, and Regulation

Blood Consists of Plasma and Formed Elements

- Plasma is the liquid portion of blood
- Stem cells give rise to the formed elements
- Platelets are cell fragments essential to blood clotting
- White blood cells help defend the body against disease
- Red blood cells transport oxygen
- The effects of blood cell disorders depend on the type of blood cell affected

Blood Types Are Determined by Antigens on the Surface of Red Blood Cells

Blood Clotting Occurs in a Regulated Sequence of Events

ENVIRONMENTAL ISSUE Lead Poisoning

11

Blood

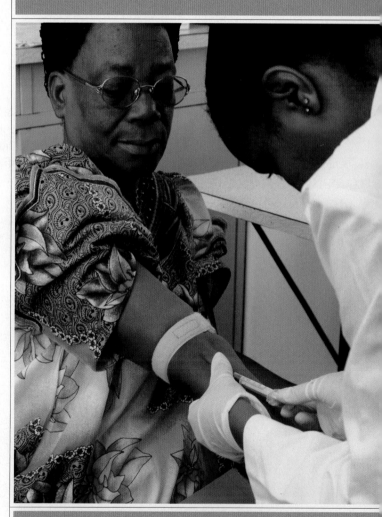

Blood, which brings oxygen and nutrients to each of our cells and carries waste away, is at all times a sensitive indicator of our well-being. A check-up at the doctor's office frequently includes the drawing of blood samples for various kinds of analyses.

Allie settled Mr. Daniels into a chair in the nurse's station and prepared to draw a specimen of blood from the gentleman's arm. Before palpating the inside of his elbow and selecting a vein, she glanced again at the laboratory order form to double check the kind and number of blood tests Dr. Mejia had ordered. In the four long rows of boxes on the form, only the ones pertaining to a cholesterol profile had checks inside them. There were at least 80 other choices on the form—levels of plasma proteins, the various blood cells, glucose, electrolytes, minerals, waste products, enzymes, cardiac risk factors, thyroid hormones, insulin, and on and on—all ways in which blood could be examined to provide information about specific organs and the overall health of the body. Allie drew the blood, carefully labeled the tube, showed Mr. Daniels how to keep pressure on the puncture site until any bleeding had stopped, and promptly took the specimen to the lab.

You probably know that blood is a critical life-sustaining fluid, but you are probably not aware of all the reasons it is so vital. We'll encounter many of those reasons in this chapter as we consider the functions and composition of blood. In addition, we will see what causes blood type and why it is so important. Finally, we will consider how blood forms clots to prevent blood loss from a wound. ∎

Blood Functions in Transportation, Protection, and Regulation

Blood is sometimes referred to as the river of life. The comparison is apt because, like many a river, blood serves as a transportation system. It carries vital materials to all the cells of the body and carries away the wastes that cells produce. But blood does more than passively move its precious cargoes. Its white blood cells help protect us against disease-causing organisms, and its clotting mechanisms help protect us from blood loss when a vessel is damaged. In addition, buffers in the blood help regulate the acid-base balance of body fluids. Blood also helps regulate body temperature by absorbing heat produced in metabolically active regions and distributing it to cooler regions and to the skin, where the heat is dissipated. We see, then, that the diverse functions of blood can be grouped into three categories: transportation, protection, and regulation.

Blood Consists of Plasma and Formed Elements

Blood *is* thicker than water. The reason for the difference is that blood contains cells suspended in its watery fluid. In fact, a single drop of blood contains more than 250 million blood cells. You may recall from Chapter 4 that blood is classified as a connective tissue because it contains cellular elements suspended in a matrix. The liquid matrix is called plasma, and the cellular elements are collectively called the formed elements (Figure 11.1).

■ Plasma is the liquid portion of blood

Plasma is a straw-colored liquid that makes up about 55% of blood. Plasma serves as the medium in which materials are transported by the blood. Almost every substance that is transported by the blood is dissolved in the plasma. These include nutrients (such as simple sugars, amino acids, lipids, and vitamins), ions (such as sodium, potassium, and chloride), dissolved gases (including carbon dioxide, nitrogen, and a small amount of oxygen), and every hormone. In addition to transporting materials to the cells, the plasma carries away cellular wastes. For example, urea from protein breakdown and uric acid from nucleic acid breakdown are carried to the kidneys, where they can be removed from the body.

In spite of the amount and variety of substances transported by the blood, most of the dissolved substances (solutes) in the blood are *plasma proteins*, which make up 7% to 8% of plasma. Plasma proteins help balance water flow between the blood and the cells. You may recall from Chapter 3 that water moves by osmosis across biological membranes from an area of lesser solute concentration to an area of greater solute concentration. Without the plasma proteins, water would be drawn out of the blood by the proteins in cells. As a result, fluid would accumulate in the tissues, causing swelling.

Most of the 50 or so types of plasma proteins fall into one of three general categories: albumins, globulins, and clotting proteins.

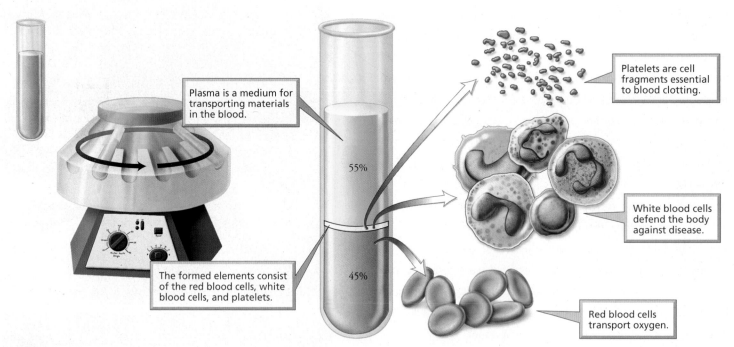

Plasma is a medium for transporting materials in the blood.

The formed elements consist of the red blood cells, white blood cells, and platelets.

55%

45%

Platelets are cell fragments essential to blood clotting.

White blood cells defend the body against disease.

Red blood cells transport oxygen.

WEB TUTORIAL 11.1 FIGURE **II.I** Whole blood consists of a straw-colored liquid, called plasma, in which cellular elements, called formed elements, are suspended. Blood can be separated into its major components when it is placed in a test tube with a substance that prevents coagulation and then spun in a centrifuge. The uppermost layer consists of plasma. The formed elements are found in two layers below the plasma. Just below the plasma is a thin layer consisting of platelets and white blood cells (leukocytes). The red blood cells (erythrocytes) are packed at the bottom of the test tube.

The albumins make up more than half of the plasma proteins. They are most important in the blood's water-balancing ability. The globulins have a variety of functions. Some globulins transport lipids, including fats and some cholesterol, as well as fat-soluble vitamins. Other globulins are antibodies, which provide protection against many diseases. An example of the third category of plasma protein, the clotting proteins, is fibrinogen, whose role we discuss toward the chapter's end.

■ Stem cells give rise to the formed elements

Among the substances transported by the plasma are the **formed elements**—platelets, white blood cells, and red blood cells. These substances perform some of the key functions of the blood. The descriptions and functions of the formed elements of the blood are summarized in Table 11.1.

Red bone marrow, a porous connective tissue that fills the cavities within many bones, is the birthplace and nursery for the formed elements. Its spongelike framework supports fat cells, but in addition it supports the undifferentiated cells called **stem cells** that divide and give rise to all the formed elements (Figure 11.2).

■ Platelets are cell fragments essential to blood clotting

Platelets, sometimes called thrombocytes (*thromb*, clot; *cyte*, cell) are essential to blood clotting. They are actually fragments of larger precursor cells called megakaryocytes and are formed in the red bone marrow when these precursor cells break apart. The fragments are released into the blood at the astounding rate of about 200 billion a day. They then mature during the course of a week, after which they circulate in the blood for about 10 to 12 days. Platelets contain several substances important in stopping the loss of blood through damaged blood vessels. This vital function of platelets will be considered later in the chapter.

■ White blood cells help defend the body against disease

White blood cells, or **leukocytes** (*leuk-*, white; *-cyte*, cell), perform certain mundane housekeeping duties—such as removing wastes, toxins, and damaged or abnormal cells—and serve as warriors in the body's fight against disease (see Chapter 13). Although leukocytes represent less than 1% of whole blood, we simply could not live without them. We would succumb to the microbes that surround us. Because the number of white blood cells increases when the body responds to microbes, white blood cell counts are often used as an index of infection.

Like the other formed elements, white blood cells are produced in the red bone marrow[1]. Unlike platelets and red blood cells, however, white blood cells are nucleated. Moreover, although they circulate in the bloodstream they are not confined there. By squeezing between the cells that form the walls of blood vessels, white blood cells can leave the circulatory system and move to a site

of infection, tissue damage, or inflammation (Figure 11.3). Having slipped out of the capillary into the fluid bathing the cells, white blood cells roam through the tissue spaces. Chemicals released by invading microbes or damaged cells attract the white blood cells and cause them to gather in areas of tissue damage or infection. Certain types of white blood cells may then engulf the "offender" in a process called **phagocytosis** (*phago-*, to eat; *-cyt-*, cell; see Chapter 3). Our consideration of white blood cells in this chapter will be brief; Chapter 13 contains a more detailed discussion of their many tactics for defending our bodies.

TYPES OF WHITE BLOOD CELLS

Each of the five types of white blood cells is assigned to one of two groups—the granulocytes and the agranulocytes—based on cytoplasmic differences. **Granulocytes** have granules in their cytoplasm. The granules are sacs containing chemicals that are used as weapons to destroy invading pathogens, especially bacteria. The **agranulocytes** lack cytoplasmic granules or have very small granules.

GRANULOCYTES: NEUTROPHILS, EOSINOPHILS, AND BASOPHILS Depending on the color of the granules after they have been stained for microscopic study, the granulocytes are classified as neutrophils, eosinophils, or basophils (see Table 11.1).

- **Neutrophils**, the most abundant of all white blood cells, are the blood cell soldiers on the front lines. Arriving at the site of infection before the other types of white blood cells, neutrophils immediately begin to engulf microbes by phagocytosis, thus curbing the spread of the infection. After engulfing a dozen or so bacteria, a neutrophil dies. But, even in death, it helps the body's defense by releasing chemicals that attract more neutrophils to the scene. Dead neutrophils, along with bacteria and cellular debris, make up pus, the yellowish liquid we usually associate with infection.

- **Eosinophils** contain substances that are important in the body's defense against parasitic worms, such as tapeworms and hookworms. They also lessen the severity of allergies by phagocytizing antibody-antigen complexes (described toward the end of the chapter) and inactivating inflammatory chemicals.

- **Basophils** release histamine, a chemical that attracts other white blood cells to the site of infection and causes blood vessels to dilate (widen), thereby increasing blood flow to the affected area. They also play a role in some allergic reactions.

AGRANULOCYTES: MONOCYTES AND LYMPHOCYTES The agranulocytes, which lack visible granules in the cytoplasm, are classified as monocytes or lymphocytes.

- **Monocytes**, the largest of all formed elements, leave the bloodstream and enter various tissues, where they develop into macrophages. Macrophages are phagocytic cells that engulf invading microbes, dead cells, and cellular debris.

- **Lymphocytes** are classified into two types: B lymphocytes and T lymphocytes. The *B lymphocytes* give rise to plasma cells, which, in turn, produce antibodies. Antibodies are proteins

---

[1]One type, the lymphocytes, also may be produced in the lymph nodes and other lymphoid tissues.

Use this table to learn more about the formed elements of blood.

TABLE II.I THE FORMED ELEMENTS OF BLOOD

TYPE OF FORMED ELEMENT	CELL FUNCTION	DESCRIPTION	NO. OF CELLS/MM^3	LIFE SPAN
Platelets	Play role in blood clotting	Fragments of a megakaryocyte; small, purple-stained granules in cytoplasm	250,000–500,000	5–10 days
White blood cells (WBCs; leukocytes)				
Granulocytes				
Neutrophils	Consume bacteria by phagocytosis	Multilobed nucleus, clear-staining cytoplasm, inconspicuous granules	3000–7000	6–72 hours
Eosinophils	Consume antibody-antigen complex by phagocytosis; attack parasitic worms	Large, pink-staining granules in cytoplasm, bilobed nucleus	100–400	8–12 days
Basophils	Release histamine, which attracts white blood cells to the site and widens blood vessels	Large, purple-staining cytoplasmic granules; bilobed nucleus	20–50	3–72 hours
Agranulocytes				
Monocytes	Give rise to macrophages, which consume bacteria, dead cells, and cell parts by phagocytosis	Gray-blue cytoplasm with no granules; U-shaped nucleus	100–700	Several months
Lymphocytes	Attack damaged or diseased cells or produce antibodies	Round nucleus that almost fills the cell	1500–3000	Many years
Red blood cells (RBCs; erythrocytes)	Transport oxygen and carbon dioxide	Biconcave disk, no nucleus	4–6 million	About 120 days

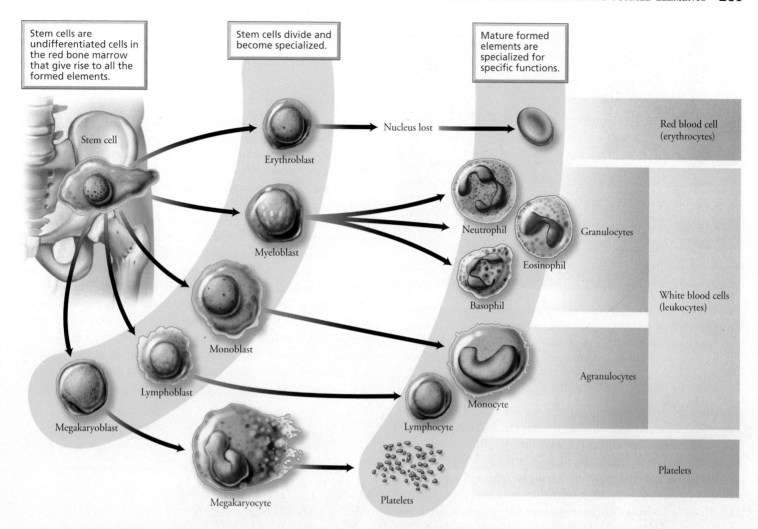

Stem cells are undifferentiated cells in the red bone marrow that give rise to all the formed elements.

Stem cells divide and become specialized.

Mature formed elements are specialized for specific functions.

Stem cell

Erythroblast

Nucleus lost

Red blood cell (erythrocytes)

Myeloblast

Neutrophil

Eosinophil

Granulocytes

Basophil

Monoblast

White blood cells (leukocytes)

Lymphoblast

Monocyte

Agranulocytes

Megakaryoblast

Lymphocyte

Megakaryocyte

Platelets

Platelets

FIGURE **11.2** All formed elements originate in the red bone marrow from an undifferentiated cell called a stem cell. Stem cells divide and differentiate, giving rise to the various types of blood cells.

that recognize specific molecules—called antigens—on the surface of invading microbes or other foreign cells. After recognizing the foreign cell by its antigens, the antibodies help prevent it from harming the body. There are several types of *T lymphocytes*, specialized white blood cells that play roles in the body's defense mechanisms. We will discuss lymphocytes further in Chapter 13.

■ Red blood cells transport oxygen

Red blood cells, also called **erythrocytes** (*erythro-*, red; *-cyt-*, cell), pick up oxygen in the lungs and ferry it to all the cells of the body. Red blood cells also carry about 23% of the blood's total carbon dioxide, a metabolic waste product. They are by far the most numerous cells in the blood. Indeed, they number 4–6 million/mm^3 of blood and constitute approximately 45% of the total blood volume.

RED BLOOD CELLS AND HEMOGLOBIN

The shape of red blood cells, as shown in Figure 11.4, is marvelously suited to their function of picking up and transporting oxygen. Red blood cells are quite small, and each is shaped like a biconcave disk. That is, it is a flattened cell indented on each side.

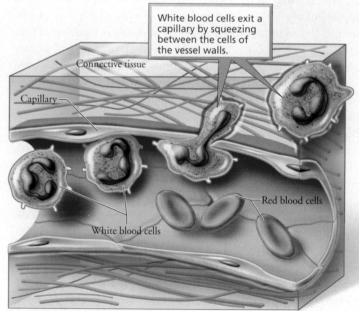

White blood cells exit a capillary by squeezing between the cells of the vessel walls.

Connective tissue

Capillary

Red blood cells

White blood cells

FIGURE **11.3** White blood cells can squeeze between the cells that form the wall of a capillary. They then enter the fluid surrounding body cells and, attracted by chemicals released by microbes or damaged cells, gather at the site of infection or injury.

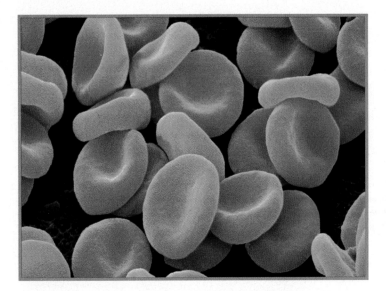

FIGURE **II.4** Red blood cells serve to ferry oxygen from the lungs to the needy tissues. Each red blood cell is a small biconcave (indented on both sides) disk. This design maximizes the surface area for gas exchange. Lacking a nucleus and other organelles, a red blood cell is essentially a bag packed with the oxygen-binding pigment hemoglobin.

The biconcave shape maximizes the surface area of the cell. Because of the greater surface area, oxygen can enter the red blood cell more rapidly than if the disk were flat. A red blood cell is also unusually flexible and thus able to squeeze through capillaries (the smallest blood vessels, those in which gas exchange occurs), even those with a diameter much smaller than red blood cells. Each red blood cell is packed with **hemoglobin**, the oxygen-binding pigment. As a red blood cell matures in the red bone marrow, it loses its nucleus and most organelles. Thus, it is scarcely more than a sack of hemoglobin molecules. Each red blood cell is packed with approximately 280 million molecules of hemoglobin. As can be seen in Figure 11.5, each hemoglobin molecule is made up of four subunits. Each subunit consists of a polypeptide chain and a heme group. The heme group includes an iron ion that actually binds to the oxygen. Therefore, each hemoglobin molecule can carry up to four molecules of oxygen. The compound formed when hemoglobin binds with oxygen is called, logically enough, *oxyhemoglobin*. Body cells use the oxygen and produce carbon dioxide. Most of the carbon dioxide travels to the lungs dissolved in plasma, but some of it binds to hemoglobin (at a site other than that where the iron atom binds oxygen).

As wonderfully adapted as it is for carrying oxygen, the hemoglobin molecule binds 200 times more readily to carbon monoxide, a product of the incomplete combustion of any carbon-containing fuel. In other words, if concentrations of carbon monoxide and oxygen were identical in inhaled air, for every 1 molecule of hemoglobin that picked up an oxygen molecule, 200 molecules of hemoglobin would bind to carbon monoxide. This is the reason carbon monoxide can be deadly. When it binds to the oxygen-binding sites on hemoglobin, it prevents the blood from carrying life-giving oxygen to the cells. As a result, the cells cannot carry out cellular respiration, and a person exposed to carbon monoxide can die. Carbon monoxide is a particularly insidious poison because it is odorless

and tasteless. Its primary source is automobile exhaust, but it can also come from indoor sources, including improperly vented heaters and leaky chimneys.

LIFE CYCLE OF RED BLOOD CELLS

The creation of a red blood cell, which takes about 6 days to complete, entails many changes in the cell's activities and structure. First, the very immature cell becomes a factory for hemoglobin molecules. After the cell is packed with hemoglobin, its nucleus is pushed out. Then a structural metamorphosis occurs, culminating in a mature red blood cell with a typical biconcave shape. At this point, the cell leaves the bone marrow and enters the bloodstream. Red marrow produces roughly 2 million red blood cells a second, for a cumulative total of more than half a ton in a lifetime.

A red blood cell lives for only about 120 days. During that time, it travels through approximately 100 km (62 mi) of blood vessels, being bent, bumped, and squeezed along the way. Its life span is probably limited by the lack of a nucleus that would otherwise maintain it and direct needed repairs. Without a nucleus, for instance, protein synthesis needed to replace key enzymes cannot take place, so the cell becomes increasingly rigid and fragile.

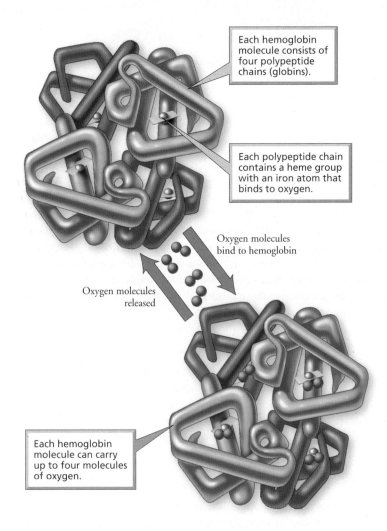

Each hemoglobin molecule consists of four polypeptide chains (globins).

Each polypeptide chain contains a heme group with an iron atom that binds to oxygen.

Oxygen molecules bind to hemoglobin

Oxygen molecules released

Each hemoglobin molecule can carry up to four molecules of oxygen.

FIGURE **II.5** The structure of hemoglobin, the pigment in red blood cells that transports oxygen from the lungs to the cells

The liver and spleen are the "graveyards" where worn-out red blood cells are removed from circulation. The old, inflexible red blood cells tend to become stuck in the tiny circulatory channels of these organs. Macrophages then engulf and destroy the dying cells. The demolished cells release their hemoglobin, which the liver degrades into its protein (globin) component and heme. The protein is digested to amino acids, which can be used to make other proteins. The iron from the heme is salvaged and sent to the red marrow for recycling.

The remaining part of the heme is degraded to a yellow pigment, called bilirubin, which is excreted by the liver in bile. Bile is released into the small intestine, where it assists in the digestion of fats. It is then carried along the digestive system to the large intestine with undigested food and becomes a component of feces. The color of feces is partly due to bilirubin that has been broken down by intestinal bacteria.

Products formed by the chemical breakdown of heme also create the yellowish tinge in a bruise that is healing. A bruise, or black-and-blue mark, results when tiny blood vessels or capillaries are ruptured and blood leaks into the surrounding tissue. As the tissues use up the oxygen, the blood becomes darker and, viewed through the overlying tissue, looks black or blue. Gradually, the red blood cells degenerate, releasing hemoglobin. The breakdown products of hemoglobin then make the bruise appear yellowish.

stop and think

Hepatitis is an inflammation of the liver that can be caused by certain viruses or exposure to certain drugs. It impairs the liver's ability to handle bilirubin properly. A symptom of hepatitis is jaundice, a condition in which the skin develops a yellow tone. Explain why hepatitis causes jaundice.

A negative feedback mechanism regulates red blood cell production according to the needs of the body, especially the need for oxygen (Figure 11.6). Most of the time, red blood cell production matches red blood cell destruction. However, there are circumstances, blood loss for instance, that trigger a homeostatic mechanism that speeds up the rate of red blood cell production. This mechanism is initiated by a decrease in the oxygen supply to the body's cells. Certain cells in the kidney sense the reduced oxygen, and they respond by producing the hormone **erythropoietin**. Erythropoietin then travels to the red marrow, where it steps up both the division rate of stem cells and the maturation rate of immature red blood cells. When maximally stimulated by erythropoietin, the red marrow can increase red blood cell production 10-fold—to 20 million cells per second! The resulting increase in red blood cell numbers is soon adequate to meet the oxygen needs of body cells. The increased oxygen-carrying capacity of the blood then inhibits erythropoietin production.

■ The effects of blood cell disorders depend on the type of blood cell affected

Disorders of red and white blood cells have many different causes. The problems associated with each disorder depend on the type of blood cells affected, because red and white blood cells have different functions.

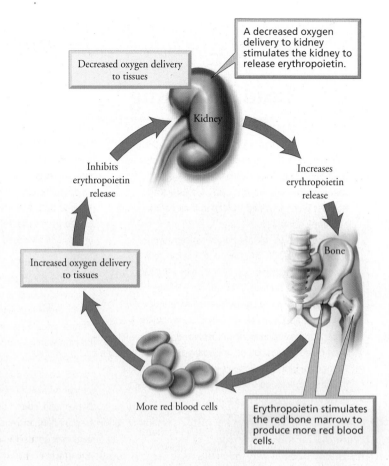

FIGURE **11.6** The production of red blood cells is regulated by a negative feedback relationship between the oxygen-carrying capacity of the blood and the production of erythropoietin.

DISORDERS OF RED BLOOD CELLS

Anemia, a condition in which the blood's ability to carry oxygen is reduced, can result from too little hemoglobin, too few red blood cells, or both. The symptoms of anemia include fatigue, headaches, dizziness, paleness, and breathlessness. In addition, an anemic person's heart often beats faster to help compensate for the blood's decreased ability to carry oxygen. The accelerated pumping can cause heart palpitations—the uncomfortable awareness of one's own heartbeat. Although anemia is not usually life threatening, it does increase susceptibility to illness, and it stresses the heart and lungs. It can also affect the quality of life because lack of energy and low levels of productive activity often go hand in hand.

Worldwide, the most common cause of anemia is an insufficiency of iron in the body, which leads to inadequate hemoglobin production. *Iron-deficiency anemia* can be caused by a diet that contains too little iron, by an inability to absorb iron from the digestive system, or from blood loss, such as might occur because of menstrual flow or peptic ulcers. Treatment for iron-deficiency anemia includes dealing with whatever is the cause of the iron depletion, as well as restoring iron levels to normal by eating foods that are rich in iron—such as meat, leafy green vegetables, and fortified cereals—or by taking pills that contain iron.

Blood loss will obviously lower red blood cell counts, but so will any condition that causes the destruction of red blood cells at

Lead Poisoning

What do children in old, rundown buildings in America today have in common with the ancient upper-class Romans? Lead poisoning. Today, some children are poisoned by the lead in paint chips. Centuries ago, Romans were poisoned by lead in their eating and drinking vessels.

The types of cells that are most sensitive to lead are nerve cells and the bone marrow cells that give rise to red blood cells. The best-known effects of lead on human health are probably those on the nervous system—mental retardation, lowered IQs, reading disabilities, irritability, hyperactivity, and even death. However, the effects of lead on the blood, including anemia and changes in blood enzymes, can also be devastating. Indeed, a blood test is the only way to positively diagnose blood poisoning.

Lead causes a two-pronged attack on the blood's ability to carry oxygen. First, it interferes with the absorption of iron from the digestive system. Second, it inhibits one of the essential enzymes leading to hemoglobin synthesis.

Lead has been a common pollutant for centuries, encountered in a surprising number of substances. The major causes of lead poisoning in the United States are as follows:

1. **House paint** Although lead has not been added to house paint for years, this remains the most important cause of poisoning. Young children tend to put almost anything they find into their mouths, and lead-containing paint chips are no exception. Although most cases of lead poisoning occur in children of poor families living in big cities, lead poisoning can also affect the affluent. The entire family—adults, children, and even pets—can be exposed to lead during the renovation of old homes. For example, lead dust is created as lead paint is sanded from surfaces. That dust can then be inhaled and absorbed into the bloodstream.

2. **Drinking water** One out of every five Americans drinks tap water containing excess levels of lead, according to Environmental Protection Agency estimates. Most of these people are unaware that their drinking water is contaminated with lead because the water still tastes and smells normal. Lead leaches into the water from sources within the homes—copper pipes that contain lead solder, lead connectors in the plumbing, or fancy faucets made of brass or bronze. The leaching process is hastened by heat, acidity, and soft water.

3. **Air and soil pollution** The primary source of lead in the atmosphere has been leaded gasoline. As early as the 1920s, tetraethyl lead was added to motor fuel because it slowed combustion and therefore reduced engine knock and wear. Tiny particles of lead are released into the atmosphere when leaded fuel is burned. Thus, as the use of leaded motor fuel increased, so did the concentration of atmospheric lead.

4. **Lead solder on food cans** The lead in the solder used to seal the seams in food and beverage cans can leach into the contents, especially if the contents are acidic foods such as tomatoes or citrus juices. Health officials estimate that this source still accounts for over 20% of low-level lead poisoning.

5. **Lead-containing dishware** China, ceramic, and earthenware dishes may be coated with lead-containing glaze. When the glazes are not properly fired, lead can leach into the food or beverage contained in the dish. Lead is most likely to leach into foods or beverages that are hot or acidic. Thus, it is probably best to avoid daily use of a ceramic mug for coffee or tea. Wine should not be stored in lead crystal decanters, and if a wine bottle is sealed with a lead foil capsule, clean the rim of the bottle before removing the cork.

6. **Fruits and vegetables grown on lead-contaminated soil** Soil can be contaminated by lead from the atmosphere or by leakage from lead-containing solid wastes, such as lead batteries in automobiles that are dumped into landfills. Important sources of atmospheric lead are the combustion of lead motor fuel, the smelting of ores and other industrial processes, and the incineration of refuse. Fruits and vegetables can then incorporate the lead or be coated in lead dust. When the contaminated food is eaten, the lead is absorbed into the bloodstream.

---

an amount that exceeds the production of red blood cells. For example, in hemolytic anemias, red blood cells are ruptured because of infections, defects in the membranes of red blood cells, transfusion of mismatched blood, or hemoglobin abnormalities. *Sickle cell anemia* is an example of a hemolytic anemia caused by abnormal hemoglobins. This abnormal hemoglobin (hemoglobin S) causes the red blood cells to become deformed to a crescent (or sickle) shape when the blood's oxygen content is low. The distorted cells are fragile and rupture easily, clogging small blood vessels and promoting clot formation. These events prevent oxygen-laden blood from reaching the tissues and can cause episodes of extreme pain.

Red blood cell numbers also drop when the production of red blood cells is halted or impaired, as occurs in *pernicious anemia*. The production of red blood cells depends on a supply of vitamin B_{12}. The small intestine absorbs vitamin B_{12} from the diet with the aid of a chemical called intrinsic factor, which is produced by the stomach lining. People with pernicious anemia do not produce intrinsic factor and are, therefore, unable to absorb vitamin B_{12}. They are treated with injections of B_{12}. Lead poisoning can also cause anemia. (See the Environmental Issue essay, *Lead Poisoning*.)

DISORDERS OF WHITE BLOOD CELLS

Infectious mononucleosis, or simply "mono," is a viral disease of the lymphocytes caused by the Epstein-Barr virus, a virus that is common in humans but usually asymptomatic. In mononucleosis, the infection causes an increase in lymphocytes that have an atypi-

what would you do?

cal appearance. Because mono often is spread from person to person by oral contact, it is sometimes called the kissing disease; however, it can also be spread by sharing eating utensils or drinking glasses. Mono is most common among teenagers and young adults, particularly those living in dormitories. It often strikes at stressful times, such as during final exams, when resistance is low.

The initial symptoms of mono are similar to those of influenza: fever, chills, headache, sore throat, and an overwhelming sense of being ill. Within a few days, the glands in the neck, armpits, and groin become painfully swollen. Mono must simply run its course. The major symptoms generally subside within a few weeks, but fatigue may linger much longer.

Leukemia is a cancer of the white blood cells that causes their uncontrolled multiplication, so that their number increases greatly. The cancerous cells—all descendants of a single abnormal cell—remain unspecialized and are therefore unable to defend the body against infectious organisms. Because they divide more rapidly and live longer than do normal cells, the abnormal cells "take over" the bone marrow, preventing the development of normal blood cells, including red blood cells, white blood cells, and platelets.

Symptoms of leukemia generally result either from the insufficient number of normal blood components or from the invasion of organs by abnormal white blood cells. The increased number of white blood cells crowds out the other formed elements. Insufficient numbers of platelets cause gum bleeding and frequent bruising. Reduced levels of red blood cells lead to anemia, which in turn causes chronic fatigue, breathlessness, and pallor. Because their white blood cells do not function properly, leukemia patients may suffer from repeated respiratory or throat infections, herpes, or skin infections. Also, bone tenderness may be experienced because the immature white blood cells pack the red marrow. Headaches, another symptom of leukemia, may be caused by anemia or by the effects of abnormal white blood cells in the brain.

The treatment of leukemia usually includes radiation therapy and chemotherapy to kill the rapidly dividing cells. In addition, transfusions of red blood cells and platelets may be given to alleviate anemia and prevent excessive bleeding. Today, many children with acute leukemia are cured with bone marrow transplants. Someone, most often a family member, whose tissue type closely matches that of the patient, must be located and agree to serve as a donor. Then the bone marrow in the leukemia patient must be destroyed by irradiation and drugs. Next, the donor's bone marrow is given to the patient intravenously, just like a blood transfusion. The marrow cells find their way to the patient's marrow and begin to grow. Treatment with stem cells from umbilical cord blood seems to be as effective as bone marrow transplant in helping some children go into remission (the signs and symptoms of leukemia go away), as discussed in Chapter 19a, Special Topic: *Stem Cells—the Body's Repair Kit.*

Blood Types Are Determined by Antigens on the Surface of Red Blood Cells

Human blood is classified into different **blood types**, depending on the presence or absence of certain molecules, mostly proteins, on the surface of the person's red blood cells (Table 11.2). As we will learn in Chapter 13, each of your body cells is labeled as "self" (that is, as belonging to your body) by proteins on its surface. If a cell that lacks these self markers enters the body, the body's defense system recognizes that the foreign cell is "nonself" and does not belong. The foreign cell will have different proteins on its surface. To the body's defense system, these proteins are antigens (mentioned above in the discussion of lymphocytes), identifying the cell as foreign and marking it for destruction. As was mentioned earlier, one way the body attacks the foreign cell is by producing proteins called antibodies that specifically target the antigen on the foreign cell's surface. Let's consider the role of antigens and antibodies in blood types and transfusions.

When asked about your blood type, you are probably used to responding by indicating one of the types in the ABO series: A, B, AB, or O. Red blood cells with only the antigen A on their surface are type A. When only the B antigen is on the red blood cell surface, the blood is type B. Blood with both A and B antigens on the red blood cell surface is designated type AB. When neither A nor B antigens are present, the blood is type O.

Normally, a person's plasma contains antibodies against those antigens that are not on his or her own red blood cells. Thus, individuals with type A blood have antibodies against the B antigen (anti-B antibodies[2]), and those with type B blood have antibodies against A (anti-A antibodies). Because individuals with type AB blood have both antigens on their red blood cells, they have neither antibody. Those with type O blood have neither antigen, so they have both anti-A and anti-B antibodies in their plasma.

In a typical test for blood type, technicians mix a drop of a person's blood with a solution containing anti-A antibodies and mix another drop of the blood with a solution containing anti-B antibodies. If clumping occurs in one of the mixtures, it means the antigen corresponding to the antibody in that mixture is present (Figure 11.7).

[2]It is not certain why these antibodies form without exposure to red blood cells bearing the foreign antigen. It may be that either the bacteria that invade our bodies or the food we eat contains a small amount of A and B antigens, enough to stimulate antibody production.

TABLE 11.2 TRANSFUSION RELATIONSHIPS AMONG BLOOD TYPES

BLOOD TYPES	ANTIGENS ON RED BLOOD CELLS	ANTIBODIES IN PLASMA	BLOOD TYPES (RBCs) THAT CAN BE RECEIVED IN TRANSFUSIONS	INCIDENCE OF BLOOD TYPE IN U.S.
A	A	Anti-B	A, O	Caucasian, 40% African American, 27% Asian, 28% Native American, 8%
B	B	Anti-A	B, O	Caucasian, 10% African American, 20% Asian, 27% Native American, 1%
AB (universal recipient)	A and B	None	A, B, AB, O	Caucasian, 5% African American, 4% Asian, 5% Native American, 0%
O (universal donor)	None	Anti-A, Anti-B	O	Caucasian, 45% African American, 49% Asian, 40% Native American, 91%

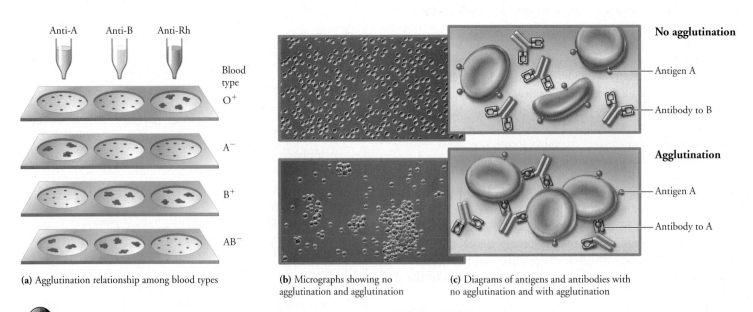

(a) Agglutination relationship among blood types

(b) Micrographs showing no agglutination and agglutination

(c) Diagrams of antigens and antibodies with no agglutination and with agglutination

WEB TUTORIAL 11.2

FIGURE **11.7** Blood is typed by mixing it with serum known to contain antibodies specific for a certain antigen. If blood containing that antigen is mixed with the serum, the blood will agglutinate (clump). Thus, one drop of blood is mixed with serum containing anti-A and another drop with serum containing anti-B. A third drop of blood is mixed with serum containing antibodies to Rh. Agglutination in response to an antibody reveals the presence of the antigen.

When a person is given a blood transfusion with donor blood containing foreign antigens, the antibodies in the recipient's blood will cause the donor's cells to clump, or **agglutinate**. This clumping of the donor's cells is damaging, and perhaps even fatal. The clumped cells get stuck in small blood vessels and block blood flow to body cells. Or they may break open, releasing their cargo of hemoglobin. The hemoglobin then clogs the filtering system in the kidneys, causing death.

It is important, therefore, to be sure that the blood types of a donor and recipient are compatible, which means that the recipient's blood does not contain antibodies to antigens on the red blood cells of the donor. The questions to ask in this case are (1) what, if any, antigens are on the donor's cells and (2) what, if any, antibodies are in the recipient's blood. For example, if a person with blood type A is given a transfusion of blood type B or of type AB, the naturally occurring anti-B antibodies in the recipient's blood will cause the red blood cells of the donor to clump because they have the B antigen. The transfusion relationships among blood types in the ABO series are shown in Table 11.2. Type O blood is sometimes called the "universal donor" because its red blood cells have neither A nor B antigens. Therefore, type O blood will not clump if there are anti-A or anti-B antibodies in the recipient's plasma. Type AB blood is called the "universal recipient" because its plasma lacks antibodies to both A and B antigens. It will not cause clumping of donated blood that has A or B antigens.

The A and B antigens are not the only important antigens found on the surface of red blood cells. The presence or absence of an **Rh factor** is also an important component of blood type. The name Rh comes from the beginning of the name of the Rhesus monkey, in which an Rh antigen (actually a group of related antigens) was first discovered. People who have any of the Rh antigens on their red blood cells are considered Rh-positive (Rh$^+$). When Rh antigens are missing from the red blood cell surface, the individual is considered Rh-negative (Rh$^-$).

An Rh-negative person will not form anti-Rh antibodies unless he or she has been exposed to the Rh antigen. For this reason, an Rh-negative individual should be given only Rh-negative blood in a transfusion. If he or she is mistakenly given Rh-positive blood, it will stimulate the production of anti-Rh antibodies. A transfusion reaction will not occur after the first such transfusion, because it takes time for the body to start making anti-Rh antibodies. After a second transfusion of Rh-positive blood, however, the antibodies in the recipient's plasma will react with the antigens on the red blood cells of the donated blood. This reaction may lead to the death of the patient.

The Rh factor can also be of medical importance in pregnancies in which the mother is Rh-negative and the fetus is Rh-positive, a situation that may occur if the father is Rh-positive (see Chapter 20; Figure 11.8). Ordinarily, the maternal and fetal blood supplies do not mix during pregnancy. However, some mixing may occur during a miscarriage or delivery as a result of blood vessel damage. If the baby's red blood cells, which bear Rh antigens, accidentally pass into the bloodstream of the mother, she will produce anti-Rh antibodies. There are usually no ill effects associated with the first introduction of the Rh antigen. However, if antibodies are

present in the maternal blood from a previous pregnancy with an Rh-positive child or from a transfusion of Rh-positive blood, the anti-Rh antibodies may pass into the blood of the fetus. This transfer can occur because anti-Rh antibodies, unlike red blood cells, can cross the placenta (a structure that forms during pregnancy to allow the exchange of selected substances between the maternal and fetal circulatory systems). These anti-Rh antibodies may destroy the fetus's red blood cells. As a result, the child may be stillborn or very anemic at birth. This condition is called *hemolytic disease of the newborn.*

The incidence of hemolytic disease of the newborn has decreased in recent years because of the development of a means of destroying any Rh-positive fetal cells in the maternal blood supply before they can stimulate the mother's cells to produce her own anti-Rh antibodies. The Rh-positive cells are killed by injecting RhoGAM, a serum containing antibodies against the Rh antigens, at about the seventh month of pregnancy and shortly after delivery

An Rh$^+$ male and an Rh$^-$ female have either a 50% or a 100% chance of having an Rh$^+$ baby (depending on the genetic makeup of the father).

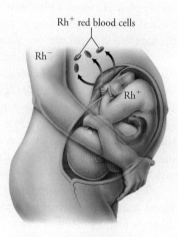

First pregnancy: At birth some of the Rh$^+$ blood of the fetus may enter the mother's circulation.

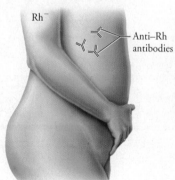

The mother forms anti-Rh antibodies over the next few months.

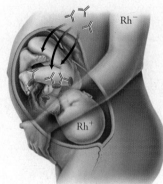

Second pregnancy with an Rh$^+$ fetus: Anti-Rh antibodies may pass into the fetus's blood, causing its blood cells to burst.

FIGURE **11.8** Rh incompatibility can result when an Rh-negative (Rh$^-$) woman is pregnant with an Rh-positive (Rh$^+$) baby if the woman has been previously exposed to Rh$^+$ blood.

if the baby is Rh-positive. Rh antigens are thus prevented from being "set" in the memory of the mother's immune system. The injected antibodies disappear after a few months. Therefore, no antibodies linger to affect the fetus in a subsequent pregnancy.

Blood Clotting Occurs in a Regulated Sequence of Events

When a blood vessel is cut, a number of responses are triggered that stop the bleeding. They are similar to the ways you might respond if the garden hose you are using springs a leak. Your initial response might be to squeeze the hose in hopes of stopping the water flow. Likewise, the body's immediate response to blood vessel injury is for the vessel to constrict (squeeze shut).

The next response is to plug the hole (Figure 11.9). Your thumb might do the job on your garden hose; in an injured blood vessel, platelets form a plug that seals the leak. The *platelet plug* is formed when platelets cling to cables of collagen, a protein fiber on the torn blood vessel surface. When the platelets attach to collagen, they swell, form many cellular extensions, and stick together. Platelets also produce a chemical that attracts other platelets to the wound and makes them stick together. Aspirin prevents the formation of the chemical and, therefore, inhibits clot formation. For this reason, a daily dose of aspirin is sometimes prescribed to prevent the formation of blood clots that could block blood vessels nourishing heart tissue and thus cause the death of heart cells (a heart attack). It is also why aspirin can cause excessive bleeding.

The next stage in stopping blood loss through a damaged blood vessel is the formation of the clot itself. There are more than 30 steps in the process of clot formation, but here we will describe only the key events. Clot formation begins when clotting factors

are released from injured tissue and from platelets. At the site of the wound, the clotting factors convert an inactive blood protein to **prothrombin activator**,[3] which then converts **prothrombin**, a plasma protein produced by the liver, to an active form, **thrombin**. Thrombin then causes a remarkable change in another plasma protein produced by the liver, **fibrinogen**. The altered fibrinogen forms long strands of **fibrin**, which make a web that traps blood cells and forms the clot. The clot is a barrier that prevents further blood loss through the wound in the vessel.

stop and think

Thromboplastin, a chemical important in the initiation of clot formation, is released from both damaged tissue and activated platelets. How does this fact explain why a scrape, which causes a great deal of tissue damage, generally stops bleeding more quickly than a clean cut, such as a paper cut or one that might occur with a razor blade?

If even one of the many factors needed for clotting is lacking, the process can be slowed or completely blocked. Vitamin K is needed for the liver to synthesize prothrombin and three other clotting factors. Thus, without vitamin K, clotting does not occur. We have two sources of vitamin K. One is the diet. Vitamin K is found in leafy green vegetables, tomatoes, and vegetable oils. The second source is bacteria living in our intestines and producing vitamin K, some of which we absorb for our own use. Antibiotic treatment for serious bacterial infections can kill gastrointestinal bacteria and lead to a vitamin K deficiency in as few as 2 days. Vitamin K is used rapidly by body tissues, so both sources are needed for proper blood clotting.

Hemophilia is an inherited condition in which the affected person bleeds excessively owing to a fault in a gene involved in pro-

[3]Prothrombin activator is an enzyme called prothrombinase.

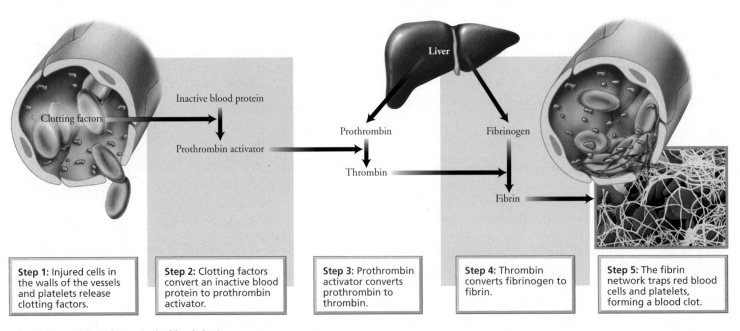

Step 1: Injured cells in the walls of the vessels and platelets release clotting factors.

Step 2: Clotting factors convert an inactive blood protein to prothrombin activator.

Step 3: Prothrombin activator converts prothrombin to thrombin.

Step 4: Thrombin converts fibrinogen to fibrin.

Step 5: The fibrin network traps red blood cells and platelets, forming a blood clot.

FIGURE **II.9** Selected steps in the blood-clotting process

ducing one of the clotting factors. Because of the way hemophilia is inherited, the condition usually occurs in males (see Chapter 20). Symptoms appear when the affected child first becomes physically active. Crawling, for instance, causes bruises on the elbows and knees, and cuts tend to bleed longer than is normal. Excessive internal bleeding can damage nerves or, when it occurs in joints, permanently cripple the hemophiliac.

Treatment for hemophilia involves restoring the missing clotting factor. Bleeding episodes can be controlled with repeated transfusions of fresh plasma or with injections of concentrated clotting factor. Concentrated clotting factor is made by combining the plasma donations of many people. In fact, each injection requires 2000 to 5000 individual donations. Fortunately, two of the clotting factors (factors VIII and IX) are now being manufactured using recombinant DNA technology.

The failure of blood to clot can shorten the life of a person with hemophilia, but in people without hemophilia the formation of blood clots when they are not needed can have much more im-mediate health consequences, because they can disrupt blood flow. A blood clot lodged in an unbroken blood vessel is called a *thrombus*. A blood clot that drifts through the circulatory system is called an *embolus*. Emboli can drift through the circulatory system until they become stuck in a narrow vessel. When the tiny vessels that nourish the heart or brain become clogged with a clot, the consequences can be severe—disability or even death. After a wound has healed, clots are normally dissolved by an enzyme called **plasmin**, which is formed from an inactive protein, *plasminogen*. Plasmin dissolves clots by digesting the fibrin strands that form the framework of the clot.

stop and think

Heparin is a drug that inactivates thrombin. It is sometimes administered to patients for the purpose of inhibiting the clotting response. How would heparin act to achieve these ends?

HIGHLIGHTING THE CONCEPTS

Blood Functions in Transportation, Protection, and Regulation (p. 198)

1. Blood transports vital material to cells and carries wastes away from cells. White blood cells defend against disease. Blood clotting prevents excessive blood loss. Blood also helps regulate body temperature.

Blood Consists of Plasma and Formed Elements (pp. 198–205)

2. Blood is a type of connective tissue that contains formed elements in a liquid (plasma) matrix.

3. Plasma is the liquid portion of the blood, consisting of water and certain dissolved components, mostly proteins. (The substances transported by the blood are also dissolved in the plasma. These include nutrients, ions, gases, hormones, and waste products.) Plasma proteins aid in the balance of water flow between blood and cells. These proteins are grouped into three categories: albumins, globulins, or clotting proteins (such as fibrinogen).

4. Stem cells are undifferentiated cells that divide and give rise to all of the formed elements, namely, platelets, white blood cells, and red blood cells.

5. Platelets play an important role in blood clotting. They are fragments of a larger cell called a megakaryocyte.

6. White blood cells (leukocytes) help the body fight off disease and help remove wastes, toxins, and damaged cells.

7. There are five types of leukocytes: neutrophils, eosinophils, basophils, monocytes, and lymphocytes. Neutrophils are the most abundant and immediately phagocytize foreign microbes. Eosinophils help the body defend against parasitic worms and play a role in allergic reactions. Basophils can increase the flow of blood by releasing histamines. Monocytes are the largest leukocytes and, after maturing into macrophages, actively fight chronic infections. Lymphocytes take part in body defense responses.

8. Red blood cells (erythrocytes) are flexible cells packed with hemoglobin, an oxygen-binding pigment. Originating in the red bone marrow, erythrocytes have a life span of about 120 days. Red blood cell production is controlled by erythropoietin, a hormone produced by the kidney in response to low oxygen. Worn and dead red blood cells are removed from circulation and broken down in the liver and spleen.

9. Anemia is a reduction in the blood's ability to carry oxygen. There are several forms, including iron-deficiency anemia, hemolytic anemias, such as sickle cell anemia, and pernicious anemia, caused by the stomach's failure to produce intrinsic factor, which is needed to absorb the vitamin B_{12} necessary for red blood cell production.

10. Infectious mononucleosis is a viral disease of lymphocytes. Leukemia is a cancer of the white blood cells. There are high numbers of white blood cells, but they do not function properly to defend against infectious agents.

WEB TUTORIAL 11.1 Blood

Blood Types Are Determined by Antigens on the Surface of Red Blood Cells (pp. 205–208)

11. Blood types are determined by the presence of certain proteins (antigens) on the surface of red blood cells: the ABO and Rh groups. The plasma contains antibodies against A and B antigens if the antigens are not present on the red blood cells. Antibodies to Rh antigens are formed only after exposure to Rh-positive blood. If blood containing foreign antigens is introduced during a transfusion, a reaction between the antibodies and antigens can cause red blood cells to clump and clog the recipient's bloodstream, which can lead to death.

12. Rh incompatibility becomes important to an Rh-negative woman who has been previously exposed to Rh-positive blood and who is pregnant with an Rh-positive fetus. Antibodies to the Rh antigens can cross the placenta and destroy the red blood cells of the fetus, a condition called hemolytic disease of the newborn.

WEB TUTORIAL 11.2 Blood Types

Blood Clotting Occurs in a Regulated Sequence of Events (pp. 208–209)

13. The prevention of blood loss involves three mechanisms: blood vessel constriction, platelet plug formation, and clotting. Clotting is initiated when platelets and damaged tissue release clotting factors that lead to the production of a chemical called prothrombin activator, which converts the blood protein prothrombin to thrombin. Thrombin then converts the blood protein fibrinogen to fibrin. Strands of fibrin make a mesh that traps red blood cells and forms the clot. Hemophiliacs are missing an important clotting factor and, therefore, bleed excessively.

KEY TERMS

plasma *p. 198*
formed elements *p. 199*
stem cells *p. 199*
platelets *p. 199*
white blood cells *p. 199*
leukocytes *p. 199*
phagocytosis *p. 199*
granulocytes *p. 199*

agranulocytes *p. 199*
neutrophils *p. 199*
eosinophils *p. 199*
basophils *p. 199*
monocytes *p. 199*
lymphocytes *p. 199*
red blood cells *p. 201*
erythrocytes *p. 201*

hemoglobin *p. 202*
erythropoietin *p. 203*
anemia *p. 203*
leukemia *p. 205*
blood type *p. 205*
agglutinate *p. 207*
Rh factor *p 207*
prothrombin activator *p. 208*

prothrombin *p. 208*
thrombin *p. 208*
fibrinogen *p. 208*
fibrin *p. 208*
hemophilia *p. 208*
plasmin *p. 209*

REVIEWING THE CONCEPTS

1. What is plasma? What are its functions? *pp. 198–199*
2. What are the three categories of plasma proteins? *pp. 198–199*
3. List the three types of formed elements and describe the function of each. *pp. 199–201*
4. Compare the size, structure, and numbers of leukocytes with the size, structure, and numbers of erythrocytes. *pp. 199–201*
5. Leukocytes function in defending the body against foreign invaders. List the five types of white blood cells and describe the role each plays in body defense. *pp. 199–201*
6. Describe the characteristics of red blood cells that make them specialized for delivering oxygen to the tissues. *pp. 201–202*
7. Describe the structure of hemoglobin. *p. 202*
8. Where are red blood cells produced? How is the production of red blood cells controlled? *pp. 202–203*
9. Describe what happens to worn or damaged red blood cells and their hemoglobin. *p. 203*
10. Why would pernicious anemia be treated with regular injections of vitamin B_{12} instead of by dietary supplements of this vitamin? *p. 204*
11. How are blood types determined? *pp. 205–207*
12. What happens if a person is given a blood transfusion with blood of an incompatible type? *pp. 205–207*
13. What blood type(s) can a person with type B blood receive? Explain. *p. 207*
14. What is hemolytic disease of the newborn? What causes it? *pp. 207–208*
15. After a blood vessel is cut, what mechanisms prevent blood loss? *p. 208*

16. Describe the steps involved in blood clotting. *p. 208*
17. What is the difference between the blood clotting that occurs after an injury and the clumping that occurs after a mismatched transfusion? *pp. 205–209*
18. Type B blood contains which antibodies?
    a. A
    b. B
    c. both A and B
    d. neither A nor B
19. An important function of white blood cells is
    a. blood clotting.
    b. transportation of oxygen.
    c. fighting infection.
    d. maintaining blood pressure.
20. A condition in which the oxygen carrying capacity of the blood is low is
    a. infection.
    b. anemia.
    c. leukemia.
    d. a kidney problem.
21. The primary function of red blood cells is to _____.
22. Hemolytic disease of the newborn may result if the mother is Rh _____ and the fetus is Rh _____.
23. _____ is a protein in red blood cells that transports oxygen.
24. The protein that forms a net that traps red blood cells and platelets and forms blood clots is _____.

APPLYING THE CONCEPTS

1. Erin has leukemia, and her white blood cell count is elevated. Why does Erin have an elevated risk of infection?
2. John is a runner from New York City who is on his college track team. The championship meet is in Denver, which is called the Mile-High City because of its elevation above sea level. There is less oxygen available for breathing as elevation increases. Because this meet is so important, John plans to arrive in Denver several weeks before the meet. Why? Would you expect John's red blood cell count to be higher or lower than normal when he returns home to New York City, which is at sea level?
3. Indira is a 25-year-old woman who has uterine polyps that cause heavy vaginal bleeding. She complains that she tires easily with physical ac-

tivity and always feels fatigued. The doctor orders that a blood test be done to determine what percentage of the blood is red blood cells. Why? Would you expect the percentage to be higher or lower than normal? The doctor suggests that she take iron supplements. Why?
4. Raul is in a car accident and is taken to the emergency room. He has type AB blood. Which blood types can he receive?
5. Elizabeth has type Rh-negative blood. She is pregnant for the second time. What information would the doctor want to know about the first and second children to know whether to expect a problem with this pregnancy?
6. Sarala was given an antibiotic that caused her platelet count to fall. What symptoms would be expected?

Additional questions can be found on the companion website.

The Cardiovascular System Consists of the Blood Vessels and the Heart

The Blood Vessels Conduct Blood in Continuous Loops

- Arteries carry blood away from the heart
- Capillaries are sites of exchange with body cells
- Veins return blood to the heart

The Heart Is a Muscular Pump

- Blood flows through the heart in two circuits
- Coronary circulation serves the heart muscle
- The cardiac cycle is the sequence of heart muscle contraction and relaxation
- The rhythmic contraction of the heart is due to its internal conduction system
- An electrocardiogram is a recording of the electrical activities of the heart
- Blood pressure is the force blood exerts against blood vessel walls

Cardiovascular Disease Is a Major Killer in the United States

- High blood pressure can kill without producing symptoms
- Atherosclerosis is a buildup of lipids in the artery walls
- Coronary artery disease is atherosclerosis in the coronary arteries
- Heart attack is the death of heart muscle

The Lymphatic System Functions in the Circulatory and Immune Systems

HEALTH ISSUE The Cardiovascular Benefits of Exercise

12
The Circulatory System

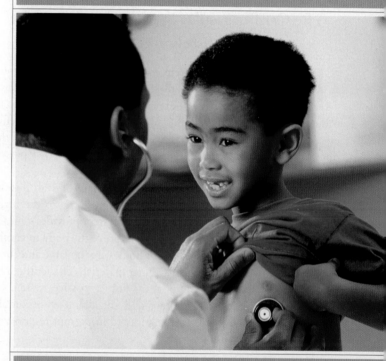

Small abnormalities in the sound of a beating heart can tell doctors a great deal about the structure and functioning of that important muscular pump.

It was past midnight, and Susan, a 38-year-old air traffic controller, had come to the emergency room because of severe pains in her chest. It was a completely frustrating puzzle to her that she would be having these symptoms night after night and occasionally during the day for the past week. At that hour of the night, she fortunately did not have to wait for treatment. The doctor gave her supplemental oxygen, performed a number of blood tests, and decided to call in a heart specialist.

The specialist performed more tests, including a coronary angiogram (a procedure to examine blood flow through blood vessels) and informed Susan that an important coronary artery was almost completely blocked. The reason for her chest pain—he called it angina—was that the blocked artery was not allowing enough oxygen to be delivered to her heart. He recommended a coronary angioplasty, a procedure that uses a balloon to stretch open the blockage in the artery and restore the supply of oxygen to the heart muscle.

Susan had the angioplasty done. Afterward, during her several-month period of cardiac rehabilitation, she was given a diet and exercise plan that would help reduce the future possibility of heart problems.

In this chapter, you will learn about the structure and function of the heart and grow to appreciate even more its marvelous design. You will also see why Susan was right to be concerned about her chest pains. An understanding of how the heart functions may help you better care for your own and take steps to maintain its health. The chapter also examines the network of vessels and other structures that make up the lymphatic system. Together, the heart and blood vessels and the lymphatic system constitute the *circulatory system*. ■

The Cardiovascular System Consists of the Blood Vessels and the Heart

The **cardiovascular system** consists of the **heart**—a muscular pump that contracts rhythmically and provides the force that moves the blood—and the blood vessels—a system of tubules through which blood flows (Figure 12.1). The blood delivers a continuous supply of oxygen and nutrients to the cells of the body and carries away metabolic waste products before they poison the cells.

Why is the cardiovascular system so critical to survival? It is the body's transportation network, similar in some ways to the highways within a country. The cardiovascular system provides a means for distributing vital chemicals from one part of the body to another quickly enough to sustain life. Our bodies are too large and complex for diffusion alone to distribute materials efficiently. The cardiovascular system is more than just a passive system of pipelines, however. The heart rate and the diameter of certain blood vessels are continually being adjusted in prompt response to the body's changing needs.

The Blood Vessels Conduct Blood in Continuous Loops

Once every minute, or about 1440 times each day, the blood moves through a life-sustaining circuit of blood. The system of blood vessels is extensive. Indeed, if all the vessels in an adult's body were placed end to end, they would stretch about 100,000 km (60,000 mi), long enough to circle the earth's equator more than twice!

The blood vessels do not form a single long tube. Instead, they are arranged in branching networks. With each circuit through the body, blood is carried away from the heart in an *artery*, which branches to give rise to narrower vessels called *arterioles*. Arterioles lead into networks of microscopic vessels called *capillaries*, which allow the exchange of materials between the blood and body cells.

The capillaries eventually merge to form *venules*, which eventually join to form larger tubes called *veins*. The venules and veins return the blood to the heart. We see, then, that the path of blood is:

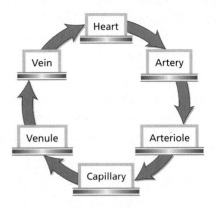

All blood vessels share some common features, but each type also has its own traits and is marvelously adapted for its specific function (Figure 12.2). The hollow interior of a blood vessel, through which the blood flows, is called the *lumen*. The inner lining that comes into contact with the blood flowing through the lumen is composed of simple squamous epithelium (the flattened, tight-fitting cells we encountered in Chapter 4). This lining, called the endothelium, provides a smooth surface that minimizes friction so as not to impede the flow. The lumen and endothelium are characteristic of all blood vessels.

■ Arteries carry blood away from the heart

Arteries are muscular tubes that transport blood *away* from the heart, delivering it rapidly to the body tissues. We have seen that the innermost layer of an arterial wall is the endothelium. Immediately outside the endothelium is a middle layer that contains elastic fibers and circular layers of smooth muscle (Figure 12.2). The elastic fibers allow an artery to stretch and then return to its original shape. The smooth muscle enables the artery to contract. The outer layer of an arterial wall is a sheath of connective tissue that contains elastic fibers and collagen. This layer adds strength to the arterial wall and anchors the artery to surrounding tissue.

The elastic fibers have two important functions: (1) They help the artery tolerate the pressure shock caused by blood surging into it when the heart contracts; and (2) they help maintain a relatively even pressure within the artery, despite large changes in the volume of blood passing through it. Consider, for instance, what happens when the heart contracts and sends blood into the *aorta*, the body's main artery. Each beat of the heart causes 70 ml (about one-fourth cup) of blood to pound against the wall of the aorta like a tidal wave. A rigid pipe could not withstand the repeated pressure surges, but the elastic walls of the artery stretch with each wave of blood and return to their original size when the surge has moved along, resulting in the intermittent waves being dampened into a continuous stream.

The alternate expansion and recoiling of arteries create a pressure wave, called a **pulse**, that moves along the arteries with each heartbeat. You can feel the pulse by slightly compressing with your

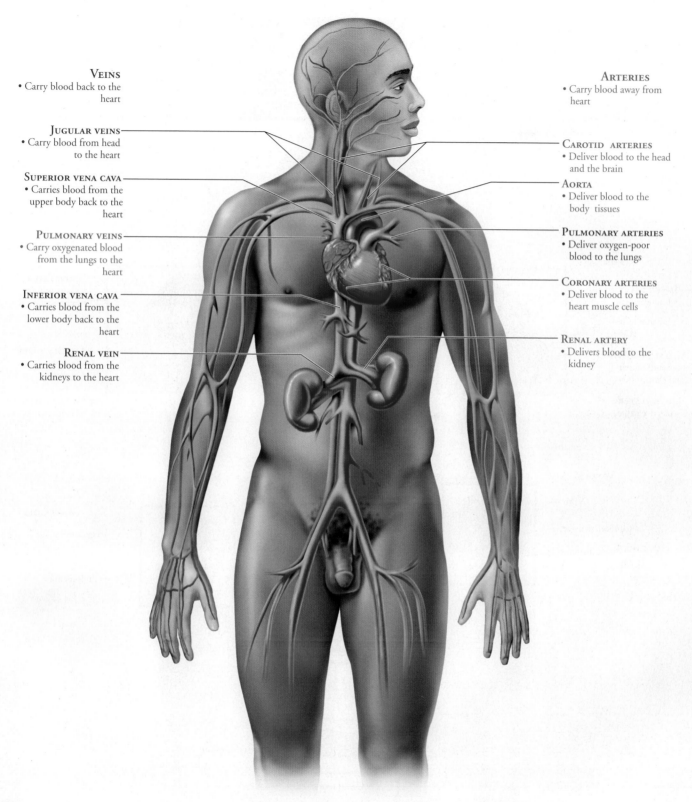

VEINS
* Carry blood back to the heart

JUGULAR VEINS
* Carry blood from head to the heart

SUPERIOR VENA CAVA
* Carries blood from the upper body back to the heart

PULMONARY VEINS
* Carry oxygenated blood from the lungs to the heart

INFERIOR VENA CAVA
* Carries blood from the lower body back to the heart

RENAL VEIN
* Carries blood from the kidneys to the heart

ARTERIES
* Carry blood away from heart

CAROTID ARTERIES
* Deliver blood to the head and the brain

AORTA
* Deliver blood to the body tissues

PULMONARY ARTERIES
* Deliver oxygen-poor blood to the lungs

CORONARY ARTERIES
* Deliver blood to the heart muscle cells

RENAL ARTERY
* Delivers blood to the kidney

FIGURE **12.1** A diagrammatic view of the cardiovascular system (heart and blood vessels). Red indicates blood that is high in oxygen. Blue indicates blood that is low in oxygen.

fingers any artery that lies near the body's surface, such as the one at the wrist or the one under the angle of the jaw.

As previously mentioned, the middle layer of an artery wall also contains smooth muscle that enables the artery to contract. When the muscle contracts, the diameter of the lumen becomes narrower, and blood flow through the artery is reduced. On the other hand, when the smooth muscle relaxes, the arterial lumen increases in diameter, which *increases* blood flow through the artery.

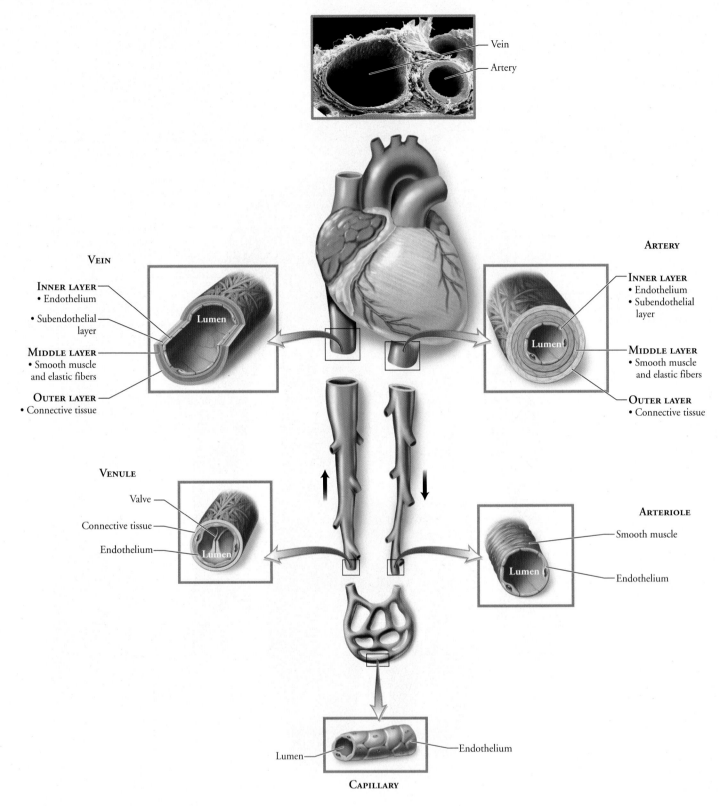

FIGURE **12.2**   The structure of blood vessels

The smooth muscle is best developed in small- to medium-sized arteries. These arteries serve to regulate the distribution of blood, adjusting flow to suit the needs of the body.

Disease, inflammation, injury, or a defect existing at birth sometimes weakens the wall of an artery. When the arterial wall be-

comes weakened, the pressure of the blood flowing through the weakened area may cause the wall to swell outward like a balloon, forming an *aneurysm*. Most aneurysms do not cause symptoms, but the condition can be threatening just the same. The primary risk is that the aneurysm will burst, causing blood loss. The tissues serviced by that vessel will then be deprived of oxygen and nutrients, a situation that can be fatal. Even if the aneurysm does not rupture, it can cause the formation of life-threatening blood clots. A clot can break free from the site of formation and float through the circulatory system until it lodges in a small vessel, where it can block blood flow and cause tissue death beyond that point. In some cases, an aneurysm can be repaired surgically.

The smallest arteries, called **arterioles**, are barely visible to the unaided eye. Their walls have the same three layers found in arteries, but the middle layer is primarily smooth muscle with only a few elastic fibers.

Arterioles have two extremely important regulatory roles. First, they are the prime controllers of **blood pressure**, which is the pressure of blood against the vessel walls (discussed later in this chapter). Second, they serve as gatekeepers to the capillary networks. A capillary network can be open or closed, depending on whether the smooth muscle in the walls of the arteriole leading to it allows blood through. In this way, metabolically active cells receive more blood than metabolically inactive ones. Minute by minute, arterioles respond to input from hormones, the nervous system, and local conditions, constantly modifying blood pressure and flow to meet the body's changing needs.

■ Capillaries are sites of exchange with body cells

Capillaries are microscopic blood vessels that connect arterioles and venules. The capillaries are well suited to their primary function: the exchange of materials between the blood and the body cells, as shown in Figure 12.3. Because capillary walls are only one cell layer thick, substances move easily between the blood and the fluid surrounding the cells outside the capillary. The plasma membrane of the capillary's endothelial cells is an effective selective barrier that determines which substances can cross. Some substances that cross the capillary walls do not pass *through* the endothelial cells. Instead, these substances filter through small slits *between* adjacent endothelial cells. The slits between the cells are just large enough for some fluids and small dissolved molecules to pass through.

The design of the capillary networks allows blood flow through capillaries to be adjusted to the body's metabolic needs. The network of capillaries servicing a particular area is called a **capillary bed** (Figure 12.4). The number of capillaries in a bed generally ranges from 10 to 100, depending on the type of tissue. A ring of smooth muscle called a *precapillary sphincter* surrounds the capillary where it branches off the arteriole and regulates blood flow into it. Contraction of the precapillary sphincter squeezes the opening to the capillary shut.

The precapillary sphincters act as valves that open and close the capillary beds. For instance, while you are resting on the beach after finishing a picnic lunch, the capillary beds servicing the digestive organs will be open, and nutrients from the food will be absorbed. If you dive into the water and start swimming, the capillary beds of the digestive organs will close down, and those in the skeletal muscles will open.

Collectively, the capillaries provide a tremendous surface area for the rapid exchange of materials between body and blood. Cap-

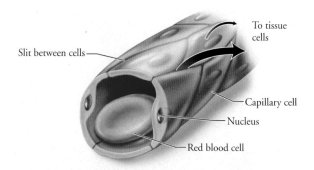

(a) Substances are exchanged between the blood and tissue fluid across the plasma membrane of the capillary or through slits between capillary cells.

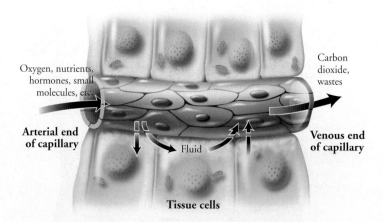

(b) At the arterial end of a capillary, blood pressure forces fluid out of the capillary to the fluid surrounding tissue cells. At the venous end, fluid is drawn back into the capillary by osmotic pressure.

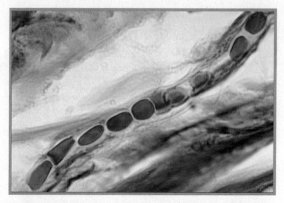

(c) Capillaries are so narrow that red blood cells must travel through them in single file.

FIGURE **12.3** Capillaries are the sites where materials are exchanged between the blood and the body cells.

illary beds bring capillaries very close to nearly every cell. Your fingernails provide windows that allow you to appreciate the efficiency with which capillary networks reach all parts of the body. You may have noticed that the tissue beneath a fingernail normally has a pink tinge. The color results from blood flowing through numerous capillaries there. Gentle pressure on the nail causes the tissue to turn white as the blood is pushed from those capillaries.

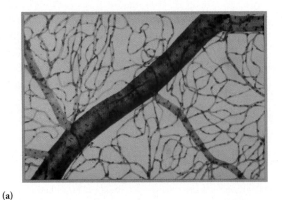

(a)

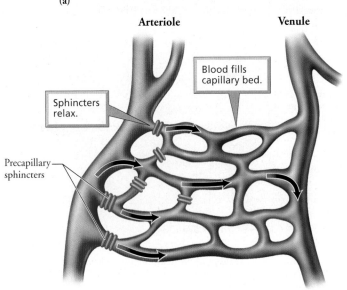

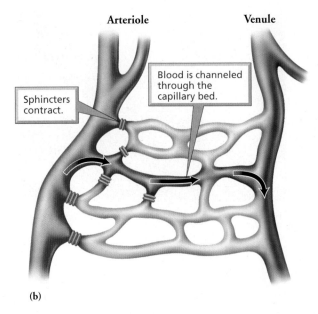

(b)

FIGURE **12.4** (a) A capillary bed is a network of capillaries. (b) The entrance to each capillary (the arterial side) is guarded by a ring of muscle called a precapillary sphincter. Blood flow through a capillary bed is regulated according to the body's metabolic needs.

Although a single capillary is scarcely wide enough for a single red blood cell to squeeze through it, there are so many capillaries that their *combined* cross-sectional area is enormous, much greater than that of the arteries or veins. Because of the large cross-sectional area of the capillaries, the blood flows much more slowly through them than through the arteries or veins. The slower rate of flow in the capillaries provides more time for the exchange of materials (Figure 12.5).

■ Veins return blood to the heart

On the side of the capillary bed that is away from the arteriole, capillaries merge to form the smallest kind of vein, a **venule**. Venules then join to form larger veins. **Veins** are blood vessels that return the blood to the heart.

Although veins share some structural features with arteries, there are also some important differences. The walls of veins have the same three layers found in arterial walls, but the walls of veins are thinner and the lumens of veins are larger than those of arteries of equal size (Figure 12.2). The thin walls and large lumens allow veins to hold a large volume of blood. Veins serve as blood reservoirs, holding up to 65% of the body's total blood supply.

The same amount of blood that is pumped out of the heart must be conducted back to the heart, but it must be moved through the veins without assistance from the high pressure generated by the heart's contractions. In the head and neck, of course, gravity helps move blood toward the heart. But how is it possible to move blood against the force of gravity—from the foot back to the heart, for instance (unless, by chance, your foot was in your mouth)?

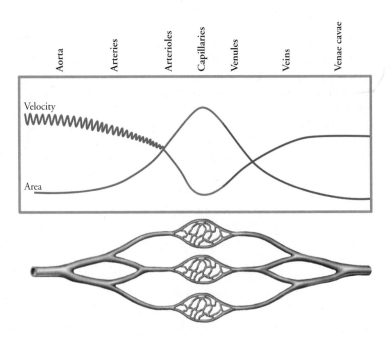

FIGURE **12.5** The capillaries are so numerous that their total cross-sectional area is much greater than that of arteries or veins. Thus, blood pressure drops and blood flows more slowly as it passes through a capillary bed. The slower rate of flow allows time for the exchange of materials between the blood and the tissues.

Three mechanisms move blood from lower parts of the body toward the heart:

1. **Valves in veins prevent backflow of blood.** Veins often contain valves that act as one-way turnstiles, allowing blood to move toward the heart but preventing it from flowing backward. These valves are pockets of connective tissue projecting from the lining of the vein, as shown in Figure 12.6a.

   A simple experiment can demonstrate the effectiveness of venous valves. Allow your hand to hang by your side until the veins on the back of your hand become distended. Place two fingertips from the other hand at the end of one of the distended veins nearest to the knuckles. Then, leaving one fingertip pressed on the end of the vein, move the other toward the wrist, pressing firmly and squeezing the blood from the vein. Lift the fingertip near the knuckle and notice that blood immediately fills the vein. Repeat the procedure, but this time lift the fingertip near the wrist. You will see the vein remain flattened, because the valves prevent the backward flow of blood.

2. **Contraction of skeletal muscle squeezes veins.** Virtually every time a skeletal muscle contracts, it squeezes nearby veins. This pressure pushes blood past the valves toward the heart. The mechanism propelling the blood is not unlike the one that causes toothpaste to squirt out of the uncapped end of the tube regardless of where the tube is squeezed. When skeletal muscles relax, any blood that moves backward fills the valves. As the valves fill with blood, they extend further into the lumen of the vein, closing the vein and preventing the flow of blood from reversing direction (Figure 12.6b). Thus, the skeletal muscles are always squeezing the veins and driving blood toward the heart.

3. **Breathing causes pressure changes that move blood toward the heart.** The thoracic (chest) cavity increases in size when we inhale (see Chapter 14). The expansion reduces pressure within the thoracic cavity and increases pressure in the abdominal cavity. Blood moves toward regions of lower pressure. Thus, reduced pressure in the thoracic cavity pulls blood back toward the heart. In addition, increased pressure in the abdominal cavity squeezes veins, also forcing blood back toward the heart.

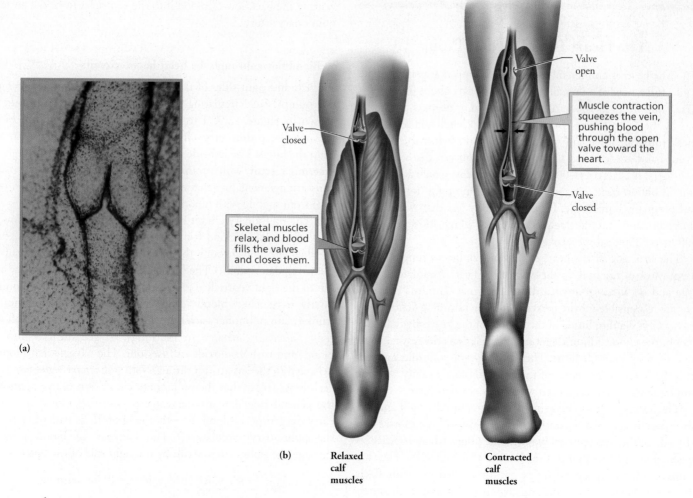

(a)

(b)

Valve closed

Skeletal muscles relax, and blood fills the valves and closes them.

Valve open

Muscle contraction squeezes the vein, pushing blood through the open valve toward the heart.

Valve closed

Relaxed calf muscles

Contracted calf muscles

FIGURE **12.6** (a) A micrograph of a vein showing a valve. (b) Pocketlike valves on the inner surface of veins assist the return of blood to the heart against gravity by preventing backflow.

stop and think

VARICOSE VEINS

Varicose veins are veins that have become distended because the blood in them is prevented from flowing freely and therefore accumulates, or "pools." As you may recall, the walls of veins are relatively thin, so they are easily stretched, especially if the vein is near the body surface, where there is little surrounding tissue to support it. Stretching is more likely to occur in veins that must support a great column of blood, such as those in the legs. Once a section of a vein has been stretched, extra blood accumulates inside it, adding to the stress on the valve below. Eventually, that valve can become stretched and ineffective, so the blood the valve once held falls back to the next valve. The resulting varicose veins may appear as bulging veins or as bluish purple, spiderlike lines.

stop and think

The Heart Is a Muscular Pump

The heart is an incredible muscular pump that generates the force needed to circulate blood. It beats about 72 times a minute every hour of every day. To appreciate the work done by the heart, form a fist and alternately clench and relax it 70 times a minute. How many minutes does it take before the muscles of your hand are too tired to continue? In contrast, the heart does not fatigue. It beats over 100,000 times each day, which adds up to about 2 billion beats over a lifetime. The pumping done by the heart is equally remarkable. It pumps slightly less than 5 liters (10 pt) of blood a minute through its chambers, which adds up to over 9400 liters (2500 gal) per day.

The structure of the heart explains its ability to function as a pump. Most of the bulk of the wall of the heart is cardiac muscle tissue and is called the **myocardium**. The myocardium's contractions are responsible for the heart's incredible pumping action. The *endocardium* is a thin lining in the cavities of the heart. By reducing friction, the endocardium's smooth surface lessens the resistance to blood flow through the heart. The *pericardium* is a fibrous sac that holds the heart in the center of the chest (thoracic) cavity without hampering its movements, even when they are vigorous.

The heart appears to be a single structure, but the right and left halves function as two separate pumps. As we will see shortly, the right side of the heart pumps blood to the lungs, where it picks up oxygen. The left side pumps the blood to the body cells. The two pumps are physically separated by a partition called a septum. Each side of the heart consists of two chambers: an upper chamber, called an **atrium** (plural, atria), and a lower chamber, called a **ventricle** (Figure 12.7). The two atria function as receiving chambers for the blood returning to the heart. The two ventricles function as the main pumps of the heart. The contraction of the ventricles forces blood out of the heart under great pressure. When we think about the work of the heart, we are, in fact, thinking about the work of the ventricles. It should not be surprising, then, that the ventricles are much larger chambers than the atria and have thicker, more muscular walls.

Two pairs of valves ensure that the blood flows in only one direction through the heart. The first pair is the **atrioventricular (AV) valves**, each leading from an atrium to a ventricle, as shown in Figure 12.8. The AV valves are connective tissue flaps, called cusps, anchored to the wall of the ventricle by strings of connective tissue called the chordae tendineae—the heartstrings. These strings prevent the AV valves from flapping back into the atria under the pressure developed when the ventricles contract. The AV valve on the right side of the heart has three flaps and is called the tricuspid valve. The AV valve on the left side of the heart has two flaps and is called the bicuspid, or mitral, valve.

For the second pair of valves, the **semilunar valves**, each is located between a ventricle and its connecting artery. The cusps of the semilunar valves are small pockets of tissue attached to the inner wall of the respective artery. When the pressure in the arteries becomes greater than the pressure in the ventricles, these valves fill with blood in a manner similar to a parachute filling with air, preventing the backflow of blood into the ventricles from the aorta or pulmonary artery.

■ Blood flows through the heart in two circuits

The left and right sides of the heart actually function as two separate pumps, each circulating the blood through a different route, as shown in Figure 12.9. The right side of the heart pumps blood through the **pulmonary circuit**, which transports blood to and from the lungs. The left side of the heart pumps blood through the **systemic circuit**, which transports blood to and from body tissues. This arrangement prevents oxygenated blood (blood rich in oxygen) from mixing with blood that is low in oxygen.

The pulmonary circuit begins in the right atrium, as veins return oxygen-poor blood from the systemic circuit. (You can trace the flow of blood through the heart in Figure 12.9 as you read the following description.) The blood then moves from the right atrium to the right ventricle. Contraction of the right ventricle pumps poorly oxygenated blood to the lungs through the pulmonary trunk (main pulmonary artery), which divides to form the left and right *pulmonary arteries*. In the lungs, oxygen diffuses into the blood, and carbon dioxide diffuses out. The oxygen-rich blood is delivered to the left atrium through four *pulmonary veins*, two from each lung. (Note that the pulmonary circulation is an *exception* to the general rule that arteries carry oxygen-rich blood and veins carry oxygen-poor blood. Exactly the opposite is true of vessels in the pulmonary circulation.) The pathway of blood pumped through the pulmonary circuit by the right side of the heart is

Right atrium → AV valve (tricuspid) → Right ventricle

Pulmonary semilunar valve → Pulmonary trunk →

Pulmonary arteries → Lungs → Pulmonary veins → Left atrium

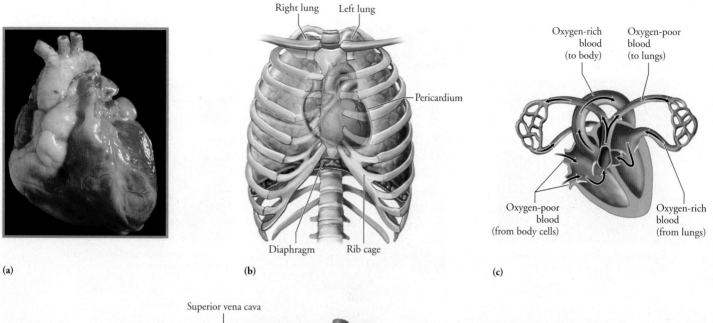

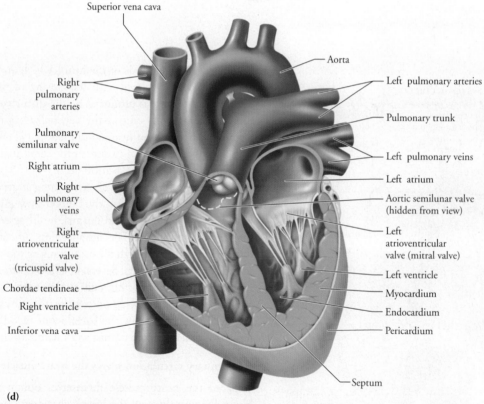

FIGURE **12.7** (a) The human heart. (b) The heart is located in the thoracic (chest) cavity. (c) Blood flows through the heart from the atria to the ventricles. (d) This diagram of a human heart shows the four chambers, the major vessels connecting to the heart, and the two pairs of heart valves.

The systemic circuit begins when oxygen-rich blood enters the left atrium (see Figure 12.9). Blood then flows to the left ventricle. When the left ventricle contracts, oxygenated blood is pushed through the largest artery in the body, the aorta. The aorta arches over the top of the heart and gives rise to the smaller arteries that eventually feed the capillary beds of the body tissues. The venous system collects the oxygen-depleted blood and eventually culminates in veins that return the blood to the right atrium. These veins

are the *superior vena cava*, which delivers blood from regions above the heart, and the *inferior vena cava*, which returns blood from regions below the heart. Thus, the pathway of blood through the systemic circuit pumped by the left side of the heart is

Left atrium → AV (bicuspid or mitral) valve → Left ventricle →

Aortic semilunar valve → Aorta → Body tissues →

Inferior vena cava or superior vena cava → Right atrium

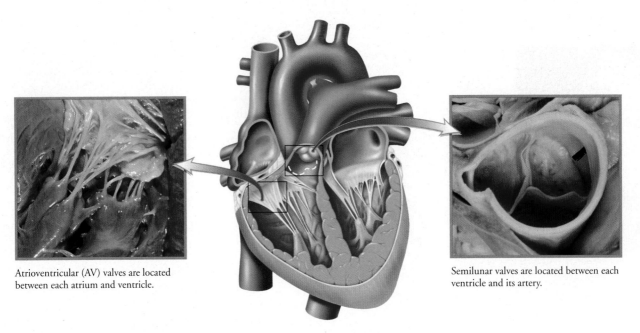

Atrioventricular (AV) valves are located between each atrium and ventricle.

Semilunar valves are located between each ventricle and its artery.

FIGURE **12.8** The valves of the heart keep blood flowing in one direction.

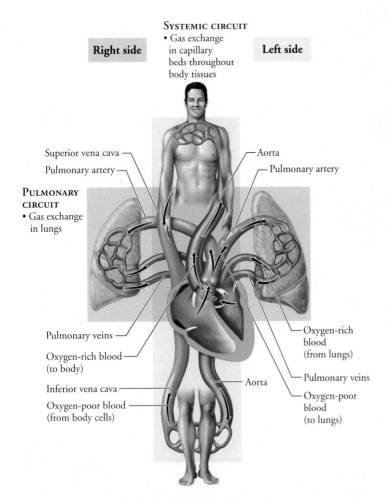

SYSTEMIC CIRCUIT
• Gas exchange in capillary beds throughout body tissues

Right side

Left side

PULMONARY CIRCUIT
• Gas exchange in lungs

Superior vena cava

Pulmonary artery

Aorta

Pulmonary artery

Pulmonary veins

Oxygen-rich blood (to body)

Oxygen-rich blood (from lungs)

Inferior vena cava

Aorta

Pulmonary veins

Oxygen-poor blood (from body cells)

Oxygen-poor blood (to lungs)

FIGURE **12.9** Circuits of blood flow. The right side of the heart pumps blood through the pulmonary circuit, which carries blood to and from the lungs. The left side of the heart pumps blood through the systemic circuit, which conducts blood to and from the body tissues.

The familiar sounds of the heart, which are often described as "lub-dup," are associated with the closing of the valves. The first heart sound ("lub") is produced by the disturbed blood flow when the AV valves snap shut as the ventricles begin to contract. The higher-pitched second heart sound ("dup") is produced by the disturbed blood flow when closure of the semilunar valves and the beginning of ventricular relaxation occur.

Heart murmurs, which are swooshing heart sounds other than lub-dup, are created by disturbed blood flow. Although heart murmurs are sometimes heard in normal, healthy people, they can indicate a heart problem. For instance, malfunctioning valves often disturb blood flow through the heart, causing the swishing or gurgling sounds of heart murmurs. Several conditions can cause valves to malfunction. In some cases, thickening of the valves narrows the opening and impedes blood flow. In other cases, the valves do not close properly and, therefore, allow the backflow of blood. In either case, the heart is strained because it must work harder to move the blood.

■ Coronary circulation serves the heart muscle

Cells of the heart muscle themselves obtain little nourishment blood flowing through the heart's chambers. Instead, an extensive network of vessels, known as the **coronary circulation**, services the tissues of the heart. The first two arteries that branch off the aorta are the coronary arteries (Figure 12.10). These arteries give rise to numerous branches, ensuring that the heart receives a rich supply of oxygen and nutrients. After passing through the capillary beds that nourish the heart tissue, blood enters cardiac veins and eventually flows into the right atrium.

■ The cardiac cycle is the sequence of heart muscle contraction and relaxation

Although the two sides of the heart pump blood through different circuits, they work in tandem. The two atria contract simultaneously, and then the two ventricles contract simultaneously.

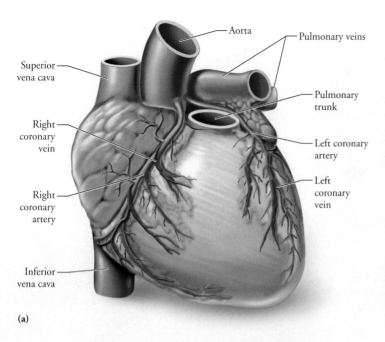

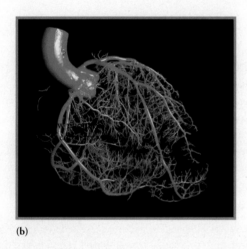

(b)

FIGURE **12.10** Coronary circulation. (a) The coronary vessels deliver a rich supply of oxygen and nutrients to the heart muscle cells and remove the metabolic wastes. (b) This cast of the coronary blood vessels reveals the complexity of the coronary circuit.

We see, then, that a heartbeat is not a single event. Each beat involves contraction, which is called **systole** (*sis*-to-lee) and relaxation, which is called **diastole** (di-*as*-to-lee). All the events associated with the flow of blood through the heart chambers during a single heartbeat are collectively called the **cardiac cycle**, as illustrated in Figure 12.11. First, all chambers relax and blood passes through the atria and enters the ventricles. When the ventricles are about 70% filled, the atria contract and push their contents into the ventricles. The atria then relax and the ventricles begin their contraction phase. Upon completion of this contraction, the whole heart again relaxes. If we were to add the contraction time of the heart during a day and compare it with the relaxation time during a day, the heart's workday might turn out to be equivalent to yours. In 24 hours, the heart spends a total of about 8 hours working (con-

tracting) and 16 hours relaxing. However, unlike your workday, the heart's day is divided into repeating cycles of work and relaxation.

■ The rhythmic contraction of the heart is due to its internal conduction system

If a human heart is removed, as in a transplant operation, and placed in a dish, it will continue to beat, keeping a lonely and useless rhythm until its tissues die. In fact, if a few cardiac muscle cells are grown in the laboratory, they too will beat on their own, each twitch a reminder of the critical role the intact organ plays. Clearly, then, the heart muscle does not require outside stimulation to beat. Instead, the tendency is intrinsic, within the heart muscle itself.

Another remarkable observation has been made of heart muscle cells grown in a laboratory dish. Although isolated heart cells twitch independently of others, if two cells should touch, they will begin beating in unison. This, too, is an inherent property of the cells, but it is partly due to the type of connections between heart muscle cells. The cell membranes of adjacent cardiac muscle cells interweave with one another at specialized junctions called *intercalated disks*. Cell junctions in the intercalated disks mechanically and electrically couple the connected cells. Adjacent cells are held together so tightly that they do not rip apart during contraction but instead transmit the pull of contraction from one cell to the next. At the same time, the junctions permit electrical communication between adjacent cells, allowing the electrical events responsible for contraction to spread rapidly over the heart by passing from cell to cell. Yet even though heart cells contract automatically and in a coordinated manner, they still need some outside control to contract at the proper rate.

The tempo of the heartbeat is set by a cluster of specialized cardiac muscle cells, called the **sinoatrial (SA) node**, located in the right atrium near the junction of the superior vena cava (Figure 12.12). Because the SA node sends out impulses that initiate each heartbeat, it is often referred to as the **pacemaker**. About 70 to 80 times a minute, the SA node sends out an electrical signal that spreads through the muscle cells of the atria, causing them to contract. The signal reaches another cluster of specialized muscle cells called the **atrioventricular (AV) node**, located in the partition between the two atria, and stimulates it. The AV node then relays the stimulus by means of a bundle of specialized muscle fibers, called the *atrioventricular bundle,* that runs along the wall between the ventricles. The bundle forks into right and left branches and then divides into many other specialized cardiac muscle cells, called *Purkinje fibers,* which penetrate the walls of the ventricles. The rapid spread of the impulse through the ventricles ensures that they contract smoothly.

When the heart's conduction system is faulty, cells may begin to contract independently. Such cellular independence can result in rapid, irregular contractions of the ventricles, called *ventricular fibrillation,* which render the ventricles useless as pumps and stop circulation. With the brain no longer receiving the blood it needs to function, death will occur unless an effective heartbeat is restored quickly. A method for stopping ventricular fibrillation is to subject the heart to an electric shock that goes through the heart; in

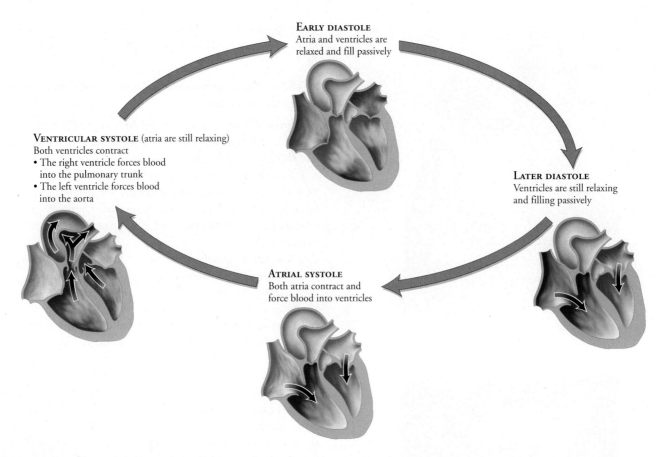

EARLY DIASTOLE
Atria and ventricles are
relaxed and fill passively

LATER DIASTOLE
Ventricles are still relaxing
and filling passively

VENTRICULAR SYSTOLE (atria are still relaxing)
Both ventricles contract
• The right ventricle forces blood
  into the pulmonary trunk
• The left ventricle forces blood
  into the aorta

ATRIAL SYSTOLE
Both atria contract and
force blood into ventricles

FIGURE **12.11** The cardiac cycle—the sequence of heart muscle relaxation and contraction. The atria contract together and the ventricles contract together. Red indicates blood high in oxygen. Blue indicates blood low in oxygen.

many cases, the SA node will once again begin to function normally. Although costly, an implantable defibrillator (a device that electrically shocks the heart) can provide a life-saving "jump-start" when needed.

Problems with the conduction system of the heart can sometimes be treated with an artificial pacemaker, a small device that monitors the heart rate and rhythm and responds to abnormalities if they occur. For instance, if the heart rate becomes too slow, the pacemaker will send electrical signals to the heart through an electrode.

REGULATION OF HEART RATE

The pace or rhythm of the heartbeat changes constantly in response to activity or excitement. The autonomic nervous system and certain hormones make the necessary adjustments so that the heart rate suits the body's needs. During times of stress, the sympathetic nervous system increases the rate and force of heart contractions. As part of this response, the adrenal medulla produces the hormone epinephrine, which can prolong the sympathetic nervous system's effects. In contrast, when restful conditions prevail, the parasympathetic nervous system dampens heart activity, in keeping with the body's more modest metabolic needs.

what would you do?

In a heart transplant operation, a person's diseased heart is replaced with a healthy heart from a person who recently died. Not surprisingly, there are thousands more people in need of a heart transplant than there are donor hearts available. An artificial heart, AbioCor, is now approved by the Food and Drug Administration. To qualify as a recipient, a person must have a 70% chance of dying within the next 30 days. Some recipients have died within a day of receiving the artificial heart. Most lived several months. As of 2006, the longest-living recipient survived for 17 months. If you were in need of a transplant, would you volunteer to participate in the testing of an artificial heart? What criteria would you use to decide?

■ An electrocardiogram is a recording of the electrical activities of the heart

The electrical events that spread through the heart with each heartbeat actually travel throughout the body, because body fluids are good conductors. Electrodes placed on the body surface can detect these electrical events, transmitting them so that they cause deflections (movements) in the tracing made by a recording device. An **electrocardiogram** (**ECG** or **EKG**) is an image of the electrical activities of the heart generated by such a recording device.

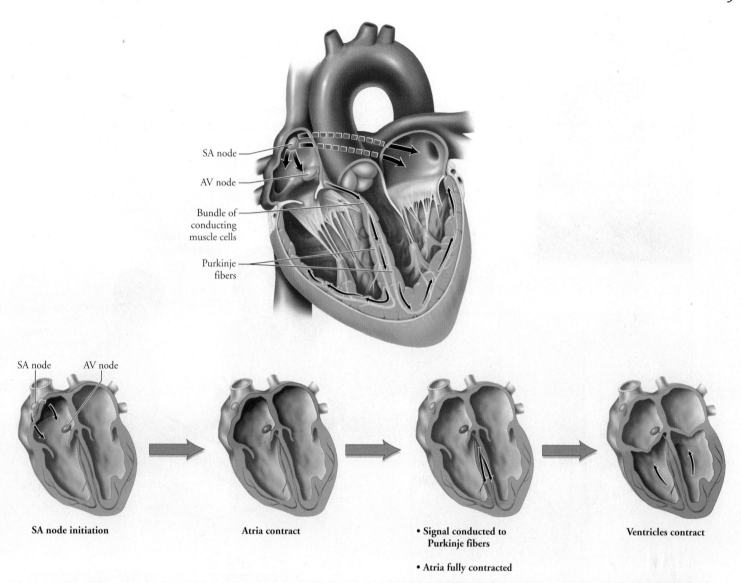

SA node

AV node

Bundle of conducting muscle cells

Purkinje fibers

SA node   AV node

SA node initiation

Atria contract

• **Signal conducted to Purkinje fibers**

• **Atria fully contracted**

Ventricles contract

FIGURE **12.12** The conduction system of the heart consists of specialized cardiac muscle cells that speed electrical signals through the heart. The sinoatrial (SA) node serves as the heart's internal pacemaker that determines the heart rate. Electrical signals from the SA node spread through the walls of the atria, causing them to contract. The signals then stimulate the atrioventricular (AV) node, which in turn sends the signals along the atrioventricular AV bundle to its forks and finally to the many Purkinje fibers that penetrate the ventricular walls. The Purkinje fibers distribute the signals to the walls of the ventricles, causing them to contract. Electrical activity is indicated by the color blue in the bottom row of drawings.

A typical ECG consists of three distinguishable deflection waves, as shown in Figure 12.13. The first wave, called the P wave, accompanies the spread of the electrical signal over the atria and the atrial contraction that follows. The next wave, the QRS wave, reflects the spread of the electrical signal over the ventricles and ventricular contraction. The third wave, the T wave, represents the return of the ventricles to the electrical state that preceded contraction (ventricular repolarization). Because the pattern and timing of these waves are remarkably consistent in a healthy heart, abnormal patterns can indicate heart problems.

■ **Blood pressure is the force blood exerts against blood vessel walls**

We hear a lot about blood pressure, usually when someone worries about having a high blood pressure reading or brags about having a low one. Specifically, blood pressure is the force exerted by the blood against the walls of the blood vessels. When the ventricles contract, they push blood into the arteries under great pressure. This pressure is the driving force that moves blood through the body, but it also pushes outward against vessel walls. Ideally, a person's blood pressure

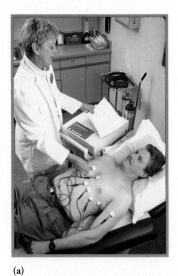

(a)

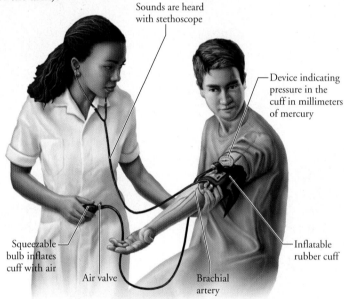

(b)

FIGURE **12.13** (a) A person having an electrocardiogram (ECG) recorded. (b) The electrical activity that accompanies each heartbeat can be visualized in an ECG tracing. The P wave is generated as the electrical signals from the SA node spread across the atria and cause them to contract. The QRS wave represents the spread of the signal through the ventricles and ventricular contraction. The T wave occurs as the ventricles recover and return to the electrical state that preceded contraction.

should be great enough to circulate the blood but not so great that it stresses the heart and blood vessels.

Blood pressure in the arteries varies predictably during each heartbeat. It is highest during the contraction of the ventricles (ventricular systole), when blood is being forced into the arteries. In a typical, healthy adult, this **systolic pressure**, the highest pressure in the artery during each heartbeat, is less than 120 mm of mercury (mm Hg).[1] Blood pressure is lowest when the ventricles

[1]Pressure is measured as the height to which that pressure could push a column of mercury (Hg).

are relaxing (diastole). In a healthy adult, the lowest pressure, or **diastolic pressure**, is less than 80 mm Hg. A person's blood pressure is usually expressed as two values—the systolic followed by the diastolic. For instance, normal adult blood pressure is said to be less than 120/80. (Do you know what your blood pressure is?)

Blood pressure is measured with a device called a *sphygmomanometer* (sfig-mo-mah-*nom*-eter), which consists of an inflatable cuff that wraps around the upper arm and is attached to a device that can measure the pressure within the cuff. Figure 12.14 shows how a manually operated sphygmomanometer uses the easily measured pressure of the air pumped into the cuff to measure the blood pressure in the brachial artery (which runs along the inner surface of the arm).

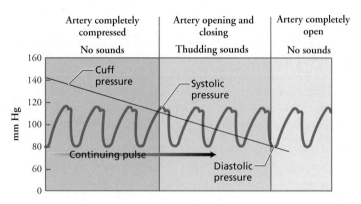

FIGURE **12.14** Blood pressure is measured with a sphygmomanometer, which consists of an inflatable cuff and a means of measuring the pressure within the cuff. The cuff is placed around the upper arm and inflated so that it compresses the brachial artery. The pressure in the cuff is slowly released, and as it descends it reaches a point where blood is able to spurt through the constricted artery only at the moments of highest blood pressure. This pressure, at which "tapping" sounds are first heard, is the systolic pressure, the blood pressure when the heart is contracting. As the pressure in the cuff continues to drop, a point is reached where the sounds disappear. The blood is now flowing continuously through the brachial artery. The pressure in the cuff when the sounds first disappear is the diastolic pressure, the blood pressure when the heart is relaxing.

Cardiovascular Disease Is a Major Killer in the United States

Cardiovascular disease is the single biggest killer of men and women in the United States. It affects slightly more men than women because, until menopause, women receive some natural protection from the female hormone estrogen. Ironically, although slightly fewer women than men have heart attacks, women who do have a heart attack are twice as likely to die within the following weeks than are men. Because heart attacks are commonly thought of as a male problem, women and their physicians often fail to recognize the symptoms, thus delaying treatments that could be lifesaving. (See the Health Issue essay, *The Cardiovascular Benefits of Exercise*.)

■ High blood pressure can kill without producing symptoms

High blood pressure, or **hypertension**, is often called the silent killer. It is *silent* because it does not produce any telltale symptoms. It is a *killer* because it can cause fatal problems, usually involving the heart, brain, blood vessels, or kidneys. Hypertension damages the heart in a number of ways but primarily by causing the heart to work harder to keep the blood moving. In response, the heart muscle thickens, and the heart enlarges. The enlarged heart works less efficiently and has difficulty keeping up with the body's needs. At the same time, the increased workload increases the heart's need for oxygen and nutrients. If these cannot be delivered rapidly enough, a heart attack can result.

High blood pressure can also damage the kidneys, reducing the blood flow through them. In response, the kidneys make matters worse by secreting renin, a chemical that leads to further increases in blood pressure in an ever-escalating cycle.

Although about 90% of the cases of hypertension have no *known* cause, many contributing factors have been identified. Between 10% and 40% of cases have a genetic basis. In other cases, the kidneys have an impaired ability to handle sodium, and the resulting fluid retention increases blood pressure by increasing blood volume. In still other people, the sympathetic nervous system reacts too strongly to stress, constricting the blood vessels and increasing heart rate. Thus, more blood per minute is pumped through vessels that provide a greater resistance to flow.

Most physicians would agree that a blood pressure of 160/90 is high and should be treated. But uncertainty clouds the treatment issue when the person's diastolic pressure is between 80 and 89. Although drug treatment may help in borderline cases, it usually must be continued for life. A diagnosis of high blood pressure may influence other aspects of a person's life, such as life insurance premiums. So, sometimes, only lifestyle changes are recommended for mild cases of hypertension.

When the diagnosis of hypertension is clear, one or more of various kinds of drugs can be prescribed, each type combating a different mechanism that contributes to high blood pressure. Some drugs, the diuretics for instance, decrease blood volume by increasing the excretion of sodium and fluids. Some other drugs reduce the pressure by causing the blood vessels to dilate (become wider).

A number of changes in lifestyle are recommended to treat or prevent hypertension.

1. **Control weight.** Maintaining normal body weight can help control blood pressure. Many overweight people with high blood pressure benefit from shedding just a few extra pounds. The best way to lose weight is to eat a moderate, balanced diet, reduce fat intake, and increase physical activity.

2. **Exercise regularly.** Aerobic exercise, such as brisk walking, jogging, swimming, or cycling, performed for at least 20 minutes three times a week, helps lower blood pressure and keep it low.

3. **Do not smoke.** Cigarette smoke contains nicotine, a drug that increases heart rate and constricts blood vessels; both of those effects increase blood pressure.

4. **Limit dietary salt.** Some people with hypertension can lower their blood pressure by lowering the amount of salt in their diet. Salt can affect fluid retention and, therefore, blood volume.

5. **Limit alcohol intake.** Alcohol consumption can elevate blood pressure. It can also interact with medications prescribed to treat hypertension.

■ Atherosclerosis is a buildup of lipids in the artery walls

Atherosclerosis (*ather-*, yellow, fatty deposit; *sclerosis*, a hardening) is a buildup of fatty deposits in the walls of arteries, fueled by an inflammatory response. In some cases, the deposits narrow the artery (Figure 12.15). Such narrowing causes problems because it reduces blood flow through the vessel, choking off the vital supply of oxygen and nutrients to the tissues served by that vessel. But contrary to the beliefs held just a few years ago, atherosclerosis is more than just a plumbing problem, a clog in a passive pipeline.

The inflammatory response thought to cause atherosclerosis is the same process that wards off infection when you scrape your knee (discussed in Chapter 13). In this case, it begins with an injury to the wall of an artery. The injury may be caused by some kind of blood-borne irritant (such as the chemicals inhaled in cigarette smoke), by cholesterol deposits in the artery lining, or by infection. Perhaps the excessively rapid or turbulent blood flow caused by high blood pressure can also produce such arterial damage. The damaged cells begin to pick up low-density lipoproteins (LDLs), the so-called bad form of cholesterol. This accumulation of LDLs is most likely to occur when the LDL concentration in the blood is high. The LDLs undergo chemical changes that stimulate the cells of the arterial lining to enlist the body's defense responses: inflammatory chemicals and defense cells. Growth factors produced by defense cells stimulate smooth muscle in the arterial wall to divide, thickening the wall. Other defense cells engulf LDLs, become enlarged, and form fatty streaks on the arterial lining. As these defense cells continue to scavenge lipids, the fatty streak increases in size and forms plaque, a bumpy, fatty layer in the artery wall. The plaque can bulge into the artery channel, blocking the blood flow, or it can expand outward into the artery wall. Although the plaque has a fibrous cap that initially keeps pieces from breaking away, the fat-filled cells secrete inflammatory substances that weaken the cap. A small break in the cap can allow the plaque to rupture, which triggers the formation of a blood clot. In any case,

HEALTH ISSUE

The Cardiovascular Benefits of Exercise

What would you say if you were told there is a simple way to reduce your risk of heart attack, stroke, diabetes, and cancer while controlling your weight, strengthening your bones, relieving anxiety and tension, and improving your memory? "Impossible!" you might say. "What's the catch?" There is none. This seemingly magical key to life is regular aerobic exercise that uses large muscle groups rhythmically and continuously and elevates heart rate and breathing rate for at least 15 to 20 minutes.

Although exercise has many beneficial effects on the body, here we will consider only the benefits to the cardiovascular system. Exercise benefits the heart by making it a more efficient pump, thus reducing its workload. A well-exercised heart beats more slowly than the heart of a sedentary person—during both exercise and rest. The lower heart rate gives the heart more time to rest between beats. Yet at the same time, the well-exercised heart pumps more blood with each beat.

Exercise also increases the oxygen supply to the heart muscle by widening the coronary arteries, thereby increasing blood flow to the heart. Moreover, because the capillary beds within the heart muscle become more extensive with regular exercise, oxygen and nutrients can be delivered to the heart cells and wastes can be removed more quickly.

Furthermore, exercise helps to ensure continuous blood flow to the heart. One way it accomplishes this benefit is by increasing the body's ability to dissolve blood clots that can lead to heart attacks or strokes. Exercise stimulates the release of a natural enzyme that prevents blood clotting and remains effective for as long as $1\frac{1}{2}$ hours after you stop exercising. Besides this, exercise stimulates the development of collateral circulation, that is, additional blood vessels that provide alternative pathways for blood flow. As a result, blood flows continuously through the heart, even if one vessel becomes blocked.

Exercise affects the blood in ways that allow more oxygen to be delivered to the cells. The amount of hemoglobin, the oxygen-binding protein in red blood cells, increases. In addition, the blood volume and the numbers of red blood cells increase.

Exercise lowers the risk of coronary artery disease by lowering blood pressure and by shifting the balance of lipids in the blood. High-density lipoproteins (HDLs), which are the "good" form of cholesterol-carrying particles that remove cholesterol from the arterial walls, increase with exercise.

To reap cardiovascular benefits, you must exercise hard enough, long enough, and often enough. The exercise must be vigorous enough to elevate your heart rate to the so-called target zone, which is between 70% and 85% of your maximal attainable heart rate. The target zone can be determined by subtracting your age in years from 222 beats per minute. The exercise must continue for at least 20 minutes and be performed at least 3 days a week, with no more than 2 days between sessions.

Doing *something* active on a regular basis can improve your quality of life. Moderate activity helps you feel better emotionally and physically. The difference has been likened to traveling first class instead of coach. 👫

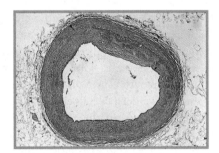

(a) A normal artery

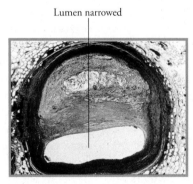

Lumen narrowed

(b) An artery partially obstructed with plaque

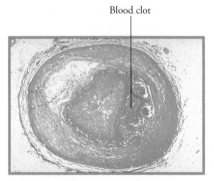

Blood clot

(c) An artery completely obstructed with plaque and a blood clot

FIGURE **12.15** Atherosclerosis, a low-level inflammatory response in the wall of an artery, is associated with the formation of fat-filled plaques. Plaque can obstruct blood flow through the artery, thus depriving the cells that would be fed life-sustaining blood by the artery. Plaque can also rupture, causing a blood clot to form. The clot may then completely clog the vessel and cause the death of tissue downstream.

the restriction of the blood flow can starve and kill the cells lying downstream. Such an occurrence in the heart (leading to a heart attack) or brain (leading to a stroke) can be fatal. Because plaques do not necessarily bulge into the artery and cause symptoms of atherosclerosis before they rupture, heart attacks often occur in patients without previous symptoms.

stop and think

C-reactive protein (CRP) is an inflammatory chemical released by injured cells in the artery lining. Why might CRP prove to be a better predictor of atherosclerosis than blood cholesterol level?

■ Coronary artery disease is atherosclerosis in the coronary arteries

Coronary artery disease is a condition in which the fatty deposits associated with atherosclerosis form within coronary arteries, the arteries that nourish the heart muscle. Coronary artery disease is the underlying cause of the vast majority of heart attacks.

A temporary shortage of oxygen to the heart is accompanied by angina pectoris—chest pain, usually experienced in the center of the chest or slightly to the left. The name *angina* comes from the Latin word *angere*, meaning "to strangle." The name is apt, since the pain of angina is often described as suffocating, viselike, or choking. Typically, the pain begins during physical exertion or emotional stress, when the demands on the heart are increased and the blood flow to the heart muscle can no longer meet the needs. The pain stops after a period of rest. Angina serves as a warning that part of the heart is receiving insufficient blood through the coronary arteries, but it does not cause permanent damage to the heart. The warning should be taken seriously, however, because each year up to 15% of those people who have angina die suddenly from a heart attack.

Although coronary artery disease is usually diagnosed from the symptoms of angina and a physical examination, a procedure called *coronary angiography* may be used to spot areas in the coronary arteries that have become narrowed by atherosclerosis. In this procedure, a contrast dye that is visible to x-rays is released in the heart, allowing the coronary vessels to be seen on film. A catheter (a slender, flexible tube) is inserted into an artery in the arm or leg and then threaded through the blood vessels until it reaches the heart. The dye is then squirted into the openings of the coronary arteries, and its movement through the arteries is then recorded in a series of high-speed x-rays.

Coronary artery disease can be treated with medicines or with surgery. Among the medicines commonly used are some that dilate (widen) blood vessels, such as nitroglycerin. Certain other drugs specifically dilate the coronary arteries. Wider blood vessels make it easier for the heart to pump blood through the circuit. When the coronary arteries dilate, more blood is delivered to the heart muscle. Also used are drugs that dampen the heart's response to stimulation from the sympathetic nervous system, thus decreasing its need for oxygen.

Two surgical operations for treating coronary artery disease are balloon angioplasty and coronary artery bypass. In *angioplasty*, the channel of an artery narrowed by soft, fatty plaque is widened by inflating a tough, plastic balloon inside the artery (Figure 12.16). First the tiny, uninflated balloon is attached to the end of a catheter and inserted through an artery in the arm or upper thigh. It is then pushed to the blocked spot in a coronary artery; x-rays are used to track its progress. After it reaches the blockage, the balloon is inflated under pressure, stretching the artery and pressing the soft plaque against the wall to widen the lumen.

After angioplasty, physicians commonly insert a metal-mesh tube called a stent into the treated arteries. The stent prevents the arteries from collapsing and keeps loose pieces of plaque from being swept into the bloodstream. Although stents boost the percentage of arteries that stay open, they sometimes trigger an inflammatory response or become clogged again. In 2006, the Food and Drug Administration approved a stent that slowly releases

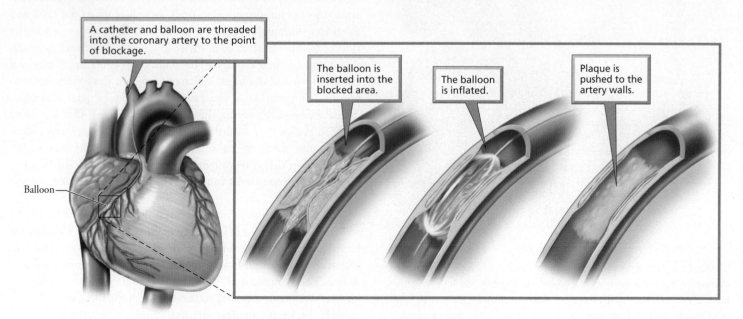

A catheter and balloon are threaded into the coronary artery to the point of blockage.

The balloon is inserted into the blocked area.

The balloon is inflated.

Plaque is pushed to the artery walls.

Balloon

FIGURE **12.16** Balloon angioplasty opens a partially blocked artery.

sirolimus, a drug that blocks cell division, preventing scar tissue from forming and blocking the artery again.

A *coronary bypass* is a procedure in which a segment of a leg vein is removed and grafted so that it provides an alternate pathway that bypasses a point of obstruction between the aorta and a coronary artery (Figure 12.17).

stop and think

In a coronary bypass operation, why is it important for the surgeon to insert the grafted vein in the correct orientation? What would happen if the piece of vein were inserted backwards?

■ Heart attack is the death of heart muscle

In a heart attack, technically known as a *myocardial infarction*, a part of the heart muscle dies because of an insufficient blood supply. (*Myocardial* refers to heart muscle; *infarct* refers to dead tissue.) Heart muscle cells begin to die if they are cut off from their essential blood supply for more than 2 hours. Depending on the extent of damage, the effects of a heart attack can spread quickly throughout the body: the brain receives insufficient oxygen; the lungs fill with fluid; and the kidneys fail. Within a short time, white blood cells swarm in to remove the damaged heart tissue. Then, over the next 8 weeks or so, scar tissue replaces the dead cardiac muscle (Figure 12.18). Because scar tissue cannot contract, part of the heart permanently loses its pumping ability.

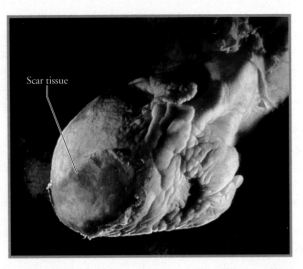

FIGURE **12.18** The ravages of a prior heart attack are visible as scar tissue at the bottom of this lifeless heart. Scar tissue replaced cardiac muscle when the blood supply to the heart muscle was shut down. Because scar tissue cannot contract, that part of the life-sustaining pump becomes ineffective.

Heart attacks can be caused in many ways. The most common type of heart attack is a coronary thrombosis, which means the attack is caused by a blood clot blocking a coronary artery. Coronary thrombosis is unlikely to happen unless the artery already contains plaques of atherosclerosis, as occurs in coronary artery disease. In some instances, the blood clot is formed elsewhere in the body but is swept along in the bloodstream until it lodges in a coronary artery. In other instances, the blockage is temporary, caused by constriction of a coronary artery, called a coronary artery spasm.

Chest pain is a common indication that a heart attack is in progress, especially in men. Unlike angina, the pain is not necessarily brought on by activity and it doesn't go away with rest. In some cases, the victim feels a severe, crushing pain that begins in the center of the chest and often spreads down the inside of one or both arms (most commonly the left one), as well as up to the neck and shoulders. Although the pain is usually severe enough to cause the victim to stop whatever he or she is doing, it is not always so strong that it is recognized as a sign of heart attack. A heart attack may also cause nausea and dizziness, which can prompt the victim to interpret the symptoms as an upset stomach. Oddly, persons who experience severe pain may be the lucky ones, because they are more likely to realize they are having a heart attack and seek immediate help. Doubt about the cause of the symptoms is unfortunate because it often delays treatment, and treatment within the first few hours after a heart attack can make the difference between life and death.

If a large enough section of heart muscle is damaged by a heart attack, the heart may not be able to continue pumping blood to the lungs and the rest of the body at an adequate rate. A condition in which the heart is no longer an efficient pump is known as *heart failure*. The symptoms of heart failure include shortness of breath, fatigue, weakness, and fluid accumulation in the lungs or limbs. Although the heart can never be restored to its former health, the symptoms of heart failure can be treated with drugs. For instance,

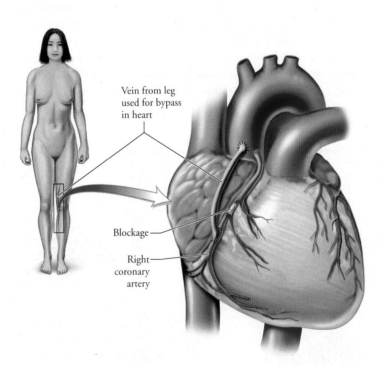

FIGURE **12.17** In coronary bypass surgery, a section of a leg vein is removed. One end of the vein is attached to the heart's main artery, the aorta, and the other to a coronary artery, bypassing the obstructed region. The grafted vein provides a pathway through which blood can reach the previously deprived region of heart muscle.

digitalis can increase the strength of heart contractions, and diuretics reduce fluid accumulation, thus lessening the heart's workload. Other drugs relax constricted arteries, thereby reducing resistance to blood flow and thus lowering the blood pressure. Together, these drugs help a weakened heart pump more efficiently.

Sometimes, drugs and bypass surgery cannot halt progressive heart failure. In this case, heart transplant surgery may provide hope for some patients, but generally only for those younger than 60 years of age. First, a donor heart must be found whose tissue is an acceptable match with that of the recipient. During transplant surgery, a heart-lung machine takes over the circulation while the weakened heart is removed, and the new heart is sewn in place. The patient is then treated with drugs that lessen the chances of organ rejection (discussed in Chapter 13).

The Lymphatic System Functions in the Circulatory and Immune Systems

The **lymphatic system** consists of **lymph**, which is a fluid identical to interstitial fluid (the fluid that bathes all the cells of the body); of **lymphatic vessels** through which the lymph flows; and of various lymphoid tissues and organs scattered throughout the body. Together, the cardiovascular system and the lymphatic system make up the **circulatory system**.

The functions of the lymphatic system are as diverse as they are essential to life.

1. **Return excess interstitial fluid to the bloodstream.** The lymphatic system maintains blood volume by returning excess interstitial fluid to the bloodstream. Only 85% to 90% of the fluid that leaves the blood capillaries and bathes the body tissues as interstitial fluid is reabsorbed by the capillaries. The rest of that fluid is absorbed by the lymphatic system and then returned to the circulatory system. This job is important. If the surplus interstitial fluid were not drained, it would cause the tissue to swell; the volume of blood would drop to potentially fatal levels; and the blood would become too viscous (thick) for the heart to pump.

    A dramatic example of the importance of returning fluid to the blood is provided by elephantiasis, a condition in which parasitic worms block lymphatic vessels (Figure 12.19). The blockage can cause a substantial buildup of fluid in the affected body region, followed by the growth of connective tissue. Elephantiasis is so named because it results in massive swelling and the darkening and thickening of the skin in the affected region, making the region resemble the skin of an elephant. Elephantiasis is a tropical disease, transmitted by mosquitoes, that affects over 400 million people.

2. **Transport products of fat digestion from the small intestine to the bloodstream.**

3. **Help defend against disease-causing organisms.** The lymphatic system helps protect against disease and cancer. We will learn more about its role in body defense in Chapter 13.

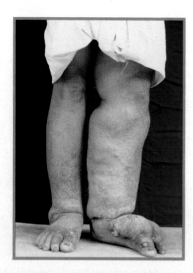

FIGURE **12.19** The leg of a person with elephantiasis. In this condition, parasitic worms plug lymphatic vessels and prevent the return of fluid from the tissues to the circulatory system.

The structure of the lymphatic vessels is central to their ability to absorb the tissue fluid not carried away by capillaries. The extra fluid enters a branching network of microscopic tubules, called *lymphatic capillaries*, that penetrate between the cells and the capillaries in almost every tissue of the body (except teeth, bones, bone marrow, and the central nervous system) (Figure 12.20). The lymphatic capillaries differ from the blood capillaries in two ways. First, lymphatic capillaries end blindly, like the fingers of a glove. In essence, they serve as drainage tubes. Fluid enters at the "fingertips" and moves through the system in only one direction. Second, lymphatic capillaries are much more permeable than blood capillaries, a feature that is key to their ability to absorb the digestive products of fats as well as excess interstitial fluid. The lymphatic capillaries drain into larger lymphatic vessels, which merge into progressively larger tubes with thicker walls. Lymph is eventually returned to the circulatory system through one of two large ducts that join with the large veins at the base of the neck.

With no pump to drive it, lymph flows slowly through the lymphatic vessels, propelled by the same forces that move blood through the veins. That is, the contractions of nearby skeletal muscles compress the lymphatic vessels, pushing the lymph along. One-way valves similar to those in veins prevent backflow. Pressure changes in the thorax (chest cavity) that accompany breathing also help pull the lymph upward from the lower body. Gravity assists the flow from the upper body.

The lymphatic vessels are studded with **lymph nodes**, small bean-shaped structures that cleanse the lymph as it slowly filters through. The lymph nodes contain macrophages and lymphocytes, white blood cells that play an essential role in the body's defense system. Macrophages engulf bacteria, cancer cells, and other debris, clearing them from the lymph. Lymphocytes serve as the surveillance squad of the immune system. They are continuously on the lookout for specific disease-causing invaders, as we will see in Chapter 13. Swollen and painful lymph nodes are a symptom of infection.

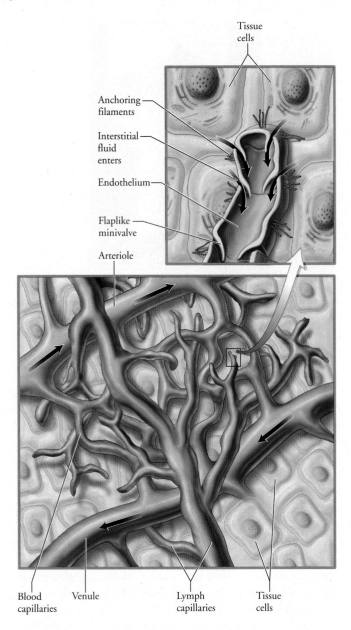

FIGURE **12.20**   The lymphatic capillaries are microscopic, blind-ended tubules through which surplus tissue fluid enters the lymphatic system to be returned to the bloodstream.

Besides the lymph nodes, there are several other *lymphoid organs* (Figure 12.21). Among these are the *tonsils*, which form a ring around the entrance to the throat, where they help protect against disease organisms that are inhaled or swallowed. The *thymus gland*, located in the chest, is another lymphoid organ. It plays its part during early childhood by processing certain lymphocytes that protect us from specific disease-causing organisms. On the left side of the abdominal region is the largest lymphoid organ, the *spleen*. In addition to serving as a birthplace for lymphocytes, the spleen clears the blood of old and damaged red blood cells and platelets. Finally, isolated clusters of lymph nodules along the small intestine, known as *Peyer's patches*, keep bacteria from breaching the intestinal wall. The *red bone marrow*, where white blood cells and other formed elements are produced, is a lymphoid organ found in the ends of long bones, ribs, sternum, and vertebrae.

stop and think

Cancer cells that break loose from their original site in a process called metastasis, have easy access to the highly permeable lymphatic capillaries. The lymphatic vessels then provide a route by which the cancer cells may spread to nearly every part of the body. Explain why the lymph nodes are often examined to determine whether cancer has spread. Why are the lymph nodes near the original cancer site often removed?

HIGHLIGHTING THE CONCEPTS

The Cardiovascular System Consists of the Blood Vessels and the Heart (p. 212)

1. The heart serves as a pump that pushes blood through blood vessels.
   WEB TUTORIAL 12.1   The Cardiovascular System

The Blood Vessels Conduct Blood in Continuous Loops (pp. 212–218)

2. Blood circulates through a branching network of blood vessels in a path that travels from the heart to arteries to arterioles to capillaries to venules to veins and then back to the heart.

3. Arteries are elastic, muscular tubes that carry blood away from the heart. By stretching and then returning to their original shape, they withstand the high pressure of blood as it is pumped from the heart. These changes help maintain a relatively even blood pressure within the arteries in spite of large changes in the volume of blood within them.

4. The pressure change along an artery as it expands and recoils is called a pulse.

5. Arteries branch to form narrower tubules called arterioles. Arterioles are important in the regulation of blood pressure, and they regulate blood flow through capillary beds.

6. The exchange of materials between the blood and tissues takes place across the thin walls of capillaries. Capillaries are arranged in highly branched networks that provide a tremendous surface area for the exchange. Each network of capillaries is called a capillary bed. Rings of muscle called precapillary sphincters determine whether blood flows into a capillary bed or is channeled through it.

7. Capillaries merge to form venules, and these, in turn, merge to form veins, which conduct blood back to the heart. Blood is returned to the heart against gravity by nearby skeletal muscles that contract and

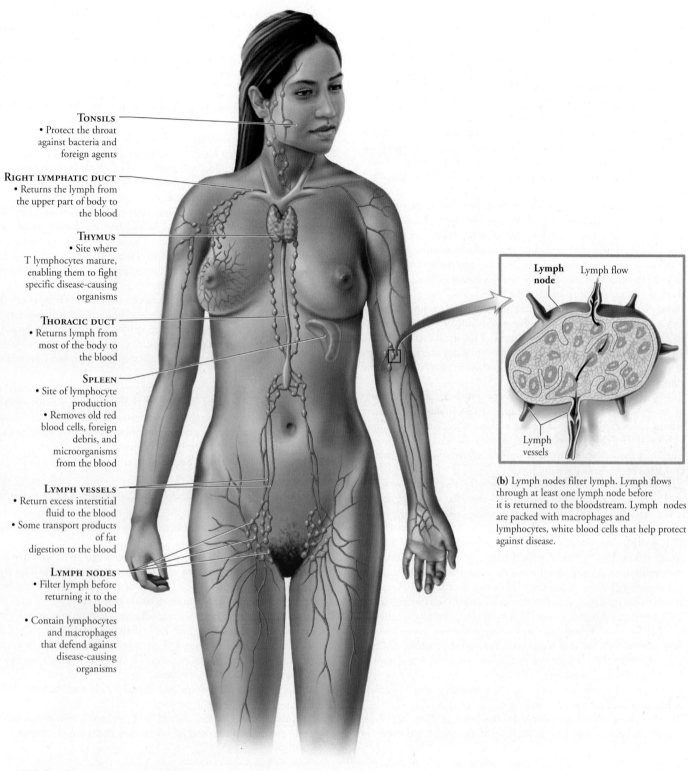

Tonsils
- Protect the throat against bacteria and foreign agents

Right lymphatic duct
- Returns the lymph from the upper part of body to the blood

Thymus
- Site where T lymphocytes mature, enabling them to fight specific disease-causing organisms

Thoracic duct
- Returns lymph from most of the body to the blood

Spleen
- Site of lymphocyte production
- Removes old red blood cells, foreign debris, and microorganisms from the blood

Lymph vessels
- Return excess interstitial fluid to the blood
- Some transport products of fat digestion to the blood

Lymph nodes
- Filter lymph before returning it to the blood
- Contain lymphocytes and macrophages that defend against disease-causing organisms

Lymph node   Lymph flow

Lymph vessels

(b) Lymph nodes filter lymph. Lymph flows through at least one lymph node before it is returned to the bloodstream. Lymph nodes are packed with macrophages and lymphocytes, white blood cells that help protect against disease.

(a) The lymphatic system returns the fluid to the bloodstream that previously left the capillaries to bathe the cells, protects against disease-causing organisms, and transports products of fat digestion from the small intestine to the bloodstream.

FIGURE **12.21** The lymphatic system is a system of lymphatic vessels containing a clear fluid, called lymph, and various lymphatic tissues and organs located throughout the body.

push the blood along within the veins. Valves within the veins prevent blood from flowing backward when the skeletal muscles are relaxed. Pressure differences generated by breathing also help draw blood toward the heart from the lower torso.

The Heart Is a Muscular Pump (pp. 218–224)

8. Every minute, the heart beats about 72 times and moves slightly less than 5 liters (10 pt) of blood through its chambers.

9. Most of the wall of the heart is composed of cardiac muscle and is called the myocardium. The heart is enclosed in a double-layered sac called the pericardium, which allows the heart to beat while still confining it near the midline of the thoracic cavity. The endocardium is a thin inner lining.

10. The right and the left halves of the heart function as two separate pumps. Each side consists of two chambers: an upper chamber called the atrium and a lower, thick-walled chamber called the ventricle. The smaller, thin-walled atria function primarily as receiving chambers that accept blood returning to the heart and pump it a short distance to the ventricles. When the ventricles contract, they push blood through the arteries to all parts of the body.

11. Blood circulates in one direction through the heart because of the action of two pairs of valves. The atrioventricular (AV) valves are located between each atrium and ventricle. The semilunar valves are located between each ventricle and its connecting artery. The heart sounds, lub-dup, are caused by blood turbulence associated with the closing of the heart valves.

12. The right side of the heart pumps blood to the lungs through a loop of vessels called the pulmonary circuit. The left side of the heart pumps blood to all parts of the body except the lungs through a loop of vessels called the systemic circuit.

13. The heart has its own network of vessels, called the coronary circuit, that services the heart tissue itself.

14. Each heartbeat consists of contraction (systole) and relaxation (diastole). The atria contract in unison, and then the ventricles do the same. The events associated with each heartbeat are collectively called the cardiac cycle.

15. The rhythmic contraction of the heart is produced by its internal conduction system. A cluster of specialized cardiac muscle cells, called the sinoatrial (SA) node, usually sets the tempo of the heartbeat and is therefore called the pacemaker. When the electrical signal reaches another cluster of specialized muscle cells, called the atrioventricular (AV) node, the stimulus is quickly relayed along the atrioventricular bundle that runs through the wall between the two ventricles and then fans out into the ventricular walls through the Purkinje fibers.

16. The rate of heartbeat is regulated by the autonomic nervous system to suit the body's level of activity.

17. An electrocardiogram (ECG or EKG) is a recording of the electrical events associated with each heartbeat.

18. Blood pressure, the force created by the heart to drive the blood around the body, is measured as the force of blood against the walls of blood vessels. The blood pressure in an artery peaks when a ventricle contracts. This is called the systolic pressure. In contrast, the lowest blood pressure of each cardiac cycle, the diastolic pressure, occurs while the heart is relaxing between contractions. Blood pressure is measured with a device called a sphygmomanometer.

Cardiovascular Disease Is a Major Killer in the United States (pp. 225–229)

19. Cardiovascular diseases are those that affect the heart and blood vessels.

20. Hypertension (high blood pressure) is called a silent killer because it does not always produce symptoms and can have fatal consequences, involving the heart, brain, blood vessels, or kidneys. The heart enlarges, but it works less efficiently.

21. Atherosclerosis is an inflammatory response associated with a buildup of fatty deposits within artery walls. The fatty material is covered with a fibrous cap and is called plaque. The plaque can bulge into the artery channel, restricting blood flow. Plaque that does not bulge into the artery may still be vulnerable to rupture, triggering blood clot formation that can block blood flow.

22. When the vessel affected by atherosclerosis is a coronary artery that supplies blood to heart tissues, the condition is known as coronary artery disease. A temporary reduction in blood supplied to the heart (ischemia) is accompanied by chest pain, called angina pectoris. Angina is a warning of a heart problem, but it does not cause permanent damage to the heart.

23. A region of heart muscle dies during a heart attack (myocardial infarction). In a surviving patient, the dead cells are gradually replaced by scar tissue. A heart attack may be caused by blocked blood flow in a coronary artery due to a blood clot or atherosclerosis. Another cause of a heart attack is a coronary artery spasm.

24. In heart failure, the heart becomes an inefficient pump.

The Lymphatic System Functions in the Circulatory and Immune Systems (pp. 229–231)

25. The lymphatic system consists of lymph, lymphatic vessels, lymphoid tissue, and lymphoid organs.

26. Three vital functions of the lymphatic system are to return interstitial fluid to the bloodstream, to transport products of fat digestion from the digestive system to the bloodstream, and to defend the body against disease-causing organisms or abnormal cells.

27. Tissue fluid enters lymphatic capillaries—microscopic tubules that end blindly and are more permeable than blood capillaries. The fluid, then called lymph, is moved along larger lymphatic vessels by contraction of nearby skeletal muscles. The lymphatic vessels have valves to prevent the backflow of lymph.

28. Lymph nodes filter lymph and contain cells that actively defend against disease-causing organisms.

29. Lymphoid organs include the red bone marrow, lymph nodes, tonsils, thymus gland, spleen, and Peyer's patches of the small intestine.

KEY TERMS

cardiovascular system *p. 212*
heart *p. 212*
artery *p. 212*
pulse *p. 212*
arteriole *p. 215*
blood pressure *p. 215*
capillary *p. 215*
capillary bed *p. 215*
venule *p. 216*

vein *p. 216*
myocardium *p. 218*
atrium *p. 218*
ventricle *p. 218*
atrioventricular (AV) valve *p. 218*
semilunar valves *p. 218*
pulmonary circuit *p. 218*
systemic circuit *p. 218*
coronary circulation *p. 220*

systole *p. 221*
diastole *p. 221*
cardiac cycle *p. 221*
sinoatrial (SA) node *p. 221*
pacemaker *p. 221*
atrioventricular (AV) node *p. 221*
electrocardiogram (ECG or EKG) *p. 222*
systolic pressure *p. 224*

diastolic pressure *p. 224*
hypertension *p. 225*
coronary artery disease (CAD) *p. 227*
lymphatic system *p. 229*
lymph *p. 229*
lymphatic vessel *p. 229*
circulatory system *p. 229*
lymph nodes *p. 229*

REVIEWING THE CONCEPTS

1. Trace the flow of blood from the heart and back to it by naming, in order, the general types of vessels through which the blood flows. *p. 212*
2. What is a pulse? *pp. 215–213*
3. What is an aneurysm? Whay is it dangerous? *pp. 214–215*
4. What are two important functions of arterioles? *p. 215*
5. What determines whether blood flows through any particular capillary bed? *p. 215*
6. Compare the structure of arteries, capillaries, and veins. Explain how the structure is suited to the function of each type of vessel. *pp. 212–216*
7. Explain how blood is returned to the heart from the lower torso against the force of gravity. *p. 217*
8. Describe the structure of the heart valves. Explain how they function. *p. 218*
9. Describe the structure of the heart. Explain how it functions as two separate pumps. *pp. 218–219*
10. Trace the path of blood from the left ventricle to the left atrium, naming each major vessel associated with the heart and the heart chambers in the correct sequence. *pp. 218–219*
11. Describe the cardiac cycle. *pp. 220–221*
12. Explain how clusters or bundles of specialized cardiac muscle cells coordinate the contraction associated with each heartbeat. *pp. 221–222*
13. What is hypertension? In what ways does it damage the body? *p. 225*
14. What is atherosclerosis? Why is it harmful? How does it develop? *pp. 225–227*
15. Explain the relationships between coronary artery disease, angina pectoris, and heart attack. *pp. 227–229*
16. What is a heart attack? What are the common causes? *pp. 228–229*
17. List three important functions of the lymphatic system. *p. 229*
18. Compare the structure of lymphatic capillaries with the structure of blood capillaries. How does the structure of lymphatic capillaries allow them to absorb tissue fluid? *p. 229*
19. What is the function of lymph nodes? *p. 229–230*
20. The semilunar valves prevent the backflow of blood from the
    a. arteries to the ventricles.
    b. veins to the atria.
    c. ventricles to the atria.
    d. arteries to the atria
21. If you cut all the nerves to the heart but kept the heart alive,
    a. the heart would stop beating.
    b. the heart would continue beating.
    c. only systole would occur.
    d. only diastole would occur.

For questions 22 and 23, imagine that you have been miniaturized and are riding through the circulatory system using a red blood cell as a life raft.

22. You are in the big toe traveling toward the heart. The last vessel you pass through before entering the heart is the
    a. aorta.
    b. inferior vena cava.
    c. coronary artery.
    d. pulmonary vein.
23. You are nearly deafened by the first heart sound, which is caused by
    a. the opening of the atrioventricular valves.
    b. the closing of the atrioventricular valves.
    c. the opening of the semilunar valves.
    d. the closing of the semilunar valves.
24. The _____ is a cluster of specialized heart muscle cells that determines the heart rate by initiating each cardiac cycle.
25. A weakened area on an artery wall that can balloon outward is called _____.
26. Oxygen and nutrients move through the walls of _____ to reach the body cells.

APPLYING THE CONCEPTS

1. Ying is a 75-year-old woman who broke her hip and had difficulty walking for several weeks. She passed the time watching television. After her hip healed, she suddenly had a heart attack while food shopping. The doctor said that the heart attack was caused by a blood clot in a leg vein that was swept away with the flow of blood. Explain the connection between these events.
2. Amelia is a friend of yours. When you call her to ask about dinner plans, she tells you that she is not feeling well. She aches all over and has swollen "glands" in her neck. You explain that these are not really glands, because they do not secrete anything. What are they? Why are they swollen?

Additional questions can be found on the companion website.

3. Nicotine, a substance in cigarette smoke, causes vasoconstriction (narrowing of certain blood vessels) and increases heart rate. Explain how these effects lead to high blood pressure. Explain why cigarette smokers are more likely to die of cardiovascular diseases than are nonsmokers.
4. Abnormally short (or long) chordae tendineae of the mitral (bicuspid) valve can cause a condition known as mitral valve prolapse, in which the mitral valves do not close properly. Why would mitral valve prolapse cause the heart to produce abnormal heart sounds?

13

Body Defense Mechanisms

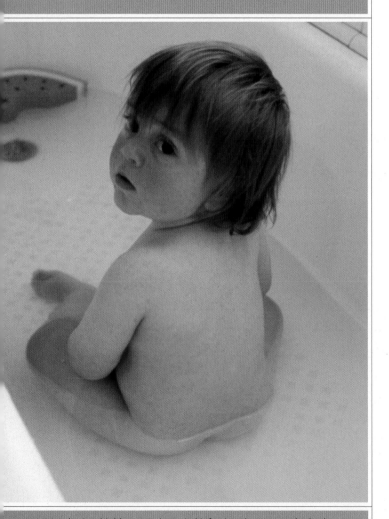

Measles is a highly contagious viral infection that enters the body through the respiratory system. There is no specific treatment for measles (although it can be prevented with a vaccine), but if this child has a well-functioning immune system, she will almost certainly recover.

The Body's Defense System Targets Pathogens and Cancerous Cells

The Body Has Three Lines of Defense

- The first line of defense consists of physical and chemical barriers that prevent entry of pathogens
- The second line of defense includes defensive cells and proteins, inflammation, and fever
- The third line of defense, the immune response, has specific targets and memory

The Immune System Distinguishes Self from Nonself

The Immune System Mounts Antibody-Mediated Responses and Cell-Mediated Responses

The Cell-Mediated Immune Response and the Antibody-Mediated Immune Response Have the Same Steps

- B cells mount an antibody-mediated immune response against antigens free in the blood or bound to a cell surface
- Cytotoxic T cells mount a cell-mediated defense against antigen-bearing cells
- Immunological memory permits a more rapid response on subsequent exposure to the antigen
- Suppressor T cells turn off the immune response

Immunity Can Be Active or Passive

Monoclonal Antibodies Are Used in Research, Clinical Diagnosis, and Disease Treatment

The Immune System Can Cause Problems

- Autoimmune disorders occur when the immune system attacks the body's own cells
- Allergies are immune responses to harmless substances

HEALTH ISSUE The Fear That Vaccines Cause Autism

HEALTH ISSUE Rejection of Organ Transplants

Sarah was eating dinner with her family when she started to feel slightly nauseous. "Feeling like this on top of my headache is all I need," she moaned. Her mother, placing her palm on Sarah's forehead, said, "Oh dear, you have a fever. Let's get you into bed." On the way up the stairs, Sarah felt weak, as if her muscles were protesting even this slight exertion. She briefly wondered if she might have the flu. The next day her symptoms worsened and a rash appeared on her arm. It was clearly time for a visit to the doctor. Fortunately, the family physician was well aware that Colorado had been experiencing an increase in cases of West Nile virus that summer. She ordered diagnostic tests, which determined that Sarah had indeed contracted the disease. The virus had entered Sarah's body by way of a bite from a mosquito that had previously fed on the blood of an infected bird. Sarah's symptoms disappeared within a couple of days, as is usually the case with West Nile virus patients who exhibit symptoms. (Most do not show symptoms.) In the worst cases, the brain becomes inflamed and the result may be death, particularly in people over 60 years of age.

In this chapter, we will study how the body reacts to an invasion such as the West Nile virus and many other organisms and substances that it perceives as threats. We will see that there are three lines of defense. We will also learn that the body can acquire long-lasting resistance to a microbe by becoming ill or by being immunized. Finally, we will consider some potential problems caused by the immune system. ■

The Body's Defense System Targets Pathogens and Cancerous Cells

Your body generally defends you against anything that it does not recognize as being part of or belonging inside you. Common targets of your defense system include organisms that cause disease or infection and body cells that have turned cancerous.

The bacteria, viruses, protozoans, fungi, parasitic worms, and prions (infectious proteins) that cause disease are called **pathogens** (discussed in Chapter 13a). Note that this term does not apply to most of the microorganisms we encounter. Many bacteria, for example, are actually beneficial. They flavor our cheese, help rid the planet of corpses through decomposition, and help keep other, potentially harmful bacteria in check within our bodies.

Cancerous cells also threaten our well-being. A cancer cell was once a normal body cell, but because of changes in its genes, it can no longer regulate its cell division. These renegade cells can multiply until they take over the body, upsetting its balance, choking its pathways, and ultimately causing great pain and sometimes death.

The Body Has Three Lines of Defense

The body's "strategy" for defending against foreign organisms, cells, or molecules has three components.

1. **Keep the foreign cells or molecules out of the body.** This is accomplished by the first line of defense—*chemical and physical surface barriers.*

2. **Attack *any* foreign cell or molecule that enters the body.** The second line of defense consists of *internal cellular and chemical defenses* that become active if the surface barriers are penetrated.

3. **Destroy the *specific* type of foreign cell or molecule that enters the body.** The third line of defense is the *immune response*, which destroys *specific* targets, usually disease-causing organisms, and remembers those targets so that a quick response can be mounted if that target enters the body again.

Thus, the first and second lines of defense consist of nonspecific mechanisms that are effective against *any* foreign organisms or substances. The third line of defense, the immune response, is a specific mechanism of defense. The three lines of defense against pathogens are summarized in Figure 13.1.

■ **The first line of defense consists of physical and chemical barriers that prevent entry of pathogens**

The skin and mucous membranes that form the first line of defense are physical barriers that help keep foreign substances from entering the body. In addition, they produce several protective chemicals.

PHYSICAL BARRIERS

Like a suit of armor, unbroken skin helps shield the body from the hostile world by providing a barrier to foreign substances (Figure 13.2). A layer of dead cells forms the tough outer layer of skin. These cells are filled with the fibrous protein keratin, which waterproofs the skin and makes it resistant to the disruptive toxins (poisons) and enzymes of most would-be invaders. Some of the strength of this barrier results from the tight connections binding the cells together. What is more, the dead cells are continuously shed and replaced, at the rate of about a million cells every 40 minutes. As dead cells flake off, they take with them any microbes that have somehow managed to latch on. Another physical barrier, the

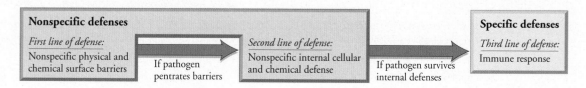

FIGURE **13.1** The body's three lines of defense against pathogens.

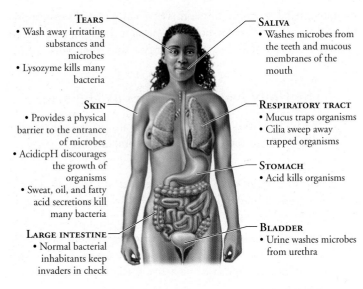

TEARS
- Wash away irritating substances and microbes
- Lysozyme kills many bacteria

SKIN
- Provides a physical barrier to the entrance of microbes
- AcidicpH discourages the growth of organisms
- Sweat, oil, and fatty acid secretions kill many bacteria

LARGE INTESTINE
- Normal bacterial inhabitants keep invaders in check

SALIVA
- Washes microbes from the teeth and mucous membranes of the mouth

RESPIRATORY TRACT
- Mucus traps organisms
- Cilia sweep away trapped organisms

STOMACH
- Acid kills organisms

BLADDER
- Urine washes microbes from urethra

FIGURE **13.2** The body's first line of defense consists of physical and chemical barriers that serve as nonspecific defenses against *any* threats to our well-being. Collectively, they prevent many invading organisms and substances from entering the body, or confine them to a local region, kill them, remove them, or slow their growth.

mucous membranes lining the digestive and respiratory passages, produce sticky mucus that traps many microbes.

CHEMICAL BARRIERS

The skin also provides chemical protection against invaders. Sweat and oil produced by glands in the skin wash away microbes. Moreover, the acidity of the secretions slows bacterial growth, and the oils contain chemicals that kill some bacteria.

There are other chemical barriers. The lining of the stomach produces hydrochloric acid and protein-digesting enzymes that destroy many pathogens. The acidity of urine slows bacterial growth. (Urine is also a physical barrier, periodically flushing microbes from the lower urinary tract.) Saliva and tears contain an enzyme called *lysozyme* that kills some bacteria by disrupting their cell walls.

■ The second line of defense includes defensive cells and proteins, inflammation, and fever

The second line of defense consists of nonspecific internal defenses against any pathogen that breaks through the physical and chemical barriers and enters the body. It includes defensive cells and proteins, inflammation, and fever, as summarized in Table 13.1.

DEFENSIVE CELLS

PHAGOCYTES Specialized "scavenger" cells called **phagocytes** (*phage*, to eat; *cyte*, cell) engulf pathogens, damaged tissue, or dead cells by the process of phagocytosis (Chapter 3). This class of white blood cells serves as the front-line soldiers in the body's internal defense system, but also as janitors that clean up debris. When a phagocyte encounters a foreign particle, cytoplasmic extensions flow from the phagocytic cell, bind to the particle, and pull it inside the cell (Figure 13.3). Once inside the cell, the particle is enclosed within a membrane-bound vesicle and quickly destroyed by digestive enzymes.

The body has several types of phagocytes. One type, *neutrophils*, arrives at the site of attack before the other types of white blood cells and immediately begins to consume the pathogens, especially bacteria, by phagocytosis. Other white blood cells (monocytes) leave the vessels of the circulatory system and enter the tissue fluids, where they develop into large **macrophages** (*macro*, big; *phage*, to eat). Macrophages have hearty and less discriminating appetites than neutrophils and attack and consume virtually anything that is not recognized as belonging in the body—including viruses, bacteria, and damaged tissue.

EOSINOPHILS A second type of white blood cell, *eosinophils*, attack pathogens that are too large to be consumed by phagocytosis, such as parasitic worms. Eosinophils get close to the parasite and discharge enzymes that destroy the organism.

NATURAL KILLER CELLS

A third type of white blood cell, called **natural killer (NK) cells**, roams the body in search of abnormal cells and quickly orchestrate their death. In a sense, NK cells function as the body's police walk-

TABLE 13.1	THE SECOND LINE OF DEFENSE—NONSPECIFIC INTERNAL DEFENSES	
DEFENSE	**EXAMPLE**	**FUNCTION**
Defensive cells	Phagocytic cells such as neutrophils and macrophages	Engulf invading organisms
	Eosinophils	Kill parasites
	Natural killer cells	Kill many invading organisms and cancer cells
Defensive proteins	Interferons	Slow the spread of viruses in the body
	Complement system	Stimulates histamine release; promotes phagocytosis; kills bacteria; enhances inflammation
Inflammation	Widening of blood vessels and increased capillary permeability, leading to redness, heat, swelling, and pain	Brings in defensive cells and speeds healing
Fever	Abnormally high body temperature	Slows the growth of bacteria; speeds up body defenses

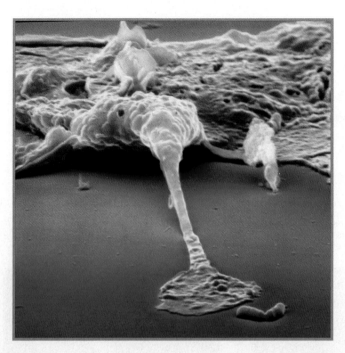

FIGURE **13.3** A macrophage ingesting bacteria (the rod-shaped structure). The bacteria will be pulled inside the cell within a membrane-bound vesicle and quickly killed.

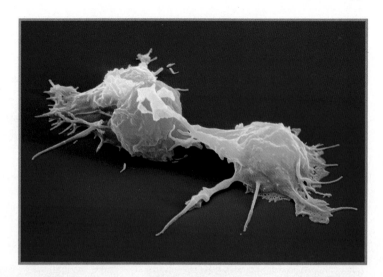

FIGURE **13.4** Natural killer cells (shown in orange) attacking a leukemia cell (shown in red). NK cells patrol the body, bumping and touching other cells as they go. When NK cells contact a cell with an altered cell surface, such as a cancer cell or a virus-infected cell, a series of events is immediately initiated. The NK cell attaches to the target cell and releases proteins that create pores in the target cell, making the membrane leaky and causing the cell to burst.

ing a beat. They are not seeking a specific villain. Instead, they respond to any suspicious character, which, in this case, is a cell whose cell membrane has been altered by the addition of proteins that are unfamiliar to the NK cell. The prime targets of NK cells are cancerous cells and cells infected with viruses. Cancerous cells routinely form but are quickly destroyed by NK cells and prevented from spreading (Figure 13.4).

As soon as it touches a cell with an abnormal surface, the NK cell attaches to the abnormal cell and delivers a "kiss of death" in the form of proteins that create numerous pores in the target cell. The pores make the target cell "leaky" so that it can no longer maintain a constant internal environment and eventually bursts.

DEFENSIVE PROTEINS

The second line of defense also includes defensive proteins. We will discuss two types of defensive proteins: interferons, which slow viral reproduction, and the complement system, which assists other defensive mechanisms.

INTERFERONS   A cell that has been infected with a virus can do little to help itself. But cells infected with a virus can help cells that are not yet infected. Before certain virally infected cells die, they secrete small proteins called **interferons** that act to slow the spread of viruses already in the body. As the name implies, interferons interfere with viral activity.

Interferons mount a two-pronged attack. First, they help rid the body of virus-infected cells by attracting macrophages and NK cells that destroy the infected cells immediately. Second, interferons protect cells that are not yet infected with the virus. When released, an interferon diffuses to neighboring cells and stimulates them to produce proteins that prevent viruses from replicating in those

cells. Because viruses cause disease by replicating inside body cells, preventing replication curbs the disease. Interferon helps protect uninfected cells from *all* strains of virus, not just the one responsible for the initial infection.

Pharmaceutical preparations of interferon have been shown to be effective against certain cancers and viral infections. For instance, interferon is often successful in combating a rare form of leukemia (hairy cell leukemia) and Kaposi's sarcoma, a form of cancer that often occurs in people with AIDS. Interferon has also been approved for treating the hepatitis C virus, which can cause cirrhosis of the liver and liver cancer; the human papilloma virus, which causes genital warts; and the herpes virus, which causes genital herpes.

Researchers have discovered a new antiviral protein secreted by certain cells of the immune system. The protein was discovered by studying elite controllers, people who are HIV$^+$ but are able to keep HIV at barely detectable levels without antiviral drugs. Researchers have not yet identified this antiviral protein, but they have narrowed the possibilities to about 80 proteins. They also suspect that elite controllers produce more interferon than healthy HIV-negative people.

THE COMPLEMENT SYSTEM   The **complement system**, or simply *complement*, is a group of at least 20 proteins whose activities enhance, or complement, the body's other defense mechanisms. These proteins enhance both nonspecific and specific defense mechanisms. The effects of complement include the following:

- **Destruction of pathogen** Complement can act *directly*, by punching holes in a target cell's membrane (Figure 13.5), so that the cell is no longer able to maintain a constant internal environment. Water enters the cell, causing it to burst, just as

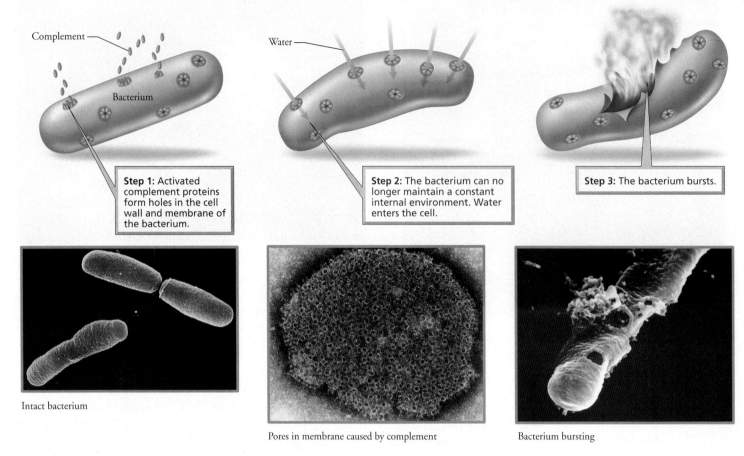

Step 1: Activated complement proteins form holes in the cell wall and membrane of the bacterium.

Step 2: The bacterium can no longer maintain a constant internal environment. Water enters the cell.

Step 3: The bacterium bursts.

Intact bacterium

Pores in membrane caused by complement

Bacterium bursting

FIGURE **13.5** Complement has a direct destructive effect on pathogens.

when NK cells secrete proteins that make a target cell's membrane leaky.

- **Enhancement of phagocytosis** Complement enhances phagocytosis in two ways. First, complement proteins attract macrophages and neutrophils to the site of infection to remove the foreign cells. Second, one of the complement proteins binds to the surface of the microbe, making it easier for macrophages and neutrophils to "get a grip" on the intruder and devour it.

- **Stimulation of inflammation** (see below) Complement also causes blood vessels to widen and become more permeable. These changes provide increased blood flow to the area and increased access for white blood cells.

INFLAMMATION

When body tissues are injured or damaged, a series of events called the **inflammatory response** or reaction occurs. This response destroys invaders and helps repair and restore damaged tissue. The four cardinal signs of inflammation that occur at the site of a wound are redness, heat (or warmth), swelling, and pain. These signs announce that certain cells and chemicals have combined efforts to contain infection, clean up the damaged area, and heal the wound. Let's consider the causes of the cardinal signs and how they are related to the benefits of inflammation (Figure 13.6).

REDNESS Redness occurs because blood vessels dilate (widen) in the damaged area, causing blood flow there to increase. The dilation is caused by **histamine**, which is released by small mobile connective tissue cells called **mast cells** and by *basophils,* a type of white blood cell, in response to chemicals from damaged cells.

The increased blood flow to the site of injury delivers phagocytes, blood-clotting proteins, and defensive proteins, including complement and antibodies (discussed later in the chapter). At the same time, the increased blood flow washes away dead cells and toxins produced by the invading microbes.

HEAT The increased blood flow also elevates the temperature in the area of injury. The elevated temperature increases the metabolic rate of the body cells in the region and speeds healing. Heat also increases the activities of phagocytic cells and other defensive cells.

SWELLING The injured area swells because histamine also makes capillaries more permeable, or leakier, than usual. Fluid seeps into the tissues from the bloodstream, bringing with it many beneficial substances. Blood-clotting factors enter the injured area and begin to wall off the region, thereby helping to protect surrounding areas from injury and preventing the loss of blood.

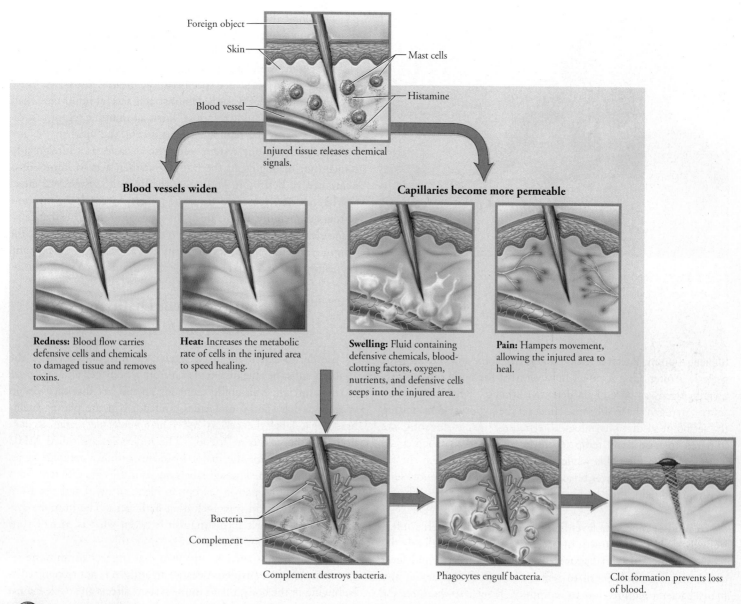

WEB
TUTORIAL
13.1 FIGURE **13.6** The inflammatory response is a general response to tissue injury or invasion by foreign microbes. It serves to defend against pathogens and to clear the injured area of pathogens and dead body cells, allowing repair and healing to occur. The four cardinal signs of inflammation are redness, heat, swelling, and pain.

The seepage also increases the oxygen and nutrient supply to the cells. If the injured area is a joint, swelling can hamper movement, an effect that might seem to be an inconvenience, but it permits the injured joint to rest and recover.

PAIN   There are several causes for the pain in an inflamed area. For example, the excessive fluid that has leaked into the tissue presses on nerves and contributes to the sensation. Some soreness might be caused by bacterial toxins. Injured cells also release pain-causing chemicals, such as prostaglandins. Pain usually causes a person to protect the area to avoid additional injury.

As a consequence of the inflammatory response, phagocytes begin to swarm to the injured site, attracted by chemicals released when tissue is damaged. Within minutes, the neutrophils squeeze through capillary walls into the fluid around cells and begin engulfing pathogens, toxins, and dead body cells. Soon macrophages arrive and continue the body's counterattack for the long term. Macrophages are also important in cleaning debris, such as dead body cells, from the damaged area. As the recovery from infection continues, dead cells, including microbes, body tissue cells, and phagocytes, may begin to ooze from the wound as pus.

FEVER

A *fever* is an abnormally high body temperature (Figure 13.7). Fevers are caused by *pyrogens* (*pyro*, fire; *gen*, producer), chemicals that set the "thermostat" in the brain (the hypothalamus) to a

FIGURE **13.7** Although a fever might make us feel uncomfortable, it can help the body fight disease.

higher set point. Bacteria release toxins that sometimes act as pyrogens. It is interesting to note, however, that the body produces its own pyrogens as part of its defensive strategy. Regardless of the source, pyrogens have the same effect on the hypothalamus, raising the set point so that physiological responses, such as shivering, are initiated to raise body temperature (as discussed in Chapter 4). Thus, we have the chills while the fever is rising. When the set point is lowered, the fever breaks and physiological responses such as perspiring reduce the body temperature until it reaches the new set point.

A mild or moderate fever helps the body fight bacterial infections by slowing the growth of bacteria and stimulating body defense responses. Bacterial growth is slowed because a mild fever causes the liver and spleen to remove iron from the blood, and many bacteria require iron to reproduce. Fever also increases the metabolic rate of body cells; the higher rate speeds up defensive responses and repair processes. On the other hand, a very high fever (over 105°F, 40.6°C) is dangerous. It can inactivate enzymes needed for biochemical reactions within body cells.

■ **The third line of defense, the immune response, has specific targets and memory**

When the body's first and second lines of defense fail, the body's specific defenses respond and target the particular pathogen or foreign molecule that has entered the body. The third line of defense, the **immune system**, provides the specific responses. The organs of the lymphatic system (see Chapter 12) are important components of the immune system because they produce the various cells responsible for immunity. However, the immune system is not an organ system in an anatomical sense. Instead, the immune system is defined by its *function*: the recognition and destruction of specific pathogens or foreign molecules. The body's specific defenses working together are called an *immune response*.

Let's look at an example. You may have noticed that after you recover from a certain disease, say measles, you are not likely to get it again. You might get chickenpox or the flu afterward, but you will not get measles. You are *immune* to measles, meaning that you have long-lasting resistance specific to the measles virus.

This simple observation demonstrates several important characteristics of an immune response. First, an immune response is directed at a particular pathogen, in this case, the virus that causes measles. The immune system recognizes measles as a foreign substance (not belonging in the body) and then acts to immobilize, neutralize, or destroy it. You know the immune system was effective because you recovered from the measles. Second, the immune system has memory. If you are again exposed to the virus that causes measles 20 years after you recovered from the original illness, the immune system remembers the virus and attacks it so quickly and vigorously that you will not become ill with measles a second time.

The Immune System Distinguishes Self from Nonself

In order to defend against a foreign organism or molecule, the body must be able distinguish it from a body cell and recognize it as foreign. This ability depends on the fact that each cell in your body has special molecules embedded in the plasma membrane that label the cell as *self*. These molecules serve as flags declaring the cell as a "friend." The molecules are called **MHC markers**, named for the major histocompatibility complex genes that code for them. The self labels on your cells are different from those of any other person (except an identical twin) and also from those of other organisms, including pathogens. The immune system uses these labels to distinguish between what is part of your body and what is not (Figure 13.8).

A nonself substance or organism that triggers an immune response is called an **antigen**. Because an antigen is not recognized as belonging in the body, the immune system directs an attack against it. Typically, antigens are large molecules, such as proteins, polysaccharides, or nucleic acids. Often, antigens are found on the surface of an invader—embedded in the plasma membrane of a unwelcome bacterial cell, for instance, or part of the protein coat of a virus. However, pieces of invaders and chemicals secreted by invaders, such as bacterial toxins, can also serve as antigens. Each antigen is recognized by its shape.

Certain white blood cells, called lymphocytes, are responsible for both the specificity and the memory of the immune response. There are two principal types of lymphocytes: **B lymphocytes**, or more simply *B cells*, and **T lymphocytes**, or *T cells*. Both types form in the bone marrow, but they mature in different organs of the body. It is thought that B cells mature in the bone marrow. The T cells, on the other hand, mature in the thymus gland, which overlies the heart.

As the T lymphocytes mature, they develop the ability to distinguish between cells that belong in the body and those that do not. The T cells must be able to recognize the specific MHC self markers of that person and *not* respond vigorously to cells bearing

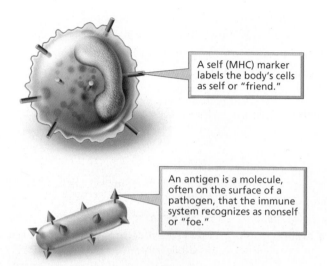

A self (MHC) marker labels the body's cells as self or "friend."

An antigen is a molecule, often on the surface of a pathogen, that the immune system recognizes as nonself or "foe."

FIGURE **13.8** All nucleated cells in the body have molecular MHC markers on their surface that label them as self. Foreign substances, including potential disease-causing organisms, have molecules on their surfaces that are not recognized as belonging in the body. Foreign molecules that are capable of triggering an immune response are called antigens.

that MHC self marker. If T cells do respond to cells with those self markers, they are destroyed. Mature T lymphocytes then circulate through the body, bumping into other cells and checking to be sure those cells have the correct self (MHC) marker. Cells with proper MHC markers are accepted. However, any substance or organism that lacks a proper self label is considered to be nonself, or "foe," and the immune system will target it for destruction.

In addition, both T and B lymphocytes, as they mature, are programmed to recognize one particular type of antigen (that is, each cell learns to recognize a different antigen). This recognition is the basis of the specificity of the immune response. Each lymphocyte develops its own particular receptors—molecules having a unique shape—on its surface. Thousands of *identical* receptor molecules pepper the surface of each lymphocyte, and they are unlike the receptor molecules on other lymphocytes. When an antigen fits into a lymphocyte's receptors, the body's defenses target that particular antigen. Because of the tremendous diversity of receptor molecules, each type occurring on a different lymphocyte, a few of the billions of lymphocytes in your body are able to respond to each of the thousands of different antigens that you will be exposed to in your lifetime.

When an antigen is detected, B cells and T cells bearing receptors able to respond to that particular invader are stimulated to divide repeatedly, forming two lines of cells. One line of descendant cells is made up of *effector cells*, which carry out the attack on the enemy. Effector cells generally live for only a few days. Thus, after the invader has been eliminated from the body, the number of effector cells declines. The other line of descendant cells is composed of *memory cells*, long-lived cells that "remember" that particular invader and mount a rapid, intense response to it if it should ever appear again. The quick response of memory cells is the mechanism that prevents you from getting the same illness twice.

The Immune System Mounts Antibody-Mediated Responses and Cell-Mediated Responses

There are some striking similarities between the body's immune defenses and a nation's military defense system. The military has scouts who look for invaders. If an invader is found, the scout alerts the commander-in-chief of the military forces and provides an exact description of the villain. The scout must also provide the appropriate password so that the commander knows it is not a spy planting misinformation. The body has scouts, called macrophages, that are part of the nonspecific defenses. Macrophages roam the tissues, looking for any invader. The cells that act as the immune system's commander-in-chief are a subset of T cells called helper T cells. When properly alerted, helper T cells call out the body's specific defensive forces and the immune responses begin.

A nation's military may have two (or more) branches. For example, it may consist of an army and a navy. Specialized to respond to slightly different forms of enemy invasion, each branch is armed with certain types of weapons. Either branch can be activated to combat a particular threat, say little green people with purple hair. The navy may be called into action if the enemy is encountered at sea, whereas the army will come to the defense if the enemy is on land.

The body, similarly, has two types of specific defense. They recognize and destroy the same antigens but do so in different ways.

- *Antibody-mediated immune responses* defend primarily against antigens found traveling freely in intercellular and other body fluids—for example, toxins or extracellular pathogens such as bacteria or free viruses. The warriors of this branch of immune defense are the effector B cells (also called plasma cells), and their weapons are Y-shaped proteins called *antibodies*, which neutralize and remove potential threats from the body. Antibodies are programmed to recognize and bind to the antigen posing the threat; they help eliminate the antigen from the body (discussed in greater detail later in the chapter).

- *Cell-mediated immune responses* protect against foreign or abnormal cells, including body cells that have become infected with viruses or other pathogens and cancer cells. The lymphocytes responsible for cell-mediated immune responses are a type of T cell called cytotoxic T cell (discussed at greater length later in the chapter). Once activated, cytotoxic T cells quickly destroy the infected or cancerous cells by causing them to burst.

Now that the various defenders have been introduced, let's see how they work together to produce your body's highly effective immune response. Table 13.2 summarizes the functions of the cells participating in the immune response, and Table 13.3 summarizes the steps in the immune response. You can refer to these tables as you read the following description.

TABLE 13.2 CELLS INVOLVED IN THE IMMUNE RESPONSE

CELL	FUNCTIONS
Macrophage	**An antigen-presenting cell** • Engulfs and digests antigens • Places a piece of consumed antigen on its plasma membrane • Presents the antigen to a helper T cell • Activates the helper T cell
T Cells	
Helper T cell	**The "on" switch for both lines of immune response** • After activation by macrophage, it divides, forming effector helper T cells and memory helper T cells • Helper T cells activate B cells and T cells
Cytotoxic T cell (effector T cell)	**Responsible for cell-mediated immune responses** • When activated by helper T cell, it divides to form effector cytotoxic T cells and memory cytotoxic T cells • Destroys cellular targets, such as infected body cells, bacteria, and cancer cells
Suppressor T cell	**The "off" switch for immune responses** Suppresses the activity of B cells and T cells after the foreign cell or molecule has been successfully destroyed
B Cells	**Involved in antibody-mediated responses** When activated by helper T cell, it divides to form plasma cells and memory cells
Plasma cell	**Effector in antibody-mediated response** Secretes antibodies specific to extracellular antigens, such as toxins, bacteria, and free viruses
Memory cells	**Responsible for memory of immune system** • Generated by B cells or any type of T cell during an immune response • Enable quick and efficient response on subsequent exposures of the antigen • May live for years

TABLE 13.3 STEPS IN THE IMMUNE RESPONSE

1. **Threat**	Foreign cell or molecule enters the body
2. **Detection**	Macrophage detects foreign cell or molecule and engulfs it
3. **Alert**	• Macrophage puts antigen from the pathogen on its surface and finds the helper T cell with correct receptors for that antigen • Macrophage presents antigen to the helper T cell • Macrophage alerts the helper T cell that there is an invader that "looks like" the antigen • Macrophage activates the helper T cell
4. **Alarm**	Helper T cell activates both lines of defense to fight that specific antigen
5. **Build specific defense (clonal selection)**	• Antibody-mediated defense—B cells are activated and divide to form plasma cells that secrete antibodies specific to the antigen • Cell-mediated defense—T cells divide to form cytotoxic T cells that attack cells with the specific antigen
6. **Defense**	• Antibody-mediated defense—antibodies specific to antigen eliminate the antigen • Cell-mediated defense—cytotoxic T cells cause cells with the antigen to burst
7. **Continued surveillance**	Memory cells formed when helper T cells, cytotoxic T cells, and B cells were activated remain to provide swift response if the antigen is detected again
8. **Withdrawal of forces**	Once the antigen has been destroyed, suppressor T cells shut down the immune response to that antigen

The Cell-Mediated Immune Response and the Antibody-Mediated Immune Response Have the Same Steps

Although the cell-mediated immune response and the antibody-mediated immune response use different mechanisms to defend against pathogens or foreign molecules (nonself), the general steps in these responses are the same (Figure 13.9).

STEP 1: Threat The immune response begins when a molecule or organism lacking the self (MHC) marker manages to evade the first two lines of defense and enters the body (Figure 13.10).

STEP 2: Detection Recall that macrophages are phagocytic cells that roam the body, engulfing any foreign material or organisms they may encounter. Within the macrophage, the engulfed material is digested into smaller pieces.

STEP 3: Alert Acting like a scout in the military, the macrophage then alerts the immune system's commander-in-chief, a *helper T cell*, that an antigen is present. The macrophage accomplishes this task by transporting some of the digested pieces to its own surface, where they bind to the MHC self markers on the macrophage membrane. The self marker acts as a secret password that identifies the macrophage as a "friend." However, the pieces of antigen bound to the self markers function as a kind of wanted poster, telling the lymphocytes that there is an invader and revealing what the invader looks like. The displayed antigens trigger the immune response. Thus, the macrophage is an important type of **antigen-presenting cell (APC)**. (B cells and dendritic cells—cells with long extensions found in lymph nodes—are two other kinds of antigen-presenting cells.)

The macrophage presents the antigen to a **helper T cell**, the kind of T cell that serves as the main switch for the entire immune response. However, the macrophage must alert the *right* kind of

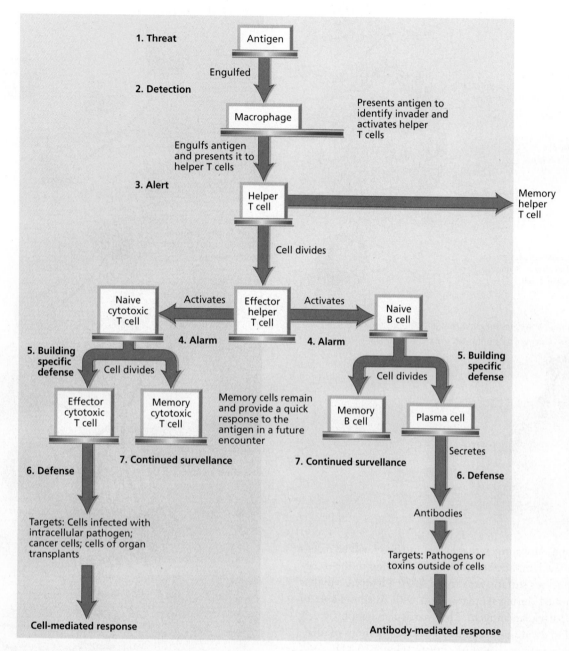

FIGURE **13.9** An overview of the immune response.

helper T cell, a helper T cell bearing receptors that recognize the specific antigen being presented. These specific helper T cells constitute only a tiny fraction of the entire T cell population. Finding the right helper T cell is like looking for a needle in a haystack. The macrophage wanders through the body until it literally bumps into an appropriate helper T cell. The encounter most likely occurs in one of the lymph nodes, because these bean-shaped structures, discussed in Chapter 12, contain huge numbers of lymphocytes of all kinds. When the antigen-presenting macrophage meets the appropriate helper T cell and binds to it, the macrophage secretes a chemical that activates the helper T cell.

STEP 4: Alarm Within hours, an activated helper T cell begins to secrete its own chemical messages. The helper T cell's message calls

into active duty the appropriate B cells and T cells, the ones with the ability to bind to the particular antigen that triggered the response.

STEP 5: Build Specific Defenses When the appropriate "naive"[1] B cells or T cells are activated, they begin to divide repeatedly. The result is a clone (a population of genetically identical cells) that is specialized to protect against the particular target antigen.

The process by which this highly specialized clone is produced, called **clonal selection**, underlies the entire immune response. We have seen that each lymphocyte is equipped to recognize an antigen

[1]A "naive" cell is one that has been preprogrammed to respond to a particular antigen but has not been previously activated to respond.

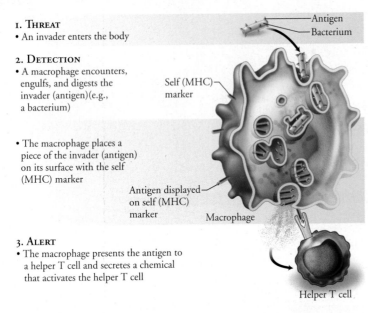

1. THREAT
• An invader enters the body

2. DETECTION
• A macrophage encounters, engulfs, and digests the invader (antigen)(e.g., a bacterium)

• The macrophage places a piece of the invader (antigen) on its surface with the self (MHC) marker

3. ALERT
• The macrophage presents the antigen to a helper T cell and secretes a chemical that activates the helper T cell

FIGURE **13.10** A macrophage is an important antigen-presenting cell. It presents the antigen, attached to a self (MHC) marker to a helper T cell and activates the helper T cell.

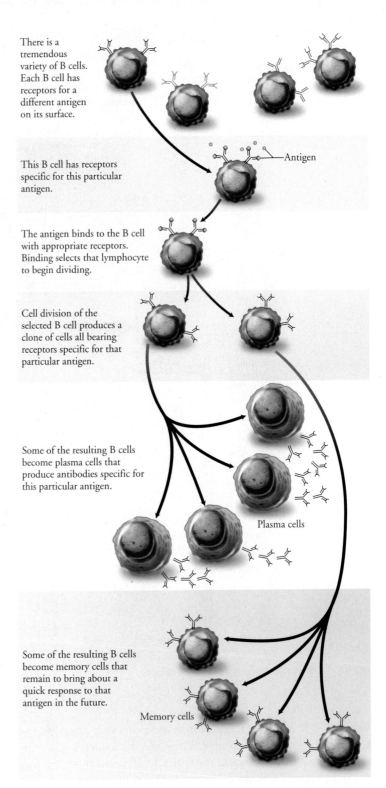

There is a tremendous variety of B cells. Each B cell has receptors for a different antigen on its surface.

This B cell has receptors specific for this particular antigen.

The antigen binds to the B cell with appropriate receptors. Binding selects that lymphocyte to begin dividing.

Cell division of the selected B cell produces a clone of cells all bearing receptors specific for that particular antigen.

Some of the resulting B cells become plasma cells that produce antibodies specific for this particular antigen.

Some of the resulting B cells become memory cells that remain to bring about a quick response to that antigen in the future.

FIGURE **13.11** Clonal selection is the process by which an immune response to a specific antigen becomes amplified. This figure shows clonal selection of B cells, but a similar process occurs with T cells. As a B cell or T cell matures, it develops receptors capable of recognizing and binding to an antigen of a specific shape. When an antigen enters the body, it binds to a B cell or T cell that bears an appropriate receptor, and the reaction causes the cell to proliferate. In this way, millions of exact copies (a clone) of cells specialized to recognize that particular antigen are created. There are two types of cells in the clone: effector cells (in the case of B cells, these are called plasma cells) and memory cells.

of a specific shape. Any antigen that enters the body will be recognized by only a few lymphocytes at most. By binding to the receptors on a lymphocyte's surface, an antigen *selects* a lymphocyte that was preprogrammed during its maturation with receptors able to recognize that particular antigen. That particular lymphocyte is then stimulated to divide and produces a clone of millions of identical cells able to recognize that same antigen (Figure 13.11).

The following analogy may be helpful for understanding clonal selection. Consider a small bakery with only sample cookies on display. A customer chooses a particular cookie and places an order for many cookies of that type. The cookies are then prepared especially for that person. The sample cookies do not take a lot of space, so a wide selection can be available. The baker does not waste energy making unnecessary cookies. Your body prepares samples of many kinds of lymphocytes. When an antigen selects the appropriate lymphocyte, the body produces many additional copies of the lymphocyte chosen by that particular antigen.

stop and think

A primary target of HIV, the human immunodeficiency virus that leads to AIDS, is the helper T cell. Why does the virus's preference for the helper T cell impair the immune system more than if another type of lymphocyte were targeted?

■ **B cells mount an antibody-mediated immune response against antigens free in the blood or bound to a cell surface**

We have already mentioned that two types of cells are produced in step 5: memory cells and effector cells. Before turning to the role of memory cells, let's look more closely at exactly *how* the effector cells protect us.

STEP 6: **Defense—The Antibody-Mediated Response** In the **antibody-mediated immune response** activated B cells divide. The effector cells they produce, which are called **plasma cells**, secrete antibodies into the bloodstream to defend against antigens free in the blood or bound to a cell surface (Figure 13.12). **Antibodies** are Y-shaped proteins that recognize a specific antigen by its shape. Each antibody is specific for one particular antigen. The specificity results from the shape of the proteins that form the tips of the Y (Figure 13.13). Because of their shapes, the antibody and antigen fit together like a lock and a key. Each antibody can bind to two identical antigens, one at the tip of each arm on the Y.

Antibodies can bind only to antigens that are free in body fluids or attached to the surface of a cell. Their main targets are toxins and extracellular microbes, including bacteria, fungi, and protozoans. Antibodies help defend against these pathogens in several ways that can be remembered with the acronym PLAN.

- **P**recipitation: The antigen-antibody binding causes antigens to clump together and precipitate, enhancing phagocytosis by making the antigens easier to capture and engulf.
- **L**ysis (bursting): Certain antibodies activate the complement system, which then pokes holes through the membrane of the target cell and causes it to burst.
- **A**ttraction of phagocytes: Antibodies also attract phagocytic cells to the area. Phagocytes then engulf and destroy the foreign material.
- **N**eutralization: Antibodies bind to toxins and viruses, neutralizing them and preventing them from causing harm.

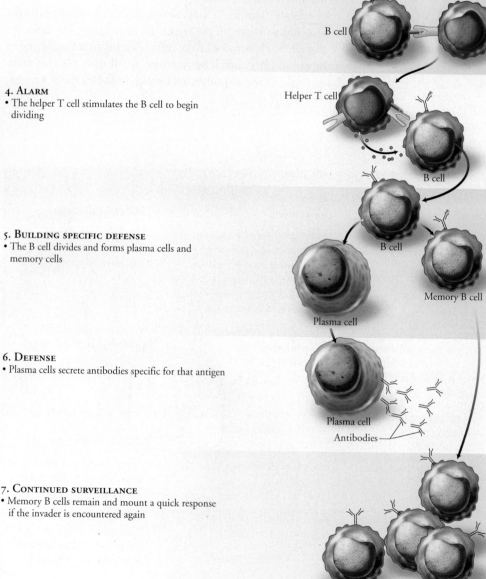

4. **ALARM**
- The helper T cell stimulates the B cell to begin dividing

B cell

Helper T cell

B cell

5. **BUILDING SPECIFIC DEFENSE**
- The B cell divides and forms plasma cells and memory cells

B cell

Memory B cell

Plasma cell

6. **DEFENSE**
- Plasma cells secrete antibodies specific for that antigen

Plasma cell

Antibodies

7. **CONTINUED SURVEILLANCE**
- Memory B cells remain and mount a quick response if the invader is encountered again

Memory B cells

FIGURE **13.12** Antibody-mediated immune response

There are five classes of antibodies, each with a special role to play in protecting against invaders. Antibodies are also called **immunoglobulins** (Ig), and each class is designated with a letter: IgG, IgM, IgE, IgA, and IgD. As you can see in Table 13.4, the antibodies of some classes exist as single Y-shaped molecules (monomers), in one class they exist as two attached molecules (dimers), and in one class they exist as five attached molecules (pentamers) radiating outward like the spokes of a wheel.

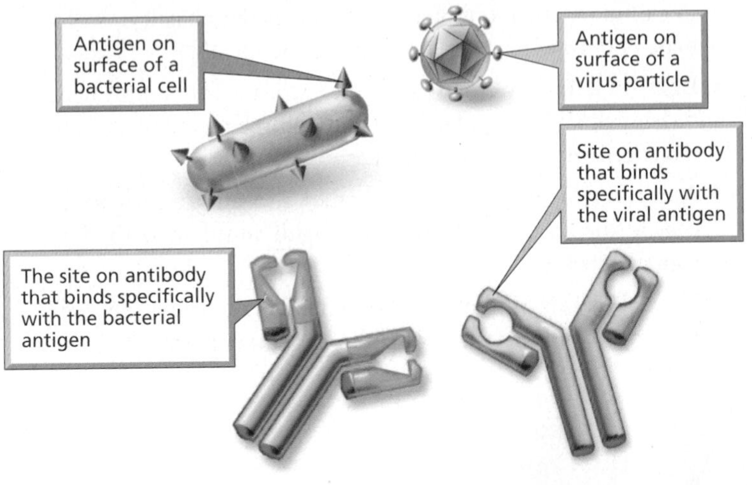

■ Cytotoxic T cells mount a cell-mediated defense against antigen-bearing cells

In a cell-mediated response, T cells divide to form cytotoxic T cells.

STEP 6: Defense—The Cell-Mediated Response The **cytotoxic T cells** are the effector T cells responsible for the **cell-mediated immune response**, which destroys antigen-bearing cells. Each cytotoxic T cell is programmed to recognize a particular antigen bound to MHC (self) markers on the surface of a cellular pathogen, an infected or cancerous body cell, or on cells of a tissue or organ transplant. A cytotoxic T cell becomes activated to destroy a target cell when two events occur simultaneously, as shown in Figure 13.14. First, the cytotoxic T cell must encounter an antigen-presenting cell, such as a macrophage. Second, a helper T cell must release a chemical to activate the cytotoxic T cell. When sufficiently activated, the cytotoxic T cell divides, producing memory cells and effector cytotoxic T cells.

An effector cytotoxic T cell releases chemicals called **perforins** that cause holes to form in the target cell membrane. The holes are large enough to allow some of the cell's contents to leave the cell so that the cell disintegrates. The cytotoxic T cell then detaches from the target cell and seeks another cell having the same type of antigen.

FIGURE **13.13** An antibody is a Y-shaped protein designed to recognize an antigen having a specific shape. The recognition of a specific antigen occurs because of the shape of the tips of the Y in the antibody molecule.

Use this table to learn more about the classes of antibodies

TABLE 13.4 CLASSES OF ANTIBODIES

CLASS	STRUCTURE	LOCATION	CHARACTERISTICS	PROTECTIVE FUNCTIONS
IgG	Monomer	Blood, lymph, and the intestines	Most abundant of all antibodies in body; involved in primary and secondary immune responses; can pass through placenta from mother to fetus and provides passive immune protection to fetus and newborn	Enhances phagocytosis; neutralizes toxins; triggers complement system
IgA	Dimer or monomer	Present in tears, saliva, and mucus as well in secretions of gastrointestinal system and excretory systems; present in breast milk	Levels decrease during stress, raising susceptibility to infection	Prevents pathogens from attaching to epithelial cells of surface lining
IgM	Pentamer	Attached to B cell where it acts as a receptor for antigens; free in blood and lymph	First Ig class released by plasma cell during primary response	Powerful agglutinating agent (10 antigen-binding sites); activates complement
IgD	Monomer	Surface of many B cells; blood and lymph	Life span of about 3 days	Thought to be involved in recognition of antigen and in activating B cells
IgE	Monomer	Secreted by plasma cells in skin, mucous membranes of gastrointestinal and respiratory systems	Become bound to surface of mast cells and basophils	Involved in allergic reactions by triggering release of histamine and other chemicals from mast cells or basophils

stop and think

Rejection of an organ transplant occurs when the recipient's immune system attacks and destroys the cells of the transplanted organ. Why would this attack occur? Which branch of the immune system would be most involved?

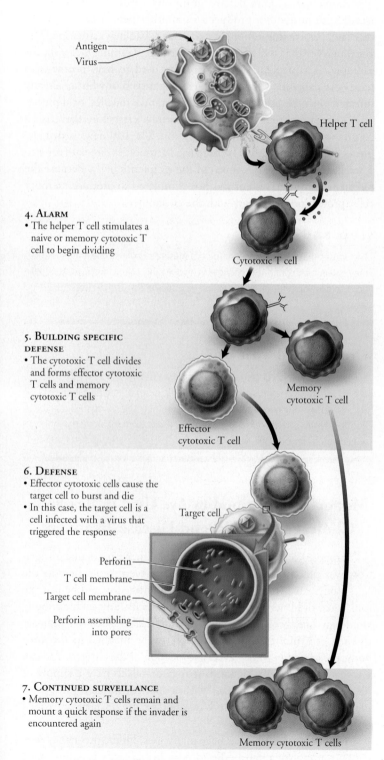

4. ALARM
- The helper T cell stimulates a naive or memory cytotoxic T cell to begin dividing

5. BUILDING SPECIFIC DEFENSE
- The cytotoxic T cell divides and forms effector cytotoxic T cells and memory cytotoxic T cells

6. DEFENSE
- Effector cytotoxic cells cause the target cell to burst and die
- In this case, the target cell is a cell infected with a virus that triggered the response

7. CONTINUED SURVEILLANCE
- Memory cytotoxic T cells remain and mount a quick response if the invader is encountered again

FIGURE **13.14** Cell-mediated immune response

■ **Immunological memory permits a more rapid response on subsequent exposure to the antigen**

In the last step of the immune response, memory cells remain to provide a swift defense. They may linger for years.

STEP 7: Continued Surveillance The first time an antigen enters the body, only a few lymphocytes can recognize it. Those lymphocytes must be located and stimulated to divide in order to produce an army of lymphocytes ready to eliminate that particular antigen. As a result, the *primary response*, the one that occurs during the body's first encounter with a particular antigen, is relatively slow. A lapse of several days occurs before the antibody concentration begins to rise, and the concentration does not peak until 1 to 2 weeks after the initial exposure to the antigen (Figure 13.15).

Following subsequent exposure to the antigen, the *secondary response* is strong and swift. Recall that when naive B cells and T cells were stimulated to divide, not only did they produce effector cells that actively defended against the invader, but they also produced memory cells. These memory B cells and T cells live for years or even decades. As a result, the number of lymphocytes programmed to respond to that particular antigen is greater than it was before the first exposure. When the antigen is encountered again, each of those memory cells divides and produces new effector cells and memory cells specific for that antigen. Therefore, the number of effector cells rises quickly during the secondary response, and within 2 or 3 days reaches a higher peak than it did during the primary response. This process is the reason we do not get measles twice.

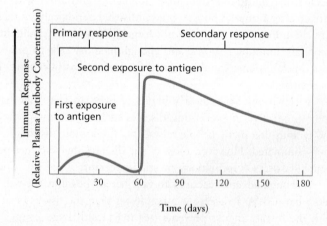

FIGURE **13.15** The primary and secondary immune responses. In the primary response, which occurs after the first exposure to an antigen, there is a delay of several days before the concentration of circulating antibodies begins to increase. It takes 1 to 2 weeks for the antibody concentration to peak because the few lymphocytes programmed to recognize that particular antigen must be located and activated. (The T cells show a similar pattern of response.) The secondary response following a subsequent exposure to an antigen is swifter and stronger than the primary response. The difference is due to the long-lived memory cells produced during the primary response; these are a larger pool of lymphocytes programmed to respond to that particular antigen.

■ **Suppressor T cells turn off the immune response**

There is one last step to a successful immune response.

STEP 8: Withdrawal of Forces As the immune system begins to conquer the invading organism and the level of antigens declines, another type of T cell, the **suppressor T cell**, releases chemicals that dampen the activity of both B cells and T cells. Suppressor T cells turn off the immune response when the antigen no longer poses a threat.

Immunity Can Be Active or Passive

In **active immunity** the body actively defends itself by producing memory B cells and T cells following exposure to an antigen. Active immunity happens naturally whenever a person gets an infection. Fortunately, active immunity can also develop through *vaccination*, a procedure that introduces a harmless form of a disease-causing microbe into the body to stimulate immune responses against that antigen. In some kinds of vaccination—whooping cough and typhoid fever, for instance—the microbe is inactivated before the vaccine is prepared. Other vaccines must be made from live organisms in order to be effective. In these cases, the microbes are first weakened so that they can no longer cause disease. Still other vaccines, including the one against smallpox, are prepared from microbes that cause related but milder diseases.

Because it leads to the production of memory cells, active immunity, even when produced by vaccination, is relatively long lived. The first dose of a vaccine causes the primary immune response, and antibodies and some memory cells are generated. In certain cases, especially when inactivated antigens are used in the vaccine, the immune system may "forget" its encounter with the antigen after a time. A booster is administered periodically to make sure the immune system does not forget. The booster results in a secondary immune response and enough memory cells to provide for a quick response should a potent form of that pathogen ever be encountered.

Vaccinations have saved millions of lives. In fact, they have been so effective in preventing diseases such as whooping cough and tetanus that many people mistakenly think those diseases have been eliminated. However, most of the diseases that vaccines prevent still exist, so vaccinations are still important.

Although adverse reactions to vaccines are possible, the risk of being harmed by a vaccine is much lower than the risk associated with the disease that it prevents (see the Health Issue essay, *The Fear That Vaccines Cause Autism*). For example, a severe paralytic illness called Guillain-Barré syndrome may develop in one or two people out of a million receiving the influenza (flu) vaccine. But the flu kills tens of thousands of people each year. In most cases, a reaction to a vaccine is mild: warmth, redness, and tenderness at the site of injection.

Passive immunity is protection that results when a person receives antibodies produced by another person or animal. For instance, some antibodies produced by a pregnant woman can cross the placenta and give the growing fetus some immunity. These maternal antibodies remain in the infant's body for as long 3 months, at which point the infant is old enough to produce its own antibodies. Antibodies in breast milk also provide passive immunity to nursing infants, especially against pathogens that might enter through the intestinal lining. The mother's antibodies are a temporary yet critical blanket of protection. Most of the pathogens that would otherwise threaten the health of a newborn have already been encountered by the mother's immune system.

People can acquire passive immunity medically by being injected with antibodies produced in another person or animal. In this case, passive immunity is a good news–bad news situation. The good news is that the effects are immediate. Gamma globulin, for example, is a preparation of antibodies used to help people who have been exposed to diseases such as hepatitis B or who are already infected with the microbes that cause tetanus, measles, or diphtheria. Gamma globulin is often given to travelers before they visit a country where viral hepatitis is common. The bad news is that the protection is short lived. The borrowed antibodies circulate for 3 to 5 weeks before being destroyed in the recipient's body. Because the recipient's immune system was not stimulated to produce memory cells, protection disappears with the antibodies.

stop and think

The viruses that cause influenza (the flu) mutate rapidly, so the antigens in the protein coat continually change. Why does this characteristic make it difficult to develop a flu vaccine that will be effective for several consecutive years?

what would you do?

Some parents are so concerned about vaccine safety that they refuse to have their children vaccinated. Most schools require proof of immunization against certain diseases for admission. Do you think this question should be decided by the parents or the government? If you were (or are) a parent, would you have your child vaccinated?

Monoclonal Antibodies Are Used in Research, Clinical Diagnosis, and Disease Treatment

Suppose you wanted to determine whether a particular antigen was present in a solution, tissue, or even somewhere in the body. An antibody specific for that antigen would be just the tool you would need. Because of its specificity, any such antibody would go directly to the target antigen. If a label (such as a radioactive tag or a molecule that fluoresces) were attached to the antibody, the antibody could reveal the location of the antigen. You can see that for a test of this kind, it is desirable to have a supply of identical antibodies that react with a specific antigen. Groups of identical antibodies that bind to one specific antigen are called **monoclonal antibodies**.

Monoclonal antibodies have many uses. Home pregnancy tests contain monoclonal antibodies produced to react with a hormone (human chorionic gonadotropin; see Chapter 18) secreted by

HEALTH ISSUE

The Fear That Vaccines Cause Autism

One of the major medical success stories has been the development of vaccines to prevent potentially deadly diseases, including measles, mumps, rubella (German measles), diphtheria, tetanus, and polio. However, concerns about a possible link between certain vaccines and autism, a neurological developmental disorder that causes impaired social interaction and communication skills, have been debated for nearly a decade.

The first hypothesis that such a link existed arose out of a 1998 study of the records of 12 children who had both autism and irritable bowel disease. The latter was thought to be caused by a persistent measles-virus infection in the intestines. The parents or pediatricians of 9 of the children suspected that the MMR (measles, mumps, and rubella) vaccine the children had received had caused the irritable bowel disease and also contributed to the children's autism. Because there were too few cases in the study to establish a causal link between the MMR vaccine and autism, 10 of the 13 authors of the paper later retracted the interpretation of the data that had been presented in the paper.

Numerous other studies have since explored the possible link between the MMR vaccine and autism. The Institute of Medicine (IOM) of the National Academies reviewed all these studies and issued a report in 2001 stating that the current evidence does not support a link between the MMR vaccine and autism. The recommendation of the IOM was to keep the immunization schedule the same.

The next hypothesis linking autism to vaccines suggested that autism may be caused by a mercury-containing organic compound called thimerosal. Thimerosal had been used to prevent the growth of microorganisms in vaccines from the 1930s until the early 2000s. As you may recall from Chapter 8, exposure to mercury is harmful to the nervous system, so it is reasonable to think that thimerosal could cause problems in the developing nervous system that might result in autism. In 2004, an IOM report concluded that it is biologically plausible that the cumulative amount of mercury in the series of recommended vaccinations for infants and toddlers could exceed the maximum federal safety standards. However, the standards pertain to methylmercury, and thimerosal contains ethylmercury, which is handled differently in the body. The IOM therefore concluded that the current evidence is inadequate for either accepting or rejecting the hypothesis that exposure to thimerosal in childhood vaccines can lead to autism. Since 2004, several studies have failed to show a link between thimerosal and autism. Even so, childhood vaccines are no longer treated with thimerosal, or they contain only trace amounts. There are, however, some remaining supplies of thimerosal-containing vaccines. The IOM recommends that these supplies not be used.

Measles was declared to be eradicated in the United States in 2000, due to the effectiveness of the MMR vaccine. For that reason and because of the fear that vaccines are linked to autism, some people did not have their children vaccinated in the years following. Measles outbreaks began to occur again as a result. For example, in 2006 a Romanian girl returned to the United States with measles after a visit to her homeland. She subsequently attended a church gathering where three other people became infected. Measles then spread to an additional 34 people in the community who had not been vaccinated. The families of those who became ill said that they did not have their children vaccinated because of fear that thimerosal would cause autism, but the MMR vaccine has never contained thimerosal. 👫

---

membranes associated with the developing embryo. A monoclonal antibody drug that will detect the early stages of inhalation anthrax infection is now being tested. Monoclonal antibodies have also proved useful in screening for certain diseases, including Legionnaire's disease, hepatitis, certain sexually transmitted diseases, and certain cancers, including those of the lung and prostate. Some monoclonal antibodies are used in cancer treatment. The radioactive material or chemical treatment is attached to a monoclonal antibody that targets the tumor cells but has little effect on other cells.

The Immune System Can Cause Problems

The immune system protects us against myriad threats from agents not recognized as belonging in the body. However, sometimes the defenses are misguided. In autoimmune disease, the body's own cells are attacked. Allergies result when the immune system protects us against substances that are not harmful. Tissue rejection following organ transplant is also caused by the immune system (see the Health Issue essay, *Rejection of Organ Transplants*).

■ Autoimmune disorders occur when the immune system attacks the body's own cells

Autoimmune disorders occur when the immune system fails to distinguish between self and nonself and attacks the tissues or organs of the body. If the immune system can be called the body's military defense, then autoimmune disease is the equivalent of "friendly fire."

A number of autoimmune disorders occur because portions of disease-causing organisms resemble antigens found on normal body cells. If the immune system mistakes the body's antigens for the foreign antigens, it may attack them. For instance, the body's attack on certain streptococcal bacteria that cause a sore throat may result in the production of antibodies that target not only the streptococcal bacteria but also similar molecules that are found in the valves of the heart and joints. The result is an autoimmune disorder known as rheumatic fever (Figure 13.16).

HEALTH ISSUE

Rejection of Organ Transplants

Each year, tens of thousands of people receive a gift of life in the form of a transplanted kidney, heart, lungs, liver, or pancreas. Although these transplants seem commonplace today, they have been performed for only about 30 years. Before organ transplants could be successful, physicians had to learn how to prevent the effector T cells of the immune system from attacking and killing the transplanted tissue because it lacked self markers. When transplanted tissue is killed by the host's immune system, we say that the transplant has been rejected.

The success of a transplant depends on the similarity between the host and transplanted tissues. The most successful transplants, then, are those in which tissue is taken from one part of a person's body and transplanted to another part. In cases of severe burns, for example, healthy skin from elsewhere on the body can replace

badly burned areas of skin. Because identical twins are genetically identical, their cells have the same self markers, and organs can be transplanted from one twin to another with little fear of tissue rejection. But few of us have an identical twin. The next best source for tissue for a transplant, and the most common, is a person whose cell surface markers closely match those of the host. Usually the transplanted tissue comes from a person who has recently died. The donor is usually brain-dead, but his or her heart is kept beating by life-support equipment. Some organs—primarily kidneys—can be harvested from someone who has died and whose heart has stopped beating. In some cases, living people can donate organs; one of two healthy kidneys can be donated to a needy recipient, as can sections of liver.

Even though the odds in favor of a successful transplant are always improving, the waiting

list of patients in need of an organ from a suitable donor has outpaced the supply. Some researchers believe that in the future, organs from nonhuman animals may fill the gap between the supply of organs and the demand. So far, however, attempts to transplant animal organs into people have failed. The biggest obstacle is hyperacute rejection. Within minutes to hours after transplant, the animal organ dies because its blood supply is choked off by the human immune system.

Other dangers may remain, even if the rejection problem is solved. Animals carry infectious agents that are harmless to their hosts but that might "jump species" and spread from person to person. If that happened, we would have to ask whether it is ethical to expose a third party to risk.

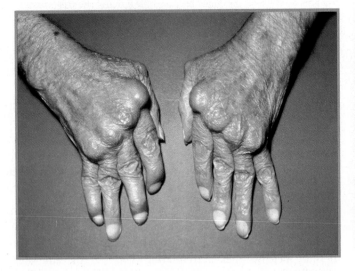 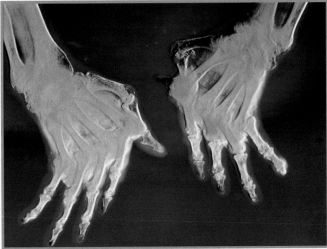

FIGURE **13.16** In autoimmune diseases, the body's lymphocytes attack the body's own cells. One example is rheumatoid arthritis, in which lymphocytes attack collagen fibers in joints.

Treatment of autoimmune disorders is usually two pronged. First, any deficiencies caused by the disorder are corrected. We saw in Chapter 10, for example, that in diabetes mellitus type 1, the immune system destroys insulin-producing cells in the pancreas. Treatment for diabetes would therefore include replacement of insulin. Second, immune system activity is suppressed with drugs.

■ Allergies are immune responses to harmless substances

An **allergy** is an overreaction by the immune system to an antigen, in this case called an *allergen*. The immune response in an allergy is considered an overreaction because the allergen usually is not harmful to the body (Table 13.5). The most common allergy is hay fever, which, by the way, is not caused by hay and does not cause a

TABLE 13.5 COMMON ALLERGIES

TYPE OF ALLERGIC RESPONSE	COMMON CAUSES	LOCATION OF REACTIVE MAST CELLS	SYMPTOMS
Hay fever (allergic rhinitis)	Pollen, mold spores, animal dander (bits of skin and hair), feces of dust mites	Lining of nasal cavity	Sneezing, nasal congestion
Asthma	Pollen, mold spores, animal dander	Airways of lower respiratory tract	Difficulty breathing
Food allergy	Chicken, eggs, fish, milk, nuts (especially peanuts), shellfish, soybeans, and wheat	Lining of digestive system	Nausea, vomiting, abdominal cramps, and diarrhea
Hives	Foods (especially shellfish, strawberries, chocolate, nuts, and tomatoes); insect bites; certain drugs (especially penicillin and aspirin); chemicals such as food additives, dyes, and cosmetics	Skin	Patches of skin become red and swollen
Anaphylactic shock	Insect stings (especially from bees, wasps, hornets, yellow jackets, fire ants); medicines (especially penicillin and tetracycline); certain foods (especially eggs, seafood, nuts, and grains)	Throughout the body	Widening of blood vessels, causing blood to pool in capillaries and resulting in dizziness, nausea, diarrhea, and unconsciousness; death

fever. Hay fever is more correctly known as allergic rhinitis (*rhino*, nose; *-itis*, inflammation of). The symptoms of hay fever—sneezing and nasal congestion—occur when an allergen is inhaled, triggering an immune response in the respiratory system. Mucous membranes of the eyes may also respond, causing red, watery eyes. Common causes of hay fever include pollen, mold spores, animal dander, and the feces of dust mites, microscopic creatures that are found all over your house (Figure 13.17). The same allergens, however, can cause asthma. During an asthma attack, the small airways in the lung (bronchioles) constrict, making breathing difficult. In food allergies, the immune response occurs in the digestive system and may cause nausea, vomiting, abdominal cramps, and diarrhea. Food allergies can also cause hives, a skin condition in which patches of skin temporarily become red and swollen.

Anaphylactic shock is an extreme allergic reaction that occurs within minutes after exposure to the substance a person is allergic to. It can cause pooling of blood in capillaries, which leads to dizziness, nausea, and sometimes unconsciousness as well as extreme difficulty in breathing. Anaphylactic shock can be fatal, but most people survive. Common triggers of anaphylactic shock include certain foods; medicines, including antibiotics such as penicillin; and insect stings, especially stings from bees, wasps, yellow jackets, and hornets.

An immediate allergic response begins when a person is exposed to an allergen and a primary immune response is launched (Figure 13.18). Soon, plasma cells churn out a certain type of antibody known as IgE, which binds to either basophils or mast cells. In subsequent exposures to that allergen, the allergen binds to IgE antibodies on the surface of basophils or mast cells and causes granules containing histamine to release their contents.

Histamine then causes the swelling, redness, and other symptoms of an allergic response. The blood vessels widen, slowing blood flow and causing redness. At the same time, the blood vessels become leaky, allowing fluid to flow from the vessels into spaces between tissue cells, swelling the tissues. Histamine also causes the release of large amounts of mucus, so the nose begins to run. In addition, histamine can cause smooth muscles of internal organs to

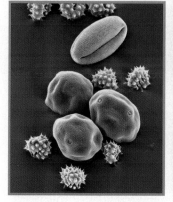
Pollen grains

Dust mite

FIGURE **13.17** Common causes of allergies are pollen grains and the feces of dust mites, such as the mite shown here.

First exposure

STEP 1
- The invader (allergen) enters the body

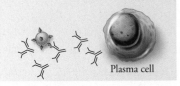

Allergen

STEP 2
- Large amounts of class IgE antibodies against the allergen are produced by plasma cells

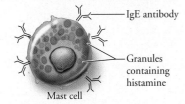

Plasma cell

STEP 3
- IgE antibodies attach to mast cells, which are found in body tissues

IgE antibody

Granules containing histamine

Mast cell

Subsequent (secondary) response

STEP 4
- More of the same allergen invades the body

STEP 5
- The allergen combines with IgE attached to mast cells
- Histamine and other chemicals are released from mast cell granules

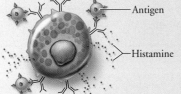

Antigen

Histamine

STEP 6: HISTAMINE
- Causes blood vessels to widen and become leaky
- Fluid enters the tissue, causing edema

- Stimulates release of large amounts of mucus

- Causes smooth muscle in walls of air tubules in lungs to contract

FIGURE **13.18** Steps in an allergic reaction

contract. If the allergen spreads from the area where it entered the body, these effects can be widespread. The result can be anaphylactic shock.

People with allergies often know which substances cause their problems. When the culprits are not known, doctors can identify them using a crude but effective technique in which small amounts of suspected allergens are injected into the skin. If the person is allergic to one of the suspected allergens, a red welt will form at the site of injection.

The simplest way to avoid the miseries of allergies is to avoid exposure to the substances that cause problems. During pollen season, spend as much time as possible indoors, using an air conditioner to filter pollen out of the incoming air. Unfortunately, spores from molds growing in air conditioners and humidifiers are also common triggers of allergies. Some foods, for instance, strawberries, may be easy to avoid. Others, such as peanut oil, can show up in some unlikely dishes, including stew, chili, or meat patties.

Certain drugs may reduce allergy symptoms. As their name implies, antihistamines block the effects of histamine. Antihistamines are most effective if they are taken before the allergic reaction begins. Unfortunately, allergies tend to become less susceptible to antihistamines over time, and most antihistamines cause drowsiness, which can impair performance on the job or in school and can make driving a car extremely hazardous.

Allergies can be treated by gradually desensitizing the person to the offending allergens. Allergy shots containing gradually increasing amounts of a known allergen are injected into the person's bloodstream. During this treatment, the allergen causes the production of another class of antibodies—IgG. Afterward, when the allergen enters the body, IgG antibodies bind to it and prevent it from binding to IgE antibodies on mast cells and triggering an allergic reaction.

exploring further . . .

In this chapter, we have considered the ways that our bodies protect us from foreign agents that could cause us harm. In the next chapter, we will learn more about the pathogens that can make us ill. We will consider the mechanisms by which they cause harm and the ways they are transmitted.

HIGHLIGHTING THE CONCEPTS

The Body's Defense System Targets Pathogens and Cancerous Cells (p. 235)

1. The targets of the body's defense system include anything that is not recognized as belonging in the body, such as disease-causing organisms and cancerous cells.

The Body Has Three Lines of Defense (pp. 235–240)

2. The first line of defense is nonspecific physical barriers, such as skin and mucous membranes, and chemical barriers, such as sweat, oil, tears, and saliva, that prevent entry of pathogens.

3. The second line of defense includes defensive cells and proteins, inflammation, and fever. Defensive cells include phagocytes, eosinophils, and natural killer cells. Two types of defensive proteins are antiviral interferons and complement, which causes cells to burst.

4. The inflammatory response occurs as a result of tissue injury or invasion by foreign microbes. It begins when mast cells in the injured area release histamine, which increases blood flow by dilating blood vessels to the region and by increasing the permeability of capillaries there. Increased blood flow causes redness and warmth in the region. Fluid leaking from the capillaries causes swelling.

5. Fever, an abnormally high body temperature, helps the body fight invading microbes by enhancing several body defense mechanisms and slowing the growth of many pathogens.

6. The third line of defense, the immune system, has specific targets and memory.
   WEB TUTORIAL 13.1 The Inflammatory Response

The Immune System Distinguishes Self from Nonself (pp. 240–241)

7. All body cells are labelled with proteins called major histocompatibility complex (MHC) proteins, which serve as self markers. Cells that lack self markers (MHC) are considered nonself and are attacked. A nonself substance or organism triggers an immune response and is called an antigen.

8. Lymphocytes are white blood cells that are responsible for immune responses. Both B lymphocytes (B cells) and T lymphocytes (T cells) develop in the bone marrow. The B cells are thought to mature in the bone marrow, but the T cells mature in the thymus gland. During maturation, B cells and T cells develop receptors on their surfaces that allow each of those cells to recognize an antigen of a different shape.

9. When an antigen is detected, B cells and T cells with receptors that respond to that antigen divide repeatedly, forming effector cells that destroy the antigen and forming memory cells that provide a quick response on subsequent exposure to that antigen.

The Immune System Mounts Antibody-Mediated Responses and Cell-Mediated Responses (pp. 241–242)

10. The antibody-mediated immune response and the cell-mediated immune response simultaneously defend against the same antigen.
    WEB TUTORIAL 13.2 Antibody- and Cell-Mediated Responses

The Cell-Mediated Immune Response and the Antibody-Mediated Immune Response Have the Same Steps (pp. 242–248)

11. Macrophages are phagocytic cells that engulf any foreign material or organism they encounter. After engulfing the material, the macrophage places a part of it on its own surface to serve as an antigen that alerts lymphocytes to the presence of an invader and reveals what the invader looks like. Macrophages also have molecular (MHC) markers on their membranes that identify them as belonging in the body, that is, as self.

12. A macrophage then presents the antigen to a helper T cell, which serves as the main switch to the entire immune response. When this encounter occurs, the macrophage secretes a chemical that activates the helper T cell. The helper T cell, in turn, secretes a chemical that activates the appropriate B cells and T cells (those specific for the antigen that the macrophage engulfed).

13. B cells are responsible for antibody-mediated immune responses, which defend against antigens that are free in body fluids, including bacteria, free virus particles, and toxins. When called into action by a helper T cell, a B cell divides repeatedly, forming two lines of descendant cells: effector cells that transform into plasma cells and memory B cells. Plasma cells secrete Y-shaped proteins called antibodies into the bloodstream. Antibodies bind to the particular antigen and inactivate it or help remove it from the body.

14. Cytotoxic T cells are responsible for cell-mediated immune responses, which are effective against cellular threats, including infected body cells and cancer cells. When a T cell is properly activated, it divides, forming two lines of descendant cells: effector cells, called cytotoxic T cells, and memory T cells. Cytotoxic T cells secrete perforins that poke holes in the foreign or infected cell, causing it to burst and die.

15. After the first encounter with a particular antigen, the primary response is initiated, which may take several weeks to become effective against the antigen. However, because of memory cells, a subsequent exposure to the same antigen triggers a quicker response, called a secondary response.

16. Suppressor T cells dampen the activity of B cells and T cells when antigen levels begin to fall.

Immunity Can Be Active or Passive (p. 248)

17. In active immunity, the body actively participates in forming memory cells to defend against a particular antigen. Active immunity may occur when an antigen infects the body, or it may occur through vaccination, a procedure that introduces a harmless form of an antigen into the body. Passive immunity results when a person receives antibodies that were produced by another person or animal. Passive immunity is short lived.

Monoclonal Antibodies Are Used in Research, Clinical Diagnosis, and Disease Treatment (pp. 248–249)

20. Monoclonal antibodies are identical antibodies. They are useful in research and in the diagnosis and treatment of diseases.

The Immune System Can Cause Problems (pp. 249–252)

21. Autoimmune disorders occur when the immune system mistakenly attacks the body's own cells.

22. An allergy is a strong immune response against an antigen (called an allergen). An allergy occurs when the allergen binds to IgE antibodies on the surface of mast cells or basophils, causing them to release histamine. Histamine, in turn, causes the redness, swelling, itching, and other symptoms of an allergic response.

KEY TERMS

pathogen   *p. 235*

phagocyte   *p. 236*

macrophage   *p. 236*

natural killer cells   *p. 236*

interferon   *p. 237*

complement system   *p. 237*

inflammatory response   *p. 238*

histamine   *p. 238*

mast cell   *p. 238*

immune system   *p. 240*

MHC marker   *p. 240*

antigen   *p. 240*

B lymphocyte   *p. 240*

T lymphocyte   *p. 240*

antigen-presenting cell (APC)
   p. 242

helper T cell   *p. 242*

clonal selection   *p. 243*

antibody-mediated immune
   response   *p. 245*

plasma cell   *p. 245*

antibody   *p. 245*

immunoglobin   *p. 246*

cytotoxic T cell   *p. 246*

cell-mediated immune response
   p. 246

perforins   *p. 246*

suppressor T cells   *p. 248*

active immunity   *p. 248*

passive immunity   *p. 248*

monoclonal antibody   *p. 248*

autoimmune disease   *p. 249*

allergy   *p. 250*

REVIEWING THE CONCEPTS

1. Explain the difference between nonspecific and specific defense mechanisms.   *p. 235*
2. List seven nonspecific defense mechanisms. Explain how each helps protect us against disease.   *pp. 235–240*
3. How does a natural killer cell kill its target cell?   *pp. 236–237*
4. What are interferons? What type of cell produces them? How do they help protect the body?   *p. 237*
5. What are the complement proteins? Explain how they act directly and indirectly to protect the body against disease.   *pp. 237–238*
6. Signs of inflammation include redness, warmth, swelling, and pain. What causes these symptoms? How does inflammation help defend against infection?   *pp. 238–240*
7. What does an antigen-presenting cell do? What is the most common type of antigen-presenting cell? How do other cells recognize the antigen-presenting cell as a "friend"?   *p. 242–243*
8. What cells are responsible for antibody-mediated immune responses? What are the targets of antibody-mediated immune responses?   *pp. 245–246*
9. Describe an antibody. How do antibodies inactivate or eliminate antigens from the body?   *pp. 245–246*
10. What is responsible for cell-mediated immune responses? What are the targets of cell-mediated immune responses?   *p. 246*
11. How does a natural killer cell differ from a cytotoxic T cell?   *pp. 236, 246*
12. Why does a secondary response occur more quickly than the primary response?   *p. 247*
13. Differentiate between active and passive immunity.   *p. 248*
14. What are monoclonal antibodies? What are some medical uses for them?   *pp. 248–249*
15. What is an autoimmune disorder?   *pp. 249–250*
16. What is an allergy? What causes the symptoms?   *pp. 250–252*
17. Indicate the *correct* statement.
    a. An antibody is specific to one particular antigen.
    b. Antibodies are held within the cell that produces them.
    c. Antibodies are produced by macrophages.
    d. Antibodies can be effective against viruses that are inside the host cell.
18. An antigen is a
    a. cell that produces antibodies.
    b. receptor on the surface of a lymphocyte that recognizes invaders.
    c. memory cell that causes a quick response to an invader when it is encountered a second time.
    d. large molecule on the surface of an invader that triggers an immune response.
19. Indicate the choice with the *incorrect* pairing of cell type and function.
    a. Helper T cell—serves as "main switch" that activates both the cell-mediated immune responses and the antibody-mediated immune responses
    b. Cytotoxic T cells—present antigen to the helper T cell
    c. Macrophage—roams the body looking for invaders, which are engulfed and digested when they are found
    d. Suppressor T cells—shut off the immune response when the invader has been removed
20. When the doctors say they are looking for a suitable donor for a kidney transplant, they are looking for someone
    a. whose tissues have self markers similar to those of the recipient.
    b. who lacks antibodies to the recipient's tissues.
    c. who has suppressor T cells that will suppress the immune response against the donor kidney.
    d. who lacks macrophages.
21. The piece of the antigen displayed on the surface of a macrophage
    a. stimulates the suppressor T cells to begin dividing.
    b. attracts other invaders to the cell, causing them to accumulate and making it easier to kill the invaders.
    c. informs the other cells in the immune system of the exact nature of the antigen they should be looking for (what the antigen "looks like").
    d. has no function in the immune response.
22. A cell that kills any unrecognized cell in the body and is part of the nonspecific body defenses is the _____.
23. _____ is a chemical released by mast cells and basophils that produces most of the symptoms of an allergy.
24. Antibodies are produced by _____.
25. _____ are important antigen-presenting cells.
26. _____ diseases are those in which the body attacks its own tissues.

APPLYING THE CONCEPTS

1. After being exposed to the hepatitis B virus, Barbara goes to the doctor and asks to be vaccinated against it. Instead, the doctor gives her an injection of gamma globulin (a preparation of antibodies). Why wasn't she given the vaccine?
2. More than 100 viruses can cause the common cold. How does this fact explain why you can catch a cold from Ramond immediately after recovering from a cold you caught from Jessica?
3. HIV is a virus that kills helper T cells. This virus is not the direct cause of death in people who are infected with it. Instead, people die of diseases caused by organisms that are common in the environment. Explain why HIV-infected persons are susceptible to these diseases.

Additional questions can be found on the companion website.

4. Steve found a deer tick attached to the back of his leg. He knows that deer ticks can transmit the bacterium that causes Lyme disease and that untreated Lyme disease can cause arthritis and fatigue. He immediately went to the doctor to get tested, which would involve drawing blood to look for antibodies to the bacterium. The doctor refused to test Steve for Lyme disease. Why?
5. Rashon has leukemia, a cancer in which the number of white blood cells increases dramatically. The doctors decide that a bone marrow transfer might help by replacing defective bone stem cells with healthy ones. His girlfriend offered to be a donor, but the doctors chose his brother instead. Why? Why was Rashon given drugs to suppress his immune system after the transplant?

13a

Infectious Disease

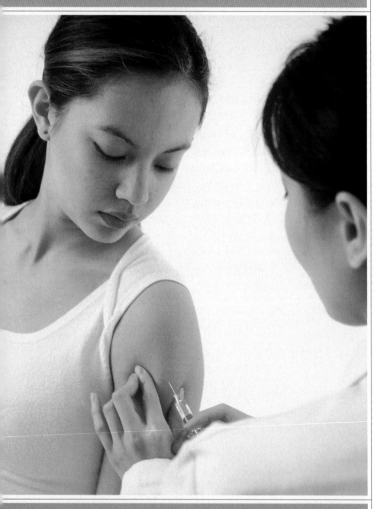

Mass vaccination campaigns have slowed or stopped the spread of a number of communicable diseases.

Pathogens Are Disease-Causing Organisms

- Certain bacteria produce toxins that cause disease
- Viruses can damage the host cell when they leave the cell after replication or when they are incorporated into the cell's chromosomes
- Protozoans cause disease by producing toxins and enzymes
- Fungi often cause disease by secreting enzymes that digest cells
- Parasitic worms cause disease by releasing toxins, feeding off blood, or competing with the host for food
- Prions induce disease by causing normal proteins to become misfolded and clump together

Disease Is Spread When a Pathogen Enters the Body through Contact, Consumption, or a Vector

Infectious Diseases Remain a Cause for Concern

- New diseases are emerging and some old diseases are reappearing
- Epidemiologists track diseases

Dani loved her boarding school. One of the best things about it was that the students came from all over the world. Living together in dormitories, they learned to accept and respect each other's differences and to recognize the more important ways they were all alike.

Unfortunately, during Dani's second year there, a minor drawback of boarding school life became apparent. Some of the students came from countries where children are not habitually immunized against childhood diseases. The school required immunization but often granted exceptions for religious or philosophical reasons. As a result, when one student developed measles a few days after returning from winter break, three others in the same dormitory became infected before anyone realized what was happening, and by the time the staff had been mobilized into action, a total of 10 students (including some who had been vaccinated when they were younger) had fallen ill. The sick ones had to stay in the health center or were hospitalized, the dormitories they had been sleeping in were quarantined, and any nonimmunized students who had not come down with measles were removed from the campus for a short time. No one became seriously ill, and in a few weeks, school life was back to normal, except that all had acquired a new appreciation for the challenges of disease control in the modern world.

More serious infectious diseases than measles have left and continue to leave their marks on humanity. In the past, the plague, cholera, and diphtheria decimated human populations. Today, tuberculosis (a lung disease) is making a comeback. HIV, the human immunodeficiency virus, has caused the deaths of millions of people worldwide.

In this chapter, we discuss the most important categories of disease-causing organisms, called *pathogens*, how they cause harm, how they are transmitted from person to person, and how they are studied so that steps may be taken to hold them in check. ■

Pathogens Are Disease-Causing Organisms

There are different types of pathogens and a wide range of differences within each type. As a result, each pathogen has specific effects on the body, and some pathogens are a greater menace than others. In this chapter, we will look at bacteria, viruses, protozoans, fungi, parasitic worms, and prions. We will consider the general means by which these different types of pathogen attack the body and cause symptoms. Keep in mind, however, as we do so, that some of the symptoms are caused not by the pathogen itself but by the immune responses our body uses to protect us (Chapter 13).

Virulence is the relative ability of a pathogen to cause disease. Some of the factors that contribute to this ability are the ease with which the pathogen invades tissues and the degree and type of damage it does to body cells. An organism that always causes disease, the typhoid bacterium, for instance, is highly virulent. On the other hand, the yeast *Candida albicans*, which *sometimes* causes disease, is moderately virulent.

■ Certain bacteria produce toxins that cause disease

Bacterial cells differ greatly from the cells that make up our bodies. Recall from Chapter 3 that our bodies are made up of eukaryotic cells that contain a true nucleus and membrane-bound organelles. Bacteria, in contrast, are prokaryotes, which means they lack a nucleus and other membrane-bound organelles (Figure 13a.1). Nearly all bacteria have a semirigid cell wall composed of a strong mesh of peptidoglycan, a type of polymer consisting of sugars and amino acids. The cell wall endows most types of bacteria with one of three common shapes: a sphere (a spherical bacterium is called a coccus) that can occur singly, in pairs, or in chains; a rod (bacillus) that usually occurs singly; or a spiral or corkscrew shape (spirilla).

Bacteria can reproduce rapidly. This rapid growth rate is a matter of concern because the greater the number of bacteria, the greater harm they can potentially do. Rapid reproduction is possible because bacteria reproduce asexually in a type of cell division called *binary fission* (Figure 13a.2), in which the bacterial genetic material (DNA) is copied, the cell is pinched in half, and each new cell contains a complete copy of the original genetic material. Under ideal conditions, certain bacteria can divide every 20 minutes. Thus, if every descendant lived, a single bacterium could result in a massive infection of trillions of bacteria within 24 hours.

Bacteria have defenses or other adaptive mechanisms that affect their virulence. Some bacteria have long, whiplike structures called flagella that allow them to move and spread through tissues. Bacteria may also have filaments called pili that help them attach to the cells they are attacking. Outside the bacterial cell, there is often a capsule that provides a means of adhering to a surface and prevents scavenger cells of the immune system (phagocytes; see Chapter 13) from engulfing them.

BACTERIAL ENZYMES AND TOXINS

Destructive enzymes and toxins (poisons) are among the offensive mechanisms of certain bacteria. Some of these bacteria secrete enzymes that directly damage tissue and cause lesions, allowing the bacteria to push through tissues like a bulldozer. An example is *Clostridium*, the bacterium that causes gas gangrene, a condition in which tissue dies because its blood supply is shut off. The bacterium secretes an enzyme that dissolves the material holding muscle cells together, permitting the bacteria to spread with ease. When this bacterium digests muscle cells for energy, a gas is produced that presses against blood vessels and shuts off the blood supply. In addition, *Clostridium* causes anemia by secreting an enzyme that bursts red blood cells.

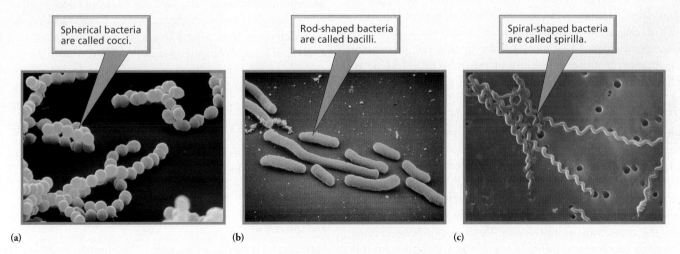

(a) (b) (c)

FIGURE **13a.1** Bacteria have three basic shapes: (a) round (coccus), (b) rod-shaped (bacillus), and (c) corkscrew-shaped (spirilla). All bacteria are prokaryotic cells, meaning they lack a nucleus and membrane-bound organelles.

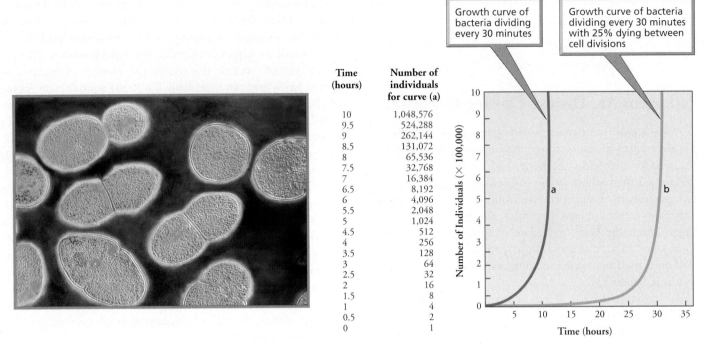

Time (hours)	Number of individuals for curve (a)
10	1,048,576
9.5	524,288
9	262,144
8.5	131,072
8	65,536
7.5	32,768
7	16,384
6.5	8,192
6	4,096
5.5	2,048
5	1,024
4.5	512
4	256
3.5	128
3	64
2.5	32
2	16
1.5	8
1	4
0.5	2
0	1

FIGURE **13a.2** Bacteria reproduce by binary fission; they copy their genetic information and pinch in half. This is one reason they reproduce very rapidly.

Most bacteria, however, do their damage by releasing toxins (poisons) into the bloodstream or the surrounding tissues. If the toxins enter the bloodstream, they can be carried throughout the body and disturb body functions.

The disease symptoms depend on which body tissues are affected by the toxin. Thus, the bacteria that cause various types of food poisoning have different effects. The most common of these bacteria, *Staphylococcus*, which is often found contaminating poultry, meat and meat products, and creamy foods such as pudding or salad dressing. These bacteria multiply when food is undercooked or unrefrigerated. The toxins they produce stimulate cells in the immune system to release chemicals that result in inflammation, vomiting, and diarrhea. Another type of food poisoning is caused by *Salmonella*, often encountered in undercooked contaminated chicken or eggs. In this case, the toxin causes changes in the permeability of intestinal cells, leading to diarrhea and vomiting. Contaminated meat, particularly ground meat, is often the cause of *Escherichia coli* (*E. coli*) food poisoning. Several major outbreaks have been traced to ground beef in hamburgers. In addition to causing vomiting and diarrhea, *E. coli* toxin can cause kidney failure in children and the elderly. The toxin that causes botulism, a type of food poisoning often brought on by eating improperly canned food, is one of the most toxic substances known. Produced by the bacterium *Clostridium botulinum*, it interferes with nerve functioning, especially motor nerves that cause muscle contraction. Death occurs because muscle paralysis prevents breathing. If enough of it is consumed, this toxin is almost always fatal.

BENEFICIAL BACTERIA

Many bacteria are not harmful except in the wrong place or when they reproduce too rapidly. Certain bacteria are important in food production, especially dairy products and alcoholic beverages. Others are important in the environment, serving as decomposers or driving the biogeochemical cycles (see Chapter 23). Yet other bacteria are important in genetic engineering (see Chapter 21). Some bacteria are normal residents in the body that keep potentially harmful microorganisms in check. For example, both group B *Streptococcus* and *E. coli* live harmlessly in most people's intestines, along with many other types of bacteria. The population of bacteria, especially strep bacteria, hold the *E. coli* in check.

Although harmless in the intestines, *E. coli* can have devastating, sometimes fatal, effects in the bloodstream. Moreover, both *E. coli* and group B *Streptococcus* can spread to a woman's vagina. If the woman is pregnant, the bacteria can be passed to the newborn during birth and cause serious problems. Beginning in 1990s, women who test positive for group B *Streptococcus* have been given the antibiotic ampicillin during delivery to prevent the spread of strep to the vulnerable newborn. However, killing the strep bacteria allowed the *E. coli* to flourish. As a result, strep infections in newborns declined, but *E. coli* infections rose. Unfortunately, *E. coli* can be more deadly than strep.

ANTIBIOTICS

Fortunately, bacteria can be killed. The human body has its own array of defenses. But when the body needs help, we can call on *antibiotics*, chemicals that inhibit the growth of microorganisms. Antibiotics work by reducing the number of bacteria or by slowing the growth rate of the population, allowing time for body defenses to conquer the bacteria. Some antibiotics kill bacteria directly by causing them to burst. Others slow bacterial growth by preventing the synthesis of bacterial cell walls. Recall that our body cells lack cell walls (see Chapter 3). Thus, our cells are unaffected by antibiotics that target cell walls. Some antibiotics block protein synthesis by bacteria but do so without interfering with cell protein synthesis in human body cells. This selective action is possible because the structure of ribosomes, the organelles on which proteins are synthesized, is slightly different in bacteria and humans. Unfortunately, some bacteria have become resistant to antibiotics, so we are losing our most powerful weapon against bacterial diseases.

ANTIBIOTIC RESISTANCE When antibiotics were introduced during the 1940s, they were considered to be miracle drugs. For the first time, there was a cure for devastating bacterial diseases such as pneumonia, bacterial meningitis, tuberculosis, and cholera. Today, there are over 160 antibiotics. These life-saving drugs have become so commonplace that we take them for granted.

Unfortunately, antibiotics are losing their power. Infections that were once easy to cure can now turn deadly as bacteria gain resistance to the drugs. Several bacterial species capable of causing life-threatening illnesses have produced strains that are resistant to every antibiotic available today.[1]

Contradictory as it may seem, the use of antibiotics can actually promote the development of resistant bacteria. When a strain of bacteria is exposed to an antibiotic, the bacteria that are susceptible die. The more resistant bacteria may survive and multiply. If the bacteria are exposed to the antibiotic again, the selection process is repeated. With each exposure to the drug, the resistant bacteria gain a stronger foothold. Making matters worse, antibiotics kill beneficial bacteria along with the harmful ones. Normally, the beneficial bacterial strains help keep the harmful strains in check. Loss of the "good" bacteria can allow the harmful ones to dominate. We saw this previously with *Streptococcus* and *E. coli*.

The overuse and misuse of antibiotics are largely to blame for the resistance problem. An example of overuse is when physicians prescribe antibiotics for illnesses that are viral, such as a cold or flu. This is overuse because antibiotics have no effect on viruses. Patients misuse antibiotics when they stop taking their medicine as soon as they feel better instead of completing the full course of treatment. By stopping too early, they may be leaving the bacteria with greater resistance alive. Hospitals use antibiotics heavily, so it is not surprising that they are breeding grounds for antibiotic-resistant bacteria. The resistant bacteria survive, outgrow suscepti-

ble strains, and spread from person to person. Indeed, most infections by antibiotic-resistant bacteria occur in hospitals. An example is offered by *Staphylococcus aureus*, which can cause many types of infections, including blood poisoning, pneumonia, skin infections, heart infections, and nervous system infections. The strain of *S. aureus* called MRSA, **m**ethicillin-**r**esistant *Staphylococcus* **a**ureus is actually resistant to many antibiotics. For many years, MRSA existed only in hospitals, but it is now found in the community at large. For a time, vancomycin was the only antibiotic that remained effective against MRSA. Unfortunately, a vancomycin-resistant *S. aureus* (VRSA) has arisen.

Over 40% (by mass) of the antibiotics used in the United States are given to livestock to promote growth. Farmers also spray crops with antibiotics to control or prevent bacterial infections in the crops. These practices also contribute to antibiotic resistance.

What can you do to slow the spread of drug-resistant bacteria? Use antibiotics responsibly. Do not insist on a prescription for antibiotics against your doctor's advice. Take antibiotics exactly as prescribed, and be sure to complete the treatment. Also, reduce your risk of getting an infection that might require antibiotic treatment by washing your hands frequently, rinsing fruits and vegetables before eating them, and cooking meat thoroughly.

■ Viruses can damage the host cell when they leave the cell after replication or when they are incorporated into the cell's chromosomes

Viruses, which are much smaller than bacteria, are responsible for many human illnesses. Some viral diseases, such as the common cold, are usually not very serious. Others, such as yellow fever, can be deadly.

Most biologists do not consider a virus to be a living organism because, on its own, it cannot perform any life processes. To copy itself, a virus must enter a cell, called the host cell. The virus exploits the host cell's nutrients and metabolic machinery to make copies of itself that then infect other host cells.

A virus consists of a strand or strands of genetic material, either DNA or RNA, surrounded by a coat of protein, called a capsid (Figure 13a.3). The genetic material carries the instructions for making new viral proteins. Some of these proteins become structural parts of the new viruses. Some of them serve as enzymes that help carry out biochemical functions important to the virus. Some are regulatory proteins, such as the proteins that trigger specific viral genes to become active under certain sets of conditions or the proteins that convert the host cell into a virus-producing factory.

Some viruses have an envelope, an outer membranous layer that is studded with glycoproteins. In some viruses, the envelope is actually a bit of plasma membrane from the previous host cell that became wrapped around the virus as it left the host cell. The envelope of certain other viruses, those in the herpes family, for instance, comes from a previous host cell's nuclear membrane. In any case, the virus produces the glycoproteins on the envelope.

[1]Bacteria resistant to all antibiotics available today include some strains of *Staphylococcus aureus* (skin infection, pneumonia), *Mycobacterium tuberculosis* (tuberculosis), *Enterococcus faecalis* (intestinal infections), and *Pseudomonas aeruginosa* (many types of infections).

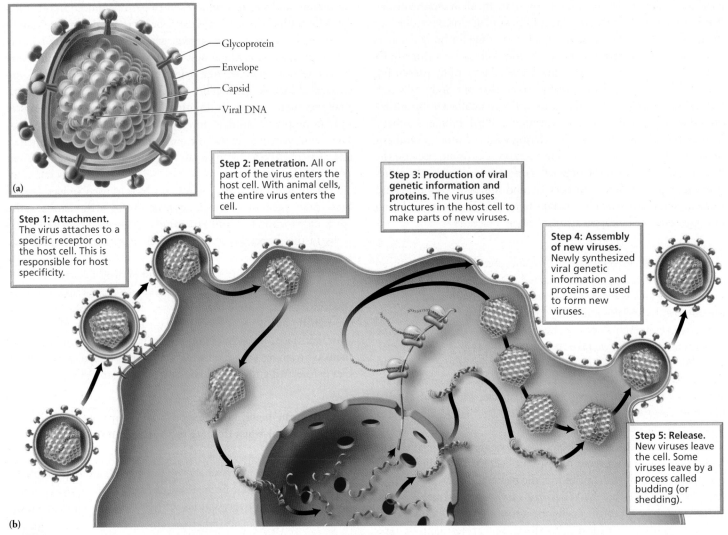

FIGURE **13a.3** (a) The structure of a typical virus. A protein coat, called a capsid, surrounds a core of genetic information made of DNA or RNA. Some viruses have an outer membranous layer, called the envelope, from which glycoproteins project. (b) Steps in viral replication.

A virus can replicate (make copies of itself) only when its genetic material is inside a host cell. Figure 13a.3 illustrates the general steps in the replication of viruses that infect animal cells:

1. **Attachment.** The virus gains entry by binding to a receptor (a protein or other molecule of a certain configuration) on the host cell surface. Such binding is possible because the viral surface has molecules of a specific shape that fit the host's receptors. The host cell receptors play a role in normal cell functioning. However, a molecule on the surface of the virus has a shape that is similar to the chemical that would normally bind to the receptor. Viruses generally attack only certain kinds of cells in certain species because a particular virus can infect only cells that bear a receptor that the virus can bind to. Whereas the virus that causes the common cold infects only cells in the respiratory system, the virus that causes hepatitis infects only liver cells.

2. **Penetration.** After a virus has bound to a receptor on an animal cell, the entire virus enters the host cell, often by phagocy-

tosis by the host cell. Once inside, the virus is stripped of its capsid, leaving only its genetic material intact.

3. **Production of viral genetic information and proteins.** Viral genes then direct the host cell machinery to make thousands of copies of viral DNA or RNA. Next, viral genes direct the synthesis of viral proteins, including coat proteins and enzymes.

4. **Assembly of new viruses.** Copies of the viral DNA (or RNA) and viral proteins then assemble to form new viruses.

5. **Release.** Some viruses—those that acquire an envelope—leave the cell in a process called budding, or shedding. The newly formed viruses push through the host cell's plasma membrane and become wrapped in the membrane, which forms the envelope. Budding need not kill the host cell. Other virus types cause the host cell membrane to rupture, releasing the newly formed viruses and killing the host cell.

Viruses can cause disease in several ways, as summarized in Table 13a.1. Some viruses cause disease when they kill the host's cells or cause the cells to malfunction. The host cell dies when viruses leave it so rapidly that it bursts, or *lyses*. In such cases, the symptoms of the disease will depend on which cells are killed. However, if viruses are shed slowly, the host cell may remain alive and continue to produce new viruses. Slow shedding causes *persistent infections* that can last a long time. Some viruses can produce *latent infections*, in which the viral genes remain in the host cell for an extended period without causing harm to the cell. At any time, however, the virus can begin replicating and cause cell death as new viruses are released.

An example of a virus that can act in all of these ways is the herpes simplex virus that causes fever blisters. The virus is spread by contact (discussed shortly) and enters the epithelial cells of the mouth, where it actively replicates. Rapid shedding kills the host cells, causing fever blisters. Slow shedding may not cause outward signs of infection, but the virus can still be transmitted. When the blisters are gone, the virus remains in a latent form within nerve

cells without causing symptoms. However, stress can activate the virus. It then follows nerves to the skin in the region of the mouth and begins actively replicating, causing new blisters.

Certain viruses can also cause cancer. Some do this when they insert themselves into the host chromosome near a cancer-causing gene and, in so doing, alter the functioning of that gene. Still other viruses bring cancer-causing genes with them into the host cell.

Unfortunately, viruses are not as easy to destroy as bacteria. One reason is the difficulty of attacking viruses inside their host cells without killing the host cell itself. Most attempts to develop antiviral drugs have failed for this reason. Nonetheless, some drugs are now available to slow viral growth, and others are being developed. Most of the antiviral drugs available today, including those against the herpes virus and HIV, work by blocking one of the steps necessary for viral replication. As mentioned in Chapter 13, interferons are proteins produced by virus-infected cells that protect neighboring cells from all strains of viruses. Interferons are not as useful as originally hoped, but they have been used for certain viral infections, including hepatitis C and the human papilloma virus that causes genital warts.

Because of these obstacles to treatment, the best way to fight viral infections is to prevent them with vaccines (discussed in Chapter 13).

■ Protozoans cause disease by producing toxins and enzymes

Protozoans are single-celled eukaryotic organisms, having a well-defined nucleus. They can cause disease by producing toxins or by releasing enzymes that prevent host cells from functioning normally. Protozoans are responsible for many diseases, including malaria (discussed shortly), sleeping sickness, amebic dysentery, and giardiasis. Giardiasis is a diarrheal disease that can last for weeks. There are frequent outbreaks of giardiasis in the United States, most of them resulting from water supplies contaminated with human or animal feces. Even clear and seemingly clean lakes and streams in the wilderness can contain *Giardia* (Figure 13a.4). Fortunately, drugs are available to treat protozoan infections. Some of these drugs work by preventing protozoans from synthesizing proteins.

TABLE 13A.1 POSSIBLE EFFECTS OF ANIMAL VIRUS ON CELLS	
Lytic infection	Rapid release of new viruses from infected cell causes cell death. Symptoms of the disease depend on which cells are killed.
Persistent infection	Slow release of new viruses causes cell to remain alive and continue to produce new viruses for a prolonged period of time.
Latent infection Primary infection Latent period Secondary infection	Delay between infection and symptoms. Virus is present in the cell without harming the cell. Symptoms begin when the virus begins actively replicating and causes cell death when new viruses exit the cell.
Transformation to cancerous cell	Certain viruses insert their genetic information into host cell chromosomes. Some carry oncogenes (cancer-causing genes) that are active in the host cell. Some disrupt the functioning of the host cell's genes that regulate cell division, causing the cell to become cancerous.

FIGURE **13a.4** *Giardia* is a protozoan that is commonly found in lakes and streams used as sources of drinking water, even those in pristine areas. It causes severe diarrhea that lasts for weeks and can be especially dangerous for children.

■ Fungi often cause disease by secreting enzymes that digest cells

Like the protozoans, fungi are also eukaryotic organisms, with a well-defined nucleus in their cells. Some fungi exist as single cells. Others are organized into simple multicellular forms, with not much difference among the cells. There are over 100,000 species of fungi, but less than 0.1% cause human ailments. Fungi obtain food by infiltrating the bodies of other organisms—dead or alive—secreting enzymes to digest the food, and absorbing the resulting nutrients. If the fungus is growing in or on a human, body cells of the human are digested, causing disease symptoms. Some fungi cause serious lung infections, such as histoplasmosis and coccidioidomycosis. Other, less-threatening fungal infections occur on the skin and include athlete's foot, ringworm, and vaginitis (discussed in Chapter 17). Most fungal infections can be cured. Fungal cell membranes have a slightly different composition from those of human cells. As a result, the membrane is a point of vulnerability. Some antifungal drugs work by altering the permeability of the fungal cell membrane. Others interfere with membrane synthesis by fungal cells. Fungal infections of the skin, hair, and nails can be combated with a drug that prevents the fungal cells from dividing.

■ Parasitic worms cause disease by releasing toxins, feeding off blood, or competing with the host for food

The parasitic worms are multicellular animals. They include flukes, tapeworms, and roundworms, such as hookworms and pinworms. They can cause illness by releasing toxins into the bloodstream, feeding off blood, or competing for food with the host. Parasitic worms cause many serious human diseases, including ascariasis, schistosomiasis, and trichinosis.

Ascariasis is caused by a large roundworm, *Ascaris*, that is about the size of an earthworm. People become infected with *Ascaris* when they consume food or drink contaminated with *Ascaris* eggs. The eggs develop into larvae (immature worms) in the person's intestine. The larvae then penetrate the intestinal wall, enter the bloodstream, and travel to the lungs. After developing further, the worms are coughed up and swallowed, thus returning to the intestine. Within 2 to 3 months, they mature into male and female worms, which live for about 2 years. During those years, female worms can produce more than 200,000 eggs a day.

As much as 25% of the world population is infected with *Ascaris*, particularly in tropical regions. Up to 50% of the children in some parts of the United States (mostly rural areas in the Southeast) are infected. Many people with ascariasis have no symptoms. However, the worms can cause lung damage and severe malnutrition. When many worms are present, they can block or perforate the intestines, leading to death.

■ Prions induce disease by causing normal proteins to become misfolded and clump together

Prions (*pree*-ons) are infectious particles of proteins—or, more simply, infectious proteins. They cause a group of diseases called transmissible spongiform encephalopathies (TSEs), which are associated with degeneration of the brain. Several of the TSEs are animal infections, notably mad cow disease, scrapie in sheep, and chronic wasting disease, which is spreading through herds of deer and elk in the United States (Figure 13a.5). Mad cow disease severely damaged the British beef industry in the 1990s. The U.S. and Canadian governments responded quickly when the first cases of mad cow disease appeared in these countries in 2003. In 2006, cases of atypical mad cow disease in Texas and in Alabama appeared to be a different strain, similar to cases in France, Italy, and Japan. This strain appears spontaneously and some researchers fear that it may be spread from cow to cow.

Prions also cause a human neurological disorder called Creutzfeldt-Jakob disease (CJD). Indeed, the prion responsible for mad cow disease is thought to cause one form of CJD. The incubation period for CJD can be months to decades. Symptoms include sensory and psychiatric problems. Once the symptoms begin, death usually occurs within a year.

Prions are misfolded versions of a harmless protein normally found on the surface of nerve cells. The host nerve cell produces a normal version of the protein. However, if a prion is present, it somehow causes the host protein to change its shape to the abnormal form. The misshapen proteins clump together and accumulate in the nerve tissue of the brain. These clumps of prions may damage the plasma membrane or interfere with molecular traffic. Spongelike holes develop in the brain, causing death. Prions cannot be destroyed by heat, ultraviolet light, or most chemical agents. Currently, there is no treatment for any disease they cause.

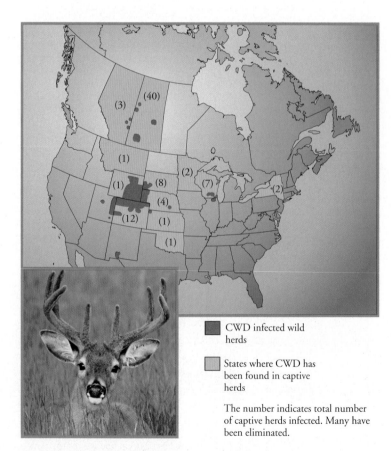

■ CWD infected wild herds

■ States where CWD has been found in captive herds

The number indicates total number of captive herds infected. Many have been eliminated.

FIGURE **13a.5** Chronic wasting disease (CWD), which is caused by a prion, is spreading through herds of deer and elk in the United States.

How does an animal become infected with prions? In the case of mad cow disease, it appears that prions eaten in contaminated food are often the source of infection. For example, prions have been passed along in the protein supplements fed to cattle to increase their growth and milk production. Those protein supplements had been prepared from the carcasses of animals considered unfit for human consumption, a practice banned in the United States in 1997. Any protein supplements prepared from animals infected with mad cow disease would have contained prions. The prions pass through the intestinal wall, enter the lymphatic system, and are then transported by nerves to the brain and spinal cord, where they destroy nerve cells. In contrast, chronic wasting disease can apparently be spread by animal-to-animal contact, including contact with body fluids such as urine or feces from infected animals. Scientists also think that the prions responsible for chronic wasting disease may remain in the soil or water for years. As a result, healthy animals may become infected from living in a region previously occupied by diseased animals. Humans can become infected with prions by eating contaminated substances, through tissue transplant, or through contaminated surgical instruments.

Disease Is Spread When a Pathogen Enters the Body through Contact, Consumption, or a Vector

Obviously, you catch a disease when the pathogen enters your body, but how do diseases travel from person to person? The answer to this question varies with the type of pathogen.

- **Direct contact** One means of transmission is direct contact of an infected person with an uninfected person, as might occur when shaking hands, hugging and kissing, or being sexually intimate. For example, sexually transmitted diseases (STDs) are spread when a susceptible body surface touches an infected body surface (see Chapter 17a). The organisms that cause STDs generally cannot remain alive outside the body for very long, so direct intimate contact is necessary. A few disease-causing organisms, HIV and the bacterium that causes syphilis, for instance, can spread across the placenta from a pregnant woman to her growing fetus.

- **Indirect contact** Indirect contact—the transfer from one person to another without their touching—can spread other diseases. Most respiratory infections, including the common cold, are spread by indirect contact (see Chapter 14). When an infected person coughs or sneezes, airborne droplets of moisture full of pathogens are carried through the air (Figure 13a.6). The droplets may be inhaled or land on nearby surfaces. When another person touches an affected surface, the organisms are transmitted. In this way, pathogens may be spread on contaminated inanimate objects, including doorknobs, drinking glasses, and eating utensils.

- **Contaminated food or water** Certain diseases are transmitted in contaminated food or water. You have read that spoiled food

FIGURE **13a.6** Pathogens can be spread through the air in droplets of moisture when an infected person sneezes or coughs.

can cause food poisoning. Other diseases transmitted by food or water include hepatitis A, an inflammation of the liver caused by a certain virus. *Legionella*, the bacterium that causes a severe respiratory infection known as Legionnaires' disease, is a common inhabitant of the water in condensers of large air conditioners and cooling towers. The disease-causing bacteria are spread through tiny airborne water droplets. Coliform bacteria come from the intestines of humans and are, therefore, an indicator of fecal contamination of water. Their numbers are monitored in drinking and swimming water. To be safe, drinking water should not have any coliform bacteria.

- **Animal vectors** Another means of transmission is by *vector*, an animal that carries a disease from one host to another. The most common vectorborne disease in the United States is Lyme disease. It is caused by a bacterium transmitted by the deer tick (the vector), an insect that is about the size of the head of a pin (Figure 13a.7). The tick larva picks up the infectious agent when it bites and sucks blood from an infected animal. When the tick subsequently feeds on a human or other

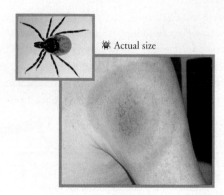

❈ Actual size

FIGURE **13a.7** A tiny tick, the deer tick, is a vector that transmits the bacterium responsible for Lyme disease. One characteristic sign of Lyme disease is a red bull's-eye rash surrounding the tick bite. The rash gradually increases in diameter.

mammalian host, the bacteria gradually move from the tick's gut to its salivary glands and then to its victim. The incubation period, during which there are no symptoms, can be as long as 6 to 8 weeks. Early symptoms include a headache, backache, chills, and fever. Often, a rash resembling a bull's eye develops, with an intense red center and border. Over a period of weeks, the circle increases in diameter. Weeks to months later, unless the disease is treated promptly, pain, swelling, and arthritis may develop. Cardiovascular and nervous system problems may follow the arthritis.

Mosquitoes transmit the protozoan parasite *Plasmodium* that causes malaria. This disease affects about 300 million to 500 million people worldwide and kills 1 million to 3 million people each year. The life cycle of this protozoan requires two hosts: a mosquito and a human. When an infected *Anopheles* mosquito feeds on human blood, the protozoans enter the

human's bloodstream (Figure 13a.8). After maturing in the liver, they return to the blood, where they infect red blood cells. They then reproduce within the red blood cells, which eventually burst and release more protozoans, which will infect additional red blood cells and gametocytes, cells that will develop into gametes. Toxins produced by the protozoans are also released when the red blood cells burst, causing chills, fevers, and vomiting. Infection and the bursting of red blood cells occur in cycles of 2 to 3 weeks. Several factors related to the loss of red blood cells can cause death. If a mosquito bites an infected person when the gametocytes are in the blood, the mosquito will ingest the gametocytes. In the gut of the mosquito, the gametocytes mature into gametes, which will fuse and develop into a new generation of protozoans. These protozoans then migrate to the mosquito's salivary glands. After about a week, the mosquito can infect another person.

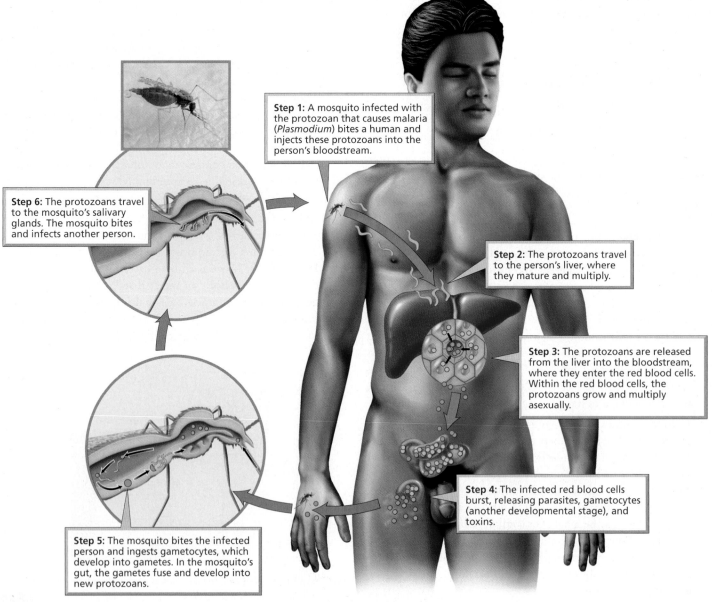

Step 1: A mosquito infected with the protozoan that causes malaria (*Plasmodium*) bites a human and injects these protozoans into the person's bloodstream.

Step 6: The protozoans travel to the mosquito's salivary glands. The mosquito bites and infects another person.

Step 2: The protozoans travel to the person's liver, where they mature and multiply.

Step 3: The protozoans are released from the liver into the bloodstream, where they enter the red blood cells. Within the red blood cells, the protozoans grow and multiply asexually.

Step 4: The infected red blood cells burst, releasing parasites, gametocytes (another developmental stage), and toxins.

Step 5: The mosquito bites the infected person and ingests gametocytes, which develop into gametes. In the mosquito's gut, the gametes fuse and develop into new protozoans.

WEB TUTORIAL 13a.2   FIGURE **13a.8**   The life cycle of *Plasmodium*, the protozoan that causes malaria

Malaria was known as far back 1000 BC, so why haven't people found a cure or means of preventing the deadly disease? Antimalarial drugs are being developed, but the creation of antimalarial vaccines has proved difficult. So far, the most effective approach has been to combat malaria by combating the mosquito. Spraying with insecticides to eliminate populations of the mosquitoes is the traditional way of controlling them. In addition, researchers have genetically modified some of the mosquitoes with a gene that prevents the development of the protozoan within them. The modified mosquitoes are unable to transmit the protozoan to mice; what is not yet known is if the same gene could prevent the transmission of malaria to humans.

Infectious Diseases Remain a Cause for Concern

An epidemic is a large-scale outbreak of an infectious disease. The most notorious epidemics—the bubonic plague, cholera, diphtheria, and smallpox—are mostly history, although new outbreaks may occur sporadically. However, outbreaks of serious new diseases continue to present problems. We discuss some of these modern-day plagues in this chapter and elsewhere in the text (Table 13a.2).

■ New diseases are emerging, and some old diseases are reappearing

Since the mid-1970s, about 30 new virulent diseases have appeared, causing tens of millions of deaths. Other diseases have reemerged that were thought to have been conquered. An *emerging disease* is a disease with clinically distinct symptoms whose incidence has increased, particularly over the last two decades. A *reemerging disease* is a disease that has reappeared after a decline in incidence (Figure 13a.9). We will consider three factors that play important roles in the emergence and reemergence of disease.

TABLE 13A.2 EXAMPLES OF MODERN-DAY INFECTIOUS THREATS

DISEASE	CAUSE	DISCUSSION
Meningitis	Bacterium or virus	Chapter 8
West Nile disease	Virus	Chapter 8
Malaria	Protozoan	Chapter 13a
Lyme disease	Bacterium	Chapter 13a
Transmissible spongiform encephalolopathies (TSEs; e.g., mad cow disease, chronic wasting syndrome, Creutzfeldt-Jakob disease)	Prion	Chapter 13a
Hantavirus pulmonary syndrome	Virus	Chapter 13a
Avian flu	Virus	Chapter 13a
Tuberculosis	Bacterium	Chapter 14
Severe acute respiratory syndrome (SARS)	Virus	Chapter 14
Influenza	Virus	Chapter 14
Hepatitis C	Virus	Chapter 15
Chlamydia	Bacterium	Chapter 17a
Gonorrhea	Bacterium	Chapter 17a
Genital herpes	Virus	Chapter 17a
Genital warts	Virus	Chapter 17a
HIV/AIDS	Virus	Chapter 17a

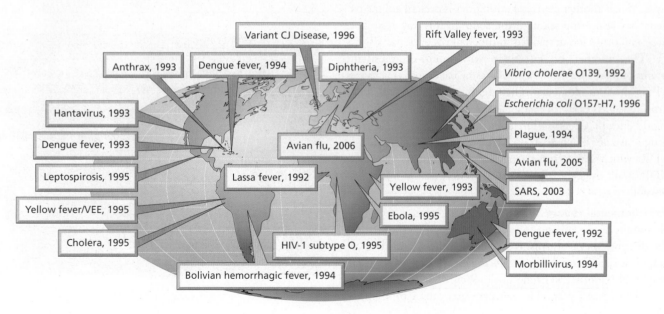

FIGURE **13a.9** Recent emerging or reemerging diseases

1. **Development of new organisms that can infect humans and of drug-resistant organisms.** Most of the time, a pathogen infects only one type or a few types of organism. Mutations are changes in genetic information that occur randomly. Some mutations allow the pathogen to "jump species" from its original host and infect another type of organism. For example, the H5N1 influenza virus that causes a particularly dangerous form of bird flu can jump from birds to humans by direct contact with infected birds or their droppings. Recall that a virus can penetrate a cell only if the virus has the appropriate molecule on its surface—one that will fit into a receptor on the host cell. Mutations—spontaneous, random changes in genetic information—have altered surface proteins on the H5N1 influenza virus so that it can now also infect humans. Another mechanism that could allow an animal virus to infect humans is if an animal virus and a human virus infected the same cell and their genetic information became mixed.

   Pathogens can also undergo changes in their response to drugs. We have seen that certain bacteria have acquired resistance to antibiotics, for example. As a result, some diseases that were once easily cured by antibiotics are now much more difficult to treat. Improper antibiotic treatment during the reemergence of tuberculosis (TB) has led to antibiotic resistant strains of TB, which, in turn, makes it more difficult to treat. In the United States, TB reached an all time low in 2005, but globally it is still the second leading cause of death.

   In some parts of the world multi-drug-resistant tuberculosis accounts for more than 10% of the new cases of tuberculosis a year. This drug resistance develops because many patients stop taking the antibiotic before the completion of 6 to 12 months of treatment. The reasons for stopping treatment range from forgetfulness to poverty. Not completing the antibiotic regimen allows the more resistant bacteria that withstood the initial drug onslaught to reproduce, creating strains of bacteria that are difficult or impossible to kill. The resistant bacteria then grow unchecked, and the patient suffers a relapse. When another course of antibiotics is prescribed, the patient may again stop taking it too soon. In this way, resistance to several drugs has developed in tuberculosis bacteria.

   The number of infections with multi-drug-resistant strains of *Mycobacterium tuberculosis*, the bacterium that causes TB, is increasing at an alarming rate. Between 2003 and 2004, drug-resistant cases of tuberculosis increased 13.3%. The World Health Organization coined a new term to describe drug resistance in a new strain of the tuberculosis bacterium—XDR, which stands for extensively drug resistant. The new XDR strain causes a tuberculosis infection that is nearly impossible to treat.

2. **Environmental change.** Changes in local climate—the annual amount of rainfall and the average temperature—can affect the distribution of organisms and change the size of the geographical region in which certain organisms can live. For example, El Niño is a warming of the Pacific Ocean that occurs in cycles of 3 to 7 years. The warmer water causes atmospheric changes that, in turn, cause climate changes. The 1991–1992 El Niño and the 1997–1998 El Niño caused wet and mild winters in the southwestern United States. Scientists think these climate changes were responsible for outbreaks of the hantavirus in the Southwest in 1993 and 1999. In each of these episodes, the change in climate led to a population explosion of adult rodents. Rodents, especially the deer mouse and the cotton rat, carry the hantavirus. The virus is transmitted to people when they inhale air contaminated by virus shed in the rodents' urine, feces, and saliva. The hantavirus causes a severe respiratory illness similar to influenza. It can result in respiratory failure and, in some cases, death.

3. **Population growth.** Another important factor in the emergence or reemergence of diseases is the increase of the human population in association with the development and growth of cities. Swelling human populations in cities cause people to move out of the city into surrounding areas, creating suburbs. If the surrounding areas were previously undeveloped, the move brings more people into contact with animals and insects that might carry infectious organisms. Indeed, wild animals serve as reservoirs for more than a hundred species of pathogens that can affect humans. The development of suburbs also destroys populations of predators, such as foxes and bobcats. In some regions of New York, the loss of predators has led to an increase in numbers of tick-carrying mice and an increase in the incidence of Lyme disease.

   Population density and mobility also enable infectious diseases to spread more easily today than in the past. Densely populated cities allow diseases to begin spreading quickly, and air travel enables them to spread over great distances, as we saw during the outbreak of severe acute respiratory syndrome (SARS) in 2003 (Figure 13a.10).

FIGURE **13a.10** Air travel is one reason that new diseases can spread rapidly.

■ **Epidemiologists track diseases**

Epidemiology is the study of patterns of disease, including rate of occurrence, distribution, and control. Most diseases can be described as having one of the following four patterns:

- *Sporadic diseases* occur only occasionally at unpredictable intervals. They affect a few people within a restricted area. Typhoid fever, an intestinal infection, is a sporadic disease in the United States. It is contracted by consuming food or water infected with the bacterium *Salmonella typhi*.
- *Endemic diseases* are always present in a population and pose little threat. The common cold provides an example.
- An *epidemic disease* occurs suddenly and spreads rapidly to many people. Outbreaks of smallpox and cholera are examples of epidemics.
- A *pandemic* is a global outbreak of disease. HIV/AIDS is considered to be a pandemic. There is grave concern that avian flu caused by the H5N1 virus may become pandemic.

Epidemiologists are "disease detectives" who try to determine why a disease is triggered at a particular time and place. The first step in answering this question is to verify that there is indeed a disease outbreak, defined as more than the expected number of cases of individuals with similar symptoms in a given area. Next, epidemiologists try to identify the cause of the disease, whether it can be transmitted to other people, and, if it can be, how the disease is transmitted. To identify the cause of an infectious disease, epidemiologists try to isolate the same infectious agent from all people showing symptoms of the condition. In addition, they try to identify factors, including age, sex, race, personal habits, and geographic location, shared by people with symptoms of the condition. These factors might provide a clue as to whether the condition can be transmitted and how. Information about whether the incidence of the disease changes with seasons can also be important in determining the means of transmission. For example, the incidence of West Nile disease increases in the summer because mosquitoes, which are prevalent during warmer months, transmit the virus responsible for the disease. On the other hand, influenza, which is spread by droplet infection, is more common in the winter, when people are indoors and in closer contact with each other. Intestinal pathogens are usually spread in food or water, so epidemiologists look for similarities in the food and drink consumed among affected people and the locations in which consumption occurred.

As an example of how epidemiologists track a disease, we will consider how the cause of SARS, the first new infectious disease of the millenium, was determined and controlled within a few months during 2003. SARS is a type of pneumonia characterized by a high fever and was first reported in China in February 2003. A Chinese physician who cared for patients with pneumonia developed symptoms himself. Nonetheless, he traveled to Hong Kong and stayed in a hotel while visiting relatives. The next day, he was admitted to the hospital and died of respiratory failure a week or so later. Shortly afterward, people displaying the symptoms that have come to be associated with SARS appeared in Hong Kong, Singapore, Vietnam, Canada, and the United States. By questioning these early victims, epidemiologists determined that the factor they all had in common was contact with the Chinese physician. Some patients were guests in the same hotel and others were close family members. This pattern of occurrence is typical of a disease caused by an infectious agent. Therefore, measures for limiting the spread of the infectious agent were put in place. These included quarantining thousands of people, restricting travel, and even checking the body temperature of people in airports. Although SARS affected more than 8000 people in 30 countries and caused more than 800 deaths in the months that followed, the restrictive measures were eventually successful in limiting its spread.

Other scientists determined that the cause of SARS was a specific type of virus called a coronavirus. It was identified as the cause of SARS because it could be isolated from all SARS patients tested. We now know that SARS is transmitted from person to person in droplets of moisture spread into the air when an infected person coughs. The virus is very stable: it can cause infection after 24 hours on a dry surface.

14

The Respiratory System

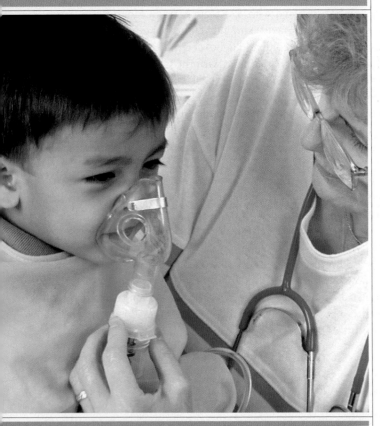

Asthma can interfere with the key objective of the respiratory system, which is the exchange of gases—oxygen and carbon dioxide—across body membranes.

In the Respiratory System, Oxygen and Carbon Dioxide Are Exchanged across a Moist Body Surface
- The nose filters and conditions incoming air and has receptors for the sense of smell
- The sinuses lighten the head and condition inhaled air
- The pharynx is a passageway for food and air
- The larynx is an adjustable entryway that produces the voice
- The trachea is the windpipe
- The bronchial tree is a system of air tubules that conducts air
- The alveoli of the lungs are surfaces for gas exchange

Pressure Changes within the Lungs Cause Breathing
- Inhalation occurs when the pressure in the lungs decreases
- Exhalation occurs when the pressure in the lungs increases
- The volume of air moved into and out of the lungs is an indication of health

Blood Transports Gases between the Lungs and the Cells
- Most oxygen is carried by hemoglobin
- Most carbon dioxide is transported as bicarbonate ions

Breathing Is Controlled Primarily by Respiratory Centers in the Brain
- Brain centers control the basic breathing pattern
- Depth and rate of breathing are affected by chemoreceptors

Respiratory Disorders Have Many Causes
- The common cold is caused by many types of viruses
- The flu is caused by three types of virus
- Pneumonia is an inflammation of the lungs
- Strep throat can have serious consequences
- Tuberculosis causes tubercles to form in the lungs
- Bronchitis is an inflammation of the bronchi
- Emphysema is caused by the destruction of alveoli

SOCIAL ISSUE Bird Flu, Will It Become a Pandemic?

HEALTH ISSUE Surviving a Common Cold

ENVIRONMENTAL ISSUE Air Pollution and Human Health

evon remembers the day he decided to be a respiratory therapist. He was 16, and his stepbrother Jordan had to be taken to the hospital because of an asthma attack. Jordan had been coughing and wheezing and seemed to be having trouble with his inhaler, so Devon's mom and Devon put him in the car, and they all went to the emergency room. The nurse at the registration desk said they were in luck because a respiratory therapist was on duty and she really knew her stuff.

While Mom was filling out the paperwork and phoning Jordan's father, an orderly wheeled Jordan to the treatment area. Devon went along to reassure Jordan, who was clearly struggling for air. Fortunately, the respiratory therapist was there when they arrived. She looked Jordan over quickly, checking his pulse and temperature and attaching a clip to his finger that she said would measure the oxygen in his blood. Next she fitted a face mask over Jordan's nose and used a powerful compressor to deliver a nebulized rescue medication into his airways. By the time Mom caught up with them, Jordan was breathing quietly and Devon was asking excited questions about the procedures he'd just observed.

This chapter explains why the respiratory system is so vital to life that certain respiratory diseases can be life threatening. It begins by following the course of inhaled air to the lungs and describing the mechanics of breathing. It then considers the transport of oxygen and carbon dioxide between the lungs and the cells and examines the control of respiration. Finally, it discusses several disorders of the respiratory system. ■

In the Respiratory System, Oxygen and Carbon Dioxide Are Exchanged across a Moist Body Surface

ithout oxygen, we would die within a few minutes. Why? To stay alive, our cells need energy, and oxygen plays an essential role extracting energy from food molecules (see Chapter 3). We store the extracted energy by producing a molecule called ATP (adenosine triphosphate), which then releases the energy as needed to do the work of the cell. Our cells can make a little ATP without oxygen but not enough to supply the energy needs of the body. Cells can make 18 times more ATP if oxygen is present.

The same chemical reactions that require oxygen for the production of ATP produce carbon dioxide as a by-product. In solution—for example, in water or blood—carbon dioxide forms carbonic acid, which can be harmful to cells.

The function of the **respiratory system** is to provide the body with oxygen and dispose of carbon dioxide, an exchange that also regulates the acidity of body fluids. Four processes play a part in respiration (Figure 14.1).

- **Breathing (ventilating):** bringing oxygen-rich air into the lungs and moving carbon dioxide-laden air out of the lungs.
- **External respiration:** the exchange of oxygen and carbon dioxide between the lungs and the blood. Oxygen moves from the lungs into the blood, and carbon dioxide moves from the blood into the lungs.
- **Gas transport:** transport of oxygen from the lungs to the cells and of carbon dioxide from the cells to the lungs.

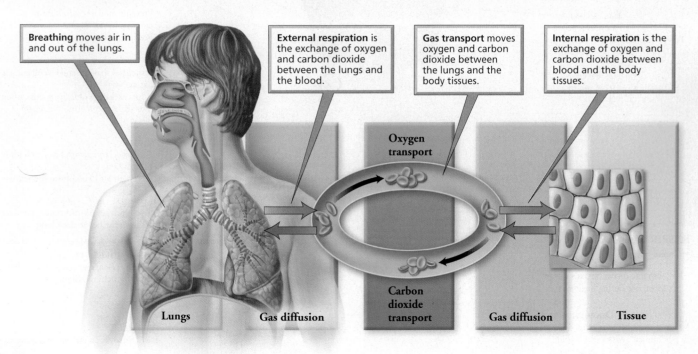

Breathing moves air in and out of the lungs.

External respiration is the exchange of oxygen and carbon dioxide between the lungs and the blood.

Gas transport moves oxygen and carbon dioxide between the lungs and the body tissues.

Internal respiration is the exchange of oxygen and carbon dioxide between blood and the body tissues.

Oxygen transport

Carbon dioxide transport

Lungs   Gas diffusion   Gas diffusion   Tissue

WEB TUTORIAL 14.1 FIGURE **14.1** An overview of respiration

- **Internal respiration:** the exchange of oxygen and carbon dioxide between the blood and the cells. Oxygen moves from the blood to the cells, where it is used in cellular respiration to produce ATP and carbon dioxide. Carbon dioxide produced by the cells moves into the blood.

We begin our exploration of how humans obtain oxygen and dispose of carbon dioxide by following the path of air from the nose to the lungs. The structures the air passes along the way are identified and described in Figure 14.2 and Table 14.1. The path the air travels is summarized in Figure 14.3. The respiratory system is generally divided into upper and lower regions. The nose (nasal cavities) and pharynx make up the *upper respiratory system*. The *lower respiratory system* consists of the larynx, trachea, bronchi, bronchioles, and lungs.

■ **The nose filters and conditions incoming air and has receptors for the sense of smell**

However large someone's nose might seem from the outside, the inside is not as roomy as you might imagine. One reason is that a thin partition of cartilage and bone called the *nasal septum* divides the inside of the nose into two **nasal cavities**. In addition, much of the space within the nasal cavities is taken up by three convoluted,

shelflike bones. These bones increase the surface area inside the nasal cavities and divide each cavity into three narrow passageways through which the air flows. Moist mucous membrane covers the entire inner surface of the nasal cavities.

We all know what a nose looks like, but what does a nose do? Your nose has three important functions: (1) filtration and cleansing, (2) conditioning the air, and (3) olfaction (smell).

FILTRATION AND CLEANSING

The nose helps clear particles from the air that moves through its passages. Considering that each of us inhales about 150,000 bacteria along with a great deal of pollen and dust each day, filtering the air is an important job. Most pollen and dust particles are removed from the air before they reach the lungs.

The nose and air tubules (described below) clean inhaled air in a variety of ways. Hairs inside the nose filter out the largest particles. In addition, certain cells in the membrane lining the surface of the nasal cavities and air tubules produce mucus, a sticky substance that catches dust particles. Cilia, tiny projections extending from the membranous lining, then sweep the mucus, trapped dirt particles, and bacteria toward the throat, where they can be swallowed and subsequently be destroyed by digestive enzymes (Figure 14.4). Particles that do not become trapped in the nasal cavities or the air tubules are deposited in the lungs.

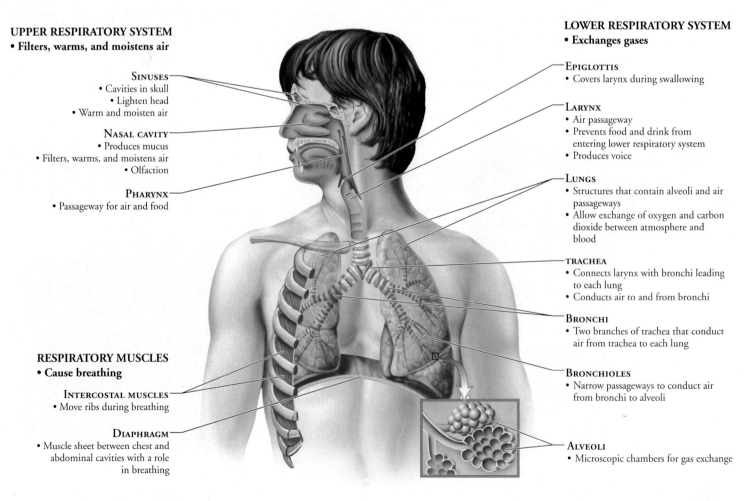

FIGURE **14.2** The respiratory system

TABLE 14.1 REVIEW OF STRUCTURES OF THE RESPIRATORY SYSTEM

STRUCTURE	DESCRIPTION	FUNCTION
Upper respiratory system		
Nasal cavity	Cavity within the nose, divided into right and left halves by nasal septum; has three shelflike bones	Filters and conditions (moistens and warms incoming air); olfaction (sense of smell)
Pharynx (throat)	Chamber connecting nasal cavities to esophagus and larynx	Common passageway for air, food, and drink
Lower respiratory system		
Larynx	Cartilaginous boxlike structure between the pharynx and trachea that contains the vocal cords and the glottis	Allows air but not other materials passage to the lower respiratory system; source of the voice
Epiglottis	Flap of tissue reinforced with cartilage	Covers the glottis during swallowing
Trachea	Tube reinforced with C-shaped rings of cartilage that leads from the larynx to the bronchi	The main airway; conducts air from larynx to bronchi
Bronchi (primary)	Two large branches of the trachea reinforced with cartilage	Conduct air from trachea to each lung
Bronchioles	Narrow passageways leading from bronchi to alveoli	Conduct air to alveoli; adjust airflow in lungs
Lungs	Two lobed, elastic structures within the thoracic (chest) cavity containing surfaces for gas exchange	Exchange oxygen and carbon dioxide between blood and air
Alveoli	Microscopic sacs within lungs, bordered by extensive capillary network	Provide immense, internal surface area for gas exchange

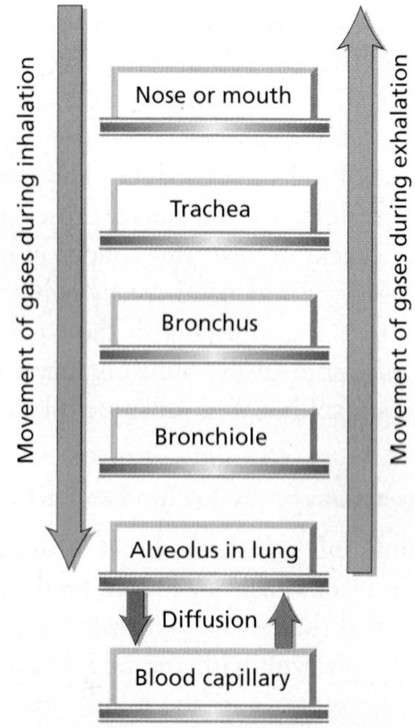

CONDITIONING THE AIR

The nose also warms and moistens the inhaled air before it reaches the delicate lung tissues. Warming the air before it reaches the lungs is extremely important in cold climates because frigid air can kill the delicate cells of the lung. Moistening the inhaled air is also essential because oxygen cannot cross dry membranes. Mucus helps moisten the incoming air so that lung surfaces do not dry out.

Tears drain from the eyes through a canal that connects to the nasal cavity. When we cry, tear production increases, causing a runny nose.

OLFACTION

Our sense of smell is due to the olfactory receptors located on the mucous membranes high in the nasal cavities behind the nose. The sense of smell is discussed in Chapter 9.

FIGURE 14.3 The path of air during inhalation and exhalation

stop and think

Very cold temperatures can slow the action of the cilia in the nasal cavities. Explain why the loss of ciliary action can cause a runny nose on a very cold day.

■ **The sinuses lighten the head and condition inhaled air**

Connected to the nasal cavities are large air-filled spaces in the bones of the face, called the **sinuses**. The sinuses make the head lighter and help warm and moisten the air we breathe. In addition, the sinuses are part of the resonating chamber that affects the quality of the voice. When you have a cold, your voice becomes muffled because the mucous membranes of the sinuses swell and produce excess fluid.

Because the air spaces of the sinuses are continuous with those of the nasal cavities, any excess mucus and fluids drain from the

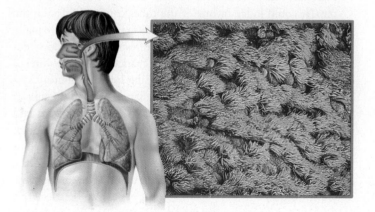

FIGURE 14.4 The respiratory passageways are lined with clumps of short hairlike structures called cilia interspersed between mucus-secreting cells. The cilia are pink in this color-enhanced electron micrograph.

sinuses into the nasal cavities. However, when the mucous membranes of the sinuses become inflamed, as they do in *sinusitis* (-*itis*, inflammation of), the swelling can block the connection between the nasal cavities and the sinuses, preventing the sinuses from draining the mucous fluid they produce. The pressure caused by the accumulation of fluids in the sinuses causes pain over one or both eyes or in the cheeks or jaws; this condition is usually called a sinus headache. Decongestant nasal sprays reduce the swelling in the tubes that connect the sinuses with the nasal cavity, allowing the sinuses to drain more easily. Sinusitis may be caused by the virus responsible for a cold or by a subsequent bacterial infection.

■ The pharynx is a passageway for food and air

The **pharynx**, commonly called the throat, is the space behind the nose and mouth. It is a passageway for air, food, and drink. Small, narrow passages, called the *auditory* (eustachian) *tubes*, connect the upper region of the pharynx with the middle ear. These passages help equalize the air pressure in the middle ear with that of the pharynx.

■ The larynx is an adjustable entryway that produces the voice

After moving through the pharynx, the air next passes through the **larynx**, which is commonly called the voice box or Adam's apple. The larynx is a boxlike structure composed primarily of cartilage (Figure 14.5).

The larynx has two main functions. It is a traffic director for materials passing through the structures in the neck, allowing air, but not other materials, to enter the lower respiratory system. The larynx is also the source of the voice. Let's consider these two functions in more detail.

1. **A selective entrance to the lower respiratory system.** The larynx provides a selective opening to the lower respiratory system: it can be opened to allow air to pass into the lungs and closed to prevent other matter, such as food, from entering the lungs. Because the esophagus (the tube leading to the stomach) is behind the larynx, food and drink must pass over the opening to the larynx to reach the digestive system. If solid material such as food were to enter the lower respiratory system, it could lodge in one of the tubes conducting air to the lungs and prevent air flow. Fluid entering the lungs is equally dangerous because it can cover the respiratory surfaces, decreasing the area available for gas exchange. Normally, foreign material is prevented from entering the lower respiratory system during swallowing because the larynx rises and causes a flap of cartilage called the **epiglottis** to move downward and form a lid over the **glottis**, the opening in the larynx through which air passes. You can feel the larynx moving if you put your fingers on your Adam's apple while swallowing. Because of this movement, you cannot breathe and swallow at the same time. (Try it!)

   If food or drink accidentally enters the trachea, we usually cough and expel it. However, if food lodges in the trachea, it may block air flow. The **Heimlich maneuver** can be used to remove the blockage and restore air flow (Figure 14.6).

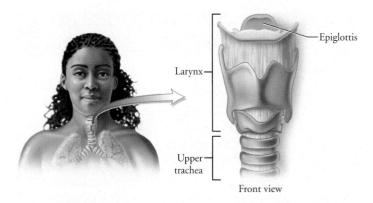

(a) The epiglottis is open during breathing but covers the opening to the larynx during swallowing to prevent food or drink from entering the trachea.

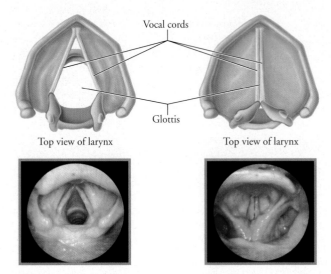

During quiet breathing, the vocal cords are near the sides of the larynx, and the glottis is open.

During speech, the vocal cords are stretched over the glottis and vibrate as air passes through them, producing the voice.

(b) The vocal cords are the folds of connective tissue above the opening of the larynx (the glottis) that produce the voice.

FIGURE **14.5** The larynx, commonly called the voice box or Adam's apple, is an adjustable entryway to the trachea and the source of the voice.

2. **Production of voice.** The voice is generated in the larynx by the vibration of the **vocal cords**, two thick strands of tissue stretched over the opening of the glottis (Figure 14.5). When you speak, muscles stretch the vocal cords across the air passageway, narrowing the opening of the glottis. Air passing between the stretched vocal cords causes them to vibrate and produce a sound, just as the edges of the neck of an inflated balloon vibrate and make noise if you stretch the balloon's neck while allowing air to escape. The vibrations of the vocal cords set up sound waves in the air spaces of the nose, mouth, and pharynx. This resonation is largely responsible for the tonal quality of your voice.

   The pitch of the voice depends on the tension of the vocal cords. When the cords are stretched, becoming thinner and more taut, the pitch of the sound when they vibrate is higher.

A person who is choking cannot speak or breathe and needs immediate help.

The **Heimlich maneuver** is a procedure intended to force a large burst of air out of the lungs and dislodge the object blocking air flow.

STEP 1: Stand behind the choking person with arms around the waist.

STEP 2: Make a fist and place the thumb of the fist beneath the victim's rib cage about midway between the navel (belly button) and the breastbone.

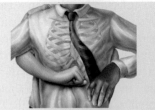

STEP 3: Grasp the fist with your other hand and deliver a rapid "bear hug" up and under the rib cage with the clenched fist. Be careful not to press on the ribs or the breastbone because doing so could cause serious injury.

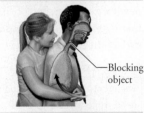

Blocking object

STEP 4: Repeat until the object is dislodged.

FIGURE **14.6** The Heimlich maneuver can be performed on a choking person who is standing or sitting. If a choking victim is lying on the ground, the same life-saving pressure changes can be generated by pushing inward and upward on the upper part of the victim's abdomen. If you begin to choke and there is no one to perform the Heimlich maneuver on you, it may be possible to dislodge the obstruction in your trachea by throwing your upper abdominal region against a table, chair, or other stationary object.

You can demonstrate the relationship between thickness and pitch for yourself by plucking a rubber band stretched between your thumb and forefinger. The more the rubber band is stretched, the higher the pitch of the twang.

When you suffer from *laryngitis*, an inflammation of the larynx, the vocal cords become swollen and thick. As a result, they cannot vibrate freely, and the voice becomes deeper and huskier. When the vocal cords are very inflamed, a person can hardly speak at all because the cords cannot vibrate in that condition.

■ **The trachea is the windpipe**

The **trachea**, or windpipe, is a tube that conducts air between the outside of the body and the lungs. It is held open by C-shaped rings of cartilage that give it the general appearance of a vacuum cleaner hose. You can feel these rings of cartilage in your neck, just below the larynx.

The support rings are necessary in the trachea and its branches to prevent these airways from collapsing during each breath when the rapid flow of air into the lungs creates a drop in pressure. Air (or fluid) passing rapidly over a surface causes a lower pressure, experienced as a "pull," on that surface. Maybe you have noticed that when you get into the shower and turn on the water, the shower curtain is drawn in toward you. The curtain moves inward because the moving water lowers the air pressure, just as the rapid movement of air through the respiratory tubules does. If the trachea were not supported open by cartilage rings, the rapid flow of air during breathing would cause it to collapse or flatten.

■ **The bronchial tree is a system of air tubules that conducts air**

The trachea divides into two air tubes called primary **bronchi**; each bronchus (singular) conducts air from the trachea to one of the lungs. The bronchi branch repeatedly within the lungs, forming progressively smaller air tubes. The smallest bronchi divide to form yet smaller tubules called **bronchioles**, which finally terminate in *alveoli*, sacs with surfaces specialized for gas exchange (discussed shortly).

The repeated branching of air tubules in the lung is reminiscent of a maple tree in winter. In fact, the resemblance is so close that the system of air tubules is often called the **bronchial tree** (Figure 14.7). All the bronchi are held open by cartilage, just as occurs in the trachea. However, the amount of cartilage decreases with the diameter of the tube. The bronchioles have no cartilage,

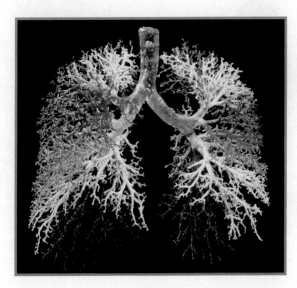

FIGURE **14.7** A resin cast of the bronchial tree of the lungs. In the body, this branching system of air tubules is hollow and serves as a passageway for the movement of air between the atmosphere and the alveoli, where gas exchange takes place.

but their walls contain smooth muscle, which is controlled by the autonomic nervous system so that air flow can be adjusted to suit metabolic needs (see Chapter 8).

Although the contraction of the muscle in bronchial walls is usually closely attuned to the body's needs, sometimes the bronchial muscles go into spasms that severely obstruct the flow of air. Such is the case with *asthma*, a chronic condition characterized by recurring attacks of wheezing and difficulty breathing. The difficult breathing is worsened by persistent inflammation of the airways. An allergy to substances such as pollen, dog or cat dander (skin particles), and the feces of tiny mites in household dust often trigger asthma attacks. However, a cold or respiratory infection, certain drugs, inhaling irritating substances, vigorous exercise, and psychological stress can also cause an attack. Some attacks start for no apparent reason. Certain inhalants prescribed to treat asthma attacks work by relaxing the bronchial muscles. Other inhalants contain steroids that reduce the inflammation of the air tubules that occurs in asthma.

■ The alveoli of the lungs are surfaces for gas exchange

Each bronchiole ends either with an enlargement called an alveolus (plural, alveoli) or, more commonly, with a grapelike cluster of alveoli. Each **alveolus** is a thin-walled, rounded chamber encased in a dense network of capillaries (Figure 14.8). Oxygen diffuses from the alveoli into the blood, which delivers the oxygen to cells. Carbon dioxide produced by the cells diffuses from the blood into the alveolar air to be exhaled.

Most of the lung tissue is composed of alveoli, making the structure of the lung much more like foam rubber than like a balloon, the image sometimes used to describe a lung. The surface area inside a simple, hollow balloon the same size as our lungs would be roughly 0.01 m^2 (about 0.2 yd^2). However, each of our lungs contains approximately 300 million alveoli, whose total surface area is about 70 to 80 m^2 (about 84 to 96 yd^2). In other words, the alveoli increase the surface area of the lung about 8500 times.

For the alveoli to function properly as a surface for gas exchange, they must be kept open. Moist membranes, such as those of the alveolar walls, are attracted to one another because of an attraction between water molecules called surface tension. If this attraction were not disrupted in the alveoli by phospholipid molecules called **surfactant**, it would pull the alveolar walls together, collapsing the air chambers.

Surfactant production usually begins during the eighth month of fetal life, so enough surfactant is present to keep the alveoli open when the newborn takes its first breath. Unfortunately, some premature babies have not yet produced a sufficient amount of surfactant to overcome the attractions between the alveolar walls. As a result, their alveoli collapse after each breath. This condition, called *respiratory distress syndrome* (RDS), makes breathing difficult for the newborn. Some newborns with RDS die as a result. However, many are saved by the use of mechanical respirators and artificial surfactant to keep them alive until their lungs mature.

stop and think

Pneumonia is a lung infection that results in an accumulation of fluid and dead white blood cells in the alveoli. Why might this result in lower blood levels of oxygen?

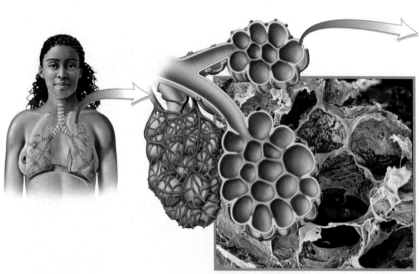

(a) Each alveolus is a cup-shaped chamber. In this section, some of the alveoli have been cut open and you can see into them.

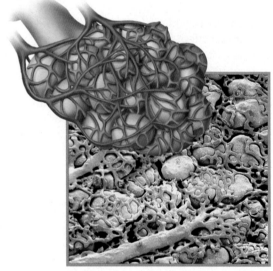

(b) Much of the surface of each alveolus is covered with capillaries. The interface provides a vast surface area for the exchange of gases between the alveoli and the blood.

FIGURE **14.8** Alveoli in the lungs create a huge surface area where oxygen and carbon dioxide are exchanged between the lungs and the blood. Oxygen diffuses from the alveoli into the blood, and carbon dioxide diffuses from the blood to the alveoli.

Pressure Changes within the Lungs Cause Breathing

Air moves between the atmosphere and the lungs in response to pressure gradients. It moves into the lungs when the pressure in the atmosphere is greater than the pressure in the lungs, and it moves out when the pressure in the lungs is greater than the pressure in the atmosphere.

The pressure changes in the lungs are created by changes in the volume of the thoracic cavity, a relationship explained by the characteristics of the pleural membrane. Each lung is enclosed in a double-layered sac of pleural membrane. One layer of membrane adheres to the wall of the thoracic cavity; the other adheres to the lung. Fluid between the layers of membrane lubricates the membrane layers and holds them together. As a result, a change in the volume of the thoracic cavity causes a similar change in the volume of the lungs. Let's consider how the changes in the size of the thoracic cavity are brought about.

■ Inhalation occurs when the pressure in the lungs decreases

Air moves into the lungs when the size of the thoracic cavity increases; this increase causes the pressure in the lungs to drop below atmospheric pressure. The increase is due to the contraction of both the **diaphragm**, a broad sheet of muscle that separates the abdominal and thoracic cavities, and the muscles of the rib cage, called the **intercostals** (*costa*, rib) (Figure 14.9a). The air pressure in the lungs decreases, and air rushes into the lungs. This process is called inhalation, or *inspiration*. The intercostals lie between the ribs, so that when those muscles contract, they pull the rib cage upward and outward. By placing your hands on your rib cage while you inhale, you can feel the rib cage move up and out. Raising the rib cage increases the size of the thoracic cavity from the front to the back. Meanwhile, the contraction of the diaphragm lengthens the thoracic cavity from top to bottom, as shown in Figure 14.9a.

■ Exhalation occurs when the pressure in the lungs increases

The process of breathing out, called exhalation, or *expiration*, is usually passive. In other words, it doesn't require work but occurs when the muscles of the rib cage and the diaphragm relax. The lungs are elastic; that is, after stretching they return to their former size. When the elastic tissues of the lung recoil, the rib cage falls back to its former lower position and the diaphragm bulges into the thoracic cavity (Figure 14.9b). The pressure within the lungs increases as the volume of the lungs decreases. When the pressure within the lungs exceeds atmospheric pressure, air moves out.

stop and think

In Victorian times, a woman often wore a corset containing whalebone that formed a band around her waist and lower chest. Corsets were laced tightly to create a wasplike waistline, and women frequently fainted. What is the most likely cause of these fainting spells?

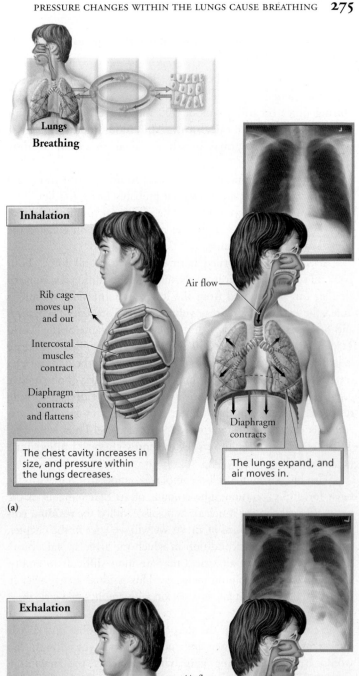

Lungs Breathing

Inhalation

Rib cage moves up and out
Intercostal muscles contract
Diaphragm contracts and flattens

Air flow
Diaphragm contracts

The chest cavity increases in size, and pressure within the lungs decreases.

The lungs expand, and air moves in.

(a)

Exhalation

Rib cage moves down and inward
Intercostal muscles relax
Diaphragm relaxes and moves upward

Air flow
Diaphragm relaxes

The chest cavity decreases in size, and pressure within the lungs increases.

The lungs recoil, and air moves out.

(b)

FIGURE 14.9 Changes in the volume of the thoracic cavity bring about inhalation and exhalation. The x-rays show the actual changes in lung volume during inhalation and exhalation.

■ **The volume of air moved into and out of the lungs is an indication of health**

The volume of air moved during each breath varies from person to person, depending largely on the person's sex, age, and height. During quiet breathing, about 500 ml, roughly 1 pint, of air moves in and out with each breath. The amount of air inhaled or exhaled during a normal breath is called the **tidal volume** (Figure 14.10).

If, after inhaling normally, you were to inhale until you could not take in any more air, you would probably bring another 1900 to 3300 ml of air into your lungs. The additional volume of air that can be brought into the lungs after normal inhalation is called the **inspiratory reserve volume**.

After you have exhaled normally, you can still force about 1000 ml of additional air from the lungs. This additional volume of air that can be expelled from the lungs after the tidal volume is called the **expiratory reserve volume**. New York City firefighters and other rescue workers at the scene of the collapse of the World Trade Center on Sept. 11, 2001 experienced a significant decrease in expiratory reserve volume in the year following the disaster. Decreases in expiratory reserve volume are characteristic of obstructive lung diseases such as bronchitis and asthma. The decline in lung function, presumably due to the inhalation of toxic dust, was equivalent to that expected from 12 years of aging. Those who were on the scene when the towers fell or shortly afterward suffered the most damage.

The lungs can never be completely emptied, even with the most forceful expiration. The amount of air that remains in the lungs after exhaling as much air as possible, called the **residual volume**, is roughly 1200 ml of air. As we will see later in the chapter, emphysema is a lung condition in which the alveolar walls break down, creating larger air spaces that are more difficult to empty. Thus, the residual volume increases. This residual air is lower in oxygen than is inhaled air, so a person with emphysema feels short of breath.

If you were to take the deepest breath possible and exhale until you could not force any more air from your lungs, you would be demonstrating your **vital capacity**, the maximum amount of air that can be moved into and out of the lungs during forceful breathing. The vital capacity, therefore, equals the sum of the tidal volume, the inspiratory reserve, and the expiratory reserve. Although average values for college-age people are about 4800 ml in men and 3400 ml in women, the values can vary tremendously depending on a person's health and fitness. One illness that affects vital capacity is pneumonia. It causes fluid to accumulate within the alveoli, taking up space that would normally be occupied by air.

Because some air is always left in the lungs, the vital capacity is not a measure of the total amount of air that the lungs can hold. The **total lung capacity**, the total volume of air contained in the lungs after the deepest possible breath, is calculated by adding the residual volume to the vital capacity. This volume is approximately 6000 ml in men and 4500 ml in women.

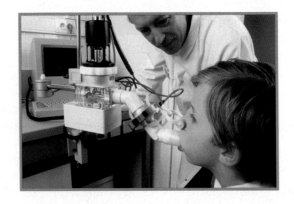

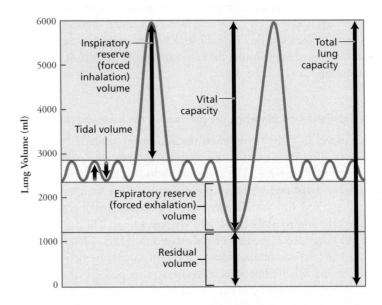

Tidal volume (~500 ml)	Amount of air inhaled or exhaled during an ordinary breath
Inspiratory reserve volume (~1900–3300 ml)	Amount of air that can be inhaled in addition to a normal breath
Expiratory reserve volume (~1000 ml)	Amount of air that can be exhaled in addition to a normal breath
Residual volume (~1100–1200 ml)	Amount of air remaining in the lungs after maximum exhalation
Vital capacity (~3400–4800 ml)	Amount of air that can be inhaled or exhaled in a single breath
Total lung capacity (4500–6000 ml)	Total amount of air in the lungs after maximal inhalation (vital capacity + residual volume)

FIGURE **14.10** A spirometer is used to measure the volumes of air in the lungs

Blood Transports Gases between the Lungs and the Cells

We have seen that breathing brings air into the lungs and expels air from the lungs. Recall that three other processes then play roles in delivering the oxygen to the cells and disposing of carbon dioxide from the cells. External respiration occurs in the alveoli of the lungs; there oxygen diffuses into the blood and carbon dioxide diffuses from the blood. Gas transport is accomplished by the blood, which carries oxygen to the cells and carbon dioxide away from the cells. Internal respiration occurs in the various tissues; there oxygen diffuses out of the blood and into the cells, and carbon dioxide diffuses out of the cells and into the blood (Figure 14.11).

■ Most oxygen is carried by hemoglobin

Oxygen is carried from the alveoli throughout the body by the blood. Almost all, about 98.5%, of the oxygen that reaches the cells is bound to hemoglobin, a protein in the red blood cells. Hemoglobin bound to oxygen is called **oxyhemoglobin** (HbO_2). The remaining 1.5% of the oxygen delivered to the cells is dissolved in the plasma. Whole blood, which consists of cells as well as plasma, carries 70 times more oxygen than an equal amount of plasma.

Hemoglobin picks up oxygen at the lungs and releases it at the cells. But what determines whether hemoglobin will bind to oxygen or release it? The most important factor deciding this question is the partial pressure of oxygen, which is directly related to its concentration. In a mixture of gases, each gas contributes only part of the total pressure of the whole mixture of gases. The pressure exerted by one of the gases in a mixture is called its partial pressure.

Recall from Chapter 3 that substances always diffuse from regions of higher concentration or pressure to regions of lower concentration or pressure. In the alveoli of the lungs, where the concentration of oxygen is high, hemoglobin in the red blood cells in nearby capillaries picks up oxygen. The oxygen is then released near the cells, where the oxygen concentration is low.

■ Most carbon dioxide is transported as bicarbonate ions

The carbon dioxide produced by cells as they use oxygen is removed by the blood. Carbon dioxide transport occurs in three principal ways:

1. **Dissolved in blood plasma.** Between 7% and 10% of the carbon dioxide is transported dissolved in the plasma as molecular carbon dioxide.

2. **Carried by hemoglobin.** Hemoglobin molecules in red blood cells carry slightly more than 20% of the transported carbon dioxide. When carbon dioxide combines with hemoglobin, it forms a compound called **carbaminohemoglobin**.

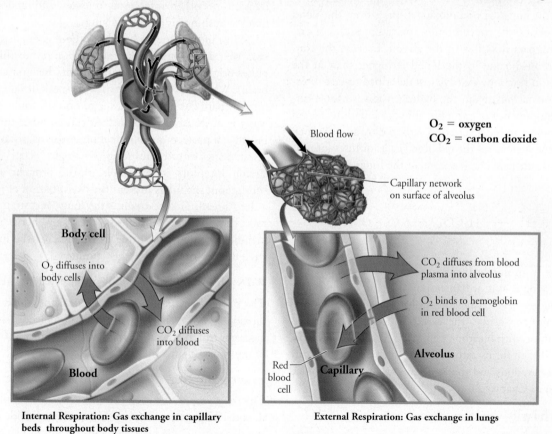

O_2 = oxygen
CO_2 = carbon dioxide

Blood flow

Capillary network on surface of alveolus

Body cell

O_2 diffuses into body cells

CO_2 diffuses into blood

Blood

CO_2 diffuses from blood plasma into alveolus

O_2 binds to hemoglobin in red blood cell

Alveolus

Red blood cell

Capillary

Internal Respiration: Gas exchange in capillary beds throughout body tissues

External Respiration: Gas exchange in lungs

FIGURE **14.11** In the lungs, oxygen diffuses from the alveoli into the blood. Oxygen is carried to the cells in red blood cells. At the cells, oxygen diffuses from the blood to the body cells, which use the oxygen and produce carbon dioxide in the process. Carbon dioxide diffuses into the blood and is carried back to the lungs, where it diffuses from the blood into an alveolus and is exhaled.

3. **As a bicarbonate ion.** By far the most important means of transporting carbon dioxide is as bicarbonate ions dissolved in the plasma. About 70% of the carbon dioxide is transported this way. Carbon dioxide (CO_2) produced by cells diffuses into the blood and into the red blood cells. In both the plasma and the red blood cells, it reacts with water (H_2O) and forms carbonic acid (H_2CO_3). Carbonic acid quickly dissociates to form hydrogen ions (H^+) and bicarbonate ions (HCO_3^-). The process of bicarbonate ion formation in the capillaries of the tissues is represented by the following formula:

$$\underset{\substack{\text{Carbon} \\ \text{dioxide}}}{CO_2} + \underset{\text{Water}}{H_2O} \xrightarrow{\substack{\text{Carbonic} \\ \text{anhydrase}}} \underset{\substack{\text{Carbonic} \\ \text{acid}}}{H_2CO_3} \longrightarrow \underset{\substack{\text{Bicarbonate} \\ \text{ion}}}{HCO_3^-} + \underset{\substack{\text{Hydrogen} \\ \text{ion}}}{H^+}$$

Although these reactions occur in the plasma as well as in red blood cells, they occur hundreds of times faster in red blood cells. The higher rate of reaction is caused by the enzyme **carbonic anhydrase** that is found within red blood cells but not in the plasma. The hydrogen ions produced by the reaction combine with hemoglobin. In this way, hemoglobin acts as a buffer, and the acidity of the blood changes only slightly as it passes through the tissues. The bicarbonate ions diffuse out of the red blood cells into the plasma and are transported to the lungs.

In the lungs, the process is reversed. When the blood reaches the capillaries of the lungs, carbon dioxide diffuses from the blood into the alveoli because the concentration (partial pressure) of carbon dioxide is comparatively low in the alveoli. Because the concentration of carbon dioxide in the blood is higher than in the alveoli, the chemical reactions we have just described reverse direction. The bicarbonate ions rejoin the hydrogen ions to form carbonic acid. In the presence of carbonic anhydrase within the red blood cells, carbonic acid is converted to carbon dioxide and water. The carbon dioxide then leaves the red blood cells, diffuses into the alveolar air, and is exhaled. The reactions in the lungs are summarized below.

$$\underset{\substack{\text{Bicarbonate} \\ \text{ion}}}{HCO_3^-} + \underset{\substack{\text{Hydrogen} \\ \text{ion}}}{H^+} \xrightarrow{\substack{\text{Carbonic} \\ \text{anhydrase}}} \underset{\substack{\text{Carbonic} \\ \text{acid}}}{H_2CO_3} \longrightarrow \underset{\substack{\text{Carbon} \\ \text{dioxide}}}{CO_2} + \underset{\text{Water}}{H_2O}$$

Besides being the form in which carbon dioxide is transported by the blood, bicarbonate ions are an important part of the body's *acid-base buffering system*. They help neutralize acids in the blood. If the blood becomes too acidic, the excess hydrogen ions are removed by combining with bicarbonate ions to form carbonic acid. The carbonic acid then forms carbon dioxide and water, which are exhaled. (Bicarbonate ions and the acid/base balance of the blood are also discussed in Chapter 2).

stop and think

Carbon monoxide binds to hemoglobin much more readily than does oxygen, and it binds in the same site as oxygen. Thus, when carbon monoxide is bound to hemoglobin, oxygen cannot bind. Explain why carbon monoxide poisoning can be fatal.

Breathing Is Controlled Primarily by Respiratory Centers in the Brain

Breathing rate influences the amount of oxygen that can be delivered to cells and the amount of carbon dioxide that can be removed from the body. Neural and chemical controls adjust breathing rate to meet the body's needs.

◼ Brain centers control the basic breathing pattern

As you sit there reading your text, your breathing is probably rather rhythmic, with about 12 to 15 breaths a minute. The basic rhythm is controlled by a breathing (respiratory) center located in the medulla of the brain (Figure 14.12). Within the breathing center are an inspiratory area and an expiratory area.

During quiet breathing, when you are calm and breathing normally, the inspiratory area shows rhythmic bouts of neural activity. While the inspiratory neurons are active, impulses that stimulate contraction are sent to the muscles involved in inhalation (the diaphragm and the intercostals). As we have seen, contraction of the diaphragm and the intercostals causes the size of the thoracic cavity to increase, thus moving air into the lungs. After about 2 seconds of inhalation, the activity of the neurons in the inspiratory center ceases for about 3 seconds. When inspiratory neurons cease activity, the diaphragm and intercostals relax, and passive exhalation occurs. During heavy breathing the expiratory center causes contraction of other intercostal muscles and abdominal muscles, quickly pushing air out of the lungs.

Most of the time we breathe without giving it a thought. However, we can voluntarily alter our pattern of breathing through impulses originating in the cerebral cortex (the "conscious" part of the brain). We control breathing when we speak or sigh, and we can voluntarily pant like a dog. Holding our breath while swimming under water is obviously a good idea. At certain other times, holding our breath can protect us from inhaling smoke or irritating gases.

During forced breathing, as might occur during exercise, stretch receptors in the walls of the bronchi and bronchioles throughout the lungs prevent the overinflation of the lungs. When a deep breath greatly expands the lungs and stretches these receptors, they send impulses over the vagus nerve that inhibit the breathing center, permitting exhalation. As the lungs deflate, the stretch receptors are no longer stimulated.

◼ Depth and rate of breathing are affected by chemoreceptors

The purpose of breathing is to control the blood levels of carbon dioxide and oxygen. We will now consider how the levels of these gases control the breathing rate, which in turn, influences the levels of the gases (Figure 14.13).

CARBON DIOXIDE

The most important chemical influencing breathing rate is carbon dioxide. The mechanism by which carbon dioxide regulates breathing depends on the hydrogen ions produced when carbon dioxide goes into solution and forms carbonic acid:

$$\underset{\substack{\text{Carbon} \\ \text{dioxide}}}{CO_2} + \underset{\text{Water}}{H_2O} \longrightarrow \underset{\substack{\text{Carbonic} \\ \text{acid}}}{H_2CO_3} \longrightarrow \underset{\substack{\text{Hydrogen} \\ \text{ion}}}{H^+} + \underset{\substack{\text{Bicarbonate} \\ \text{ion}}}{HCO_3^-}$$

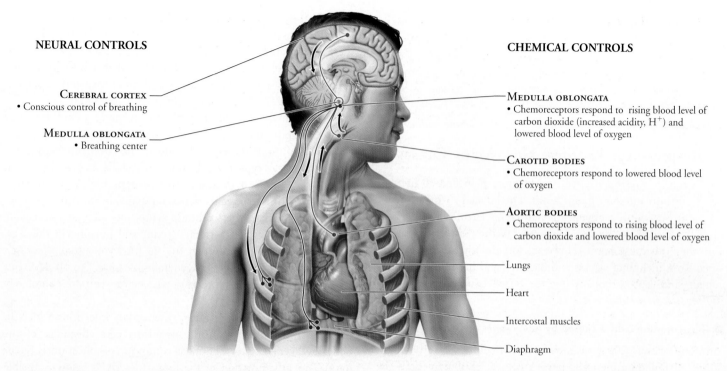

NEURAL CONTROLS

CEREBRAL CORTEX
• Conscious control of breathing

MEDULLA OBLONGATA
• Breathing center

CHEMICAL CONTROLS

MEDULLA OBLONGATA
• Chemoreceptors respond to rising blood level of carbon dioxide (increased acidity, H^+) and lowered blood level of oxygen

CAROTID BODIES
• Chemoreceptors respond to lowered blood level of oxygen

AORTIC BODIES
• Chemoreceptors respond to rising blood level of carbon dioxide and lowered blood level of oxygen

Lungs

Heart

Intercostal muscles

Diaphragm

FIGURE **14.12** Neural and chemical controls of breathing

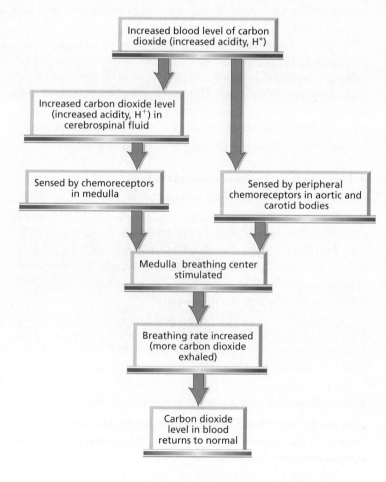

FIGURE **14.13** The role of carbon dioxide in controlling the breathing rate

Three groups of chemoreceptors respond to the changing levels of hydrogen ions in the blood (Figure 14.12). Central chemoreceptors are located in a region of the brain called the medulla, near the breathing center. Peripheral chemoreceptors are located in the aortic bodies and the carotid bodies, small structures associated with the main blood vessel leaving the heart to the body and the main blood vessels to the head. The chemoreceptors in the aortic bodies and carotid bodies also respond when the blood level of oxygen is low.

The chemoreceptors in the medulla are near its surface, where they are bathed in cerebrospinal fluid. Carbon dioxide diffuses from the blood into the cerebrospinal fluid, where it raises the hydrogen ion concentration by forming carbonic acid. When the rising hydrogen ion concentration stimulates the chemoreceptors in the medulla, breathing rate is increased and more carbon dioxide is exhaled, causing a decrease in the blood level of carbon dioxide.

OXYGEN

Because oxygen, not carbon dioxide, is essential to survival, it may be somewhat surprising to learn that oxygen does not influence the breathing rate unless its blood level falls dangerously low. Only then do oxygen-sensitive chemoreceptors in the aortic bodies and carotid bodies send a message that the blood oxygen level is at a critical point, initiating a last-minute call to the medulla to increase the breathing rate and raise oxygen levels. If the oxygen level falls much more, the neurons in the inspiratory area die from a lack of oxygen and do not respond well to impulses from the chemoreceptors. As a result, the inspiratory area begins to send fewer impulses to the muscles of inspiration, and the breathing rate decreases and may even cease completely.

stop and think

Divers sometimes take several deep and rapid breaths before diving. This decreases blood carbon dioxide levels. Why would this practice allow the diver to remain under water longer? Occasionally, a diver who hyperventilates before diving loses consciousness while under water. Why would hyperventilating cause a diver to pass out?

Respiratory Disorders Have Many Causes

Even in a restful day, you move more than 86,000 liters of air into and out of your lungs. The inhaled air may contain disease-causing organisms or noxious chemicals and particles. The respiratory system clears most of these materials out before they can cause harm. However, some of the disease-causing organisms, including viruses and bacteria, and harmful substances remain and cause problems (see the Social Issue essay, *Bird Flu, Will It Become a Pandemic?*).

■ The common cold is caused by many types of viruses

Any one of over 200 viruses can cause a cold. Not surprisingly, some 30 million Americans have a cold at this moment. (The "common cold" is indeed common.) Because there are so many cold-causing viruses, you can, and usually do, get several colds a year, each from a different virus. Most people get their first cold before they are a year old.

Typically, a cold begins with a runny nose, possibly a sore throat, and sneezing. In the beginning, the nasal discharge is thin and watery, but it becomes thicker as it fills the nasal cavity. Almost any part of the respiratory system can be affected. Sneezing and a stuffy nose indicate that the infection is in the upper respiratory system. When the pharynx is affected, a sore throat results. The infection may spread to the bronchi, causing a cough, or to the larynx, making your voice hoarse.

As miserable as you feel with a cold, you can take comfort from the fact that your suffering will not last forever (see the Health Issue essay, *Surviving a Common Cold*). A cold is self-limiting, lasting only 1 to 2 weeks. Furthermore, colds are seldom fatal, except occasionally among the very young or very old or in those people already seriously ill with another malady.

Colds are spread when the causative virus is transmitted from an infected person. The viruses are plentiful in nasal secretions. However, transmission of the virus is not usually the result of direct inhalation of infected droplets from a cough or sneeze. It is more likely to be the result of handling an object that is contaminated with the virus. Cold viruses may remain alive on the skin or on an object for several hours, waiting for an unsuspecting person to touch them and contaminate his or her fingers. Subsequently, when the virus-laden fingers are touched to the mucous membranes of the nose, the transfer is completed. The best ways to prevent a cold are to wash your hands frequently and avoid being around people who have colds.

■ The flu is caused by three types of virus

Flu is an abbreviation of *influenza*, another viral disease. The kinds of viruses that cause the flu are few compared to the number of different viruses that can cause a cold. In fact, the viruses that cause the flu in humans are all variants of three major types—A, B, and C (but there are hundreds of variants of these three basic types). Influenza of the A type is often more serious than B in that it is frequently accompanied by severe complications and more frequently results in death. Influenza C causes a mild illness with cold symptoms.

Symptoms of the flu are similar to those of a cold, but they appear to start suddenly and are more severe. (Actually, by the time flu symptoms emerge, the disease has been incubating for several days.) A typical flu begins with chills and a high fever, about 103°F (39°C) in adults and perhaps higher in children. Many flu victims experience aches and pains in the muscles, especially in the back. Other common symptoms include a headache, sore throat, dry cough, weakness, pain and burning in the eyes, and sensitivity to light. When the flu hits, you usually feel sick enough to go to bed. The flu generally lasts for 7 to 10 days, but an additional week or more may pass before you are completely back on your feet.

The flu is often complicated by secondary infections, which set in when other disease-causing organisms take advantage of the body's weakened state. The most common complication is pneumonia, an inflammation of the lungs (described below). Possible secondary infections caused by bacteria include bronchitis, sinusitis, and ear infections.

One way to prevent the flu is by getting a flu shot—a vaccine made from the strains of viruses that scientists anticipate will cause the next outbreaks of the illness. Flu shots are only about 60% to 70% effective because the viruses they target mutate rapidly, causing new strains to appear. The new strains are not recognized by the immune system defenses that were programmed by the latest vaccine. Because each flu season brings new strains of flu viruses, the effectiveness of the vaccine lasts only as long as that season's most prevalent strains. As a result, new vaccines must be developed continually to protect us, and flu shots must be repeated each year.

■ Pneumonia is an inflammation of the lungs

Pneumonia is an inflammation of the lungs that causes fluid to accumulate in the alveoli, thus reducing gas exchange. It also causes swelling and narrowing of the bronchioles, making breathing difficult. Pneumonia is usually caused by infection with bacteria or viruses, but fungi and protozoans can also cause it. Many cases of pneumonia develop after a common cold or influenza. Radiation, chemicals, and allergies can also bring it on.

Symptoms of pneumonia often begin suddenly. They include fever and chills, chest pain, cough, and shortness of breath. The severity of pneumonia varies from mild to life threatening. Treatment depends on the cause of the illness.

■ Strep throat can have serious consequences

Strep throat, a sore throat caused by *Streptococcus* bacteria, is a problem mainly in children 5 to 15 years old. The soreness is usually accompanied by swollen glands and a fever.

Although the pain of strep throat may be so mild that a doctor is never consulted, ignoring a strep infection can have serious consequences. If untreated, the *Streptococcus* bacteria can spread to

Bird Flu, Will It Become a Pandemic?

According to the World Health Organization (WHO) a pandemic, a global outbreak of a severe infectious disease, can be expected to occur three or four times each century. The last pandemic, Spanish flu, occurred in 1918 to 1919 and killed more than 40 million people. Public health officials around the world are fearful that bird influenza virus H5N1 may cause the next pandemic.

Migratory waterfowl, wild ducks in particular, become infected with a mild form of the bird influenza virus, but the virus does not usually kill the waterfowl. Within domestic flocks, however, the H5N1 virus mutates to a more deadly form that kills domestic birds. Scientists are increasingly convinced that infected migrant birds transmit the virus to domestic flocks of poultry, such as chickens and turkeys, along their migration route.

If the bird flu virus infected only birds, it would cause huge economic losses but would not be a major health concern. However, bird influenza virus H5N1 can jump from infected birds to humans. The first strain of H5N1 able to make humans sick showed up in Vietnam, Cambodia, and Thailand in 2003 and 2004. A second strain appeared in Indonesia in 2005 and has since spread to Europe. The virus spreads to people who come into close contact with the feces of infected birds or who handle sick birds. There is no evidence that bird flu can be transmitted by eating cooked poultry.

Bird influenza H5N1 virus can also infect certain other mammals—cats, stone martins (a weasel-like animal), and civets (in the mongoose family). Some health officials worry that migratory birds infected with the H5N1 strain of avian flu may die and be eaten by cats that are family pets. Most humans are in more direct contact with their family pets than they are with wild or domestic birds. Scientists do not yet know whether infected pets could transmit the virus to humans.

In humans, bird flu begins with a high fever and influenza-like symptoms, including watery diarrhea, chest and abdominal pain, and bleeding from the nose and gums. Many bird flu patients have symptoms of respiratory distress. Some develop brain infections, and others have failure of multiple organ systems.

Organisms that cause pandemics usually share three characteristics: they mutate rapidly, cause severe illness in humans, and are easily transmitted from person to person. Bird flu virus H5N1 already has the first two characteristics. It can spread from infected bird to humans, and it kills more than half of those people infected. Bird flu virus H5N1 could gain the ability to spread among humans by mutation (a spontaneous change in its genetic information) or by a mixing of viral genes during a dual infection of H5N1 and an influenza virus that normally infects humans. Bird flu virus H5N1 has a history of being able to acquire genes from other animals. As the number of human infections increases, so does the likelihood that a person could be simultaneously infected with human influenza and bird flu influenza. A double infection could allow mixing of the genetic material of the viruses, creating a new bird flu virus that could spread from person to person. Because people have little natural immunity against it, the H5N1 virus would spread rapidly, possibly reaching pandemic proportions. Concerned about this possibility, WHO is watching H5N1 closely.

Scientists are working on developing a vaccine against the H5N1 virus, but none is commercially available now. In August 2006, the Chinese announced that they have a vaccine to protect against bird flu that passed the first phase of clinical trials, indicating that it is safe and effective for humans. The vaccine must pass two more phases of clinical trials before it can be commercially available. To be effective, a vaccine must be tailored to the pandemic virus itself, so if a bird flu pandemic does occur, a commercial vaccine is not likely to be available for the first few months of it. However, currently available antiviral drugs such as Tamiflu and Relenza reduce the severity and duration of seasonal flu. If a person with bird flu takes one of these drugs early in the infection, it may improve their prospects of survival (but scientists are not yet certain that it will).

Many governments around the world are preparing for the possibility of a pandemic. Most of their plans begin with containment and surveillance. Infected flocks are destroyed. Beginning in the spring of 2006, the number of migratory birds tested for the H5N1 virus was increased. In addition, health-care professionals are being trained to diagnose and treat bird flu, and hospitals are preparing for a possible surge in the need for care. Communication lines are being established within communities and among local, state, and federal governments to keep people and agencies informed.

---

other parts of the body and cause rheumatic fever or kidney problems. The main symptoms of rheumatic fever are swollen, painful joints and a characteristic rash. About 60% of rheumatic fever sufferers develop disease of the heart valves. Another possible consequence of a streptococcal infection is kidney disease (glomerulonephritis). The kidney damage is due to a reaction from the body's own protective mechanisms. The body produces antibodies that destroy the bacteria; but if these antibodies persist after the bacteria have been killed, they can cause the kidneys to become inflamed. The inflamed kidneys may be unable to filter the blood, and blood may leak into the urine.

Symptoms of strep throat include a sore throat and two of three of the following: fever of 101°F (38.33°C), white or yellow coating on tonsils, or swollen glands in the neck. Because many viruses can cause sore throats that look like strep infections, the only way to identify strep throat is to test for the causative organism. If *Streptococcus* bacteria are found, an antibiotic, usually penicillin, is prescribed to prevent rheumatic fever and kidney disease.

Surviving a Common Cold

The only thing more common than the cold is advice on how to treat it. Here we will examine the validity of some frequently suggested treatments for a cold.

1. **Take large doses of vitamin C?** Vitamin C will not prevent a cold unless you are malnourished or under extreme physical stress. Nonetheless, some people who take vitamin C may experience less severe or shorter colds than do people who do not take vitamin C.

2. **Suck on a zinc lozenge?** Some evidence exists that zinc slows viral replication, prevents cold viruses from adhering to nasal membranes, and boosts the immune system. But it is not clear whether zinc lozenges can alleviate cold symptoms. Some people who begin taking zinc within the first day or two of the onset of symptoms and continue taking a lozenge every two or three hours for several days do experience some relief from cold symptoms.

3. **Take echinacea?** Commonly known as the purple coneflower, echinacea has been used for centuries by cold sufferers. However, a study published in the *New England Journal of Medicine* in 2005 showed that the herbal medication was not effective in preventing or treating the common cold.

4. **Take an antibiotic?** Antibiotics are *not* effective against viruses and cannot cure a cold. However, a doctor may prescribe antibiotics to control secondary bacterial infections—such as a middle ear infection, bronchitis, or sinusitis—that may accompany a cold. Unnecessary use of an antibiotic may cause side effects such as diarrhea and can lead to the development of bacterial resistance to the drug. The rise of strains of bacteria that are resistant to antibiotics is likely to be a major threat to public health in the coming years, as discussed in Chapter 13a.

5. **Go to bed?** Bed rest enables the body to muster its resources and fight secondary infections. Staying at home with a cold is also socially responsible, because it helps prevent the spread of the virus. But if you are too busy to spend a few days in bed because of a cold, you probably are not hurting yourself. Bed rest will not cure your cold or shorten its duration.

6. **Have some chicken soup?** Grandmothers have long prescribed chicken soup to treat a cold, and doctors finally agree that the advice has some merit. You should always consume plenty of fluids when you have a cold. They help loosen secretions in the respiratory tract and thus reduce congestion. As a result, the patient can breathe more freely, and the amount of time the cold viruses are in contact with the cells lining the respiratory system is reduced. Hot fluids, such as chicken soup, are more effective than cold ones for increasing the flow of nasal mucus.

Although a cold cannot be cured, there are ways to make it more bearable. Some ways to relieve the misery of a cold are as follows.

For Nasal Congestion

The congestion is caused when the mucous membranes of the nasal cavities become swollen and produce increased amounts of mucus because of viral infection. The best ways to relieve the congestion are to drink plenty of extra fluids and to inhale moist air from a hot bath, shower, or vaporizer.

By constricting small blood vessels in the nose, decongestants may reduce the accumulation of fluid causing nasal congestion. Unfortunately, if taken orally, decongestants constrict small blood vessels throughout the body; in the process, they raise blood pressure. Other possible side effects of oral decongestants include nervousness, sleeplessness, and dryness of the mouth. Therefore, if you must use a decongestant, it is wise to choose one in the form of nose sprays or drops. Use these at the recommended dosage and only for a few days. Although antihistamines are present in many cold remedies, there is no evidence that they are effective in reducing nasal congestion caused by colds. Furthermore, antihistamines may cause blurred vision, retention of urine, and dizziness. Drowsiness is a major side effect, so if you do take an antihistamine, avoid driving or other hazardous activities.

For Coughs

Coughing is a protective reflex controlled by a cough center in the brain. Besides clearing the respiratory tubules of foreign material, coughing loosens and removes phlegm and mucus. Therefore, it is not always a good idea to suppress a cough. On the other hand, if a cough persists for more than a week, you should see a physician.

The safest and cheapest cough remedies are substances such as hard candy or honey, which coat and soothe the throat, and drinking extra fluids, which loosens the mucus and helps relieve some irritation.

Cough medications fall into two general classes: suppressants and expectorants. Cough suppressants work by reducing the activity of the brain's cough center. These are helpful in easing dry, hacking coughs. The most effective nonprescription cough suppressant is dextromethorphan. Codeine also suppresses coughs, but it is available only by prescription. A cough that brings up sputum is performing a useful function and should not be suppressed. Instead, removal of the irritating substances should be assisted with an expectorant, which increases the flow of respiratory tract secretions.

For Fever and Pain Relief

The most common medicines for relief of fever and pain are aspirin, acetaminophen, and ibuprofen. Children and teenagers should not take aspirin because it is linked to Reye's syndrome, a deadly disease that affects all body organs. Keep in mind that fever can be a good thing. The additional body heat slows the growth and reproduction (or replication) of many disease-causing organisms (discussed in Chapter 13). Unless the fever is very high (above 102°F, 40°C, in an adult), you may want to let it run its course.

■ Tuberculosis causes tubercles to form in the lungs

Tuberculosis (TB) is caused by a rod-shaped bacterium, *Mycobacterium tuberculosis*. It is spread when the cough of an infected person sends bacteria-laden droplets into the air and the bacteria are inhaled into the lungs of an uninfected person. Because the bacteria are inhaled, the lungs are usually the first sites attacked, but the bacteria can spread to any part of the body, especially to the brain, kidneys, or bone.

As a defense against the bacteria, the body forms fibrous connective tissue casings, called tubercles, that encapsulate the bacteria (hence, the name of the disease). Although the formation of tubercles slows the spread of the disease, it does not actually kill the bacteria. The immune system destroys at least some of the walled-off bacteria and may, in fact, kill them all. But pockets of bacteria may persist undetected for many years. Later, the disease may progress to the secondary stage as pockets of bacteria become reactivated. Furthermore, bacteria may escape from the tubercles and be carried by the bloodstream to other parts of the body. As a result, whenever the victim becomes weak, ill, or poorly nourished, the disease may flare up.

The initial symptoms of tuberculosis, if they occur, are similar to those of the flu. In the secondary stage, the patient usually develops a fever, loses weight, and feels tired. If the infection is in the lungs, as is usual, it causes a dry cough that eventually produces pus-filled and blood-streaked phlegm. TB can be fatal, especially if it is caused by a multi-drug-resistant strain of bacteria; and resistant strains are becoming increasingly common (discussed in Chapter 13a).

what would you do?

DOTS (Directly Observed Therapy, short course) is the World Health Organization's recommended treatment for tuberculosis. DOTS, which mandates that someone witness the TB patient swallowing medication each day, is in place in most major cities in the United States. The treatment ranges from 6 months to 2 years (the latter, if a multi-drug-resistant strain is present). Patients' failure to complete TB treatment is what has led to the drug-resistant strains of the TB bacterium. Is DOTS a fair balance of protection of the public with personal rights? What do you think?

■ Bronchitis is an inflammation of the bronchi

Viruses, bacteria, or chemical irritation may cause the mucous membrane of the bronchi to become inflamed, a condition called *bronchitis*. The inflammation results in the production of excess mucus, which triggers a deep cough that produces greenish yellow phlegm.

There are two types of bronchitis: acute and chronic. Acute bronchitis, which often follows a cold, is usually caused by the cold virus itself, but it may be caused by bacteria that take advantage of the body's lowered resistance and invade the trachea and bronchi. An antibiotic will hasten recovery if the cause is bacterial.

When a cough that brings up phlegm is present for at least 3 months in each of 2 consecutive years, the condition is called chronic bronchitis, a more serious problem that is usually associated with cigarette smoking or air pollution. (See the Environmental

Issue essay, *Air Pollution and Human Health*.) Some people with chronic bronchitis may lack an enzyme that normally protects the air passageways from such irritants. As the disease progresses, breathing becomes increasingly difficult, partly because the linings of the air tubules thicken, narrowing the passageway for air. Contraction of the muscles in bronchiole walls and excessive secretion of mucus further obstruct the air tubules.

Chronic bronchitis can have serious consequences. The degenerative changes in the lining of the air tubules make removal of mucus more difficult. As a result, the patient is more likely to develop lung infections such as pneumonia, which can be fatal, and degenerative changes in the lungs, such as emphysema.

■ Emphysema is caused by the destruction of alveoli

Emphysema is a common consequence of smoking, although it can have other causes as well. In *emphysema*, the alveoli break down and merge, thereby becoming fewer and larger (Figure 14.14). This change has two major effects: a reduction in the surface area available for gas exchange and an increase in the volume of residual, or "dead," air in the lungs. Exhalation, you may recall, is a passive process that depends on the elasticity of lung tissue. In emphysema, the lungs lose that elasticity, and air becomes trapped inside them. As the dead air space increases, adequate ventilation requires more forceful inhalation. Forcing the air in causes more alveolar walls to rupture, further increasing the dead air space. Lung size gradually increases as the residual volume of air becomes greater, giving a person with emphysema a characteristic barrel chest; but, gas exchange continues to become more difficult. To get an idea of what poor lung ventilation caused by increased dead air space feels like, take a

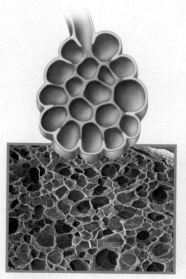

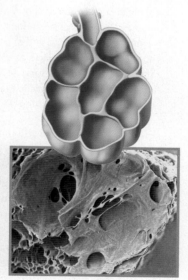

(a) Normal alveoli

(b) Emphysema causes breakdown of alveolar walls.

FIGURE **14.14** A comparison of (a) normal alveoli and (b) alveoli in an individual with emphysema. Notice that in emphysema, the alveolar walls rupture and there is a decrease in the surface area for gas exchange, an increase in the dead air space, and a thickening of the alveolar walls.

ENVIRONMENTAL ISSUE

Air Pollution and Human Health

Caution: The air you breathe may be hazardous to your health. It may even kill you—especially if you have heart or respiratory problems. In most major cities, poor air quality can be obvious, signaled by a brown haze. Even in remote national parks, air pollution is often significant enough to reduce visibility. Yet the consequences of air pollution can be subtle and slow to make themselves felt. They include damage to the environment as well as to human health.

Two major human sources of air pollution are motor vehicle exhaust and industrial emissions. The exhaust pipes on our cars and the smokestacks on factories spew oxides of sulfur, nitrogen, and carbon; a mixture of hydrocarbons; and many particulates (such as soot and smoke) into the air. In the atmosphere, sulfur dioxide and nitrogen dioxide dissolve in water vapor and form an aerosol of strong acids—sulfuric acid and nitric acid, respectively. Some of this aerosol may drift upward, forming acidic clouds that can be blown hundreds of miles by the winds and then fall as acid rain. On the other hand, the acidic aerosol may remain close to its source as a component of the haze created by air pollution. In addition, sunlight can cause hydrocarbons and nitrogen dioxide to react with one another and form a mix of hundreds of substances called photochemical smog. The most harmful component of photochemical smog is ozone, which attacks cells and tissues, irritates the respiratory system, damages plants, and even erodes rubber. Paradoxically, the ozone

that is so damaging when it is found at ground level in photochemical smog is the same compound that is beneficial in the upper layers of the atmosphere, where it prevents much of the sun's ultraviolet radiation from reaching the earth's surface. Unfortunately, the ozone found in smog does not make its way to the ozone layer of the upper atmosphere, because it is converted to oxygen within a few days.

When we breathe polluted air, the respiratory system is, not surprisingly, the first to be affected. As air containing toxic substances fills the lungs, cells lining the airways and within the lungs are injured. Damaged cells release histamine (Chapter 13), which causes nearby capillaries to widen and become more permeable to fluid. As a result, fluid leaks from the capillaries and accumulates within the tissues. Even a brief exposure to oxides of sulfur (5 parts per million, or ppm, for a few minutes), the oxides of nitrogen (2 ppm for 10 minutes), or ozone leads to fluid accumulation, increased mucus production, and spasms (intense involuntary contractions) of the bronchioles. These effects obstruct air flow and reduce gas exchange.

Long-term irritation of bronchi by pollutants is a cause of chronic bronchitis and emphysema. The process begins when irritation causes fluid accumulation that, in turn, stimulates mucus production and coughing. The cough and mucus, signs of chronic bronchitis, irritate the lungs even more. As the bronchitis continues, the air passageways become narrower, trapping air in the lungs. When the increased pressure ac-

companying a cough causes the overinflated alveoli to rupture, emphysema begins.

Some pollutants can cause cancer. They overload or in other ways damage the mechanisms that would normally cleanse the respiratory system so that any cancer-causing chemicals in the inhaled air are no longer effectively removed. Within the airways these chemicals may then bring about changes in genetic material that can lead to cancer.

With so many documented health consequences of air pollution, you may wonder why the world's great minds have not developed the technology to solve the problem. In fact, they have. But unfortunately, air pollution is not just a scientific problem; it is also a social, political, and economic problem. The technology to prevent air pollution is available, but it costs money. Are you willing to pay more for goods and services to offset those costs? How much more are you willing to pay for low-sulfur fuels, for instance? Some of the strategies for improving air quality are likely to impose a certain amount of inconvenience. Are you willing to carpool, take a bus or, better yet, walk instead of drive to improve the air quality in your community? Considering that the pollutants in air are often blown hundreds of miles from their source, are you willing to make the same sacrifices to improve a stranger's air quality? Who should make the laws for controlling pollution when the pollutants are likely to cross state or national borders? Who should enforce those laws? ⧓

---

deep breath, then exhale only slightly, and repeat this process several times. Notice how quickly you feel an oxygen shortage if you continue taking very shallow breaths that keep the lungs almost completely filled with air.

Shortness of breath, the main symptom of emphysema, has several causes. Two causes are the decreased surface area for gas exchange and the increased dead air space. As the disease progresses, gas exchange becomes even more difficult because the alveolar walls thicken with fibrous connective tissue. The oxygen that does reach the alveoli has difficulty crossing the connective tissue to enter the blood. Thus, a person with emphysema constantly gasps for air.

exploring further . . .

In this chapter, we have seen how the respiratory system provides an extensive surface for gas exchange between inhaled air and the blood, moves air into and out of the lungs, and protects the gas exchange surfaces in the lungs. We have also considered some of the disorders of the respiratory system that result from inhaling disease-causing organisms or noxious substances. We cannot help inhaling many potentially harmful substances. However, we can usually control whether we inhale cigarette smoke, a very harmful substance. In the next chapter, we will discuss some of the health effects of smoking cigarettes.

HIGHLIGHTING THE CONCEPTS

In the Respiratory System, Oxygen and Carbon Dioxide Are Exchanged across a Moist Body Surface (pp. 269–274)

1. The oxygen that we breathe is needed to maximize the number of energy-storing ATP molecules formed from food energy. Exhaling carbon dioxide, a waste product formed by the same reactions, helps regulate the acid-base balance of body fluids. The role of the respiratory system is to exchange oxygen and carbon dioxide between the air and the blood.

2. The first structure that inhaled air usually passes through is the nose, which serves to clean, warm, and moisten the incoming air. Olfactory, or smell, receptors are located in the nasal cavities. The sinuses are air-filled spaces in the facial bones that also help warm and moisten the air. After leaving the nose, the inhaled air passes through the pharynx, or throat, and then the larynx, or voice box. Reflex movements of the larynx prevent food from entering the airways and lungs. The larynx is the source of the voice. The air passageways include the trachea, which branches to form bronchi, which branch various times in the lungs and eventually form progressively smaller branching tubules called bronchioles. The bronchioles terminate at the lungs' gas-exchange surfaces, the alveoli. Each alveolus is a thin-walled air sac encased in a capillary network.
   WEB TUTORIAL 14.1 The Human Respiratory System

Pressure Changes within the Lungs Cause Breathing (pp. 275–276)

3. Pressure changes within the lungs caused by changes in the size of the thoracic cavity move air into and out of the lungs. Inspiration occurs when the size of the thoracic cavity increases, causing the pressure in the lungs to drop below atmospheric pressure. Expiration occurs when the size of the thoracic cavity decreases and pressure in the lungs rises above atmospheric pressure.

Blood Transports Gases between the Lungs and the Cells (pp. 277–278)

4. Oxygen and carbon dioxide are exchanged between the alveolar air and the capillary blood by diffusion along their concentration (partial pressure) gradients. Oxygen diffuses from the alveoli into the blood, where it binds to hemoglobin within the red blood cells and is delivered to the body cells. A small amount of carbon dioxide is carried to the lungs dissolved in the blood plasma or bound to hemoglobin. Most, however, is transported to the lungs as bicarbonate ions.

Breathing Is Controlled Primarily by Respiratory Centers in the Brain (pp. 278–280)

5. The basic rhythm of breathing is controlled by the inspiratory area within the medulla of the brain. The neurons within this center undergo spontaneous bouts of activity. When they are active, messages are sent, causing contraction of the diaphragm and the muscles of the rib cage. As a result, the thoracic cavity increases in size and air is drawn into the lungs. When the inspiratory neurons are inactive, the diaphragm and rib cage muscles relax, and exhalation occurs passively. Also in the medulla is an expiratory area that causes forceful exhalation during heavy breathing.

6. The most powerful stimulant to breathing is an increased number of hydrogen ions in the blood, formed from carbonic acid when carbon dioxide dissolves in plasma. Extremely low levels of oxygen also increase breathing rate.

Respiratory Disorders Have Many Causes (pp. 280–284)

7. Two common respiratory diseases, the common cold and the flu, are caused by viruses. Pneumonia is an inflammation of the lungs, usually caused by infection, that causes fluid to fill the alveoli and narrows bronchioles. Strep throat, a sore throat caused by *Streptococcus* bacteria, can lead to rheumatic fever (and consequently to disease of the heart valves) or to kidney disease. Tuberculosis is a bacterial lung disease in which tubercles form in the lung. Acute bronchitis is caused by either bacteria or a virus. Chronic bronchitis is a persistent irritation of the bronchi. Emphysema is a breakdown of the alveolar walls and thus a reduction in the gas-exchange surfaces. Chronic bronchitis and emphysema are usually caused by smoking or air pollution.

KEY TERMS

respiratory system *p. 269*	Heimlich maneuver *p. 272*	surfactant *p. 274*	residual volume *p. 276*
nasal cavities *p. 270*	vocal cords *p. 272*	diaphragm *p. 275*	vital capacity *p. 276*
sinus *p. 271*	trachea *p. 273*	intercostal muscles *p. 275*	total lung capacity *p. 276*
pharynx *p. 272*	bronchi (sing. bronchus) *p. 273*	tidal volume *p. 276*	oxyhemoglobin *p. 277*
larynx *p. 272*	bronchioles *p. 273*	inspiratory reserve volume	carbaminohemoglobin *p. 277*
epiglottis *p. 272*	bronchiole tree *p. 273*	*p. 276*	carbonic anhydrase *p. 278*
glottis *p. 272*	alveoli (sing. alveolus) *p. 274*	expiratory reserve volume *p. 276*	

REVIEWING THE CONCEPTS

1. Why must we breathe oxygen? *p. 269*
2. Trace the path of air from the nose to the cells that use the oxygen. *pp. 270–274*
3. How are most particles and disease-causing organisms removed from the inhaled air before it reaches the lungs? *p. 270*
4. Describe the reflex that normally prevents food from entering the lower respiratory system. *p. 272*
5. How is human speech produced? *pp. 272–273*

6. What is the function of the cartilage rings in the trachea? *p. 273*
7. What is the bronchial tree? *p. 273*
8. How are the pressure changes in the thoracic cavity that are responsible for breathing created? *p. 275*
9. Is tidal volume or the vital capacity a larger volume of air? Explain. *p. 276*
10. How is most oxygen transported to the body cells? *pp. 276–277*

11. How is most carbon dioxide transported from the cells to the lungs? *pp. 277–278*

12. What region of the brain causes the basic breathing rhythm? *p. 278*

13. Explain how blood carbon dioxide levels regulate the breathing rate. *p. 278*

14. What are the causes of the shortness of breath experienced by people with emphysema? *pp. 283–284*

15. Choose the *correct* statement.
    a. During quiet breathing, expiration does not usually involve the contraction of muscles.
    b. Expiration occurs when the diaphragm and the rib muscles contract.
    c. Expiration occurs as the chest (thoracic) cavity enlarges.
    d. The larynx acts like a suction pump to pull air into the lungs.

16. You should be able to hold your breath longer than normal after you hyperventilate (breathe rapidly for a while) because hyperventilating
    a. decreases your blood oxygen levels.
    b. decreases blood carbon dioxide levels.
    c. increases blood oxygen levels.
    d. increases blood carbon dioxide levels.

17. The structure specialized to produce the sound of your voice is the
    a. trachea.
    b. larynx.
    c. bronchiole.
    d. epiglottis.

18. In a healthy person, most of the particles that are inhaled into the respiratory system
    a. are trapped in the mucus and moved by cilia to the pharynx (toward the digestive system).
    b. pass through the alveoli into the circulatory system, where they are engulfed by white blood cells.
    c. are caught on the vocal cords.
    d. are trapped in the sinuses.

19. In emphysema
    a. the number of alveoli is reduced.
    b. cartilage rings in the trachea break down.
    c. the diaphragm is paralyzed.
    d. the epiglottis becomes less mobile.
    e. the pharynx is constricted.

20. The _____ is the flap that covers the trachea to prevent food from entering during swallowing.

21. The enzyme in red blood cells that reversibly converts carbonic acid to bicarbonate ions and hydrogen ions is _____.

APPLYING THE CONCEPTS

1. Tatyana is a young woman with iron-deficiency anemia, so her blood does not carry enough oxygen. Would you expect this condition to affect her breathing rate or tidal volume? Why or why not?

2. Cigarette smoke destroys the cilia in the respiratory system. Explain why the loss of these cilia is a reason that cigarette smokers tend to lose more workdays because of illness than do nonsmokers.

3. Rosa has a 4-year-old son, Juan, who threatens to hold his breath until she gives him a candy bar. Should she be worried? Why?

4. Vincent is having an asthma attack. During an asthma attack the bronchioles constrict (get narrower in diameter). Does Vincent have more difficulty inhaling or exhaling? Why?

Additional questions can be found on the companion website.

Smoking Is the Leading Cause of Death in the United States

Cigarette Smoke Contains Poisons and Cancer-Causing Substances

Smoking Causes Several Deadly Diseases
- Smoking causes lung disease
- Smoking causes cancer
- Smoking causes heart disease
- Smoking causes other health problems

Smoking Poses Additional Health Risks for Women

Passive Smoking Causes Serious Health Problems

No Cigarette Is Safe

The Health Benefits of Quitting Smoking Are Numerous

14a

SPECIAL TOPIC

Smoking and Disease

H elvi was both hopeful and upset. She had spent the past 2 hours surfing the Web to learn more about emphysema and chronic bronchitis because she was anxious about her favorite aunt, Ilta. Helvi's mother had sent an e-mail earlier in the day describing Ilta's trip to the doctor, the tests, the prognosis, and the new lifestyle that Ilta was going to have to embrace. Helvi couldn't help but feel angry and disappointed that Ilta had not quit smoking 30 years ago when Helvi's mother did. Instead, Ilta had continued smoking in spite of constant coughing, shortness of breath, general poor health, and other family members' pleas to stop. During her Web search of authoritative sites, Helvi learned that one couldn't state with 100% accuracy that Ilta's lung diseases arose from smoking, but she learned that 80% to 90% of patients with lung disease were smokers. Her Web search also yielded some hope that there are treatments to alleviate the symptoms, permitting some short-term relief, but the fact remains that there is no cure. She gathered her strength for the painful call to her aunt.

In this chapter, we will consider the reasons why cigarette smoking leads to lung disease, heart disease, and cancer. We will see that no cigarette is safe, but much of the harm done to a smoker's body can be repaired after the person quits smoking. ∎

Since 1965, federal law has required warning labels on cigarette packages.

Smoking Is the Leading Cause of Death in the United States

Smoking is the greatest single preventable cause of disease, disability, and death in our society (Figure 14a.1a). In fact, every cigarette pack and cigarette advertisement in the United States must bear a warning from the Surgeon General. Yet tobacco, which causes bodily harm when used exactly as intended, is legal to sell to anyone at least 18 years old.

Let's put it this way. Each cigarette a person smokes shortens the smoker's life by about 5 to 7 minutes, a little less than the time it takes to smoke the cigarette. On average, smokers shorten their lives by 10 years (Figure 14a.1b). Consequently, each year, more than 3 million lives around the world—more than 400,000 of them American—are snuffed out prematurely because of smoking. In spite of statistics like these warning of the dangers of smoking, in 2006 21% of U.S. adults and 22% of high school students smoked cigarettes regularly.[1]

The economic consequences of lost workdays, premature death, and health-care expenditures due to cigarette smoke are also staggering. This is one reason nonsmokers want smokers to quit. (Another reason is that sometimes they love them.) According to the most recent (2004) report of the Surgeon General on the health consequences of smoking, smoking-related illnesses cost the United States $157.7 billion a year.

Cigarette Smoke Contains Poisons and Cancer-Causing Substances

The damage begins the instant the smoke arrives at a smoker's lips. Cigarette smoke harms every living tissue it touches—mouth, tongue, throat, esophagus, air passageways, lungs, and stomach. When autopsied, even light smokers (those who smoke less than a pack a day) show lung damage. Substances in the smoke are metabolized (broken down) in the liver, but even the breakdown products can injure the bladder, pancreas, and kidneys. Smoking a pipe or cigar is somewhat safer than smoking cigarettes because the pipe or cigar smoke generally is not inhaled. However, the risk of cancer of the lips, mouth, and tongue of a pipe or cigar smoker is still greater than a nonsmoker's, and so is the risk of lung cancer or heart disease. The effects of chewing smokeless tobacco are similar to those of smoking a pipe.

The average American smoker consumes a pack and a half of cigarettes each day—equal to about 300 puffs a day and over 109,000 puffs a year—year after year. With such exposure, we should certainly ask ourselves, what is in the smoke? The answer is alarming: smoke contains at least 4700 substances, and scientists are still testing the adverse health effects of each. So far, at least 50 have been shown to cause cancer. Also present are some well-known poisons: hydrogen cyanide (the poisonous gas used in gas chambers), carbon monoxide (common in auto exhaust), and

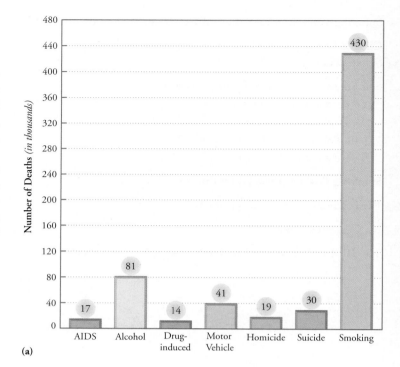

(a)

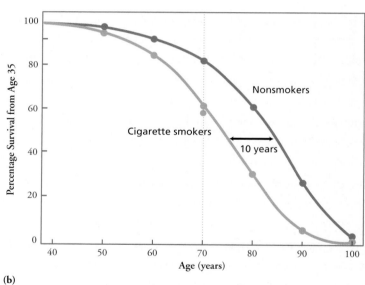

(b)

FIGURE **14a.1** Cigarette smoking reduces life expectancy. (a) Smoking causes more than 400,000 deaths each year in the United States, more people each year than the combined number of deaths due to AIDS, alcohol, drugs, car accidents, murder, and suicide. These data were published by the Centers for Disease Control and Prevention in 2000. Deaths from cigarette smoking have increased since then, but the relative importance of each cause of death remains similar. (b) In a study that tracked nearly 35,000 male British doctors for 50 years, cigarette smoking cut an average of 10 years off smokers' lives. A quarter of the lifetime smokers died before age 70.

Source: Doll, R., R. Peto, et al. (2004). "Mortality in relation to smoking: 50 years' observations on male British doctors." *British Medical Journal* 328: 1519–153.

[1]National Institutes of Health State-of-the-Science Statement, June 2006.

cresols (chemicals similar to the tars used to preserve telephone poles). In addition, tobacco smoke contains significant amounts of radioactive substances (thorium-228, radium-226, and polonium-210). This exposure to radiation may account for as much as 40% of maximum permissible annual exposure to radiation.

The three most dangerous substances in smoke are nicotine, carbon monoxide, and tar. Nicotine, which is a psychoactive drug, is perhaps the most insidious of the three because it creates pleasurable feelings of relaxation and yet has many harmful effects on the body. Moreover, it is nicotine that causes smokers to become "hooked" on cigarettes, ensuring continued exposure to the other injurious substances in the smoke. Depending on the brand, a cigarette contains 0.5 to 2.0 mg of nicotine. Each puff delivers about 0.2 mg of the drug to the bloodstream, and it reaches the brain within 6 to 7 seconds—twice as fast as injected heroin!

Although smoking can be relaxing, as noted above, nicotine is actually a stimulant of the brain and circulatory system. Under the influence of nicotine, the heart beats as many as 33 more beats per minute. At the same time, the blood vessels constrict, so not only is the heart beating faster, but it is forcing blood through a less receptive circulatory system. The result is an increase in blood pressure. Nicotine also affects the platelets in the blood (these are described in Chapter 11 as cell fragments containing chemicals that initiate clotting). Nicotine causes the platelets to become sticky, increasing the likelihood of abnormal clots forming that may lead to heart attacks or strokes. A heart attack is the death of heart muscle cells and a stroke is the death of nerve cells in the brain. Either can be caused by blockage of a blood vessel by a clot.

As mentioned, nicotine is powerfully addictive. Ninety-five percent of smokers are physiologically dependent on it. Nicotine causes dopamine to be released in the brain's reward center, causing a pleasurable feeling. You may recall from Chapter 8a that drugs that stimulate the brain's reward center, such as cocaine and heroin, are highly addictive. Thus, smokers experience withdrawal symptoms when they try to quit. Indeed, some opium addicts have reported that it was easier to do without opiates than nicotine. When a smoker tries to quit, nerve cells become hyperactive, causing withdrawal symptoms that include irritability, anxiety, headache, nausea, constipation or diarrhea, craving for tobacco, and insomnia. The majority of smokers continue smoking to avoid these unpleasant symptoms. Those who quit find that most of the withdrawal symptoms begin to lessen after a week without nicotine but some may continue for weeks or even months. Certain symptoms, such as drowsiness, difficulty concentrating, and craving for a cigarette, seem to worsen about 2 weeks after quitting. Consequently, many smokers return to their habit and resume their exposure to the harmful substances in smoke.

The amount of carbon monoxide in cigarette smoke is 1600 ppm (parts per million), which greatly exceeds the 10 ppm considered dangerous in industry. Furthermore, carbon monoxide from smoking a cigarette lingers in the bloodstream for up to 6 hours. You may recall from Chapter 11 that carbon monoxide is a poison that prevents red blood cells from transporting oxygen. In fact, carbon monoxide from smoking a cigarette lowers the oxygen-carrying capacity of the blood by about 12%, reducing oxygen delivery to every part of the body, including the brain and heart. The diminished oxygen supply to the brain can impair judgment, vision, and attentiveness to sounds. For these reasons, smoking can be hazardous for drivers.

Tar is a mixture of thousands of substances in the smoke that settles out as a brown sticky substance when the smoke cools within the body. A pack-a-day smoker coats his or her respiratory system with about 50 mg of tar each day. Besides the cancer-causing chemicals in tar, other chemicals it contains destroy the elasticity of the lung.

Smoking Causes Several Deadly Diseases

Most of the health problems associated with cigarette smoking are caused by increases in the risk of three diseases, all of which can kill: lung disease, cancer, and heart disease. We will now examine these three diseases. But keep in mind that smoking affects health in many other ways as well.

■ Smoking causes lung disease

Because the objective of smoking is to bring smoke into the lungs, we can expect some of its most damaging effects to be seen there. Even teenage smokers show damage to airways and lungs, but because teenagers are young and more resilient than adults, the damage may be apparent only in breathing tests or during athletic activities. Young smokers are more likely to be short-winded than their nonsmoking friends. If they continue to smoke, the effects become much more noticeable. By age 60, most smokers have significant changes in their airways and lungs.

The damage to the respiratory system of smokers is gradual and progressive. It begins as the smoke hampers the actions of two of the lungs' cleansing mechanisms—cilia and macrophages. Recall from Chapter 14 that the cells lining the airways are covered with hairlike cilia. Even the first few puffs from a cigarette slow the movement of the cilia, making them less effective in sweeping debris from the air passageways. Smoking an entire cigarette prevents the cilia from moving for an hour or longer. With continued smoking, the nicotine and sulfur dioxide in the smoke paralyze the cilia, and the cyanide destroys the ciliated cells (Figure 14a.2).

Cigarette smoke causes a smoker's lungs to be chronically inflamed. The inflammation summons macrophages, the wandering cells that engulf foreign debris, to the lungs for the purpose of cleaning the lung surfaces. But, just as smoke paralyzes the cilia, it also paralyzes the macrophages, further hampering the cleansing efforts. As the cilia and macrophages become less effective, greater quantities of tar and disease-causing organisms remain within the respiratory system. As a result, cigarette smokers spend more time sick in bed and lose more workdays each year than do nonsmokers.

At the same time that the cilia and macrophages are being slowed, the smoke stimulates the mucus-secreting cells in the linings of the respiratory passageways. In consequence, the smaller airways become plugged with mucus, making breathing more difficult. At this point, if not before, "smoker's cough" begins. Coughing is a protective reflex, and initially the smoker coughs simply because smoke irritates air passageways. However, as smoking

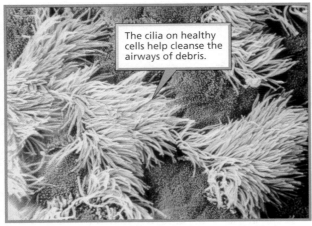

(a)

(b)

FIGURE **14a.2** Changes in the ciliated linings of the air passageways accompanying smoking. (a) The cilia on the cells lining the airways of a healthy nonsmoker cleanse the airways of debris. (b) Cigarette smoke first paralyzes and then destroys the cilia. As a result, hazardous materials can accumulate on the surfaces of the air passageways.

continues and the cilia become increasingly less able to remove mucus and debris, the only way to remove the material from the passageways is to cough. The cough is generally worse in the morning as the body attempts to clear away the mucus that accumulated overnight.

Gradually, the inflammation and congestion within the lungs, along with the constant irritation from smoke, lead to chronic bronchitis, a disease characterized by a persistent deep cough that brings up mucus. The air passageways become narrow because of the thickening of their linings caused by repeated infection, the accumulation of mucus, and the contraction of the smooth muscle in their walls. Air flow is restricted, resulting in breathlessness and wheeziness. Bronchial infections become more common because the body has lost the ability to clear disease-causing organisms from the passageways. Bacterial infections may be treated with antibiotics, bringing slight temporary relief, but chronic bronchitis will continue as long as smoke irritates the airways.

Emphysema, in which the walls of alveoli are destroyed, is often the next stage in the progressive damage to the lungs (Chapter 14). Airways and alveolar walls lose elasticity as tar causes body defense cells to secrete destructive enzymes. Consequently, the lung tissues can no longer absorb the increase in pressure that accompanies a cough, and the delicate alveolar walls break like soap bubbles. With more and more alveoli destroyed, the surface area for gas exchange is reduced, so less oxygen is delivered to the body. Furthermore, as the alveolar walls break down, the alveoli become larger, increasing the amount of air that stays trapped in the lungs during exhalation. Exhalation thus becomes more difficult, and greater amounts of cyanide, formaldehyde, and carcinogens from the smoke remain in the lungs, killing even more cells.

As the alveoli are damaged, the small blood vessels going to and from the alveoli rupture. If these blood vessels sustain enough damage, stress on the right ventricle of the heart increases, because it must push the same amount of blood to the lung tissues through fewer vessels.

■ Smoking causes cancer

Smoking is the major single cause of lung cancer, and it causes other cancers as well (Table 14a.1). In fact, smoking is responsible for 30% of all cancer deaths. Sadly, cancer-causing chemicals in tobacco damage every tissue they touch.

Cancers can develop because the smoke contains carcinogens, or cancer-causing chemicals, that cause cells to lose control over cell division. Some of the carcinogens change the structure of the genetic material, DNA. Others, such as formaldehyde, cause enzyme changes that allow cells to become cancerous. Still other components of the smoke work as co-carcinogens, chemicals that enhance the action of other carcinogens or promote cell division and, therefore, tumor growth once the cancer has begun.

It is sobering to realize that between 85% and 90% of all cases of lung cancer are caused by smoking and are, therefore, preventable (Figure 14a.3). Unfortunately, nearly 90% of individuals diagnosed as having lung cancer that has spread die within 5 years.

TABLE 14A.1	TYPES OF INCREASED CANCER RISK DUE TO SMOKING
TYPE	**INCREASED RISK**
Lung	13–23 times
Mouth and lips	4 times
Larynx	5 times in light smokers (less than 1 pack per day)
	20–30 times in heavy smokers (more than 1 pack per day)
Esophagus	2–9 times
Kidney and bladder	2–10 times
Pancreas	2–5 times (especially if alcohol is also consumed)

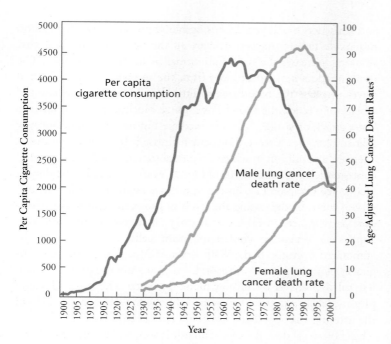

FIGURE **14a.3** Tobacco use and lung cancer in the United States. Lung cancer usually takes about 20 years to develop. Notice that the number of deaths from lung cancer increases and decreases with tobacco use but with about a 20-year delay. The lung cancer death rate of females is lower than that of males because there are fewer female smokers than there are male smokers. The data are age-adjusted to 2000 U.S. standard population.

Source: http://www.cancer.org/downloads/STT/Cancer_Statistics_2006_Presentation.ppt. Death rates: U.S. Mortality Public Use Tapes, 1960–2000, U.S. Mortality Volumes, 1930–1959, National Center for Health Statistics, Centers for Disease Control and Prevention, 2002. Cigarette consumption: U.S. Department of Agriculture, 1900–2000.

Lung cancer usually has no symptoms until it is quite advanced. Therefore, it is not usually detected in time for a cure.

Admittedly, not every smoker gets lung cancer, but smokers are 13 to 23 times more likely to get it than are lifetime nonsmokers. The likelihood of a smoker getting lung cancer depends on several factors, such as the number of cigarettes smoked, the number of years of smoking, the age at which smoking began, how deeply the smoke is inhaled, and the amount of tar and nicotine in the brand of cigarette smoked. There are also individual differences in genetic and biological makeup that influence cancer risk.

The progression to lung cancer is marked by changes in the cells of the airway linings of smokers (Figure 14a.4). In a non-smoker, the lining of the air passageways has a basement membrane underlying basal cells and a single layer of ciliated columnar cells. In a smoker, one of the first signs of damage is an increase in the number of layers of basal cells. Next, the ciliated columnar cells die and disappear. The nuclei of the basal cells then begin to change as mutations accumulate, and the cells become disorganized. This is the beginning of cancer. Eventually, the uncontrolled cell division forms a tumor (Figure 14a.5). When cancer cells break through the basement membrane, they can spread to other parts of the lung and to the rest of the body, a process called metastasis (see Chapter 21a).

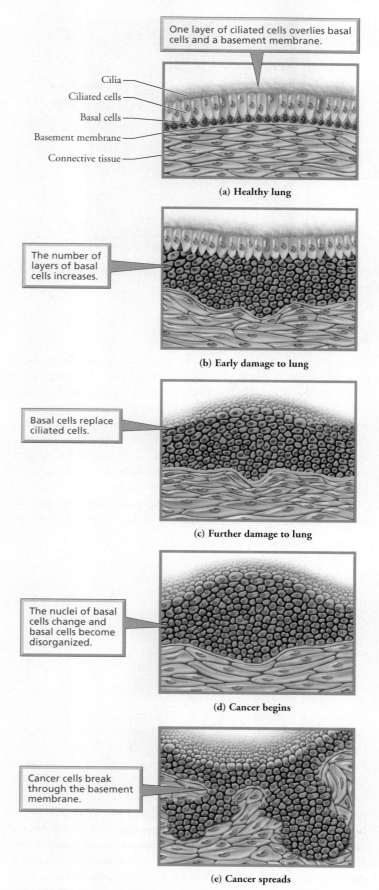

FIGURE **14a.4** Lung cancer developing in the lining of a bronchus, an air passageway in the lung

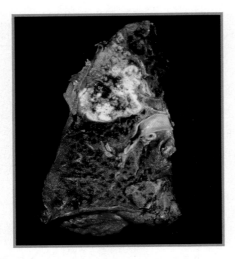

FIGURE **14a.5** Lung cancer. The tumor is the light-colored solid mass shown in the upper region of the lung.

■ Smoking causes heart disease

When we think about the hazards of tobacco smoke, lung cancer generally leaps to mind. But the increased risk of cardiovascular (heart and blood vessel) disease is even more significant. Each year, cardiovascular disease kills many more people than does lung cancer, and smokers have a twofold to threefold increase in the risk of heart disease (Figure 14a.6). The American Heart Association estimates that about 25% of all fatal heart attacks are caused by cigarette smoke. This translates to roughly 200,000 heart attacks a year in the United States that could have been prevented by not smoking.

Smoking stresses the heart and blood vessels in many ways. We saw earlier in the chapter that nicotine makes the heart beat faster

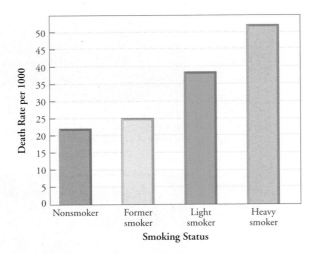

FIGURE **14a.6** Death rates due to heart disease among nonsmokers and smokers. Notice that the death rate from heart disease increases with the number of cigarettes smoked per day. People who smoke over a pack per day (heavy smokers) have more than twice the risk of death due to heart disease than do people who have never smoked. In any case, a smoker who successfully quits is much less likely to die of heart disease than if the smoking habit continues.

and constricts blood vessels, raising blood pressure, and that carbon monoxide reduces oxygen delivery to the heart. We also saw that smoking increases the risk of abnormal blood clot formations. The resulting clots may break loose from the site where they form and travel through the bloodstream until they lodge in a small vessel where they block the blood flow. These blockages may result in a heart attack or stroke. A less immediate but no less important way that smoking leads to cardiovascular disease is by increasing atherosclerosis, a condition in which lipid deposits, primarily composed of cholesterol, form in the walls of blood vessels, restricting the flow of blood (see Chapter 12). Smoking influences atherosclerosis in two ways. One is by decreasing the levels of protective cholesterol-transport particles, called HDLs, that carry cholesterol to the liver, perhaps even removing cholesterol from cells, so that it can be eliminated from the body. With fewer HDLs, more cholesterol begins to clog the arteries. A second way that smoking promotes cholesterol deposits is by raising blood pressure. The elevated blood pressure causes rapid, turbulent blood flow that damages the walls of the arteries, making them more susceptible to cholesterol deposit. The cholesterol deposits cause inflammation, which leads to atherosclerosis. The narrowing of blood vessels caused by atherosclerosis leads to starvation of tissue downstream and an increase in blood pressure. When atherosclerotic deposits form in the arteries that supply blood to the heart, as they often do, the blood supply to the heart may be reduced or shut down completely, causing heart cells to die.

■ Smoking causes other health problems

The reduction in the amount of oxygen reaching tissues impairs the body's ability to make collagen, a main supportive protein in certain connective tissues, including skin and bones. One result is wrinkles. The extent of wrinkling increases with cigarette consumption and the duration of the smoking habit. Heavy smokers are 3.5 times more likely than nonsmokers to show crow's feet (wrinkles around the eyes). In smokers, impaired collagen production also slows down the healing of wounds in bones and skin.

Smoking can dim your vision. In the United States, smoking is responsible for 20% of all cases of cataracts, a clouding of the lens of the eye that leads to blindness if not corrected surgically. Unfortunately, ex-smokers continue to have a higher risk of cataracts. So, unlike many of the other effects of smoking, this is a problem that quitting might not solve.

Urinary incontinence (lack of urinary control) is more common among smokers than nonsmokers. The hacking smoker's cough weakens the muscle that normally holds urine in the bladder. Nicotine also causes the muscles of the bladder to contract, allowing urine to leak.

Smoking Poses Additional Health Risks for Women

More than a quarter of women of child-bearing age smoke, and almost half of them have tried to quit. Unfortunately, the hazards of smoking may be greater for women than for men. For instance, some studies suggest that women smokers have a greater chance of getting lung cancer than do male smokers. Although the reasons for this difference are not clear, genes may play a role. Women are more than three times more likely than men to carry a certain genetic mutation, the *K-ras* mutation. This mutation helps the lung tumor to grow, particularly in response to the female hormone estrogen. Another gene that promotes lung cancer growth—gastrin-releasing peptide receptor—is activated by nicotine. This gene is located on the X chromosome in a region that escapes inactivation;[2] because women have two X chromosomes and men have only one, women have twice as many copies of the gene than do men, making women more susceptible to the cancer-causing effects of smoking.

Furthermore, some risks are unique to women. Women who smoke are three times more likely than nonsmokers to develop cervical cancer. In addition, women who smoke reach menopause 2 to 3 years earlier than nonsmokers. Female smokers also have higher rates of osteoporosis—a condition more common in women than men, in which bones become less dense and, therefore, weaker.

Women of reproductive age may be at special risk because the combination of smoking and oral contraceptives can be deadly; both elevate blood pressure and increase the likelihood of abnormal clot formation. The death rate among women smokers who also take birth control pills is three times that of nonsmokers who are on the pill. The increased death rate is due to a much higher incidence of strokes, heart attacks, and blood clots in the legs.

Smoking has many adverse effects on a woman's reproductive ability (Figure 14a.7). A woman smoker who would like to become pregnant will have much better luck conceiving if she quits smoking. Women who smoke more than a pack a day are half as fertile as women who do not smoke. Furthermore, if a woman smoker does become pregnant and continues to smoke, she is twice as likely to miscarry as a nonsmoker. Smoking during pregnancy increases the risk of complications such as premature separation of the placenta

FIGURE **14a.7** Cigarette smoking during pregnancy exposes the fetus to toxins and reduces its oxygen supply. Pregnant women who smoke are more likely to miscarry or to have a baby die shortly after birth.

[2]One of the X chromosomes in women is inactivated.

from the uterus, which shuts off the oxygen supply to the fetus, or poor positioning of the placenta, which may partially or completely block the birth canal. (Incidentally, exposure to someone else's cigarette smoke can also affect a fetus whose nonsmoking mother is exposed to the smoke while she is pregnant.)

In a very real sense, tobacco smoke suffocates the growing fetus, because nicotine constricts its lifeline, the blood vessels of the umbilical cord, and because carbon monoxide reduces the amount of oxygen carried by the blood. Smoking two packs a day blocks approximately 40% of the oxygen supply to the fetus. As a result, the fetus's heart rate and blood pressure increase, and the acid-base balance of its blood is shifted.

Smoking leads to an estimated 5000 preventable fetal deaths annually during the 8 weeks before or 7 days after birth. The incidence of stillbirth among pregnant women who smoke is 7.8% compared with 4.1% among nonsmoking women. Furthermore, a newborn of a woman who smoked during pregnancy has a one-third higher risk of dying soon after birth. Among the reasons for the increased infant mortality are premature birth and low birth weight. Newborns of women who smoke weigh an average of 200 grams (about $\frac{1}{2}$ lb) less than those born to nonsmoking women.

Passive Smoking Causes Serious Health Problems

It is not true that people who smoke are hurting only themselves. After all, their smoke pollutes the air of others who may not smoke. This dangerous pollution takes the form both of exhaled smoke and of sidestream smoke that comes from the burning end of the cigarette. As a result, two-thirds of a cigarette's smoke actually enters the environment. And, because it is not filtered through tobacco, this smoke contains higher levels of many hazardous materials than the smoke that is inhaled through a cigarette. For example, sidestream smoke contains more cadmium, which is associated with high blood pressure, chronic bronchitis, and emphysema. It also contains twice the tar and nicotine of inhaled smoke and five times as much carbon monoxide. Because the concentration of cancer-causing nitrosamines in sidestream smoke is 50 times greater than in inhaled smoke, a nonsmoker who spends an hour in a smoke-filled room may inhale an amount of nitrosamines equal to smoking 15 filter cigarettes.

When people are smoking during a typical campus party, the number of particulates in the air of the room is 40 times greater than the U.S. standard for air quality. After 30 minutes in a smoky room, a nonsmoker's heartbeat and blood pressure begin to increase. The carbon monoxide level in the blood begins to rise enough to hamper one's ability to distinguish time intervals or to perceive the relative brightness of two lights. (This impairment could be of critical significance if a person needs to distinguish between the headlights of oncoming cars.) After a single hour in a very smoky room, a nonsmoker's blood levels of carbon monoxide and nicotine match those of someone who smoked a cigarette. For 3 to 4 hours after leaving the room, a nonsmokers' blood still contains carbon monoxide.

Ironically, because we generally spend the most time with people we care about the most, the loved ones of smokers are often the people who suffer the greatest harm from the effects of passive smoking. Passive exposure to smoke most often occurs at home. Many other people encounter secondhand smoke on a regular basis in the workplace.

We have seen that the major health risks for smokers are increased risk of lung disease, cancer, and heart disease. Long-term exposure to secondhand tobacco smoke jeopardizes the health of nonsmokers in the same ways.

No Cigarette Is Safe

There is no such thing as a safe cigarette. True, filters do reduce certain risks by trapping some tar, nicotine, carbon monoxide, and other poisonous gases. As a result, smokers who smoke cigarettes with filters have a slightly lower risk of lung cancer than those who smoke cigarettes without filters. However, the filters do not trap everything, and these smokers still have a risk of developing lung cancer that is 6.5 times that of a nonsmoker.

Because the tars and nicotine cause many of the harmful effects of cigarette smoke, the risks associated with low-tar, low-nicotine cigarettes might be expected to be lower than those associated with other brands. However, the low-tar, low-nicotine products are not necessarily safer. The craving for nicotine often causes people to adjust their smoking habits so that the blood level of nicotine remains constant regardless of the type of cigarette smoked. For example, there is evidence that smokers who switch to low-tar, low-nicotine cigarettes inhale more deeply or puff more often. They may even smoke more cigarettes.

A definite downside to cigarettes with low tar and low nicotine is that they are easier for novices to smoke. Such cigarettes have made it easier for a whole new group—young girls and women—to become hooked. Females are more sensitive than males to unpleasant side effects of nicotine, but now they can smoke because they tolerate the low levels of nicotine more easily than the levels in regular cigarettes.

The Health Benefits of Quitting Smoking Are Numerous

Three of four smokers say they want to quit, and 60% say they have tried. In any one serious try, chances of success are only one in five. However, the chances of success go up to three in five with repeated attempts. Despite the difficulty of quitting, more than 30 million Americans have stopped smoking.

The health benefits if you quit smoking are enormous. The major payoff is a longer life for you, your friends and loved ones, and your coworkers. A recent study (depicted in Figure 14a.1b) followed almost 35,000 British male doctors for 50 years. On average, lifelong smokers in this study cut about 10 years off their lives, but those who stopped smoking at age 50 lost only 5 years. Smokers who quit by age 30 had the same average life expectancy as nonsmokers did.

Much of the damage caused by smoke is reversible once you quit. Blood pressure and heart rate begin to decrease toward normal levels. The risk of heart attack begins to drop a year after quitting. After 5 years, the risk of heart attack for someone who smoked a pack a day is the same as for someone who never smoked. The risk of lung cancer also drops, but it never falls as low as the level for people who have never smoked (Figure 14a.8).

If you are a smoker who would like to break the habit, here are some suggestions from the Centers for Disease Control and Prevention:

1. **Prepare.** Set a date to quit. Throw away all cigarettes and ashtrays in your home, car, and workplace. Do not allow other people to smoke inside your home. If this is not the first time you are trying to quit, review the actions that helped your efforts in the past and those that did not help.

2. **Do not have even a puff of any kind of cigarette.** Low tar, low nicotine cigarettes do not help you quit.

3. **Enlist support and encouragement from your family, friends, and health-care provider.**

4. **Make a list of the reasons you want to quit.** Do you want to gain control of your life? Do you want better health? Do you want to protect your family and friends from the smoke from your cigarette? Is there a special person who makes you want a longer life? Focusing on why you want to quit increases motivation to quit.

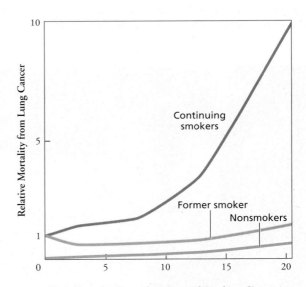

FIGURE **14a.8** A comparison of lung cancer deaths among lifetime nonsmokers, ex-smokers, and continuing smokers. Quitting results in dramatic reduction of a smoker's risk of death due to lung cancer.

5. **Learn new skills and behaviors.** Distract yourself from smoking by walking, talking, or keeping busy with a task. Change your routine to avoid behaviors or places that you associate with smoking. Find ways to reduce your stress; you could try taking a hot bath, exercising, or reading a book.

6. **Recognize that quitting will be difficult.** You are addicted to nicotine. There are medications to help you quit. Most of these medications provide nicotine in a form other than smoking: gum, patch, inhaler, or nasal spray. Bupropion SR is an antidepressant drug available by prescription that can help if you feel depressed after quitting.

7. **Be prepared to face difficult situations.** Try to avoid drinking alcohol, because drinking lowers the likelihood of success. Try to avoid places where people will be smoking, because the smoke will increase your cravings. Be aware that you are likely to gain weight. Many people gain weight when they quit smoking, but the gain is usually less than 10 pounds. Do not let weight gain discourage your attempt to quit smoking.

15

The Digestive System

The Digestive System Consists of a Long Tube That Runs through the Body, Along with Accessory Glands

The Digestive System Is Divided into Specialized Compartments for Food Processing

- The mouth begins mechanical digestion and the chemical digestion of starch
- The pharynx is shared by the digestive and respiratory systems
- The esophagus conducts food from the pharynx to the stomach
- The stomach stores and liquefies food and begins protein digestion
- The small intestine is the primary site of digestion and absorption
- The pancreas, liver, and gallbladder are accessory organs that aid the processes of digestion and absorption within the small intestine
- The large intestine absorbs water and other useful substances

Nerves and Hormones Control Digestive Activities

HEALTH ISSUE Heartburn and Peptic Ulcers—Those Burning Sensations

This baby must put the spaghetti in her mouth and chew and swallow it. After that, the digestion of her meal will proceed automatically.

Balamani, an expert computer programmer, was perplexed. The programming task itself was simple. The problem was that part of the animation seemed all too real. The pharmaceutical company had asked for an instructional animation that would show how the gastrointestinal (GI) tract worked, along with examples of specific malfunctions or diseases for which they had products. Of immediate concern to Balamani was a malfunction along the GI tract described in the storyboards. It was called gastroesophageal reflux. The symptoms matched his own. He often experienced heartburn—an occasional backing up of stomach content—when he could taste acidic fluid, and once in a while his stomach hurt. Antacids didn't seem to help. He had assumed that the symptoms were due to the drastic changes in his diet since his move from India to the United States and that they would go away after a while. The storyboards, however, said that "acid reflux" is caused by a problem at the base of the esophagus where it empties into the stomach.

The next day, Balamani consulted his doctor, who gave him a number of instructions for dealing with his condition: elevate the head of the bed to allow gravity to work for his esophagus, (keeping fluids down); cut back on fatty foods; cut out coffee and orange juice; reduce the size of meals while increasing the frequency; and by all means pay fewer visits to fast food restaurants. Within a few weeks, Balamani was feeling much better, and at about the same time, he finalized the animation.

We will learn in this chapter that food usually travels in one direction through the digestive system, unless there is a problem such as gastroesophageal reflux. We will see how the digestive system is organized for breaking food molecules down and making them available for use as a source of energy or as raw material for the growth and repair of cells. Food is digested and nutrients and water are absorbed in specialized compartments along a long tube, the GI tract. Undigested materials are then eliminated from the body. ■

The Digestive System Consists of a Long Tube That Runs through the Body, Along with Accessory Glands

There is some truth to the saying "You are what you eat." Rest assured, however, that no matter how many hamburgers you eat, you will never become one. Instead, the hamburger becomes *you*. The transformation is possible largely because of the activities of the digestive system. Like an assembly line in reverse, the digestive system takes the food we eat and breaks the complex organic molecules into their chemical subunits. The subunits are molecules small enough to be absorbed into the bloodstream and delivered to body cells. Food molecules ultimately meet one of two fates: they may be used to provide energy for daily activities or they may provide materials for growth and repair of the body. Imagine that you just ate a hamburger on a bun. The starch in the bun may fuel a jump for joy; the protein in the beef may be used to build muscle; and the fat may become myelin sheaths that insulate nerve fibers. Without the digestive system, this food would be useless to us, because it could never reach the cells of our body.

The **digestive system** consists of a long, hollow tube, called the **gastrointestinal (GI) tract**, into which various accessory glands release their secretions (Figure 15.1). The hollow area of the tube that food and fluids travel through is called the *lumen*. Along most of its length, the walls of the GI tract have four basic layers, as shown in Figure 15.2.

- **Mucosa.** The innermost layer is the moist, mucus-secreting layer called the mucosa. The mucus helps lubricate the tube, allowing food to slide through easily. Mucus also helps protect the cells in the lining from rough substances in the food and from digestive enzymes. In some regions of the digestive system, cells in the mucosa also secrete digestive enzymes. In addition, the mucosa in some parts of the digestive tract is highly folded, which increases the surface area for absorption.
- **Submucosa.** The next layer, the submucosa, consists of connective tissue containing blood vessels, lymph vessels, and nerves. The blood supply maintains the cells of the digestive system and, in some regions, picks up and transports the products of digestion. The nerves are important in coordinating the contractions of the next layer.
- **Muscularis.** The next layer, the muscularis, is responsible for movement of materials along the GI tract and for mixing materials with digestive secretions. In most sections of the tube, the muscularis is a double layer of smooth muscle. (The stomach, as you will read later, is an exception; it has three layers of muscle.) In the inner, "circular" layer of muscle, the muscle cells encircle the tube. In the outer, longitudinal layer, the muscle cells are arranged parallel to the digestive tube. The muscle layers churn the food until it is liquefied, mix the resulting liquid with enzymes, and propel the food along the GI tract in a process called *peristalsis*, which you will read about later in the chapter.
- **Serosa.** The serosa, a thin layer of epithelial tissue supported by connective tissue, wraps around the GI tract. It secretes a fluid that reduces friction with contacting surfaces of the intestine and other abdominal organs.

Regions of the GI tract are specialized to process food in particular ways. One aspect of that processing is *mechanical digestion*, the physical breaking of food into smaller pieces, and another is *chemical digestion*, the breaking of chemical bonds so that complex molecules are taken apart into smaller subunits. Chemical digestion produces molecules that can be absorbed into the bloodstream and used by the cells. We will trace the path food travels along the GI tract to see how it is processed and absorbed.

The Digestive System Is Divided into Specialized Compartments for Food Processing

As food moves along the GI tract, it passes through the mouth, pharynx, esophagus, stomach, small intestine, and large intestine. The salivary glands, liver, and pancreas add secretions along the way. Most nutrients are absorbed from the small intestine. Additional water is absorbed in the large intestine. Undigested and indigestible materials pass out the anus. Table 15.1 identifies the structures of the digestive system and describes their roles in both mechanical and chemical digestion. You can refer to the table as you read about each of the digestive structures below.

■ The mouth begins mechanical digestion and the chemical digestion of starch

The entryway to the digestive tract and the first stop on food's journey through the digestive tract is the *mouth*, also called the oral cavity. The roof of the mouth is called the *palate*. The region of the palate closest to the front of the face, the hard palate, is reinforced with bone. Farther to the back of the mouth is the soft palate, which consists only of muscle and prevents food from entering the nose during swallowing. The mouth serves several functions: (1) mechanical and, to some extent, chemical digestion begin; (2) food quality is monitored; and (3) food is moistened and manipulated so that it can be swallowed. The teeth, salivary glands, and tongue all contribute to these functions.

TEETH AND MECHANICAL DIGESTION

As we chew, our teeth break solid foods into smaller fragments that are easier to swallow and digest. The sharp, chisel-like incisors in

ORGANS

ACCESSORY STRUCTURES

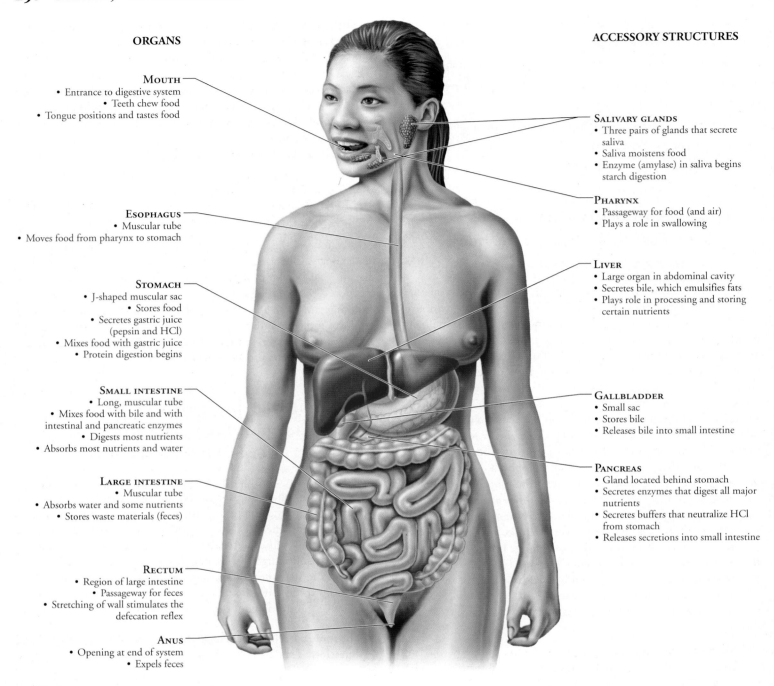

MOUTH
- Entrance to digestive system
- Teeth chew food
- Tongue positions and tastes food

SALIVARY GLANDS
- Three pairs of glands that secrete saliva
- Saliva moistens food
- Enzyme (amylase) in saliva begins starch digestion

PHARYNX
- Passageway for food (and air)
- Plays a role in swallowing

ESOPHAGUS
- Muscular tube
- Moves food from pharynx to stomach

LIVER
- Large organ in abdominal cavity
- Secretes bile, which emulsifies fats
- Plays role in processing and storing certain nutrients

STOMACH
- J-shaped muscular sac
- Stores food
- Secretes gastric juice (pepsin and HCl)
- Mixes food with gastric juice
- Protein digestion begins

SMALL INTESTINE
- Long, muscular tube
- Mixes food with bile and with intestinal and pancreatic enzymes
- Digests most nutrients
- Absorbs most nutrients and water

GALLBLADDER
- Small sac
- Stores bile
- Releases bile into small intestine

PANCREAS
- Gland located behind stomach
- Secretes enzymes that digest all major nutrients
- Secretes buffers that neutralize HCl from stomach
- Releases secretions into small intestine

LARGE INTESTINE
- Muscular tube
- Absorbs water and some nutrients
- Stores waste materials (feces)

RECTUM
- Region of large intestine
- Passageway for feces
- Stretching of wall stimulates the defecation reflex

ANUS
- Opening at end of system
- Expels feces

FIGURE **15.1** The digestive system consists of a long tube, called the gastrointestinal tract, into which accessory glands release their secretions.

the front of the mouth (see Figure 15.3a) slice the food as we bite into it. At the same time, the pointed canines to the sides of the incisors tear the food. Then the food is ground, crushed, and pulverized by the premolars and molars, which lie along the sides of the mouth.

Teeth are alive. In the center of each tooth is the pulp, which contains the tooth's life support systems—blood vessels that nourish the tooth and nerves that sense heat, cold, pressure, and pain (Figure 15.3b). Surrounding the pulp is a hard, bonelike substance, called dentin. The crown of the tooth (the part visible above the

gum line) is covered with enamel, a nonliving material that is hardened with calcium salts. The root of the tooth (the part below the gum line) is covered with a calcified, yet living and sensitive connective tissue called cementum. The roots of the teeth fit into sockets in the jawbone. Blood vessels and nerves reach the pulp through a tiny tunnel through the root called the root canal.

Tooth decay is caused by acid produced by bacteria living in the mouth. When you eat, food particles become trapped between the teeth and in the small spaces where the teeth meet the gums. Bacteria in the mouth are nourished by the sugar in these food par-

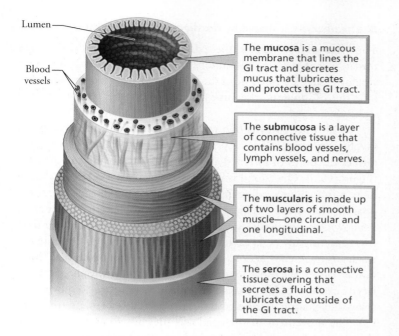

Lumen

Blood vessels

The **mucosa** is a mucous membrane that lines the GI tract and secretes mucus that lubricates and protects the GI tract.

The **submucosa** is a layer of connective tissue that contains blood vessels, lymph vessels, and nerves.

The **muscularis** is made up of two layers of smooth muscle—one circular and one longitudinal.

The **serosa** is a connective tissue covering that secretes a fluid to lubricate the outside of the GI tract.

FIGURE **15.2** Along most of its length, the wall of the digestive system has four basic layers: the mucosa, the submucosa, the muscularis, and the serosa.

disease, occurs when plaque that has formed along the gum line causes the gums to become inflamed and swollen. The swollen gums can bleed and do not fit as tightly around the teeth as they should. The pocket that forms between the tooth and the gum traps additional plaque. The bacteria in the plaque can then attack the bone and soft tissues around the tooth, a condition called *periodontitis* (*peri*, around; *dont*, teeth; *itis*, inflammation of). As the tooth's bony socket and the tissues that hold the tooth in place are eroded, the tooth becomes loose.

what would you do?

For the past 50 years, some communities whose water is not naturally fluoridated have added low doses of fluoride to the public water supply to reduce tooth decay. Nearly 98% of Americans have, or have had, tooth decay. Over 6 million teeth are removed each year. Proponents of fluoridation point out that the incidence of tooth decay has declined after the addition of fluoride to drinking water. Opponents of the practice argue that (1) fluoridation of the water supply is a form of forced medication; (2) the dosage cannot be properly controlled because of differences in people's body weight and the amount of water people consume; and (3) excessive fluoride intake causes teeth to become brown and mottled. If you had to vote on fluoridation of your public water supply, what additional information would you want before voting? How would you vote at present? Explain your reasons.

ticles. As bacteria digest the sugar, acid is produced that dissolves the enamel and causes a cavity to form. When the enamel has been penetrated, bacteria can invade the softer dentin beneath. If a dentist does not fill the cavity, the bacteria can infect the pulp (Figure 15.4). *Plaque*, an invisible film of bacteria, mucus, and food particles, promotes tooth decay because it holds the acid against the enamel. Daily brushing and flossing helps remove plaque, reducing the chance of tooth decay.

Gum disease, which affects two of three middle-aged people in the United States, is a major cause of tooth loss in adults. *Gingivitis* (*gingiv-*, the gums; *-itis*, inflammation of), an early stage of gum

SALIVARY GLANDS AND CHEMICAL DIGESTION

Three pairs of salivary glands—the sublingual (below the tongue), submandibular (below the jaw), and parotid—release their secretions, collectively called **saliva**, into the mouth (Figure 15.5). As we chew, food is mixed with saliva. Water in saliva moistens food, and mucus binds food particles together.

Saliva also contains an enzyme, called **salivary amylase**, that begins to chemically digest starches into shorter chains of sugar. You will notice the result of salivary amylase activity if you chew a piece of bread for several minutes: The bread will begin to taste sweet. Try it.

TABLE 15.1 REVIEW OF STRUCTURES OF THE GASTROINTESTINAL TRACT			
STRUCTURE	**DESCRIPTION/FUNCTIONS**	**MECHANICAL DIGESTION**	**CHEMICAL DIGESTION**
Mouth	Receives food; contains teeth and tongue; tongue manipulates food and monitors quality	Teeth tear and crush food into smaller pieces	Digestion of carbohydrates begins
Pharynx	Area that both food and air pass through	None	None
Esophagus	Tube that transports food from mouth to stomach	None	None
Stomach	J-shaped muscular sac for food storage	Churning of stomach mixes food with gastric juice, creating liquid chyme	Protein digestion begins
Small intestine	Long tube where digestion is completed and nutrients are absorbed	Segmental contractions mix food with intestinal enzymes, pancreatic enzymes, and bile	Carbohydrate, protein, and fat digestion completed
Large intestine	Final tubular region of GI tract; absorbs water and ions; houses bacteria; forms and expels feces	None	Some digestion is carried out by bacteria
Anus	Terminal outlet of digestive tract	None	None

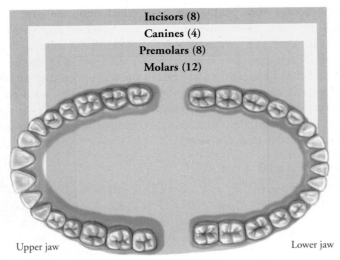

Incisors (8)
Canines (4)
Premolars (8)
Molars (12)

Upper jaw   Lower jaw

(a) The teeth slice, tear, and grind food until it can be swallowed.

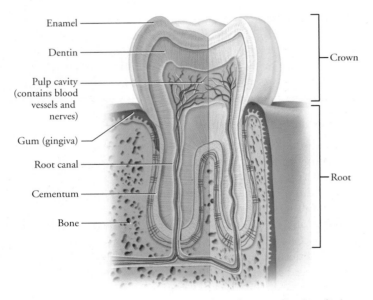

Enamel

Dentin

Pulp cavity
(contains blood
vessels and
nerves)

Gum (gingiva)

Root canal

Cementum

Bone

Crown

Root

(b) The structure of the human tooth is suited for its function of breaking food into smaller pieces.

FIGURE **15.3** Adult human teeth

THE TONGUE: TASTE AND FOOD MANIPULATION

The *tongue* is a large skeletal muscle studded with taste buds. Our ability to control the position and movement of the tongue is critical to both speech and the manipulation of food within the mouth. Once food is dissolved in saliva, the chemicals in the food can stimulate receptors in taste buds located primarily on the tongue. Information from the taste buds, along with input from the olfactory receptors in the nose, helps us to monitor the quality of food. For instance, spoiled or poisonous food usually tastes bad, so we can spit it out before swallowing. The tongue moves food into position for crushing and grinding by the teeth, to mix it with saliva, and to shape it into a small, soft mass, called a *bolus*, that is easily swallowed. The tongue also initiates swallowing by pushing the bolus to the back of the mouth.

■ **The pharynx is shared by the digestive and respiratory systems**

The **pharynx**, which is the passageway commonly called the throat, is shared by the respiratory and digestive systems. When we swallow, food is pushed from the mouth, through the pharynx, and into the **esophagus**, the tube that connects the pharynx to the stomach.

Swallowing consists of a voluntary component followed by an involuntary one. When a person begins to swallow, the tongue pushes the bolus of softened and moistened food into the pharynx (Figure 15.6). Once food is in the pharynx, it is too late to change one's mind about swallowing. Sensory receptors in the wall of the pharynx detect the presence of food and stimulate the involuntary swallowing reflex. Reflex movements of the soft palate prevent food from entering the nasal cavity or respiratory pathways. Other involuntary muscle contractions push the larynx (the voice box, commonly called the Adam's apple) upward. The movement of the larynx causes a cartilaginous flap called the *epiglottis* to move, covering the opening to the airways of the respiratory system (the glottis). The movement of the epiglottis prevents food from entering the airways. Instead, food is pushed into the esophagus.

■ **The esophagus conducts food from the pharynx to the stomach**

The esophagus is a muscular tube that conducts food from the pharynx to the stomach. Food is moved along the esophagus and all the rest of the digestive tract by rhythmic waves of muscle contraction called **peristalsis** (Figure 15.7). In the esophagus, small intestine, and large intestine, peristalsis is produced by the two layers of muscle in the muscularis. The muscles of the inner layer circle the tube, causing a constriction when they contract. The muscles in the outer layer run lengthwise, causing a shortening of the region where they contract. The following analogy may help you understand the effects of muscle contraction. Imagine a partially inflated, long balloon. Squeezing the balloon in its center extends the ends. This extension is the equivalent of the contraction of circular muscles. Compressing the ends of the balloon causes its center to bulge outward. This compression is the equivalent of the longitudinal muscle contracting. Stretching of the esophagus by the food bolus causes the circular muscles behind the bolus to contract, pushing food forward. The longitudinal muscles in front of the bolus then contract, shortening that region.

■ **The stomach stores and liquefies food and begins protein digestion**

The **stomach** is a muscular sac that is well designed to carry out its three important functions: (1) storage of food, (2) liquefaction of food, and (3) the initial chemical digestion of proteins.

STORAGE OF FOOD

Like any good storage compartment, the stomach is expandable and has openings that can close to seal the contents within or can open for filling and emptying. When empty, the stomach is a small J-shaped sac, about the size of a sausage, that can hold only about

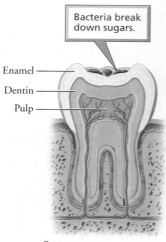
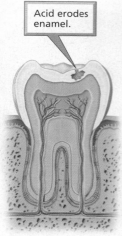
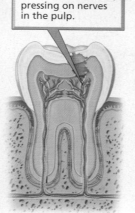
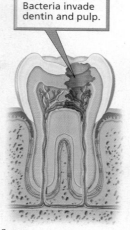
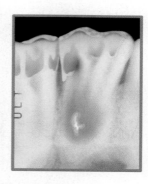

Bacteria break down sugars.

Acid erodes enamel.

Blood vessels widen, pressing on nerves in the pulp.

Bacteria invade dentin and pulp.

Enamel

Dentin

Pulp

STEP 1
Tooth decay is caused by acid produced when bacteria living on the tooth surface break down sugars in food particles adhered to teeth.

STEP 2
The acid erodes the tooth's enamel, causing a cavity to form.

STEP 3
The body responds by widening blood vessels in the pulp, to increase delivery of white blood cells to fight the infection. When widened blood vessels press on nerves within the pulp, a toothache results.

STEP 4
Bacteria can then infect the softer dentin beneath the enamel and later the pulp at the heart of the tooth.

A dental cavity can be seen as a orange area in this artificially colored x-ray of teeth.

FIGURE **15.4** The development of a dental cavity

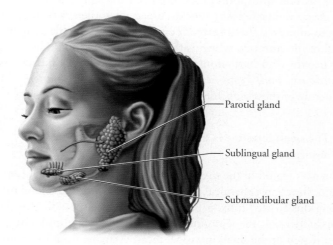

Parotid gland

Sublingual gland

Submandibular gland

FIGURE **15.5** Three pairs of salivary glands release their secretions into the mouth. These secretions, collectively called saliva, make food easier to swallow, dissolve substances so they can be tasted, and begin the chemical digestion of starch.

50 ml (a quarter of a cup) without stretching. However, the wall of the empty stomach has folds that can spread out, allowing the stomach to expand as it fills. When fully expanded, as after a large meal, the stomach can hold several liters of food.

Bands of circular muscle called sphincters guard the openings at each end of the stomach. Contraction of a sphincter closes the opening, and relaxation of a sphincter allows material to pass through.

LIQUEFACTION OF FOOD

Food is generally stored and processed within the stomach for 3 to 5 hours. Mechanical digestion occurs as the food is churned and mixed with secretions produced by the glands of the stomach until it is a soupy mixture called **chyme**. The stomach wall has three layers of smooth muscle, each oriented in a different direction. The coordinated contractions of these layers twist, knead, and compress the stomach contents, physically breaking food into smaller pieces.

CHEMICAL DIGESTION OF PROTEIN

In an adult, chemical digestion in the stomach is limited to the initial breakdown of proteins. The lining of the stomach has millions of gastric pits, within which are **gastric glands** containing several types of secretory cells. Certain secretory cells produce hydrochloric acid (HCl), which kills most of the bacteria swallowed with food or drink. Hydrochloric acid also breaks down the connective tissue of meat and activates pepsinogen, which is secreted by other cells in the gastric glands. Pepsinogen is the inactive form of **pepsin**, a protein-digesting enzyme. When the mixture of pepsin and HCl, called *gastric juice*, is released into the stomach, the pepsin begins the chemical digestion of the protein in food. Still other cells within the gastric glands secrete mucus, which helps protect the stomach wall from the action of gastric juice. Although not related to the digestive function of the stomach, a very important material secreted by the gastric glands is *intrinsic factor*, a protein necessary for the absorption of vitamin B_{12} from the small intestine.

The stomach wall is composed of the same materials that gastric juice is able to attack, so various protections are in place to keep the stomach from digesting itself. One, mentioned above, is the

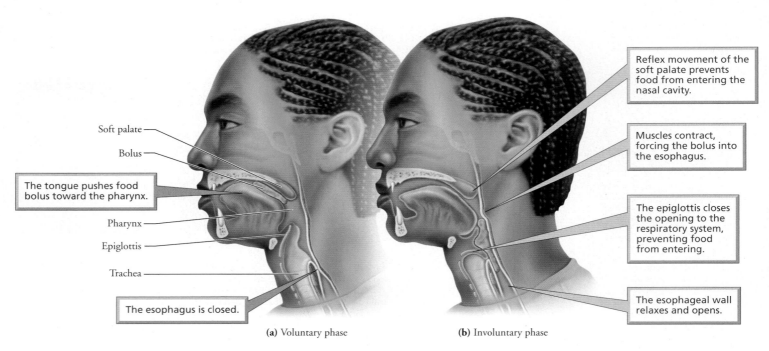

Soft palate

Bolus

The tongue pushes food bolus toward the pharynx.

Pharynx

Epiglottis

Trachea

The esophagus is closed.

Reflex movement of the soft palate prevents food from entering the nasal cavity.

Muscles contract, forcing the bolus into the esophagus.

The epiglottis closes the opening to the respiratory system, preventing food from entering.

The esophageal wall relaxes and opens.

(a) Voluntary phase

(b) Involuntary phase

FIGURE **15.6**   Swallowing consists of (a) voluntary and (b) involuntary phases.

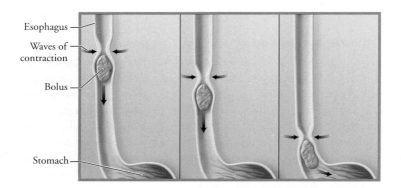

Esophagus

Waves of contraction

Bolus

Stomach

FIGURE **15.7**   Peristalsis is a wave of muscle contraction that pushes food along the esophagus and the entire remaining GI tract. When circular muscles contract, the tube is narrowed and food is pushed forward. The longitudinal muscles in front of the bolus contract, shortening that region.

presence of mucus. Mucus forms a thick protective coat that prevents gastric juice from reaching the cells of the stomach wall. The alkalinity of mucus helps to neutralize the HCl. Another protection is that pepsin is produced in an inactive form that cannot digest the cells that produce it. In addition, neural and hormonal reflexes regulate the production of gastric juice so that little is released unless food is present to absorb and dilute it. Finally, if the stomach lining is damaged, it is quickly repaired. The high rate of cell division in the stomach lining replaces a half million cells every minute. As a result, you have a new stomach lining every 3 days!

Very little absorption of food materials occurs in the stomach because food simply has not been broken down into molecules small enough to be absorbed. Notable exceptions are alcohol and aspirin. The absorption of alcohol from the stomach is the reason its effects can be felt very quickly, especially if there is no food present to dilute it. The absorption of aspirin can cause bleeding of the wall of the stomach, which is the reason aspirin should be avoided by people who have stomach ulcers (see the Health Issue essay, *Heartburn and Peptic Ulcers—Those Burning Sensations*). The structure of the stomach is shown in Figure 15.8.

■ The small intestine is the primary site of digestion and absorption

The next region of the digestive tract, the **small intestine**, has two major functions: chemical digestion and absorption. As food moves along this twisted tube, it passes through three specialized regions: the duodenum, the jejunum, and the ileum. The **duodenum**, the first region of the small intestine, receives chyme from the stomach and digestive juices from the pancreas and liver. However, most chemical digestion and absorption occur in the jejunum and the ileum.

CHEMICAL DIGESTION WITHIN THE SMALL INTESTINE

Within the small intestine, a battery of enzymes completes the chemical digestion of virtually all the carbohydrates, proteins, fats, and nucleic acids in food. Although both the small intestine and the pancreas contribute enzymes, most of the digestion that occurs in the small intestine is performed by pancreatic enzymes (Table 15.2).

Fats present a special digestive challenge because they are insoluble in water. You have observed this when oil (a fat) quickly separates from the vinegar (a water solution) in your salad dressing. In water, droplets of fat tend to coalesce into large globules. The problem is that lipase, the enzyme that chemically breaks down fats, is

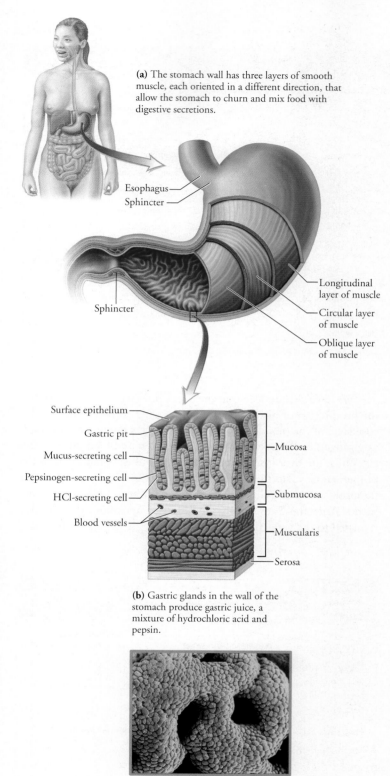

(a) The stomach wall has three layers of smooth muscle, each oriented in a different direction, that allow the stomach to churn and mix food with digestive secretions.

Esophagus

Sphincter

Sphincter

Longitudinal layer of muscle

Circular layer of muscle

Oblique layer of muscle

Surface epithelium

Gastric pit

Mucus-secreting cell

Pepsinogen-secreting cell

HCl-secreting cell

Blood vessels

Mucosa

Submucosa

Muscularis

Serosa

(b) Gastric glands in the wall of the stomach produce gastric juice, a mixture of hydrochloric acid and pepsin.

(c) The holes seen in this electron micrograph are the gastric pits, openings in the stomach wall through which gastric glands release their secretions.

FIGURE **15.8** The structure of the stomach is well suited to its functions of churning food and digestive secretions, storing food, and beginning protein digestion.

soluble in water and not in fats. As a result, lipase can work only at the surface of a fat globule. Large fat globules have less combined surface area than do smaller droplets, so their digestion by lipase proceeds more slowly.

Bile, a mixture of water, ions, cholesterol, bile pigments, and bile salts, plays an important role in the digestion of fats. The bile salts emulsify fats; that is, they keep fats separated into small droplets. This separation exposes a larger combined surface area to lipase, making the digestion and absorption of fats faster and more complete. Bile is produced by the liver, is stored in the gallbladder, and acts in the small intestine.

We will have more to say about the pancreas, liver, and gallbladder shortly.

STRUCTURE OF THE SMALL INTESTINE

The small intestine, the primary site of absorption in the digestive system, is extremely effective at its task because it is long and has several structural specializations that vastly increase its surface area (Figure 15.9). First, the entire lining of the small intestine is pleated, like an accordion, into circular folds. These circular folds increase the surface area for absorption and cause chyme to flow through the small intestine in a spiral pattern. The spiral flow helps mix the chyme with digestive enzymes and increases its contact with the absorptive surfaces. Covering the entire lining surface are tiny 1-mm projections called **villi** (singular, villus). The villi give the lining a velvety appearance and, like the pile on a bath towel, increase the absorptive surface. Indeed, the villi increase the surface area of the small intestine 10-fold. In addition, thousands of microscopic projections, called *microvilli*, cover the surface of each villus, increasing the surface area of the small intestine by another 20 times. The microvilli form a fuzzy surface, known as a *brush border*, on the absorptive epithelial cells. The circular folds, villi, and microvilli create a surface area of 300 to 600 m^2—greater than the size of a tennis court!

The core of each villus is penetrated by a network of capillaries and a **lacteal**, which is a lymphatic vessel. Thus, as substances are absorbed from the small intestine, they cross only two cell layers: the epithelial cells of the villi and the wall of either a capillary or a lacteal. Most materials enter the epithelial cells by active transport, facilitated diffusion, or simple diffusion (Figure 15.10; see Chapter 3). Monosaccharides, amino acids, water, ions, vitamins, and minerals diffuse across the capillary wall into the bloodstream and are delivered to body cells. The products of fat digestion combine with bile salts in the small intestine, creating particles called *micelles*. When a micelle contacts an epithelial cell of a villus, the products of fat digestion easily diffuse into the cell. Within an epithelial cell, glycerol and fatty acids are reassembled into triglycerides, mixed with cholesterol and phospholipids, and coated with special proteins, thus becoming part of a complex known as a *chylomicron*. The protein coating makes the fat soluble in water, allowing it to be transported throughout the body. The chylomicrons leave the epithelial cell by exocytosis (see Chapter 3). Chylomicrons are too large to pass through capillary walls. However, they easily diffuse into the more porous lacteal and enter the lymphatic system, which carries them to the bloodstream.

≋ **Use this table to find more information on digestives enzymes.**

TABLE 15.2 MAJOR DIGESTIVE ENZYMES

SITE OF PRODUCTION	ENZYME	SITE OF ACTION	SUBSTRATE AND PRODUCTS
Salivary glands	Salivary amylase	Mouth	Polysaccharides into shorter molecules
Stomach	Pepsin	Stomach	Proteins into protein fragments (polypeptides)
Pancreas	Trypsin	Small intestine	Proteins and polypeptides into smaller fragments
	Chymotrypsin	Small intestine	Proteins and polypeptides into smaller fragments
	Amylase	Small intestine	Polysaccharides into disaccharides
	Carboxypeptidase	Small intestine	Polypeptides into amino acids
	Lipase	Small intestine	Triglycerides (fats) into fatty acids and glycerol
	Nucleases (deoxyribonuclease and ribonuclease)	Small intestine	DNA or RNA into nucleotides
Small intestine	Maltase	Small intestine	Maltose into glucose units
	Sucrase	Small intestine	Sucrose into glucose and fructose
	Lactase	Small intestine	Lactose into glucose and galactose
	Aminopeptidase	Small intestine	Peptides into amino acids

■ **The pancreas, liver, and gallbladder are accessory organs that aid the processes of digestion and absorption within the small intestine**

The pancreas, liver, and gallbladder are not part of the GI tract. However, they play vital roles in digestion by releasing their secretions into the small intestine. The functions of the secretions of the accessory organs are summarized in Table 15.3.

THE PANCREAS

The **pancreas** is an accessory organ that lies behind the stomach, extending toward the person's left from the small intestine. In addition to enzymes, pancreatic juice contains water and ions, including bicarbonate ions that are important in neutralizing the acid in chyme when it emerges from the stomach. Neutralization is essential for optimal enzyme activity in the small intestine. Pancreatic juice drains from the pancreas into the pancreatic duct, which fuses with the common bile duct from the liver just before entering the duodenum of the small intestine (Figure 15.11).

Collectively, the pancreatic enzymes and intestinal enzymes break nutrients into their component building blocks: proteins to amino acids, carbohydrates to monosaccharides, and triglycerides (a type of lipid) to fatty acids and glycerol.

THE LIVER

The nutrient-laden blood from the capillaries in the villi travels through the hepatic portal vein to the **liver**, the largest internal organ in the body, which has a variety of metabolic and regulatory roles (Figure 15.12). A portal system consists of the blood vessels that link two capillary beds (we encountered one associated with the pituitary gland of the brain, described in Chapter 10). The hepatic portal system delivers blood from a capillary bed in the small intestine to a second capillary bed in the liver.

We have already seen that the liver's primary role in digestion is the production of bile. One of its other roles is to control the glucose level of the blood, either removing excess glucose or storing it as glycogen or breaking down glycogen to raise blood glucose levels. Thus, the liver keeps the glucose levels of the blood within the proper range. The liver also packages lipids with protein carrier molecules, forming lipoproteins, which transport lipids in the blood. After the liver adjusts the blood composition, the blood is returned to the general circulation through the hepatic veins.

TABLE 15.3 REVIEW OF ACCESSORY STRUCTURES OF THE DIGESTIVE SYSTEM

STRUCTURE	SECRETIONS/FUNCTIONS	SITE OF ACTION OF CHEMICAL SECRETIONS
Salivary glands (sublingual, submandibular, parotid)	Secrete saliva, a liquid that moistens food and contains an enzyme (amylase) for digesting carbohydrates	Mouth
Pancreas	Digestive secretions include bicarbonate ions that neutralize acidic chyme and enzymes that digest carbohydrates, proteins, fats, and nucleic acids	Small intestine
Liver	Digestive function is to produce bile, a liquid that emulsifies fats, making chemical digestion easier and facilitating absorption	Small intestine
Gallbladder	Stores bile and releases it into small intestine	Small intestine

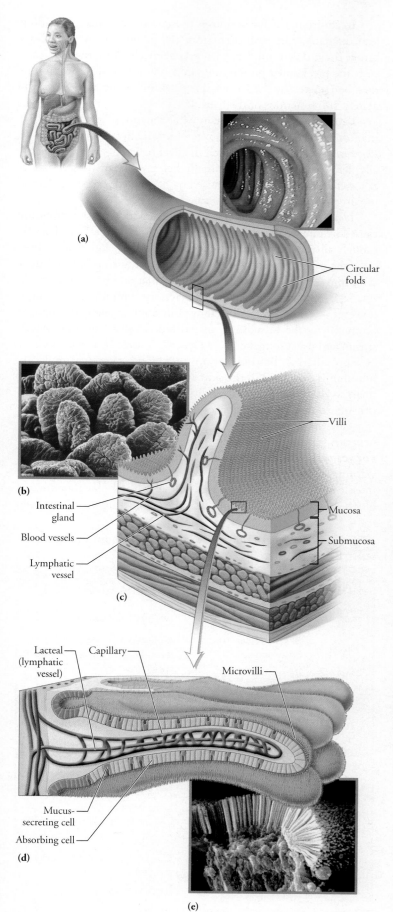

FIGURE **15.9** The small intestine is specialized for the absorption of nutrients by structural modifications that increase its surface area. (a) Its wall contains accordion-like pleats called circular folds. (b) This electron micrograph shows the intestinal villi, the numerous fingerlike projections on the intestinal lining (the mucosa in part c). (d) The surface of each villus bristles with thousands of microscopic projections (of cell membranes) called microvilli. In the center of each villus, a network of blood capillaries, which carries away absorbed products of protein and carbohydrate digestion as well as ions and water, surrounds a lacteal (a small vessel of the lymphatic system) that carries away the absorbed products of fat digestion. (e) An electron micrograph of the microvilli that cover each villus.

Circular folds

Villi

Intestinal gland

Blood vessels

Lymphatic vessel

Mucosa

Submucosa

Lacteal (lymphatic vessel)

Capillary

Microvilli

Mucus-secreting cell

Absorbing cell

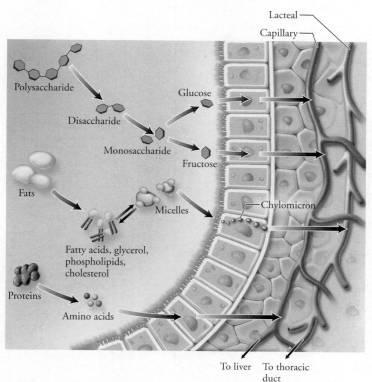

Lacteal

Capillary

Polysaccharide

Disaccharide

Monosaccharide

Glucose

Fructose

Fats

Micelles

Fatty acids, glycerol, phospholipids, cholesterol

Chylomicron

Proteins

Amino acids

To liver   To thoracic duct

FIGURE **15.10** The small intestine is the primary site for chemical digestion and absorption. The digestive products, such as monosaccharides, amino acids, fatty acids, and glycerol, enter absorptive epithelial cells of villi by active transport, facilitated diffusion, or diffusion. Monosaccharides and amino acids, along with water, ions, and vitamins, then enter the capillaries within the villus and are carried to body cells by the bloodstream.

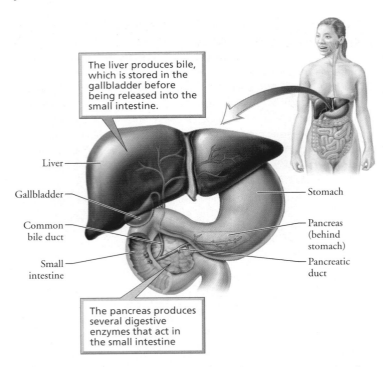

The liver produces bile, which is stored in the gallbladder before being released into the small intestine.

Liver

Gallbladder

Common bile duct

Small intestine

Stomach

Pancreas (behind stomach)

Pancreatic duct

The pancreas produces several digestive enzymes that act in the small intestine

FIGURE **15.11** The pancreas, liver, and gallbladder are accessory organs of the digestive system.

Inferior vena cava

Hepatic veins

Liver

Stomach

Large intestine

Small intestine

Step 4: Hepatic veins deliver blood to the circulatory system.

Step 3: The liver monitors blood contents.

Step 2: Digested food molecules then travel through hepatic portal veins to the liver.

Step 1: Products of digestion are absorbed into the capillaries within the villi of the small intestine.

FIGURE **15.12** A portal system transports blood from one capillary bed to another. In the hepatic portal system the hepatic portal vein carries blood from the capillaries' network of the villi of the small intestine to the capillary beds of the liver. The liver monitors blood content and processes nutrients before they are delivered to the bloodstream.

In addition, the liver removes poisonous substances, including lead, mercury, and pesticides, from the blood and in some cases breaks them down into less harmful chemicals. Also, the liver converts the breakdown products of amino acids into urea, which can then be excreted by the kidney. These are but a few of the approximately 500 functions of the liver.

Considering the liver's many vital functions, it is no surprise that diseases of the liver can be serious and life threatening. *Cirrhosis* is a condition in which the liver becomes fatty and gradually deteriorates, its cells eventually being replaced by scar tissue. Since cirrhosis is sometimes caused by prolonged, excessive alcohol use, it is discussed in Chapter 8a.

Hepatitis is inflammation of the liver. It is most commonly caused by one of six viruses, designated as A, B, C, D, E, and G. Although all the hepatitis viruses attack the liver and destroy liver cells, there are differences in their means of transmission and their symptoms. Table 15.4 summarizes information about hepatitis A, B, and C. All forms of hepatitis do have one symptom in common, however. Healthy liver cells remove bilirubin, a yellowish pigment produced by the breakdown of red blood cells, from the bloodstream and use it to make bile. Liver cells injured by hepatitis viruses stop filtering bilirubin from the blood. The accumulating bilirubin is deposited in the skin and the whites of the eyes, giving them a yellowish tint. This condition, called jaundice, is characteristic of any disease that damages the liver.

Currently, about 4 million people in the United States have hepatitis C, and most of them have no idea that they are infected. For years, the disease has spread silently because it has no outward warning signs or very mild symptoms—vague fatigue or flulike muscle and joint pain. It is spread primarily through contaminated blood. Hundreds of thousands of intravenous drug users have been

Use this table to find more information about hepatitis A, B, and C.

TABLE 15.4	FORMS OF VIRAL HEPATITIS	
HEPATITIS VIRUS	**MEANS OF TRANSMISSION**	**SYMPTOMS AND COURSE**
A	Sewage-contaminated water or food	Acute infection: fatigue, jaundice, fever, abdominal pain, loss of appetite
		Recovery within several months
B	Sexual contact, contact with contaminated blood, mother to newborn at birth	Acute infection: fatigue, jaundice, fever, abdominal pain, loss of appetite, joint pain
		Chronic, with possible liver failure and death
		Can lead to liver cancer
C	Contaminated blood or body fluids	Acute infection: fatigue, jaundice, dark urine, abdominal pain, nausea
		Chronic, with possible liver failure and death
		Can lead to liver cancer

HEALTH ISSUE

Heartburn and Peptic Ulcers—Those Burning Sensations

Heartburn is a burning sensation behind the breastbone that occurs when acidic gastric juice backs up into the esophagus. At least 30% of all Americans suffer from heartburn on a monthly basis and 10% on a daily basis. Heartburn occurs when the pressure of the stomach contents overwhelms the sphincter at the lower end of the esophagus. In some people, this sphincter is weak and unable to keep stomach contents out of the esophagus. A person with a normal esophageal sphincter will experience heartburn when the stomach contents exert a greater pressure than usual against the sphincter, as might occur after a large meal, during pregnancy, when lying down, or when constipated. Acidic foods, such as tomatoes and citrus fruits, and spicy, fatty, or caffeine-containing products can aggravate the problem by causing the sphincter to relax and open. Tight clothing also worsens heartburn.

There are several treatments for heartburn. Antacids will provide temporary relief. Unfortunately, long-term use of certain antacids can have undesirable side effects. The fizzing bicarbonate types are high in sodium and should be avoided by people with high blood pressure. Others are high in calcium, which might promote the formation of kidney stones and also stimulate the production of stomach acid. Newer medications, such as Tagamet, Zantac, Pepcid, Axid, Prilosec, and Nexium, decrease the production of stomach acid.

People who have chronic heartburn—at least one attack per week—have an increased risk of developing cancer of the esophagus. The risk increases with the frequency of attacks and the number of years of experiencing heartburn. Those who have one attack per week have a risk eight times higher than normal. Esophageal cancer is particularly aggressive and has become more common in recent years. It is detected using a test called endoscopy, in which a thin, lighted tube is snaked down the throat.

At some point in their lives, nearly 13% of Americans experience a failure of the mechanisms that protect the stomach and duodenum from their acidic contents, so that the lining of some region in the GI tract becomes eroded. The resulting sore resembles a canker sore of the mouth and is called a peptic ulcer (Figure 15.A). Although a peptic ulcer may form in the esophagus or the stomach, the most common site is the duodenum of the small intestine. Ulcers are usually between 10 and 25 mm (0.33 and 1 in.) in diameter and may occur singly or in multiple locations.

The symptoms of an ulcer are variable. A common symptom is abdominal pain, which can be quite severe. Vomiting, loss of appetite, bloating, indigestion, and heartburn are other common symptoms. However, some people with ulcers, especially those who are taking nonsteroidal anti-inflammatory drugs (NSAIDs), have no pain. Unfortunately, the degree of pain is a poor indicator of the severity of ulceration. Often, people with no symptoms are unaware of their ulcers until serious complications develop. Gastric juice can erode the lining of the GI tract until it bleeds. In some cases, the ulcer can eat a hole completely through the gut wall (a condition called perforated ulcer). Recurrent ulcers can cause scar tissue to form, which may narrow or block the lower end of the stomach or duodenum.

Although acidic gastric juice is the direct cause of peptic ulcers, factors that interfere with the mechanisms that normally protect the lining of the GI tract from the acid are considered to be the real causes. NSAIDs, which include as-

pirin, ibuprofen, and naproxen, can cause ulcers because they slow the production of chemicals called prostaglandins, which normally help protect the lining of the GI tract from damage by acid.

The leading cause of peptic ulcers, however, is thought to be infection with the bacterium *Helicobacter pylori*. More than 80% of persons with ulcers in the stomach or duodenum are infected with *H. pylori*. These corkscrew-shaped bacteria live in the layer of mucus that protects the lining of the GI tract. Here, partially protected from gastric juice, the bacteria attract body defense cells—specifically, macrophages and neutrophils—that cause inflammation leading to ulcer formation. Toxic chemicals produced by the bacteria also contribute to ulcers.

H. pylori infections may last for years. These bacteria affect more than a billion people throughout the world and approximately 50% of the people in the United States who are over 60 years of age. For some reason, however, only about 10% to 15% of those who are infected actually develop peptic ulcers. Besides ulcers, an *H. pylori* infection is a risk factor for esophageal and stomach cancer. However, it may soon be possible to vaccinate children against this bacterium, thereby preventing both peptic ulcers and stomach cancer.

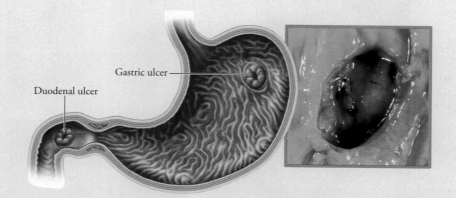

Gastric ulcer

Duodenal ulcer

FIGURE **15.A** A peptic ulcer is a raw area that forms when gastric juice erodes the lining of the esophagus, stomach, or, most commonly, the duodenum. The ulcer shown in this stomach wall is bleeding. The most common symptom is abdominal pain that occurs when the stomach is empty.

infected by sharing contaminated needles. Hepatitis C can also be spread through contaminated needles used in body piercing or tattooing. Before there was a way to test for the virus, many people became infected when they received transfusions with contaminated blood.

what would you do?

Egypt has one of the world's highest hepatitis C infection rates—about 15% to 25% of the population. The high infection rate is the result of a government-sponsored campaign to vaccinate people against schistosomiasis, a disease caused by a parasite. Improperly disinfected needles were used to vaccinate many people. Treatment for hepatitis C is expensive and sometimes debilitating, and the government does not have enough money to pay for treatment for all those infected. Can you think of sources of funds to treat hepatitis C in Egypt? Should this be considered a global concern? What would you do to help?

THE GALLBLADDER

After it is produced by the liver, bile is stored, modified, and concentrated in a muscular, pear-shaped sac called the **gallbladder**. When chyme enters the small intestine, a hormone causes the gallbladder to contract, squirting bile through the common bile duct into the duodenum of the small intestine.

Bile is rich in cholesterol. Sometimes, if the balance of dissolved substances in bile becomes upset, a tiny crystalline particle precipitates out of solution. Cholesterol and other substances can then build up around the particle to form a *gallstone* (Figure 15.13). Many people develop more than one gallstone.

■ The large intestine absorbs water and other useful substances

Now that we have discussed the accessory organs that assist digestion and absorption in the small intestine, we will go back to following the movement of ingested substances through the GI tract. The material that was not absorbed in the small intestine moves into the final major structure of the digestive system, the **large intestine**. The principal functions of the large intestine are to absorb most of the water remaining in the indigestible food residue, thereby adjusting the consistency of the waste material, or feces; to store the feces; and to eliminate them from the body. The large intestine is home to many types of bacteria, some of which produce vitamins that may be absorbed for use by the body.

stop and think

In a gastric bypass operation, the esophagus is detached from the stomach and reattached farther along the GI tract. Which part of the GI tract would you want to bypass to decrease food absorption?

The large intestine has four regions: the cecum, colon, rectum, and anal canal, as shown in Figure 15.14. The cecum is a pouch that hangs below the junction of the small and large intestines. Extending from the cecum is another slender, wormlike pouch, called the **appendix**. The appendix has no digestive function. Some scientists believe the appendix plays a role in the immune system, which protects the body against disease.

Each year, about 1 of 500 people develops *appendicitis*, inflammation of the appendix. Appendicitis is usually caused by an infection that arises in the appendix after it becomes blocked by a piece

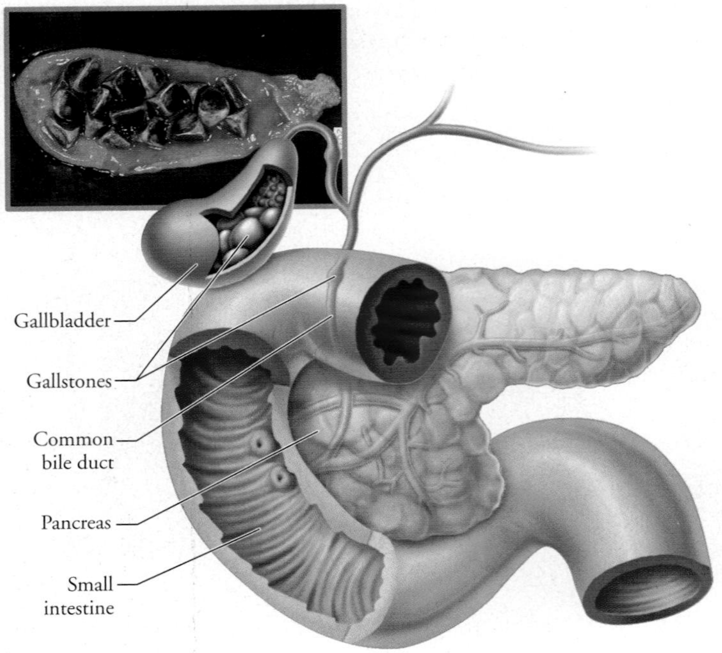

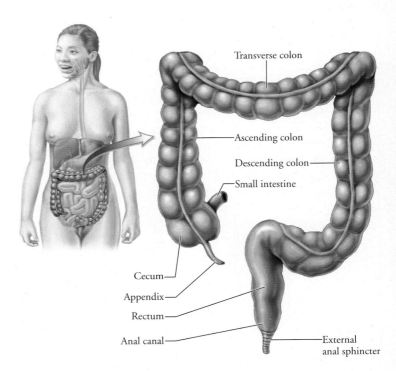

FIGURE **15.13** Gallstones consist primarily of cholesterol that has precipitated from bile during storage in the gallbladder. A gallstone can intermittently or continuously block the ducts that drain bile into the small intestine. The photograph shows a gallbladder and several gallstones.

FIGURE **15.14** The large intestine consists of the cecum, colon, rectum, and anal canal. It absorbs water from undigested material, forming the feces, and houses bacteria.

of hardened stool, food, or a tumor. At first, appendicitis is usually experienced as vague bloating, indigestion, and a mild pain in the region of the navel (bellybutton). As the condition worsens, the pain becomes more severe and is localized in the region of the appendix, the lower right abdomen. The pain is typically accompanied by fever, nausea, and vomiting. Untreated infection in the appendix usually causes the appendix to rupture, allowing its contents to spill into the abdominal cavity. Spillage from the appendix generally leads to infection and inflammation throughout the abdomen (a condition called peritonitis), which is potentially fatal.

The largest region of the large intestine, the **colon**, is composed of the ascending colon on the right side of the abdomen, the transverse colon across the top of the abdominal cavity, and the descending colon on the left side (see Figure 15.14). Although much of the water that was originally in chyme was absorbed in the small intestine, the material entering the colon is still quite liquid. As it passes through the colon, 90% of the remaining water and sodium and potassium ions are absorbed. The material left in the large intestine after passing through the colon is called feces and consists primarily of undigested food, sloughed-off epithelial cells, water, and millions of bacteria. The brown color of feces comes from bile pigments.

The bacteria are not normally disease causing and, in fact, are beneficial. Intestinal bacteria produce several vitamins that we are unable to produce on our own, including vitamin K and some of the B vitamins. Some of the vitamins are then absorbed from the colon for our own use. There are roughly 50 species of bacteria, including the well-known *Escherichia coli* (*E. coli*), living in the healthy colon. They are nourished by undigested food and by material that we are unable to digest, including certain components of plant cells. When the intestinal bacteria use the undigested food for their own nutrition, their metabolic processes liberate gas that sometimes has a foul odor. Although most of the gas is absorbed through the intestinal walls, the remaining gas can produce some embarrassing moments when released as flatus.

stop and think

Certain foods, beans for instance, are notorious for producing intestinal gas. Beans contain large amounts of certain short-chain carbohydrates that our bodies are unable to digest. Explain how the nutritional content of beans is related to flatulence.

Periodic peristaltic contractions move material through the large intestine, but these contractions are slower than the contractions in the small intestine. Eventually, the feces are pushed into the **rectum**, stretching the rectal wall and initiating the *defecation reflex*. Nerve impulses from the stretch receptors in the rectal wall travel to the spinal cord, which sends motor impulses back to the rectal wall, stimulating muscles there to contract and propel the feces into the **anal canal**. Two rings of muscles, called sphincters, must relax to allow *defecation*, the expulsion of feces. The internal sphincter relaxes automatically as part of the defecation reflex. The external sphincter is under voluntary control, allowing the person to decide whether or not to defecate. Conscious contraction of the abdominal muscles can increase abdominal pressure and help expel the feces.

The water absorption that occurs in the colon adjusts the consistency of feces. When material passes through the colon too rapidly, as might occur when colon contractions are stimulated by toxins from microorganisms or by excess food or alcoholic drink, too little water is absorbed. As a result, the feces are very liquid. This condition, which results in frequent loose stools, is called *diarrhea*. Persistent diarrhea can be dangerous, especially in an infant or young child, because it can lead to dehydration. It is one of the major causes of death worldwide.

On the other hand, if material passes through the colon too slowly, too much water is absorbed, resulting in infrequent, hard stools, a condition called *constipation*. People who are constipated may need to strain during bowel movements. Straining increases pressure within veins in the rectum and anus, causing them to stretch and enlarge. The resulting varicose veins are known as hemorrhoids. The wall of the large intestine also experiences great pressure during a strained bowel movement. Then, like the inner tube of an old tire, the weaker spots in the intestinal wall can begin to bulge outward, forming small outpouchings called diverticula (singular, diverticulum) (Figure 15.15). Diverticula are very common in people older than age 50. When diverticula do not cause problems or symptoms, the condition is called diverticulosis. But if the diverticula become infected with bacteria and inflamed, the condition is then called diverticulitis, which can cause abrupt, cramping abdominal pain, a change in bowel habits, fever, and rectal bleeding.

stop and think

Stimulant laxatives enhance peristalsis in the large intestine but do not affect the small intestine. Some people with eating disorders use laxatives to speed the movement of food through the digestive system, thinking that the calories in the food will not be absorbed. Are they correct in their thinking? After purging in this way, the person may weigh less on the bathroom scale. What accounts for the weight loss? Could laxative use help a person lose body fat?

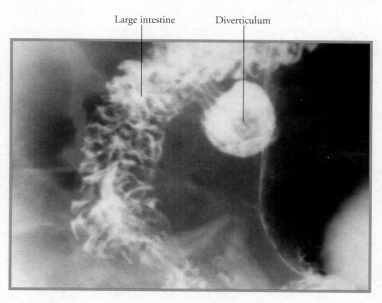

Large intestine   Diverticulum

FIGURE **15.15** A diverticulum is a small pouch that forms in the wall of the large intestine that is usually caused by repeated straining during bowel movements. A high fiber diet results in softer, bulkier stools that are easier to pass. Thus fiber makes it less likely that diverticula will form.

Cancers of the colon or rectum are common and can be deadly. Colorectal cancer is the second leading cause of cancer deaths. Early detection and treatment cut the risk of death dramatically.

Colorectal cancer begins with a small noncancerous growth called a polyp. Most polyps stop growing, but a small percentage continues to grow. As they do, genetic mutations accumulate in them that can transform a cell into a cancerous tumor. Polyps may take as long as 10 years to grow and turn cancerous, allowing plenty of time for detection. Sometimes polyps bleed, so one sign of colorectal polyps and perhaps cancer is blood in the stool, although the blood is not usually visible and must be detected with a diagnostic test. However, because many polyps do not bleed, screening methods that allow a direct view of the wall of the rectum and colon are generally more effective. A long, flexible fiber optic tube is threaded through the first third of the colon in a procedure known as a sigmoidoscopy or through the entire colon in a procedure called a colonoscopy. If a polyp is detected, it can be removed and biopsied to determine whether it is cancerous (see Chapter 21a).

Nerves and Hormones Control Digestive Activities

As we have seen, ingested material moves along the gastrointestinal tract, stopping for specific kinds of treatment along the way. For digestion to occur, enzymes must be present in the right place at the right time. However, because the body is composed of many of the same substances found in food, digestive enzymes should not be released until food is present (otherwise the enzymes might start digesting the digestive system). Both nerves and hormones play a role in orchestrating the release of digestive secretions, timing the release of each to the presence of food at each stop.

Food spends little time in the mouth, so to be effective, saliva must be secreted quickly. Because nervous stimulation is faster than hormonal stimulation, it is not surprising to learn that the nervous system controls salivation. Some saliva is released before food even enters the mouth, which may begin to "water" simply at the *thought* of food—and certainly begins at the sight or smell of food. The major trigger for salivation, however, is the presence of food in the mouth—its flavor and pressure. Salivary juices continue to flow for some time after the food is swallowed, helping to rinse out the mouth.

While food is still being chewed, neural reflexes stimulate the stomach lining to begin secreting gastric juice and mucus. The distention of the stomach, along with the presence of partially digested proteins, stimulates cells in the stomach lining to release the hormone **gastrin**. Gastrin enters the bloodstream and circulates throughout the body and back to the stomach, where it increases the production of gastric juice.

The presence of acidic chyme in the small intestine is the most important stimulus for the release of enzymes from both the small intestine and the pancreas, as well as bile from the gallbladder. Local nerve reflexes that are triggered by acidic chyme are the most important factor regulating the movement of material through the small intestine and the release of intestinal secretions. Acid chyme also causes the small intestine to release several hormones that, in turn, are responsible for the release of digestive enzymes and bile. For instance, one hormone, *vasoactive intestinal peptide*, is released from the small intestine into the bloodstream and is carried back to the small intestine, where it causes the release of intestinal juices. At the same time, the small intestine releases a second hormone, *secretin*, which stimulates the release of sodium bicarbonate from the pancreas into the small intestine to help neutralize the acidity of chyme. A third hormone from the small intestine is *cholecystokinin*, which causes the pancreas to release its digestive enzymes and causes the gallbladder to contract and release bile. The neural and hormonal controls of the digestive system are summarized in Tables 15.5 and 15.6.

TABLE 15.5 EXAMPLES OF NEURAL CONTROLS ON DIGESTIVE ACTIVITY

STIMULUS	EFFECT
Sight of food, thought of food, presence of food in mouth	Release of saliva from salivary glands
Chewing food	Release of gastric juice (enzymes from stomach and HCl) and mucus from cells of stomach lining
Presence of acidic chyme in small intestine	Release of enzymes from small intestine into the small intestine; increased motility in small intestine

exploring further . . .

In this chapter, we have seen how the digestive system is organized to break down food molecules for energy or building materials. In the next chapter, we will learn about your body's use of specific nutrients and consider ways in which the composition of the diet influences health.

TABLE 15.6 EXAMPLES OF HORMONAL CONTROL ON DIGESTIVE ACTIVITY

HORMONE	STIMULUS	ORIGIN	TARGET	EFFECTS
Gastrin	Distention of stomach by food; presence of partially digested proteins in stomach	Stomach	Stomach	Release of gastric juice (enzymes from stomach and HCl)
Vasoactive intestinal peptide	Presence of acidic chyme in small intestine	Small intestine	Small intestine	Release of enzymes from small intestine
Secretin	Presence of acidic chyme in small intestine	Small intestine	Pancreas	Release of sodium bicarbonate into small intestine to neutralize acidic chyme
Cholecystokinin	Arrival of chyme containing lipids	Small intestine	Pancreas	Release of enzymes from pancreas
			Gall bladder	Contraction of gallbladder and release of bile

HIGHLIGHTING THE CONCEPTS

The Digestive System Consists of a Long Tube That Runs through the Body, Along with Accessory Glands (p. 297)

1. The digestive system consists of a long tube called the gastrointestinal (GI) tract (mouth, pharynx, esophagus, stomach, small intestine, and large intestine) and several accessory organs (salivary glands, pancreas, liver, and gallbladder).

    WEB TUTORIAL 15.1 Physical and Chemical Digestion

The Digestive System Is Divided into Specialized Compartments for Food Processing (pp. 297–310)

2. The mouth serves several functions. Teeth tear and grind food, making it easier to swallow. The salivary glands produce salivary amylase, which is released into the mouth to begin the chemical breakdown of starches. Taste buds help monitor the quality of food. Finally, the tongue manipulates food so that it can be swallowed.

3. The pharynx is shared by the digestive and respiratory systems.

4. The esophagus, a tube that leads to the stomach, conducts food from the pharynx to the stomach. When we swallow, food is pushed from the mouth, and waves of muscle contraction called peristalsis push the food along the esophagus.

5. The stomach stores food, liquefies it by mixing it with gastric juice, begins the chemical digestion of proteins, and regulates the release of chyme into the small intestine. Gastric juice consists of hydrochloric acid (HCl) and pepsin, a protein-splitting enzyme. Pepsin is produced in an inactive form, called pepsinogen, that is activated by HCl.

6. The small intestine is the primary site of digestion and absorption, where enzymes produced by the small intestine and pancreas work to chemically digest carbohydrates, proteins, and fats into their component subunits. Bile emulsifies fat (breaks it into tiny droplets) in the small intestine, thereby increasing the combined surface area of fat droplets. Bile's action makes fat digestion by water-soluble lipase faster and more complete.

7. The small intestine's surface area for absorption is increased by circular folds in its lining, fingerlike projections called villi, and microscopic projections covering the villi, called microvilli.

8. Products of digestion are absorbed into the epithelial cells of villi by active transport, facilitated diffusion, or simple diffusion. Most materials, including monosaccharides, amino acids, water, and ions, then enter the capillary blood network in the center of each villus. However, fatty acids and glycerol are resynthesized into triglycerides, combined with cholesterol, and covered in protein, forming droplets called chylomicrons. The chylomicrons enter a lymphatic vessel called a lacteal in the core of each villus. They are then delivered to the bloodstream by way of lymphatic vessels.

9. The pancreas, liver, and gallbladder are accessory organs that aid in digestion and absorption within the small intestine. The pancreas secretes enzymes to digest most nutrients. The liver produces bile, which is stored in the gallbladder.

10. The large intestine consists of the cecum, colon, rectum, and anal canal. The large intestine absorbs water, ions, and vitamins. It is home to millions of beneficial bacteria that live on undigested material that has passed from the small intestine. The bacteria produce several vitamins, some of which we then absorb for our own use.

11. Material left in the large intestine after passing through the colon is called feces. Feces consist of undigested or indigestible material, bacteria, sloughed-off cells, and water.

Nerves and Hormones Control Digestive Activities (pp. 310–311)

12. Neural and hormonal mechanisms regulate the release of digestive secretions. Neural reflexes trigger the release of saliva, initiate the secretion of some gastric juice, and are the most important factors regulating the release of intestinal secretions. Gastrin, vasoactive intestinal peptide, secretin, and cholecystokinin are hormones that regulate digestive activities.

KEY TERMS

digestive system   *p. 297*
gastrointestinal tract   *p. 297*
saliva   *p. 299*
salivary amylase   *p. 299*
pharynx   *p. 300*
esophagus   *p. 300*
peristalsis   *p. 300*

stomach   *p. 300*
chyme   *p. 301*
gastric glands   *p. 301*
pepsin   *p. 301*
small intestine   *p. 302*
duodenum   *p. 302*
bile   *p. 303*

villi (sing. villus)   *p. 303*
lacteal   *p. 303*
pancreas   *p. 304*
liver   *p. 304*
gallbladder   *p. 308*
large intestine   *p. 308*
appendix   *p. 308*

colon   *p. 309*
rectum   *p. 309*
anal canal   *p. 309*
gastrin   *p. 310*

REVIEWING THE CONCEPTS

1. List the structures of the gastrointestinal tract in the order in which food passes through them.   *pp. 297–310*
2. Describe how food is processed in the mouth. What are the functions of the teeth and tongue?   *pp. 297–300*
3. Describe the structure of a tooth. What causes tooth decay?   *pp. 297–299*
4. What are the functions of the stomach? What are the digestive functions of gastric juice?   *pp. 300–302*
5. Why are so few substances absorbed from the stomach?   *p. 302*
6. How does bile assist the digestion and absorption of fats?   *pp. 302–303*
7. Where are carbohydrates digested? Proteins? Fats?   *pp. 299, 301–302*
8. Describe the structural features that increase the surface area for absorption in the small intestine.   *p. 303*
9. Which structures produce the digestive enzymes that act in the small intestine?   *pp. 304–306, 308*
10. What are the functions of the large intestine?   *pp. 308–310*
11. Describe the neural or hormonal mechanisms that regulate the release of digestive juices from each structure involved in digestion.   *p. 310*
12. The villi in the wall of the small intestine function to _____.
    a. increase the surface area for absorption.
    b. help mix the food with digestive enzymes.
    c. secrete bile.
    d. secrete digestive enzymes.
13. It is possible to swallow while standing on one's hands because
    a. valves in the digestive system keep food from moving backward.
    b. peristalsis pushes food along the digestive tract in the right direction.
    c. bacteria clog the digestive tube and prevent food from moving in the wrong direction.
    d. of a wave of muscle contraction called emulsification.
14. The most important function of the stomach is
    a. absorption of nutrients.
    b. chemical digestion.
    c. mucus secretion.
    d. storage of food.
15. You go out with your friends to celebrate your birthday and share a sausage pizza. The digestion of the oil *begins* in the
    a. mouth.
    b. esophagus.
    c. stomach.
    d. small intestine.
16. _____ is a liquid mixture of food and gastric juices found in the stomach.
17. The _____ is the organ that produces bile.
18. Saliva contains an enzyme that begins the chemical digestion of _____.
19. _____ is a hormone produced by the small intestine that causes the gallbladder to contract and release bile.

APPLYING THE CONCEPTS

1. Barbara is a 20-year-old woman with cystic fibrosis, a genetic disease in which abnormally thick mucus is produced. The thick mucus sometimes blocks the pancreatic duct that allows pancreatic juice to enter the small intestine. Explain why cystic fibrosis has caused Barbara to be malnourished.
2. Dehydration occurs when too much body fluid is lost. It can be a serious, even fatal, condition. Doctors worry about dehydration if a person, especially an infant, has severe diarrhea lasting several days. Why?
3. A 35-year-old man sporting a tattoo that says "Janet" visits the dermatologist to have the tattoo removed. When asked why he wants the tattoo removed, he replies that Janet was his first wife. He is getting remarried, and the new wife does not want to see "Janet" on her husband's bicep. Noticing that the patient's skin has a yellowish tint, the dermatologist strongly recommends that the patient visit a physician. Why? What is the dermatologist concerned about?
4. You spend the day with your Aunt Sally on her 55th birthday. She is overweight, so she has a salad for lunch and suffers no ill effects. However, she indulges in fried chicken and French fries for dinner. A few hours later, she is rushed to the emergency room with a severe pain on the right side of her abdomen. She reports that she has had bouts of pain over the last month or so. What do you think is Aunt Sally's problem? Why isn't the pain continuous?

Additional questions can be found on the companion website.

MyPyramid Is a Food Guide for Planning a Healthy Diet

Nutrients Provide Energy or Have a Structural or Functional Role in the Body

- Lipids include fats, oils, and cholesterol
- Carbohydrates include sugars, starches, and dietary fiber
- Proteins are chains of amino acids
- Vitamins are needed in small amounts to promote and regulate the body's chemical reactions
- Minerals play structural and functional roles in the body
- Water is critical and needed in large amounts

Food Labels Help Us Make Wise Food Choices

For Body Energy Balance, Calories Gained in Food Must Equal Calories Used

Obesity Is Body Weight 20% or More above the Body Weight Standard

Successful Weight-Loss Programs Usually Involve Reducing Calorie Intake, Increasing Calorie Use, and Changing Behavior

Anorexia Nervosa and Bulimia Are Eating Disorders That Create Calorie Deficits

15a

SPECIAL TOPIC

Nutrition and Weight Control

Amber was excited about spending the weekend with her former college roommate. She hadn't seen Dana for over a year and looked forward to some long talks and laughter-filled reminiscences. Secretly, Amber had to admit, she also couldn't wait to show off her newly slimmed figure. Dana responded to the honk of her horn by rushing to the car door, pulling it open, and giving Amber a long, tight hug. "It's so good to see you!" she exclaimed, followed immediately by, "My, my, haven't we lost weight! You look great."

Amber couldn't help but blush with pride. "It's nothing—just one of those low-carbohydrate diets." An hour later, though, after they'd been talking a while, Amber admitted that although she had lost more than 20 pounds, she wasn't feeling as well as she'd expected. She was frequently constipated, and she had developed intense cravings for fruit and any kind of bread product. Dana remembered taking some nutrition courses in college that had stressed the importance of a balanced diet. She pointed out that the lack of fiber was probably causing Amber's constipation. She also warned that the next phase of her weight-loss program could be even more difficult if she wanted to keep the weight off, because she would have to focus on nutritional balance while restricting carbohydrates. Amber nodded and told Dana that she was going to make an appointment with a nutritionist when she returned home. They toasted her commitment, enjoyed the dinner Dana had prepared, and spent the evening discussing the fun and perils of their lives since graduating from college.

Eating a balanced diet and maintaining a healthy weight are important ways of safeguarding one's health.

In this chapter, we will consider carbohydrates as well as the other nutrients in a balanced diet—proteins, fats, vitamins, minerals, and water. We will see how the body uses these nutrients and examine ways to combine the foods we eat to promote health and fitness. ∎

MyPyramid Is a Food Guide for Planning a Healthy Diet

Planning a healthy diet requires more than simply eating certain foods to avoid deficiencies of particular nutrients. Choosing the right balance of foods can help improve health and reduce the risk of serious chronic diseases, such as heart disease, cancer, and diabetes.

MyPyramid is a food guide released by the U.S. Department of Agriculture (USDA) in 2005. Its purpose is to help any person plan a well-balanced diet even if he or she does not understand the reasons for eating certain foods (Figure 15a.1). The easy-to-use website www.MyPyramid.gov personalizes the nutritional information, guiding each user to one of 12 pyramid-shaped diagrams that summarizes the eating pattern considered optimal for someone of his or her age, gender, and level of activity. *MyPyramid* follows the recommendations in the USDA's Dietary Guide for Americans 2005. This guide promotes a healthy lifestyle that includes wise decisions about food and physical activity. All 12 pyramids share certain features. The human figure shown running up the steps on the side of each pyramid reminds us that daily physical activity is a key to healthy living. There are 12 pyramids because a healthy dietary plan should be personalized, not treated as "one size fits all." Dietary plans ought to be individualized because people differ in the number of calories they require. Thus, each of the 12 pyramids is based on the number of calories required by a person in a particular category. Each food group is color coded, corresponding to colored bands on the pyramid that remind us to eat a variety of foods, while the width of each band suggests what proportion of our daily diets should come from that food group. The bands are wider at the bottom, suggesting that most of the food we eat in each food group should be nutrient dense (packed with nutrients) and contain few "empty" calories (a concept we explain later in this chapter). The pyramid's slogan, *Steps to a Healthier You*, reminds us that improvements in diet should happen gradually. Besides being personalized, *MyPyramid* differs from the old food pyramid by specifying quantities to be consumed in cups or ounces instead of in servings. The 2005 Dietary Guide also allows discretionary calories. These are calories you are free to obtain from any food group after you have met your basic nutrient needs with nutrient-dense foods. The flexibility represented by discretionary calories allows you to consume some foods and drinks that contain added fats, added sugar, or alcohol.

The 2005 Dietary Guide makes three general recommendations for a healthy lifestyle: (1) choose food wisely, (2) stay within your calorie needs, and (3) engage in physical activity. *MyPyramid* can help you plan a nutritionally balanced diet to satisfy the first recommendation. The remainder of this chapter explains some of the biology on which the dietary guidelines are based.

Nutrients Provide Energy or Have a Structural or Functional Role in the Body

The daily decisions we make about what to eat are based on many and varied factors, including personal preferences, family and peer influences, our ethnic and religious background, media advertising, the money at our disposal, and how hungry we are at the time. The choices we make can influence how healthy we will be in the future. A high-fat diet, for instance, promotes certain cancers as well as diseases of the heart and blood vessels. On the other hand, a high-fiber diet has many benefits.

What does the body do with the food you eat? Food provides fuel, building blocks, molecules needed to carry out chemical processes in the body, and water.

- Food is needed as an energy source for all cellular activities. (Energy is measured in a unit called a calorie, which is the amount needed to raise the temperature of 1 g of water 1°C. In discussions of biochemical reactions, gains and losses of energy are usually reported in kilocalories. A kilocalorie is 1000 calories of energy. In nonscientific discussions, however, the *kilo* is dropped, and kilocalories are referred to simply as "calories." We will follow this tradition because it is the way most nutritional values are reported to the public.[1])

- Building blocks are needed for cell division, maintenance, and repair.

- Molecules, such as vitamins, are needed to coordinate life's processes.

- Water is necessary for maintaining the proper cellular environment and for certain cellular reactions.

The digestive system breaks complex molecules of carbohydrates, proteins, fats, and nucleic acids into their component subunits. Most cells of the body, especially those of the liver, are able to use these subunits as building blocks, by joining them together, and sometimes by converting one type of molecule into another. Your body, however, cannot synthesize all of the required kinds of amino and fatty acids, at least not in quantities sufficient to meet body needs. These substances that the body cannot synthesize are called essential amino acids and essential fatty acids. *Essential* means that the substances must be included in the diet.

A nutrient is a substance in food that provides energy, becomes part of a structure, or performs a function in growth, maintenance, or repair. Three nutrients—fats (lipids), carbohydrates, and proteins—can provide energy. Although proteins are included in this list, they are usually used to build cell structures or regulatory molecules, such as enzymes or certain hormones. Vitamins, minerals, water, and fiber do not provide energy, but they are essential to cellular functioning, as we will see.

∎ Lipids include fats, oils, and cholesterol

To weight-conscious Americans, *fat* is a dirty word. What adds insult to injury is that you do not have to eat fat to get fat. When you

---

[1]In technical writing, kilocalories are written as Calories.

GRAINS Make half your grains whole	VEGETABLES Vary your veggies	FRUITS Focus on fruits	MILK Get your calcium-rich foods	MEAT & BEANS Go lean with protein
Eat at least 3 oz. of whole-grain cereals, breads, crackers, rice, or pasta everyday 1 oz. is about 1 slice of bread, about 1 cup of breakfast cereal, or 1/2 cup of cooked rice, cereal, or pasta Eat 6 oz. every day	Eat more dark-green veggies like broccoli, spinach, and other dark leafy greens Eat more orange vegetables like carrots and sweet potatoes Eat more dry beans and peas like pinto beans, kidney beans, and lentils Eat 2.5 cups every day	Eat a variety of fruit Choose fresh, frozen, canned, or dried fruit Go easy on fruit juices Eat 2 cups every day	Go low-fat or fat-free when you choose milk, yogurt, and other milk products If you don't or can't consume milk, choose lactose-free products or other calcium sources such as fortified foods and beverages Eat 3 cups every day	Choose low-fat or lean meats and poultry Bake it, broil it, or grill it Vary your protein routine-choose more fish, beans, peas, nuts, and seeds Eat 5.5 oz. every day

Find your balance between food and physical activity

- Be sure to stay within your daily calorie needs.
- Be physically active for at least 30 minutes most days of the week.
- About 60 minutes a day of physical activities may be needed to prevent weight gain.
- For sustaining weight loss, at least 60 to 90 minutes a day of physical activities may be required.
- Children and teenagers should be physically active for 60 minutes every day, or most days.

Know the limits on fats, sugars, and salt (sodium)

- Make most of your fat sources from fish, nuts, and vegetables oils.
- Limit solid fats like butter, stick margarine, shortening, and lard, as well as foods that contain these.
- Check the Nutrition Facts label to keep saturated fats, *trans* fats, and sodium low.
- Choose food and beverages low in added sugars. Added sugars contribute calories with few, if any, nutrients.

FIGURE **15a.1** The food guide pyramid can helps you plan a well-balanced diet, even if you are not an expert nutritionist. The amounts specified in this figure are for a 2000 calorie per day diet.

regularly consume more calories than you use for energy, your body makes its own fat.

Lipid is the more technical name for what we have been calling fat. You may recall from Chapter 2 that there are several types of lipids. They include fats, oils, and cholesterol. Even so, 95% of the lipids found in food are triglycerides—the neutral fats that we commonly think of when we hear the term *fat*. A triglyceride is a molecule made from three fatty acids (hence, *tri*) attached to a molecule of glycerol (hence, *glyceride*). The fatty acids in the triglyceride give the molecule its characteristics.

An important way in which fatty acids can differ is their degree of saturation, or the extent to which each carbon in the fatty acid is bonded to as many hydrogen atoms as possible. Recall from Chapter 2 that a saturated fatty acid contains all the hydrogen it can hold. A polyunsaturated fatty acid could hold four or more additional hydrogens, and a monounsaturated fatty acid could hold two more hydrogens. In general, *saturated fats* are solid at room temperature, and they usually come from animal sources. In contrast, *unsaturated fats* are liquid at room temperature. Oils are unsaturated fats. The oils in our diet usually come from plant sources.

FUNCTIONS OF LIPIDS

Although cultural attitudes regarding body fat have varied throughout history, the biological need for some fat in the diet has never changed. Fat is an excellent medium for storing energy. It holds 9 calories of energy per gram, compared with the 4 calories per gram stored in carbohydrate or protein. It builds up between muscle cells, including those of the heart, to provide a ready source of energy. In addition, fat is a poor conductor of heat, so the layer of fat stored just beneath the skin provides some insulation against extreme heat or cold. Fat deposits also cushion and protect many vital organs, including the kidneys and the eyeballs.

Certain lipids are essential components of all cell membranes; some are used in the construction of myelin sheaths that insulate nerve fibers. Lipids are also needed for the absorption of the fat-soluble vitamins A, D, E, and K (discussed later in this chapter). These vitamins are absorbed from the intestines along with the products of fat digestion. Lipids then carry these vitamins in the bloodstream to the cells that use them. Glands in our skin produce oils that keep the skin soft and prevent dryness. Certain other lipids provide the raw materials for the construction of chemicals used in communication between cells. Cholesterol is the structural basis of the steroid hormones, including the sex hormones.

DIETARY FAT, OBESITY, AND ATHEROSCLEROSIS

Unfortunately for most of us, dietary fat quickly becomes too much of a good thing. The daily need for dietary fat is a mere tablespoon, yet the average American adult consumes 6 to 8 tablespoons of fat (78 to 196 g) a day, which can quickly add inches to the waistline! By itself, obesity is associated with health problems such as high blood pressure and increased risk of diabetes. In addition, the high consumption of fat is related to certain cancers, including cancer of the colon, prostate gland, lungs, and perhaps the breast.

The clearest health risk of a high-fat diet is the possibility of developing atherosclerosis, a condition in which fatty deposits form in the walls of blood vessels. The deposits promote an inflammatory response in the artery wall, thereby increasing the risk of heart attack and stroke (see Chapter 12). The risk of atherosclerosis increases with the blood level of cholesterol. In general, blood cholesterol levels under 200 mg/dl[2] are recommended for adults. The bad news is that the average blood cholesterol level for a middle-aged adult in the United States is 215 mg/dl. The good news is that people who lower their blood cholesterol levels can slow, or even reverse, atherosclerosis and, therefore, their risk of heart attack.

Total blood cholesterol, however, provides only a partial picture of a person's risk of atherosclerosis. Before cholesterol can be transported in the blood or lymph, it is combined with protein and triglycerides to form a lipoprotein, which makes it soluble in water. Low-density lipoproteins (LDLs) are considered to be a *bad* form of cholesterol because, although they bring cholesterol to the cells that need it to sustain life, they also deposit cholesterol in artery walls. In contrast, high-density lipoproteins (HDLs) help the body eliminate cholesterol. Thus, HDLs are considered to be a *good* form of cholesterol and to be protective against heart disease.

Because of the different roles the two kinds of lipoproteins play in transporting cholesterol, the ratio of HDLs to LDLs is considered a more important indicator of risk than is the total blood cholesterol (HDL + LDL) level alone. The ratio of total cholesterol to HDLs should not be greater than 4:1. HDL levels higher than 60 mg/dl are considered to reduce the risk of heart disease.

Blood cholesterol comes from one of two sources: the diet or the liver. Of the two, cholesterol production by the liver is more significant because *most* of the cholesterol in blood comes from the liver and not from the food we eat. Diet is still important, and, surprisingly, saturated fat in the diet raises blood levels of cholesterol more than dietary cholesterol does. Moreover, different people's bodies handle cholesterol differently depending on their genetic makeup (as well as the amount of cholesterol consumed in a meal). As a result, there is not a clear relationship between dietary and blood cholesterol.

The types of fat in the diet can influence blood cholesterol level and, more importantly, the ratio of LDLs to HDLs. Saturated fats and *trans* fats are bad for the heart. Saturated fat consistently boosts blood levels of harmful LDL cholesterol, both by directly stimulating the liver to step up its production of LDLs and by slowing the rate at which LDLs are cleared from the bloodstream. *Trans* fatty acids may behave like saturated fatty acids and raise LDL levels. In addition, they lower good HDLs. *Trans* fatty acids are formed when hydrogens are added to unsaturated fats (oils) in such a way as to stabilize them or to solidify them, as when margarine is formed from vegetable oil. *Trans* fats are also found in many packaged foods (Figure 15a.2).

But, not all fats are bad. Monounsaturated fats and polyunsaturated fats, such as omega-3 fatty acids and omega-6 fatty acids,[3] are good fats; they lower total blood cholesterol and LDLs. Monounsaturated fats are found in olive, canola, and peanut oils and in nuts. Omega-3 fatty acids are found in the oils of certain fish, such as Atlantic mackerel, lake trout, herring, tuna, and salmon. The most important omega-6 fatty acid is linoleic acid, an essential fatty acid found in corn and safflower oils (Figure 15a.2).

RECOMMENDATIONS FOR DIETARY FATS

The dietary goals for fat are (1) to limit all fats in the diet, so as to reduce obesity, and (2) to balance the types of fats consumed, so as

---

[2] dl = deciliter (100 milliliters).

[3] The number in omega-3 and omega-6 refers to the location of the first double bond in the carbon chain of the fatty acid.

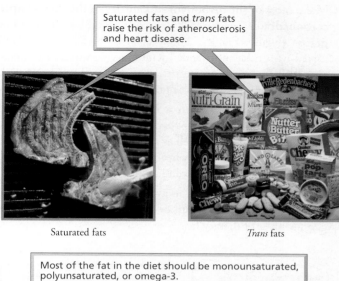

Saturated fats and *trans* fats raise the risk of atherosclerosis and heart disease.

Saturated fats

Trans fats

Most of the fat in the diet should be monounsaturated, polyunsaturated, or omega-3.

Monounsaturated fats

Polyunsaturated fats

Omega-3 fats

FIGURE **15a.2** Types of dietary fats. The total fat content of the diet should be moderate. The types of fats in the diet can influence the risk of heart attack and stroke.

to lower blood cholesterol and therefore reduce the risk of atherosclerosis. The strategy for balancing the types of fats is as follows.

- **Minimize saturated fats and *trans* fats.** These bad fats raise LDL levels and lower HDL levels.

- **Make most dietary fats monounsaturated, polyunsaturated, or omega-3 fats.** These fats are considered to be heart healthy because they lower LDL levels and raise HDL levels, and they should represent most of the fat calories in the diet.

For most of us, achieving this balance of fats would require a drastic reduction in the amount of saturated fat in the diet. Foods high in saturated fat include meat (especially red meat), butter, cheese, whole milk, and other dairy products.

■ **Carbohydrates include sugars, starches, and dietary fiber**

Carbohydrates in the diet include sugars, starches, and roughage (dietary fiber). The basic unit of a carbohydrate molecule is a monosaccharide. Sugars are *simple carbohydrates,* meaning they are

generally monosaccharides or disaccharides (two monosaccharides linked together). Sugars taste sweet and are naturally present in whole foods such as fruit and milk. In whole foods, sugars are present with other nutrients, including vitamins and minerals. However, so-called refined sugars are sugars that have been removed from their plant sources and purified, so they are no longer mixed with other nutrients. The calories from refined sugars are therefore described as "empty." They provide energy but have no other nutritive value. Refined sugars are found in candies, cookies, cakes, and pies.

Starches and fiber are *complex carbohydrates,* which means they are polysaccharides. Plants store energy in starches—long, sometimes branched chains of hundreds or thousands of linked molecules of the monosaccharide glucose. Common sources of starches include wheat, rice, oats, corn, potatoes, and legumes. *Fiber* is the indigestible part of edible plants. Complex carbohydrates are usually accompanied by other nutrients.

FUNCTIONS OF CARBOHYDRATES

The most important function of carbohydrates is to provide fuel for the body. The monosaccharide glucose is the carbohydrate that cells use for fuel most of the time. Even a temporary shortage of blood glucose can impair brain function and kill nerve cells. One gram of carbohydrate yields 4 calories—less than half of the energy provided by one gram of fat.

DIETARY FIBER

Dietary fiber is found in all plants that are eaten for food, including fruits, vegetables, dried beans, and whole grains. It is a form of carbohydrate that humans cannot digest into its component monosaccharides. Because only monosaccharides can be absorbed from the small intestine, dietary fiber is passed along to the large intestine. Some of the fiber is digested by the bacteria living in the large intestine, and the remaining fiber gives bulk to feces.

There are two types of dietary fiber. Water-soluble fiber is found in the fluids around or inside of plant cells; for example, it is found in pectins and gums. Oats, apples, and beans are good sources of soluble fibers. Insoluble fiber comes from structural parts of plant cells; an example is cellulose from plant cell walls (discussed in Chapter 2). Wheat bran and rye bran are good sources of insoluble fibers.

Although fiber cannot be digested or absorbed, it is still an important part of a healthful diet. Water-soluble fiber is good for the heart and blood vessels. It lowers LDLs and total cholesterol but does not lower the beneficial HDL cholesterol levels.

Intestinal disorders such as constipation and hemorrhoids improve when the amount of fiber in the diet is increased (discussed in Chapter 15). Soluble fiber can absorb an amazing amount of water, thereby softening stools and making them easier to pass.

RECOMMENDATIONS FOR CARBOHYDRATES IN THE DIET

Nutritionists recommend that you limit your intake of simple carbohydrates, especially refined sugars, and instead choose sources of carbohydrates that contain more than just calories. For example, fruits, vegetables, grains, and milk are food groups that supply many other nutrients besides carbohydrates.

Sugar consumption in the United States averages 125 pounds per person per year. Even people who faithfully read food labels may not be aware of the sugar content in processed food, because simple sugars have so many different names, including honey, brown sugar, dextrose, maltose, molasses, fructose, levulose, and corn syrup. Foods that taste sweet often contain whopping amounts of sugar. A single glazed donut, for example, contains 6 teaspoons (24 g) of sugar. Be aware that sugar may also be present in foods that do not taste sweet. Who would expect that 1 tablespoon of ketchup contains 1 teaspoon (4 g) of sugar?

Even complex carbohydrates are not all healthful because they differ in the way they affect blood sugar. Recall that the digestive system breaks down all carbohydrates (except fiber) to simple sugars, primarily glucose, which is absorbed into the bloodstream. A measure called the glycemic response describes how quickly a serving of food is converted to blood sugar and how much the level of blood sugar is affected. The glycemic index is a numerical ranking of carbohydrates based on their glycemic response. The scale of the glycemic index ranges from 0 to 100, with pure glucose serving as a reference point of 100. Sugar and starchy foods such as white bread, potatoes, and white rice have a high value on the glycemic index because they cause the blood sugar level to rise sharply. Foods with a low value on the glycemic index, including whole fruit, whole-grain foods, brown rice, and barley, cause a more modest and gradual increase in blood sugar. Generally, foods with a low value on the glycemic index are high in fiber (Table 15a.1).

The glycemic response is important because it influences how the body reacts to different foods. After you eat a carbohydrate with a high glycemic index value, your blood sugar rises. As discussed in Chapter 10, a rise in blood glucose level causes the pancreas to release more of the hormone insulin, which lowers blood glucose (thus inducing hunger) by converting excess glucose to fat. In short, in a healthy person, consumption of a high glycemic food is followed by fat formation and by increased hunger. In people with diabetes mellitus, rising blood-sugar levels due to high-glycemic foods may cause medical problems, because these people are unable to produce or to use insulin (see Chapter 10.)

- **Make most of the carbohydrate calories in the diet complex carbohydrates, especially, fruits, vegetables, and whole grains** (Figure 15a.3). Unlike sugars, starches are generally mixed with other nutrients, including vitamins and minerals, and with dietary fiber. Nutritionists generally agree that the diet should emphasize whole grains, vegetables, and fruits.

- **Avoid carbohydrates with a high glycemic response.** The glycemic response of carbohydrates in your diet can affect weight management, as well as the risk of heart disease and diabetes (see Chapter 10). A healthful diet contains more foods with low values on the glycemic index than with high values.

- **Increase dietary fiber.** The average American diet does not contain an adequate amount of fiber. Figure 15a.4 shows some good sources of dietary fiber.

Simple carbohydrates are empty calories, because they provide only energy.

Complex carbohydrates provide energy along with other nutrients.

FIGURE **15a.3** Types of carbohydrates. Simple carbohydrates, especially refined sugars, should be minimized in the diet. Most of the carbohydrates in the diet should be complex carbohydrates.

Use this table for additional information on the glycemic indexes of selected carbohydrates.

TABLE 15A.1 THE GLYCEMIC INDEX VALUES OF SELECTED CARBOHYDRATES			
FOOD	**GLYCEMIC INDEX**	**FOOD**	**GLYCEMIC INDEX**
Peanuts	14	Macaroni and cheese	64
Lowfat yogurt	33	White rice	64
Apple	38	Table sugar	68
Spaghetti	42	White bread	70
Brown rice	55	Popcorn	72
Oatmeal	58	Baked potato	85
Ice cream	61		

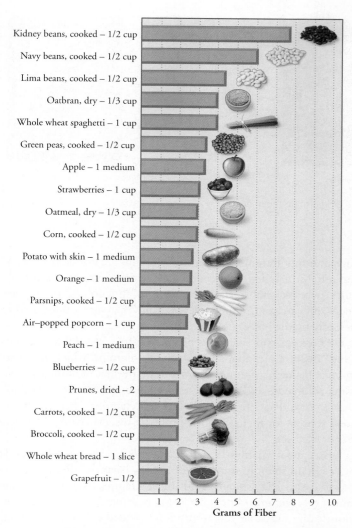

FIGURE **15a.4** Good sources of dietary fiber include fruits, vegetables, dried beans, and whole grains.

When your body makes a protein, the amino acids are strung together in a specific order, depending on which protein is being made. If a particular amino acid is needed for a certain protein but is not available, then that protein cannot be synthesized. Consider this analogy: A sign maker with a bag containing only 1 copy of the letter *R* and 100 copies of the other 25 letters in the alphabet could make only one NO PARKING sign. There is simply no way to substitute for the limiting letter, *R*. The same principle applies to protein synthesis; if a needed amino acid is lacking, protein production stops or the body breaks down existing proteins to get that amino acid.

In short, the pool of amino acids available for protein synthesis must always contain sufficient amounts of all the essential amino acids. Dietary proteins described as *complete proteins* contain ample amounts of all of the essential amino acids. Animal proteins are generally complete. Plant proteins are generally *incomplete proteins* and are low in one or more of the essential amino acids. Eating certain combinations of incomplete proteins from two or more plant sources can ensure that the pool of amino acids available for protein synthesis will contain ample amounts of all the essential amino acids (Figure 15a.5). Such combinations are called *complementary proteins* because if the combinations are correct, they supply enough of all the essential amino acids (Figure 15a.6). A vegetarian must be sure to consume complementary proteins.

FUNCTIONS OF PROTEIN

Protein is an essential structural component of every cell in your body. Indeed, if the water were removed from your body, half of your remaining body weight would be protein. Protein forms the framework of bones and teeth. It forms the contractile part of muscle and helps red blood cells carry oxygen. Dietary protein provides the raw materials for replacing or repairing cells that are stressed by everyday wear and tear and for forming new tissues as you grow.

Proteins regulate body processes. Certain proteins are enzymes that help the chemical reactions of the body take place at

■ Proteins are chains of amino acids

Protein seems to have a privileged status in our culture. When we think of protein, images of vitality, strength, and muscularity often come to mind. Proteins were named from the Greek word *proteios*, which means "of prime importance." Before we consider why proteins are so important, we should review their composition.

A protein consists of one or more chains of amino acids. Human proteins contain 20 different kinds of amino acids. The protein in the food you eat is digested into its component amino acids, which are then absorbed into the bloodstream and delivered to the cells, creating a pool of available amino acids. Your cells then draw the amino acids needed to build proteins in your body from those available in the pool. In addition, the body is able to synthesize 11 of the amino acids from nitrogen and molecules derived from carbohydrates, fats, or other amino acids. The 9 remaining amino acids that the body cannot synthesize—called *essential amino acids*—must be supplied by the diet.

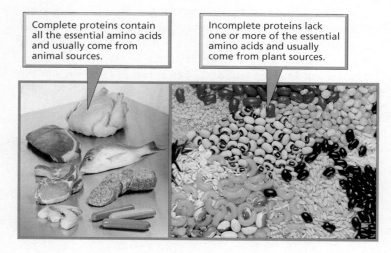

Complete proteins contain all the essential amino acids and usually come from animal sources.

Incomplete proteins lack one or more of the essential amino acids and usually come from plant sources.

FIGURE **15a.5** Types of protein in the diet

- Hummus (chickpeas and sesame seeds)
- Tofu and cashew stir-fry

- Trail mix (roasted soybeans and nuts)
- Tahini (sesame seeds) and peanut sauce

- Rice and beans
- Black-eyed peas and corn bread
- Bean burrito in corn tortilla

- Peanut butter on bread
- Rice and tofu
- Rice and lentils

Nuts and seeds

Legumes

Grains

FIGURE **15a.6** Complementary proteins are combinations of two or more incomplete proteins that together supply all the essential amino acids.

body temperature and at rates fast enough to support life. Other proteins are hormones, which are chemical messengers that enable cells in distant parts of the body to communicate with one another.

Proteins known as antibodies help the body defend itself against foreign invaders, such as bacteria. Also, proteins dissolved in the blood help maintain the water balance in the body.

Proteins can also be used for energy. When 1 gram of protein is oxidized, it yields 4 calories. However, protein is not the preferred fuel for the body. Protein is used for energy only if the supply of carbohydrates and fats is insufficient or if dietary protein exceeds the body's needs.

RECOMMENDATIONS FOR PROTEIN IN THE DIET

Meat, poultry, fish, beans, milk, and milk products are good sources of protein.

- **Choose lean, low-fat, or fat-free sources of protein.** Sources of animal protein usually contain a lot of fat, whereas plant protein sources are usually good sources of carbohydrates, too. The tendency in the United States to make meat the focal point of every meal is the main reason that the typical diet here contains excess fat. When choosing among animal sources of protein, reduce the amount of red meat in favor of chicken (with the fatty skin removed) or fish.

- **Combine plant sources of protein that contain complementary amino acids.** Although the diet must include protein choices that ensure an adequate supply of all the essential amino acids, the proteins need not come from animals. Eating a variety of plant proteins will supply all the essential amino acids and will help reduce the percentage of fat calories in the diet as well as boost the percentage of calories from complex carbohydrates.

■ Vitamins are needed in small amounts to promote and regulate the body's chemical reactions

A vitamin (*vita*, life) is an organic (carbon-containing) compound that, although essential for health and growth, is needed only in minute quantities—milligrams or micrograms. All the vitamins you need in a day would fill only an eighth of a teaspoon. These tiny amounts are sufficient because vitamins are not destroyed during use. Most function as coenzymes, which are nonprotein molecules necessary for certain enzymes to function. Enzymes and coenzymes are continuously recycled and, therefore, can be used repeatedly by the body.

There are two categories of vitamins: water-soluble vitamins, which dissolve in water, and fat-soluble vitamins, which are stored in fat. Of the 13 vitamins known to be needed by humans, 9 are water soluble (C and the various B vitamins) and 4 are fat soluble (A, D, E, and K). Table 15a.2 lists the vitamins, their functions, good sources for them, and the problems associated with deficiencies or excesses.

Except for vitamin D, our cells cannot make vitamins, so we must obtain them in our food. A varied, balanced diet is the best way to ensure an adequate supply of all vitamins. No one food contains every vitamin, but most contain some. Vitamins are often more easily available for absorption when the foods containing them are cooked. Cooked carrots, for instance, are a better source of vitamin A than are raw carrots. However, water-soluble vitamins are just that—water soluble—so they are likely to be lost if the vegetables containing them are cooked by being boiled in water. Steaming those vegetables is a better way to preserve their vitamin content.

Certain vitamins have to be consumed in adequate amounts every day. Folic acid, one of the B vitamins that is particularly abundant in dark leafy greens, plays a role in preventing birth defects, such as spina bifida, that involve the brain and spinal cord. It now seems that folic acid, along with vitamins B_6 and B_{12}, may also help prevent heart disease. Five daily servings of fruits and vegetables should provide enough of these vitamins to protect the heart. Vitamins that act as antioxidants, including vitamin C, vitamin E, and beta-carotene, are thought by some researchers to slow the aging process and protect against cancer, atherosclerosis, and macular degeneration (the leading cause of irreversible blindness in people over age 65). These, too, should be ingested every day. Spinach, collard greens, and carrots are good sources of antioxidant vitamins.

Can vitamin supplements be "too much of a good thing"? It depends. The answer to whether you personally should take vitamin supplements depends partly on the quality of your everyday

Use this table to learn more about the functions of vitamins.

TABLE 15A.2 VITAMINS

VITAMIN	GOOD SOURCES	FUNCTION	EFFECTS OF DEFICIENCY	EFFECTS OF EXCESS
Fat-soluble vitamins				
A	Liver, egg yolk, fat-containing and fortified dairy products; formed from carotene (found in deep yellow and deep green leafy vegetables)	Components of rhodopsin, the eye pigment responsible for black-and-white vision; maintains epithelia; cell differentiation	Night-blindness; dry, scaly skin; dry hair; skin sores; increased respiratory, urogenital, and digestive infections; xerophthalmia (the leading cause of preventable blindness worldwide); most common vitamin deficiency in world	Drowsiness; headache; dry, coarse, scaly skin; hair loss; itching; brittle nails; abdominal and bone pain
D	Fortified milk, fish liver oil, egg yolk; formed in skin when exposed to ultraviolet light	Increases absorption of calcium; enhances bone growth and calcification	Bone deformities in children, rickets, bone softening in adults	Calcium deposits in soft tissues, kidney damage, vomiting, diarrhea, weight loss
E	Whole grains, dark green vegetables, vegetable oils, nuts, seeds	May inhibit effects of free radicals; helps maintain cell membranes; prevents oxidation of vitamins A and C in gut	Rare; possible anemia and nerve damage	Muscle weakness, fatigue, nausea
K	Primary source from bacteria in large intestine; leafy green vegetables, cabbage, cauliflower	Important in forming proteins involved in blood clotting	Easy bruising, abnormal blood clotting, severe bleeding	Liver damage and anemia
Water-soluble vitamins				
C (ascorbic acid)	Citrus fruits, cantaloupe, strawberries, tomatoes, broccoli, cabbage, green pepper	Collagen synthesis; may inhibit free radicals; improves iron absorption	Scurvy, poor wound healing, impaired immunity	Diarrhea, kidney stones; may alter results of certain diagnostic lab tests
Thiamin (B_1)	Pork, legumes, whole grains, leafy green vegetables	Coenzyme in energy metabolism; nerve function	Water retention in tissues, nerve changes leading to poor coordination, heart failure, beriberi	None known
Riboflavin (B_2)	Dairy products such as milk; whole grains, meat, liver, egg whites, leafy green vegetables	Coenzyme used in energy metabolism	Skin lesions	None known
Niacin (B_3)	Nuts, green leafy vegetables, potatoes; can be formed from tryptophan found in meats	Coenzyme used in energy metabolism	Contributes to pellagra (damage to skin, gut, nervous system)	Flushing of skin on face, neck, and hands; possible liver damage
B_6	Meat, poultry, fish, spinach, potatoes, tomatoes	Coenzyme used in amino acid metabolism	Nervous, skin, and muscular disorders; anemia	Numbness in feet, poor coordination
Pantothenic acid	Widely distributed in foods, animal products, and whole grains	Coenzyme in energy metabolism	Fatigue, numbness and tingling of hands and feet, headaches, nausea	Diarrhea, water retention
Folic acid (folate)	Dark green vegetables, orange juice, nuts, legumes, grain products	Coenzyme in nucleic acid and amino acid metabolism	Anemia (megaloblastic and pernicious), gastrointestinal disturbances, nervous system damage, inflamed tongue, neural tube defects	High doses mask vitamin B_{12} deficiency
B_{12}	Poultry, fish, red meat, dairy products except butter	Coenzyme in nucleic acid metabolism	Anemia (megaloblastic and pernicious), impaired nerve function	None known
Biotin	Legumes, egg yolk; widely distributed in foods; bacteria of large intestine	Coenzyme used in energy metabolism	Scaly skin (dermatitis), sore tongue, anemia	None known

diet. Even though you may know that eating a well-balanced diet, including lots of vegetables, is the best way to obtain all the necessary nutrients, the pressures of time or force of habit may prevent you from eating well. In that case, a daily multivitamin supplement is a good idea. Other factors influencing whether vitamin supplements will be beneficial or harmful are the amount and type of vitamin consumed. Excess water-soluble vitamins are usually excreted in the urine. In contrast, excess fat-soluble vitamins are stored in fat and can accumulate in the body, causing serious problems if they reach high levels.

■ Minerals play structural and functional roles in the body

The minerals needed in our diet are inorganic substances essential to a wide range of life processes. We need fairly large amounts, although not what would be described as "megadoses," of seven minerals: calcium, phosphorus, potassium, sulfur, sodium, chloride, and magnesium. In addition, we need trace amounts of about a dozen others (Table 15a.3).

We can obtain the necessary minerals from the foods we eat, as long as we prepare them in ways that do not reduce their mineral

 Use this table to learn more about the functions of minerals.

TABLE 15A.3 MINERALS

MINERAL	GOOD SOURCES	FUNCTION	EFFECTS OF DEFICIENCY	EFFECTS OF EXCESS
Major minerals				
Calcium	Milk, cheese, dark green vegetables, legumes	Hardness of bones, tooth formation, blood clotting, nerve and muscle action	Stunted growth, loss of bone mass, osteoporosis, convulsions	Impaired absorption of other minerals, kidney stones
Phosphorus	Milk, cheese, red meat, poultry, whole grains	Bone and tooth formation; component of nucleic acids, ATP, and phospholipids; acid-base balance	Weakness, demineralized bone	Impaired absorption of some minerals
Magnesium	Whole grains, green leafy vegetables, milk, dairy products, nuts, legumes	Component of enzymes	Muscle cramps, neurologic disturbances	Neurologic disturbances
Potassium	Available in many foods including meats, fruits, vegetables, and whole grains	Body water balance, nerve function, muscle function, role in protein synthesis	Muscle weakness	Muscle weakness, paralysis, heart failure
Sulfur	Protein-containing foods including meat, legumes, milk, and eggs	Component of body proteins	None known	None known
Sodium	Table salt	Body water balance, nerve function	Muscle cramps, reduced appetite	High blood pressure in susceptible people
Chloride	Table salt, processed foods	Formation of hydrochloric acid in stomach, role in acid-base balance	Muscle cramps, reduced appetite, poor growth	High blood pressure in susceptible people
Trace minerals				
Iron	Meat, liver, shellfish, egg yolk, whole grains, green leafy vegetables, nuts, dried fruit	Component of hemoglobin, myoglobin, and cytochrome (transport chain enzyme)	Iron-deficiency anemia, weakness, impaired immune function	Liver damage, heart failure, shock
Iodine	Marine fish and shellfish, iodized salt, dairy products	Thyroid hormone function	Enlarged thyroid	Enlarged thyroid
Fluoride	Drinking water, tea, seafood	Bone and tooth maintenance	Tooth decay	Digestive upsets, mottling of teeth, deformed skeleton
Copper	Nuts, legumes, seafood, drinking water	Synthesis of melanin, hemoglobin, and transport chain components; collagen synthesis; immune function	Rare; anemia, changes in blood vessels	Nausea, liver damage
Zinc	Seafood, whole grains, legumes, nuts, meats	Component of digestive enzymes; required for normal growth, wound healing, and sperm production	Difficulty in walking, slurred speech, scaly skin, impaired immune function	Nausea, vomiting, diarrhea, impaired immune function
Manganese	Nuts, legumes, whole grains, leafy green vegetables	Role in synthesis of fatty acids, cholesterol, urea, and hemoglobin; normal neural function	None known	Nerve damage

content. Like certain vitamins, many minerals are water soluble and can be lost during food preparation.

Sodium, a component of table salt, is essential to health, but most Americans consume too much of it. High salt intake causes high blood pressure in some people. The Dietary Recommendations for Americans advises that salt intake should not exceed 2400 mg (slightly more than 1 teaspoon) a day. Processed foods are especially high in salt, which is added to preserve food and enhance the taste. Salt is found in nearly every processed food product, including canned vegetables, cheese, bread, and processed meats. To reduce your salt intake, use salt sparingly when cooking fresh food and read the labels on prepared food.

■ **Water is critical and needed in large amounts**

Water is perhaps the most important nutrient. We can live without food for about 8 weeks but without water for only about 3 days. Although you look solid, the environment within your body is a virtual sea. Even bone is about one-fourth water! A newborn is, in fact, 85% water. The water content gradually decreases with age to between 55% to 65% in an adult female and between 65% to 75% in an adult male. Females contain less water than males do because their bodies contain a greater proportion of fat, which does not hold water as well as muscle does.

Water is so common—found in nearly every food and beverage we consume—that we rarely think about its importance. It is an excellent solvent. Thus, in the blood and lymph, it transports a wide variety of materials, including nutrients, metabolic wastes, and hormones. Water also provides a medium in which chemical reactions can take place; in addition, it participates in many of those reactions. One example is hydrolysis reactions (*hydro-,* water; *lysis,* splitting), in which chemical bonds are broken by the addition of water (Chapter 2). For instance, the food molecules we eat are split into their component subunits during digestion by hydrolysis. Water is also a lubricant. It keeps your joints from creaking and the soft tissues of your body from sticking together. It also forms a protective cushion within the eyes and around the brain and spinal cord. During pregnancy, water in the amniotic fluid protects the fetus. In addition, water plays an important role in the regulation of body temperature. Perspiration, for instance, cools the body.

Nutritionists recommend that we consume eight 8-ounce glasses (2 quarts) of water a day. Water in fruits and vegetables makes up about half of that requirement for the average adult. The rest of the requirement does not have to be consumed as plain water, but it should not come from carbonated sweet drinks, caffeinated beverages, or alcoholic beverages. Carbonation interferes with water absorption, and sugar adds empty calories. Caffeine and alcohol increase water loss in urine.

Food Labels Help Us Make Wise Food Choices

Grocery stores generally offer several choices for essentially the same product—whole wheat crackers, for instance. Food labels and the knowledge you now have about nutrients can help you make healthy choices.

Figure 15a.7, containing a food label from a box of baked whole wheat crackers, provides some useful general pointers for reading food labels. First, note the serving size described on the label, and remember that the amount you actually eat, whether larger or smaller than what is described on the label, will be what determines the number of calories and amount of nutrients you consume.

Next, notice the number of calories reported per serving and the number of calories from fat. In the example shown in the figure, 40 of the 120 calories in a serving, or 34% of the calories, come from fat. Dietary guidelines recommend that you consume less than 30% of your total calories as fat. However, if you crave these crackers, you can make up for the amount of fat in them, to some extent, by eating low-fat food at another time during the day. Also note the amount of saturated fat and *trans* fat reported on the label, because these kinds of fat should be minimized in your diet.

In considering the percent daily values of various nutrients reported on the label, keep two things in mind. First, these values are based on a 2000-calorie diet. Your own diet may require more or fewer calories. Second, as becomes clear from the daily values provided at the bottom of the label, you should be attempting to keep dietary fat and sodium consumption *below* those amounts.

Food labels can also help you increase your intake of nutrients that are important to consume. Be sure to get enough vitamin A, vitamin C, calcium, iron, and phosphorus. The daily values reported for carbohydrates and dietary fiber will also help you consume enough of those substances.

For Body Energy Balance, Calories Gained in Food Must Equal Calories Used

We have seen how to choose foods wisely to meet our nutritional needs. Let's now focus on why we have to meet our nutritional needs without exceeding our caloric needs. As mentioned earlier, our bodies obtain energy from the carbohydrates, proteins, and fats that we eat. Carbohydrates and proteins provide 4 calories per gram, and fats provide 9 calories per gram.

Although energy can be neither created nor destroyed, it can be changed from one form to another. Such energy conversions take place in the biochemical reactions that occur in the body. They are the reason that the energy in a potato you eat can fuel muscle contraction, allowing you to run to catch the bus. One characteristic of a healthy lifestyle is a dynamic balance between the energy the body takes in (in food) and the energy it uses to function. Food energy that is not used for the body's various activities is stored as fat or glycogen, explaining why you gain weight when you consume more calories than you use.

The body requires energy for maintenance of basic body functions, for physical activity, and for processing the food that is eaten. The energy that is needed strictly for maintenance is called the *basal metabolic rate (BMR)*; it is the minimum energy needed to keep an awake, resting body alive, and it generally represents between 60% and 75% of the body's energy needs. A male usually has

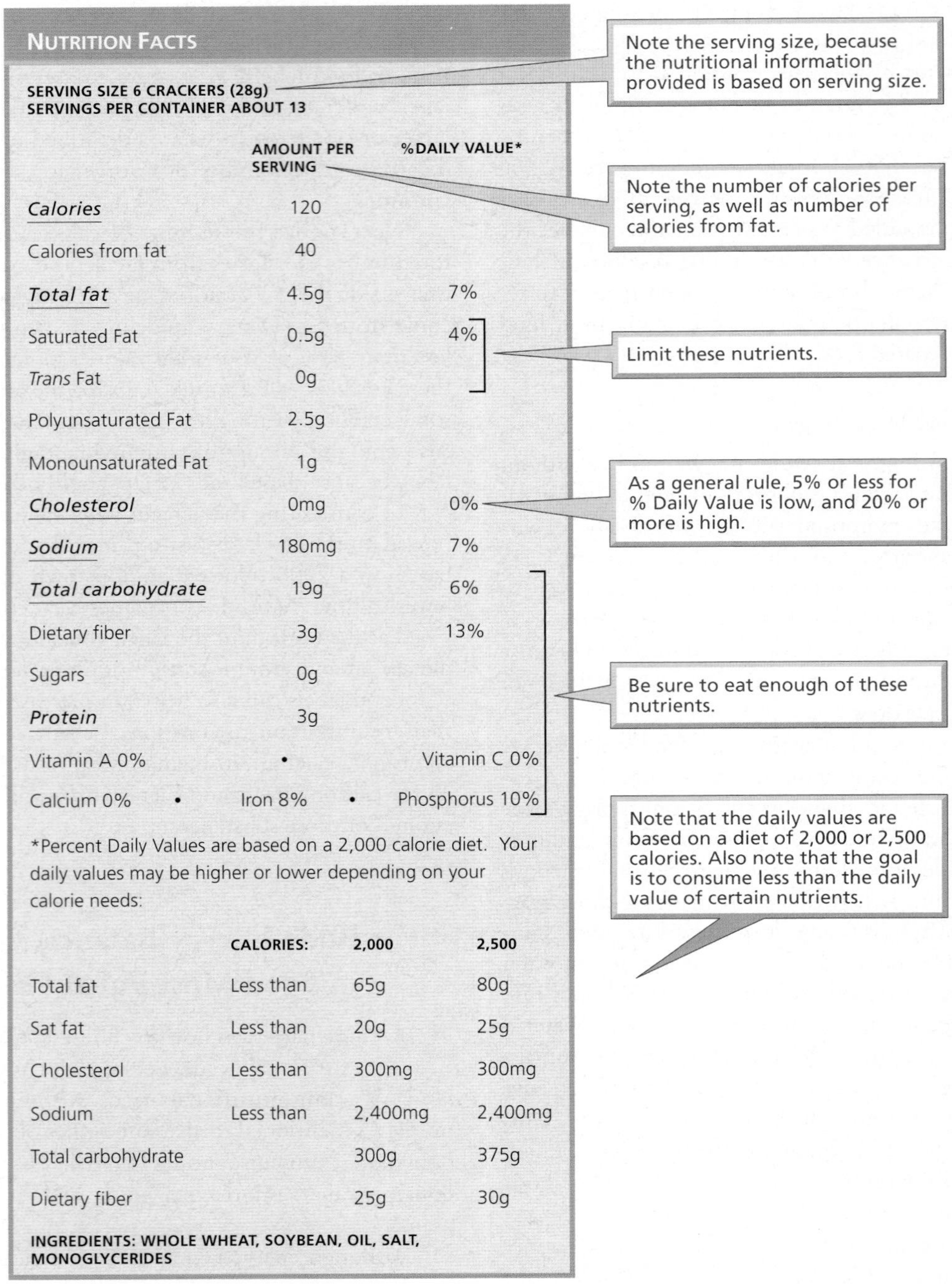

NUTRITION FACTS

SERVING SIZE 6 CRACKERS (28g)
SERVINGS PER CONTAINER ABOUT 13

	AMOUNT PER SERVING	%DAILY VALUE*
Calories	120	
Calories from fat	40	
Total fat	4.5g	7%
Saturated Fat	0.5g	4%
Trans Fat	0g	
Polyunsaturated Fat	2.5g	
Monounsaturated Fat	1g	
Cholesterol	0mg	0%
Sodium	180mg	7%
Total carbohydrate	19g	6%
Dietary fiber	3g	13%
Sugars	0g	
Protein	3g	

Vitamin A 0%	•	Vitamin C 0%
Calcium 0% •	Iron 8% •	Phosphorus 10%

*Percent Daily Values are based on a 2,000 calorie diet. Your daily values may be higher or lower depending on your calorie needs:

	CALORIES:	2,000	2,500
Total fat	Less than	65g	80g
Sat fat	Less than	20g	25g
Cholesterol	Less than	300mg	300mg
Sodium	Less than	2,400mg	2,400mg
Total carbohydrate		300g	375g
Dietary fiber		25g	30g

INGREDIENTS: WHOLE WHEAT, SOYBEAN, OIL, SALT, MONOGLYCERIDES

Note the serving size, because the nutritional information provided is based on serving size.

Note the number of calories per serving, as well as number of calories from fat.

Limit these nutrients.

As a general rule, 5% or less for % Daily Value is low, and 20% or more is high.

Be sure to eat enough of these nutrients.

Note that the daily values are based on a diet of 2,000 or 2,500 calories. Also note that the goal is to consume less than the daily value of certain nutrients.

FIGURE **15a.7** Tips for reading food labels

a higher metabolic rate than does a female because a male's body has more muscle and less fat than a female's. Muscles use more energy than fat does. So, while a man and a woman of equal size sit on the couch and watch television together, he burns 10% to 20% more calories than she does. As you age, muscle mass and metabolic rate both decline. Together, these factors reduce caloric needs. If other adjustments in lifestyle are not made to compensate, these metabolic changes can add extra pounds each year after age 35.

The second largest use of energy is physical activity (Table 15a.4). Exercise is an excellent way to burn calories. It not only boosts your energy needs during the activity but also speeds up metabolic rate for a while afterward. The calories you habitually use in exercise are added to those needed to maintain your BMR when you determine your calorie needs.

Exercise helps keep the body in good working order. The *Dietary Guide for Americans 2005* encourages adults to engage in at least 30 minutes of moderate-intensity physical activity on most days of the week. Aerobic exercise reduces the risk of diseases of the heart and blood vessels and lowers blood pressure (discussed in Chapter 12). Weight-bearing exercise reduces the risk of osteoporo-

TABLE 15A.4 APPROXIMATE NUMBER OF CALORIES BURNED PER HOUR BY VARIOUS ACTIVITIES			
ACTIVITY	100-LB PERSON	150-LB PERSON	200-LB PERSON
Bicycling, 6 mph	160	240	312
Bicycling, 12 mph	270	410	534
Jogging, 5.5 mph	440	660	962
Jogging, 10 mph	850	1280	1664
Jumping rope	500	750	1000
Swimming, 25 yd/min	185	275	358
Swimming, 50 yd/min	325	500	650
Walking, 2 mph	160	240	312
Walking, 4.5 mph	295	440	572
Tennis (singles)	265	400	535

Source: American Heart Association

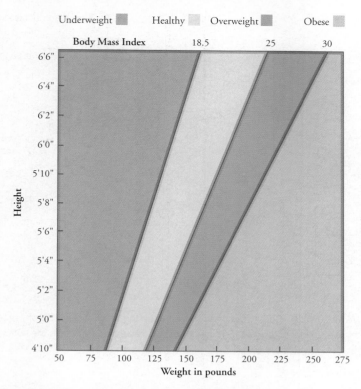

FIGURE **15a.8** The Body Mass Index (BMI) evaluates body weight relative to height. A BMI over 30 is usually considered to be unhealthy and a sign of obesity.

sis, a loss of bone density (discussed in Chapter 5). In general, regular physical exercise reduces stress and the risk of certain chronic diseases, including diabetes. Regular exercise also manages body weight, helping to prevent the unhealthy weight gain that can occur in adulthood.

Obesity Is Body Weight 20% or More above the Body Weight Standard

Although *overweight* and *obese* are both terms used to describe people who have excess body weight, they do not have exactly the same meaning. An obese person is overweight because of excess fat. An athletic person who because of well-developed muscle weighs more than the weight listed as desirable on height-weight tables is not obese.

The Body Mass Index (BMI) is a number that provides a reliable indicator of body fat because it evaluates your weight in relation to your height (Figure 15a.8). A BMI greater than 30 is generally considered unhealthy and an indication of obesity. However, just as it is possible for a very muscular person to have a BMI above 30 and not be considered obese, it is possible for a person in the healthy weight range to have too much fat and little muscle. The Centers for Disease Control and Prevention estimates that 64% of Americans are overweight and that almost 31% are obese. The number of obese people in the United States has been steadily rising since 1985 (Figure 15a.9).

Although most dieters are motivated to slim down for cosmetic reasons, more important reasons are the health risks associated with obesity. For instance, obesity leads to disease of the heart and blood vessels. Even individuals who are just slightly overweight are at increased risk of having a heart attack. Obesity also raises total cholesterol levels in the blood while lowering levels of the beneficial

HDL cholesterol. In addition, it increases the risk of high blood pressure, which can lead to death from heart attack, stroke, or kidney disease. Obesity has harmful effects besides those on the heart and blood vessels. It can induce diabetes, which results in elevated blood glucose levels; it is a major cause of gallstones; and it can worsen degenerative joint diseases.

Successful Weight–Loss Programs Usually Involve Reducing Calorie Intake, Increasing Calorie Use, and Changing Behavior

Successful weight-loss programs generally have three components: (1) a reduction in the number of calories consumed, without departing from the recommended nutritional guidelines; (2) an increase in energy expenditure; and (3) behavior modification. Gradual changes in eating habits are most likely to lead to permanent lifestyle changes.

To determine the number of calories needed each day to maintain a desirable weight, a person must take activity level and age into account. (Figure 15a.10 provides a way of estimating that number.) A pound of fat contains approximately 3500 calories, so to lose 1 pound a week, a person should reduce calorie consumption by 500 calories a day (500 calories × 7 days = 3500 calories), or increase calorie use by 500 calories a day, or any

OBESITY TRENDS* AMONG U.S. ADULTS
1991–2004

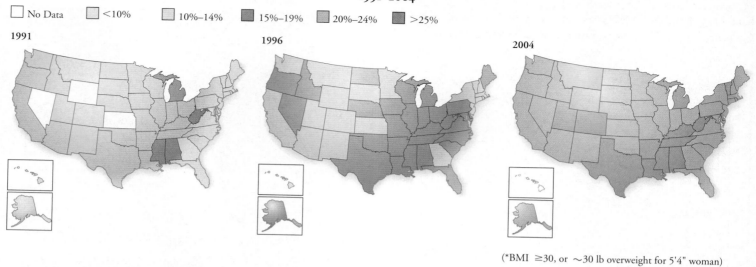

(*BMI ≥30, or ~30 lb overweight for 5'4" woman)

FIGURE **15a.9** Obesity trends among adults in the United States. Obesity is defined as a high amount of body fat relative to total body weight. There has been an alarming increase in the number of obese people over the last decade. (Centers for Disease Control and Prevention)

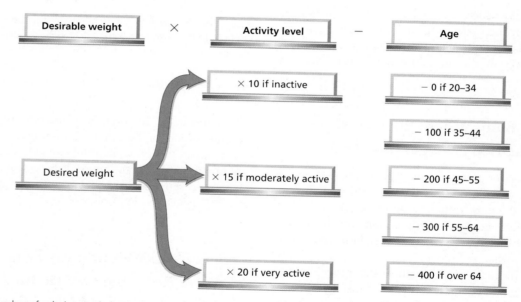

FIGURE **15a.10** The number of calories needed to maintain a desirable weight is influenced by one's activity level and age. The formula shown here provides a way of estimating that number.

equivalent combination. The lowest daily caloric intake recommended for a female is 1200 calories and for a male, 1500 calories, unless they are in a medically supervised program.

LOW-CALORIE, LOW-FAT DIETS

The easiest way to reduce calorie intake while continuing to eat healthily is to cut back on fatty foods, especially those that contain saturated fat, because fat contains more than twice as many calories as an equivalent weight of carbohydrate or protein. In addition, it is easier for the body to store the fat in foods as body fat than as protein or carbohydrate. Indeed, most of us do end up wearing the fat we eat.

Another recommended way of reducing calories is to avoid sugar. As we have seen, foods that taste sweet are packed with sugar and, therefore, calories.

A third healthful diet tip is to increase the amount of fiber in the diet. High-fiber foods, such as fruits and vegetables, tend to be low in calories and fat but also high in vitamins and minerals. Because fiber is bulky, these foods are also filling.

LOW-CARBOHYDRATE DIETS

Low-carbohydrate diets, such as the well-publicized Atkins diet, have helped many people lose weight. The first stage of the four-stage Atkins diet severely restricts carbohydrates to force the body to burn fat for energy instead of carbohydrates. Proponents of low-carbohydrate diets point out that carbohydrates are digested to glucose, which is absorbed into the blood. When blood glucose levels rise, the pancreas releases the hormone insulin, which allows cells to take in glucose. Blood glucose levels plummet as glucose is taken into cells, causing the brain to send out hunger signals. Insulin also stimulates fat storage. In short, they believe that eating carbohydrates stimulates appetite and promotes an increase in body fat.

Factors other than the control of insulin levels probably play at least some part in helping people on a low-carbohydrate diet to lose weight. One factor might be that a low-carbohydrate diet is also usually a low-calorie diet. Snack foods, desserts, and soda are high in both carbohydrates and calories. Another factor might be that a *moderate* amount of fat (which is allowed in low-carbohydrate diets) makes food more satisfying and filling, which helps the dieter eat less and consume fewer calories.

An important criticism of the Atkins diet is that it doesn't distinguish between sources of protein, permitting dieters to indulge heavily in animal sources of protein. Animal sources of protein (meat) are usually high in saturated fats, which nearly all nutrition experts agree are bad for health.

THE YO-YO EFFECT

Approximately 60% to 90% of dieters who lose weight will later regain all the weight they lost (Figure 15a.11). Often, the weight is regained because the weight loss was achieved by drastically cutting back calories, which can be unhealthy and is very difficult to sustain for long periods. As old eating habits return, so do the pounds.

When the determination to shed some pounds returns, the diet begins again. This process is commonly known as "the yo-yo effect."

What causes the yo-yo effect? When there is a severe restriction in calories, as occurs in the typical crash diet, the body adopts a calorie-sparing defense that reduces the resting metabolic rate by as much as 45%. This response conserves energy and evolved long ago to help our ancestors survive during times of food scarcity. Today, it makes weight loss from successive diets progressively more difficult. Furthermore, with each diet, a person usually loses both lean and fat tissue, especially if increased exercise is not part of the weight-loss program. When the weight is regained, most of it is fat. Fat is less metabolically active than lean muscle tissue, so the person's metabolic rate drops even lower. Thus, repeated crash dieting may be a "no-lose situation."

Anorexia Nervosa and Bulimia Are Eating Disorders That Create Calorie Deficits

Obesity can be considered an eating disorder associated with overeating. The obese and overweight would be wise to follow a nutritionally sound weight-loss program. It is important to remember, however, that dieting can be taken too far. Most people with the eating disorders anorexia nervosa and bulimia began their descent into these disorders by dieting. Anorexia nervosa is a deliberate self-starvation. A person whose body weight is 85% or less than expected for his or her height and weight is considered to be anorexic. In contrast, bulimia is marked by binge eating large quantities of food and then purging by self-induced vomiting, enemas, laxatives, or diuretics.

The changes in eating habits associated with these eating disorders are thought to be the result of psychological, social, and

(a) 1986

(b) 1988

(c) 2006

FIGURE **15a.11** Typical dieters change their energy balance temporarily, and, when old habits are resumed, the weight is regained. The cycle is then repeated. The yo-yo effect makes weight loss more difficult with successive diets. The popular talk show host Oprah Winfrey has been candid about her struggle with fluctuating weight.

physiological factors. Both disorders are associated with a preoccupation with body size and shape. Anxious depression also seems to play a role. In addition, people with anorexia often have a low tolerance for change. The problem often begins when the person leaves home for college.

The behavior patterns associated with anorexia and bulimia differ, but both result in a severe deficit in calories. A person with anorexia eats very little food and, therefore, consumes few calories. But, anorexia, which means "lack of appetite," is misnamed. Although people with anorexia deny feeling hunger, their refusal to eat stems from an intense fear of becoming fat—*not* from lack of appetite. Indeed, they are often preoccupied with food and may develop strange rituals around eating. For instance, food may be measured out in extremely small amounts or cut into tiny pieces. Excessive exercise is also typical in anorexia. No matter how much weight is lost, however, it can never be enough, because an anorexic person has a distorted body image. He or she perceives the body as fat even when emaciated (Figure 15a.12).

In contrast, a person with bulimia eats a huge amount of food but then eliminates it from the body. During a bulimic binge, which may last as long as 8 hours, as many as 20,000 calories may be consumed. Several shorter binges may occur in a single day. Each binge is followed by attempts to purge the body of the calories, usually by self-induced vomiting or by laxatives. Some bulimic people take as many as 200 laxatives a week, expecting to rid the body of unwanted calories. It is estimated, however, that laxatives can eliminate only 10% of the calories consumed.

FIGURE **15a.12**   Anorexia is a form of self-starvation. No matter how emaciated the person becomes, the individual still perceives the body as being fat.

Eating disorders have many negative effects on the body. One major side effect of anorexia is a severe decrease in bone health. Although the excessive exercise engaged in by most anorexic people may have a slight strengthening effect on bones, many other factors work to weaken the bones. For example, amenorrhea (cessation of menstruation), malnutrition, and low body weight, particularly low body fat, are all consequences of anorexia, and they can contribute to poor bone health. There is a loss of bone mass, particularly from vertebrae and the long bones of the arms and legs. This makes the bones fragile and increases the risk of fractures, particularly hip fractures.

A bulimic person's practice of purging by either self-induced vomiting or overuse of laxatives creates special problems. Purging can disturb the balance of electrolytes, especially sodium and potassium, which in turn can lead to muscle cramping, an irregular heartbeat, or cardiac arrest. Laxatives and vomiting also cause dehydration, resulting in neurological problems and kidney damage. Frequent self-induced vomiting can cause other serious problems: glands in the face and neck may become swollen; repeated regurgitation of acidic stomach contents can cause tooth decay, chronic heartburn, and sores in the mouth or lips. In addition, some of the techniques used to induce vomiting can tear the throat or esophagus.

The effects of anorexia are not so different from those of starvation. Without treatment, up to 20% of people with serious eating disorders die. Even with treatment, 2% to 3% die. In the early phase of the illness, an anorexic person typically chooses a diet that is low in energy-dense foods but rather high in proteins and other essential nutrients. Dietary protein, combined with the high activity levels characteristic of a person with anorexia nervosa, has a nitrogen-sparing effect. As a result, the initial weight loss is almost entirely due to loss of fat tissue. However, when fat reserves are exhausted and refusal of food becomes more severe, the body begins to break down its own proteins to use as an energy source. The primary sources of these proteins are skeletal and heart muscle; therefore, skeletal and heart muscle mass decreases. At the same time, water loss is accelerated, especially from *within* the body cells. This water loss leads to disturbances in metabolism and electrolyte balance.

Heart problems are the most common cause of death in people with anorexia. Starvation, dehydration, and electrolyte disturbances cause the heartbeat to slow (bradycardia) and blood pressure to fall (hypotension). The shrinkage of the heart muscle, which occurs as the patients lose muscle mass as well as body fat, can cause heart murmurs as well as abnormal blood flow through the heart. In addition, the person can develop congestive heart failure if fluids are replaced too quickly during treatment.

A potential cause of death associated with either of these eating disorders is hypoglycemia, an abnormally low blood glucose level. Because the brain depends entirely on glucose for its metabolism, hypoglycemia can cause unconsciousness and death.

Organs from Several Body Systems Eliminate Waste

By Producing Urine, the Kidneys Maintain Homeostasis
- Each kidney has three regions
- Nephrons are the functional units of the kidneys
- The kidneys help maintain acid-base balance
- The kidneys help conserve water
- Hormones influence kidney function
- The kidneys help produce red blood cells and activate vitamin D

Dialysis and Transplant Surgery Help When Kidneys Fail
- Dialysis cleanses the blood
- Renal function can be restored with a kidney transplant

Urination Has Involuntary and Voluntary Components

Bacteria Can Enter the Urethra and Cause Urinary Tract Infections

HEALTH ISSUE: Kidney Stones and Their Shocking Treatment

HEALTH ISSUE: Urinalysis: What Your Urine Says about You

16
The Urinary System

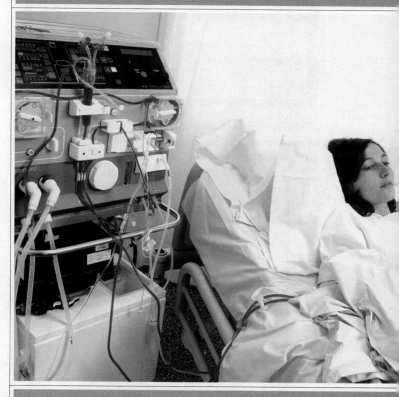

The kidneys are the principal organs of the urinary system. They filter the blood and help maintain its proper chemistry and composition. When kidneys fail, a dialysis machine can perform some of their functions.

Amanda was feeling depressed. She was 20 years old and doing well in college, but the rest of her life was on hold as she waited for a kidney donor. Her best friend was taking the extended European vacation they had planned to take together this summer, but Amanda remained tethered to her home by her need for a dialysis machine.

The total number of candidates on the waiting list for organs in the United States is updated minute by minute. At 4:56 P.M. on March 21, 2006, that number totaled 91,492. The vast majority of these candidates—65,807—were waiting for kidneys. Why did kidneys top the list? What important roles do kidneys play in our bodies? What medical options—transplantation or others—are available when such critical organs fail? Perhaps you have considered carrying an organ donor card that says you wish to donate organs or tissues at the time of your death (Figure 16.1). Your kidney could save someone else's life, or perhaps someone else's kidney could save you.

In this chapter we will examine kidney structure and consider the critical roles kidneys play in maintaining homeostasis by removing wastes from the body and regulating the volume and solute concentration of blood. We then describe disorders of the kidneys and various ways to maintain kidney function. We also consider what the characteristics of our urine can say about us and why medical personnel so frequently ask us to provide a urine sample. We finish the chapter with a discussion of urination and urinary tract infections. ∎

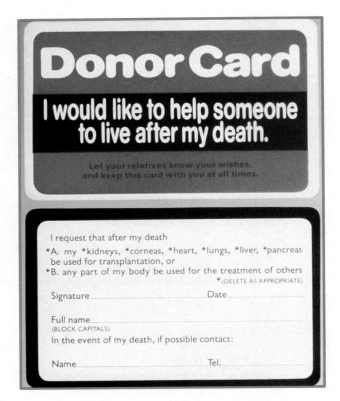

FIGURE **16.1** An organ donor card available through http://www.organdonor.gov. In addition to carrying this card, it is important to let family members know of your decision to donate your organs or tissues. Your intent to donate also can be indicated on your driver's license.

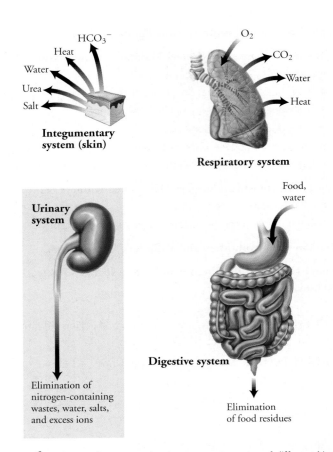

FIGURE **16.2** Organs from several systems remove wastes of different kinds from the body. The kidneys, organs of the urinary system, play a major role in the excretion of nitrogen-containing wastes, water, salts, and excess ions.

Organs from Several Body Systems Eliminate Waste

Without the kidneys and other organs that eliminate waste substances, the accumulation of those wastes in the body would severely upset homeostasis. Consider that our cells, like tiny laboratories, conduct the multitude of chemical reactions that constitute metabolism. Just like the day-to-day activities of actual chemical laboratories, the cells' ongoing synthesis and breakdown of molecules produce wastes. Metabolic wastes include carbon dioxide, water, heat, salts, and nitrogen-containing molecules such as ammonia, urea, and uric acid. Our bodies must diligently dispose of them all.

The organs that eliminate wastes and excess essential ions—such as hydrogen (H^+), sodium (Na^+), potassium (K^+), and chloride (Cl^-)—are shown in Figure 16.2. Lungs and skin eliminate heat, water, and carbon dioxide (as the gas CO_2 from the lungs and as bicarbonate ions, HCO_3^-, from the skin). Skin also excretes salts and urea. Organs of our gastrointestinal tract eliminate solid wastes and many of the other metabolic wastes. Our focus in this chapter is the kidneys, which are the organs of the urinary system responsible for the formation of urine. In urine, the body excretes nitrogen-containing wastes, as well as water, carbon dioxide (as HCO_3^-), inorganic salts, and hydrogen ions.

The **urinary system** consists of two kidneys, two ureters, one urinary bladder, and one urethra (Figure 16.3). The main function

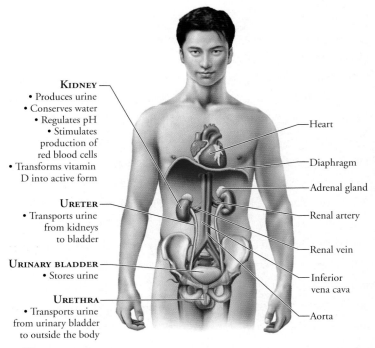

KIDNEY
• Produces urine
• Conserves water
• Regulates pH
• Stimulates production of red blood cells
• Transforms vitamin D into active form

URETER
• Transports urine from kidneys to bladder

URINARY BLADDER
• Stores urine

URETHRA
• Transports urine from urinary bladder to outside the body

Heart
Diaphragm
Adrenal gland
Renal artery
Renal vein
Inferior vena cava
Aorta

FIGURE **16.3** Organs of the urinary system (labeled on the left side) and their relation to major blood vessels (identified on the right side). Digestive organs are not shown.

of this system is to regulate the volume, pressure, and composition of the blood. The **kidneys** are the organs of the urinary system that accomplish this task by regulating the amount of water and dissolved substances that are removed from and returned to the blood. Wastes and excess materials removed from the blood form **urine**, the yellow fluid produced by each kidney.

Urine from the kidneys travels down tubes called **ureters** to the urinary bladder. The **urinary bladder** is a muscular organ that temporarily stores urine until it is excreted from the body. Urine leaves the body through the **urethra**, a tube that transports urine from the urinary bladder to the outside of the body through an external opening. The female urethra transports only urine. The male urethra, however, carries urine and reproductive fluids (but not simultaneously). The reproductive function of the male urethra is discussed in Chapter 17.

By Producing Urine, the Kidneys Maintain Homeostasis

Our kidneys are red in color and shaped like kidney beans (see Figure 16.3). Each one is about the size of a fist. They are located just above the waist, sandwiched between the parietal peritoneum (the membrane that lines the abdominal cavity) and the muscles of the dorsal body wall. The slightly indented, or concave, border of each kidney faces the midline of the body. Perched on top of each kidney is an adrenal gland.

The kidneys are covered and supported by several layers of connective tissue (Figure 16.4). The outermost layer is a tough fibrous layer that anchors each kidney and its adrenal gland to the abdominal wall and surrounding tissues. Beneath this layer is a protective cushion of fat (the middle connective tissue layer also called the fat capsule). The innermost layer covering the kidneys is a layer of collagen fibers. This layer protects the kidneys from trauma and infection. The numerous protective barriers and cushions surrounding our kidneys highlight the importance of these organs to our daily existence.

■ Each kidney has three regions

The ureter leaves the kidney at a notch in the concave border, as shown in Figure 16.4. This notch is also the area where blood vessels enter and exit the kidney. The renal arteries branch off the aorta and carry blood to the kidneys. The renal veins carry filtered blood away from the kidneys to the inferior vena cava, which transports the blood to the heart.

Each kidney has three regions: an outer region, the **renal cortex**; a region enclosed by the cortex, the **renal medulla**; and an inner chamber, the **renal pelvis** (see Figure 16.4). The renal cortex begins at the outer border of the kidney, and portions of it, called renal columns, extend between the pyramid-shaped subdivisions (renal pyramids) of the renal medulla. The narrow end of each renal pyramid joins a cuplike extension of the renal pelvis. As we will soon see, urine produced by the kidneys eventually drains into the renal pelvis and out the ureter to the urinary bladder.

■ Nephrons are the functional units of the kidneys

Nephrons are the microscopic functional units of the kidneys and are responsible for the formation of urine. Each kidney contains 1 million to 2 million nephrons.

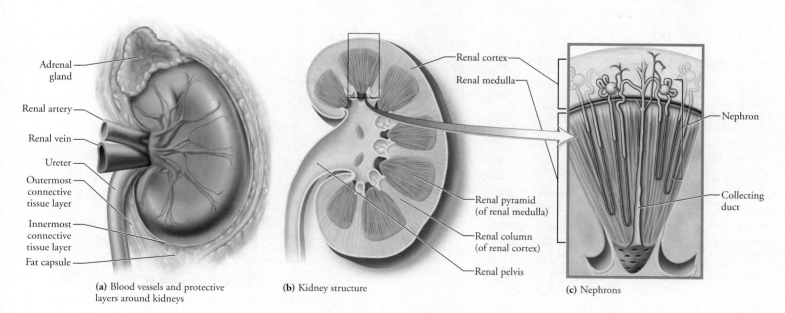

(a) Blood vessels and protective layers around kidneys

(b) Kidney structure

(c) Nephrons

FIGURE **16.4** Structure of the kidney. These views of the left kidney show (a) the blood vessels and three protective connective tissue layers and (b) the basic internal structure of the kidney. (c) The location of nephrons within these regions is also shown.

STRUCTURE OF NEPHRONS

A nephron has two basic parts: the renal corpuscle and the renal tubule, as shown in Figure 16.5. The **renal corpuscle** is the portion of the nephron where fluid is filtered from the blood. It consists of a tuft of capillaries, the **glomerulus**, and a surrounding cuplike structure, the **glomerular capsule** (sometimes called *Bowman's capsule*). The glomerular capsule has two layers, as would the end of the finger of a glove if it were pushed inward. There is a space between the two layers. Blood enters a glomerulus by an afferent (incoming) arteriole. Inside the glomerular capillaries, water and small solutes move from the blood into the glomerular capsule and then into the renal tubule, where they are considered filtrate. The blood then leaves the glomerulus by an efferent (outgoing) arteriole. The **renal tubule** is the site where substances are removed from and added to the filtrate. This tubule has three sections: the **proximal convoluted tubule**, the **loop of the nephron** (sometimes called the *loop of Henle*), and the **distal convoluted tubule**. The loop of the nephron resembles a hairpin turn, having a descending limb and an ascending limb. The distal convoluted tubules of several nephrons empty into a single **collecting duct** that eventually drains into the renal pelvis. The renal pelvis is connected to the ureter. Urine exits the kidney by way of the ureter and moves into the urinary bladder.

About 80% of the nephrons in our kidneys have short loops and are confined almost entirely to the renal cortex. The remaining 20% have long loops that extend from the cortex into the renal medulla. Once in the medulla, the loops of these nephrons turn abruptly upward, back into the cortex, where they lead into distal convoluted tubules. As we will see, the nephrons whose loops extend deep into the medulla play an important role in water conservation.

FUNCTIONS OF NEPHRONS

The kidneys are crucial for maintaining homeostasis. For one thing, they filter wastes and excess materials from the blood. They also assist the respiratory system in the regulation of blood pH. Finally, the kidneys maintain fluid balance by regulating the volume and composition of blood and urine.

To understand what occurs in the kidneys, we must examine the work of nephrons. Nephrons perform three functions: (1) glomerular filtration, (2) tubular reabsorption, and (3) tubular secretion. We can compare these functions to the steps you might take during a selective cleaning of small items from your bedroom closet. First, you might remove almost all of the small items—the valuable along with the unwanted. This activity is analogous to glomerular filtration, which removes from the blood all materials

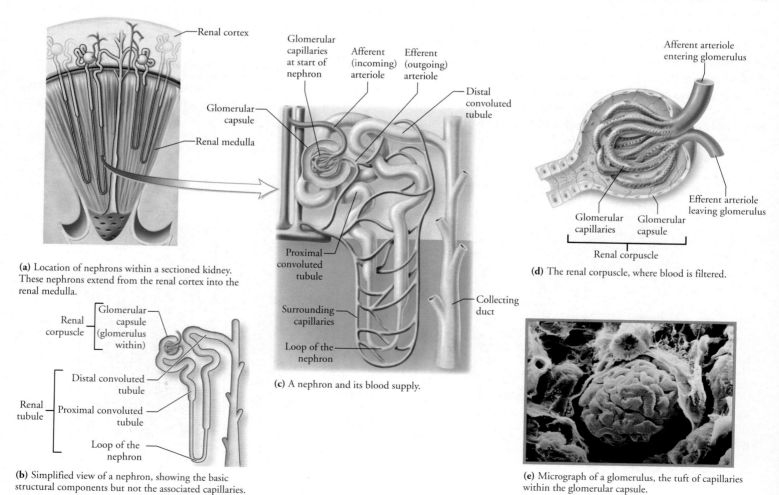

(a) Location of nephrons within a sectioned kidney. These nephrons extend from the renal cortex into the renal medulla.

(b) Simplified view of a nephron, showing the basic structural components but not the associated capillaries.

(c) A nephron and its blood supply.

(d) The renal corpuscle, where blood is filtered.

(e) Micrograph of a glomerulus, the tuft of capillaries within the glomerular capsule.

FIGURE **16.5** Structure of the nephron

small enough to fit through the pores of the kidney's filter (see below). The next step in cleaning your closet might be to look through the various items you have removed and put back "the good stuff," the items worth saving. Returning valuable materials to the closet is analogous to tubular reabsorption, which returns useful materials to the blood. The final step in cleaning the closet might be to once again scan what you have in your closet and to selectively remove items, such as those in excess. This last step is analogous to tubular secretion, in which wastes and excess materials are removed from the blood and added to the filtrate that will eventually leave the body as urine. Here are the three steps in more detail.

1. **Glomerular filtration** occurs as blood pressure forces water and small solutes from the blood in the glomerulus to the inside of the glomerular capsule. The filter consists of two cell layers: the lining of the glomerular capillaries and the inner lining of the glomerular capsule. In between the two cell layers is the basement membrane present in all epithelial tissues (Figure 16.6). Large molecules, such as plasma proteins, and blood cells cannot pass through the pores of the filter. However, water and small solutes, such as ions, nutrients, and metabolic wastes, pass through the filter into the glomerular capsule.

These substances are known collectively as glomerular filtrate. The concentrations of the molecules dissolved in the glomerular filtrate are approximately the same as in the blood plasma.

Several things can change the rate of filtration by the glomerulus. An increase in the diameter of afferent (incoming) arterioles brings more blood into the glomerulus and produces higher pressure in the glomerular capillaries. Higher pressure results in higher filtration rates (more filtrate produced by the kidneys each minute). A reduction in the diameter of efferent (outgoing) arterioles also produces higher pressure in glomerular capillaries and higher filtration rates. General (systemic) increases in blood pressure can also produce higher filtration rates.

2. **Tubular reabsorption** is the process that removes useful materials from the filtrate and returns them to the blood. This process occurs in the renal tubule, primarily in the proximal convoluted tubule. The characteristics of the cells lining the proximal convoluted tubule make it an ideal location for reabsorption (Figure 16.7). These epithelial cells have numerous microvilli (projections of the plasma membrane) that reach into the lumen (passageway) of the tubule. Similar in function to the microvilli in

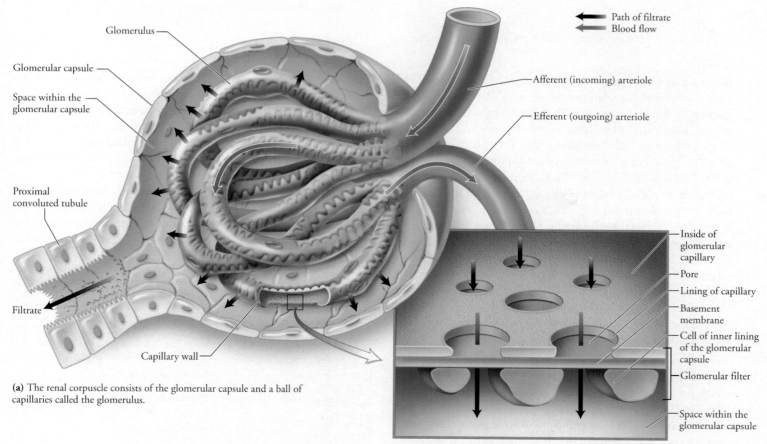

Path of filtrate
Blood flow

Glomerulus

Glomerular capsule

Space within the glomerular capsule

Proximal convoluted tubule

Filtrate

Capillary wall

Afferent (incoming) arteriole

Efferent (outgoing) arteriole

Inside of glomerular capillary

Pore

Lining of capillary

Basement membrane

Cell of inner lining of the glomerular capsule

Glomerular filter

Space within the glomerular capsule

(a) The renal corpuscle consists of the glomerular capsule and a ball of capillaries called the glomerulus.

(b) Cross section through the wall of a glomerular capillary showing how waste and dissolved substances in the blood move first through the pores in the lining of the capillary, then through the basement membrane, and finally through slits in the inner lining of the glomerular capsule into the space within the glomerular capsule.

FIGURE **16.6** The renal corpuscle is the site of glomerular filtration.

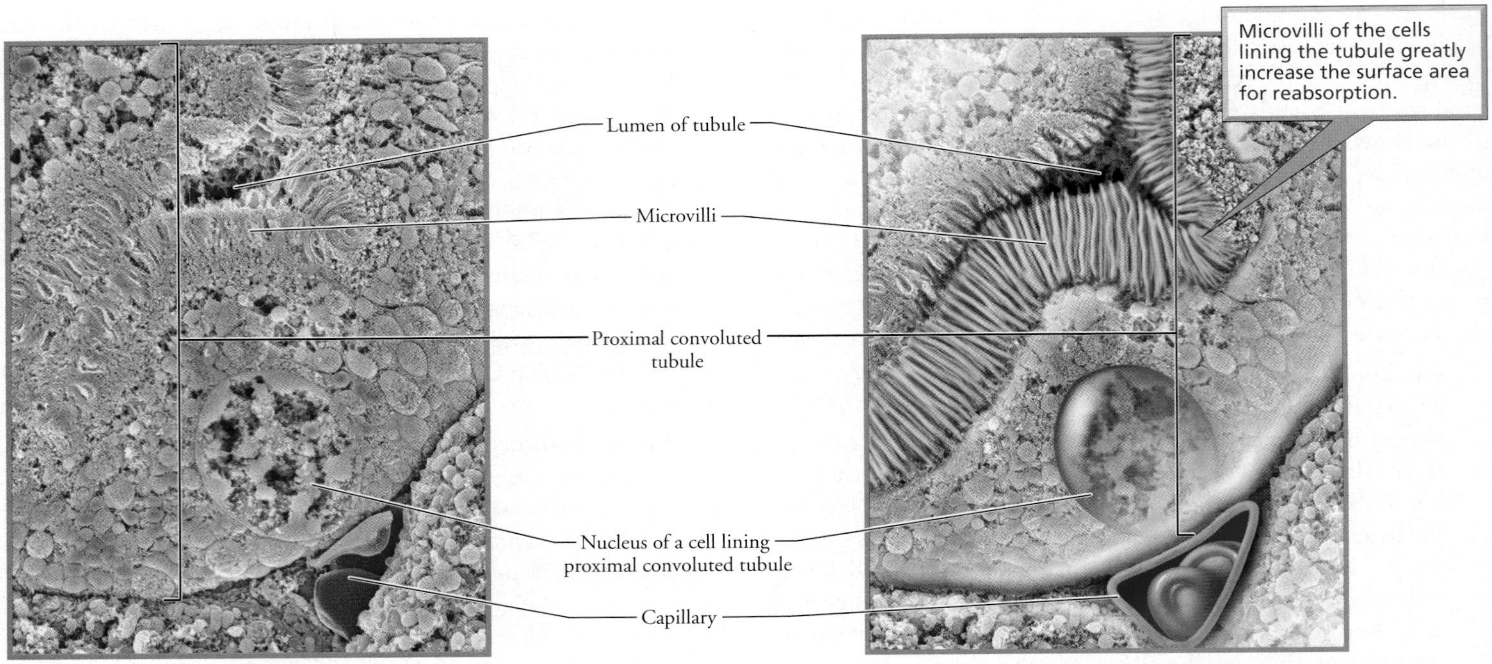

Lumen of tubule

Microvilli

Proximal convoluted tubule

Nucleus of a cell lining proximal convoluted tubule

Capillary

Microvilli of the cells lining the tubule greatly increase the surface area for reabsorption.

(a) Micrograph showing a cross section through a proximal convoluted tubule

(b) Diagram showing a cross section through a proximal convoluted tubule. Substances move from the filtrate through the cells of the proximal convoluted tubule and into the surrounding fluid. Eventually, the substances move into the capillaries nearby.

FIGURE **16.7** The proximal convoluted tubule is the site of tubular reabsorption.

the small intestine, these microvilli dramatically increase the surface area for the reabsorption of materials. Remarkably, as the glomerular filtrate passes through the renal tubule, about 99% of it is returned to the blood in the surrounding capillaries. Thus, only about 1% of the glomerular filtrate is eventually excreted as urine. Put another way, of the approximately 180 liters (48 gal) of filtrate that enter the glomerular capsule each day, between 178 and 179 liters (47 gal) are returned to the blood by reabsorption. The remaining 1 to 2 liters (approximately 1 gal) are excreted as urine. Reabsorption returns water, essential ions, and nutrients to the blood. Imagine how much water and food we would have to consume if we did not have reabsorption to offset the losses from glomerular filtration! Some wastes are not reabsorbed at all; others, such as urea, are partially reabsorbed. We will see that **antidiuretic hormone (ADH)**, released by the posterior pituitary gland, regulates the amount of water reabsorbed in parts of the renal tubule.

3. **Tubular secretion** removes additional wastes and excess ions from the blood. It also removes drugs such as penicillin. These wastes are added to the filtered fluid that will become urine. For example, hydrogen ions (H^+), potassium ions (K^+), and ammonium ions (NH_4^+) in the blood are actively transported into the renal tubule, where they become part of the filtrate to be excreted. Tubular secretion occurs along the proximal and distal convoluted tubules and collecting duct.

The regions of the nephron and their roles in filtration, reabsorption, and secretion are shown in Figure 16.8 and reviewed in

Table 16.1. When reviewing these, keep in mind the directions that substances move in each of the three processes: (1) during glomerular filtration, substances move from the blood into the nephron to form filtrate; (2) during tubular reabsorption, useful substances move from the filtrate within the nephron back into the blood; (3) during tubular secretion, drugs and substances in excess move from the blood into the filtrate.

By the end of glomerular filtration, tubular reabsorption, and tubular secretion, blood leaving the kidneys contains most of the water, nutrients, and essential ions that it contained upon entering the kidneys. Wastes and excess materials have been removed, leaving

TABLE 16.1 REVIEW OF NEPHRON REGIONS AND THEIR ROLES	
REGION OF NEPHRON	**ROLE**
Renal corpuscle (glomerular capsule and glomerulus)	Filters the blood, removing water, glucose, amino acids, ions, nitrogen-containing wastes, and other small molecules
Proximal convoluted tubule	Reabsorbs water, glucose, amino acids, some urea, Na^+, Cl^-, and HCO_3^- Secretes drugs, H^+, NH_4^+
Loop of the nephron	Reabsorbs water, Na^+, Cl^-, and K^+
Distal convoluted tubule	Reabsorbs water, Na^+, Cl^-, and HCO_3^- Secretes drugs, H^+, K^+, NH_4^+

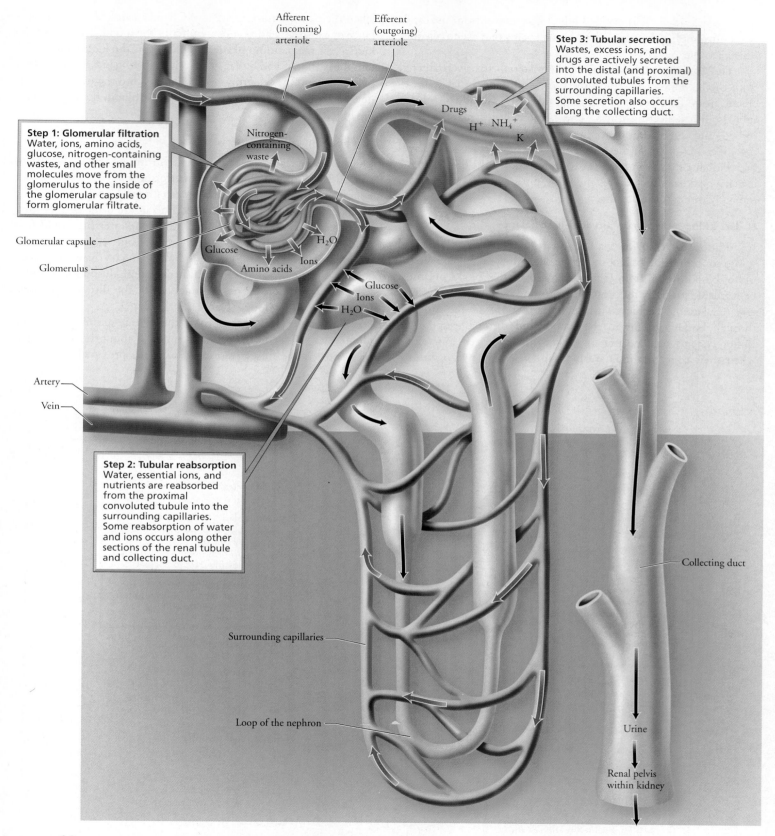

Afferent (incoming) arteriole

Efferent (outgoing) arteriole

Step 3: Tubular secretion
Wastes, excess ions, and drugs are actively secreted into the distal (and proximal) convoluted tubules from the surrounding capillaries. Some secretion also occurs along the collecting duct.

Drugs

H^+ NH_4^+

K

Step 1: Glomerular filtration
Water, ions, amino acids, glucose, nitrogen-containing wastes, and other small molecules move from the glomerulus to the inside of the glomerular capsule to form glomerular filtrate.

Nitrogen-containing waste

Glomerular capsule

Glucose

H_2O

Amino acids

Ions

Glomerulus

Glucose
Ions
H_2O

Artery

Vein

Step 2: Tubular reabsorption
Water, essential ions, and nutrients are reabsorbed from the proximal convoluted tubule into the surrounding capillaries. Some reabsorption of water and ions occurs along other sections of the renal tubule and collecting duct.

Collecting duct

Surrounding capillaries

Loop of the nephron

Urine

Renal pelvis within kidney

FIGURE **16.8** Summary of glomerular filtration, tubular reabsorption, and tubular secretion along the nephron

the blood cleansed. This purified blood now moves from the capillaries surrounding the nephron into small veins that eventually join the renal vein.

The urine now contains all of the materials that were filtered from the blood and not reabsorbed, plus substances that were secreted from the blood. It empties from the distal convoluted tubules into collecting ducts. There, more water may be reabsorbed, and some additional removal of wastes also occurs. From the collecting ducts, urine moves into the renal pelvis and leaves each kidney through a ureter. It then travels down the ureters to the urinary bladder, which stores the urine until it is eliminated from the body through the urethra.

■ The kidneys help maintain acid-base balance

In addition to removing wastes and regulating the volume and solute concentration of blood plasma, the kidneys help regulate the pH of blood. Recall from Chapters 2 and 14 that blood pH must be regulated precisely for proper functioning of the body. This precise regulation is achieved through the actions of the kidneys, through buffer systems in the blood, and through respiration. Buffer systems regulate pH by picking up hydrogen ions (H^+) when their concentrations are high and releasing hydrogen ions

when their concentrations are low. Chapter 2 described the importance of carbonic acid as such a buffer in the blood. The role of the kidneys in maintaining pH is twofold. First, the kidneys help sustain the carbonic acid buffer system by returning bicarbonate to the blood. Second, by secreting hydrogen ions into the urine, the kidneys remove excess hydrogen ions from the blood, thereby increasing blood pH.

■ The kidneys help conserve water

Our kidneys enable us to conserve water through the production of concentrated urine. This task is performed by the 20% of our nephrons with long loops that dip deep into the renal medulla. The special ability of these nephrons to concentrate urine derives from an increasing concentration of solutes in the interstitial fluid (the fluid that fills the spaces between cells) from the cortex to the medulla of the kidneys. The most important of these solutes are sodium chloride (NaCl) and urea. Let's explore this mechanism for urine concentration by tracing the path of the filtrate as it flows through these long loops (Figure 16.9).

The solute concentration of the filtrate passing from the glomerular capsule to the proximal tubule is about the same as that of blood. As the filtrate moves through the proximal convoluted

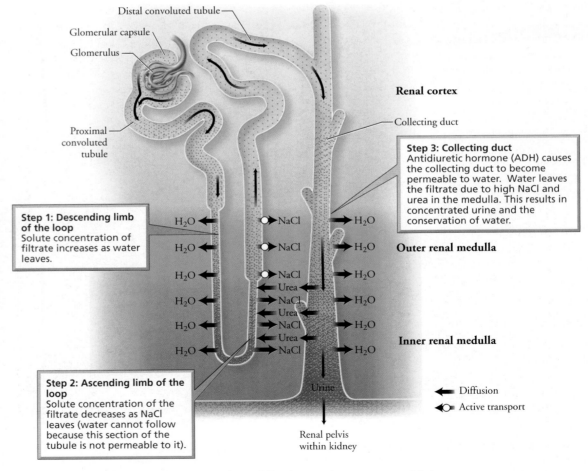

Step 1: Descending limb of the loop
Solute concentration of filtrate increases as water leaves.

Step 2: Ascending limb of the loop
Solute concentration of the filtrate decreases as NaCl leaves (water cannot follow because this section of the tubule is not permeable to it).

Step 3: Collecting duct
Antidiuretic hormone (ADH) causes the collecting duct to become permeable to water. Water leaves the filtrate due to high NaCl and urea in the medulla. This results in concentrated urine and the conservation of water.

Distal convoluted tubule
Glomerular capsule
Glomerulus
Renal cortex
Collecting duct
Proximal convoluted tubule
Outer renal medulla
Inner renal medulla
Urine
Renal pelvis within kidney

← Diffusion
← Active transport

FIGURE **16.9** Some nephrons have loops that extend deep into the medulla. These nephrons are responsible for water conservation. The steps by which these nephrons concentrate urine and conserve water are shown here. Stippling indicates solute concentration.

tubule, large amounts of water *and* salt are reabsorbed. This reabsorption produces dramatic reductions in the volume of filtrate but little change in its solute concentration (because salt is reabsorbed, too). However, when the filtrate enters the descending limb of the loop, major changes in solute concentration begin (Figure 16.9). This path takes it from the cortex to the medulla. Along the descending limb, water leaves the filtrate by osmosis (a special type of diffusion described in Chapter 3). The departure of water creates an increase in the concentration of solutes, including sodium chloride, within the filtrate. The concentration of salt in the filtrate peaks at the curve of the loop, setting the stage for the next step in the process of urine concentration. Now the filtrate moves up the ascending limb of the loop. As it does so, large amounts of sodium chloride are actively transported out of the filtrate into the interstitial fluid of the medulla. Water, however, remains in the filtrate because the ascending limb is not permeable to water. When the filtrate reaches the distal convoluted tubule in the cortex, it is therefore quite dilute. In fact, the filtrate is hypotonic to body fluids because it has a lower solute concentration than body fluids (see Chapter 3). The filtrate then moves into a collecting duct and begins to descend once again toward the medulla. This pathway is one of increasing salt concentration in the interstitial fluid because of all the salt that was transported out of the filtrate as it ascended the loop. Collecting ducts are permeable to water but not to salt (ADH increases the permeability of the collecting duct to water; see below). Thus, as the filtrate encounters increasing concentrations of salt in the fluid of the inner medulla, water leaves the filtrate by osmosis. With this departure of large amounts of water, urea is now concentrated in the filtrate. In the lower regions of the collecting duct, some of the urea moves into the interstitial fluid of the medulla. This leakage of urea contributes to the high solute concentration of the inner medulla and thus aids in concentrating the filtrate. The remaining urea is excreted.

At its most concentrated, urine is hypertonic to blood and interstitial fluid from any other part of the body except the inner medulla, where it is isotonic. (Recall from Chapter 3 that *hypertonic* means having a greater solute concentration than another fluid. *Isotonic* means having the same solute concentration as another fluid.) Together, the loop of the nephron and collecting duct maintain extraordinarily high solute concentrations in the interstitial fluid of the kidneys, making possible the concentration of urine and conservation of water by the kidneys.

■ Hormones influence kidney function

Our health depends on our keeping the salt and water levels in our body near certain optimum values. This, as we have seen, is an important job of the kidneys. It is also a challenging job, because our activities produce constant fluctuations in those levels. For example, on a hot day or after exercise, we may lose body water and salts through perspiration. In contrast, eating a tub of salted popcorn at the movies can boost our salt intake. The kidneys must deal with these challenges and adjust the concentration of solutes in the urine to keep water and salt levels in our body relatively constant.

Three hormones—ADH, aldosterone, and atrial natriuretic peptide—play important roles in adjusting kidney function to meet the body's needs. ADH is manufactured by the hypothalamus and then travels to the posterior pituitary for storage and release (see Chapter 10). This hormone regulates the amount of water reabsorbed by the collecting ducts. The hypothalamus responds to changes in the concentration of water in the blood by increasing or decreasing secretion of ADH. Decreases in the concentration of water in the blood stimulate increased secretion of ADH, as shown in Figure 16.10. Higher levels of ADH in the bloodstream then increase the permeability to water of the collecting ducts of nephrons, resulting in more water being reabsorbed from the filtrate. The movement of increased amounts of water from the filtrate back into the blood results in increased blood volume and pressure and production of small amounts of concentrated urine. Just the opposite occurs when the concentration of water in the blood increases. In this case, the release of ADH is inhibited, thereby reducing water reabsorption from the filtrate. Reduced water reabsorption causes reduced blood volume and pressure and the production of large amounts of dilute urine. Alcohol, too, inhibits the secretion of ADH, causing reduced water reabsorption by the kidneys and production of large amounts of dilute urine. Thus, it makes little sense to try to quench your thirst and restore body fluids by drinking an alcoholic beverage on a hot day. Substances such as alcohol that promote urine production are called **diuretics**. *Diabetes insipidus,* a disease characterized by excretion of large amounts of dilute urine, is caused by a deficiency of ADH.

Aldosterone is released by the adrenal cortex (see Chapter 10). It increases reabsorption of sodium by the distal convoluted tubules and collecting ducts. This process is important because water follows sodium. As more sodium is transported out of the nephron into the capillaries, increased amounts of water go with it, resulting in increased blood volume and pressure and the production of small

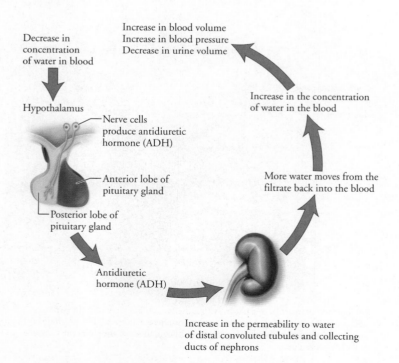

FIGURE **16.10** The pathway by which ADH influences kidney function and blood volume and blood pressure

amounts of concentrated urine. Caffeine, a familiar chemical found in coffee, tea, and many soft drinks, has just the opposite effect. Caffeine decreases the blood's reabsorption of sodium, thus decreasing the amount of water moving out of tubules back into the bloodstream. Ingestion of caffeine causes decreased blood volume and pressure and large amounts of dilute urine. Because caffeine, like alcohol, increases urine output, it too is considered a diuretic.

What stimulates the release of aldosterone? Ultimately, it is the blood pressure in the afferent (incoming) arteriole carrying blood to the glomerulus. The blood pressure in this arteriole is monitored by a part of the nephron called the **juxtaglomerular apparatus**, a group of cells located where the distal convoluted tubule contacts the afferent arteriole (Figure 16.11). Ultimately, this group of cells, through the renin-angiotensin system (see below), influences the secretion of aldosterone. When the blood pressure in the afferent arteriole drops, so does the glomerular filtration rate. In turn, this drop in filtration rate reduces the volume of filtrate within nephrons. Cells within the juxtaglomerular apparatus respond by releasing the enzyme renin. **Renin** converts angiotensinogen, a protein produced by the liver, into another protein, *angiotensin I*. Angiotensin I is then converted by yet another enzyme into angiotensin II. *Angiotensin II* is the active form of the protein that stimulates the adrenal gland to release aldosterone. Aldosterone increases reabsorption of sodium and water by the distal convoluted tubules and collecting ducts of nephrons, resulting in increased blood volume and pressure. These changes increase the filtration rate within the glomerulus, resulting in an increased volume of filtrate within the nephron.

A final hormone that influences kidney function is **atrial natriuretic peptide (ANP)**, released by cells in the right atrium of the heart. The cells release ANP in response to stretching of the heart caused by increased blood volume and pressure. Atrial natriuretic peptide decreases water and Na^+ reabsorption by the kidneys, causing declines in blood volume and pressure and the production of large amounts of dilute urine. ANP also influences kidney function by inhibiting secretion of aldosterone and renin. Table 16.2 reviews the effects of ADH, aldosterone, and ANP on kidney function.

stop and think

Estrogens are female sex hormones that are chemically similar to aldosterone. As a result, the effects of estrogens on the distal convoluted tubules and collecting ducts of the kidneys are similar to those of aldosterone. How might this similarity explain the water retention experienced by many women as their estrogen levels rise during the menstrual cycle?

■ The kidneys help produce red blood cells and activate vitamin D

The kidneys have two additional functions that are important to homeostasis but are not directly related to the urinary system. First, the kidneys release **erythropoietin,** a hormone that travels to the red bone marrow, where it stimulates production of red blood cells. Second, the kidneys have an effect on vitamin D. Vitamin D is a substance provided by certain foods in our diet or produced by the skin in response to sunlight. The kidneys transform vitamin D into its active form, calcitriol, which promotes the absorption and use of calcium and phosphorus by the body. Table 16.3 reviews the components of the urinary system and their functions.

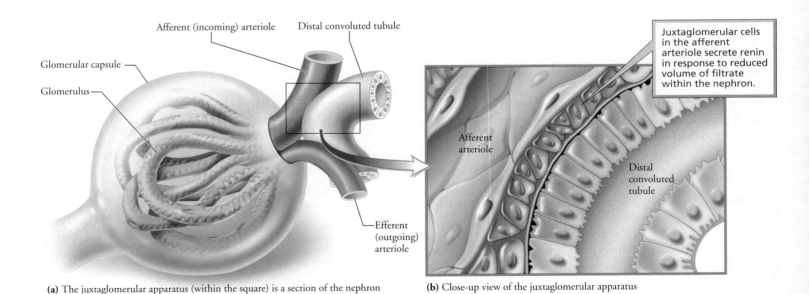

(a) The juxtaglomerular apparatus (within the square) is a section of the nephron where the distal convoluted tubule contacts the afferent arteriole. The nearby renal corpuscle is shown in ghosted view to reveal its components, the glomerular capsule and the glomerulus.

(b) Close-up view of the juxtaglomerular apparatus

FIGURE **16.11** The juxtaglomerular apparatus

TABLE 16.2 REVIEW OF SOME HORMONES THAT INFLUENCE KIDNEY FUNCTION

HORMONE	EFFECT ON WATER AND SOLUTE REABSORPTION IN NEPHRON	EFFECT ON BLOOD VOLUME AND PRESSURE	URINE PRODUCED
Antidiuretic hormone (ADH)	Increases permeability to water of collecting ducts, resulting in more water moving from filtrate to blood	Increases	Concentrated
Aldosterone	Increases reabsorption of Na^+ by distal convoluted tubules and collecting ducts, resulting in more water following Na^+ as it moves from filtrate to blood	Increases	Concentrated
Atrial natriuretic peptide (ANP)	Decreases reabsorption of Na^+ by distal convoluted tubules and collecting ducts, resulting in more Na^+ and water remaining in filtrate	Decreases	Dilute

TABLE 16.3 REVIEW OF URINARY SYSTEM COMPONENTS

COMPONENT	FUNCTION
Kidneys	Filter wastes and excess material from the blood
	Help regulate blood pressure and pH
	Maintain fluid balance by regulating the volume and composition of blood and urine
	Release erythropoietin, which stimulates production of red blood cells
	Transform vitamin D into its active form
Ureters	Transport urine from kidneys to urinary bladder
Urinary bladder	Stores urine
	Contracts and expels urine into urethra
Urethra	Transports urine from urinary bladder to outside the body
	In males, transports semen to outside the body

Dialysis and Transplant Surgery Help When Kidneys Fail

Renal failure is a decrease or complete cessation of glomerular filtration. In other words, the kidneys stop working. The failure can be acute or chronic. *Acute renal failure* is an abrupt, complete or nearly complete, cessation of kidney function. It typically develops over a few hours or days and is characterized by little output of urine. Causes of acute renal failure include nephrons damaged by severe inflammation, certain drugs, and poison. Acute renal failure also may be caused by low blood volume due to profuse bleeding or obstruction of urine flow by kidney

stones (see the Health Issue essay, *Kidney Stones and Their Shocking Treatment*). *Chronic renal failure*, on the other hand, is a progressive and often irreversible decline in the rate of glomerular filtration over a period of months or years. Kidney disease may destroy nephrons and cause a progressive decline of this type. Nephrons lost to kidney disease cannot be replaced. Polycystic disease, for example, is an inherited and progressive condition in which fluid-filled cysts and tiny holes form throughout kidney tissue. Few symptoms are apparent at the beginning of chronic renal failure because the remaining nephrons enlarge and take over for those that have been destroyed. With time, however, the loss of nephrons becomes so severe that symptoms of decreased glomerular filtration rate appear. For example, increased levels of nitrogen-containing wastes are found in the blood. By end-stage renal failure, about 90% of the nephrons have been lost, making necessary a kidney transplant or the use of an artificial kidney machine.

Renal failure, whether acute or chronic, has many consequences. These include (1) acidosis, a decrease in blood pH caused by the inability of the kidneys to excrete hydrogen ions; (2) anemia, low numbers of red blood cells caused by the failure of damaged kidneys to release erythropoietin; (3) edema, the buildup of fluid in the tissues because of water and salt retention; (4) hypertension, an increase in blood pressure caused by failure of the renin-angiotensin system and salt and water retention; and (5) accumulation of nitrogen-containing wastes in the blood. In short, failure of the kidneys severely disrupts homeostasis. Untreated kidney failure will lead to death within a few days.

■ Dialysis cleanses the blood

A common way of coping with failure or severe impairment of the kidneys is **hemodialysis**, the use of artificial devices to cleanse the blood. Often this is done with an artificial kidney machine (Figure 16.12). A tube is inserted into an artery of the patient's arm. Blood

HEALTH ISSUE

Kidney Stones and Their Shocking Treatment

Kidney stones are small, hard crystals. They form when substances such as calcium or uric acid precipitate out of urine because of higher-than-normal concentrations. In the majority of cases, dehydration is the reason for these higher-than-normal concentrations of stone-forming substances. Most people's urine also contains substances that prevent the formation of crystals, but these substances may not function in people with a history of kidney stones.

Kidney stones often begin as extremely tiny grains in the renal pelvis. They grow larger over a period of years as additional material is deposited on them (Figure 16.A). Some stones may attain diameters of 25 mm (1 in.). Before reaching anything near that size, however, many stones are flushed out of the kidneys and down the ureters to the bladder, and from there they are expelled

with urine out the urethra. These tiny stones rarely cause problems, and people may not even be aware of their passing. However, somewhat larger stones may cause considerable pain during their journey through the urinary tract. The sharp edges of large stones may gouge into the walls of the ureters and urethra. The pain comes in waves, and each wave may cause the person to double over. Nausea, vomiting, and chills accompany the pain, and blood may appear in the urine. Even more serious problems arise when large stones become lodged in the kidneys or ureters, blocking the flow of urine. The high internal pressure that results from the obstruction can damage the nephrons and impair kidney function.

Kidney stones usually are diagnosed with x-rays of the kidneys, ureters, and bladder. Blood and urine tests also are performed to determine what might be causing the problem. Today, a

large stone in a kidney may be surgically removed through an incision in the back that extends into the kidney. To locate a stone lodged in a ureter, doctors may pass a small fiberoptic instrument through the urethra and bladder and then use another instrument to remove the stones without making an incision. Certain stones can be dissolved with drugs. Another option is extracorporeal shock wave lithotripsy, a relatively painless technique in which high-energy shock waves are created outside the body and then directed at the stones within. During the procedure, the patient lies in a water bath or on a soft cushion. The shock waves pulverize the stones into tiny pieces that can be passed painlessly in the urine. This technique also does not require an incision and recovery is rapid. Nevertheless, recent data suggest that lithotripsy may increase the risk of developing diabetes and hypertension later in life.

flows through the tube and into the kidney machine, where it is filtered and then returned to the patient's body. Within the machine, the blood flows through tubing made of a selectively permeable membrane. The tubing is immersed in a dialysis solution called the dialysate. The selectively permeable membrane permits wastes and excess small molecules to move from the blood into the dialysate.

At the same time, the membrane prevents passage into the dialysate of blood cells and most proteins because they are too large. Nutrients are sometimes provided in the dialysate for absorption into the blood. The composition of the dialysate is precisely controlled to maintain proper concentration gradients between the dialysate and the blood. Once the blood has completed the circuit through the

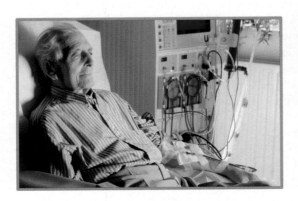

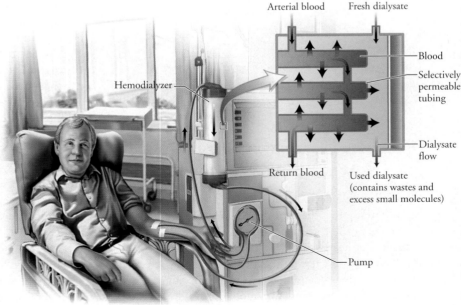

FIGURE **16.12** An artificial kidney machine can be used to cleanse the blood when the kidneys fail.

Because the pancreas is in the "blast path" of the shock waves, it may be damaged during treatment for kidney stones, resulting in diabetes. Hypertension may be caused by scarring of the kidneys from the shock waves; it is possible that such scarring influences production of renin, the protein produced by the kidneys that influences blood pressure.

A simple way to prevent the formation of kidney stones is to drink large quantities of water, about 2.4 to 2.8 liters (5 to 6 pt) a day. Drinking this amount of liquid helps to dilute the urine and flushes any existing stones from the urinary tract. For people who are prone to developing certain types of kidney stone, dietary changes may be recommended. Finally, people with a history of developing kidney stones usually can take specific medications to prevent formation of new stones once the old ones have been removed. 👫

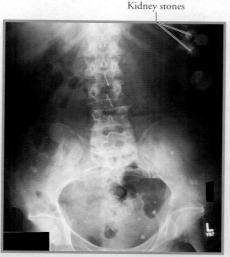

Kidney stones

(a) An x-ray showing several kidney stones in the left kidney

(b) Although some kidney stones may be 25 mm (1 in.) across, most are much smaller. Here is an example of a large kidney stone that was removed through surgery.

FIGURE **16.A** Kidney stones

---

tubing, it is returned, free of wastes, to a vein in the arm. Patients requiring hemodialysis typically undergo the procedure about three times a week. Each session lasts several hours.

Another option for removing wastes from the blood is continuous ambulatory peritoneal dialysis (CAPD). In this procedure, the peritoneum, one of the body's own selectively permeable membranes, is used as the dialyzing membrane (Figure 16.13). The peritoneum lines the abdominal cavity and covers the internal organs.

Dialyzing fluid held in a plastic container suspended over the patient flows down a tube inserted into the abdomen. As the fluid bathes the peritoneum, wastes move from the blood vessels that line the abdomen across the peritoneum and into the solution. The solution is then returned to the plastic container and discarded. Typically, fluid is passed into the abdomen, left for a few hours, and then removed and replaced with new fluid. CAPD requires about three or four fluid changes each day. However, the patient is

Dialysate bag

Fresh dialyzing fluid

Fresh dialyzing fluid flows into the abdominal cavity, where wastes in nearby blood vessels move across the peritoneum and into the dialyzing fluid.

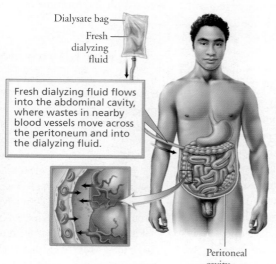

Peritoneal cavity

Dialyzing fluid containing wastes is drained from the abdominal cavity and discarded.

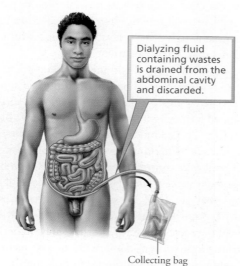

Collecting bag

FIGURE **16.13** CAPD uses the peritoneum (the membrane that lines the abdominal cavity and covers the organs within) as the dialyzing membrane.

free to move around between fluid changes while the dialysis proceeds internally.

CAPD is gentler in its waste removal and less costly than hemodialysis. However, CAPD requires several daily changes of dialysis fluid, and each change requires hooking up a new container of dialysate. Thus, CAPD provides more opportunities for bacteria to move down the tube and into the abdomen, where they may cause peritonitis (inflammation of the peritoneum). Even the most meticulous patients suffer at least one episode of peritonitis a year, and every episode requires hospitalization and treatment with an antibiotic. Eventually, the peritoneum may become so scarred from bouts of peritonitis that CAPD is no longer possible.

■ Renal function can be restored with a kidney transplant

The ultimate hope for many people whose kidneys fail is to receive a healthy kidney from another person. The kidney was one of the first organs to be successfully transplanted. The availability of dialysis was a crucial part of this success, because dialysis keeps people alive until a suitable donor organ can be found. Progress in liver and heart transplantation has been slower because of the difficulty of providing temporary mechanical replacements for those organs.

What makes a suitable donor organ? The main obstacle to successful acceptance of a transplanted organ is the rejection of foreign tissue by the patient's own immune system (see Chapter 13). The most suitable kidney would come from a patient's identical twin. It is not surprising, then, that the first successful transplant occurred in 1955 between identical twins. A kidney from a close relative, such as a parent or sibling, would be the next best choice. Removal of a healthy kidney is a safe operation, and the donor who gives a kidney away can get along fine with the remaining kidney. About 90% of kidneys donated by close relatives are still functioning 2 years after transplantation. However, most donated kidneys come from individuals who are unrelated to the patient but who agreed before their death (usually a sudden death in an accident) to donate their organs. About 75% of kidneys donated from unrelated individuals, matched as closely as possible to the patient's tissue and blood type, are still functioning 2 years after transplantation. Even if a transplant fails after a few years, successful second and third transplants may extend a patient's life.

In most situations, the kidneys of the patient needing a transplant are not removed. The donor kidney is simply transplanted into a protected area within the pelvis (Figure 16.14). The ureter of the donated kidney is attached to the recipient's bladder, and blood vessels of the donated kidney are attached to the recipient's vessels. The recipient's own kidneys would be removed if they were infected or if they were causing problems such as high blood pressure.

The high success rate of kidney transplants is linked to the use of drugs that suppress the recipient's immune system. Typically, transplant recipients must take two or more of these medications for the rest of their lives. Recently, a team of researchers developed a way to perform kidney transplants without the recipients' needing to permanently remain on immune-suppressing drugs. The suppressing drugs were given to recipients after transplant surgery. Next, the recipients received several radiation treatments, which were directed at reducing the number of cells in their immune system capable of

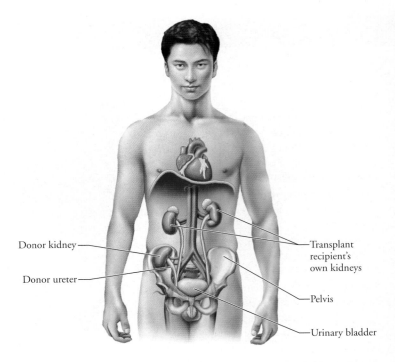

FIGURE **16.14** In most transplant situations, the donor organ is located in a safe region within the pelvis, and the recipient's own kidneys are left in place. The ureter from the transplanted kidney is attached to the recipient's bladder. Blood vessels from the transplanted kidney are attached to the recipient's vessels.

attacking the new organ. Finally, the recipients received blood stem cells from the person who donated the transplanted kidney. These new stem cells moved to the recipient's bone marrow and generated new blood cells and immune cells. In essence, each recipient now had a hybrid immune system, a mix of his or her own cells and those of the donor. The hybrid immune systems did not recognize the transplanted kidneys as foreign, and recipients were gradually weaned off the immune-suppressing medications.

what would you do?

The constant shortage of donor organs has led some people to suggest a change in the laws regarding organ donation. At present, a person must give permission to use his or her organs after death. If a potential donor is incapable of giving such permission, then medical personnel may consult close relatives. In recent years, people concerned about organ shortages have suggested changing the laws so that a person would have to provide written proof that they did *not* want their organs made available for transplantation after death. In other words, without anything in writing, the law would presume that the person had no objection to organ donation. Many healthy people probably have no objections to organ donation but simply do not think about death and the future use of their organs. On the other hand, some might worry that doctors and nurses may let certain people die because other people needed their organs. If you were given the responsibility of deciding whether to change the laws regarding organ donation, what would you do?

Despite major advances in kidney transplantation, there is still room for improvement. For example, donor kidneys can be kept alive and healthy for only about 24 to 48 hours (Figure 16.15). This short window of opportunity necessitates rapid location of a recipient, shipment of the donor kidney, and transplantation. Perhaps most important, kidneys available for transplantation are always in short supply. Many patients wait for months or years for a new kidney.

Urination Has Involuntary and Voluntary Components

Urination is the process by which the urinary bladder is emptied. This process includes both involuntary and voluntary actions and is summarized in Figure 16.16. The kidneys produce urine around the clock. The urine trickles into the ureters and down to the urinary bladder where it is temporarily stored. When at least 200 ml (0.42 pt) of urine have accumulated, stretch receptors in the wall of the bladder send impulses along sensory neurons to the lower part of the spinal cord. From there, impulses are sent along motor neurons back to the bladder, where they

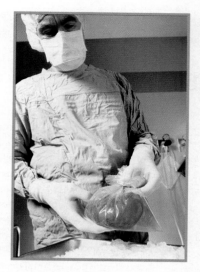

FIGURE **16.15** Before transplantation, donor kidneys must be kept in a cool salt solution under sterile conditions. Even under such conditions, kidneys will deteriorate after about 1 or 2 days.

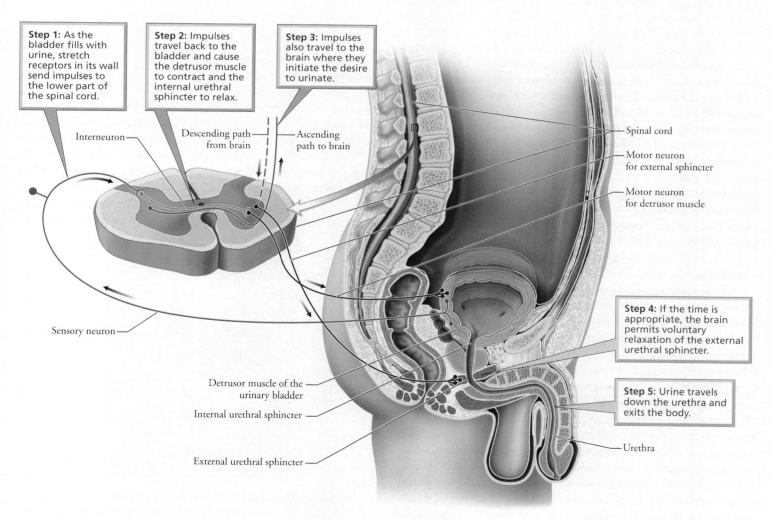

Step 1: As the bladder fills with urine, stretch receptors in its wall send impulses to the lower part of the spinal cord.

Step 2: Impulses travel back to the bladder and cause the detrusor muscle to contract and the internal urethral sphincter to relax.

Step 3: Impulses also travel to the brain where they initiate the desire to urinate.

Step 4: If the time is appropriate, the brain permits voluntary relaxation of the external urethral sphincter.

Step 5: Urine travels down the urethra and exits the body.

Interneuron

Descending path from brain

Ascending path to brain

Spinal cord

Motor neuron for external sphincter

Motor neuron for detrusor muscle

Sensory neuron

Detrusor muscle of the urinary bladder

Internal urethral sphincter

External urethral sphincter

Urethra

FIGURE **16.16** The steps involved in urination

Urinalysis: What Your Urine Says about You

The kidneys are our body's filtering system, so the urine they produce contains substances that originate from almost all of our different organs. A detailed analysis of the urine, therefore, tells us not only how the organs of the urinary tract are functioning but also about the general health of our other organ systems. In other words, a complete urinalysis including measures of volume, microorganisms, and the physical and chemical properties of urine provides an overview of a person's basic health.

Healthy urine exiting the body contains no microorganisms. The presence of bacteria in a properly collected urine sample usually signals infection of the urinary system. Bacteria found in a sample may be cultured to determine their identity, which can help in diagnosing the infection. Urine also may be screened for fungi or protozoans that cause inflammation within the urinary tract.

Sometimes a urine sample is contaminated because of improper collection. For example, if the perineal area has not been cleaned and the collection cup or urine is allowed to contact that area, bacteria may be introduced. (In males, this is the area between the anus and scrotum; in females, it is between the anus and vulva.) It is therefore important to be careful when collecting a urine sample.

The following physical characteristics are routinely checked in urine analyses: color, turbidity, pH, and specific gravity. The color of urine comes from urochrome, a yellow pigment produced as a waste product by the liver during the breakdown of hemoglobin in red blood cells. The urochrome travels in the bloodstream from the liver to the kidneys, where it is filtered from the blood and excreted with urine. The color of urine varies somewhat according to diet. Beets and blackberries, for example, lend urine a red color, and asparagus causes a green tinge. The color of urine also varies with concentration. More concentrated urine, such as that collected first thing in the morning, is darker than more dilute urine. An abnormal color of urine—particularly red, when followed by microscopic confirmation of the presence of red blood cells—can indicate kidney

---

cause a smooth muscle in the wall of the bladder, called the *detrusor muscle*, to contract. The impulses also cause the *internal urethral sphincter*, a thickening of smooth muscle located at the junction of the bladder and urethra, to relax. These combined actions push stored urine into the urethra. Upon arrival of sensory impulses in the lower spinal cord, information also travels up to the brain and initiates a desire to urinate. If the circumstances are appropriate, the brain permits voluntary relaxation of the *external urethral sphincter*, a small band of skeletal muscle farther down the urethra. When the external sphincter relaxes, urine exits the body. If the circumstances are not appropriate for urination, the brain does not permit the external sphincter to relax, and the bladder continues to store the urine until a better time. Urine will exit the body later, when the person consciously allows the external sphincter to relax.

Thus, although urination is a reflex—a relatively rapid response to a stimulus that is mediated by the nervous system—it can nevertheless be started and stopped voluntarily because of the conscious control exerted by the brain over the external urethral sphincter. However, not everyone can control his or her external urethral sphincter. Lack of voluntary control over urination is called *urinary incontinence*. Incontinence is the norm for infants and children younger than 2 or 3 years of age, because nervous connections to the external urethral sphincter are incompletely developed. As a result, infants and young children void whenever their bladder fills with enough urine to activate its stretch receptors. Toilet training occurs when toddlers learn to bring urination under conscious control (Figure 16.17). This step is made possible by the development of complete neural connections to the external sphincter.

Damage to the external sphincter may cause incontinence in adults. In men, such damage may occur during surgery on the prostate gland. The prostate gland surrounds the male urethra just below the urinary bladder and contributes substances to semen (see Chapter 17). Incontinence also may occur because of bladder muscles that contract before the bladder is full. This condition is often described as an overactive or spastic bladder. Infection of the urinary system (see below) can also cause incontinence, in any age group.

Mild incontinence, especially the kind called stress incontinence, is common in adults. *Stress incontinence* is characterized by the escape of small amounts of urine when sudden increases in abdominal pressure force urine past the external sphincter. Laughing, sneezing, or coughing may cause these sudden increases in pressure.

FIGURE **16.17**   Conscious control over urination is usually acquired by the age of 3, when neural connections to the external urethral sphincter are fully developed.

stones and trauma to urinary organs. Because menstrual blood can contaminate urine samples, women should always inform their doctor if they are menstruating at the time of urine collection. Freshly voided urine is usually transparent. Cloudy, or turbid, urine may indicate a urinary tract infection, particularly if white blood cells are detected. Healthy urine has a pH of about 6, although considerable variation may occur in response to diet. High-protein diets produce acidic urine, and vegetarian diets produce alkaline urine. An alkaline pH also is associated with some bacterial infections. The specific gravity, or density, of urine (the ratio of its weight to the weight of an equal volume of distilled water), is a measure of the concentration of solutes in urine and therefore of the concentrating ability of the nephrons within our kidneys. When urine becomes highly concentrated for

long periods, substances such as calcium and uric acid may precipitate out and form kidney stones.

Quantitative chemical tests are run to assess the levels of specific chemical constituents of urine. The major constituent is water, making up about 95% of the total volume. The remaining 5% consists of solutes from outside sources (such as drugs) or from the metabolic activities of our cells. Tests run on urine specimens look for certain abnormal constituents of urine (such as glucose, red blood cells, or white blood cells) and for normal constituents present in abnormal amounts (such as increases or decreases in calcium, potassium, or sodium).

Although daily urine volume varies considerably, most people expel between 1000 and 2000 ml (about 1 to 2 qt) of urine each day. Factors such as diet, temperature, and blood volume and

pressure influence urine volume. For example, high blood volume and pressure cause decreased reabsorption of water in the renal tubules of nephrons, resulting in increased urine output. Thus, in conjunction with other tests, an analysis showing high urine volume may indicate high blood pressure. Typically, volume is not measured at the time of providing a routine urine sample. It is only when a problem with the urinary tract is suspected that people are asked to monitor their urine output over a 24-hour period. Such monitoring may reveal polyuria, which is excessive urine production (well over 2 liters or 2.1 qt per day), or oliguria, an inadequate urine production (less than 500 ml, or 0.53 qt per day). Failure to produce any urine is an extreme health crisis called anuria.

Stress incontinence is common in women, particularly after childbirth. Giving birth may stretch or damage the external sphincter, making it less effective in controlling the flow of urine. Such damage can be minimized or repaired by exercises designed to strengthen the external sphincter and muscles along the floor of the pelvic cavity.

Urinary retention is the failure to expel urine from the bladder to a normal degree. This condition may result from a lack of the sensation that one must urinate, as might occur temporarily after general anesthesia. It may also result from contraction or obstruction of the urethra. For example, enlargement of the prostate gland in men may obstruct the urethra. Immediate treatment for retention usually involves the use of a tube called a urinary catheter to drain urine from the bladder.

Bacteria Can Enter the Urethra and Cause Urinary Tract Infections

Healthy urinary systems contain no microorganisms. Nevertheless, microorganisms that make their way into the urinary system can thrive there and cause a **urinary tract infection (UTI)**. Some bacteria arrive at the kidneys by way of the bloodstream. Most bacteria, however, enter the urinary system by moving up the urethra from outside the body. Bacteria called *Escherichia coli* normally inhabit the colon and often cause UTIs. Sometimes, microorganisms called *Chlamydia* and *Mycoplasma* cause UTIs. In contrast to *E. coli*, these microorganisms are sexually transmitted (see Chapter 17a). Many disorders of the urinary system can be diagnosed by a detailed examination of the urine in a

medical laboratory. See the Health Issue essay, *Urinalysis: What Your Urine Says about You.*

The urethras of males and females differ in length, as shown in Figure 16.18. The urethra in females is 3 to 4 cm (1.5 in.) long and lies in front of the vagina. The urethra in males is about 20 cm (8 in.) long and opens to the outside at the tip of the penis. Urinary tract infections are more common in females than in males, perhaps because of the shorter urethras. In females, the bacteria need to travel only a short distance from the external urethral orifice (opening) through the urethra to the bladder. In addition, the external urethral orifice is closer to the anus in females than it is in males. Thus, improper (from back to front) wiping after defecation can easily carry fecal bacteria to the female urethra. Bacteria also may enter the urethra during sexual intercourse. Women are advised to urinate soon after sex to flush bacteria from the lower urinary tract.

To avoid UTIs, men and women alike are advised to drink plenty of water each day and to urinate frequently throughout the day. Other tips to prevent UTIs include avoiding wet or tight clothing and nylon underwear. Such clothing allows the opening of the urethra to stay moist and creates an environment where bacteria can thrive. Showers are recommended over baths. Bubble baths, in particular, may irritate the urethra and increase susceptibility to UTIs. Finally, some physicians recommend drinking plenty of cranberry juice. This juice acidifies the urine and inhibits the growth of some bacteria.

Bacteria may infect only the urethra and cause inflammation there, a type of infection called **urethritis**. Sometimes, however, the bacteria travel farther into the urinary system. These bacteria may cause **cystitis**, infection of the urinary bladder. Bacteria may also be

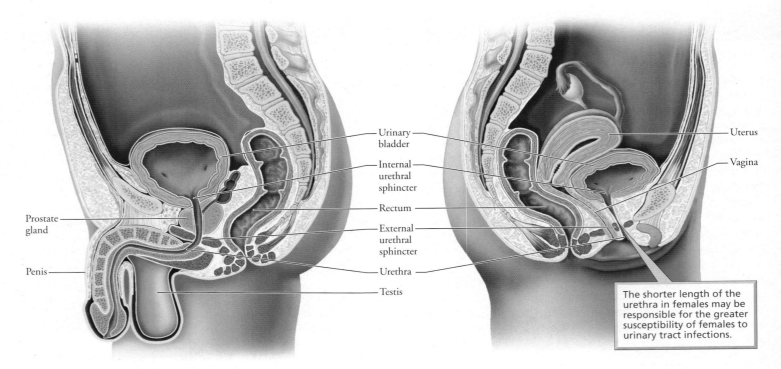

FIGURE **16.18** The urinary bladder and urethra in a male (left) and a female (right)

introduced into the bladder when a catheter is used. Although the insertion of a catheter does pose a risk of infection, use of these devices can be lifesaving. Sometimes bacteria do not stop at the bladder and move through the ureters and into the kidneys. An infection of the kidneys is called **pyelonephritis**.

Symptoms of urinary tract infections include fever, blood in the urine, and painful and frequent urination. Bed wetting may be a symptom in young children. In addition, there may be lower abdominal pain (if the bladder is involved) or back pain (if the kidneys are involved). Physicians usually diagnose such infections by having the urine checked for bacteria and blood cells. People who

have recurrent UTIs may wish to purchase nonprescription dipsticks to check for infection. The dipstick is held in the stream of the first morning urine and changes color when it detects nitrite, a compound made by the bacteria. These dipsticks can detect about 90% of UTIs.

Urinary tract infections are treated with antibiotics. Infections of the lower urinary tract (the urethra and bladder) should be treated immediately and the prescribed antibiotics taken for their full term. These steps are critical to prevent the spread of infection to the kidneys, where very serious damage can occur.

HIGHLIGHTING THE CONCEPTS

Organs from Several Body Systems Eliminate Waste (pp. 330–331)

1. Several organs from different body systems remove waste from our bodies. Lungs and skin eliminate carbon dioxide, heat, and water. The skin also excretes salt and urea. Organs of the digestive system eliminate solid waste, and those of the urinary system excrete nitrogen-containing wastes, water, carbon dioxide, inorganic salts, and hydrogen ions.

2. Two kidneys, two ureters, the urinary bladder, and the urethra make up the urinary system. The main function of this system is to regulate the volume, pressure, and composition of the blood. The kidneys filter excess materials and wastes from the blood and produce urine. Urine travels down the ureters to the urinary bladder, where it is

stored until excreted from the body through the urethra. The male urethra transports reproductive fluid in addition to urine (but not at the same time). The female urethra transports only urine.

By Producing Urine, the Kidneys Maintain Homeostasis (pp. 331–339)

3. The kidneys are located just above the waist against the back wall of the abdominal cavity. Each kidney has an outer portion (the renal cortex), an inner region (the renal medulla), and an inner chamber (the renal pelvis).

4. Properly functioning kidneys are critical to maintaining homeostasis. These organs filter wastes and excess materials from the blood, help

regulate blood pH, and maintain fluid balance by regulating the volume and composition of blood and urine.

5. Nephrons are tiny tubules responsible for the formation of urine. There are two basic parts to a nephron: (1) the renal corpuscle, which consists of a tuft of capillaries called the glomerulus and a surrounding cuplike structure called the glomerular capsule; and (2) the renal tubule. The renal tubule has three sections: the proximal convoluted tubule, the loop of the nephron, and the distal convoluted tubule. The distal convoluted tubules of several nephrons empty into a single collecting duct. Urine in collecting ducts moves into a large chamber within the kidneys called the renal pelvis. From the renal pelvis, urine leaves the kidneys by traveling down the ureters to the urinary bladder.

6. Glomerular filtration is the process by which only certain substances are allowed to pass out of the blood and into the nephron. High pressure within glomerular capillaries forces water and small solutes in the blood across a filter into the glomerular capsule. Large proteins and blood cells normally cannot cross the filter.

7. From the glomerular capsule, the filtrate—consisting of water, ions, and small molecules filtered from the blood—moves into the renal tubule of the nephron. As it passes through the renal tubule, about 99% of it is returned to the blood in a process known as tubular reabsorption. Almost all the water, ions, and nutrients in the filtrate are returned to the blood. Wastes are either partially reabsorbed or not reabsorbed at all. Most reabsorption occurs in the proximal convoluted tubule.

8. During tubular secretion, drugs and any wastes or excess essential ions that may have escaped glomerular filtration are removed from the blood in the capillaries surrounding nephrons and added to the filtered fluid that will become urine. This process occurs primarily along the proximal and distal convoluted tubules and collecting ducts.

9. The kidneys work with the lungs and buffer systems to regulate pH. Specifically, tubular secretion removes excess hydrogen ions from the blood.

10. Most nephrons have short loops that are restricted almost entirely to the renal cortex. About 20% of the nephrons have long loops that extend deep into the renal medulla. These nephrons are responsible for the extreme water-conserving ability of the kidneys because they maintain high solute concentrations in the interstitial fluid within the kidneys. Maintenance of this gradient between the interstitial fluid and the filtrate enables large amounts of water to move out of collecting ducts by osmosis and leads to the production of concentrated urine.

11. Several hormones influence kidney function. ADH is produced by the hypothalamus and released by the posterior lobe of the pituitary gland. ADH increases water reabsorption in the collecting ducts of nephrons. Aldosterone is secreted by the adrenal cortex. Aldosterone increases sodium reabsorption by the distal convoluted tubules and collecting ducts. Because water follows sodium, aldosterone increases water reabsorption by nephrons. ANP is a hormone released by cells of the right atrium of the heart. ANP causes decreased water and sodium reabsorption by the kidneys.

12. The kidneys release the hormone erythropoietin, which is necessary for red blood cell production. The kidneys also convert vitamin D into its active form that promotes the body's absorption of calcium and phosphorus.

WEB TUTORIAL 16.1 The Urinary System

Dialysis and Transplant Surgery Help When Kidneys Fail (pp. 339–343)

13. Renal failure is a decrease or complete cessation of glomerular filtration. Failure of the kidneys severely disrupts homeostasis. Conditions such as acidosis, anemia, edema, hypertension, and a toxic buildup of nitrogen-containing wastes in the blood may result when the kidneys stop working.

14. Treatments for renal failure include hemodialysis (using artificial devices to cleanse the blood), continuous ambulatory peritoneal dialysis (using the patient's own peritoneum as the dialyzing membrane), and kidney transplantation.

Urination Has Involuntary and Voluntary Components (pp. 343–345)

15. Urination, the process by which the urinary bladder is emptied, entails both involuntary and voluntary actions. Moderate filling of the bladder stimulates stretch receptors in its wall that send impulses along sensory neurons to the lower spinal cord. From the spinal cord, impulses move along motor neurons back to the bladder, where they cause the muscular wall of the bladder to contract. The impulses also cause the internal urethral sphincter, a thickening of smooth muscle at the junction of the bladder and urethra, to relax. These actions push stored urine into the urethra. If conditions are appropriate, the brain then permits voluntary relaxation of the external urethral sphincter, a band of skeletal muscle farther down the urethra, causing urine to exit the body.

Bacteria Can Enter the Urethra and Cause Urinary Tract Infections (pp. 345–346)

16. Urinary tract infections (UTIs) are more common in females than in males, possibly because bacteria can more readily move from outside the body up the shorter urethra of females to the bladder. Infections may occur in the urethra, bladder, and kidneys, with infections of the kidneys being the most serious.

KEY TERMS

urinary system *p. 330*	renal corpuscle *p. 332*	tubular reabsorption *p. 333*	erythropoietin *p. 338*
kidney *p. 331*	glomerulus *p. 332*	antidiuretic hormone (ADH) *p. 334*	renal failure *p. 339*
urine *p. 331*	glomerular capsule *p. 332*		hemodialysis *p. 339*
ureter *p. 331*	renal tubule *p. 332*	tubular secretion *p. 334*	urination *p. 343*
urinary bladder *p. 331*	proximal convoluted tubule *p. 332*	diuretic *p. 337*	urinary tract infection (UTI) *p. 345*
urethra *p. 331*		aldosterone *p. 337*	
renal cortex *p. 331*	loop of the nephron *p. 332*	juxtaglomerular apparatus *p. 338*	urethritis *p. 345*
renal medulla *p. 331*	distal convoluted tubule *p. 332*	renin *p. 338*	cystitis *p. 345*
renal pelvis *p. 331*	collecting duct *p. 332*	atrial natriuretic peptide (ANP) *p. 338*	pyelonephritis *p. 346*
nephron *p. 331*	glomerular filtration *p. 333*		

REVIEWING THE CONCEPTS

1. List the components of the urinary system and their functions. *pp. 330–331*
2. Describe the ways in which kidneys maintain homeostasis. *pp. 330–331*
3. Describe the structure of a nephron. *p. 332*
4. Explain glomerular filtration, tubular reabsorption, and tubular secretion by nephrons. Where in the nephron does each process occur? *pp. 333–335*
5. How do nephrons contribute to the regulation of blood pH? *p. 336*
6. How do the kidneys conserve water? *pp. 336–337*
7. Explain how each of the following influences kidney function: ADH, aldosterone, and ANP. *pp. 337–338, 339*
8. What treatments are available for persons with severely impaired kidneys? *p. 339–343*
9. Describe the process of urination. *pp. 343–344*
10. How do the urethras of males and females differ? What are the clinical implications of these differences? *pp. 345–346*
11. Glomerular filtration
    a. occurs at the renal corpuscle.
    b. returns useful substances to the blood.
    c. occurs along the renal tubule.
    d. is the last step of three in cleansing the blood.
12. Which of the following organs does *not* play a role in eliminating metabolic wastes?
    a. skin
    b. lungs
    c. esophagus
    d. kidneys
13. Diabetes insipidus is caused by a deficiency in
    a. aldosterone.
    b. antidiuretic hormone.
    c. erythropoietin.
    d. atrial natriuretic peptide.
14. Which of the following is *not* performed by the kidneys?
    a. regulation of pH
    b. conservation of water
    c. transformation of vitamin D into active form
    d. contraction to initiate urination
15. Tubular reabsorption
    a. allows only certain substances to pass from the blood into the nephron.
    b. returns useful substances from the filtrate to the blood.
    c. removes wastes and excess essential ions that escaped filtration from the blood.
    d. occurs at the renal corpuscle.
16. _____ are the functional units of the kidneys.
17. _____ is the use of artificial devices to cleanse the blood.
18. The _____ urethral sphincter is made of _____ muscle and is involuntary. The _____ urethral sphincter is made of _____ muscle and allows voluntary control over urination.

APPLYING THE CONCEPTS

1. Rachel has noticed that since the birth of her daughter, a small amount of urine escapes whenever she sneezes. What condition might Rachel have? What would explain this condition, and what can she do to improve it?
2. After his surgery, a urinary catheter was inserted into Miguel's bladder. A few days later, he had a fever, pain in his lower abdomen, and painful urination. What do you think Miguel might be suffering from? How would you diagnose and treat his condition?
3. Sean has just finished an exhausting game of basketball on a very hot day. He thinks that a cold beer might be just the thing to quench his thirst and restore body fluids. What do you think, and why?
4. Anne has polycystic disease, an inherited and progressive condition that is destroying her kidneys. What will be the health consequences of her chronic renal failure? What are her options for restoring kidney function?

Additional questions can be found on the companion website.

The Gonads Produce Gametes and Sex Hormones

A Male's Reproductive Role Differs from a Female's

The Male Reproductive System Delivers Sperm to the Egg

- The testes produce sperm and male hormones
- The duct system stores and transports sperm
- The accessory glands produce most of the volume of semen
- The penis transfers sperm to the female
- As sperm develop, changes occur in the number of chromosomes and the structure and function of cells
- The interplay of hormones controls male reproductive processes

The Female Reproductive System Produces the Eggs and Nurtures the Embryo and Fetus

- The ovaries produce eggs and female hormones
- The oviducts transport the immature egg, zygote, and early embryo
- The uterus supports the growth of the developing embryo
- External genitalia lie outside the vagina
- The breasts produce milk to nourish the baby
- The events of the ovarian cycle lead to release of an egg
- The interplay of hormones coordinates the ovarian and menstrual cycles
- Menopause ends a woman's reproductive cycles

Problems with the Female Reproductive System Vary in the Severity of Health Consequences

The Human Sexual Response Has Four Stages

Birth Control Is the Prevention of Pregnancy

- Abstinence is refraining from intercourse
- Sterilization may involve cutting and sealing gamete transport tubes
- Hormonal contraception interferes with regulation of reproductive processes
- Intrauterine devices prevent the union of sperm and egg and implantation
- Barrier methods of contraception prevent the union of sperm and egg
- Spermicidal preparations kill sperm
- Fertility awareness is the avoidance of intercourse when fertilization could occur
- The morning-after pill is emergency contraception

HEALTH ISSUE: Breast Cancer

ENVIRONMENTAL ISSUE: Environmental Estrogens

17

Reproductive Systems

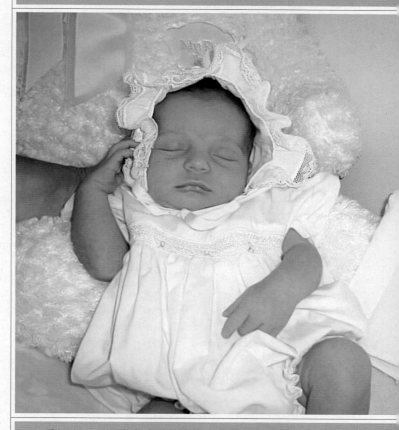

This rosy, resting baby looks comfortable in her new surroundings, although they can hardly be as sheltering and secure as her mother's uterus.

Tim and Andrea had attempted for 2 years to get Andrea pregnant, trying everything from advice they read in medical columns to ideas suggested by old wives' tales. Nothing worked. Finally, they went to a highly recommended fertility clinic for help. After many tests and examinations, and thorough questioning about their medical history, Dr. Perry told them that Tim was likely to be the source of the problem. He explained that Tim had "poor–performing sperm—slow swimmers and too few of them." He went on to explain that these problems with Tim's sperm were probably caused by varicoceles, enlargements of the veins that drain blood from the testes. The varicoceles create a slightly warmer environment for sperm, resulting in a reduced sperm count and a high percentage of deformed sperm. Tim had surgery to correct the problem, and 18 months later, Andrea gave birth to their beautiful baby girl.

In this chapter, we will gain an understanding of why Tim and Andrea had difficulty conceiving a child. We will look at the structures of the reproductive systems and how they function. We will also consider the sex hormones and the ways in which they regulate reproductive processes. ∎

The Gonads Produce Gametes and Sex Hormones

In both males and females, the **gonads**—testes or ovaries—are the most important structures in the reproductive system. The gonads serve two important functions: (1) they produce the **gametes**, meaning the eggs and sperm—the cells that will fuse and develop into a new individual; and (2) they produce the sex hormones. The male gonads are the **testes**, and the gametes they produce are the **sperm**. The testes also produce the male sex hormone **testosterone**. The **ovaries** are the female gonads. The gametes they produce are the **eggs**. The female sex hormones they produce are **estrogen** and **progesterone**.

A Male's Reproductive Role Differs from a Female's

A male's reproductive role differs from that of a female. It is true that they each make the same genetic contribution to the next generation; that is, they each contribute one copy of each chromosome to their offspring. However, they have different "reproductive strategies" to help ensure that their DNA (packaged in their chromosomes) is passed along to a new generation. A male's strategy is to produce millions of sperm, deliver them to the female reproductive system, and hope that one sperm reaches an egg. In contrast, a female usually produces only one egg approximately once a month. If sperm delivery is appropriately timed with egg production, the sperm and egg may fuse in a process called **fertilization**. The cell created by fertilization, called a zygote, will develop into a new individual (see Chapter 18). A male's role in the reproductive process ends with delivery of sperm, but a female's role continues after egg production; she nourishes and protects the offspring until birth. The egg is packed with nutrients to provide for early development, and the uterus serves as a nourishing, protective environment for the developing offspring.

Sperm and eggs are specialized for their reproductive roles in various ways. One specialization they share is a reduction in the number of chromosomes. Each of your body cells contains 23 *pairs* of chromosomes (for a total of 46 chromosomes). One member of each pair came from your father's sperm; the other member came from your mother's egg. Each of the 23 kinds of chromosome contains a different portion of the instructions for making and maintaining your body, and two of each kind is necessary for health and normal development. It is important, therefore, that one, but only one, member of each pair be present in a gamete. The development of gametes involves two types of cell division. The first is *mitosis*, a process in which a cell with 23 pairs of chromosomes replicates its chromosomes and divides into two identical daughter cells, each having 23 pairs of chromosomes. The alterations in genetic content occur during the second type of cell division, called *meiosis*, which begins with a cell that has two copies of each of the 23 kinds of chromosome (called a *diploid* cell) and ends with up to four cells that each contain only one of each kind of chromosome (called *haploid* cells). Meiosis entails two rounds of cell division. (The details of mitosis and meiosis are described in Chapter 19.)

The Male Reproductive System Delivers Sperm to the Egg

The male reproductive system consists of the testes, a system of ducts through which the sperm travel, the penis, and various accessory glands. Accessory glands produce secretions that help protect and nourish the sperm as well as provide a transport medium that aids the delivery of sperm to the outside of the male's body. The structure of the male reproductive system is shown in Figure 17.1 and summarized in Table 17.1.

TABLE 17.1 REVIEW OF MALE REPRODUCTIVE SYSTEM	
STRUCTURE	**FUNCTION**
Testes	Produce sperm and testosterone
Epididymis	Location of sperm storage and maturation
Vas deferens	Conducts sperm from epididymis to urethra
Urethra	Tube through which sperm or urine leaves the body
Prostate gland	Produces secretions that make sperm mobile and that counteract the acidity of the female reproductive tract
Seminal vesicles	Produce secretions that make up most of the volume of semen
Bulbourethral glands	Produce secretions just before ejaculation; may lubricate; may rinse urine from urethra
Penis	Delivers sperm to female reproductive tract

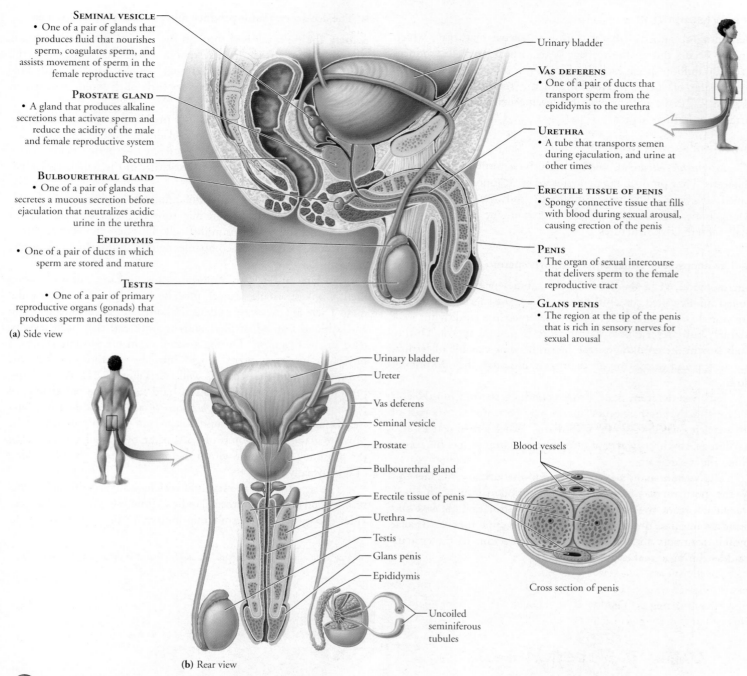

SEMINAL VESICLE
• One of a pair of glands that produces fluid that nourishes sperm, coagulates sperm, and assists movement of sperm in the female reproductive tract

PROSTATE GLAND
• A gland that produces alkaline secretions that activate sperm and reduce the acidity of the male and female reproductive system

Rectum

BULBOURETHRAL GLAND
• One of a pair of glands that secretes a mucous secretion before ejaculation that neutralizes acidic urine in the urethra

EPIDIDYMIS
• One of a pair of ducts in which sperm are stored and mature

TESTIS
• One of a pair of primary reproductive organs (gonads) that produces sperm and testosterone

(a) Side view

Urinary bladder

VAS DEFERENS
• One of a pair of ducts that transport sperm from the epididymis to the urethra

URETHRA
• A tube that transports semen during ejaculation, and urine at other times

ERECTILE TISSUE OF PENIS
• Spongy connective tissue that fills with blood during sexual arousal, causing erection of the penis

PENIS
• The organ of sexual intercourse that delivers sperm to the female reproductive tract

GLANS PENIS
• The region at the tip of the penis that is rich in sensory nerves for sexual arousal

Urinary bladder
Ureter
Vas deferens
Seminal vesicle
Prostate
Bulbourethral gland
Erectile tissue of penis
Urethra
Testis
Glans penis
Epididymis
Uncoiled seminiferous tubules

Blood vessels

Cross section of penis

(b) Rear view

FIGURE **17.1** The male reproductive system

WEB TUTORIAL 17.1

■ The testes produce sperm and male hormones

The male reproductive system has two testes (singular, testis). Each is located externally to the body in a sac of skin called the **scrotum**. The temperature in the scrotum is several degrees cooler than within the abdominal cavity. The lower temperature is important in the production of healthy sperm.

Reflexes in the scrotum help keep the temperature within the testes fairly stable. Cooling of the scrotum, such as might occur when a male jumps into frigid water, triggers contraction of a muscle that pulls the testes closer to the warmth of the body. In a hot shower, however, the muscle relaxes, and the testes hang low, away

from the heat of the body. The skin of the scrotum is also amply supplied with sweat glands that help cool the testes.

Testicular cancer is the most common form of cancer among men between the ages of 15 and 35 years, although it affects only 1% of all men. Testicular cancer is more likely to occur in men whose testes did not descend into the scrotum or descended after 6 years of age. Because this cancer does not usually cause pain, it is important for every man to examine his testes each month to feel for a lump or a change in consistency. The cure rate for tumors caught in the early stages is nearly 100%.

SPERM PRODUCTION

Beginning at puberty (the transition to sexual maturity), which usually occurs during the teenage years, a healthy male produces over 100 million sperm each day for the rest of his life. The microscopic sites of sperm production are the **seminiferous tubules** (Figure 17.2). We will consider the development of sperm in more detail later in this chapter.

HORMONE PRODUCTION

The *interstitial cells* are located between the seminiferous tubules of the testis. They produce the male steroid sex hormones, collectively called *androgens*. The most important androgen is *testosterone*, which is needed for sperm production and the maintenance of male reproductive structures.

■ The duct system stores and transports sperm

Sperm produced in the seminiferous tubules next enter a highly coiled tubule called the **epididymis**, where the sperm mature and are stored (see Figure 17.1). Sperm that enter the epididymis look mature, but they cannot yet function as mature sperm. During their stay in the epididymis, the sperm become capable of fertilizing an egg and of moving on their own, although they do not yet do so.

Each **vas deferens** is a tube that conducts sperm from the epididymis to the urethra. Some sperm may be stored in the part of the vas deferens closest to the epididymis. When a man reaches sexual climax, rhythmic waves of muscle contraction propel the sperm along the vas deferens.

The *urethra* conducts urine (from the urinary bladder) or sperm (from the vas deferens) out of the body through the *penis*, the male organ of sexual intercourse and urination. Sperm and urine do not pass through the urethra at the same time. A circular muscle contracts and pinches off the connection to the urinary bladder during sexual excitement.

■ The accessory glands produce most of the volume of semen

Semen, the fluid released through the urethra at sexual climax, is composed of sperm and the secretions of the accessory glands: the prostate gland, the paired seminal vesicles, and paired bulbourethral glands. Interestingly, very little of the volume of semen is made up of sperm.

About the size of a walnut, the **prostate gland** surrounds the urethra, just beneath the urinary bladder. Prostate secretions are slightly alkaline and serve both to activate the sperm, making them fully motile, and to counteract the acidity of the female reproductive tract.

The prostate often begins enlarging when a man reaches middle age. The enlarged prostate may squeeze the urethra and restrict urine flow, making urination difficult. This benign enlargement of the prostate gland that accompanies aging is not related to prostate cancer.

Prostate cancer is an important cause of cancer deaths among men. Unlike testicular cancer, prostate cancer usually affects older men. There are two ways of detecting prostate cancer: a rectal exam and a blood test. Most physicians recommend that both tests be used in the diagnosis. During a rectal exam for prostate cancer, a physician inserts a gloved finger into the rectum and feels the prostate through the rectum's wall (see Figure 17.1). A cancerous prostate is firm or may contain a hard lump. The blood test for prostate cancer measures the amount of a protein called prostate-specific antigen (PSA). Because this protein is produced only by the prostate gland, the amount of PSA in the blood reflects the size of the prostate. As a tumor within the prostate grows, the blood level of PSA usually rises.

The secretions of the **seminal vesicles** contain citric acid, fructose, amino acids, and prostaglandins.[1] Fructose is a sugar that provides energy for the sperm's long journey to the egg. Some of the

---

[1]Prostaglandins are chemicals secreted by one cell that alter the activity of other cells.

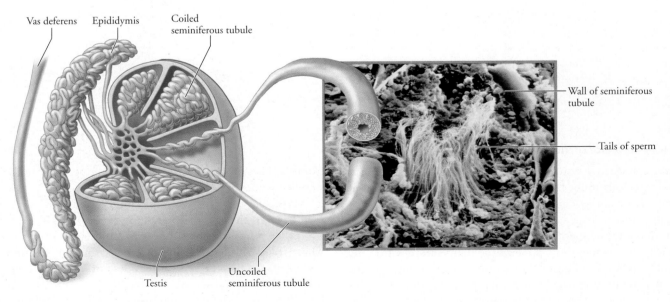

Vas deferens | Epididymis | Coiled seminiferous tubule

Wall of seminiferous tubule

Tails of sperm

Testis | Uncoiled seminiferous tubule

FIGURE **17.2** The internal structure of the testis and epididymis

amino acids thicken the semen. This thickening helps keep the sperm within the vagina and also protects the sperm from contacting the acidic environment of the vagina. The prostaglandins serve to cut the viscosity of cervical mucus (that could otherwise slow the movement of sperm) and to cause uterine contractions that assist the movement of sperm.

The **bulbourethral glands** release a clear, slippery liquid immediately before ejaculation. This fluid may rinse the slightly acidic urine remnants from the urethra before the sperm pass through.

■ The penis transfers sperm to the female

The **penis** is a cylindrical organ whose role in reproduction is to deliver the sperm to the female reproductive system. The tip of the penis is enlarged, forming a smooth rounded head known as the *glans penis*, which has many sensory nerve endings and is important in sexual arousal. When a male is born, a cuff of skin called the *foreskin* covers the glans penis. The foreskin can be pulled back to expose the glans penis. The surgical removal of the foreskin is called circumcision.

The transfer of sperm to the female reproductive system usually requires the penis to be erect. An erection consists of increases in the length, width, and firmness of the penis, due to changes in the blood supply to the organ. Within the penis are three columns of spongy *erectile tissue,* which is a loose network of connective tissue with many empty spaces (see Figure 17.1). During sexual arousal, the arterioles that pipe blood into the spongy tissue dilate (widen), and the spongy tissue fills with blood, causing the penis to become larger. At the same time, the spongy tissue squeezes shut the veins that drain the blood from the penis. As a result, the blood flows into the penis faster than it can leave, causing the spongy tissue to fill with blood and press against a connective tissue casing. This makes the penis firm (like a water balloon), larger, and erect.

Erectile dysfunction (ED, also called impotence) is a male's inability to achieve or maintain an erection (and thus to have sexual intercourse). ED is not unusual for a man to experience at some point in his life. There are a number of psychological causes for it, including worry, stress, a quarrel with the partner, and depression. However, there are also many physical causes. Nerve damage, such as that which often accompanies chronic alcoholism and sometimes diabetes, can be responsible. Because an erection depends on adequate blood supply, fatty deposits in the arteries serving the penis (as in atherosclerosis) can also cause ED. Medications, especially certain drugs used to treat high blood pressure, antihistamines, antinausea and antiseizure drugs, antidepressants, sedatives, and tranquilizers may cause the problem. Cigarette smoking, excessive alcohol consumption, or marijuana use can also cause ED.

The first step in treating ED is to eliminate the cause of the problem, if possible. Should impotency continue, drugs for erectile dysfunction—Viagra, Levitra, Cialis—may be prescribed. These drugs can help a man achieve and maintain an erection when he is sexually aroused. They work by prolonging the effect of a chemical (nitric oxide) that is released when the man becomes sexually aroused and that causes the widening of the arterioles in the penis. As the arterioles widen, blood flow increases, and an erection results.

stop and think

Explain why Viagra, Levitra, or Cialis cannot cause an erection in a man who is not sexually aroused.

■ As sperm develop, changes occur in the number of chromosomes and the structure and function of cells

The sequence of events within the seminiferous tubules that leads to the development of sperm is called *spermatogenesis*. This process reduces the number of chromosomes in the resulting gametes to one member of each pair and changes the sperm cells' shape and functioning to make the sperm efficient chromosome-delivery vehicles.

The process of spermatogenesis (Figure 17.3) begins in the outermost layer of each seminiferous tubule, where undifferentiated diploid cells called *spermatogonia* (singular, spermatogonium)

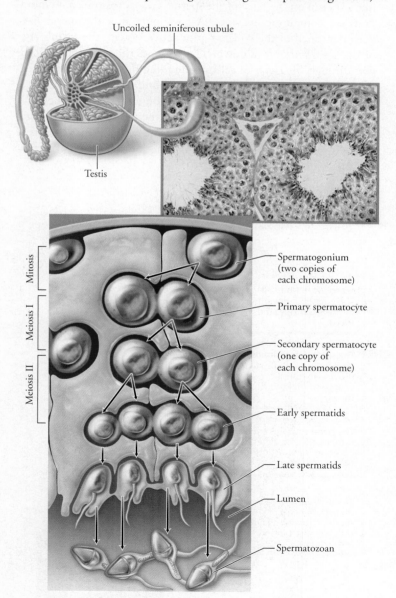

FIGURE **17.3** The stages of spermatogenesis in the wall of a seminiferous tubule. As the cells that will become sperm develop, they are pushed from the outer wall of the tubule to the lumen, or central canal.

develop. Each spermatogonium divides by mitosis to produce two new diploid spermatogonia. One of these spermatogonia pushes deeper into the wall of the tubule, where it will enlarge to form a *primary spermatocyte,* which is also a diploid cell. Primary spermatocytes undergo the two divisions of meiosis to form *secondary spermatocytes,* which mature into *spermatids.* Although the spermatids are haploid and have the correct set of chromosomes for joining with the egg during fertilization, numerous structural changes must still occur to create cells capable of swimming to the egg and fertilizing it. These changes convert spermatids to streamlined *spermatozoa,* or sperm.

The mature sperm cell has three distinct regions: the head, the midpiece, and the tail (Figure 17.4). The head of the sperm is a flattened oval that contains little else besides the 23 densely packed chromosomes (remember that the sperm is haploid). Positioned like a ski cap on the head of the sperm is the **acrosome**, a membranous sac containing enzymes. A few hours after the sperm have been deposited in the female reproductive system, the membranes of the acrosomes break down. The enzymes then spill out and digest through the layers of cells surrounding the egg, assisting fertilization. Within the midpiece, mitochondria are arranged in a spiral. The mitochondria are the powerhouses of the cell. They provide energy in the form of ATP to fuel the movements of the tail. The tail contains contractile filaments. The whiplike movements of the tail propel the sperm during its long journey through the female reproductive system.

■ The interplay of hormones controls male reproductive processes

Testosterone, secreted by the interstitial cells of the testes, is important for sperm production as well as many other male characteristics. At puberty, testosterone turns boys into men (literally). It is responsible for the growth spurt that accompanies puberty, stimulating the growth of the long bones of the arms and legs and also making the male reproductive organs, including the testes and penis, larger. In addition, testosterone is responsible for the development and maintenance of the male secondary sex characteristics, which are features associated with "masculinity" but not directly related to reproductive functioning. For example, the growth of muscles and the skeleton tends to result in wide shoulders and narrow hips. Pubic hair develops, as does hair under the arms. A beard appears, perhaps accompanied by hair on the chest. Meanwhile, the voice box enlarges and the vocal cords thicken, causing a deepening of the male voice. And, importantly, testosterone is responsible for sex drive.

The level of testosterone in the body remains relatively steady because its production is regulated by a negative feedback loop (see Chapter 10) involving hormones from the hypothalamus, the anterior pituitary gland, and the testes. The hormones that regulate male reproductive processes are summarized in Figure 17.5 and Table 17.2. Testosterone levels rise when a region of the brain called the hypothalamus releases **gonadotropin-releasing hormone**, or **GnRH**. This hormone stimulates the anterior pituitary gland, a pea-sized endocrine gland on the underside of the brain, to secrete **luteinizing hormone (LH)**. In turn, LH stimulates the production of testosterone by the interstitial cells of the testis. (For this reason, LH is sometimes called *interstitial cell–stimulating hormone,* or

The whiplike movements of the **tail** propel the sperm.

The **midpiece** contains mitochondria that will provide metabolic energy to fuel the trip to the egg.

The **head** contains the father's chromosomes, his genetic contribution to the next generation.

The **acrosome**, a sac that covers the head of the sperm, contains enzymes that will assist in fertilization.

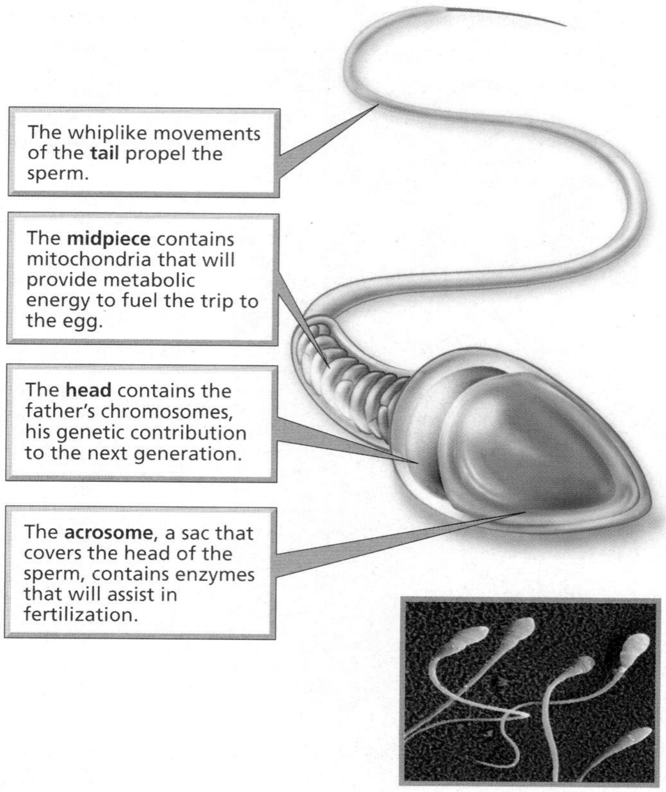

FIGURE **17.4** The structure of a mature sperm (spermatozoan)

TABLE 17.2 REVIEW OF HORMONES IMPORTANT IN THE REGULATION OF MALE REPRODUCTIVE PROCESSES

HORMONE	SOURCE	EFFECTS
Testosterone	Interstitial cells in testes	Sperm production; development and maintenance of male reproductive structures, male secondary sex characteristics; sex drive
Gonadotropin-releasing hormone (GnRH)	Hypothalamus (in brain)	Stimulates the anterior pituitary gland to release LH
Luteinizing hormone (LH)	Anterior pituitary gland (in brain)	Stimulates interstitial cells of testis to produce testosterone
Follicle-stimulating hormone (FSH)	Anterior pituitary gland (in brain)	Enhances sperm formation
Inhibin	Seminiferous tubules in testes	Inhibits FSH secretion by anterior pituitary gland, causing a decrease in sperm production and testosterone production

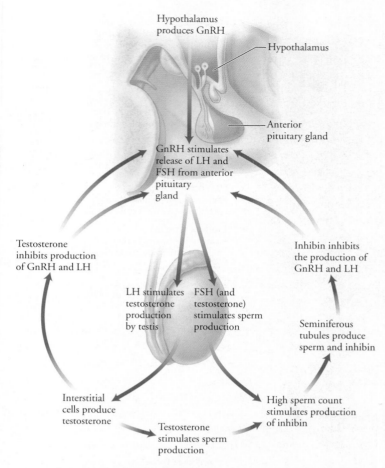

FIGURE **17.5** The feedback relationships among the hypothalamus, anterior pituitary, and testes control the production of both sperm and testosterone.

ICSH.) The rising testosterone level inhibits the release of GnRH from the hypothalamus. As a result, the amount of LH produced by the anterior pituitary drops. This decrease in the LH level then lowers the amount of testosterone secretion, removing the inhibition on the hypothalamus. These interactions keep testosterone levels constant.

Sperm production is regulated by another negative feedback loop. Besides secreting LH, the anterior pituitary gland produces **follicle-stimulating hormone**, or **FSH**. Follicle-stimulating hormone stimulates sperm production by making the cells that will become sperm more sensitive to the stimulatory effects of testosterone. FSH works by causing certain cells within the seminiferous tubules to secrete a protein that binds and concentrates testosterone.

The seminiferous tubules produce a hormone called *inhibin* in addition to sperm. Inhibin is named for its ability to *inhibit*, or lessen, the production of FSH from the anterior pituitary. It also may inhibit the hypothalamic secretion of GnRH. Rising levels of inhibin cause a decline in testosterone level and sperm production. As the sperm count falls, so does the level of inhibin. Released from inhibition, the anterior pituitary then increases production of FSH, and the hypothalamus produces GnRH, which increases sperm production again.

stop and think

Anabolic steroids are taken by some male athletes to build muscles. A side effect of steroid abuse is a reduction in testis size. Considering that the anabolic steroids mimic the effect of testosterone, how can the shrinking testis size be explained?

The Female Reproductive System Produces the Eggs and Nurtures the Embryo and Fetus

The female reproductive system consists of the ovaries, the oviducts, the uterus, and the vagina. Its structure is shown in Figure 17.6, and its functions are summarized in Table 17.3.

■ **The ovaries produce eggs and female hormones**

The ovaries have two important functions: (1) to produce eggs in a process called *oogenesis* and (2) to produce the female hormones, estrogen and progesterone. These two functions are intimately related and occur in cycles, as we will see.

■ **The oviducts transport the immature egg, zygote, and early embryo**

Two **oviducts**, also known as fallopian tubes or uterine tubes, extend from the uterus. They transport the immature egg, called an *oocyte*, from the ovary to the uterus. The end of each oviduct nearest the ovary is open and funnel shaped. Its many ciliated, finger-like projections drape over the ovary but are rarely in direct contact with it. At about the time that the egg is released from the ovary, the projections from the tubes begin to wave. The currents they create and those caused by the cilia lining the tubes help draw the oocyte into the oviduct. If fertilization occurs, it usually takes place in the oviduct, near the ovary. The resulting zygote begins its development into an embryo in the oviduct. The beating cilia and the rhythmic muscular contractions of the oviduct sweep the egg or the early embryo along the oviduct toward the uterus.

■ **The uterus supports the growth of the developing embryo**

The **uterus** receives and nourishes the developing baby (first called an embryo, but after 8 weeks called a fetus). During pregnancy, as the fetus grows, the uterus expands to about 60 times its original

TABLE 17.3	REVIEW OF FEMALE REPRODUCTIVE SYSTEM
STRUCTURE	**FUNCTION**
Ovary	Produces eggs and the hormones estrogen and progesterone
Oviducts	Transport ovulated egg (or embryo if fertilization occurred) from ovary to uterus; the usual site of fertilization
Uterus	Receives and nourishes embryo
Vagina	Receives penis during intercourse; serves as birth canal
Clitoris	Contributes to sexual arousal
Breasts	Produce milk

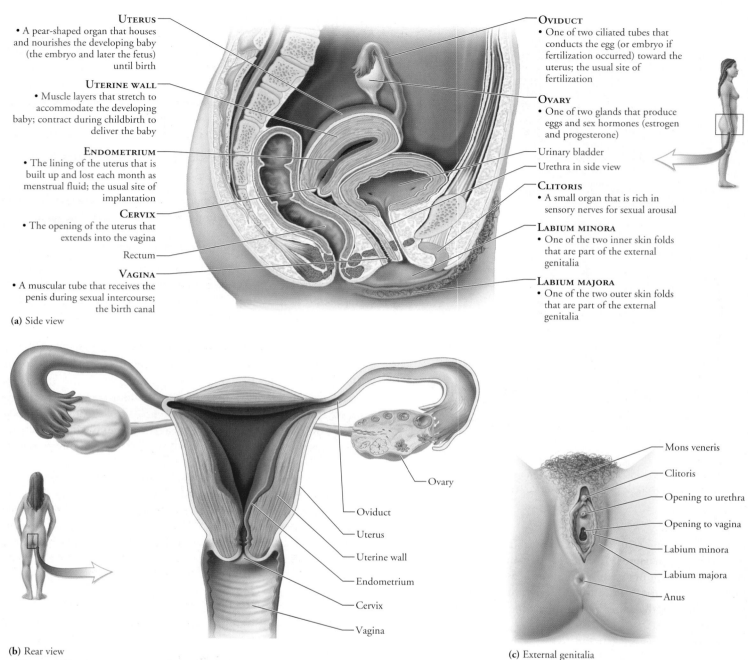

UTERUS
- A pear-shaped organ that houses and nourishes the developing baby (the embryo and later the fetus) until birth

UTERINE WALL
- Muscle layers that stretch to accommodate the developing baby; contract during childbirth to deliver the baby

ENDOMETRIUM
- The lining of the uterus that is built up and lost each month as menstrual fluid; the usual site of implantation

CERVIX
- The opening of the uterus that extends into the vagina

Rectum

VAGINA
- A muscular tube that receives the penis during sexual intercourse; the birth canal

(a) Side view

OVIDUCT
- One of two ciliated tubes that conducts the egg (or embryo if fertilization occurred) toward the uterus; the usual site of fertilization

OVARY
- One of two glands that produce eggs and sex hormones (estrogen and progesterone)

Urinary bladder

Urethra in side view

CLITORIS
- A small organ that is rich in sensory nerves for sexual arousal

LABIUM MINORA
- One of the two inner skin folds that are part of the external genitalia

LABIUM MAJORA
- One of the two outer skin folds that are part of the external genitalia

Ovary

Oviduct

Uterus

Uterine wall

Endometrium

Cervix

Vagina

(b) Rear view

Mons veneris

Clitoris

Opening to urethra

Opening to vagina

Labium minora

Labium majora

Anus

(c) External genitalia

FIGURE 17.6 The female reproductive system

size. After childbirth, the uterus never quite returns to its prepregnancy size.

The wall of the uterus has two main layers: a muscular layer and a lining called the **endometrium** (*endo-*, within; *metr-*, uterus; *-ium*, region). The smooth muscle of the uterine wall contracts rhythmically in waves during childbirth and forces the infant out. The thickness of the endometrium varies over a cycle of approximately 1 month (this cycle is discussed later in this chapter). If an embryo forms, it implants (embeds) in the endometrium, which provides nourishment during early pregnancy. If an embryo does not form, the endometrium completes its cycle by being lost as menstrual flow.

The narrow neck of the uterus, called the **cervix**, projects into the **vagina**, a muscular tube that opens to the outside of the body. The vagina receives the penis during sexual intercourse. Sperm that are deposited in the vagina can enter the uterus through an opening in the cervix and then swim to the oviduct to meet the egg. At birth, the infant is pushed through the cervix and then the vagina on its way to greet the world.

If an embryo implants in an area other than the uterus, the condition is described as an ectopic pregnancy (*ect-*, outside). The most common type of ectopic pregnancy is a tubal pregnancy, in which the embryo implants in an oviduct. A tubal pregnancy must be surgically terminated because it places the life of the mother in danger. If the embryo is permitted to continue growing, it will eventually rupture the oviduct, and the rupture can cause the mother to bleed to death internally.

■ External genitalia lie outside the vagina

The female reproductive structures that lie outside the vagina are collectively known as the external genitalia or the *vulva*. Two hair-covered folds of skin surround the vaginal opening. The *labia majora* ("big lips") enclose two thinner skin folds, the *labia minora* ("little lips"). The anterior portions of the labia minora form a hood over the *clitoris*. Like the penis, the clitoris has many nerve endings sensitive to touch. During tactile stimulation, the erectile tissue in the clitoris becomes swollen with blood and contributes to a woman's sexual arousal.

■ The breasts produce milk to nourish the baby

The *breasts*, or mammary glands, are present in both sexes, but they produce milk to nourish a newborn only in females. Inside the breast are 15 to 25 groups of milk-secreting glands. A milk duct drains each group through the nipple. Interspersed around the glands and ducts is fibrous connective tissue that supports the breast (Figure 17.7).

The monthly changes in level of the ovarian hormones estrogen and progesterone prepare the breast for the possibility of preg-nancy. Early in each cycle, estrogen stimulates the growth of the glands, ducts, and fibrous tissue of the breast. Progesterone then triggers the initial steps of the milk-secreting process (although milk production does not begin for 2 to 3 days after the newborn begins suckling; see Chapter 18.). Breast cancer is the second lead-ing cause of cancer deaths among women and appears to be influ-enced by the changing levels of estrogen (see the Health Issue essay, *Breast Cancer*).

■ The events of the ovarian cycle lead to release of an egg

The changes in the ovary that produce the egg occur in a cycle about 1 month long called the **ovarian cycle**. Although the ovaries do not produce mature eggs until a female reaches puberty, the preparations for egg production begin before she is born. Recall that the egg is a haploid cell formed by meiosis, the type of cell di-vision that forms gametes. During fetal development, *oogonia*, which are diploid cells in the ovaries, enlarge and begin to store nu-trients. Some begin meiosis by making a copy of their chromo-somes (but the copies of each chromosome remain attached to one another). The cells that prepare for meiosis are called *primary oocytes*. (They are immature eggs.) A single layer of flattened cells, called follicle cells, surrounds each primary oocyte. The entire structure is called a primary **follicle**.

Long before birth, all of a woman's potential eggs have formed. In fact, she may have formed as many as 2 million primary follicles, but only about 700,000 remain when she is born, and the number continues to decrease after birth. The eggs remain in this immature state until she reaches puberty, usually at 10 to 14 years of age. By the time of puberty, a female's lifetime supply of potential eggs has dwindled to between 200,000 and 400,000.

Beginning at puberty, some primary follicles, usually one each month, resume their development (Figure 17.8, page 360). The follicle's outer cells begin dividing, forming layers of cells and se-creting a fluid that contains estrogen. The ovarian cycle that con-verts a primary follicle into an egg each month consists of the following steps:

- **Follicle maturation.** The follicle cells continue dividing, and fluid begins to accumulate between them. As the fluid accu-mulates, the wall of follicle cells splits into two layers. The inner layer of follicle cells directly surrounds the primary oocyte. The outer layer forms a balloonlike sphere enclosing the fluid and the oocyte. This structure grows rapidly.

- **Formation of a mature follicle.** Within about 10 to 14 days after its development began, the follicle assumes its mature form, called a secondary or Graafian follicle. The primary oocyte then completes the first meiotic division, which it pre-pared for years earlier. When the primary oocyte divides, it forms two cells of unequal size: a large cell that contains most of the cytoplasm and nutrients, called a *secondary oocyte*, and a tiny cell, called the first *polar body*. The polar body is essential-ly a garbage bag for one set of chromosomes. It has very few cellular constituents and plays no further role in reproduction. The secondary oocyte is a much larger cell packed with nutri-ents that will nourish the embryo until it reaches the uterus.

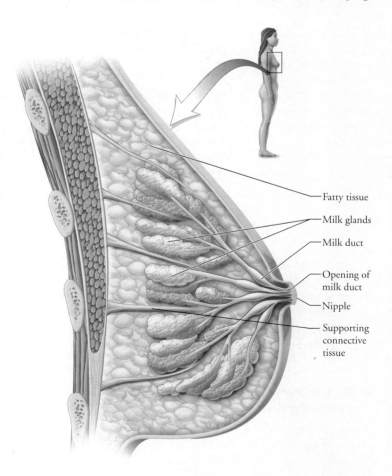

Fatty tissue

Milk glands

Milk duct

Opening of milk duct

Nipple

Supporting connective tissue

FIGURE **17.7** Breast structure

HEALTH ISSUE

Breast Cancer

Breast cancer usually begins with abnormal growth of the cells lining the milk ducts of the breast, but it sometimes begins in the milk glands themselves. Some breast tumors grow quite large without spreading. Other, more life-threatening types of breast cancer aggressively invade surrounding tissues. Typically, cancerous cells begin to spread when the tumor is about 20 mm (about 3/4 in.) in diameter. At this point, they break through the membranes of the ducts or glands where they initially formed and move into the connective tissue of the breast. They may then move into the lymphatic vessels or blood vessels permeating the breast or into both; the vessels may transport the cells throughout the body.

Detecting Breast Cancer

Early detection is a woman's best defense against breast cancer. A monthly breast self-exam (BSE) is helpful in detecting a lump early (Figure 17.A). If a woman begins doing regular breast self-exams in early adulthood, she becomes familiar with the consistency of her breast tissue. Later in life, it is easy to notice changes that might be signs of breast cancer.

Mammograms, which are x-rays of breast tissue, can also help detect early breast cancer, because they can reveal a tumor too small to be felt as a lump.

Early detection can make the difference between life and death. A tumor large enough to be felt contains a billion or more cells—a few of which may have already spread from the tumor to other tissues of the body. After cancer cells spread, the woman's chance of survival decreases dramatically. An added benefit of mammograms is that they can detect tumors that are small enough to be removed by a type of surgery called lumpectomy that removes the lump but spares the breast. At later stages of breast cancer, the entire breast may have to be removed, a type of surgery called mastectomy.

Risk Factors

Many of the factors that increase a woman's risk of breast cancer cannot be altered. One of these is the genes she inherited from her parents. At least two genes increase a woman's risk of breast cancer. Although only 5% to 10% of all breast cancers are related to these genes, the unlucky women who inherit either of them have an 85% chance of developing breast cancer at some point in life.

Exposure to estrogen is a common thread running through the tapestry of risk factors associated with breast cancer. During each menstrual cycle, estrogen stimulates breast cells to begin dividing in preparation for milk production, in case the egg is fertilized. Excessive estrogen exposure, therefore, may be a factor that pushes cell division to a rate characteristic of cancer.

One factor that influences estrogen levels is the number of times a woman ovulates during her lifetime, because estrogen is produced by both the maturing ovarian follicle and the corpus luteum. The number of times a woman ovulates is, in turn, affected by factors such as the following.

1. **Age when menstruation begins.** Ovulation usually occurs in each menstrual cycle. Thus, the younger a woman is when menstruation begins, the more opportunities there are for ovulation and the greater her exposure to estrogen.

2. **Menopause after age 55.** The older a woman is at menopause, the more menstrual cycles she is likely to have experienced. Estrogen levels are low and ovulation ceases after menopause.

3. **Childlessness and late age at first pregnancy.** Ovulation does not occur during pregnancy. Thus, pregnancy gives the ovaries a rest. Furthermore, the hormonal patterns of pregnancy appear to transform breast tissue in a way that protects against cancer. As a result, delaying pregnancy until after age 30 or remaining childless increases a woman's risk of developing breast cancer later in life.

4. **Breast-feeding.** Women who breast-feed their infants have a 20% lower risk of developing breast cancer before they reach menopause than do women who bottle-feed their infants. Nursing may guard against breast cancer by blocking ovulation or by

---

- **Ovulation.** About 12 hours after the secondary oocyte has formed, the mature follicle pops, like a blister, releasing the oocyte mass, which is about the size of the head of a pin. The release of the oocyte from the ovary is called **ovulation**. If a sperm penetrates the secondary oocyte, meiosis advances to completion: the replicate chromosomes are separated. This time, too, the cytoplasm divides unequally. One of the resulting sets of chromosomes goes into a small cell, called the second polar body, and the other set goes into a large cell, the mature *ovum*, or egg. If fertilization does not occur, the egg does not complete its meiosis.

- **Formation of the corpus luteum.** The cells that made up the outer sphere of the mature Graafian follicle remain in the ovary, and luteinizing hormone (LH) transforms them into an endocrine structure called the **corpus luteum** (which means yellow body). The corpus luteum secretes both estrogen and progesterone. Unless pregnancy occurs, the corpus luteum degenerates. If pregnancy occurs, the corpus luteum will be maintained by a hormone from the embryo called human chorionic gonadotropin, as we will see later in this chapter.

■ The interplay of hormones coordinates the ovarian and menstrual cycles

Because of the events we have been describing, a woman's fertility is cyclic. At approximately monthly intervals, an egg matures and is released from an ovary. Simultaneously, the uterus is readied to receive and nurture the young embryo. These two processes must be

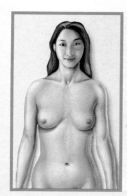
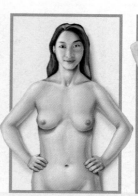

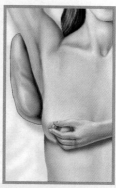

Step 1: Stand in front of the mirror and look at each breast to see if there is a lump, a depression, a difference in texture, or any other change in appearance.

Step 2: Get to know how your breasts look. Be especially alert for any changes in the nipples' appearance.

Step 3: Raise both arms and check for any swelling or dimpling in the skin of your breasts.

Step 4: Lie down with a pillow under your right shoulder and put your right arm behind your head. Perform a manual breast examination. With the nipple as the center, divide your breast into imaginary quadrants.

Step 5: With the pads of the fingers of the left hand, make firm circular movements over each quadrant, feeling for unusual lumps or areas of tenderness. When you reach the upper, outer quadrant of your breast, continue toward your armpit. Press down in all directions.

Step 6: Feel your nipple for any change in size and shape. Squeeze your nipple to see if there is any discharge. Repeat from step 4 on the other breast.

FIGURE **17.A** A monthly breast self-exam can help a woman find lumps in her breast before they have spread to surrounding tissue. It also makes a woman familiar with the consistency of her breast tissue and therefore able to detect changes in consistency that might indicate cancer.

causing physiological changes that leave the breast tissue more resistant to cancer-producing chemicals in the environment.

5. **Exercise.** Even moderate exercise can suppress ovulation in adolescents and women in their twenties (but, unfortunately, exercise is not as likely to suppress ovulation in older women).

Other factors may also influence estrogen levels. For instance, obese women have higher estrogen levels than do thin ones because estrogen is produced in fat cells as well as by the ovaries.

Indeed, obesity, especially if the fat is carried above the waist, translates to a threefold elevated risk of breast cancer. A weight gain as small as 10 pounds when a woman is in her twenties also increases her risk of breast cancer later in life.

coordinated. If fertilization does not occur, the uterine provisions are discarded as menstrual flow. The ovaries and uterus will prepare again during the next cycle. The events in the ovary, known as the ovarian cycle, must be closely coordinated with those in the uterus, known as the **uterine cycle** (or the *menstrual cycle*). The length of the uterine cycle may vary from cycle to cycle and from woman to woman. The timing of events within a uterine cycle will vary with the cycle's length. We will describe the events as they might occur in a 28-day cycle.

A female's fertility is governed by hormones (Table 17.4). Events in both the uterus and the ovary are coordinated by interactions between hormones from the anterior pituitary gland and from the ovary (Figure 17.9). The anterior pituitary gland produces follicle-stimulating hormone (FSH) and luteinizing hormone

(LH). In the female, these hormones cause the ovary to release its hormones, estrogen and progesterone. As in the male, the release of FSH and LH is regulated by a hormone from the hypothalamus called GnRH. Progesterone and estrogen (except at a very high level) exert negative feedback on the anterior pituitary, which causes the levels of FSH and LH to decline. As a result, the levels of estrogen and progesterone decline, reducing the inhibition on the anterior pituitary gland: and so, the hormones cycle.

MENSTRUATION Traditionally, the first day of menstrual flow is considered day 1 of the uterine cycle because it is the most easily recognized event in the cycle (Table 17.5). The bleeding usually lasts an average of 5 days. During menstruation, the ovarian hormones, estrogen and progesterone, are at their lowest levels,

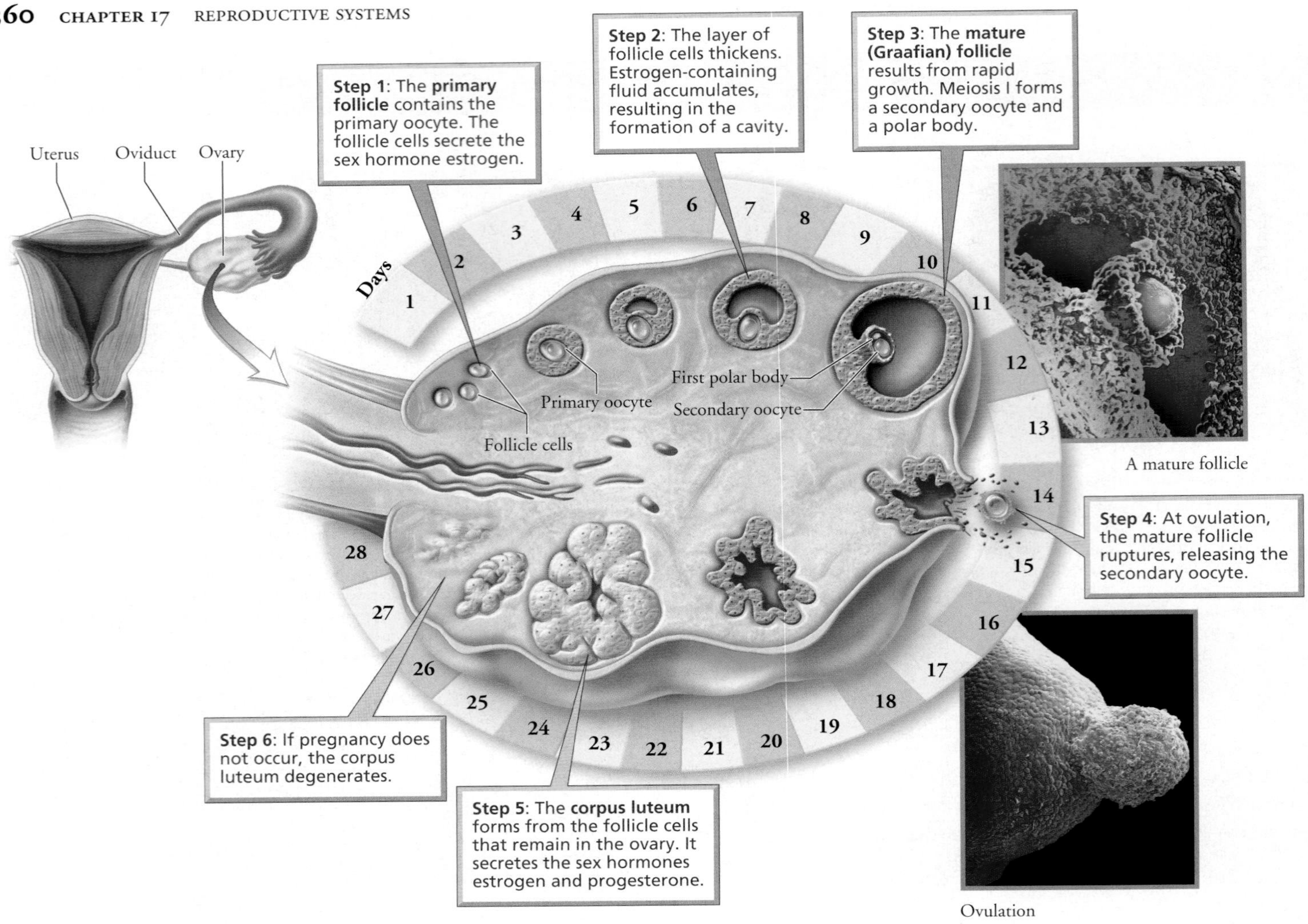

Step 1: The **primary follicle** contains the primary oocyte. The follicle cells secrete the sex hormone estrogen.

Step 2: The layer of follicle cells thickens. Estrogen-containing fluid accumulates, resulting in the formation of a cavity.

Step 3: The **mature (Graafian) follicle** results from rapid growth. Meiosis I forms a secondary oocyte and a polar body.

Step 4: At ovulation, the mature follicle ruptures, releasing the secondary oocyte.

Step 5: The **corpus luteum** forms from the follicle cells that remain in the ovary. It secretes the sex hormones estrogen and progesterone.

Step 6: If pregnancy does not occur, the corpus luteum degenerates.

Uterus   Oviduct   Ovary

Days

Primary oocyte

Follicle cells

First polar body

Secondary oocyte

A mature follicle

Ovulation

FIGURE **17.8** The ovarian cycle. A follicle does not move around the ovary during its development, as depicted here. The steps indicate the sequence of events that occurs in a single place in the ovary during the approximately 28-day cycle.

allowing the anterior pituitary gland to produce its hormones, especially FSH. In turn, FSH causes a number of egg follicles in the ovary to develop and produce estrogen. (Hence its name, follicle-stimulating hormone.) Thus, even as the uterus loses the endometrial lining it prepared during the previous cycle in which no egg was fertilized, the egg that will be released in the next cycle is developing.

ENDOMETRIUM THICKENS   As the egg follicle develops, it produces an increasing amount of estrogen. Estrogen causes the cells in the endometrial lining of the uterus to divide. These cells store nutrients to nourish a future embryo during its early stages of development. In addition, estrogen inhibits the release of FSH through a negative feedback cycle (see Chapter 10). When the egg and follicle are nearly mature, the estrogen level rises rapidly and causes a sudden and spectacular release of LH (and FSH) from the anterior pituitary gland.[2]

[2]Above a critical level, estrogen exerts a positive feedback on the brain and anterior pituitary gland.

TABLE 17.4 REVIEW OF HORMONES INVOLVED IN THE REGULATION OF FEMALE REPRODUCTIVE PROCESSES		
HORMONE	**SOURCE**	**EFFECTS**
Estrogen	Ovaries (follicle cells and corpus luteum)	Maturation of the egg; development and maintenance of female reproductive structures, secondary sex characteristics; thickens endometrium of uterus in preparation for implantation of embryo; cell division in breast tissue
Progesterone	Ovaries (corpus luteum)	Further prepares uterus for implantation of embryo; maintains endometrium
Follicle-stimulating hormone (FSH)	Anterior pituitary gland (in brain)	Stimulates development of a follicle in the ovary
Luteinizing hormone (LH)	Anterior pituitary gland (in brain)	Triggers ovulation; causes formation of the corpus luteum

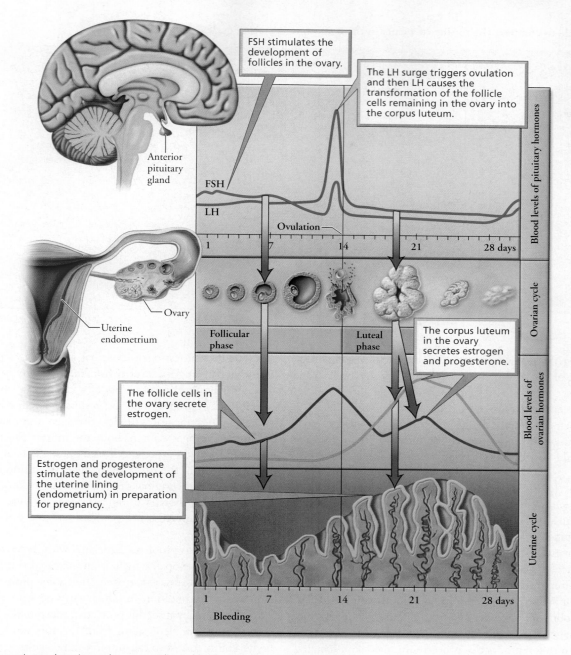

FSH stimulates the development of follicles in the ovary.

The LH surge triggers ovulation and then LH causes the transformation of the follicle cells remaining in the ovary into the corpus luteum.

Anterior pituitary gland

FSH

LH

Ovulation

1   7   14   21   28 days

Blood levels of pituitary hormones

Ovary

Uterine endometrium

Follicular phase

Luteal phase

The corpus luteum in the ovary secretes estrogen and progesterone.

Ovarian cycle

The follicle cells in the ovary secrete estrogen.

Estrogen and progesterone stimulate the development of the uterine lining (endometrium) in preparation for pregnancy.

Blood levels of ovarian hormones

Uterine cycle

1   7   14   21   28 days

Bleeding

FIGURE **17.9** The ovarian and uterine cycles are coordinated by the interplay of hormones from the anterior pituitary gland and the ovary. The timing of events shown is for a 28-day cycle.

OVULATION AND FORMATION OF CORPUS LUTEUM (IN OVARY) The LH surge causes several important events. It causes the egg to undergo its first meiotic division. Next, LH triggers ovulation, and the egg bursts out of the ovary to begin its journey along the oviduct. Continued secretion of LH then transforms the remaining follicle cells into the corpus luteum. The corpus luteum continues the estrogen secretion begun by the follicle cells and, importantly, also secretes progesterone.

ENDOMETRIUM FURTHER PREPARED FOR IMPLANTATION Together, estrogen and progesterone make the endometrial lining of the uterus a hospitable place for the embryo. The blood supply to the endometrium increases. Uterine glands develop that secrete a mucous material that can nourish the young embryo when it arrives in the uterus.

The rising estrogen and progesterone levels inhibit pituitary secretion of FSH and LH. As FSH levels decline, the development of new follicles is inhibited.

CORPUS LUTEUM DEGENERATES If fertilization does not occur, the corpus luteum will degenerate within about 2 weeks (14 days plus or minus 2 days, regardless of the length of the menstrual cycle). The degeneration of the corpus luteum results in falling levels of estrogen and progesterone.

Use this table to compare the timing of events in the ovarian and uterine cycles.

TABLE 17.5 OVARIAN AND UTERINE CYCLES

OVARIAN CYCLE		UTERINE CYCLE	
APPROXIMATE TIMING IN 28-DAY CYCLE	EVENTS	APPROXIMATE TIMING IN 28-DAY CYCLE	EVENTS
Days 1–13	Follicle develops, caused by FSH	Day 1	Onset of menstrual flow (breakdown of endometrium)
	Follicle cells produce estrogen	Day 6	Endometrium begins to get thicker
Day 14	Ovulation is triggered by LH surge		
Days 15–21	Corpus luteum forms and secretes estrogen and progesterone	Days 15–23	Endometrium is further prepared for implantation of the embryo by estrogen and progesterone
Days 22–28	Corpus luteum degenerates, causing estrogen and progesterone level to decline	Days 24–28	Endometrium begins to degenerate owing to declining maintenance by progesterone

MENSTRUATION BEGINS AGAIN Progesterone is essential to the maintenance of the endometrium. As progesterone levels drop because of the degeneration of the corpus luteum, the blood vessels nourishing the endometrial cells collapse. The cells of the endometrium then die and are sloughed off, along with mucus and blood, as menstrual flow. No longer inhibited by estrogen and progesterone, FSH and LH levels begin to climb. Thus, the cycle begins again.

If the egg is fertilized, a hormone from the embryo, called *human chorionic gonadotropin* (*HCG*) maintains the corpus luteum. If pregnancy occurs, HCG will prevent the degeneration of the corpus luteum, keeping estrogen and progesterone levels high enough to prevent endometrial shedding. HCG is detectable in the mother's blood within 7 to 9 days after fertilization and in her urine less than 2 weeks after fertilization. Between the second and third month of development, the placenta has developed sufficiently to take over the production of estrogen and progesterone during the rest of the pregnancy. The corpus luteum then degenerates.

We have seen that reproductive processes in males and females are controlled by the interplay of hormones. Unfortunately, we humans have polluted our environment with chemicals that can interfere with the normal effects of reproductive hormones (see Environmental Issue, *Environmental Estrogens*).

■ Menopause ends a woman's reproductive cycles

A woman's fertility usually peaks when she is in her twenties and then gradually declines. By the time she reaches 45 to 55 years of age, few of the potential eggs prepared before birth remain in the ovaries. Those that do remain become increasingly less responsive to FSH and LH, the hormones that cause egg development and ovulation. The ovaries, therefore, gradually stop producing eggs, and the levels of estrogen and progesterone fall. During this time, menstrual cycles become increasingly irregular. Eventually, ovulation and menstruation stop completely, a transition called **menopause**.

The drop in estrogen levels has a number of physiological effects that can range in severity from annoying to life threatening. At one end of the scale is the loss of a layer of fat that was formerly promoted by estrogen. This loss results in a reduction in breast size and the appearance of wrinkles. Estrogen also plays an important role in regulating a woman's body thermostat. So, as estrogen levels fall, many women experience hot flashes, waves of warmth spreading upward from the trunk to the face. In addition, the absence of estrogen can cause vaginal dryness that can make sexual intercourse painful. Furthermore, without estrogen, the male hormones produced by the adrenal gland predominate and can cause facial hair to grow.

More serious problems associated with the estrogen deficit that accompanies menopause include an increased risk of diseases of the heart and blood vessels. Estrogen offers some protection against atherosclerosis, a condition in which fatty deposits clog the arteries. Thus, after menopause, this protective effect on the heart is lost. Another serious problem stems from the role estrogen plays in the body's ability to absorb calcium from the digestive system and to deposit it in bone. Without calcium, bone becomes weak and porous, a condition known as osteoporosis (discussed in Chapter 5).

Problems with the Female Reproductive System Vary in the Severity of Health Consequences

*P*remenstrual syndrome (PMS) is a collection of symptoms that appear in some women 7 to 10 days before their period begins. These symptoms include depression, irritability, fatigue, and headaches.

Some researchers suggest that a progesterone deficiency is to blame for the symptoms of PMS. Progesterone, for example, has a calming effect on the brain. It also decreases fluid retention. As

Environmental Estrogens

According to the International Union of Pure and Applied Chemistry (IUPAC), certain pollutants are causing gender changes in certain animals, including alligators, polar bears, seals, and some fish. Some of these pollutants, such as the pesticide DDT and bisphenol A (a breakdown product of plastics), mimic the effects of estrogen. Others, such as certain PCBs (polychlorinated biphenyls) that were once used in coolants and plastics, block the effects of natural estrogens. Some pollutants bind to receptors for androgens and can feminize a male by blocking the action of natural androgens. There are at least 45 environmental contaminants reported to cause changes in reproductive systems.

There is already evidence that chemicals with estrogenic properties, consumed as contaminants in food, can accumulate in humans. For example, seal and beluga whale blubber, which are loaded with DDT and PCBs, are a common component of the diet of certain Inuits living in northern Quebec. The breast milk of Inuit women who eat blubber is four to seven times higher in estrogenic pesticides than the breast milk of women in Quebec who do not eat blubber.

Sex hormones choreograph so many biological activities—development, anatomy, physiology, and behavior—that there is a myriad of ways that environmental estrogens could possibly play havoc with human reproduction. Some researchers argue that synthetic estrogenic chemicals cause cancers of the reproductive system, such as breast cancer and testicular cancer, as well as a variety of noncancer effects, including the development of female characteristics in males, decreased sperm counts, higher probability of endometriosis in women, and weakened immune systems. Although evidence of effects on other animal species is mounting, studies of effects on humans remain contradictory.

The developing fetus is probably the most sensitive of all biological entities, and as its organ systems take shape, the proper balance of sex hormones is critical. Too much estrogen at the wrong time can cause a genetically male fetus to have the physical appearance of a female, including feminized sex organs, or to exhibit female behaviors later in life.

It is clear that we have polluted our environment with chemicals that have the potential of altering reproductive anatomy, physiology, and behavior. It is also clear that this pollution is already affecting various animal species. What is not clear is how great a threat these chemicals are to humans or how we humans will decide to cope with this kind of self-induced problem. IUPAC recommends research on these gender-bending pollutants and global monitoring of their levels in the environment. Implementing these recommendations would be costly. As a taxpayer, are you willing to pay for these precautions? Would you vote for a political candidate who supported their implementation?

---

progesterone levels plummet in the days just before menstruation, the nervous system may be stimulated and fluids may be retained. Fluid retention causes an uncomfortable, bloated feeling.

Treatments for PMS are varied. For women whose symptoms are severe, drugs that elevate the levels of serotonin—a neurotransmitter in the brain (Chapter 7)—are administered. Some women with milder symptoms are helped by changes in diet. Caffeine, alcohol, fat, and sodium should be avoided. On the other hand, foods high in calcium, potassium, manganese, and magnesium should be increased. Aerobic exercise for at least half an hour stimulates the release of enkephalins and endorphins (brain chemicals that produce pleasure), which may provide some relief.

Prostaglandins, chemicals used in communication between cells in many parts of the body, are the primary cause of *menstrual cramps*. Endometrial cells produce prostaglandins that, among other things, make the smooth muscle cells of the uterus contract, causing cramps. High levels of prostaglandins can cause sustained contractions, called muscle spasms, of the uterus. These muscle spasms may cut down blood flow and, therefore, the oxygen supply to the uterine muscles, resulting in pain.

Endometriosis is a condition in which tissue from the lining of the uterus is found outside the uterine cavity—commonly in the oviducts, on the ovaries, or on the outside surface of the uterus, the bladder, or the rectum. In these cases, the endometrial tissue has moved out the open ends of the oviducts to the abdominal cavity. Endometrial tissue, wherever it is, grows and breaks down with the hormonal changes that occur during each menstrual cycle. These cyclic changes can cause extreme pain. Moreover, women with endometriosis often have difficulty becoming pregnant.

Vaginitis is an inflammation of the vagina that is most commonly caused by one of three types of organisms: yeast (*Candida albicans*), a protozoan (*Trichomonas vaginalis*), or a bacterium (*Gardnerella vaginalis*). Whereas a yeast infection is generally not sexually transmitted, infections with *Gardnerella* or *Trichomonas* are usually transmitted through sexual intimacy with an infected person. Although vaginal discharge is always the primary sign of vaginitis, the nature of the discharge varies with the causative organism. The discharge accompanying a yeast infection, for instance, is white and curdy, like cottage cheese. It causes intense itching and sometimes burning and itching of the outer lips of the vagina. An infection with *Trichomonas* causes a watery, foamy, greenish or yellowish discharge, vaginal itching, and pain. In contrast, the discharge caused by *Gardnerella* is gray or white and has a fishy odor. It does not usually cause itching. It is important to identify the cause of a vaginal infection because each has a different treatment.

Pelvic inflammatory disease (PID), which affects more than a million women each year in the United States alone, is a catch-all term used to refer to an infection of the pelvic organs. The symptoms of PID—abdominal pain or tenderness, lower-back pain, pain during intercourse, abnormal vaginal bleeding or discharge, fever, or chills—can come on gradually or suddenly and vary in intensity from mild to severe.

Pelvic inflammatory disease can be caused by a number of organisms, but the most common culprits are the sexually transmitted bacteria that cause chlamydia and gonorrhea (discussed in Chapter 17a). The bacteria ascend from the vagina through the cervix to the uterus. There, they may infect the endometrium and then spread to the wall of the uterus and to the oviducts. Near the ovary, the oviduct opens to the abdominal cavity, allowing the infection to spread into the abdominal cavity. The infection can be curbed by treatment with antibiotics, which kill the bacteria.

If PID is not treated quickly, its consequences can be long lasting and severe: reduced fertility and an increased risk of ectopic pregnancy. As the infection spreads through the oviducts, the body attempts to defend itself by secreting fibrous material to wall off the area. The resulting scar tissue can block the tubes. If both tubes are completely blocked, the woman is sterile (unable to have children), because the sperm cannot reach the egg. If a tube is partially blocked, the woman may have difficulty conceiving a child. If pregnancy does occur, the scarring also increases the odds that the pregnancy will be ectopic. An egg is many times larger than sperm. Thus, sperm may be able to swim past the scar tissue and fertilize the egg. The resulting embryo, however, being too big to pass the scar tissue and reach the uterus, may implant in the oviduct.

The Human Sexual Response Has Four Stages

In both men and women, sexual arousal and sexual intercourse involve two basic physiological changes: certain tissues fill with blood (vasocongestion) and certain muscles undergo sustained or rhythmic contractions (myotonia). However, men and women differ in which tissues fill with blood and which muscles contract. The sequence of events that accompanies sexual arousal and intercourse is called the sexual response cycle. Let's consider the four stages of the human response cycle in men and in women.

1. **Excitement.** During the excitement phase of the sexual response cycle, sexual arousal increases. This stage occurs during foreplay, the activities that precede sexual intercourse, and prepares the penis and vagina for intercourse. In a male, blood fills the spongy tissue of the penis, which becomes erect, as described earlier. In a female, blood flow to the breasts, nipples, labia, and clitoris causes these structures to swell. Partly due to an increase in blood flow, the vaginal walls seep fluid that serves as a lubricant, which makes the insertion of the penis easier. In both men and women, muscle contraction can cause the erection of nipples and tension in the arms and legs. Breathing rate and heart rate increase.

2. **Plateau.** During the plateau stage, sexual arousal is maintained at a high level. The changes that began during the first phase will continue through the plateau stage until orgasm. In men, increased blood flow causes the testes to enlarge, and muscle contraction pulls them closer to the body. In women, increased blood flow causes the outer third of the vagina to enlarge and the clitoris to retract under the clitoral hood. In both men and women, breathing rate and heart rate continue to increase. Muscle spasms may begin in the hands, feet, and face.

3. **Orgasm.** The peak of the sexual response cycle, orgasm, is brief, but usually intensely pleasurable. In both sexes, orgasm is characterized by muscle contractions of the reproductive structures. In men, ejaculation occurs during orgasm. Ejaculation takes place in two stages. In the first stage, contractions of glands and ducts in the reproductive system force sperm and the secretions of the accessory glands to the urethra. During the second stage of ejaculation, the urethra contracts and semen is forcefully expelled from the penis. Orgasm in women involves rhythmic muscle contractions in the vagina and uterus. Some women experience orgasm rarely or not at all, and others can experience orgasm repeatedly. The absence of orgasm does not affect a woman's ability to reproduce.

4. **Resolution.** During the resolution stage, the body slowly returns to its normal level of functioning. Men usually experience an interval during the resolution stage when it is impossible for them to achieve another erection or orgasm. Resolution usually takes longer in women than it does in men.

Birth Control Is the Prevention of Pregnancy

Sexual activity is actually a double-edged sword: One edge is the possibility of pregnancy, and the other is the possibility of contracting a sexually transmitted disease (STD; many STDs are discussed in Chapter 17a). The method couples use (if any) to dodge one of the edges of the sword influences the chances of being cut by the other edge. For this reason, as we discuss various means of contraception, we will also consider whether they reduce the risk of spreading sexually transmitted infections.

There are many means of preventing pregnancy available. Different methods interrupt the reproductive process at different points, have varying degrees of effectiveness, and provide varying degrees of protection against STDs.

■ Abstinence is refraining from intercourse

Abstinence, that is, not having intercourse at all, is the most reliable way to avoid both pregnancy and the spread of STDs. However, abstinence is not always used as a means of contraception.

■ Sterilization may involve cutting and sealing gamete transport tubes

Well before the end of their reproductive life spans, most people have had all the children they want. Other than abstinence, *sterilization* is the most effective way to ensure that pregnancy does not occur. However, unlike abstinence, sterilization offers no protection against STDs.

Sterilization in men usually involves an operation called a *vasectomy* in which the vas deferens on each side is cut to prevent sperm from leaving the man's body. The procedure, which usually takes about 20 minutes, can be performed in a physician's office under local anesthesia. The physician simply makes small openings in the scrotum through which each vas deferens can be pulled and cut; a small segment is removed, and at least one end is sealed shut.

Because sperm make up only about 1% of the semen, the volume of semen the man ejaculates is not noticeably reduced. His interest in sex is not lessened, because testosterone, which is responsible for the sex drive, is still released from the interstitial cells of the testis and carried around the body in the blood.

The most common method of female sterilization, *tubal ligation,* involves blocking the oviducts to prevent the egg and sperm from meeting. Commonly, the tubes are cut and the ends seared shut or mechanically blocked with clips or rings. Tubal ligation is frequently done using a procedure called a laparoscopy, in which two small incisions are made in the abdominal cavity. Laparoscopy is generally performed in a hospital under general anesthesia. Because the abdominal cavity is opened, the risk of infection is greater after tubal ligation than it is after a vasectomy. A woman continues to menstruate after a tubal ligation, because the ovarian hormones continue to be produced in a cyclic manner.

Sterilization should be considered permanent even though it is sometimes possible to reverse the procedure. At best, the reversal procedure is expensive and requires that the surgeon have special training in microsurgical techniques. Even then, success is not guaranteed.

■ **Hormonal contraception interferes with regulation of reproductive processes**

Hormonal contraception is currently available only to females. There are two basic types: the methods that combine estrogen and progesterone and the progesterone-only means of contraception.

THE COMBINATION OF ESTROGEN AND PROGESTERONE

Several forms of birth control combine synthetic forms of estrogen and progesterone. Each of these methods works by mimicking the effects of natural hormones that would ordinarily be produced by the ovaries. Among these effects is the suppression of the release of FSH and LH from the anterior pituitary gland. Without these pituitary hormones, the egg does not mature and is not released from the ovary.

The oldest and most familiar means of combined hormonal birth control is "the pill." A hormone-containing birth control pill is taken daily for 3 weeks, followed by a week of daily pills without hormones.

Over the past few years, additional methods of combined hormonal contraception have been developed. There is a vaginal ring (NuvaRing) and a skin patch (Ortho Evra) that slowly release hormones for the 3 weeks they are in place. When a woman uses the birth control pill, the vaginal ring, or the patch, she menstruates during the week in which hormones are not administered. Women who have difficulty remembering to take a pill each day may prefer these newer options. In addition, there is regimen of hormonal contraception in which estrogen and progesterone pills are taken for 84 days, followed by a week of low-dose estrogen pills. A woman menstruates only four times a year, while taking the estrogen-only pills.

When most people speak about "the pill" they are referring to the combination birth control pill, so named because it contains synthetic forms of both estrogen and progesterone. For most healthy, nonsmoking women younger than 35 years of age, the pill is an extremely safe means of contraception. Nonetheless, a small number of pill users do die each year from a pill complication, so the possibility of such complications should be recognized. The risk increases with age and with cigarette smoking. The pill and cigarette smoking can both increase blood pressure and the risk of abnormal blood clot formation. Together, they have a greater effect than either does alone. Problems with the circulatory system are the most important of the serious (sometimes fatal) complications of pill use. Most pill-associated deaths are caused by heart attack or stroke caused when the blood supply to the heart or a region of the brain is blocked. The pill also increases the risk of abnormal blood clot formation.

Pill users have a greater risk of catching certain STDs from an infected partner than do women who use other forms of birth control or no birth control at all. One way the pill increases the likelihood of transmission of these diseases is by making the vaginal environment more alkaline, a condition that favors the growth of bacteria that cause gonorrhea and chlamydia. Another reason that women on the pill have a greater risk of getting STDs is that it makes the user's cervix more vulnerable to disease-causing organisms.

PROGESTERONE-ONLY CONTRACEPTION

The most popular form of progesterone-only contraception is an injection (Depo-Provera) given every 3 months. Progesterone injections are an extremely effective means of contraception. There is also a progesterone-only pill (POP). As with the combined birth control pill, a woman must take this pill faithfully every day. POPs are generally less effective than the combined pill. A progesterone-containing implant, Implanon, was approved by the Food and Drug administration in mid-2006. It is a slender, flexible silicone capsule about the size of a matchstick. Implanon inserted under the skin in a woman's upper arm are barely visible and provide extremely effective protection for three years as the progesterone slowly diffuses out of the capsules.

There are several mechanisms through which different progesterone-only contraceptives may prevent pregnancy. All types may prevent ovulation, but they vary in their ability to do so. They all cause thickening of cervical mucus, which hampers the passage of sperm to reach the egg. In addition, because estrogen is not used in this contraceptive, the endometrium is not prepared properly for the implantation of an embryo should fertilization occur. An important drawback is that, as with all hormonal contraceptives, progesterone-only contraceptives provide no protection against sexually transmitted diseases.

what would you do?

Clinical trials are underway to test the safety and effectiveness of various forms of hormonal birth control designed to be used by males. The hormones reduce sperm levels enough to make a man infertile, and sperm production resumes within a few months without the hormones. The implants do not protect against STDs. If you were a male in a long-term relationship, would you volunteer for this study? If you were the female in this relationship, would you want your partner to participate in the study? What criteria would you use to decide?

■ Intrauterine devices prevent the union of sperm and egg and implantation

An *intrauterine device (IUD)*, is a small device that is inserted into the uterus by a physician to prevent pregnancy (Figure 17.10a). It can be left in place for several years and should be removed by a physician. An IUD is highly effective in preventing pregnancy. It can interfere with both fertilization and implantation. However, IUDs offer no protection against the spread of sexually transmitted diseases. In the future, there may be a male equivalent of an IUD, called an Intra Vas Device, which is a plug inserted into the vas deferens to block the passage of sperm.

A risk associated with the use of an IUD is pelvic inflammatory disease. The risk is greatest in the weeks following IUD insertion, because disease-causing organisms, usually sexually transmitted organisms, can enter the uterus from the vagina at this time. The risk of pelvic infection due to an IUD is low, less than 1%. However, if it does occur, it can lead to sterility or ectopic pregnancy.

■ Barrier methods of contraception prevent the union of sperm and egg

Barrier methods of contraception include the diaphragm, cervical cap, contraceptive sponge, and male and female condoms. Methods in this category work, as their name suggests, by creating a barrier between the egg and sperm.

A *diaphragm* is a dome-shaped soft rubber cup containing a flexible ring. It is inserted into the vagina before intercourse so that it covers the cervix, preventing sperm from passing through (Figure 17.10b). A *cervical cap* is smaller than a diaphragm and fits snugly over the cervix. Before the diaphragm or cervical cap is inserted, spermicidal cream or jelly should be added to the inner surface. Because a diaphragm covers the cervix and the top of the vagina, it offers some, though limited, protection to the woman against important STDs, including chlamydia and gonorrhea. By protecting the cervix from the virus that causes genital warts, it also lessens the risk of cervical cancer. A cervical cap, being smaller, offers no protection against STDs.

Unlike a diaphragm or cervical cap, a contraceptive sponge or condom can be purchased without a prescription. A *contraceptive sponge* is a small sponge that contains a spermicide. One side of the sponge has an indentation into which the cervix fits. The other side has a strap that makes removal easier. The sponge must be thoroughly moistened before use to activate the spermicide. It is then inserted and offers protection against pregnancy for the next 24 hours.

The *male condom* is a thin sheath of latex, polyurethane, or natural membranes ("skin") that is rolled onto an erect penis, where it fits like a glove (Figure 17.10c). The sperm are trapped within the condom and cannot enter the vagina. The effectiveness of condoms in preventing pregnancy depends largely on how consistently they are used.

Other than complete abstinence from sexual activity, the latex condom is the best means of preventing the spread of STDs available today. The skin condoms may offer greater sensitivity, but they do allow some microorganisms, particularly viruses, to pass through. Latex has no pores, so microorganisms cannot pass through it. Health advisors highly recommend that people who might be exposed to STDs use a condom for disease protection, even if they are already using another means of birth control to prevent pregnancy. Keep in mind, however, that a condom can prevent disease transmission only between the body surfaces it separates. Diseases can still be spread by contact between other, unprotected body surfaces that are vulnerable to infection.

The *female condom* is a loose sac of polyurethane, a clear plastic that resembles the type used in a food-storage bag. At each end of the sac is a flexible ring that helps hold the device in place (Figure 17.10d). The female condom does reduce the risk of spreading sexually transmitted infections. Female condoms offer a barrier against STDs and pregnancy to women who cannot count on their male partner to use condoms.

stop and think

A latex condom provides an impenetrable barrier to disease-causing organisms. Yet, studies indicate that women who use a diaphragm or vaginal sponge routinely have lower rates of STDs than do women who rely on their male partners to use a condom. What might explain this seemingly illogical difference?

■ Spermicidal preparations kill sperm

Spermicidal preparations consist of a sperm-killing chemical (nonoxynol-9) in some form of carrier, such as foam, cream, jelly, film, or tablet. When spermicide is used without any other means

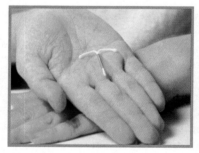

(a) Intrauterine device (IUD)

(b) Diaphragm and spermicidal cream or jelly

(c) Male latex condom

(d) Female latex condom

FIGURE **17.10** Selected methods of birth control

of contraception, foams are the most effective option because they act immediately and disperse more evenly than the others. The sperm-killing effect of any spermicide lasts for about 1 hour after the product has been activated. Laboratory tests have shown that nonoxynol-9 kills the organisms responsible for many STDs. However, this chemical also damages the cells lining the vagina, and this damage could increase a woman's susceptibility to STDs.

■ Fertility awareness is the avoidance of intercourse when fertilization could occur

Fertility awareness, which also goes by the names of *natural family planning* and the *rhythm method*, is a way to reduce the risk of pregnancy by avoiding intercourse on all days on which sperm and egg might meet. This sounds easier than it is. Sperm can live in the female reproductive tract for 2 to 5 days, but the lower value is most likely. An egg lives only 12 to 24 hours after ovulation. Consequently, there are only 4 days in each cycle during which fertilization might occur. But which 4 days? Therein lies the problem. It is difficult enough to pinpoint when ovulation occurs, much less predict it several days in advance. The situation is not entirely hopeless, however. In almost all women, regardless of their cycle length, ovulation occurs 14 (plus or minus 2) days *before* the onset of flow in the next cycle. Thus, if the woman has a stable cycle length, she can count back 14 days from the date she expects her next period and add 2 days on either side to determine the most likely days of ovulation. Intercourse should be avoided for 3 days before the earliest date on which ovulation might occur, because sperm may remain alive that long. It should also be avoided for 1 day after the latest date on which ovulation is expected, because the egg survives for 24 hours. Irregularities in cycle length add complications, of course.

■ The morning-after pill is emergency contraception

The so-called morning-after pill is a means of emergency contraception that can actually be used in the *first few days* after unprotected intercourse. One type (Preven) combines estrogen and progesterone. The second type (Plan B) contains only progesterone. With either type, the first dose should be taken within 3 to 5 days of unprotected intercourse and the second 12 hours later. Scientists do not fully understand the precise mechanisms of action; possible explanations for the pill's efficacy include inhibition or delay of ovulation, prevention of fertilization, thickening of cervical mucus, and alteration of the endometrium, making it an inhospitable place for implantation of the young embryo. Plan B can be purchased over-the-counter in a pharmacy by women 18 years of age or older.

exploring further . . .

In this chapter, we have seen how the male and female reproductive systems function to produce offspring. Unfortunately, during sexual intimacy, the vulnerable surfaces of certain parts of the reproductive system come into contact, potentially allowing sexually transmitted organisms to spread, as we will see in the next chapter.

HIGHLIGHTING THE CONCEPTS

The Gonads Produce Gametes and Sex Hormones (p. 350)

1. The gonads (testes and ovaries) are the reproductive structures that produce the gametes (sperm and eggs), as well as sex hormones. The male sex hormone is testosterone. The female sex hormones are estrogen and progesterone. A sperm and egg fuse at fertilization, producing a cell called a zygote that will develop into a new individual.

A Male's Reproductive Role Differs from a Female's (p. 350)

2. A male produces a large number of sperm and delivers them to the female reproductive system. A female usually produces only one nutrient-filled egg during each approximately month-long interval. The female also nourishes and protects the developing baby (the embryo and later the fetus) in her uterus.

The Male Reproductive System Delivers Sperm to the Egg (pp. 350–355)

3. The male reproductive system is composed of the testes, a series of ducts (the epididymis, vas deferens, and urethra), accessory glands (the prostate, seminal vesicles, and bulbourethral glands), and the penis.
4. The testes are located outside the body cavity in a sac called the scrotum, which helps regulate the temperature of the testes to ensure proper sperm development.
5. Within the seminiferous tubules of the testis, the production of gametes, called spermatogenesis, is a continuous process. Sperm are haploid gametes produced by meiosis.
6. After leaving the seminiferous tubules, sperm mature and are stored in the epididymis. At ejaculation, sperm travel through the vas deferens and the urethra to leave the body.

7. Semen is the fluid released when a man ejaculates. Most of the semen consists of the secretions of the accessory glands.
8. The penis transfers sperm to the female. Erectile dysfunction is the inability to achieve an erection. There are several options for treatment.
9. Each sperm has three distinct regions. The head of the sperm contains the male's genetic contribution to the next generation. The midpiece is packed with mitochondria, which produce ATP to power movement. The whiplike tail propels the sperm to the egg.
10. The male hormone testosterone is produced by the interstitial cells, which are located within the testes between the seminiferous tubules.
11. Male reproductive processes are regulated by an interplay of hormones from the anterior pituitary gland in the brain (LH and FSH), from the hypothalamus (GnRH), and from the testis (testosterone and inhibin).

WEB TUTORIAL 17.1 The Male Reproductive System

The Female Reproductive System Produces the Eggs and Nurtures the Embryo and Fetus (pp. 355–362)

12. The female reproductive system consists of the ovaries, oviducts, uterus, vagina, and external genitalia.
13. The ovaries produce eggs and the female hormones estrogen and progesterone.
14. The oviducts transport the immature egg, zygote, and then the early embryo to the uterus. Fertilization usually occurs in the oviduct.
15. The uterus supports the growth of the developing baby (embryo and then fetus). The embryo implants in the lining of the uterus, the endometrium, which provides nourishment during development. The muscular wall of the uterus allows the uterus to stretch as the fetus grows, and contractions of the muscle force the baby out of the uterus during childbirth. The cervix is the opening of the uterus.

16. The breasts produce milk to nourish the baby.

17. The events of the ovarian cycle lead to the release of an egg. A female is born with a finite number of primary follicles. A primary follicle is a primary oocyte (an immature egg) surrounded by a single layer of follicle cells. The primary oocytes will remain in this state until puberty, when (usually) one each month will resume development and be ovulated as a secondary oocyte.

18. Hormones from the pituitary gland (FSH and LH) and from the ovary (estrogen and progesterone) regulate the ovarian cycle (which prepares an egg for fertilization) and the uterine or menstrual cycle (which prepares the endometrium of the uterus for implantation of the embryo). However, if the egg is fertilized, the young embryo produces human chorionic gonadotropin (HCG), which maintains the corpus luteum. HCG is the hormone that pregnancy tests detect.

19. At menopause, which usually occurs between the ages of 45 and 55, menstruation and ovulation stop. As a result, the levels of estrogen and progesterone drop. Menopause has many psychological and physiological effects on a woman's body.

WEB TUTORIAL 17.2 The Female Reproductive System
WEB TUTORIAL 17.3 Preparation of the Endometrium for Implantation

Problems with the Female Reproductive System Vary in the Severity of Health Consequences (pp. 362–364)

20. Premenstrual syndrome (PMS), a collection of symptoms that occur in some women several days before their period begins, may be caused by low levels of progesterone or of the brain chemical serotonin. Menstrual cramps are caused by prostaglandins, which make muscles contract. Endometriosis is a condition in which endometrial tissue is found outside the uterus. Vaginitis is an inflammation of the vagina caused by yeast, a protozoan, or a bacterium. Pelvic inflammatory disease is a bacterial infection of the pelvic organs that can increase a woman's risk of becoming infertile or of having an ectopic pregnancy.

The Human Sexual Response Has Four Stages (p. 364)

21. The human sexual response is the sequence of events that occur during sexual intercourse. This cycle consists of four phases: excitement (increased arousal), plateau (continued arousal), orgasm (climax), and resolution (return to a normal level of functioning).

Birth Control Is the Prevention of Pregnancy (pp. 364–367)

22. Abstinence (refraining from intercourse) and sterilization (vasectomy or tubal ligation) are the most effective ways to prevent pregnancy. Hormonal contraception (combined estrogen plus progesterone or progesterone-only methods) interferes with the regulation of reproductive processes. An intrauterine device prevents the union of sperm and egg and implantation. Barrier methods of contraception (the diaphragm, cervical cap, contraceptive sponge, and male or female condom) prevent the union of sperm and egg. Spermicidal preparations kill sperm. Natural family planning (the rhythm method) consists of abstinence at times when fertilization could occur. Morning-after pills are emergency contraception that can reduce the risk of an unwanted pregnancy resulting from unprotected intercourse.

WEB TUTORIAL 17.4 Ovulation and Hormonal Birth Control Methods

KEY TERMS

gonad p. 350	scrotum p. 351	gonadotropin-releasing hormone (GnRH) p. 354	vagina p. 356
gamete p. 350	seminiferous tubule p. 352	luteinizing hormone (LH) p. 354	ovarian cycle p. 357
testes p. 350	epididymis p. 352		follicle p. 357
sperm p. 350	vas deferens p. 352	follicle-stimulating hormone (FSH) p. 355	ovulation p. 358
testosterone p. 350	semen p. 352	oviduct p. 355	corpus luteum p. 358
ovary p. 350	prostate gland p. 352	uterus p. 355	uterine cycle (menstrual cycle) p. 359
egg p. 350	seminal vesicles p. 352	endometrium p. 356	menopause p. 362
estrogen p. 350	bulbourethral gland p. 353	cervix p. 356	
progesterone p. 350	penis p. 353		
fertilization p. 350	acrosome p. 354		

REVIEWING THE CONCEPTS

1. Name the male and female gonads. What are the functions of these organs? *p. 350*

2. How is the temperature maintained in the testes? Why is temperature control important? *p. 351*

3. Trace the path of sperm from their site of production to their release from the body, naming each tube the sperm pass through. *p. 352*

4. Name the male accessory glands and give their functions. *pp. 352–353*

5. What is the function of the penis? Describe the process by which the penis becomes erect. *p. 353*

6. Name and describe the functions of the three regions of a sperm cell. *p. 354*

7. List the hormones from the hypothalamus, the anterior pituitary gland, and the testes that are important in the control of sperm production. Explain the interactions between these hormones. *pp. 354–355*

8. What are the two main layers in the wall of the uterus? *pp. 355–356*

9. List the major structures of the female reproductive system and give their functions. *pp. 355–357*

10. What is an ectopic pregnancy? Why is it dangerous to the mother's health? *pp. 356–357*

11. Describe the structure of breasts. What is their function? *p. 357*

12. Describe the ovarian cycle. Include in your description primary oocytes, primary follicles, mature Graafian follicles, and the corpus luteum. *pp. 357–358*

13. Describe the interplay of hormones from the anterior pituitary and from the ovaries that is responsible for the menstrual cycle. *pp. 358–362*

14. What is menopause? Why does menopause lower estrogen levels? What are some effects of lowered estrogen levels? *p. 362*

15. What are the stages of the human sexual response? *p. 364*

16. Describe a vasectomy and a tubal ligation. *pp. 364–365*
17. How does the combination birth control pill reduce the chances of pregnancy? *p. 365*
18. What health risks are associated with use of the birth control pill? *p. 365*
19. How do progesterone-only means of contraception reduce the risk of pregnancy? *p. 365*
20. What is an IUD? *p. 366*
21. How do the diaphragm, male condom, and female condom prevent pregnancy? *p. 366*
22. The interstitial cells
    a. are found in the seminal vesicles.
    b. produce a secretion that makes up most of the volume of semen.
    c. secrete testosterone.
    d. store sperm.
23. Choose the *incorrect* statement about semen.
    a. It helps lubricate passageways in the male reproductive system, so sperm travel through them more easily.
    b. Sperm cells make up most of the volume of semen.
    c. It helps reduce the acidity of the female reproductive system, thereby increasing sperm survival.
    d. It contains nourishment for the sperm.

24. After sperm are released in the female reproductive system, they can swim to the egg and fertilize it. Which is the correct order of structures through which the *sperm* will pass?
    a. Vagina, cervix, body of uterus, oviduct
    b. Ovary, cervix, body of uterus, oviduct
    c. Endometrium, cervix, oviduct, ovary
    d. Ovary, oviduct, endometrium, cervix
25. Which is the *correct pairing* of a structure with its function?
    a. Endometrium, the usual site of fertilization
    b. Corpus luteum, production of estrogen and progesterone
    c. Epididymis, production of testosterone
    d. Seminal vesicles, store sperm while they mature
26. The controversial "abortion pill" RU486 works by preventing progesterone from acting. This effect would cause an abortion because progesterone is needed to
    a. trigger ovulation.
    b. cause the formation of the corpus luteum.
    c. maintain the endometrium.
    d. increase the levels of LH.
27. Sperm are stored and mature in the _____.
28. Fertilization usually occurs in the _____.

APPLYING THE CONCEPTS

1. Katie is a 25-year-old woman who had pelvic inflammatory disease 2 years ago. She was treated with antibiotics and cured. She and her husband have been trying unsuccessfully to conceive a child for the last year. What is the most likely reason that she is having difficulty conceiving a child?
2. You are a health-care provider in a family planning clinic. Your first client is a woman who is 22 years old and in excellent health. She is about to marry her high school sweetheart. She and her fiancé have never dated anyone else. They want a very effective means of contraception because they would like to wait until he finishes graduate school to start a family. She does not smoke. Which means of birth control would you recommend for this client? Why?
3. Endometriosis, a condition in which endometrial tissue implants on pelvic organs outside the uterus, causes pain in the pelvic area. Why would the pain be greatest around the time of menstruation?

Additional questions can be found on the companion website.

4. Doctors in fertility clinics sometimes use FSH as a fertility drug to treat women who are having difficulty conceiving a child. Women who use FSH as a fertility drug are more likely than women who conceive naturally to have several babies at once. Why would FSH increase the incidence of multiple births?
5. A male friend of yours and his wife would like to have a baby. They are having difficulty conceiving a child. The doctor told him that the reason for this difficulty is a low sperm count. Since further tests are costly and would take time to complete, the doctor told your friend that as a first step, he might try wearing loose clothing and switching to loose boxer shorts, rather than his customary tight briefs. Why do you think the doctor made this suggestion?

17a

SPECIAL TOPIC

Sexually Transmitted Diseases and AIDS

STDs can be dangerous and highly infectious. You would be wise to learn about modes of transmission, signs and symptoms, and how to protect yourself from contracting STDs.

STDs Are Extremely Common and Can Have Long-Lasting Effects

STDs Caused by Bacteria Can Be Cured with Antibiotics

- Chlamydia can cause pain during urination, PID, or no symptoms
- Gonorrhea can cause pain during urination, PID, or no symptoms
- Syphilis can progress through three stages when untreated

STDs Caused by Viruses Can Be Treated but Not Cured

- Genital herpes can cause painful, fluid-filled blisters
- Genital warts can lead to cervical, penile, or anal cancer

An HIV Infection Progresses to AIDS

- HIV/AIDS is a global epidemic
- HIV consists of RNA and enzymes encased in a protein coat
- HIV enters the cell, rewrites its RNA as DNA, inserts DNA into the host chromosome, and replicates
- Most HIV is transmitted through sexual contact, intravenous drug use, or from a pregnant woman to her fetus
- Sites of HIV infection include the immune system and the brain
- An HIV infection progresses through several stages as helper T cell numbers decline
- Treatments for HIV infection are designed to block specific steps in HIV's replication cycle

Kathie, 19, felt she was in perfect health. Nonetheless, she had recently experienced both an urge to urinate frequently and pain during urination, so she made an appointment with her physician. Dr. Spradley took a urine sample and gave it to a technician for testing. While awaiting the results, she asked Kathie many questions, including whether Kathie was sexually active. Kathie replied that she was not and asked why information about her sexual history was important. Dr. Spradley replied that Kathie's symptoms were those of a urinary tract infection, and a common cause is a bacterium called *Chlamydia trachomatis*. In Kathie's case, they could rule out *Chlamydia* as a cause of the infection because this bacterium is transmitted through sexual contact. The doctor also explained that infections caused by *Chlamydia* could make it difficult for a woman to become pregnant in the future. Dr. Spradley left to get the results of Kathie's urine test. "Sure enough," she said. "You do have a urinary tract infection. Another bacterium, called *E. coli*, is the most likely cause, because you are not sexually active. We will culture your urine to be sure, and I'll prescribe an antibiotic. You should feel better in a day or two, but be sure to continue taking the antibiotic until the entire prescription is used."

This chapter discusses sexually transmitted diseases (STDs), diseases that are transmitted by sexual contact. It first describes three STDs that are caused by bacteria: chlamydia, gonorrhea, and syphilis. Then it describes three STDs that are caused by viruses: genital herpes, genital warts, and HIV/AIDS. HIV is discussed in this chapter because sexual contact is the most common means of transmission, although the virus can be transmitted in other ways as well. ■

STDs Are Extremely Common and Can Have Long-Lasting Effects

Sexually transmitted diseases (STDs) take a serious toll on humanity, both in terms of their direct effects on the victims and because of the cost to society in general. Thirteen million infections occur each year in the United States. Two-thirds of the U.S. residents infected are younger than 25. The cost of treating their infections is in the millions of dollars.

We should also mention another trait of STDs: they are sexist. They affect women more severely than men, causing sterility more often in women than in men, ectopic pregnancy (a pregnancy in which the embryo begins development outside the uterus), and cervical cancer.

One of the reasons that STDs are so rampant is that people are often unaware of being infected. Many cases of STDs lack symptoms. Indeed, the prevalence of asymptomatic cases has led to use of the term sexually transmitted *infection* (STI) rather than sexually transmitted *disease*. Also, the symptoms of some STDs, such as syphilis, disappear without treatment, leading the person to believe that he or she is cured. Unfortunately, the person would be mistaken, as we will see.

STDs Caused by Bacteria Can Be Cured with Antibiotics

We will consider three STDs that are caused by bacteria: chlamydia, gonorrhea, and syphilis.

■ Chlamydia can cause pain during urination, PID, or no symptoms

Chlamydia is the most frequently reported infectious disease in the United States. Chlamydial infections are rapidly becoming epidemic because they are highly contagious and because they do not necessarily cause noticeable symptoms that would prompt the infected person to seek treatment (Figure 17a.1). In fact, most people who have chlamydia do not have symptoms, and many people learn that they have it only because a responsible partner diagnosed with the infection informs them. Any symptoms that do develop may take weeks or months to appear. In the meantime, even without outward signs of infection, the disease can be passed along to others.

Chlamydia is caused by a very small bacterium (*Chlamydia trachomatis*) that cannot grow outside a human cell. It is completely

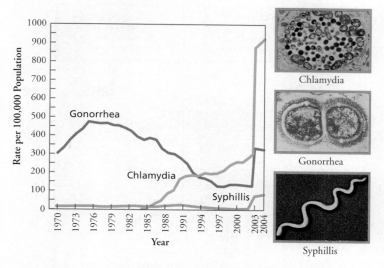

FIGURE **17a.1** The relative rates of chlamydia, gonorrhea, and syphilis from 1984 to 2004. The bacteria that cause these STDs are shown at the right.

dependent on another cell for energy, because it cannot produce its own ATP molecules.

Symptoms of chlamydia depend largely on the site of the infection. The bacteria can infect the cells of mucous membranes, including those of the urethra, vagina, cervix, oviducts, throat, anus, and eyes. Worldwide, the major cause of blindness is chlamydial infection.

The most common symptoms of chlamydia are those of urinary tract infection. Inflammation of the urethra causes a burning sensation during urination, itching or burning around the opening of the urethra, and a clear or white discharge from the urethra. The burning pain experienced while urinating spurs the person to seek medical attention. In males, the most common site of chlamydial infection is the urethra, because this is the mucous membrane most likely to be exposed to the bacteria during sexual intimacy.

Chlamydial infections in women, however, are more likely to affect the pelvic organs than the urethra, because the vagina and cervix are a woman's primary sites of contact during sexual intimacy. The general term for an infection of the pelvic organs is *pelvic inflammatory disease* (PID; see Chapter 17). Symptoms of PID include vaginal discharge, bleeding between periods (which is actually from the cervix), and pain during intercourse. Unfortunately, pelvic infections are less likely to produce symptoms than are urinary tract infections, and the untreated infection may permanently damage a female's reproductive system. Therefore, it is important that a man who is diagnosed with chlamydia inform everyone with whom he has been sexually intimate.

Chlamydia can have long-term reproductive consequences because it often causes scar tissue to form in the tubes through which gametes travel. If the vas deferens or oviducts are completely blocked by scar tissue, the man or woman is sterile. Sterility is more likely to occur in women than in men, because women are less likely to have symptoms that prompt treatment. In a woman, partial blockage of the oviduct may allow the tiny sperm to reach the egg and fertilize it, but the blockage will not allow the much larger embryo to move past the scar tissue to reach the uterus (Figure 17a.2). The embryo may then implant in the oviduct, resulting in an ectopic pregnancy that places the woman's life in danger.

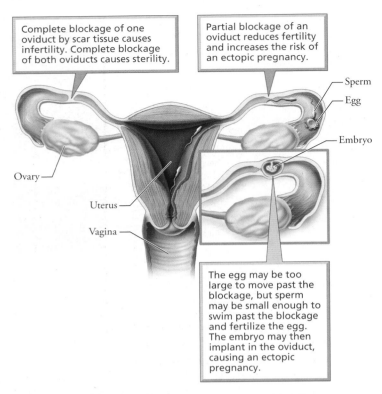

Complete blockage of one oviduct by scar tissue causes infertility. Complete blockage of both oviducts causes sterility.

Partial blockage of an oviduct reduces fertility and increases the risk of an ectopic pregnancy.

Sperm

Egg

Embryo

Ovary

Uterus

Vagina

The egg may be too large to move past the blockage, but sperm may be small enough to swim past the blockage and fertilize the egg. The embryo may then implant in the oviduct, causing an ectopic pregnancy.

FIGURE **17a.2** Chlamydia and gonorrhea can cause pelvic inflammatory disease, which can cause scar tissue to form in the oviducts.

Chlamydial infections during pregnancy place the fetus at risk. Specifically, chlamydia can cause the protective membranes around the fetus to rupture early, which can kill the fetus. In addition, newborns can be infected as they pass through an infected cervix or vagina at birth. Chlamydial infections picked up during birth can affect the mucous membrane of the eye (the conjunctiva), the throat, vagina, rectum, or lungs and can be fatal to an infant.

Diagnosis of chlamydia can now be done in a physician's office with quick, accurate tests, including a urine test that detects the DNA of *Chlamydia*. Considering the suffering that chlamydia can cause, it is almost ironic that it can be so easily cured with antibiotics.

■ Gonorrhea can cause pain during urination, PID, or no symptoms

One of the oldest known sexually transmitted diseases, gonorrhea, is caused by the bacterium *Neisseria gonorrhoeae*. Gonorrhea, like chlamydia, primarily infects the mucous membranes of the genital or urinary tract, throat, or anus. The bacteria that cause gonorrhea cannot survive long if exposed to air, so gonorrhea is generally transferred by direct contact between an infected mucous membrane and an uninfected one during sexual intimacy. A woman who has unprotected intercourse with a man who has gonorrhea has a 50% chance of becoming infected. If she is taking birth control pills, however, she will almost certainly get it. The pill creates an alkaline environment in which the bacteria thrive. On the other hand, a man who has intercourse with a woman who has gonorrhea has only about a 22% chance of becoming infected. The difference

in transfer rates is probably due to differences in the surface area of the uninfected membrane exposed during intercourse—the small opening of the urethra in a male versus the entire vaginal lining and cervix in a female. Gonorrhea can also spread to the lining of the anus during anal intercourse or to the throat during oral sex. Furthermore, it can infect the eyes if an eye is touched by a finger that has just touched an infected area. In the case of infants, eye infection can occur as the baby passes through an infected vagina or cervix during the birth process.

The symptoms of gonorrhea are similar to those described for chlamydia. In males, symptoms of urethritis (inflammation of the urethra) usually begin within 2 to 20 days of exposure. Typically, there is a discharge from the urethra that is thin and watery at first. Several days to a week later, the discharge becomes thicker and turns to yellow or white (Figure 17a.3). Urination can produce a painful, burning sensation. In a manner similar to that of chlamydia, gonorrhea can spread to the vas deferens, epididymis, prostate, or bladder. In women, it may infect the urethra, causing a discharge and burning sensation. However, it is more likely to infect the vagina and cervix and eventually result in PID. In the long run, gonorrhea can cause sterility in either sex and increase the risk of ectopic pregnancy in women. Rarely, gonorrhea can cause arthritis or fatal heart and liver infections. Keep in mind that gonorrhea does not always produce symptoms. Gonorrhea can be diagnosed in several ways. A urine test can detect the DNA in *Neisseria* to confirm gonorrhea. Gonorrhea can also be diagnosed by observing the bacterium in a smear of cells taken from the infected area or grown in a laboratory.

At one time, gonorrhea could be cured easily with antibiotics such as penicillin or ceftriaxone. Unfortunately, because of the misuse of antibiotics, some strains of the gonococcal bacteria are now drug resistant. Therefore, every patient undergoing treatment must be retested for gonorrhea when the antibiotic regimen has been completed to make sure the infection is cured. Immunity to gonorrhea does not develop, so you can become infected with each new exposure. Also, if all partners are not treated and cured, the disease can "Ping-Pong" among them.

FIGURE **17a.3** A yellowish white discharge from the urethra caused by gonorrhea

■ Syphilis can progress through three stages when untreated

Syphilis rates decreased steadily during the 1990s, reaching their lowest point in 2000. Since then, the frequency of syphilis has risen slightly, primarily among men who have sex with men.

Syphilis is caused by a corkscrew-shaped bacterium (a spirochete) called *Treponema pallidum*. As terrible as these bacteria are, they are extremely delicate and cannot survive drying or even minor temperature changes. Thus, a person cannot contract syphilis from toilet seats, wet towels, or drinking glasses. You become infected only by direct contact with an infected sexual partner. The bacteria can invade any mucous membrane or enter through a break in the skin. Unfortunately, the bacteria can also cross the placenta and infect the growing fetus if a pregnant woman is infected.

If untreated, syphilis progresses through three stages. The first stage is characterized by a painless bump, called a chancre (*shang* ker), that forms at the site of contact, usually within 2 to 8 weeks of the initial contact (Figure 17a.4). Because syphilis is sexually transmitted, the chancre usually appears on the genitals. The chancre can, however, form anywhere on the body, because minute scratches on the skin will allow the bacteria to enter. The chancre appears as a hard, reddish-brown bump with raised edges that make it resemble a crater. Unfortunately, a chancre is not always noticed, because it can form in places that are difficult to see and because it may be mistaken for something else. The chancre lasts for one to a few weeks. During this time, it ulcerates, becomes crusty, and disappears. The disease, however, has not disappeared.

During the primary stage of syphilis, diagnosis is made by identifying the bacterium in the discharge from a chancre. Treatment at this stage usually involves an antibiotic such as penicillin.

If syphilis is not detected and treated in the first stage, it can progress to the second stage, which is characterized by a reddish-brown rash covering the entire body, including the palms of the hands and soles of the feet. The rash usually appears within a few weeks to a few months after the disappearance of the chancre. The infected person may confuse the rash with an ordinary skin rash, such as might be caused by an allergy or German measles. The rash does not hurt or itch. Each bump eventually breaks open, oozes fluid, and becomes crusty. The ooze contains millions of bacteria. Therefore, the secondary stage is the most contagious stage of syphilis.

Secondary syphilis usually has symptoms besides the rash, and these may produce the first pain associated with the disease. At first, there may be flulike symptoms, including a slight temperature, chills, a sore throat, and gastrointestinal upset. In addition, warty growths may appear in the genital area. Patches of hair may fall out from the head, eyebrows, and eyelashes.

Syphilis in this stage is diagnosed by a blood test to detect antibodies for *Treponema* bacteria. Second-stage syphilis *can* be cured with antibiotics such as penicillin, but the disease becomes increasingly difficult to treat as it progresses.

Again, the symptoms go away whether or not they are treated. However, in some people, the symptoms recur periodically. Other people may have no outward signs of syphilis for years.

The third stage brings the drama to a grisly conclusion. At this stage, lesions called gummas may appear on the skin or certain internal organs. Gummas often form on the aorta, the major artery that delivers blood from the heart to the rest of the body. As a result, the artery wall may be weakened and may burst, causing the person to bleed to death internally. The bacteria that cause syphilis can also infect the nervous system, damaging the brain and spinal cord, so that the person may have difficulty walking, become paralyzed, or become insane. The disease may also cause blindness by affecting the optic nerve, the iris, and other parts of the eye.

In its third stage, syphilis is more difficult to diagnose and very difficult to treat. General blood tests may be negative, but specialized blood tests are usually positive. Treatment requires massive doses of antibiotics over a prolonged period. The damage that has already been done to the body cannot be repaired.

Characteristics of bacterial STDs (their symptoms, diagnosis, treatment, and effects) are summarized in Table 17a.1.

STDs Caused by Viruses Can Be Treated but Not Cured

Unlike STDs caused by bacteria, viral STDs cannot be cured with antibiotics. One can treat the symptoms, but one can never be certain that the virus has been eliminated. Therefore, it is always important to take precautions not to pass the virus on to others.

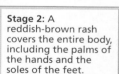
Stage 1: The first stage of syphilis is characterized by a chancre, a hard, painless, crater-shaped bump at the place in the body where the bacteria entered, usually the genitals.

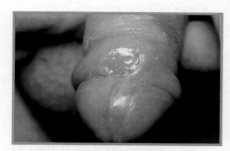

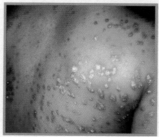
Stage 2: A reddish-brown rash covers the entire body, including the palms of the hands and the soles of the feet.

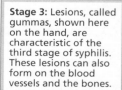
Stage 3: Lesions, called gummas, shown here on the hand, are characteristic of the third stage of syphilis. These lesions can also form on the blood vessels and the bones.

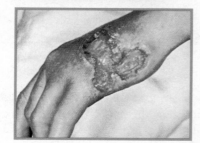

FIGURE **17a.4** Untreated syphilis goes through three stages.

Use this table as a reference for information on bacterial STDs.

TABLE 17A.1 OVERVIEW OF BACTERIAL STDs

DISEASE	SYMPTOMS	DIAGNOSIS AND TREATMENT	EFFECTS
Chlamydia	First symptoms occur 7–21 days after contact Up to 75% of women and 50% of men show no symptoms *Women:* Vaginal discharge Vaginal bleeding between periods Pain during urination and intercourse Abdominal pain accompanied by fever and nausea *Men:* Urethral discharge Pain during urination	*Diagnosis:* Urine test for chlamydial DNA *Treatment:* Antibiotics	Long-term reproductive consequences such as sterility Infection can pass to infant during childbirth Can cause rupture of the protective membrane surrounding the fetus
Gonorrhea	First symptoms occur 2–21 days after contact About 30%–40% of men and women show no symptoms *Women:* Vaginal discharge Pain during urination and bowel movement Cramps and pain in lower abdomen More pain than usual during menstruation *Men:* Thick yellow or white discharge from penis Inflammation of the urethra Pain during urination and bowel movements	*Diagnosis:* Examination of penile discharge or cervical secretions Urine test for DNA of the bacterium that causes gonorrhea Cell culture *Treatment:* Antibiotics	Can cause long-term reproductive consequences such as sterility Infection can pass to infant during childbirth Can cause heart trouble, arthritis, and blindness
Syphilis	*Stage 1:* Occurs 2–8 weeks after contact. Chancre forms at site of contact. Lymph nodes in groin area swell *Stage 2:* Occurs 6 weeks to 6 months after contact Reddish brown rash appears anywhere on the body Flulike symptoms present Ulcers or warty growth may appear Patches of hair may be lost *Stage 3:* Lesions appear on skin and internal organs May affect nervous system Blindness Brain damage	*Diagnosis:* Identification of the bacterium from a chancre Blood test to detect antibodies to the bacterium that causes syphilis *Treatment:* Large doses of antibiotics over a prolonged period of time	Infection can pass to fetus during pregnancy Can cause heart disease, brain damage, blindness, and death

■ **Genital herpes can cause painful, fluid-filled blisters**

Genital herpes is caused by herpes simplex viruses (each abbreviated HSV). There are actually two types of HSV. They cause similar sores but tend to be active in different parts of the body. Type 1 (HSV-1) is most commonly found above the waist, where it causes fever blisters, or cold sores. Type 2 (HSV-2) is more likely to be found below the waist, where it causes genital herpes on the genitals, buttocks, or thighs. As a result of oral-genital sex with an infected person, however, HSV-1 can cause sores on the genitals, and HSV-2 can cause cold sores (Figure 17a.5).

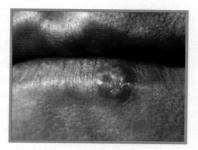

(a) Oral herpes: Cold sores (also called fever blisters) are most often caused by HSV-1.

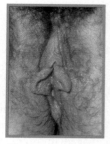

(b) Genital herpes, shown here on the external genitalia of a female, is usually caused by HSV-2.

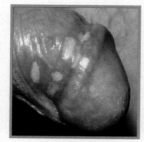

(c) Genital herpes, shown here on the penis of a male, is usually caused by HSV-2.

FIGURE **17a.5** Herpes simplex viruses cause blisters to form at the site of infection.

Herpes simplex virus is quite contagious and can be spread by direct contact with viruses that are being shed from an infected surface or that are in the fluid on the blisters. Mucous membranes are most susceptible. Skin is a good barrier, unless there is a cut, abrasion, or burn; acne; eczema; or other break in the skin.

The first hints of genital herpes infection begin about 2 to 20 days (an average of 6 days) after the initial contact. The initial bout may be severe, often accompanied by fever, aching muscles, and swollen glands in the groin. Soon blisters appear, accompanied by local swelling, itching, and possibly burning, especially if the blisters get wet during urination. The blisters form at the site of contact and last about 2 days before they ulcerate, leaving small, painful sores.

When the symptoms subside, the virus is not gone from the body. During dormant periods, the virus retreats to ganglia (clusters of nerve cells) near the spinal cord. Then, at times of stress (emotional or physical), the virus may be reactivated and blisters may re-form.

Unfortunately, most people with herpes apparently never develop symptoms, or the symptoms are so mild that they go unnoticed. Nonetheless, these asymptomatic individuals can unwittingly transmit the virus. In fact, *most* cases of genital herpes are transmitted by partners who never had symptoms or whose symptoms were atypical and undiagnosed!

Genital herpes is most contagious when active sores are present, so sexual contact should be avoided during an outbreak. Moreover, unprotected contact should be avoided as soon as there is a

hint of a recurrence. As we have seen, genital herpes can also be spread when there are no symptoms.

A herpes infection can cause problems during pregnancy. Most women with herpes have successful pregnancies and normal deliveries. In some cases, however, the infection spreads to the fetus as it is growing in the uterus and can cause miscarriage or stillbirth. HSV can also be transmitted to the newborn during vaginal delivery, especially if active sores are present. To avoid exposing the baby to HSV, physicians generally recommend delivery by cesarean section (surgical removal from the uterus) if the mother has active sores.

Clinicians diagnose herpes by examining the sores or by testing the fluid from sores for the presence of the virus. It is also possible to identify the DNA of the herpes virus in the material wiped off when the genital area is gently swabbed. Blood tests can also detect antibodies to the herpes viruses. However, such blood tests may not be useful during an initial attack because the antibodies may not show up for several months. Newer tests can distinguish HSV-2 from HSV-1. This distinction is helpful because most people have had a cold sore and will test positive for HSV-1. However, HSV-2 is usually the cause of genital herpes.

Although there is no cure for HSV, there are three antiviral drugs—Zovirax, Famvir, and Valtrex—that help ease symptoms during the initial outbreak and reduce the frequency of recurrences. Unfortunately, strains of HSV that are resistant to antiviral drugs are beginning to crop up.

New therapies that are intended to stimulate immune responses against HSV-2 are now being tested. One is a vaccine. Although the vaccine's effectiveness is limited to a select group of women, there is hope that it could reduce genital herpes in the general population if girls were vaccinated before they are exposed to herpes viruses. In addition, a gel is available that boosts the immune response. When applied to the affected area, this gel lengthens the amount of time between recurrences.

■ Genital warts can lead to cervical, penile, or anal cancer

Genital warts may be caused by any of several human papilloma viruses (HPVs). These are not the same viruses that cause warts on the hands and feet. Although chlamydia holds the title of being the most common of all STDs in the United States, genital warts is the most common of the viral STDs.

Several factors contribute to the prevalence of genital warts. HPVs are slow-growing viruses. Thus, there may be a long delay, an average of 2 to 3 months (although longer time periods are possible), between exposure to the virus and the formation of a wart. From the time of infection, months before warts actually appear, the virus can be spread, even though the person has no idea that he or she has been infected. Moreover, although warts can form in a visible location (Figure 17a.6), they often form in locations where they are not likely to be discovered—the vagina, cervix, or anus. In any location, the warts are highly contagious. Genital warts are usually diagnosed on the basis of their appearance. In women, a Pap test, which looks for precancerous cells on the cervix, is also helpful. A Pap test is a painless test in which the cervix is gently swabbed to collect cells that are then examined for abnormalities under a microscope.

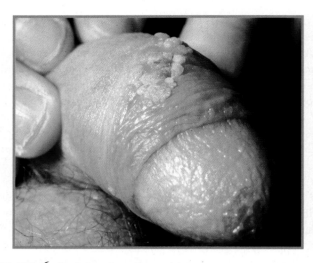

FIGURE **17a.6** Genital warts on the penis

Treatments for genital warts are intended to kill the cells that contain the virus. Methods for removing genital warts include (1) freezing (cold cautery), (2) burning with an electrical instrument (hot cautery), (3) laser (high-intensity light), (4) surgery, and (5) podophyllin (a chemical that is painted onto the warts and washed off after the prescribed time period, before it burns the skin). Creams are also available that are applied by the patient to shrink genital warts. The creams work by boosting the body's defense mechanisms against the virus. These treatments may destroy visible warts, but HPV may remain nearby in normal-looking tissue, which can cause new warts to form weeks or months after old ones have been destroyed.

HPV is closely linked to both cervical cancer in women and penile cancer in men. It is thought to be responsible for one-third of all cases of penile cancers in the United States, and it can be isolated in 90% of women with cervical cancer. For this reason, any woman who has ever had genital warts should have a Pap test at least once a year. HPV can also cause anal cancer. In mid-2006, the Food and Drug Administration approved a vaccine that is effective against several HPVs, including two that are responsible for most cases of cervical cancer. Health advisors recommend that girls be vaccinated when they are 10 to 12 years old.

The characteristics of genital herpes and genital warts, both caused by viruses, are summarized in Table 17a.2.

An HIV Infection Progresses to AIDS

In the 25 years since its appearance, *a*cquired *i*mmune *d*eficiency *s*yndrome, or AIDS, has left its mark on many aspects of society, including medicine, science, law, economics, and education. The name of the condition is apt, because the symptoms are those associated with a damaged immune system, but unlike many other immune deficiencies, AIDS is not inherited but acquired. A syndrome is a set of symptoms that tend to occur together, and people with AIDS experience a devastating set of symptoms.

We now know that a virus, the human immunodeficiency virus (HIV), causes AIDS. However, as we will learn shortly, AIDS is just the final stage in an HIV infection during which the immune system—the body's defense system—is slowly weakened. One of the primary targets of HIV is helper T cells, which serve as the main switch for the entire immune response (see Chapter 13).

≋ **Use this table as a reference for information on viral STDs.**

TABLE 17A.2 OVERVIEW OF VIRAL STDS

DISEASE	SYMPTOMS	DIAGNOSIS AND TREATMENT	EFFECTS
Genital herpes	First symptoms appear 2–20 days after contact Some people have no symptoms Flulike symptoms present Small, painful blisters that can leave painful ulcers appear Blisters go away but the virus remains Symptoms recur periodically	*Diagnosis:* Examination of blisters Laboratory tests on the fluid from the sore to detect the presence of the virus Blood test for antibodies *Treatment:* Antiviral drugs can ease symptoms	Cannot be cured Recurrences of blisters Infection can pass to fetus, causing miscarriage or stillbirth Can cause brain damage in newborns
Genital warts	First symptoms appear 1–6 months after exposure Small warts appear on sex organs May cause itching, burning, irritation, discharge, bleeding	*Diagnosis:* Appearance of growth In women, Pap test may help *Treatment:* For removal: freezing, burning, laser surgery	Formation of additional warts Closely associated with cervical and penile cancer Infection can pass to infant during childbirth

Once HIV enters a helper T cell, the T cell stops functioning well, although this cessation of function is not immediately apparent. HIV will, in the end, kill the infected helper T cell. The infection and eventual death of helper T cells cripple the immune system, giving disease-causing organisms that routinely surround us the opportunity to cause infection. These are the opportunistic infections that characterize AIDS and, in time, cause death.

■ HIV/AIDS is a global epidemic

According to the World Health Organization (WHO), at the end of 2005, 36.4 to 46 million people globally were living with an HIV infection. Africa, with more than 25.8 million people living with HIV/AIDS in 2005, is currently the area hardest hit by HIV.

■ HIV consists of RNA and enzymes encased in a protein coat

The genetic material in HIV is RNA, rather than DNA. The RNA, along with several virus-specified enzymes, is encased in a protein coat. The RNA and protein coat constitute the core of the virus (Figure 17a.7).

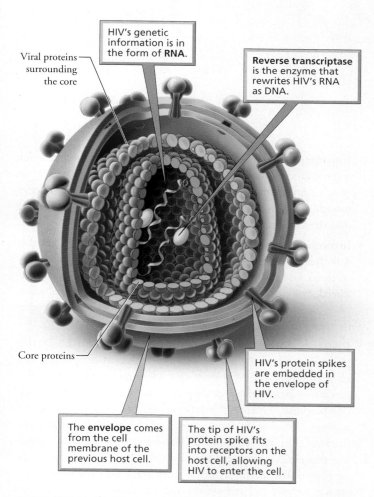

Viral proteins surrounding the core

HIV's genetic information is in the form of **RNA**.

Reverse transcriptase is the enzyme that rewrites HIV's RNA as DNA.

Core proteins

HIV's protein spikes are embedded in the envelope of HIV.

The **envelope** comes from the cell membrane of the previous host cell.

The tip of HIV's protein spike fits into receptors on the host cell, allowing HIV to enter the cell.

FIGURE **17a.7** The structure of HIV (human immunodeficiency virus)

Surrounding the protein coat is an outer covering, or envelope, consisting of spikes of protein embedded in a lipid membrane. The lipid membrane is actually a piece of plasma membrane stolen from the previous host cell and altered for use by the virus. The protein spikes of the viral envelope are responsible for binding the virus to the host cell.

■ HIV enters the cell, rewrites its RNA as DNA, inserts DNA into the host chromosome, and replicates

HIV binds to an uninfected cell when the spherical region at the end of a protein spike fits into a receptor on the host cell surface like a key in a lock. The host cell receptor is a surface protein called CD4. Helper T cells are the predominant cell types with these CD4 receptors and, therefore, the most common targets of HIV. When an HIV spike is properly docked in a CD4 receptor, the contents of the virus enter the host cell (Figure 17a.8).

Once within the host cell, the RNA of HIV undergoes a process called reverse transcription. During this process, the viral RNA is rewritten as double-stranded DNA. Because going from RNA to DNA is the reverse of the usual genetic information transfer, viruses that work this way are known as retroviruses (*retro-*, backward). The backward copying of genetic information from RNA into DNA is performed by an enzyme called reverse transcriptase, which is inserted into the host cell along with viral RNA.

The newly formed viral DNA, which contains all the instructions necessary for producing thousands of new viruses, is then spliced into the host DNA. Once HIV's DNA is integrated into the host cell chromosome, that DNA is called the HIV provirus. After it has been incorporated into the host cell DNA, the host cell treats the HIV genes as it would its own. Each time the cell reproduces, the viral DNA is copied along with the cell's DNA. When HIV is transmitted to another person, it is usually in the form of a latent—that is, inactive—HIV provirus when the infected cell enters the person's body.

The HIV genes can reside in a helper T cell chromosome for years, until the cell is activated to respond to some foreign antigen, such as another kind of virus, a fungus, or a parasite. At this point, the virus begins making copies of itself instead of allowing the cell to fight the invader. The viral genome is activated and turns the cell into a virus factory. Some of the newly produced HIV RNA will become genetic material for new viruses, and some will be used to produce viral proteins. The viral components gather at the cell membrane and bud off from the host cell. The HIV RNA and proteins then self-assemble into new, mature viruses. The mature viruses can move through the bloodstream to infect new cells (Figure 17a.9).

■ Most HIV is transmitted through sexual contact, intravenous drug use, or from a pregnant woman to her fetus

HIV is found in bodily fluids—blood, semen, vaginal secretions, breast milk, saliva, tears, urine, cerebrospinal fluid, and amniotic fluid. However, HIV is not easily transmitted. Only the first four of the bodily fluids listed contain HIV concentrations high enough to cause infection in another person.

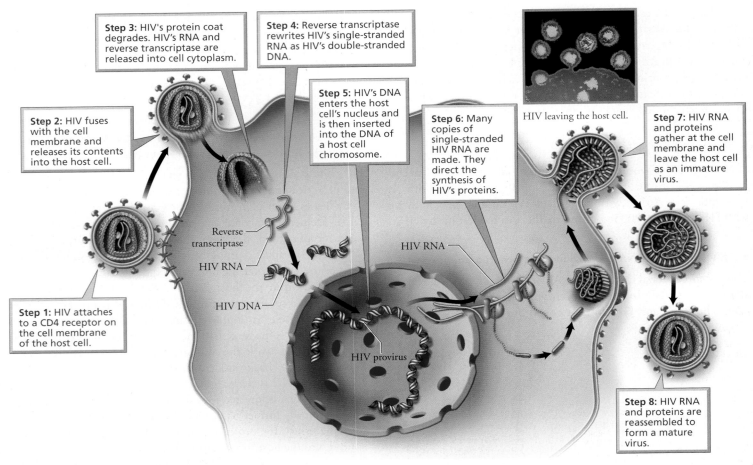

Step 2: HIV fuses with the cell membrane and releases its contents into the host cell.

Step 3: HIV's protein coat degrades. HIV's RNA and reverse transcriptase are released into cell cytoplasm.

Step 4: Reverse transcriptase rewrites HIV's single-stranded RNA as HIV's double-stranded DNA.

Step 5: HIV's DNA enters the host cell's nucleus and is then inserted into the DNA of a host cell chromosome.

Step 6: Many copies of single-stranded HIV RNA are made. They direct the synthesis of HIV's proteins.

Step 7: HIV RNA and proteins gather at the cell membrane and leave the host cell as an immature virus.

Step 1: HIV attaches to a CD4 receptor on the cell membrane of the host cell.

Reverse transcriptase

HIV RNA

HIV DNA

HIV provirus

HIV RNA

HIV leaving the host cell.

Step 8: HIV RNA and proteins are reassembled to form a mature virus.

WEB TUTORIAL 17a.1

FIGURE **17a.8** The life cycle of HIV

Today, the major modes of transmission are sexual contact without a latex condom and contact with contaminated blood through intravenous drug use, but HIV can also be transmitted from a pregnant woman to her fetus. HIV *cannot* be transmitted by casual contact.

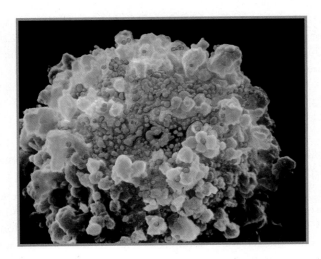

FIGURE **17a.9** HIV particles (red) on the surface of a lymphocyte (green)

- **Unprotected sexual activity.** HIV can be transmitted through unprotected sexual activity of any kind—that is, activity in which skin or mucous membranes of the vagina, vulva, penis, mouth, or anus of the partners are not separated by a barrier that the virus cannot penetrate.

- **Intravenous drug use.** Contact with infected blood is another way that HIV can be transmitted. Since 1985, all blood donated to blood banks is tested for HIV; this testing has greatly reduced the risk of getting HIV from a blood transfusion. However, intravenous drug users who share needles *do* have an increased risk of HIV infection.

- **Infected mother to child before or during birth.** During pregnancy, maternal and fetal blood supplies are brought into close proximity (see Chapter 18), allowing nutrients and oxygen to be exchanged, but they are separated by membranes in the placenta. During the last months of fetal development, however, small tears may occur in the placenta and may allow the infection to spread to the bloodstream of the fetus. Bleeding during the birth process may also cause infection in the newborn. Treatment with the drug AZT and delivery by cesarean section can reduce the risk of transmission. HIV can also be passed to the infant in breast milk.

■ Sites of HIV infection include the immune system and the brain

HIV can infect any cell that has a CD4 receptor. The most important of these cells is the helper T cell, which turns on the entire immune response (see Chapter 13). The hallmark of an HIV infection is a progressive decline in helper T cell numbers. This loss is devastating because it leaves the body increasingly defenseless against other infections.

HIV can also infect the brain, killing nerve cells. When the brain is infected, the symptoms can include forgetfulness, impaired speech, inability to concentrate, depression, seizures, and personality changes. Roughly 60% of people with AIDS have signs of dementia.

■ An HIV infection progresses through several stages as helper T cell numbers decline

An HIV infection usually progresses through a series of stages: the initial infection, an asymptomatic stage, initial disease symptoms, early immune failure, and AIDS. The progress of an HIV infection is usually monitored by following both the decline in the number of helper T cells and the increase in viral load (the number of HIV free in the blood). The changes in T cell number and viral load are summarized in Figure 17a.10.

INITIAL INFECTION

During the initial stages of infection, the virus actively replicates, and the circulating level of HIV rises. The body's immune system produces antibodies against the virus in an attempt to eliminate it. Antibodies can usually be detected within 8 weeks or so, but in some cases they are not detected for many months or even years. An HIV test looks for the presence of antibodies to HIV in the blood. If antibodies are found, the person is said to be HIV-positive (HIV$^+$).

Many people have no symptoms when they first become infected; others experience some mild disease symptoms during the initial infection. Early symptoms might include enlarged lymph nodes throughout the body, fatigue, and fever. If the brain becomes infected, the initial symptoms may include headaches, fever, and difficulty concentrating, remembering, or solving problems.

ASYMPTOMATIC STAGE

Weeks to months after the initial stage of infection, the person usually feels well again, often for several years. This is the period of asymptomatic infection. During this stage, the immune system mounts a defense strong enough to control, but not conquer, the infection. The rate of T cell production increases. For a while, T cell production matches destruction.

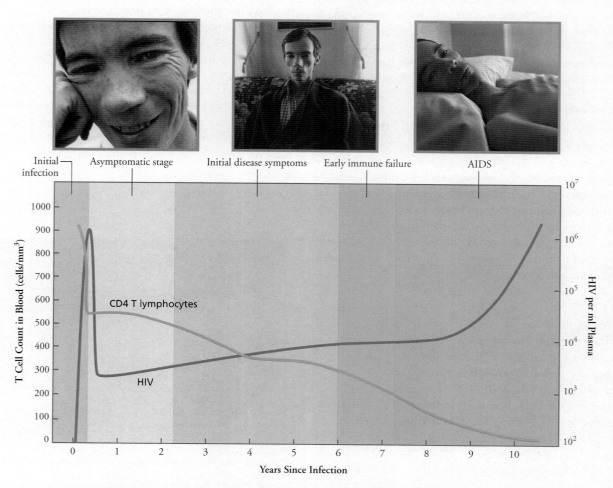

FIGURE **17a.10** An HIV infection can be monitored by following the decline in the number of T cells and the rise in viral load (number of viruses in the bloodstream).

During the asymptomatic stage, HIV is far from idle. It is "hiding out" in certain lymphatic organs, including the spleen, tonsils, and adenoids, but most importantly in the lymph nodes, where it replicates. Consequently, the spaces between cells in the lymph nodes become packed with virus particles. Most helper T cells reside in the lymph nodes, and many of them have the DNA form of HIV genes incorporated into their own chromosomes. Because helper T cells frequently circulate through the lymph nodes, this is an excellent place for HIV to encounter healthy T cells. Thus, millions of T cells become infected as they move through the lymph nodes on their way to other parts of the body.

INITIAL DISEASE SYMPTOMS

Eventually, the immune system begins to falter as the virus gains the upper hand. As helper T cell numbers gradually drop, symptoms set in again, signaling the beginning of the end. The initial disease symptoms fall into three classes.

- *Wasting syndrome* is characterized by an otherwise unexplained loss in body weight of more than 10%, often accompanied by diarrhea. The weight loss is similar to that of a cancer patient.

- *Swollen lymph nodes* occur in the neck, armpits, and groin. These structures may remain swollen.

- *Neurological symptoms* may appear, either because HIV has spread to the brain or because other organisms have caused a brain infection. Among the neurological symptoms are dementia, weakness, and paralysis caused by spinal cord damage. Other symptoms include pain, burning, or a tingling sensation, usually in hands or feet, caused by peripheral nerve damage.

EARLY IMMUNE FAILURE

As T cell numbers continue their gradual decline, the body becomes increasingly vulnerable to infection. Early signs of immune failure include thrush and shingles. Thrush is a yeast (fungus) infection in the mouth caused by *Candida*, which is normally held in check by other microbes and the immune system. Thrush causes white, sore, furry-feeling areas to form in the mouth. Shingles is a painful rash caused by the same virus that causes chickenpox (*Varicella zoster*). The virus can remain dormant in nerve cells long after a person recovers from chickenpox, and it may become active again when the immune system is compromised.

AIDS

AIDS (sometimes called frank, or full-blown, AIDS) is the final stage of an HIV infection. The time from HIV infection to diagnosis of AIDS can be 10 or more years, depending on the number of other infections the person is exposed to and the health of the immune system before infection with HIV. Now, although the body has fought long and hard, it is beginning to lose the battle. The immune system is crippled, making the person vulnerable to certain characteristic fungal, protozoan, bacterial, and viral infections, as well as cancers.

A diagnosis of AIDS is made when an HIV$^+$ person develops one of the following conditions: (1) a helper T cell count below 200/mm^3 of blood; (2) 1 of 26 opportunistic infections, the most common of which are *Pneumocystis carinii* pneumonia and Kaposi's sarcoma, a cancer of connective tissue that primarily affects the skin; (3) a greater than 10% loss of body weight (wasting syndrome); or (4) dementia (mental incapacity such as forgetfulness or inability to concentrate).

Pneumocystis carinii, mentioned above, is a fungus that infects the lungs and is a leading cause of death in people with AIDS. It causes a dry cough and progressive shortness of breath.

Protozoans are single-celled organisms that can also cause serious illness when HIV weakens the immune system. One (*Cryptosporidium*) can infect the lining of the intestinal tract and cause long and severe bouts of diarrhea. Another (*Toxoplasma gondii*) is an intracellular parasite that can infect many organs. When the brain becomes infected, symptoms such as convulsions, disorientation, and dementia develop.

Important among the bacterial infections that commonly plague people with AIDS is tuberculosis. People with AIDS usually get an atypical form of tuberculosis that may infect both the lungs and the bone marrow.

Cytomegalovirus (CMV), a member of the herpes family, is a common virus that infects many people in childhood, but it can be reactivated in AIDS patients. For healthy people, infection with CMV does not cause major problems. When CMV infection recurs in people with AIDS, however, it can cause blindness, an adrenal hormone imbalance, or pneumonia.

Kaposi's sarcoma and lymphomas are cancers more common in people with AIDS than in the general population. In Kaposi's sarcoma, tumors form in connective tissue, especially in blood vessels, and appear as pink, purple, or brown spots on the skin (Figure 17a.11). The tumors spread and eventually may affect all linings of the body. A virus in the herpes family, Kaposi sarcoma-associated herpesvirus, is thought to be the most probable cause of Kaposi's sarcoma. This virus is almost always present in lesions caused by Kaposi's sarcoma but is rare in other lesions.

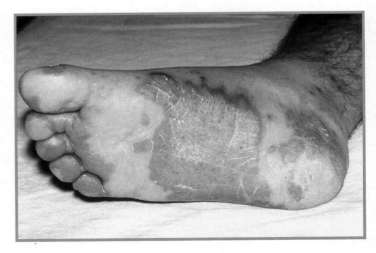

FIGURE **17a.11**   Kaposi's sarcoma, a cancer common among persons living with AIDS, characterized by purple, pink, or brown spots on the skin.

■ Treatments for HIV infection are designed to block specific steps in HIV's replication cycle

Death in a person with AIDS is usually caused by one of the opportunistic infections. Thus, treating opportunistic infections, or preventing their onset, can improve the quality of life and extend the length of life for people with AIDS. However, because AIDS is just the final stage in an HIV infection, treatment is also aimed at slowing the progress of the infection. The current prevailing strategy is to slow the rate at which HIV can make new copies of itself. At present, there are three classes of drugs that slow viral replication: reverse transcriptase inhibitors, protease inhibitors, and fusion inhibitors. Each class of drugs prevents HIV from replicating by blocking a different step in the viruss life cycle.

The original class of antiviral drugs used to slow the progression of an HIV infection consisted of drugs that act on reverse transcriptase (AZT, ddI, ddC). Recall that reverse transcriptase is the enzyme necessary for converting the RNA of HIV to DNA that can then be inserted into the DNA of the host cell chromosome.

A second class of antiviral drugs consists of protease inhibitors. Recall that after the HIV genetic information has been inserted into the DNA of the host cell, it can begin producing proteins needed to form new copies of HIV. The proteins initially produced are too big to be used and must be cut to the proper size by an enzyme called a protease. Protease inhibitors block the action of this enzyme, preventing a necessary step in the preparation of proteins for the assembly of new viruses.

The newest class of antiviral drugs consists of fusion inhibitors. These block the proteins on the surface of HIV that bind to the CD4 receptor on host cells, thereby blocking HIV from entering the host cell.

A treatment called highly active antiretroviral therapy (HAART), also known as a drug cocktail, consists of combinations of drugs. Typically, the cocktail contains two inhibitors of reverse transcriptase and a protease inhibitor. The idea behind this drug combination is that a protease inhibitor will slow the rate of formation of new viruses so that the development of resistance to the other drugs will be minimized. Combination drug treatments are prolonging the lives of HIV$^+$ people.

HAART can reduce the viral load to undetectable levels, but the treatment is not a cure. If treatment is stopped, HIV can leave the latent state and begin actively replicating again.

A variety of other approaches to combating an HIV infection are under active investigation. Clearly, though, the best way to fight the devastation caused by HIV is to prevent it from establishing the initial infection. Theoretically, this could be accomplished by the development of a vaccine similar to those for other viral infections such as small pox and measles. Recall from Chapter 13 that vaccines work by causing a person's body to produce antibodies against the organisms that cause a disease, without actually causing the disease. Unfortunately, there are several characteristics of HIV that have thwarted efforts to develop a vaccine:

1. The mutation rate in HIV is very high, especially for the proteins in its outer coat, which are the proteins that an antibody would have to recognize.

2. There are two types of HIV and many strains of each type. Antibodies would have to recognize the exact strain entering the body.

3. An HIV infection is latent when its genetic material has become incorporated into the host DNA (as a provirus). Antibodies cannot attack the virus while it is hidden within the host cell.

4. HIV can pass directly from the interior of an infected cell to the interior of an adjacent uninfected cell. Antibodies can attack viruses outside the cell only, they cannot prevent the spread of the virus by this means of transmission.

Development throughout Life

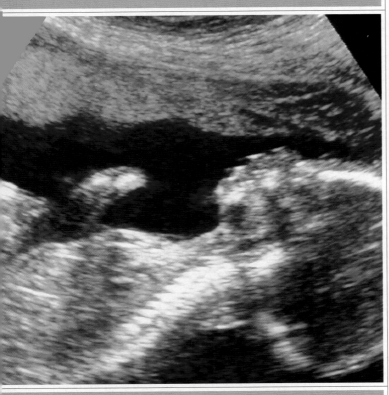

Human development begins with fertilization and continues until death. This ultrasound image shows a fetus 20 weeks into a life that may last 70 to 80 years.

Human Life Has Two Main Periods of Development

The Prenatal Period Begins at Fertilization and Ends at Birth
- Fertilization, cleavage, and implantation characterize the pre-embryonic period
- Tissues and organs form during the embryonic period
- Rapid growth characterizes the fetal period

Birth Is the Transition from Prenatal to Postnatal Development
- True labor has three stages
- The placenta may send the signal to initiate labor
- Drugs and breathing techniques can relieve pain during childbirth

Environmental Disruptions during the Embryonic Period Cause Major Birth Defects

The Mother's Mammary Glands Produce Milk

The Postnatal Period Begins with Birth and Continues into Old Age
- Possible causes of aging
- Medical advances and a healthy lifestyle can help achieve a high-quality old age

HEALTH ISSUE Making Babies, but Not the Old-Fashioned Way

"Hey, want to see a picture of my baby?" Maritza held up a black-and-white image that seemed to show what a tiny Martian alien would look like on the radar screen of a star cruiser. Her coworkers hadn't even known she was pregnant (though some now claimed they had suspected it). She told them that because she was under treatment for hypertension and had miscarriages in the past, she'd waited to make the announcement. Her family physician had sent her to an obstetrician who specialized in complicated pregnancies, and the obstetrician had done an ultrasound examination of her developing baby at 6 weeks and again at 20 weeks, to see if its development was proceeding normally. Her friends peered through the "snow" on the photographed ultrasound image as Maritza pointed to the baby's head (the largest and most easily recognizable feature); the chambers of the heart, which she had been able to see beating during the procedure; and the edge of the placenta, which according to the doctor was exactly where it belonged. Various measurements of the developing baby had confirmed its age and indicated a normal rate of growth. Relieved that her baby had been given a clean bill of health so far, Maritza was ready to share her excitement with everyone she knew.

Development is the process of growth and change that a living organism experiences throughout its life. Maritza's baby is still in the earliest stages of its development. Maritza, though a fully grown adult, has many decades of health and new stages in her own life to look forward to. In this chapter, we will describe the major milestones of human development, from conception to the transition into old age. ∎

Human Life Has Two Main Periods of Development

A newborn enters the world outside his mother's womb and rests (Figure 18.1). How did this tiny human develop? Consider that he—like you—began as a fertilized egg, no bigger than the period at the end of this sentence. The miracle of human development begins with **fertilization**, the union of a sperm and an egg, and the early stages take place within the female reproductive system (see Chapter 17). Birth occurs about 266 days after fertilization, marking the transition to development outside the mother's body.

The period of development before birth is the prenatal period. The period after birth is the postnatal period. Relative to prenatal development, postnatal development is a lengthy process, taking from 20 to 25 years to reach adulthood. Bodily changes do not cease once adulthood is reached, however. Our bodies continue to change throughout adulthood as part of the aging process.

The Prenatal Period Begins at Fertilization and Ends at Birth

W e start our discussion of human development at the moment it all begins—when sperm meets egg. In this section, which presents the major events of prenatal development, we consider what normally happens and what can happen when things go wrong.

From the standpoint of human development, the prenatal period is divided into three periods: (1) the **pre-embryonic period**, from fertilization through the second week (during which the de-

veloping human is called a **pre-embryo**); (2) the **embryonic period**, from week 3 through week 8 (during which the developing human is called an **embryo**); and (3) the **fetal period**, from week 9 until birth (during which the developing human is called a **fetus**). We will consider the developmental events that define each stage.

∎ Fertilization, cleavage, and implantation characterize the pre-embryonic period

The pre-embryonic period begins with fertilization, the union between the nucleus of an egg and the nucleus of a sperm. Fertilization takes about 24 hours from start to finish and usually occurs in a widened portion of the oviduct, not far from the ovary, as shown in Figure 18.2. Fertilization is a beautifully orchestrated performance by egg and sperm, aided by the actions of the oviducts and uterus.

FERTILIZATION

Let's begin our discussion of the pre-embryonic period with an overview of how egg and sperm meet. At ovulation, the ovary releases the egg, still a secondary oocyte (see Chapter 17). The egg is swept into the oviduct by fingerlike projections called fimbriae (singular, *fimbria*; see Figure 18.2). Cilia and waves of peristalsis move the oocyte slowly along the oviduct toward the uterus. Not far into its journey, the oocyte pauses in the widened portion of the oviduct.

The trip made by sperm is a considerably more competitive and precarious venture than the movements of the typically lone oocyte. During sexual intercourse, from 200 million to 600 million sperm are deposited in the vagina and on the cervix, the neck of the uterus that extends into the vagina. Many sperm become trapped at the boundary of the vagina and cervix. Indeed, less than 1% of deposited sperm actually enter the uterus. Sperm that escape entrapment use their whiplike tails to swim into the uterus and eventually the oviduct. Along the way, the sperm are aided by small uterine contractions stimulated by chemicals (prostaglandins) in the semen. Of the vast number of sperm originally deposited in the female reproductive tract, only about 200 reach the site of fertilization in the widened portion of the oviduct.

Timing is everything when it comes to fertilization. Typically, an egg cannot be fertilized more than 24 hours after its release from the ovary. Most eggs are fertilized within 12 hours. In addition, most sperm survive no more than 2 days in the female reproductive tract, though some may survive 5 days.

Two critical processes must precede fertilization (Figure 18.3). First, secretions from the uterus or oviducts must alter the surface of the acrosome, the enzyme-containing cap on the head of a sperm. Alteration of the acrosome requires about 6 to 7 hours and appears to involve the removal of cholesterol and possibly other substances from the surface of the sperm. These changes destabilize the sperm's plasma membrane. The second process occurs when sperm with unstable membranes contact the corona radiata, a layer of cells surrounding the secondary oocyte (Figure 18.3). Once such contact occurs, perforations develop in the sperm's plasma membrane and outer acrosomal membrane. Enzymes then spill out of the acrosome and disrupt the attachments between cells of the corona radiata. This disruption enables sperm to pass through the corona radiata to the layer below called the zona pellucida. The

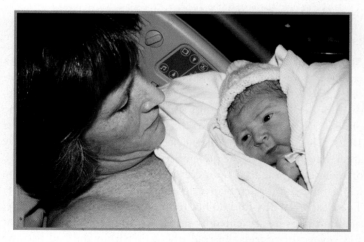

FIGURE **18.1** A newborn and his mother. Birth is the transition from prenatal to postnatal development.

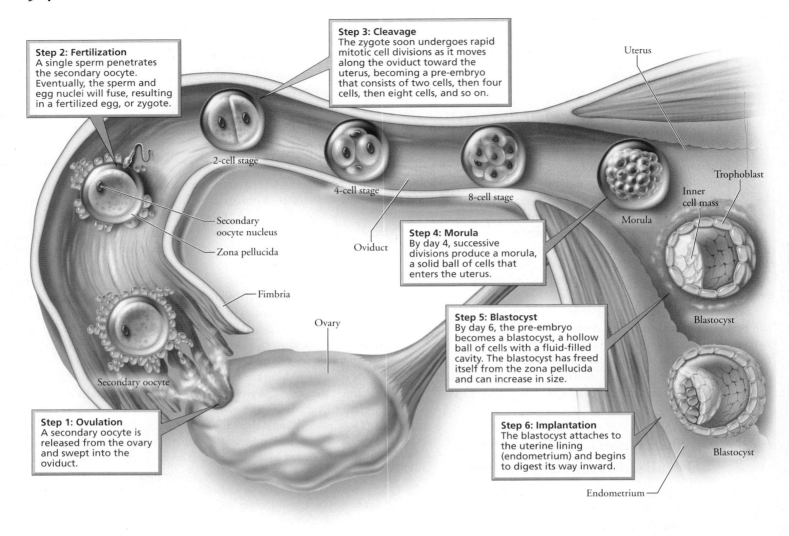

Step 2: Fertilization
A single sperm penetrates the secondary oocyte. Eventually, the sperm and egg nuclei will fuse, resulting in a fertilized egg, or zygote.

Step 3: Cleavage
The zygote soon undergoes rapid mitotic cell divisions as it moves along the oviduct toward the uterus, becoming a pre-embryo that consists of two cells, then four cells, then eight cells, and so on.

Step 4: Morula
By day 4, successive divisions produce a morula, a solid ball of cells that enters the uterus.

Step 5: Blastocyst
By day 6, the pre-embryo becomes a blastocyst, a hollow ball of cells with a fluid-filled cavity. The blastocyst has freed itself from the zona pellucida and can increase in size.

Step 6: Implantation
The blastocyst attaches to the uterine lining (endometrium) and begins to digest its way inward.

Step 1: Ovulation
A secondary oocyte is released from the ovary and swept into the oviduct.

2-cell stage

4-cell stage

8-cell stage

Secondary oocyte nucleus

Zona pellucida

Fimbria

Ovary

Oviduct

Secondary oocyte

Uterus

Morula

Inner cell mass

Trophoblast

Blastocyst

Blastocyst

Endometrium

FIGURE **18.2** Early stages in the reproductive process

zona pellucida is a thick noncellular layer that immediately surrounds the secondary oocyte. Enzymes released from the acrosome also break apart the zona pellucida, creating a pathway to the oocyte.

Several sperm may arrive at the secondary oocyte at about the same time, although usually only one sperm crosses the zona pellucida. The plasma membrane of the successful sperm and that of the secondary oocyte fuse, triggering two important events. First, enzymes released by granules near the plasma membrane of the oocyte cause the zona pellucida to harden and thereby prevent passage of other sperm. This block to entry by additional sperm ensures equal genetic contributions from each parent and prevents abnormal numbers of chromosomes in the resulting embryo. Second, the oocyte undergoes its second meiotic division (Chapter 17) and is now considered an ovum. The nucleus of the ovum and the nucleus of the sperm fuse (Figure 18.3). The fertilized ovum is called a **zygote**. Just visible to the unaided eye, the zygote contains genetic material from both the mother (23 chromosomes) and father (23 chromosomes), and thus has 46 chromosomes.

CLEAVAGE

About 1 day after fertilization, the zygote undergoes **cleavage**, a rapid series of mitotic cell divisions. During cleavage, the single-celled zygote becomes a pre-embryo, first consisting of two cells, then four cells, then eight cells, and so on. Cleavage occurs as the pre-embryo moves along the oviduct toward the uterus (see Figure 18.2). The cleaving pre-embryo becomes a solid ball of cells by about day 3. By day 4, the pre-embryo is a *morula*, a solid ball of 12 or more cells produced by successive divisions of the zygote. The name reflects its resemblance to the fruit of the mulberry tree. These early cell divisions do not result in an overall increase in size; such increases are prevented by the tight-fitting zona pellucida that still covers the morula. Instead, the cells within the ball become progressively smaller as divisions occur.

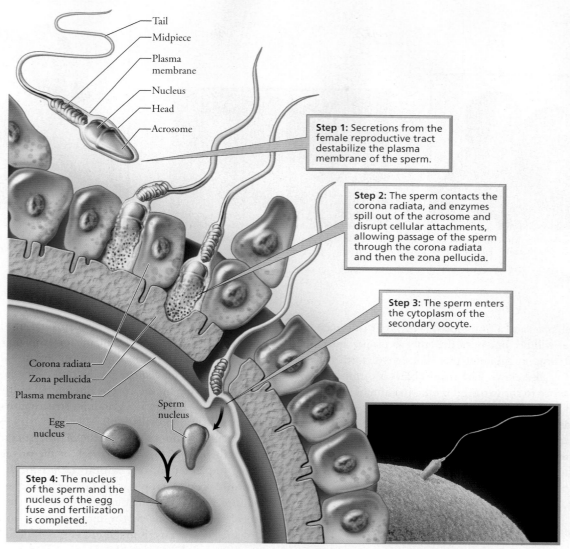

Tail
Midpiece
Plasma membrane
Nucleus
Head
Acrosome

Corona radiata
Zona pellucida
Plasma membrane
Egg nucleus
Sperm nucleus

Step 1: Secretions from the female reproductive tract destabilize the plasma membrane of the sperm.

Step 2: The sperm contacts the corona radiata, and enzymes spill out of the acrosome and disrupt cellular attachments, allowing passage of the sperm through the corona radiata and then the zona pellucida.

Step 3: The sperm enters the cytoplasm of the secondary oocyte.

Step 4: The nucleus of the sperm and the nucleus of the egg fuse and fertilization is completed.

(a) Diagram showing the processes that occur before and during fertilization

(b) Micrograph of a sperm contacting a secondary oocyte

FIGURE **18.3** Fertilization

Sometimes during an early stage of cleavage, the mass of cells splits into two, and two pre-embryos are formed. *Identical twins*, also called monozygotic twins ("from one zygote"), develop in this way. Such twins have identical genetic material and always are the same gender. In rare cases, the splitting of the pre-embryo is incomplete and *conjoined twins* result. Such twins may be surgically separated after birth once doctors have evaluated which structures are shared and the likelihood of successful separation. *Fraternal twins* occur when two secondary oocytes are released from the ovaries and fertilized by different sperm. Such twins also are called dizygotic twins ("from two zygotes"). Fraternal twins may or may not be the same gender and are no more genetically similar than siblings who are not twins. Identical twins (separate and conjoined) and fraternal twins are shown in Figure 18.4.

stop and think

Given what you know about the events necessary to produce identical and fraternal twins, how might a set of triplets form in which two are identical and one fraternal?

The morula enters the uterus about 5 days after fertilization. As cell division continues, a cavity begins to form at the morula's center. Cells lining the cavity flatten and compact as the zona pellucida still prevents increases in overall size. Fluid from the uterine cavity accumulates within the forming cavity, and the morula is converted into a **blastocyst**, a ball of cells with an inner fluid-filled cavity. As shown in Figure 18.2, the blastocyst is made up of two

(a) Identical twins result when a single fertilized egg splits in two very early in development.

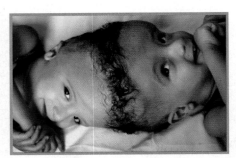

(b) Rarely, the splitting is incomplete and conjoined twins result.

(c) Fraternal twins result from the fertilization of two oocytes by different sperm.

FIGURE **18.4** Twins

parts: an inner cell mass and a trophoblast. The **inner cell mass** is a group of cells that will become both the embryo proper and some of the extraembryonic membranes that will extend from or surround the embryo (see below). The **trophoblast** is a thin layer of cells that will give rise to the extraembryonic membrane that is the embryo's contribution to the placenta. The **placenta** is the organ that delivers oxygen and nutrients to the embryo and carries carbon dioxide and other wastes away. Before the blastocyst becomes implanted in the uterine wall and the placenta develops, the blastocyst floats freely in the uterus. Increases in overall size of the blastocyst become possible when it frees itself from the zona pellucida, which then degenerates.

IMPLANTATION

The blastocyst attaches to the endometrium (the lining of the uterus) about 6 days after fertilization. Then, in a process called **implantation**, it begins to digest its way inward until it is firmly embedded in the endometrium (Figure 18.5). Implantation normally occurs high up on the back wall of the uterus. Sometimes, however, a blastocyst implants outside the uterus, and an *ectopic pregnancy* results. The vast majority of implantations outside the uterus occurs in the oviducts, usually because passage of the dividing pre-embryo through the oviduct is impaired in some way. Factors that hinder passage of the pre-embryo include structural abnormalities and scar tissue from surgery or pelvic inflammatory disease (see Chapters 17 and 17a). The embryo is surgically removed (and hence the pregnancy terminated), because rupture of the oviduct and hemorrhage can be fatal to the mother.

For successful early development and implantation, conditions must be just right. Indeed, an estimated one-third to one-half of all zygotes fail to become blastocysts and to implant. The failed zygotes and early pre-embryos are either reabsorbed by cells of the endometrium or expelled from the uterus in an early spontaneous abortion (such an end to a pregnancy is described as spontaneous because it was not medically induced). Causes of early spontaneous

abortion include chromosomal abnormalities in the zygote and an inhospitable uterine environment for implantation. The latter might result from the presence of an intrauterine device (IUD; Figure 17.10) or inadequate production of the hormones estrogen and progesterone by the corpus luteum. (Recall from Chapter 17 that the corpus luteum is a glandular structure that forms from the ovarian follicle after ovulation.)

When conditions are right, the blastocyst completes implantation by the end of the second week of the pre-embryonic period. During implantation, cells within the blastocyst produce **human chorionic gonadotropin (HCG)**, a hormone that enters the bloodstream of the mother and is excreted in her urine. The presence of this hormone in the mother's urine forms the basis for many pregnancy tests. Enough HCG is produced by the end of the second week to be detected by the pregnancy test and yield a positive result. Its usefulness in pregnancy tests aside, the physiological function of HCG is to maintain the corpus luteum and to stimulate it to continue producing the hormone progesterone. Progesterone is essential for maintenance of the endometrium, the very lining in which the embryo is implanting.

Infertility is the inability of a female to conceive (become pregnant) or of a male to cause conception. Implantation is a major hurdle in the series of steps leading to a successful pregnancy. The Health Issue essay *Making Babies, But Not the Old-Fashioned Way* describes many of the options available to infertile couples who wish to have children.

EXTRAEMBRYONIC MEMBRANES

Toward the end of the pre-embryonic period, four membranes—the amnion, yolk sac, allantois, and chorion—begin to form around the pre-embryo (Figure 18.6). Formation of these membranes extends into the embryonic period, when the developing human is called an embryo. The membranes lie outside the embryo and are called **extraembryonic membranes**. These membranes protect and nourish the embryo and later the fetus. The **amnion**

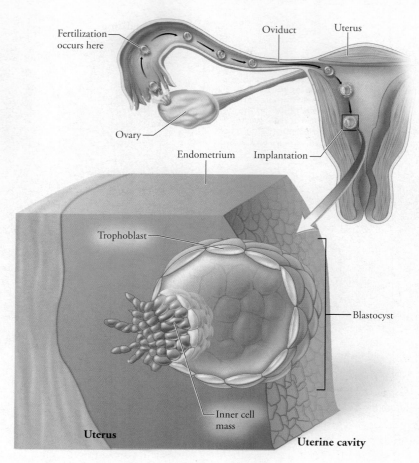

FIGURE **18.5** Implantation. About 6 days after fertilization, the blastocyst attaches to the endometrium of the uterus and begins to digest its way inward.

surrounds the entire embryo, enclosing it in a fluid-filled space called the amniotic cavity. Amniotic fluid forms a protective cushion around the embryo that later can be examined as part of prenatal testing in a procedure known as amniocentesis (see Chapter 20).

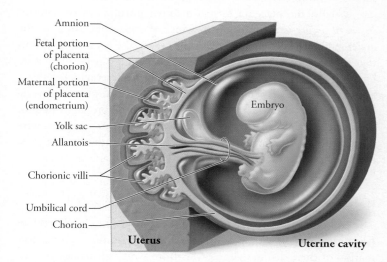

FIGURE **18.6** Extraembryonic membranes. The amnion, yolk sac, chorion, and allantois begin to form during the second to third week after fertilization.

The **yolk sac** is the primary source of nourishment for embryos in many species of vertebrates. In embryonic humans, however, it remains quite small and does not provide nourishment. Human embryos receive nutrients from the placenta. The yolk sac in humans is a site of early blood cell formation. It also contains *primordial germ cells* that migrate to the gonads (testes or ovaries), where they differentiate into immature cells that will eventually become sperm or oocytes. The **allantois** is a small membrane whose blood vessels become part of the **umbilical cord**, the ropelike connection between the embryo and the placenta. The umbilical cord consists of blood vessels and supporting connective tissue. Finally, the **chorion** is the outermost membrane. It develops largely from the trophoblast and becomes the embryo's major contribution to the placenta. The amnion, yolk sac, and allantois develop from the inner cell mass.

THE PLACENTA

We mentioned that developing humans rely on the placenta, rather than the yolk sac, for nourishment. Indeed, the placenta orchestrates all interactions between the mother and the fetus. The placenta forms from the chorion of the embryo and a portion of the endometrium of the mother (specifically, the endometrium in the area where implantation occurred; Figure 18.7).

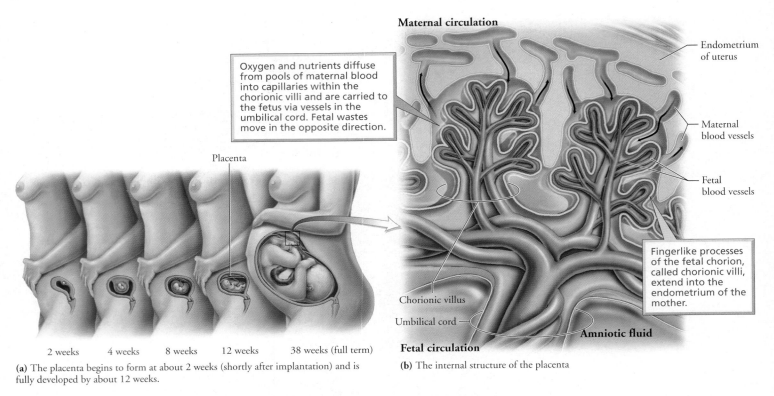

Oxygen and nutrients diffuse from pools of maternal blood into capillaries within the chorionic villi and are carried to the fetus via vessels in the umbilical cord. Fetal wastes move in the opposite direction.

Placenta

Maternal circulation

Endometrium of uterus

Maternal blood vessels

Fetal blood vessels

Fingerlike processes of the fetal chorion, called chorionic villi, extend into the endometrium of the mother.

Chorionic villus

Umbilical cord

Amniotic fluid

Fetal circulation

2 weeks   4 weeks   8 weeks   12 weeks   38 weeks (full term)

(a) The placenta begins to form at about 2 weeks (shortly after implantation) and is fully developed by about 12 weeks.

(b) The internal structure of the placenta

FIGURE **18.7** The placenta is formed from the chorion of the embryo and the endometrium of the mother.

A major function of the placenta is to allow oxygen and nutrients to diffuse from maternal blood into embryonic blood. Wastes such as carbon dioxide and urea diffuse from embryonic blood into maternal blood. The placenta also produces hormones such as HCG, estrogen, and progesterone. Placental hormones, as mentioned earlier, are essential for the continued maintenance of pregnancy. High levels of HCG from the placenta may be responsible for *morning sickness*, the nausea and vomiting experienced by some women early in pregnancy. (Morning sickness is not restricted to the morning, by the way.)

Formation of the placenta begins shortly after implantation. During implantation, cells of the outer layer of the trophoblast rapidly divide and invade the endometrium. Cavities form that fill with blood from maternal capillaries severed by the invading cells. Soon, the inner layer of the trophoblast invades the endometrium, and fingerlike processes of the chorion, called **chorionic villi**, grow into the endometrium (refer again to Figure 18.7). These chorionic villi grow, divide, and continue their invasion of maternal tissue, causing ever larger cavities and pools of maternal blood to form. The placenta is fully developed by the third month of pregnancy. At this time, it weighs about 680 g (1.5 lb).

Chorionic villi contain blood vessels connected to the developing embryo. Oxygen and nutrients in the pools of maternal blood diffuse through the capillaries of the villi into the umbilical vein and travel to the embryo. Wastes leave the embryo by the umbilical arteries, move into capillaries within the chorionic villi, and diffuse into maternal blood. Thus, the chorionic villi provide exchange surfaces for diffusion of nutrients, oxygen, and wastes. Some chorionic villi also anchor the embryonic portion of the placenta to maternal tissue.

Under normal circumstances, there is no direct mixing of maternal and fetal blood; all exchanges occur across capillary walls. Nevertheless, the closeness of the maternal and fetal blood vessels within the placenta causes the fetus to share in maternal nutrition, habits, and lifestyle. Substances that cross the placenta include drugs, alcohol, caffeine, toxins in cigarette smoke, and HIV (human immunodeficiency virus, the virus that causes AIDS).

Usually, the placenta forms in upper portions of the uterus. *Placenta previa* is a condition in which the placenta forms in the lower half of the uterus and covers the cervix, the neck of the uterus that projects into the vagina. Placenta previa may cause premature birth or maternal hemorrhage and usually makes vaginal delivery impossible. About 75% of women with diagnosed placenta previa have their baby delivered by cesarean section before labor begins. **Cesarean section**, often shortened to C-section, is the procedure by which the fetus and placenta are removed surgically from the uterus through an incision in the abdominal wall and uterus.

■ Tissues and organs form during the embryonic period

The embryonic period extends from the third to the eighth week of development. It is a time of great change, beginning with the formation of three distinct germ layers from which all tissues and organs develop. We first describe the general process by which germ layers form, and then we consider development of the central nervous system and reproductive system as examples of tissue and organ

Making Babies, but Not the Old-Fashioned Way

Some couples desperately want a child, but they cannot conceive one. The inability to become pregnant or to cause a pregnancy is called infertility. Couples are considered infertile if conception does not occur after 1 year of unprotected sexual intercourse. Infertility is not uncommon, striking about one in six American couples. In some of these couples, the woman is infertile. Hormonal imbalances may disrupt ovulation or implantation, or scarring from disease may block the oviducts and prevent passage of gametes. Some women have had their uterus removed because of cervical cancer. In women older than 40, aging eggs may be the problem. In other couples, it is the man who is infertile. Male infertility often is caused by production of few or sluggish sperm or sperm with structural abnormalities. Also, in a strange twist of fate, otherwise compatible couples may be incompatible at the cellular level. Women, for example, are sometimes allergic to their partner's sperm and produce antibodies that kill the sperm before they are able to fertilize an egg. Finally, in many of the most frustrating cases, the cause of infertility is unknown. Can anything be done to help those who are infertile? For roughly half of the couples seeking help the answer is yes, although the road to reproduction may be long, expensive, and filled with emotional ups and downs. The procedures to treat infertility are collectively called assisted reproductive techniques (ARTs).

Some cases of female infertility can be treated by administering hormones. For example, hormones that trigger ovulation may be given to women whose ovaries fail to properly release eggs. Women who tend to miscarry (spontaneously abort) may be given progesterone to enhance the receptivity of their uterine lining. However, some causes of infertility, such as scarred oviducts in women and low sperm counts in men, do not respond to hormone therapy. What can be done in these cases?

Artificial insemination is one option available to couples whose infertility is caused by a low sperm count. In this procedure, sperm that have been donated and stored at a sperm bank are deposited (usually with a syringe) in the woman's cervix or vagina at about the time of ovulation. Sperm from the male member of the couple may be concentrated and then used. Alternatively, couples may use semen from an anonymous donor. (Artificial insemination with donor sperm also is frequently used by lesbian couples seeking to have children.) Another possibility for couples in which the man has few sperm, or sperm that lack the strength or enzymes necessary to penetrate an egg, is intracytoplasmic sperm injection (ICSI). In this procedure, a tiny needle is used to inject a single sperm into an egg (Figure 18.A).

Some cases of infertility can be treated with in vitro fertilization. Literally meaning fertilization in glass, IVF involves placing eggs and sperm together in a dish in the laboratory. First, the woman is treated with hormones to trigger superovulation, the ovulation of several eggs, rather than the typical lone egg. (An abundance of eggs is the key to many infertility treatments. Large numbers of eggs can be fertilized to produce many zygotes, which, in turn, will develop into pre-embryos. Having several pre-embryos available allows selection of the healthiest ones for immediate use and freezing of other healthy ones for future use.) Next, the eggs are removed from the woman's ovaries with a suction device and placed into a dish that contains a sample of the man's sperm. If fertilization occurs, the zygotes are transferred to a solution that will support further development. A few days later, when the pre-embryos are at an early blastocyst stage (about 3 to 5 days after fertilization), one or more of them is transferred to the woman's uterus. The uterus has been primed with estrogen and progesterone, and it is hoped that the pre-embryos will implant and complete development there. Women with blocked oviducts may conceive with IVF because the technique bypasses the oviducts altogether. If the woman lacks a uterus, the couple can hire a surrogate mother to gestate their baby. The egg and sperm may come from the couple or from other individuals.

Failure of pre-embryos to implant is in part responsible for the relatively low success rate of IVF. (Implantation is also a major hurdle for fertile couples, who lose about one in three pregnancies at this time.) The problem with IVF seems to be the rather violent squirt of the pre-

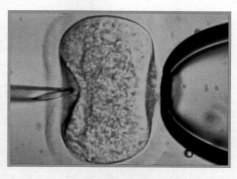

FIGURE **18.A** Intracytoplasmic sperm injection. An egg, held in place with a blunt-nosed pipette (right), is injected with a single sperm.

embryo into a reproductive tract that is already traumatized from hormone treatments and retrieval of eggs. Two new procedures avoid the problem of the pre-embryo's abrupt arrival in the uterus. In gamete intrafallopian transfer (GIFT), eggs and sperm are collected from a couple and inserted into the woman's oviduct, where fertilization is then expected to occur. Afterward, any resulting pre-embryos drift naturally (and gently) into the uterus. The second procedure is zygote intrafallopian transfer (ZIFT). In this procedure, eggs and sperm are collected and brought together in a dish in the laboratory. If fertilization occurs, the resulting zygotes are inserted into the woman's oviducts, where they continue to develop and travel on their own to the uterus. Normal, healthy oviducts are needed for GIFT and ZIFT to work.

At the beginning of this essay we stated that about half of couples who seek treatment for infertility are eventually rewarded with the birth of a child. Success rates of ARTs range from about 20% to 28% live births per egg retrieved. Although costs of the techniques vary somewhat, each attempt typically costs between $8000 and $12,000, and multiple attempts often are needed. Many health insurance plans do not cover these expenses because the procedures are considered experimental or elective. And sometimes assisted reproductive techniques result in exceptionally large families with exceptionally large demands and responsibilities (Figure 18.B).

continued →

HEALTH ISSUE (CONTINUED)

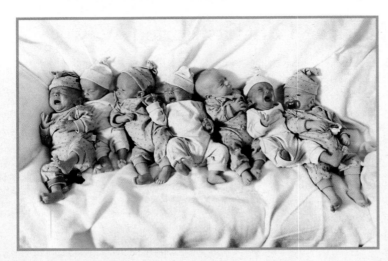

FIGURE 18.B Some assisted reproductive techniques may result in multiple births. Shown here are the McCaughey septuplets, conceived as a result of fertility drugs.

What about the couples who do not conceive with ARTs? Is there hope that some day they might also conceive? The answer is a decided yes. Advances in reproductive research and technology are rapidly providing potential treatments for infertility. Someday, a uterine transplant may be an option. In March 2002, doctors in Saudi Arabia transplanted a uterus from a 46-year-old woman undergoing a hysterectomy into a 26-year-old woman. (The recipient's uterus had previously been removed because of excessive bleeding following a cesarean section.) The transplanted uterus had to be removed after 99 days because of blood clotting. Nevertheless, Saudi doctors viewed the transplant as a success and were hopeful of better outcomes in the future.

Scientists are also working on techniques with mice in which spermatogonia, the stem cells deep within the testes that produce sperm, are transferred from a fertile animal to a sterile one. The once-sterile mouse can then sire offspring. Some day it might be possible to remove stem cells from a man with a low sperm count, encourage the cells to multiply under laboratory conditions, and then return them—now in much greater numbers—to their rightful owner, who might then father children.

What about research with eggs? Although sperm and pre-embryos are routinely frozen, eggs quickly become inviable when manipulated outside the body and are thus difficult to store. However, new techniques for freezing eggs are

being developed. Research also is under way on ovarian tissue transplantation as a way to produce functioning ovaries. Tissue from ovaries rich in potential eggs could be removed from a young female, frozen, and then placed back in her body years later. It might also be possible, though more difficult, to transplant ovarian tissue between females. Other research focuses on using eggs from fetuses. Fetal ovaries are a rich source of eggs, containing about 2 million as compared with the estimated 400 eggs ovulated by most women over the course of their reproductive years. Scientists are investigating the possibility of transplanting into infertile women tissue from the ovaries of fetuses made available from elective or spontaneous abortions. Fetal eggs might one day be stored at an egg bank, where they could be used by infertile women. The babies produced from such eggs would be in the truly unusual position of having a biological mother who had been aborted. Indeed, the aborted fetus with her wealth of eggs could potentially become the biological mother of many children.

Obviously, current and future reproductive technologies raise many ethical questions. Among them, do people who donate sperm or eggs have a claim to their biological offspring? What happens to viable pre-embryos that are created in the laboratory but turn out to be "extra"? Do they have a right to life? And who gets the pre-embryos when couples divorce or die? Finally, is it in our long-term interest to help obviously defective sperm fertilize eggs or to have women old enough to be grandmothers bearing children?

---

formation during the embryonic period. By the end of the embryonic period, all organs have formed, and the embryo has a distinctly human appearance. Three interrelated processes produce this tiny human: cell division (which continues from the pre-embryonic period); **cell differentiation** (the process by which cells become specialized with respect to structure and function); and **morphogenesis** (the development of overall body organization and shape).

GASTRULATION

Morphogenesis begins during the third week after fertilization. Just before the start of morphogenesis, during implantation, the inner cell mass moves away from the surface of the blastocyst, and the amniotic cavity forms. The amniotic cavity is filled with amniotic fluid and lined by the amnion, one of the four extraembryonic

membranes. The inner cell mass, which now becomes a flattened platelike structure called the *embryonic disk* (Figure 18.8), is destined to become the embryo proper. But first, the cells within the embryonic disk must differentiate and migrate, forming three **primary germ layers** known as ectoderm, mesoderm, and endoderm. (In anatomical drawings, these germ layers have traditionally been depicted with the following colors: blue for ectoderm, red for mesoderm, yellow for ectoderm. We follow this tradition in our figures.) The cell movements that establish the primary germ layers are called **gastrulation**, and the embryo during this period is called a *gastrula* (Figure 18.8).

All tissues and organs develop from the primary germ layers. **Ectoderm** covers the surface of the embryo and forms the outer layer of skin and its derivatives such as hair, nails, oil glands, sweat

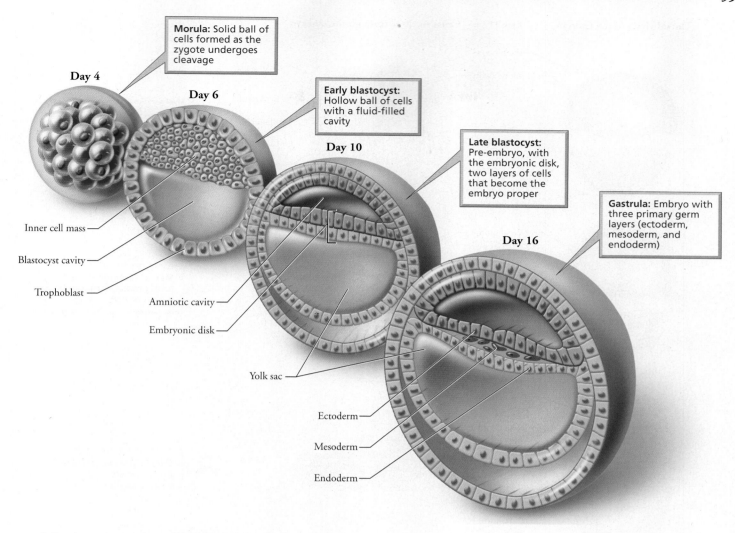

Day 4

Morula: Solid ball of cells formed as the zygote undergoes cleavage

Day 6

Early blastocyst: Hollow ball of cells with a fluid-filled cavity

Day 10

Late blastocyst: Pre-embryo, with the embryonic disk, two layers of cells that become the embryo proper

Day 16

Gastrula: Embryo with three primary germ layers (ectoderm, mesoderm, and endoderm)

Inner cell mass

Blastocyst cavity

Trophoblast

Amniotic cavity

Embryonic disk

Yolk sac

Ectoderm

Mesoderm

Endoderm

FIGURE **18.8** Early stages of development in cross section

glands, and mammary glands. Ectoderm also forms the nervous system. Another germ layer, the **endoderm**, turns inward to form the lining of digestive tract. Endoderm also forms the lining of the urinary and respiratory tracts, as well as other organs and glands (for example, the pancreas, liver, thyroid gland, and parathyroid glands). **Mesoderm** fills some of the space between ectoderm and endoderm, and gives rise to muscle; bone and other connective tissue; and various organs, including the heart, kidneys, ovaries, and testes. Because gastrulation initiates the processes by which the developing embryo's body takes shape and becomes organized, it is considered a key part of morphogenesis.

As gastrulation proceeds, a flexible rod of mesoderm tissue called the **notochord** develops where the vertebral column will form. The notochord defines the long axis of the embryo and gives the embryo some rigidity. The notochord also prompts overlying ectoderm to begin formation of the central nervous system (CNS), as discussed below. Vertebrae eventually form around the notochord, and the notochord degenerates. The pulpy, elastic material in the center of intervertebral disks (pads that help cushion the bones of the vertebral column) is all that remains of our notochord.

DEVELOPMENT OF THE CENTRAL NERVOUS SYSTEM

A major milestone in embryonic development is the formation of the central nervous system (brain and spinal cord) from ectoderm, as shown in Figure 18.9. The developing notochord induces the overlying ectoderm to thicken. This thickened region of ectoderm above the notochord is called the neural plate. The neural plate folds inward, forming a groove that extends the length of the embryo, along its back surface. The raised sides of the groove, known as neural folds, grow upward and eventually meet and fuse to form the **neural tube**, a fluid-filled tube that will become the central nervous system. The process by which the neural tube is formed is called **neurulation**, and the embryo during this period is called a *neurula*. The anterior portion of the neural tube develops into the brain, and the posterior portion forms the spinal cord. Alongside the neural tube, mesoderm cells organize into blocks called somites. **Somites** eventually form skeletal muscles of the neck and trunk, connective tissues, and vertebrae.

Failure of the neural tube to develop and close properly results in neural tube defects. *Spina bifida* ("split spine") is a type of neural tube defect in which part of the spinal cord develops abnormally, as

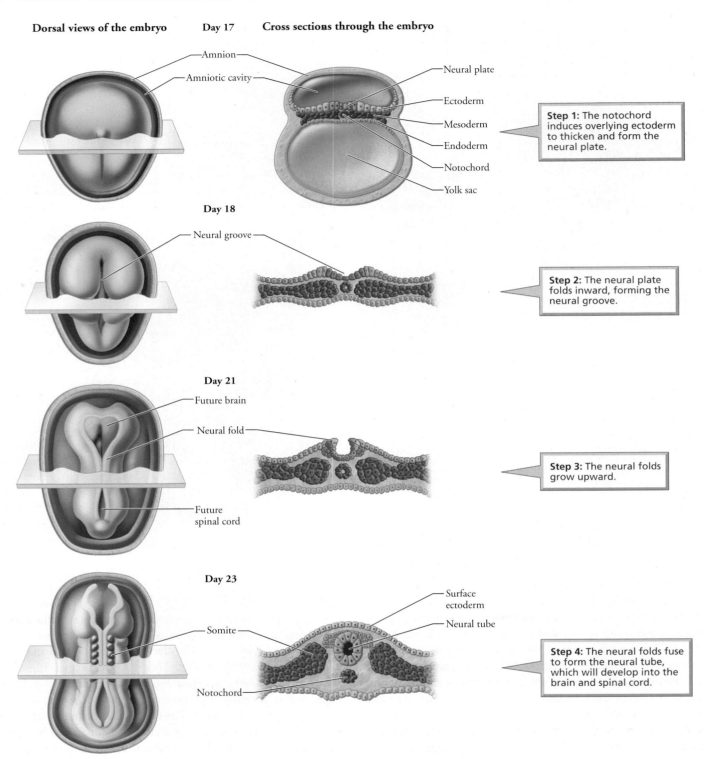

FIGURE **18.9** Formation of the central nervous system from ectoderm. Dorsal views of a human embryo (left) and corresponding cross-sectional views or parts of such views (right) are shown at 4 days during development of the brain and spinal cord.

does the adjacent area of the spine. The severity of spina bifida varies greatly, and some cases can be improved through surgery. *Anencephaly* is a neural tube defect that involves incomplete development of the brain and results in stillbirth or death shortly after birth.

DEVELOPMENT OF THE REPRODUCTIVE SYSTEM

Among the 23 chromosomes provided by the mother's egg is an X chromosome. Among the 23 chromosomes provided by the father's sperm is an X or a Y chromosome. The X and Y chromosomes are called sex chromosomes, and they determine the gender of human embryos. More specifically, the gender of an embryo is determined at fertilization by the type of sperm that fertilizes the egg. If an X-bearing sperm fertilizes the egg, then the zygote is XX and will normally develop as a female. If a Y-bearing sperm fertilizes the egg, then the zygote is XY and will normally develop as a male.

what would you do?

Methods are currently available to separate X-bearing sperm from Y-bearing sperm. During the procedure, sperm are treated with a fluorescent dye that binds to DNA. X-bearing and Y-bearing sperm glow differently because the X chromosomes have almost 3% more DNA than Y chromosomes. Thus X- and Y-bearing sperm can be sorted by an automated machine. Such "sperm sorting" makes possible the selection of a child's gender. Couples could decide whether to use X-bearing sperm to produce a female baby or Y-bearing sperm to produce a male baby. Some couples want to select the gender of their child to avoid X-linked genetic diseases (see Chapter 20); others want to balance their families with respect to numbers of sons and daughters. The chosen sperm are deposited within the woman's vagina or cervix at a time when pregnancy is likely to occur. (The process of depositing semen within the female reproductive tract is called artificial insemination.) Should parents be allowed to choose the gender of their children? What implications would gender selection have for families and for society? If you were given the responsibility of deciding whether such technology should continue to be made available to parents, what would you do?

Male and female embryos have the same external and internal anatomy for the first few weeks of development; their reproductive organs have not yet differentiated. The embryos at this time are described as being "sexually indifferent." However, about 6 weeks after fertilization, a region of the Y chromosome—called SRY for sex-determining region of the Y chromosome—initiates development of testes in XY embryos. The testes soon begin producing testosterone, the sex hormone that directs development of male reproductive organs. Female embryos lack the Y chromosome, and, in its absence, ovaries develop. The development of female reproductive structures is not influenced by hormones, and female structures will develop even if ovaries are absent. Thus, it is the absence of the Y chromosome and testosterone that leads to female development. We will focus on development of the external genitalia, but keep in mind that internal reproductive organs also are undergoing differentiation.

Both male and female embryos of about 6 weeks of age have a small bud of tissue between their legs (Figure 18.10). Just below

this bud is a shallow depression called the urogenital groove. The urogenital groove is surrounded by urogenital folds, which, in turn, are surrounded by labioscrotal swellings. In the male embryo, testosterone produced by the newly differentiated testes causes the urogenital groove to elongate and completely close. The bud then becomes the glans penis; the urogenital folds develop into the shaft of the penis; and the labioscrotal swellings become the scrotum (into which the testes will later descend). In the female, the urogenital groove also elongates, but instead of closing, as in the male, it remains open. The bud becomes the clitoris in females, and the urogenital folds develop into the labia minora. The labioscrotal swellings of females become the labia majora.

■ Rapid growth characterizes the fetal period

The fetal period extends from the ninth week after fertilization until birth. Although tissues and organs continue to differentiate during this period, the most notable change in most body parts is rapid growth, made possible by the placenta. The placenta completes its development early in the fetal period, and through its efficient functioning, oxygen and nutrients are carried to cells of the fetus and wastes such as carbon dioxide and urea are carried away. We first describe growth during the fetal period and then consider the circulatory pathway that makes growth possible.

GROWTH

Growth during the fetal period is extremely rapid, reflected in phenomenal increases in weight (from 8 g to 3400 g, or from about 0.3 oz to 120 oz) and substantial increases in length (crown-rump length changes from 50 to 360 mm, or from about 2 in. to 14 in.), when the pregnancy continues for the full 38 weeks. (Also known as "sitting height," crown-rump length is a measurement of the distance from the head to the rump.) One striking change is that the rate of growth of the head slows relative to the rate of growth of other regions of the body, altering the way the body is proportioned (Figure 18.11). Difference in the relative rates of growth of various parts of the body is called **allometric growth**. Such growth continues after birth and helps to shape developing humans and other organisms.

FETAL CIRCULATION

The fetal circulatory system differs from circulation after birth because several organs—the lungs, kidneys, and liver, to name a few—do not perform their postnatal functions in the fetus. Before birth, most blood is shunted past these organs through temporary vessels or openings. Recall that the fetus receives its oxygen and nutrients from maternal blood and gives up its carbon dioxide and wastes to maternal blood at the placenta. We first describe the path of blood flow from the fetus to the placenta and back to the fetus. Then we consider the circulatory bypasses unique to life before birth. We end with a description of the changes in circulation that occur at birth when the organs of the newborn begin to function. This discussion is summarized in Figure 18.12.

Fetal blood must travel to the placenta to rid itself of carbon dioxide and wastes and to pick up oxygen and nutrients. It does so

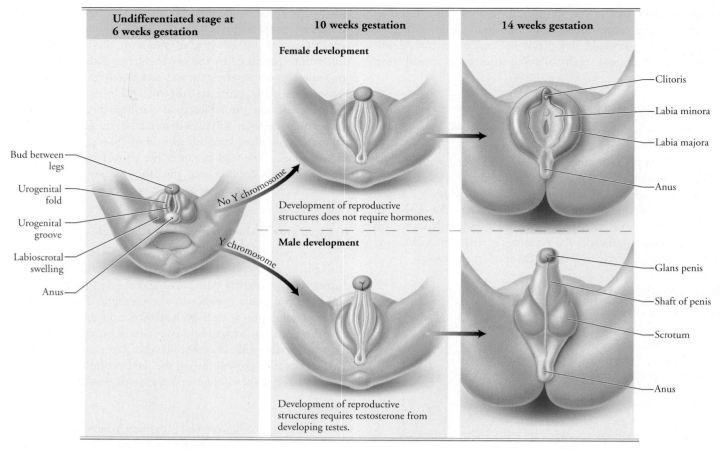

FIGURE **18.10** Development of external genitalia

by flowing through umbilical arteries that branch off large arteries in the legs of the fetus. The two umbilical arteries run through the umbilical cord to the placenta. Once exchanges between fetal and maternal blood have occurred at the placenta, blood rich in oxygen and nutrients travels back to the fetus in the umbilical vein (within the umbilical cord; Figure 18.12a). Some of this blood flows to the fetal liver, which produces red blood cells but does not yet function in digestion. Most of the blood from the placenta, however, bypass-

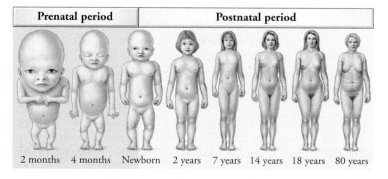

FIGURE **18.11** Allometric growth. Changes in body proportions occur throughout prenatal and postnatal development. For comparison, all stages are drawn to the same total height.

es the liver by means of a temporary vessel called the *ductus venosus* and heads to the heart.

Like the liver, fetal lungs are not yet functional and need only a small amount of blood for growth and removal of wastes from their cells. Once blood enters the right atrium of the heart, some of it passes to the right ventricle and on to the lungs, as it does in circulation after birth. Most of the blood, however, moves into the left atrium through a small hole in the wall between the atria, called the *foramen ovale*. The blood in the left atrium moves into the left ventricle and out to the body of the fetus. Another shunt, called the *ductus arteriosus*, connects the pulmonary trunk to the aorta and also functions to divert blood away from the lungs. Thus, blood flowing out of the right ventricle into the pulmonary trunk is shunted away from the lungs and into the aorta, where it travels to all areas of the fetus. Another glance at Figure 18.12a reveals that much of the blood traveling through the fetus is moderate to low in oxygen. Indeed, the umbilical vein is the only fetal vessel that carries fully oxygenated blood.

At birth, when the lungs, liver, and other organs begin their postnatal functions, fetal circulation converts to the postnatal pattern (Figure 18.12b). Blood flow to the placenta ceases when the umbilical cord is tied and cut off. The scar left by the cord becomes the baby's navel, or belly button. Within the infant, the umbilical arteries, umbilical vein, ductus venosus, and ductus arte-

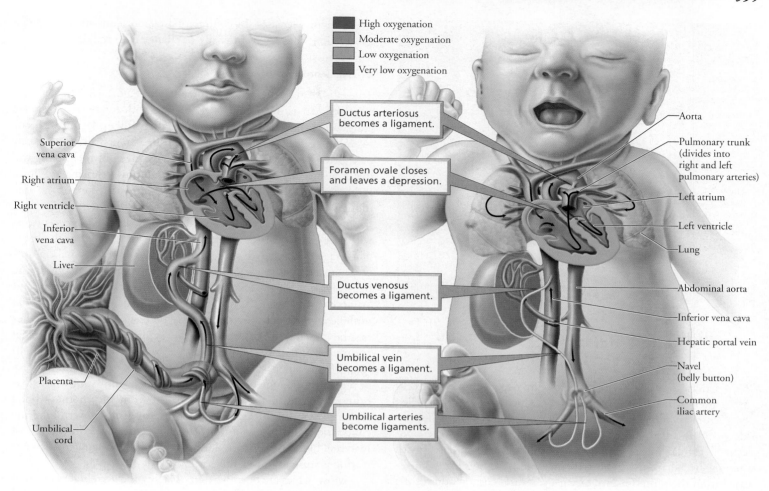

High oxygenation
Moderate oxygenation
Low oxygenation
Very low oxygenation

Ductus arteriosus
becomes a ligament.

Foramen ovale closes
and leaves a depression.

Ductus venosus
becomes a ligament.

Umbilical vein
becomes a ligament.

Umbilical arteries
become ligaments.

Superior
vena cava

Right atrium

Right ventricle

Inferior
vena cava

Liver

Placenta

Umbilical
cord

Aorta

Pulmonary trunk
(divides into
right and left
pulmonary arteries)

Left atrium

Left ventricle

Lung

Abdominal aorta

Inferior vena cava

Hepatic portal vein

Navel
(belly button)

Common
iliac artery

(a) Fetal circulation is characterized by a connection to the placenta (the umbilical cord) and several bypasses around organs that do not yet perform their postnatal functions.

(b) At birth, the umbilical cord is tied off and cut, leaving the navel (belly button). The bypasses close, allowing more blood to reach the now functional organs of the newborn.

FIGURE **18.12** Fetal circulation and changes at birth

riosus constrict, shrivel, and form ligaments. The foramen ovale normally closes shortly after birth, leaving a small depression in the wall between the two atria. In some newborns, the foramen ovale fails to close. Known as *blue babies*, these infants have a bluish appearance because much of their blood still bypasses the

lungs and is low in oxygen. Fortunately, this defect can be corrected with surgery.

The major events that occur during prenatal development are summarized by developmental period in Table 18.1. Several stages of prenatal human development are shown in Figure 18.13.

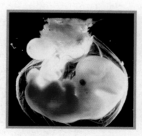

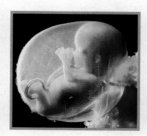

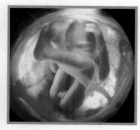

(a) A zygote

(b) A blastocyst implanting into the wall of the uterus

(c) An embryo about 7 weeks old

(d) A fetus about 4 months old

(e) A fetus about 5 months old

FIGURE **18.13** The developing human

TABLE 18.1 REVIEW OF MAJOR EVENTS DURING PRENATAL DEVELOPMENT

PERIOD	MAJOR EVENTS
Pre-embryonic (fertilization–week 2)	Fertilization
	Cleavage
	Formation and implantation of the blastocyst
	Beginning of formation of extraembryonic membranes and placenta
Embryonic (week 3–week 8)	Gastrulation
	Formation of tissues, organs, and organ systems
Fetal (week 9–birth)	Continued differentiation and growth of tissues and organs
	Increase in crown-rump length
	Increase in weight

Birth Is the Transition from Prenatal to Postnatal Development

Birth, also called parturition, usually occurs about 38 weeks after fertilization. The process by which the fetus is expelled from the uterus and moved through the vagina to the outside world is called **labor**. Uterine contractions during true labor occur at regular intervals, are usually painful, and intensify with walking. Recall from Chapter 10 that the hormone oxytocin released from the posterior pituitary gland causes contractions of the smooth muscle of the uterus. These contractions, in turn, stimulate further release of oxytocin in a positive feedback loop. As true labor proceeds, contractions become more intense, and the interval be-

tween them decreases. In contrast, false labor consists of irregular contractions that fail to intensify or change with walking.

■ True labor has three stages

True labor can be divided into three stages (Figure 18.14). The first stage, known as the *dilation stage*, begins with the onset of regular contractions and ends when the cervix has fully dilated to 10 cm (4 in.). This stage usually lasts about 6 to 7 hours, although it can be shorter in women who previously have given birth. During the dilation stage, the amniotic sac usually ruptures, releasing amniotic fluid. Known commonly as *breaking the water*, rupture of the amniotic sac is done deliberately by a doctor or midwife if it does not happen spontaneously. The second stage, the *expulsion stage*, begins with full dilation of the cervix and ends with delivery of the baby. This stage may last 1 or more hours. Some physicians make an incision to enlarge the vaginal opening, just before passage of the baby's head. This procedure, an *episiotomy*, facilitates delivery and avoids ragged tearing. Once delivery of the baby is complete, the umbilical cord is clamped and cut. The newborn takes his or her first breath, and the conversion of fetal circulation to the postnatal pattern begins. Labor, however, is not over. The third (and final) stage of true labor is the *placental stage*, which begins with delivery of the newborn and ends when the afterbirth, consisting of the placenta and fetal membranes, is expelled from the mother's body about 15 minutes after the baby. Continuing and powerful uterine contractions aid in expulsion of the placenta and constriction of maternal blood vessels torn during delivery. Any vaginal incision or tearing is sutured once labor has ended.

Most babies are born head first, facing the vertebral column of their mother. Some babies, however, are born buttocks first, and their delivery is called *breech birth*. Breech births are associated with difficult labors and umbilical cord accidents, such as compression

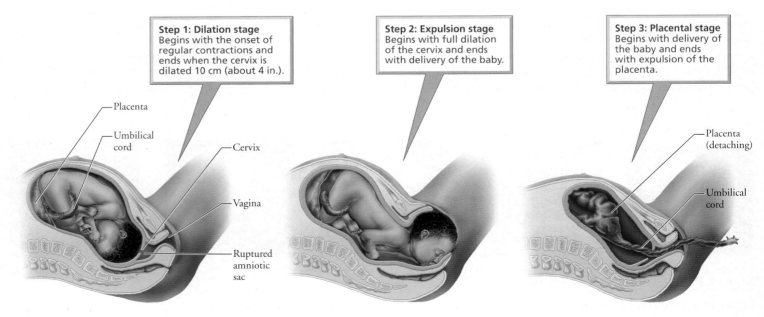

Step 1: Dilation stage
Begins with the onset of regular contractions and ends when the cervix is dilated 10 cm (about 4 in.).

Step 2: Expulsion stage
Begins with full dilation of the cervix and ends with delivery of the baby.

Step 3: Placental stage
Begins with delivery of the baby and ends with expulsion of the placenta.

Placenta

Umbilical cord

Cervix

Vagina

Ruptured amniotic sac

Placenta (detaching)

Umbilical cord

FIGURE **18.14** The stages of true labor

or looping of the cord around the baby's neck. Thus, many physicians attempt to turn breech babies into the headfirst position before delivery. The baby may be delivered by cesarean section if turning is not successful.

Babies born at least 38 weeks after fertilization are called *full-term infants*. These babies weigh, on average, 3400 g (7.5 lb). Some babies are born 1 or more weeks late. Because the placenta begins to deteriorate with time, labor is typically induced if a pregnancy continues for 41 or more weeks. Babies born before 37 weeks of gestation are called *premature infants* (*gestation* refers to the time a pre-embryo, embryo, or fetus is carried in the female reproductive tract). About 12% of babies born in the United States are born prematurely. Their chances for survival increase with time spent in the female reproductive tract. Infants born at only 22 weeks of age usually are not revived because major disabilities are almost inevitable. Such babies weigh about 630 g (1.4 lb). Amazingly, infants born at 25 weeks gestation and weighing about 1 kg (2.2 lb) have a 50% to 80% chance of surviving, provided they receive intensive care. Intensive medical care is needed because the organ systems of these infants have not matured sufficiently to take over the functions normally performed by the mother's body. Care for premature infants often includes numerous blood transfusions, tube feeding, and maintenance on a respirator (Figure 18.15). Even with excellent intensive care, health problems may persist well beyond infancy.

what would you do?

Babies born before 25 weeks gestation typically have a low chance of survival, even with excellent medical care. In addition, they have a very high probability of suffering from severe disabilities, if they do survive. Should efforts be made to save all such infants? Or should criteria be used to decide which infants the doctors should try to save and which they should let die? If you were given the responsibility of developing such criteria, what would you suggest? More to the point, if your infant were born before 25 weeks, what would you ask the doctor to do?

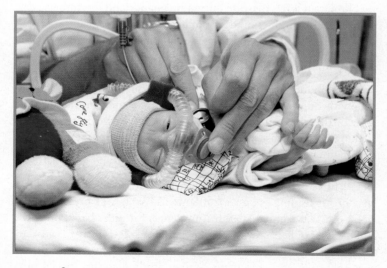

FIGURE **18.15** Many premature infants require intensive care because their organs are not yet functional.

■ The placenta may send the signal to initiate labor

In humans, the signal to initiate labor appears to come from the placenta in the form of *corticotropin-releasing hormone (CRH)*. Levels of CRH begin to rise in fetal and maternal bloodstreams beginning about 12 weeks into gestation. Sharp increases in CRH occur during the last few weeks of gestation. This placental hormone stimulates the anterior pituitary gland of the fetus to release adrenocorticotropic hormone, which, in turn, prompts the fetal adrenal glands to release a hormone that is converted by the placenta into estrogen. As a result of this conversion, levels of estrogen increase relative to progesterone, causing a cascade of events that initiates labor. These events include coordinated uterine contractions, dilation of the cervix, and release of oxytocin. Some recent studies support CRH as the initiating signal. In these studies, the level of CRH between weeks 16 and 20 of gestation predicted the timing of birth. Women with the highest levels of CRH were likely to deliver prematurely. Those with the lowest levels were likely to deliver late. Prevention of premature deliveries and the consequent health risks to infants is a major goal for understanding the events that initiate labor.

■ Drugs and breathing techniques can relieve pain during childbirth

Most women experience pain during childbirth. The level of pain varies greatly and seems to depend on factors such as degree of fear and tension, the physical surroundings, and the support received. Typically, the more prepared and relaxed a woman is, the less pain she feels. Reduced stress is one reason health-care workers encourage expectant mothers and their partners to enroll in childbirth classes. These classes provide information on the stages of labor and medical procedures associated with delivery. Childbirth classes also teach breathing and relaxation techniques.

Drugs may be administered to reduce the pain of childbirth. Analgesics—substances that relieve pain—may be administered intramuscularly or intravenously. Regional anesthetics, substances that produce loss of sensation in specific areas of the body, also may be used. For example, in epidural anesthesia, an anesthetic is injected in the lower back into the space between the dura mater (the outer membrane covering the spinal cord) and the vertebral canal. This injection blocks pain and motor impulses to the abdomen. Because motor impulses are blocked, epidural anesthesia makes pushing out the baby impossible. Epidural procedures are performed fairly early during labor and not toward the end, when pushing by the mother is critical for successful vaginal delivery. Epidural anesthesia is associated with slowing the fetal heartbeat. Thus, its use necessitates continuous fetal monitoring.

Many women rely on breathing or relaxation techniques rather than medications to relieve the pain of childbirth. Childbirth without the use of pain-killing drugs is often called *natural childbirth*. Whether to use medication during childbirth is a decision that is best made by weighing the costs and benefits to both mother and baby. Also, because it is impossible to predict with certainty what will occur during labor and delivery, all expectant mothers should keep an open mind.

Environmental Disruptions during the Embryonic Period Cause Major Birth Defects

The development of a fertilized egg into a baby is an incredibly wondrous yet complex process. In view of its complexity, it is not difficult to imagine that things might go wrong along the way. Indeed, the potential for mistakes seems enormous. Thus, we should not be surprised to learn that not every zygote develops into a baby, and not all babies are born healthy. Developmental defects present at birth are called **birth defects**. They may involve structure, function, behavior, or metabolism. The study of birth defects is called teratology (*teratos*, monster), reflecting the unfortunate term used in earlier times to describe severely deformed infants.

The causes of birth defects may be genetic or environmental. Genetic causes include mutant genes or changes in the number or structure of chromosomes. Environmental causes include drugs, chemicals, radiation, deficiencies in maternal nutrition, and certain viruses, such as herpes simplex and rubella (the cause of German measles). Birth defects resulting from genetic causes will be discussed in Chapter 19. Some of the birth defects associated with psychoactive drugs were considered in Chapter 8a. Here we focus on *when* in development disruptive agents have their greatest impact.

Environmental agents that disrupt development have their greatest effects during periods of rapid differentiation, when cells and tissues acquire specific functions and form organs. Recall that these events largely occur during the embryonic period (week 3 through week 8). Exposure to disruptive agents during the embryonic period can cause major birth defects (Figure 18.16). Such defects, characterized by major structural abnormalities, occur in 2% to 3% of newborns. Spina bifida is an example of a major birth defect. Disruptive agents produce more minor defects during the fetal period, when most organs are growing rather than differentiating. Minor birth defects such as birthmarks occur in about 4% to 5% of newborns.

We have said that all organs and organ systems develop primarily during the embryonic period. Nevertheless, different organs develop at somewhat different times and rates. Such variation results in each organ system having a critical period in which it is most susceptible to disruption by environmental agents. The CNS has a critical period extending from week 3 through week 16. During this time, the developing nervous system is highly sensitive, and disruptive agents can have major consequences. By comparison, exposure to such agents after week 16 tends to induce relatively minor defects. The CNS has the longest critical period. It extends beyond the embryonic period and into the fetal period. This long critical period may explain why the brain is the organ in which defects in structure and function are most commonly found. The critical period for development of the upper and lower limbs is relatively short, extending from about week 4 to week 6. It is during this short window of time that certain environmental agents cause major abnormalities such as tiny limbs or complete absence of limbs.

Another look at Figure 18.16 reveals that exposure to harmful agents during the first 2 weeks after fertilization is not known to cause birth defects. Development during this period (the pre-embryonic period) centers on formation of structures outside the embryo, such as the extraembryonic membranes described earlier. However, the developing pre-embryo is not immune to environmental influences. Disruptive agents in the pre-embryonic period may interfere with cleavage or implantation and cause a spontaneous abortion. Chromosomal abnormalities also account for a large proportion of spontaneous abortions at this time.

what would you do?

During the late 1950s and early 1960s, the drug thalidomide was widely used in Europe as a sedative and as relief for morning sickness. Previous animal testing had indicated that thalidomide was safe; although the testing reportedly did not include pregnant animals. Unfortunately, in humans, the drug caused major defects of the limbs such as complete absence of limbs or abnormally small limbs lacking elbows and knees. Thalidomide also caused abnormalities of the heart (some of which caused death shortly after birth), ears, and urinary and digestive systems. About 12,000 infants whose mothers had taken thalidomide were affected. In later years, thalidomide was found to be an effective treatment for conditions such as leprosy and cancer of the bone marrow and also appears to relieve some of the conditions associated with HIV infection. Today, thalidomide is available by prescription in the United States and is sold over the counter in South America, where "thalidomide babies" are still being born. Should a drug known to cause severe birth defects be made available for treatment of seriously ill adults? How can we ensure that pregnant women do not mistakenly take thalidomide? If you were responsible for developing a policy regarding the use of thalidomide, what would you do?

The Mother's Mammary Glands Produce Milk

Lactation is the production and ejection of milk from the mammary glands. Recall from Chapter 10 that the hormone prolactin from the anterior pituitary gland promotes milk production. Recall also that oxytocin from the posterior pituitary gland stimulates milk ejection. The structure of the breast was described in Chapter 17.

Prolactin levels increase during pregnancy and reach their peak at birth. Milk production does not begin during pregnancy because high levels of estrogen and progesterone inhibit the actions of prolactin. Thus, prolactin cannot initiate milk production until birth, when levels of estrogen and progesterone decline. Milk usually is available about 3 days later. The breasts produce *colostrum*, a cloudy yellowish fluid of different composition than milk, in the interval after birth, when milk is not yet available. After birth, the infant's sucking on the breast stimulates the release of prolactin, promoting and maintaining the production of milk. When sucking ends, so too does milk production.

Which is best—bottle or breast? Most people today believe that breast-feeding is best for infants. Advocates of breast-feeding point out that compared with commercial formulas based on cow's milk, breast milk is more digestible and much less likely to cause an allergic reaction, constipation, or obesity in infants. In addition, maternal milk and colostrum contain antibodies and special proteins that boost the immune system. Finally, obtaining milk from a

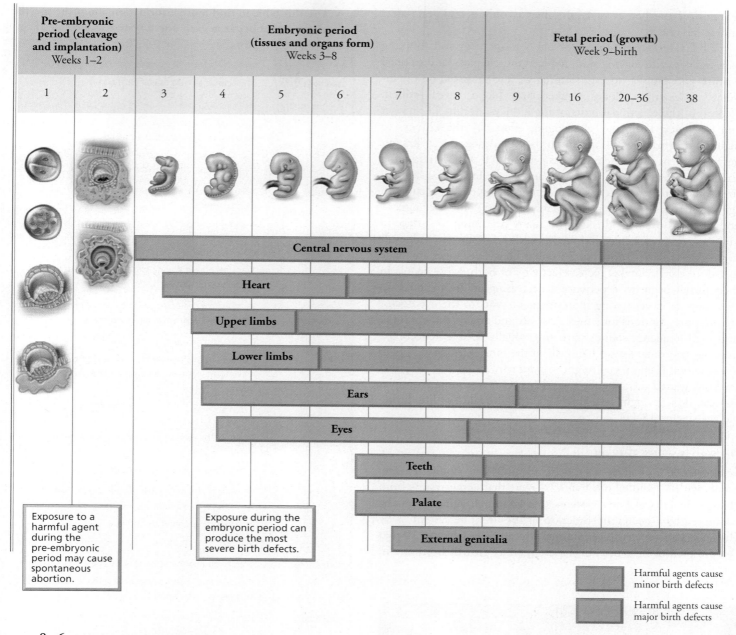

Pre-embryonic period (cleavage and implantation) Weeks 1–2		Embryonic period (tissues and organs form) Weeks 3–8						Fetal period (growth) Week 9–birth			
1	2	3	4	5	6	7	8	9	16	20–36	38

Central nervous system

Heart

Upper limbs

Lower limbs

Ears

Eyes

Teeth

Palate

External genitalia

Exposure to a harmful agent during the pre-embryonic period may cause spontaneous abortion.

Exposure during the embryonic period can produce the most severe birth defects.

Harmful agents cause minor birth defects

Harmful agents cause major birth defects

FIGURE **18.16** Critical periods in development

breast requires more effort than sucking on a bottle. Health professionals think this greater effort promotes optimum development of the infant's jaws, teeth, and facial muscles.

Breast-feeding also is beneficial to the mother. Nursing helps the mother's uterus return toward its prepregnant size. When oxytocin is released in response to an infant's sucking, it helps the uterus shrink. Perhaps most important, breast-feeding establishes prolonged periods of skin-to-skin contact between mother and infant.

Despite its benefits, breast-feeding is not for everyone. Some women simply prefer bottle-feeding. Others are warned against breast-feeding because the medications they take pass into breast milk. Finally, mothers who are HIV positive should not breast-feed because the virus can be transmitted in breast milk.

The Postnatal Period Begins with Birth and Continues into Old Age

The postnatal period—the period of growth and development after birth—includes the following stages: infancy (from birth to 12 months), childhood (from 13 months to 12 or 13 years), puberty (from 12 to 15 years in girls and from 13 to 16 years in boys), adolescence (from puberty to late teens), and adulthood (generally reached by around 20 or 21 years of age). The start of adulthood is imprecise because of the many factors—physical, behavioral, societal—that influence how people define an adult. Growth and formation of bone usually are completed by age 25. **Aging**, the normal but progressive decline in the structure and function of the bodies of adults, begins only a few years after growth has ceased.

Observable characteristics of old age—graying and thinning hair, wrinkles and sagging skin, reduced muscle mass, and stooped posture—are familiar to us. However, we also grow old on the inside. For example, bones weaken and blood vessels stiffen. Nephrons, the functional units of the kidneys, decline in number and efficiency, ultimately challenging efforts by the kidneys to maintain the balance of fluids in our bodies. Aging also strikes our nervous system and sense organs, as evidenced by our deteriorating ability to remember, see, hear, taste, and smell. Table 18.2 summarizes some of the changes to organ systems that typically occur with aging. Of course, we do not suddenly become "old." Changes to body systems occur gradually, with many beginning to surface in middle age. For example, declines in basal metabolic rate will result in weight gain after about age 35 unless lifestyle changes are made to reduce caloric intake, increase activity, or both. Most people in their forties begin to experience a decreasing ability to focus on close objects, necessitating more frequent visits to the eye doctor and new prescriptions for lenses. Also, around age 50, our ability to hear high-frequency sounds starts to gradually decline, reflecting a slow but continual loss of hair cells in the inner ear that begins in our twenties! Even so, you might wonder why we discuss aging in a textbook geared toward college students. Our reasoning is simple; you can do a lot right now to ensure a high quality of life during old age (Figure 18.17).

■ Possible causes of aging

Scientists do not agree about what causes our bodies to age. Some, with a "whole body" view, suggest that aging results from changes in critical body systems. For example, aging might be prompted by a decline in function of the immune system or by changing levels of certain hormones. Recall from Chapter 10 that the production of some hormones, such as growth hormone in

 Use this table to find more information about the specific changes that occur in each organ system as a result of aging.

ORGAN SYSTEM	SOME CHANGES THAT OCCUR WITH AGING
Integumentary	Wrinkles appear as skin becomes thinner and less elastic.
	Sweat glands decrease in number, making regulation of body temperature more challenging.
	Hair thins owing to death of hair follicles and turns gray as pigment-producing cells die.
Skeletal	Bones become lighter and more brittle, especially in women after menopause.
	About 7.6 cm (3 in.) of height are lost as the intervertebral disks deteriorate and the vertebrae move closer together.
	Joints become stiff and painful because of decreased production of synovial fluid.
Muscular	Muscle mass decreases owing to loss of muscle cells and decreased size of remaining muscle cells.
Nervous	Brain mass decreases.
	Movements and reflexes slow as conduction velocity of nerve fibers decreases and release of neurotransmitters slows.
	Hearing becomes less acute as hair cells in the inner ear are lost.
	Ability of the eye to focus declines as the lens of the eye stiffens.
	Smell and taste become less acute.
Endocrine	In women, production of estrogen and progesterone decreases with menopause.
	In men, production of testosterone decreases.
	In both sexes, production of growth hormone decreases.
Circulatory	Cardiac output decreases as walls of the heart stiffen.
	Blood pressure rises as arteries become less elastic and clogged by fatty deposits.
Respiratory	Lung capacity decreases as alveoli break down and lung tissue becomes less elastic.
Digestive	Basal metabolic rate declines.
	Ability of the liver to detoxify substances declines.
Urinary	In both sexes, kidney mass declines, as does the rate of filtration of the blood by nephrons.
	Particularly in women, the external urethral sphincter weakens, causing incontinence.
	In men, the prostate gland enlarges, causing painful and frequent urination.
Reproductive	In men, fewer viable sperm are produced.
	In women, ovulation and menstruation cease at menopause.

TABLE 18.2 CHANGES IN ORGAN SYSTEMS AS WE AGE

FIGURE **18.17** Wise and informed lifestyle decisions made when young can lead to a healthy old age.

both sexes and estrogen in females, declines as we age. Perhaps such declines induce the changes in bodily structure and function that characterize aging. Other scientists seek explanations for the causes of aging at the cellular and molecular levels, and we consider these ideas next.

CESSATION OF CELL DIVISION

Ongoing cell division is necessary to replace cells that die. Without new cells, most tissues and organs could not continue to function effectively. Yet cell division appears to slow in aging animals. Supporting the idea that cessation of cell division is genetically programmed is the observation that cells grown in culture in the laboratory do not divide indefinitely. In fact, they divide only a certain number of times. The number of divisions seems to be correlated with the age of the individual that donated the cells (be the donor a roundworm, mouse, or human) and the life span of the particular species. Telomeres, protective pieces of DNA at the tips of chromosomes, may help cells keep track of how many times they have divided (see Chapter 21a). A tiny piece of each telomere is sliced off each time the DNA is copied before cell division. After a certain number of cell divisions, the telomere is gone, and the chromosome is no longer protected from the slicing. Chromosomes that sustain such damage can no longer participate in cell division. Cells grown in culture demonstrate this phenomenon, yet a clear link to human longevity has not been established.

DAMAGE TO DNA AND OTHER MACROMOLECULES

Other researchers postulate that highly reactive molecules known as free radicals disrupt cell processes and lead to aging. Free radicals are by-products of normal cellular activities. They have an unpaired electron and readily combine with and damage DNA, proteins, and lipids. Mitochondrial DNA seems particularly susceptible to such damage. Recall that mitochondria play a critical role in processing energy for cells (discussed in Chapter 3). Thus, damage to mitochondria threatens the supply of energy needed to sustain cellular activities. Aging also is associated with a decline in the ability of cells to repair damaged DNA. Such a decline might lead to an accumulation of gene mutations and ultimately to a decline in cell function.

Scientists are also studying *Werner's syndrome*, a rare inherited disease characterized by premature aging. People with Werner's syndrome develop the appearance and ailments of the elderly while still in their twenties. They typically die in their late forties. An understanding of this condition, which appears to be related to impaired DNA replication, may provide information on the normal aging process.

Finally, some scientists have suggested that glucose—our main source of fuel—is the culprit. Glucose changes proteins such as collagen by causing cross-linkages to form between their molecules. This shackling of protein molecules may cause the stiffening of connective tissue and heart muscle associated with aging.

Is there a primary cause of aging? Most scientists believe that aging is not caused by a single factor but rather by several processes that interact to produce our eventual deterioration and demise.

■ Medical advances and a healthy lifestyle can help achieve a high-quality old age

The news today is full of stories about antiaging products. For example, human growth hormone (HGH) is billed as being able to reverse many of the effects of aging. (Recall from Chapter 10 that growth hormone is secreted by the anterior pituitary gland.) Administration of HGH seems to increase lean body mass and the density of some bones. It also appears to reduce thinning of the skin and the amount of adipose (fat) tissue. However, the administration of HGH has side effects such as fluid accumulation, increased blood pressure, and diabetes. In addition, some animal studies have suggested that administering growth hormone might actually shorten life.

Aging is a normal biological process that, at present, cannot be slowed, stopped, or reversed. While this news might depress you, especially after reading the long list of declines in organ systems outlined in Table 18.2, the good news is that we can stave off much of the disease and disability associated with aging. The keys to a healthy old age lie in medical advances and lifestyle. Physicians are now able to treat many conditions of old age. For example, worn-out hip or knee joints can be replaced with artificial joints. The clouding of vision caused by cataracts is now treated readily by replacing the old lens of the eye with an artificial lens. Drugs are being tested that may break the cross-links that glucose creates in proteins and thereby restore elasticity to our arteries. In addition, research is under way to determine if antioxidants (substances that inhibit the formation of free radicals) are effective in delaying certain aspects of aging.

Today, the maximum documented life span for humans is 122 years, a record recently established by Madame Jeanne Calment of France (Figure 18.18). Most of us will likely never reach her age. In

FIGURE **18.18** The longest documented human life span is 122 years; this record was established by Madame Jeanne Calment, shown here on her last birthday.

fact, life expectancy for babies born in the United States today is about 77 years. Nevertheless, we can do a lot to promote an enjoyable and illness-free old age. The human life span appears to be determined by genes, environment, and lifestyle. Lifestyle is the factor over which we have the most personal control. Some components of a healthy lifestyle include proper nutrition, plenty of exercise and sleep, no smoking, and routine medical checkups. Indeed, the lifestyle choices we make when young can delay some aspects of aging. For example, we can delay wrinkling and aging of the skin

by avoiding excessive exposure to the sun. The importance of a healthy lifestyle continues into our later years, when exercise, social and intellectual stimulation, and good nutrition can help to prevent the physical and mental declines associated with aging. Although it is never too late to change lifestyle habits, the earlier we begin to develop healthy habits, the better. Finally, aging is not only about declining organ systems; it has positive aspects as well. Not the least of these, as the years pass, are the gains we make in experience, perspective, and wisdom.

HIGHLIGHTING THE CONCEPTS

Human Life Has Two Main Periods of Development (p. 383)

1. The period of development before birth is the prenatal period. The period of development after birth is the postnatal period.

The Prenatal Period Begins at Fertilization and Ends at Birth (pp. 383–396)

2. The prenatal period can be subdivided into the pre-embryonic period, embryonic period, and fetal period. The pre-embryonic period runs from fertilization through the second week and is characterized by formation and implantation of the blastocyst. The extraembryonic membranes and placenta also begin to form at this time. The embryonic period extends from week 3 through week 8 and is characterized by gastrulation and the formation of organs and organ systems. The fetal period runs from the ninth week until birth and is a period of intense growth.

3. Fertilization is the union of an egg and a sperm to form a single cell called the zygote.

4. Cleavage is a rapid series of mitotic cell divisions that transforms the zygote into a morula (a solid ball of cells) and then a blastocyst (a hollow ball of cells with a fluid-filled cavity). Within the blastocyst are the inner cell mass, which will become the embryo and some of the extraembryonic membranes, and the trophoblast, which will form part of the placenta. Toward the end of the first week, the blastocyst begins implantation, the process by which the embryo becomes embedded in the endometrium of the uterus.

5. Extraembryonic membranes—the amnion, yolk sac, chorion, and allantois—lie outside the embryo and begin to form during the second or third week after fertilization. The placenta, formed from the chorion of the embryo and the endometrium of the mother is the organ that delivers nutrients and oxygen to the embryo (and later the fetus) and carries wastes away. The placenta also produces hormones such as estrogen, progesterone, human chorionic gonadotropin (HCG), and corticotropin-releasing hormone (CRH).

6. Gastrulation includes the cell movements by which the primary germ layers—ectoderm, mesoderm, and endoderm—are established. Ectoderm forms the nervous system and the top layer of skin and its derivatives (hair, nails, oil glands, sweat glands, and mammary glands). Mesoderm forms muscle, bone, connective tissue, and organs such as the heart, ovaries, and testes. Endoderm forms organs such as the liver, some endocrine glands, and the lining of the urinary, respiratory, and digestive tracts.

7. Neurulation is the process by which the neural tube forms from ectoderm. The anterior portion of the neural tube forms the brain, and the posterior portion forms the spinal cord. Somites, blocks of mesoderm organized alongside the neural tube, eventually form vertebrae, connective tissue, and skeletal muscles of the neck and trunk.

8. The gender of a human embryo is determined at fertilization by the type of sperm (X-bearing or Y-bearing) that fertilizes the egg (which is X-bearing). An XX zygote will develop into a female, and an XY zygote into a male. Internal reproductive organs and external anatomy begin to differentiate about 6 weeks after fertilization when a region on the Y chromosome initiates development of testes in male embryos. Absence of the Y chromosome results in female development.

9. Growth during the fetal period is extremely rapid. The rate of growth of the head slows relative to that of other parts of the body, producing changes in the way the body is proportioned. Such change in the relative rates of growth of different parts of the body is called allometric growth.

10. Fetal circulation differs from circulation after birth in that most blood is shunted through temporary vessels or openings past organs such as the lungs and liver, which do not yet perform their postnatal functions. At birth, when organs begin their postnatal functions, the bypasses of fetal circulation begin to close.

WEB TUTORIAL 18.1 Embryonic Development

Birth Is the Transition from Prenatal to Postnatal Development (pp. 396–397)

11. Labor is the process by which the fetus is expelled from the uterus and moved through the vagina into the outside world. Parturition (birth) usually occurs about 38 weeks after fertilization and marks the transition from the prenatal period of development to the postnatal period.

Environmental Disruptions during the Embryonic Period Cause Major Birth Defects (p. 398)

12. Birth defects are developmental defects present at birth. Such defects may be caused by genetic factors (mutant genes or changes in the number or structure of chromosomes) or environmental agents such as drugs or radiation. Disruptive agents have their greatest effects on the developing embryo during periods of rapid differentiation when organs and organ systems are forming.

The Mother's Mammary Glands Produce Milk (pp. 398–399)

13. Lactation is the production and ejection of milk from the mammary glands. At birth, when levels of estrogen and progesterone drop, prolactin initiates milk production. Oxytocin causes milk ejection.

The Postnatal Period Begins with Birth and Continues into Old Age (pp. 399–402)

14. Stages in postnatal development include infancy, childhood, puberty, adolescence, and adulthood. Growth stops in adulthood, and a few years later aging begins. Aging is the progressive decline in the structure and function of the body. Changes occur in all organ systems. Possible causes of aging include genetically programmed cessation of cell division; damage to DNA, protein, and lipids caused by free radicals; the inability of cells to repair damaged DNA; alteration of proteins by glucose such that cross-linkages form and produce a stiffening in tissues; and declines in the functioning of key organ systems.

KEY TERMS

fertilization   p. 383
pre-embryonic period   p. 383
pre-embryo   p. 383
embryonic period   p. 383
embryo   p. 383
fetal period   p. 383
fetus   p. 383
zygote   p. 384
cleavage   p. 384
blastocyst   p. 385
inner cell mass   p. 386

trophoblast   p. 386
placenta   p. 386
implantation   p. 386
human chorionic gonadotropin (HCG)   p. 386
infertility   p. 386
extraembryonic membranes   p. 386
amnion   p. 386
yolk sac   p. 387
allantois   p. 387

umbilical cord   p. 387
chorion   p. 387
chorionic villi   p. 388
cesarean section   p. 388
cell differentiation   p. 390
morphogenesis   p. 390
primary germ layers   p. 390
gastrulation   p. 390
ectoderm   p. 390
endoderm   p. 391
mesoderm   p. 391

notochord   p. 391
neural tube   p. 391
neurulation   p. 391
somite   p. 391
allometric growth   p. 393
labor   p. 396
birth defect   p. 398
lactation   p. 398
aging   p. 399

REVIEWING THE CONCEPTS

1. List the three stages of prenatal development. What ages and major developmental milestones are associated with each stage?   pp. 383–396
2. What effect do secretions from the female reproductive tract have on sperm?   p. 383
3. How is fertilization of an oocyte by more than one sperm prevented?   p. 384
4. Describe implantation. Where does it usually occur? What happens if it occurs elsewhere?   p. 386
5. What are the functions of the four extraembryonic membranes?   pp. 386–387
6. What is the placenta, and how does it form? Explain how nutrients, oxygen, and wastes are exchanged between fetal and maternal blood.   pp. 387–388
7. What is gastrulation? What tissues and organs are formed from each of the three primary germ layers?   pp. 390–391
8. Describe formation of the central nervous system.   pp. 391–393
9. How is the gender of a human embryo determined?   p. 393
10. How does fetal circulation differ from circulation after birth?   pp. 393–395
11. What are the three stages of labor? What initiates labor?   pp. 396–397
12. Explain the relationship between the effects of environmental agents that disrupt development and embryonic age.   pp. 398, 399
13. Why is breast-feeding considered better than bottle-feeding for most infants and mothers?   pp. 398–399
14. What are some possible causes of aging? What can you do today to improve the quality of your old age?   pp. 399–402
15. Fertilization
    a. typically occurs in the uterus.
    b. often involves more than one sperm uniting with an egg.
    c. is the process by which the zygote becomes embedded in the uterine lining.
    d. typically occurs in an oviduct.
16. The yolk sac
    a. is a primary source of nourishment for human embryos.
    b. is the embryo's major contribution to the placenta.
    c. contains primordial germ cells.
    d. encloses the embryo in a fluid-filled sac.

17. Which of the following does *not* characterize the placenta?
    a. Produces estrogen, progesterone, and human chorionic gonadotropin
    b. Is the site where fetal and maternal blood directly mix
    c. Appears to send the signal to initiate labor in humans
    d. Is expelled after birth
18. Neurulation
    a. is the process by which the neural tube forms.
    b. is the process by which the notochord forms.
    c. occurs just before gastrulation.
    d. involves endoderm.
19. The fetal period
    a. is the time when tissues and organs form.
    b. is characterized by rapid growth.
    c. runs from week 2 to week 8.
    d. is when exposure to harmful environmental agents is most likely to produce major birth defects.
20. Which of the following statements regarding childbirth is *true*?
    a. Most babies are born feet first.
    b. Delivery of the baby follows delivery of the placenta.
    c. Full-term babies are those born at 32 weeks gestation.
    d. Infants born prematurely at very low birth weight require intensive neonatal care.
21. Within the blastocyst, the _____ becomes the embryo proper and some extraembryonic membranes, and the _____ becomes part of the placenta.
22. Soon after fertilization, the zygote undergoes _____, a series of rapid mitotic divisions without an increase in overall size.
23. The placenta forms from the _____ of the embryo and the _____ of the mother.
24. _____ is the process by which primary germ layers form.
25. The hormone _____ stimulates milk production, and the hormone _____ stimulates milk ejection.
26. _____ are highly reactive molecules produced by normal cellular metabolism that damage DNA and may lead to aging.

APPLYING THE CONCEPTS

1. Tanya had pelvic inflammatory disease 2 years ago. Now she is a few weeks pregnant and is experiencing severe abdominal pain on the right side. What condition might Tanya have? What is typically done for women with this condition?

2. Juan was described as a "blue baby" at birth and underwent corrective surgery. What did the surgeons correct and why?

3. Why might exposure to thalidomide during weeks 4 to 6 of gestation produce more severe birth defects than a similar level of exposure toward the end of gestation?

4. Dora is expecting her baby in about a month. She is deciding whether to breast-feed or bottle-feed. You are her physician. First describe how milk is produced and ejected from the mammary glands. Then, present the advantages and disadvantages of both methods of feeding.

5. Your grandmother is taking vitamin E and explains to you that it is a potential antioxidant. What is an antioxidant and what "condition" is she hoping to treat? Is the vitamin E likely to work?

Additional questions can be found on the companion website.

The Human Life Cycle Has Two Types of Cell Division

Chromosomes Consist of DNA and Protein

Our Cells Divide in a Characteristic Cyclic Pattern

- Interphase is a period of growth and preparation for cell division
- Division of body cells entails division of the nucleus and the cytoplasm
- Mitosis has four stages
- Cytokinesis occurs toward the end of mitosis

Meiosis Forms Haploid Gametes

- Meiosis keeps the chromosome number constant over generations and increases genetic variability in the population
- Meiosis involves two cell divisions
- Crossing over and independent assortment cause genetic recombination during meiosis
- Failure of chromosomes to separate during meiosis creates cells with extra or missing chromosomes

HEALTH ISSUE Turner Syndrome and Klinefelter Syndrome

19
Chromosomes and Cell Division

Down syndrome, which results from an error in cell division, is the most frequent inherited cause of mild to moderate retardation.

As her little brother walked across the stage to receive his high school diploma, Lynn had never been prouder. When he stopped the graduation procession, looked around, held up his diploma, and waved at her with that huge infectious smile of his, people in the auditorium cheered. Everyone in school knew Zach. He wore the mascot suit at home games and encouraged everyone to root for the teams. He designed the huge Picasso-like murals used as a backdrop for the junior play and helped paint them and anchor them into place. It was the most original set *The Music Man* had ever had.

Zach's life hadn't been easy. He had Down syndrome, which means he was born with an extra chromosome 21. He had endured several heart surgeries as a young child and spent countless hours with speech therapists. People had not always been kind to him either, partly because he was slower to learn in some ways than most people. Sometimes his feelings were hurt. Lynn nevertheless envied the people who would meet Zach for the first time next fall when he went to community college. He was going to study art, but Lynn knew that his most precious gift was his genius for teaching. He would teach other students to value and appreciate their own unique gifts.

In this chapter, we will examine why it is important for each cell to have the correct number of chromosomes. We begin by considering the physical basis of heredity: the chromosomes. Then we consider how the chromosomes are parceled out during two types of cell division: mitosis and meiosis. ■

The Human Life Cycle Has Two Types of Cell Division

We begin life as a single cell called a *zygote,* formed by the union of an egg and a sperm. By adulthood, our bodies consist of trillions of cells. What happened in the intervening years? How did we go from a single cell to the multitude of cells that make up the tissues of a fully functional adult? To put it simply, cell division happened—over and over again. Even in adults, many cells continue to divide for growth and repair of body tissues. With very few exceptions, each of those cells carries the same genetic information as its ancestors. The type of nuclear division that results in identical body cells is called *mitosis.*

In Chapter 17 you learned that males and females produce specialized reproductive cells called gametes (eggs or sperm). *Meiosis* is a special type of nuclear division that gives rise to gametes. In females, meiosis occurs in the ovaries and produces eggs. In males, meiosis occurs in the testes and produces sperm. Meiosis is important because through it the gametes end up with half the amount of genetic information (half the number of chromosomes) in the original cell. When an egg and sperm unite (fertilization), the chromosome number is restored to that of the original cell. As a result, the number of chromosomes in body cells remains constant from one generation to the next.

The roles of mitosis (which produces new body cells) and meiosis (which forms gametes) are summarized in the diagram of the human life cycle in Figure 19.1. You will learn more about both mitosis and meiosis later in this chapter.

Chromosomes Consist of DNA and Protein

A chromosome is a combination of a DNA molecule (which contains genetic information for the organism) and specialized proteins called histones. Chromosomes are found in the cell nucleus. The information contained in the DNA molecules in chromosomes directs the development and maintenance of the body. The proteins combined with the DNA are for support and control of gene expression. A **gene** is a specific segment of the DNA that directs the synthesis of a protein, which in turn plays a structural or functional role within the cell. By coding for a specific protein, a gene determines the expression of a particular characteristic, or trait. Each chromosome in a human cell contains a specific assortment of genes. Like beads on a string, genes are arranged in a fixed sequence along the length of specific chromosomes.

In the human body, **somatic cells**, that is, all cells except for eggs or sperm, have 46 chromosomes. Those 46 chromosomes are actually two sets of 23 chromosomes—one set that came from the mother's egg and another set that came from the father's sperm. Thus, each cell contains 23 **homologous pairs** of chromosomes, a pair being two chromosomes (one from the mother and one from the father) with genes for the same traits. Homologous pairs are called homologues for short. Any cell with two sets of chromosomes is described as being **diploid** (written as 2n). In diploid cells, then, genes, too, occur in pairs, and the members of each pair are located at the same position on homologous chromosomes.

One of the 23 pairs of chromosomes consists of the **sex chromosomes** that determine whether a person is male or female. There

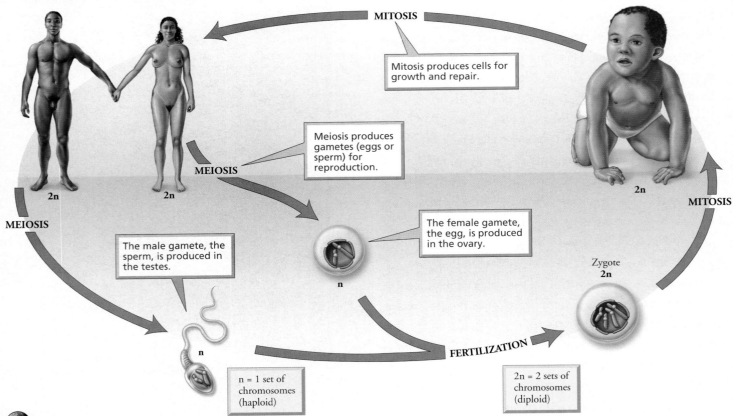

MITOSIS

Mitosis produces cells for growth and repair.

Meiosis produces gametes (eggs or sperm) for reproduction.

MEIOSIS

MEIOSIS

The male gamete, the sperm, is produced in the testes.

The female gamete, the egg, is produced in the ovary.

Zygote
2n

MITOSIS

FERTILIZATION

2n

2n

2n

n

n

n = 1 set of chromosomes (haploid)

2n = 2 sets of chromosomes (diploid)

WEB TUTORIAL 19.1

FIGURE **19.1** The human life cycle

are two types of sex chromosomes, X and Y. A person who has two X chromosomes is described as XX and is genetically female; a person who has an X and a Y chromosome is described as XY and is genetically male. The other 22 pairs of chromosomes are called the **autosomes**. The autosomes determine the expression of most of a person's inherited characteristics.

Our Cells Divide in a Characteristic Cyclic Pattern

In **mitosis**, one nucleus divides into two daughter nuclei containing the same number and kinds of chromosomes. But mitosis is only one phase during the life of a dividing cell. The entire sequence of events that a cell goes through from its origin in the division of a parent cell through its own division into two daughter cells is called the **cell cycle** (Figure 19.2). The cell cycle consists of two major phases: interphase and cell division.

■ Interphase is a period of growth and preparation for cell division

Interphase is the period of the cell cycle between cell divisions. It usually accounts for most of the time that elapses during a cell cycle. During active growth and divisions (depending on the type of cell), an entire cell cycle might take about 16 to 24 hours to complete, and only 1 to 2 hours are spent in division. Interphase is not a "resting period" as once thought. Instead, interphase is a time when the

cell carries out its functions and grows. If the cell is going to divide, interphase is a time of intense preparation for cell division. It is a time when the DNA and organelles are duplicated. These preparations ensure that when the cell divides, each of its resulting cells, called daughter cells, will receive the essentials for survival.

Interphase consists of three parts: G_1 (first "gap"), S (DNA synthesis), and G_2 (second "gap"). All three parts of interphase are times of cell growth, characterized by the production of organelles and the synthesis of proteins and other macromolecules. There are, however, some events specific to certain parts of interphase.

- G_1: A time of major growth before DNA synthesis begins
- S: The time during which DNA is synthesized
- G_2: A time of growth after DNA is synthesized and before mitosis begins

The details of DNA synthesis (replication) are described in Chapter 21. Our discussion here introduces some basic terminology pertaining to the cell cycle.

During G_1 at the start of interphase, each chromosome consists of a DNA molecule and proteins. Throughout interphase, the chromosomes are long, thin threads that are often called *chromatin*. They twist randomly around one another like tangled strands of yarn. In this state, DNA can be synthesized (replicated) and genes can be active. When the chromosomes are being replicated during the S phase, the chromosome copies remain attached. The two copies, each an exact replicate of the original chromosome, stay attached to one another at a region called the **centromere**. As long

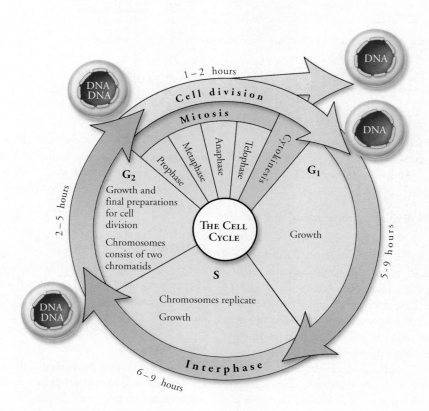

FIGURE **19.2** The cell cycle

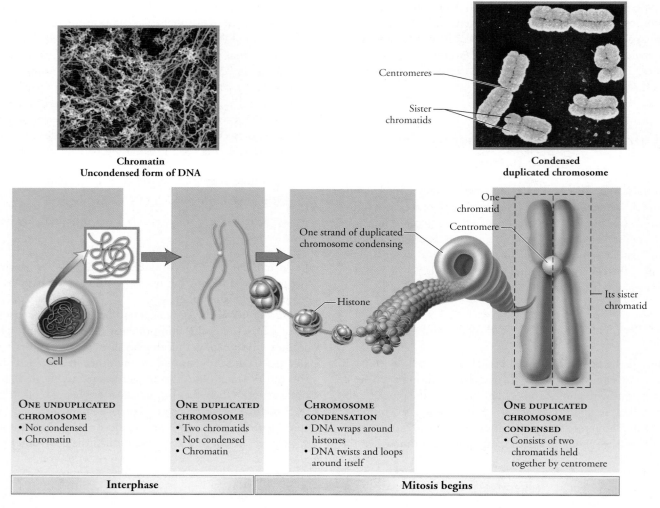

Chromatin
Uncondensed form of DNA

Condensed
duplicated chromosome

Centromeres

Sister
chromatids

One strand of duplicated
chromosome condensing

One
chromatid

Centromere

Histone

Its sister
chromatid

Cell

ONE UNDUPLICATED CHROMOSOME
• Not condensed
• Chromatin

ONE DUPLICATED CHROMOSOME
• Two chromatids
• Not condensed
• Chromatin

CHROMOSOME CONDENSATION
• DNA wraps around histones
• DNA twists and loops around itself

ONE DUPLICATED CHROMOSOME CONDENSED
• Consists of two chromatids held together by centromere

Interphase

Mitosis begins

FIGURE **19.3** Changes in chromosome structure because of DNA replication during interphase and preparation for nuclear division in mitosis

as the replicate copies remain attached, each copy is called a **chromatid**. The two attached chromatids are genetically identical and are called *sister chromatids*, as shown in Figure 19.3.

■ **Division of body cells entails division of the nucleus and the cytoplasm**

Body cells divide continually in the developing embryo and fetus. Such division also plays an important role in the growth and repair of body tissues in children. In the adult, some cells, such as nerve cells, lose their ability to divide. Late in G_1 of interphase, these cells enter the G_0 stage; they are metabolically active but do not divide. Other adult cells, such as liver cells, stop dividing but retain the ability to undergo cell division should the need for tissue repair and replacement arise. Still other cells actively divide throughout life. Skin cells, for example, continue to divide in adults. The ongoing cell division in skin cells serves to replace the enormous numbers of cells worn off each day.

The division of body cells (after interphase) consists of two processes that overlap somewhat in time. The first process, division of the nucleus, is called *mitosis*. The second process is **cytokinesis**, which is the division of the cytoplasm that occurs toward the end of mitosis (Figure 19.4).

■ **Mitosis has four stages**

For the purpose of discussion, mitosis is usually divided into four stages: prophase, metaphase, anaphase, and telophase. The major events of each stage are depicted in Figure 19.5 and are described later in the chapter.

PROPHASE

Mitosis begins with **prophase**, a time when changes occur both in the nucleus and in the cytoplasm. In the nucleus, the chromosomes undergo packing for easy separation. They begin to condense as the DNA wraps around histones. The DNA then loops and twists to

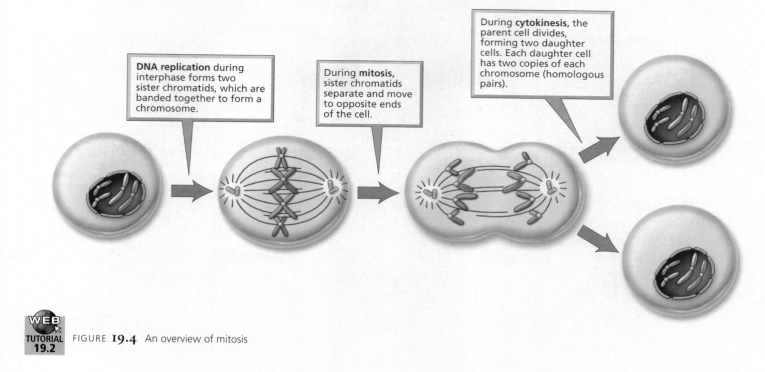

DNA replication during interphase forms two sister chromatids, which are banded together to form a chromosome.

During mitosis, sister chromatids separate and move to opposite ends of the cell.

During cytokinesis, the parent cell divides, forming two daughter cells. Each daughter cell has two copies of each chromosome (homologous pairs).

FIGURE **19.4** An overview of mitosis

form a tightly compacted structure (see Figure 19.3). When DNA is in this condensed state, it cannot be replicated, and gene activity is shut down. These thicker, shorter chromosomes become visible under a light microscope when stained. In this condensed state, the sister chromatids are easier to separate without breaking. At about this time, the nucleolus disappears, and the nuclear membrane also begins to break down.

Outside the nucleus, in the cytoplasm, the mitotic spindle forms. The mitotic spindle is made of microtubules associated with the centrioles (discussed in Chapter 3). During prophase, the centrioles, duplicated during interphase, move away from each other toward opposite ends of the cell.

METAPHASE

During the next stage of mitosis, **metaphase**, the chromosomes attach to the mitotic spindles, forming a line at the equator (center of the mitotic spindles). This alignment ensures each daughter cell receives one chromatid from each chromosome when the chromosomes separate at the centromere. Thus, each daughter cell receives a complete set of the parent cell's chromosomes.

ANAPHASE

Anaphase begins when the sister chromatids of each chromosome begin to separate, splitting at the centromere. Now separate entities, the sister chromatids are considered chromosomes in their own right. The spindle fibers pull the chromosomes toward opposite poles of the cell. By the end of anaphase, equivalent collections of chromosomes are located at the two poles of the cell.

TELOPHASE

During **telophase**, a nuclear envelope forms around each group of chromosomes at each pole. The mitotic spindle disassembles, and nucleoli reappear. The chromosomes also become more threadlike in appearance.

stop and think

Cancer cells divide rapidly and without end. One type of drug used in cancer chemotherapy inhibits the formation of spindle fibers. Why is this an effective anticancer treatment?

KARYOTYPES As we have seen, one of the major features of cell division is the shortening and thickening of the chromosomes. In this state, the chromosomes are visible with a light microscope and can be used for diagnostic purposes, such as when potential parents want to check their own chromosomal makeup for defects. One often-used method takes white blood cells from a blood sample and grows them for a while in a nourishing medium. The culture then is treated with a drug that destroys the mitotic spindle, thus preventing separation of the chromosomes and halting cell division at metaphase. Next the cells are fixed, stained, and photographed so that the images of the chromosomes can be arranged in pairs on the basis of physical characteristics such as location of the centromere and overall length. This arrangement of chromosomes is called a **karyotype** (Figure 19.6). Karyotypes can be checked for irregularities in number or structure of chromosomes.

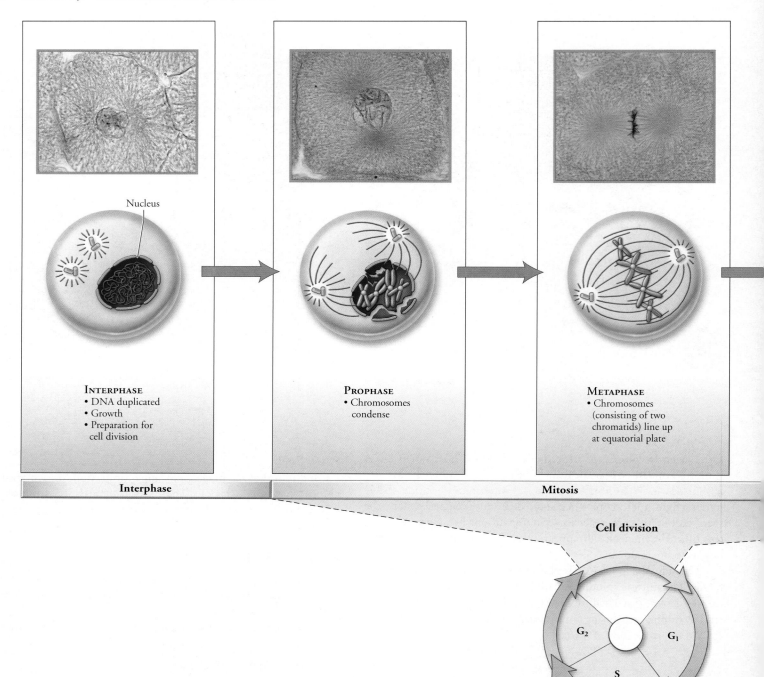

FIGURE **19.5** The stages of cell division (mitosis and cytokinesis) captured in light micrographs and depicted in schematic drawings

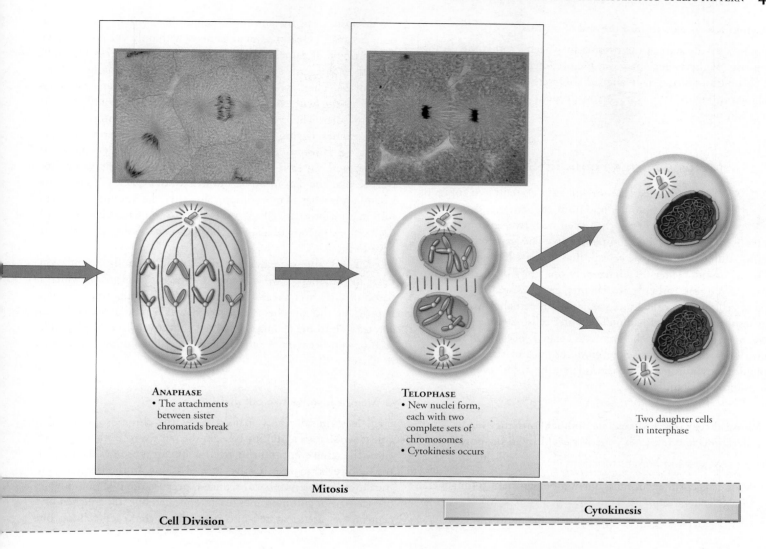

ANAPHASE
* The attachments between sister chromatids break

TELOPHASE
* New nuclei form, each with two complete sets of chromosomes
* Cytokinesis occurs

Two daughter cells in interphase

Mitosis

Cytokinesis

Cell Division

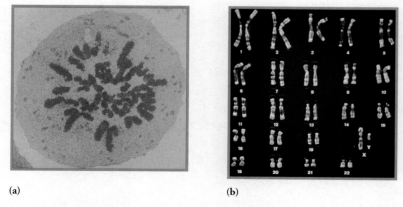

(a)

(b)

FIGURE **19.6** Chromosomes in dividing cells can be examined for defects in number or structure. (a) Photomicrograph of metaphase chromosomes from a human white blood cell. (b) A karyotype constructed by arranging the chromosomes from photographs like the one in part a. The chromosomes have been arranged on the basis of location of centromere and size.

■ **Cytokinesis occurs toward the end of mitosis**

Cytokinesis—division of the cytoplasm—begins sometime during telophase. At this time, a band of microfilaments in the area where the chromosomes originally aligned contracts and forms a furrow, as shown in Figure 19.7. The furrow deepens, eventually pinching the cell in two.

Meiosis Forms Haploid Gametes

We have seen that the somatic cells contain two copies of every chromosome, one from the father and one from the mother. Recall that a cell with two sets of chromosomes is described as being diploid, 2n. The gametes—eggs or sperm—differ from somatic cells in that they are **haploid**, indicated by n, meaning that they have only one copy of each chromosome. As you read earlier in the chapter, gametes are produced by a type of cell division called **meiosis,** which is actually two divisions that result in up to four haploid daughter cells. When a sperm fertilizes an egg, a new cell, the zygote, is created. Because the egg and sperm both contribute a set of chromosomes to the zygote, it is diploid. Many mitotic cell divisions later, the zygote eventually develops into a new individual.

■ **Meiosis keeps the chromosome number constant over generations and increases genetic variability in the population**

Meiosis serves two important functions:

- Meiosis keeps the chromosome number constant from generation to generation.
- Meiosis increases genetic variability in the population.

Meiosis keeps the chromosome number constant over generations because it creates haploid gametes with only one copy of each chromosome. If gametes were produced by mitosis, they would be diploid; each sperm and egg would contain 46 chromosomes instead of 23. Then, when a sperm containing 46 chromosomes fertilized an egg with 46 chromosomes, the zygote would have 92 chromosomes. The zygote created by that process would have 184 chromosomes, having been formed by an egg and sperm each containing 92 chromosomes. The next generation would have 368 chromosomes in each cell, and the next one 736—and so on. You can see that the chromosome number would quickly become unwieldy and, what is more important, alter the amount of genetic information in each cell. As we will see toward the chapter's end, even one extra copy of a single chromosome usually causes an embryo to die.

Meiosis also increases genetic variability in the population. Later in this chapter we will consider the mechanisms by which it accomplishes this increase. Genetic variability is important because it provides the raw material through which natural selection can act, leading to the changes described collectively as evolution. The relationship between genetic variability and evolution is discussed in Chapter 22.

■ **Meiosis involves two cell divisions**

First, let's consider how meiosis keeps the chromosome number constant. Meiosis and mitosis begin the same way. Both are preceded by the same event—the replication of chromosomes. Unlike mitosis, however, meiosis involves *two* divisions. In the first division, the chromosome number is reduced, because the two homologues of each pair of chromosomes (each replicated but still attached by a centromere) are separated into two cells so that each cell has one

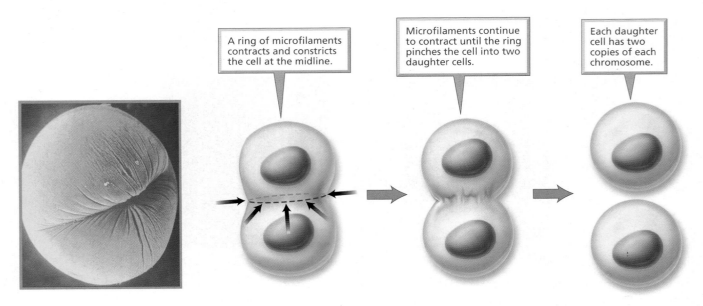

A ring of microfilaments contracts and constricts the cell at the midline.

Microfilaments continue to contract until the ring pinches the cell into two daughter cells.

Each daughter cell has two copies of each chromosome.

FIGURE **19.7** Cytokinesis is the division of the cytoplasm to form two daughter cells.

complete set of chromosomes. In the second division, the replicated copies of the single set of chromosomes are separated. We see, then, that meiosis begins with one diploid cell and, two divisions later, produces four haploid cells. The orderly movements of chromosomes during meiosis ensure that each haploid gamete produced contains one copy of each chromosome. The stages in meiosis are summarized in Figure 19.8. Although not shown in the summary figure, each of the two meiotic divisions has four stages similar to those in mitosis: prophase, metaphase, anaphase, and telophase.

Subsequent changes in the shape and functioning of the four haploid cells result in functional gametes. In males, one diploid cell results in four functional sperm. In contrast, in females, meiotic divisions of one diploid cell result in only one functional egg and up to three nonfunctional polar bodies. As a result, the egg contains most of the nutrients found in the original diploid cell, and these nutrients will nourish the early embryo (Figure 19.9).

MEIOSIS I

The first meiotic division—meiosis I—produces two cells, each with 23 chromosomes. Note that the daughter cells do not contain a random assortment of any 23 chromosomes. Instead, each daughter cell contains *one complete set of chromosomes* (one member of each homologous pair), with each chromosome consisting of two sister chromatids.

It is important that each daughter cell receive a complete set of chromosomes during meiosis I. It would not do if one of the daughter cells had two copies of chromosome 3 and no copy of chromosome 6. Although there would still be 23 chromosomes present, part of the instructions for the structure and function of the body (chromosome 6) would be missing. The separation of chromosomes into complete sets occurs reliably during meiosis I because, during prophase I, members of homologous pairs line up next to one another, a phenomenon called **synapsis** ("bringing together"). For example, the chromosome 1 that was originally from your father would line up with the chromosome 1 from your mother. Paternal chromosome 2 would pair with maternal chromosome 2, and so on. During metaphase I, matched homologous pairs become positioned at the midline of the cell and attach to spindle fibers. The pairing of homologous chromosomes helps ensure that the daughter cells will receive one copy of each homologous pair. Consider the following analogy. Pairing your socks before putting

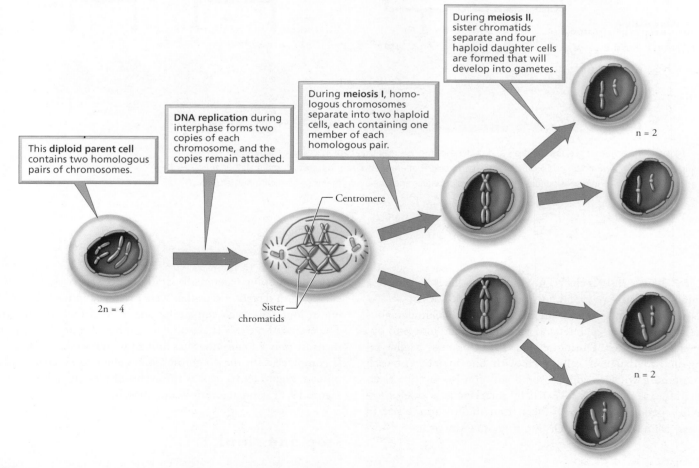

This **diploid parent cell** contains two homologous pairs of chromosomes.

DNA replication during interphase forms two copies of each chromosome, and the copies remain attached.

During **meiosis I**, homologous chromosomes separate into two haploid cells, each containing one member of each homologous pair.

During **meiosis II**, sister chromatids separate and four haploid daughter cells are formed that will develop into gametes.

Centromere

Sister chromatids

2n = 4

n = 2

n = 2

WEB TUTORIAL 19.3 FIGURE **19.8** Overview of meiosis. Meiosis reduces the chromosome number from the diploid number to the haploid number. Meiosis involves two cell divisions.

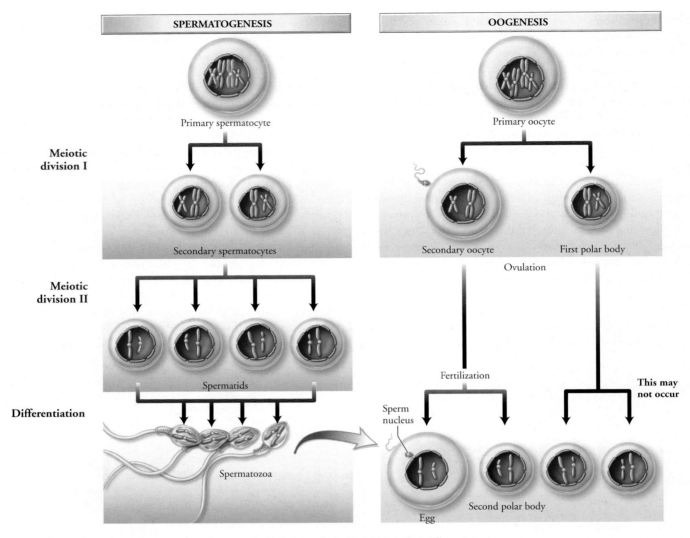

FIGURE 19.9 Comparison of spermatogenesis and oogenesis. Meiosis results in haploid cells that differentiate into mature gametes. Spermatogenesis produces four sperm cells that are specialized to transport the male's genetic information to the egg. Oogenesis produces up to three polar bodies and one ovum that is packed with nutrients to nourish the early embryo.

them in a drawer ensures that you will be more likely to put on a matching pair than if you randomly pulled out two socks.

Next, during anaphase I, the members of each homologous pair of chromosomes separate, and each homologue moves to opposite ends of the cell. During telophase I, cytokinesis begins, resulting in two daughter cells, each with one member of each chromosome pair. Each chromosome still consists of two sister chromatids. Telophase I is followed by interkinesis, a brief interphase-like period. *Interkinesis* differs from mitotic interphase in that there is no replication of DNA during interkinesis.

MEIOSIS II

During the second meiotic division—meiosis II—each chromosome lines up in the center of the cell independently (as occurs in mitosis), and the sister chromatids (attached replicates) making up

each chromosome separate. Separation of the sister chromatids occurs in both daughter cells that were produced in meiosis I, resulting in four cells, each containing one copy of each chromosome. The events of meiosis II are similar to those of mitosis, except that there are only 23 chromosomes lining up independently in meiosis II compared with the 46 chromosomes aligning independently in mitosis. Figure 19.10 depicts the events of meiosis. Table 19.1 and Figure 19.11 compare mitosis and meiosis.

stop and think

If you were examining dividing cells under a microscope, how could you determine whether a particular cell was in metaphase of mitosis or metaphase I of meiosis?

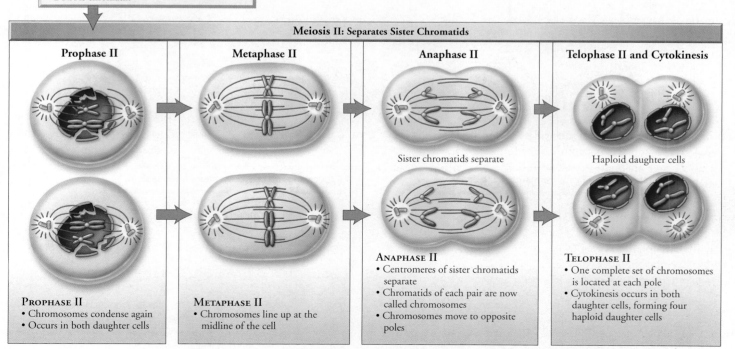

Interphase

Centriole pairs — — DNA

PRE-MEIOTIC INTERPHASE
• DNA replicates
• Copies remain attached to one another by centromere
• Each copy is called a chromatid

Meiosis I: Separates Homologues

Prophase I

Spindle apparatus — Synapsis — Sister chromatids

PROPHASE I
• Chromosomes condense
• Synapsis occurs (homologous chromosomes pair and become perfectly aligned with one another)
• Crossing over takes place

Metaphase I

Centromere

METAPHASE I
• Homologous pairs of chromosomes line up at the midline of the cell
• Spindle fiber from one pole attaches to one member of each pair while spindle fiber from opposite pole attaches to the homologue

Anaphase I

Homologues separate

ANAPHASE I
• Homologous pairs of chromosomes separate and move to opposite ends of the cell
• Each homologue still consists of two chromatids

Telophase I and Cytokinesis

TELOPHASE I
• One member of each homologous pair is at each pole
• Cytokinesis occurs and forms two haploid daughter cells
• Each chromosome still consists of two chromatids

Two important sources of genetic variation

Recombination due to crossing over occurs during prophase I.
Parts of nonsister chromatids are exchanged.

Independent assortment occurs during metaphase I.
Maternal and paternal members of homologous pairs align randomly at the equatorial plate, creating a random assortment of maternal and paternal chromosomes in the daughter cells.

INTERKINESIS
• A brief interphase-like period
• No DNA replication occurs during interkinesis
• DNA is chromatin

Meiosis II: Separates Sister Chromatids

Prophase II

PROPHASE II
• Chromosomes condense again
• Occurs in both daughter cells

Metaphase II

METAPHASE II
• Chromosomes line up at the midline of the cell

Anaphase II

Sister chromatids separate

ANAPHASE II
• Centromeres of sister chromatids separate
• Chromatids of each pair are now called chromosomes
• Chromosomes move to opposite poles

Telophase II and Cytokinesis

Haploid daughter cells

TELOPHASE II
• One complete set of chromosomes is located at each pole
• Cytokinesis occurs in both daughter cells, forming four haploid daughter cells

FIGURE **19.10** Stages of meiosis

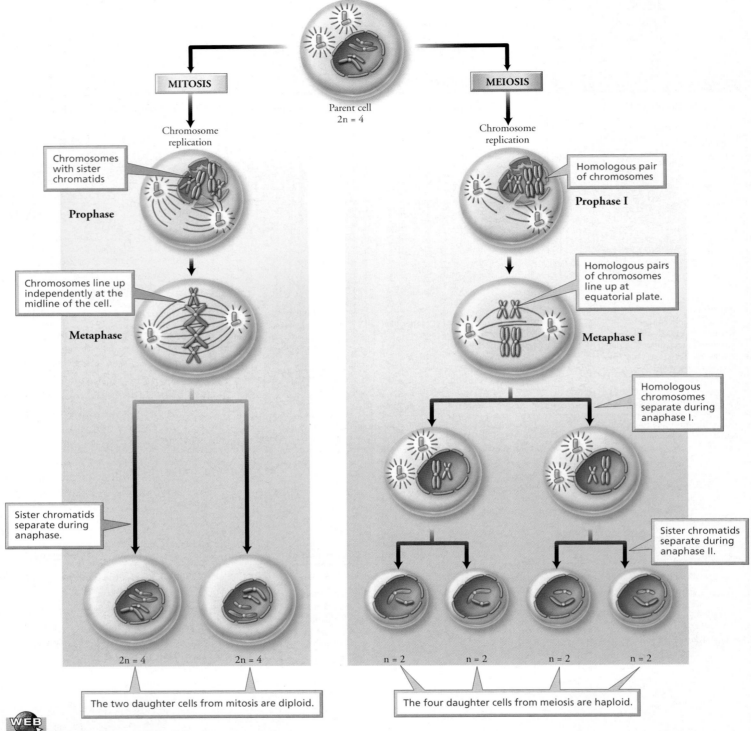

FIGURE **19.11** A comparison of meiosis and mitosis

TABLE 19.1 MITOSIS AND MEIOSIS COMPARED	
MITOSIS	**MEIOSIS**
Involves one cell division	Involves two cell divisions
Produces two diploid cells	Produces up to four haploid cells
Occurs in most somatic cells	Occurs only in ovary and testes during the formation of gametes (egg and sperm)
Results in growth and repair	Results in gamete (egg and sperm) production
No exchange of genetic material	Parts of chromosomes are exchanged in crossing over
Daughter cells are genetically similar	Daughter cells are genetically dissimilar

FIGURE **19.12** Each child inherits a unique combination of maternal and paternal genetic characteristics because of the shuffling of chromosomes that occurs during meiosis. This photograph shows Mary and Eric Goodenough with their four sons: Derick, Stephen, David, and John.

■ Crossing over and independent assortment cause genetic recombination during meiosis

At the moment of fertilization, when the nuclei of an egg and a sperm fuse, a new, *unique* individual is formed. Although certain family characteristics may be passed along, each child bears its own assortment of genetic characteristics (Figure 19.12).

Genetic variation arises largely because of the shuffling of maternal and paternal forms of genes during meiosis. One way this mixing occurs is through a process called **crossing over**, in which corresponding pieces of chromatids of maternal and paternal homologues (nonsister chromatids) are exchanged during synapsis when the homologues are aligned side by side (Figure 19.13). After crossing over, the affected chromatids have a mixture of DNA from the two parents. Because the homologues align gene by gene during synapsis, the exchanged segments contain genetic information for the same traits. However, because the genes of the mother and those of the father may direct different expressions of the trait—attached or unattached earlobes, for instance—the chromatids that are affected by crossing over have a new, novel combination of genes. Thus, crossing over increases the genetic variability of gametes.

Independent assortment is a second way that meiosis provides for the shuffling of genes between generations (Figure 19.14).

Recall that the homologous pairs of chromosomes line up at the equator (center) of the mitotic spindle during metaphase I. However, the orientation of the members of the pair is random with respect to which member is closer to which pole. Thus, like the odds that a flipped coin will come up heads, there is a fifty-fifty chance that a given daughter cell will receive the maternal chromosome. Each of the 23 pairs of chromosomes orients independently during metaphase I. The orientations of all 23 pairs will determine the assortments of maternal and paternal chromosomes in the daughter cells. Thus, each child (other than identical siblings) of the same parents has a unique genetic makeup.

■ Failure of chromosomes to separate during meiosis creates cells with extra or missing chromosomes

Most of the time, meiosis is a precise process that results in the chromosomes being distributed evenly to gametes. Nonetheless, meiosis is not foolproof, and occasionally a mistake is made. A pair of chromosomes or sister chromatids may adhere so tightly to one

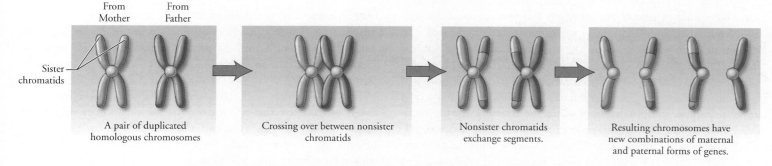

From Mother From Father

Sister chromatids

A pair of duplicated homologous chromosomes

Crossing over between nonsister chromatids

Nonsister chromatids exchange segments.

Resulting chromosomes have new combinations of maternal and paternal forms of genes.

FIGURE **19.13** Crossing over. During synapsis, when the homologous chromosomes of the mother and the father are closely aligned, corresponding segments of nonsister chromatids are exchanged. Each of the affected chromatids has a mixture of maternal and paternal genetic information.

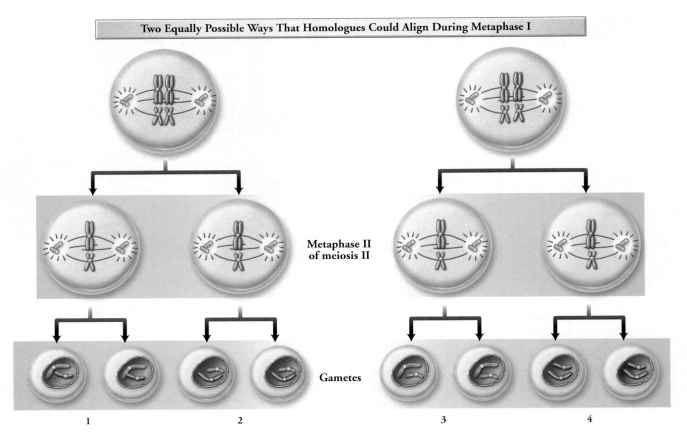

FIGURE 19.14 Independent assortment. The relative positioning of homologous maternal and paternal chromosomes with respect to the poles of the cell is random. The members of each homologous pair orient independently of the other pairs. Notice that with only two homologous pairs, there are four possible combinations of chromosomes in the resulting gametes.

another that they do not separate during anaphase. As a result, both go to the same daughter cell, and the other daughter cell receives none of this type of chromosome (Figure 19.15). The failure of homologous chromosomes to separate during meiosis I or of sister chromatids to separate during meiosis II is called **nondisjunction**.

What happens if nondisjunction creates a gamete with an extra or a missing chromosome and that gamete is then united with a normal gamete during fertilization? The resulting zygote will have an excess or deficit of chromosomes. For instance, if the abnormal gamete has an extra chromosome, the resulting zygote will have three copies of one type of chromosome and two copies of the rest. This condition, in which there are three representatives of one chromosome, is called **trisomy**. If, on the other hand, a gamete that is missing a representative of one type of chromosome joins with a normal gamete during fertilization, the resulting zygote will have only one copy, rather than the normal two copies, of one type of chromosome. The condition in which there is only one representative of a particular chromosome in a cell is called **monosomy**. The imbalance of chromosome numbers usually causes abnormalities in development. Most of the time, the resulting malformations are severe enough to cause the death of the fetus, which will result in a miscarriage. Indeed, in about 70% of miscarriages, the fetus has an abnormal number of chromosomes.

When a fetus inherits an abnormal number of certain chromosomes—for instance, chromosome 21 or the sex chromosomes—the resulting condition is usually not fatal. The upset in chromosome balance does, however, cause a specific syndrome. (A syndrome is a group of symptoms that generally occur together.) The Health Issue essay, *Turner Syndrome and Klinefelter Syndrome*, considers the consequences of the failure of sex chromosomes to separate.

DOWN SYNDROME

One in every 800 to 1000 infants born has three copies of chromosome 21 (trisomy 21), a condition known as *Down syndrome*. The symptoms of Down syndrome include moderate to severe mental retardation, short stature or shortened body parts due to poor skeletal growth, and characteristic facial features (Figure 19.16). Individuals with Down syndrome typically have a flattened nose, a forward-protruding tongue that forces the mouth open, upward-slanting eyes, and a fold of skin at the inner corner of each eye. Approximately 50% of all infants with Down syndrome have heart defects, and many of them die as a result of this defect. Blockage in the digestive system, especially in the esophagus or small intestine, is also common and may require surgery shortly after birth.

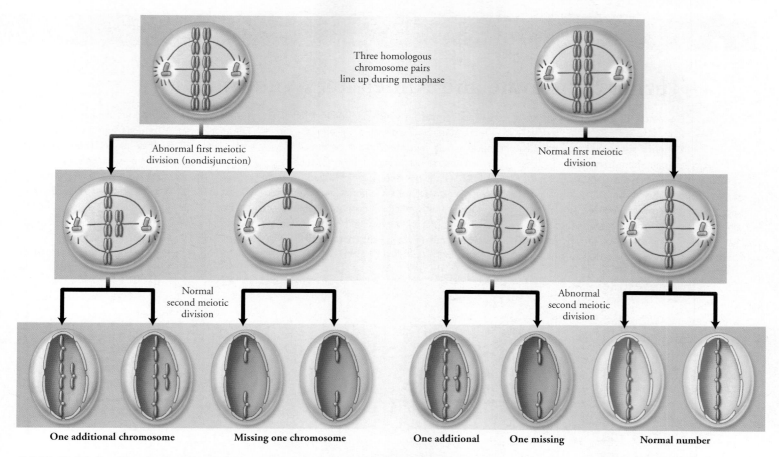

Three homologous
chromosome pairs
line up during metaphase

Abnormal first meiotic
division (nondisjunction)

Normal first meiotic
division

Normal
second meiotic
division

Abnormal
second meiotic
division

One additional chromosome   **Missing one chromosome**   **One additional**   **One missing**   **Normal number**

FIGURE **19.15** Nondisjunction is a mistake that occurs during cell division in which homologous chromosomes or sister chromatids fail to separate during anaphase. One of the resulting daughter cells will have an extra copy of one chromosome, and the other will be missing that type of chromosome.

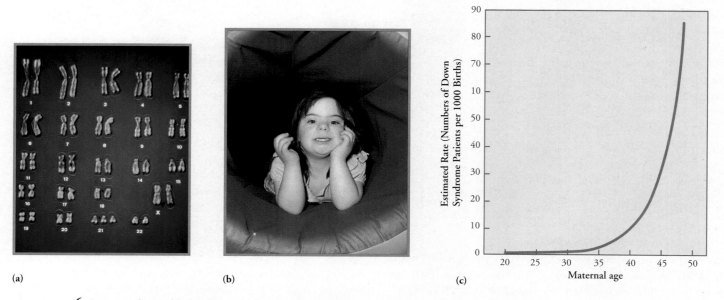

(a)   (b)   (c)

Maternal age

Estimated Rate (Numbers of Down Syndrome Patients per 1000 Births)

FIGURE **19.16** Down syndrome. (a) Three copies of chromosome 21 (trisomy 21), as seen here, cause Down syndrome. (b) A person with Down syndrome is moderately to severely mentally retarded and has a characteristic appearance. (c) The incidence of Down syndrome increases with the age of the mother, especially after age 35.

HEALTH ISSUE

Turner Syndrome and Klinefelter Syndrome

Like autosomes, sex chromosomes may fail to separate during anaphase. This error can occur during either egg or sperm formation. A male is chromosomally XY, so when the X and Y separate during anaphase, equal numbers of X-bearing and Y-bearing sperm are produced. However, if nondisjunction occurs during sperm formation, half of the resulting sperm will carry both X and Y chromosomes, whereas the other resulting sperm

will not contain any sex chromosome. A female is chromosomally XX, so each of the eggs she produces should contain a single X chromosome. When nondisjunction occurs, however, an egg may contain two X chromosomes or none at all. When a gamete with an abnormal number of sex chromosomes is joined with a normal gamete during fertilization, the resulting zygote has an abnormal number of sex chromosomes (Figure 19.Aa).

Turner syndrome occurs in individuals who have only a single X chromosome (XO). Approximately 1 in 2000 female infants is born with Turner syndrome, but this represents only a small percentage of the XO zygotes that are formed. Most of these XO zygotes are lost as miscarriages. A person with Turner syndrome has the external appearance of a female (Figure 19.Ba). Before puberty, she looks like a normal, XX female. The only hint of Turner syndrome may be a thick fold of

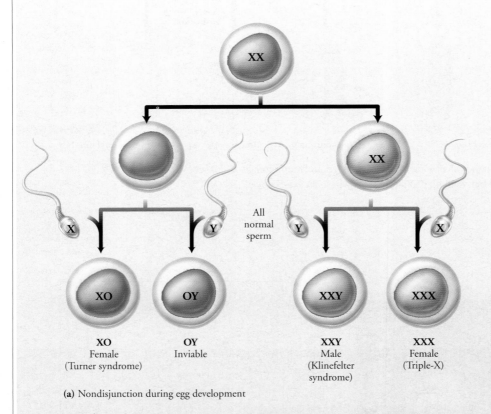

(a) Nondisjunction during egg development

(b) Nondisjunction during sperm development

FIGURE **19.A**   The sex chromosomes may fail to separate during formation of either eggs or sperm. When a gamete with an abnormal number of sex chromosomes joins a normal gamete in fertilization, the resulting zygote has an abnormal number of sex chromosomes. Imbalances of sex chromosomes upset normal development of reproductive structures.

The risk of having a baby with Down syndrome increases with the age of the mother. A 30-year-old woman is twice as likely to give birth to a child with Down syndrome as is a 20-year-old woman. After age 30, the risk rises dramatically, so at age 45 a mother is 45 times as likely to give birth to a Down syndrome infant as is a 20-year-old woman.

what would you do?

It is possible to determine whether a 2-month-old fetus has Down syndrome. If you were a 30-year-old woman who became pregnant (or the father of the child), would you want to know whether the child would have Down syndrome? What criteria would you use to decide?

skin on the neck. As she ages, however, she generally is noticeably shorter than her peers. Her chest is wide, and her breasts underdeveloped. In 90% of the women with Turner syndrome, the ovaries are also poorly developed, leading to infertility. Pregnancy may be possible through *in vitro* fertilization (see Chapter 18), in which a fertilized egg from a donor is implanted in her uterus.

Klinefelter syndrome is observed in males who are XXY. Although the extra X chromosome can be inherited as a result of nondisjunction during either egg or sperm formation, it is twice as likely to come from the egg. Increased maternal age may increase the risk slightly.

Klinefelter syndrome is fairly common. Approximately 1 in 500 to 1 in 1000 of all newborn

males is XXY. However, not all XXY males display the symptoms of having an extra X chromosome. In fact, many of them live their lives without ever suspecting that they are XXY. Signs that a male has Klinefelter syndrome do not usually show up until puberty. During the teenage years, the testes of an XY male gradually increase in size. In contrast, the testes of many XXY males remain small and do not produce an adequate amount of the male sex hormone, testosterone. As a result of the testosterone insufficiency, they may grow taller than average. Secondary sex characteristics, such as facial and body hair, fail to develop fully. The arms and legs may be exceptionally long, and the breasts may develop slightly (Figure 19.Bb). The penis is usually of normal size, but the testes

do not produce sperm, so men with Klinefelter syndrome are usually sterile.

Nondisjunction can also result in a female with three X chromosomes (XXX, triple X syndrome) or a male with two Y chromosomes (XYY, Jacob syndrome, produced when the chromatids of a replicated Y chromosome fail to separate). Women with triple X syndrome (XXX) may have menstrual disturbances and reach menopause earlier in life than do women with two X chromosomes. Males with two Y chromosomes (XYY) are often taller than normal, and some have slightly lower than normal intelligence. ♂♀

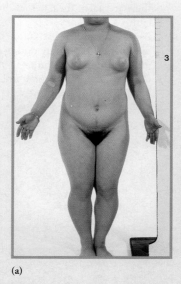
(a)

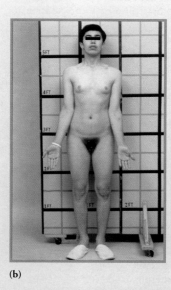
(b)

FIGURE **19.B** Missing or extra sex chromosomes cause Turner syndrome and Klinefelter syndrome. (a) A female with Turner syndrome has one rather than the normal two X chromosomes. Because the ovaries are underdeveloped, she is usually infertile. (b) A male with Klinefelter syndrome is XXY. Klinefelter syndrome is characterized by small testes and sometimes breast enlargement.

exploring further . . .

In this chapter we considered cell division: mitosis, which gives rise to new body cells for growth and repair, and meiosis, which gives rise to the gametes (eggs and sperm). In the next chapter, we will consider mitosis further and explore stem cells, which are undifferentiated cells that can divide continuously and develop into different tissue types.

HIGHLIGHTING THE CONCEPTS

The Human Life Cycle Has Two Types of Cell Division (p. 406)

1. The human life cycle requires two types of nuclear division—mitosis and meiosis. Mitosis creates cells that are exact copies of the original cell. Mitosis occurs in growth and repair. Meiosis creates cells with half the number of chromosomes as were in the original cell. Gamete production requires meiosis.
WEB TUTORIAL 19.1 The Human Life Cycle

Chromosomes Consist of DNA and Protein (pp. 406–407)

2. A chromosome contains DNA and proteins called histones. A gene is a segment of DNA that codes for a protein that plays a structural or functional role in the cell. Genes are arranged along a chromosome in a specific order. Each of the 23 different kinds of chromosomes in human cells contains a specific sequence of genes.

3. Somatic cells (all cells except for eggs and sperm) are diploid; that is, they contain two sets of chromosomes, one from each parent. Homologous chromosomes carry genes for the same traits. In humans, the diploid number of chromosomes is 46, two sets of 23 homologous pairs. One pair of chromosomes, the sex chromosomes, determines gender. Males are XY, and females are XX. The other 22 pairs of chromosomes are called autosomes. Eggs and sperm are haploid; they contain only one set of chromosomes.

Our Cells Divide in a Characteristic Cyclic Pattern (pp. 407–412)

4. The cell cycle consists of two major phases: interphase and cell division. Interphase is the period between cell divisions.

5. During interphase, the cell carries out the activities it is specialized to perform. DNA and organelles become replicated in preparation for the cell to divide and produce two identical daughter cells. Somatic cell division consists of mitosis (division of the nucleus) and cytokinesis (division of the cytoplasm).

6. In mitosis, the original cell, having replicated its genetic material, distributes it equally between its two daughter cells. There are four stages of mitosis: prophase, metaphase, anaphase, and telophase.

7. Cytokinesis, division of the cytoplasm, usually begins sometime during telophase. A band of microfilaments at the midline of the cell contracts and forms a furrow. The furrow deepens and eventually pinches the cell in two.
WEB TUTORIAL 19.2 Mitosis

Meiosis Forms Haploid Gametes (pp. 412–421)

8. Meiosis, a special type of nuclear division that occurs in the ovaries or testes, begins with a diploid cell and produces four haploid cells that will become gametes (egg or sperm).

9. Meiosis is important because it halves the number of chromosomes in gametes, thereby keeping the chromosome number constant between generations. When a sperm fertilizes an egg, a diploid cell called a zygote is created. After many mitotic divisions, the zygote will develop into a new individual.

10. Before meiosis begins, the chromosomes are replicated, and the copies remain attached to one another by centromeres. The attached replicated copies are called sister chromatids. There are two cell divisions in meiosis. During the first meiotic division (meiosis I), members of homologous pairs are separated. Thus, the daughter cells contain only one member of each homologous pair (although each chromosome still consists of two sister chromatids). During the second meiotic division (meiosis II), the sister chromatids are separated.

11. Genetic recombination during meiosis results in variation among offspring from the same two parents. One cause of genetic recombination is crossing over, in which corresponding segments of DNA are exchanged between maternal and paternal homologues, creating new combinations of alleles in the resulting chromatids.

12. A second cause of genetic recombination is the independent assortment of maternal and paternal homologues into daughter cells during meiosis I. The orientation of the members of the pair relative to the poles of the cell determines whether a daughter cell will receive the maternal or the paternal chromosome. Each pair aligns independently of the others.

13. Nondisjunction is the failure of homologous chromosomes or sister chromatids to separate during cell division. It results in an abnormal number of chromosomes in the resulting cells. Nondisjunction of chromosome 21 can result in Down syndrome.
WEB TUTORIAL 19.3 Meiosis
WEB TUTORIAL 19.4 Comparing Mitosis and Meiosis

KEY TERMS

chromosome p. 406
gene p. 406
somatic cells p. 406
homologous pair p. 406
diploid p. 406
sex chromosomes p. 406
autosomes p. 407

mitosis p. 407
cell cycle p. 407
interphase p. 407
centromere p. 407
chromatid p. 408
cytokinesis p. 408
prophase p. 408

metaphase p. 409
anaphase p. 409
telophase p. 409
karyotype p. 409
haploid p. 412
meiosis p. 412
synapsis p. 413

crossing over p. 417
independent assortment p. 417
nondisjunction p. 418
trisomy p. 418
monosomy p. 418

REVIEWING THE CONCEPTS

1. Explain the relationship between genes and a chromosome. *p. 406*
2. Define mitosis and cytokinesis. *pp. 407–412*
3. Why is meiosis important? *p. 412*
4. Describe the alignment of chromosomes at the midline during meiosis I and meiosis II. Explain the importance of these alignments in creating haploid gametes from diploid cells. *pp. 412–417*
5. Explain how crossing over and independent assortment result in genetic recombination that causes variability among offspring from the same two parents. *pp. 417–418*
6. Define nondisjunction. Explain how nondisjunction can result in abnormal numbers of chromosomes in a person. *pp. 417–418*
7. What causes Down syndrome? What are the usual characteristics of the condition? *pp. 418–420*
8. The process of mitosis results in
   a. two haploid cells.
   b. two diploid cells.
   c. four haploid cells.
   d. four diploid cells.

9. DNA is synthesized (replicated) during
   a. interphase.
   b. prophase.
   c. metaphase.
   d. anaphase.
10. Crossing over occurs during which stage of meiosis?
    a. Prophase I
    b. Metaphase I
    c. Prophase II
    d. Metaphase II
11. During meiosis, the processes of _____ and _____ increase genetic diversity.
12. _____ chromosomes carry genes for the same traits.
13. _____ is the pairing of chromosomes during meiosis.
14. The stage of *mitosis* during which sister chromatids separate is _____.
15. The stage of *meiosis* during which sister chromatids separate is _____.

APPLYING THE CONCEPTS

1. Cathy is a 35-year-old woman who just finished graduate school. She and her husband, Bob, decided to start a family. When Cathy became pregnant, the doctor performed an amniocentesis so as to create a karyotype of the fetus's genes. The karyotype revealed that the fetus had three copies of chromosome 21. How could this condition occur? How will it affect the child?

Additional questions can be found on the companion website.

2. A cell biologist is studying the cell cycle. She is growing the cells in culture, and they are actively dividing mitotically. One particular cell has half as much DNA as most of the other cells. Which stage of mitosis is this cell in? How do you know?
3. What would happen if the spindle fibers failed to form during mitosis?

19a

Stem Cells—The Body's Repair Kit

Dolly, the first animal cloned from an adult cell, was created using a technique called somatic cell nuclear transplant. Related techniques, described in this chapter, could potentially be used to produce stem cells for therapeutic cloning.

Stem Cells Are Unspecialized Cells That Divide Continually and Can Be Made to Develop into Different Tissue Types

There Are Several Sources of Human Stem Cells

- Embryonic stem cells can be harvested from unused embryos created for *in vitro* fertilization
- Adult stem cells exist in many tissues
- Stem cells can be harvested from umbilical cord blood
- Embryonic stem cells can be created by somatic cell nuclear transfer

Embryonic Stem Cell Research Raises Ethical and Political Controversy

Robyn is taking a course in scientific and medical journalism. This week, her assignment for the course is to report on any controversial medical issue about which she has yet to form an opinion. Robyn immediately thinks of human embryonic stem cell research. She's aware that it has been widely covered in the news media, and yet she has never paid enough attention to the reports to know what the research is used for or to understand the arguments on different sides of the debate. She thinks she has heard that stem cells must all be taken from human embryos. She'll need to find out if this is true. She also remembers her friend Taylor, who has diabetes, saying that someday stem cells might make his pancreas produce its own insulin, so he would no longer have to give himself injections every day. Why would that be something stem cells could do? Starting to feel excited about the assignment, Robyn hurries home to do some reading.

In this chapter we will learn what stem cells are, where they are found, and the promises they may hold for the future. ■

Stem Cells Are Unspecialized Cells That Divide Continually and Can Be Made to Develop into Different Tissue Types

Most of the trillions of cells in your body have become specialized to perform a particular job. Muscle cells are specialized to contract, and neurons are specialized to conduct nerve impulses. The specialization of cells during embryonic development is directed by biological cues in each cell's immediate environment. The cues are usually provided by neighboring cells and include growth factors, surface proteins, salts, and touch. Because all cells in the body have the same genetic makeup, the specialization of different cells to perform different functions means that particular genes in each type of cell have been turned on or off.

Stem cells, in contrast, are relatively unspecialized cells that divide continually, creating a pool of undifferentiated cells for possible use. If stem cells are given the correct signals—exposure to a particular growth factor or hormone, for example—they can be coaxed into differentiating into a particular specialized cell type. The specialized cell types can then be used to treat diseases or regenerate injured tissues. As we will see, stem cells from different sources differ in the number of specialized cell types they can develop into.

The possibilities for therapeutic use of stem cells are mind boggling. Imagine growing new heart cells to repair damage from a heart attack, curing Parkinson's disease by restoring the dopamine-producing neurons that are lacking in such patients, or generating insulin-producing pancreatic islet cells to cure diabetes. Think of finding a cure for multiple sclerosis, Alzheimer's disease, Lou Gehrig's disease, and spinal cord injuries. These advances may someday be possible through the use of stem cells, the body's own repair kit.

Scientists hope that stem cells can also be used to regenerate important immune system cells that form the foundation of the body's defense system. This ability would open the door for new cancer therapies, especially for leukemia. Research is currently under way for developing ways of using stem cells to treat certain autoimmune diseases; that is, diseases in which the body's immune system mistakenly attacks a part of the body (discussed in Chapter 13). Two autoimmune diseases that have already been treated with stem cells are rheumatoid arthritis, an inflammation of the joints, and diabetes type 1, a condition in which too little of the hormone insulin is produced. Diabetes type 1 develops when the body's own T lymphocytes attack the islet cells in the pancreas, the cells that produce insulin.

Although the manipulation of stem cells is still in its infancy, the basic therapy technique for using them will probably consist of harvesting them from some tissue source and providing them with developmental cues that will make them differentiate into the type of cell needed to do a certain job. Perhaps it will be possible to coax a single kind of stem cell into forming any of various different types of cells depending on the environment in which it is placed. In some cases, the proper environment will be one created in a laboratory culture dish. In other cases, stem cells may be transplanted into the organ where they are needed. In those cases, the neighboring cells would provide the environmental triggers that transform the stem cells into the desired cell types. Almost all of the research done so far has made use of nonhuman animals. Only a few pioneer studies have been done using humans. However, scientists are hopeful that the day is near when stem cell technology will be used to improve the quality of human life.

There Are Several Sources of Human Stem Cells

There are several sources of stem cells, including early embryos, certain adult tissues, and umbilical cord blood. The general steps for obtaining and using stem cells—embryonic stem cells and adult stem cells—are depicted in Figure 19a.1 and described throughout this section.

■ Embryonic stem cells can be harvested from unused embryos created for *in vitro* fertilization

The early embryo is a rich source of stem cells because the cells it consists of would ultimately produce all the cells of the new individual. When stem cells are extracted, the embryo is about the size of the head of a pin and is called a blastocyst (see Chapter 18; Figure 19a.2). It is a sphere of cells with a cluster of about 20 to 30 cells, called the inner cell mass, adhering to one side. The stem cells make up the inner cell mass.

Embryonic stem cells are fairly easy to harvest, but their use has generated much ethical and political turmoil. The ones used for stem cell research come from embryos that were created for reproductive purposes but were not used. Only a few days old, these embryos left in fertility clinics are destined for destruction. Nevertheless, many people object to the fact that removing the stem cells destroys such an embryo, putting an end to its ability to become a fetus.

■ Adult stem cells exist in many tissues

Adults also have stem cells, but they are more difficult to locate than those in an embryo. Waiting in the bone marrow, brain, skin, liver, and other organs, adult stem cells remain ready to generate new cells for repair or replacement of old ones in many parts of the body. Diseased or dying cells provide biological cues that summon stem cells and prompt them to differentiate into the needed type of cell.

Unfortunately, adult stem cells do not seem to be as versatile as stem cells from an embryo. Whereas embryonic stem cells can transform into any cell type in the body, adult stem cells usually form cells of a particular lineage only. Adult stem cells in the bone marrow form blood cells. Those in the skin form epithelial cells. However, as scientists learn more about the signals that direct differentiation, they may be able to induce adult stem cells to form many more types of cells than is normal for them.

There have been some successes. In 2004, a German man had his jaw removed because of cancer. Doctors created a mesh mold in the shape of the jaw bone that had been removed, and they implanted the mold into the muscle below the man's shoulder blade,

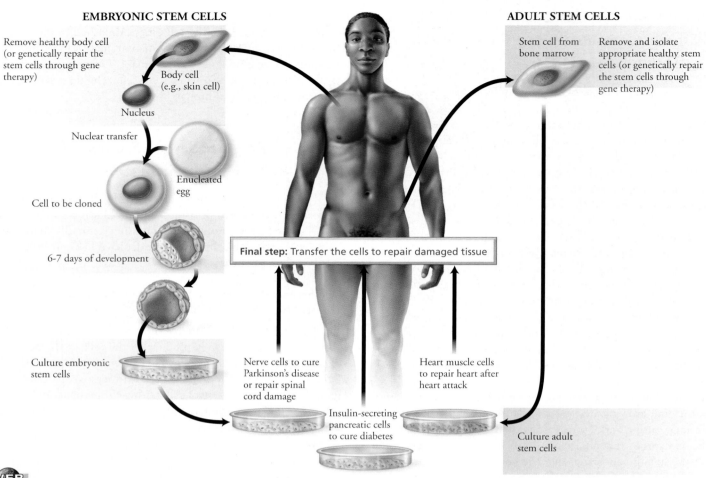

EMBRYONIC STEM CELLS

Remove healthy body cell (or genetically repair the stem cells through gene therapy)

Body cell (e.g., skin cell)

Nucleus

Nuclear transfer

Enucleated egg

Cell to be cloned

6-7 days of development

Culture embryonic stem cells

ADULT STEM CELLS

Stem cell from bone marrow

Remove and isolate appropriate healthy stem cells (or genetically repair the stem cells through gene therapy)

Final step: Transfer the cells to repair damaged tissue

Nerve cells to cure Parkinson's disease or repair spinal cord damage

Heart muscle cells to repair heart after heart attack

Insulin-secreting pancreatic cells to cure diabetes

Culture adult stem cells

FIGURE **19a.1** The proposed procedures for using adult stem cells and embryonic stem cells

WEB TUTORIAL 19a.1

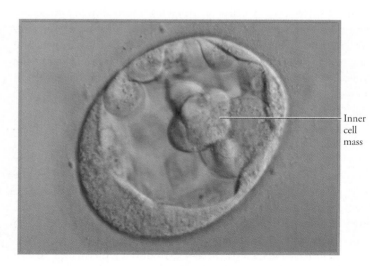

Inner cell mass

FIGURE **19a.2** Embryonic stem cells come from the inner cell mass, a cluster of 20 to 30 cells, from a 6–7 day old embryo. At this age the embryo is about the size of a head of a pin.

along with some cow-derived bone mineral, a growth factor to stimulate bone growth, and blood extracted from the man's bone marrow, which contains stem cells (Figure 19a.3). The mold was left in place until bone scans showed that new bone had formed around it, at which point the new jaw was removed, along with some muscle and blood vessels, and implanted in the correct location. Four weeks later, the man could chew solid food—a bratwurst sandwich! Two years later, doctors grew new urinary bladders for seven patients using the patients' own cells. In this case, doctors extracted muscle and bladder cells from the patient and grew them in a petri dish. As the cells formed layers of tissue, the tissue was shaped into new bladders, which were then implanted into the patients. Within a few weeks, the bladders grew to normal size and performed the required functions.

■ Stem cells can be harvested from umbilical cord blood

Blood from the umbilical cord is also a good source of stem cells. Stem cells do not settle into the bone marrow until a few days after birth. Before that, during fetal development, they circulate in the bloodstream and therefore travel through vessels in the umbilical cord when circulating to and from the placenta. Umbilical cord blood collected at birth is therefore likely to be rich in stem cells.

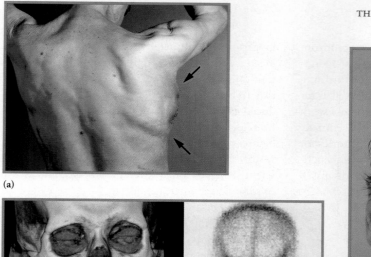

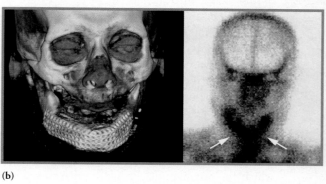

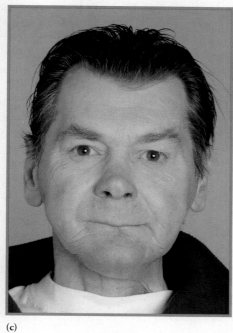

FIGURE **19a.3** Adult stem cells from bone marrow were used to grow a new jaw to replace one that had been removed because of cancer. A mold in the shape of the missing jaw was created out of mesh and seeded with bone marrow cells and bone growth factor. The mold was placed in the muscle of the man's back until bone filled the mold. The new jaw was then implanted to replace the missing jaw.

Blood stem cells can be harvested from bone marrow, but they are more easily extracted from umbilical cord blood. Less than one in every thousand marrow cells is a stem cell, although techniques have been developed to identify those stem cells and separate them from the other marrow cells. Another advantage of taking stem cells from cord blood is that they are less likely to be rejected by the recipient than are stem cells from bone marrow. Lastly, stem cells from cord blood are less likely to carry infections and, importantly, can differentiate into a greater variety of specialized cells.

The use of umbilical cord blood as a source of stem cells raises fewer ethical (and legal) problems than does obtaining stem cells from embryos. Although stem cells from embryos are more plentiful and can specialize into more cell types than those from umbilical cord blood, the embryo must be destroyed to obtain the cells. Furthermore, the federal government currently funds embryonic stem cell research only when the research is restricted to using existing lines of stem cells. We will revisit the issue of using embryonic stem cells at the end of this chapter.

Most of the stem cells in cord blood are blood stem cells that give rise to red blood cells, white blood cells, and platelets. Within the body, blood stem cells produce about 260 billion new cells (about an ounce of blood) every day. However, umbilical cord blood also contains some stem cells that can turn into other kinds of cells—bone, cartilage, heart muscle, brain tissue, and liver tissue.

Because blood stem cells are able to produce an endless supply of blood cells, they hold promise for treating a host of blood conditions. For instance, some patients with the inherited disorders sickle cell anemia and beta-thalassemia, in which abnormal forms of hemoglobin are produced, have already been treated successfully with stem cell therapy. Stem cells that produce normal hemoglobin can be transplanted into a person with the inherited anemia. Normal red blood cells then begin replacing the abnormal ones.

Most of the medical benefits that may be achieved using umbilical cord blood have yet to be realized. Nonetheless, there are already umbilical cord blood banks that store cord blood from newborn babies. Some parents pay a set-up fee and an annual service fee to store their baby's cord blood just in case the blood is needed some day. Some banks collect and store thousands of samples of cord blood in hopes of having a tissue match for most individuals in the population. Researchers have developed ways to amplify umbilical cord blood stem cells by growing them in the laboratory.

The existence of umbilical cord blood banks raises certain social issues. Should stored blood be reserved only for the possible future need of the donor, or should it be made available for anyone in need whose tissue type matches? Should cord blood banks be privately owned or funded by the government? These are examples of the questions that remain to be addressed.

■ Embryonic stem cells can be created by somatic cell nuclear transfer

To *clone* means to produce one or more genetically identical copies of a cell or organism. *Somatic cell nuclear transplant*, one form of cloning, produces multiple copies of a cell by coaxing an adult cell into dividing repeatedly. Born on July 5, 1996, Dolly was the first sheep—indeed, the first mammal—to be cloned *from an adult cell*. She was created using somatic cell nuclear transplant, shown in

Figure 19a.4. In Dolly's case, the nucleus of a mammary cell from an adult sheep's udder was placed into an egg from which the nucleus had been removed. (Dolly was named after Dolly Parton, a country singer who is known for her mammary "cells" as well as for her voice.) The goal of Dr. Ian Wilmut, the Scotsman whose work produced Dolly, was to develop techniques that would eventually lead to the production of animals that could be used as factories for manufacturing proteins, such as hormones, that would be beneficial to humans. Cloning could make genetic engineering (discussed in Chapter 21) more efficient. The ability to clone an adult would mean a scientist need engineer an animal with a particular desired trait only once. Then, when the animal was old enough for scientists to be sure it had the desired trait, the animal could be reproduced exactly, in multiple copies. Cloning from an adult instead of from an embryo is advantageous because the scientist can be certain before cloning that the animal has the desired traits. In other words, "What you see is what you get." Since the birth of Dolly, scientists have cloned several kinds of animals using the technique of somatic cell nuclear transfer.

Discussions about cloning often become heated when the possibility of human cloning is raised. In these discussions, a distinction is often drawn between reproductive cloning and therapeutic cloning. *Therapeutic cloning* would produce a clone of "replacement cells" having the same genetic makeup as a given patient so that they could be used in the patient's medical therapy without being rejected by the immune system. The cells produced would be embryonic stem cells created using somatic cell nuclear transfer, the technique used to create Dolly. A cell would be taken from the patient, and its nucleus would be put into a donor egg cell from which the original nucleus had been removed. The newly created cell would then begin development into an embryo (Figure 19a.4). (If the patient's disease were genetic, a normal form of the relevant gene would be substituted for the defective gene before the nucleus was put in the enucleated egg. Replacement of a specific gene has been accomplished in mice but has not yet been done in humans.) The creation of a human embryo by somatic cell nuclear transfer has not yet been accomplished in humans. In 2004 and 2005 South Korean scientists claimed to have created a human embryo in this way and extracted embryonic stems cells. Later in 2005, the South Koreans' work was declared to be fraudulent (see the Social Issue essay in Chapter 1, *Scientific Misconduct*).

Once an embryo was created by somatic cell transfer, stem cells could be removed and treated with growth factors that make them develop into the type of cells needed for treatment. The stem cells could then be transplanted into the patient to replace faulty cells. For example, neurons might be transplanted to cure Parkinson's disease or to repair spinal cord damage, and insulin-producing cells might be transplanted to cure diabetes. Because the embryonic stem cells would have the same genetic makeup as the patient, they would not be attacked as foreign by the patient's body-defense responses.

Reproductive cloning, in contrast, produces a new individual with a known genetic makeup. The cloning of nonhuman animals described above is reproductive cloning. Currently, reproductive cloning is an extremely inefficient process. Hundreds of cloned embryos must be used in order to obtain one newborn mammal. When cloned embryos do proceed to develop into a new individual, they grow faster and larger than normal. There is also a suspicion that cloned mammals age prematurely. For example, there are indications that Dolly was genetically older than her birth age: one is that Dolly had arthritis, a painful joint condition usually found only in older animals. For reasons such as these, experts agree that scientists are still far from being able to produce a cloned human baby.

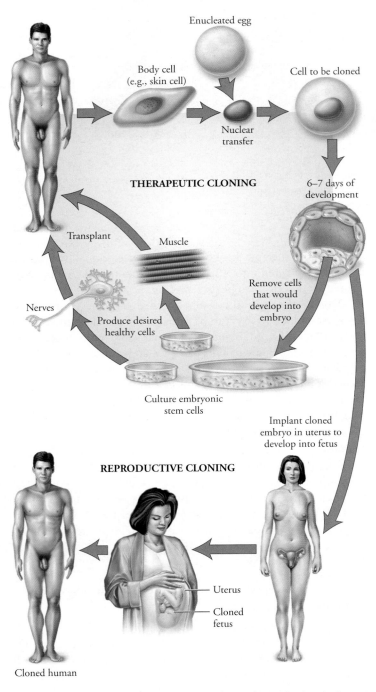

Enucleated egg

Body cell
(e.g., skin cell)

Cell to be cloned

Nuclear transfer

THERAPEUTIC CLONING

6–7 days of development

Transplant

Muscle

Nerves

Produce desired healthy cells

Remove cells that would develop into embryo

Culture embryonic stem cells

Implant cloned embryo in uterus to develop into fetus

REPRODUCTIVE CLONING

Uterus

Cloned fetus

Cloned human

FIGURE **19a.4** Methods for human therapeutic and reproductive cloning

If reproductive cloning of humans does become possible some-day, what might this technology mean to our society? How would it affect human life? One thing it would *not* do is give us the power to create an exact copy of a specific person. Environment *and* genes determine personality characteristics. Although cloning could exactly replicate the genetic makeup of a specific person, there would be no way to replicate all the past experiences in the life of the person being cloned, so the clone would be a different person from the original one. Thus, we could not replace a loved one who has died or achieve immortality of our own in the body of a clone. We could, however, produce an identical twin of a person.

Reproductive cloning could offer a solution to the kinds of medical problems that could be cured by a transplant—of, say, bone marrow or a kidney—by providing an organ or other body part or material that has a matching tissue type. A family with a child in desperate need of a bone marrow transplant to cure leukemia might be glad to raise a second, genetically identical child to save the life of the first. The second child would be an identical twin to the first child.

Embryonic Stem Cell Research Raises Ethical and Political Controversy

The use of embryonic stem cells for therapeutic purposes has been controversial from the start. Some people argue that destroying an embryo to create embryonic stem cell lines is immoral. They suggest that research should focus on adult stem cells. Others argue that the unused embryos in fertility clinics would be destroyed anyway, so why not put them to good use? They view the use of embryonic stem cells as taking a life to save a life. They point out that adult stem cells are not as useful as embryonic stem cells because they cannot replicate indefinitely and cannot differentiate into as many types of cells.

The controversy flared in August 2006 when researchers announced that they could create embryonic stem cell lines without destroying the embryo. The researchers used a technique that is used to diagnosis severe genetic defects is an embryo created in a fertility clinic. The technique is called preimplantation genetic diagnosis (PGD). In PGD a single cell is removed from an embryo that consists of only 8 to 10 cells. That cell is analyzed for genetic

defects. If the cell is normal, the embryo is then implanted in the woman's uterus to continue development. Approximately 1500 PGD procedures have been done in the United States and have led to normal development. The researchers argue that the cell removed from the embryo could instead be used to establish an embryonic stem cell line, and the embryo would be unharmed. Thus, this technique could be used to create "ethical" embryonic stem cell lines.

Opponents of the creation of new embryonic stem cell lines argue that the new procedure is still unethical. They suggest that the single cell could develop into a new individual, and so the procedure takes a life. They also point out that because the researchers who developed the procedure did not implant the embryos into a woman, they destroyed those embryos. Opponents also suggest that it is always immoral to use an embryo for research, especially if the embryo will not benefit from the procedure.

The controversy has extended into the political realm. In August 2001, President George W. Bush announced that embryonic stem cell research cannot receive federal funding unless the research is limited to using only previously existing stem cell lines. (Use of federal funds is also prohibited for the creation of human embryos to be used in research or for reproductive cloning.) Today, there are only 21 approved embryonic stem cell lines, which means the genetic diversity of the allowed lines is limited. Furthermore, the cell lines are old and don't grow well.

Since then, some scientists have left the United States to do stem cell research in other countries. Other U.S. scientists continue research on embryonic stem cells in privately funded research institutions. Several states have voted to fund their own stem cell research. In mid-2006, the United States Congress voted to ease restrictions on stem cell research, but the decision was vetoed by President Bush.

Where do you stand on this thorny issue? Does the saving of a person's life justify the use of a human embryo? The embryos will still be destroyed even if they are not used for research. Should the harvesting of embryonic stem cells be considered equivalent to salvaging organs from the dead for transplant to the living? If so, should it be legal to create an embryo using somatic cell nuclear transplant for therapeutic cloning? Or should we avoid using embryonic stem cells altogether and shift our focus entirely to research using adult stem cells?

20

Genetics and Human Inheritance

The genetic makeup of these twins is identical. They look slightly different from one another because of environmental influences. Other than identical siblings, each person has a unique genetic inheritance.

Principles of Inheritance Help Us Predict How Simple Traits Are Passed to the Next Generation

- During gamete formation, alleles segregate and assort independently
- Mendelian genetics considers patterns of inheritance
- Pedigrees help us to ascertain genotype
- A dominant allele often produces a protein that the recessive allele does not
- Codominant alleles both produce functioning gene products
- In incomplete dominance, the heterozygote has an intermediate phenotype
- In pleiotropy, one gene has many effects
- Certain genes have multiple alleles in a population
- Most traits are controlled by many genes
- Genes on the same chromosome are usually inherited together
- Sex-linked genes are located on the sex chromosomes
- Sex-influenced genes are autosomal genes whose expression is influenced by sex hormones

Breaks in Chromosomes Change Chromosomal Structure and Function

Certain Genetic Disorders Can Be Detected by Laboratory Tests

 **SOCIAL ISSUE** Gene Testing

Steve's brother, Dan, is a risk taker. Two years ago, he hiked through the Andes in Chile. Four years ago, he learned to scuba dive. Recently, he took up sky diving.

Dan hasn't always been so daring. Six years ago, he was finishing law school and looking forward to a career in politics. Then their father was diagnosed with Huntington's disease, a disorder caused by the degeneration of brain cells. This devastating disease usually appears in middle age with symptoms of clumsiness and forgetfulness. Now Dad is still able to work, but his handwriting has become illegible, and he cannot sit for long periods without twitching.

Huntington's disease is genetic. Dan and Steve were each told they had a 50% chance of also carrying the gene. If they do carry it, they will surely develop the disease in middle age, because the gene for Huntington's disease is dominant. After Dad's diagnosis, Dan decided to have his genes analyzed to see if he carried the gene that affected their father. Unfortunately, he did. Dan then started to live the life of a risk taker—what did he have to lose, he thought? Steve, on the other hand, elected not to have his genes tested but to continue to live the satisfying life he and his wife Jessica had always enjoyed. But now Jessica was pregnant. She wanted to have the child, but not if it would be born with the gene for Huntington's disease. She insisted that the baby's genes be tested. Fortunately for Steve, Jessica, and the baby, the test was negative for the Huntington's disease gene.

The genes we receive at the moment of conception influence all the biochemical reactions taking place inside our cells, our susceptibility to disease, our behavior patterns, and even our life span. Our environment is also an important influence, but our genes provide the basic blueprint for our possibilities and limitations. In this chapter, you will learn more about the genetic foundation that has been so important in shaping who you are. You will learn how heritable traits are passed to new generations and how to predict the distribution of traits from one generation to the next. ■

Principles of Inheritance Help Us Predict How Simple Traits Are Passed to the Next Generation

An understanding of meiosis helps us answer important questions about inheritance. Why is your brother the only sibling with Mom's freckles and widow's peak (a hairline that comes to a point on the forehead)? How can you have blue eyes when both your parents are brown eyed? Will you be bald at 40, like Dad? Let's look again at how chromosomes, meiosis, and heredity are related.

Before beginning, you may want to review some of the terms that were introduced in Chapter 19 and are summarized in Figure 20.1. Recall that somatic (body) cells have two copies of every chromosome. Thus, human cells have 23 pairs of chromosomes. One member of each pair was inherited from the female parent, and the other from the male parent. The chromosomes that carry genes for the same traits are a *homologous pair of chromosomes,* or homologues. Chromosomes are made of DNA and protein. Certain segments of the DNA of each chromosome function as genes. A **gene** directs the synthesis of a specific protein that can play either a structural or a functional role in the cell.[1] (In some cases, the gene directs the synthesis of a polypeptide that forms part of a protein.) In this way, the gene-determined protein can influence whether a certain **trait**, or characteristic, will develop. For instance, the formation of your brother's widow's peak was directed by a protein coded for by a gene that he inherited from Mom. Genes for the same trait are found at the same specific location on homologous chromosomes.

Different forms of a gene are called **alleles**. Alleles produce different versions of the trait they determine. For example, there is a gene that determines whether freckles will form. One allele of this gene causes freckles to form; the other does not. Somatic cells carry two alleles for each gene, one allele on each homologous chromosome. If a person has at least one allele for freckles, melanin will be deposited, and freckles will form. If neither homologue bears the

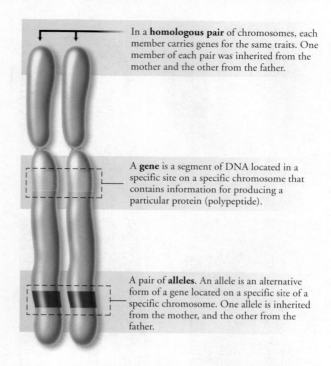

In a **homologous pair** of chromosomes, each member carries genes for the same traits. One member of each pair was inherited from the mother and the other from the father.

A **gene** is a segment of DNA located in a specific site on a specific chromosome that contains information for producing a particular protein (polypeptide).

A pair of **alleles**. An allele is an alternative form of a gene located on a specific site of a specific chromosome. One allele is inherited from the mother, and the other from the father.

FIGURE **20.1** Important terms in genetics

freckle allele that leads to melanin deposition, for example, the person will not have freckles.

Individuals with two copies of the same allele of a gene are said to be **homozygous** (*homo,* same; *zygo,* joined together) for that trait. Those with different alleles of a given gene are said to be **heterozygous** (*hetero,* different; *zygo,* joined together). When the effects of a certain allele can be detected, regardless of whether an alternative allele is also present, the allele is described as **dominant**. Dimples and freckles are human traits dictated by dominant alleles. An allele whose effects are masked in the heterozygous condition is described as **recessive**. Because of this masking, only homozygous recessive alleles are expressed. Conditions such as cystic fibrosis, in which excessive mucus production impairs lung and pancreatic function, and albinism, in which the pigment melanin is missing in the hair, skin, and eyes, result from recessive alleles. By convention, a dominant allele is designated with a capital letter, and a recessive allele with a lowercase letter—*A* and *a,* for example.

We observe the dominant form of the trait whether the individual is homozygous dominant (*AA*) or heterozygous (*Aa*) for that trait. As a consequence, we cannot always tell exactly which alleles are present. It is important to remember that an individual's genetic makeup is not always revealed by the individual's appearance. A **genotype** is the precise set of alleles a person possesses for a given trait or traits. It tells us whether the individual is homozygous or heterozygous for a given gene. A **phenotype**, on the other hand, is the observable physical trait or traits of an individual. Figure 20.2 shows the genotype or genotypes for several phenotypes in humans (for example, the freckled phenotype has two genotypes: *FF* and *Ff).* Table 20.1 summarizes terms commonly used in genetics as they apply to the inheritance of freckles, which is a dominant trait.

---

[1]This is a simplified definition of a gene. As we will see in Chapter 21, some genes contain regulatory regions of DNA within their boundaries. Also, some genes code for RNA molecules that are needed for the production of a protein but are not part of it.

Freckles: *FF* or *Ff*   No freckles: *ff*

Widow's peak: *WW* or *Ww*   Straight hairline: *ww*

Unattached earlobes: *EE* or *Ee*   Attached earlobes: *ee*

tt   *Tt*   *TT* or *Tt*
Tongue rolling

FIGURE **20.2** Genotype or genotypes for selected human phenotypes

TABLE 20.1 REVIEW OF COMMON TERMS IN GENETICS

GENOTYPE: THE ALLELES THAT ARE PRESENT	DESCRIPTION	PHENOTYPE: THE OBSERVABLE TRAIT
FF	Homozygous dominant: • Two dominant alleles present. • Dominant phenotype expressed.	Freckles
Ff	Heterozygous: • Different alleles present. • Dominant phenotype expressed.	Freckles
ff	Homozygous recessive: • Two recessive alleles present. • Recessive phenotype expressed.	No freckles

■ **During gamete formation, alleles segregate and assort independently**

Recall from Chapter 19 that the members of each homologous pair of chromosomes segregate during meiosis I, with each homologue going to a different daughter cell. Thus, an egg or a sperm has only one member of each homologous pair. Consider what this fact means for inheritance. Because the alleles for each gene segregate during gamete formation, half the gametes bear one allele, and half bear the other. This principle, known as the law of segregation, explains why, for every gene in our chromosomes, one of the alleles comes from our mother and one comes from our father.

Furthermore, each pair of homologous chromosomes lines up at the midline of the cell during meiosis I independently of the other pairs. The orientation of the paternal and maternal homologues relative to the poles of the cell (that is, with regard to which homologue is closer to which pole) is entirely random. What this fact means for inheritance is that when pairs of alleles are on different chromosomes, they segregate into gametes independently. This principle, known as the law of independent assortment, explains why the mixture of genes that came from the mother and genes that came from the father is different in every gamete.

■ **Mendelian genetics considers patterns of inheritance**

During the nineteenth century, Gregor Mendel, a monk who grew up in a region of Austria that is now part of the Czech Republic, worked out much of what we know today about the laws of heredity by performing specific crosses of pea plants. Although Mendel knew nothing about chromosomes, his ideas about the inheritance of traits are consistent with what we now know about the movement of chromosomes during meiosis. Mendel's ideas are used today to predict the outcome of hereditary crosses.

ONE-TRAIT CROSSES

We will begin by using Mendel's ideas to consider the inheritance of a single trait, using the example of freckles. The inheritance of freckles, a characteristic determined by a dominant allele. Suppose a freckled female who is homozygous dominant (*FF*) mated with a homozygous recessive male with no freckles (*ff*) (Figure 20.3). The following sequence of steps allows us to predict the degree of likelihood that their child will have freckles.

1. **Identify the possible gametes that each parent can produce.** We know that alleles segregate during meiosis, but in this example, both parents are homozygous. Therefore, each can produce only one type of gamete (as far as freckles are concerned). The female produces gametes (eggs) with the dominant allele (*F*), and the male produces gametes (sperm) with the recessive allele (*f*).

2. **Use a Punnett square to determine the probable outcome of the genetic cross.** In a Punnett square, columns are set up and labeled to represent each of the possible gametes of one parent, let's say the father. Remember, the alleles of each gene segregate during meiosis. In this case, then, there would be two columns to represent the gametes of the male without freckles, each labeled with a recessive allele, *f*. In a similar manner, rows are established across the columns and labeled to represent all the possible gametes formed by the other parent, in this case, the woman with freckles. There would be two rows in this Punnett square, each labeled *F*. Each square in the resulting table is then filled in by combining the labels of the corresponding rows and columns. The squares represent the possible offspring of these parents. In this first case, all the children would have freckles and would be heterozygous for the trait (*Ff*). Figure 20.4 shows the Punnett square for this example.

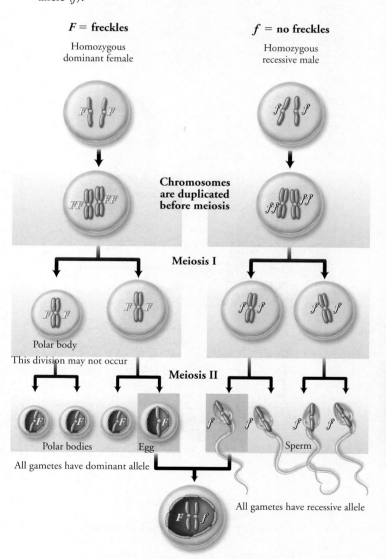

Fertilization produces heterozygous offspring

FIGURE **20.3** Gamete formation by a female who is homozygous dominant for freckles (*FF*), a male who is homozygous recessive for no freckles (*ff*), and the heterozygous (*Ff*) individual resulting from the union of these gametes.

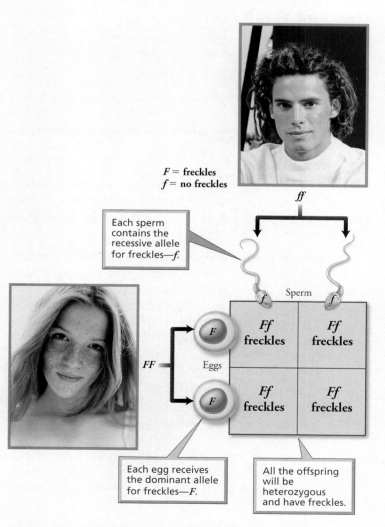

FIGURE **20.4** This Punnett square illustrates the probable offspring from a cross between a homozygous dominant female with freckles (*FF*) and a homozygous recessive male without freckles (*ff*). The columns are labeled with the possible gametes the male could produce (one gamete per column). The rows are labeled with the possible gametes the female could produce (one gamete per row). Combining the labels on the corresponding rows and columns yields the genotype of possible offspring.

Now let's consider why it is possible for parents who are both heterozygous for the freckle trait (*Ff*) to have a child without freckles. The *F* and *f* alleles segregate during meiosis. So half the gametes bear the *F* allele, and half bear the *f* allele (Figure 20.5a). This time the two rows in the Punnett square are labeled with *F* and *f*, and so are the two columns (Figure 20.5b). Filling the boxes according to the labels on rows and columns, we see that the probable genotypes of the children would be homozygous dominant (*FF*), heterozygous (*Ff*), and homozygous recessive (*ff*), in a genotypic ratio of 1 *FF*:2 *Ff*:1 *ff*. The ratio of phenotypes of children with freckles (*FF* and *Ff*) compared with children without freckles (*ff*) would be 3:1. This ratio means that each child born to this couple will have a 75% (3/4) chance of having freckles. But there is a 1 in 4, or 25%, chance that each child will not have freckles. A cross in which both parents are heterozygous for one trait of interest is called a *monohybrid cross.*

stop and think

Toward the beginning of the chapter, a question was raised about why only one sibling, your hypothetical brother, inherited your mother's freckles. Implicit in the way the question was phrased was that there is at least one other sibling (you), who, like your father, does not have freckles. Use a Punnett square to explain why the mother in this example must be heterozygous for the freckle trait.

TWO-TRAIT CROSSES

We use the same steps to predict the probable outcome for inheritance of two traits of interest that we used for inheritance of a single trait—as long as the genes for the two traits are on different chromosomes.

1. **Identify the possible gametes that each parent can produce.** Like freckles, a widow's peak is controlled by a dominant allele. What would the children look like if the parents were a woman who is homozygous for both freckles and a widow's peak (*FFWW*) and a man with no freckles and a straight hairline (*ffww*)? To answer that question, we begin by determining the possible gametes each mate can produce. Since they are homozygous for both traits, the woman can produce only gametes bearing dominant alleles (*FW*), and the man can produce only gametes with the recessive alleles (*fw*).

2. **Use a Punnett square to determine the probable outcome of the genetic cross.** In this case, each parent can produce only one type of gamete, so all the children will have the same genotype and the same phenotype. All the children will be heterozygous for each trait (*FfWw*) and will have both freckles and a widow's peak.

Suppose one of these children mates with someone who is also heterozygous for freckles and a widow's peak. This would be a *dihybrid cross,* a mating of individuals who are both heterozygous for two traits of interest. Would it be possible for this couple to have a child that had neither freckles nor a widow's peak? Yes, but the chances are only 1 in 16. Let's see why. First, determine the possible gametes that each parent could produce (Figure 20.6).

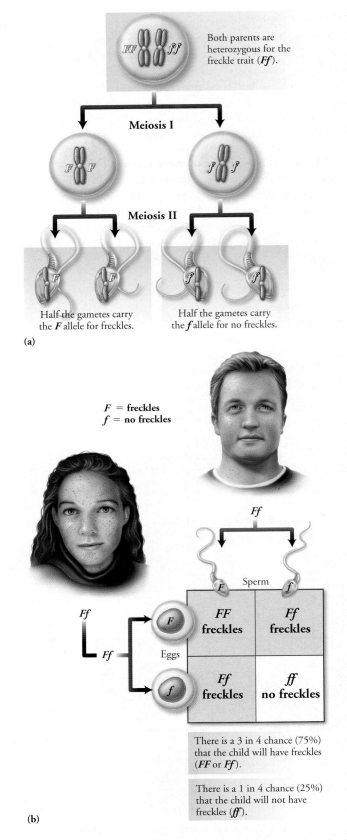

FIGURE **20.5** (a) Gamete formation by a person who is heterozygous for the freckle trait (*Ff*). (b) A Punnett square showing the probable outcome of a mating between two people who are heterozygous for the freckle trait (*Ff*).

Keep in mind that alleles for a gene segregate and that genes for different traits that are located on different chromosomes will segregate independently of one another. There are, therefore, four possible allele combinations in gametes produced by a person who is heterozygous for these two different genes. The gametes are *FW, Fw, fW,* and *fw.* We construct a Punnett square with *four* columns, because this time there are four possible male gametes, and four rows labeled with the possible female gametes. Each type of gamete has an equal chance of joining with any other type at fertilization. The squares are filled by combining the labels of the corresponding columns and rows, giving us the possible genotypes of the offspring. Notice in Figure 20.7 that the relative numbers of phenotypes possible for the next generation are be 9 freckled, widow's peak: 3 freckled, straight hairline: 3 no freckles, widow's peak: 1 no freckles, straight hairline. We see, then, that the expected phenotypic ratio resulting from a dihybrid cross is 9:3:3:1.

■ Pedigrees help us to ascertain genotype

It is sometimes important to find out the genotype of a human with a dominant phenotype for a particular trait. We can often deduce the unknown genotype by looking at the expression of the trait in the person's ancestors or descendants. A chart showing the genetic connections between individuals in a family is called a **pedigree**. Family or medical records are used to fill in the pattern of expression of the trait in question for as many family members as possible. Pedigrees are useful not just in determining an unknown genotype but also in predicting the chances that one's offspring will display the trait. Pedigrees for the inheritance of autosomal-dominant disorders and autosomal-recessive disorders are shown in Figure 20.8 (recall that autosomes are non-sex chromosomes; we will discuss genes on sex chromosomes later in the chapter).

Marfan syndrome is an autosomal-dominant disorder (like the neurological disorder called Huntington's disease, which was discussed in the chapter's opening scenario). Marfan syndrome is a connective tissue disorder in which a dominant allele for the production of an elastic connective tissue protein, fibrillin, produces a nonfunctional protein. Because connective tissue is widespread in the body, so are the symptoms of Marfan syndrome. Connective tissue is an important component in blood vessel walls, heart valves, tendons, ligaments, and cartilage. A serious problem caused by Marfan syndrome is weak walls of large blood vessels, because they can tear or burst.

stop and think

If the pedigree in Figure 20.8a were depicting the inheritance of Marfan syndrome, what is the genotype of the male in the third generation who has Marfan syndrome? How do you know?

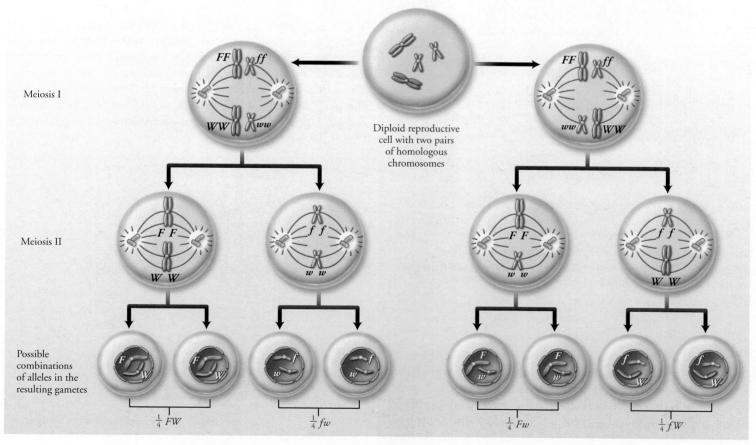

FIGURE **20.6** A person who is heterozygous for two genes that are located on different chromosomes can produce four different types of gametes.

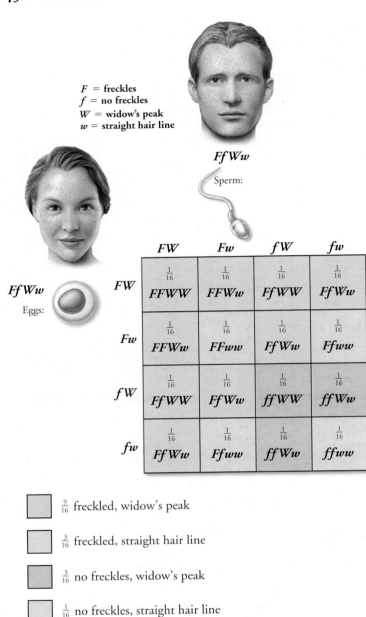

F = freckles
f = no freckles
W = widow's peak
w = straight hair line

FfWw

Sperm:

FfWw

Eggs:

	FW	**Fw**	**fW**	**fw**
FW	$\frac{1}{16}$ *FFWW*	$\frac{1}{16}$ *FFWw*	$\frac{1}{16}$ *FfWW*	$\frac{1}{16}$ *FfWw*
Fw	$\frac{1}{16}$ *FFWw*	$\frac{1}{16}$ *FFww*	$\frac{1}{16}$ *FfWw*	$\frac{1}{16}$ *Ffww*
fW	$\frac{1}{16}$ *FfWW*	$\frac{1}{16}$ *FfWw*	$\frac{1}{16}$ *ffWW*	$\frac{1}{16}$ *ffWw*
fw	$\frac{1}{16}$ *FfWw*	$\frac{1}{16}$ *Ffww*	$\frac{1}{16}$ *ffWw*	$\frac{1}{16}$ *ffww*

$\frac{9}{16}$ freckled, widow's peak

$\frac{3}{16}$ freckled, straight hair line

$\frac{3}{16}$ no freckles, widow's peak

$\frac{1}{16}$ no freckles, straight hair line

FIGURE **20.7** A dihybrid cross is a mating of individuals heterozygous for two traits, each governed by a gene on a different chromosome. Analyzed on a Punnett square, it illustrates the law of independent assortment, that is, that each allele pair is inherited independently of others found on different chromosomes. The phenotypes of the offspring of a dihybrid cross would be expected to occur in a 9:3:3:1 ratio.

Genetic disorders are often caused by recessive alleles, so knowledge of whether the parents carry the allele will help predict whether their child could be homozygous for the trait and therefore born with the condition (Figure 20.8b). For example, cystic fibrosis (CF), a disorder in which abnormally thick mucus is produced, is controlled by a recessive allele. Approximately 1 in every 2000 infants in the United States is born with CF. It is a leading cause of childhood death. The average length of life is 24 years.

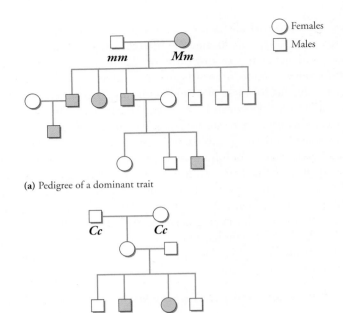

○ Females
□ Males

mm   *Mm*

(a) Pedigree of a dominant trait

Cc   *Cc*

(b) Pedigree of a recessive trait

FIGURE **20.8** Pedigrees showing the inheritance of (a) a dominant autosomal trait and (b) a recessive autosomal trait. A pedigree is constructed so that each generation occupies a different horizontal line, numbered from top to bottom, with the most ancestral at the top. Males are indicated as squares, and females as circles. A horizontal line connects the partners in a marriage. An affected individual is indicated with a solidly colored symbol.

The three primary signs of CF are salty sweat, digestive problems, and respiratory problems. Many children with CF suffer from malnutrition because mucus clogs the pancreatic ducts, preventing pancreatic digestive enzymes from reaching the small intestine, where they would otherwise function in digestion. Thick mucus also plugs the respiratory passageways, making breathing difficult and increasing susceptibility to lung infections such as pneumonia (Figure 20.9).

FIGURE **20.9** Cystic fibrosis, controlled by a recessive gene, is a condition in which abnormally thick mucus is produced causing serious digestive and respiratory problems. This child with cystic fibrosis is using a therapeutic toy to enhance airflow. When she exhales with enough force, the ribbons wave.

Because CF is inherited as a recessive allele, children with CF are usually born to normal, healthy parents who had no idea they carried the trait. A **carrier** is someone who displays the dominant phenotype but is heterozygous for a trait and can, therefore, pass the recessive allele to descendants. Approximately 1 in 22 Caucasians in the United States is a carrier for CF. (CF is less common among Asians and rare among African Americans.)

stop and think

If the pedigree in Figure 20.8b were depicting the inheritance of CF, what is the genotype of the male in the third generation who has CF? How do you know? Which individuals are carriers for the trait? How do you know?

■ A dominant allele often produces a protein that the recessive allele does not

You may wonder what makes an allele dominant or recessive. In many cases, the dominant allele produces a normal, functional protein, but the recessive allele does not produce any protein or it produces the protein in an altered form that does not function properly. Consider the inheritance of the most common form of albinism, the inability to produce the brown pigment melanin that normally gives color to the eyes, hair, and skin. Because of the lack of melanin, a person with albinism has pale skin and white hair (Figure 20.10). A child with the trait has pink eyes, but the eye color darkens to blue in an adult. Because there is no melanin in the skin to protect against sunlight's ultraviolet rays, a person with albinism is very vulnerable to sunburn and skin cancer.

The ability to produce melanin depends on the enzyme tyrosinase. The dominant allele that results in normal skin pigmentation produces a functional form of tyrosinase. A single copy of the dominant allele can produce all of this enzyme that is needed. The recessive allele that causes albinism produces a nonfunctional form of tyrosinase, and melanin cannot be formed.

■ Codominant alleles both produce functioning gene products

The examples we have described so far represent **complete dominance**, a situation in which a heterozygote exhibits the trait associated with the dominant allele but not that of the recessive allele. In other words, the dominant allele produces a functional protein, and the protein's effects are apparent; but the recessive allele produces a less functional protein or none at all, and its effects are not apparent. However, complete dominance is not the only possibility for a heterozygous genotype. In some cases, both alleles produce functional proteins. In this situation, described as **codominance**, the effects of *both* alleles are separately apparent in a heterozygote.

The inheritance of type AB blood is an example of codominance. Two alleles, I^A and I^B, result in the production of two different polysaccharides on the surface of red blood cells. In persons with type AB blood, both alleles are expressed, and their red blood cells have both A and B polysaccharides on their surface. Figure 20.11 illustrates the inheritance of blood types as an example of codominance. (We will return to the example of blood groups shortly.)

■ In incomplete dominance, the heterozygote has an intermediate phenotype

In **incomplete dominance**, the expression of a trait in a heterozygous individual is somewhere between the expression of the trait in a homozygous dominant individual and the expression of the trait in a homozygous recessive individual. The dominant allele produces a functional protein product. As a result, a heterozygote has only one "dose" of the protein product—half the amount in a homozygous dominant person. The recessive allele does not produce that product.

Sickle cell hemoglobin provides an example of incomplete dominance. Recall from Chapter 11 that hemoglobin is the pigment in red blood cells that carries oxygen. A red blood cell filled with normal hemoglobin (Hb^A) is a biconcave disk. The allele for sickling hemoglobin (Hb^S) produces an abnormal form of hemoglobin that is less efficient in binding oxygen. In the homozygous sickling condition ($Hb^S Hb^S$), called sickle cell anemia, the red blood cells contain only the abnormal form of

FIGURE **20.10** A person with albinism lacks the brown pigment melanin in the skin, hair, and irises of the eyes.

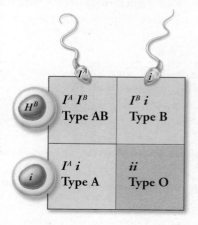

FIGURE **20.11** The inheritance of ABO blood type is an example of codominance. The alleles I^A and I^B are codominant; they are both expressed. Both these alleles are dominant over i.

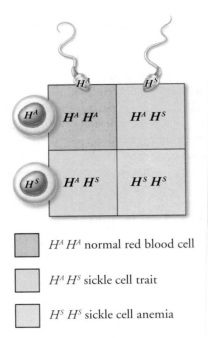

☐ $H^A H^A$ normal red blood cell

☐ $H^A H^S$ sickle cell trait

☐ $H^S H^S$ sickle cell anemia

FIGURE **20.12** The inheritance of sickle cell trait ($H^A H^S$) is an example of incomplete dominance.

hemoglobin. When the oxygen content of the blood drops below a certain level, as might occur during excessive exercise or respiratory difficulty, the red blood cells become sickle-shaped and tend to clump together. The clumped cells can clog capillaries and break open, causing great pain. Vital organs may be damaged by lack of oxygen. People who are homozygous for sickle cell anemia usually die at a young age.

The normal allele for the sickle cell gene shows incomplete dominance, so people who are heterozygous $Hb^A Hb^S$ have the *sickle cell trait*. They are generally healthy, but sickling and clumping of red cells may occur if there is a prolonged drop in the oxygen content of the blood, as might happen during travel at high elevations. (Sickle cell anemia was discussed in more detail in Chapter 11. The maintenance of the sickle cell trait in a population is discussed in Chapter 22.) Figure 20.12 shows the inheritance of the sickle cell trait as an example of incomplete dominance.

stop and think

Straight hair shows incomplete dominance over curly hair. A homozygous dominant person has straight hair; a heterozygote has wavy hair; and a homozygous recessive person has curly hair. What is the probability that a curly-haired person and a wavy-haired person would have a child with wavy hair?

■ In pleiotropy, one gene has many effects

In addition to providing an example of incomplete dominance, sickle cell anemia is an example of **pleiotropy**: one gene having many effects. As you can see in Figure 20.13, the sickling of red blood cells caused by the abnormal hemoglobin has effects throughout the body.

■ Certain genes have multiple alleles in a population

Many genes have more than two alleles. When three or more forms of a given gene exist, they are referred to as **multiple alleles**. Keep in mind, however, that one individual has only two alleles for a given gene (one on each homologue), even if multiple alleles exist in the population. These alleles segregate independently during meiosis in the same manner as alleles for any other gene.

The ABO blood types (discussed in Chapter 11) provide an example of multiple alleles. Blood type is determined by the presence of certain polysaccharides (sugars) on the surface of red blood cells. Type A blood has the A polysaccharide; type B has the B polysaccharide; type AB has both A and B polysaccharides; and type O has neither. The synthesis of each kind of polysaccharide is directed by a specific enzyme: one enzyme produces A; a different enzyme produces B. Each enzyme is specified by a different allele of the gene.

The gene controlling ABO blood types therefore has three alleles, I^A, I^B, and i. Alleles I^A and I^B specify the A and B polysaccharides, respectively. When both these alleles are present, both polysaccharides are produced. I^A and I^B are, therefore, codominant. Allele i, which produces no enzyme, is recessive to both I^A and I^B. The possible combinations of these alleles and the resulting blood types are shown in Table 20.2.

TABLE 20.2 THE RELATIONSHIP BETWEEN GENOTYPE AND ABO BLOOD TYPES	
GENOTYPE	**BLOOD TYPE**
$I^A I^A$, $I^A i$	A
$I^B I^B$, $I^B i$	B
$I^A I^B$	AB
ii	O

stop and think

A man who has blood type AB is accused of fathering a child with blood type O. The mother has blood type B. Is it possible for the accused man to be the father of this child? (*Hint:* What are the possible genotypes of the man, woman, and child? Given each possible genotype of this man and woman, what gametes could each produce? Use Punnett squares to determine whether any combination of these genotypes could produce a child with the same genotype as this child.)

■ Most traits are controlled by many genes

So far, we have discussed traits governed by single genes, although the genes may have multiple alleles. When a single gene controls a trait, the trait is usually either present or it is not, even though incomplete dominance or environmental factors may modify its expression. In the case of multiple alleles, there may be several distinct classes of phenotypes, as in A, B, AB, and O blood types.

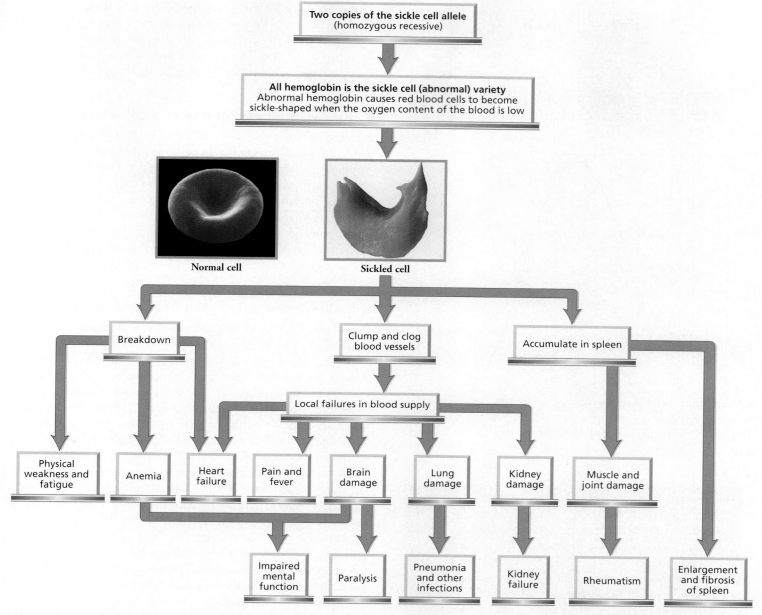

FIGURE **20.13** Sickle cell anemia is an example of pleiotropy, a condition in which a single gene has many effects.

The expression of most traits is much more variable, however. Many traits, including height, skin color, and eye color, vary almost continuously from one extreme to another. Environment can play a role in creating such a smooth continuum. For instance, diet and disease influence adult height, and exposure to sunlight darkens skin color. But even when all environmental factors are equal, there is still considerable variation in the expression of certain traits. Such variation results from **polygenic inheritance**, that is, the involvement of two or more genes in producing a trait. The more genes involved, the smoother the gradations and the greater the extremes of trait expression.

Although human height is probably controlled by more than three genes, we will pretend it is determined by only three, partly to simplify our discussion, but also to show how much variation in expression is possible even with as few as three genes—A, B, and C—interacting to determine a trait. Let's assume that the dominant alleles (A, B, C) of each gene add height, and the recessive alleles (a, b, c) do not. How tall would we expect the children to be if both parents were of medium height and heterozygous for all three genes? As you can see in Figure 20.14, there would be seven genetic height classes, ranging from very short to very tall. The probability of the children having either extreme of stature is slim, 1/64. The most likely probability (a 20/64 chance) is that they will be of medium height, like their parents.

Skin color is also determined by several genes. The allele for the kind of albinism described above prevents melanin production,

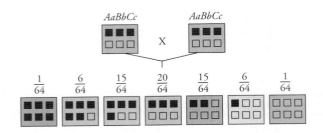

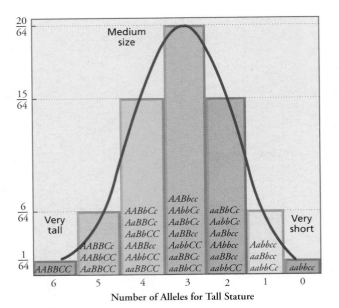

FIGURE 20.14 Human height varies along a continuum. (a) One reason is that height is determined by more than one gene (polygenic inheritance). This figure shows the distribution of alleles for tallness in children of two parents of medium height, assuming that three genes are involved in the determination of height. The top line of boxes shows the parental genotypes, and the second line of boxes indicates the possible genotypes of the offspring. Alleles for tallness are indicated with dark squares. (b) Students organized according to height.

so if a person is homozygous recessive for this allele, no melanin can be deposited in the skin. In addition, there are probably at least four other genes involved in determining the amount of melanin deposited in the skin. Two alleles for each of four genes would create nine classes of skin color, ranging from pale to dark.

■ Genes on the same chromosome are usually inherited together

Scientists estimate that there are about 20,000 to 25,000 human genes distributed among the 23 pairs of chromosomes. Thus, each chromosome bears a great number of genes. Genes on the same chromosome tend to be inherited together because an entire chromosome moves into a gamete as a unit. Genes that tend to be inherited together are described as being *linked*. We see, then, that linked genes do not *usually* assort independently. *Usually* is emphasized here because there is a mechanism that can unlink genes on the same chromosome: crossing over (discussed in Chapter 19).

■ Sex-linked genes are located on the sex chromosomes

Recall from Chapter 19 that one pair of chromosomes is called the sex chromosome. There are in fact two kinds of sex chromosomes, X and Y. They are not truly homologous because the Y chromosome is much smaller than the X chromosome, and they do not carry all of the same genes.[2] The Y chromosome carries very few genes, but it is more important than the X chromosome in determining gender. If a particular gene on the Y chromosome is present, an embryo will develop as a male. In the absence of that gene, an embryo will develop as a female. In contrast, most of the genes on an X chromosome have nothing to do with sex determination. For instance, genes for certain blood-clotting factors and for the pigments in cones (the photoreceptors responsible for color vision) are found on the X chromosome but not on the Y chromosome. Furthermore, the X chromosome has about as many genes as a typical autosome, but the Y chromosome has relatively few. Thus, most genes on the X chromosome have no corresponding alleles on the Y chromosome and are known as **X-linked genes**.

Because most X-linked genes have no homologous allele on the Y chromosome, they have a different pattern of inheritance than do autosomes. A male is XY and therefore will express virtually all of the alleles on his single X chromosome, even those that are recessive. A female, on the other hand, is XX, so she does not always express recessive X alleles. As a result, the recessive phenotype is much more common in males than in females (Figure 20.15). Furthermore, a son cannot inherit an X-linked recessive allele from his father. To be male, a child must have inherited his father's Y chromosome, not his X chromosome. Consequently, a son can inherit an X-linked recessive allele only from his mother. A daughter, however, can inherit an X-linked recessive allele from either parent. She must be homozygous for the recessive allele to show the reces-

[2]X and Y are considered to be a homologous pair because they each have a small region at one end that carries some of the same genes. During meiosis, the tiny homologous region on X and Y will pair in synapsis. As a result, they segregate into gametes in the same manner as autosomes do.

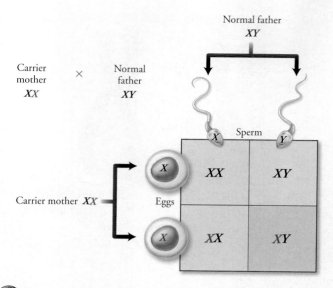

Normal father
XY

Carrier
mother   ×   Normal
XX   father
                    XY

Sperm
X   **Y**

Carrier mother **XX**

Eggs

	XX	**XY**
X	XX	XY
X	XX	XY

FIGURE **20.15** Genes that are X linked have a different pattern of inheritance than do genes on autosomes, as seen in this cross between a carrier mother and a father who is normal for the trait. The recessive allele is indicated in red. Notice that each son has a 50% chance of displaying the recessive phenotype. All daughters will appear normal, but each daughter has a 50% chance of being a carrier.

sive phenotype. If she is heterozygous for the trait, she will have a normal phenotype but be a carrier for that trait.

Among the disorders caused by X-linked recessive alleles are red-green color blindness, two forms of hemophilia, and Duchenne muscular dystrophy. Red-green color blindness, the inability to distinguish red and green, was discussed in Chapter 9. Hemophilia (discussed in Chapter 11) is a bleeding disorder caused by a lack of a blood-clotting factor: Hemophilia A is a lack of blood-clotting factor VIII, and hemophilia B is a lack of clotting factor IX. Duchenne muscular dystrophy is discussed in Chapter 6.

stop and think

Explain why Duchenne muscular dystrophy is inherited from one's mother but is usually expressed only in sons. (*Hint:* Consider its mode of inheritance.)

■ **Sex-influenced genes are autosomal genes whose expression is influenced by sex hormones**

The expression of certain autosomal genes, those that are not located on the sex chromosomes, is powerfully influenced by the presence of sex hormones, so their expression differs in males and females. These traits are described as *sex-influenced* traits.

Male pattern baldness, premature hair loss on the top of the head but not on the sides, is an example of a sex-influenced trait. Male pattern baldness is much more common in men than in women because its expression depends on both the presence of the allele for baldness and the presence of testosterone, the male sex hormone. The allele for baldness, then, acts as a dominant allele in

males because of their high level of testosterone and as a recessive allele in females because females have a much lower testosterone level. A male will develop pattern baldness whether he is homozygous or heterozygous for the trait. However, only women who are homozygous for the trait will develop pattern baldness. When a woman does develop pattern baldness, it usually appears later in life than it does in a man. The allele is expressed in women because the adrenal glands produce a small amount of male hormone. After menopause, when the supply of estrogen declines, adrenal male hormones may cause the expression of the baldness gene. However, in many women, balding may be merely thinning of hair.

stop and think

Male pattern baldness was passed from father to son through at least four generations of the Adams family. John Adams (1735–1826), the second U.S. president, passed it to his son, John Quincy Adams (1767–1848), the sixth U.S. president. He, in turn, passed the gene to his son, Charles Frances Adams (1807–1886), a diplomat, who passed it to his son, Henry Adams (1838–1918), a historian. Explain why father-to-son transmission of the trait rules out X-linked inheritance.

Breaks in Chromosomes Change Chromosomal Structure and Function

Chromosomes can break, which may lead to other alterations in structure. Breakage can be caused by certain chemicals, radiation, or viruses. Breakage also occurs as an essential part of crossing over. Although it does not happen often, chromosomes can be misaligned when crossing over occurs. Then, when the pieces reattach, one chromatid will have lost a segment, and the other will have gained a segment.

The loss of a piece of chromosome is called a *deletion*. The most common type of deletion occurs when the tip of a chromosome breaks off and then, during cell division, does not move into the same daughter cell as the rest of the chromosome. Deletion of more than a few genes on an autosome is usually lethal, and the loss of even small regions causes disorders.

In humans, the most common deletion, the loss of a small region near the tip of chromosome 5, causes cri-du-chat syndrome (meaning *cry of the cat*). An infant with this syndrome has a high-pitched cry that sounds like a kitten meowing. The unusual sound of the cry is caused by an improperly developed larynx (voice box). Infants with this syndrome have a round face; wide-set, downward-sloping eyes with a fold of skin at the corner of each eye; and misshapen ears (Figure 20.16). Although the condition is not usually fatal, it does cause severe mental retardation.

The addition of a piece of chromosome is called a *duplication*. The effects of a duplication depend on its size and position. In general, however, a small duplication is less harmful than a deletion of comparable size. There is a small region of chromosome 9 that is sometimes duplicated, resulting in cells containing three copies of this segment. The result is mental retardation, accompanied by facial characteristics that may include a bulbous nose, wide-set squinting eyes, and a lopsided grin.

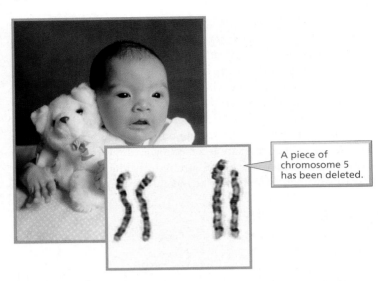

A piece of chromosome 5 has been deleted.

FIGURE **20.16** Occurring in 1 out of every 50,000 live births, cri-du-chat syndrome is the most common genetic deletion found in humans. It is caused by the loss of a small region near the tip of chromosome 5. The name of the syndrome, which means *cry of the cat*, describes the sound of the cry of affected babies. The infants have a characteristically round face and wide-set eyes.

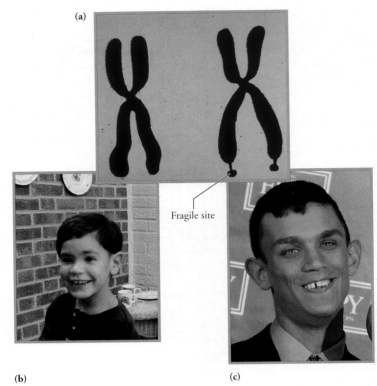

Fragile site

FIGURE **20.17** Fragile X syndrome. (a) Duplication of a region on the X chromosome makes the chromosome fragile and easily broken. (b) A child with fragile X syndrome appears normal. (c) Characteristics of an adult with fragile X syndrome include a long face and large ears. In addition, fragile X syndrome causes mental retardation.

Genetic disorders also occur when certain sequences of DNA are duplicated multiple times. The fragile X syndrome provides an example. The syndrome is so named because an abnormally long sequence of repeats makes the X chromosome fragile and easily broken. Besides making the chromosome fragile, the repeated DNA can shut down the activity of the entire chromosome. Fragile X syndrome is the most common form of inherited mental retardation, affecting roughly 1 in 1250 males and 1 in 2500 females. It is not known, however, exactly how fragile X syndrome causes the retardation. Other characteristics may include attention deficit, hyperactivity, large ears, long face, and flat feet (Figure 20.17).

what would you do?

The degree of mental retardation accompanying fragile X syndrome is variable. Nearly all males who inherit a fragile X chromosome exhibit retardation. About a quarter of the girls with fragile X syndrome will develop mental retardation, although others will only have learning problems. Some cities screen children with mental retardation and children with learning difficulties for fragile X chromosomes. People who endorse this practice argue that the children with fragile X syndrome are then benefited because specialists such as speech therapists, counselors, and physicians can be called on to help. Genetic counselors can help the families of these children decide whether to have additional children. Others oppose the screening, arguing that there is no special treatment for children with fragile X syndrome other than that available to all children with learning problems. After diagnosis, insurance companies may discriminate against the children and drop their health coverage. In your opinion, do the benefits of screening for fragile X syndrome outweigh the disadvantages?

Certain Genetic Disorders Can Be Detected by Laboratory Tests

More than 4000 disorders have their roots in our genes. Tests are now available to look for predispositions to many of these genetic disorders. Some tests can even confirm the presence of a suspected disease-related allele in a particular person. Knowledge about the presence or absence of a faulty gene can be very helpful to couples who are planning a family. Normal, healthy parents can carry recessive alleles for disorders such as CF and Tay-Sachs disease (a disorder of lipid metabolism that causes death, usually between the ages of 1 and 5; discussed in Chapter 3).

When specific tests are unavailable, pedigree analysis can help prospective parents determine whether they *might* be carriers of recessive alleles. It cannot answer the question with certainty, but it does allow people to weigh the risks of passing a lethal allele to their children versus the possibility of having a child who is not affected. You can read more about the pros and cons of genetic testing in the Social Issue essay, *Gene Testing*.

Prenatal testing is usually recommended when a defective gene runs in the family or when the mother is older than 35 (because age increases the risk of problems due to failure of the homologous chromosomes to separate; see Chapter 19). There are two available

Gene Testing

Genetic screening is the practice of testing people who have no symptoms, to determine whether they carry genes that will influence their chances of developing certain genetic diseases. Genetic screening technologies are advancing rapidly, and their use is gaining momentum. However, genetic screening raises many ethical questions.

Among the advantages of genetic testing is that it enables people who discover they are at risk for a treatable or preventable condition to take steps to reduce their risk. By informing people that they carry a recessive allele they were unaware of or a dominant allele that is not expressed until late in life, it can also help reduce the incidence of serious genetic disorders in future generations. Consider Tay-Sachs disease, which causes the death of children, usually by the age of 5. Tay-Sachs disease is especially prevalent in descendants of Jewish people from eastern Europe. As a result of voluntary screening programs, the number of children born with Tay-Sachs disease has decreased 10-fold in many communities.

Genetic testing also has a dark side. The psychological consequences of test results can be devastating. Many genetic disorders cannot be prevented or treated. How does a person who may have one of these disorders prepare for the consequences of knowing *now* what will cause his or her death? Huntington's disease is caused by a dominant allele that provides no hints of its existence until relatively late in life, usually past child-bearing years. About 60% of the people with Huntington's disease are diagnosed between the ages of 35 and 50. The gene causes degeneration of the brain, leading to muscle spasms, personality disorders, and death, usually within 10 to 15 years. Because Huntington's disease is caused by a dominant allele, a bearer has a 50% chance of passing it to his or her children. Thus, a person whose parent died of Huntington's disease might well be tested and receive the good news that the test did not detect the allele. But, it is equally likely that the allele *will* show up in the test. Many persons at risk for Huntington's disease prefer to live without the knowledge of their possible fate.

A predictive gene test confirming the risk of a serious disease can cause anxiety and depression in carriers and feelings of guilt in siblings who do not carry the allele. Yet many of these tests deal only in probabilities, not in certainties. Thus the carrier's worry will often prove unnecessary. A mutation in the *BRCA1* gene on chromosome 17, for example, has been found to increase a woman's risk of developing breast cancer and ovarian cancer. Carriers have an 85% chance of developing breast cancer by age 65. A woman who learns she is a carrier must make medical decisions that could influence the length of her life. One option would be to have frequent mammograms in the hope of discovering breast cancer, if it should develop, early enough for treatment. Of course, there is a chance that cancer would not be detected before it spread, making effective treatment difficult. A second option would be to have a mastectomy quickly, except that there is a 15% chance that a person with the gene will not develop breast cancer.

There is also concern that the results of gene tests will not remain private information but instead be used by employers as well as by life and health insurers. If you were an employer who had genetic information about prospective employees, would you choose to invest time and money in training a person who carried an allele that increased the risk of cancer, heart disease, Alzheimer's disease, or alcoholism? As an insurer, would you knowingly cover such a carrier?

The results of gene testing can have both positive and negative consequences for those being tested and for their families. Who, then, should decide whether screening should be done, for which genes, on whom, and in which communities? A flippant answer might be, "There oughta be a law!" Should we leave ethical issues to judges and legislators? Should moral matters be decided by society or clergy? Or should they be personal decisions?

If gene testing *is* done, should the person being tested be told the results no matter what? If the affected person is an infant, should the parents always be told the results, even if the condition is poorly understood? How do we balance helping such children with the possibility of stigmatizing them?

Another issue to consider is that we live in a world of limited resources. In addition to deciding who should be tested, we must decide who should pay the bill. Both testing and treatment are expensive. Should testing be done only when treatment or preventive measures are available? How much say should the agent that pays for the procedure have in who is tested and who receives medical treatment?

These are not easy questions to answer, or even to think about, yet if we do not take part in the debate, we will be allowing others to decide these crucial issues for us.

procedures for diagnosing genetic problems in the fetus: amniocentesis and chorionic villi sampling (Figure 20.18). Although it is possible to look for more than 100 disorders with these procedures, tests are run only for those disorders that are common as well as any that are of particular concern in that pregnancy. Reassuring results from either form of prenatal testing cannot guarantee a healthy baby, despite the high degree of accuracy with which they can detect genetic disorders.

In **amniocentesis**, a needle is inserted through the lower abdomen into the uterus, and a small amount of amniotic fluid, 10 to 20 ml (about 2 to 4 tsp), is withdrawn. (Ultrasound is performed first, to find the safest spot for insertion, away from the fetus, umbilical cord, and placenta.) Floating in the amniotic fluid are living cells that have sloughed off the fetus. These cells are grown in the laboratory for a week or two and then are examined for abnormalities in the number of chromosomes and for the presence of certain alleles that are likely to cause specific diseases. Biochemical tests are also done on the fluid to look for certain chemicals that are indicative of problems. For example, a high level of alpha-fetoprotein, a substance produced by the fetus, suggests a problem with the development of

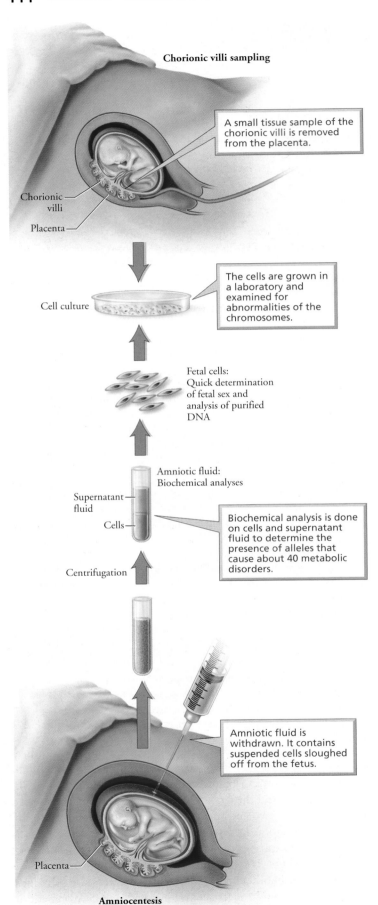

Chorionic villi sampling

A small tissue sample of the chorionic villi is removed from the placenta.

Chorionic villi

Placenta

Cell culture

The cells are grown in a laboratory and examined for abnormalities of the chromosomes.

Fetal cells:
Quick determination of fetal sex and analysis of purified DNA

Amniotic fluid:
Biochemical analyses

Supernatant fluid

Cells

Biochemical analysis is done on cells and supernatant fluid to determine the presence of alleles that cause about 40 metabolic disorders.

Centrifugation

Amniotic fluid is withdrawn. It contains suspended cells sloughed off from the fetus.

Placenta

Amniocentesis

FIGURE **20.18** Amniocentesis and chorionic villi sampling (CVS) are procedures available for prenatal genetic testing.

the central nervous system (a neural tube defect). Amniocentesis is generally safe for both the mother and the fetus, but there is a small risk of triggering a miscarriage or of the needle injuring the mother and causing infection or bleeding.

Amniocentesis is usually done between 15 and 20 weeks after the woman's last menstrual period, when there is enough amniotic fluid, about 250 ml (about 1 cup), to minimize the risk of injuring the fetus. (A few medical centers are now able to perform amniocentesis at 12 weeks of pregnancy.)

Chorionic villi sampling (CVS) removes and analyzes a small amount of tissue containing chorionic villi, the small, fingerlike projections of the part of the placenta called the chorion (see Chapter 18). Cells of the chorion have the same genetic material as those of the fetus. With the help of ultrasound, a small tube is inserted through the vagina and cervix to where the villi are located. Gentle suction is then used to remove a small tissue sample, which can be analyzed for genetic abnormalities.

There are pros and cons to CVS. Advantages are that it can be performed several weeks earlier in the pregnancy than amniocentesis, and the results are available within a few days. More than 95% of the high-risk women who opt for prenatal tests receive good news, and an early diagnosis saves weeks of worry over the health of the fetus. Also, if there is a genetic problem in the fetus and the couple wishes to terminate the pregnancy, the procedure can be performed earlier in the pregnancy, when it is safer for the mother. When abortion is not chosen, early diagnosis allows more time to plan the safest time, location, and method of delivery. A disadvantage of CVS is that it has a slightly greater risk of triggering miscarriage than does amniocentesis.

If detected early enough, certain birth defects can be prevented. Congenital adrenal hyperplasia (an overgrowth of the adrenal glands), for instance, will cause a female fetus to develop abnormal genitalia, unless she is treated with hormones from week 10 to week 16 of gestation. Early diagnosis with CVS can tell a physician whether hormone treatment is needed.

A simple blood test is now used routinely to screen newborns for phenylketonuria (PKU), an inherited metabolic disorder. People with PKU produce a defective enzyme that prevents them from converting phenylalanine (an amino acid in food) to tyrosine. As a result, they have too much phenylalanine in their bodies and too little tyrosine, an imbalance that somehow causes brain damage. Although nothing can be done to correct the enzyme, brain damage can be prevented with a strict diet that excludes most phenylalanine. Nearly all proteins contain phenylalanine, so it is often necessary to substitute a specially prepared mixture of amino acids for most protein-containing foods, such as meat. It is also necessary to avoid foods containing the artificial sweetener NutraSweet, which contains aspartame, consisting of the amino acids phenylalanine and aspartic acid. When aspartame is digested, phenylalanine is separated from aspartic acid and can reach dangerous levels.

Many predictive genetic tests are now available or are being developed. These tests identify people who are at risk of getting a disease but do not yet have any symptoms. They are simple tests that can usually be done with a small blood sample. When steps can be taken to prevent the disease, predictive gene tests can be lifesaving. For instance, colon cancer will develop in nearly everyone who has the alleles for familial adenomatous polyposis, a condition in which thousands of benign polyps grow in the intestine. If the colon is routinely inspected for polyps and any polyps are then removed, cancer can be prevented. Genetic counselors help people understand their risks of developing inherited disorders and the medical consequences of such disorders.

Other predictive gene tests look for alleles that might predispose a person to a disorder. A protein that transports cholesterol in the blood, called ApoE, comes in three forms, each specified by a different allele. Having two alleles for ApoE-2, one form of the protein, causes catastrophically high blood cholesterol levels, which can lead to heart attack and stroke. Knowing that a person has this genetic makeup, a physician could prescribe medication to lower blood cholesterol.

what would you do?

Having two copies of another *ApoE* allele, *ApoE-4*, increases a person's risk of heart disease by 30% to 50%. It also nearly guarantees that the person will develop Alzheimer's disease by 80 years of age. Alzheimer's disease is an untreatable condition in which brain tissue degenerates (see Chapter 7), gradually robbing the victim of memories, of the ability to function normally in society, and eventually of life itself. Suppose that a gene test that is performed out of concern for a person's risk of heart disease reveals the presence of two copies of *ApoE-4*. In your opinion, is a physician with this knowledge morally obligated to tell the patient about the inevitability of Alzheimer's disease, or should factors in the patient's life be considered in making the decision of whether to tell? If your genes were tested and found to have two copies of *ApoE-4*, would you want to be told? Why or why not?

HIGHLIGHTING THE CONCEPTS

Principles of Inheritance Help Us Predict How Simple Traits Are Passed to the Next Generation (pp. 431–441)

1. Different forms of a gene are called alleles. An individual who has two of the same alleles is said to be homozygous. An individual with two different alleles for a gene is said to be heterozygous. The allele that is expressed in the heterozygous condition is described as being dominant. The allele that is masked in a heterozygote is described as recessive.

2. The genotype for one or more traits consists of the specific alleles present in an individual. The phenotype refers to the observable expression of the trait or traits.

3. The alleles for each gene separate during gamete formation so that half the gametes receive one allele and the other half receive the other allele.

4. Each pair of alleles located on one kind of chromosome separate into gametes independently of the alleles of a gene pair located on a different kind of chromosome.

5. If a homozygous dominant individual (*AA*) mates with a homozygous recessive one (*aa*), the genotypes of all the offspring would be heterozygous (*Aa*) and the dominant phenotype. If heterozygous individuals (*Aa*) mate, the probable genotypes of children will be 1 in 4 (25%) homozygous dominant (*AA*):2 in 4 (50%) heterozygous dominant (*Aa*):1 in 4 (25%) homozygous recessive (*aa*). Their phenotypes will be 3 dominant:1 recessive.

6. Pedigrees, which are diagrams constructed to show the genetic relationships among individuals in an extended family, are often useful in determining the unknown genotypes of humans showing dominant phenotypes.

7. In many cases, the dominant allele produces a functional protein and the recessive allele produces a nonfunctional protein or no protein at all.

8. In complete dominance, the dominant allele completely masks the recessive allele. When two alleles are codominant, both are separately apparent in the phenotype. In incomplete dominance, the expression of the trait in a heterozygous individual is somewhere in between the expression of the trait in a homozygous dominant individual and the expression of the trait in a homozygous recessive individual.

9. Three or more alleles for a particular gene in a population are called multiple alleles. ABO blood types are determined by three alleles: I^A, I^B, and i. Blood type refers to the presence of certain polysaccharides on the surface of red blood cells.

10. Many traits, including height, skin pigmentation, and eye color, are determined by more than one gene (polygenic inheritance). Such traits show a wide range of variation. Moreover, when many genes are involved in determining a trait, the variation in expression can be continuous.

11. An X-linked gene is one that is located on the X chromosome and that has no corresponding allele on the Y chromosome. A recessive X-linked allele will always be expressed in a male, but it will be expressed only in the homozygous condition in a female. Red-green color blindness, hemophilia, and Duchenne muscular dystrophy are disorders caused by X-linked recessive alleles.

12. The expression of a sex-influenced trait depends on both the presence of the allele and the presence of sex hormones. Therefore, the expression of the allele depends on the person's sex. For example, the expression of the allele for male pattern baldness requires the presence of testosterone.

WEB TUTORIAL 20.1   One- and Two-Trait Crosses
WEB TUTORIAL 20.2   Codominance and Incomplete Dominance
WEB TUTORIAL 20.3   Sex-Linked Traits

Breaks in Chromosomes Change Chromosomal Structure and Function (pp. 441–442)

13. The loss of a piece of a chromosome is called a deletion. The gain of a piece of chromosome is called a duplication. Either chromosome abnormality can cause a genetic disorder.

Certain Genetic Disorders Can Be Detected by Laboratory Tests (pp. 442–445)

14. Tests such as amniocentesis and chorionic villi sampling (CVS) are available to determine whether a fetus is likely to develop a genetic disease. In amniocentesis, a sample of amniotic fluid is taken. The fluid and the fetal cells in the fluid are analyzed for genetic problems. CVS samples cells from the chorion of the placenta and analyzes them for genetic disorders.

KEY TERMS

gene   *p. 431*
trait   *p. 431*
allele   *p. 431*
homozygous   *p. 431*
heterozygous   *p. 431*
dominant   *p. 431*

recessive   *p. 431*
genotype   *p. 431*
phenotype   *p. 431*
pedigree   *p. 435*
carrier   *p. 437*
complete dominance   *p. 437*

codominance   *p. 437*
incomplete dominance   *p. 437*
pleiotropy   *p. 438*
multiple alleles   *p. 438*
polygenic inheritance   *p. 439*

X-linked genes   *p. 440*
amniocentesis   *p. 443*
chorionic villi sampling (CVS)
        p. 444

REVIEWING THE CONCEPTS

1. What is the ratio of genotypes and phenotypes in the offspring resulting from a cross between a homozygous dominant individual and a homozygous recessive one?   *pp. 433–434*
2. What is a pedigree? What can a family pedigree reveal about the inheritance of a trait?   *pp. 435–436*
3. Using an example, explain what is meant by codominance.   *p. 437*
4. Differentiate between multiple alleles and polygenic inheritance. *pp. 438–440*
5. What are linked genes? Why are they usually inherited together? What can cause such genes to become unlinked?   *p. 440*
6. Explain why the pattern of inheritance for recessive X-linked genes is different from the pattern for recessive autosomal alleles.   *pp. 440–441*
7. What two procedures are used for prenatal testing? How do they differ?   *pp. 442–444*
8. Which of the following crosses could produce offspring with the recessive phenotype?
   a. *AA × aa*
   b. *Aa × Aa*
   c. *AA × Aa*
   d. *AA × AA*

9. All of the following crosses will have a 50% probability of producing 50% of the offspring with a heterozygous genotype *except*
   a. *AA × aa*
   b. *Aa × Aa*
   c. *AA × Aa*
   d. *Aa × aa*
10. A trait controlled by many genes is described as being _____.
11. The _____ of an individual is the physical expression of one or more genes of interest, and the _____ is the set of alleles the person possesses for the gene or genes of interest.
12. The genotype of a person with two copies of the same allele is _____, the genotype of a person with two different alleles for a trait is _____.
13. Genes for different traits that are located on the same chromosome are described as being _____.

APPLYING THE CONCEPTS

1. Klaus has red-green color blindness, which is a sex-linked recessive trait. His wife, Helen, is homozygous for normal color vision. (Normal color vision is dominant to red-green color blindness.) What is the probability of their having a color-blind daughter? What is the probability of their having a color-blind son?
2. George and Sue have an infant, Sammy, who is lethargic, is vomiting, and has liver disease. Sammy is diagnosed with galactosemia, an autosomal recessive disorder in which the affected individuals are unable to metabolize the milk sugar galactose. Sammy is placed on a diet free of lactose and galactose and slowly recovers. George and Sue would like to have a second child. What are the chances that the second child will have galactosemia? (*Hint:* What are the genotypes of George and Sue? Use a Punnett square to determine the expected results of a cross with those genotypes.)

3. The parents of first cousins are siblings. Explain why the offspring of first cousins are more likely to have harmful recessive traits than are offspring of unrelated individuals.
4. A woman with blood type AB accuses a man with blood type O of fathering her child. The child has blood type AB. Could the man be the father? Why or why not?
5. Straight hair shows incomplete dominance over curly hair. A homozygous dominant person has straight hair; a heterozygote has wavy hair; and a homozygous recessive person has curly hair. The first child of a curly-haired person and a wavy-haired person has wavy hair. What is the probability that a second child would have wavy hair?

Additional questions can be found on the companion website.

DNA Is a Double Helix Consisting of Two Strings of Nucleotides

During Replication of DNA, Each Original Strand Serves as a Template for a New Strand

DNA Codes for RNA, Which Codes for Protein
- Transcription is RNA synthesis
- Translation is protein synthesis

Mutations Result from Nucleotide Substitution, Insertion, or Deletion

Gene Activity Can Be Turned On or Off
- Coiling and uncoiling of chromosomes regulate gene activity at the chromosome level
- Certain genes regulate the activity of other genes
- Chemical signals regulate gene activity

Genetic Engineering Is the Manipulation of DNA for Human Purposes
- Recombinant DNA is made of DNA from different sources
- Genetic engineering produces proteins of interest or transgenic organisms with desirable traits
- Gene therapy replaces faulty genes with functional genes

Genomics Can Be Used to Study How Genes Function and How Diseases Are Inherited
- The Human Genome Project sequenced a representative human genome
- Microarray analysis is a useful tool in genomics
- Comparing genomes of different species can be useful

HEALTH ISSUE Genetically Modified Food

SOCIAL ISSUE Forensic Science, DNA, and Personal Privacy

21
DNA and Biotechnology

A DNA fingerprint is based on the sequence of bases in a person's DNA. Thus, a DNA fingerprint is unique to each person (except for those who happen to have identical siblings).

"Can DNA evidence really be used to solve crimes that occurred decades ago?" asked Sarah, a student in the high school class that Detective Jim Young was speaking to.

"Yes, it's possible," Det. Young explained. "In the lab, DNA can be extracted from archived biological samples such as blood on clothing, saliva on a cigarette butt, or semen from a swab in a rape kit. Using methods that mimic natural DNA replication, we can multiply the sample and then characterize it. The tests focus on several sites in the DNA that are highly variable among individuals. If a suspect's DNA matches those sites, it is almost certain that he or she was the source of the original biological sample."

"But if a crime was committed years ago, who is considered a suspect?" asked Sean.

"Good question," Det. Young said. "Each case is unique, but one approach is to compare DNA evidence to a database of DNA samples collected from convicted criminals. A number of "cold" cases have been solved because the guilty party was later convicted of a different crime and became a part of the database."

Another hand went up. "But I thought that DNA evidence was used to get people out of jail, not to put them in jail," Erin commented.

Det. Young explained, "Those are the cases we hear about. Many innocent people have been released from jail after serving years for crimes they didn't commit. Because DNA evidence can help convict or to clear a person, DNA evidence can be the most powerful tool in my arsenal against crime."

In this chapter we will learn that DNA fingerprinting is possible because the DNA of each person (except for identical twins) is unique. We will become familiar with the structure and function of DNA and discover how this molecule is able to serve as the basis of our genetic inheritance as well as the source for the diversity of life on earth. We will learn that the importance of DNA on a personal level is that it directs the synthesis of specific polypeptides (proteins) that play structural or functional roles in our bodies. We will then consider the technology that our understanding of DNA has already made available and what possibilities such technology may hold for the future. ∎

DNA Is a Double Helix Consisting of Two Strings of Nucleotides

DNA is sometimes referred to as the thread of life—and a very slender thread it is. When DNA is unwound, it measures a mere 50-trillionths of an inch in diameter. Even so, if all the DNA strands in a single cell were fastened together end to end, the thread would stretch over 5 ft in length. DNA might also be considered the thread that ties all life together, because the DNA of organisms ranging from bacteria to humans is built from the same kinds of subunits. The order of these subunits encodes the information needed to make the proteins that build and maintain life.

Deoxyribonucleic acid, or **DNA**, is a double-stranded molecule resembling a ladder that is gently twisted to form a spiral called a double helix, as shown in Figure 21.1. Each side of the ladder, including half of each rung, is made from a string of repeating subunits called nucleotides. You may recall from Chapter 2 that a nucleotide is composed of one sugar (deoxyribose, in DNA), one phosphate, and one nitrogenous base. DNA contains four types of nitrogenous bases: adenine (A), guanine (G), thymine (T), and cytosine (C). The sides of the ladder are composed of alternating sugars and phosphates; the rungs consist of paired bases. Following the rules of **complementary base pairing**, adenine pairs *only* with thymine (an A–T pair), and cytosine pairs *only* with guanine (a C–G pair). Each base pair is held together by weak hydrogen bonds. The pairing of complementary bases is specific because of the shapes of the bases and the number of hydrogen bonds that can form between them. You may also recall from Chapter 2 that a molecule formed by the joining of nucleotides is called a nucleic acid. Thus, DNA is a nucleic acid.

Because of the specificity of base pairing, the bases on one strand of DNA are *always* complementary to the bases on the other strand. Thus, the order of bases on one strand determines the sequence of bases on the other strand. For instance, if the sequence of bases on one strand were CATATGAG, what would the complementary sequence be? Remember, cytosine (C) always pairs with guanine (G), and adenine (A) always pairs with thymine (T). As a result, the complementary sequence on the opposite strand would be GTATACTC.

The DNA within each human cell has an astounding 3 billion base pairs. Although the pairing of adenine with thymine and cytosine with guanine is specific, the sequence of bases throughout the length of different DNA molecules can vary in myriad ways. As we will see, genetic information is encoded in the exact sequence of bases.

During Replication of DNA, Each Original Strand Serves as a Template for a New Strand

For DNA to be the basis of inheritance, its genetic instructions must be passed from one generation to the next. Moreover, for DNA to direct the activities of each cell, its instructions must be present in every cell. These requirements dictate that DNA be copied before both mitotic and meiotic cell division (see Chapter 19). It is important that the copies be exact. The key to the precision of the copying process lies in the complementarity of the bases.

The copying process, or **DNA replication**, begins when an enzyme breaks the hydrogen bonds that hold together the two nucleotide strands of the double helix, thereby "unzipping" and unwinding the strands. As a result, the nitrogenous bases on the separated regions of each strand are exposed. Individual nucleotide bases, which are always present within the nucleus, can then attach to complementary bases on the open DNA strands. Enzymes called *DNA polymerases* link the sugars and phosphates of the newly

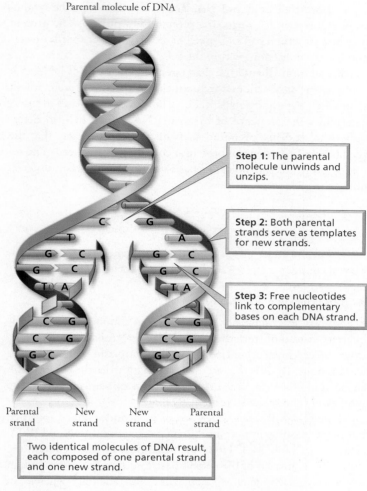

Parental molecule of DNA

Step 1: The parental molecule unwinds and unzips.

Step 2: Both parental strands serve as templates for new strands.

Step 3: Free nucleotides link to complementary bases on each DNA strand.

| Parental strand | New strand | New strand | Parental strand |

Two identical molecules of DNA result, each composed of one parental strand and one new strand.

FIGURE **21.2** DNA replication is called semiconservative because each daughter molecule consists of one "old" strand and one "new" strand.

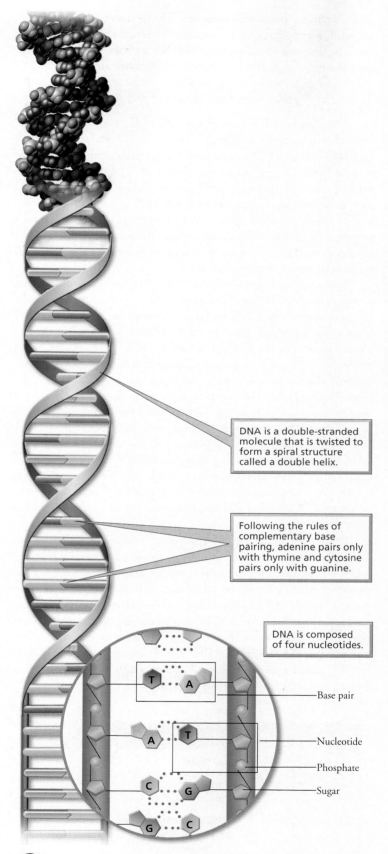

DNA is a double-stranded molecule that is twisted to form a spiral structure called a double helix.

Following the rules of complementary base pairing, adenine pairs only with thymine and cytosine pairs only with guanine.

DNA is composed of four nucleotides.

Base pair

Nucleotide

Phosphate

Sugar

FIGURE **21.1** DNA is a double-stranded molecule that twists to form a spiral structure called a double helix.

attached nucleotides together to form a new strand. As each of the new double-stranded DNA molecules forms, it twists into a double helix (Figure 21.2).

Each strand of the original DNA molecule serves as a template for the formation of a new strand. This process is called **semiconservative replication**, because in each of the new double-stranded DNA molecules, one original strand is saved (conserved) and the other strand is new. Look at Figure 21.2. Notice the original (parental) strand of nucleotides in each new molecule of DNA. Complementary base pairing creates two new DNA molecules that are identical to the parent molecule.

DNA Codes for RNA, Which Codes for Protein

The process of replication ensures that genetic information is passed accurately from a parent cell to daughter cells and from generation to generation. The next obvious question is, "How does DNA issue commands that direct cellular activities?" The answer is that DNA directs the synthesis of another nucleic acid, **ribonucleic acid**, or **RNA**. RNA, in turn, directs the synthesis of a polypeptide (a part of a protein) or a protein. The protein may be a structural part of the cell or an enzyme that speeds up certain chemical reactions within the cell.

Recall from Chapter 20 that a **gene** is a segment of DNA that contains the instructions for producing a specific protein (or in some cases, a specific polypeptide).[1] The sequence of bases in DNA determines the sequence of bases in RNA, which in turn determines the sequence of amino acids of a protein. We say that the gene is *expressed* when the protein it codes for is produced. The resulting protein is the molecular basis of the inherited trait; it determines the phenotype.

Gene expression:

DNA $\longrightarrow$ RNA $\longrightarrow$ Protein

To gain a fuller appreciation for how gene expression works, we will consider each step in slightly more detail.

■ Transcription is RNA synthesis

Just as the CEO of a major company issues commands from headquarters instead of from the factory floor, DNA issues instructions from the cell nucleus and not from the cytoplasm where the cell's work is done. RNA is the intermediary that carries the information encoded in DNA from the nucleus to the cytoplasm and directs the synthesis of the specified protein.

Like DNA, RNA is composed of nucleotides linked together, but there are some important differences between DNA and RNA, as shown in Table 21.1. First, the nucleotides of RNA contain the sugar ribose, instead of the deoxyribose found in DNA. Second, in RNA the nucleotide uracil (U) pairs with adenine, whereas in DNA thymine pairs with adenine. Third, most RNA is single-stranded. Recall that DNA is a double-stranded molecule.

The first step in converting the DNA message to a protein is to copy the message as RNA, a process called **transcription**.

DNA $\xrightarrow{\text{transcription}}$ RNA $\longrightarrow$ Protein

Three types of RNA are produced in cells. Each plays a different role in protein synthesis (Table 21.2). **Messenger RNA (mRNA)** carries DNA's instructions for synthesizing a particular protein from the nucleus to the cytoplasm. The order of bases in mRNA specifies the sequence of amino acids in the resulting protein, as we will see. Each **transfer RNA (tRNA)** molecule is specialized to bring a specific amino acid to where it can be added to a polypeptide that is under construction. **Ribosomal RNA (rRNA)** combines with proteins to form ribosomes, which are the structures on which protein synthesis occurs.

Transcription begins with the unwinding and unzipping of the specific region of DNA to be copied; these actions are performed by an enzyme. The DNA message is determined by the order of bases in the unzipped region of the DNA molecule. One of the unwound strands of the DNA molecule serves as the template. RNA nucleotides present in the nucleus pair with their complementary bases on the template—cytosine with guanine and uracil with adenine (Figure 21.3). The signal to start transcription is given by a specific sequence of bases on DNA, called the **promoter**. An enzyme called **RNA polymerase** binds with the promoter on DNA and then moves along the DNA strand, aligning the appropriate RNA nucleotides and linking them together. Another sequence of bases on the DNA serves as a stop signal. When RNA polymerase reaches this stop signal, transcription gradually ceases, and the newly formed strand of RNA, called the *RNA transcript*, is released from the DNA.

Messenger RNA usually undergoes certain modifications before it leaves the nucleus (Figure 21.4). Most stretches of DNA between a promoter and the stop signal include regions that do not contain codes that will be translated into protein. These unexpressed regions of DNA are called *introns*, short for *intervening sequences*. The regions of mRNA corresponding to the introns are snipped out of the newly formed mRNA strand by enzymes before the strand leaves the nucleus. The remaining segments of DNA or mRNA, called *exons* for *expressed sequences*, are the sequences that direct the synthesis of a protein.

TABLE 21.1 COMPARISON OF DNA AND RNA

	DNA	RNA
Similarities	Are nucleic acids	
	Are composed of linked nucleotides	
	Have a sugar-phosphate backbone	
	Have four types of bases	
Differences	Is a double-stranded molecule	Is a single-stranded molecule
	Has the sugar deoxyribose	Has the sugar ribose
	Contains the bases adenine, guanine, cytosine, and thymine	Contains the bases adenine, guanine, cytosine, and uracil (instead of thymine)
	Functions primarily in the nucleus	Functions primarily in the cytoplasm

TABLE 21.2 REVIEW OF THE FUNCTIONS OF RNA

MOLECULE	FUNCTIONS
Messenger RNA (mRNA)	Carries DNA's information in the sequence of its bases (codons) from the nucleus to the cytoplasm
Transfer RNA (tRNA)	Binds to a specific amino acid and transports it to be added, as appropriate, to a growing polypeptide chain
Ribosomal RNA (rRNA)	Combines with protein to form ribosomes (structures on which polypeptides are synthesized)

[1]We will develop this concept as the chapter proceeds. Some genes code for a polypeptide that is only part of a functional protein. A gene can also code for RNA that forms part of a ribosome or that transports amino acids during protein synthesis.

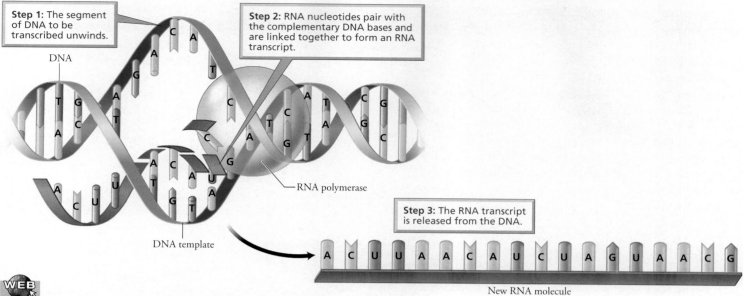

Step 1: The segment of DNA to be transcribed unwinds.

Step 2: RNA nucleotides pair with the complementary DNA bases and are linked together to form an RNA transcript.

DNA

RNA polymerase

DNA template

Step 3: The RNA transcript is released from the DNA.

ACUUAACAUCUAGUAACG

New RNA molecule

FIGURE **21.3** Transcription is the process of producing RNA from a DNA template.

■ Translation is protein synthesis

The newly formed mRNA carries the genetic message (transcribed from DNA) from the nucleus to the cytoplasm, where it is translated into protein at the ribosomes. Just as we might translate a message written in Spanish into English, **translation** converts the nucleotide language of mRNA into the amino acid language of a protein.

$$ \text{DNA} \xrightarrow{\text{transcription}} \text{RNA} \xrightarrow{\text{translation}} \text{Protein} $$

Before examining the process of translation, we should become more familiar with the language of mRNA.

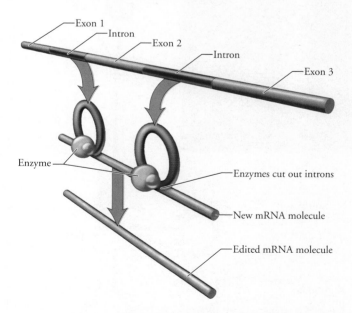

FIGURE **21.4** Newly formed messenger RNA is modified before it leaves the nucleus. Noncoding regions of DNA called introns are snipped out of the corresponding regions of mRNA molecule. Segments of mRNA that code for protein are then spliced together.

THE GENETIC CODE

To use any language, you must know what the words are and what they mean, as well as where sentences begin and end. The **genetic code** is the "language" of genes that translates the sequence of bases in DNA ultimately into the specific sequence of amino acids in a protein. We have seen that the sequence of bases in DNA determines the sequence of bases in mRNA through complementary base pairing. The "words" in the genetic code, called **codons**, are sequences of three bases on mRNA that specify 1 of the 20 amino acids or the beginning or end of the protein chain. All the codons of the genetic code are shown in Table 21.3. For instance, the codon UUC on mRNA specifies the amino acid phenylalanine. (The complementary sequence on DNA would be AAG.)

stop and think

Look at the strand of mRNA in Figure 21.3. Notice that the codon at the end of the mRNA strand is ACG. Which amino acid does this specify? (Use Table 21.3.)

The four bases in RNA (A, U, C, and G) could form 64 combinations of three-base sequences. The number of possible codons, therefore, exceeds the number of amino acids. As Table 21.3 indicates, however, there are several sets of codons that code for the same amino acid. Note, too, that the codon AUG can either serve as a start signal to initiate translation or can specify the addition of the amino acid methionine to the growing protein chain, depending on where it occurs in the mRNA molecule. In addition, three codons (UAA, UAG, and UGA) are stop codons that specify the end of a protein. If we think of the codons as genetic words, then a stop codon functions as the period at the end of the sentence.

TRANSFER RNA

A language interpreter translates a message from one language to another. Transfer RNA (tRNA) serves as an interpreter that converts

TABLE 21.3 THE GENETIC CODE

	Second base				
First base	**U**	**C**	**A**	**G**	**Third base**
U	UUU ⎤ Phenylalanine UUC ⎦ UUA ⎤ Leucine UUG ⎦	UCU ⎤ UCC ⎥ Serine UCA ⎥ UCG ⎦	UAU ⎤ Tyrosine UAC ⎦ UAA Stop UAG Stop	UGC ⎤ Cysteine UGC ⎦ UGA Stop UGG Tryptophan	U C A G
C	CUU ⎤ CUC ⎥ Leucine CUA ⎥ CUG ⎦	CCU ⎤ CCC ⎥ Proline CCA ⎥ CCG ⎦	CAU ⎤ Histidine CAC ⎦ CAA ⎤ Glutamine CAG ⎦	CGU ⎤ CGC ⎥ Arginine CGA ⎥ CGG ⎦	U C A G
A	AUU ⎤ AUC ⎥ Isoleucine AUA ⎦ AUG Methionine Start	ACU ⎤ ACC ⎥ Threonine ACA ⎥ ACG ⎦	AAU ⎤ Asparagine AAC ⎦ AAA ⎤ Lysine AAG ⎦	AGU ⎤ Serine AGC ⎦ AGA ⎤ Arginine AGG ⎦	U C A G
G	GUU ⎤ GUC ⎥ Valine GUA ⎥ GUG ⎦	GCU ⎤ GCC ⎥ Alanine GCA ⎥ GCG ⎦	GAU ⎤ Asparagine GAC ⎦ GAA ⎤ Glutamic Acid GAG ⎦	GGU ⎤ GGC ⎥ Glycine GGA ⎥ GGG ⎦	U C A G

the genetic message carried by mRNA into the language of protein, which is a particular sequence of amino acids. To accomplish this conversion, a tRNA molecule must be able to recognize both the codon on mRNA and the amino acid that the codon specifies (Figure 21.5).

There are many kinds of tRNA—at least one for each of the 20 amino acids. Each type of tRNA molecule binds to a particular amino acid. Enzymes ensure the specificity of this binding. The tRNA then ferries the amino acid to the correct location along a strand of mRNA.

How does the tRNA know the correct location? The location is determined by a sequence of three nucleotides on the tRNA called the **anticodon**. In a sense, the anticodon "reads" the language of mRNA by binding to a codon on the mRNA molecule according to the complementary base-pairing rules. When the tRNA's anticodon binds to the mRNA's codon, a specific amino acid is brought to the growing polypeptide chain. For example, a tRNA molecule with the anticodon AAA binds to the amino acid phenylalanine and ferries it to the mRNA molecule, where the codon UUU is presented for translation. Phenylalanine will then be added to the growing amino acid chain.

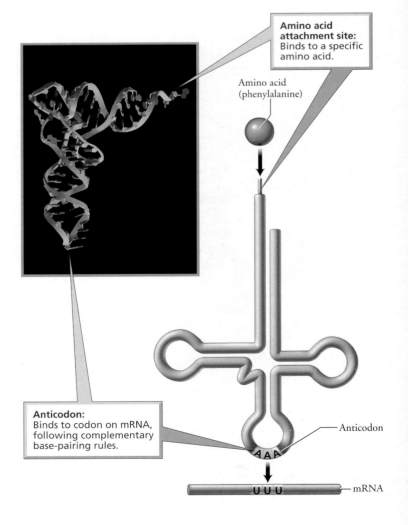

Amino acid attachment site: Binds to a specific amino acid.

Amino acid (phenylalanine)

Anticodon: Binds to codon on mRNA, following complementary base-pairing rules.

Anticodon

AAA

UUU — mRNA

FIGURE **21.5** A tRNA molecule is a short strand of RNA that twists and folds on itself. The job of tRNA is to ferry a specific amino acid to the ribosome and insert it in the appropriate position in the growing peptide chain.

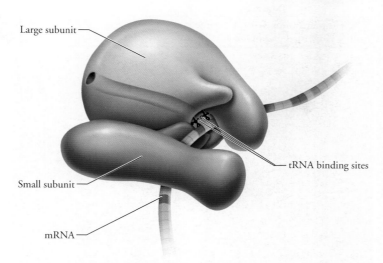

Large subunit

tRNA binding sites

Small subunit

mRNA

FIGURE **21.6** A ribosome consists of two subunits of different sizes. When the two subunits join together to form a functional ribosome, a groove for mRNA is formed. The ribosome has two binding sites for tRNA molecules. It also contains an enzyme that promotes the formation of a peptide bond between the amino acids that are attached to the tRNAs in the binding sites.

RIBOSOMES

Ribosomes function as the workbenches on which proteins are built from amino acids. A ribosome consists of two subunits (small and large), each composed of ribosomal RNA (rRNA) and protein (Figure 21.6). The subunits form in the nucleus and are shipped to the cytoplasm. They remain separate except during protein synthesis. The role of the ribosome in protein synthesis is to bring the tRNA bearing an amino acid close enough to the mRNA to interact. As you can see in Figure 21.6, when the two subunits fit together to form a functional ribosome, a groove for mRNA is formed. Two binding sites position tRNA molecules so that an enzyme in the ribosome can cause bonds to form between their amino acids.

PROTEIN SYNTHESIS

Translation—essentially, protein synthesis—can be divided into three stages: initiation, elongation, and termination (Figure 21.7).

1. **During initiation, the major players in protein synthesis (mRNA, tRNA, and ribosomes) come together.** First, the small ribosomal subunit attaches to the mRNA strand at the start codon, which is usually AUG. The tRNA with the complementary anticodon, UAC, quickly pairs with the start codon. The larger ribosomal subunit then joins the smaller one to form a functional, intact ribosome with mRNA positioned in a groove between the two subunits.

2. **Elongation of the protein occurs as additional amino acids are added to the chain.** With the start codon positioned in one binding site, the next codon is aligned in the other binding site. The tRNA bearing an anticodon that will pair with the exposed codon slips into place at the binding site, and the amino acid it carries binds to the previous amino acid with the assis-

tance of enzymes. The first tRNA, no longer attached to its amino acid, then leaves the ribosome. The ribosome moves along the mRNA molecule, positioning the next codon in the open site. An appropriate tRNA slips into the open site, and its amino acid binds to the previous one. The tRNA in the first binding site leaves the ribosome. This process is repeated many times, adding one amino acid at a time to the growing polypeptide chain.

Numerous ribosomes may glide along a given mRNA strand at the same time, each producing its own copy of the protein directed by that mRNA (Figure 21.8). As soon as one ribosome moves past the start codon, another ribosome can attach. A cluster of ribosomes simultaneously translating the same mRNA strand is called a **polysome**.

3. **Termination occurs when a stop codon moves into the ribosome.** There are no tRNA anticodons that pair with the stop codons, so when a stop codon moves into the ribosome, protein synthesis is terminated. The newly synthesized polypeptide, the mRNA strand, and the ribosomal subunits then separate from one another.

stop and think

Streptomycin is an antibiotic, a drug taken to slow the growth of invading bacteria and allow body defense mechanisms more time to destroy them. Streptomycin works by binding to the bacterial ribosomes and preventing an accurate reading of mRNA. Why would this process slow bacterial growth?

Mutations Result from Nucleotide Substitution, Insertion, or Deletion

DNA is remarkably stable, and the processes of replication, transcription, and translation generally occur with amazing precision. However, sometimes DNA is altered, and the alterations can change its message. Changes in DNA are called **mutations**. One type of mutation occurs when whole sections of chromosomes become rearranged, duplicated, or deleted, as discussed in Chapter 20. Now that we are familiar with the chemical structure of DNA and how it directs the synthesis of proteins, we can consider another type of mutation—a gene mutation. A gene mutation results from changes in the order of nucleotides in DNA. Although a gene mutation can occur in any cell, the only way it can be passed on to offspring is if it is present in a cell that will become an egg or a sperm. A mutation that occurs in a body cell can affect the functioning of that cell and the cells produced by that cell, sometimes with disastrous effects, but it cannot be transmitted to a person's offspring.

One type of gene mutation is a point mutation, which is the replacement of one nucleotide pair by a different nucleotide pair in the DNA double helix. During DNA replication, bases may accidentally pair incorrectly. For example, adenine might mistakenly pair with cytosine instead of thymine. Repair enzymes normally replace the incorrect base with the correct one. However, sometimes the enzymes recognize that the bases are incorrectly paired but mistakenly replace the original base (the one on the old strand) rather

Initiation: Translation Begins

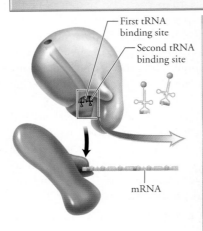

First tRNA binding site
Second tRNA binding site

mRNA

STEP 1
• The small ribosomal subunit joins to mRNA at the start codon, AUG

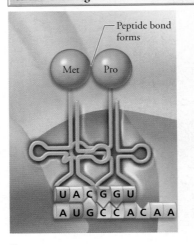

Pro

Met

GGU

UAC

AUGCCACAA

STEP 2
• A tRNA with a complementary anticodon pairs with the start codon
• Ribosomal subunits join to form a functional ribosome

Elongation: Amino Acids Are Added to the Growing Chain One at a Time

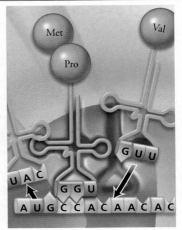

Peptide bond forms

Met Pro

UACGGU

AUGCCACAA

STEP 3
• A tRNA with the appropriate anticodon pairs with the next codon on mRNA
• Enzymes link the amino acids

Met

Pro

Val

GUU

UAC

GGU

AUGCCACAACAC

STEP 4
• The first tRNA leaves the ribosomes
• The ribosome moves along the mRNA, exposing the next codon
• Enzymes link the amino acids

FIGURE **21.7** Translation occurs in three steps: initiation, elongation, and termination.

than the new, incorrect one. The result is a complementary base pair consisting of the wrong nucleotides (Figure 21.9).

Other types of gene mutations are caused by the insertion or deletion of one or more nucleotides. Generally, a mutation of this kind has more serious effects than does a mutation caused by substitution. Recall that the mRNA is translated in units of three nucleotides (codons). If one or two nucleotides are inserted or deleted, *all* the triplet codons that follow the insertion or deletion are likely to change. As a consequence, mutations due to the insertion or deletion of one or two nucleotides can greatly change the resulting protein. A sentence consisting of three-letter words (rep-

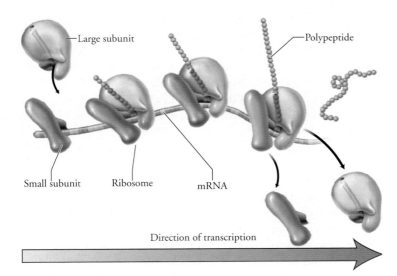

Large subunit

Polypeptide

Small subunit   Ribosome   mRNA

Direction of transcription

FIGURE **21.8** A polysome is a group of ribosomes reading the same mRNA molecule.

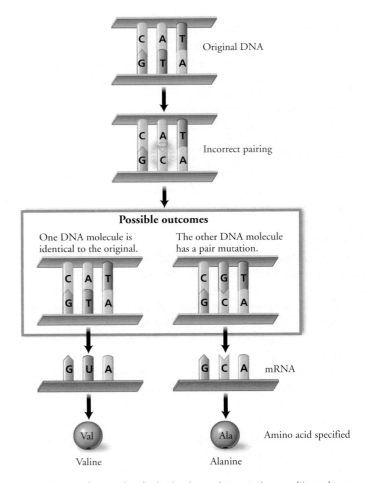

C A T
G T A
Original DNA

C A T
G C A
Incorrect pairing

Possible outcomes

One DNA molecule is identical to the original.

The other DNA molecule has a pair mutation.

C A T
G T A

C G T
G C A

G U A   mRNA

G C A   mRNA

Val

Ala   Amino acid specified

Valine

Alanine

FIGURE **21.9** A base-pair substitution is a point mutation resulting when a base is paired incorrectly. This may change the amino acid specified by the mRNA and alter the structure of the protein.

Termination: The Newly Synthesized Protein Is Released

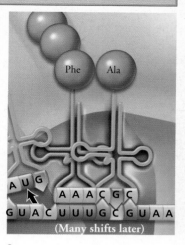

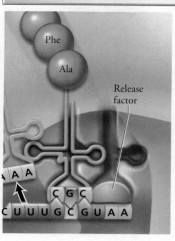

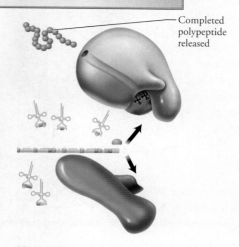

Completed polypeptide released

STEP 5
- The tRNA in the first binding site leaves the ribosome
- The ribosome moves along the mRNA, exposing the next codon
- Enzymes link the amino acids
- The process is repeated many times

STEP 6
- The stop codon moves into the ribosome

STEP 7
- Release factors cause the release of the newly formed polypeptide and the separation of the ribosomal subunits and the mRNA

resenting codons) illustrates what can happen. Deleting a single letter from the sentence "The big fat dog ran" renders the sentence nonsensical:

Original: THE BIG FAT DOG RAN
After deletion of the E in THE: THB IGF ATD OGR AN

Gene Activity Can Be Turned On or Off

At the time of your conception, you received one set of chromosomes from your father and one set from your mother. The resulting zygote then began a remarkable series of cell divisions—some of which continue in many of your body cells to this day. With each division, the genetic information was faithfully replicated, and exact copies were parceled into the daughter cells. Thus, every nucleated cell you possess, except eggs or sperm, contains a complete set of identical genetic instructions for making every structure and performing every function in your body.

How, then, can liver, bone, blood, muscle, and nerve cells look and act so differently from one another? The answer is deceptively simple: Only certain genes are active in a certain type of cell; most genes are turned off. Indeed, as cells become specialized for specific jobs, even the timing of the activity of specific genes is critical. The active genes produce specific proteins, which determine the structure and function of that particular cell. For example, genes for producing the hormone insulin are present in every cell, but those genes are active only in certain cells of the pancreas.

As is often the case in biology, the answer to one question prompts other questions. What controls gene activity? How are genes turned on or off? The answers to those questions are a bit more complex, because gene activity is controlled in several ways.

■ Coiling and uncoiling of chromosomes regulate gene activity at the chromosome level

At the chromosome level, gene activity is affected by the coiling and uncoiling of the DNA. When the DNA is tightly coiled, or condensed, the genes are not expressed. When a particular protein is needed in a cell, the region of the chromosome containing the necessary gene unwinds, allowing transcription to take place. Presumably, the uncoiling allows enzymes responsible for transcription to reach the DNA in that region of the chromosome. Other regions of the chromosome remain tightly coiled and are not expressed.

■ Certain genes regulate the activity of other genes

Gene activity can also be regulated by "master genes" that turn on a battery of other genes; the activation of those genes, in turn, leads to the production of proteins that cause the development of a particular cell type. For instance, a master gene in an immature muscle cell produces a protein called myoD1. MyoD1 then turns on other genes that produce the muscle-specific proteins needed to transform the immature cell into a muscle cell.

If we compare master genes to the ignition switch in a car, then other regions of DNA, called enhancers, are analogous to the accelerator. Enhancers are segments of DNA that increase the *rate* of transcription of certain genes and, therefore, the amount of a specific protein that is produced. Enhancers also specify the timing of expression and a gene's response to external signals and developmental cues that affect gene expression.

■ Chemical signals regulate gene activity

Chemical signals can also regulate gene activity. You may recall from Chapter 10 that one of the ways certain hormones bring about their effects is by turning on specific genes. Steroid hormones, for instance, bind to receptors within a target cell. The hormone-receptor complex then finds its way to the nuclear chromatin and turns on specific genes. One such complex turns on the genes in cells that produce facial hair—explaining why your father may have a beard but your mother probably does not, even though she has the necessary genes to grow one. In this case, the male hormone testosterone binds to a receptor and turns on hair-producing genes. Facial hair follicle cells of both men and women have the necessary testosterone receptors. However, women usually do not produce enough testosterone to activate the hair-producing genes, so bearded women are rare.

stop and think

Explain why female athletes who inject themselves with testosterone to stimulate muscle development can develop facial hair.

In addition to hormones that turn on specific genes, regulatory proteins can activate a gene. An activator protein, for instance, can bind to DNA somewhere near the promoter for a gene. The activator protein then helps to position and activate RNA polymerase so that transcription begins.

Genetic Engineering Is the Manipulation of DNA for Human Purposes

The manipulation of genetic material for human purposes, a practice called **genetic engineering**, began almost as soon as scientists began to understand the language of DNA. It is part of the broader endeavor of **biotechnology**, a field in which scientists make controlled use of living cells to perform specific tasks. Genetic engineering has been used to produce pharmaceuticals and hormones, improve diagnosis and treatment of human diseases, increase food production from plants and animals, and gain insight into the growth processes of cells.

■ Recombinant DNA is made of DNA from different sources

The basic idea behind genetic engineering is to put a gene of interest, one that produces a useful protein or trait, into another piece of DNA to create **recombinant DNA**, which is DNA combined from two or more sources. The recombinant DNA, carrying the gene of interest, is then placed into a rapidly multiplying cell that quickly produces many copies of the gene. The final harvest may consist of large amounts of the gene product or many copies of the gene itself. Let's take a closer look at the procedure one step at a time.

1. **The gene of interest is sliced out of its original organism and spliced into vector DNA.** Both the DNA originally containing the gene of interest and the **vector** DNA, which receives the transferred genes, are cut at specific places by a **restriction enzyme**. This is a type of enzyme that makes a staggered cut between specific base pairs in DNA, leaving several unpaired bases on each side of the cut. There are many kinds of restriction enzymes; each kind recognizes and cuts a different sequence of DNA. The stretch of unpaired bases produced on each side of the a cut is called a *sticky end* because of its tendency to pair with the single-stranded stretches of complementary base sequences on the ends of other DNA molecules that were cut with the same restriction enzyme (Figure 21.10).

The sticky ends are the secret to splicing the gene of interest and the vector DNA. The sticky ends of DNA from different sources will be complementary and stick together as long as they have been cut with the same restriction enzyme. The initial attachment between sticky ends is temporary, but the ends can be "pasted" together permanently by another enzyme, DNA ligase. The resulting recombinant DNA contains DNA from two sources.

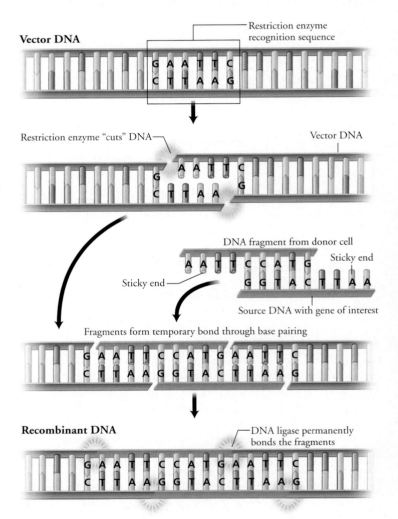

FIGURE **21.10** DNA from different sources can be spliced together using a restriction enzyme to make cuts in the DNA. A restriction enzyme makes a staggered cut at a specific sequence of DNA, leaving a region of unpaired bases on each cut end. The region of single-stranded DNA at the cut end is called a sticky end, because it tends to pair with the complementary sticky end of any other piece of DNA that has been cut with the same restriction enzyme, even if the pieces of DNA came from different sources.

2. **The recombinant DNA is transferred to a new host cell by the vector.** Biological carriers that ferry the recombinant DNA to a host cell are called vectors. For example, some bacteria have **plasmids**, small, circular pieces of self-replicating DNA that exist separately from the bacterial chromosome,[2] and they are commonly used as vectors to carry a gene of interest into bacteria. The source DNA (that is, the source of the gene of interest) and the plasmid (vector) DNA are both treated with the same restriction enzyme. Afterward, fragments of source DNA, some of which will contain the gene of interest, will be incorporated into plasmids when their sticky ends join. The recombinant DNA is mixed with bacteria in a test tube. Under the right conditions, some of the bacterial cells will then take up the recombined plasmids.

Although the basic strategy is usually the same, there are many variations on this theme of transporting a gene into a new host. For instance, the gene of interest is sometimes combined with viral DNA. The viruses are then used as vectors to insert the recombinant DNA into a host cell. Hosts other than bacteria, including yeast or animal cells, can also be used.

3. **The recombinant organism containing the gene of interest is identified and isolated from the mixture of recombinants.** Each recombinant plasmid is introduced into a single bacterial cell, and each cell is then grown into a colony. Each colony contains a different recombinant plasmid. The bacteria containing the gene of interest must be identified and isolated.

4. **The gene is amplified through bacterial cloning or by use of a polymerase chain reaction.** After the colony containing the gene of interest has been identified, researchers usually amplify (that is, replicate) the gene, producing numerous copies. Gene amplification is accomplished using one of two techniques: bacterial cloning or a polymerase chain reaction.

Bacteria containing the plasmid with the gene of interest can be grown in huge numbers by cloning. Each bacterium divides many times to form a colony. Thus, each colony is a clone—a group of genetically identical organisms all descended from a single cell. In this case, all the members of the clone carry the same recombinant DNA. The general genetic engineering procedure, including bacterial cloning, is summarized in Figure 21.11. Later, the plasmids can be separated from the bacteria, a process that partially purifies the gene of interest. The plasmids then can be taken up by other bacteria that will thus become capable of performing a service deemed useful by humans, such as cleaning up oil spills. Alternatively, the plasmids can be transferred into plants or animal cells—creating transgenic organisms.

The second technique for increasing the amount of specific DNA is the **polymerase chain reaction (PCR)** (Figure 21.12). In a PCR, the DNA of interest is unzipped, by gentle heating, to form single strands. The single strands, which will serve as templates, are then mixed with primers—special short pieces of nucleic acid—

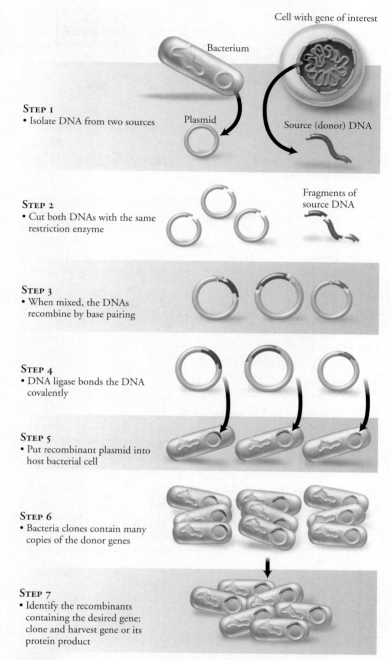

STEP 1
• Isolate DNA from two sources

Cell with gene of interest

Bacterium

Plasmid

Source (donor) DNA

STEP 2
• Cut both DNAs with the same restriction enzyme

Fragments of source DNA

STEP 3
• When mixed, the DNAs recombine by base pairing

STEP 4
• DNA ligase bonds the DNA covalently

STEP 5
• Put recombinant plasmid into host bacterial cell

STEP 6
• Bacteria clones contain many copies of the donor genes

STEP 7
• Identify the recombinants containing the desired gene; clone and harvest gene or its protein product

FIGURE **21.11** An overview of genetic engineering using plasmids

one primer with bases complementary to each strand. The primers serve as start tags for DNA replication. Nucleotides and a special heat-resistant DNA polymerase, which promotes DNA replication, are also added to the mixture, which is then cooled to allow base pairing. Through base pairing, a complementary strand forms for each single strand. The procedure is then repeated many times, and each time the number of copies of the DNA of interest is doubled. In this way, billions of copies of the DNA of interest can be produced in a short time.

[2]Plasmids seem to have evolved as a means of moving genes between bacteria. A plasmid can replicate itself and pass, with its genes, into another bacterium.

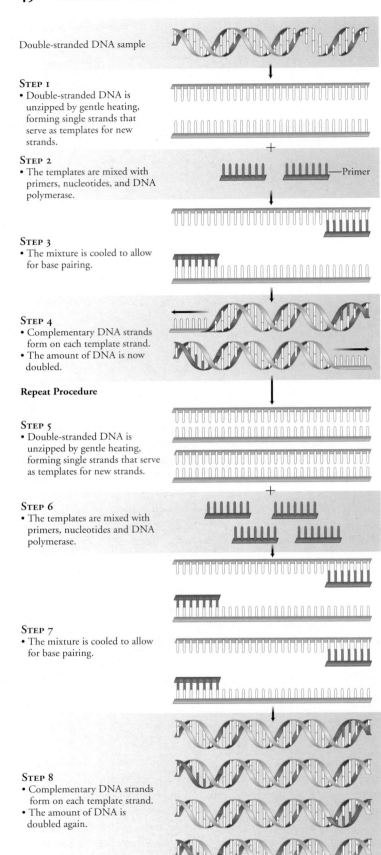

Double-stranded DNA sample

STEP 1
- Double-stranded DNA is unzipped by gentle heating, forming single strands that serve as templates for new strands.

STEP 2
- The templates are mixed with primers, nucleotides, and DNA polymerase.

Primer

STEP 3
- The mixture is cooled to allow for base pairing.

STEP 4
- Complementary DNA strands form on each template strand.
- The amount of DNA is now doubled.

Repeat Procedure

STEP 5
- Double-stranded DNA is unzipped by gentle heating, forming single strands that serve as templates for new strands.

STEP 6
- The templates are mixed with primers, nucleotides and DNA polymerase.

STEP 7
- The mixture is cooled to allow for base pairing.

STEP 8
- Complementary DNA strands form on each template strand.
- The amount of DNA is doubled again.

The procedure is repeated many times, doubling the amount of DNA with each round.

FIGURE **21.12** The polymerase chain reaction (PCR) rapidly produces a multitude of copies of a single gene or of any desired segment of DNA. PCR amplifies DNA more quickly than does bacterial cloning. It has many uses besides genetic engineering, including DNA fingerprinting.

what would you do?

Genetic engineering involves altering an organism's genes—adding new genes and traits to microbes, plants, or even animals. Do you think we have the right to "play God" and alter life forms in this way? The U.S. Supreme Court has approved patenting of genetically engineered organisms, first microbes but now mammals as well. If you were asked to decide whether it is ethical to patent a new life form, how would you respond?

■ **Genetic engineering produces proteins of interest or transgenic organisms with desirable traits**

Genetic engineering has been used in two general ways.

- **Genetic engineering provides a way to produce large quantities of a particular gene product.** The useful gene is transferred to another cell, usually a bacterium or a yeast cell, that can be grown easily in large quantities. The cells are cultured under conditions that cause them to express the gene, after which the gene product is harvested. For example, genetically engineered bacteria have been used to produce large quantities of human growth hormone (Figure 21.13). Treatment with growth hormone allows children with an underactive pituitary gland to grow to nearly normal height.

- **Genetic engineering allows a gene for a trait considered useful by humans to be taken from one species and transferred to another species.** The transgenic organism, an organism containing genes from another species, then exhibits the desired trait. For example, scientists have endowed salmon with a gene from an eel-like fish. This gene causes the salmon to produce growth hormone year-round. As a result, the salmon grow faster than normal.[3]

ENVIRONMENTAL APPLICATIONS

Genetic engineering also has environmental applications. In sewage treatment, genetically engineered microbes lessen the amount of phosphate and nitrate discharged into waterways. Phosphate and nitrate can cause excessive growth of aquatic plants, which could choke waterways and dams, and of algae, which can produce chemicals that are poisonous to fish and livestock. Microorganisms are also being genetically engineered to modify or destroy chemical wastes or contaminants so that they are no longer harmful to the environment. For instance, oil-eating microbes that can withstand the high salt concentrations and low temperatures of the oceans have proven useful in cleaning up after marine oil spills.

LIVESTOCK

Genetic engineering has also been used on livestock. Genetically engineered vaccines have been created to protect piglets against a

---

[3]In mid-2006, the Food and Drug Administration was considering approval for genetically modified salmon to be marketed.

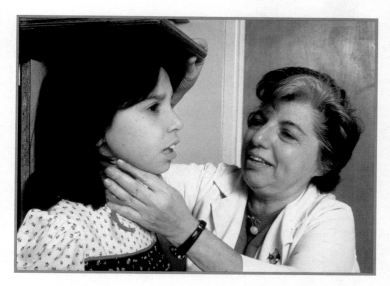

FIGURE **21.13** Genetic engineering is used to produce large quantities of a desired protein or to create an organism with a desired trait. This girl has an underactive pituitary gland. Its undersecretion of growth hormone would have caused her to be very short, even as an adult. However, growth hormone from genetically engineered bacteria has helped her grow to almost a normal height.

form of dysentery called scours, sheep against foot rot and measles, and chickens against bursal disease (a viral disease that is often fatal). Genetically engineered bacteria produce bovine somatotropin (BST), a hormone naturally produced by a cow's pituitary gland that enhances milk production. Injections of BST can boost milk production by nearly 25%.

Transgenic animals have been created by injecting a fertilized egg with the gene of interest in a petri dish. The goals of creating transgenic animals include making animals with leaner meat, sheep with softer wool, cows that produce more milk, and animals that mature more quickly than normal.

PHARMACEUTICALS

Genes have been put into a variety of cells, ranging from microbes to mammals, to produce proteins for treating allergies, cancer, heart attacks, blood disorders, autoimmune disease (see Chapter 13), and infections. Among these products are hormones, including insulin for treating diabetes and human growth hormone.

Genetically engineered bacteria have also been used to create vaccines for humans. You may recall from Chapter 13 that a vaccine typically uses an inactivated bacterium or virus to stimulate the body's immune response to the active form of the organism. The idea is that the body will learn to recognize proteins on the surface of the infectious organism and mount defenses against any organism bearing those proteins. Because the organism used in the vaccine was rendered harmless, the vaccine cannot trigger an infection. Scientists produce genetically engineered vaccines by putting the gene that codes for the surface protein of the infectious organism into bacteria. The bacteria then produce large quantities of that protein, which can be purified and used as a vaccine. The vaccine cannot cause infection because only the surface protein is used.

Plants have also been used to produce therapeutic proteins. Engineered bananas that produce an altered form of the hepatitis B virus surface protein are being developed as an edible vaccine against the liver disease hepatitis B. Someday, you may eat a banana in order to be vaccinated, instead of receiving an injection against hepatitis B. Plants are also being engineered to produce "plantibodies," antibodies made by plants. For example, soybeans are being cultivated that contain human antibodies to the herpes simplex virus that causes genital herpes. A human gene for an antibody that binds to tumor cells has been transplanted into corn. The antibodies can then deliver radioisotopes to cancer cells, selectively killing them.

Pharming is a word that comes from the combination of the words farming and pharmaceuticals. In gene pharming, transgenic animals are created that produce a protein with medicinal value in their milk, eggs, or blood. The protein is then collected and purified for use as a pharmaceutical. When the pharm animal is a mammal, the gene is expressed in mammary glands. The desired protein is then extracted and purified from milk (Figure 21.14). For example, the gene for the protein alpha-1-antitrypsin (AAT) has been inserted into sheep that then secrete AAT in their milk. People with an inherited, potentially fatal form of emphysema (a lung disease) take AAT as a drug. It is also being tested as a drug to prevent lung damage in people with cystic fibrosis. The first drug made from a transgenetic goat, an anticlotting drug called ATryn, gained approval for use in 25 European countries in mid-2006. It is given to people with a blood-clotting deficiency who must undergo surgery. At about the same time researchers announced that they had created a transgenic goat to produce milk containing lysozyme, an antibacterial agent. Lysozyme can be used to treat intestinal infections that kill millions of children in underdeveloped countries.

Researchers are currently working on a malaria vaccine that will be expressed in goat's milk. Malaria affects 300 million to 500 million people a year (discussed in Chapter 13a). If the researchers are successful, eight goats could produce enough vaccine to inoculate 20 million people.

what would you do?

Some ecologists are concerned about the harm to the environment or humans that could occur if genetically modified plants or animals escaped from the places where they were bred. The U.S. Department of Agriculture (USDA) has approved tests of plants that are genetically altered to produce pharmaceuticals. Most of these tests have been conducted in Hawaii. Environmental groups, worried that tradewinds might carry pollen from some of these biopharm plants out of the test area and the altered genes might spread to native plants and animals. This spread could threaten Hawaii's fragile biodiversity and harm humans. In 2006, The U.S. District judge in Hawaii ruled the USDA should have determined whether the gene-altered crops posed a threat to any of Hawaii's plants or animals before approving the tests. The genetically modified corn and sugar cane had already been harvested, because the experiments using them were finished. What should be done now? Do you think that the potential benefits of genetically engineered plants and animals outweigh the harm that might occur if the genetically altered organisms escaped? Would you object if a biopharm were developed near your home? What are your reasons?

RAW MATERIALS

Genetic engineering is also used to produce useful new materials. For example, spider silk is five times stronger than steel and still lightweight. Attempts have been made to farm spiders for their silk, but spiders are too aggressive to live close together. It is hoped that when goats that have been genetically engineered to possess the gene for spider silk reproduce, they will create a herd of goats that will secrete spider silk in their milk (Figure 21.15). The spider silk

FIGURE **21.15** This transgenic goat has the gene for making spider silk, one of the strongest substances known. The spider silk protein can be extracted from the goat's milk and spun into threads that can be used for products in which strength and light weight are important qualities.

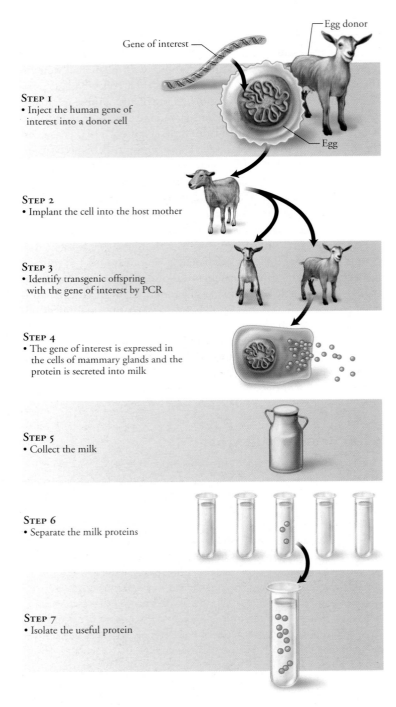

STEP 1
• Inject the human gene of interest into a donor cell

Gene of interest

Egg donor

Egg

STEP 2
• Implant the cell into the host mother

STEP 3
• Identify transgenic offspring with the gene of interest by PCR

STEP 4
• The gene of interest is expressed in the cells of mammary glands and the protein is secreted into milk

STEP 5
• Collect the milk

STEP 6
• Separate the milk proteins

STEP 7
• Isolate the useful protein

FIGURE **21.14** The procedure for creating a transgenic animal that will produce a useful protein in its milk

proteins could then be spun into a fine thread called BioSteel, which could be used for bullet-proof clothing, thinner surgical thread, and stronger yet lighter racing cars and aircraft.

AGRICULTURE

We experience some of the results of genetic engineering at our dinner tables (see the Health Issue essay, *Genetically Modified Food*). The most common traits that have been genetically engineered into crops are resistance to pests and resistance to herbicides. Scientists also developed two virus-resistant strains of papaya and distributed them to papaya growers in Hawaii, saving the industry from ruin. In addition, different strains of rice have been genetically engineered to resist disease-causing bacteria and to withstand flooding of the paddy. Other plants have been genetically engineered to be more nutritious. For example, golden rice is a strain of rice that has been genetically engineered to produce high levels of beta-carotene, which is in short supply in certain parts of the world. Other crops have also been created that grow faster, produce greater yields, and have longer shelf lives.

■ Gene therapy replaces faulty genes with functional genes

The problems associated with many genetic diseases arise because a mutant gene fails to produce a normal protein product. The goal of **gene therapy** is to cure genetic diseases by putting normal, functional genes into the body cells that were affected by the mutant gene. The functional gene would then produce the needed protein.

METHODS OF DELIVERING A HEALTHY GENE

One way a healthy gene can be transferred to a target cell is by means of viruses. Viruses generally attack only one type of cell. For instance, an adenovirus, which causes the common cold, typically attacks cells of the respiratory system. A virus consists largely of genetic material, commonly DNA, surrounded by a protein coat (see Chapter 13a). Once inside a cell, the viral DNA uses the cell's

Genetically Modified Food

From dinner tables to diplomatic circles, people are now discussing genetically modified (GM) food. This sudden interest is somewhat ironic, considering that people in the United States have been eating GM food since the mid-1990s. More than 70% of processed foods sold in the United States contain genetically modified ingredients. Yet many people vehemently object to GM food.

Why is something as common as GM food controversial? The concerns about it can generally be divided into three categories: health issues, social issues, and environmental issues (Table 21.A). Let's explore these categories one at a time.

Health concerns A panel of the National Academy of Sciences (NAS) has issued a report saying that genetically engineered crops do not pose health risks that cannot also be caused by crops created by conventional breeding. However, because genetic engineering could produce unintended harmful changes in food, the NAS panel recommends scrutiny of GM foods before they can be marketed. Currently, the U.S. Department of Agriculture, the Food and Drug Administration (FDA), and the Environmental Protection Agency regulate genetically modified foods. The NAS panel concluded that the GM foods already on the market are safe.

A common safety concern is that GM foods may contain allergens (substances that cause allergies). After a protein is produced, the cell modifies it in various ways. The protein may be modified in the genetically modified plant differently from the way it would in an unmodified cell, and the modification could produce an allergen. The likelihood that this may occur is reduced by rigorous testing. Most known allergens share certain properties. They are proteins, relatively small molecules, and resistant to heat, acid, and digestion in the stomach. If a protein produced by a GM plant has any of the properties typical of an allergen or is structurally similar to a known allergen, the FDA considers it to be a potential allergen and requires that the protein undergo additional allergy testing.

Bacterial resistance to antibiotics is likely to be a major threat to public health in this decade (see Chapter 13a). When bacteria are resistant to an antibiotic, the drug will not kill them and thus will no longer cure the human disease for which they are the cause. Some people worry about the scientific practice of putting genes for resistance to an antibiotic into GM crops as markers to identify the plants with the modified genes. Plant seedlings thought to be genetically modified are grown in the laboratory in the presence of an antibiotic. Only those seedlings with the gene for resistance will survive. Because of the way they were engineered, the surviving plants also contain the "useful" gene.

What worries some people is that the genes for resistance to antibiotics could be transferred to bacteria, making the bacteria resistant to antibiotics. The receiving bacteria might be those that normally live in the human digestive system, or they might be bacteria ingested with food. It is not known whether genes can be transferred from a plant to a bacterium. However, it is known that bacteria can easily and quickly transfer genes for antibiotic resistance to one another. Thus, a harmless bacterium in the gut could transfer the gene for antibiotic resistance to a disease-causing bacterium.

The transfer of antibiotic-resistance genes from GM plants to bacteria could have serious consequences. For this reason, antibiotic-resistance marker genes are being phased out in favor of other marker genes, such as a green fluorescent protein. Scientists also have developed a way to inactivate the antibiotic-resistance gene if it were to be transferred to bacteria.

Environmental concerns Proponents of GM foods argue that herbicide-resistant and pesticide-resistant crops reduce the need for spraying with herbicides and pesticides. So far, experience has shown that the validity of this argument depends on the crop. Pest-resistant cotton has substantially reduced the use of pesticides, but pest-resistant corn probably has not. Farmers who grow herbicide-resistant crops still spray with herbicides, but they change the type of herbicide they use to a type that is less harmful to animals.

Unfortunately, engineered crops containing insecticides could have undesirable effects on insects. Genetically engineering insecticides into plants could hasten the development of insect resistance to that insecticide, making the insecticide ineffective—not just for the genetically modified crop but for all crops.

Another concern is that genetically modified organisms could harm other organisms. One example is that pollen from some pest-resistant corn has been shown to harm monarch butterfly caterpillars. Fortunately, monarch caterpillars rarely encounter enough pollen to be harmed, and most of the pest-resistant corn grown in the United States today does not produce pollen that is harmful to monarch butterflies. A second example of a genetically modified organism that may have the potential to harm other organisms is the salmon engineered to produce more growth hormone and thus to grow several times faster than wild

TABLE 21.A CONCERNS ABOUT GENETICALLY MODIFIED FOOD

Health concerns	Is GM food safe for human consumption? Could GM crops increase bacterial resistance to antibiotics?
Environmental concerns	What effects will GM crops have on the level of use of pesticides? Can GM crops harm other organisms? Will GM crops become "superweeds"?
Social concerns	Are GM foods the proper approach to reducing world hunger?

continued →

salmon do. The GM salmon would be raised on fish farms. If the FDA grants approval, the gene-altered salmon could dramatically cut costs for fish farmers and consumers. However, when the genetically modified salmon are grown in tanks with ordinary salmon and food is scarce, the genetically modified salmon eat most of the food *and* some of their ordinary companions. What would happen if the genetically modified salmon escaped from their pens on the fish farm? They might mate with the wild salmon and create less healthy offspring or outcompete wild salmon for food, which could eventually cause the extinction of the wild salmon. During the past few years, hundreds of thousands of fish have escaped from fish farms when their floating pens were damaged by storms or sea lions. To minimize the risk that genetically modified salmon could destroy the population of wild salmon, scientists plan to breed the fish inland, sterilize the offspring, and ship only sterile fish to coastal pens. The sterilization procedure is effective in small batches of fish but has not yet been proved completely effective in large batches of fish.

Critics of GM foods also fear that crops genetically engineered to resist herbicides could become "superweeds" that could not be controlled with existing chemicals. In Canada, canola crops in several areas have been genetically modified, each for resistance to a different herbicide. Because pollen from canola plants can spread about 8 km (5 mi.), neighboring crops have cross-pollinated, producing seeds that are resistant to many herbicides. Could seeds left behind after a harvest grow and begin to invade other areas? Indeed, in mid-2006, genetically-engineered grass was found growing in the wild, near the location where a field test on the grass was performed a few years earlier.

Social concerns   Proponents of GM food claim that GM food can assist in the battle against world hunger. We have seen that genetic engineering can produce crops that resist pests and disease. It can also produce crops with greater yields and crops that will grow in spite of drought, depleted soil, or excess salt, aluminum, or iron. Foods can also be genetically modified to contain higher amounts of specific nutrients. One example is the "golden rice" mentioned in this chapter. More than 100 million children worldwide suffer from vitamin A deficiency, and 500,000 of them go blind every year because of that deficiency. Although golden rice cannot supply a complete recommended daily dose of vitamin A, the amount it contains could be helpful to a person whose diet is very low in vitamin A.

Critics of using GM food to battle world hunger argue that the problem of hunger has nothing to do with an inability to produce enough food. The problem, they say, is a social one of distributing food so that it is available to the people who need it.

Do you consider GM food to be a blessing or a danger to the world? What are your reasons? 👥

---

metabolic machinery to produce viral proteins. If a healthy gene is spliced into the DNA of a virus that has first been rendered harmless, the virus will deliver the healthy gene to the host cell and cause the desired gene product to be produced (Figure 21.16).

Another type of virus used in gene therapy is a retrovirus, a virus whose genetic information is stored as RNA rather than DNA. Once inside the target cell, a retrovirus rewrites its genetic information as double-stranded DNA and inserts the viral DNA into a chromosome of the target cell.

EARLY RESULTS OF GENE THERAPY RESEARCH

More than 4000 human diseases have been traced to defects in single genes. Although the Food and Drug Administration has not yet approved a gene therapy for any of these conditions, hundreds of clinical trials of gene therapies are currently under way, including trials of possible therapies for cystic fibrosis and cancer (discussed in Chapter 21a).

The first condition to be treated experimentally with gene therapy was a disorder referred to as severe combined immunodeficiency disease (SCID). The immune system of children with SCID is nonfunctional, leaving them vulnerable to infections. The cause of the problem is a mutant gene that prevents the production of an enzyme called adenosine deaminase (ADA). Without ADA, white blood cells never mature but die while still developing in the bone marrow. The first gene therapy trial began in 1990, when white blood cells of a 4-year-old SCID patient, Ashanthi DeSilva, were genetically engineered to carry the ADA gene and then returned to her tiny body. Her own gene-altered white blood cells began producing ADA, and her body defense mechanisms were strengthened. Ashanthi's life began to change. She was not ill as often as she had been before. She could play with other children. However, the life span of white blood cells is measured in weeks, and when the number of gene-altered cells declined, new gene-altered cells had to be infused. Ashanthi is now a teenager with a reasonably healthy immune system. However, she still needs repeated treatments.

French scientists believe they have cured 10 children with X-SCID using gene therapy (Figure 21.17). X-SCID is a severe combined immunodeficiency syndrome caused by a mutant gene on the X-chromosome. It is still too soon to know for certain whether all 10 children will require treatment in the future. Three children in the French studies developed leukemia from the therapy.

RISKS OF GENE THERAPY

Unfortunately, gene therapy using viruses has resulted in the death of an 18-year-old boy. Jessie Gelsinger died from a reaction to gene therapy treatment while participating in a clinical trial in 1999. Also, as mentioned above, three toddlers in France developed leukemia, as a result of gene therapy that used retroviruses to treat X-SCID.

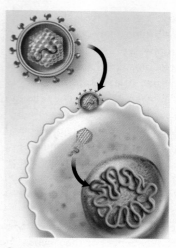

STEP 1
Incorporate a healthy form of the gene into the virus.

STEP 2
Remove bone marrow stem cells from the patient.

STEP 3
Infect the patient's stem cells with the virus that is carrying the healthy form of the gene.

STEP 4
Return the genetically engineered stem cells to the patient. The gene is expressed to produce the needed protein.

FIGURE **21.16** Gene therapy using a virus. In gene therapy, a healthy gene is introduced into a patient who has a genetic disease caused by a faulty gene.

what would you do?

Without treatment, children with SCID die at a young age. The gene therapy treatment for X-SCID that caused leukemia in three French boys does appear to have cured this deadly disorder in other patients. If you had a child with X-SCID, would you want him to have this gene therapy treatment? Why or why not? What factors would you consider in making your decision?

FIGURE **21.17** Rhys Evans is the first person to be cured of X-linked severe combined immunodeficiency disease (X-SCID) by gene therapy. The immune system of a person with X-SCID is nonfunctional. Rhys Evans' immune system was strengthened, and he can now go to public places without fearing contact with people who might carry germs. He can play with other children.

Genomics Can Be Used to Study How Genes Function and How Diseases Are Inherited

A **genome** is the entire set of genes carried by one member of a species, in this case one person. **Genomics** is the study of entire genomes and the interactions of the genes with one another and the environment.

■ The Human Genome Project sequenced a representative human genome

One goal of genomics is to determine the location and sequences of genes. Researchers have developed supercomputers that automatically sequence (determine the order of bases in) DNA. These were put to use on a massive scale in the Human Genome Project, a worldwide research effort, completed in 2003, to sequence the human genome. As a result, we now have some idea of the locations of genes along the 23 pairs of chromosomes and the sequence of the estimated 3 billion base pairs that make up those chromosomes. Although the exact number of human genes is still not known for certain, scientists now estimate that the human genome consists of 20,000 to 25,000 genes, not 100,000 as originally thought. One reason for the smaller number of actual genes is that many gene families have related or redundant functions and therefore are able to share certain genes, so that fewer are needed to carry out all the body's functions. A second reason is that many genes are now known to code for parts of more than one protein.

In addition, researchers have identified and mapped to specific locations on specific chromosomes genes for more than 1400 genetic diseases. It is hoped that this information will give scientists a greater ability to diagnose, analyze, and eventually treat many of the 4000 diseases of humans that have a genetic basis. Researchers have already cloned the genes responsible for many genetic diseases, including Duchenne muscular dystrophy, retinoblastoma, cystic fibrosis, and neurofibromatosis. These isolated genes can now be used to test for the presence of the same disease-causing genes in specific individuals. As we saw in Chapter 20, some gene tests can be used to identify people who are carriers for certain genetic diseases such as cystic fibrosis, allowing families to make choices based on known probabilities of bearing an affected child; similar tests can be used for prenatal diagnosis and for diagnosis before symptoms of the disease begin. After a disease-related gene has been identified, scientists can study it to learn more about the protein it codes for and perhaps discover ways to correct the problem.

We have also learned from the Human Genome Project that humans are identical in 99.9% of the sequences of their genes. As scientists gain greater understanding of the 0.1% of DNA that differs from person to person, they expect to learn more about why some people develop heart disease, cancer, or Alzheimer's disease and others do not.

■ **Microarray analysis is a useful tool in genomics**

A second goal of genomics is to understand the mechanisms that control gene expression. More than 95% of human DNA does not code for protein; however, some of these noncoding DNA sequences function as regulatory regions that determine when, where, and how much of certain proteins are produced. Because gene activity plays a role in many diseases, the study of how these regions turn genes on or off may lead to advances in diagnosis and treatment.

One of the tools researchers use in this effort is the microarray, which consists of thousands of DNA sequences stamped onto a single glass slide called a *DNA chip*. Researchers use microarrays to monitor large numbers of DNA segments in order to discover which genes are active and which are turned off under different conditions, such as in different tissue types, different stages of development, or in health and disease. For example, they may use microarrays to identify genes that are active in cancerous cells, but not in healthy cells. Presumably, the genes that are active in cancerous cells play a role in the development of cancer.

In addition to identifying gene activity in health and disease, microarray analysis is useful in identifying genetic variation in the members of a population. Some of these genetic differences are in the form of single-nucleotide polymorphisms (SNPs, or *snips*). These are DNA sequences that can vary by one nucleotide from person to person and the differences in their protein products are thought to influence how we respond to stress and diseases, among other things. As researchers learn more about SNPs, they may be able to develop treatments tailored to the genetic makeup of each

individual. These are the kinds of discoveries that can open the door for gene therapy. Although gene therapy may be useful in the future, we already use individual differences in DNA in DNA fingerprinting (see the Social Issue essay, *Forensic Science, DNA, and Personal Privacy*).

what would you do?

Some people worry that once we know the location and function of every gene and have perfected the techniques of gene therapy, we will no longer limit gene manipulation to repairing faulty genes but will begin to modify genes to enhance human abilities. Should people be permitted to design their babies by choosing genes that they consider superior? What do you think? Where should the line be drawn? Who should draw that line? Who should decide which genes are "good?"

The International HapMap Project is a scientific consortium whose purpose is to describe genetic variation between populations. When SNPs are located near one another on a chromosome, they tend to be inherited together. A group of SNPs in a region of a chromosome is called a haplotype. Researchers collaborating on this project will compare haplotype frequencies in groups of people who have a certain disease to those of a group without the disease, hoping to identify genes associated with the disease.

■ **Comparing genomes of different species can be useful**

The DNA of certain widely studied organisms, including the mouse, the fruit fly, a roundworm, and yeast, are also being mapped, with the hope of gaining some insight into basic biology, including basic principles of the organization of genes within the genome, gene regulation, and molecular evolution. Humans share many genes with other organisms. For example, we share half of our genes with the fruit fly and 90% with the mouse. These genetic similarities are evidence of our common evolutionary past. The genes and genetic mechanisms we share with other organisms are likely to be important in determining body plan and also to influence development and aging.

exploring further . . .

In Chapter 19, we learned about the cell cycle. In Chapters 20 and 21, we learned about genes, their inheritance, and their regulation, and also considered how mutations affect gene functions. In Chapter 21a, we will use this information to understand cancer, a family of diseases in which mutations in genes that regulate the cell cycle cause a loss of control over cell division.

SOCIAL ISSUE

Forensic Science, DNA, and Personal Privacy

So-called DNA fingerprints, like the more conventional prints left by fingers, can help identify the individuals they belong to out of a large population. DNA fingerprinting refers to techniques of identifying individuals on the basis of unique features of their DNA. DNA fingerprints are possible because there are many regions of DNA that are composed of small, specific sequences of DNA that are repeated many times. Most commonly used are repeated units of 1 to 5 bases, which are called short tandem repeats (STRs). The number of times these sequences are repeated varies considerably from person to person, from few to 100 repeats. Because of these differences, the segments can be used to match a sample of DNA to the person whose cells produced the sample.

The first step in preparing a DNA fingerprint is to extract DNA from a tissue sample. The type of tissue does not matter. Commonly used sources include blood, semen, skin, and hair follicles, because they are not too painful to remove or are readily available or because they are left at a crime scene.

First, the amount of DNA is greatly increased using PCR, as described in Figure 21.12. The primers used are sequence-specific for the regions on either side of the repeating region. This produces many copies of the repeating region, which are then analyzed to determine the number of repeats present.

The FBI uses 13 STRs as a core set for forensic analysis. The resulting DNA fingerprint is unique to the person who produced the DNA. Moreover, any DNA sample taken from the same person would always be identical. *But* the fingerprint profile resulting from the DNA of *different* people are always different (except for identical siblings), because the number and sizes of the fragments are determined by the unique sequence of bases in each person's DNA.

DNA fingerprinting has many applications, but the most familiar is probably its use in crime investigations. In these cases, the DNA fingerprint is usually created from a sample of tissue such as blood or hair follicles collected at the crime scene. A fingerprint can be produced from tissue left at the scene years before. This fingerprint is then compared with the DNA fingerprints of various suspects. A match reveals, with a high degree of certainty, the person who was the source of the sample from the crime scene (Figure 21.A).

Of course, the degree of certainty of a match between DNA fingerprints depends on how carefully the analysis was done. Because DNA fingerprints are being used as evidence in an increasing number of court cases each year, it is important that national standards be set to ensure the reliability of these molecular witnesses.

It is generally easier to declare with certainty that two DNA fingerprints do *not* match than it is to be sure that they do. Many convicts, including more than 175 on death row, have been found innocent through DNA testing since 1990. Do you think everyone arrested of a crime should have the right to DNA fingerprinting to prove his or her innocence. If so, who should pay for the process?

Has DNA testing gone too far? All states in the United States collect DNA samples from people convicted of sex crimes and murder. Several other states also collect DNA samples from people convicted of other felonies, such as robbery. As of mid-2006, 28 states collect DNA samples from people accused of misdemeanors, including loitering, shoplifting, or vandalism. In many cases, the DNA is stored in a database, even if the person is found innocent of the crime. People who are simply cooperating with the investigation may also provide DNA samples. These too are added to a national database. Is this an invasion of privacy? Under what conditions do you think DNA samples should be obtained?

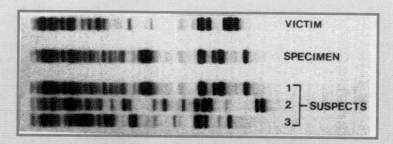

FIGURE **21.A** The pattern of banding in a DNA fingerprint is determined by the sequence of bases in a person's DNA and is, therefore, unique to each person. A match between DNA fingerprints can identify the source of a tissue sample from a crime scene with a high degree of certainty. Which suspect's DNA fingerprint matches this specimen found at a crime scene?

HIGHLIGHTING THE CONCEPTS

DNA Is a Double Helix Consisting of Two Strings of Nucleotides (p. 448)

1. DNA consists of two strands of covalently linked nucleotides twisted together to form a double helix. Each nucleotide consists of a phosphate, a sugar called deoxyribose, and one of four nitrogenous bases: adenine, thymine, cytosine, or guanine. The sugar and phosphate components of the nucleotides alternate along the two sides of the molecule. Pairs of bases meet and form hydrogen bonds in the interior of the double helix; these pairs resemble the rungs on a ladder.

2. According to the rules of complementary base pairing, adenine binds only with thymine, and cytosine binds only with guanine.

WEB TUTORIAL 21.1 The Structure of DNA

During Replication of DNA, Each Original Strand Serves as a Template for a New Strand (pp. 448–449)

3. DNA replication is semiconservative. Each new double-stranded DNA molecule consists of one old and one new strand. The enzyme DNA polymerase "unzips" the two strands of a molecule (the parent molecule), allowing each strand to serve as a template for the formation of a new strand. Complementary base pairing ensures the accuracy of replication.

DNA Codes for RNA, Which Codes for Protein (pp. 449–453)

4. Genetic information is transcribed from DNA to RNA and then translated into a protein.

5. Transcription is the synthesis of RNA by means of base pairing on a DNA template. RNA differs from DNA in that the sugar ribose replaces deoxyribose, and the base uracil replaces thymine. Most RNA is single-stranded.

6. Messenger RNA (mRNA) carries the DNA genetic message to the cytoplasm, where it is translated into protein. The genetic code is read in sequences of three RNA nucleotides; each triplet is called a codon. Each of the 64 codons specifies a particular amino acid or indicates the point where translation should start or stop.

7. Transfer RNA (tRNA) interprets the genetic code. At one end of the tRNA molecule is a sequence of three nucleotides called the anticodon that pairs with a codon on mRNA in accordance with base-pairing rules. The tail of the tRNA binds to a specific amino acid.

8. Each of the two subunits of a ribosome consists of ribosomal RNA (rRNA) and protein. A ribosome brings tRNA and mRNA together for protein synthesis.

9. Translation of the genetic code into protein begins when the two ribosomal subunits and an mRNA assemble, with the mRNA sitting in a groove between the ribosome's two subunits. The mRNA attaches to the ribosome at the mRNA's start codon. Then the ribosome slides along the mRNA molecule, reading one codon at a time. Molecules of tRNA ferry amino acids to the mRNA and add them to a growing protein chain. Translation stops when a stop codon is encountered. The protein chain then separates from the ribosome.

WEB TUTORIAL 21.2 Transcription
WEB TUTORIAL 21.3 The Genetic Code
WEB TUTORIAL 21.4 Translation

Mutations Result from Nucleotide Substitution, Insertion, or Deletion (pp. 453–455)

10. A point mutation is a change in one or a few nucleotides in the sequence of a DNA molecule. When one nucleotide is mistakenly substituted for another, the function of the resulting protein may or may not affect the function of the protein. The insertion or deletion of a nucleotide always changes the resulting protein.

Gene Activity Can Be Turned On or Off (pp. 455–456)

11. Gene activity is regulated at several levels. Usually, most of the DNA is folded and coiled. For a gene to be active, the region of DNA in which it is located must be uncoiled. Gene activity can be affected by other segments of DNA. Master genes can turn other genes on. Regions of DNA called enhancers can increase the amount of RNA produced. Chemical signals such as regulatory proteins or hormones can also affect gene activity.

Genetic Engineering Is the Manipulation of DNA for Human Purposes (pp. 456–463)

12. Genetic engineering is the purposeful manipulation of genetic material by humans. It can be used to produce large quantities of a particular gene product or to transfer a desirable genetic trait from one species to another or to another member of the same species.

13. Genetic engineering uses restriction enzymes to cut the source DNA, which contains the gene of interest, and the vector DNA at specific places, creating sticky ends composed of unpaired complementary bases that allow the cut segments to recombine. The recombinant vector is then used to transfer the recombinant DNA to a host cell. Common vectors include bacterial plasmids and viruses. The host cell is often a type of cell that reproduces rapidly, such as a bacterium or yeast cell. Each time a host cell divides, both daughter cells receive a copy of the gene of interest. Introducing a gene from one species into a different species results in a transgenic plant or animal.

14. Genetic engineering has had many applications in plant and animal agriculture, environmental science, and medicine.

15. In gene therapy, a healthy form of a gene is introduced into body cells to correct problems caused by a defective gene.

WEB TUTORIAL 21.5 Recombinant DNA
WEB TUTORIAL 21.6 The Polymerase Chain Reaction (PCR)

Genomics Can Be Used to Study How Genes Function and How Diseases Are Inherited (pp. 463–465)

16. A genome consists of all the genes in a single organism. Genomics is the study of genomes and the interaction genes and the interactions with one another and the environment. It is now believed that the human genome consists of 20,000 to 25,000 genes.

17. Scientists use microarrays to analyze gene activity under different conditions. It uses DNA chips, glass slides with thousands of DNA segments stamped on them. The information may be helpful in treating genetic diseases. Microarray analysis also allows scientists to discover small differences in the gene sequences of people. Scientists hope to use this information to develop individualized treatments.

18. Large portions of our DNA are the same as the DNA in other organisms, the closer the evolutionary relationship, the greater portion of DNA in common.

KEY TERMS

deoxyribonucleic acid (DNA)
  p. 448
complementary base pairing
  p. 448

DNA replication *p. 448*
semiconservative replication
  p. 449
ribonucleic acid (RNA) *p. 450*

gene *p. 450*
transcription *p. 450*
messenger RNA (mRNA) *p. 450*
transfer RNA (tRNA) *p. 450*

ribosomal RNA (rRNA) *p. 450*
promoter *p. 450*
RNA polymerase *p. 450*
translation *p. 451*

REVIEWING THE CONCEPTS

1. Describe the structure of DNA.   *p. 448*
2. Explain why complementary base pairing is crucial to exact replication of DNA.   *p. 448*
3. Why is DNA replication described as semiconservative?   *pp. 448–449*
4. Explain the roles of transcription and translation in converting the DNA message to a protein.   *pp. 449–451*
5. In what ways does RNA differ from DNA?   *p. 450*
6. What roles do mRNA, tRNA, and rRNA play in the synthesis of protein?   *p. 450*
7. Define codon. What role do codons play in protein synthesis?   *p. 451*
8. What is the role of an anticodon?   *p. 452*
9. Describe the events that occur during the initiation of protein synthesis, the elongation of the protein chain, and the termination of synthesis.   *p. 453*
10. What is a point mutation?   *pp. 453–455*
11. How is gene activity regulated?   *pp. 455–456*
12. Define genetic engineering. Explain the roles of restriction enzymes and vectors in genetic engineering.   *pp. 456–457*
13. Describe some of the ways in which genetic engineering has been used in farming and medicine.   *pp. 459–460*
14. What is gene therapy?   *p. 460*
15. The complementary base for thymine is
    a. adenine.
    b. cytosine.
    c. guanine.
    d. uracil.

16. Although the amount of any particular base in DNA will vary among individuals, the amount of guanine will always equal the amount of
    a. thymine.
    b. adenine.
    c. cytosine.
    d. uracil.
17. A codon is located on
    a. DNA.
    b. mRNA.
    c. tRNA.
    d. rRNA.
18. Translation produces
    a. a polypeptide chain.
    b. mRNA complementary to a template strand of DNA.
    c. tRNA complementary to mRNA.
    d. rRNA.
19. The anticodon is located on a molecule of _____.
20. In RNA, the nucleotide _____ binds with adenine.
21. In genetic engineering, the staggered cuts in DNA that allow genes to be spliced together are made by _____.

APPLYING THE CONCEPTS

Use the genetic code in Table 21.3 (p. 452) to answer questions 1 and 2.

1. What would be the amino acid sequence in the polypeptide resulting from a strand of mRNA with the following base sequence?

   AUG ACA UAU GAG ACG ACU

2. Below are base sequences in four mRNA strands: one normal and three with mutations. (Keep in mind that translation begins with a start codon and ends with a stop codon.) Which of the mutated sequences is likely to have the most severe effects? Why? Which would have the least severe effects? Why?

   Normal mRNA: AUG ACA UAU GAG ACG ACU
   Mutation 1:   AUG ACC UAC GAA ACG ACC
   Mutation 2:   AUG ACU UAA GAG ACG ACA
   Mutation 3:   AUG ACG UAU GAG ACG ACG

3. You are a crime scene investigator asked to testify at a murder trial. The DNA fingerprints shown on the right are those of a bloodstain at the murder scene (not the victim's blood) and those of seven suspects (numbered 1 through 7). Which suspect's blood matches the bloodstain from the crime scene?

Additional questions can be found on the companion website.

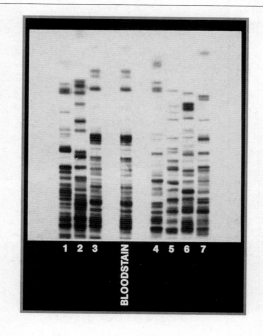

1 2 3   BLOODSTAIN   4 5 6 7

21a

SPECIAL TOPIC
Cancer

Scheduling appropriate screening exams on a regular basis is one of several things you can do to reduce your risk from cancer.

Cancer Is Uncontrolled Cell Division
- Tumors can be benign or malignant
- Tumor development progresses through stages

Cancer Begins with a Single Cell That Escapes Normal Control Mechanisms
- Cancer cells lose restraints on cell division
- Cancer cells do not self-destruct when their DNA is damaged
- Cancer cells divide indefinitely
- Cancer cells attract a blood supply
- Cancer cells do not adhere to neighboring cells
- Body defense cells destroy cancer cells

Viruses, Certain Chemicals, and Radiation Can Cause Cancer
- Certain viruses can disrupt genetic control of cell division
- Some chemicals can cause mutations
- Radiation can cause mutations

Certain Lifestyle Habits Reduce the Risk of Cancer

There Are Several Ways to Diagnose Cancer

Surgery, Radiation, and Chemotherapy Are Conventional Ways to Treat Cancer
- Surgery is used to remove tumors
- Radiation therapy is used to kill localized cancer cells
- Chemotherapy is used to kill cancer cells throughout the body
- Immunotherapy boosts immune responses against cancer cells
- Inhibition of blood vessel formation may slow the spread of cancer cells
- Gene therapy may someday help fight cancer in several ways

When Margaret turned 50, her doctor insisted that she schedule a screening sigmoidoscopy, a procedure that uses a short, flexible, lighted tube to look inside the lower part of the descending colon. He explained that the test was important to make sure that Margaret had no signs of colon cancer. It would also help him monitor any future changes in the lining of her colon that might signal precancerous conditions. Margaret reluctantly agreed.

During the sigmoidoscopy, Dr. Greer found a small growth, or polyp. He removed the polyp and sent it to a laboratory for testing. He explained to Margaret that her risk of colon cancer was very low. He later called and explained that the polyp was not cancerous, but she should schedule another sigmoidoscopy in 5 years. Margaret realized she had been foolish to be reluctant to have the procedure and was thankful that her doctor had insisted on it. Now she encourages all of her friends to have a sigmoidoscopy when she sends them a birthday card on their fiftieth birthday.

Cancer, the "Big C," is perhaps the disease that people in the industrialized world dread the most—and with good reason. Cancer touches the lives of nearly everyone. One of every three people in the United States will develop cancer at some point in life, and the other two are likely to have a friend or relative with cancer.

In this chapter, we will consider how cancer cells escape the normal controls over cell division. We will then learn about some causes of cancer and how we can reduce our risk of developing the disease. Finally, we will identify some means of diagnosing and treating cancer. ■

Cancer Is Uncontrolled Cell Division

All forms of cancer share one characteristic—uncontrolled cell division. Cancer cells behave like aliens taking over our bodies, but they are actually traitorous body cells.

■ Tumors can be benign or malignant

An abnormal growth of cells can form a mass of tissue called a *tumor* or a *neoplasm* (meaning *new growth*). However, not all tumors are cancerous: tumors can be either benign or malignant. A *benign tumor* is an abnormal mass of tissue that is surrounded by a capsule of connective tissue and that usually remains at the site where it forms. Its cells do not invade surrounding tissue or spread to distant locations. In most cases, a benign tumor does not threaten health, because it can be removed completely by surgery. Benign tumors can be harmful when they press on nearby tissues enough to interfere with the functioning of those tissues. If a "benign" tumor of that type is also inoperable, as may occur with a tumor in the brain, it can be life threatening. Even so, only *malignant tumors*, tumors that can invade surrounding tissue and spread to multiple locations throughout the body, are properly called cancerous. The spread of cancer cells from one part of the body to another is called *metastasis*.

■ Tumor development progresses through stages

Cells on their way to becoming cancerous are accumulating genetic damage. As a result, precancerous cells typically look different from normal cells. *Dysplasia* is the term used to describe the changes in shape, nuclei, and organization within tissues of precancerous cells. Their ragged edges give precancerous cells an abnormal shape. Their nuclei become unusually large and atypically shaped and may contain increased amounts of DNA. We can see the differences by comparing the chromosomes of a normal cell with those from a cancer cell. Notice in Figure 21a.1 that the cancer cell has extra copies of some chromosomes, is missing parts of some chromosomes, and has extra parts of other chromosomes. In a group, precancerous cells form a disorganized clump and, significantly, have an unusually high percentage of cells in the process of dividing.

Eventually, the tumor will reach a critical mass consisting of about a million cells. Although the tumor is still only a millimeter or two in diameter (smaller than a BB), the cells in the interior cannot get a sufficient supply of nutrients, and their own waste is poi-

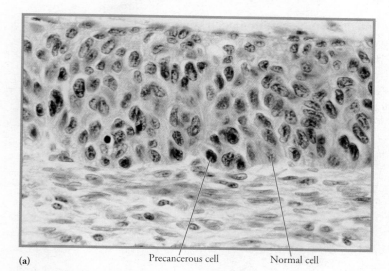

(a)   Precancerous cell   Normal cell

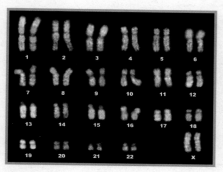

(b)

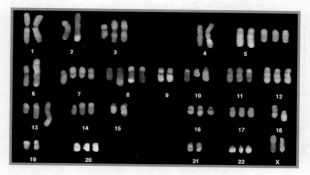

(c)

FIGURE **21a.1** Cancer cells have an abnormal appearance. (a) In dysplasia, precancerous cells have large, irregularly shaped nuclei that contain increased amounts of DNA. A comparison of karyotypes from (b) a normal cell and (c) a cancerous cell shows that cancer cells contain extra chromosomes and chromosomes with extra pieces. (A karyotype is an arrangement of photographed chromosomes in their pairs, identified by physical features.)

soning them. This tiny mass is now called carcinoma *in situ*, which literally means *cancer in place*.

Unless the tumor is removed, a time will come, perhaps years later, when some of its cells will start to secrete chemicals that cause blood vessels to invade the tumor. This process marks an ominous point of transition, because the tumor cells now have supply lines bringing in nutrients to support continued growth and carry away waste. Of equal importance, the tumor cells have an escape route; they can enter the blood or nearby lymphatic vessels and travel throughout the body. Like cancerous seeds, the cancer cells that spread, or metastasize, can begin to form tumors in their new locations (Figure 21a.2).

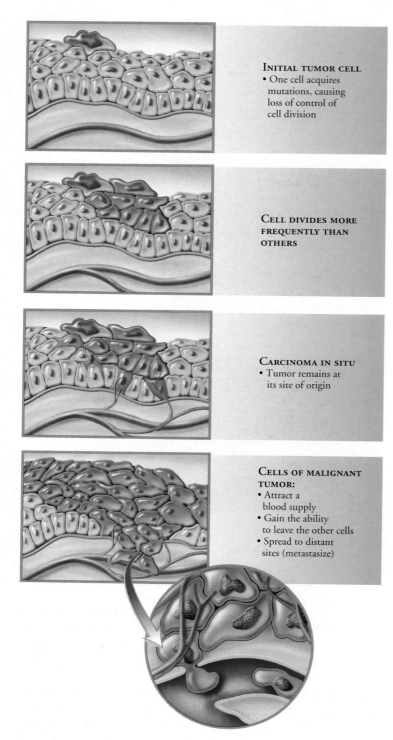

INITIAL TUMOR CELL
• One cell acquires mutations, causing loss of control of cell division

CELL DIVIDES MORE FREQUENTLY THAN OTHERS

CARCINOMA IN SITU
• Tumor remains at its site of origin

CELLS OF MALIGNANT TUMOR:
• Attract a blood supply
• Gain the ability to leave the other cells
• Spread to distant sites (metastasize)

FIGURE **21a.2** The progression of cancer from the initial cancer cell to a malignant tumor

As long as a tumor stays in place, it can grow quite large and a surgeon would still be able to remove it (depending on the location). However, once cancer cells leave the original tumor, they usually spread to so many locations that a surgeon's scalpel is no longer an effective weapon. At this point, chemotherapy or radiation is generally used to kill the cancer cells wherever they are hiding (these treatments are discussed later in the chapter). The original tumor is rarely a cause of death. Instead, the tumors that form in distant sites in the body are responsible for 90% of the deaths of people with cancer.

Once cancer cells have separated from the original tumor, they usually enter the circulatory or lymphatic system, which carries the renegade cells to distant sites. Thus, circulatory pathways in the body often explain the patterns of metastasis. For example, cancer cells escaping from tumors in most parts of the body, including the skin, encounter the next capillary bed in the lungs. Consequently, many cancers spread to the lungs. However, blood leaving the intestine travels directly to the liver, so colon cancer typically spreads to the liver.

The simplest explanation of how cancer causes death is that it interferes with the ability of body cells to function normally. For instance, cancer cells are greedy. They deprive normal cells of nutrients, thereby weakening them, sometimes to the point of death. Cancer cells can also prevent otherwise healthy cells from performing their usual functions. In addition, tumors can block blood vessels or air passageways in the respiratory system or press on vital nerve pathways in the brain.

Cancer Begins with a Single Cell That Escapes Normal Control Mechanisms

The 30 trillion to 50 trillion cells in the human body generally work cooperatively, much like the members of any organized society. There are "rules," or controls, that tell a cell when and how often to divide, when to self-destruct, and when to stay in place. However, cancer cells are outlaws. They evade the many controls that would normally maintain order in the body. As a result, cancer cells can become a threat. Let's discuss some of the normal systems of checks and balances that regulate healthy cells and see how cancer cells are able to get around these safeguards to divide indefinitely and spread out of control.

Recall from Chapter 19 that the cell cycle is the life cycle of the cell. During the cell cycle, there are checkpoints at which the cell determines whether conditions are favorable for moving on to the next stage. This is the cell's system of damage control. If a healthy cell detects damage, such as a mutated gene, it stops the cell cycle, assesses the damage, and begins repair (Figure 21a.3). If the repair is successful, the cell cycle resumes. If the damage is recognized as too severe to repair, a program of cell death is initiated, as will be discussed shortly. Unfortunately, if the damage repair is unsuccessful or incomplete, genetic damage accumulates and can lead to cancer.

Tumor-suppressor genes are an important part of the cell's system of damage control. Some tumor-suppressor gene products de-

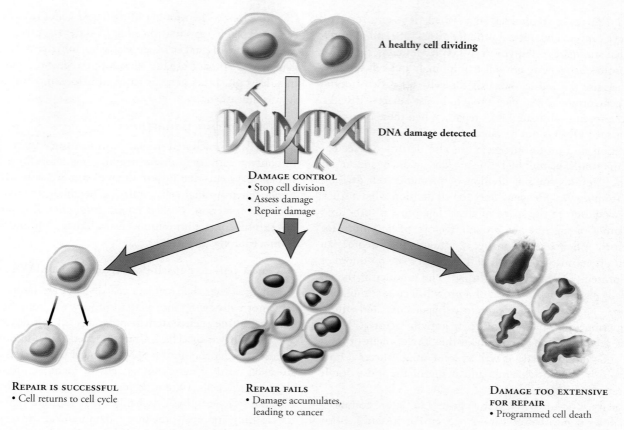

A healthy cell dividing

DNA damage detected

DAMAGE CONTROL
• Stop cell division
• Assess damage
• Repair damage

REPAIR IS SUCCESSFUL
• Cell returns to cell cycle

REPAIR FAILS
• Damage accumulates, leading to cancer

DAMAGE TOO EXTENSIVE FOR REPAIR
• Programmed cell death

FIGURE **21a.3** Steps in controlling DNA damage during the cell cycle

tect damaged DNA.[1] When they do, other tumor-suppressor gene products serve as "managers of the cell's repair shop." These gene products assess the damage and coordinate the activities of other genes whose products serve as "mechanics" and repair the damage. If the damage turns out to be too extensive, the manager activates still other genes whose products cause cell death. A particularly important tumor-suppressor gene is *p53*. We will consider some of the activities of the p53 protein as we discuss the relationship between genes and cancer.

 ■ **Cancer cells lose restraints on cell division**

When genes that regulate cell division are mutated, they usually do not function properly, and the cell loses control over cell division. Cancer, we will see, is fundamentally a disease in which certain genes mutate and produce proteins that malfunction, or produce proteins in abnormal amounts or in inappropriate locations.

MUTATIONS AND CANCER

Two types of genes usually regulate cell division: proto-oncogenes and the tumor-suppressor genes mentioned above. Proto-oncogenes stimulate cell division in a variety of ways, including by producing

[1]Recall that genes exert their effect through the proteins they code for. It is really the proteins that are acting.

growth factors or affecting their function or by producing proteins that affect the activity of certain genes. In contrast, tumor-suppressor genes inhibit or stop cell division. Thus, tumor-suppressor gene products act like brakes on cell division. The combined activities of these two types of genes allow the body to control cell division so as to divide and develop normally, repair defective cells, and replace dead cells.

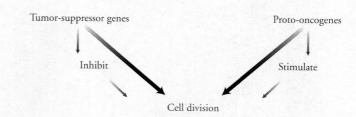

Tumor-suppressor genes

Proto-oncogenes

Inhibit

Stimulate

Cell division

When mutations affect the functioning of these gene products, the normal system of checks and balances that regulates cell division goes awry, and the disruption can result in the unrestrained cell division that characterizes cancer. A mutation in a tumor-suppressor gene can promote cancer by taking the brakes off cell division. The tumor-suppressor gene *p53* produces a protein that regulates another gene whose job is to produce a protein that keeps cells in a nondividing state. When *p53* mutates, cell division is no longer curbed. Mutant *p53* seems to be an important factor in more than half of all cancers. Mutation in another tumor-suppressor gene, called *RB*, causes retinoblastoma, which is a rare form of childhood eye cancer. The normal product of *RB* turns off a proto-oncogene that stimulates cell division. When *RB* mutates, the activity of the proto-oncogene is no longer inhibited, and cell division continues because the cell cannot exit the cell cycle. Two tumor-suppressor gene products that play a role in breast cancer—BRCA1 and BRCA2 proteins—initiate DNA repair. When these two genes mutate, damaged DNA is not repaired.

A mutation in a proto-oncogene can also destroy the regulation of cell division. The mutated gene is called an oncogene, and it increases the stimulus for cell division or promotes cell division without a stimulus. An oncogene does to cell division what a stuck accelerator would do to the speed of a car. The proteins produced by many proto-oncogenes are growth factors or receptors for growth factors. When a proto-oncogene mutates and becomes an oncogene, it often causes too much of its protein to be produced or makes the protein more active than usual. The product of the *ras* gene normally signals the presence of a growth factor, stimulating cell division. The *ras* oncogene protein is hyperactive and stimulates cell division even in the absence of growth factors. The *ras* oncogene is thought to be important in the development of most pancreatic and colon cancers, as well as some lung cancers. Other oncogenes play a role in leukemia and many of the most deadly forms of breast and ovarian cancer.

Damage must occur in *at least* two genes (and more commonly more than six genes) before cancer occurs. For instance, colon cells must accumulate damage in at least one proto-oncogene and three tumor-suppressor genes before they become cancerous

(Figure 21a.4). The number of damaged genes needed to produce cancer explains why it is possible to inherit a predisposition to a certain form of cancer. A person who inherits only one mutant gene may be predisposed to cancer, but a second event, a mutation in at least one other gene, is required before uncontrolled cell division is unleashed.

LOSS OF CONTACT INHIBITION

Most cells are anchored in place within tissues and organs and are in contact with appropriate neighboring cells. Normal cells have signaling systems that inform them of contact with other cells. As a result, when normal cells contact a neighbor, they stop dividing. This phenomenon is called *contact inhibition*. But cancer cells do not exhibit contact inhibition. Instead, they continue to divide and form a tumor.

■ Cancer cells do not self-destruct when their DNA is damaged

When the genes that regulate cell division are faulty, backup systems normally swing into play to protect the body from the renegade cell. One such system is programmed cell death, also called *apoptosis*, in which cells activate a genetic suicide program in response to a biochemical or physiological signal. Activation of the so-called death genes prompts cells to manufacture proteins that then kill the cells. Often, the condemned cells go through a predictable series of physical changes that indicate that the cell will die. During apoptosis, the outer membrane of the condemned cell produces bulges, called blebs, that are pinched off the cell (Figure 21a.5).

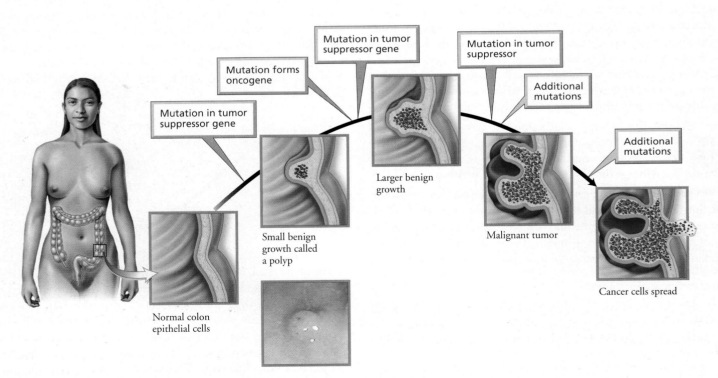

FIGURE **21a.4** Multiple mutations must occur in a single cell before it becomes cancerous. At least one proto-oncogene and three tumor-suppressor genes must mutate in colon cells before they become cancerous.

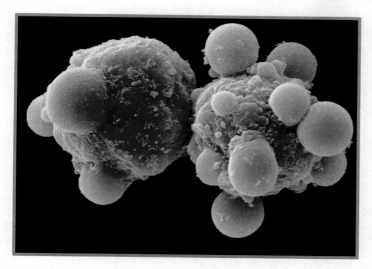

FIGURE **21a.5** Programmed cell death is a backup system that protects the body from a cell in which the genes regulating cell division have been damaged. DNA damage that is too extensive to repair normally triggers a genetic suicide program that causes the cell to self-destruct, as these cells are doing. As the cell goes through programmed cell death, its plasma membrane forms bulges called blebs. Cancer cells are able to evade this protective mechanism.

Cancer cells often fail to trigger apoptosis. Although not the only way that cancer cells evade this safeguard, a faulty tumor-suppressor gene *p53* is often at least partly responsible. Besides producing a protein that inhibits cell division, the p53 protein normally prevents the replication of damaged DNA. If damage is detected, the p53 protein halts cell division until the DNA can be mended. If the damage is beyond repair, the p53 protein triggers the events that lead to programmed cell death.

In a cancer cell, however, a faulty p53 protein fails to initiate the events leading to cellular self-destruction, so the cells are free to divide uncontrollably. Tumors containing cells with damaged *p53* grow aggressively and spread easily and quickly to new locations in the body. Cancer cells containing mutations in *p53* are difficult to kill with radiation or chemotherapy, because these techniques are intended to damage the DNA of the cancer cell and trigger programmed self-destruction. Many cancer cells are simply unable to self-destruct in response to DNA damage.

■ Cancer cells divide indefinitely

Healthy cells have yet another safeguard against unrestrained cell division: a mechanism that limits the number of times a cell can divide during its lifetime. When grown in the laboratory, most human cells divide only about 50 or 60 times before entering a nondividing state called senescence. Like sand running through an hourglass, cell division has a predetermined end.

How does a cell "count" the number of times it has divided? The answer might lie in telomeres—pieces of DNA at the tips of chromosomes that protect the ends of the chromosomes like the plastic pieces on the ends of shoelaces protect the shoelace—or in telomerase, the enzyme that constructs the telomeres (Figure 21a.6). Soon after an embryo is fully developed, most cells stop producing telomerase, putting an end to the maintenance of telomere length. Each time DNA is copied in preparation for cell division, a tiny piece of every telomere in the cell is shaved off, shortening the chromosomes slightly. When the telomeres are completely gone, the chromosome tips can fuse together, disrupting the genetic message and causing the cell to die. Telomeres, then, may be the cell's way of limiting the number of times division can occur. When the telomeres are gone, time is up for that cell. Thus, telomere length may serve both as a gauge of a cell's age and as an indicator of how long that line of cells will continue to divide.

Some scientists currently suspect that the "fountain of youth" that bestows immortality on cancer cells, allowing them to divide indefinitely, may be their unceasing production of telomerase. This enzyme reconstructs the telomeres after each cell division, stabilizing telomere length and protecting the important genes at chromosome tips. Although most types of mature human cells can no longer produce telomerase, the genes for telomerase apparently become turned on in most cancer cells. Telomerase is present in nearly 90% of biopsies of human tumors.

■ Cancer cells attract a blood supply

We have seen that cancer cells have escaped the normal cellular controls on cell division and are unable to issue the orders that would lead to their own death. Instead, they multiply and form a tumor.

As mentioned earlier, when a tumor reaches a critical size, about a million cells, its growth will stop unless it can attract a blood supply to deliver the nutrients it needs to support its growth and to remove waste. Cancer cells release special growth factors that cause

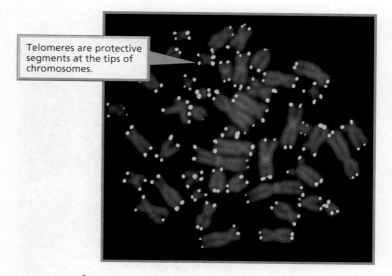

Telomeres are protective segments at the tips of chromosomes.

FIGURE **21a.6** Telomeres, shown here in yellow, protect the tips of chromosomes in much the way that the small plastic pieces protect the ends of shoelaces. Telomeres may serve as molecular counting mechanisms that limit the number of times a cell can divide. Cancer cells retain the ability to construct new telomeres to replace the bits that are shaved off.

capillaries to invade the tumor (Figure 21a.7). These tiny blood vessels become the lifeline of the tumor, removing wastes and delivering fresh nutrients and additional growth factors that will spur tumor growth. They also serve as pathways by which the cancer cells are able to leave the tumor and spread to other sites in the body.

In a healthy person, blood vessel formation is uncommon and is usually limited to repairing cuts or other wounds. Abnormal invasion of blood vessels into tissues can cause damage. For instance, when blood vessels invade the eye's light-sensitive retina, blindness can result. When vessels invade joints, they can cause arthritis. To avoid such damage, cells produce a protein that prevents new blood vessels from forming in tissues. The gene that normally produces this protein is by now familiar—*p53*. Mutations in *p53* can block the production of the protein that prevents the attraction of blood vessels, allowing blood vessels to invade the tumor.

■ Cancer cells do not adhere to neighboring cells

With access to blood vessels, the cancer cells can begin to spread. Their ability to travel through the body is yet another example of their freedom from normal cellular control mechanisms. Normal cells are "glued" in place by special molecules on their surfaces called cellular adhesion molecules (CAMs). When most normal cells become "unglued" from other cells, they stop dividing, and their genetic program for self-destruction is activated.

Cancer cells must become 'unglued' from other cells to travel through the body. One way cancer cells break loose is by secreting enzymes that break down the CAMs that hold them and their neighbors in place. In this way, their anchors are broken, and mechanical barriers, such as basement membranes, that would prevent metastasis are breached. The reason unanchored cancer cells continue dividing and evade self-destruction is that their oncogenes send a false message to the nucleus saying that the cell is properly attached (Table 21a.1).

FIGURE **21a.7** Tumor cells release growth factors that cause blood vessels to grow into the tumor mass. These vessels bring nutrients and additional growth factors and remove wastes. They also provide a pathway by which cancer cells can leave the tumor and spread to other locations in the body. Here, blood vessels have formed in a tumor on the spinal cord.

TABLE 21A.1 REVIEW OF CONTROL MECHANISMS THAT FAIL IN CANCER	
MECHANISM THAT PROTECTS CELLS FROM CANCER	**METHOD OF EVASION USED BY CANCER CELLS**
Genetic controls on cell division	
Proto-oncogenes stimulate cell division through effects on growth factors and certain other cell-signalling mechanisms	Oncogenes promote cell division
Tumor-suppressor genes inhibit cell division	Mutations in tumor-suppressor genes take the "brakes" off cell division
Programmed cell death	
A genetic program that initiates events that lead to the death of the cell when damaged DNA is detected or another signal is received	Mutations in tumor-suppressor genes: Mutant gene p53 no longer triggers cell death when damaged DNA is detected
Limitations on the number of times a cell can divide	
Telomeres protect the ends of chromosomes, but a fraction of each is shaved off each time the DNA is copied; when the telomeres are gone, the chromosome tips can stick together, causing the cell to die	Genes to produce telomerase, the enzyme that reconstructs telomeres, are turned on in cancer cells so telomere length is stabilized
Controls that prevent the formation of new blood vessels	
These controls are normally in effect except in a few instances such as wound healing	Cancer cells produce growth factors that attract new blood vessels and proteins that counter the normal proteins that inhibit blood vessel formation
Controls that keep normal cells in place	
Cellular adhesion molecules (CAMs) hold cells in place; unanchored cells stop dividing and self-destruct	Cancer cell oncogenes send a false message to the nucleus that the cell is properly anchored

■ Body defense cells destroy cancer cells

Despite all the safeguards that prevent cells from becoming cancerous, cancer cells develop in our bodies every day. Fortunately, certain body defense cells—natural killer cells and cytotoxic T cells (see Chapter 13)—usually kill those cancer cells. The processes that lead to the creation of cancer cells also produce new and slightly different proteins on cancer cell membranes. The defense cells recognize these proteins as nonself and destroy the cancer cells. But sometimes cancer cells evade destruction. Some types of cancer cells actively inhibit the defense cells, and some simply multiply so quickly that the defense cells cannot destroy them all. In either case, the tumor is able to grow and spread.

Viruses, Certain Chemicals, and Radiation Can Cause Cancer

We have seen that many of the "tactics" a cancer cell uses to evade normal cellular safeguards are consequences of changes in genes. Those genetic changes are often brought about by viruses or by mutations caused by certain chemicals or radiation.

■ Certain viruses can disrupt genetic control of cell division

Viruses cause some human cancers (Table 21a.2). It is estimated that viruses cause about 5% of the cancers in the United States. Some of the viruses that cause cancer have oncogenes among their genes. Once inside the host cell, the viral oncogene behaves as a host oncogene would, taking the cell one step closer to becoming cancerous. A viral oncogene is partly why the human papilloma virus that causes genital warts can cause cervical and penile cancer. If the viral genetic information is in the form of RNA, the enzyme reverse transcriptase uses viral RNA to synthesize viral DNA, which then inserts into a host cell chromosome. Proteins produced by the viral DNA may then drive the cellular proto-oncogene to be expressed in abnormal levels or in the wrong place or time. RNA viruses can also pick up a proto-oncogene from one host cell and introduce it to a new host cell, thus promoting cell division. In other cases, a virus causes cancer because viral DNA becomes inserted into the host DNA in a location that disrupts the functioning of a gene that influences cell division. The viral DNA could, for example, be inserted into a regulatory gene that controls a proto-oncogene, breaking the switch that turns the gene off. Some viruses interfere with the function of the immune system, lessening its ability to find and destroy cancer cells as they arise.

■ Some chemicals can cause mutations

A carcinogen is an environmental agent that fosters the development of cancer. Some chemicals, especially certain organic chemicals, cause cancer by causing mutations. As we saw in Chapter 21, a change in as little as one nucleotide in a DNA sequence can alter a gene's message. Thus, even a small alteration in DNA can wreak havoc with a cell's regulatory mechanisms and lead to cancer.

Chemical carcinogens are around us most of the time—in air, food, water, and other substances in our surroundings. We can attempt to avoid contact with some of them. For instance, tobacco smoke contains a host of chemical carcinogens. Among the carcinogens in tobacco smoke is one that specifically mutates the tumor-suppressor gene *p53* and another that specifically mutates one of the *ras* oncogenes. Tobacco smoke is responsible for 30% of all cancer deaths and may contribute to as many as 60% of all cancer deaths in the United States. Excessive alcohol consumption is another cancer risk factor that can be avoided. Other chemical carcinogens are more difficult to avoid. These include benzene, formaldehyde, hydrocarbons, certain pesticides, and chemicals in some dyes and preservatives.

Some chemicals contribute to the development of cancer by stimulating cell division, which increases the chance of additional mutations arising. If a cancerous cell has already formed, this stimulation will cause the cancer to progress. Certain hormones can promote cancer in this way. For example, the female hormone estrogen stimulates cell division in the tissues of the breast and in the lining of the uterus (the endometrium). Sustained high levels of estrogen are linked with breast cancer;[2] and the incidence of endometrial cancer is elevated in postmenopausal women whose hormone replacement therapy consists entirely of estrogen. Estrogen does not seem to promote endometrial cancer when taken in combination with progesterone. However, since 2002, several studies have shown that estrogen taken in combination with progesterone does slightly increase a woman's risk of developing breast cancer. Furthermore, breast tumors in women taking combined hormone pills were larger and more likely to have spread than were tumors in women not taking hormone pills.

■ Radiation can cause mutations

Radiation, too, can cause cancer by causing mutations in DNA. It is impossible to avoid exposure to radiation from natural sources, such as the ultraviolet light from the sun, cosmic rays, radon, and uranium. However, we can take reasonable precautions to minimize our risks. For example, sunlight's ultraviolet rays cause skin cancer. Although we probably would not choose to spend our lives entirely indoors to reduce the risk of skin cancer, it would be wise to avoid sunbathing, as well as tanning lamps and tanning parlors. It would also be a good idea to use a sunscreen whenever exposure to the sun is unavoidable.

Certain Lifestyle Habits Reduce the Risk of Cancer

Although we tend to think of cancer as one disease, it is in fact a family of more than 200 diseases, usually named for the organ in which the tumor arises. Cancers of the epithelial tissues are carcinomas. Leukemias are cancers of the bone marrow. Sarcomas are cancers of the muscle, bone, cartilage, or

TABLE 21A.2 SOME VIRUSES LINKED TO HUMAN CANCERS

VIRUS	TYPES OF CANCER
Human papilloma viruses (HPVs)	Cervical, penile, and other anogenital cancers in men and women
Hepatitis B and C viruses	Liver cancer
Epstein–Barr virus	B cell lymphomas, especially Burkitt's lymphoma; nasopharyngeal carcinoma
Human T cell leukemia virus (HTLV–1)	Adult T cell leukemia
Cytomegalovirus (CMV)	Lymphomas and leukemias
Kaposi sarcoma-associated virus	Kaposi's sarcoma

[2]The link between estrogen and breast cancer is discussed in Chapter 17 and in Chapter 10's Heath Issue essay, *Is It Hot in Here, or Is It Me? Hormone Replacement Therapy and Menopause.*

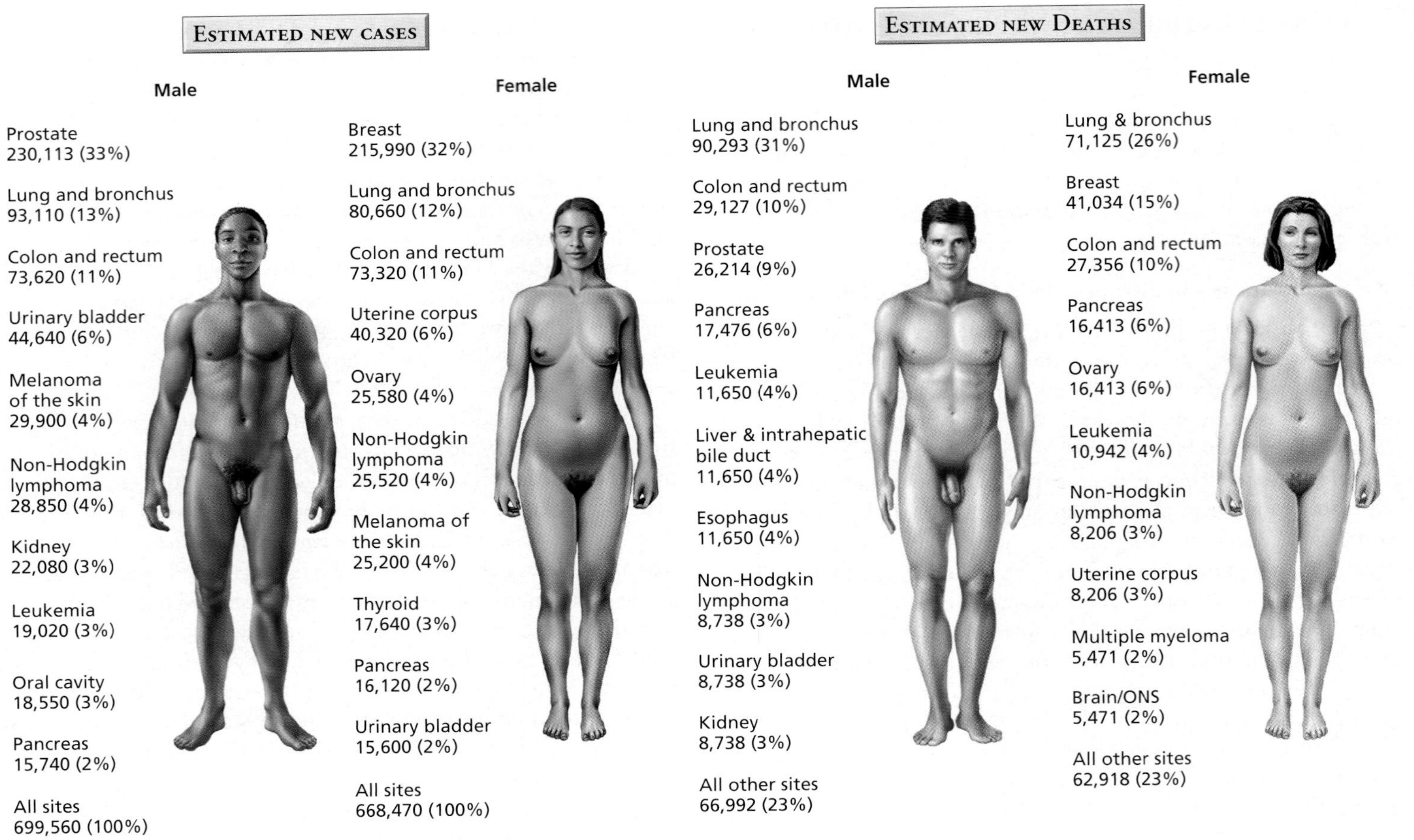

ESTIMATED NEW CASES

Male

Prostate
230,113 (33%)

Lung and bronchus
93,110 (13%)

Colon and rectum
73,620 (11%)

Urinary bladder
44,640 (6%)

Melanoma
of the skin
29,900 (4%)

Non-Hodgkin
lymphoma
28,850 (4%)

Kidney
22,080 (3%)

Leukemia
19,020 (3%)

Oral cavity
18,550 (3%)

Pancreas
15,740 (2%)

All sites
699,560 (100%)

Female

Breast
215,990 (32%)

Lung and bronchus
80,660 (12%)

Colon and rectum
73,320 (11%)

Uterine corpus
40,320 (6%)

Ovary
25,580 (4%)

Non-Hodgkin
lymphoma
25,520 (4%)

Melanoma of
the skin
25,200 (4%)

Thyroid
17,640 (3%)

Pancreas
16,120 (2%)

Urinary bladder
15,600 (2%)

All sites
668,470 (100%)

ESTIMATED NEW DEATHS

Male

Lung and bronchus
90,293 (31%)

Colon and rectum
29,127 (10%)

Prostate
26,214 (9%)

Pancreas
17,476 (6%)

Leukemia
11,650 (4%)

Liver & intrahepatic
bile duct
11,650 (4%)

Esophagus
11,650 (4%)

Non-Hodgkin
lymphoma
8,738 (3%)

Urinary bladder
8,738 (3%)

Kidney
8,738 (3%)

All other sites
66,992 (23%)

Female

Lung & bronchus
71,125 (26%)

Breast
41,034 (15%)

Colon and rectum
27,356 (10%)

Pancreas
16,413 (6%)

Ovary
16,413 (6%)

Leukemia
10,942 (4%)

Non-Hodgkin
lymphoma
8,206 (3%)

Uterine corpus
8,206 (3%)

Multiple myeloma
5,471 (2%)

Brain/ONS
5,471 (2%)

All other sites
62,918 (23%)

FIGURE **21a.8** The American Cancer Society's 2006 estimates for the leading types of cancer in terms of new cases and deaths. The figures do not include basal and squamous cell skin cancer or *in situ* carcinomas other than those of the urinary bladder. The percentages may not total 100% because of rounding.

(Source: American Cancer Society.)

connective tissues. Lymphomas are cancers of the lymphatic tissues. Adenocarcinomas are cancers of the glandular epithelia. Figure 21a.8 shows the estimated number of cases of cancer and cancer deaths for various types of cancer, and Table 21a.3 indicates where in the text certain types of cancer are discussed.

Cancer is the second leading cause of death in industrialized countries, but the good news is that some lifestyle changes can

TABLE 21A.3 SOME DISCUSSIONS OF CANCER IN THIS BOOK

CANCER	CHAPTER
Skin	Chapter 4 (p. 75)
Leukemia	Chapter 11 (p. 205)
Lung	Chapter 14a (pp. 290-291)
Colon, stomach, esophagus	Chapter 15 (pp. 304, 310)
Testis, prostate	Chapter 17 (pp. 351, 352)
Breast	Chapter 17 (pp. 358-359)
Cervical	Chapter 17a (p. 380)
Karposi sarcoma	Chapter 17a (p. 380)

greatly decrease your risk of developing cancer (Table 21a.4). To-bacco use and unhealthy diet are responsible for two-thirds of all cancer deaths in the United States. Tobacco smoke is the leading carcinogen, and it is obvious how to modify that risk (which, in any case, was the focus of Chapter 14a). Here we will focus on diet as a way to reduce your cancer risk.

A few simple diet changes may reduce your risk of developing cancer. The best rules are to eat a well-balanced diet and to eat all foods in moderation. For instance, a high-fat diet is linked to colon and breast cancers. Most people in the United States consume far too much fat. Thus, it is wise to reduce fat intake, especially saturated fat, which comes from animal sources such as red meat. Consuming large quantities of smoked, salt-cured, and nitrite-cured foods, such as ham, bologna, and salami, increases the risk of cancers of the esophagus and stomach.

A diet rich in fruits and vegetables can reduce your cancer risk because they are high in fiber, which can dilute the contents of the intestines, bind to carcinogens, and reduce the amount of time the carcinogens spend in the intestine by speeding passage of intestinal contents. The so-called colorful vegetables—vegetables having colors other than green—are usually high in antioxidant vitamins, and these vitamins may play a role in protecting against cancer. As their name implies, antioxidants interfere with oxidation, a process that

TABLE 21A.4 TIPS FOR REDUCING YOUR CANCER RISK

1. Do not use tobacco. If you do, quit. Avoid exposure to secondhand smoke.

2. Reduce the amount of saturated fat in your diet, especially the fat from red meat.

3. Minimize your consumption of salt-cured, pickled, and smoked foods.

4. Eat at least five servings of fruits and vegetables every day.

5. Avoid excessive alcohol intake. If you consume alcohol, one or two drinks a day should be the maximum.

6. Watch your caloric intake, and maintain a healthy body weight.

7. Avoid excessive exposure to sunlight. Wear protective clothing. Use sunscreen.

8. Avoid unnecessary medical x-rays.

9. Have the appropriate screening exams on a regular basis. Women should have PAP tests and mammograms. Men should have prostate tests. All adults should have tests for colorectal cancer.

can result in the formation of molecules called free radicals that can damage DNA and lead to cancer. The three major antioxidants are beta-carotene, vitamin E, and vitamin C. The first two are common in red, yellow, and orange fruits and vegetables, and the last abounds in citrus fruits, among other sources.

There Are Several Ways to Diagnose Cancer

Early detection is critical to cancer survival because treatment is much more likely to be successful if the cancer has not yet spread. You know your own body better than anyone else does. The American Cancer Society suggests that it is wise for you to be aware of cancer's seven warning signs, the first letters of which spell the word CAUTION:

Change in bowel or bladder habit or function

A sore that does not heal

Unusual bleeding or bloody discharge

Thickening or lump in breast or elsewhere

Indigestion or difficulty swallowing

Obvious change in wart or mole

Nagging cough or hoarseness

There are additional ways to diagnose cancer—some more involved than others:

- **Routine screening.** Many routine tests can detect cancer in people who do not have symptoms. You can perform some of the tests on yourself; others require a visit to a medical professional (Table 21a.5).

- **Imaging.** Many imaging techniques allow physicians to look inside the body and identify tumors. These include x-rays, computerized tomography (CT) scans, magnetic resonance

 Use this table as a reference for the appropriate ages and frequencies to have cancer screening tests

TABLE 21A.5 RECOMMENDED CANCER SCREENING TESTS

Guidelines suggested by the American Cancer Society for the early detection of cancer in people without symptoms, age 20 to 40

Cancer-related checkup every 3 years

Should include the procedures listed below plus health counseling (such as tips on quitting cigarette smoking) and examinations for cancers of the thyroid, testes, prostate, mouth, ovaries, skin, and lymph nodes. Some people are at higher than normal risk for certain cancers and may need to have tests more frequently.

Breast	• Exam by doctor every 3 years • Self-exam every month • One baseline breast x-ray ages 35–40 Higher risk for breast cancer: Personal or family history of breast cancer; never had children; had first child after 30
Uterus	• Pelvic exam every 3 years
Cervix	• Yearly PAP test beginning at age 18 or when sexual activity begins Higher risk for cervical cancer: Early age at first intercourse; multiple sex partners

Guidelines suggested by the American Cancer Society for the early detection of cancer in people without symptoms, age 40 and over

Cancer-related checkup every year

Should include the procedures listed below plus health counseling (such as tips on quitting cigarette smoking) and examinations for cancers of the thyroid, testes, prostate, mouth, ovaries, skin, and lymph nodes. Some people are at higher than normal risk for certain cancers and may need to have tests more frequently.

Breast	• Exam by doctor every year • Self-exam every month • Breast x-ray every year after 40 Higher risk for breast cancer: Personal or family history of breast cancer; never had children; had first child after 30
Uterus	• Pelvic exam every year
Cervix	• Yearly PAP test Higher risk for cervical cancer: Early age at first intercourse; multiple sex partners
Endometrium	• Endometrial tissue sample at menopause if at risk Higher risk for endometrial cancer: Infertility, obesity, failure of ovulation, abnormal uterine bleeding, estrogen therapy
Prostate	• Yearly prostate-specific antigen (PSA) blood test and digital rectal exam after age 50
Colon and rectum	• Fecal occult blood test every year after age 50 • Flexible sigmoidoscopy beginning at age 50 and every 5 years thereafter Higher risk for colorectal cancer: Personal or family history of colon or rectal cancer; personal or family history of polyps in the colon or rectum; ulcerative colitis

imaging (MRI), and ultrasound. *Biopsy* is the removal and analysis of a small piece of tissue suspected to be cancerous. A biopsy is often done using a needle instead of surgery. In either case, cells are then examined under a microscope to see whether they have the characteristic appearance of cancer cells.

- **Tumor marker tests.** When cancer is suspected, certain blood tests can be used to look for tumor markers, which are chemicals produced either by the cells of the tumor or by body cells in response to a tumor. Prostate cells, for example, produce prostate-specific antigen (PSA). Thus, elevated blood PSA levels suggest the presence of prostate cancer. Currently, PSA is the only tumor marker that is useful in the original diagnosis of a cancer, but other tumor markers may reveal whether certain cancers have spread or returned. Blood levels of a marker called TA–90 can help determine whether melanoma (a type of skin cancer) has spread. The tumor marker CA 125 can identify ovarian cancer. CA–15–3 indicates a recurrence of breast cancer, and CEA indicates a recurrence of colon cancer.

- **Genetic tests.** DNA analysis of certain substances can identify gene mutations associated with certain cancers: sputum is examined for signs of lung cancer, urine for signs of bladder cancer, and feces for signs of colon cancer. There are other signs of cancer that can be detected by still other tests. For instance, the enzyme telomerase is produced by cancer cells but rarely by normal ones. A test for telomerase appears to be helpful in diagnosing certain cancers, but this test is still in an experimental stage.

Surgery, Radiation, and Chemotherapy Are Conventional Ways to Treat Cancer

The conventional cancer treatments—surgery, radiation therapy, and chemotherapy—are still the mainstays of cancer treatment. But many new treatments hold the promise of a brighter future.

■ Surgery is used to remove tumors

When a cancerous tumor is accessible and can be removed without damaging vital surrounding tissue, surgery is usually performed to try to eradicate the cancer or remove as much as possible. If every cancer cell is removed, as can be done with early tumors (carcinoma *in situ*), a complete cure is possible. However, if cancer cells have begun to invade surrounding tissue or have spread to distant locations, surgery alone cannot "cure" the cancer. If the cancer has spread, surrounding tissue and perhaps even nearby lymph nodes may also be removed. Unfortunately, more than half of all tumors have already metastasized by the time of diagnosis, so further treatment is necessary.

■ Radiation therapy is used to kill localized cancer cells

If cancer has spread from the initial site but is still localized, surgery is usually followed by radiation therapy (Figure 21a.9). In some cases, such as cancer of the larynx (voice box), localized tumors that

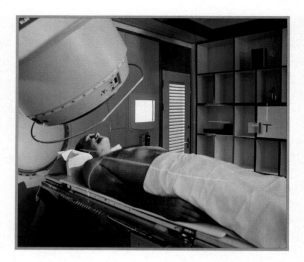

FIGURE **21a.9** Radiation therapy is often used to treat cancer that is localized to one area.

are difficult to remove surgically without damaging surrounding tissue may be treated with radiation alone.

As we have seen, radiation damages DNA, and extensive DNA damage triggers programmed cell death. The greatest damage is done to cells that are dividing rapidly. The intended targets of radiation, cancer cells, are dividing actively, but so are the cells of several types of tissues, called renewal tissues, whose cells normally continue dividing throughout life. These tissues include cells of the reproductive system, cells that replace layers of skin or the lining of the stomach, cells that give rise to blood cells, or cells that give rise to hair. Unfortunately, radiation cannot distinguish between cancer cells and renewal tissue, so good cells are sacrificed in killing the harmful ones. The destruction of renewal tissues leads to the side effects of radiation such as temporary sterility, nausea, anemia, and hair loss.

■ Chemotherapy is used to kill cancer cells throughout the body

When cancer is thought to have spread by the time of diagnosis, chemotherapy is often used. In general, the drugs used in chemotherapy reach all parts of the body and kill all rapidly dividing cells, just as radiation does. Some of the drugs block DNA synthesis, others damage DNA, and a few others prevent cell division by interfering with other cellular processes. The side effects are similar to those that accompany radiation therapy.

As we have seen, the idea underlying radiation and chemotherapy is that the damage they do to DNA in rapidly dividing cells will cause the cells to self-destruct. However, *p53*, the gene that detects DNA damage and initiates programmed cell death, is mutant in more than half of all cancers. As a result, even though the treatment succeeds in damaging the DNA in cancer cells, the cells do not self-destruct, and treatment fails.

■ Immunotherapy boosts immune responses against cancer cells

Cytotoxic T cells of the immune system continually search the body for abnormal cells, such as cancer cells, and kill any that they find (see Chapter 13). The goal of immunotherapy, then, is to boost the patient's immune system so that it becomes more effective in destroying cancer cells. One form of immunotherapy involves the administration of factors normally secreted by lymphocytes, including interleukin-2 (which stimulates lymphocytes that attack cancer cells), interferons (which stimulate the immune system and also have direct effects on the tumor cells), and tumor necrosis factor (which has direct effects on cancer cells).

Two types of vaccines can be used in the battle against cancer. One type is a vaccine against a virus that causes cancer. For example, a vaccine against four of the human papilloma viruses that cause cervical cancer was approved in the summer of 2006. Two of these viruses are HPV16 and HPV18, which cause 70% of the cases of cervical cancer. Federal health officials now recommend that all girls 10 to 12 years of age receive this vaccination.

The other type of vaccine is being tested in clinical trials with promising results. It is designed to work by stimulating T cells to attack and kill the cancer cells. Unlike most vaccines, these vaccines cannot *prevent* disease (cancer); they can only treat it. Cancer vaccines contain dead cancer cells, parts of cells, or proteins from cancer cells. Recall that vaccines cause the immune system to fight cells having the same characteristics as the cells in the vaccine. The vaccine strategy is now being tested against melanoma (a deadly form of skin cancer) and leukemia, as well as cancers of the kidney, prostate, colon, and lung.

■ Inhibition of blood vessel formation may slow the spread of cancer cells

In recent years, there has been a major change in the way doctors and scientists think about cancer. Gone are the days when the only aim was to kill the cancer cells. As researchers have learned more about the molecular biology of cancer, new ways of slowing its progression or dealing with it as a genetic disease have been developed.

The formation of blood vessels is a critical step in the life of a tumor, because they bring nourishment and provide a pathway for cell migration. Researchers are working on ways to cut off these lifelines and starve the tumor. Many drugs have been developed for this purpose, and some are nearing the end of clinical trials. For example, when a certain drug that blocks blood vessel formation is combined with chemotherapy, the survival rate of patients with colorectal cancer has been shown to improve.

■ Gene therapy may someday help fight cancer in several ways

There are currently over 400 clinical trials using gene therapy to treat cancer, but no means of gene therapy has been approved by the Food and Drug Administration (FDA) for cancer treatment. Nearly all these studies are in very early stages. One of the treatment strategies being tested is to insert normal tumor-suppressor genes into the cancerous cells. For instance, you may recall that gene *p53* normally triggers programmed cell death when DNA damage is detected but that this gene is often faulty in cancer cells. Researchers hope that inserting a healthy form of *p53* into cancer cells will lead to their death, causing the tumor to shrink.

Another strategy is to insert into a cancer cell a piece of DNA that will prevent an oncogene from exerting its effects. The inserted DNA is called antisense DNA because it is complementary to the mRNA produced by the oncogene. The antisense DNA would be expected to bind to the mRNA produced by the oncogene and prevent it from being translated into protein, thereby preventing the oncogene from exerting its effects. Currently, this is being tried as a means of treating leukemia.

One of the most promising approaches using gene therapy is to insert a gene into tumor cells that makes them sensitive to a drug that will kill them. This method is now being evaluated as treatment for brain, ovarian, and prostate cancers. A viral gene for the enzyme thymidine kinase is inserted into the tumor cells, and when the gene is expressed, the resulting enzyme makes the cell sensitive to a drug called ganciclovir. The drug is inactive except inside the cells that produce the enzyme coded for by the inserted gene. Thus, only the tumor cells are killed.

In the fall of 2006, researchers announced that they had used gene therapy to cure two patients (out of 17 patients who were treated) of the aggressive skin cancer melanoma. The researchers used a virus to transfer a gene into body defense cells called T cells (discussed in Chapter 13). The inserted gene codes for a protein called T-cell receptor or TCR. With this gene the T cell can find tumor cells and destroy them.

22

Evolution and Our Heritage

An understanding of human biology requires an understanding of our evolutionary heritage and of the lives of our early ancestors.

Life Evolved on the Earth About 3.8 Billion Years Ago

- Inorganic molecules formed small organic molecules
- Small organic molecules joined to form larger molecules
- Macromolecules aggregated into droplets

The Scale of Evolutionary Change May Be Small or Large

- Microevolution is small-scale evolutionary change over short periods of time
- Macroevolution is large-scale evolutionary change over long periods of time

Evidence of Evolution Comes from Diverse Sources

- The fossil record provides evidence of evolution
- Geographic distributions reflect evolutionary history
- Comparative anatomy and embryology reveal common descent
- Comparative molecular biology also reveals evolutionary relationships

Human Roots Trace Back to the First Primates

- Primates have distinct characteristics
- Discussions of human origins often evoke controversy
- Misconceptions distort the picture of human evolution
- Walking on two feet was a critical step in hominid evolution

SOCIAL ISSUE   Conducting Research on Our Relatives

The year is 1831. The country is England. A young man, 22 years of age, walks to the ship he is to board and upon which he will live for the next several years. On this day, he has no thoughts about revolutionizing the prevailing view of the earth and its inhabitants and forever changing the course of scientific thinking. Charles Darwin, an avid naturalist and somewhat disenchanted student of medicine and theology, left Great Britain in 1831 to spend 5 years aboard the H.M.S. *Beagle* (Figure 22.1). The purpose of the voyage was to chart the coastline of South America—a goal that allowed Darwin to examine, describe, and collect plants and animals from the South American coast and nearby islands. In 1836, Darwin returned to Great Britain, where he began to mull over all that he had seen and read. Twenty-three years later, in 1859, his book *On the Origin of Species* was published. In it, Darwin proposed that the incredible diversity of life on the earth was the product of evolution. Broadly defined, **evolution** is descent with modification from a common ancestor. It is the process by which life forms on the earth have changed from their earliest beginnings to today.

How did life arise and evolve on the earth? How does evolution work? How has evolution shaped species, including our own? What were our ancestors like? We address those questions in this chapter. ■

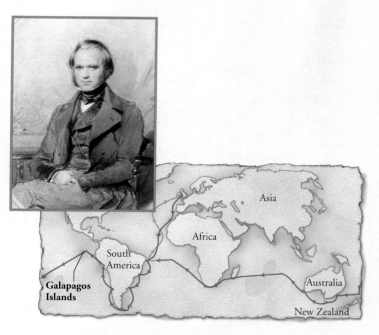

FIGURE **22.1** Charles Darwin shortly after his return to England from his voyage on the H.M.S. *Beagle*

Life Evolved on the Earth About 3.8 Billion Years Ago

The earth is estimated to be 4.5 billion years old. Evidence from physical and chemical changes in the earth's crust and atmosphere suggests that life has existed on the earth for about 3.8 billion years. The environment of the early earth was very different from that of today and would have been an extremely hostile place for most organisms (Figure 22.2). The earth's crust was hot and volcanic. There were intense lightning and ultraviolet radiation, and there was almost no gaseous oxygen (O_2) in the atmosphere. The early atmosphere likely consisted of the following gases: hydrogen (H_2), nitrogen (N_2), carbon dioxide (CO_2),

methane (CH_4), and carbon monoxide (CO). How could life have evolved under such conditions? In the sections that follow, we present a plausible sequence of events that many scientists believe explains the origin of life. The sequence, known as **chemical evolution,** suggests that life evolved from chemicals slowly increasing in complexity over a period of perhaps 300 million years.

■ Inorganic molecules formed small organic molecules

Scientists hypothesize that conditions of the early earth favored the synthesis of small organic molecules from inorganic molecules. Specifically, the low-oxygen atmosphere of the primitive earth encouraged the joining of simple molecules to form complex molecules. The low-oxygen atmosphere was important because oxygen attacks chemical bonds. Scientists further hypothesize that the energy required for the joining of simple molecules could have come from the lightning and intense ultraviolet (UV) radiation striking the primitive earth. (UV radiation was likely more intense during those times than it is now because young suns emit more UV radiation than do old suns, and the early earth lacked an ozone layer to shield it from much of the UV radiation.)

Stanley Miller and Harold Urey of the University of Chicago tested the hypothesis that organic molecules could be synthesized from inorganic ones in 1953. In their laboratory, these scientists recreated conditions presumed to be similar to those of the early earth (Figure 22.3). Miller and Urey discharged electric sparks (meant to simulate lightning) through an atmosphere that contained some of the gases thought to be present in the early atmosphere. They generated a variety of small organic molecules, supporting the hypothesis that organic molecules could be synthesized from inorganic ones.

■ Small organic molecules joined to form larger molecules

Scientists hypothesize that these small organic molecules accumulated in the early oceans and over a long period formed a complex mixture. Somehow, perhaps by being washed onto clay or hot sand or lava, the small molecules joined to form macromolecules (Chapter 2). Some scientists suggest that deep-sea vents were

The atmosphere of the early earth contained little gaseous oxygen (O_2) and probably consisted of gases such as hydrogen (H_2), methane (CH_4), carbon dioxide (CO_2), and carbon monoxide (CO).

FIGURE **22.2** Representation of the early earth

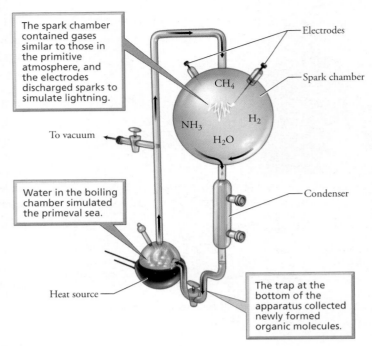

The spark chamber contained gases similar to those in the primitive atmosphere, and the electrodes discharged sparks to simulate lightning.

Electrodes

Spark chamber

CH_4

NH_3 H_2

H_2O

To vacuum

Condenser

Water in the boiling chamber simulated the primeval sea.

Heat source

The trap at the bottom of the apparatus collected newly formed organic molecules.

(a) Using this apparatus, Stanley Miller (shown) and Harold Urey demonstrated that simple organic molecules could be synthesized from inorganic ones in the low-oxygen environment of the primitive earth.

(b) Diagram of the apparatus

FIGURE **22.3** Testing the hypothesis that organic molecules could be synthesized from inorganic ones

important locations for the synthesis of small organic molecules and their joining to form larger molecules.

Which macromolecule led to the formation of the first cells? Some scientists suggest that RNA was the critical macromolecule, because it can act as an enzyme and because its self-replicating abilities allowed the transfer of information from one generation to the next. Cells today store their genetic information as DNA. Thus, if RNA was the first genetic material, then DNA is likely to have evolved later from an RNA template. Other scientists suggest that proteins were the macromolecules that led to the first cells.

■ Macromolecules aggregated into droplets

Scientists hypothesize that the newly formed organic macromolecules aggregated into droplet-like structures that were the precursors of cells. The early droplets displayed some of the same properties as living cells, such as the ability to maintain an internal environment different from surrounding conditions. Fossil evidence indicates that the earliest cells were prokaryotes. Recall from Chapter 3 that eukaryotic cells, such as those in your body, differ from prokaryotic cells in many ways. Eukaryotic cells have an extensive system of internal membranes, a cytoskeleton, and membrane-bound organelles such as the nucleus.

So the question arises, "How did more complex cells arise from the earliest prokaryotic cells?" It is possible that some of the organelles within eukaryotic cells appeared when other, smaller organisms became incorporated into the early, primitive cells. Mitochondria, for example, may be descendants of once free-living bacteria. These bacteria either invaded or were engulfed by the ancient cells and formed a mutually beneficial relationship with them. This idea is called the **endosymbiont theory.** It is generally accepted by the biological community to explain the origin of some features of eukaryotic cells, including mitochondria. At present, scientists are not certain whether endosymbiosis was a factor in the origin of other eukaryotic features such as the cytoskeleton or membrane-bound nucleus. An alternative possibility is that infolding of the plasma membrane of some ancient prokaryote produced some of the organelles (endoplasmic reticulum, Golgi apparatus) found in eukaryotic cells of today.

Fossils of prokaryotic cells have been dated at 3.5 billion years old, and scientists estimate that the first prokaryotic cells probably arose around 3.8 billion years ago. Eukaryotic cells evolved some 2 billion years later. Multicellularity then evolved in eukaryotes and eventually led to organisms such as plants, fungi, and animals. Figure 22.4 summarizes the possible steps responsible for the origin of life on the earth.

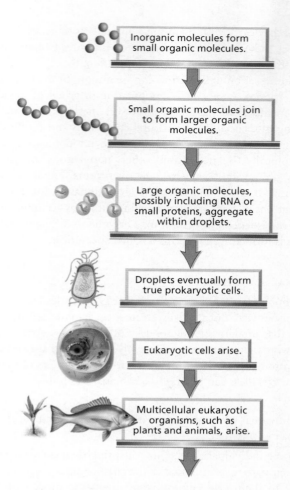

FIGURE **22.4** Possible steps in the origin of life on the earth

The Scale of Evolutionary Change May Be Small or Large

Evolution occurs on two levels. One level, microevolution, is small; and the other, macroevolution, is large. **Micro-evolution** occurs through changes in allele frequencies within a population over a few generations. **Macroevolution**, on the other hand, consists of larger scale evolutionary change over longer periods of time, such as the origin of groups of species (for example, mammals) and mass extinctions (the catastrophic disappearance of large numbers of species). We begin with microevolution, examining genetic variation within populations and describing the causes of microevolution. We then turn to the larger-scale phenomena of macroevolution.

■ Microevolution is small-scale evolutionary change over short periods of time

Look around. You will see variation in almost any group of individuals belonging to the same species. Consider your classmates. They do not all look alike, and, unless you are an identical twin, you probably do not precisely resemble your brothers or sisters. Before we look at what makes individuals in such a group different, let's define some basic terminology. A **population** is a group of individuals of the same species living in a particular area. A population of bluegill sunfish inhabits a pond and a population of deermice inhabits a small tract of forest. A **gene pool** consists of all the alleles of all of the genes of all individuals in a population. (Recall from Chapter 20 that a *gene* is a segment of DNA on a chromosome that directs the synthesis of a specific protein, whereas *alleles* are different forms of a gene.) Now let's take a closer look at what makes individuals in a population different and examine some of the ways that variation can appear in populations.

Sexual reproduction and mutation produce variation in populations. Sexual reproduction shuffles alleles already present in the population. As discussed in Chapter 20, the gametes (eggs or sperm) of any one individual show substantial genetic variation due to crossing over and independent assortment. Also, the combination of gametes that unite at fertilization is a chance event. Of the millions of sperm produced by a male, only one fertilizes the egg. This union produces a new individual with a new combination of alleles.

The only source of new characteristics in a population is *mutation,* a change in the DNA of genes. Mutations occur at a low rate in any set of genes. For example, the overall human mutation rate is estimated to be 1×10^{-6} per gene per generation. Because mutations occur at such a low rate, their contribution to genetic diversity in large populations is quite small. Mutations can appear spontaneously from mistakes in DNA replication, or they can be caused by outside sources such as radiation or chemical agents.

Recall that microevolution involves changes in the frequency of certain alleles relative to others within the gene pool of a population. Some of the processes that produce those changes are genetic drift, gene flow, natural selection, and mutation.

GENETIC DRIFT

Genetic drift occurs when allele frequencies within a population change randomly because of chance alone. This process is usually negligible in large populations. However, in small populations (fewer than about 100 individuals), chance events can cause allele frequencies to drift randomly from one generation to the next. Two mechanisms that facilitate genetic drift in natural populations are the bottleneck effect and the founder effect.

THE BOTTLENECK EFFECT   Sometimes, dramatic reductions in population size occur because of disasters such as fires and earthquakes that kill many individuals at random. Consider a population that experiences a flood in which most members die. The genetic makeup of the survivors may not be representative of the original population. This change in the gene pool is the

bottleneck effect, so named because the population experiences a dramatic decrease in size much as the size of a bottle decreases at the neck. Certain alleles may be more or less common in the survivors of the flood than in the original population simply by chance alone. In fact, some alleles may be completely lost, resulting in reduced overall genetic variability among survivors.

THE FOUNDER EFFECT   Genetic drift also occurs when a few individuals leave their population and establish themselves in a new, somewhat isolated place. By chance alone, the genetic makeup of the colonizing individuals is probably not representative of the population they left. Genetic drift in new, small colonies is called the **founder effect**. This effect typically results in a relatively high frequency of inherited disorders.

The relatively high incidence of Ellis–van Creveld syndrome among the Amish in Lancaster County, Pennsylvania, has been linked to the founder effect. This inherited condition is associated with dwarfism and extra fingers or toes (Figure 22.5). Its incidence in the general population of the United States is about 1 in 60,000 live births. Its incidence in the Old Order Amish is a striking 300 in 60,000 live births. About 200 Amish moved to Pennsylvania from Switzerland between 1722 and 1770. Among the new arrivals, either a Mr. Samuel King or his wife probably carried the recessive allele for Ellis–van Creveld syndrome. The couple and their descendants were apparently quite prolific, and the frequency of the deleterious allele drifted higher in subsequent generations.

GENE FLOW

Another cause of microevolution is **gene flow**, which occurs when individuals move into and out of populations. As individuals come and go, they carry with them their unique sets of genes. Gene flow occurs if these individuals successfully interbreed (mate and produce offspring) with the resident population.

The cessation of gene flow can be important to the formation of new species. A **species** is a population or group of populations whose members are capable of successful interbreeding. Such interbreeding must occur under natural conditions and produce fertile offspring. But consider what happens when a population becomes geographically isolated from other populations of the same species—for example, when a new river forms. The isolated population may begin to take a separate evolutionary route as distinctly different sets of allele frequencies and mutations accumulate. Eventually, the isolated population may become so different that a new species is formed. This process is called **speciation**.

NATURAL SELECTION

In his book *On the Origin of Species* (1859), Charles Darwin argued that species were not specially created, unchanging forms. He suggested that modern species are descendants of ancestral species. In other words, present-day species evolved from past species. Darwin also proposed that evolution occurred by the process of natural selection. His ideas can be briefly summarized as follows:

1. Individual variation exists within a species. Some of this variation is inherited.

2. Some individuals have more offspring than do others because their particular inherited characteristics make them better suited to their local environment; this is the process of **natural selection**.

3. Evolutionary change occurs as the traits of individuals that survive and reproduce become more common in the population. Traits of less successful individuals become less common.

According to Darwin's ideas, an individual's evolutionary success can be measured by fitness (sometimes called Darwinian fitness). **Fitness** compares the number of reproductively viable offspring among individuals. Individuals with greater fitness have more of their genes represented in future generations. To succeed in terms of evolution, one must reproduce (Figure 22.6). Indeed, you could live more than 100 years, but your individual fitness would be zero if you did not reproduce. Some of the diseases or conditions discussed in earlier chapters are associated with zero fitness because they cause death before reproduction (such as Tay-Sachs disease) or sterility (such as Turner's syndrome).

One result of natural selection is that populations become better suited to their particular environment. This transformation of the population toward better fitness in their environment is called **adaptation.** If the environment changes, however, the individuals who have become most finely attuned to the initial environment could lose their advantage, and other individuals with different alleles might be selected by nature to leave more offspring. If the environment once again stabilizes, the population reestablishes itself around a new set of allele frequencies that better meet the new conditions.

FIGURE **22.5** An Amish woman holding her child with Ellis–van Creveld syndrome, a disorder characterized by dwarfism and extra digits. The founder effect (genetic drift in new, small colonies) may explain the relatively high incidence of this genetic disorder among the Amish in Lancaster County, Pennsylvania.

FIGURE **22.6** Fitness is the number of offspring left by an individual. These parents clearly have high fitness.

The evolution of antibiotic resistance in bacteria provides an example of natural selection in action. Consider what can happen when you take an antibiotic to treat a bacterial infection. Many bacteria are killed by the treatment, but some may survive. Those that survive have genes that confer resistance to the antibiotic. The survivors reproduce and pass on the trait for resistance to future generations of bacteria. Over time, antibiotic resistance becomes more common in bacterial populations than before. The evolution of drug-resistant strains may prove to be one of the greatest threats to humankind (Chapter 13a).

Natural selection does not lead to perfect organisms. It can act on available variation only, and the available variation may not include the traits that would be ideal. Also, natural selection can only modify existing structures; it cannot produce completely new and different structures from scratch. Finally, organisms have to do many different things in order to survive and reproduce—such as escaping from predators and finding food, shelter, and mates—and it simply is not possible to be perfect at all of them. Indeed, adaptations often are compromises between the many competing demands the organism faces.

stop and think

Some people have suggested that natural selection no longer operates within the human species because modern medicine levels the playing field. Medical technology, they argue, helps "fit" and "unfit" individuals alike to survive and reproduce. Others believe that natural selection continues to operate in modern society and that some present-day conditions do act as selective agents. Air pollution, for example, might select for individuals with genetic resistance to respiratory disease. Which argument do you favor and why?

MUTATION

Mutations, you will recall, are rare changes in the DNA of genes; they are the fourth way (we have already considered genetic drift, gene flow, and natural selection) that the frequencies of certain alleles can change relative to others within gene pools. Mutations produce new alleles that, when transmitted in gametes, cause an immediate change in the gene pool. Essentially the new (mutant) allele is substituted for another allele. If the frequency of the mutant allele increases in a population, it is not because mutations are suddenly occurring with greater frequency. Instead, possession of the mutant allele might confer some advantage that enables individuals with the mutant allele to produce more offspring than those without it. In other words, the increased frequency of the mutant allele relative to others results from natural selection.

■ Macroevolution is large-scale evolutionary change over long periods of time

Large-scale evolutionary change is macroevolution. Whereas microevolution involves changes in the frequencies of alleles within populations, macroevolution produces changes in groups of species, such as might occur with major changes in climate. Our discussion of macroevolution begins with a description of how species are named. We then consider how their evolutionary histories can be analyzed and diagrammed.

SCIENTIFIC NAMES

Systematic biology deals with the naming, classification, and evolutionary relationships of organisms. A universal system for naming and classifying organisms is essential for communicating information about them. Scientists use the naming system that Swedish naturalist Carl Linnaeus developed more than 200 years ago. Each organism is given a Latin binomial, or two-part name, consisting of the genus name followed by the specific epithet. For example, humans are in the genus *Homo*, and the specific epithet is *sapiens*, so our binomial is *Homo sapiens*. By convention, the genus name and specific epithet are italicized, the first letter of the genus name is always capitalized, and the specific epithet is all lowercase. Sometimes, the genus name is abbreviated: *H. sapiens*.

Linnaeus also developed a system for classifying organisms using a series of increasingly broad categories: species, genus, family, order, class, phylum, and kingdom. Similar species were placed in the same genus; similar genera (the plural of *genus*) were placed in the same family; similar families in the same order, and so on. Above the kingdom level, scientists have added the category domain to Linnaeus's scheme (the three domains are Archaea, Bacteria, and Eukarya; Chapter 1). Figure 22.7 presents the categories to which humans belong.

PHYLOGENETIC TREES

Phylogenetic trees are branching diagrams used by scientists to depict hypotheses about evolutionary relationships among species or groups of species. Such trees can illustrate in simple graphic form concepts that are difficult to express in words.

Scientists developing hypotheses about these relationships may begin by creating a character matrix, such as the example in

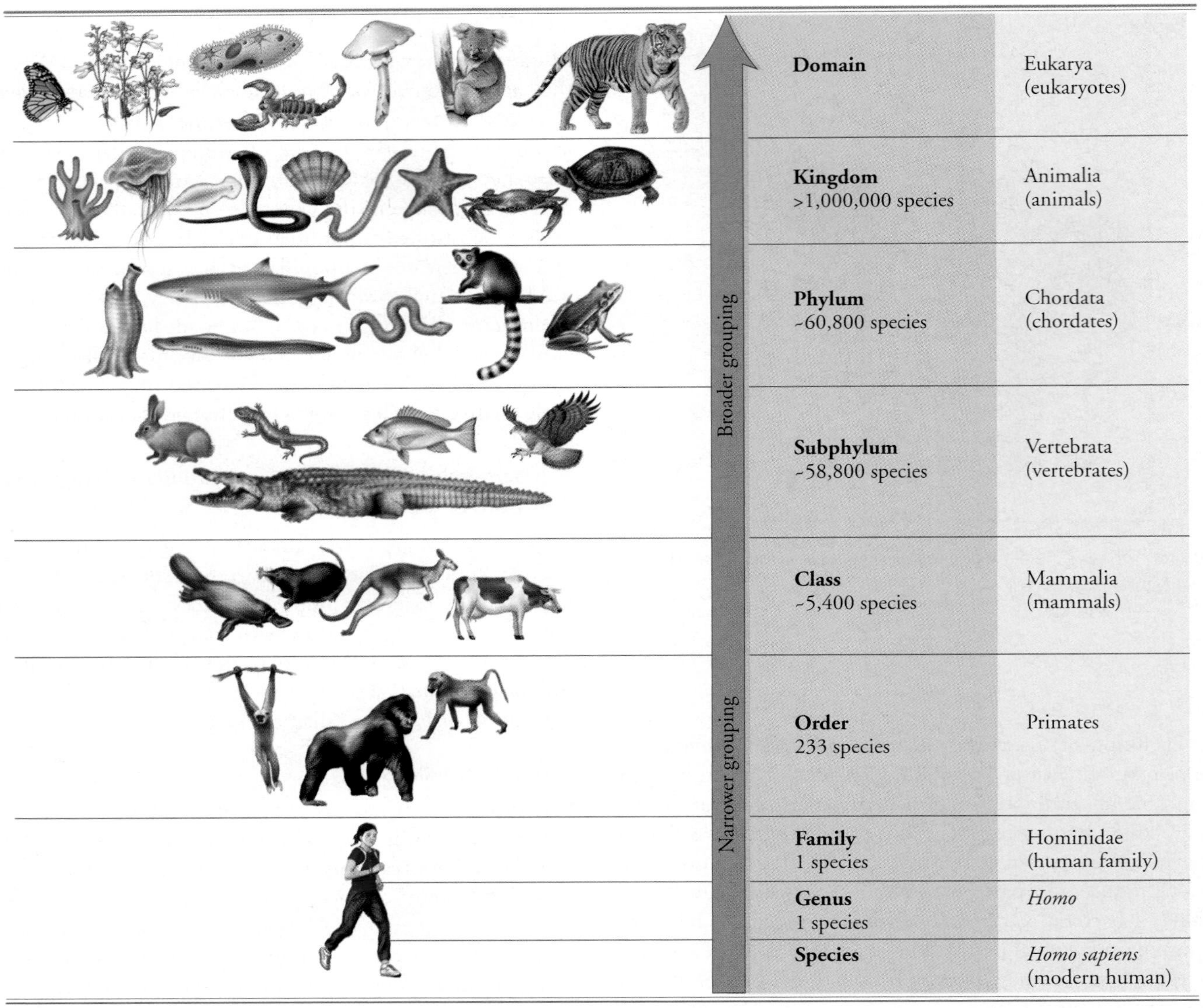

Domain	Eukarya (eukaryotes)
Kingdom >1,000,000 species	Animalia (animals)
Phylum ~60,800 species	Chordata (chordates)
Subphylum ~58,800 species	Vertebrata (vertebrates)
Class ~5,400 species	Mammalia (mammals)
Order 233 species	Primates
Family 1 species	Hominidae (human family)
Genus 1 species	*Homo*
Species	*Homo sapiens* (modern human)

Broader grouping — Narrower grouping

FIGURE **22.7** Categories in the classification of living organisms. Any organism, including a human being, can be classified using a hierarchy of increasingly general categories. Modern humans are called *Homo sapiens*. We are the only living species in our genus (*Homo*) and in our family (Hominidae).

Figure 22.8a. The character matrix can then be used to construct a phylogenetic tree. Typical matrices consist of vertical columns representing species or other groups and horizontal rows representing "characters" that are either present or absent in those species or groups. Figure 22.8a shows a simple matrix with a shark (fish), frog (amphibian), and human (mammal) being compared in terms of the presence or absence of three characters: two paired appendages (fins or limbs), individual digits, and hair. A phylogenetic tree based on the results from the matrix is shown in Figure 22.8b. This tree suggests that humans and frogs have more in common with one another (two paired appendages and individual digits) than they have with sharks (two paired appendages), and thus humans and frogs are likely to be more closely related (share a more recent common ancestor). The tree also shows that humans differ from frogs in having hair.

Evidence of Evolution Comes from Diverse Sources

We know that evolution has occurred because the physical evidence of evolution surrounds us. Such evidence comes from many sources, including the fossil record, biogeography, the comparison of anatomical and embryological structures, and molecular biology.

■ The fossil record provides evidence of evolution

The earth is littered with silent relics of organisms that lived long ago (Figure 22.9). We find, for example, tiny spiders preserved in resin that dripped as sticky sap from some ancient tree. Also, we find mineralized bones and teeth, hardened remains that tell us

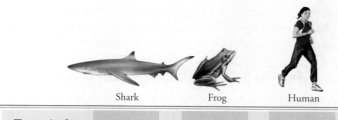

	Shark	Frog	Human
Two paired appendages	Yes	Yes	Yes
Digits	No	Yes	Yes
Hair	No	No	Yes

(a) A character matrix is used to construct a phylogenetic tree.

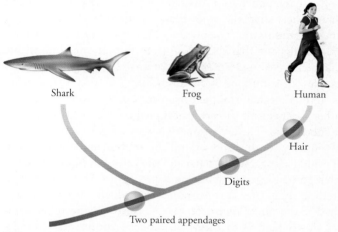

(b) A sample tree developed from the above character matrix

FIGURE **22.8** A phylogenetic tree depicts hypotheses about evolutionary relationships among organisms.

much about the ancestry of today's vertebrates (animals with a backbone). These preserved remnants and impressions of past organisms are **fossils**. Most fossils occur in sedimentary rocks, such as limestone, sandstone, shale, and chalk. These rocks form when sand and other particles settle to the bottom of rivers, lakes, and oceans; accumulate in layers; and harden.

Fossilization is any process by which fossils form (Figure 22.10). In a typical case, an organism dies and settles to the bottom of a body of water. If not destroyed by scavengers, the organism is buried under accumulating layers of sediment. Soft parts usually decay. Hard parts such as bones, teeth, and shells may be preserved if they become impregnated with minerals from surrounding water and sediments. As new layers of sediment are added, the older (lower) layers solidify under the pressure generated by overlying sediments. Eventually, if the sediments are uplifted by geological processes and the water disappears, weather may erode the surface of the rock formation and expose the fossil.

Fossils provide strong evidence of evolution. Fossils of extinct organisms show both similarities to and differences from living species. Similarities to other fossil and modern species are used to assess degrees of evolutionary relationships. Often, fossils reveal combinations of features not seen in any living forms. Such combinations help us understand how major new adaptations arose. Sometimes we are lucky enough to find transitional forms that closely link ancient organisms to modern species. Consider a fossil whale discovered in the last few years that had hind limbs and an aquatic lifestyle. This transitional form links older terrestrial forms that had hind limbs to modern whales, which are aquatic and have no hind limbs. The relative ages of fossils can be determined because fossils found in deeper layers of rock are typically older than those found in layers closer to the surface. These kinds of observations enable scientists to piece together the chronological emergence of different kinds of organisms. For example, fishes are the first fossil vertebrates to appear in deep (old) layers of rock. Above those

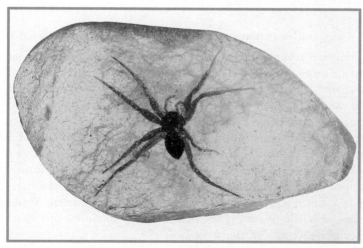

(a) A spider preserved in amber

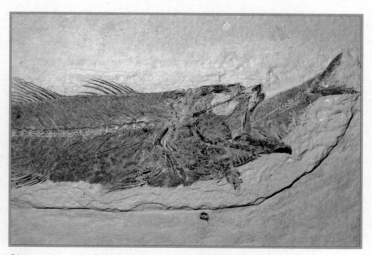

(b) An ancient act of predation preserved in sedimentary rocks

FIGURE **22.9** A sampling of past life in fossils

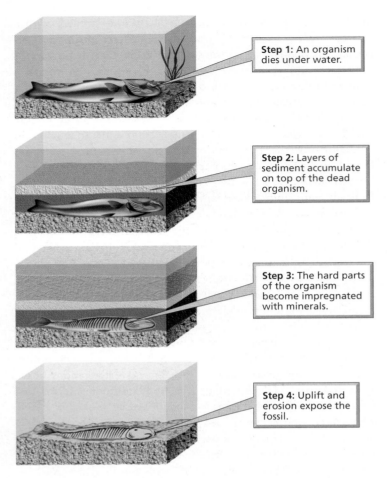

Step 1: An organism dies under water.

Step 2: Layers of sediment accumulate on top of the dead organism.

Step 3: The hard parts of the organism become impregnated with minerals.

Step 4: Uplift and erosion expose the fossil.

FIGURE **22.10** A typical sequence for fossilization

layers, fossils of amphibians appear, then reptiles, then mammals, and finally birds. This chronological sequence for the appearance of the major groups of vertebrates has been supported by other lines of evidence, some of which we consider later in this section. Other large-scale evolutionary events revealed by the fossil record include changes in diversity over time and changes in faunas (animal life) and floras (plant life) as continents moved (geologists have mapped major changes in the sizes and locations of continents over time).

Although the fossil record tells us much about past life, it has limitations. First, fossils are relatively rare. When most animals or plants die, their remains are eaten by predators or scavengers or are broken down by microorganisms, chemicals, or mechanical processes. Even if a fossil should form, the chances are small that it will be exposed by erosion or other forces and not be destroyed by those same forces before it is discovered. Second, the fossil record represents a biased sampling of past life. Aquatic plants and animals have a much higher probability of burial in deep sediment than do terrestrial organisms. Thus, preservation of aquatic organisms is more likely. Large animals with a hard skeleton are far more likely to be preserved than are small animals with soft parts. Organisms from large, enduring populations are more likely to be represented in the fossil record than are those from small, quickly disappearing populations. Despite these limitations, fossils document that life on the earth has not always been the same as it is today. The simple fact of these changes is potent evidence of evolution.

■ Geographic distributions reflect evolutionary history

Biogeography is the study of the geographic distribution of organisms. Geographic distributions often reflect evolutionary history and relationships because related species are more likely than are unrelated species to be found in the same geographic area. A careful comparison of the animals in a given place with those occurring elsewhere can yield clues about the relationship of the groups. If groups of animals have been separated, biogeography can tell us how long ago the separation occurred.

For example, today, we find many species of marsupials—mammal species such as opossums and kangaroos, in which the female often has a pouch—in Australia but only a few in North and South America. Australia is an island, remote from other major continental landmasses. The presence of so many species of marsupials there suggests that they arose from distant ancestors whose descendants were not replaced by animals arriving from other places. New distributions of organisms occur by two basic mechanisms. In one mechanism, the organisms disperse to new areas. In the other mechanism, the areas occupied by the organisms move or are subdivided. The evolutionary history of the Australian marsupials involves both dispersal and the movement of continents. Fossil evidence suggests that marsupials evolved in North America and that some dispersed southward into South America, then to Antarctica, and later to Australia, to which Antarctica was attached at the time (Figure 22.11). As Australia and other landmasses slowly shifted to form the modern continental arrangement, the ancestors of today's marsupials were carried away from their place of origin to evolve in isolation in Australia. As the continental landmasses separated, they carried populations of other living things with them as well. Trees, freshwater fish, and even birds are quite different in Australia than in other parts of the world, reflecting Australia's long history of isolation and continental movement.

■ Comparative anatomy and embryology reveal common descent

Comparative anatomy, as its name suggests, is the comparison of the anatomies (physical structures) of different species. Common, or shared, traits among different species have long been considered a measure of relatedness. Put simply, two species with more shared traits are considered more closely related than are two species without shared traits (this principle underlies the use of character matrices to construct phylogenetic trees). For example, the vast majority of vertebrates alive today have jaws and paired fins or limbs. All vertebrates with these features are likely to share a common ancestor. As another example, many very different vertebrates share similarities in the bones of their forelimbs that show they too have an ancestor in common (Figure 22.12). Structures that are similar and that probably arose from a common ancestry are called **homologous structures**. The forelimbs that support bird wings and bat wings are homologous structures. Sometimes,

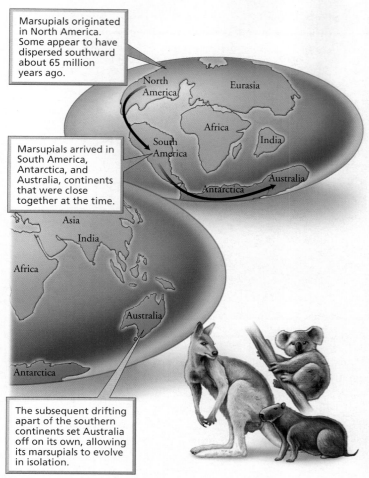

Marsupials originated in North America. Some appear to have dispersed southward about 65 million years ago.

Marsupials arrived in South America, Antarctica, and Australia, continents that were close together at the time.

The subsequent drifting apart of the southern continents set Australia off on its own, allowing its marsupials to evolve in isolation.

WEB TUTORIAL 22.3 FIGURE **22.11** The story of marsupials and Australia involves both dispersal and movement of continents.

however, similarities are not inherited from a common ancestor. For example, the wings of birds and those of insects both permit flight, but they are made from entirely different structures. Whereas bird wings consist of forelimbs, the wings of insects are not true appendages; they are extensions of the insect's cuticle (exoskeleton). Thus, the wings of birds and insects do not reflect common ancestry. Instead, birds and insects independently evolved wings because of similar ecological roles and selection pressures, in a process known as **convergent evolution**. Structures that are similar because of convergent evolution are called **analogous structures.**

Homologous structures arise from the same kind of embryonic tissue. Hence, comparative embryology, the comparative study of early development, also can be a useful tool for studying evolution. Common embryological origins can be considered evidence of common descent. For example, 4-week-old human embryos closely resemble embryos of other vertebrates, including fish. Indeed, human embryos at 4 weeks gestation come complete with a tail and gill pouches, as shown in Figure 22.13. As development proceeds, the gill pouches of fish develop into gills. The gill pouches of hu-

mans develop into other structures such as the auditory tubes connecting the middle ear and throat. Nevertheless, the fact that human, fish, and all other vertebrate embryos look very similar at early stages of development indicates common descent from an ancient ancestor.

■ Comparative molecular biology also reveals evolutionary relationships

Just as visible traits such as forelimbs can be compared, so can the molecules that are the basic building blocks of life. For example, scientists can compare the sequences of amino acids in proteins or the nucleotide sequences in DNA. As described in Chapter 21, the Human Genome Project has provided information on the location of genes along our chromosomes and the order of the base pairs that make up our chromosomes. More recently, the genome of the chimpanzee has been described and compared with that of humans. Such comparison reveals that the genomes of humans and chimps are strikingly similar; in fact, we share about 96% of our DNA sequence.

The study of comparative molecular biology suggests the existence of a **molecular clock,** a constant rate of divergence of macromolecules (such as proteins) over time as species evolve. The molecular clock hypothesis is based on the notion that single nucleotide changes in DNA, called point mutations, and consequent changes in the amino acids of proteins occur with steady, clocklike regularity. If this is true, then scientists can compare molecular sequences in two species as a way of estimating the amount of time that has passed since the species' lineages diverged from a single ancestor. The more differences in sequence, the more time that has elapsed since the common ancestor. For example, a comparison of such sequences tells us that humans and chimpanzees diverged from a common ancestor about 6 million years ago. (See the Social Issue essay, *Conducting Research on Our Relatives.*)

We have learned about microevolution, macroevolution, and the evidence for evolution. Now let's look at our own past and see how humans evolved.

Human Roots Trace Back to the First Primates

We begin our discussion of human evolution with the primates, an order of mammals that includes humans, apes, monkeys, and related forms (such as lemurs). Scientists do not agree about how long ago the first primates evolved. Paleontologists (scientists who study fossils) believe that primates evolved about 65 million years ago. In contrast, molecular biologists have suggested that primates evolved about 90 million years ago.

stop and think

In the following discussion of our ancestry, we will be describing the species that led to us. At some point the line *became* us. Before you proceed to read about our ancestry, what qualities would you now say are necessary for a species to qualify as human?

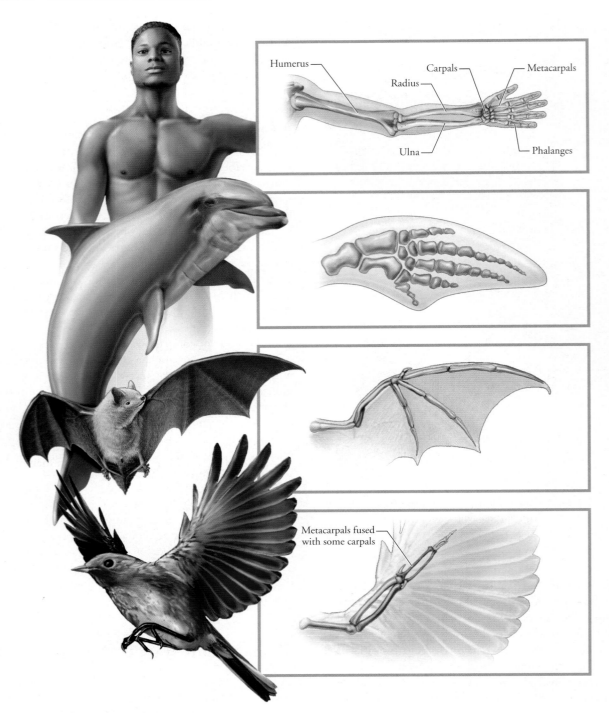

Humerus · Carpals · Metacarpals · Radius · Ulna · Phalanges

Metacarpals fused with some carpals

FIGURE **22.12** Homologous structures. The similarity of the bones of the forelimbs of humans, dolphins, bats, and birds suggests that these organisms share a common ancestry.

■ Primates have distinct characteristics

Despite debate over when primates evolved, most scientists agree that the first primates probably arose from an insect-eating mammal that lived in trees. This ancestor to primates might have looked something like a modern tree shrew (Figure 22.14). Primates share many characteristics that reflect an arboreal lifestyle specialized for the visual hunting and manual capture of insects. For example, primates have flexible, rotating shoulder joints and exceptionally mobile digits with sensitive pads on their ends. Flattened nails replace claws. In many primate species, the big toe is separated from the other toes, and thumbs are opposable to other fingers. Thus, primates have grasping feet *and* hands, features that help in the pursuit and capture of insects along branches. In addition, a complex visual system and a large brain relative to body size provide the

Development

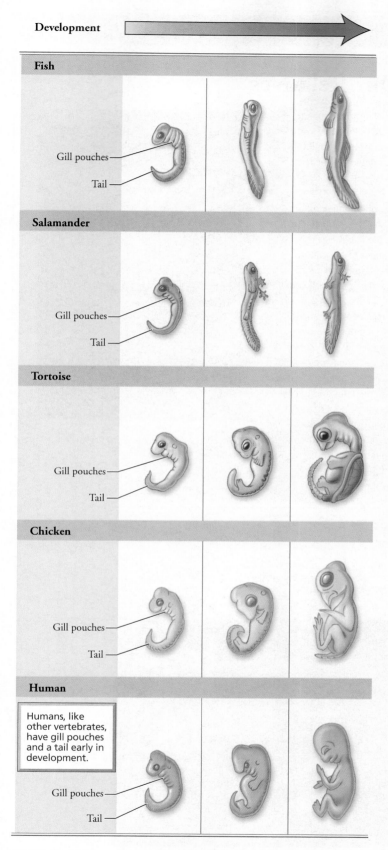

Fish

Gill pouches

Tail

Salamander

Gill pouches

Tail

Tortoise

Gill pouches

Tail

Chicken

Gill pouches

Tail

Human

Humans, like other vertebrates, have gill pouches and a tail early in development.

Gill pouches

Tail

FIGURE **22.13** Resemblance early in development indicates common descent. The embryos in the different representative vertebrates have been drawn to the same approximate size to permit comparison.

FIGURE **22.14** A tree shrew. These modern animals resemble the arboreal, insect-eating mammals from which primates probably evolved over 65 million years ago.

well-developed depth perception, hand-eye coordination, and neuromuscular control needed by arboreal insectivores. Also, most primates give birth to only one infant at a time (even two is quite rare) and provide extensive parental care—characteristics that may reflect the difficulty of carrying and rearing infants in trees (Figure 22.15). Humans are the most terrestrial of primates, yet we retain many of the basic characteristics that evolved in the trees. The characters that distinguish primates from other mammals are summarized in Table 22.1.

Modern primates are divided into two main groups. The first group contains lemurs, lorises, and pottos (grouped together because they retain ancestral primate features such as small body size and nocturnal habits). Figure 22.16 shows representatives of this first group. The second group contains monkeys, apes, and hu-

FIGURE **22.15** This female loris with young illustrates many of the features characteristic of primates and associated with an arboreal lifestyle, including grasping hands and feet, forward-facing eyes, and small litter size.

SOCIAL ISSUE

Conducting Research on Our Relatives

Chimpanzees look and behave somewhat differently than we do. Nevertheless, it is hard to look into their eyes and not see something of ourselves (Figure 22.A). Chimpanzees are our closest living relatives, sharing about 96% of our DNA sequence. Is it ethical to keep chimpanzees in laboratory cages and use them for invasive scientific research that might benefit us? What about other primate species to which we are more distantly related?

An estimated 55,000 nonhuman primates were used in research in the United States during the year 2004. Despite the controversy and cost of using nonhuman primates, they often are preferred as subjects in certain experiments precisely because they are so similar to humans. For example, human and nonhuman primates have brains with similar organization, develop comparable plaques in their arteries, and experience many of the same changes in anatomy, physiology, and behavior with age. In some cases, Nobel Prize–winning research has resulted from the contributions of research with nonhuman primates, including development of vaccines for yellow fever (1951) and polio (1954)

and insight into how visual information is processed in the brain (1981). Research with nonhuman primates has also led to significant advances in our understanding of Alzheimer's disease, AIDS, and SARS (severe acute respiratory syndrome).

The care and use of nonhuman primates (and other animals) in research is regulated by federal agencies such as the Public Health Service and the U.S. Department of Agriculture. Animal research also is regulated at the local level. Each college, university, or research center has an Institutional Animal Care and Use Committee whose members include veterinarians, teachers, researchers, and members from the public. In addition to federal and local oversight, scientists and animal care personnel are making substantial efforts to provide appropriate housing for captive nonhuman primates and to consider their psychological well being. Even with such efforts, however, controversy and questions remain. Should we ban the use of nonhuman primates in medical research? And if we do, will progress in fighting diseases such as AIDS and Alzheimer's be slowed? ✷

FIGURE **22.A** Our closest living relative, the chimpanzee

(**a**) Lemurs   (**b**) Loris   (**c**) Potto

FIGURE **22.16** Examples of modern primates that retain ancestral primate features and form the first major group of primates

TABLE 22.1 REVIEW OF CHARACTERISTICS OF PRIMATES
Flexible shoulder joints
Grasping forelimbs; some species with fully opposable thumbs
Grasping feet, with big toe separated from other toes (except humans)
Sensitive digits with flattened nails instead of claws
Forward-facing eyes
Small litter size
Complex social behavior, including extensive parental care
Reliance on learned behavior

mans (Figure 22.17). The term **hominid** is traditionally used for members of the human family, such as those in the genera *Australopithecus* and *Homo* (discussed later in this chapter). Figure 22.18 shows hypothesized relationships among living primates. As described earlier, molecular evidence suggests that the lines leading to modern humans and chimpanzees diverged about 6 million ago. Molecular data further indicate that after chimpanzees, gorillas are our next closest living relatives, followed by orangutans and then gibbons.

■ Discussions of human origins often evoke controversy

Paleoanthropology is the scientific study of human origins and evolution. This field combines the disciplines of paleontology (the study of fossils) and anthropology (the study of humans). Many of the debates concerning human evolution arose because of practices followed by earlier paleoanthropologists. For example, in the past, paleoanthropologists tended to give each new fossil discovery a new name and to consider it an example of a separate line. More recent studies indicate that many of these "finds" are probably of the same lineage. The resulting confusion was compounded by hasty conclu-

sions some paleoanthropologists extrapolated from sketchy information, such as a partial skull or a few teeth. In addition, the early paleoanthropologists failed to recognize the inevitability of variation in any given population. Two specimens that differed slightly in some traits were likely to be interpreted as belonging to different lines when in fact they may have simply been examples of variation within a single species. Then, there is the problem of the overall paucity of fossils. There just have not been enough hominid remains found to reconstruct a solid, unequivocal history of our species. Finally, many arguments have arisen because of certain popular misconceptions among nonscientists, as we see in the next section.

■ Misconceptions distort the picture of human evolution

One misconception expressed by many nonscientists is the idea that we descended from chimpanzees or any of the other modern apes. Humans and chimpanzees represent separate phylogenetic branches that diverged about 6 million years ago. Thus, the common ancestor of humans and chimpanzees was different from any modern species of ape.

Another misconception is that modern humans evolved in an orderly, stepwise fashion. We often see such a stepwise progression depicted in drawings, sometimes humorously, and its appeal lies in its simplicity. However, as is so often the case, the real story is less tidy and far more complex. The path to modern humans has been fraught with unsuccessful phenotypes leading to dead end after dead end. In fact, the path looks more like a family "bush" than an orderly progression from primitive to modern. (Note, for example, in Figure 22.22, that more than one species of hominid existed simultaneously at some points in the past.)

A final misconception is that, over the course of human evolution, the various bones and organ systems evolved simultaneously and at the same rate. They did not. There is no reason to believe that the human brain evolved at the same rate as, say, the appendix or the foot. Instead, different traits evolved at different times and rates, a phenomenon known as **mosaic evolution.**

(a) Spider monkey   **(b)** Baboons   **(c)** Orangutans   **(d)** Gorilla

FIGURE **22.17** Monkeys (such as spider monkeys and baboons) and apes (such as orangutans and gorillas) are placed with humans in the second major group of primates.

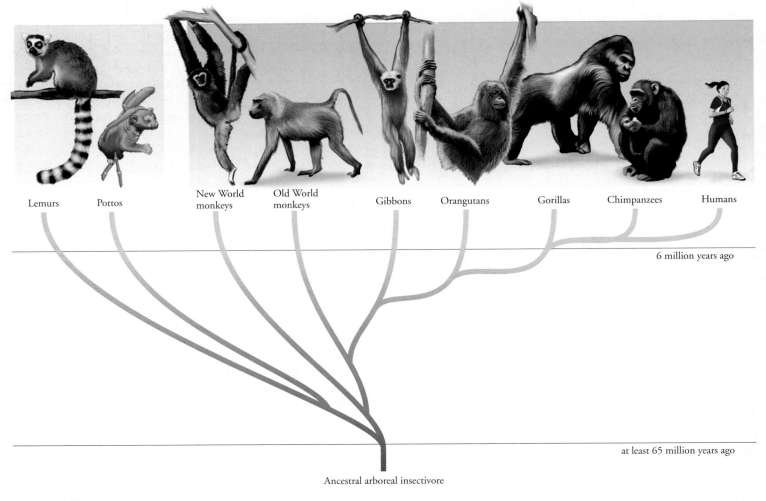

| Lemurs | Pottos | New World monkeys | Old World monkeys | Gibbons | Orangutans | Gorillas | Chimpanzees | Humans |

6 million years ago

at least 65 million years ago

Ancestral arboreal insectivore

FIGURE **22.18** Summary of primate evolution. (Figures are not drawn to scale.)

■ Walking on two feet was a critical step in hominid evolution

Several evolutionary trends are apparent in the history of hominids. **Bipedalism**—walking on two feet—evolved early and probably set the stage for the evolution of other characteristics, such as increases in brain size. Cultural developments, such as tool use and language, are linked to increases in brain size. Once the hands were freed from the requirements of locomotion, they could be used for tasks such as making tools. Evidence that bipedalism preceded increases in brain size and cultural developments comes from fossilized hominid footprints found in Tanzania, Africa. These footprints, estimated to be about 3.6 million years old, were apparently made by two adults and a child (Figure 22.19). The footprints clearly predate the oldest stone tools, from 2.6 million years ago.

Other changes associated with upright posture include the S-shaped curvature of the vertebral column (the lumbar curve); modifications to the bones and muscles of the pelvis, legs, and feet; and positioning of the skull on top of the vertebral column. The faces of hominids also changed (see Figure 22.22). For example, the forehead changed from sloping to vertical, and sites of muscle

FIGURE **22.19** These hominid footprints from Laetoli, Tanzania, predate the oldest known tools and thus provide evidence that bipedalism preceded increases in brain size and cultural trends such as tool making. The larger prints were made by two individuals, one following in the footsteps of the other. The smaller prints may have been made by a child walking with the two individuals.

TABLE 22.2 REVIEW OF MAJOR TRENDS IN HOMINID EVOLUTION
• Bipedalism
• Increasing brain size and associated cultural trends such as tool use, language, and behavioral complexity
• Shortening of the jaw and flattening of the face
• Reduced difference in size between males and females

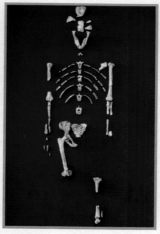

(a) Lucy is a remarkably complete skeleton of a young female *Australopithecus afarensis*, a hominid in existence over 3 million years ago.

(b) Reconstruction of Lucy on display at the St. Louis Zoo

FIGURE **22.20** Lucy

attachment, such as the brow ridges and crests on the skull, became smaller. The jaws became shorter, and the nose and chin more prominent. The overall size difference between males and females decreased. Males of our early ancestors appear to have been 1.5 times the size of females. Modern human males weigh about 1.2 times as much as females. A difference in appearance between the males and females of a species is called *sexual dimorphism.* Table 22.2 summarizes major trends in hominid evolution.

When considering our ancestors, we will focus on the hominids that we know the most about—those in the genus *Australopithecus* and the genus *Homo.* We will also mention some recent finds of apparently older hominids. Keep in mind that as new hominid fossils are found and genetic studies conducted, the dates for the origin of many species will likely be pushed back.

AUSTRALOPITHICINES

The first hominid remains discovered were given the genus name *Australopithecus,* meaning *southern ape.* Species within *Australopithecus* are sometimes collectively termed australopithecines. *Australopithecus anamensis,* considered the earliest species of the australopithecine line, is known from a small number of fossils found in Kenya and Ethiopia and dated at between 4.2 and 3.9 million years old. The most spectacular australopithecine fossil found to date is that of a young adult female of the species *Australopithecus afarensis* (because she was found in the Afar region of Ethiopia). She was named Lucy because the Beatles' song "Lucy in the Sky with Diamonds" was playing the night Donald Johanson and his coworkers celebrated her discovery. When more than 60 pieces of Lucy's bones were arranged, scientists estimated that, at death, she was about 1 m (3 ft) tall and weighed about 30 kg (66 lb) (Figure 22.20). Her bones were determined to be 3.2 million years old. As more remains were found, it became apparent that the males of Lucy's species were somewhat taller (about 1.5 m, or about 5 ft) and heavier (about 45 kg, or 99 lb). The brain of *A. afarensis* was similar in size to that of modern chimpanzees or gorillas, about 430 cm^3 (26 in.^{3}). Although many aspects of the anatomy of *A. afarensis* suggest adaptations for living in trees, the remains also indicate bipedalism. In *A. afarensis,* we see an example of mosaic evolution—bipedalism evolving before substantial increases in brain size.

Since the mid-1990s, researchers have found hominid remains older than those of *A. anamensis.* Some remains from Ethiopia were dated at 4.4 million years old and assigned to the species *Ardipithecus ramidus.* Some researchers think *A. ramidus* displayed bipedalism, although that point is debated. Another uncertainty is whether *A. ramidus* is the ancestor of *Australopithecus* or a separate line that became extinct.

stop and think

Brain size and shape can be estimated by making an endocast, a model of the inside of the skull. Such reconstructions are possible even when only a few parts of the skull are discovered and reassembled. Endocasts reveal general brain anatomy and size, allowing scientists to draw tentative conclusions about the cognitive capabilities of the skull's owner. Endocasts do not, however, capture the folds or convolutions on the surface of a brain. Do you think endocasts and brain volumes make reliable indicators of intelligence? Why or why not?

About 3 million years ago, when *A. afarensis* had been in existence for nearly 1 million years, several new hominid species appeared in the fossil record. Scientists believe that *Australopithecus africanus,* one of these new species, was a hunting-and-gathering omnivore. *A. africanus,* like *A. afarensis,* was a gracile, or slender, hominid. Three more "robust" hominids (previously called *Australopithecus aethiopicus, A. robustus,* and *A. boisei,* but now considered by many researchers to represent a group deserving of placement in a separate genus, *Paranthropus*) also appeared and are thought to have been savanna-dwelling vegetarians. The robust hominids had a massive skull, heavy facial bones, a pronounced brow, and huge teeth. Whatever they ate, it required a lot of chewing. It is unclear whether Lucy's species, *A. afarensis,* gave rise to these other species or simply lived at the same time as they did. The robust hominids appear to have been evolutionary dead ends. Descendants of *A. afarensis* may have lead to our genus, the genus *Homo.*

HOMO HABILIS

Homo habilis ("handy man"), the first member of the modern genus of humans, appeared in the fossil record about 2.5 million years ago. The remains classified as *H. habilis* are highly variable, causing some researchers to question their classification. Some researchers believe the remains are varied enough to represent more than one species. *H. habilis* differed from *A. afarensis* primarily in brain size. The cranial capacity of *H. habilis* has been estimated at between 500 and 800 cm^3 (30 and 49 in.^{3}). Some scientists hypothesize that *H. habilis* was the first hominid to use stone tools. Simple stone tools, dating from 2.5 million to 2.7 million years ago have been found in Africa. Whether these tools were used by *H. habilis* or one of the species of robust hominids is unclear. It also is possible that *H. habilis* was capable of rudimentary speech. Casts of one brain made from reassembled skull fragments indicate a bulge in Broca's area, the area of the brain important in speech (see Chapter 8).

HOMO ERGASTER AND HOMO ERECTUS

A new hominid, *Homo ergaster* ("working man"), appeared in the fossil record about 1.9 million years ago. The name reflects the many tools found with the remains. Traditionally, scientists had classified these remains as *Homo erectus* ("upright man"), but they now differentiate them from *H. erectus*. *H. ergaster* appears to have originated in East Africa and coexisted there for several thousand years with some of the robust hominids.

H. erectus is thought to have diverged from *H. ergaster* around 1.6 million years ago. *H. erectus* was a wanderer, believed by many to be the first hominid to migrate out of Africa, spreading to Asia. *H. erectus* differed from earlier hominids in being larger (up to 1.85 m, or 6 ft tall, and weighing at least 65 kg, or 143 lb) and less sexually dimorphic. *H. erectus* had a brain volume of about 1000 cm^3 (61 in.^{3}). Evidence indicates that *H. erectus* used so-phisticated tools and weapons and in addition may have used fire. *H. erectus* disappeared from most locations about 400,000 years ago, but some remains from Java have been dated at only 50,000 years old. The Java remains suggest that at least one population of *H. erectus* existed at the same time as modern humans (*Homo sapiens*).

HOMO HEIDELBERGENSIS, HOMO SAPIENS, AND HOMO NEANDERTHALENSIS

Traditionally, fossils that did not quite resemble modern humans were classified as Archaic *Homo sapiens*. Scientist now place these fossils in the species *Homo heidelbergensis;* the name refers to Heidelberg, Germany, where a fossil lower jaw intermediate to those of earlier forms and *H. sapiens* was found. These intermediate specimens range from about 800,000 years old to about 130,000, the age of the first modern human remains. Many scientists believe that *H. heidelbergensis* evolved from *H. ergaster* and not *H. erectus.*

We now come to our own species, *H. sapiens* ("thinking man"); the oldest fossil evidence for modern humans comes from Africa and is about 130,000 years old. *H. sapiens* differs from earlier humans in having a larger brain (1300 cm^3, or 79 in.^{3}), flat forehead, absent or very small brow ridges, prominent chin, and a very gracile body form.

Neanderthals, close evolutionary relatives of ours, are known from Europe and Asia from about 200,000 years ago to 30,000 years ago. Neanderthals had distinct features apparently adapted for life in a cold climate. Some Neanderthals lived in caves ("the cave men"), and others made temporary camps (Figure 22.21a). Neanderthal burial sites have been discovered, making them the first people known to have buried their dead. Also, the discovery of 50,000-year-old remains of sick, injured, and elderly individuals suggests that Neanderthals cared for the less fortunate among them.

(a) The Neanderthals had an appearance adapted for life in a cold climate. Some anthropologists consider Neanterthals to be a subspecies of our own species; others consider them a species separate from our own.

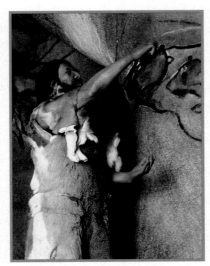

(b) The Cro-Magnons were quite similar to us in appearance, and some anthropologists believe they were responsible for the disappearance of the Neantherthals.

FIGURE **22.21** Relatively recent representatives of the genus *Homo*

Interestingly, Neanderthals had a larger brain case than do *H. sapiens* and a slightly larger brain volume (about 1450 cm³ or 88 in.³). However, these features may not have been correlated with intelligence but rather with the Neanderthals' more massive body. They had larger bones, suggesting heavier musculature, and rather short legs. They also had a thick brow ridge, large nose, broad face, and well-developed incisors and canines. Some anthropologists consider Neanderthals to be a subspecies of *H. sapiens*, calling them *H. sapiens neanderthalensis* (according to this scheme modern humans are known as *H. sapiens sapiens*). Most anthropologists, however, assign Neanderthals species status and call them *Homo neanderthalensis*.

Neanderthals vanished from the fossil record some 30,000 years ago for still mysterious reasons. Some scientists suggest they were outcompeted or simply killed outright by a form of *H. sapiens* called Cro-Magnons (Figure 22.21b). Other scientists suggest that interbreeding between anatomically modern humans and Neanderthals might have resulted in the loss of the Neanderthal phenotype. The extent and type of interactions between Neanderthals and Cro-Magnons are still debated. It is clear, however, that the Cro-Magnons had superior tools and weapons, including bows and arrows, knives, and stone-tipped spears. It is generally believed that Cro-Magnons were quite similar in appearance to humans of today and lived in small to moderate-sized groups. They also were accomplished hunters and artists.

Major milestones in human evolution are summarized in Table 22.3, and the major hominid species are shown in Figure 22.22.

TABLE 22.3 REVIEW OF SOME MILESTONES IN HUMAN EVOLUTION

HOMINID	YEARS AGO	MILESTONE
Ardipithecus ramidus	4.4 million	Bipedalism?
Homo habilis	2.4–1.6 million	Tool use, speech
Homo erectus	1.6 million–50,000	Fire, migration
*Homo neanderthalensis**	200,000–30,000	Buried their dead

*Some scientists consider Neanderthals to be a subspecies of *Homo sapiens* and call them *Homo sapiens neanderthalensis*.

what would you do?

Human remains can tell scientists much about the diets, diseases, lifestyles, and genetic relationships of our ancestors, and such information can help piece together our evolutionary past. Sometimes, however, keeping human remains for scientific study conflicts with the wishes of modern-day descendants who wish to have their ancestors' remains returned to them for reburial. In the United States, for example, many museums and universities are developing policies regarding the treatment and disposition of Native American and Native Hawaiian remains. How should human remains be treated, and, in the end, who should get them? If you were director of a museum and were asked to develop a policy for the treatment of human remains, what would you do?

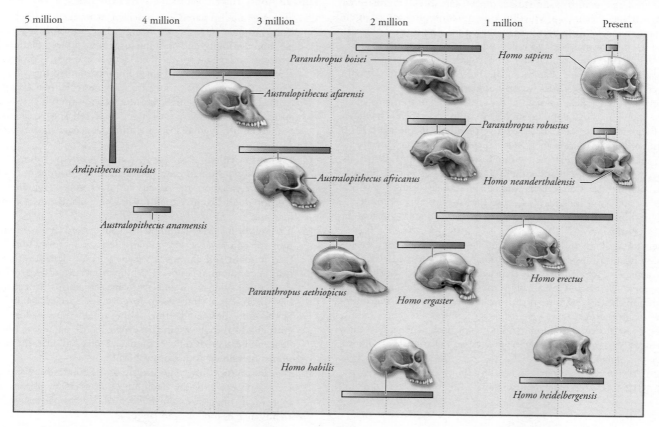

FIGURE **22.22** The major hominid species

Did modern humans evolve several times in different regions? The idea that they did is known as the **multiregional hypothesis.** It suggests that modern humans evolved independently in locations such as Europe, Asia, and Africa from distinctive local populations of earlier humans. The alternative idea is known as the **Out of**

Africa hypothesis. It suggests a single origin for all *H. sapiens.* According to the Out of Africa hypothesis, anatomically modern humans evolved from earlier humans in Africa and only later migrated to Europe, Asia, and other locations, where they replaced earlier human populations.

HIGHLIGHTING THE CONCEPTS

Life Evolved on the Earth About 3.8 Billion Years Ago (pp. 481–483)

1. The earth is estimated to be 4.5 billion years old, and life is thought to have originated about 3.8 billion years ago. A plausible sequence of events known as chemical evolution suggests that life arose from chemicals that grew increasingly complex over a long period of time. A critical step in this process was the origin of genetic material (possibly RNA and not DNA), which allowed the transfer of information from one generation to the next. Eventually, large organic molecules aggregated into droplets that became the precursors to cells.

2. The earliest cells were prokaryotic. According to the endosymbiont theory, more complex cells formed as smaller organisms were incorporated into the cells, forming organelles such as mitochondria. Multicellularity evolved, leading to the appearance on the earth of organisms such as plants and animals.

The Scale of Evolutionary Change May Be Small or Large (pp. 483–486)

3. Microevolution involves change in the frequencies of alleles within populations over a few generations. Macroevolution is large-scale evolution such as the origin or extinction of groups of species over long periods of time. Speciation is the formation of new species.

4. Populations are groups of individuals of the same species that live in a particular area. Sexual reproduction and mutation produce variation in populations. With respect to sexual reproduction, sources of variation include crossing over and independent assortment during meiosis. Also, fertilization produces a zygote with new combinations of alleles. Mutations, rare changes in the DNA of genes, are the only source of new characteristics within populations.

5. A gene pool is a collection of all of the alleles of all of the genes of all of the individuals in a population. Microevolution involves change in the frequency of certain alleles relative to others in the gene pool. Causes of microevolution include genetic drift (random change in allele frequencies due to chance within a small population), gene flow (movement of alleles as individuals move into and out of populations), natural selection (mechanism of evolutionary change by which traits of successful individuals—those that survive and reproduce—become more common in a population), and mutation (change in the DNA of genes).

6. Systematic biology deals with the naming, classification, and evolutionary relationships of organisms. Organisms are named according to a scheme developed by Carl Linnaeus in which each organism is given a two-part name consisting of the genus name followed by the specific epithet. In addition, organisms may be classified using a hierarchy of increasingly more general categories: species, genus, family, order, class, phylum, kingdom, and domain. A phylogenetic tree is a schematic diagram depicting hypothesized relationships among species or higher categories.
   WEB TUTORIAL 22.1 Principles of Evolution
   WEB TUTORIAL 22.2 Agents of Change

Evidence of Evolution Comes from Diverse Sources (pp. 486–489)

7. Fossils are the preserved remnants and impressions of past organisms. The fossil record documents that life on the earth has not always been the same as it is today, and thus provides evidence of evolution.

8. Biogeography is the study of the geographic distribution of organisms. New distributions of organisms occur either when the organisms disperse to a new location or when the areas they occupy move or are subdivided. In general, related species are more likely than are unrelated species to be found in the same geographic area.

9. Comparative anatomy also provides evidence of evolution. In general, species with more shared traits are considered more likely to be related. Structures that have arisen from common ancestry are called homologous structures and usually arise from the same embryonic tissue. Thus, common embryological origins can also be viewed as evidence of common descent.

10. Molecules that are the basic building blocks of life can be compared for evidence of evolution. For example, scientists compare the sequences of amino acids in proteins or the nucleotide sequences in DNA of different species to gauge relatedness and to estimate the time of divergence from a most recent common ancestor.
    WEB TUTORIAL 22.3 Biogeography and Continental Movement

Human Roots Trace Back to the First Primates (pp. 489–498)

11. Humans are primates, an order of mammals that also includes lemurs, monkeys, and apes. Primates have forward-facing eyes, flexible shoulder joints, and grasping forelimbs and feet. Flattened nails—rather than claws—cover their sensitive digits. Primates provide extensive parental care to a small number of offspring.

12. There are two groups of modern primates. The first group includes lemurs, lorises, and pottos, and the second group includes monkeys, apes, and humans. The term *hominid* refers to members of the human family, such as species within the genera *Australopithecus* and *Homo.*

13. Paleoanthropology is the study of human origins and evolution. Argument about human evolution stems, in part, from questions of intraspecific variation, the paucity of fossils, and the tendency of some anthropologists to give each fossil a new name and to consider it a new line.

14. The following three common misconceptions about human evolution stir controversy: (1) humans descended from chimpanzees (in fact, humans and chimps represent separate branches that diverged from a common ancestor); (2) human evolution occurred in an orderly progression from ancient to modern forms (in actuality, there have been several times when two or more species of hominids lived at the same time, and some of those lines were evolutionary dead ends); and (3) traits of humans evolved at the same rate (instead, evidence indicates that traits of humans evolved at different rates, a phenomenon known as mosaic evolution).

15. Bipedalism probably evolved early on in hominids. Walking on two feet set the stage for the evolution of other characteristics such as an increase in brain size, which, in turn, was associated with cultural trends such as tool use and language.

16. The oldest hominid remains found to date are those of *Ardipithecus ramidus*, estimated to be at least 4.4 million years old. Few fossils have been found of *Australopithecus anamensis,* the most primitive australopithecine; this species is believed to have existed between 4.2 and 3.9 million years ago. Excellent specimens have been found of *Australopithecus afarensis,* a species about 3.8 million years old. About 3 million years ago, several new hominid species appeared (*Australopithecus africanus* and three more robust species now placed in the genus *Paranthropus*).

17. About 2.5 million years ago, *Homo habilis* remains appear in the fossil record; these were the first hominids to use tools. About 1.9 million years ago, *Homo ergaster* arose, and *Homo erectus* diverged from this species 1.6 million years ago. Fossils of *H. erectus* are not restricted to Africa, indicating that this was the first hominid to disperse from Africa; this species may also have used fire. The oldest remains of modern humans, *Homo sapiens,* are dated at 130,000 years old. *Homo neanderthalensis* was a cold-adapted species that coexisted with modern humans until about 30,000 years ago.

18. The multiregional hypothesis suggests that *H. sapiens* evolved independently in different regions from distinctive populations of early humans. The Out of Africa hypothesis suggests that modern humans evolved from early humans in Africa and then dispersed to other regions, where they replaced existing hominid species.

KEY TERMS

evolution *p. 480*
chemical evolution *p. 481*
endosymbiont theory *p. 482*
microevolution *p. 483*
macroevolution *p. 483*
population *p. 483*
gene pool *p. 483*

genetic drift *p. 483*
bottleneck effect *p. 484*
founder effect *p. 484*
gene flow *p. 484*
species *p. 484*
speciation *p. 484*
natural selection *p. 484*

fitness *p. 484*
adaptation *p. 484*
phylogenetic trees *p. 485*
fossil *p. 487*
biogeography *p. 488*
homologous structure *p. 488*
convergent evolution *p. 489*

analogous structure *p. 489*
molecular clock *p. 489*
hominid *p. 493*
mosaic evolution *p. 493*
bipedalism *p. 494*
multiregional hypothesis *p. 498*
Out of Africa hypothesis *p. 498*

REVIEWING THE CONCEPTS

1. How might life have evolved from inorganic molecules to complex cells? *pp. 481–483*
2. Distinguish between microevolution and macroevolution. *p. 483*
3. What are four sources of variation within populations? *pp. 483–485*
4. How does genetic drift lead to microevolution? *pp. 483–484*
5. Define speciation and relate it to gene flow. *p. 484*
6. Define natural selection. How can variation within populations be maintained in the face of natural selection? *pp. 484–485*
7. Describe the binomial system by which organisms are named and the hierarchical system by which they are classified. *pp. 485–486*
8. What is a phylogenetic tree? *pp. 485–487*
9. What is a fossil? Describe the process of fossilization and relate it to limitations of the fossil record. *pp. 486–488*
10. How do new distributions of organisms arise? *p. 488*
11. Distinguish between homologous and analogous structures. *pp. 488–489*
12. Describe how comparative embryology provides evidence of evolution. *pp. 489, 491*
13. What is a molecular clock? *p. 489*
14. What characters set primates apart from other mammals? *pp. 490–491, 493*
15. Why do discussions of human evolution inspire controversy? *p. 493*
16. What are three popular misconceptions about human evolution? *p. 493*
17. Describe the major species of hominids and trends in hominid evolution. *pp. 494–497*
18. Which of the following does *not* produce variation in populations?
    a. Bottleneck effect
    b. Mutation
    c. Crossing over
    d. Independent assortment

19. Which of the following occurs when fertile individuals move into and out of populations?
    a. Genetic drift
    b. Speciation
    c. Mutation
    d. Gene flow
20. Which of the following was in very short supply in the environment of the early earth?
    a. Lightning
    b. Volcanoes
    c. Gaseous oxygen (O_2)
    d. UV radiation
21. The earliest cells
    a. were prokaryotic.
    b. were eukaryotic.
    c. evolved over 10 billion years ago.
    d. were part of multicellular organisms.
22. Which of the following primates is our closest living relative?
    a. Gorilla
    b. Orangutan
    c. Chimpanzee
    d. Gibbon
23. Which of the following features of an organism would promote fossilization?
    a. Terrestrial existence
    b. Member of a small population
    c. Soft body parts
    d. Aquatic existence
24. Which hominid was probably the first to migrate out of Africa?
    a. *Australopithecus afarensis*
    b. *Homo sapiens*
    c. *Homo erectus*
    d. *Australopithecus africanus*

25. The major events of human evolution occurred in
    a. Africa.
    b. South America.
    c. Europe.
    d. North America.
26. Which of the following hominid characters evolved the earliest?
    a. Bipedalism
    b. Large brain
    c. Language
    d. Use of fire

27. _____ is differential survival and reproduction.
28. _____ is when different traits evolve at different rates.
29. The _____ model of human evolution suggests that modern humans evolved independently in several places from local populations of earlier humans.

APPLYING THE CONCEPTS

1. A friend of yours believes that all organisms were specially created. You believe in evolution and want to present your case to your friend. What would you say?
2. Can life arise on the earth from inanimate material today? Why or why not?
3. You are a biologist exploring the dense tropical rain forests of Brazil. You come upon an unfamiliar medium-sized mammal moving above you in the trees. How would you determine whether this animal is a primate?
4. Tay-Sachs disease has zero fitness because it causes death before the individual reaches reproductive age. How does a trait with zero fitness persist in a population?

5. Did modern humans evolve from chimpanzees? Why or why not?
6. Apes have a fairly easy time giving birth, compared with humans. A female ape, has a relatively large pelvis, and the small head of her infant passes easily through a roomy birth canal (the channel formed by the cervix, vagina, and vulva). Birth in humans is decidedly more difficult and risky. With its large head, the typical human fetus may spend hours making its way through a narrow birth canal. What trends in human evolution might have produced the tight squeeze of human birth?

Additional questions can be found on the companion website.

The Earth Is a Closed Ecosystem with Energy as the Only Input

The Biosphere Is the Part of the Earth Where Life Exists

Ecological Succession Is the Change in Species Occupying a Given Location over Time

Energy Flows through Ecosystems from Producers to Consumers
- Food chains and food webs depict feeding relationships
- Energy is lost as it is transferred through trophic levels
- Ecological pyramids depict energy or biomass at each trophic level
- Ecological pyramids have health and environmental ramifications

Chemicals Cycle through Ecosystems
- Water cycles between the atmosphere and land
- Carbon cycles between the environment and living bodies
- Nitrogen cycles through several nitrogenous compounds
- Phosphorus cycles between rocks and living organisms

Humans Can Upset Biogeochemical Cycles
- Humans sometimes cause shortage or pollution of water
- Atmospheric levels of carbon dioxide affect global temperatures
- Disruptions to the nitrogen and phosphorus cycles can cause eutrophication

 **ENVIRONMENTAL ISSUE**   Global Warming

23

Ecology, the Environment, and Us

The best legacy we can give our children is a world in ecological balance.

John stopped the car next to the river. He knew that the eagles had started to arrive in Illinois a couple of days ago. Bald eagles were once plentiful throughout North America. Three feet tall, with a majestic white head and impressive 7-foot wing span, they once nested and fished along most major waterways. Unfortunately, human settlers killed many eagles, and as the human population grew, wild places where the eagles could nest began to disappear. Moreover, when natural waters became polluted with industrial chemicals and pesticides, these substances eventually found their way into fish. Every time an eagle ate a contaminated fish, it was at risk. On July 4, 1976, America's national bird, the bald eagle, was officially listed as an Endangered Species. But now the eagles were coming back. John was eager to show them to his 3-year-old son, Jason. "Look. There's a huge eagle up on the bluff!"

"Where?" asked Jason, followed by "Wow! He's as tall as I am!"

In this chapter, we will see that humans are but one of a host of species that share our small planet. We have many of the same needs as other species, and we face many of the same threats. If we are ever to understand the world around us and our place in it, we must have some knowledge of its physical characteristics and of our dependence on the other species with which we share it.

Our focus in this chapter is ecology, a science whose name comes from the Greek words *oikos*, meaning home, and *logos*, meaning to study. Thus, ecology is the study of our home—the earth—encompassing both its living (biotic) components and nonliving (abiotic) components. More precisely, ecology is the study of the interactions between organisms and between organisms and the environment. The key word here is *interaction*. Ecologists are the scientists who study these interactions. ■

The Earth Is a Closed Ecosystem with Energy as the Only Input

We can make two important observations when we view the earth from a distance (Figure 23.1). First, we see that the earth is isolated. With respect to materials, this observation is accurate. Aside from an occasional meteorite and other bits of debris from space, there is no source of new materials for our planet. Many of the materials that came together to form it some 4.5 billion years ago cycle repeatedly from organism to organism and between the living and nonliving components of our world. A carbon atom in your body may once have been part of a dinosaur or of Aristotle. Second, the light reflected from the earth's surface reminds us that the earth receives one very important contribution from without—energy in the form of sunlight. As we will see, this energy is captured by green plants and transferred from organism to organism to sustain nearly all life on the earth.

FIGURE **23.1** A view of the earth from space shows us that the earth is isolated. Because there is no regular input of materials, important elements must be cycled from organism to organism and between the living and nonliving components of the earth. The only input to the system is energy from the sun, which sustains nearly all life on the earth.

The Biosphere Is the Part of the Earth Where Life Exists

The part of the earth where life exists is the **biosphere**. The biosphere consists of many **ecosystems**, each consisting of the organisms in a specific geographic area and their physical environment. All the living species in an ecosystem that can potentially interact form a **community**. A **population** is all the individuals of the *same species* that can potentially interact. Thus, populations of different species form a community.

When ecologists consider an individual species, they describe it according to its niche (sometimes called *ecological niche*). The **niche** is the organism's role in the ecosystem, defined by all the physical, chemical, and biological factors that keep the organism healthy and allow it to reproduce. Such factors include the nature of the organism's food and how it obtains food, its predators, and its specific needs for shelter, even extending to the temperature, light, water, and oxygen required for survival. Thus, the organism's **habitat**—the place (physical area) where it lives—is also a part of its niche. You may want to think of an organism's habitat as its address and its niche as its profession. For example, the habitat of an African lion is semi-open plains. Its niche includes the habitat, when it is active (any time of day), how it obtains food (predation on large animals), how it relates to other animals (cubs are preyed on by hyenas and leopards; lives in social group called pride; marks territory by roaring, urinating, and patrolling), and how it reproduces (gives birth to litters after 110 to 199 days of pregnancy).

Ecological Succession Is the Change in Species Occupying a Given Location over Time

Ecosystems change over long time periods (in fact, everything changes on the earth). The sequence of changes in the kinds of species making up a community is called ecological **succession**. There are two types of ecological succession: primary and secondary.

Primary succession occurs where no community previously existed (Figure 23.2). Such places may be found on rocky outcroppings, or where lava has solidified into a new surface, or where an island has appeared. When primary succession begins, no soil exists.

The first living things to invade such an area are called *pioneer species*. Among the most prominent of these are lichens, which are actually two species—a fungus and a photosynthetic organism, usually an alga—combined in a mutually beneficial relationship. The fungus provides the attachment to the barren surface, retains water, and releases minerals from the rock to be used by the alga; the alga provides the food. The lichens secrete acid, which helps break down the rock, beginning soil formation. In time, the lichens die and their remains mix with the rock particles, furthering soil formation in cracks and crevices.

Soil building is a slow process, but once soil is in place, succession occurs more quickly. Plants begin to appear. Their roots push into every crevice, and their leaves fall and decompose, accelerating the pace of soil building and new plant colonization. In some areas, trees eventually become the dominant plant form.

This thin mat of lichen is helping break down the bare rock, beginning soil formation.

After soil has begun to accumulate, plant species appear that often include shrubs and dwarf trees.

Trees later become the dominant plant form.

Each species that moves into the area changes the environmental conditions slightly, thus changing the available resources in ways that favor some species and hamper others. The community that eventually forms and remains, if no disturbances occur, is called the *climax community* (Figure 23.3). The nature of a climax community is determined by many factors, including temperature, rainfall, nutrient availability, and exposure to sun and wind.

When an existing ecological community is cleared away—either by natural means or by human activity—and is then left alone by humans, it undergoes a sequence of changes in species composition known as *secondary succession*. Secondary succession occurs where soil is already in place. Such areas include old deserted fields and farms (originally cleared and planted by humans but later abandoned) or in areas damaged by catastrophic fire, flood, wind, or grazing (Figure 23.4). The initial invaders are likely to be grasses, weeds, and shrubs, but these are gradually outcompeted as other plants move in from the surrounding community. The area finally "heals," and the community that forms often resembles the one that existed before the disturbance.

 Energy Flows through Ecosystems from Producers to Consumers

WEB TUTORIAL 23.1

Virtually all the energy that propels life on this small planet comes from the sun. Only a small fraction of the sun's energy that reaches the earth's surface worldwide is captured by living organisms. Nonetheless, the life that abounds on the earth owes its existence to that captured energy, which is shifted, shuffled, channeled, and scattered through various systems from one level to the next.

■ Food chains and food webs depict feeding relationships

The flow of energy through the living world begins when light from the sun is absorbed by photosynthetic organisms, such as plants, algae, and cyanobacteria (a group of photosynthetic bacteria that usually live in water). *Photosynthesis* is a chemical process that essentially captures light energy and transforms it into the chemical energy of the sugar glucose, which is manufactured from carbon dioxide and water. The photosynthesizers themselves use some of the energy stored in these glucose molecules to fuel their own metabolic activities. Any remaining energy can be used for growth and reproduction. Once the photosynthesizer uses the energy in glucose to make new organic molecules, these molecules may become food for an animal.

The photosynthesizers, called the **producers**, form the lowest **trophic level** (*trophic* means feeding). All other organisms are **consumers** and use the energy that producers store (Figure 23.5).

FIGURE **23.2** Primary succession is the sequence of changes in the species composition of a community that begins where no life previously existed.

Temperate deciduous forest. These forests receive 75–125 cm (30–50 in.) of rainfall per year. Summers are hot, and winters are cold. Trees lose their leaves in the winter so as to avoid water loss when it is too cold to photosynthsize. Insects, mice, squirrels, and many species of birds are common in these forests.

Temperate grasslands. These grasslands receive 25–75 cm (10–30 in.) of rainfall per year. Long dry periods and fire are important factors in maintaining grasslands. Grazing animals such as antelope and burrowing animals such as prairie dogs are common.

Desert. Lack of water defines the desert community. Deserts receive less than 25 cm (10 in.) of rain each year. Most deserts are hot, but some are cold. Both plants and animals must be able to conserve water. Many desert plants are succulents with leaves that retain and store water. Animals may tend to avoid the sun by foraging at night.

Taiga. The taiga is composed of evergreen forests with highly variable rainfall of 50–100 cm (20–40 in.) per year. Winters are long and cold, and summers are short. The needles on evergreen trees help save water by providing little surface through which water can leave. Animals such as the grizzly bear, moose, wolf, and snowshoe hare are found in the taiga.

Tropical rain forest. Tropical rain forests may receive 200–1000 cm (80–400 in.) of rain each year. It is hot throughout the year. Tropical rain forests have a tremendous diversity of life.

FIGURE **23.3** Selected climax communities of the earth. Environmental conditions, particularly the temperature and the availability of water, affect the distribution of organisms. Each species is adapted to certain conditions.

Consumers belong to higher trophic levels and are grouped on the basis of their food source.

- **Herbivores**, or **primary consumers**, eat plants.
- **Carnivores**, or **secondary consumers**, feed on herbivores. Some carnivores eat other carnivores, forming still higher trophic levels—tertiary and quaternary consumers.
- **Omnivores** eat both plants and animals.
- **Decomposers**, such as bacteria, fungi, and worms consume dead organic material for energy and release inorganic material that can then be reused by producers.

At one time, the pattern of feeding relationships responsible for the flow of energy through an ecological system was described as a **food chain**—a linear sequence in which A eats B, which eats C, which eats D, and so on. However, now we know that the chain analogy is too simplistic because many organisms eat at several trophic levels. To illustrate, consider the number of trophic levels on which humans feed. Whereas eating chicken makes us secondary consumers, eating tuna (a predatory fish) makes us tertiary consumers (or quaternary consumers if the tuna has eaten a predatory fish). But we also can be considered primary consumers because we eat vegetables (Figure 23.6). More realistic patterns of feeding relationships, consisting of many interconnected food chains, are described as **food webs**. Figure 23.7 describes part of a food web in a community where land meets water.

Immediately after a fire there are no visible signs of life. Dead trees stand as ghostly reminders of the forest that had existed, and gray ash covers the forest floor.

The following spring young plants, such as fireweed and heartleaf arnica, appear and begin the stages of secondary succession.

FIGURE **23.4** Secondary succession following the 1988 fires in Yellowstone National Park

Tertiary consumers are carnivores that feed on secondary consumers.

Secondary consumers are carnivores that feed on herbivores.

Primary consumers (herbivores) consume producers.

Producers use the energy of the sun to produce organic molecules.

FIGURE **23.5** Trophic levels

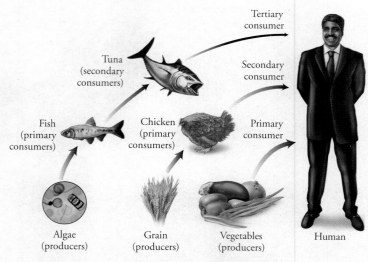

FIGURE **23.6** Humans eat at several trophic levels.

■ Energy is lost as it is transferred through trophic levels

Energy is lost as it is transferred from one trophic level to the next. Although the efficiency of transfer can vary greatly depending on the organisms in the system, on average only 10% of the energy available at one trophic level is transferred to the next higher level. As a result, ecosystems rarely have more than four or five trophic levels.

As we consider why energy transfer between trophic levels is so inefficient, keep in mind that only the energy that is converted to **biomass** (the dry weight of the organism) is available to the next higher trophic level. One reason for energy loss between trophic levels is that an animal must first expend energy in order to obtain its food, usually by grazing or hunting (the chicken in Figure 23.8 must do a lot of pecking). Furthermore, not all the food available at a given trophic level is captured and consumed. In addition, some of the food eaten cannot be digested and is lost as feces. The energy in the indigestible material is unavailable to the next higher trophic level. Lastly, roughly two-thirds of the energy in the food that is digested is used by the animal for cellular respiration. However, the

remaining energy can be converted to biomass and will be available to the next higher trophic level. These losses become multiplied as the energy is transferred to successively higher trophic levels, until a point is reached where there simply is not enough food or energy to feed another level. For example, for every 100 calories of solar energy captured by producers, 10 calories are transferred to herbivores and only 1 calorie is transferred to carnivores.

■ Ecological pyramids depict energy or biomass at each trophic level

An **ecological pyramid** is a diagram that compares certain properties in a series of related trophic levels. The *pyramid of energy*, for example, in Figure 23.9 shows the amount of energy available at each trophic level. The base of the pyramid reflects *primary productivity*—the amount of light energy converted to chemical energy in organic molecules, primarily by photosynthetic organisms. The tiers representing successively higher trophic levels grow smaller, forming a pyramid, because only about 10% of the energy contained in one trophic level is transferred to the next. *Pyramids of biomass* describe the number of individuals at each trophic level multiplied by their biomass (dry weight). Because the energy available at a trophic level determines the biomass that can be supported, a pyramid of energy and a pyramid of biomass generally have the same shape.

■ Ecological pyramids have health and environmental ramifications

Ecological pyramids convey important lessons. Let's take look at two of these lessons: (1) nondegradable substances accumulate in organisms living at higher trophic levels; and (2) more humans could be nourished on a vegetarian diet than on a diet containing meat.

BIOLOGICAL MAGNIFICATION

Chemicals that are essential to life—carbon, hydrogen, oxygen, nitrogen, and phosphorus—are passed from one trophic level to the

FIGURE **23.7** In this simplified food web, the arrows show the direction of flow of energy in the form of food.

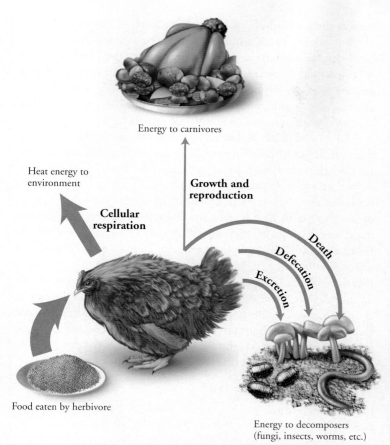

Energy to carnivores

Heat energy to environment

Cellular respiration

Growth and reproduction

Death

Defecation

Excretion

Food eaten by herbivore

Energy to decomposers (fungi, insects, worms, etc.)

FIGURE **23.8** As energy flows through a food web, only a small amount of it is stored as body mass and becomes available to the next higher trophic level. Some of the food energy is simply not captured; some is undigested and is lost in fecal waste; and some is used for cellular respiration. Only the remaining energy becomes available to the next higher trophic level.

next, thus being continuously recycled from organism to organism. Organic molecules are broken down and then either metabolized for energy or put together to form the biomass of another individual.

However, certain substances—including chlorinated hydrocarbon pesticides such as DDT, heavy metals such as mercury, and radioactive isotopes—are broken down or excreted very slowly. Substances such as these tend to stay in the body, often stored in fatty tissues such as liver, kidneys, and the fat around the intestines. As more of these types of substances are consumed, they accumulate and become concentrated in the animal's body.

The concentrations of such substances become magnified at each higher trophic level. Remember that, on average, only 10% of the energy available at one trophic level transfers to the next. As a consequence, consumers at one trophic level must eat many organisms from the previous trophic level to obtain enough energy to support life. If the organisms that are eaten are contaminated, the pollutant will accumulate in the consumer. Consider a simplified example. Assume that a nondegradable substance such as mercury pollutes the water in a certain aquatic ecosystem. Phytoplankton

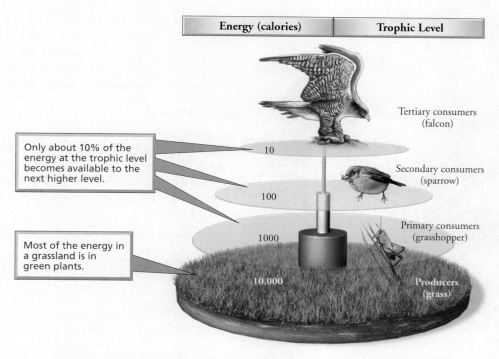

Energy (calories)	Trophic Level

Only about 10% of the energy at the trophic level becomes available to the next higher level.

Most of the energy in a grassland is in green plants.

10 — Tertiary consumers (falcon)

100 — Secondary consumers (sparrow)

1000 — Primary consumers (grasshopper)

10,000 — Producers (grass)

FIGURE **23.9** An ecological pyramid. Note that each level contains less energy and less biomass than the level below it.

(minute photosynthetic organisms that are abundant in aquatic environments) will absorb the mercury, and mercury levels in the phytoplankton will be in some dilute concentration that we will represent as 1. An herbivore such as a zooplankton (a minute non-photosynthetic organism) feeds on large numbers of phytoplankton. The mercury thus becomes more concentrated in the zooplankton's tissues than it was in the phytoplankton—let's say about 10 times more. Large numbers of zooplankton may in turn be eaten by a small fish. Once in the fish's body, the mercury stays there and accumulates, let's say another 10 times, to reach a relative concentration of 100. When these mercury–containing fish are eaten by a tuna, the mercury passes to the tuna's body and accumulates. Because many small fish must be consumed to keep a tuna alive, the mercury concentration in the tuna's body might be 1000 times greater than that in water. (Our estimate that the mercury will be 10 times more concentrated at each successive level on the trophic scale is a realistic one.) This tendency of a nondegradable chemical to become more concentrated in organisms as it passes along the food chain is known as **biological magnification** (Figure 23.10).

If humans are wise, we will learn an important lesson from observing the effects of biological magnification. We are often top carnivores, and we continue to pollute our environment with nondegradable, potentially harmful substances. Mercury, for example, enters water through volcanic activity but also through coal combustion and improper disposal of medical waste. Bacteria convert the mercury to a particularly toxic form, methyl mercury, which then works its way through food webs, as described previously, and accumulates in fish, particularly long-lived predators

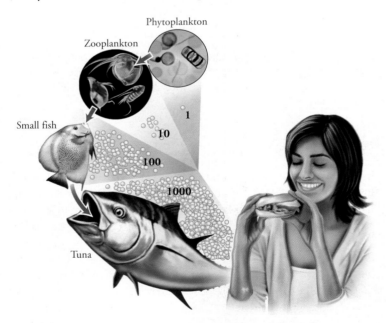

FIGURE **23.10** Substances such as mercury, that are broken down slowly or excreted slowly, tend to accumulate in the body. The concentration becomes magnified at each successive trophic level, because a consumer must eat many individuals from a lower trophic level to stay alive. As a result of this biomagnification, the mercury concentration in the tuna eaten by a human at the end of this food chain is about 1000 times greater than the mercury concentration in the phytoplankton.

such as tuna, swordfish, or shark. High levels in humans affect the nervous system, causing muscle tremors, personality disorders, and birth defects (see Chapter 7). For this reason, the Food and Drug Administration now recommends that pregnant women, breast-feeding women, women of child-bearing age, and children stop eating shark, marlin, and swordfish, and limit their intake of albacore (white) tuna to one 6-oz. serving a week.

Polychlorinated biphenyls (PCBs) are industrial chemicals with various uses. Unfortunately they are suspected of causing cancer and nervous system damage. They have been called environmental estrogens because they mimic the effects of the female sex hormone estrogen or enhance estrogen's effects. Scientists, recognizing that PCBs have had feminizing effects on certain male animals, are investigating the possibility that PCBs may be reducing human fertility (for example by reducing human sperm count) and affecting the rates of cancer in certain reproductive structures (breasts, ovaries, prostate gland, and testes). Like mercury, PCBs are nondegradable and accumulate in the food web. PCBs can be found in chicken, beef, and dairy products. In addition, some farm-raised Atlantic salmon have PCB levels high enough to trigger health warnings from the Environmental Protection Agency. (Wild salmon have lower levels of PCBs.)

what would you do?

DDT is a pesticide that was widely used in the 1950s. During the 1960s and 1970s, it was blamed for the alarming decline in certain species of birds. The United States banned the use of DDT during the 1970s, and many other countries followed suit.

Malaria is a disease that infects 300 million people a year and kills about 200 million of them. It is transmitted by a mosquito (see Chapter 13a), and DDT is an affordable and effective way to kill mosquitoes. However, neither the United States nor the World Health Organization will fund the use of DDT to control malaria-causing mosquitoes. Some of the areas that are hit the hardest by malaria—Africa, Asia, and Latin America—are too poor to afford other pesticides, which cost four to six times more than DDT. What do you think should be done to curb the transmission of malaria? Do you think the use of DDT should be permitted? What are your reasons?

WORLD HUNGER

The world's human population has grown at an alarming rate (a problem discussed in Chapter 24). The pyramid of energy suggests a way to feed the growing population more efficiently: persuade people to eat lower on the food chain. Because only about 10% of the energy available on one trophic level is transferred to the next one, we see that

10,000 calories of corn energy → 1000 calories of beef energy → 100 calories of human energy

However, if humans began to eat one level lower on the food chain, about 10 times more energy would be available to them:

10,000 calories of grain (corn, wheat, rice, and so on) → 1000 calories of human energy

In short, more people could be fed and less land would have to be cultivated if we adopted a largely, or exclusively, vegetarian diet (Figure 23.11). This fact is one reason why people in densely pop-

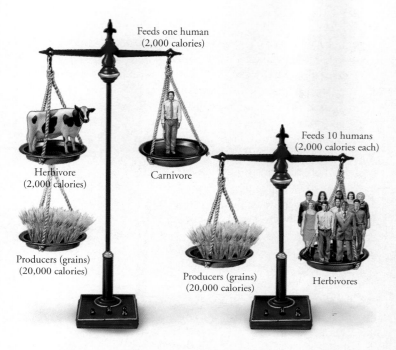

FIGURE **23.11** Energy pyramids may hold an important lesson for humans. Because only approximately 10% of the energy available at one trophic level is available at the next higher level, a greater number of people could be fed if they ate a vegetarian diet rather than a diet containing meat.

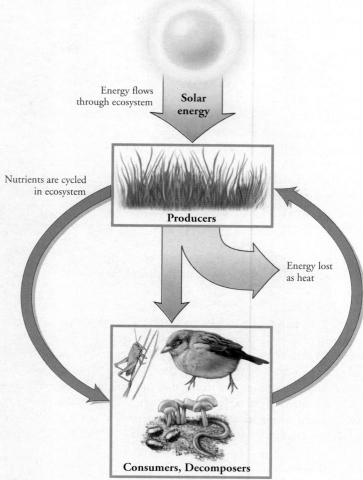

FIGURE **23.12** Through biogeochemical cycles, matter cycles between living organisms and the physical environment. This cycling of matter (blue arrows) is in contrast to the path of energy (yellow arrows), which flows through the ecosystem in one direction.

ulated regions of the world, including China and India, are primarily vegetarians. As the human population continues to expand, meat is likely to become even more of a luxury throughout the world than it is today.

stop and think

The traditional diet of the Inuit Eskimos, one of the native peoples of the North American coast, is part of a relatively long food chain:

Diatoms (producers) → Tiny marine animals → Fish → Seals → Inuits

Explain how the length of this food chain may be one factor contributing to the small population size of Inuit groups.

Seal blubber contains high levels of DDT and PCBs. Explain why the levels of DDT and PCBs are four to seven times higher in the breast milk of Inuit women who eat blubber than in the breast milk of women in Quebec who eat lower on the food chain, on fish for example.

Chemicals Cycle through Ecosystems

Resources on the earth are limited. Life on the earth is demanding. Many of the earth's reserves would be quickly depleted if it were not for nature's cycling. Materials move through a series of transfers, from living to nonliving systems and back again (Figure 23.12). Let's look at some of the more important of these **biogeochemical cycles**, the recurring pathways certain materials travel between living and nonliving systems.

■ Water cycles between the atmosphere and land

Each drop of rain reminds us that water recycles continuously, precipitating from the atmosphere and falling to the earth; collecting in ponds, lakes, or oceans; and then evaporating back into the atmosphere. Most of the rain or snow that falls over land returns to the sea at some point. This cycle provides us with a renewable source of drinking water. Because water is so critical to life, large amounts of it pause for a time in the bodies of living things. In living cells, water helps regulate temperature and acts as a solvent for biological reactions. The very oxygen we breathe is produced from water through the reactions of photosynthesis. Water also cycles back to the environment from living things; plants return 99% of the water they absorb to the atmosphere in the process of transpiration (the evaporation of water from the leaves and stems of plants). The water cycle is shown in Figure 23.13.

■ Carbon cycles between the environment and living bodies

If you have ever seen a dead opossum beside the road, you know something about carbon cycling. In the carbon cycle, carbon moves

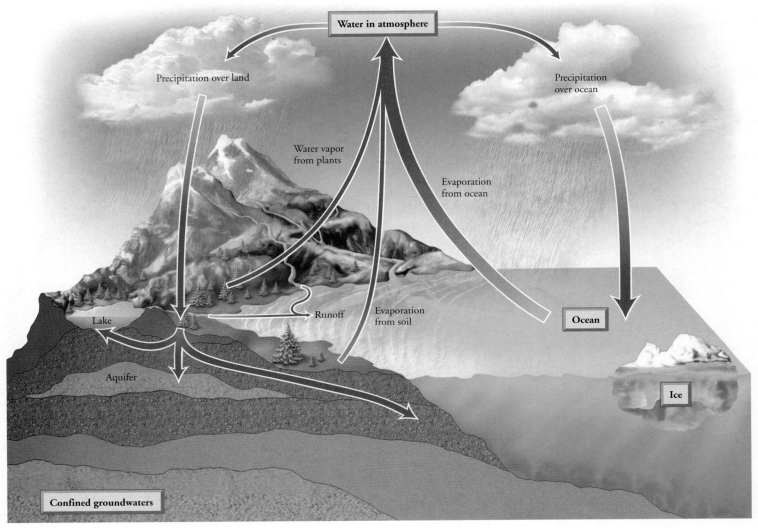

FIGURE 23.13 The water cycle. Water, which is essential to life, cycles from the atmosphere to the land as precipitation, collects in oceans and other bodies of water, and evaporates back to the atmosphere. Water also returns to the atmosphere in the form of vapor lost from the leaves of plants. This cycling provides us with a renewable source of drinking water.

from the environment, into the bodies of living things, and back to the environment (Figure 23.14). Living organisms need carbon to build the molecules that give them life: proteins, carbohydrates, fats, and nucleic acids. Conversely, certain processes of living organisms—photosynthesis and respiration—are an integral part of the carbon cycle.

The primary movement of carbon from the environment into living organisms occurs during photosynthesis, as plants, algae, and cyanobacteria use carbon dioxide (CO_2) to produce sugars and other organic molecules. When photosynthesizers are eaten by herbivores, these organic molecules serve as a carbon source for the herbivores. The herbivores use that carbon to produce their own organic molecules, which then serve as a carbon source for carnivores. When an organism dies, like the opossum beside the road, the organic molecules in its body will serve as a carbon source for

decomposers. However, while alive, all organisms cycle carbon back to the environment through the reactions of cellular respiration, which breaks down organic molecules to CO_2.

Some carbon is significantly delayed before being cycled back into the environment. For example, carbon may remain tied up in the wood of some trees for hundreds of years. Most of the carbon that has left the carbon cycle is thought to be stored in limestone, a type of sedimentary rock formed from the shells of marine organisms that sank to the bottom of the ocean floor and were covered and compressed by newer sediments. Other vast carbon stores are the fossil fuels (oil, coal, and natural gas), so named because they formed from the remains of organisms that lived millions of years ago.

Three processes release carbon from long-term storage and return it to the environment: decomposition, erosion, and combus-

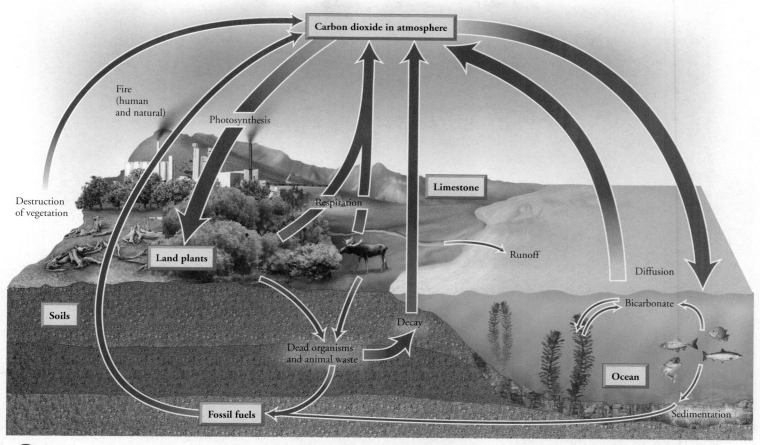

WEB TUTORIAL 23.3

FIGURE **23.14** The carbon cycle. Carbon cycles between the environment and living organisms. Carbon dioxide (CO_2) is removed from the environment as producers use it to synthesize organic molecules by photosynthesis. The carbon in those organic molecules then moves through the food web, serving as a carbon source for herbivores, carnivores, and decomposers. Carbon is returned to the atmosphere as CO_2 when organisms use the organic compounds in cellular respiration.

tion. When the trees eventually die, the natural process of decomposition will make the carbon available for new organisms, which will respire and release CO_2 to the atmosphere. The carbon in limestone is recycled through erosion. Millions of years after it forms, sedimentary rock containing limestone can be lifted to the earth's surface, where it is eroded by chemical and physical weathering. These changes make the carbon available to cycle through the food web once again. Combustion, or burning, returns the carbon in fossil fuels to the environment. Today, fossil fuels such as coal, oil, and natural gas are being burned in huge amounts, and the carbon they contain is being returned to the atmosphere as CO_2.

■ **Nitrogen cycles through several nitrogenous compounds**

Nitrogen is a principal constituent of a number of molecules needed for life, including proteins and nucleic acids. Nitrogen is often in short supply to living systems, so its cycling is of particular importance (Figure 23.15).

The largest reservoir of nitrogen is the atmosphere, which is about 79% nitrogen gas (N_2). However, nitrogen gas cannot inter-

act with life directly. (As you sit reading, you are bathed in nitrogen gas, but you do not interact with it.) Before it can be used, nitrogen gas must be converted to a form that living organisms can use—ammonium (NH_4^+). The process of converting nitrogen gas to ammonium is called **nitrogen fixation**. This is performed by nitrogen-fixing bacteria, many of which live in nodules on the roots of leguminous plants such as peas and alfalfa. Ammonium is then converted to nitrite (NO_2^-) and then to nitrate (NO_3^-) by nitrifying bacteria living in the soil, in a process called **nitrification**. The ammonium or nitrates are first absorbed by plants, which use the nitrogen to form plant proteins and nucleic acids. The nitrogen then passes through the food web and is incorporated into the nitrogen-containing compounds of animals.

When plants and animals die, decomposers such as bacteria break down the waste products and dead bodies of plants and animals, producing ammonium (NH_4^+). Much of the ammonium is converted to nitrate by nitrifying bacteria. Other bacteria balance the nitrogen cycle by performing a process called denitrification, in which the nitrates that are not assimilated into living organisms are

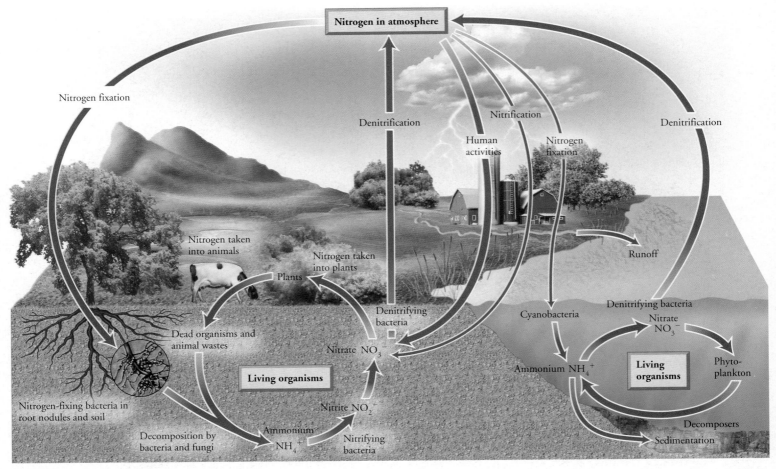

FIGURE **23.15** The nitrogen cycle. Atmospheric nitrogen can be converted to ammonium by nitrogen-fixing bacteria, which then convert the ammonium to nitrate, the main form of nitrogen absorbed by plants. Plants use nitrate to produce proteins and nucleic acids. Animals eat the plants and use the plant's nitrogen-containing chemicals to produce their own proteins and nucleic acids. When plants and animals die, their nitrogen-containing molecules are converted to ammonium by bacteria. Denitrifying bacteria return nitrogen to the atmosphere.

converted to nitrogen gas. Denitrifying bacteria are found in wet soil, swamps, and estuaries.

■ Phosphorus cycles between rocks and living organisms

Phosphorus is an important component of many biological molecules, including the genetic material DNA, energy-transfer molecules such as adenosine triphosphate (ATP), and phospholipids found in membranes. Phosphorus is also essential in vertebrate bones and teeth.

Unlike the water, carbon, and nitrogen cycles, the phosphorus cycle does not have an atmospheric component (Figure 23.16). Instead, the reservoir for phosphorus is sedimentary rock, where it is found as phosphate ions. The cycle begins when erosion caused by rainfall or runoff from streams dissolves the phosphates in the rocks. The dissolved phosphate is readily absorbed by producers and incorporated into their biological molecules. When animals eat the plants, the phosphates are passed through

food webs. Decomposers return the phosphates to the soil or water, where they become available to plants and animals once again. Much of the phosphate is lost to the sea. Although some of this phosphate may cycle through marine food webs, most of it becomes bound in sediment. The phosphates in sediment become unavailable to the biosphere unless geological forces bring the sediment to the surface.

Humans Can Upset Biogeochemical Cycles

For many millions of years, biogeochemical cycles have worked just as described in the preceding pages. However, human activity, particularly since the time of the Industrial Revolution, which began around the mid-1800s, has disrupted the normal patterns of cycling. Some of these disturbances result from new technologies, but most are caused by the demands of an ever-increasing human population.

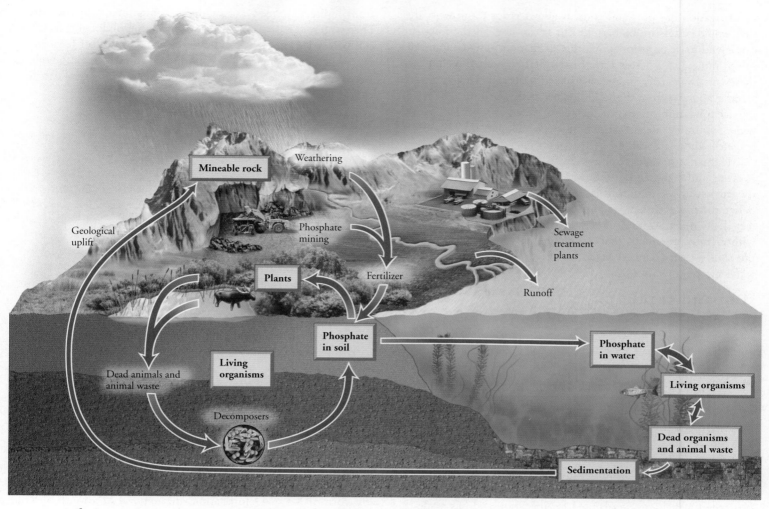

FIGURE **23.16** The phosphorus cycle. Phosphates dissolve from rocks in rainwater. They are then absorbed by producers and passed to other organisms in food webs. Decomposers return phosphates to the soil. Some phosphates are carried to the oceans and eventually deposited in marine sediments.

■ Humans sometimes cause shortage or pollution of water

The world's water supply is threatened because of overuse by humans. Both water shortages and water pollution can result from human activities.

WATER SHORTAGE

The earth has been called the Water Planet because water is so plentiful here. Still, relatively little is available for human use because most of the earth's water is in the oceans. You might think that at least all the fresh water would be available, but much of it is tied up in ice or clouds, hidden in underground rivers, or otherwise inaccessible for immediate human use.

In some places, water shortages are caused by drought, or lack of rain. A drought has affected the western United States for more than a decade. In some regions, farming and ranching have been severely affected. For example, in 2006, ranchers in Colorado had to sell off parts of their herds because there was no grass to feed them. Drought also increases the risk of wildfires that kill plants, wildlife,

and humans, and destroy property. (At the same time, an interesting new study blames climate change—discussed later in the chapter—for the increase in numbers of wildfires in recent years, pointing out that the average fire season in the western United States is now 2 months longer than in 1970.)

In most regions, the underlying cause of water shortage is too many people drawing from a limited supply. However, North Americans use an exceptionally large amount per person. In the United States, each individual uses about 7500 liters (1950 gal) of water a day, whereas a person in a less-developed country uses less than 75 liters (19.5 gal) of water a day.

You might respond to these numbers defensively and claim that *you* certainly do not use that much water. In a sense you would be right. Personal use accounts for only a small percentage of total U.S. water use. Most of the water is used in agriculture. In an effort to feed a growing human population, our country increasingly relies on irrigation to make arid areas fruitful. The rest of the water is used primarily in industry for steam generation or the cooling of power plants.

Irrigation has pros and cons. It allows crops to grow in areas that would otherwise be barren. But it can also, in the long run, cause land to be unfit for agriculture. Irrigation water contains dissolved minerals. Whereas runoff from natural rainfall would carry these salts away, irrigation water soaks into the soil. Then, when the water evaporates from the soil, the salts are left behind. The resulting accumulation of salts in the soil is called *salinization*. Worldwide, salinization destroys the fertility of 5000 km² (1930 mi²) of irrigated land each year.

Humans use water from two main sources: surface water, such as rivers and lakes, and groundwater, which is water found under the earth's surface in porous layers of rock, such as sand or gravel. When too much surface water is used in an area, ecosystems are affected. As much as 30% of a river's flow can usually be removed without affecting the natural ecosystem. However, in some parts of the southwestern United States, as much as 70% of the surface water has been removed.

stop and think

Conservationists sometimes buy wetlands, hoping to preserve the habitats of endangered plants and wildlife. Despite this conservation strategy, the wetlands sometimes dry up. Explain why it would also be important to buy the rights to the water that feeds the wetland.

Human use of groundwater is depleting the aquifers (porous layers of underground rock where groundwater is found). Water is added to aquifers by rainfall or melting snow. In some areas, however, water is being removed from aquifers faster than it can be replenished. The Ogallala Aquifer underlies eight states in the Great Plains region of the United States. Much of the water pumped up from the Ogallala Aquifer is used for irrigation and has made this region some of the most productive farmland in the country. However, the water table in the aquifer is dropping by 6 ft a year. The water tables beneath the highly populated northeastern United States are also dwindling dangerously. What steps can be taken to reduce water shortages?

1. **Reduce water use.** Each of us could help by individual efforts. However, because most water is used in agriculture, the biggest benefit would come from improved irrigation methods. Between 1950 and 1980, for example, Israel reduced its water wastage from 83% to a mere 5%, primarily by changing from spray irrigation to drip irrigation.

2. **Raise the price of water.** The cost of water in the United States is substantially below its cost in European countries. Experience has shown that the consumer's interest in conserving water is directly related to its cost. Clearly, economic and legal incentives for conserving water are needed.

WATER POLLUTION

Another threat to the water supply is that much of the water we rely on is polluted, often by our own activities. Factories, refineries, and waste treatment plants release fluids of varying quality directly into our water supplies. The effects of mercury and PCBs in water were discussed earlier in the chapter. Water can also become polluted

from indirect sources. For example, contaminants in the atmosphere, such as gaseous emissions from automobiles and factories, can be carried into water supplies by rain. Runoff of fertilizer or pesticides into storm drains, groundwater, and streams also pollutes water supplies. We will consider some of the effects of fertilizer runoff later in this chapter.

what would you do?

Because of water shortages in Colorado, some farmers are giving up their farms. In 2004, some farmers in Colorado's Arkansas Valley sold their water rights to Aurora, a suburb of Denver. About half of Aurora's water is used to irrigate the landscaping around private homes. With these sales, the valley lost 2600 acres of farmland. Those who fear that the loss of farmland will harm the local economy are lobbying the state legislature to prevent people who hold water rights from selling them to buyers outside the local area. The farmers holding water rights, their lobbying neighbors, and the people in Aurora have conflicting interests. What do you think would be an equitable way to reduce the conflict over water use?

■ Atmospheric levels of carbon dioxide affect global temperatures

The level of atmospheric carbon dioxide (CO_2) is increasing (Figure 23.17). Two human activities—burning fossil fuels and deforestation—are important causes of this increase. Fossil fuels were formed from deposits of dead plants and animals that were buried in sediments and escaped decomposition between 345 million and 280 million years ago. High temperatures and pressure over millions of years converted the deposits to coal, oil, and natural gas. The burning of these fuels returns carbon to the environment in the form of CO_2. Deforestation, the removal of a forest without adequate replanting, as is occurring in areas of the Amazon rain forest and the U.S. Pacific Northwest, increases atmospheric CO_2 in two ways. First, it *reduces the removal* of CO_2 from the atmosphere through the trees' photosynthesis. Second, the trees are often

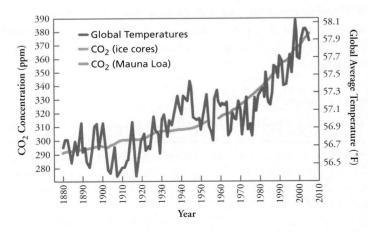

FIGURE **23.17** The concentration of carbon dioxide (CO_2) in the atmosphere has been increasing for many years. Two major causes are the burning of fossil fuels, which releases CO_2, and deforestation, which leaves fewer trees to remove CO_2 from the atmosphere. Global temperatures have also risen during this period.

burned after cutting, and burning *adds* CO_2 to the atmosphere. About three-quarters of the CO_2 added to the atmosphere comes from cars and factories, and the remaining quarter is largely from deforestation.

The rise in atmospheric CO_2 raises concerns because CO_2 is one of the greenhouse gases that play a role in warming our atmosphere. Other greenhouse gases include methane, nitrous oxide, chlorofluorocarbons (CFCs), and ozone in the lower atmosphere. These gases are accumulating in the atmosphere, largely because of human activity. Like the glass on a greenhouse, these gases allow the sunlight to pass through to the earth's surface, where it is absorbed and radiated back to the atmosphere as heat, long-wave infrared radiation. Greenhouse gases then absorb the infrared radiation, thus trapping the heat. Because of this effect, there is a concern that the increase in atmospheric CO_2 is leading to a rise in temperatures throughout the world—global warming.

Is the earth getting warmer? According to a 2006 report of the National Academy of Science, global average surface temperature has increased by 0.7°C (1.4° F) during the 20th century. As of this writing, ten of the hottest years in history have occurred since 1989. Although the rate and extent of the rise are not clear, scientists believe that global temperatures will continue to rise with the level of greenhouse gases (see the Environmental Issue essay, *Global Warming*). What is not known for certain, is *how much* the temperatures will rise. The range of the prediction for temperature increase is large because of uncertainties concerning the effects of many of the interacting factors, such as cloud cover due to evaporation, cooling effects of atmospheric particulates such as sulfates, possible increased primary productivity due to the increase in CO_2, and the ability of the ocean to absorb CO_2.

Unfortunately, the greater amounts of CO_2 being absorbed by the ocean are already having a noticeable effect on the water itself. They have lowered the normally alkaline pH of ocean water to a pH that is actually acidic. Scientists fear this change will have far-reaching consequences for marine life. For one thing, it could seriously reduce the numbers of certain plankton, on which so many food webs in the ocean depend.

Some people do not believe that rising atmospheric CO_2 is the cause of increasing global temperature. Some point out that rising temperatures could be the result of a natural climate cycle. Others suggest that domesticated animals, such as cattle, sheep, and goats, are to blame. Bacteria in the guts of these animals produce methane, a greenhouse gas whose heat-trapping ability is 20 times greater than that of carbon dioxide. The collective belches and flatulence of all the cattle on the earth accounts for 20% of the methane gas released into the atmosphere.

stop and think

Scientists are working on a diet for cattle and sheep that is low in nitrogen and will reduce the amount of methane produced by these animals. How might such a diet affect global warming? On what factors might the outcome depend?

■ Disruptions to the nitrogen and phosphorus cycles can cause eutrophication

Human activity is also disturbing the nitrogen and phosphorus cycles. Inadequate nitrogen and phosphorus in the soil can slow the growth of plants. For this reason, many crops are fertilized with synthetic preparations containing both nitrogen (as ammonium) and phosphates. The nitrogen in these fertilizers was "fixed" artificially by an industrial process. The demand for synthetic fertilizers has increased with the demand for food to feed a growing human population. One problem it creates is that natural processes of denitrification cannot keep pace with the rate of industrial nitrogen fixation and the amount of nitrogen added to the land in commercial fertilizers. The result is an excess of fixed nitrogen.

Nitrogen and phosphate from fertilizer can wash into nearby streams, rivers, ponds, or lakes in runoff. Fertilizer runoff is one cause of *eutrophication*, the enrichment of water in a lake or pond by nutrients. The nutrient boost can lead to an explosive growth of photosynthetic organisms, such as algae, and in shallow areas, weeds. When these organisms die, their remains accumulate at the bottom of the lake. Decomposers then thrive, depleting the water of dissolved oxygen. Gradually, fish, such as pike, sturgeon, and whitefish, which require higher oxygen concentrations in the water, are replaced by less desirable species, such as catfish and carp, which can tolerate lower levels of dissolved oxygen.

HIGHLIGHTING THE CONCEPTS

The Earth Is a Closed Ecosystem with Energy as the Only Input (p. 502)

1. Ecology is the study of the interactions between organisms and between organisms and their environment.
2. Materials on the earth are recycled among living organisms and the environment. The only input to the earth is energy from the sun.

The Biosphere Is the Part of the Earth Where Life Exists (p. 502)

3. The biosphere consists of all the ecosystems on the earth. An ecosystem consists of a community of organisms and their physical environment. An organism's niche is the specific role it plays in the community; its habitat is the place where it lives.

Ecological Succession Is the Change in Species Occupying a Given Location over Time (pp. 502–503)

4. Ecological succession is the sequence of changes in communities over long periods. Primary succession occurs where no community previously existed. The first invaders are called pioneer species. A climax community eventually forms and remains as long as no disturbances occur. Secondary succession describes the sequences of changes that occur when an existing community is disturbed by human or natural means.

ENVIRONMENTAL ISSUE

Global Warming

Why is there such commotion over global warming? Why should anyone care whether the atmosphere becomes 1° to 2°C (1.8° to 3.6°F) warmer? The Intergovernmental Panel on Climate Change (IPCC) is coordinating efforts to predict the effects of global warming. The task is not easy, because global warming affects and is affected by ongoing changes in the degree to which people use fossil fuels compared with the degree to which they use renewable energy. The Kyoto Protocol, an agreement between more than 163 nations to reduce CO_2 emissions to pre-1990 levels through technology and a reduction in energy use, was ratified in early 2005. In mid-2006, believing that technology can solve the problem of global warming, the United States and the European Union agreed to share information on technological advances. These agreements are intended to lessen the rise of carbon dioxide levels in the future. The IPCC is therefore preparing predictions based on several possible scenarios. Their report will be published in 2007.

Let's look at the current extent of global warming and possible consequences suggested in the 2001 IPCC report.

1. **Melting ice caps, rising sea level, and warmer oceans.** The polar ice caps have shrunk by 20% since 1979 and are currently shrinking by 9% per decade. The largest single block of ice in the Arctic, the Ward Hull ice cap, began splitting in 2000, had split in half by 2002, and is now breaking into pieces. Nearly all mountain glaciers are melting, as are the glaciers in Greenland. When ice on land melts, sea level rises. The resulting reduction of white covering means less of the sun's energy is reflected back to space. Less reflection increases warming, and the ocean water expands as it warms. During the last century, sea level rose about 10 to 20 cm (4 to 8 in.), and the temperature of the oceans increased. According to the IPCC, the sea level will continue to rise.

2. **Coastal cities could flood.** The rising sea level would have the greatest effect on low-lying coastal countries such as Bangladesh, Egypt, Vietnam, and Mozambique. Seawater could cover significant areas in these densely populated lands. For example, according to the IPCC, flooding in Bangladesh could increase by 40% in this century. If the global temperature increased by 6°C (10.8°F)—the worst-case scenario of the report—more than half of Bangladesh would be submerged for most of the year. In the United States, major cities such as New York, Miami, Jacksonville, Boston, San Francisco, and Los Angeles could be largely under water.

3. **Changing weather patterns.** Global warming would cause changes in both temperature and rainfall patterns. Not all researchers agree on why the oceans are warmer—global warming or a natural cycle—but most agree that warmer oceans can fuel hurricanes. A study published in mid-2006 identifies warmer oceans as a major factor in the 2005 Atlantic hurricanes, including Katrina, Rita, and Wilma.

Recall that the distribution of climax communities, each with its characteristic plant and animal life, is largely determined by temperature and rainfall. If global warming occurs, some species will thrive and others may become extinct. There will be a shift in the location of agricultural regions, meaning food production in the central plains of the United States and Canada could drop. Areas that now produce enough food for export may not be able to produce enough for the local population. Meanwhile, other regions—parts of India and Russia, for instance—might receive increased rainfall that would benefit agriculture. These changes will alter the distribution of "haves" and "have-nots." How will the world community deal with the shifts in economic and political power?

4. **Human health.** The IPCC predicts that global warming will have various negative effects on human health. The change in climate is likely to increase the number of deaths due to heat waves, flooding, and droughts leading to starvation. Climate change could also increase air pollution. In addition, because mosquitoes and ticks, as well as other vectors of disease, would thrive in a warmer climate, we may see increases in malaria, West Nile virus encephalitis (inflammation of the brain), Lyme disease, and the like.

What can individuals do to slow global warming? An important step, which would also have other benefits, would be to reduce our production of greenhouse gases. It would be wise, for example, to walk, ride a bicycle, or take public transportation instead of driving. When driving is necessary, use car pools whenever possible. When you purchase appliances, choose those that are energy efficient, and wear clothing that minimizes the use of heating or cooling systems in your home. In short, conserve energy. The environment will benefit; the exercise will improve your health; and you will save money. ✹

Energy Flows through Ecosystems from Producers to Consumers (pp. 503–509)

5. Most of the energy in living systems comes from solar energy absorbed and stored in the molecules of photosynthetic organisms, called producers. The energy stored in producers' molecules enters the animal world through plant-eating herbivores (primary consumers), which may be eaten by carnivores (secondary consumers) or by animals that eat other carnivores (tertiary consumers). Decomposers consume dead organic material, releasing inorganic compounds that can be used by producers. The position of an organism in these feeding relationships is referred to as a trophic level.

6. The feeding relationships that result in the one-way flow of energy through the ecosystem and in the cycling of materials among organisms are called food chains or food webs.

7. Ecological pyramids depict the amount of energy or biomass at each trophic level. Pyramids of energy show the loss of energy from one trophic level to another. On average, only 10% of the energy available at one trophic level is available at the next higher level. Ecosystems

generally have only four or five trophic levels. Pyramids of biomass have the same shape as pyramids of energy, because the available energy determines the biomass that can be formed.

8. Biological magnification—the tendency for certain nondegradable substances to become more concentrated in organisms as it passes through a food web—is a consequence of the energy loss between trophic levels. Thus, top carnivores are most likely to be poisoned by nondegradable toxic substances in the environment.

9. Because energy is lost with each transfer in the trophic scale, one way to feed more people would be for humans to adopt a largely vegetarian diet.

WEB TUTORIA 23.1	Energy Flow and Food Webs
WEB TUTORIA 23.2	The Water Cycle
WEB TUTORIA 23.3	The Carbon Cycle and Global Warming
WEB TUTORIA 23.4	The Nitrogen Cycle

Chemicals Cycle through Ecosystems (pp. 509–512)

10. In biogeochemical cycles, materials cycle between organisms and the environment.

11. The water cycle is the pathway of water as it falls as precipitation; collects in ponds, lakes, and seas; and returns to the atmosphere through evaporation.

12. The carbon cycle is the worldwide circulation of carbon from the abiotic environment (carbon dioxide in air) to the biotic environment (carbon in organic molecules of living organisms) and back to the air. Carbon enters living systems when photosynthetic organisms incorporate carbon dioxide into organic materials. Carbon dioxide is formed again when the organic molecules are used by the living organism for cellular respiration.

13. The nitrogen cycle is the worldwide circulation of nitrogen from nonliving to living systems and back again. Atmospheric nitrogen (N_2) cannot enter living systems. Nitrogen-fixing bacteria living in nodules on the roots of leguminous plants convert N_2 to ammonium (NH_4^+), which is converted to nitrites (NO_2^-) and then to nitrates (NO_3^-) by nitrifying bacteria. The ammonium and nitrates are then available to plants to use in their proteins and nucleic acids. Next, the nitrogen is transferred to organisms that consume the plants. Nitrates that are not assimilated into living organisms can be converted to nitrogen gas by denitrifying bacteria.

14. Phosphorus in the form of phosphates is washed from sedimentary rock by rainfall. The dissolved phosphates are used by producers to produce important biological molecules, including DNA and ATP. When animals eat producers or other animals, phosphates are passed through the food webs. Decomposers release phosphates into the soil or water from dead organisms.

Humans Can Upset Biogeochemical Cycles (pp. 512–515)

15. The water cycle can be disturbed because of humans causing shortages through overuse of water supplies. Most water use is for agriculture, primarily irrigation. Irrigation can cause salts to accumulate in the soil (salinization), which can make the land unusable for agriculture.

16. Humans have disturbed the carbon cycle through activities that increase carbon dioxide levels in the atmosphere: burning fossil fuels, which directly adds carbon dioxide, and deforestation, which decreases the removal of carbon dioxide from the atmosphere. Carbon dioxide is a greenhouse gas (along with methane, nitrous oxide, CFCs, and ozone) that traps heat in the earth's atmosphere and may cause global warming.

17. Humans have disrupted the nitrogen and phosphorus cycles through the industrial fixation of nitrogen to produce fertilizer and the addition of phosphates to fertilizer. Some of the fixed nitrogen and phosphates wash into nearby waterways. Fertilizer runoff is one cause of eutrophication.

KEY TERMS

biosphere *p. 502*	succession *p. 502*	carnivore *p. 504*	biomass *p. 505*
ecosystem *p. 502*	producers *p. 503*	secondary consumer *p. 504*	ecological pyramid *p. 505*
community *p. 502*	trophic level *p. 503*	omnivore *p. 504*	biological magnification *p. 508*
population *p. 502*	consumer *p. 503*	decomposers *p. 504*	biogeochemical cycle *p. 509*
niche *p. 502*	herbivore *p. 504*	food chain *p. 504*	nitrogen fixation *p. 511*
habitat *p. 502*	primary consumer *p. 504*	food web *p. 504*	nitrification *p. 511*

REVIEWING THE CONCEPTS

1. What is ecological succession? How does primary succession differ from secondary succession? *pp. 502–503*

2. Explain how energy flows through an ecosystem. *pp. 503–504*

3. Define the following terms: producer, primary consumer, secondary consumer, and decomposer. Explain the role each plays in cycling nutrients through an ecosystem. *pp. 503–504*

4. Explain why the feeding relationships in a community are more realistically portrayed as a food web than as a food chain. *p. 504*

5. Explain the reasons for the inefficiency of energy transfer from one trophic level to the next higher one. Why does this loss of energy limit the number of possible trophic levels? *p. 505*

6. Define a pyramid of energy. What causes the pyramidal shape? *p. 505*

7. What is meant by a pyramid of biomass? *p. 505*

8. Define biological magnification. Explain how biological magnification is a consequence of the energy loss between trophic levels. Why should humans be concerned about biological magnification? *pp. 505–508*

9. Explain why more people could be fed on a vegetarian diet than on a diet in which meat provides most of the protein calories. *pp. 508–509*

10. Describe the water cycle, the carbon cycle, the nitrogen cycle, and the phosphorus cycle. *pp. 509–512*

11. Explain some of the ways that humans are disturbing the water cycle. *pp. 513–514*

12. Which human activities are primarily responsible for the rising level of atmospheric carbon dioxide? *pp. 514–515*

13. What is meant by the greenhouse effect? Explain why there is concern that the rising level of atmospheric carbon dioxide could lead to global warming.   *p. 515*

14. Which of the following is *not* constantly recycled in the biosphere?
    a. Energy
    b. Water
    c. Carbon
    d. Nitrogen

15. The greenhouse effect is due in part to
    a. phosphorus.
    b. nitrogen.
    c. carbon dioxide.
    d. carbon monoxide.

16. Secondary succession is most likely to be found
    a. on rocky outcroppings.
    b. in areas that were clear cut for timber.
    c. where an island has appeared.
    d. in areas where estuaries have filled in to form land.

17. An organism's role in a community is its _____.

18. An animal that eats an herbivore is called a _____.

APPLYING THE CONCEPTS

1. In a particular grassland ecosystem, energy flows in the following path:

    Grass → Crickets → Frogs → Herons

    On the assumption that the efficiency of energy transfer is typical, how many calories will the herons receive if there were 100,000 calories of grass?

Additional questions can be found on the companion website.

2. Explain why the mercury levels in a lake may be low enough for the water to be safe to drink yet fish from the same lake may be poisonous.

3. You are a crime scene investigator at a murder scene. You find a leaf unlike any others in the vicinity of the crime stuck to the bottom of the victim's shoe. What clues could you get from the leaf?

Populations Change in Size over Time

- Rates of addition and subtraction determine population growth rate
- The age structure of a population influences future growth
- Immigration and emigration affect population size

The Human Population Is Increasing, but Its Rate of Growth Is Slowing

Environmental Factors Regulate Population Size

The Growing Human Population Has Altered the Earth's Carrying Capacity

- The human population may be reaching the earth's carrying capacity
- Human activities cause pollution
- Human activities deplete the earth's resources
- Human activities are reducing biodiversity

Our Future Depends on the Decisions We Make Today

SOCIAL ISSUE Maintaining Our Remaining Biodiversity

24

Human Population, Limited Resources, and Pollution

The demands of the growing human population, combined with the effectiveness of large-scale commercial fishing methods, have begun to deplete the oceans of many species of fish.

Brad and Krista walked toward the boat, ready for their parasailing adventure. Dale, the driver, met them along the way. Krista asked nervously, "How long have you been taking people parasailing, Dale?'

"About three years now," Dale explained, "ever since the fishing gave out around here and I had to give it up. The big trawlers have dangerously overexploited the waters. I guess it's just as well because we would have had to stop fishing anyway. Some company dumped a load of electronic equipment—mostly cell phones—off the coast a while back, and now they've found lead and some other metals in the shellfish catches."

As they reached the boat, Krista said, "I love the ocean. I didn't realize that it could be depleted of fish or polluted. It seems so huge! I thought nothing could harm it."

Dale replied, "Overfishing by trawlers and pollution have already cost large numbers of people their traditional livelihood. For this part of the ocean to become a reliable fishery again, many countries will have to work together to make conservation and clean-up a top priority."

In this chapter, we will consider overfishing and other problems that stem from the growth of the human population. We will start by reviewing general principles governing population size. Then we will consider how the rapid growth of the human population has led to pollution, depletion of the earth's resources, and reductions in biodiversity. Lastly, we will discuss how the choices we make today can affect the quality of life on earth in the future. ■

Populations Change in Size over Time

There is no way to discuss human populations without arousing concern. The worries arise from two kinds of information. First, the increase in our numbers is startling, and we just do not know how long it can continue. Second, humans are using the world's natural resources much faster than they can be replaced (*if* they can be replaced). We will review the evidence and possible consequences of each of these trends—and also consider some solutions.

■ Rates of addition and subtraction determine population growth rate

Population size changes when individuals are added and removed at different rates. A population grows when more individuals are added than are subtracted. Some individuals join a population or leave it by immigration or emigration, but most of a population's members are added by birth and removed by death. Most likely, your own family has added to the population growth over the last few generations. How many living descendants do your grandparents have (Figure 24.1)?

Births and deaths are usually expressed as rates. Usually, the rate is reported as the number of births or deaths per 1000 persons per year. In 2006, the estimated *birth rate* of the world's population was 20 births per 1000 individuals per year, and the *death rate* was 9 deaths per 1000 individuals per year. Thus the growth rate

of the world's population, an increase of 11 individuals per 1000, was 1.1%.

$$\frac{\text{Birth} - \text{Deaths}}{1000} = \frac{20 - 9}{1000} = \frac{11}{1000} = \frac{1.1}{100} = 1.1\%$$

The age at which the females have their first offspring has a dramatic impact on the birth rate of a population. The younger the women are when they begin to reproduce, the faster the population grows. In fact, the age when reproduction begins is the most important factor influencing a female's overall reproductive potential. For example, the increasing proportion of women who are postponing childbirth until their later reproductive years has been helping to slow the rate of population growth of the United States. The birth rate in the United States is declining for women of all ages, but the largest drop in birth rate has occurred among women in their twenties. One reason may be that increasing numbers of women want to establish careers before beginning motherhood. Regardless of the reasons, this trend is contributing to slower U.S. population growth.

■ The age structure of a population influences future growth

When we want to predict future growth, a population's **age structure**—the relative number of individuals of each age—can be helpful. The age structure of the population is important because only individuals within a certain age range reproduce. Among humans, for instance, toddlers and octogenarians are counted as members of the population, but they do not reproduce. The ages are often grouped into prereproductive, reproductive, and postreproductive categories. Generally, only individuals of reproductive age add to the size of the population. However, the prereproductive group (those who are currently too young to reproduce) usually get older, and when they enter the reproductive class, they have children, adding additional members to the population. Even if the birth rate remains the same, the overall size of the population will grow more rapidly as the size of the reproductive class increases.

stop and think

Health care has improved greatly in the last century and has increased population growth. Why would measures that decrease child mortality (death) contribute more to population growth than measures that increase life expectancy?

FIGURE **24.1** Gladys Davis is surrounded by her six great-grandchildren. She and her husband had four children, two of whom had children of their own, producing three grandchildren. The three grandchildren produced these six great-grandchildren. By mid-2006, two of the children shown here had started their own families, producing six great-great-grandchildren. Although the size of each individual family was small, there are currently 15 living descendants of Jack and Gladys Davis. How much has your family grown in the last few generations?

We see, then, that the relative size of the base of a population's age structure (the prereproductive age category) determines how quickly numbers will be added to the population in the future. A wide base to any age structure reflects a growing population, whereas a narrow base is characteristic of a population that is getting smaller (Figure 24.2). A large base to any such pyramid, as exists in India, spells trouble in terms of future demands on the country's commodities. Moderately stable nations, such as the United States, show a more consistent age distribution, with some fluctuation but

Postreproductive
(46-100 years)

Reproductive
(15-45 years)

Prereproductive
(0-14 years)

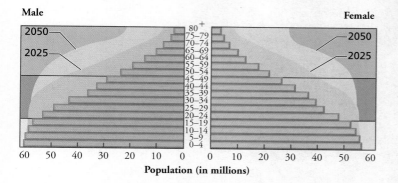

Expanding: India, 2007

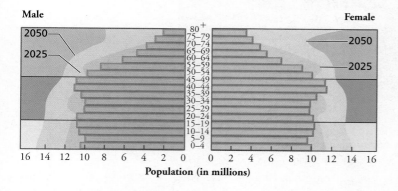

Moderately stable: United States, 2007

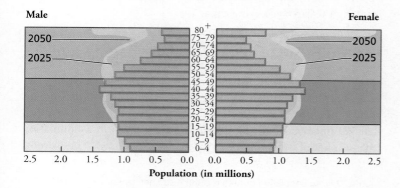

Declining: Canada, 2007

 FIGURE **24.2** Age structures of India, the United States, and Canada. The age structures shown are estimates for 2007, 2025, and 2050. (Source: U.S. Census Bureau, International Data Base.)

with a small prereproductive population. The prereproductive base of the age structure of Canada is smaller than the reproductive base. Canada's population will probably decline in the future.

■ Immigration and emigration affect population size

If we consider the human population of the world as a whole, the effects of immigration and emigration on population size are eliminated. However, immigration and emigration can have dramatic effects on the population size of countries, cities, and regions. For example, in 2006, there were an estimated 35 million legal immigrants and 10 to 12 million illegal immigrants living in the United States. In 1950, 75% of residents identified as minorities in the United States were African Americans. However, by 2001, Hispanics were the largest minority. In 2006, immigrant and native-born Hispanics made up 14% of the U.S. population and represented half of the growth in population during the previous year. Such shifts in population due to immigration can have social, political, and economic effects in the communities where they occur.

what would you do?

The United States is currently struggling with several issues related to immigration and its cost in public funds. Should an illegal immigrant be able to obtain a driver's license? Should our schools be required to provide bilingual education? Should a student who is an illegal immigrant be eligible for financial aid to attend college? Should certain illegal immigrants be given the opportunity to become legal immigrants? What do you think? What are your reasons?

The Human Population Is Increasing, but Its Rate of Growth Is Slowing

For most of the period following the Stone Age, the human population grew steadily but relatively slowly. Now, however, the rapidity of its growth is disturbing (Figure 24.3). It did not reach 1 billion until 1804, but since then we have been adding additional billions at a much faster rate. In mid-2006, the world population was approximately 6.4 billion people. It is expected that, by 2016, our numbers will reach nearly 7.3 billion people.

Unrestrained growth can occur only as long as the environment offers plenty of resources and adequate removal of waste, but no environment on earth can support unlimited numbers of individuals of any species. The number of individuals of a given species that a particular environment can support for a prolonged period is called its **carrying capacity** (Figure 24.4). It is determined by such factors as availability of resources, including food, water, and space; a means of cleaning away wastes; disease; and predation pressure.

Where resources are plentiful, populations grow rapidly at first but then grow increasingly slowly as the population size approaches the environment's carrying capacity. Eventually, environmental pressures such as limited resources or accumulating wastes cause growth to level off and the population size to remain more or less stable, fluctuating slightly around the carrying capacity. The growth rate for human populations in some parts of the world is

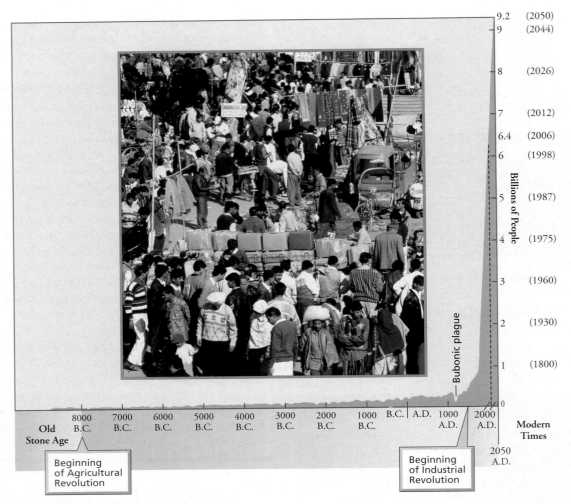

9.2 (2050)
9 (2044)
8 (2026)
7 (2012)
6.4 (2006)
6 (1998)
5 (1987)
4 (1975)
3 (1960)
2 (1930)
1 (1800)
0

Billions of People

Bubonic plague

8000 B.C. | 7000 B.C. | 6000 B.C. | 5000 B.C. | 4000 B.C. | 3000 B.C. | 2000 B.C. | 1000 B.C. | B.C. | A.D. | 1000 A.D. | 2000 A.D. | Modern Times

Old Stone Age

2050 A.D.

Beginning of Agricultural Revolution

Beginning of Industrial Revolution

WEB TUTORIAL 24.2

FIGURE **24.3** The human population has grown steadily throughout most of human history. It skyrocketed after the Industrial Revolution. The growth rate is now showing signs of slowing.

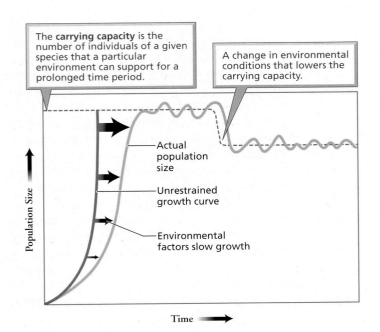

The **carrying capacity** is the number of individuals of a given species that a particular environment can support for a prolonged time period.

A change in environmental conditions that lowers the carrying capacity.

Actual population size

Unrestrained growth curve

Environmental factors slow growth

Population Size

Time

greater than in other parts. The growth rate in less developed countries—those in which the majority of people live in poverty—greatly exceeds the growth rate in more developed countries. Nonetheless, the growth rate of the world population has been slowing steadily during the past decade, suggesting that we may be reaching the earth's carrying capacity.

Environmental Factors Regulate Population Size

The factors that reduce the size of populations can be described as density-dependent or density-independent. *Density-independent regulating factors* are causes of death that are not related to the density of individuals in a population (the number of individuals in a given area of habitat). These events

FIGURE **24.4** Population growth is often restrained by environmental factors, including the availability of food, water, and space and the accumulation of wastes. Environmental factors such as these determine the environment's carrying capacity. Eventually, growth levels off and the population size fluctuates around the carrying capacity. If a change in the environment lowers the carrying capacity, the population will stabilize at a smaller size.

FIGURE **24.5** Food shortages may reflect density-dependent population controls, because they have a greater impact as population density increases.

include such natural disasters as floods, mudslides, earthquakes, hurricanes, and fires. For example, hurricane Katrina killed more than 1800 people (with many more still unaccounted for) when it hit the Gulf Coast of the United States in the summer of 2005. Such events are density independent because adding or removing people from the population would not alter the mortality rate of the event itself.

Density-dependent regulating factors are events that have a greater impact on the population as conditions become more crowded. For example, famine can be a density-dependent factor. If only so much food is available, an increasing population would mean there was less food for each individual until, finally, some individuals would begin to starve (Figure 24.5). The greater the population density, the greater the effects of food scarcity on the population. Because disease-causing organisms and parasites spread more easily in crowded conditions than in uncrowded conditions, these too are density-dependent factors that regulate population sizes.

The Growing Human Population Has Altered the Earth's Carrying Capacity

Whatever your personal predictions for the future of humankind, one thing should be clear: The world is becoming an increasingly crowded place. You may recall from Chapter 23 that there is no input of new resources on the earth. Therefore, we run the risk of using up our resources and making our world unlivable because of our waste. In fact, the size of the human population contributes to most of the problems we face today.

■ The human population may be reaching the earth's carrying capacity

The earth's ability to support people depends not only on natural constraints, such as limited resource availability, but also on human activities and choices. These, in turn, are influenced by economics, politics, technology, and values. Thus, the earth's carrying capacity for humans is uncertain and constantly changing. For instance, technological advances in agriculture and pollution control act to increase the earth's carrying capacity. At the same time, however, we have lowered the carrying capacity by consuming the earth's resources faster than they can be replaced.

Not surprisingly, estimates of the earth's carrying capacity vary widely, ranging from 5 billion to 20 billion people. Those who accept the lower values point to resource depletion and pollution as evidence that we have already exceeded the earth's carrying capacity with a world population of more than 6.4 billion.

World hunger is one consequence of human population growth. Most of us know what *hungry* means, but not *hunger*. Hungry is the feeling we get when lunch is late; hunger is a condition in which lack of food causes abnormal physiological changes within our body. Hunger can be brought on by two conditions: *undernourishment*, when not enough food is eaten, and *malnourishment*, when the diet is not balanced because certain foods are missing from it.

Hunger has its severest effects on children, pregnant women, and nursing mothers. Children are particularly susceptible to two very dangerous nutritional diseases. *Marasmus* is caused by a diet low in proteins and calories. The result is wrinkled skin, a startling appearance of old age, and thin limbs caused by wasting muscles. *Kwashiorkor* also results from a diet low in proteins and calories. It often appears when a child is weaned from breast milk. One of the many symptoms of the disease is fluid retention (edema), typically seen as a pronounced swelling of the abdomen. Scientists are not sure why a diet low in proteins and calories progresses to marasmus in some people and kwashiorkor in others.

Chronic undernourishment (starvation) is associated with irreversible changes in growth and brain development. If children receive less than 70% of the required daily calories, their growth and activity fall below normal levels, the number of brain cells decreases, and brain chemistry is altered. Thus, undernourishment can lower intelligence.

One reason for hunger is that food is unequally distributed throughout the world. North Africa, for example, produces very little food because of a combination of drought, poor farming practices, and warfare. On the other hand, both food abundance and scarcity may exist in the same country. For example, southern Brazil produces a great deal of food, but the northern part of the country does not, so the people living in the north may suffer from malnutrition. The United States is an important exporter (our population has not yet caught up with our food production), but even here we find hunger in certain areas.

Agricultural production has increased in many countries, largely because of the so-called *Green Revolution* of the 1970s and 1980s—the development of high-yield varieties of crops and the use of modern cultivation methods, including the use of farm machinery, fertilizers, pesticides, and irrigation. For example, Indonesia at one time imported more rice than any other country, but now it has enough rice to feed its people, as well as some to export.

Nonetheless, the Green Revolution has a big price tag: high energy costs and environmental damage. Although crop yields may be four times greater than with more traditional methods, modern farming practices use as much as a hundred times more energy and mineral resources. It takes energy—usually obtained from fossil fuels—to produce fertilizers, to power tractors and combines, and to install and operate irrigation systems.

Overfishing is another practice that is lowering the carrying capacity of the earth. Fishing has long provided humans with an abundant food supply, but now many fish populations are being depleted to such levels that some fish can no longer be caught in certain areas. Recent examples include Atlantic cod in New England and salmon in the northwestern United States. The problem is that we have not given the fish enough time to replace their numbers at their reduced population size. As numbers of fish have dwindled, the human response has been to fish harder, using more boats and new techniques, including electronic searches. Some species, the Atlantic bluefin tuna for instance, may already be so overfished that they will become extinct.

■ Human activities cause pollution

We discussed water pollution in Chapter 23, the health consequences of air pollution in Chapter 14, and acid rain caused by air pollution in Chapter 2. Here we will focus on another consequence of air pollution: the destruction of the ozone layer.

Ozone, a gas whose molecules are composed of three atoms of oxygen (O_3), can be a good thing or a bad thing. At the earth's surface, say anywhere in the first few miles of atmosphere, it is generally a bad thing. In fact, ozone is the primary component of photochemical smog, the noxious gas produced largely by sunlight interacting with air pollutants. Ozone irritates the eyes, skin, lungs, nose, and throat. (The effects of photochemical smog on the respiratory system are discussed in the Environmental Issue essay in Chapter 14.) Ozone can also damage forests and crops and dissolve rubber.

However, naturally produced ozone is an essential part of the stratosphere, a layer of the lower atmosphere that encircles the earth about 10 to 45 km (6 to 28 mi) above its surface. The concentration of ozone in the stratosphere reaches levels of only about 1 molecule in 100,000. If all this ozone were compressed from top to bottom, its depth would be equal to only the diameter of a pencil lead. Still, the ozone is able to shield the earth from excessive ultraviolet (UV) radiation from the sun; without it, most terrestrial forms of life could not exist. UV radiation causes cataracts, aging of the skin, sunburn, and snow blindness; and it is the primary cause of skin cancer that kills nearly 11,000 Americans each year. In addition, UV radiation inhibits the immune system and can interfere with plant growth.

In 1985, British scientists reported a sharp drop in the concentration of ozone over the Antarctic. This "hole" appears on a yearly cycle—at the beginning of the Antarctic spring (early September through mid-October; Figure 24.6). The primary cause was discovered to be the action of chlorofluorocarbons (CFCs)—chemicals once important in cooling systems, as aerosol propellants, in the manufacture of plastic foam such as styrofoam, and as solvents in the electronic industry. The CFCs drift up to the stratosphere, where UV radiation breaks them down into chlorine, fluorine, and carbon. Then, under the conditions found in the stratosphere, chlorine can react with ozone, converting it to oxygen. Because chlorine is not altered in this reaction, a single chlorine molecule can destroy thousands of ozone molecules.

A number of laws are now in place to prevent the production or release of CFCs. In 1990 and 1992, industrialized countries signed agreements to reduce the production of CFCs. These countries have now phased out CFC production. It is expected that developing countries will phase out CFC production by 2010. Substitutes for CFCs have been found for aerosol cans, refrigerators, and air conditioners. There is evidence that the reduction in CFC production is paying off. In 1996, 2 years after CFC use peaked, chlorine levels in the lower atmosphere were slightly lower than before. By 1999, chlorine levels in the stratosphere were reduced. In 2006, scientists predicted that the ozone hole will recover by 2068.

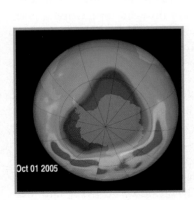

A satellite image of the hole in the ozone layer over Antarctica. The ozone hole is depicted in blue.

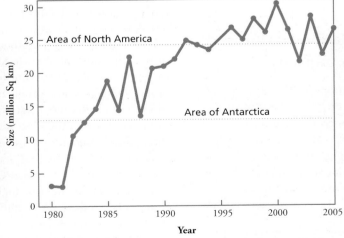

The recovery of the ozone later is occuring more slowly than originally expected. Its recovery is expected by 2068.

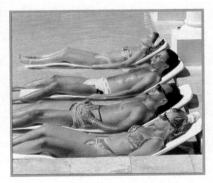

The ozone layer absorbs ultraviolet (UV) radiation. A thinner ozone layer allows more UV radiation to reach the earth, increasing the cancer risk.

FIGURE **24.6** The thinning of the ozone layer

■ Human activities deplete the earth's resources

An intact natural ecosystem usually retains soil and remains fertile. However, overuse and misuse of land can destroy its fertility.

SOIL EROSION

As the human population grows, land is cultivated, grazed, and stripped of vegetation faster than it can recover. Wind and rain then carry away the topsoil, and the productivity of the area declines. In this way, overfarming and overgrazing transform marginal farmland to desert, a process called **desertification** (Figure 24.7). According to the United Nations, a fifth of the world's population is currently threatened by desertification.

DEFORESTATION

People often do not value what they have until it is gone. This may be true for forests. Some of us live in areas where trees are so common that we take them for granted. However, trees, especially when they are grouped together to form forests, are essential to the global ecosystems. As we learned in Chapter 23, forests play an important role in water, carbon, and nitrogen cycles. Tree roots also reduce erosion by holding soil in place. If trees are cut down, rainwater runs off the land, carrying away soil and causing floods. In addition, ecosystems containing trees support an incredible variety of life. Trees also influence local and global climate, including temperature and rainfall. And, in hot weather, their shade is pleasant.

Why are forests disappearing? The answer is simple: The human population is expanding. People need space in which to live, and they need wood for homes and furniture. Livestock cannot graze in forests, so trees are cut. Most are not replanted.

Deforestation is the removal of trees from an area without replacing them. Deforestation is taking place in many regions of the world (Figure 24.8), including the United States. Before the colonization of North America, forests covered most of the eastern seacoast. Cities have now replaced many of those forests. In the U.S. Pacific Northwest, 80% of the forests are slated for logging.

Deforestation is removing trees from an area and not replacing them.

The countries in which the most deforestation is occurring are shown in red and orange.

Rate of deforestation:
Red = 2,000–14,280 km² per year
Orange = 100–1900 km² per year

FIGURE **24.8** Deforestation damages soil quality.

Nonetheless, tropical forests are falling the fastest. Why? The primary reason the land is being cleared is so native families can feed themselves. The second reason (in terms of impact) is commercial logging. The third reason is cattle ranching. Cattle can graze for about 6 to 10 years after trees have been cleared from a tropical forest before shrubs take over and make the area unsuitable for rangeland. Foreign companies own most of these cattle ranches, and the beef is often exported to fast-food restaurant chains. Thus, the decisions we make about what to eat for lunch may have an indirect influence on the rate of tropical deforestation. Other reasons include mining and the development of hydroelectric dams.

Whatever the reason for tropical deforestation, soil fertility declines rapidly when these forests are cut. As a result, native people who depend on the soil for their living become even poorer.

■ Human activities are reducing biodiversity

Biodiversity can be described as species richness. It is the number of species living in a given area. Recently, scientists have raised an

FIGURE **24.7** Desertification is the conversion of farmland, or in this case rangeland, to desert through overuse or other harmful practices.

Maintaining Our Remaining Biodiversity

In the face of powerful evidence of a marked reduction in the earth's biodiversity, some experts express guarded optimism that we may be able to reverse the trend. Various measures are being suggested.

First, both developed (richer) and developing (poorer) countries must take a careful census of their life forms to produce inventories of the species they harbor. This need is especially acute in tropical countries, which, unfortunately, are often the least likely to be able to afford such programs. Advocates therefore argue that the effort must be worldwide, with developed countries subsidizing the research in less affluent areas.

Second, nations must cooperate in linking the economic development of impoverished regions, particularly the tropics, to conservation. Lending agencies must stipulate that certain areas must be set aside and left undeveloped. The decisions regarding which locations should be spared development should be made with the advice of ecologists so that the effects on the world's biodiversity will be minimized. (One cannot help wondering what effect such rules would have had on the development of the United States.)

Third, educating the people whose decisions have the most immediately effect on biodiversity should be a priority. For example, recent studies have shown that native people are able to make more money through sustainable use of a forest (as in gathering nuts and fruits) than would be possible if the forest were cleared for agriculture or ranching. Many people in developing countries are also learning that scientists worldwide are interested in the medicines and healing knowledge they have, so in a sense, the educational opportunities are reciprocal.

Fourth, biologists worldwide must work more closely with zoning and land-use personnel to introduce multiple land-use planning so as to maximize the use of areas being cleared or cultivated. Such planning includes, for example, rotation of crops to maintain healthy soil and increase biodiversity or replacement of single-species forests planted by paper companies with more diverse forests that could harbor more species of animals. Of course, the paper companies may ask why they should be required to reduce their profits to increase biodiversity. This is the type of issue that will have to be negotiated in the process of bringing people to an understanding that what is good for the environment is ultimately good for everyone. ✶

alarm worldwide because there are indications that globally, and especially in certain critical areas, biodiversity is decreasing. We are losing species, perhaps as many as 100 species each day, in large part because of human activity. *Mass extinctions*, the loss of many species from the earth, have occurred in previous eras, but not because of humans. Today, many species are being forced to compete with us as we change the environment to suit ourselves, rendering it unsuitable for them.

How many species are there? The answer is that no one knows. Scientific estimates range from 8 million to 30 million species now in existence, with about 90% of them living on land. However, only about 1.7 million of these species have been identified, and only 3% of those have been studied.

Most terrestrial species live in the tropics. Although tropical rain forests cover only 7% of the earth's surface, they are home to between 50% and 80%—estimated as 7 million to 8 million—of the earth's species (see the Social Issue essay, *Maintaining Our Remaining Biodiversity*).

Habitat destruction is largely responsible for the loss of biodiversity. The primary reasons habitats are destroyed are to create living space for humans and to provide room or resources for economic development. Calculations indicate that most of the original tropical rain forest will be destroyed by 2040. This means that in the next 35 years, we could lose a quarter of all the species on the earth. Evolutionary biologists tell us that, currently, we are losing species at the rate of 1000 to 10,000 times the average rate for the last 65 million years, since the extinction of the dinosaurs.

Rain forest is not the only type of habitat being destroyed. Lumbering and acid rain are destroying northern temperate forests. Also, marine ecosystems are being destroyed by pollution, overfishing, and coastal development. For example, shark populations have declined by more than 70% in the last 15 years. Commercial fishing is largely to blame for the reduction of shark populations. People eat shark meat and use the fins to make soup. Moreover, some sharks, hammerheads for example, are accidentally caught when they try to prey on the herring and squid being used as bait by people fishing for tuna or swordfish. In the United States, there are federal regulations that restrict shark fishing; however, shark fishing is not regulated in Europe. If shark numbers continue to fall, we may not be able to save them from extinction, because sharks grow slowly, mature late in life, have few young at a time, and females do not reproduce every year.

Perhaps you are wondering why we should care about the loss of species, especially those we did not even know existed. Several reasons immediately come to mind. The first is that biodiversity preserves the genetic diversity encoded in the chromosomes of species. This genetic diversity is useful for cross-breeding. The cross-breeding of two strains of a species, each possessing different desirable traits, can combine those desirable traits in a single new strain. For example, a new kind of corn was found on a Mexican hillside. (The farmer had to be persuaded not to plow it under to plant his usual corn crop.) The newly discovered corn was a perennial and proved to be resistant to a disease called corn smut, a fungal disease that causes the kernels to turn black and become oddly shaped. Plans were made to cross the perennial corn with common forms of agricultural corn, and these sorts of experiments continue.

Interestingly, the usefulness of genetic diversity as a reservoir of genetic stores has increased dramatically with modern technology. Owing to recent advances in genetic engineering, specific genes can be isolated from their source organisms and moved into other kinds of organisms. Thus, the maintenance of genetic stores becomes even more important as we use genetic engineering to maximize crop yields or otherwise create new forms of organisms that might be useful to society (see Chapter 21).

The second benefit to maintaining biodiversity lies in developing new kinds of medicines. About 25% of all known drugs come from plants. The rosy periwinkle from Madagascar, as discussed in Chapter 1, is a source of two anticancer drugs, vincristine and vinblastine, and the Mexican yam was once a source of oral contraceptives. Most of the plants with medicinal value have been found in the tropics, which, as we said, are being rapidly destroyed. A third benefit is the many services that biodiversity provides in functioning ecosystems—cleansing water and air, enriching soil, cycling minerals, and pollinating crops for example.

Our Future Depends on the Decisions We Make Today

In many ways, our grandmothers were right: older ideas *were* better—at least, some were. Today's economy is wasteful. Many of us behave as if the world is here for humans alone.

In Chapter 23 and this chapter, we have seen that humans are part—a *small* part—of the biosphere and that it must remain healthy if we are to remain healthy. We must realize that we are subject to the same principles that govern all other life, indeed, all of the earth, and that instead of trying to conquer nature, we must work with natural laws. Above all, we must recognize that resources on the earth are limited and must be shared with all living organisms.

This dramatic turnabout in thinking and behavior will not be easy to accomplish. It will require changing the way governments and businesses operate, as well as the ingrained habits of individuals. However, as was said earlier, humans are, by nature, problem solvers. We now have the opportunity to approach with intelligence one of the greatest problems ever to confront humankind.

Jane Goodall, famed animal behavior researcher and founder of the Jane Goodall Institute, a center for environmental studies, is optimistic about our future. In her words, "My reasons for hope are fourfold: (1) the human brain; (2) the resilience of nature; (3) the energy and enthusiasm that is found or can be kindled among young people worldwide; and (4) the indomitable human spirit.[1] We have to be optimistic that a solution is possible and work together to achieve it.

---

[1]Jane Goodall (with Phillip Berman), *Reasons for Hope. A Spiritual Journey* (New York: Warner Books, 1999), 233.

HIGHLIGHTING THE CONCEPTS

Populations Change in Size over Time (pp. 520–521)

1. Population dynamics describes how populations change in size. Population size changes when the number of individuals added differs from the number of individuals removed. Individuals are added through births and immigration. Individuals are removed through death and emigration.

2. The change in population is usually expressed as a rate: the number of births or deaths for a certain number of individuals during a certain time period (usually per 1000 individuals per year). If we rule out immigration and emigration, we can say the growth rate of a population equals the birth rate minus the death rate.

3. The age at which females have their first offspring is the most important factor in determining the rate at which a population grows.

4. The age structure of a population is an important determinant of future changes in population size. The age structure of a population consists of the relative numbers of individuals in each age category: prereproductive, reproductive, and postreproductive. A growing population has a large base of prereproductive individuals. A stable population has approximately equal numbers of individuals in each age category. A population that is getting smaller has a large proportion of individuals who are past reproductive age.

5. Immigration (movement into a population) and emigration (movement out of a population) can influence local population size.
   WEB TUTORIAL 24.1 Age Structure and Population Growth

The Human Population Is Increasing, but Its Rate of Growth Is Slowing (pp. 521–522)

6. When it is not limited by finite resources, a population grows without restraint.

7. The carrying capacity of the environment is the number of individuals (of a particular species) it can support over a long period of time. Carrying capacity is determined by the amount of available resources.

8. Generally, as a population grows and its size approaches the carrying capacity of the environment, the growth rate slows and the population size levels off, fluctuating around the carrying capacity.
   WEB TUTORIAL 24.2 Human Population Growth

Environmental Factors Regulate Population Size (pp. 522–523)

9. Population size can be regulated by density-independent regulating factors, events such as natural disasters, whose effects are not influenced by the size of the population, and by density-dependent regulating factors, which are factors that have a greater impact as the population becomes more crowded. Density-dependent factors include food availability and disease.

The Growing Human Population Has Altered the Earth's Carrying Capacity (pp. 523–527)

10. Humans have raised the carrying capacity of the earth through technological advances, but they have also decreased it through resource depletion and pollution. The size of the human population contributes to many of the problems we face today.

11. Hunger can be caused by undernourishment (starvation) or by malnourishment (a diet that does not contain enough of the right kinds of nutrients). Hunger has its greatest effects on children, pregnant women, and nursing mothers.

12. The Green Revolution has increased agricultural production through the use of high-yield varieties of crops and modern cultivation methods. The Green Revolution also has high energy costs, depletes natural resources, and causes environmental damage. Many fishing areas are becoming depleted because of overfishing.

13. Air pollution in the form of chlorofluorocarbons (CFCs) has depleted the ozone layer in the lower atmosphere (the stratosphere). The ozone layer traps ultraviolet (UV) light, protecting life on the earth from the harmful effects of UV radiation.

14. Overgrazing and overfarming are converting marginal farmlands to deserts, a process called desertification.

15. Deforestation—removing trees from an area without replacing them—is a growing problem throughout the world, especially in the tropical rain forests.

16. Biodiversity, the number and variety of living things, is being reduced dramatically, largely because of human activity. Most of the loss is occurring in the tropics and is due to habitat destruction. Two practical reasons for concern over the loss of biodiversity are that the disappearing species could have genes that would someday prove useful or that they could be found to produce chemicals with medicinal qualities.

Our Future Depends on the Decisions We Make Today (p. 527)

17. Humans are a small part of a large ecosystem. We must find ways to work within natural laws to preserve ecosystem earth and its biodiversity.

KEY TERMS

age structure   *p. 520*

carrying capacity   *p. 521*

desertification   *p. 525*

deforestation   *p. 525*

biodiversity   *p. 525*

REVIEWING THE CONCEPTS

1. How is the growth rate of a population expressed?   *p. 520*
2. What is meant by the age structure of a population? How is a population's age structure related to predictions of its future growth? *pp. 520–521*
3. What is the carrying capacity of the environment?   *pp. 521–522*
4. Differentiate between density-independent and density-dependent regulating factors. Give examples of each.   *pp. 522–523*
5. Describe some of the health consequences of hunger.   *p. 523*
6. What determines whether ozone is helpful or harmful? What is causing the destruction of the ozone layer?   *p. 524*
7. What causes desertification?   *p. 525*
8. Define deforestation. Where is it occurring most rapidly? What are the primary reasons for deforestation?   *p. 525*
9. Define biodiversity. Where is it greatest? Why should we be concerned about the loss of biodiversity?   *pp. 525–527*

10. Which of the following is an example of a density-dependent regulating factor?
    a. Earthquake
    b. Starvation
    c. Flood
    d. Drought
11. If a population reaches the carrying capacity of the environment,
    a. unrestrained growth will occur.
    b. the population will decline rapidly.
    c. food and other resources will increase.
    d. the population size will fluctuate around this level.
12. The ozone layer of the atmosphere absorbs
    a. carbon dioxide.
    b. chlorine.
    c. nitrogen dioxide.
    d. ultraviolet radiation.
13. The number of individuals an environment can support is the _____.

APPLYING THE CONCEPTS

1. The populations of Kenya and Italy (shown at the right) have the same number of people, and the growth rate of the two populations is the same. Assuming that the growth rate stays the same, will the populations have the same number of people in 50 years? Explain.

2. Around 1993, four species of Asian carp escaped from fish farms in Mississippi and Arkansas, and by 1997 they had spread from Arkansas to Louisiana. By 2003, the number of Asian carp in Louisiana had increased dramatically. Asian carp are very large fish that eat plankton, which is also consumed by many other freshwater fish.

   a. Explain why the Asian carp population was able to increase extremely rapidly in a new environment (Louisiana).

   b. Explain why scientists are concerned that Asian carp will reduce the population size of other freshwater fish, even though the carp do not eat these other fish.

3. When soldiers returned home after World War II, many of them began having children, leading to the birth of the baby boomers, the largest generation in United States history. Born between 1946 and 1964, the baby boomers began to turn 60 in 2006. Many will soon reach retirement age. How would you expect this change in population structure to affect the way in which tax money will be spent?

4. Explain why population growth is usually expressed as a *rate*.

Additional questions can be found on the companion website.

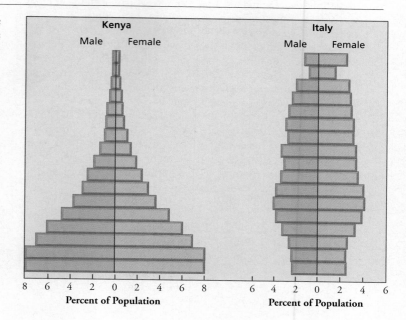

Accommodation A change in the shape of the lens of the eye brought about by contraction of the smooth muscle of the ciliary body that changes the degree to which light rays are bent so that an image can be focused on the retina.

Acetylcholine A neurotransmitter found in both the central nervous system and the peripheral nervous system. It is the neurotransmitter released at neuromuscular junctions and causes muscle contraction.

Acetylcholinesterase An enzyme that breaks the neurotransmitter acetylcholine into its inactive component parts, acetate and choline. Acetylcholinesterase stops the action of acetylcholine at a synapse.

Acid Any substance that increases the concentration of hydrogen ions in solution.

Acinar cells Exocrine cells of the pancreas that secrete digestive enzymes into ducts that empty into the small intestine.

Acromegaly A condition characterized by thickened bones and enlarged soft tissues caused by overproduction of growth hormone in adulthood.

Acrosome A membranous sac at the tip of a sperm cell that contains enzymes that facilitate sperm penetration into the egg during fertilization.

Actin The contractile protein that makes up the major portion of the thin filaments in muscle cells. An actin (thin) filament is composed of actin, troponin, and tropomyosin. In muscle cells, contraction occurs when actin interacts with another protein called myosin.

Actin filaments The thin filaments in muscle cells composed primarily of the protein actin and essential to muscle contraction. In addition to actin, thin filaments contain two proteins important in the regulation of muscle contraction: tropomyosin and troponin.

Action potential A nerve impulse. An electrochemical signal conducted along an axon. A wave of depolarization caused by the inward flow of sodium ions followed by repolarization caused by the outward flow of potassium ions.

Active immunity Immune resistance in which the body actively participates by producing memory B cells and T cells after exposure to an antigen, either naturally or through vaccination.

Active site A specific location on an enzyme where the substrate binds.

Active transport The movement of molecules across the plasma membrane, usually against a concentration gradient (from a region of lower concentration to one of higher concentration) with the aid of a carrier protein and energy (usually in the form of adenosine triphosphate, or ATP) supplied by the cell.

Acute renal failure An abrupt, complete or nearly complete, cessation of kidney function.

Adaptation The process by which organisms become better attuned to their particular environments as a result of natural selection.

Adaptive trait A characteristic (structure, function, or behavior) of an organism that makes an individual better able to survive and reproduce in its natural environment. Adaptive traits arise through natural selection.

Addison's disease An autoimmune disorder characterized by fatigue, loss of appetite, low blood pressure, and increased skin pigmentation resulting from undersecretion of cortisol and aldosterone.

Adenosine triphosphate (ATP) A nucleotide that consists of the sugar ribose, the base adenine, and three phosphate groups. ATP is the energy currency of all living cells.

Adipose tissue A type of loose connective tissue that contains cells specialized for storing fat.

Adolescence The stage in postnatal development that begins with puberty and ends by about age 17. It is a period of rapid physical and sexual maturation during which the ability to reproduce is achieved.

Adrenal cortex The outer region of the adrenal gland that secretes glucocorticoids, mineralocorticoids, and gonadocorticoids.

Adrenal glands The body's two adrenal glands are located on top of the kidneys. The outer region of each adrenal gland secretes glucocorticoids, mineralocorticoids, and gonadocorticoids, and the inner region secretes epinephrine and norepinephrine.

Adrenaline See epinephrine.

Adrenal medulla The inner region of the adrenal gland that secretes epinephrine and norepinephrine.

Adrenocorticotropic hormone (ACTH) The anterior pituitary hormone that controls the synthesis and secretion of glucocorticoid hormones from the cortex of the adrenal glands.

Adulthood The stage in postnatal development that is generally reached somewhere between 18 and 21 years of age. Growth and formation of bone are usually completed by age 25, and thereafter developmental changes occur quite slowly.

Afferent (sensory) neuron A nerve cell specialized to conduct nerve impulses from the sensory receptors *toward* the central nervous system.

Age structure Of a population, the number of males and females of each age. The ages are often grouped into prereproductive, reproductive, and postreproductive categories. Generally, only individuals of reproductive age add to the size of the population.

Agglutinate To clump together.

Aging The normal and progressive alteration in the structure and function of the body. Aging is possibly caused by declines in critical body systems, disruption of cell processes by free radicals, slowing or cessation of cell division, and decline in the ability to repair damaged DNA.

Agranulocytes The white blood cells without granules in their cytoplasm, including monocytes and lymphocytes.

AIDS Acquired immune deficiency syndrome. A diagnosis of AIDS is made when an HIV-positive person develops one of the following conditions: (1) a helper T cell count below 200/mm³ of blood; (2) one of 26 opportunistic infections, the most common of which are *Pneumocystis carinii* pneumonia and Kaposi's sarcoma, a cancer of connective tissue that affects primarily the skin; (3) a loss of more than 10% of body weight (wasting syndrome); or (4) dementia (mental incapacity such as forgetfulness or inability to concentrate).

Albinism A genetic inability to produce the brown pigment melanin that normally gives color to the eyes, hair, and skin.

Aldosterone A hormone (the primary mineralocorticoid) released by the adrenal cortex that stimulates the reabsorption of sodium within kidney nephrons.

Allantois The extraembryonic membrane that serves as a site of blood cell formation and later becomes part of the umbilical cord, the ropelike connection between the embryo and the placenta.

Allele An alternative form of a gene. One of two or more slightly different versions of a gene that codes for different forms of the same trait.

Allergen An antigen that stimulates an allergic response.

Allergy A strong immune response to an antigen (an allergen) that is not usually harmful to the body.

Allometric growth The change in the relative rates of growth of various parts of the body. Such growth helps shape developing humans and other organisms.

Alveolus (plural, alveoli) A thin-walled rounded chamber. In the lungs, the alveoli are the surfaces for gas exchange. They form clusters at the end of each bronchiole that are surrounded by a vast network of capillaries. The alveoli greatly increase the surface area for gas exchange.

Amino acid The building blocks of proteins consisting of a central carbon atom bound to a hydrogen atom, an amino group (NH_2), a carboxyl group (COOH), and a side chain designated by the letter *R*. There are 20 amino acids important to human life; some can be synthesized by our bodies (nonessential amino acids), whereas others cannot be synthesized and must be obtained from the foods we eat (essential amino acids).

Amniocentesis A method of prenatal testing for genetic problems in a fetus in which amniotic fluid is withdrawn through a needle so that the fluid can be tested biochemically and the cells can be cultured and examined for genetic abnormalities.

Amnion The extraembryonic membrane that encloses the embryo in a fluid-filled space called the amniotic cavity. Amniotic fluid forms a protective cushion around the embryo that later can be examined as part of prenatal testing in a procedure known as amniocentesis.

Ampulla A wider region in a canal or duct. In the inner ear, the ampulla is found at the base of a semicircular canal.

Anabolic steroids Synthetic hormones that mimic testosterone and stimulate the body to build muscle and increase strength. Steroid abuse can have many dangerous side effects.

Anabolism The building (synthetic) chemical reactions within living cells, as when cells build complicated molecules from simple ones. Compare with catabolism.

Anal canal The canal between the rectum and the anus. Feces pass through the anal canal.

Analgesic A substance, such as Demerol, that relieves pain.

Analogous structure A structure of one organism that is similar to that of another organism because of convergent evolution and not because the organisms arose from a common ancestry. Compare with homologous structures.

Anaphase In mitosis, the phase when the chromatids of each chromosome begin to separate, splitting at the centromere. Now separate entities, the chromatids are considered chromosomes, and they move toward opposite poles of the cell.

Anaphylactic shock An extreme allergic reaction that occurs within minutes after exposure to a substance to which a person is allergic and that can cause pooling of blood in capillaries, which causes dizziness, nausea, and sometimes unconsciousness as well as extreme difficulty in breathing. Anaphylactic shock can cause death.

Androgen A steroid sex hormone secreted by the testes in males and produced in small quantities by the adrenal cortex in both sexes.

Anemia A condition in which the blood's ability to carry oxygen is reduced. It can result from too little hemoglobin, too few red blood cells, or both.

Anencephaly A neural tube defect that involves incomplete formation of the brain and results in stillbirth or death shortly after birth.

Anesthesia The drug-induced loss of the sensation of pain. It may be general or regional.

Aneuploidy A general term describing the condition whereby gametes or cells contain too many or too few chromosomes compared with the normal number.

Aneurysm A blood-filled sac in the wall of an artery caused by a weak area in the artery wall.

Angina pectoris Choking or strangling chest pain, usually experienced in the center of the chest or slightly to the left, that is caused by a temporary insufficiency of blood flow to the heart. It begins during physical exertion or emotional stress, when the demands on the heart are increased and the blood flow to the heart muscle can no longer meet the needs.

Angioplasty A procedure that widens the channel of an artery obstructed because of atherosclerosis. It involves inflating a tough, plastic balloon inside the artery.

Angiotensin I Renin converts angiotensinogen into this protein.

Angiotensin II A protein that stimulates the adrenal gland to release aldosterone.

Anorexia nervosa An eating disorder characterized by deliberate self-starvation, a distorted body image, and low body weight.

Antagonistic pairs Muscles arranged in pairs so that the actions of the members of the pair are opposite to one another. This arrangement is characteristic of most skeletal muscles.

Antibody A Y-shaped protein produced by plasma cells during an immune response that recognizes and binds to a specific antigen because of the shape of the molecule. Antibodies defend against invaders in a variety of ways, including neutralization, agglutination and precipitation, or activation of the complement system.

Antibody-mediated immune responses Immune system responses conducted by B cells that transform into plasma cells and produce antibodies, and that defend primarily against enemies that are free in body fluids, including toxins or extracellular pathogens, such as bacteria or free viruses.

Anticodon A three-base sequence on transfer RNA (tRNA) that binds to the complementary base pairs of a codon on the mRNA.

Antidiuretic hormone (ADH) A hormone manufactured by the hypothalamus but stored in and released from the posterior pituitary. It regulates the amount of water reabsorbed by the distal convoluted tubules and collecting ducts of nephrons. ADH causes water retention at the kidneys and elevates blood pressure. It is also called vasopressin.

Antigen A substance that is recognized as foreign by the immune system. Antigens trigger an immune response.

Antigen-presenting cell (APC) A cell that presents an antigen to a helper T cell, initiating an immune response toward that antigen. An important type of antigen-presenting cell is a macrophage.

Aorta The body's main artery that conducts blood from the left ventricle toward the cells of the body. The aorta arches over the top of the heart and gives rise to the smaller arteries that feed the capillary beds of the body tissues.

Apoptosis A series of predictable physical changes in a cell that is undergoing programmed cell death. Apoptosis is sometimes used as a synonym for programmed cell death.

Appendicular skeleton The part of the skeleton comprising the pectoral girdle (shoulders), the pelvic girdle (pelvis), and the limbs (arms and legs).

Appendix A slender closed pouch that extends from the large intestine near the juncture with the small intestine.

Aqueous humor The fluid within the anterior chamber of the eye. It supplies nutrients and oxygen to the cornea and lens and carries away their metabolic wastes.

Arachnoid The middle layer of the meninges (the connective tissue layers that protect the central nervous system).

Areolar connective tissue A type of loose connective tissue composed of cells in a gelatinous matrix. It serves as a universal packing material between organs and anchors skin to underlying tissues and organs.

Arrector pili The tiny, smooth muscles attached to the hair follicles in the dermis.

Arteriole A small blood vessel located between an artery and a capillary. Arterioles serve to regulate blood flow through capillary beds to various regions of the body. They also regulate blood pressure. Arterioles are barely visible to the naked eye.

Artery A large-diameter muscular tube (blood vessel) that transports blood away from the heart toward the cells of body tissues. Arteries conduct blood low in oxygen to the lungs and blood high in oxygen to the body tissues. Arteries typically have thick muscular and elastic walls that dampen the blood pressure pulsations caused by heart contractions.

Arthritis An inflammation of a joint.

Artificial insemination A treatment for infertility in which sperm are deposited in the woman's cervix or vagina at about the time of ovulation.

Association neuron An interneuron. These neurons are located within the central nervous system between sensory and motor neurons and serve to integrate information.

Asthma A condition marked by spasms of the muscles of bronchioles, making air flow difficult. It is often triggered by allergy.

Astigmatism Irregularities in the curvature of the cornea or lens that cause distortions of a visual image because the irregularities cause light rays to converge unevenly.

Atherosclerosis A narrowing of the arteries caused by thickening of the arterial walls and a buildup of lipid (primarily cholesterol) deposits. Atherosclerosis reduces blood flow through the vessel, choking off the vital supply of oxygen and nutrients to the tissues served by that vessel.

Atom A unit of matter that cannot be further broken down by chemical means; it is composed of subatomic particles, which include protons (positively charged particles), neutrons (with no charge), and electrons (with negative charges).

Atomic number The number of protons in the nucleus of an atom.

Atrial fibrillation Rapid, ineffective contractions of the atria of the heart.

Atrial natriuretic peptide (ANP) The hormone released by cells in the right atrium of the heart in response to stretching of the heart caused by increased blood volume and pressure. ANP decreases water and solute reabsorption by the kidneys, resulting in the production of large amounts of urine.

Atrioventricular (AV) bundle A tract of specialized cardiac muscle cells that runs along the wall between the ventricles of the heart and conducts an electrical impulse that originated in the sinoatrial (SA) node and was conducted to the atrioventricular node to the ventricles. The bundle forks into right and left branches and then divides into many other specialized cardiac muscle cells, called Purkinje fibers, that penetrate the walls of the ventricles.

Atrioventricular (AV) node A region of specialized cardiac muscle cells located in the partition between the two atria. It receives an electrical signal that spreads through the atrial walls from the sinoatrial node and relays the stimulus to the ventricles by means of a bundle of specialized muscle fibers, called the atrioventricular bundle, that runs along the wall between the ventricles.

Atrioventricular (AV) valves Heart valves located between the atria and the ventricles that keep blood flowing in only one direction, from the atria to the ventricles. The right AV valve consists of three flaps of tissue and is also called the tricuspid valve. The left AV valve consists of two flaps of tissue and is also called the bicuspid or the mitral valve.

Atrium (plural, atria) An upper chamber of the heart that receives blood from the veins and pumps it to the ventricles.

Auditory tubes Small tubes that join the upper region of the pharynx (throat) with the middle ear. They help to equalize the air pressure between the middle ear and the atmosphere. Also called eustachian tubes.

Autoimmune disorder An immune response misdirected against the body's own tissues.

Autonomic nervous system The part of the peripheral nervous system that governs the involuntary, unconscious activities that maintain a relatively stable internal environment. The autonomic nervous system has two branches: the sympathetic and the parasympathetic.

Autosomes The 22 pairs of chromosomes (excluding the pair of sex chromosomes) that determine the expression of most of the inherited characteristics of a person.

Autotroph An organism that makes its own food (organic compounds) from inorganic substances. The autotrophs include photoautotrophs, which use the energy of light, and chemoautotrophs, which use the energy in chemicals.

Axial skeleton The part of the skeleton comprising the skull, the vertebral column, the breastbone (sternum), and the rib cage.

Axon A long extension from the cell body of a neuron that carries an electrochemical message away from the cell body toward another neuron or effector (muscle or gland). The tips of the axon release a chemical called a neurotransmitter that can affect the activity of the receiving cell. Typically, there is one long axon on a neuron.

Axon terminal The tip of a branch of an axon that releases a chemical (neurotransmitter) that alters the activity of the target cell. A synaptic knob.

B cell See B lymphocyte.

B lymphocyte B cell. A type of white blood cell important in antibody-mediated immune responses that can transform into a plasma cell and produce antibodies.

Balanced polymorphism A phenomenon in which natural selection maintains two or more alleles for a trait in a population from one generation to the next. It occurs when the environment changes frequently or when the heterozygous condition is favored over either homozygous condition.

Ball-and-socket joint A joint that allows motion in all directions, such as the shoulder and hip joints.

Barr body A structure formed by a condensed, inactivated X chromosome in the body cells of female mammals.

Basal body The structure that anchors the microtubules of a cilium or flagellum to a cell. It contains nine triplets of microtubules arranged in a ring.

Basal cell carcinoma The most common type of skin cancer, occurring in the rapidly dividing cells of the basal layer of the epidermis.

Basal ganglia A region of the cerebrum consisting of collections of cell bodies located deep within it called the white matter. They are important in coordinating movement and may also play a role in cognition and memory of learned skills.

Basal metabolic rate (BMR) A measure of the minimum energy required to keep an awake, resting body alive. It generally represents between 60% and 75% of the body's energy needs.

Base Any substance that reduces the concentration of hydrogen ions in solution.

Basement membrane A noncellular layer beneath epithelial tissue that binds the epithelial cells to underlying connective tissue. It helps epithelial tissue resist stretching and forms a boundary.

Basophil A white blood cell that releases histamine, a chemical that both attracts other white blood cells to the site and causes blood vessels to widen during an inflammatory response.

Benign tumor An abnormal mass of tissue that usually remains at the site where it forms.

Bicuspid valve A heart valve located between the left atrium and ventricle. It is also called the mitral valve or the left atrioventricular (AV) valve.

Bile A mixture of water, ions, cholesterol, bile pigments, and bile salts that emulsifies fat (keeps fat as small globules), facilitating digestion. Bile is produced by the liver, stored in the gallbladder, and acts in the small intestine.

Bilirubin A red pigment produced from the breakdown of the heme portion of hemoglobin by liver cells. It is excreted by the liver in bile.

Binary fission A type of asexual reproduction in which the genetic information is replicated and then a cell divides into two equal parts.

Biodiversity The number and variety of all living things in a given area. It includes genetic diversity, species diversity, and ecological diversity.

Biofeedback The use of artificial signals to provide feedback about unconscious visceral and motor activity, particularly that associated with stress.

Biogeochemical cycle The recurring process by which materials (for example, carbon, water, nitrogen, and phosphorus) cycle between living and nonliving systems and back again.

Biogeography The study of the geographic distribution of organisms. New distributions of organisms occur when organisms move to new areas (dispersal) and when areas occupied by the organisms move or are subdivided.

Biological magnification The tendency of a nondegradable chemical to become more concentrated in organisms as it passes along the food chain.

Biomass In ecosystems, the dry weight of the body mass of a group of organisms in a particular habitat.

Biopsy The removal and examination, usually microscopic, of a piece of tissue to diagnose a disease, usually cancer.

Biosphere The part of the earth in which life is found. It encompasses all of the earth's living organisms.

Biotechnology The industrial or commercial use or alteration of living organisms, cells, or molecules to achieve specific useful goals.

Bipedalism Walking on two feet. The trait that evolved early in hominid evolution and set the stage for the evolution of other characteristics, such as increases in brain size.

Bipolar neuron A neuron that has only two processes. The axon and the dendrite extend from opposite sides of the cell body. Bipolar neurons are receptor cells found only in some of the special sensory organs, such as in the retina of the eye and in the olfactory membrane of the nose.

Birth defects Developmental defects present at birth. Such defects involve structure, function, behavior, or metabolism and may or may not be hereditary.

Birth rate The number of births per a specified number of individuals in the population during a specific time.

Bladder A muscular saclike organ that receives urine from the two ureters and temporarily stores it until release into the urethra.

Blastocyst The stage of development consisting of a hollow ball of cells. It contains the inner cell mass, a group of cells that will become the embryo, and the trophoblast, a thin layer of cells that will give rise to part of the placenta.

Blind spot The region of the retina where the optic nerve leaves the eye and on which there are no photoreceptors. Objects focused on the blind spot cannot be seen.

Blood Connective tissue that consists of cells and platelets suspended in plasma, a liquid matrix.

Blood-brain barrier A mechanism that protects the central nervous system by selecting the substances permitted to enter the cerebrospinal fluid from the blood. The barrier results from the relative impermeability of the capillaries in the brain and spinal cord.

Blood pressure The force exerted by the blood against the walls of the blood vessels. It is caused by the contraction of the ventricles and is influenced by vasoconstriction.

Blood type A characteristic of a person's red blood cells determined on the basis of large molecules, primarily proteins, on the surface of the plasma membrane.

Blue babies Newborns whose foramen ovale, the fetal opening between the right and left atria of the heart, fails to close. As a result, much of their blood still bypasses the lungs and is low in oxygen. The condition can be corrected with surgery.

BMR See basal metabolic rate.

Bolus A small, soft ball of food mixed with saliva.

Bone marrow The soft material filling the cavities in bones. Yellow bone marrow serves as a fat-storage site. Red bone marrow is the site where blood cells are produced.

Bone Strong connective tissue with specialized cells in a hard matrix composed of collagen fibers and mineral salts.

Bone remodeling The ongoing process of bone deposition and absorption in response to hormonal and mechanical factors.

Bottleneck effect The genetic drift associated with dramatic, random reductions in population size so that by chance alone the genetic makeup of survivors is not representative of the original population.

Brain The organ composed of neurons and glial cells that receives sensory input and integrates, stores, and retrieves information.

Brain waves The patterns recorded in an EEG (electroencephalogram) that reflect the electrical activity of the brain and are correlated with the person's state of alertness.

Breast The front of the chest, especially either of the two protuberant glandular organs (mammary glands) that in human females and other female mammals produce milk to nourish newborns.

Breathing center A region in the medulla of the brain that controls the basic breathing rhythm.

Breech birth Delivery in which the baby is born buttocks first rather than head first. It is associated with difficult labors and umbilical cord accidents.

Bronchi (singular, bronchus) The respiratory passageways between the trachea and the bronchioles that conduct air into the lungs.

Bronchial tree The term given to the air tubules in the respiratory system because their repeated branching resembles the branches of a tree.

Bronchioles A series of small tubules branching from the smallest bronchi inside each lung.

Bronchitis Inflammation of the mucous membranes of the bronchi, causing excess mucus and a deep cough.

Brush border A fuzzy border of microvilli on the surface of absorptive epithelial cells of the small intestine.

Buffer A substance that prevents dramatic changes in pH by removing excess hydrogen ions from solution when concentrations increase and adding hydrogen ions when concentrations decrease.

Bulbourethral glands Cowper's glands. Male accessory reproductive glands that release a clear slippery liquid immediately before ejaculation.

Bulimia An eating disorder characterized by binge eating followed by purging by means of self-induced vomiting, enemas, laxatives, or diuretics.

Bursa (plural, bursae) A flattened sac containing a thin film of synovial fluid that surrounds and cushions certain synovial joints. Bursae are common in locations where ligaments, muscles, skin, or tendons rub against bone.

Bursitis Inflammation of a bursa (a sac in a synovial joint that acts as a cushion). Bursitis causes fluid to build up within the bursa, resulting in intense pain that becomes worse when the joint is moved and cannot be relieved by resting in any position.

Calcitonin (CT) A hormone secreted by the thyroid gland when blood calcium levels are high. It stimulates the removal of calcium from the blood and inhibits the breakdown of bone.

Callus A mass of repair tissue formed by collagen fibers secreted from fibroblasts or woven bone that forms around and links the ends of a broken bone.

Capillary A microscopic blood vessel between arterioles and venules with walls only one cell layer thick. It is the site where the exchange of materials between the blood and the tissues occurs.

Capillary bed A network of capillaries servicing a particular area. Precapillary sphincters regulate blood flow through the capillary bed. When the sphincters relax, blood fills the capillary bed and materials can be exchanged between the blood and the tissues. When the sphincters contract, blood flows directly from an arteriole to a venule, bypassing the capillaries.

Carbaminohemoglobin The compound formed when carbon dioxide binds to hemoglobin.

Carbohydrate An organic molecule that provides fuel for the human body. Carbohydrates, which we know as sugars and starches, can be classified by size into the monosaccharides, disaccharides, and polysaccharides.

Carbonic anhydrase An enzyme in the red blood cells that catalyzes the conversion of unbound carbon dioxide to carbonic acid.

Carcinogen A substance that causes cancer.

Carcinoma *in situ* A tumor that has not spread; "cancer in place."

Cardiac cycle The events associated with the flow of blood through the heart during a single heartbeat. It consists of systole (contraction) and diastole (relaxation) of the atria and then of the ventricles of the heart.

Cardiac muscle tissue A contractile tissue that makes up the bulk of the walls of the heart. Cardiac muscle cells are cylindrical and have branching interconnections between cells. Cardiac muscle cells are striped (striated) and have a single nucleus. Contraction of cardiac muscle is involuntary.

Cardiovascular Pertaining to the heart and blood vessels.

Cardiovascular system The organ system composed of the heart and blood vessels. The cardiovascular system distributes blood, delivers nutrients, and removes wastes.

Carnivore An animal that obtains energy by eating other animals. A secondary, tertiary, or quaternary consumer.

Carpal tunnel syndrome A condition of the wrist and hand whose symptoms may include numbness or tingling in the affected hand, along with pain in the wrist, hand, and fingers that is caused by repeated motion in the hand or wrist, causing the tendons to become inflamed and press against the nerve.

Carrier An individual who displays the dominant phenotype but is heterozygous for a trait and can therefore pass the recessive allele to descendants.

Carrying capacity The number of individuals of a given species that a particular environment can support for a prolonged time period. The carrying capacity of the environment is determined by such factors as availability of resources, including food, water, and space; ability to clean away wastes; and predation pressure.

Cartilage A type of specialized connective tissue with a firm gelatinous matrix containing protein fibers for strength. The cartilage cells (chondrocytes) lie in small spaces (lacunae) within the matrix.

Catabolism Chemical reactions within living cells that break down complex molecules into simpler ones, releasing energy from chemical bonds. Compare with anabolism.

Cataract A cloudy or opaque lens. Cataracts reduce visual acuity and may be caused by glucose accumulation associated with type I diabetes, excessive exposure to sunlight, and exposure to cigarette smoke.

CD4 cell See helper T cell.

Cell The smallest structure that shows all the characteristics of life.

Cell adhesion molecule (CAM) A molecule that pokes through the plasma membranes of most cells and helps hold cells together to form tissues and organs.

Cell body The part of a neuron that contains the organelles and nucleus needed to maintain the cells.

Cell cycle The entire sequence of events that a cell goes through from its origin in the division of a parent cell through its own division into two daughter cells. The cell cycle consists of two major phases: interphase and cell division.

Cell-mediated immune responses Immune system responses conducted by T cells that protect against cellular threats, including body cells that have become infected with viruses or other pathogens and cancer cells.

Cellular respiration The oxygen-requiring pathway by which glucose is broken down by cells to yield carbon dioxide, water, and energy.

Cell differentiation The process by which cells become specialized with respect to structure and function.

Cell theory This fundamental organizing principle of biology states that (1) a cell is the smallest unit of life; (2) cells make up all living things, from unicellular to multicellular organisms; and (3) new cells arise from preexisting cells.

Cellulose A structural polysaccharide found in the cell walls of plants. Humans lack the enzymes necessary to digest cellulose, and thus it passes unchanged through our digestive tract. Although cellulose has no value as a nutrient, it is an important form of dietary fiber known to facilitate the passage of feces through the large intestines.

Cementum A calcified but sensitive part of a tooth that covers the root.

Central nervous system (CNS) The brain and the spinal cord.

Centriole A structure, found in pairs, within a centrosome. Each centriole is composed of nine sets of triplet microtubules arranged in a ring.

Centromere The region of a replicated chromosome at which sister chromatids are held together until they separate during cell division.

Centrosome The region near the nucleus that contains centrioles. It forms the mitotic spindle during prophase.

Cerebellum A region of the brain important in sensory-motor coordination. It is largely responsible for posture and smooth body movements.

Cerebral cortex The extensive area of gray matter covering the surfaces of the cerebrum. It is often referred to as the conscious part of the brain. The cerebral cortex has sensory, motor, and association areas.

Cerebral white matter A region of the cerebrum beneath the cortex consisting primarily of myelinated axons that are grouped into tracts that allow various regions of the brain to communicate with one another.

Cerebrospinal fluid The fluid bathing the internal and external surfaces of the central nervous system. It serves as a shock absorber, supports the brain, nourishes the brain, delivers chemical messengers, and removes waste products.

Cerebrovascular accident See stroke.

Cerebrum The largest and most prominent part of the brain, composed of the cerebral hemispheres. It is responsible for thinking, sensory perception, originating most conscious motor activity, personality, and memory.

Cervical cap A barrier means of contraception consisting of a small rubber dome that fits snugly over the cervix and is held in place partly by suction. It prevents sperm from reaching the egg.

Cervix The narrow neck of the uterus that projects into the vagina whose opening provides a passageway for materials moving between the vagina and the body of the uterus.

Cesarean section A procedure by which the baby and placenta are removed from the uterus through an incision in the abdominal wall and uterus. The term is often shortened to *C-section*.

Chancre A painless bump that forms during the first stage of syphilis at the site of contact, usually within 2 to 8 weeks of the initial contact.

Chemical digestion A part of the digestive process that involves breaking chemical bonds so that complex molecules are broken into their component subunits. Chemical digestion produces molecules that can be absorbed into the bloodstream and used by the cells.

Chemical evolution The sequence of events by which life evolved from chemicals slowly increasing in complexity over perhaps 300 million years.

Chemistry The branch of science concerned with the composition and properties of material substances, including their abilities to change into other substances.

Chemoreceptor A sensory receptor specialized to respond to chemicals. We describe the input from the chemoreceptors of the mouth as taste (gustation) and from those of the nose as smell (olfaction). Other chemoreceptors monitor levels of chemicals in body fluids, such as carbon dioxide, oxygen, and glucose.

Childhood The stage in postnatal development that runs from about 13 months to 12 or 13 years of age. Growth during this period is initially rapid and then slows. Toward the end of childhood there is a dramatic increase in growth known as a *growth spurt*.

Chitin A structural polysaccharide found in the exoskeletons (hard outer coverings) of animals such as insects, spiders, and crustaceans.

Chlamydia A genus of bacteria. In this text, it is an infection (usually sexually transmitted) caused by *Chlamydia trachomatus*, commonly causing urethritis and pelvic inflammatory disease.

Cholecystokinin A hormone secreted by the small intestine that stimulates the pancreas to release its digestive enzymes and the gallbladder to contract and release bile.

Chordae tendineae Strings of connective tissue that anchor the atrioventricular valves to the wall of the heart, preventing the backflow of blood.

Chorion The extraembryonic membrane that becomes the embryo's major contribution to the placenta.

Chorionic villi Fingerlike projections of the chorion that grow into the uterine lining of the mother during formation of the placenta and become part of the placenta.

Chorionic villi sampling (CVS) A procedure for screening for genetic defects of a fetus by removing a piece of chorionic villi and examining the cells for genetic abnormalities.

Choroid The pigmented middle layer of the eyeball that contains blood vessels.

Chromatid One of the two identical replicates of a duplicated chromosome. The two chromatids that make up a chromosome are held together by a centromere and are referred to as sister chromatids. During cell division, the two strands separate and each becomes a chromosome in one of the two daughter cells.

Chromatin DNA and associated proteins in a dispersed, rather than condensed, state.

Chromosomal mutation A change in DNA in which a section of a chromosome becomes rearranged, duplicated, or deleted.

Chromosome DNA (which contains the genetic information of a cell) and specialized proteins, primarily histones.

Chronic renal failure A progressive and often irreversible decline in the rate of glomerular filtration.

Chylomicron A particle formed when proteins coat the surface of the products of lipid digestion, making lipids soluble in water and allowing them to be transported throughout the body.

Chyme The semifluid, creamy mass created during digestion once the food has been churned and mixed with the gastric juices of the stomach.

Cilia Extensions of the plasma membrane found on some cells, such as those lining the respiratory tract, that move in a back-and-forth motion. They are usually shorter and much more numerous than flagella but have the same 9 + 2 arrangement of microtubules at their core.

Ciliary body A portion of the middle coat of the eyeball near the lens that consists of smooth muscle and ligaments. Contractions of the smooth muscle of the ciliary body change the shape of the lens, which then focuses images on the retina.

Circulatory system An organ system composed of the cardiovascular system (heart and blood vessels).

Circumcision The surgical removal of the foreskin of the penis, usually performed when the male is an infant.

Cirrhosis A chronic disease of the liver in which the liver becomes fatty and the liver cells are gradually replaced with scar tissue.

Citric acid cycle The cyclic series of chemical reactions that follows the transition reaction and yields two molecules of adenosine triphosphate (one from each acetyl CoA that enters the cycle) and several molecules of nicotine adenine dinucleotide (NADH) and flavin adenine dinucleotide ($FADH_2$), carriers of high-energy electrons that enter the electron transport chain. This phase of cellular respiration occurs inside the mitochondrion and is sometimes called the Kreb's cycle.

Cleavage A rapid series of mitotic cell divisions in which the zygote first divides into two cells, and then four cells, and then eight cells, and so on. Cleavage usually begins about one day after fertilization as the zygote moves along the oviduct toward the uterus.

Climax community The relatively stable community that eventually forms at the end of ecological succession and remains if no disturbances occur.

Clitoris A small, erectile body in the female that plays a role in sexual stimulation and is developed from the same embryological structure that develops into the glans penis in a male.

Clonal selection The hypothesis that, by binding to a receptor on a lymphocyte surface, an antigen selectively activates only those lymphocytes able to recognize that antigen and programs that lymphocyte to divide, forming an army of cells specialized to attack the stimulating antigen.

Clone A population of identical cells descended from a single ancestor.

Cochlea The snail-shaped portion of the inner ear that contains the actual organ of hearing (the spiral organ [of Corti]).

Codominance The condition in which the effects of both alleles are separately expressed in a heterozygote.

Codon A three-base sequence on messenger RNA (mRNA) that specifies one of the 20 common amino acids or the beginning or end of the protein chain.

Coenzyme An organic molecule such as a vitamin that functions as a cofactor and helps enzymes convert substrate to product.

Cofactor A nonprotein substance such as zinc, iron, and vitamins that helps enzymes convert substrate to product. It may permanently reside at the active site of the enzyme or may bind to the active site at the same time as the substrate.

Collagen fibers Strong insoluble protein fibers common in many types of connective tissues.

Collecting ducts Within the kidneys, the tubes that receive filtrate from the distal convoluted tubules of many nephrons and that eventually drain into the renal pelvis. Some tubular secretion occurs along collecting ducts.

Colon The division of the large intestine composed of the ascending colon, the transverse colon, and the descending colon.

Colostrum A cloudy yellowish fluid produced by the breasts in the interval after birth when milk is not yet available. Its composition is different from that of milk.

Columnar epithelium A type of epithelial tissue composed of tall, rectangular cells that are specialized for secretion and absorption.

Coma An unconscious state caused by trauma to neurons in regions of the brain responsible for stimulating the cerebrum, particularly those in the reticular activating system or thalamus. Coma can be caused by mechanical shock, such as might be caused by a blow to the head, tumors, infections, drug overdose (from barbiturates, alcohol, opiates, or aspirin), or failure of the liver or kidney.

Combination birth control pill A means of hormonal contraception that consists of a series of pills that contain synthetic forms of estrogen and progesterone. The hormones in the pills mimic the effects of natural hormones ordinarily produced by the ovaries and inhibit FSH and LH secretion by the anterior pituitary gland and, therefore, prevent the development of an egg and ovulation.

Common cold An upper respiratory infection caused by one of the adenoviruses.

Community An assortment of organisms of various species interacting in a defined habitat.

Compact bone Very dense, hard bone, containing internal spaces of microscopic size and narrow channels that contain blood vessels and nerves. It makes up the shafts of long bones and the outer surfaces of all bones.

Complementary base pairing The process by which specific bases are matched: adenine with thymine (in DNA) or with uracil (in RNA) and cytosine with guanine. Each base pair is held together by weak hydrogen bonds.

Complementary proteins A combination of foods each containing incomplete proteins that provide ample amounts of all essential amino acids when combined.

Complement system A group of about 20 proteins that enhances the body's defense mechanisms. The complement system destroys cellular pathogens by creating holes in the plasma membrane, making the cell leaky, enhancing phagocytosis, and stimulating inflammation.

Complete dominance In genetic inheritance, the dominant allele in a heterozygote completely masks the effect of the recessive allele. Complete dominance often occurs because the dominant allele produces a functional protein and the recessive allele produces a less functional protein or none at all.

Complete protein A protein that contains ample amounts of all the essential amino acids. Animal sources of protein are generally complete proteins.

Compound A molecule that contains two or more different elements.

Computed tomography (CT scanning) A method of visualizing body structures, including the brain, using an x-ray source that moves in an arc around the body part to be imaged, thereby providing different views of the structure.

Concentration gradient A difference in the number of molecules or ions between two adjacent regions. Molecules or ions tend to move away from an area where they are more concentrated to an area where they are less concentrated. Each type of molecule or ion moves in response to its own concentration gradient.

Condom, female A barrier means of contraception used by a female that consists of a loose sac of polyurethane held in place by flexible rings (one at each end). It is used to prevent sperm from entering the female tract.

Condom, male A barrier means of contraception consisting of a thin sheath of latex or animal intestines that is rolled onto an erect penis, where it prevents sperm from entering the vagina. A latex condom also helps prevent the spread of sexually transmitted diseases.

Cone One of the three types of photoreceptors responsible for color vision.

Confounding variable In a controlled experiment, it is a second variable that differentiates between the control group and the experimental group.

Conjoined twins Individuals that develop from a single fertilized ovum that fails to completely split in two at an early stage of cleavage. Such twins have identical genetic material and thus are always the same gender. They may be surgically separated after birth.

Connective tissue Tissue that binds together and supports other tissues of the body. All connective tissues contain cells in an extracellular matrix, which consists of protein fibers and a noncellular ground substance.

Constipation Infrequent, difficult bowel movements of hard feces. Constipation occurs when feces move through the large intestine too slowly and too much water is reabsorbed.

Consumer In ecosystems, an organism that obtains energy and raw materials by eating the tissues of other organisms.

Contact inhibition The phenomenon whereby cells placed in a dish in the presence of growth factors stop dividing once they have formed a monolayer. This may result from competition among the cells for growth factors and nutrients. Cancer cells do not display density-dependent contact inhibition and continue to divide, piling up on one another until nutrients run out.

Continuous ambulatory peritoneal dialysis (CAPD) A method of hemodialysis whereby the peritoneum, one of the body's own selectively permeable membranes, is used as the dialyzing membrane. It is an alternative to the artificial kidney machine during kidney failure.

Contraceptive sponge A barrier means of contraception consisting of a sponge containing spermicide.

Controlled experiment An experiment in which the subjects are divided into two groups, usually called the control group and the experimental group. Ideally, the groups differ in only the factor(s) of interest.

Convergence In vision, the process by which the eyes are directed toward the midline of the body as an object moves closer. Convergence is necessary to keep the image focused on the fovea of the retina.

Convergent evolution The process by which two species become more alike because they have similar ecological roles and selection pressures. For example, birds and insects independently evolved wings.

Core temperature The temperature in body structures below the skin and subcutaneous layers.

Cornea A clear, transparent dome located in the front and center of the eye that both provides the window through which light enters the eye and helps bend light rays so that they focus on the retina.

Coronary artery bypass A technique for bypassing a blocked coronary blood vessel to restore blood flow to the heart muscle. Typically, a segment of a leg vein is removed and grafted so that it provides a shunt between the aorta and a coronary artery past the point of obstruction.

Coronary artery disease (CAD) A condition in which fatty deposits associated with atherosclerosis form on the inside of coronary arteries, obstructing the flow of blood. It is the underlying cause of the vast majority of heart attacks.

Coronary circulation The system of blood vessels that services the tissues of the heart itself.

Coronary sinus A vessel that returns deoxygenated blood collected from the heart muscle to the right atrium of the heart. The coronary sinus is formed from the merger of cardiac veins.

Corpus callosum A band of myelinated axons (white matter) that connects the two cerebral hemispheres so they can communicate with one another.

Corpus luteum A structure in the ovary that forms from the follicle cells remaining in the ovary after ovulation. The corpus luteum functions as an endocrine structure that secretes estrogen and progesterone.

Corticotropin-releasing hormone (CRH) A hormone released from the placenta that seems to initiate labor.

Covalent bond A chemical bond formed when outer shell electrons are shared between atoms.

Cowper's glands See bulbourethral glands.

Cranial nerves Twelve pairs of nerves that arise from the brain and service the structures of the head and certain body parts such as the heart and diaphragm. Cranial nerves can be sensory, motor, or mixed.

Cranium The skeletal portion of the skull that forms the cranial case. It is formed from eight (sometimes more) flattened bones including the frontal bone, two parietal bones, the occipital bone, two temporal bones, the sphenoid bone, and the ethmoid bone.

Creatine phosphate A compound stored in muscle tissue that serves as an alternative energy source for muscle contraction.

Cretinism A condition characterized by dwarfism and delayed mental and sexual development. It is caused by undersecretion of thyroid hormone during fetal development or infancy.

Cristae Infoldings of the inner membrane of a mitochondrion.

Cross-bridges Myosin heads. Club-shaped ends of a myosin molecule that bind to actin filaments and can swivel, causing actin filaments to slide past the myosin filaments, which causes muscle contraction.

Crossing over The breaking and rejoining of nonsister chromatids of homologous pairs of chromosomes during meiosis (specifically at prophase I when homologous chromosomes pair up side by side). Crossing over results in the exchange of corresponding segments of chromatids and increases genetic variability in populations.

Cross-tolerance The development of tolerance for a drug that is not used, caused by the development of tolerance to another, usually similar, drug.

Crown The part of a tooth that is visible above the gum line. It is covered with enamel, a nonliving material that is hardened with calcium salts.

CT scanning See computed tomography.

Cuboidal epithelium A type of epithelial tissue composed of cube-shaped cells that are specialized for secretion and absorption.

Culture Social influences that produce an integrated pattern of knowledge, belief, and behavior.

Cupula A pliable gelatinous mass covering the hair cells within the ampulla of the semicircular canals of the inner ear and whose movement bends hair cells, triggering the generation of nerve impulses that are interpreted by the brain as movement of the head.

Cushing's syndrome A condition characterized by accumulation of fluid in the face and redistribution of body fat caused by prolonged exposure to cortisol.

Cutaneous membrane The skin. It is thick, relatively waterproof, and dry.

Cystitis Inflammation of the urinary bladder caused by bacteria.

Cytokinesis The division of the cytoplasm and organelles into two daughter cells during cell division. Cytokinesis usually occurs during telophase.

Cytomegalovirus Any of a group of highly host-specific herpes viruses that produce an infection characterized by unique large cells with nuclear inclusions. Cytomegalovirus infects many people in childhood but can be reactivated in people with an HIV infection.

Cytoplasm The part of a cell that includes the aqueous fluid within the cell and all the organelles with the exception of the nucleus.

Cytoskeleton A complex network of protein filaments within the cytoplasm that gives the cell its shape, anchors organelles in place, and functions in the movement of entire cells or certain organelles or vesicles within cells. The cytoskeleton includes microtubules, microfilaments, and intermediate filaments.

Cytotoxic T cell A type of T lymphocyte that directly attacks infected body cells and tumor cells by releasing chemicals called perforins that cause the target cells to burst.

Darwinian fitness See fitness.

Decomposer An organism that obtains energy by consuming the remains or wastes of other organisms. Decomposers release inorganic materials that can then be used by producers. Bacteria and fungi are important decomposers.

Deductive reasoning A logical progression of thought proceeding from the general to the specific. It involves making specific deductions based on a larger generalization or premise. The statement is usually in the form of an "if . . . then" premise.

Deforestation Removing trees from an area without replacing them.

Dehydration synthesis The process by which polymers are formed. Monomers are linked together through the removal of a water molecule.

Deletion Pertaining to chromosomes, the loss of a nucleotide or segment of a chromosome.

Denaturation The process by which changes in the environment of a protein, such as increased heat or changes in pH, cause it to unravel and lose its three-dimensional shape. Change in the shape of a protein results in loss of function.

Dendrite A process of a neuron specialized to pick up messages and transmit them toward the cell body. There are typically many short branching dendrites on a neuron.

Dense connective tissue Connective tissue that contains many tightly woven fibers and is found in ligaments, tendons, and the dermis.

Density-dependent regulating factor One of many factors that have a greater impact on the population size as conditions become more crowded. Such factors include disease and starvation.

Density-independent regulating factor One of many factors that regulate population size by causing deaths that are not related to the density of individuals in a population. Such factors include natural disasters.

Dentin A hard, bonelike substance that forms the main substance of teeth. It is covered by enamel on the crown and by cementum on the root.

Deoxyribonucleic acid (DNA) The molecular basis of genetic inheritance in all cells and some viruses. A category of nucleic acids that usually consists of a double helix of two nucleotide strands. The sequence of nucleotides carries the instructions for assembling proteins.

Depolarization A change in the difference in electrical charge across a membrane that moves it from a negative value toward 0 mV. During a nerve impulse (action potential), depolarization is caused by the inward flow of positively charged sodium ions.

Depo-Provera A means of hormonal contraception that consists of an injection of progesterone every 3 months.

Dermis The layer of the skin that lies just below the epidermis and is composed of connective tissue. The dermis contains blood vessels, oil glands, sensory structures, and nerve endings. The dermis does not wear away.

Desertification The process by which overfarming and overgrazing transform marginal farmlands and rangelands to deserts.

Desmosome A type of junction between cells that anchors adjacent cells together.

Detrital food web Energy flow begins with detritus (organic material from the remains of dead organisms) that is eaten by a primary consumer.

Detritivore An organism that obtains energy by consuming the remains or wastes of other organisms. Detritivores release inorganic materials that can then be used by producers.

Detrusor muscle A layer of smooth muscle within the walls of the urinary bladder. It plays a role in urination.

Diabetes insipidus A condition characterized by excessive urine production caused by inadequate ADH production.

Diabetes mellitus A condition characterized by excessive urine production in which large amounts of glucose are lost as a result of insulin deficiency.

Diaphragm A broad sheet of muscle that separates the abdominal and thoracic cavities. When the diaphragm contracts, inhalation occurs.

Diaphragm, contraceptive A barrier means of contraception that consists of a dome-shaped soft rubber cup on a flexible ring. A diaphragm is used in conjunction with a spermicide to prevent sperm from reaching the egg.

Diarrhea Abnormally frequent, loose bowel movements. Diarrhea occurs when feces pass through the large intestine too quickly and too little water is reabsorbed. Diarrhea sometimes leads to dehydration.

Diastole Relaxation of the heart. Atrial diastole is the relaxation of the atria. Ventricular diastole is the relaxation of the ventricles.

Diastolic pressure The lowest blood pressure in an artery during the relaxation of the heart. In a typical, healthy adult, the diastolic pressure is about 80 mm Hg.

Digestive system The organ system that digests and absorbs food. The digestive system includes the mouth, esophagus, stomach, small intestine, and large intestine. Associated structures include the teeth, tongue, salivary glands, liver, gallbladder, and pancreas.

Dilation stage The first stage of true labor. It begins with the onset of contractions and ends when the cervix has fully dilated to 10 cm (4 in.).

Diploid The condition of having two sets of chromosomes in each cell. Somatic (body) cells are diploid.

Disaccharide A molecule formed when two monosaccharides covalently bond to each other through dehydration synthesis. It is known as a double sugar.

Distal convoluted tubule The section of the renal tubule where reabsorption and secretion occur.

Diuretic A substance that promotes urine production. Alcohol is a diuretic.

DNA See deoxyribonucleic acid.

DNA fingerprint The pattern of DNA fragments that have been cut by a restriction enzyme and sorted by size. Each person has a characteristic, individual DNA fingerprint.

DNA library A large collection of cloned recombinant DNA fragments containing the entire genome of an organism.

DNA ligase An enzyme that catalyzes the formation of bonds between the sugar and the phosphate molecules that form the sides of the DNA ladder during replication, repair, or the creation of recombinant DNA.

DNA polymerase Any one of the enzymes that catalyze the synthesis of DNA from free nucleotides using one strand of DNA as a template.

DNA replication Copying DNA. During replication each DNA strand serves as a template for a new strand.

Dominant allele The allele that is fully expressed in the phenotype of an individual who is heterozygous for that gene. The dominant allele usually produces a functional protein, whereas the recessive allele does not.

Dopamine A neurotransmitter in the central nervous system thought to be involved in regulating emotions and in the brain pathways that control complex movements.

Dorsal nerve root The portion of a spinal nerve that arises from the back (posterior) side of the spinal cord and contains axons of sensory neurons. It joins with the ventral nerve root to form a single spinal nerve, which passes through the opening between the vertebrae.

Doubling time The number of years required for a population to double in size at a given, constant growth rate.

Down syndrome A collection of characteristics that tend to occur when an individual has three copies of chromosome 21. It is also known as trisomy 21.

Drug cocktail A combination of drugs used to treat people who are HIV positive. The combination usually includes a drug that blocks reverse transcription and a protease inhibitor.

Ductus arteriosus A small vessel in the fetus that connects the pulmonary artery to the aorta. It diverts blood away from the lungs.

Ductus venosus A small vessel in the fetus through which most blood from the placenta flows, bypassing the liver.

Duodenum The first region of the small intestine. The duodenum receives chyme from the stomach and digestive juices from the pancreas and liver.

Duplication Pertaining to chromosomes, the duplication of a region of a chromosome that often results from fusion of a fragment from a homologous chromosome.

Dura mater The tough, leathery outer layer of the meninges that protects the central nervous system. Around the brain, the dura mater has two layers that are separated by a fluid-filled space containing blood vessels.

Dysplasia The changes in shape, nuclei, and organization of adult cells. It is typical of precancerous cells.

Eardrum See tympanic membrane.

Ecological pyramid A diagram in which blocks represent each trophic (feeding) level.

Ecological succession The sequence of changes in the species making up a community over time.

Ecology The study of the interactions among organisms and between organisms and their environment.

Ecosystem All the organisms living in a certain area that can potentially interact, together with their physical environment.

Ectoderm The primary germ layer that forms the nervous system, epidermis, and epidermal derivatives such as hair, nails, and mammary glands.

Ectopic pregnancy A pregnancy in which the embryo (blastocyst) implants and begins development in a location other than the uterus, most commonly in the oviducts (a tubal pregnancy).

Edema Swelling caused by the accumulation of interstitial fluid.

EEG See electroencephalogram.

Effector A muscle or a gland that brings about a response to a stimulus.

Effector cells Lymphocytes that are responsible for the attack on cells or substances not recognized as belonging in the body.

Efferent (motor) neuron A neuron specialized to carry information *away from* the central nervous system to an effector, either a muscle or a gland.

Egg A mature female gamete. An ovum. The egg contains the mother's genetic contribution to the next generation and nutrients.

Elastic cartilage The most flexible type of cartilage because of an abundance of wavy elastic fibers in its matrix.

Elastic fibers Coiled proteins found in connective tissues that allow the connective tissue to be stretched and recoil. Elastic fibers are common in tissues that require elasticity.

Electrocardiogram (ECG or EKG) A graphic recording of the electrical activities of the heart.

Electroencephalogram (EEG) A graphic record of the electrical activity of the brain.

Electron transport chain A series of carrier proteins embedded in the inner membrane of the mitochondrion that receives electrons from the molecules of nicotine adenine dinucleotide (NADH) and flavin adenine dinucleotide ($FADH_2$) produced by glycolysis and the citric acid cycle. During the transfer of electrons from one molecule to the next, energy is released and this energy is then used to make adenosine triphosphate (ATP). Oxygen is the final electron acceptor in the chain.

Element Any substance that cannot be broken down into simpler substances by ordinary chemical means.

Embolus A blood clot that drifts through the circulatory system and can lodge in a small blood vessel and block blood flow.

Embryo The developing human from week 3 through week 8 of gestation (the embryonic period).

Embryonic disk The flattened platelike structure that will become the embryo proper. It develops from the inner cell mass.

Embryonic period The period of prenatal development that extends from week 3 through week 8 of gestation. It is when tissues and organs form.

Emigration The departure of individuals from a population for some other area.

Emphysema A condition in which the alveolar walls break down, thicken, and form larger air spaces, making gas exchange difficult. This change results in less surface area for gas exchange and an increase in the volume of residual air in the lungs.

Enamel A nonliving material that is hardened with calcium salts and covers the crown of a tooth.

Encephalitis An inflammation of the meninges around the brain.

Endocardium A thin layer that lines the cavities of the heart.

Endocrine gland A gland that lacks ducts and releases its products (hormones) into the fluid just outside the cells.

Endocytosis The process by which large molecules and single-celled organisms such as bacteria enter cells. It occurs when a region of the plasma membrane surrounds the substance to be ingested, then pinches off from the rest of the membrane, enclosing the substance in a saclike structure called a vesicle that is released into the cell. Two types of endocytosis are phagocytosis ("cell eating") and pinocytosis ("cell drinking").

Endoderm The primary germ layer that forms some organs and glands (for example, the pancreas, liver, thyroid gland, and parathyroid glands) and the epithelial lining of the urinary, respiratory, and gastrointestinal tracts.

Endometriosis A painful condition in which tissue from the lining of the uterus (the endometrium) is found outside the uterine cavity.

Endometrium The inner layer of the uterus consisting of connective tissue, glands, and blood vessels. The endometrium thickens and develops with each menstrual cycle and is then lost as menstrual flow. It is the site of embryo implantation during pregnancy.

Endoplasmic reticulum The network of internal membranes within eukaryotic cells. Whereas rough endoplasmic reticulum (RER) has ribosomes attached to its surface, smooth endoplasmic reticulum (SER) lacks ribosomes and functions in the production of phospholipids for incorporation into cell membranes.

Endorphin A chemical released by nerve cells that binds to the so-called opiate receptors on the pain-transmitting neurons and quells the pain. The term is short for *endo*genous mor*phine*-like substance.

Endosymbiont theory The theory that organelles such as mitochondria were once free-living prokaryotic organisms that either invaded or were engulfed by primitive eukaryotic cells with which they established a symbiotic (mutually beneficial) relationship.

Endothelium The lining of the heart, blood vessels, and lymphatic vessels. It is composed of flattened, tight-fitting cells. The endothelium forms a smooth surface that minimizes friction and allows the blood or lymph to flow over it easily.

Enkephalin A chemical released by nerve cells that binds to the so-called opiate receptors on the pain-transmitting neurons and quells the pain.

Enzyme A substance (usually a protein, but sometimes an RNA molecule) that speeds up chemical reactions without being consumed in the process.

Enzyme-substrate complex A complex formed when a substrate binds to an enzyme at the active site.

Eosinophil The type of white blood cell important in the body's defense against parasitic worms. It releases chemicals that help counteract certain inflammatory chemicals released during an allergic response.

Epidermis The outermost layer of the skin, composed of epithelial cells.

Epididymis A long tube coiled on the surface of each testis that serves as the site of sperm cell maturation and storage.

Epiglottis A part of the larynx that forms a movable lid of cartilage covering the opening into the trachea (the glottis).

Epinephrine Adrenaline. A hormone secreted by the adrenal medulla, along with norepinephrine, in response to stress. They initiate the physiological fight-or-flight reaction.

Epiphyseal plate A plate of cartilage that separates the head of the bone from the shaft, permitting the bone to grow. In late adolescence, the epiphyseal plate is replaced by bone, and growth stops. The epiphyseal plate is commonly called the growth plate.

Episiotomy An incision made to enlarge the vaginal opening, just before passage of the baby's head at the end of the second stage of labor.

Epithelial tissue One of the four primary tissue types. The tissue that covers body surfaces, lines body cavities and organs, and forms glands.

Equatorial plate A plane at the midline of a cell where chromosomes line up during mitosis or meiosis.

Erythrocyte A red blood cell. A nucleus-free biconcave cell in the blood that is specialized for transporting oxygen to cells and assists in transporting carbon dioxide away from cells.

Erythropoietin A hormone released by the kidneys when the oxygen content of the blood declines that stimulates red blood cell production.

Esophagus A muscular tube that conducts food from the pharynx to the stomach using peristalsis.

Essential amino acid Any of the eight amino acids that the body cannot synthesize and, therefore, must be supplied in the diet.

Estrogen A steroid female sex hormone produced by the follicle cells and the corpus luteum in the ovary. Estrogen helps oocytes mature, stimulates cell division in the endometrium and the breast with each uterine cycle, and maintains secondary sex characteristics. The adrenal cortex also secretes estrogen.

Eukaryotic cell A cell with a nucleus and extensive internal membranes that divide it into many compartments and enclose organelles. Eukaryotes include cells in plants, animals, and all other organisms except bacteria and archaea.

Eustachian tubes Auditory tubes. Small tubes that join the upper region of the pharynx (throat) with the middle ear. They help to equalize the air pressure between the middle ear and the atmosphere.

Eutrophication The enrichment of water in a lake or pond by nutrients. Eutrophication is often caused by nitrogen or phosphate that washes into bodies of water.

Evolution Descent with modification from a common ancestor. It is the process by which life forms on the earth have changed over time.

Excitatory synapse A synapse in which the response of the receptors for that neurotransmitter on the postsynaptic membrane increases the likelihood that an action potential will be generated in the postsynaptic neuron. The postsynaptic cell is excited because it becomes less negative than usual (slightly depolarized), usually because of the inflow of sodium ions.

Exhalation Breathing out (expiration) involves the movement of air out of the respiratory system into the atmosphere.

Exhaustion phase The final phase of the general adaptation syndrome that occurs in response to extreme and prolonged stress. During the exhaustion phase, the heart, blood vessels, and adrenal glands begin to fail after the taxing metabolic and endocrine demands of the resistance phase.

Exocrine glands Glands that secrete their product through ducts onto body surfaces, into the spaces within organs, or into a body cavity. Examples include the salivary glands of the mouth and the oil and sweat glands of the skin.

Exocytosis The process by which large molecules leave cells. It occurs when products packaged by cells in membrane-bound vesicles move toward the plasma membrane. Upon reaching the plasma membrane, the membrane of the vesicle fuses with it, spilling its contents outside the cell.

Exon The nucleotide sequences of DNA that code for a protein or the corresponding sequences of a newly synthesized messenger RNA (mRNA). Exons are spliced together to form the mature mRNA that is ultimately translated into protein.

Exophthalmos A condition characterized by protruding eyes that is caused by the accumulation of interstitial fluid due to oversecretion of thyroid hormone.

Expiration The process by which air is moved out of the respiratory system into the atmosphere. It is also called exhalation.

Expiratory reserve volume The additional volume of air that can be forcefully expelled from the lungs after normal exhalation.

Expulsion stage The second stage of true labor. It begins with full dilation of the cervix and ends with delivery of the baby.

External auditory canal The canal leading from the pinna of the ear to the eardrum (tympanic membrane).

External respiration The exchange of oxygen and carbon dioxide between the lungs and the blood.

External urethral sphincter A sphincter made of skeletal muscle that surrounds the urethra. This voluntary sphincter helps stop the flow of urine down the urethra when we wish to postpone urination.

Exteroceptor A sensory receptor that is located near the surface of the body and that responds to changes in the environment.

Extracellular fluid The watery solution outside cells. It is also called interstitial fluid.

Extraembryonic membranes Membranes that lie outside the embryo, where they protect and nourish the embryo and later the fetus. They include the amnion, yolk sac, chorion, and allantois.

Facial bones The bones that form the face. They are composed of 14 bones that support several sensory structures and serve as attachments for most muscles of the face.

Facilitated diffusion The movement of a substance from a region of higher concentration to a region of lower concentration with the aid of a membrane protein that either transports the substance from one side of the membrane to the other or forms a channel through which it can move.

Farsightedness A condition in which distant objects are seen more clearly than near ones. Farsightedness occurs either because the eyeball is too short or the lens is too thin, causing the image to be focused behind the retina.

Fascicle A bundle of skeletal muscle fibers (cells) that forms a part of a muscle. Each fascicle is wrapped in its own connective tissue sheath.

Fast-twitch cells Muscle fibers that contract rapidly and powerfully, with little endurance. They have few mitochondria and large glycogen reserves. They depend on anaerobic pathways to generate adenosine triphosphate (ATP) during muscle contraction.

Fatigue A state in which a muscle is physiologically unable to contract despite continued stimulation. Muscle fatigue results from a relative deficit of adenosine triphosphate (ATP).

Fat-soluble vitamin A vitamin that does not dissolve in water and is stored in fat. This group includes vitamins A, D, E, and K.

Feces Waste material discharged from the large intestine during defecation. Feces consist primarily of undigested food, sloughed-off epithelial cells, water, and millions of bacteria.

Fermentation A pathway by which cells can harvest energy in the absence of oxygen. It nets only 2 molecules of ATP as compared with the approximately 36 molecules produced by cellular respiration.

Fertilization The union between an egg (technically a secondary oocyte) and a sperm. It takes about 24 hours from start to finish and usually occurs in a widened portion of the oviduct, not far from the ovary.

Fetal alcohol syndrome A group of characteristics in children born to mothers who consumed alcohol during pregnancy. It can include certain facial features including an eye fold near the bridge of the nose, mental retardation, and slow growth.

Fetal period The period of prenatal development that extends from week 9 of gestation until birth. It is when rapid growth occurs.

Fetus The developing human from week 9 of gestation until birth (the fetal period).

Fever An abnormally elevated body temperature. Fever helps the body fight disease-causing invaders in a number of ways.

Fibrillation Rapid, ineffective contraction of the heart.

Fibrin A protein formed from fibrinogen by thrombin that forms a web that traps blood cells, forming a blood clot.

Fibrinogen A plasma protein produced by the liver that is important in blood clotting. It is converted to fibrin by thrombin.

Fibroblasts Cells in connective tissue that secrete the protein fibers that are found in the matrix of the connective tissue. Fibroblasts also secrete collagen fibers for the repair of body tissues.

Fibrocartilage Cartilage with a matrix containing many collagen fibers. Fibrocartilage is found around the edges of joints and the intervertebral disks.

Fibrous joints Joints that are held together by connective tissue and lack a joint cavity. Most fibrous joints do not permit movement.

Fight-or-flight response The body's reaction to stress or threatening situations by the stimulation of the sympathetic division of the autonomic nervous system.

Fitness Compares the number of reproductively viable offspring among individuals. It is sometimes called Darwinian fitness.

Flagellum A whiplike appendage of cells that moves in an undulating manner. It is composed of an extension of the plasma membrane containing microtubules in a 9 + 2 array. It is found on sperm cells.

Floating ribs The last two ribs that do not attach directly to the sternum (breastbone).

Fluid mosaic A term used to describe the structure of the plasma membrane. Proteins interspersed throughout the lipid molecules give the membrane its mosaic quality. The ability of some proteins to move sideways gives the membrane its fluid quality.

Follicle A spherical structure in the ovary that contains an oocyte and is surrounded by follicle cells.

Follicle-stimulating hormone (FSH) A hormone secreted by the anterior pituitary gland that in females stimulates the development of the follicles in the ovaries, resulting in the development of ova (eggs) and the production of estrogen, and in males stimulates sperm production.

Fontanels The "soft spots" in the skull of a newborn. The membranous areas in the skull of a newborn that hold the skull bones together before and shortly after birth. The fontanels are gradually replaced by bone.

Food chain The successive series of organisms through which energy (in the form of food) flows in an ecosystem. Each organism in the series eats or decomposes the preceding one. It begins with the photosynthesizers and flows to herbivores and then to carnivores.

Food web The interconnection of all the feeding relationships (food chains) in an ecosystem.

Foramen magnum The opening at the base of the skull (in the occipital bone) through which the spinal cord passes.

Foramen ovale In the fetus, a small hole in the wall between the right atrium and left atrium of the heart that allows most blood to bypass the lungs.

Formed elements Cells or cell fragments found in the blood. They include platelets, white blood cells, and red blood cells.

Fossilization The process by which fossils form. Typically an organism dies in water and is buried under accumulating layers of sediments. Hard parts become impregnated with minerals from surrounding water and sediments. Eventually the fossil may be exposed when sediments are uplifted and wind erodes the rock formation.

Fossils The preserved remnants or impressions of past organisms. Most fossils occur in sedimentary rocks.

Founder effect Genetic drift associated with the colonization of a new place by a few individuals so that by chance alone the genetic makeup of the colonizing individuals is not representative of the population they left.

Fovea A small region on the retina that contains a high density of cones but no rods. Objects are focused on the fovea for sharp vision.

Frameshift mutation A mutation that occurs when the number of nucleotides inserted or deleted is not a multiple of three, causing a change in the codon sequence on the mRNA molecule as well as a change in the resulting protein.

Fraternal twins Individuals that develop when two oocytes are released from the ovaries and fertilized by different sperm. Such twins may or may not be the same gender and are as similar genetically as any siblings. They are also called dizygotic twins.

Free radicals Molecular fragments that contain an unpaired electron.

FSH See follicle-stimulating hormone.

Full-term infant Baby born 38 weeks after fertilization.

Gallbladder A muscular pea-sized sac that stores, modifies, and then concentrates bile. Bile is released from the gallbladder into the small intestine.

Gamete A reproductive cell (sperm or egg) that contains only one copy of each chromosome. A sperm and egg fuse at fertilization, producing a zygote.

Gamete intrafallopian transfer (GIFT) A procedure in which eggs and sperm are collected from a couple and then inserted into the woman's oviduct, where fertilization may occur. If fertilization occurs, resulting embryos drift naturally into the uterus.

Ganglion (plural, ganglia) A collection of nerve cell bodies outside the central nervous system.

Gap junction A type of junction between cells that links the cytoplasm of adjacent cells through small holes, allowing physical and electrical continuity between cells.

Gastric glands Any one of several glands in the stomach mucosa that contribute to the gastric juice (hydrochloric acid and pepsin).

Gastric juice The mixture of hydrochloric acid (HCl) and pepsin released into the stomach.

Gastrin A hormone released from the stomach lining in response to the presence of partially digested proteins. Gastrin triggers the production of gastric juice by the stomach.

Gastrointestinal (GI) tract A long tubular system specialized for the processing and absorption of food that begins at the mouth and continues to the esophagus, stomach, intestines, and anus. Several accessory glands empty their secretions into the GI tract to assist digestion.

Gastrula The embryo during gastrulation, when cells move to establish primary germ layers.

Gastrulation Cell movements that establish the primary germ layers of the embryo. The embryo during this period is called a gastrula.

Gated ion channel An ion channel that is opened to allow ions to pass through or closed to prevent passage by changes in the shape of a protein that functions as a gate.

Gene A segment of DNA on a chromosome that directs the synthesis of a specific polypeptide that will play a structural or functional role in the cell. Some genes have regulatory regions of DNA within their boundaries. Also, some genes code for RNA molecules that are needed for the production of the polypeptide but are not part of it.

Gene flow Movement of alleles between populations as a result of the movement of individuals. It is a cause of microevolution.

Gene pool All of the alleles of all of the genes of all individuals in a population.

General adaptation syndrome (GAS) A series of physiological adjustments made by our bodies in response to prolonged, extreme stress.

General senses The sensations that arise from receptors in the skin, muscles, joints, bones, and internal organs; they include touch, pressure, vibration, temperature, a sense of body and limb position, and pain.

Gene therapy Treating a genetic disease by inserting healthy functional genes into the body cells that are affected by the faulty gene.

Genetic code The base triplets in DNA that specify the amino acids that go into proteins or that function as start or stop signals in protein synthesis. It is used to convert the linear sequence of bases in DNA to the sequence of amino acids in proteins.

Genetic drift The random change in allele frequencies within a population due to chance alone. It is a cause of microevolution that is usually negligible in large populations.

Genetic engineering The manipulation of genetic material for human practical purposes.

Genital herpes A sexually transmitted disease caused by the herpes simplex virus (HSV) that is usually characterized by painful blisters on the genitals. It is usually caused by HSV-2 but can be caused by HSV-1.

Genital warts Warts that form in the genital area caused by the human papilloma viruses (HPV). These viruses also cause cervical cancer and penile cancer.

Genome The complete set of DNA of an organism, including all of its genes.

Genomics The study of entire genomes and the interactions of the genes with one another and the environment.

Genotype The genetic makeup of an individual. It refers to precise alleles that are present.

Giantism A condition characterized by rapid growth and unusual height caused by abnormally high levels of growth hormone in childhood.

Gingivitis An inflammation of the gums.

Gland Epithelial tissue that secretes a product.

Glaucoma A condition in which the pressure within the anterior chamber of the eye increases as a result of the buildup of aqueous humor. It can cause blindness.

Glycogen The storage polysaccharide in animals.

Glial cells Nonexcitable cells in the nervous system that are specialized to support, protect, and insulate neurons. Also called neuroglial cells.

Global warming A long-term increase in atmospheric temperatures caused by a buildup of carbon dioxide (CO_2) and other greenhouse gases in the atmosphere. Greenhouse gases trap heat in the atmosphere.

Glomerular (Bowman's) capsule A cuplike structure surrounding the glomerulus of a nephron.

Glomerular filtration The process by which water and dissolved substances move from the blood in the glomerulus to the inside of the glomerular (Bowman's) capsule.

Glomerulus A tuft of capillaries within the renal corpuscle of a nephron.

Glottis The opening to the airways of the respiratory system from the pharynx into the larynx.

Glucagon The hormone secreted by the pancreas that elevates glucose levels in the blood.

Glucocorticoids Hormones secreted by the adrenal cortex that affect glucose homeostasis, thereby influencing metabolism and resistance to stress.

Gluconeogenesis The conversion of noncarbohydrate molecules to glucose.

Glycolysis The splitting of glucose, a six-carbon sugar, into two three-carbon molecules called pyruvate. Glycolysis takes place in the cytoplasm of a cell and is the starting point for cellular respiration and fermentation.

GnRH See gonadotropin-releasing hormone.

Goiter, simple An enlarged thyroid gland caused by iodine deficiency.

Golgi complex An organelle consisting of flattened membranous disks that functions in protein processing and packaging.

Golgi tendon organs The highly branched nerve fibers located in the tendons that sense the degree of muscle tension.

Gonad An ovary in a female or a testis in a male. The gonads produce gametes (eggs or sperm) and sex hormones.

Gonadocorticoids The male and female sex hormones, androgens and estrogens, secreted by the adrenal cortex.

Gonadotropin-releasing hormone (GnRH) A hormone produced by the hypothalamus that causes the secretion of luteinizing hormone and follicle-stimulating hormone from the anterior pituitary gland.

Gonorrhea A sexually transmitted disease caused by the bacterium *Neisseria gonorrhoeae* that commonly causes urethritis and pelvic inflammatory disease. It may not cause symptoms, especially in women.

Graafian follicle A mature follicle in the ovary.

Graded potential A temporary local change in the membrane potential that varies directly with the strength of the stimulus.

Granulocytes White blood cells with granules in their cytoplasm, including neutrophils, eosinophils, and basophils.

Graves' disease An autoimmune disorder caused by undersecretion of thyroid hormone. It is characterized by increased heart and metabolic rates, sweating, nervousness, and exophthalmos.

Gray matter Regions of the central nervous system that contain neuron cell bodies and unmyelinated axons. These regions are gray because they lack myelin. Gray matter is important in neural integration.

Greenhouse effect An increase in atmospheric carbon dioxide (CO_2) that could lead to a rise in temperatures throughout the world. Greenhouse gases trap heat in the atmosphere.

Growth factor A type of signaling molecule that stimulates growth by stimulating cell division in target cells.

Growth hormone (GH) An anterior pituitary hormone with the primary function of stimulating growth through increases in protein synthesis, cell size, and rates of cell division. GH stimulates growth in general, especially bone growth.

Growth plate (epiphyseal plate) A plate of cartilage that separates each end of a bone from its shaft that permits bone to grow.

Gumma A large sore that forms during the third stage of syphilis.

Habitat The natural environment or place where an organism, population, or species lives.

Hair cells A type of mechanoreceptor that generates nerve impulses when bent or tilted. Hair cells in the inner ear are responsible for hearing and equilibrium.

Hair root plexus The nerve endings that surround the hair follicle and are sensitive to touch.

Haploid The condition of having one set of chromosomes, as in eggs and sperm.

HCG See human chorionic gonadotropin.

Heart A muscular pump that keeps blood flowing through an animal's body. The human heart has four chambers: two atria and two ventricles.

Heart attack The death of heart muscle cells caused by an insufficient blood supply; a myocardial infarction.

Heartburn A burning sensation behind the breastbone that occurs when acidic gastric juice backs up into the esophagus.

Heart murmur Heart sounds other than "lub dup" that are created by turbulent blood flow. Heart murmurs can indicate a heart problem, such as a malfunctioning valve.

Heimlich maneuver A procedure intended to force a large burst of air out of the lungs to dislodge material lodged in the trachea.

Helper T cell The kind of T lymphocyte that serves as the main switch for the entire immune response by presenting the antigen to B cells and by secreting chemicals that stimulate other cells of the immune system. It is also known as a T4 cell or a CD4 cell, after the receptors on its surface.

Hematocrit The percentage of red blood cells in blood by volume. It is a measure of the oxygen-carrying ability of the blood.

Hemodialysis The use of artificial devices, such as the artificial kidney machine, to cleanse the blood during kidney failure.

Hemoglobin The oxygen-binding pigment in red blood cells. It consists of four subunits, each made up of an iron-containing heme group and a protein chain.

Hemolytic disease of the newborn A condition in which the red blood cells of an Rh-positive fetus or newborn are destroyed by anti-Rh antibodies previously produced in the bloodstream of an Rh-negative mother.

Hemophilia An inherited blood disorder in which there is insufficient production of blood-clotting factors. Hemophilia results in excessive bleeding in joints, deep tissues, and elsewhere. Hemophilia usually occurs in males.

Hepatitis An inflammation of the liver. Hepatitis is usually caused by one of six viruses.

Herbivore An animal that eats primary producers (green plants or algae). A primary consumer.

Heterotroph An organism that cannot make its own food from inorganic substances and instead consumes other organisms or decaying material.

Heterozygote advantage The phenomenon in which the heterozygous condition is favored over either homozygous condition. It maintains genetic variation within a population in the face of natural selection.

Heterozygous The condition of having two different alleles for a particular gene.

Hinge joint A joint that permits motion in only one plane, such as the knee joint or the elbow.

Hippocampus The part of the limbic system of the brain that plays a role in converting short-term memory into long-term memory.

Histamine A substance released by basophils and mast cells during an inflammatory response that causes blood vessels to widen (dilate) and become more permeable.

Homeostasis The ability of living things to maintain a relatively constant internal environment in all levels of body organization.

Hominid A member of the human family, such as members of the genera *Australopithecus* and *Homo*.

Homologous chromosomes A pair of chromosomes that bears genes for the same traits. One member of each pair came from each parent. Homologous chromosomes are the same size and shape and line up with one another during meiosis I.

Homologous structures Structures that have arisen from a common ancestry. Compare with analogous structure.

Homozygous The condition of having two identical alleles for a particular gene.

Hormonal implants A means of hormonal contraception consisting of silicon rods containing progesterone that are implanted under the skin in the upper arm and prevent pregnancy for up to 5 years.

Hormone A chemical messenger released by cells of the endocrine system that travels through the circulatory system to affect receptive target cells.

Host (vector) DNA The DNA that is recombined with pieces of DNA from another source (that might contain a desirable gene) in the formation of recombinant DNA.

HSV-1 Herpes simplex virus 1. HSV-1 usually infects the upper half of the body and causes cold sores (fever blisters), but it can cause genital herpes if contact is made with the genital area.

HSV-2 Herpes simplex virus 2. HSV-2 causes genital herpes, but it can cause cold sores if contact is made with the mouth.

Human chorionic gonadotropin (HCG) A hormone produced by the cells of the early embryo (blastocyst) and the placenta that maintains the corpus luteum for approximately the first 3 months of pregnancy. HCG enters the bloodstream of the mother and is excreted in her urine. HCG forms the basis for many pregnancy tests.

Human papilloma virus (HPV) One of the group of viruses that commonly cause genital warts.

Hyaline cartilage A type of cartilage with a gel-like matrix that provides flexibility and support. It is found at the end of long bones as well as parts of the nose, ribs, larynx, and trachea.

Hydrocephalus A condition resulting from the excessive production or inadequate drainage of cerebrospinal fluid.

Hydrogen bond A weak chemical bond formed between a partially positively charged hydrogen atom in a molecule and a partially negatively charged atom in another molecule or in another region of the same molecule.

Hydrolysis The process by which polymers are broken apart by the addition of water.

Hyperglycemia An elevated blood glucose level.

Hypertension High blood pressure. A high upper (systolic) value usually suggests that the person's arteries have become hardened, a condition known as arteriosclerosis, and are no longer able to dampen the high pressure of each heartbeat. The lower (diastolic) value is generally considered more important because it indicates the pressure when the heart is relaxing.

Hyperthermia Abnormally elevated body temperature.

Hypertonic solution A solution with a higher concentration of solutes than plasma, for example.

Hypodermis The layer of loose connective tissue below the epidermis and dermis that anchors the skin to underlying tissues and organs.

Hypoglycemia Depressed levels of blood glucose often resulting from excess insulin.

Hypophysis See pituitary gland.

Hypothalamus A small brain region located below the thalamus that is essential to maintaining a stable environment within the body. The hypothalamus influences blood pressure, heart rate, digestive activity, breathing rate, and many other vital physiological processes. It acts as the body's "thermostat"; regulates food intake, hunger, and thirst; coordinates the activities of the nervous system and the endocrine (hormonal) system; is part of the circuitry for emotions; and functions as a master biological clock.

Hypothermia Abnormally low body temperature.

Hypothesis A testable explanation for a specific set of observations that serves as the basis for experimentation.

Hypotonic solution A solution with a lower concentration of solutes than plasma, for example.

Identical twins Individuals that develop from a single fertilized ovum that splits in two at an early stage of cleavage. Such twins have identical genetic material and thus are always the same gender. They are also called monozygotic twins.

Immigration Movement of new individuals from other populations into an area.

Immune response The body's response to specific targets not recognized as belonging in the body.

Immune system The system of the body directly involved with body defenses against specific targets—pathogens, cells, or chemicals not recognized as belonging in the body.

Immunoglobulin (Ig) Any of the five classes of proteins that constitute the antibodies.

Implantation The process by which a blastocyst (pre-embryo) becomes embedded in the lining of the uterus. It normally occurs high up on the back wall of the uterus.

Impotence The inability to achieve or maintain an erection long enough for sexual intercourse.

Incomplete dominance In genetic inheritance, expression of the trait in a heterozygous individual is somewhere in between expression of the trait in a homozygous dominant individual and homozygous recessive individual.

Incomplete proteins Proteins that are deficient in one or more of the essential amino acids. Plant sources of protein are generally incomplete proteins.

Incontinence A condition characterized by the escape of small amounts of urine when sudden increases in abdominal pressure, perhaps caused by laughing, sneezing, or coughing, force urine past the external sphincter. This condition is common in women, particularly after childbirth, an event that may stretch or damage the external sphincter, making it less effective in controlling the flow of urine.

Incus The middle of three small bones in the middle ear that transmit information about sound from the eardrum to the inner ear. The incus is also known as the anvil.

Independent assortment The process by which homologous chromosomes and the alleles they carry segregate randomly during meiosis, creating mixes of maternal and paternal chromosomes in gametes. This is an important source of genetic variation in populations.

Independent assortment, law of In genetic inheritance, a principle that states that the alleles of unlinked genes (those that are located on different chromosomes) are randomly distributed to gametes.

Inductive reasoning A logical progression of thought proceeding from the specific to the general. It involves the accumulation of facts through observation until the sheer weight of the evidence forces some general statement about the phenomenon. A conclusion is reached on the basis of a number of observations.

Infancy The stage in postnatal development that roughly corresponds to the first year of life. It is a time of rapid growth when total body length usually increases by one-half and weight may triple.

Infertility The inability to conceive (become pregnant) or to cause conception (in the case of males).

Inflammatory response A nonspecific body response to injury or invasion by foreign organisms. It is characterized by redness, swelling, heat, and pain.

Influenza The flu. Influenza is a viral infection caused by any of the variants of influenza A or influenza B viruses.

Informed consent A document that a person must sign before participating in an experiment that lists all possible harmful consequences that might result from participation.

Inguinal canal A passage through the abdominal wall through which the testes pass in their descent to the scrotum and that contains the testicular artery and vein and the vas deferens.

Inhalation Breathing in. Inhalation (inspiration) involves the movement of air into the respiratory system.

Inhibin A hormone produced in the testes that increases with sperm count and inhibits follicle-stimulating hormone secretion and, therefore, inhibits sperm production.

Inhibiting hormone A hormone that inhibits secretion of another hormone.

Inhibitory synapse A synapse in which the response of the receptors for that neurotransmitter on the postsynaptic membrane decreases the likelihood that an action potential will be generated in the postsynaptic neuron. The postsynaptic cell is inhibited because its resting potential becomes more negative than usual.

Inner cell mass A group of cells within the blastocyst that will become the embryo proper and some extraembryonic membranes.

Inner ear A series of passageways in the temporal bone that houses the organs for hearing (cochlea) and the sense of equilibrium (vestibular apparatus).

Insertion The end of the muscle that is attached to the bone that moves when the muscle contracts.

Insoluble fiber A fiber that does not easily dissolve in water. These fibers include cellulose, hemicellulose, and lignin.

Inspiration Inhalation. The movement of air into the respiratory system.

Inspiratory reserve volume The volume of air that can be forcefully brought into the lungs after normal inhalation.

Insulin The hormone secreted by the pancreas that reduces glucose levels in the blood.

Insulin shock A condition that results from severely depressed glucose levels in which brain cells fail to function properly, causing convulsions and unconsciousness.

Integral proteins Proteins embedded in the plasma membrane, either completely or incompletely spanning the bilayer.

Integumentary system The skin.

Intercalated disks Thickenings of the plasma membranes of cardiac muscle cells that strengthen cardiac tissue and promote rapid conduction of impulses throughout the heart.

Intercostals The muscles between the ribs that raise and lower the rib cage during breathing.

Interferon A type of defensive protein produced by T lymphocytes that slows the spread of viruses already in the body by interfering with viral replication. Interferons also attract macrophages and natural killer cells, which kill the virus-infected cell.

Interleukin 1 A chemical secreted by a macrophage that activates helper T cells in an immune response.

Interleukin 2 A chemical released by a helper T cell that activates both B cells and T cells.

Intermediate filament A component of the cytoskeleton made from fibrous proteins. It functions in the maintenance of cell shape and anchoring organelles such as the nucleus.

Internal respiration Movement of oxygen from the blood to the tissues, and movement of carbon dioxide from the tissues to the blood.

Internal urethral sphincter A thickening of smooth muscle at the junction of the bladder and the urethra. The action of this sphincter is involuntary and keeps urine from flowing into the urethra while the bladder is filling.

Interneuron An association neuron. Neurons located within the central nervous system between sensory and motor neurons that serve to integrate information.

Interoceptor A sensory receptor located inside the body, where it monitors conditions. Interoceptors play a vital role in maintaining homeostasis. They are an important part of the feedback loops that regulate blood pressure, blood chemistry, and breathing rate. Interoceptors may also cause us to feel pain, hunger, or thirst, thereby prompting us to take appropriate action.

Interphase The period between cell divisions when the DNA, cytoplasm, and organelles are duplicated and the cell grows in size. Interphase is also the time in the cell's life cycle when the cell carries out its functions in the body.

Interstitial cells Cells located between the seminiferous tubules in the testes that produce the male steroid sex hormones, collectively called androgens.

Intervertebral disks Pads of cartilage that help cushion the bones of the vertebral column.

Intracytoplasmic sperm injection (ICSI) A procedure in which a tiny needle is used to inject a single sperm cell into an egg. It is an option for treating infertility when the man has few sperm or sperm that lack the strength or enzymes necessary to penetrate the egg.

Intrauterine device (IUD) A means of contraception consisting of a small plastic device that either is wrapped with copper wire or contains progesterone. It is inserted into the uterus to prevent pregnancy.

Intrinsic factor A protein secreted by the stomach necessary for the absorption of vitamin B_{12} from the small intestine.

Intron The unexpressed regions of DNA or of a newly formed mRNA. Introns are cut out of the mature mRNA molecule and are not expressed (used to form a protein).

In vitro fertilization A procedure in which eggs and sperm are placed together in a dish in the laboratory. If fertilization occurs, early-stage embryos are then transferred to the woman's uterus, where it is hoped they will implant and complete development. It is a common treatment for infertility resulting from blocked oviducts.

Ion An atom or group of atoms that carries an electric charge resulting from the loss or gain of electrons.

Ion channel A protein-lined pore or channel through a plasma membrane through which one type or a few types of ions can pass. Nerve cell ion channels are important in the generation and propagation of nerve impulses.

Ionic bond A chemical bond that results from the mutual attraction of oppositely charged ions.

Iris The colored portion of the eye. The iris regulates the amount of light that enters the eye.

Iron-deficiency anemia A reduction in the ability of the blood to carry oxygen due to an insufficient amount of iron in the diet, an inability to absorb iron from the digestive system, or blood loss.

Ischemia A temporary reduction in blood supply caused by obstructed blood flow. It causes reversible damage to heart muscle.

Islets of Langerhans See pancreatic islets.

Isotonic solution A solution with the same concentration of solutes as plasma, for example.

Isotopes Atoms that have the same number of protons but different numbers of neutrons.

Jaundice A condition in which the skin develops a yellow tone caused by the buildup of bilirubin in the blood and its deposition in certain tissues such as the skin. It is an indication that the liver is not handling bilirubin adequately.

Joint A point of contact between two bones; an articulation.

Junctional complexes Membrane specializations that attach adjacent cells to each other to form a contiguous sheet. There are three kinds of junctions between cells: tight junctions, desmosomes, and gap junctions.

Juxtaglomerular apparatus The region of the kidney nephron where the distal convoluted tubule contacts the afferent arteriole bringing blood into the glomerulus. Cells in this area secrete renin, an enzyme that triggers events eventually leading to increased reabsorption of sodium and water by the distal convoluted tubules and collecting ducts of nephrons.

Kaposi's sarcoma A cancer that forms tumors in connective tissue and manifests as pink or purple marks on the skin. It is common in people with a suppressed immune system, such as people living with HIV/AIDS, and is thought to be associated with a new Kaposi's sarcoma-related virus.

Karyotype The arrangement of chromosomes based on physical characteristics such as length and location of the centromere. Karyotypes can be checked for defects in number or structure of chromosomes.

Keratinocytes Cells of the epidermis that undergo keratinization, the process in which keratin gradually replaces the contents of maturing cells.

Ketoacidosis A lowering of blood pH resulting from the accumulation of breakdown by-products of fats.

Kidneys Reddish, kidney bean–shaped organs responsible for filtering wastes and excess materials from the blood, assisting the respiratory system in the regulation of blood pH, and maintaining fluid balance by regulating the volume and composition of blood and urine.

Kidney stones Small, hard crystals formed in the urinary tract when substances such as calcium (the most common constituent), uric acid, or magnesium ammonium phosphate precipitate from urine as a result of higher than normal concentrations. They are also called renal calculi.

Klinefelter syndrome A genetic condition resulting from nondisjunction of the sex chromosomes in which a person inherits an extra X chromosome that results in an XXY genotype. The person has a male appearance.

Labia majora One of two elongated skin folds lateral to the labia minora. They are part of the female external genitalia.

Labia minora One of two small skin folds on either side of the vagina and interior to the labia majora. They are part of the external genitalia of a female.

Labor The process by which the fetus is expelled from the uterus and moved through the vagina and into the outside world. During true labor, uterine contractions occur at regular intervals, are often painful, and intensify with walking. Labor is usually divided into the dilation stage, expulsion stage, and placental stage. False labor is characterized by irregular contractions that fail to intensify or change with walking.

Lactation The production and ejection of milk from the mammary glands. The hormone prolactin from the anterior pituitary gland promotes milk production, and the hormone oxytocin released from the posterior pituitary gland makes milk available to the suckling infant by stimulating milk ejection, or let-down.

Lacteal A lymphatic vessel in an intestinal villus that aids in the absorption of lipids.

Lactic acid fermentation The process by which glucose is broken down by muscle cells when oxygen is low during strenuous exercise.

Lacuna (plural, lacunae) A tiny cavity. It contains osteocytes (bone cells) in the matrix of bone and cartilage cells in the matrix of cartilage.

Lanugo Soft, fine hair that covers the fetus beginning about the third or fourth month after conception.

Large intestine The final segment of the gastrointestinal tract, consisting of the cecum, colon, rectum, and anal canal. The large intestine helps absorb water, forms feces, and plays a role in defecation.

Laryngitis An inflammation of the larynx in which the vocal cords become swollen and can no longer vibrate and produce sound.

Larynx The voice box or Adam's apple. A boxlike cartilaginous structure between the pharynx and the trachea held together by muscles and elastic tissue.

Latent period Pertaining to muscle contraction, the interval between the reception of the stimulus and the beginning of muscle contraction.

Latin binomial The two-part scientific name that consists of the genus name followed by the specific epithet.

Lens A transparent, semispherical body of tissue behind the iris and pupil that focuses light on the retina.

Leukemia A cancer of the blood-forming organs that causes white blood cell numbers to increase. The white blood cells are abnormal and do not effectively defend the body against infectious agents.

Leukocytes White blood cells. They are cells of the blood including neutrophils, eosinophils, basophils, monocytes, B lymphocytes, and T lymphocytes. Leukocytes are involved in body defense mechanisms and the removal of wastes, toxins, or damaged, abnormal cells.

LH See luteinizing hormone.

Life expectancy The average number of years a newborn is expected to live.

Ligament A strong band of connective tissue that holds the bones together, supports the joint, and directs the movement of the bones.

Limbic system A collective term for several structures involved in emotions and memory.

Linkage The tendency for a group of genes located on the same chromosome to be inherited together.

Lipid A compound, such as a triglyceride, phospholipid, or steroid, that does not dissolve in water.

Liver A large organ that functions mainly in the production of plasma proteins, the excretion of bile, the storage of energy reserves, the detoxification of poisons, and the interconversion of nutrients.

Local signaling molecules Chemical messengers, such as neurotransmitters, prostaglandins, growth factors, and nitric oxide, that act locally rather than traveling to distant sites within the body.

Locus The point on a chromosome where a particular gene is found.

Longitudinal fissure A deep groove that separates the cerebrum into two hemispheres.

Long-term memory Memory that stores a large amount of information for hours, days, or years.

Loop of the nephron (Loop of Henle) A section of the renal tubule that resembles a hairpin turn. Reabsorption occurs here.

Loose connective tissue Connective tissue, such as areolar and adipose tissue, that contains many cells in which the fibers of the matrix are fewer in number and more loosely woven than those found in dense connective tissue.

Lower esophageal sphincter A ring of muscle at the juncture of the esophagus and the stomach that controls the flow of materials between them. It relaxes to allow food into the stomach and contracts to prevent too much food from moving back into the esophagus.

Lumen The hollow cavity or channel of a tubule through which transported material flows.

Luteinizing hormone (LH) A hormone secreted by the anterior pituitary gland that in females stimulates ovulation and the formation of the corpus luteum (which produces estrogen and progesterone) and prepares the mammary glands for milk production. In males, it stimulates testosterone production by the interstitial cells within the testes.

Lymph The fluid within the vessels of the lymphatic system. It is derived from the fluid that bathes the cells of the body (interstitial fluid).

Lymphatic system A body system consisting of lymph, lymphatic vessels, and lymphatic tissue and organs. The lymphatic system helps return interstitial fluid to the blood, transports the products of fat digestion from the digestive system to the blood, and assists in body defenses.

Lymphatic vessels The vessels through which lymph flows. A network of vessels that drains interstitial fluid and returns it to the blood supply and transports the products of fat digestion from the digestive system to the blood supply.

Lymph nodes Small nodular organs found along lymph vessels that filter lymph. The lymph nodes contain macrophages and lymphocytes, cells that play an essential role in the body's defense system.

Lymphocyte A type of white blood cell important in nonspecific and specific (immune) body defenses. The lymphocytes include B lymphocytes (B cells) that transform into plasma cells and produce antibodies and T lymphocytes (T cells) that are important in defense against foreign or infected cells.

Lymphoid organs Various organs that belong to the lymphatic system, including the tonsils, spleen, thymus, and Peyer's patches.

Lymphoid tissue The type of tissue that predominates in the lymphoid organs except the thymus. The organs in which lymphoid tissue is found are the lymph nodes, tonsils, and spleen. Lymphoid tissue is an important component of the immune system.

Lymphoma Cancer of the lymphoid tissues. In people with AIDS, it commonly affects the B cells.

Lysosomal storage diseases Disorders such as Tay-Sachs disease caused by the absence of lysosomal enzymes. In these disorders, molecules that would normally be degraded by the missing enzymes accumulate in the lysosomes and interfere with cell functioning.

Lysosome An organelle that serves as the principal site of digestion within the cell.

Lysozyme An enzyme present in tears, saliva, and certain other body fluids that kills bacteria by disrupting their cell walls.

Macroevolution Large-scale evolutionary changes (affecting one or more species) such as those that might result from long-term changes in the climate or position of the continents. Examples include mass extinctions and the evolution of mammals.

Macromolecule A giant molecule of life such as a nucleic acid, protein, or polysaccharide. A macromolecule is formed by the joining together of smaller molecules.

Macrophage A large phagocytic cell derived from a monocyte that lives in loose connective tissue and engulfs anything detected as foreign.

Magnetic resonance imaging (MRI) A means of visualizing a region of the body, including the brain, in which the picture results from differences in the way the hydrogen nuclei in the water molecules within the tissues vibrate in response to a magnetic field created around the area to be pictured.

Malignant tumor A cancerous tumor. An abnormal mass of tissue that can invade surrounding tissue and spread to multiple locations throughout the body.

Malleus The first of three small bones in the middle ear that transmit information about sound from the eardrum to the inner ear. The malleus is also known as the hammer.

Malnourishment A form of hunger that occurs when the diet is not balanced and the right foods are not eaten.

Mammary glands The milk-producing glands in the breasts.

Mass The measure of how much matter is in an object.

Mast cells Small, mobile connective tissue cells often found near blood vessels. In response to injury, mast cells release histamine, which dilates blood vessels and increases blood flow to an area, and heparin, which prevents blood clotting.

Matter Anything that takes up space and has mass. Matter is made up of atoms.

Mechanical digestion A part of the digestive process that involves physically breaking food into smaller pieces.

Mechanoreceptor A sensory receptor that is specialized to respond to distortions in the receptor itself or in nearby cells. Mechanoreceptors are responsible for the sensations we describe as touch, pressure, hearing, and equilibrium.

Medulla oblongata The part of the brain stem containing reflex centers for some of life's most vital physiological functions: the pace of the basic breathing rhythm, the force and rate of heart contraction, and blood pressure. It connects the spinal cord to the rest of the brain.

Medullary cavity The cavity in the shaft of long bones that is filled with yellow marrow.

Medullary rhythmicity center The region of the brain stem controlling the basic rhythm of breathing.

Meiosis A type of cell division that occurs in the gonads and gives rise to gametes. As a result of two divisions (meiosis I and meiosis II), haploid gametes are produced from diploid germ cells.

Meissner's corpuscles Encapsulated nerve cell endings that are common on the hairless, sensitive areas of the skin, such as the lips, nipples, and fingertips, and tell us exactly where we have been touched.

Melanin A pigment produced by the melanocytes of the skin. There are two forms of melanin: black-to-brown and yellow-to-red.

Melanocytes Spider-shaped cells located at the base of the epidermis that manufacture and store melanin, a pigment involved in skin color and absorption of ultraviolet radiation.

Melanocyte-stimulating hormone (MSH) The anterior pituitary hormone that stimulates melanin production.

Melanoma The least common and most dangerous form of skin cancer. It arises in the melanocytes, the pigment-producing cells of the skin.

Melatonin A hormone secreted by the pineal gland that influences daily rhythms and sleep.

Membranous epithelium A type of epithelial tissue that forms linings and coverings. Depending on its location, it may be specialized to protect, secrete, and absorb.

Memory cell A lymphocyte (B cell or T cell) of the immune system that forms in response to an antigen and that circulates for a long period of time; such cells are able to mount a quick immune response to a subsequent exposure to the same antigen.

Meninges Three protective connective tissue membranes that surround the central nervous system: the dura mater, the pia mater, and the arachnoid.

Meningitis An inflammation of the meninges (protective coverings of the brain and spinal cord).

Menopause The end of a female's reproductive potential when ovulation and menstruation cease.

Menstrual (uterine) cycle The sequence of events that occurs on an approximately 28-day cycle in the uterine lining (endometrium) that involves the thickening of and increased blood supply to the endometrium and the loss of the endometrium as menstrual flow.

Merkel cells Cells of the epidermis found in association with sensory neurons where the epidermis meets the dermis.

Merkel disk The Merkel cell–neuron combination that functions as a sensory receptor for light touch, providing information about objects contacting the skin. It is found on both the hairy and the hairless parts of the skin.

Mesoderm The primary germ layer that gives rise to muscle, bone, connective tissue, and organs such as the heart, kidneys, ovaries, and testes.

Messenger RNA (mRNA) A type of RNA synthesized from and complementary to a region of DNA that attaches to ribosomes in the cytoplasm and specifies the amino acid order in the protein. It carries the DNA instructions for synthesizing a particular protein.

Metabolism The sum of all chemical reactions within the living cells.

Metaphase In mitosis, the phase when the chromosomes, guided by the fibers of the mitotic spindle, form a line at the center of the cell. As a result of this alignment, when the chromosomes split at the centromere, each daughter cell receives one chromatid from each chromosome and thus a complete set of the parent cell's chromosomes.

Metastasize To spread from one part of the body to another part not directly connected to the first part. Cancerous tumors metastasize and form new tumors in distant parts of the body.

MHC markers Molecules on the surface of body cells that label the cell as *self*.

Microevolution Changes in allele frequencies within a population over a few generations. The causes include genetic drift, gene flow, natural selection, and mutation.

Microfilament A component of the cytoskeleton made from the globular protein actin. Microfilaments form contractile units in muscle cells and are responsible for amoeboid movement.

Microtubule A component of the cytoskeleton made from the globular protein tubulin. Microtubules are responsible for the movement of cilia and flagella and serve as tracks for the movement of organelles and vesicles.

Microvilli Microscopic cytoplasm-filled extensions of the cell membrane that serve to increase the absorptive surface area of the cell.

Midbrain A region of the brain stem that coordinates reflex responses to auditory and visual stimuli.

Middle ear An air-filled space in the temporal bone that includes the tympanic membrane (eardrum) and three small bones (the hammer, anvil, and stapes). It serves as an amplifier of sound pressure waves.

Mineralocorticoids Hormones secreted by the adrenal cortex that affect mineral homeostasis and water balance.

Minipill A birth control pill that contains only progesterone.

Mitochondrion (plural, mitochondria) An organelle within which most of cellular respiration occurs in a eukaryotic cell. Cellular respiration is the process by which oxygen and an organic fuel such as glucose are consumed and energy is released and used to form ATP.

Mitosis A type of cell division occurring in somatic cells in which two identical cells, called daughter cells, are generated from a single cell. The original cell first replicates its genetic material and then distributes a complete set of genetic information to each of its daughter cells. Mitosis is usually divided into prophase, metaphase, anaphase, and telophase.

Mitral valve A heart valve located between the left atrium and left ventricle that prevents the backflow of blood from the ventricle to the atrium. It is also called the bicuspid valve or the left atrioventricular (AV) valve.

Molecular clock The idea that there is a constant rate of divergence of macromolecules (such as proteins) from one another. It is based on the notion that single nucleotide changes in DNA (point mutations) and the amino acid changes in proteins that can be produced by some point mutations occur with steady, clocklike regularity. It permits comparison of molecular sequences to estimate the time of separation between species (the more differences in sequence, the more time that has elapsed since the common ancestor).

Molecule A chemical structure composed of atoms held together by covalent bonds.

Monoclonal antibodies Defensive proteins specific for a particular antigen secreted by a clone of genetically identical cells descended from a single cell.

Monocyte The largest white blood cell. Monocytes are active in fighting chronic infections, viruses, and intracellular bacterial infections. A monocyte can transform into a phagocytic macrophage.

Monohybrid cross A genetic cross that considers the inheritance of a single trait from individuals differing in the expression of that trait.

Monomer A small molecule that joins with identical molecules to form a polymer.

Mononucleosis A viral disease caused by the Epstein-Barr virus that results in an elevated level of monocytes in the blood. It causes fatigue and swollen glands, and there is no available treatment.

Monosaccharide The smallest molecular unit of a carbohydrate. Monosaccharides are known as simple sugars.

Monosomy A condition in which there is only one representative of a chromosome instead of two representatives.

Mons veneris A round fleshy prominence over the pubic bone in a female. Part of the female external genitalia.

Morning sickness The nausea and vomiting experienced by some women early in pregnancy. It is not restricted to the morning and may be caused, in part, by high levels of the hormone human chorionic gonadotropin (HCG).

Morphogenesis The development of body form that begins during the third week after fertilization.

Morula A solid ball of 12 or more cells produced by successive divisions of the zygote. The name reflects its resemblance to the fruit of the mulberry tree.

Mosaic evolution The phenomenon whereby various traits evolve at their own rates.

Motor (efferent) neuron A neuron specialized to carry information *away from* the central nervous system to an effector, either a muscle or a gland.

Motor unit A motor neuron and all the muscle fibers (cells) it stimulates.

MRI See magnetic resonance imaging.

mRNA See messenger RNA.

Mucosa The innermost layer of the gastrointestinal tract. It secretes mucus that helps lubricate the tube, allowing food to slide through easily.

Mucous membranes Sheets of epithelial tissue that line many passageways in the body that open to the exterior. Mucous membranes are specialized to secrete and absorb.

Mucus A sticky secretion that serves to lubricate body parts and trap particles of dirt and other secretions. It also helps protect the stomach from the action of gastric juice.

Multiple alleles Three or more alleles of a particular gene existing in a population. The alleles governing the ABO blood types provide an example.

Multiple sclerosis An autoimmune disease in which the body's own defense mechanisms attack myelin sheaths in the nervous system. As a patch of myelin is repaired, a hardened region called a sclerosis forms.

Multipolar neuron A neuron that has at least three processes, including an axon and a minimum of two dendrites. Most motor neurons and association neurons are multipolar.

Multiregional hypothesis The idea that modern humans evolved independently in several different areas from distinctive local populations of earlier humans. Compare with Out of Africa hypothesis.

Muscle tissue Tissue composed of muscle cells that contract when stimulated and passively lengthen to the resting state. There are three types of muscle tissue: skeletal, smooth, and cardiac.

Muscle fiber A muscle cell.

Muscle spindles Specialized muscle fibers with sensory nerve cell endings wrapped around them that report to the brain whenever a muscle is stretched.

Muscle twitch Contraction of a muscle in response to a single stimulus.

Muscularis The muscular layers of the gastrointestinal tract. These layers help move food through the gastrointestinal system and mix food with digestive secretions.

Mutation A change in the base sequence of the DNA of a gene. A mutation may occur spontaneously or be caused by outside sources, such as radiation or chemical agents. Mutations are the only source of new characteristics in a population.

Myelin sheath An insulating layer around axons that carry nerve impulses over relatively long distances that is composed of multiple wrappings of the plasma membrane of certain glial cells. Outside the brain and spinal cord, Schwann cells form the myelin sheath. The myelin sheath greatly increases the speed at which impulses travel and assists in the repair of damaged axons. The Schwann cells that form the myelin sheath are separated from one another by short regions of exposed axon called nodes of Ranvier.

Myocardial infarction A heart attack. A condition in which a part of the heart muscle dies because of an insufficient blood supply.

Myocardium Cardiac muscle tissue that makes up the bulk of the heart. The contractility of the myocardium is responsible for the heart's pumping action.

Myofibril A rodlike bundle of contractile proteins (myofilaments) found in skeletal and cardiac muscle cells essential to muscle contraction.

Myofilament A contractile protein within muscle cells. There are two types: myosin (thick) filaments and actin (thin) filaments.

Myoglobin An oxygen-binding pigment in muscle fibers.

Myometrium The smooth muscle layer in the wall of the uterus.

Myosin filaments The thick filaments in muscle cells composed of the protein myosin and essential to muscle contraction. A myosin molecule is shaped like a golf club with two heads.

Myosin heads Club-shaped ends of a myosin molecule that bind to actin filaments and can swivel, causing actin filaments to slide past the myosin filaments, which causes muscle contraction. They are also called cross-bridges.

Myxedema A condition characterized by swelling of the facial tissues because of the accumulation of interstitial fluids caused by undersecretion of thyroid hormone.

Nasal cavities Two chambers in the nose, separated by the septum.

Nasal conchae The three convoluted bones within each nasal cavity that increase surface area and direct airflow.

Nasal septum A thin partition of cartilage and bone that divides the inside of the nose into two nasal cavities.

Natural childbirth Childbirth without the use of pain-killing drugs.

Natural killer (NK) cells A type of cell in the immune system, probably lymphocytes, that roam the body in search of abnormal cells and quickly kill them.

Natural selection The process by which some individuals produce more offspring than other individuals because their particular inherited characteristics make them better suited to their local environment.

Nearsightedness Myopia. Nearsightedness is a visual condition in which nearby objects can be seen more clearly than distant objects. Nearsightedness occurs because either the eyeball is elongated or the lens is too thick, causing the image to be focused in front of the retina.

Negative feedback mechanism The homeostatic mechanism in which the outcome of a process feeds back on the system, shutting down the process.

Nephrons Functional units of the kidneys responsible for the formation of urine. These microscopic tubules number 1 million to 2 million per kidney and perform filtration (only certain substances are allowed to pass out of the blood and into the nephron), reabsorption (some useful substances are returned from the nephron to the blood), and secretion (the nephron directly removes wastes and excess materials in the blood and adds them to the filtered fluid that becomes urine).

Nerve A bundle of parallel axons, dendrites, or both from many neurons. A nerve is usually covered with tough connective tissue.

Nervous tissue Tissue consisting of two types of cells, neurons and neuroglia, that make up the brain, spinal cord, and nerves.

Neural tube The embryonic structure that gives rise to the brain and spinal cord.

Neuroglia (neuroglial cells) Cells of the nervous system that support, insulate, and protect nerve cells; also called glial cells.

Neuromuscular junction The area of contact between the terminal end of a motor neuron and the sarcolemma of a skeletal muscle fiber. When an action potential reaches the terminal end of the motor neuron, acetylcholine is released, triggering events that can lead to muscle contraction.

Neurons Nerve cells involved in intercellular communication. A neuron consists of a cell body, dendrites, and an axon. Neurons are excitable cells in the nervous system specialized to generate and transmit electrochemical signals called action potentials or nerve impulses.

Neurotransmitter A chemical released from the axon tip of a neuron that affects the activity of another cell (usually a nerve, muscle, or gland cell) by altering the electrical potential difference across the membrane of the receiving cell.

Neurula The embryo during neurulation, formation of the brain and spinal cord from ectoderm.

Neurulation A series of events during embryonic development when the central nervous system (brain and spinal cord) forms from ectoderm. During this period, the embryo is called a neurula.

Neutrophil A phagocytic white blood cell important in defense against bacteria and removal of cellular debris. Most abundant of white blood cells.

New World monkeys A group of monkeys whose members are arboreal, have nostrils that open to the side, and have a grasping (prehensile) tail. They occur in South and Central America and include spider monkeys, capuchins, and squirrel monkeys.

Niche The role of a species in an ecosystem. The niche includes the habitat, food, nest sites, and so on. It describes how a member of a particular species uses materials in the environment and how it interacts with other organisms.

Nitrification The conversion of ammonia to nitrate (NO_3^-) by nitrifying bacteria living in the soil.

Nitrogen fixation The process of converting nitrogen gas to ammonium (a nitrogen-containing molecule that can be used by living organisms). The process of nitrogen fixation is carried out by nitrogen-fixing bacteria living in nodules on the roots of leguminous plants such as peas and alfalfa.

Node of Ranvier A region of exposed axon between Schwann cells forming a myelin sheath. In myelinated nerves, the impulse jumps from one node of Ranvier to the next, greatly increasing the speed of conduction. This type of transmission is called saltatory conduction.

Nondisjunction Failure of the members of a pair of homologous chromosomes or the sister chromatids to separate during mitosis or meiosis. Nondisjunction results in cells with an abnormal number of chromosomes.

Norepinephrine Noradrenaline. A hormone secreted by the adrenal medulla, along with epinephrine, in response to stress. They initiate the physiological fight-or-flight reaction. Norepinephrine is also a neurotransmitter found in both the central and peripheral nervous systems. In the central nervous system, it is important in the regulation of mood, in the pleasure system of the brain, arousal, and dreaming sleep. Norepinephrine is thought to produce an energizing good feeling. It is also thought to be essential in hunger, thirst, and sex drive.

Notochord The flexible rod of tissue that develops during gastrulation and signals where the vertebral column will form. The notochord defines the axis of the embryo, gives the embryo some rigidity, and prompts overlying ectoderm to begin formation of the central nervous system. During development, vertebrae form around the notochord. The notochord eventually degenerates, existing only as the pulpy, elastic material in the center of intervertebral disks.

Nuclear envelope The double membrane that surrounds the nucleus.

Nuclear pore An opening in the nuclear envelope that permits communication between the nucleus and the cytoplasm.

Nucleolus A specialized region within the nucleus that forms and disassembles during the course of the cell cycle. It plays a role in the generation of ribosomes, organelles involved in protein synthesis.

Nucleoplasm The chromatin and the aqueous environment within the nucleus.

Nucleotide A subunit of DNA composed of one five-carbon sugar (either ribose or deoxyribose), one phosphate group, and one of five nitrogen-containing bases. Nucleotides are the building blocks of nucleic acids (DNA and RNA).

Nucleus The command center of the cell containing almost all the genetic information.

Oil glands Glands associated with the hair follicles that produce sebum. They are also called sebaceous glands.

Old World monkeys A group of monkeys whose members lack a prehensile tail, often having a short tail or none at all, and have nostrils that open downward. Some Old World monkeys are arboreal and some are terrestrial. Examples include baboons and rhesus monkeys.

Omnivore An organism that feeds on a variety of food types, such as plants and animals.

Oocyte A cell whose meiotic divisions will produce an ovum and up to three polar bodies.

Oogenesis The production of ova (eggs), including meiosis and maturation.

Oogonium (plural, oogonia) A germ cell in an ovary that divides, giving rise to oocytes.

Opsin One of several proteins that can be bound to retinal to form the visual pigments in rods and cones.

Optic nerve One of two nerves, one from each eye, responsible for bringing processed electrochemical messages from the retina to the brain for interpretation.

Organ A structure composed of two or more different tissues with a specific function.

Organelle A specialized membrane-bound compartment within a eukaryotic cell.

Organ of Corti The spiral organ. The portion of the cochlea in the inner ear that contains receptor cells that sense vibrations caused by sound. It is most directly responsible for the sense of hearing.

Organ system A group of organs with a common function.

Origin In reference to a muscle, the end of the muscle that is attached to the bone that remains relatively stationary during a movement.

Osmosis A special case of diffusion in which water moves across the plasma membrane or any other selectively permeable membrane from a region of lower concentration of solute to a region of higher concentration of solute.

Osteoarthritis An inflammation in a joint that is caused by degeneration of the surfaces of the joint caused by wear and tear.

Osteoblast A bone-forming cell.

Osteoclast A large cell responsible for the breakdown and absorption of bone.

Osteocytes Mature bone cells found in lacunae that are arranged in concentric rings around the central canal. Cytoplasmic projections from osteocytes extend tiny channels that connect with other osteocytes.

Osteon The structural unit of compact bone that appears as a series of concentric circles of lacunae. The lacunae contain bone cells around a central canal containing blood vessels and nerves.

Osteoporosis A decrease in bone density that occurs when the destruction of bone outpaces the formation of new bone, causing bone to become thin, brittle, and susceptible to fracture.

Otoliths Granules of calcium carbonate embedded in gelatinous material in the utricle and saccule of the ear. Otoliths cause the gelatin to slide over and bend sensory hair cells when the head is moved. The bending generates nerve impulses that are sent to the brain for interpretation as the position of the head.

Out of Africa hypothesis The idea that modern humans evolved from earlier humans in Africa and later migrated to Europe, Asia, and Australia. It suggests a single origin for *Homo sapiens*. Compare with multiregional hypothesis.

Outer ear The external appendage on the outside of the head (pinna) and the canal (the external auditory canal) that extends to the eardrum. It functions as the receiver for sound vibrations.

Oval window A membrane-covered opening in the inner ear (cochlea) through which vibrations from the stirrup (stapes) are transmitted to fluid within the cochlea.

Ovarian cycle The sequence of events in the ovary that leads to ovulation. The cycle is approximately 28 days long and is closely coordinated with the menstrual cycle.

Ovary One of the female gonads. The female gonads produce the ova (eggs) and the hormones estrogen and progesterone.

Oviduct One of two tubes that conduct the ova from the ovary to the uterus in the female reproductive system. It is also called a uterine tube or a fallopian tube.

Ovulation The release of the secondary oocyte from the ovary.

Ovum A mature egg; a large haploid cell that is the female gamete.

Oxygen debt The amount of oxygen required after exercise to oxidize the lactic acid formed during exercise.

Oxyhemoglobin Hemoglobin bound to oxygen.

Oxytocin (OT) The hormone released at the posterior pituitary that stimulates uterine contractions and milk ejection.

Pacemaker See sinoatrial (SA) node.

Pacinian corpuscle A large encapsulated nerve cell ending that is located deep within the skin and near body organs that responds when pressure is first applied. It is important in sensing vibration.

Pain receptor A sensory receptor that is specialized to detect the physical or chemical damage to tissues that we sense as pain. Pain receptors are sometimes classified as chemoreceptors, because they often respond to chemicals liberated by damaged tissue, and occasionally as mechanoreceptors, because they are stimulated by physical changes, such as swelling, in the damaged tissue.

Palate The roof of the mouth. The front region of the palate, the hard palate, is reinforced with bone. The tongue pushes against the hard palate while mixing food with saliva. The soft palate is farther to the back of the mouth and consists of only muscle. The soft palate prevents food from entering the nose during swallowing.

Paleoanthropology The scientific study of human origins and evolution.

Pancreas An accessory organ behind the stomach that secretes digestive enzymes, bicarbonate ions to neutralize the acid in chyme, and hormones that regulate blood sugar.

Pancreatic islets Small clusters of endocrine cells in the pancreas; also called islets of Langerhans.

Parasympathetic nervous system The branch of the autonomic nervous system that is active during restful conditions. Its effects generally oppose those of the sympathetic nervous system. The parasympathetic nervous system adjusts bodily functions so that energy is conserved during nonstressful times.

Parathormone See parathyroid hormone.

Parathyroid glands Four small, round masses at the back of the thyroid gland that secrete parathyroid hormone (parathormone).

Parathyroid hormone (PTH) A hormone released from the parathyroid glands that regulates blood calcium levels by stimulating osteoclasts to break down bone. PTH, also called parathormone, is secreted when blood calcium levels are too low.

Parental generation The parents—the individuals in the earliest generation under consideration.

Parkinson's disease A progressive disorder that results from the death of dopamine-producing neurons that lie in the heart of the brain's movement control center, the substantia nigra. Parkinson's disease is characterized by slowed movements, tremors, and muscle rigidity.

Parturition Birth, which usually occurs about 38 weeks after fertilization.

Passive immunity Temporary immune resistance that develops when a person receives antibodies that were produced by another person or animal.

Pathogen A disease-causing organism.

Pectoral girdle The bones that connect the arms to the rib cage. The pectoral girdle is composed of the shoulder blades (scapulae) and the collarbones (clavicles).

Pedigree A chart showing the genetic connections among individuals in an extended family that is often used to trace the expression of a particular trait in that family.

Pelvic girdle The bones that connect the legs to the vertebral column. The pelvic girdle is composed of the paired hipbones.

Pelvic inflammatory disease (PID) A general term for any bacterial infection of a woman's pelvic organs. PID is usually caused by sexually transmitted bacteria, especially those that cause chlamydia and gonorrhea.

Penis The cylindrical external reproductive organ of a male through which most of the urethra extends and that serves to deliver sperm into the female tract during sexual intercourse.

Pepsin A protein-splitting enzyme initially secreted in the stomach in the inactive form of pepsinogen that is activated into pepsin by hydrochloric acid.

Peptic ulcer A local defect in the surface of the stomach or small intestine characterized by dead tissue and inflammation.

Peptide A chain containing only a few amino acids.

Perforin A type of protein released by a natural killer cell that creates numerous pores (holes) in the target cell, making it leaky. Fluid is then drawn into the leaky cell because of the high salt concentration within, and the cell bursts.

Pericardium The fibrous sac enclosing the heart that holds the heart in the center of the chest without hampering its movements.

Perichondrium The layer of dense connective tissue surrounding cartilage that contains blood vessels, which supply the cartilage with nutrients.

Periodontitis An inflammation of the gums and the tissues around the teeth.

Periosteum The membranous covering that nourishes bone.

Peripheral nervous system The part of the nervous system outside the brain and spinal cord. It keeps the central nervous system in continuous contact with almost every part of the body. It is composed of nerves and ganglia. The two branches are the somatic and the autonomic nervous systems.

Peripheral proteins Proteins attached to the inner or outer surface of the plasma membrane.

Peristalsis Rhythmic waves of muscular contraction and relaxation in the walls of hollow tubular organs, such as the digestive organs, that serve to push contents through the tubes.

Peroxisome An enzyme-containing organelle within which chemical reactions occur that generate hydrogen peroxide. Such reactions have diverse functions ranging from the breakdown of fats to the detoxification of poisons and alcohol.

PET scan See positron emission tomography.

Phagocytes Scavenger cells specialized to engulf and destroy particulate matter, such as pathogens, damaged tissue, or dead cells.

Phagocytosis The process by which cells such as white blood cells ingest foreign cells or substances by surrounding the foreign material with cell membrane. It is a type of endocytosis.

Pharynx The space shared by the respiratory and digestive systems that is commonly called the throat. The pharynx is a passageway for air, food, and liquid.

Phenotype The observable physical and physiological traits of an individual. Phenotype results from the inherited alleles and their interactions with the environment.

Phospholipid An important component of cell membranes. It has a nonpolar "water-fearing" tail (made up of fatty acids) and a polar "water-loving" head (containing an R group, glycerol, and phosphate).

Photoreceptor A sensory receptor specialized to detect changes in light intensity. Photoreceptors are responsible for the sensation we describe as vision.

pH scale A scale for measuring the concentration of hydrogen ions. The scale ranges from 0 to 14, with a pH of 7 being neutral, a pH of less than 7 being acidic, and a pH of greater than 7, basic.

Phylogenetic trees Generalized descriptions of the history of life. They depict hypotheses about evolutionary relationships among species or higher taxa.

Physical dependence A condition in which continued use of a drug is needed to maintain normal cell functioning.

Pia mater The innermost layer of the meninges (the connective tissue layers that protect the central nervous system).

Piloerection Contraction of the arrector pili muscles causing hairs to stand on end and form a layer of insulation.

Pineal gland The gland that produces the hormone melatonin and is located at the center of the brain.

Pinealocyte A cell of the pineal gland that secretes melatonin.

Pinna The visible part of the ear on each side of the head that gathers sound and channels it to the external auditory canal.

Pinocytosis A type of endocytosis in which droplets of fluid and the dissolved substances therein are engulfed by cells.

Pituitary dwarfism A condition caused by insufficient growth hormone during childhood.

Pituitary gland The endocrine organ connected to the hypothalamus by a short stalk. It consists of the anterior and posterior lobes and is also called the hypophysis.

Pituitary portal system The system in which a capillary bed in the hypothalamus connects to veins that lead into a capillary bed in the anterior lobe of the pituitary gland. It allows hormones of the hypothalamus to control the secretion of hormones from the anterior pituitary gland.

Placebo In a controlled experiment to test the effectiveness of a drug, it is a substance that appears to be identical to a drug but has no known effect on the condition for which it is taken.

Placenta The organ that delivers oxygen and nutrients to the embryo and carries carbon dioxide and wastes away from it. The placenta is also called the afterbirth.

Placental stage The third (and final) stage of true labor. It begins with delivery of the baby and ends when the placenta detaches from the wall of the uterus and is expelled from the mother's body.

Placenta previa The condition in which the placenta forms in the lower half of the uterus, entirely or partially covering the cervix. It may cause premature birth or maternal hemorrhage and usually makes vaginal delivery impossible.

Plaque A bumpy layer consisting of smooth muscle cells filled with lipid material, especially cholesterol, that bulges into the channel of an artery and reduces blood flow. Another type of plaque is a buildup of food material and bacteria on teeth that leads to tooth decay.

Plasma A straw-colored liquid that makes up about 55% of blood. It serves as the medium for transporting materials within the blood. Plasma consists of water (91%–93%) with substances dissolved in it (7%–9%).

Plasma cell The effector cell, produced from a B lymphocyte, that secretes antibodies.

Plasma membrane The thin outer boundary of a cell that controls the movement of substances into and out of the cell.

Plasma proteins Proteins dissolved in plasma, including albumins, which are important in water balance between cells and the blood; globulins, which are important in transporting various substances in the blood; and antibodies, which are important in the immune response.

Plasmid A small, circular piece of self-replicating DNA that is separate from the chromosome and found in bacteria. Plasmids are often used as vectors in recombinant DNA research.

Plasmin An enzyme that breaks down fibrin and dissolves blood clots. Plasmin is formed from plasminogen.

Plasminogen A plasma protein. It is the inactive precursor of plasmin.

Platelet A cell fragment of a megakaryocyte that releases substances necessary for blood clotting. It is formed in the red bone marrow from a precursor cell called a megakaryocyte. It is sometimes called a thrombocyte.

Platelet plug A mass of platelets clinging to the protein fiber collagen at a damaged blood vessel to prevent blood loss.

Pleiotropy One gene having many effects.

PMS See premenstrual syndrome.

Point mutation A mutation that involves changes in one or a few nucleotides in DNA.

Polar body Any of three small nonfunctional cells produced during the meiotic divisions of an oocyte. The divisions also produce a mature ovum (egg).

Polygenic inheritance Inheritance in which several independently assorting or loosely linked genes determine the expression of a trait.

Polymer A large molecule formed by the joining together of many smaller molecules of the same general type (monomers).

Polymerase chain reaction (PCR) A technique used to amplify (increase) the quantity of DNA in vitro using primers, DNA polymerase, and nucleotides.

Polypeptide A chain containing 10 or more amino acids.

Polysaccharide A complex carbohydrate formed when large numbers of monosaccharides (most commonly glucose) join together to form a long chain through dehydration synthesis. Most polysaccharides store energy or provide structure.

Polysome A cluster of ribosomes simultaneously translating the same messenger RNA (mRNA) strand.

Pons A part of the brain that connects upper and lower levels of the brain.

Population A group of potentially interacting individuals of the same species living in a distinct geographic area.

Population dynamics Changes in population size over time.

Portal system A system whereby a capillary bed drains to veins that drain to another capillary bed.

Positive feedback mechanism The mechanism by which the outcome of a process feeds back on the system, further stimulating the process.

Positron emission tomography (PET) A method that can be used to measure the activity of various brain regions. The person being scanned is injected with a radioactively labeled nutrient, usually glucose, that is tracked as it flows through the brain. The radioisotope emits positively charged particles, called positrons. When the positrons collide with electrons in the body, gamma rays are released. The gamma rays can be detected and recorded by PET receptors. Computers then use the information to construct a PET scan that shows where the radioisotope is being used in the brain.

Postnatal period The period of development after birth. It includes the stages of infancy, childhood, puberty, adolescence, and adulthood.

Postsynaptic neuron The neuron located after the synapse. The receiving neuron in a synapse. The membrane of the postsynaptic neuron has receptors specific for certain neurotransmitters.

Precapillary sphincter A ringlike muscle that acts as a valve that opens and closes a capillary bed. Contraction of the precapillary sphincter squeezes the capillary shut and directs blood through a thoroughfare channel to the venule. Relaxation of the precapillary sphincter allows blood to flow through the capillary bed.

Pre-embryo The developing human from fertilization through the second week of gestation (the pre-embryonic period).

Pre-embryonic period The period during prenatal development that extends from fertilization through the second week. Cleavage and implantation follow fertilization.

Prefrontal cortex An association area of the cerebral cortex that is important in decision making.

Premature infant A baby born before 37 weeks of gestation.

Premenstrual syndrome (PMS) A collection of uncomfortable symptoms, including irritability, stress, and bloating, that appears 7 to 10 days before a woman's menstrual period and is associated with hormonal cycling.

Prenatal period The period of development before birth. It is further subdivided into the pre-embryonic period (from fertilization through the second week), the embryonic period (from the third through the eighth weeks), and the fetal period (from the ninth week until birth).

Presynaptic neuron The neuron located before the synapse. The sending neuron in a synapse. The neuron in a synapse that releases neurotransmitter from its synaptic knobs into the synaptic cleft.

Primary consumer An herbivore. An animal that eats primary producers (green plants or algae).

Primary germ layers The layers produced by gastrulation from which all tissues and organs form. They are ectoderm, mesoderm, and endoderm.

Primary motor area A band of the frontal lobe of the cerebral cortex that initiates messages that direct voluntary movements.

Primary productivity In ecosystems, the gross primary productivity is the amount of light energy that is converted to chemical energy in the bonds of organic molecules during a given period. The net primary productivity is the amount of productivity left after the photosynthesizers have used some of the energy stored in organic molecules for their own metabolic activities.

Primary response The immune response that occurs during the body's first encounter with a particular antigen.

Primary somatosensory area A band of the parietal lobe of the cerebral cortex to which information is sent from receptors in the skin regarding touch, temperature, and pain and from receptors in the joints and skeletal muscles.

Primary spermatocyte The original large cell that develops from a spermatogonium during sperm development in the seminiferous tubules. It undergoes meiosis and gives rise to secondary spermatocytes.

Primary structure The precise sequence of amino acids of a protein. This sequence, determined by the genes, dictates a protein's structure and function.

Primary succession The sequence of changes in the species making up a community over time that begins in an area where no community previously existed and ends with a climax community.

Primordial germ cells Cells that migrate from the yolk sac of the developing human to the ovaries or testes, where they differentiate into immature cells that will eventually become sperm or oocytes.

Probe A single-stranded molecule of radioactive DNA or RNA with a sequence of bases complementary to the sequence in the gene of interest that is used to locate the desirable gene.

Prodrome The symptoms that precede recurring outbreaks of a disease such as genital herpes.

Producers In ecosystems, the producers are the organisms that convert energy from the physical environment into chemical energy in the bonds of organic molecules through photosynthesis or chemosynthesis. The producers form the first trophic level.

Progesterone A female sex hormone produced by the corpus luteum in the ovary. Progesterone helps prepare the endometrium (lining) of the uterus for pregnancy and maintains the endometrium.

Programmed cell death A genetically programmed series of events that causes a cell to self-destruct.

Prokaryotic cell A cell that lacks a nucleus and other membrane-enclosed organelles. The prokaryotes include bacteria and archaea.

Prolactin (PRL) An anterior pituitary hormone that stimulates the mammary glands to produce milk.

Promoter A specific region on DNA next to the "start" gene that controls the expression of the gene.

Prophase In mitosis, the phase when the chromosomes begin to thicken and shorten, the nucleolus disappears, the nuclear envelope begins to break down, and the mitotic spindle forms in the cytoplasm.

Prostaglandins The lipid molecules found in and released by the plasma membranes of most cells. They are often called the local hormones because they exert their effects on nearby cells.

Prostate gland An accessory reproductive gland in males that surrounds the urethra as it passes from the bladder. Its secretions contribute to semen and serve to activate the sperm and to counteract the acidity of the female reproductive tract.

Proteins The macromolecules composed of amino acids linked by peptide bonds. The functions of proteins include structural support, transport, movement, and regulation of chemical reactions.

Prothrombin A plasma protein synthesized by the liver that is important in blood clotting. It is converted to an active form (thrombin) by thromboplastin that is released from platelets.

Prothrombin activator A blood protein that converts prothrombin to thrombin as part of the blood-clotting process.

Protozoans A group of single-celled organisms with a well-defined eukaryotic nucleus. Protozoans can cause disease by producing toxins or by releasing enzymes that prevent host cells from functioning normally.

Proximal convoluted tubule The section of the renal tubule where reabsorption and secretion occur.

Pseudopodia The temporary extensions of a cell formed by microfilaments and used when feeding and moving. The name means *false foot*.

Puberty The stage in postnatal development when secondary sexual characteristics such as pubic and underarm hair develop. This stage usually occurs slightly earlier in girls (from 12 to 15 years of age) than in boys (from 13 to 16 years of age).

Pulmonary arteries Blood vessels that carry blood low in oxygen from the right ventricle to the lungs, where it is oxygenated.

Pulmonary circuit (or circulation) The pathway that transports blood from the right ventricle of the heart to the lungs and back to the left atrium of the heart.

Pulmonary veins Blood vessels that carry oxygenated blood from the lungs to the left atrium of the heart.

Pulp The center of a tooth that contains the tooth's life-support systems.

Pulse The rhythmic expansion of an artery created by the surge of blood pushed along the artery by each contraction of the ventricles of the heart. With each beat of the heart, the wave of expansion begins, moving along the artery at the rate of 6 to 9 meters per second.

Punnett square A diagrammatic method used to determine the probable outcome of a genetic cross. The possible allele combinations in the gametes of one parent are used to label the columns and the possible allele combinations of the other parent are used to label the rows. The alleles of each column and each row are then paired to determine the possible genotypes of the offspring.

Pupil The small hole through the center of the iris through which light passes to enter the eye. The size of the pupil is altered to regulate the amount of light entering the eye.

Purkinje fibers The specialized cardiac muscle cells that deliver an electrical signal from the atrioventricular bundle to the individual heart muscle cells in the ventricles.

Pyelonephritis An infection of the kidneys caused by bacteria.

Pyloric sphincter A ring of muscle between the stomach and small intestine that regulates the emptying of the stomach into the small intestine.

Pyramid of biomass A diagram in which blocks represent the amount of biomass (dry body mass of organisms) available at each trophic (feeding) level.

Pyramid of energy A graphic representation in which blocks represent the decreasing amount of energy available at each trophic (feeding) level.

Pyramid of numbers A graphic representation of the number of individuals at each trophic level.

Pyrogen A fever-producing substance.

Quaternary structure The shape of an aggregate protein. It is determined by the mutually attractive forces between the protein's subunits.

Radioisotopes Isotopes that are unstable and spontaneously decay, emitting radiation in the form of gamma rays and alpha and beta particles.

Receptor A protein molecule located in the cytoplasm and on the plasma membrane of cells that is sensitive to chemical messengers.

Receptor potential An electrochemical message (a change in the degree of polarization of the membrane) generated in a sensory receptor in response to a stimulus. Receptor potentials vary in magnitude with the strength of the stimulus.

Recessive allele The allele whose effects are masked in the heterozygous condition. The recessive allele usually produces a nonfunctional protein.

Recombinant DNA Segments of DNA from two sources that have been combined in vitro and transferred to cells in which their information can be expressed.

Rectum The final section of the gastrointestinal tract. The rectum receives and temporarily stores the feces.

Red blood cell See erythrocyte.

Red marrow Blood cell–forming connective tissue found in the marrow cavity of certain bones.

Reduction division The first meiotic division (meiosis I) that produces two cells, each of which contains one member of each homologous pair (23 chromosomes with replicates attached in humans).

Referred pain Pain felt at a site other than the area of origin.

Reflex A simple, stereotyped reaction to a stimulus.

Reflex arc A neural pathway consisting of a sensory receptor, a sensory neuron, usually at least one interneuron, a motor neuron, and an effector.

Refractory period The interval following an action potential during which a neuron cannot be stimulated to generate another action potential.

Relaxin The hormone released from the placenta and the ovaries. It facilitates delivery by dilating the cervix and relaxing the ligaments and cartilage of the pubic bones.

Releasing hormone A hormone that stimulates hormone secretion by another gland.

Renal corpuscle The portion of the nephron where fluid is filtered. It consists of a tuft of capillaries, the glomerulus, and a surrounding cuplike structure called the glomerular (Bowman's) capsule.

Renal cortex The outer region of the kidney, containing renal columns.

Renal failure A decrease or complete cessation of glomerular filtration.

Renal medulla The inner region of the kidney. It contains cone-shaped structures called renal pyramids.

Renal pelvis The innermost region of the kidney; the chamber within the kidney.

Renal tubule The site of reabsorption and secretion by the nephron. It consists of the proximal convoluted tubule, the loop of the nephron, and the distal convoluted tubule.

Renin An enzyme released by cells of the juxtaglomerular apparatus of nephrons. Renin converts angiotensinogen, a protein produced by the liver and found in the plasma, into another protein, angiotensin I. These actions of renin initiate a series of hormonal events that leads to increased reabsorption of sodium and water by the distal convoluted tubules and collecting ducts of nephrons.

Rennin The gastric enzyme that breaks down milk proteins.

Replication Copying from a template, as occurs during the synthesis of new DNA from preexisting DNA.

Repolarization The return of the membrane potential to approximately its resting value. Repolarization of the nerve cell membrane during an action potential occurs because of the outflow of potassium ions.

Residual volume The amount of air that remains in the lungs after a maximal exhalation.

Resistance phase The phase of the general adaptation syndrome that follows the alarm reaction. It is characterized by anterior pituitary release of adrenocorticotropic hormone (ACTH), thyroid-stimulating hormone (TSH), and growth hormone (GH) and the sustainment of metabolic demands by the body's fat reserves.

Respiratory distress syndrome (RDS) A condition in newborns caused by an insufficient amount of surfactant, causing the alveoli of the lungs to collapse and thereby making breathing difficult.

Respiratory system The organ system that carries out gas exchange. The respiratory system includes the nose, pharynx, larynx, trachea, bronchi, and lungs.

Resting potential The separation of charge across the plasma membrane of a neuron when the neuron is not transmitting an action potential. A neuron's resting potential is the electrical potential difference across its plasma membrane when the neuron is not conducting an impulse. It is primarily caused by the unequal distributions of sodium ions, potassium ions, and large negatively charged proteins on either side of the plasma membrane. The resting potential of a neuron is about −70 mV.

Restriction enzyme An enzyme that recognizes a specific sequence of bases in DNA and cuts the DNA into two strands at that sequence. Restriction enzymes are used to prepare DNA containing "sticky ends" during the creation of recombinant DNA. Their natural function in bacteria is to control the replication of viruses that infect the bacteria.

Reticular activating system (RAS) An extensive network of neurons that runs through the medulla and projects to the cerebral cortex. It filters out unimportant sensory information before it reaches the brain and controls changing levels of consciousness.

Reticular fibers Interconnected strands of collagen in certain connective tissues that branch extensively. Networks of reticular fibers support soft tissues, including the liver and spleen.

Retina The light-sensitive innermost layer of the eye containing numerous photoreceptors (rods and cones).

Retinal The light-absorbing portion of pigment molecules in the photoreceptors. Retinal combines with one of four opsins (proteins) to form the light-absorbing pigments in rods and cones.

Retrovirus Any one of the viruses that contain only RNA and carry out transcription from RNA to DNA (reverse transcription).

Rheumatoid arthritis An inflammation of a joint caused by an autoimmune response. It is marked by inflammation of the synovial membrane and excess synovial fluid accumulation in the joints, causing swelling, pain, and stiffness.

Rh factor A group of antigens found on the surface of the red blood cells of most people. A person who has these antigens is said to be Rh$^+$. A person who lacks these antigens is Rh$^-$.

Rhodopsin The light-absorbing pigment in the photoreceptors called rods. Rhodopsin is responsible for black and white vision.

Rhythm method of birth control A method of reducing the risk of pregnancy by avoiding intercourse on all days that might result in sperm and egg meeting.

Ribonucleic acid (RNA) A single-stranded nucleic acid that contains ribose (a five-carbon sugar), phosphate, adenine, uracil, cytosine, or guanine. RNA plays a variety of roles in protein synthesis.

Ribosomal RNA (rRNA) A type of RNA that combines with proteins to form the ribosomes, structures on which protein synthesis begins. The most abundant form of RNA.

Ribosome The site where protein synthesis begins in a cell. It consists of two subunits, each containing ribosomal RNA and proteins.

RNA See ribonucleic acid.

RNA polymerase One of the group of enzymes necessary for the synthesis of RNA from a DNA template. It binds with the promoter on DNA that aligns the appropriate RNA nucleotides and links them together.

Rod One of the photoreceptors containing rhodopsin and responsible for black and white vision. Rods are extremely sensitive to light.

Root The part of a tooth that is below the gum line. It is covered with a calcified, yet living and sensitive, connective tissue called cementum.

Root canal A channel through the root of a tooth that contains the blood vessels and nerves.

Rough endoplasmic reticulum (RER) Endoplasmic reticulum that is studded with ribosomes. It processes amino acid chains made by the attached ribosomes and sends the chains to the Golgi complex.

Round window A membrane-covered opening in the cochlea that serves to relieve the pressure caused by the movements of the oval window.

rRNA See ribosomal RNA.

Rugae Folds in the mucosa of the lining of the empty stomach's walls that can unfold, allowing the stomach to expand as it fills.

Salinization An accumulation of salts in soil caused by irrigation over a long period of time that makes the land unusable.

Saliva The secretion from the salivary glands that helps moisten and dissolve food particles in the mouth, facilitating taste and digestion. An enzyme in saliva (salivary amylase) begins the chemical digestion of starch.

Salivary amylase An enzyme in saliva that begins the chemical digestion of starches, breaking them into shorter chains of sugars.

Salivary glands Exocrine glands in the facial region that secrete saliva into the mouth to begin the digestion process.

Saltatory conduction The type of nerve transmission along a myelinated axon in which the nerve impulse jumps from one node of Ranvier to the next. Saltatory conduction greatly increases the speed of impulse conduction.

Sarcomere The smallest contractile unit of a striated or cardiac muscle cell.

Sarcoplasmic reticulum An elaborate form of smooth endoplasmic reticulum found in muscle fibers. The sarcoplasmic reticulum takes up, stores, and releases calcium ions as needed in muscle contraction.

Schizophrenia A mental illness characterized by hallucinations and disordered thoughts and emotions that is caused by excessive activity at dopamine synapses in one part of the brain (the midbrain). As a result, dopamine is no longer in the proper balance with glutamate, a neurotransmitter released by neurons in another brain region (the cerebral cortex).

Schwann cell A type of glial cell in the peripheral nervous system that forms the myelin sheath by wrapping around the axon many times. The myelin sheath insulates axons, increases the speed at which impulses are conducted, and assists in the repair of damaged neurons.

Science A systematic approach to acquiring knowledge through carefully documented investigation and experimentation.

Scientific method A procedure underlying most scientific investigations that involves observation, formulating a hypothesis, making predictions, experimenting to test the predictions, and drawing conclusions. Experimentation usually involves a control group and an experimental group that differ in one or very few factors (variables). New hypotheses may be generated from the results of experimentation.

Sclera The white part of the eye that protects and shapes the eyeball and serves as a site of attachment for muscles that move the eye.

Scrotum A loose fleshy sac containing the testes.

Seasonal affective disorder (SAD) A form of depression associated with winter months when overproduction of melatonin is triggered by short day length.

Sebum An oily substance made of fats, cholesterol, proteins, and salts secreted by the oil glands.

Secondary consumer A carnivore. An animal that obtains energy by eating other animals.

Secondary response The immune response during the body's second or subsequent exposures to a particular antigen. The secondary immune response is much quicker than is the primary response because memory cells specific for the antigen are present.

Secondary spermatocyte A haploid cell formed by meiotic division of a primary spermatocyte during sperm development in the seminiferous tubules.

Secondary structure The bending and folding of the chain of amino acids of a protein to produce shapes such as coils, spirals, and pleated sheets. These shapes form as a result of hydrogen bonding between different parts of the polypeptide chain.

Secondary succession The sequence of changes in the species making up a community over time that takes place after some disturbance destroys the existing life. Soil is already present.

Second messenger A molecule in the cytoplasm of a cell that is activated when a water-soluble hormone binds to a receptor on the surface of the cell. Second messengers influence the activity of enzymes and ultimately the activity of the cell to produce the effect of the hormone.

Secretin A hormone released by the small intestine that inhibits the secretion of gastric juice and stimulates the release of bicarbonate ions from the pancreas and the production of bile in the liver.

Segregation, law of A genetic principle that states that the alleles for each gene separate (segregate) during meiosis and gamete formation, so half of the gametes bear one allele and the other half bear the other allele.

Selectively permeable A characteristic of the plasma membrane because it permits some substances to move across and denies access to others.

Semen The fluid expelled from the penis during male orgasm. Semen consists of sperm and the secretions of the accessory glands.

Semicircular canals Three canals in each ear that are oriented at right angles to one another and contain sensory receptors that precisely monitor any sudden movement of the head. They detect body position and movement.

Semiconservative replication Replication of DNA in which the two strands of a DNA molecule become separated and each serves as a template for a new double-stranded DNA. Each new double-stranded molecule consists of one new strand and one old strand.

Semilunar valves Heart valves located between each ventricle and its connecting artery that prevent the backflow of blood from the artery to the ventricle. Whereas the cusps of the atrioventricular (AV) valves are flaps of connective tissue, those of the semilunar valves are small pockets of tissue attached to the inner wall of their respective arteries.

Seminal vesicles A pair of male accessory reproductive glands located posterior to the urinary bladder. Their secretions contribute to semen and serve to nourish the sperm cells, reduce the acidity in the vagina, and coagulate sperm.

Seminiferous tubules Coiled tubules within the testes where sperm are produced.

Sensory adaptation A gradual decline in the responsiveness of a sensory receptor that results in a decrease in awareness of the stimulus.

Sensory (afferent) neuron A nerve cell specialized to conduct nerve impulses from the sensory receptors *toward* the central nervous system.

Sensory receptors The structures specialized to respond to changes in their environment (stimuli) by generating electrochemical messages that are eventually converted to nerve impulses if the stimulus is strong enough. The nerve impulses are then conducted to the brain, where they are interpreted to build our perception of the world.

Serosa A thin layer of connective tissue that forms the outer layer of the gastrointestinal tract. It secretes a fluid that reduces friction with contacting surfaces.

Serotonin A neurotransmitter in the central nervous system thought to promote a generalized feeling of well-being.

Serous membranes Sheets of epithelial tissue that line the thoracic and abdominal cavities and the organs within them. Serous membranes secrete a fluid that lubricates the organs within these cavities.

Sex chromosomes The X and Y chromosomes. The pair of chromosomes involved in determining gender.

Sex-influenced inheritance An autosomal genetic trait that is expressed differently in males and females, usually because of the presence of sex hormones.

Sex-linked gene A gene located on the X chromosome.

Sexual dimorphism A difference in appearance between males and females within a species.

Short-term memory Memory of new information that lasts for a few seconds or minutes.

Sickle cell anemia A type of anemia caused by a mutation that results in a change in one amino acid in a globin chain of hemoglobin (the iron-containing protein in red blood cells that transports oxygen). Such a change causes the red blood cell to assume a crescent (sickle) shape when oxygen levels are low. The sickle-shaped cells clog small blood vessels, leading to pain and tissue damage from insufficient oxygen.

Simple diffusion The spontaneous movement of a substance from a region of higher concentration to a region of lower concentration.

Simple epithelial tissue Epithelial tissue with only a single layer of cells.

Simple goiter An enlarged thyroid gland caused by iodine deficiency.

Sinoatrial (SA) node A region of specialized cardiac muscle cells located in the right atrium near the junction of the superior vena cava that sets the pace of the heart rate at about 70 to 80 beats a minute. It is also known as the pacemaker. The SA node sends out an electrical signal that spreads through the muscle cells of the atria, causing them to contract.

Sinuses Large, air-filled spaces in the bones of the face.

Sinusitis Inflammation of the mucous membranes of the sinuses making it difficult for the sinuses to drain their mucous fluid.

Skeletal muscle tissue A contractile tissue. One of three types of muscle in the body. Skeletal muscle cells are cylindrical, have many nuclei, and have stripes (striations). Skeletal muscle provides for conscious, voluntary control over contraction. It attaches to bones and forms the muscles of the body. It is also called striated muscle.

Skeleton A framework of bones and cartilage that functions to support and protect internal organs and to permit body movement.

Sliding filament model A model of the mechanism of muscle contraction in which the myofilaments actin and myosin slide across one another, causing the sarcomere to shorten. When enough sarcomeres shorten, the muscle contracts.

Slow-twitch cells Muscle fibers that are specialized to contract slowly but with incredible endurance when stimulated. They contain an abundant supply of myoglobin and mitochondria and are richly supplied with capillaries. They depend on aerobic pathways to generate adenosine triphosphate (ATP) during muscle contraction.

Small intestine The organ located between the stomach and large intestine responsible for the final digestion and absorption of nutrients.

Smooth endoplasmic reticulum (SER) Endoplasmic reticulum without ribosomes. It produces membrane and detoxifies drugs.

Smooth muscle tissue A contractile tissue characterized by the lack of visible striations and by unconscious control over its contraction. It is found in the walls of blood vessels and airways and in organs such as the stomach, intestines, and bladder.

Sodium-potassium pump A molecular mechanism in a plasma membrane that uses cellular energy in the form of adenosine triphosphate (ATP) to pump ions against their concentration gradients. Typically, each pump ejects three sodium ions from the cell while bringing in two potassium ions.

Soluble fiber A type of dietary fiber that either dissolves or swells in water. It includes the pectins, gums, mucilages, and some hemicelluloses. Soluble fiber has a gummy consistency.

Somatic cells All body cells except for gametes (egg and sperm). Somatic cells contain the diploid number of chromosomes, which in humans is 46.

Somatic nervous system The part of the peripheral nervous system that carries information to and from the central nervous system, resulting in voluntary movement and sensations.

Somites Blocks formed from mesoderm cells of the developing embryo that eventually form skeletal muscles of the neck and trunk, connective tissues, and vertebrae.

Source (donor) DNA DNA containing the "gene of interest" that will be combined with host DNA to form recombinant DNA.

Special senses The sensations of smell, taste, vision, hearing, and the sense of balance or equilibrium.

Speciation The formation of a new species.

Species A population or group of populations whose members are capable of successful interbreeding under natural conditions. Such interbreeding must produce fertile offspring.

Sperm A mature male gamete. A spermatozoon.

Spermatid A haploid cell that is formed by mitotic division of a haploid secondary spermatocyte and that develops into a spermatozoon.

Spermatocyte A cell developed from a spermatogonium during sperm development in the seminiferous tubules.

Spermatogenesis The series of events within the seminiferous tubules that gives rise to physically mature sperm from germ cells. It involves meiosis and maturation.

Spermatogonium (plural, spermatogonia) The undifferentiated male germ cells in the seminiferous tubules that give rise to spermatocytes.

Spermicide A means of contraception that consists of sperm-killing chemicals in some form of a carrier, such as foam, cream, jelly, film, or tablet.

Sphincter A ring of muscle between regions of a system of tubes that controls the flow of materials from one region to another past the sphincter.

Sphygmomanometer A device for measuring blood pressure. A sphygmomanometer consists of an inflatable cuff that wraps around the upper arm attached to a device that can measure the pressure within the cuff.

Spina bifida A birth defect in which the neural tube fails to develop and close properly. The mother's taking vitamins and folic acid before conception appears to reduce the chance of having a baby with spina bifida. Some cases can be improved with surgery.

Spinal cord A tube of neural tissue that is continuous with the medulla at the base of the brain and extends about 45 cm (17 in.) to just below the last rib. It conducts messages between the brain and the rest of the body and serves as a reflex center.

Spinal nerves Thirty-one pairs of nerves that arise from the spinal cord. Each spinal nerve services a specific region of the body. Spinal nerves carry both sensory and motor information.

Spiral organ (of Corti) The portion of the cochlea in the inner ear that contains receptor cells that sense vibrations caused by sound. It is most directly responsible for the sense of hearing.

Spongy bone The bone formed from a latticework of thin struts of bone with marrow-filled areas between the struts. It is found in the ends of long bones and within the breastbone, pelvis, and bones of the skull. Spongy bone is less dense than compact bone and is made of an irregular network of collagen fibers surrounded by a calcium matrix.

Sprain Damage to a ligament (a strap of connective tissue that holds bones together).

Squamous cell carcinoma The second most common form of skin cancer that arises in the newly formed skin cells as they flatten and move toward the skin surface.

Squamous epithelium A type of epithelial tissue composed of flattened cells. It forms linings and coverings.

Stapes The last of three small bones in the middle ear that transmit information about sound from the eardrum to the inner ear. The stapes is also known as the stirrup.

Starch The storage polysaccharide in plants.

Stem cell A type of cell that can give rise to other types of cells.

Steroid A lipid, such as cholesterol, consisting of four carbon rings with functional groups attached.

Steroid hormones A group of closely related hormones chemically derived from cholesterol and secreted primarily by the ovaries, testes, and adrenal glands.

Stomach A muscular sac that is well designed for the storage of food, the liquefaction of food, and the initial chemical digestion of proteins.

Stratified epithelial tissue Epithelial tissue with several layers of cells.

Strep throat A sore throat that is caused by *Streptococcus* bacteria.

Stress incontinence A mild form of incontinence characterized by the escape of small amounts of urine when sudden increases in abdominal pressure force urine past the external urethral sphincter.

Striated muscle See skeletal muscle.

Stroke A cerebrovascular accident. A condition in which nerve cells die because the blood supply to a region of the brain is shut off, usually because of hemorrhage or atherosclerosis. The extent and location of the mental or physical impairment caused by a stroke depend on the region of the brain involved.

Succession, ecological The sequence of changes in the species making up a community over time.

Summation (of muscle contraction) A phenomenon that results when a muscle is stimulated to contract before it has time to completely relax from a previous contraction. The response to each stimulation builds on the previous one.

Superovulation The ovulation of several oocytes. It is usually prompted by administration of hormones.

Suppressor T cell A type of T lymphocyte that turns off the immune response when the level of antigen falls by releasing chemicals that dampen the activity of both B cells and T cells.

Suprachiasmatic nucleus (SCN) The region of the brain in the hypothalamus where the "master clock" controlling biological rhythms is believed to reside.

Surface area-to-volume ratio The physical relationship that dictates that increases in the volume of a cell occur faster than increases in its surface area. This relationship explains why most cells are small.

Surfactant Phospholipid molecules coating the alveolar surfaces that prevent the alveoli from collapsing.

Suture An immovable joint between the bones of the skull.

Survivorship The number of individuals of a group born in the same year that is expected to be alive at any given age. It is one way to represent age-specific mortality.

Sweat glands One type of sweat gland is functional throughout life and releases sweat onto the surface of the skin. Also in the skin, another type releases its secretions into hair follicles and becomes functional at puberty.

Sympathetic nervous system The branch of the autonomic nervous system responsible for the fight-or-flight responses that occur during stressful or emergency situations. Its effects are generally opposite to those of the parasympathetic nervous system.

Synapse The site of communication between a neuron and another cell, such as another neuron or a muscle cell.

Synapsis The physical association of homologous pairs of chromosomes that occurs during prophase I of meiosis. The term literally means *bringing together.*

Synaptic cleft The gap between two cells forming a synapse, for example, two communicating nerve cells.

Synaptic knob A small bulblike swelling of an axon terminal that releases neurotransmitter. An axon terminal.

Synaptic vesicle A tiny membranous sac containing molecules of a neurotransmitter. Synaptic vesicles are located in the synaptic knobs of axon endings, and they release their contents when an action potential reaches the synaptic knob.

Synergistic muscles Two or more muscles that work together to cause movement in the same direction.

Synovial cavity A fluid-filled space surrounding a synovial joint formed by a double-layered capsule. The fluid within the cavity is called synovial fluid.

Synovial fluid A viscous, clear fluid within a synovial cavity that acts as both a shock absorber and a lubricant between the bones.

Synovial joint A freely movable joint. A synovial joint is surrounded by a fluid-filled cavity. This is the most abundant type of joint in the body.

Synovial membranes Membranes that line the cavities of freely movable joints and secrete a fluid that lubricates the joint.

Syphilis A sexually transmitted disease caused by the bacterium *Treponema pallidum.* If untreated, it can progress through three stages and cause death. The first stage is characterized by a painless craterlike bump called a chancre that forms at the site where the bacterium entered the body.

Systematic biology The discipline that deals with the naming, classification, and evolutionary relationships of organisms. It is also called systematics.

Systemic circuit (or circulation) The pathway of blood from the left ventricle of the heart to the cells of the body and back to the right atrium.

Systole Contraction of the heart. Atrial systole is contraction of the atria. Ventricular systole is contraction of the ventricles.

Systolic pressure The highest pressure in an artery during each heartbeat. The higher of the two numbers in a blood pressure reading. In a typical, healthy adult, the systolic pressure is about 120 mm Hg.

T cell See T lymphocyte.

T4 cell A helper T cell. The kind of T lymphocyte that serves as the main switch for the entire immune response by presenting the antigen to B cells and by secreting chemicals that stimulate other cells of the immune system.

T lymphocyte T cell. A type of white blood cell. Some T lymphocytes attack and destroy cells that are not recognized as belonging in the body, such as an infected cell or a cancerous cell.

Tanning The buildup of melanin in the skin in response to ultraviolet (UV) exposure.

Target cell A cell with receptors that recognize and bind a specific hormone.

Taste bud A structure consisting of receptors responsible for the sense of taste surrounded by supporting cells. Taste buds are located primarily on the surface epithelium and certain papillae of the tongue.

Taste hairs Microvilli that project into a pore at the tip of the taste bud that bear the receptors for certain chemicals found in food.

Taxon (plural, taxa) A group under consideration, such as a genus, a family, or an order.

Tectorial membrane A membrane that forms the roof of the organ of Corti (the actual organ of hearing). It projects over and is in contact with the sensory hair cells. Pressure waves caused by sound cause the sensory cells to push against the tectorial membrane and bend, resulting in nerve impulses that are carried to the brain by the auditory nerve.

Telomerase The enzyme that synthesizes telomeres.

Telomere A piece of DNA at the tips of chromosomes that protects the ends of the chromosomes.

Telophase In mitosis, the phase when nuclear envelopes form around the group of chromosomes at each pole, the mitotic spindle is disassembled, and nucleoli reappear. The chromosomes also become less condensed and more threadlike in appearance.

Temporomandibular joint (TMJ) syndrome A group of symptoms including headaches, toothaches, and earaches caused by physical stress on the mandibular joint.

Tendinitis Inflammation of a tendon caused by excessive stress on the tendon.

Tendon A band of connective tissue that connects muscle to bone.

Terminal hair Thick, strong hair found on the scalp, eyebrows, eyelashes, and after puberty, the underarms and pubic area.

Tertiary structure The three-dimensional shape of proteins formed by hydrogen, ionic, and covalent bonds between different side chains.

Testes The male gonads. The male reproductive organs that produce sperm and the hormone testosterone.

Testosterone A male sex hormone needed for sperm production and the maintenance of male reproductive structures. Testosterone is produced primarily by the interstitial cells of the testes.

Tetanus A smooth, sustained contraction of muscle caused when stimuli are delivered in such rapid succession that there is no time for muscle relaxation.

Thalamus A brain structure located below the cerebral hemispheres that is important in sensory experience, motor activity, stimulation of the cerebral cortex, and memory.

Theory A broad-ranging explanation for some aspect of the universe that is consistent with many observations and experiments.

Thermoreceptor A sensory receptor specialized to detect changes in temperature.

Threshold The degree to which the voltage difference across the plasma membrane of a neuron or other excitable cell must change to trigger an action potential.

Thrombin A plasma protein formed from prothrombin by thromboplastin that is important in blood clotting. It converts fibrinogen to fibrin, which forms a web that traps blood cells and forms the clot.

Thrombocyte See platelet.

Thrombus A stationary blood clot that forms in the blood vessels. A thrombus can block blood flow.

Thymopoietin A hormone produced by the thymus gland that promotes maturation of T lymphocytes.

Thymosin A hormone produced by the thymus gland that promotes maturation of T lymphocytes.

Thymus gland A gland located on the top of the heart that secretes the hormones thymopoietin and thymosin. It decreases in size as we age.

Thyroid gland The shield-shaped structure at the front of the neck that synthesizes and secretes thyroid hormone and calcitonin.

Thyroid hormone (TH) A hormone released by the thyroid gland that regulates blood pressure and the body's metabolic rate and production of heat. It also promotes normal development of several organ systems.

Thyroid-stimulating hormone (TSH) The anterior pituitary hormone that acts on the thyroid gland to stimulate the synthesis and release of thyroid hormones.

Tidal volume The amount of air inhaled or exhaled during a normal breath.

Tight junction A type of junction between cells in which the membranes of neighboring cells are attached, forming a seal to prevent fluid from flowing across the epithelium through the minute spaces between adjacent cells.

Tissue A group of cells that works together to perform a common function.

Tolerance A progressive decrease in the effectiveness of a drug with continued use.

Tongue The large skeletal muscle studded with taste buds that aids in speech and eating.

Total lung capacity The total volume of air contained in the lungs after the deepest possible breath. It is calculated by adding the residual volume to the vital capacity.

Trabecula (plural, trabeculae) A supporting bar or strand of spongy bone that forms an internal strut that braces the bone from within.

Trachea The tube that conducts air into the thoracic cavity toward the lungs. The trachea is reinforced with C-shaped rings of cartilage to prevent it from collapsing during inhalation and exhalation.

Trait A phenotypically expressed characteristic.

Transcription The process by which a complementary single-stranded messenger RNA (mRNA) molecule is formed from a single-stranded DNA template. As a result, the information in DNA is transferred to RNA.

Transfer RNA (tRNA) A type of RNA that binds to a specific amino acid and transports it to the appropriate region of messenger RNA (mRNA). Transfer RNA acts as an interpreter between the nucleic acid language of mRNA and the amino acid language of proteins.

Transgenic organism An organism that contains certain genes from another species that code for a desired trait. It can be created, for example, by injecting foreign DNA into an egg cell or an early embryo.

Transition reaction The phase of cellular respiration that follows glycolysis and involves pyruvate reacting with coenzyme A (CoA) in the matrix of the mitochondrion to form acetyl CoA. The acetyl CoA then enters the citric acid cycle.

Translation Protein synthesis. The process of converting the nucleotide language of messenger RNA (mRNA) into the amino acid language of a protein.

Transverse tubules T tubules. The tiny, cylindrical inpocketings of the muscle fiber's plasma membrane that carry nerve impulses to almost every sarcomere.

Tricuspid valve A heart valve located between the right atrium and right ventricle that prevents the backflow of blood from the ventricle to the atrium. It is also called the right atrioventricular (AV) valve.

Triglycerides The lipids composed of one molecule of glycerol and three fatty acids. They are known as fats when solid and oils when liquid.

Trisomy A condition in which there are three representatives of a chromosome instead of only two representatives.

tRNA See transfer RNA.

Trophic level The feeding level of one or more populations in a food web. The producers form the first trophic level. Herbivores, which eat the producers, form the second trophic level. Carnivores that eat herbivores form the third trophic level. Carnivores that eat other carnivores form the fourth trophic level.

Trophoblast A group of cells within the blastocyst that will give rise to the chorion, the extraembryonic membrane that will become part of the placenta.

Tropic hormone A hormone that influences another endocrine gland.

Tropomyosin A protein on the thin (actin) filaments in muscle cells that works with troponin to prevent actin and myosin from binding in the absence of calcium ions.

Troponin A protein on the thin (actin) filaments in muscle cells that works with tropomyosin to prevent actin and myosin from binding in the absence of calcium ions.

Tubal ligation A sterilization procedure in females in which each oviduct is cut and sealed to prevent sperm from reaching the eggs.

Tubal pregnancy An ectopic pregnancy in which the embryo implants in an oviduct. This is the most common type of ectopic pregnancy.

Tuberculosis (TB) A highly contagious disease caused by a rod-shaped bacterium, *Mycobacterium tuberculosis*. TB is spread by coughing.

Tubular reabsorption The process by which useful materials are removed from the filtrate within the renal tubule and returned to the blood.

Tubular secretion The process by which wastes and excess ions that escaped glomerular filtration are removed from the blood and added to the filtrate within the renal tubule.

Tumor A neoplasm. An abnormal growth of cells. A tumor forms from the new growth of tissue in which cell division is uncontrolled and progressive.

Turner syndrome A genetic condition resulting from nondisjunction of the sex chromosomes in which a person has 22 pairs of autosomes and a single, unmatched X chromosome (XO). The person has a female appearance.

Tympanic membrane The eardrum. A membrane that forms the outer boundary of the middle ear and that vibrates in response to sound waves. Vibrations of the tympanic membrane are transferred through the middle ear by three small bones (hammer, anvil, and stirrup) to the inner ear, where hearing occurs when neural messages are generated in response to the pressure waves caused by the vibrations.

Type 1 diabetes mellitus An autoimmune disorder characterized by abnormally high glucose in the blood due to insufficient production of insulin by cells in the pancreas.

Type 2 diabetes mellitus A condition characterized by a decreased sensitivity to insulin caused by decreased numbers of insulin receptors on target cells.

Umbilical cord The ropelike connection between the embryo (and later the fetus) and the placenta. It consists of blood vessels (two umbilical arteries and one umbilical vein) and supporting connective tissue.

Undernourishment Starvation. A form of hunger that occurs when inadequate amounts of food are eaten.

Ureters Tubular organs that carry urine from the kidneys to the urinary bladder.

Urethra The muscular tube that transports urine from the floor of the urinary bladder to the outside of the body. In males, it also conducts sperm from the vas deferens out of the body through the penis.

Urethritis Inflammation of the urethra caused by bacteria.

Urinalysis An analysis of the volume, microorganism content, and physical and chemical properties of urine.

Urinary bladder The muscular organ that temporarily stores urine until it is excreted from the body.

Urinary incontinence Lack of voluntary control over urination. Incontinence is the norm for infants and children younger than 2 or 3 years old, because nervous connections to the external urethral sphincter are incompletely developed. In adults, incontinence may result from damage to the external sphincter (often caused, in men, by surgery on the prostate gland), disease of the urinary bladder, and spinal cord injuries that disrupt the pathways along which travel impulses related to conscious control of urination. In any age group, urinary tract infection can result in incontinence.

Urinary retention The failure to completely or normally expel urine from the bladder. This condition may result from lack of the sensation to urinate, as might occur temporarily after general anesthesia, or from contraction or obstruction of the urethra, a condition caused, in men, by enlargement of the prostate gland. Immediate treatment for retention usually involves use of a urinary catheter to drain urine from the bladder.

Urinary system The system that consists of two kidneys, two ureters, one urinary bladder, and one urethra. Its main function is to regulate the volume, pressure, and composition of the blood.

Urinary tract infection (UTI) An infection caused by bacteria in the urinary system. Most bacteria enter the urinary system by moving up the urethra from outside the body.

Urination The process, involving both involuntary and voluntary actions, by which the urinary bladder is emptied. It is also called voiding or micturition.

Urine The yellowish fluid produced by the kidneys. It contains wastes and excess materials removed from the blood. Urine produced by the kidneys travels down the ureters to the urinary bladder, where it is stored until being excreted from the body through the urethra.

Uterine (menstrual) cycle The sequence of events that occurs on an approximately 28-day cycle in the uterine lining (endometrium) that involves the thickening of and increased blood supply to the endometrium and the loss of the endometrium as menstrual flow.

Uterus A hollow muscular organ in the female reproductive system in which the embryo implants and develops during pregnancy.

Vaccination A procedure that introduces a harmless form of the disease-causing organism into the body to stimulate immune responses against that antigen.

Vagina A muscular tube in the female reproductive system that extends from the uterus to the vulva and serves to receive the penis during sexual intercourse and as the birth canal.

Vaginitis An inflammation of the vagina.

Variable In a controlled experiment, the factor that differs between the control group and experimental groups.

Varicose veins Veins that have become stretched and distended because blood is prevented from flowing freely and so accumulates, or "pools," in the vein. A common cause of varicose veins is weak valves within the veins.

Vas deferens A tubule that conducts sperm from the epididymis to the urethra.

Vasectomy A sterilization procedure in men in which both of the vas deferens are cut and sealed to prevent sperm from leaving the man's body.

Vasoactive intestinal peptide (VIP) A hormone released by the small intestine into the bloodstream that triggers the small intestine to release intestinal juices.

Vasoconstriction A decrease in the diameter of blood vessels, commonly of the arterioles. Blood flow through the vessel is reduced, and blood pressure rises as a result of vasoconstriction.

Vasodilation An increase in the diameter of blood vessels, commonly of the arterioles. Blood flow through the vessels increases, and blood pressure decreases as a result of vasodilation.

Vasopressin See antidiuretic hormone.

Vector A biological carrier, usually a plasmid or a virus, that ferries the recombinant DNA to the host cell.

Vein A blood vessel formed by the merger of venules that transports blood back toward the heart. Veins have walls that are easily stretched, so they serve as blood reservoirs, holding up to 65% of the body's total blood supply.

Vellus hair Soft, fine hair that persists throughout life covering most of the body surface.

Vena cava One of two large veins that empty oxygen-depleted blood from the body to the right atrium of the heart. The superior vena cava delivers blood from regions above the heart. The inferior vena cava delivers blood from regions below the heart.

Ventilation In respiration, breathing.

Ventral nerve root The portion of a spinal nerve that arises from the front (anterior) side of the spinal cord and contains axons of motor neurons. It joins with the dorsal nerve root to form a single spinal nerve, which passes through the opening between the vertebrae.

Ventricle One of the two lower chambers of the heart that receive blood from the atria. The ventricles function as the main pumps of the heart. The right ventricle pumps blood to the lungs. The left ventricle pumps blood to body tissues.

Ventricular fibrillation Rapid, ineffective contractions of the ventricles of the heart, which render the ventricles useless as pumps and stop circulation.

Venule A small blood vessel that receives blood from the capillaries. Venules merge into larger vessels called veins. The exchange of materials between the blood and tissues across the walls of a venule is minimal.

Vertebra (plural, vertebrae) One of a series of joined bones that forms the vertebral column.

Vertebral column The backbone. It is composed of 26 vertebrae (7 cervical, 12 thoracic, 5 lumbar, 1 sacrum, and 1 coccyx) and associated tissues. The spinal cord passes through a central canal within the vertebrae.

Vesicle A membrane-bound sac formed during endocytosis.

Vestibular apparatus A closed fluid-filled maze of chambers and canals within the inner ear that monitors the movement and position of the head and functions in the sense of balance.

Vestibule A space or cavity at the entrance to a canal. In the inner ear, the vestibule is a structure consisting of the utricle and saccule.

Villi (singular, villus) Small fingerlike projections on the small intestine wall that increase surface area for absorption.

Virus A minute infectious agent that consists of a nucleic acid encased in protein. A virus cannot replicate outside a living host cell.

Vital capacity The maximal amount of air that can be moved into and out of the lungs during forceful breathing.

Vitiligo A condition in which melanocytes disappear from areas of the skin, leaving white patches in their wake.

Vitreous humor The jellylike fluid filling the posterior cavity of the eye between the lens and the retina that helps to keep the eyeball from collapsing and holds the thin retina against the wall of the eye.

Vocal cords Folds of tissue in the larynx that vibrate when air passes through them, producing sound.

Water-soluble vitamin A vitamin that dissolves in water. Water-soluble vitamins include vitamin C and the various B vitamins.

Werner's syndrome A rare inherited disease characterized by premature aging.

White blood cells See leukocytes.

White matter Regions of the central nervous system that are white owing to the presence of myelinated nerve fibers. White matter is important in neural communication over distances.

X-linked genes Genes located on the X chromosome. Most X-linked genes have no corresponding allele on the Y chromosome and will be expressed in a male and in a female if she is homozygous.

Yellow marrow A connective tissue found in the shaft of long bones that stores fat. It forms from red marrow, and, if the need arises, it can convert back to red marrow and form blood cells.

Yolk sac The extraembryonic membrane that is the primary source of nourishment for embryos in many vertebrates. In embryonic humans, however, the yolk sac does not provide nourishment and remains quite small (human embryos receive nutrients from the placenta). In humans, the yolk sac is a site of blood cell formation and contains cells, called primordial germ cells, that migrate to the gonads, where they differentiate into immature cells that will eventually become sperm or oocytes.

Zygote The diploid cell resulting from the joining of an egg nucleus and a sperm nucleus. The first cell of a new individual.

Zygote intrafallopian transfer (ZIFT) A procedure in which zygotes created by the union of egg and sperm in a dish in the laboratory are inserted into the woman's oviducts. Zygotes travel on their own from the oviducts to the uterus.

Length

1 inch (in.) = 2.54 centimeters (cm)
1 centimeter (cm) = 0.3937 inch (in.)
1 foot (ft) = 0.3048 meter (m)
1 meter (m) = 3.2808 feet (ft) = 1.0936 yard (yd)
1 mile (mi) = 1.6904 kilometer
1 kilometer (km) = 0.6214 mile (mi)

Area

1 square inch (in.^{2}) = 6.45 square centimeters (cm^2)
1 square centimeter (cm^2) = 0.155 square inch (in.^{2})
1 square foot (ft^2) = 0.0929 square meter (m^2)
1 square meter (m^2) = 10.7639 square feet (ft^2) =
   1.1960 square yards (yd^2)
1 square mile (mi^2) = 2.5900 square kilometers (km^2)
1 acre (a) = 0.4047 hectare (ha)
1 hectare (ha) = 2.4710 acres (a) = 10,000 square meters (m^2)

Volume

1 cubic inch (in.^{3}) = 16.39 cubic centimeters (cm^3 or cc)
1 cubic centimeter (cm^3 or cc) = 0.06 cubic inch (in.^{3})
1 cubic foot (ft^3) = 0.028 cubic meter (m^3)
1 cubic meter (m^3) = 35.30 cubic feet (ft^3) =
   1.3079 cubic yards (yd^3)
1 fluid ounce (oz) = 29.6 milliliters (mL) = 0.03 liter (L)
1 milliliter (mL) = 0.03 fluid ounce (oz) =
   1/4 teaspoon (approximate)
1 pint (pt) = 473 milliliters (mL) = 0.47 liter (L)
1 quart (qt) = 946 milliliters (mL) = 0.9463 liter (L)
1 gallon (gal) = 3.79 liters (L)
1 liter (L) = 1.0567 quarts (qt) = 0.26 gallon (gal)

Mass

1 ounce (oz) = 28.3496 grams (g)
1 gram (g) = 0.03527 ounce (oz)
1 pound (lb) = 0.4536 kilogram (kg)
1 kilogram (kg) = 2.2046 pounds (lb)
1 ton (tn), U.S. = 0.91 metric ton (t or tonne)
1 metric ton (t) = 1.10 tons (tn), U.S.

Metric Prefixes

Prefix	Abbreviation	Meaning
giga-	G	$10^9 = 1,000,000,000$
mega-	M	$10^6 = 1,000,000$
kilo-	k	$10^3 = 1,000$
hecto-	h	$10^2 = 100$
deka-	da	$10^1 = 10$
		$10^0 = 1$
deci-	d	$10^{-1} = 0.1$
centi-	c	$10^{-2} = 0.01$
milli-	m	$10^{-3} = 0.001$
micro-	μ	$10^{-6} = 0.000001$

Appendix 2

SELECTED ANSWERS TO REVIEWING THE CONCEPTS QUESTIONS

Chapter 1: 10. c, 11. homeostasis, 12. hypothesis, 13. adaptive

Chapter 2: 18. d, 19. b, 20. a, 21. b, 22. a, 23. d, 24. denaturation, 25. nucleotides, 26. primary, 27. phospholipids, 28. starch; glycogen, 29. ATP, 30. pH, 31. covalent, 32. dehydration synthesis; hydrolysis

Chapter 3: 12. c, 13. a, 14. d, 15. c, 16. b, 17. a, 18. a, 19. c, 20. cytoplasm, 21. pinocytosis, 22. nucleolus, 23. oxygen, 24. lactic acid fermentation

Chapter 4: 14. b, 15. a, 16. d, 17. calcium, 18. homeostasis

Chapter 5: 10. c, 11. c, 12. axial, 13. calcium, 14. osteocytes, 15. sprain

Chapter 6: 12. c, 13. a, 14. actin, myosin, 15. sarcomere

Chapter 7: 14. a, 15. c, 16. b, 17. b, 18. b, 19. myelin sheath, 20. sodium-potassium pump, 21. dopamine

Chapter 8: 16. a, 17. a, 18. d, 19. hypothalamus, 20. medulla, 21. sympathetic nervous system, 22. cerebrum

Chapter 9: 19. a, 20. b, 21. d, 22. a, 23. b, 24. cones, 25. lens, 26. cochlea, 27. sudden movement of the head, including acceleration and deceleration

Chapter 10: 17. a, 18. d, 19. c, 20. c, 21. a, 22. lipid-soluble; water-soluble, 23. giantism; acromegaly, 24. calcitonin; parathyroid hormone (parathormone), 25. adrenal cortex, 26. insulin; glucagon

Chapter 11: 18. a, 19. c, 20. b, 21. transport oxygen, 22. negative, positive, 23. hemoglobin, 24. fibrin

Chapter 12: 20. a, 21. b, 22. b, 23. b, 24. sinoatrial node, 25. an aneurysm, 26. capillaries

Chapter 13: 17. a, 18. d, 19. b, 20. a, 21. c, 22. natural killer cell, 23. histamine, 24. plasma cells, 25. macrophages, 26. autoimmune

Chapter 14: 15. a, 16. b, 17. b, 18. a, 19. a, 20. epiglottis, 21. carbonic anhydrase

Chapter 15: 12. a, 13. b, 14. d, 15. d, 16. chyme, 17. liver, 18. starch, 19. cholecystokinin

Chapter 16: 11. a, 12. c, 13. b, 14. d, 15. b, 16. nephrons, 17. hemodialysis, 18. internal; smooth; external; skeletal

Chapter 17: 22. c, 23. b, 24. a, 25. b, 26. c, 27. epididymis, 28. oviduct

Chapter 18: 15. d, 16. c, 17. b, 18. a, 19. b, 20. d, 21. inner cell mass; trophoblast, 22. cleavage, 23. chorion; endometrium, 24. gastrulation, 25. prolactin; oxytocin, 26. free radicals

Chapter 19: 8. b, 9. a, 10. a, 11. crossing over and independent assortment, 12. homologous, 13. synapsis, 14. anaphase, 15. anaphase II

Chapter 20: 8. b, 9. a, 10. polygenic, 11. phenotype, genotype, 12. homozygous, heterozygous, 13. linked

Chapter 21: 15. a, 16. c, 17. b, 18. a, 19. tRNA, 20. uracil, 21. restriction enzyme

Chapter 22: 18. a, 19. d, 20. c, 21. a, 22. c, 23. d, 24. c, 25. a, 26. a, 27. natural selection, 28. mosaic evolution, 29. multiregional

Chapter 23: 14. a, 15. c, 16. b, 17. niche, 18. secondary consumer, carnivore

Chapter 24: 10. b, 11. d, 12. d, 13. carrying capacity

HINTS FOR APPLYING THE CONCEPTS QUESTIONS

Chapter 1

1. Design an experiment with a control group and an experimental group.

2. Consider the discussion on page 10 and the list of questions on page 12.

3. Look through the newspaper or your favorite magazine.

Chapter 2

1. What functions do triglycerides perform in your body?

2. Cellulose is an important form of dietary fiber.

3. Consider the following qualities of water: polarity, heat of vaporization, and high heat capacity.

4. What harm can radiation cause to the body?

Chapter 3

1. Consider how the plasma membrane functions in cell-cell recognition.

2. Mitochondria process energy for cells, and thus they occur in large numbers in cells with a high demand for energy.

3. What cytoskeletal element is involved in cell division?

Chapter 4

1. What are the functions of skin?

2. What is a negative feedback relationship?

3. Acetaminophen changes the body's temperature set point. What physiological responses adjust body temperature when it is above the set point? What physiological responses adjust body temperature when it is below the set point?

4. What is the function of collagen?

Chapter 5

1. What kind of activity builds bone density?

2. Why would a pregnant woman need more calcium than she did before she was pregnant?

3. What structure would indicate that the bone was still capable of growth?

Chapter 6

1. What role does acetylcholine play in muscle contraction?

2. What causes tendinitis? How is it treated?

3. What role do calcium ions play in muscle contraction?

4. Are fast twitch or slow twitch fibers darker in color? How do the properties of fast twitch fibers differ from those of slow twitch fibers?

Chapter 7

1. What effect does an inhibitory neurotransmitter have on the postsynaptic neuron?

2. What would happen if a neurotransmitter were not removed from the synapse?

3. What factors cause ions to cross the membrane during an action potential? What role do potassium ions play in the action potential?

4. What role does the myelin sheath play in the conduction of an action potential?

5. Why is the resting potential important?

Chapter 8

1. Where in the brain is the language center of most people located? Which side of the brain controls movement on the right side of the body?

2. What are the functions of the spinal cord? Would loss of any of the functions due to injury have varying effects depending on the location of injury?

3. What is the function of the sympathetic nervous system?

4. What could happen if sensory nerves from the tongue were anesthetized?

5. What structure of the brain is important in transferring short term memory to long term memory?

Chapter 9

1. In what type of vision are distant objects seen more clearly than nearby objects?

2. What part of the ear is responsible for equilibrium?

3. How would loud sounds affect the organ of hearing in the inner ear?

4. What changes when you change your focus from an object in the distance to a nearby object?

Chapter 10

1. Cortisone is a glucocorticoid.

2. Matt's age can be used to make an educated guess about his type of diabetes.

3. Theresa's symptoms and the timing of their onset suggest melatonin may be involved.

4. What hormones are produced by the adrenal glands?

Chapter 11

1. Are the white blood cells that are produced in leukemia functional?

2. What regulates the circulating number of red blood cells?

3. What is the function of red blood cells? How is iron related to the ability of red blood cells to carry out their function?

4. Which antibodies against antigens on red blood cells does Raul have?

5. What could happen if either of Elizabeth's babies had Rh+ blood?

6. What are the functions of platelets?

Chapter 12

1. What would be the result if a blood clot were lodged in a small blood vessel in the heart?

2. What happens in the lymph nodes when you get sick?

3. The pressure that propels blood through the arteries is equal to the pressure against the arterial walls. What force generates this pressure? What is the relationship between high blood pressure and atherosclerosis?

4. What happens if the valves do not close properly?

Chapter 13

1. Is a vaccine effective immediately?

2. How specific is the immune response?

3. What role do helper T cells play in the immune response?

4. How long does it take antibodies to form after the first exposure to an antigen?

5. In an organ transplant, what would happen to the transplanted cells if they had very different self MHC markers from the recipient's?

Chapter 14

1. What factors regulate breathing rate and tidal volume?

2. What is the function of the cilia in the respiratory tubules?

3. What happens to Juan's blood level of carbon dioxide when he holds his breath? How would this affect his breathing?

4. Does inhalation or expiration involve muscle contraction? Which is a passive process?

Chapter 15

1. What role does the pancreas play in digestion?

2. What is the relationship between diarrhea and water reabsorption?

3. What would cause skin to develop a yellow tone? What might the cause of the yellow skin tone have to do with a tattoo?

4. What substance is released into the small intestine when fatty food enters the small intestine?

Chapter 16

1. Consider the role of the external urethral sphincter in urination.

2. Consider what might have been introduced to Miguel's urinary tract when the catheter was inserted.

3. How does alcohol affect antidiuretic hormone?

4. Consider the functions of the kidneys and the ways to replace these functions.

Chapter 17

1. What are the possible consequences of pelvic inflammatory disease?

2. When making your recommendation for a means of contraception, consider its effectiveness in preventing pregnancy, this couple's need for protection against STDs, and the health effects of the means of contraception.

3. What happens to the endometrium during menstruation?

4. What is the function of FSH?

5. What does warm temperature have to do with sperm count?

Chapter 18

1. What might pelvic inflammatory disease have done to Tanya's oviduct?

2. How does fetal circulation differ from circulation after birth?

3. Consider the concept of critical periods in development.

4. Consider the hormones involved in milk production and ejection.

5. Consider your grandmother's age.

Chapter 19

1. What is nondisjunction?

2. Look at Figure 19.

3. What is the function of the spindle fibers?

Chapter 20

1. Colorblindness is a sex-linked trait.

2. What are the genotypes of George and Sue? Use a Punnett square to determine the expected results from a cross with those genotypes.

3. Recessive alleles are not expressed.

4. A recessive trait is only expressed in the homozygous condition.

5. Use a Punnett square to determine the outcome of the cross. The probability of the phenotype for each offspring remains the same.

Chapter 21

1. What is the start codon? What is the stop codon? What amino acids do the other codons code for?

2. Translate each of the mRNA strands. Remember that more than one codon can code for the same amino acid.

3. Are any bands present in a father, but absent in both the mother and the child?

Chapter 22

1. Describe the evidence in support of evolution.

2. Consider chemical evolution.

3. What characteristics distinguish primates from other mammals?

4. Tay-Sachs disease is caused by a recessive allele.

5. Consider misconceptions about human evolution.

6. Review major changes that occurred over the course of hominid evolution.

Chapter 23

1. On average, what percentage of the energy available at one trophic level is available to the next level?

2. What happens to mercury as it moves up the food chain?

3. Average rainfall and temperature largely determine the organisms found in a given location.

Chapter 24

1. What will happen to the size of each population as the prereproductive individuals reach reproductive age?

2a. What factors regulate population size?

2b. How might Asian carp affect factors regulating the population size of other species of fish?

3. How would spending be different in a young, growing population and an aging population?

4. The size of the population influences how quickly individuals are added to the population.